Accident Prevention Manual

FOR BUSINESS & INDUSTRY

Administration & Programs
Eleventh Edition

National Safety Council Mission Statement

The mission of the National Safety Council is to educate and influence society to adopt safety, health, and environmental policies, practices, and procedures that prevent and mitigate human suffering and economic losses arising from preventable causes.

OCCUPATIONAL SAFETY AND HEALTH SERIES

The National Safety Council's OCCUPATIONAL SAFETY AND HEALTH SERIES is composed of six volumes written to help readers establish and maintain safety and health programs. The latest information on establishing priorities, collecting and analyzing data to help identify problems, and developing methods and procedures to reduce or eliminate illness and accidents, thus mitigating injury and minimizing economic loss resulting from accidents, is contained in all volumes in the series:

ACCIDENT PREVENTION MANUAL FOR BUSINESS & INDUSTRY
(4-volume set)
> Administration & Programs
> Engineering & Technology
> Environmental Management
> Security Management

OCCUPATIONAL HEALTH & SAFETY

FUNDAMENTALS OF INDUSTRIAL HYGIENE

Other safety and health references published by the Council include:
ACCIDENT FACTS
LOCKOUT/TAGOUT: THE PROCESS OF CONTROLLING
    HAZARDOUS ENERGY
SUPERVISORS' SAFETY MANUAL
OUT IN FRONT: EFFECTIVE SUPERVISON IN THE
    WORKPLACE
PRODUCT SAFETY MANAGEMENT GUIDELINES
OSHA BLOODBORNE PATHOGENS EXPOSURE CONTROL
    PLAN (National Safety Council/CRC-Lewis Publication)
COMPLETE CONFINED SPACES HANDBOOK (National Safety
    Council/CRC-LEWIS Publication)

Accident Prevention Manual

FOR BUSINESS & INDUSTRY

Administration & Programs
Eleventh Edition

Editors:
Gary R. Krieger, MD, MPH, DABT
John F. Montgomery, Ph.D.

National Safety Council
Itasca, Illinois

Editors: Gary R. Krieger, John F. Montgomery
Project Editor: Patricia M. Laing

©1997 by the National Safety Council
All Rights Reserved
Printed in the United States of America
01 00 99 98 97 5 4 3 2

Library of Congress Cataloging-in-Publication Data
Accident prevention manual for business & industry: administration & programs/edited by Gary R. Krieger, John F. Montgomery—11th ed.
p.   cm.—(Occupational safety and health series)
Includes bibliographical references and index.

ISBN: 0–87912–191–2
1. Industrial safety—United States—Handbooks, manuals, etc.
2. Accidents—United States—Prevention—Handbooks, manuals, etc.
I. Krieger, Gary R. II. Montgomery, John F. (John Franklin), 1994–.
III. Series: Occupational safety and health series (Chicago, Ill.)
T55.A333   1996
658.4'08—dc20   96–43732
CIP
4M997   Product Number: 121450000

Contents

Foreword

SAFETY AND HEALTH TRENDS
FOR THE TWENTY-FIRST CENTURY

The rate of change in occupational safety and health in the United States and throughout the world has increased substantially in recent decades. All indications are that the changes will continue to accelerate during the last years of the 1990s and into the next century. These new trends will have considerable impact on the users of this 11th Edition of the *Accident Prevention Manual for Business & Industry*.

The *Accident Prevention Manual for Business & Industry* contains information based on the most current knowledge and best practices within the safety and health field. However, as changes occur, occupational safety and health professionals must keep abreast of current and future economic, environmental, regulatory, and technical advances affecting their field. The following review of trends in safety and health will assist and focus their efforts.

In 1985 the Executive Committee of the Industrial Division (now the Business and Industry Division) of the National Safety Council asked its Research Projects Committee to conduct a study of emerging trends in the safety field to assist future safety program planning. The Research Committee ultimately identified 34 emerging trends for evaluation and ranking. A panel of 13 experts from the Research Projects Committee and Business and Industry Division Group Directors were asked to evaluate each trend on the basis of three criteria:

- importance—the trend's impact on safety and health policy and procedures and the degree to which the trend may be associated with deaths and injuries
- interest—the degree to which the public may become involved in each trend
- time frame—the period of time during which the trend will have its maximum impact.

Most trends were rated "moderate" in importance and interest, and "long term" in their impact. The trends that the panel ranked the highest on the basis of all these criteria were:

- changing composition/nature of the workforce/jobs
- occupational versus nonoccupational illnesses
- long-term effects of toxic materials in the environment
- defining acceptable risks
- development of safety standards

A second study, begun in December, 1989, and summarized in May, 1991, was completed by a 26-member panel of experts from the Research Projects Committee, the Business and Industry Division Group Directors, the Research Committee of the NSC Board of Directors, and the Chairman of the Labor Division. This group used four criteria to select emerging trends or issues:

- importance—the degree to which the trend/issue is or will become a major safety and health concern or factor
- persistence—the degree to which the trend/issue will remain throughout the decade
- significance—the degree to which the trend/issue is or should be considered a priority by the National Safety Council
- confidence—the respondents' overall confidence in their responses.

The group agreed that five trends or issues ranked highest on the basis of these criteria: (1) ergonomics; (2) consideration of safety, health, and ergonomic issues at the concept and design stage; (3) incorporating safety into the management system; (4) repetitive motion trauma, including carpal tunnel syndrome; and (5) back injuries. Only slightly less important in the rankings were trends in emergency planning and global environmental issues. In late 1996, these still rank among the highest issues, with the addition of process safety as an important consideration at the concept and design stage.

The final step in the project was to examine all the trends/issues and to form clusters of the trends that were judged to be closely or strongly related. This composite set of groups could provide a more global view of the subject areas covered by the original trends/issues submitted. The panel devised the following five cluster groups, considering them all top priority areas for the future:

- occupational health and wellness
- public safety and health issues
- competence of health and safety professionals
- changing nature of the workforce/jobs
- responses to changing technology.

An ongoing study begun in October 1995 by the National Safety Council highlights continuing concerns in these five top priority areas. In the occupational health and wellness area, a trend toward outsourcing to healthcare service providers is noted. In public safety and health

issues, concerns focus on reducing workplace violence and stress, the increasing importance and use of global safety and health standards, and increasing sensitivity to our global living environment. On the issue of the competence of health and safety professionals, concern is expressed about trends toward increased certification requirements but a shortage of academically qualified faculty. In the area of the changing nature of the workforce/jobs, trends noted include a shift in OSHA from an inspection or regulatory body to an advisory board and a shift from corporate jobs to consultant or managed healthcare systems jobs.

Based on the results of these emerging trends studies, it is possible to draw some conclusions and to make specific recommendations to support enhanced safety and health efforts during the coming decade. Specific major problem issues and corrective actions recommended include the following:

1. Occupational safety, medical, and industrial hygiene will be integrated into managed healthcare systems for comprehensive occupational health services. This is particularly attractive to small and mid-sized businesses, but for large businesses it presents the opportunity to outsource an existing department. Employee cutbacks may continue in the near future, which could mean fewer corporate safety/health/ environment professionals to perform the work necessary to achieve appropriate safety/health/environment goals within any organization. Managers should make hiring and supporting qualified safety and health professionals a top priority.

2. If there are too few qualified staff members to accomplish the jobs required, then companies will have to pay more attention to time management. Because all safety actions can be completed within a given time, companies will have to do risk management prioritization based on the loss potential of identified risks and the cost-benefit impact of available countermeasures. The process will require a reliable, accurate system of hazard identification and evaluation.

3. Many major safety and health problems existing within an organization can be traced to what managers have judged to be "acceptable risk." That is, as long as no major losses occur, managers assume that the risks they are taking are "acceptable." They need to improve their ability to define and assess risk and "acceptable risk" based on loss-potential measurement criteria instead of after-the-fact accident losses.

   [Risk is the individual likelihood of harm (one airplane crash in over nine million airline flights) from some level of exposure (110 million airline flights [1975-1985]) to a particular hazard (windshear at airports). Risk and exposure are combined to determine how much harm (e.g., 12 crashes and

400 lives lost) is caused. The term "risk" is also used to represent the sum of all individual risks, particularly when developing population estimates. Risk at the general level involves two major components: (1) the existence of a possible unwanted consequence or loss and (2) the probability such a consequence will occur. Managers can reduce risk either by decreasing the probability of occurrence or by educating the risk taker to recognize and to control the hazard.]

4. In times of financial crises organizations may decide not to spend money to correct a safety or health problem. If such a choice must be made, only problems with the lowest loss potential should remain uncorrected for a time. This approach assumes that managers are able to measure accurately the potential loss associated with various identified safety and health problems.

5. Personal safety, health, environmental safety, equipment safety, process safety, and product safety are becoming more interrelated and creating an expanded role for the safety professional. Practicing safety and health professionals must upgrade their knowledge, skills, and job performance abilities to handle these new responsibilities. Course work and content within university degree programs in the safety field should be examined to ensure adequate coverage of those subjects needed to prepare the safety professional properly, such as life-cycle analysis, ISO 14000, and design in safety.

6. Physical hazards and unsafe behaviors, as well as high-risk human-machine-environment interactions, must be included among the measurement criteria used to evaluate safety performance. The key to proper safety and health program evaluation is measurement criteria based on determining the loss potential of certain hazards and behaviors rather than on after-the-fact analysis of actual accident losses. Two examples of evaluation methodologies based on no-loss measurement criteria include behavior-environment sampling and the Critical Incident Technique.

7. Accident prevention within an organization is not the exclusive responsibility of the safety and health professional. Managers and supervisors at all levels must be involved in day-to-day safety and health activities and decisions. Managers should be properly trained to identify and assess potential accident situations before they result in actual losses (injury, death, property damage, production stoppage, etc.). Unfortunately, many managers believe that no safety or health problem exists if no accidents occur that cause immediately measurable losses. Safety and health training must be included in the education of future managers and in upgrading the qualifications and abilities of present managers.

8. Ergonomics needs more attention. Two-thirds of illnesses and one-fourth of injuries reported to OSHA are ergonomics related. The key points addressed in the proposed U.S. Occupational Safety and Health Administration ergonomic initiatives include management commitment and employee involvement in safety; worksite analysis requiring a thorough evaluation of workplace ergonomic hazards; hazard prevention and control, including the requirement to follow established procedures to control ergonomic hazards; and education and training of managers, supervisors, and employees to help them understand the ergonomic hazards associated with their jobs.

9. Management needs to place greater emphasis on the use of engineering techniques in accident prevention. For example, they can apply engineering concepts and design to make manufacturing processes safer and to minimize waste. Instead of focusing primarily on correcting workers' unsafe practices, management should examine and assess the total human-machine-environment system to appraise its loss potential. They can then apply the principles of engineering science to minimize future accidents and their subsequent losses.

10. Companies need a better, more accurate method for determining direct and indirect costs of employer accident losses on a regular, sustained basis and for assessing the impact of accident prevention programs in dollar units that can be translated into profits. Employers need a cost accounting/cost investigation procedure to identify both direct and indirect (uninsured) costs of employee accidents occurring on and off the job. Because of employers' vested interest in loss reduction, a true determination of dollar losses resulting from accidents should strongly motivate management to initiate or expand various worker safety and health programs.

11. With an increase in vehicular traffic, both on and off the job, highway traffic safety is becoming more important to employers. Although management should pay particular attention to on-the-job motor vehicle safety, off-the-job vehicle accidents are also costly to employers. As a result, managers need to focus on a systems approach when developing a comprehensive company highway traffic safety program.

12. The demand for well-qualified, competent safety and health professionals will increase significantly during the 21st century. Jobs, however, will shift from the corporate arena to independent consultant services or comprehensive occupational healthcare services. More sophisticated, well-educated, technically competent individuals will be needed to respond to the emerging trends/issues identified in the safety and health field. Safety and

health professionals of the future can help to prepare themselves by graduating from an accredited
safety degree program and acquiring the appropriate certificates and degrees.

13. The next decade will witness the continuing emergence of a world market and a greater effort
toward harmonization of safety/health/environ-
mental practices and standards. For example, formation of the European Economic Union (EU) has
brought about changes in worker safety, commercial and industrial products, and other goods and
services. The Framework Directive of the EU
(passed in 1989) encourages the formation of
on-the-job safety and health programs for workers
and management and applies to all private and
public sectors of business. As more nations join the
world economic community, solving occupational
health and safety problems and preserving the
environment must become top international priorities for all countries.

Jerry Scannell
President
National Safety Council

Preface

The 11th edition of the *Accident Prevention Manual for Business & Industry: Administration & Programs* continues a tradition begun in 1946 with the publication of the first *Accident Prevention Manual*. This Manual brings to the safety/health/environmental professional the broad spectrum of topics, specific hazards, best practices, control procedures, resources, and sources of help known in the field today. To accommodate the expansion of knowledge and topics, the Manual is now printed in three volumes: *Administration & Programs* (11th edition), *Engineering & Technology* (11th edition), and *Environmental Management* (1st edition).

We have also expanded the professional community from which the technical information is drawn. In addition to the expertise of National Safety Council volunteers and staff, we have received expert assistance in developing, writing, and reviewing from contributors representing various disciplines and from the editors, John F. Montgomery and Gary R. Krieger (see list of Contributors).

New Material

The *Accident Prevention Manuals* are intended for a wide range of users, for the student using it as a textbook, the corporate or company manager searching for solutions to safety and health problems, the new safety specialist who must plan and organize a safety and health program within a company, or the experienced safety professional seeking to improve an operating program and to learn more about advances in the field of safety and health. To increase their usefulness, we have added new topics and features, and have updated all chapters. In this edition, we have added seven completely new or rewritten chapters and review questions to the *Administration & Programs* volume:

- Chapter 4: Safety, Health, and Environmental Auditing
- Chapter 9: Computers and Information Management
- Chapter 11: Industrial Hygiene Program
- Chapter 14: Employee Assistance Programs
- Chapter 17: Retail/Service/Warehouse Facilities
- Chapter 20: Laboratory Safety
- Chapter 26: Process Safety Management
- Review Questions at the end of each chapter

• Appendix 3: Answers to Review Questions

We have added two new chapters and review questions to the *Engineering & Technology* volume:

• Chapter 1: Designing in Safety
• Chapter 9: Occupational Medical Surveillance
• Review Questions at the end of each chapter
• Appendix 4: Answers to Review Questions

Revised Material

In addition to updating technical information and references, we have added significant new sections to many chapters. For example, there are new sections on International Regulations and Multinational Regulations in Chapter 2, Regulatory History; on Aviation Safety in Chapter 18, Transportation Safety; on Contractor Safety in Chapter 21, Contractor and Nonemployee Safety; on selected motivational techniques in Chapter 22, Motivation; and on developing and delivering structured, performance-based training in Chapter 23, Safety and Health Training.

Definitions of Terms

As the concerns and responsibilities of safety/health/environmental professionals expand, so must their ability to communicate and educate. Technical terms are defined in the text where they are used and also in Appendix 3, Glossary, in the *Engineering & Technology* volume. However, the terms *incident* and *accident* deserve a special note. In the years since the original publication of this manual, many theories of accident causation and definitions of the term "accident" have been advanced. Notwithstanding the traditional title of this manual, the National Safety Council continues to work to increase awareness that an incident is a near-accident and that so-called accidents are not random events but rather preventable events. To that end, the term "incident" is used in its broadest sense to include incidents that may lead to property damage, work injuries, or both. The following definitions are generally used in this manual:

Accident. An unplanned, undesired event, not necessarily resulting in injury, but damaging to property and/or interrupting the activity in process.

Incident. An undesired event that may cause personal harm or other damage. In the United States, OSHA specifies that incidents of a certain severity be recorded.

With proper hazard identification and evaluation, management commitment and support, preventive and corrective procedures, monitoring, evaluation, and training, unwanted events can be prevented.

Design

To represent the trend toward global cooperation and understanding in safety, health, and environmental initiatives, the design of this 11th edition features a visual and universal mode of communication—the language of symbols. These symbols are adapted from Henry Dreyfuss, *Symbol Sourcebook: An Authoritative Guide to International Graphic Symbols,* © 1972 by The McGraw-Hill Companies; reproduced with permission of The McGraw-Hill Companies. See Figure 1 for symbols used in the two volumes of the 11th edition.

STAFF REVIEWERS

The National Safety Council wishes to thank members of its staff who contributed significant expertise to the development and review of this manual:

Jean Adams	Joseph Lasek
J. David Amos	BJ LoMastro
Linda Bennett	Robert J. Marecek
Dale Haskin	Amber Nicholson
Alan Hoskin	Mike Peltier
Carol Huybrecht	Tom Planek
Ron Koziol	Roseann Solak

CONTRIBUTORS

The following safety, health, and environmental professionals have contributed to the 11th edition as reviewers and/or writers of chapters or sections.

Richard Besserman, MD; President, Health Safety Engineering Management, Inc. Dr. Besserman is the founder of InfoGEN Corporation and the developer of SENTRY Occupational Health and Safety Surveillance System. Dr. Besserman is a consultant in occupational health and safety and has extensive experience with information management. Using his knowledge of medical, safety, and environmental issues, he has been a pioneer in the development of integrated OH&S computer systems that are widely used in industry. Mailing address: HSE Management, Inc., 3115 E. Marshall Avenue, Phoenix, AZ 85016.

Morley Brickman, PE, CSE; President, Morley Brickman & Associates, Ltd. Mr. Brickman has 25 years of experience in the construction industry and 15 years with the Occupational Safety and Health Administration. Mr. Brickman is a member of the National Safety Council, American Society of Safety Engineers, World Safety Organization, National Academy of Forensic Engineers, and National Society of Professional Engineers. He serves on the A10 committee of the American National Standards Institute. Mailing address: 9221 Drake Avenue, Evanston, IL 60203-1626.

John E. Brodbeck CSP: President, Mississippi Safety Services, Inc., Clinton, MS. Mr. Brodbeck has been employed in occupational safety for 21 years and is a

Administration & Programs Volume

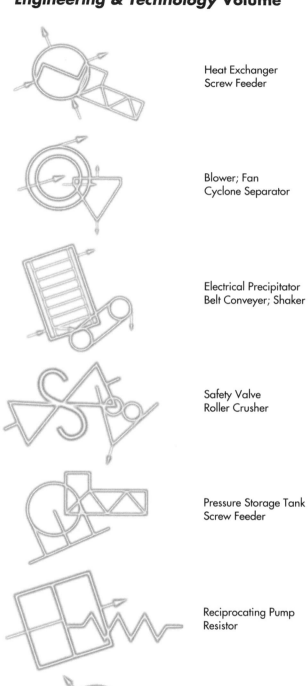

Device Control
Decision

Unit Separator
Preparation

Total
Decision

Ionizing Rays
Double Insulation

Conference
Listen

Stop If There Is a Mistake
End of Operation

Review
External Condition

Fast Forward
Movement

Engineering & Technology Volume

Heat Exchanger
Screw Feeder

Blower; Fan
Cyclone Separator

Electrical Precipitator
Belt Conveyer; Shaker

Safety Valve
Roller Crusher

Pressure Storage Tank
Screw Feeder

Reciprocating Pump
Resistor

All Control Valves
Water Cooler

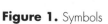

Figure 1. Symbols

member of the American Society of Safety Engineers. He has most recently served as editor for the National Safety Council's *Motor Fleet Safety Manual,* 4th edition.

Mike Carrier, MA, MS; Director of Human Resources, SuperValu, Perryman, MD. Mr. Carrier has extensive experience in the wholesale/retail industry, risk management, and human resources. He is an active member of the National Safety Council Executive Committee for Retail, Trades and Services. Mailing address: SuperValu, 504 Advantage Avenue, Perryman, MD 21130.

Somadeepti N. Chengalur; PhD; Ergonomist; Corporate Ergonomics, Eastman Kodak Company. Dr. Chengalur is responsible for evaluating existing and proposed workplaces, processes, and task designs; providing practical recommendations for improvement of the human interface; developing and conducting training programs covering a wide variety of ergonomics issues. Mailing address: Eastman Kodak Company, 1100 Ridgeway, 2/320/KP MC 26257, Rochester, NY 14652.

Maryanne DiBerto; Vice President, Arthur D. Little, Inc., and Managing Director of Environmental, Health, and Safety Consulting. Ms. DiBerto advises corporate management regarding ways to enhance the efficiency and effectiveness of their auditing activities. She has coauthored three Arthur D. Little books on EHS auditing, as well as the International Chamber of Commerce's *Guide to Effective Environmental Auditing.* Ms DiBerto has also served on the Board of Directors of the Environmental Auditing Roundtable.

Denzell B. Ekey, CHCM; President of Safety Consulting, Inc. For the past 25 years Mr. Ekey has provided consultation to industry, construction, and governmental agencies. He conducts on-site audits, and lectures nationwide on safety program development, hazard recognition, regulatory requirements, and compliance measures. In addition, he also provides accident investigations, research, and testimony in litigated cases. Mr. Ekey is past president of Veterans of Safety, President-Elect of the American Academy of Safety Education and a charter member of the National Safety Council of Construction Division. Mailing address: Safety Consulting, Inc., PO Box 2789, Topeka, Kansas 66601-2789.

J. Nigel Ellis, PhD, CSP, PE, CPE; President, Dynamic Scientific Controls (DSC), Wilmington, DE. Dr. Ellis has been a leading authority in the field of fall protection since 1970. For 26 years, he was CEO of RTC, producers of an extensive line of fall protection systems. DSC is a consulting firm that has specialized in fall hazard control planning and training since 1984. Dr. Ellis is a member of the National Safety Council, Construction Division; a former president and current member of the Lower Delaware Valley chapter of the American Society of Safety Engineers (ASSE); and a two-time winner of the National

Safety Council Cameron Award for the Construction section. He is a member of several ANSI fall protection committees and represents the United States on the International Standards Organization (ISO) fall protection committee. His publications include the *Best Safety Directory* fall protection guidelines (since 1973) and the textbook *Introduction to Fall Protection,* published by ASSE, 1988, 1993. Mailing address: DSC, PO Box 445, Wilmington, DE 19899. Web page: www.FallSafety.com. Email: NigelEllis@DOL.net.

Teresa Emig, RN, BSN, COHN; Occupational Health Nurse, Grand Canyon National Park Lodges. Editorial Review Board for *Update Series IV* AAOHN. Ms. Emig is contributing author for the National Safety Council continuing education class, *Safety Principles for Occupational Health Nurses* and Speaker National Safety Council Congress *Gender Differences—Personal Protective Equipment,* 1993 and 1994. Mailing address: Grand Canyon National Park Lodges, PO Box 699, Grand Canyon, AZ 86023.

Paul Farrow; Vice President, Arthur D. Little of Canada Limited and Director of Environmental, Health, and Safety Consulting. Using his 20 years of expertise in a broad range of industries and knowledge of environmental, health, and safety auditing and assessment approaches, Mr. Farrow assists organizations in strengthening their EHS management systems and improving efficiency.

J. Terrence Grisim, CSP, CDS, CPSM, ARM; President, Safety Management Consultants. Mr. Grisim is a product safety consultant with 28 years' experience. He also does litigation consulting in the field of product safety. He has been Chairman of the National Safety Council Product Safety Committee since 1993. He was the technical advisor for NSC's *Product Safety Management Guidelines.*

Earl Hansen, CHCM, CSE, EdD; Professor of Safety Studies, Northern Illinois University, DeKalb, IL. Professor Hansen teaches seminars and courses in occupational safety and has served as a consultant and expert witness in safety/risk management, ergonomics, hazardous waste management, product liability, and materials handling. He has contributed to the Study Guide: Accident Prevention Manuals, 9th edition and 10th edition and the review questions and answers, 11th edition.

Jessica Herzstein, MD, MPH; consultant, President of Environmental Health Resources. Dr. Herzstein is certified by the American Boards of Internal Medicine and Preventive (Occupational) Medicine. She is Associate Clinical Professor of Medicine at Temple University School of Medicine. She has over 10 years experience in the fields of occupational and environmental toxicology and corporate health and safety development and oversight. Her toxicologic expertise has also been utilized in environmental issues including risk management and risk management of waste treatment facilities, CERCLA sites, and community exposure hazards. She has written and lectured extensively on toxicology and medical surveillance in environmental and occupational medicine. Dr. Herzstein is co-editor of the textbook *Environmental Medicine* (Mosby, 1995). She is currently authoring/editing two books on International Occupational and Environmental Medicine. Mailing address: Environmental Health Resources, 7 Lothrop Circle, Lexington, MA 02173.

James T. Knorpp, PE, CSP, is a safety and health consultant specializing in OSHA compliance and providing expert witness services. For the past 24 years, Mr. Knorpp has been an Area Director for the Occupational Safety and Health Administration. Mailing address: Knorpp Safety Services, 2149 Misty's Run, Keller, Texas 76248.

Richard S. Kraus, PE, CSP, CDS, BCFE; Safety and Fire Protection consultant. Mr. Kraus has over 30 years experience in the safety and fire protection industry, and serves as Safety and Fire Protection Codes Coordinator and Marketing Department Codes Group Safety Advisor for the American Petroleum Institute. In addition, he provides technical expertise to the National Safety Council, National Fire Protection Association, and the National Committee for Motor Fleet Supervisor Training. He has written many publications and presentations, including Chapter 26 of the *Accident Prevention Manual,* 11th Edition (National Safety Council, 1992.) Mr. Kraus has been elected to several *Who's Who* volumes including *Who's Who in Finance and Industry* and *Who's Who in South and Southwest.* He has been lecture/speaker at Delaware State, Maryland, Michigan State, and other academic institutions. Mailing address: 8712 Chippendale Court, Annandale, VA 22003–3807.

Gary Krieger, MD, MPH in Occupational Medicine/Toxicology, DABT; Manager and Principal, Health Systems Group, Dames & Moore. Dr. Krieger holds simultaneous certification by the American Boards of Internal Medicine, Preventive (Occupational) Medicine, and Toxicology. He is an Assistant Professor of Toxicology at the University of Colorado, Graduate School and has served as a visiting professor at Johns Hopkins School of Hygiene and Public Health and the Mayo Clinic. He has extensive academic training in occupational medicine and toxicology as applied to hazardous waste investigations, 10 years full-time experience in the fields of public health/environmental toxicology and risk assessment; and more than seven years experience with RCRA facility investigations and CERCLA risk assessments. He has Public Forum experience as an expert witness on incinerators, cogeneration, and risk assessments for heavy metals, PCBs, dioxins, chlorinated and nonchlorinated solvents, asbestos, radioactive materials total petroleum hydrocarbons, and polyaromatic hydrocarbons; and has written many publications and presentations, including eight chapters of *Hazardous Materials Toxicology: Clinical Principles of Environmental Health,* of which he is co-editor with J. B. Sullivan (Williams & Wilkens, 1991). Dr. Krieger is also

editor of *Accident Prevention Manual for Business & Industry: Environmental Management* (National Safety Council, 1995). Mailing address: Manager, Health Services Group, Dames & Moore, First Interstate Tower North, 633 Seventeenth Street, Suite 2500, Denver, CO 80202.

Curt Lewis, CSP, PE; Manager of Flight Safety; American Airlines, Inc. Mr. Lewis has more than twenty years experience as a professional pilot, safety engineer, and air safety investigator. Mr. Lewis has published numerous safety articles, is contributing editor to the American Airlines pilots magazine *Flight Deck*, and a frequent speaker at industry meetings and seminars, community service organizations, and corporations. He has been listed in several *Who's Who* volumes, including *Who's Who in Science and Technology* and *Who's Who in Finance and Industry.* Mr. Lewis is a professional member of the American Society of Safety Engineers, the International Society of Air Safety Investigators, and the System Safety Society. Mailing address: Corporate Manager, Flight Safety, American Airlines, PO Box 619616, Dallas/Fort Worth, TX 75216-9616.

Daniel G. Liddell, CIH; Sr. Industrial Hygienist—Ground Safety; AMR Corporation/American Airlines. Mr. Liddell has over ten years experience in comprehensive industrial hygiene practice in construction, aviation, industrial, and electronics components manufacturing. He has responsibility of all Industrial Hygiene activities for over 120,000 employees both in the United States and internationally, and holds Radiation and Laser Safety Officer designation for the state of Texas Bureau of Radiation Control. In addition he has earned a technical degree in Environmental and Public Health, and is a Certified Industrial Hygienist (CIH) in comprehensive practice by the American Board of Industrial Hygienist (ABIH). He has been listed in *Who's Who in Industrial Hygiene,* ABIH Roster of Diplomats, Adjunct faculty member University of Wisconsin—Eau Claire, member of the National American Industrial Hygiene Association, North Texas Chapter American Industrial Hygiene Association.

Fred A. Manuele, CSP, PE; President, Hazards, Limited, Arlington Heights, IL. Mr. Manuele retired in 1991 from Marsh & McLennan where he was a managing director and manager of M&M Protection Consultants. He is currently serving as a member of the Executive Committee of the Business and Industry Division and as a member of the Board of Directors of the National Safety Council. He is a former member of the Board of Directors of the American Society of Safety Engineers and a member and president of the Board of Certified Safety Professionals. He was given the Distinguished Service to Safety Award by the National Safety Council and was awarded the honor of Fellow by the American Society of Safety Engineers. In November, 1992, he was inducted into the Safety and Health Hall of Fame International. His publications include papers on hazards management and a book, *On the Practice of Safety,* 1993. Mailing address: Hazards, Limited, 437 S. Fernandez, Arlington Heights, IL 60005.

Linda M. Monteiro-Hopper, MS; consultant; Arthur D. Little, Inc. As a consultant for Arthur D. Little, Ms. Monteiro-Hopper is extensively involved in a wide range of projects. She is responsible for conducting environmental, health, and safety compliance audits, developing auditing and environmental, health, and safety programs and training materials, and conducting industrial hygiene monitoring and program evaluation. She is chapter contributor to American Industrial Hygiene Association Publication, *Industrial Hygiene Auditing*, and co-author of the environmental regulatory field guide for use in Arthur D. Little's course, *Understanding the U. S. Regulatory Framework for Key Environmental Issues.* Mailing address: Environmental Management, Arthur D. Little, Inc., Acorn Park, Cambridge, MA 02140-2390.

John F. Montgomery, PhD, CSP, CHCM, CHMM; Corporate Manager, Environmental Department, AMR Corporation/American Airlines. Dr. Montgomery is a writer and speaker on industrial safety and health topics and Assistant Professor/Lecturer at Central Missouri State University, Texas A&M University and other academic institutions. He has served in several safety positions with AMR Corporation, and has been elected to several *Who's Who* volumes, including *Who's Who in the Environment* and *Who's Who in the World.* Mailing address: Manager, Environmental Safety & Health, American Airlines, PO Box 619616 M/D 5425, Dallas/Ft. Worth, TX 75261-9616.

Thomas G. Natsch, CIH, Division Health and Safety Manager. Mr. Natsch has over 16 years experience in environmental health and industrial hygiene. He has previously worked in government, insurance, and industry where he had developed a broad based background and knowledge of both environmental and occupational health issues. He has extensive technical background and has developed and administered industrial hygiene monitoring programs, conducted on-site audits, provided field and client safety consultations and directed junior industrial hygienists and site safety officers. Mr. Natsch recently managed the Health and Safety program for a large remediation project at the U. S. Department of Energy Weldon Spring Remedial Action Project in Missouri. Mailing address: Dames & Moore, 11701 Borman Dr., Suite 340, St. Louis, MO 63146.

Daniel J. Papa, Jr., MBA, MSIM, BSIE; Consulting Engineer for Crown Power and Redevelopment Corporation. For the past 23 years, Mr. Papa has been the primary engineer responsible for domestic and some foreign facilities planning/manufacturing/distribution center layouts and warehousing interface. He is also responsible for materials handling solutions. Mailing address: Crown Center-Facilities Planning & Design, 2460 Pershing Road, Lobby Level, Mail Drop #272, Kansas City, MO 64141.

Charles E. Paulson, MS; Corporate Director of Safety, MMC Corporation/Hallmark Cards Inc. Mr. Paulson has been associated with the field of safety and health for almost 20 years. Mr. Paulson is a member of the National Safety Council's Construction Division. He is also Chair for the Kansas City Mechanical Contractor's Association Safety Committee, and a professional member of the American Society of Safety Engineers, Heart of America Chapter. Mailing address: 11100 Ash, Suite 100, Leawood, Kansas 66211.

Richard G. Pearson, PhD, F. Egr. S, CPE; consultant, Professor of Industrial Engineering and Director of Graduate Programs; Director of NIOSH training grant in occupational safety and ergonomics, North Carolina State University. Dr. Pearson is a member of the Human Factors Society, Ergonomics Society (U.K.); Aerospace Medical Association, American Society of Safety Engineers, and the American Nuclear Society. Mailing address: Department of Industrial Engineering, North Carolina, Box 7906, Raleigh, NC 27695–7906.

Kent W. Peterson, MD, FACOEM, FACPM; President, Occupational Health Strategies, Inc., Charlottesville, VA. Dr. Peterson is an internationally recognized expert in preventive, occupational, and environmental medicine. His firm, founded in 1984, provides management consulting, strategic planning, health information services, education, and training to Fortune 500 employers and health care companies worldwide. Dr. Peterson is currently President of the American College of Occupational and Environmental Medicine, Vice President of the American Board of Independent Medical Examiners, and Treasurer of the Medical Review Officer Certification Council. He is a Clinical Professor of Environmental Medicine at New York University and holds faculty appointments at Georgetown University and the Federal Executive Institute. He has authored more than 250 books, chapters, articles, and scientific presentations to professional societies. These include the *Handbook of Occupational Health & Safety Software,* 10th edition, published by ACOEM, and the *Handbook of Health Risk Appraisals,* 3rd edition, published by the Society of Prospective Medicine. Mailing address: OHS, Suite 400, 901 Preston Avenue, Charlottesville, VA 22903-4491.

Joy Prescott, MS; owner/consultant; On Target Consulting. Ms. Prescott has provided consultation for the development and ongoing teaching of Hazard Communication, Bloodborne Pathogen, and HAZWOPER Training Programs. She has conducted DOT Drug and Alcohol compliance audits. She has served as Director of Medical Programs for Continental Airlines, Inc. where she was responsible for the direction of all medical matters for the corporation. Mailing address: 9820 Memorial Drive, Suite C 62, Houston, TX 77024.

Richard Reber, MIS, ASP; consultant; Minnesota Safety Council. Mr. Reber provides consulting services through the Safety Council, including physical hazard audits, program evaluations, and program development. He also provides training for management and employees on various topics including General Safety Programs

(AWAIR), Supervisor Development, Lockout/Tagout, Confined Space Entry, Personal Protective Equipment, Introduction to OSHA, Injury/Illness Recordkeeping, Electrical Safety, and Effective Safety Committees. Mailing address: Minnesota Safety Council, 474 Concordia Avenue, St. Paul, Minnesota 55103.

Bonnie Rogers, DrPH, COHN, FAAN, Associate Professor in the School of Nursing and Public Health; Director of the Occupational Health Nursing Program at the University of North Carolina, Chapel Hill, North Carolina. Dr. Rogers is the author of *Occupational Health Nursing: Concepts and Practice*, the only textbook on the subject. She is the senior author of the book *Occupational Health Nursing Guidelines for Primary Clinical Conditions*. She is a strong advocate of occupational health nursing and serves on numerous editorial panels as an ethics consultant. Mailing address: Curriculum in Public Health Nursing, School of Public Health, 263 Rosenau, CB 7400, University of North Carolina, Chapel Hill, NC 27599-7400.

John Saylor, MA, CEAP; Manager of the American Airlines Employee Assistance Program, supervising seven clinicians who provide full internal EAP services to the airline and AMR's subsidiaries. Mr. Saylor organized and now maintains American's Psychiatric and Chemical Dependency PPO Networks, as well as overseeing the management of all aspects of addiction treatment provided under the company health plans. He is a Licensed Psychologist and Certified Employee Assistance Professional who has specialized in the treatment of addictions. Before entering the EAP field six years ago, he had been Executive Director of Parkside Lodge—Westgate, an addiction treatment hospital in Denton, Texas, and had previously worked at various addiction treatment centers over a period of 15 years. He is a Senior Member of the American College of Addiction Treatment Administrators.

James H. Schaarsmith, JSD, Principal in the Strategic Environmental Management Practice of Woodward-Clyde Consultants. Mr. Schaarsmith has over 25 years of experience in the insurance industry. For the past two years, his efforts have centered on environmental planning for business and the integration of environmental management with other management systems. He is an active member of the U. S. Technical Advisory Group to ISO's Technical Committee 207 (TC 207), and is currently participating in the meetings of Sub Tag 4 on Environmental Performance Evaluation and in the ISO 14000 Ad Hoc Legal Issues Forum.

Trey Shaffer, General Manager, EmTech Environmental Services, Inc., Ft. Worth, TX. Mr. Shaffer's expertise is in emergency response, remediation, and environmental design.

Les Virgil, MA, CEAP, LCDC; experienced presenter and trainer. Mr. Virgil's experience in the health care field includes positions as director of occupational programs for the Texas Commission on Alcoholism and Drug Abuse; president of Business Health Management, Inc.; regional manager for CNR Health, Inc.; and administrator of behavioral health management for the Harris Methodist Health System. His experience in managed behavioral health includes health maintenance organizations, preferred provider organizations, point of service or triple option policies, administrative services only, and exclusive provider organizations. He is experienced with access management, utilization management, quality assurance, claims processing, and the recruitment, contracting, organization, and maintenance of provider networks. His experience with employee assistance programs includes programs in government, heavy industry, and other business settings. He is an original board member, past vice president, and current member of the Employee Assistance Society of North America and past chair of the Arlington Task Force on Alcohol and Drug Abuse. He has published articles in *The Business Press,* Fort Worth; *Employee Assistance* magazine; and the *Employee Assistance Quarterly.* Mailing address: 4144 Goodnight Circle, Fort Worth, TX 76137.

Richard L. Wade, PhD, MPH; Vice President, Environmental Health Programs, Princess Cruises. Dr. Wade's professional experience includes four years as Director, Environmental Health, City of Seattle, and State of Minnesota; four years as Chief, California OSHA, and 15 years as a consultant and expert witness on risk management for international corporations. He served as a professor for 23 years at the University of Minnesota, University of Washington, and University of California. Mailing address: Princess Cruises, 10100 Santa Monica Boulevard, Los Angeles, CA 90067-4189.

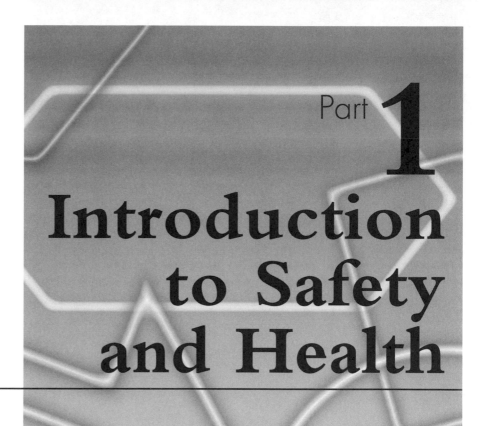

Part **1**

Introduction to Safety and Health

INTRODUCTION

The constant feature of safety and health is change. The historical overview of safety and health practices clearly illustrates that as society's perspective on the workplace has evolved, so have attitudes toward safety and health issues. The remarkable turnaround in workplace safety and health statistics in the twentieth century bears witness to this transformation in both attitudes and actual practices. The first two chapters in this manual present an overview of the historical and regulatory history of safety and health issues in the United States. As these chapters indicate, vigilance over safety and health issues has become an accepted standard of the modern workplace.

1 Historical Perspectives

Protecting life and promoting health is the basic mission of the National Safety Council. The Council's hallmarks over the years have been flexibility and the ability to create new programs to cope with ever-changing challenges. Many of the major safety and health problems that faced safety professionals at the beginning of the 20th century are virtually nonexistent today. However, occupational and environmental health, bloodborne diseases, and drug and alcohol abuse present new challenges to today's safety professionals.

In an imperfect world, there will always be risks, and the National Safety Council will continually strive to reduce the number and severity of those risks—no matter what the cause. In working to protect people from accidental death or injury, the Council seeks ways to ensure that everyone enjoys a safe and healthful environment.

In this, the eleventh edition of the *Accident Prevention Manual for Business & Industry,* the National Safety Council presents a compilation of knowledge and experience that forms part of the safety movement's general heritage. For specific information regarding occupational health and industrial hygiene, see two other National Safety Council books: *Occupational Health and Safety* and the *Fundamentals of Industrial Hygiene,* respectively.

PHILOSOPHY OF ACCIDENT PREVENTION

In medieval times, the master craftsman instructed apprentices and journeymen to work skillfully and safely because he knew the value of high-quality, uninterrupted production. However, the Industrial Revolution, which began in England during the 18th century, shifted the emphasis to faster and greater production and created the conditions that inspired the development of accident prevention as a specialized field.

The industrial safety philosophy developed because the hazardous work environment of early factories and other production and distribution sites produced an appalling rate of worker injuries and deaths. If these conditions had not been corrected to stop the waste of personnel and resources, the growing number of accidents and injuries would have staggered the imagination.

In the beginning, one way to encourage management to accept responsibility for preventing accidents was to pass workers' compensation laws. This "new" line of thinking held the employer responsible for a share of the economic loss suffered by an employee involved in an accident.

It was a rather short step from this approach to the realization that most accidents could be prevented and that the same industrial knowledge used to develop mass production methods also could be applied to accident prevention. Managers soon discovered that efficient production and safety were closely related. From this beginning grew the safety movement as it is known today.

The progress in reducing the number of accidents and injuries in the relatively short time since this movement began has exceeded the most optimistic expectations of early safety pioneers. The accidental death rate per 100,000 persons in the United States has decreased 57% over the past 82 years (*Accident Facts,* 1995).

Experience has shown that virtually any hazard can be overcome by practical safety measures. To further that belief, the National Safety Council continues its concerted efforts to prevent accidents and occupational illnesses.

In summary, here are six reasons for working hard to prevent accidents and occupational illnesses:

1. Needless destruction of life and health is morally unjustified.
2. Failure to take necessary precautions against predictable accidents and occupational illnesses makes management and workers morally responsible for those accidents and occupational illnesses.
3. Accidents and occupational illnesses severely limit efficiency and productivity.
4. Accidents and occupational illnesses produce far-reaching social harm.
5. The safety movement has demonstrated that its techniques are effective in reducing accident rates and promoting efficiency.
6. Recent state and federal legislation mandates management responsibility to provide a safe, healthful workplace.

THE INDUSTRIAL REVOLUTION

Until the 1700s, production methods were labor-intensive, with work being done by hand in cottages. Three developments were to change this way of life: (1) In England, inventors developed the spinning jenny in 1764; (2) the power loom was perfected in 1784; and (3) in America, Eli Whitney developed the cotton gin in 1792. These and other innovations ushered in what would later be called the Industrial Revolution. What began in Britain in the 18th century and spread to the European continent and the United States transformed the life of Western culture and society, and dramatically impacted traditional relationships between groups of people.

Specifically, the innovations in the processes and organization of production included:

- substitution of mechanical energy for animal sources of power, particularly steam power through the combustion of coal
- substitution of machines for human skills and strength
- invention of new methods for transforming raw materials into finished goods, particularly in iron and steel production and industrial chemicals
- organization of work into large units, such as factories or forges or mills. This made possible

Figure 1-1. Little was known in the early years of industrialization about the effects of work environment on the worker. Poorly lit and ventilated shops and accommodations that showed no awareness of ergonomic concerns were the rule rather than the exception.

direct supervision of the manufacturing process and an efficient division of labor. Paralleling these production changes were the altered technologies employed in agriculture and transportation.

Initially, this new way of organizing work was termed the "factory system." Later, however, when it reached larger and more complex scales, it was designated the Industrial Revolution by A. Toynbee (Toynbee, 1884). His nephew, Arnold J. Toynbee, is described as "the first economic historian to think of, and to set out to describe, the Industrial Revolution as a single great historical event, in which all the details come together to make an intelligent and significant picture."

Unfortunately, these changes in production methods with their need for masses of workers also created hazards never before encountered. These conditions greatly affected the history of occupational safety and health. Many health workers and industrial experts recognized the increasing need for hazard control.

The effects of the Industrial Revolution were first felt in the United States about a century after the revolution began in Great Britain. Before the 19th century, most families in the United States lived and worked on farms. Some industries had developed in the new country, namely printing, shipbuilding, quarrying, cabinetmaking, bookbinding, clock making, and the production of paper, chocolate, and cottonseed oil. However, it was the textile industry that introduced the new factory system into the United States, especially in New England where hundreds of spinning mills were built. As the Industrial Revolution continued its rapid growth, unsafe production methods exacted a heavy toll on the work force in terms of job-related injuries and deaths. (Felton, 1986). (See Figure 1-1.)

HISTORY OF U.S. SAFETY AND HEALTH MOVEMENT

During the last half of the 19th century, American factories were expanding their product lines and producing at previously unimagined rates. While the factories were far

Figure 1-2. When the safety movement began (c. 1900–1910), children commonly worked amidst heavy machinery—even without shoes or any protective equipment.

superior in terms of production to the preceding small handicraft shops, they were often vastly inferior in terms of human values, health, and safety.

In the area of human values alone, the facts make a grim picture. The 1900 census showed 1,750,178 working children between 10 and 15 years of age. Some 25,000 were employed in mines and quarries; 12,000 in making chewing tobacco and cigars; 5,000 in sawmills; 5,000 at or near steam-driven planers and lathes; 7,000 in laundries; 2,000 in bakeries; and 138,000 as servants and waiters in hotels and restaurants. These children often worked 12 to 14 hours a day; there were no health or safety guidelines in effect, even for children under age 10 (Figure 1-2).

The lag between the emergence of new working methods and the creation of health and safety standards was probably inevitable. The tools of mass production had to be invented and applied before anyone could begin to imagine the problems they might create. In turn, the problems had to be known before corrective measures could be considered, tested, and proved. Thus, for some

time deaths and injuries were accepted as part of "industrial progress"—one of the costs of doing business.

While this revolution in the work environment was taking place, the thinking of the public, management, and the law still reflected the past, when the worker was an independent craftsman or a member of the family-owned shop. Common law provided the employer with a defense that gave the injured worker little chance for compensation. The three doctrines of common law that favored the employer were:

Fellow servant rule—Employer was not liable for injury to an employee that resulted from negligence of a fellow employee.

Contributory negligence—Employer was not liable if the employee was injured due to his own negligence.

Assumption of risk—Employer was not liable because the employee took the job with full knowledge of the risks and hazards involved.

In 1906, the Pittsburgh Survey, sponsored by the Russell Sage Foundation, marked the first attempt to pinpoint the serious nature of occupational accidents and deaths. The survey team realized they did not have the resources to survey the problem throughout the United States and instead concentrated on Allegheny County in Pennsylvania. They constructed a "death calendar" of the county, showing that industrial accidents accounted for an average of nearly two deaths per day throughout the year. The number of crippling injuries was far higher. If this was the case in only one county, people asked, what must the situation be for the entire United States? The Pittsburgh Survey made it clear that the accident and death rate was serious and gave the safety movement a much-needed boost.

In large industrial centers, the ugly results of industrial accidents and poor occupational health conditions became more and more obvious. Individuals and public and private organizations raised their voices to protest these conditions. Though some employers denied the problem existed, wiser managements began to take action to improve their work environments.

As early as 1867, Massachusetts had begun to use factory inspectors. Ten years later, the state passed a law requiring employers to safeguard hazardous machinery. During 1877, Massachusetts also passed the Employer's Liability Law making employers liable for damages when a worker was injured. However, court decisions based on common law often let the employer escape liability.

From 1898 on, there were additional efforts to make the employer financially liable for accidents. In his presidential message of 1908, Theodore Roosevelt stated: "The number of accidents which result in the death or crippling of wage earners is simply appalling. In a very few years it runs up a total far in excess of the aggregate of the dead in any major war." Roosevelt's message acquired force when his social legislation passed that year in Congress. Although this first workers' compensation law covered only federal employees, it set a precedent for state laws to follow.

The first bill for workers' compensation (the Wainwright Law) was passed in New York in 1910, but it was declared unconstitutional by the New York Court of Appeals. The court ruled that the law violated both the federal and New York State constitutions, "because it took property from the employer and gave it to his employee without due process of law." On the same day the 1910 act was declared unconstitutional, March 25, 1911, a devastating fire in New York City's Triangle clothing factory killed 146 employees. This disaster, called the Triangle Fire, outraged the public and spurred demand for factory legislation and health and safety reform. After an amendment to the state constitution was approved in 1913 at the general election, a compulsory Workmen's Compensation Act finally became effective in mid–1914.

In 1911, Wisconsin passed its first effective Workers' Compensation Act, but it was declared unconstitutional by the Wisconsin Supreme Court within a few months. New Jersey and Washington also passed laws that year.

At first, the courts continued to declare such laws invalid because they conflicted with the due process of law provisions of the 14th Amendment. However, after the United States Supreme Court in 1916 declared workers' compensation to be constitutional in *New York Central Railroad Co. v White,* 243 U.S. 188, many states passed compulsory workers' compensation laws.

By the late 1800s and early 1900s, the railroads had crisscrossed the East and West but exacted a heavy toll among employees. It was said that a man was killed for each mile of track laid. By 1907, annual railroad employee deaths had reached 4,353.

Industry experts made some progress on the technical side of the problem. The railroads adopted air brakes and the automatic coupler well before the turn of the century. They also worked on guarding and fire prevention. They came to realize, however, that guarding was not the total answer. People's actions were equally important factors in creating accident situations.

At the same time, insurance companies started to relate the cost of premiums for workers' compensation insurance to the cost of accidents. Management began to understand the close relationship between successful production and safe production.

During the first decade of the 20th century, two giant industries, railroads and steel, began the first large-scale organized safety programs (Figure 1-3). From this period comes one of the historic documents of safety. In 1906, Judge Elbert Gary, president of the United States Steel Corporation, wrote:

> The United States Steel Corporation expects its subsidiary companies to make every effort practicable to prevent injury to its employees. Expenditures necessary for such purposes will be authorized. Nothing which will add to the protection of the workmen should be neglected.

The Association of Iron and Steel Electrical Engineers, organized soon after this announcement, devoted considerable attention to safety problems.

Birth of the National Safety Council

The year 1912 proved to be a landmark for accident prevention. In the previous year, the Association of Iron and Steel Electrical Engineers (which had been formed in 1907) had called for a general industrial safety conference on a national scale. The result was the First Cooperative Safety Congress, which met in 1912 in Milwaukee. The following year, at a second national meeting in New York, the delegates formed the National Council for Industrial Safety. Shortly afterward, members changed the organization's name to the National Safety Council and broadened its program to include all aspects of accident prevention. The program also includes occupational health. Yet, it must be remembered that the Council was the creation of industry and that its activities have always been heavily concentrated on industrial safety.

The group that met in Milwaukee and New York was composed of a few safety professionals, some management

Figure 1-3. Because of hazards encountered in railroad operations, the railroad industry was one of the first industries to develop organized safety programs. This photograph shows some of the hazards associated with building a bridge at Rockford, Illinois, in 1869. (Courtesy Chicago & North Western Railroad.)

leaders, public officials, and insurance specialists. Their one point in common was a desire to attack a problem which most people considered either unimportant or insoluble. The determination of these safety pioneers helped to create the safety movement as we know it today.

Actually, the members' underlying objective in forming the National Safety Council in 1913 was standardization. Thus, the primary purpose of the Council was to provide an avenue of communication, an exchange of views, and various solutions to common problems in accident prevention.

In 1918, the Council conducted the first national survey of state, federal, and municipal regulations, together with a study of insurance recommendations, technical association recommendations, and the practices of industry. The survey revealed utter chaos in industrial safety, and a clear need for industry-wide methods and practices.

Realizing its own limitations, the Council consulted the National Bureau of Standards, which agreed to call a conference to discuss establishing procedures to standardize safety methods and practices. Meeting in Washington, DC, in 1919, the attendees expressed the belief that uniform industry standards were not only desirable but essential to promote effective worker health and safety. The conference voted to formulate safety standards under the auspices and procedures of the American Engineering Standards Committee (AESC), which had been formed in 1918 by five engineering societies and three governmental departments.

American Standards Association Beginnings

In 1920, the National Safety Code Program was brought into the AESC. This resulted in the first reorganization of the Committee and marked the beginning of what later became the American Standards Association (ASA). A national code committee was organized to suggest the initial safety code projects. This later became the Safety Codes Correlating Committee, the first of ASA's 18 standards boards. Bringing manufacturing companies and trade associations into AESC membership also initiated a broader program of engineering standards. These steps launched an enlarged national standardization program.

In 1928, recognizing that the extending activities called for a more formal type of organization, the member groups reorganized the AESC as the American Standards Association, now known as the American National Standards Institute (ANSI). ASA continued to be an important partner in the safety movement. This group handled the "materials" aspect of safety while the National Safety Council focused on the "people" portion of accident and occupational illness prevention.

Accident Prevention Discoveries

As industry developed some experience in safety, it discovered that engineering could prevent accidents, that employees could be reached through education, and that safety rules could be established and enforced. Thus the "Three E's of Safety"—engineering, education, and enforcement—were developed.

Two of the many breakthroughs that safety groups made during the 1900–1990 era were the identification of occupational diseases such as mercury and lead poisoning, and the efforts to control these hazards. Other discoveries have had equally profound implications. For example, asbestos was found to be a carcinogen that causes lung cancer and another type of cancer, mesothelioma. Health professionals also studied the effects of chromium compounds and beryllium on industrial workers.

These and other discoveries led safety and health professionals to argue that savings in compensation costs and medical expenses would repay safety expenditures many times over. Thoughtful business leaders soon learned that these savings were only a fraction of the financial benefits to be derived from accident prevention work. Newer, more effective techniques have been discovered and are described elsewhere in this volume. See especially Chapters 3 and 11 through 14 in this volume, and Chapter 5 in the Engineering and Technology volume.

ACCELERATION OF THE DRIVE FOR SAFETY AND HEALTH

Industrial safety received wide acceptance in the years between World Wars I and II. During World War II the growth of safety procedures and policies intensified, particularly as the federal government began encouraging its contractors to adopt safe work practices. As industry expanded to meet the needs of the war effort, additional safety personnel were hastily trained in an effort to keep pace with the new risks and hazards in the workplace. The acceptance of safety activities as part of the industrial picture did not diminish with the end of the war. By then, the importance of safety to quality production was well established. The small handful of people dedicated to safety in 1912 had grown to tens of thousands of trained personnel. In 1948, for example, Admiral Ben Moreell, then president of Jones and Laughlin Steel Corporation, wrote:

> Although safe and healthful working conditions can be justified on a cold dollars-and-cents basis, I prefer to justify them on the basic principle that it is the right thing to do. In discussing safety in industrial operations, I have often heard it stated that the cost of adequate health and safety measures would be prohibitive and that "we can't afford it."
>
> My answer to that is quite simple and quite direct. It is this: "If we can't afford safety, we can't afford to be in business."

A discussion of current United States safety legislation follows later in this chapter under Safety and the Law, and also in Chapter 2, Regulatory History and Compliance.

One by-product of organized safety activities has been a growing interest in safety engineering on the part of colleges and universities. Many schools offer degrees and advanced courses on this subject and are contributing to a higher standard of knowledge among professionals in the field.

In addition, the World War II labor shortage dramatically brought home to management the magnitude and seriousness of the problem of off-the-job accidents to industrial employees. The wartime theme of the National Safety Council, "Save Manpower for Warpower," focused attention on the need to reduce off-the-job accidents in order to maintain efficient, safe production on the job.

An increasing number of employers are including off-the-job safety in their overall safety programs. Companies realize their operating costs and production schedules are affected almost as much when employees are injured away from work as when they are injured on the job. Off-the-job safety generally is an extension of a company's on-the-job safety program and is intended to educate the employee to follow, in outside activities, the safe practices used on the job. Companies have found that on-the-job and off-the-job programs complement each other.

From the earliest days of industrial safety, it has been difficult to make a clear distinction between illness and injury (accident hazard). Is dermatitis an injury or an illness? What about hernias, hearing loss, and heart trouble? Inevitably, safety professionals have become interested in many health problems on the borderline between illnesses and injuries. In 1939, the American Industrial Hygiene Association was established to promote the recognition, evaluation, and control of environmental stresses arising in or from the workplace (Figure 1-4).

EVALUATION OF ACCOMPLISHMENTS

Because safety factors are complex, no simple rating scale can yield all the answers to the question, "What has the safety movement accomplished?" Instead, an attempt to answer the question must be made by assembling several kinds of data.

First, the question must be asked, "Has the safety movement, in fact, really helped to prevent accidents?" The answer to that question is a clear "Yes!" (Figure 1-5). If the annual accidental death rate per 100,000 of population recorded in 1912 had continued, more than 3,700,000 additional accidental deaths would have occurred since that time. Instead, the death rate for persons of normal working age—25 to 64 years—has steadily declined by some 63%, while the rate for all ages of the entire population has declined 57%. Medical progress accounts for some of this gain, but the larger part is certainly the product of organized safety work.

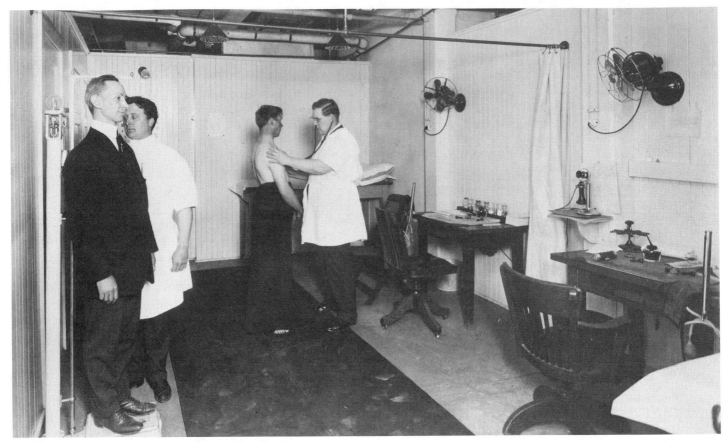

Figure 1-4. As the relationships between health and employment hazards became recognized, employers began to provide medical examinations.

Since World War II, the number of work-related deaths per 100,000 population, standardized to the age distribution of the population in 1940, also has decreased steadily. This fact indicates that the risk of on-the-job death has declined for the population as a whole. Part of the progress made in lowering the overall death rate, however, can be attributed to the rapid growth in recent years of the economy's service sector with its lower death rate, and the decline of some fairly high-risk segments of the manufacturing sector. In 1945, 43% of the nonagricultural workforce was in production-related industries (mining, construction, and manufacturing). By 1994 that proportion had declined to 21%.

A clearer picture emerges by looking at the trends on a more detailed level. Table 1-A shows a significant decline in death rates within the major industry groups. In six of the seven private sector groups, death rates have been reduced by 75% or more over the past 49 years. This clearly indicates, by one criterion, the effectiveness of the organized safety movement.

Long-term trends in nonfatal occupational injury rates cannot be examined because of a break in continuity of the historical statistical series. Until the early 1970s, injury rates were based on the voluntary American National Standards *Method of Recording and Measuring Work Injury Experience,* ANSI Z16.1. With the passage of the Occupational Safety and Health Act of 1970, it became mandatory for most private-sector employers in the United States to keep occupational injury and illness records in accordance with OSHA record-keeping requirements.

A clear trend has not yet emerged in the occupational injury and illness incidence rates published by the Bureau of Labor Statistics since 1972. Business cycles and changes in the distribution of the labor force among industries can mask any short-term changes in rates caused by more effective or more intensive safety efforts.

The Dollar Values

It has been estimated that the annual cost of occupational accidents in the United States exceeds $120 billion. If the

Table 1-A. Work Deaths per 100,000 Workers			
Industry Group	**1945**	**1994**	**Percent Change**
Agriculture, forestry, and fishing	53	26	–51
Mining and quarrying	187	27	–91
Construction	126	15	–88
Manufacturing	19	4	–79
Transportation and public utilities	52	12	–77
Wholesale and retail trade	10	2	–80
Services	20	2	–90

Source: National Safety Council, *Accident Facts,* 1946 and 1995 editions.

Figure 1-5. These women working during World War I are shown risking their lives on unsecured scaffolding. Note also the version of protective footwear of that era. (Courtesy Women's Bureau, National Archives.)

1912 accident rate had continued unchanged and if there had been no organized safety movement, this annual cost could easily have been two to three times as great, even in constant dollars.

Against such dollar savings, the relatively small expenditures for safety throughout the United States provide a striking contrast. Each dollar industry spends for safety may be returning a clear profit of up to several hundred percent.

Industry and Nonwork Accidents

Directly and indirectly, industry bears a substantial part of the cost related to nonwork accidents and their prevention. Although the National Safety Council is the creation of industry and largely supported by it, the Council, along with state and local safety organizations, plays a major role in the fight to prevent such accidents. Industry is a major supporter of the efforts to inform the general public on safety issues through the press, radio, and television.

This nonwork accident-prevention campaign is having a definite impact on public safety. From the time records on nonwork accidents were first kept in 1921, both home and public accident death rates have substantially declined.

Although industry has been a large contributor to this successful work, it has also been a major beneficiary. The reduction in nonwork injuries, illnesses, and deaths has lessened disruption of the labor force, and has reduced hardship among employees, consumers' loss of purchasing power, and tax burdens required to support hospitals and relief agencies.

RESOURCES FOR SAFETY

Safety statistics measure what has been accomplished in this area. Safety resources describe the tools, methods, and knowledge developed for safety and health professionals in their efforts to overcome future accident and occupational illness problems.

Knowledge and Experience

This *Accident Prevention Manual for Business & Industry,* for example, is an accumulation of facts and experiences that are a part of the safety and health movement. Its purpose is to present key points of knowledge to people interested in safety, whether they are students new in the field or advanced and experienced practitioners.

An individual using this Manual can find better answers to a wider range of industrial safety problems than were available to the wisest and best-trained professional safety practitioner several decades ago. Yet, even this Manual cannot contain all of the knowledge available to fight the never-ending war against accidents and occupational illnesses.

Other material may be found in numerous pamphlets, books, and periodicals published by safety and health organizations, government agencies, and insurance companies, and in the studies and directives of individual industrial concerns. The literature of various trades and professions is likewise rich in safety information. A list of handbooks is presented in Safety Tables, an appendix of the Engineering and Technology volume. At the end of this chapter is a list of the general safety books used as sources of questions for the Certified Safety Professional examination.

The National Safety Council offers a series of training courses, at both the beginning and advanced levels, for professionals. The Council also offers extensive consulting services, books, and software.

Finally, through conferences, technical seminars, newsletters, and other publications, professional safety engineers, executives, supervisors, and rank-and-file employees regularly exchange safety information. The annual National Safety Congress and Exhibition is an excellent means of enhancing professional development.

The Heritage of Cooperation

The safety movement would be far less effective if its members had concealed their discoveries from their colleagues who worked in competing companies. It was teamwork that created the safety activities of the Association of Iron and Steel Electrical Engineers. It was broadened teamwork that organized the first Milwaukee Conference, which led to the formation of the National Safety Council and other safety-related organizations.

Effective accident prevention requires cooperation at all levels of industry and government. Through the Council and other safety organizations, safety professionals meet to exchange ideas, develop safety publications, and stimulate one another in friendly competition. The tradition that there should be "no secrets in safety," no denial of help even to a competitor when it involves saving lives, is one of the great strengths in the safety movement.

Goodwill

In its early days, safety did not rank highly among management concerns. Today, a significant part of the safety professional's capital is the prestige and goodwill associated with this area that have accumulated over the years, making management more receptive to safety proposals and expenditures. Where yesterday's safety pioneers had to battle management every step of the budgetary way, today's safety professionals usually obtain a far more sympathetic hearing regarding health and safety issues.

Professionalism

Dedicated safety and health professionals continue to be the most valuable asset of accident and occupational illness prevention. Their ranks have grown. In 1995, membership in the American Society of Safety Engineers (ASSE) was over 31,000. This organization, dedicated to these professionals' interests and development, has 136 chapters in the United States and Canada. Individual membership is worldwide. Other professional societies include the National Safety Management Society, the Board of Certified Hazard Control Management, and the System Safety Society.

In 1968, the ASSE was instrumental in forming the Board of Certified Safety Professionals (BCSP). Its purpose is to certify qualified safety people as professionals once they meet strict educational and experience requirements and pass an examination. Similarly, professional certification of industrial hygienists (CIH) was sponsored by the American Industrial Hygiene Association (AIHA). Both the ASSE and the AIHA are described in the Appendix, Sources of Help.

Advancement of Knowledge

The tremendous increase in scientific knowledge and technological advancement since the 1950s has added to the complexities of safety work (Figures 1-6a and 1-6b). Prevention and control measures have oscillated between those that emphasize environmental control or engineering and those that emphasize human factors. From this, several important trends in safety work and the safety professional's development have emerged. All are discussed in subsequent chapters of this volume.

- First, more emphasis is being placed on analyzing the loss potential of any organization or projected activity. Such analysis requires the ability (1) to predict where and how loss- and injury-producing events will occur and (2) to find ways of preventing such events.
- Second, industry is developing more factual, unbiased, and objective information about loss-producing problems and accident causation to help those who are ultimately responsible for worker health and safety make sound decisions.
- Third, management is making greater use of the safety and health professional's knowledge and assistance in developing safe products. The application of the principles of accident causation and control to product manufacturing is assuming more importance because of the rise in product liability cases, the recent legal emphasis on the concept of negligent design, and the potential impact of the life-cycle analysis of a product on the environment.

To identify and evaluate the magnitude of the safety problem, safety professionals must be concerned with all facets of the problem—personal and environmental, transient and permanent. This perspective will help to determine the causes of accidents or to identify loss-producing conditions, practices, or materials. On the basis of this collected and analyzed information, safety professionals propose alternate solutions, together with recommendations founded on their specialized knowledge and experience, to those who have decision-making responsibilities.

Therefore, application of this knowledge—whether to industry, transportation, the home, or in recreation—makes it imperative that those in the safety field be trained to use scientific principles and methods to achieve adequate results. Most important, they need to have the knowledge, skill, and ability to integrate machines, equipment, and environments with people and their capabilities.

In performing these functions, safety and health professionals draw upon specialized knowledge in both the physical and social sciences. They should know how to apply the principles of measurement and analysis to evaluate safety performance and must have a fundamental knowledge of statistics, mathematics, physics, chemistry, and engineering. Students in safety and health degree programs should be certain these subjects are included. Safety and health professionals must know and understand management systems so they can properly advise line managers concerning safety and health management.

They also require training in the fields of behavior, motivation, communications, and management principles along with the theory of business and government organization. Their specialized knowledge must include a thorough understanding of the causative factors contributing to accidents and occupational illnesses as well as methods and procedures designed to control such events.

Safety professionals also need a good general education if they are to meet future challenges. The population explosion, problems of urban areas and future transportation systems, weakening of the family structure, decline of respect for authority, and an uncertain economy, coupled with the increasing complexities of everyday life, will create many problems and stretch safety professionals' creativity to its maximum. They will need all their skills if they are to provide knowledge and leadership to conserve life, health, and property.

Training for the safety and health professional of the future can no longer be solely "on-the-job" or one-on-one education. It must include specialized undergraduate courses that lead to a bachelor's or higher degree. Training courses, such as those offered by the National Safety Council, will continue to educate a large number of people who began performing safety functions as part of their jobs and who need initial or advanced training in certain specialized areas.

A large number of four-year U.S. colleges and universities offer courses in safety and health, and several dozen offer a bachelor's or higher degree in safety. Two-year community colleges offer associate degrees or certificates for courses designed for the safety technician or part-time administrator.

ACHIEVEMENTS OF THE SAFETY MOVEMENT

The safety movement has helped save more than 3 million lives. It is saving industry and its employees billions of dollars every year. It faces the future with numerous resources for preventing and controlling accidents and occupational illnesses—resources in know-how, teamwork, goodwill, and education programs that produce trained and dedicated safety workers.

Figure 1-6a. In 1919, women making small parts for telephones sat in uncomfortable chairs without back support, working at unguarded, inadequately lit machinery.

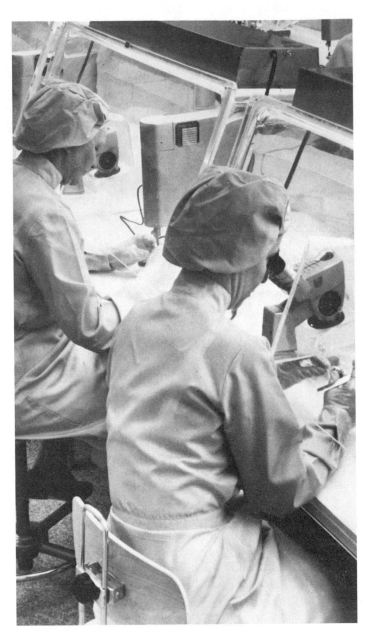

Figure 1-6b. In 1969, workstations were better lit and more comfortable. Improvements continue today. (Courtesy Western Electric Company.)

The safety movement has, therefore, done much to meet the double challenge presented to it. The movement has been able to deal with current accidents and occupational illnesses and to build a solid foundation for the long-range attack upon these problems in the future.

To answer the question, "What has the safety movement accomplished?" we look at continued growth in safety awareness and accident reduction. To answer the question, "Where does the safety movement stand?" we must look at what is wrong, as well as what is right, with the present situation. The answer can be found by comparing where the safety movement now stands to where it ought to be. The first point to be considered is simple and grim:

Accidents from all sources still cost this country more than 92,200 lives a year, cause about 18.6 million disabling injuries, and account for a total financial loss of more than $440.9 billion (*Accident Facts*, 1995). Work accidents destroy more than 5,000 lives a year. Disabling injuries in work accidents affect about 3.5 million persons annually and cost more than $120.7 billion (1994 figures).

In recent years, the ratio of off-the-job deaths to on-the-job deaths has been about 7.2 to 1; more than half of the injuries suffered by employees occurred off the job. In terms of time lost, all injuries to workers, both on and off the job, resulted in a loss of about 100 million staff days of work per year.

Accident rates vary widely from industry to industry and from company to company. The wholesale and retail trade, services, finance, insurance, and real estate industries all have occupational injury and illness incidence rates below the private sector average. On the other hand, rates are above average in construction, agriculture, manufacturing, transportation and public utilities, and mining industries.

Injury incidence rates by size of establishment are lowest for businesses with 1 to 19 employees but rise steadily until they reach a maximum in establishments of 100 to 249 employees. The rates then decline steadily as establishment size continues to increase (U.S. Bureau of Labor Statistics, 1995).

Small Establishments

As has been stated, businesses with 100 to 249 employees have proportionately more work injuries than large corporations or companies with 1 to 19 employees. As a general rule, companies—large or small—that ignore safety and health efforts will have more than their share of accidents and occupational illnesses.

The serious safety and health problems of small enterprises are widely recognized, and the National Safety Council has devoted considerable effort to reducing accident and illness rates in this sector. One way has been to establish a liaison between the National Safety Council and the trade associations representing many small companies.

Certain aspects of the small-company problem can be stated with assurance:

1. The small establishment may not be able to employ specialized safety and health personnel to deal with its accident and occupational health problems.
2. The number of accidents or the financial position of many small companies makes it difficult to convince them that spending money for proper equipment, layout, guarding, and other elements is important.
3. Managers of small operations deal with a host of problems and seldom have the expertise or time for the proper study of accidents and occupational illnesses and their causes.
4. In small units, statistical measures of performance are unreliable. As a result, it is difficult to produce clear-cut evidence of the cost of accidents versus the effectiveness of accident-prevention work. In other words, a small operation may have, by luck, a good or bad accident record over a few years, whether or not its safety program is sound.

These aspects present serious obstacles to any progress in worker health and safety. They are not, of course, excuses for failure to prevent accidents and occupational illnesses. The trade association approach offers the best hope for improvement in this sector.

Labor's Participation in Safety and Health

One of the primary goals of organized labor has always been to protect the safety and health of its members. Many of today's international unions were originally organized to deal with and improve severely hazardous situations in the workplace. They have a sincere desire to work with management on methods to prevent occupational injury and illness to all workers.

In 1949, the National Safety Council issued a policy statement declaring the common interest of labor and management in accident prevention. Even before this date, representatives of leading labor organizations served as members of the Council's governing boards. In 1955, a Council Labor Department and a Labor Conference (now known as the Labor Division) were formally established. The Labor Division serves as a vital link between industry management and the nation's labor unions.

Labor Division representatives review products, training materials, and policy statements. The Division shares information with labor leaders and nearly 500 volunteers from international and local labor unions. These groups have combined their educational efforts to help organized labor's millions of members improve the quality of their lives, both on and off the job. Some unions have done extensive safety work, published printed matter, and released films or videos that promote the safety and health movement.

The National Safety Council's Labor Division and its Industrial Division, along with other affected divisions, often prepare Council position statements on such matters as standards action, oversight testimony, publicity releases, and other areas bearing on occupational safety and health. As a result of these efforts, Council positions on health and safety issues are recognized as representing all elements of society, which gives them even greater impact on administrative agencies and legislative bodies.

As a result of a program started in January 1978 with a symposium of leaders from government, industry, and organized labor, the National Safety Council has launched an extensive inquiry to determine causal factors of occupational injuries. Because most data in the past have cataloged only types of injuries, the focus of this program is twofold: (1) to change investigatory and reporting methods and (2) to provide an information exchange bank that lists the factors actually causing injury or occupational illness.

Scope of Safety and Health Committees

The purpose of an occupational safety and health committee is to make the workplace a safer, healthier environment. Some people may want the committee to address matters not directly connected with safety and health. The committee must be careful to stay clear of these side issues, which can detract from its ability to handle its own work objectively. Other issues, such as employee relations problems, should be kept separate from safety- and health-related items or concerns.

The committee also must be careful not to assume the employer's ultimate responsibility for providing a safe and healthful workplace. The committee should be active and encourage management to develop a system for responding to committee recommendations, but this should not be interpreted as reducing the employer's liability. Many unions have included language that specifies the committee structure and duties in labor agreements with the employer in an attempt to better define some of these critical points. A similar system occurs in many European countries, e.g., France, where there are legally mandated health and safety committees.

Experience has shown that certain elements are necessary for a joint safety and health activity to be successful in any workplace:

- A sincere commitment to the safety and health effort must be displayed by both labor and management leaders in the workplace.
- Specific roles and responsibilities should be defined for all committee members.
- An effective communication link must be established between the committee members and all employees in the workplace. This includes feedback to workers from committee members.
- Measurable, realistic goals and objectives should be established for committee activity.

It is important that the committee be prepared to evaluate their progress and effectiveness on a regular basis and to modify or change their activities as required. Because of changes in the culture of many workplaces, such as the introduction of Total Quality Management, or similar programs, those involved with safety and health activities are becoming more well known as teams, rather than as committees. Regardless of the terminology used, the basic concept involves people working together to find and reduce or eliminate workplace hazards that affect employees' health and safety and cost employers in lost time and money every day. Joint safety and health programs can provide a win-win atmosphere for labor and management in most workplaces.

Unions and Safety Committees

Because assigning total responsibility for a union's safety program may be too much for one person, many unions have organized committees to seek out and provide corrections for job hazards. Workers concerned about hazards in the workplace often do not know where to turn for help and information. The union steward or president may not know much about correcting these problems, and the contract may make no provisions for filing grievances about these matters. By setting up a safety and health committee, the union establishes a regular procedure for dealing with problems. Members know where to go first to get matters resolved. Many problems can be resolved immediately without relying on government agencies.

Companies have structured their safety and health committees in as many different ways as there are different workplaces. The only constant in developing these groups is whatever works in that particular worksite. Some committees may work with management, others may be totally separate. In fact, many unions have discovered that in some workplaces, forming two occupational safety and health committees—one joint committee for coordination with management and one limited to union members—works even better.

Besides protecting workers from hazards, a union occupational safety and health committee can also encourage interested members to get involved in union activities. The union safety and health committee shows employees that the union is concerned with their safety and health. Workers will realize that by participating actively in the union's functions they are helping to improve their own work environments.

Independent Union Safety and Health Committees

A union occupational safety and health committee is totally independent of management and is not to be confused with a joint (union and management) safety and health committee. A union committee can be created without special contract language. The scope, function, and constitution are entirely matters for the union to decide.

A union safety and health committee should be elected or appointed according to the union constitution. To be effective, the committee should be fully integrated into the local union organization. It should be involved at contract negotiation time and should meet regularly with the grievance committee. If a joint union-management committee is also established, some workers should be members of both committees, which makes it easier to coordinate activities.

An independent committee can be important for several reasons:

- It allows representatives of the union to meet and consider the safety of the workplace, without any interference from management.
- It gives the union its own forum to discuss and set priorities and strategies for dealing with workplace hazards.
- It allows the committee members to gain appropriate expertise in researching hazards and seeking effective solutions. (Note that there is an important difference between gaining expertise and becoming an expert.)
- It can be used to monitor the performance of a joint union-management committee, if one exists.
- If a joint committee runs into roadblocks or becomes ineffective and the union side withdraws, the union will already have a structure for handling safety and health concerns.

A union and its members should not hesitate to form a safety and health committee. The OSHAct not only encourages worker participation in workplace safety and health efforts, but makes it illegal for an employer to punish or discriminate against a worker for job safety and health activities, such as participating in a workplace safety and health committee.

Joint Union-Management Safety and Health Committees

In addition to a local union committee, the union and management at a workplace may choose to form a joint committee to handle occupational safety and health issues. It makes sense to set up a formalized joint committee because safety and health in the workplace depend on communication with the employer. It is important, however, to be sure that the committee is set up in such a way that both parties can freely express their views and positions. This independent posture is often established through specific contract provisions.

Because every work situation is different, joint committees should be organized to meet the specific needs of their particular workplace. Both labor and management should consider several guidelines when establishing a joint committee at a workplace.

- Union and management should have an equal number of members serving on the committee.
- The union should elect or select the members who will serve as their representatives on the committee.
- The role of committee chairperson should be rotated between labor and management or the committee may choose to have co-chairs.
- Management should consider appointing committee representatives who have enough authority to make real decisions about projects or spending money to avoid creating delays or unnecessary interference.
- The committee should be able to recommend corrections or request assistance with any occupational safety and health concern or issue.
- Union and management should have an equal voice in the decision-making process and in planning committee actions and agendas.
- There should be a method or procedure established to monitor and evaluate the effectiveness of the joint committee activities.

Merely establishing a joint safety and health committee does not guarantee that it will be effective. During the planning and early development phases, it is critical that certain commitments or guarantees are obtained from the union leadership and from management if the committee is to be successful. Many times these guarantees are stated in a labor agreement and may include some of the following:

- Management should issue a policy statement directly related to safety and health. Usually these statements cover issues like implementing committee decisions

or responding quickly to employee concerns or complaints.
- The committee should be funded by the employer with committee members being compensated at their normal rate of pay for time spent on committee activities or projects.
- The committee should have the right to request assistance, such as environmental monitoring, from qualified specialized personnel.
- The committee should have access to useful information in company files, such as monitoring or exposure records, accident and injury reports, and records or lists of chemicals used in the workplace.

Statistics, Standards, and Research

Statistical data on industrial accidents have been compiled by the National Safety Council for more than 64 years. Analyses are computed annually and published in *Work Injury and Illness Rates* and *Accident Facts.*

Some industries through their trade associations have recorded accident rates for almost 60 years. In most instances, even the divisions of an industry can record specific numbers and types of accidents and compare their experiences with national averages.

A large number of standards by the American National Standards Institute (ANSI) and other professional organizations relate to safety. Continuing research over the years has kept these standards in line with current industrial developments and the creation of new products and materials.

The Council's Research and Statistical Services (RSS) group has developed several survey instruments that measure employee satisfaction with an organization's safety and health operations and management practices. Survey results are maintained in a data base that employers use to rank their employees' perceptions about the adequacy of their safety and health programs against those of a nationwide employee sample.

The RSS group also conducts consultation projects for business and industry. Recently, the group completed a seminal study focusing on the risk of occupational fatalities associated with hazardous waste site remediation. The findings showed that the fatality risks to workers engaged in remediation were at least a thousand times greater than the human cancer risks associated with hazardous waste sites. The group also conducts special studies focused on critical issues, such as a nationwide survey of employers to investigate the acceptance, makeup, activities, and perceived effectiveness of safety and health committees in industry.

Safety and the Law

Early legal action in industrial safety took the form of laws to regulate and investigate industry working conditions and death and injury rates. The next phase was largely concerned with workers' compensation payments.

In subsequent years, all governments gradually expanded their roles in regulating industry on safety matters.

The Walsh-Healey Act, which mandates safety measures in companies having supply contracts with the federal government, is an example of such regulation.

In certain industries—notably mining and transportation—U.S. government regulation and inspection have been extensive. The Construction Safety Act, which Congress passed in 1969, addresses the particular health and safety problems of that industry.

In 1970, the Williams-Steiger Occupational Safety and Health Act (OSHAct) was passed. For the first time, the United States had a comprehensive national safety law. The legislation covers every business affected by interstate commerce and employing one or more persons. Safety took on a new direction and meaning as a result of the Occupational Safety and Health Act.

But change is also happening worldwide. For example, in the United Kingdom the Health and Safety at Work Act was passed in 1974. The Act permitted ministers to make regulations to replace existing piecemeal legislation by regulations and codes of practice requiring improved standards of safety, health, and welfare. It provides for a coordinating enforcement authority and gives inspectors powers to initiate actions. Coverage was extended to anyone employed. The Act addressed discussions on the atmosphere and also dealt with certain building codes and standards. It was a major step involving legislation.

In Australia the Victoria Occupational Health and Safety Act of 1985 provides for mandatory regulation and is supplemented by codes of practice. The Act requires union, industry, and government input, and it intends for most safety and health issues to be resolved at the workplace.

In Canada the Ontario Occupational Health and Safety Act of 1978 for Industrial Establishments, Construction, Mines and Mining Plants requires the recording and posting of accident records, penalties, and inspection of workplaces. The Act's provisions were coupled with the Workers' Compensation program.

Another Canadian act developed out of provisions of OSHA. Working with all Canadian provinces, the Workplace Hazardous Management Information System (WHMIS) was created. WHMIS has three parts: (1) labels, (2) MSDS, and (3) a worker education program.

A third Canadian act, the Workers' Compensation Act, became law on January 1, 1990. However, it covers only the province of Ontario. The law deals with a number of subjects: injury, employee retention, accommodation for disabled employees, employee benefit continuation after employee injury, and disability employee benefits.

Besides the OSHAct, several other laws have impacted both industry and the safety professional, particularly those laws dealing with environmental issues. These laws include: Clean Air and Clean Water Acts (CAA, CWA), Toxic Substance Control Act (TSCA), Resource Conservation and Recovery Act (RCRA), Comprehensive Environmental Response Compensation and Liability Act

(CERCLA, "Superfund"), Superfund Amendments and Reauthorization Act (SARA or Community Right-to-Know). These acts are presented in greater detail in Chapter 12, Environmental Management, and in the *Accident Prevention Manual for Business & Industry: Environmental Management*.

Concern with the safety and health of workers is a major priority for management. This concern goes beyond the obvious benefits of less downtime, reduced costs for Workers' Compensation insurance, and lower medical and administrative expenses resulting from disability, death, and impaired productivity. Management also knows there are penalties for not adhering to the law. For example, failure to comply with health and safety requirements can mean citations, which (at the least) create administrative costs but could also lead to serious monetary penalties. The federal government can also institute criminal sanctions against employers and even against individual managers who ignore or disregard the law. The criminal action has not only come from federal and state job safety and health agencies, however. Local prosecutors have successfully convicted individual managers for murder and aggravated assault in the deaths and injuries of workers on the job.

Management must address serious emerging issues in worker health and safety law. These issues include off-the-job safety and ways to deal with the special problem of employees who are at risk in the work environment because of physical condition, language problems, or particular susceptibility to injury or disease. Another issue is the burgeoning paperwork required to comply with OSHA and other agency record-keeping regulations, with the Medical Access Standard, and with the Hazard Communication Standard. Industry accepts almost without question the concept of financial responsibility for work injuries. Not all of industry, however, is convinced of the cost effectiveness of government regulation of safety procedures.

In addition, some states have gone further and established laws requiring compulsory health and accident insurance to cover employee disabilities from diseases or accidents occurring *off* the job. This compulsory insurance might be considered either a drastic extension of the principle of workers' compensation or an extension of social security legislation. It differs from workers' compensation in that it puts a financial burden upon management for diseases and accidents brought about by conditions beyond its control.

Whatever the theory, the result of these laws is to give the employer a direct financial stake in dealing with the off-the-job accident problem. (See Chapter 4, Risk Management Programs.)

International Standards

Several factors have spurred the drive for international standardization of health and safety regulations. These include the following:

- Emerging global markets have intensified the need for international standardization.
- Worldwide technological innovations result in changes to industrial methods and organizations that threaten worker and consumer safety.
- The rapid pace of change in science and technology is outstripping standards development in most countries.
- Developing countries' efforts to industrialize means they may downplay safety and health regulations in favor of rapid economic growth.

Regulatory change has not only occurred in the United States, but also in other nations. For example, the formation of the European Union, or EU, (formerly European Economic Community) has caused changes affecting (1) worker safety, (2) products, and (3) other goods and services. The nations involved are Austria, Belgium, Denmark, Finland, France, Germany, Greece, Iceland, Italy, Luxembourg, the Netherlands, Portugal, Spain, Sweden, and the United Kingdom. Three of the more important regulatory guidelines are the Framework Directive (89/391/EEC) of June 12, 1989; the International Organization for Standardization (ISO) 9000 series; and the ISO 14000 series, issued in 1996. Although compliance with ISO's 9000 and 14000 is voluntary, companies doing business overseas, and their domestic suppliers, have found compliance necessary if their overseas venture are to succeed.

The Framework Directive

The Framework Directive covers the introduction of measures to encourage improvements in on-the-job safety and health of workers. The range of application for the Framework Directive is very broad indeed. It applies to all sectors of activity, both private and public, with the exception of specific working forces such as the armed forces and the police.

The Framework Directive lays down the obligations of employers relating to the safety and health of workers:

- taking the measures necessary for their safety and health, bearing in mind technical progress
- evaluating hazards and instructing workers accordingly
- setting up protection and prevention services—for example, the precision of safety and health practitioners—within the workplace, possibly by enlisting competent external services or persons
- organization of first aid
- evacuation of workers in the event of serious damage

The Directive also lists workers' obligations in relation to their own health and safety:

- correct use of machinery
- correct use of protective equipment

- the need to report defects in equipment, defects in procedures, and potentially dangerous situations such as near-miss accidents

To those who export goods and products to the EU countries, the materials will only have to conform to one standard rather than 15 different ones. Although exporting may be easier, conforming to EU product standards may be more difficult. Under current regulations testing and certification is required before exporting to EU countries. Companies not in EU countries will not be permitted to self-certify. Companies in countries who are EU members can self-certify.

ISO 9000 Series

The ISO 9000 series differs from the Framework Directive in that it is not a set of product standards. Instead, these standards describe a process for establishing quality management and quality assurance. It consists of five quality management standards:

- ISO 9000: Provides general guidelines for applying ISO 9001-9003.
- ISO 9001: Provides a quality systems model for quality assurance in design, development, production, installation, and servicing.
- ISO 9002: Provides a quality systems model for quality assurance in production and installation.
- ISO 9003: Offers a quality systems model for quality assurance in final inspection and testing.
- ISO 9004: Provides guidelines for quality management and quality system elements.

The basic goal of the 9000 series is to give companies guidelines for achieving consistency and uniformity of products or services through the supply chain from primary supplier to the final customers. Companies can use ISO 9000 as an international benchmark against which they can measure their own performance.

Companies do not need to follow all the 9000 standards, however. For example, organizations involved in manufacturing, shipping, research and development, and service contracts would choose ISO 9001 as their benchmark. ISO 9001 consists of 20 sections (4.1-4.20). The following list indicates how this quality standard would apply to health and safety or environmental management in a company's operations.

- 4.1 Management Responsibility: Responsibility and commitment required of management to establish, disseminate, monitor, and refine an environmental health and safety program in the company.
- 4.2 Management Systems: Description of various systems that organizations can use to define and implement health and safety activities.
- 4.3 Contract Review: Description of the company's responsibilities to safeguard the health and welfare of

all outside contractors and the general public affected by the company's operations.

- 4.4 Design Control: Defines how the company controls and implements change in terms of the effect on quality and safety.
- 4.5 Document Control: Describes procedures for writing instructions, operational procedures, and other controlled documents.
- 4.6 Purchasing: Relates to supply selection and assurance of product quality.
- 4.7 Purchaser Supplied Product: Describes quality standards for products given to a manufacturer by another supplier to complete.
- 4.8 Product Identification & Traceability: Establishes methods to trace problem materials throughout the manufacturing process.
- 4.9 Process Control: Ensures consistency of quality and performance in manufacturing processes.
- 4.10 Inspection and Testing: Establishes a quality assurance system to confirm quality, quantity, and uniformity of products against predetermined standards.
- 4.11 Inspection Measuring and Testing Equipment: Refers to calibration, maintenance, and operation of test equipment.
- 4.12 Inspection and Test Status: Establishes a system that ensures reliability and test status of all test equipment.
- 4.13 Control of Nonconforming Product: Describes emergency procedures and training to handle incidents relating to nonconforming products.
- 4.14 Corrective Action: Establishes a system to identify problems and correct them on the job.
- 4.15 Handling, Storage, Packaging, and Delivery: Establishes systems to ensure quality of product after manufacture.
- 4.16 Environmental Health and Safety Records: Refers to records needed to demonstrate compliance with standards and policies.
- 4.17 Internal Audits: Describes a verification system to ensure compliance with policies, procedures, and standards for the system being controlled.
- 4.18 Training: Ensures that workers are competent to perform their tasks and are continually trained in health and safety matters on the job.
- 4.19 Servicing: Refers to product stewardship and responsibility for safe, environmentally sound operations and waste disposal.
- 4.20 Statistical Techniques: Refers to techniques used to monitor performance or progress in safety and health areas.

ISO 14000 Series

The ISO 14000 standards are also related to quality assurance but are primarily directed toward environmental management systems. Companies have a major incentive to comply with these standards because of possible regulatory relief. For instance, the Environmental Protection Agency (EPA) has stated that it might consider up to a 100% decrease in National Pollutant Discharge Elimination System reporting requirements for companies who comply with the ISO standards. Also, penalties for violations might be reduced for certified companies. Although the first ISO 14000 standards were issued in 1996, momentum has been building since then to create another set of standards for occupational health and safety performance.

The ISO 14000 shares many management system principles with ISO 9000 quality system standards. Organizations may choose to use their management systems developed in conformity with the ISO 9000 series as a basis for environmental management. Quality management systems deal with customer needs, while environmental management systems handle issues related to environment protection. Under the ISO 14000, companies are audited by third-party registrars to earn certification from appropriate registration agencies. Companies may also declare themselves to be in conformity with the standard. In Europe, nations have adopted the ISO standard as meeting the requirements of European Community environmental standards. Public pressure in Europe may force U.S. companies to do the same.

The time frame for implementing the ISO 14000 standard is as follows:

- 1996: Environmental management systems specification ("specification" indicates company will be audited by third party to confirm progress); guidelines for auditing environmental management systems (guidelines are voluntary)
- 1997: Guidelines for environmental labeling (product claims)
- 1998: Environmental performance evaluation guidelines
- 1999: Life cycle assessment guidelines

Safety and Occupational Health

Cooperation between medical and safety personnel in accident prevention activities began during the earliest days of the safety movement. Nevertheless, interest in safety on the part of the medical profession and, conversely, interest in employee and public health on the part of the safety professional is increasing. This trend has increased since the early 1980s and has assumed increasing importance in the 1990s.

Part of the interest in workers' health results from concern with occupational disease, noise, radiation, and other problems that extend beyond the former concepts of occupational accident prevention. Interest in protecting the health of citizens in nearby communities from industrial accidents or hazardous materials releases comes partly from such infamous and well-publicized events as the Three-Mile Island accident; the Chernobyl, Ukraine, meltdown; and the Bhopal, India, disaster.

Figure 1-7. Only time, experience, and a concerted effort by all involved disciplines revealed the many relationships between physical illness and work environment. (Courtesy Library of Congress.)

Work Environment

Over time, safety and health professionals became aware of the relationship between physical illness and working conditions. They found that workers in certain industries such as mining, chemical manufacturing, and steel exhibited a higher than normal incidence of such problems as dermatitis, musculoskeletal problems, pulmonary disease, mental illness, and cancer. The improved safety and health of today's workers is the result of concerted efforts by a safety, industrial hygiene, and occupational health team working with a management that realizes an organization's primary asset is a safe and healthy work force (Figure 1–7).

For more details, refer to the Council's Fundamentals of Industrial Hygiene and Occupational Health and Safety, part of the Occupational Safety and Health Series.

Community Environments

Increasingly, the public has demanded a larger role in the management of community environmental risks. Both public and private risk managers realize that providing avenues for public participation is a necessary part of their decision-making process. The problem is how to ensure public involvement and at the same time improve the quality of safety and health decisions.

To help fill the gap in credible risk communication on environmental health and safety issues, the National Safety Council established the Environmental Health Center. This special-purpose organization is led by a Board of Governors and operates mostly through philanthropic funding from concerned corporations, foundations, labor unions, and individuals. Its goals include development of accurate and objective information on environmental and public health risks, improvement of public knowledge about these risks, and dissemination of this information to the public.

Workers with Disabilities

The employment of workers with disabilities by progressive companies and the impact of federal and state equal opportunity laws have modified the practice of preemployment examinations to screen out unfit or undesirable prospects. Medical personnel perform a preplacement examination simply to determine what physical or mental restrictions are appropriate to the prospective employee; they do not determine fitness for a specific job. The job description must specify realistic physical and mental requirements that the human resources department can match to medical restrictions. The Americans with Disabilities Act requires that the job be modified to accommodate the disabled worker. (See the discussion in Chapter 20, Workers with Disabilities.)

Psychology and "Accident Proneness"

Safety professionals who are thoughtfully looking for ways to improve their work encounter a great deal of useful information in modern psychological writing—and a great deal of careless and misleading generalizations. For example, concern about the so-called "accident-prone" individual in industry is as old as the safety movement itself. Statistical information suggests that such individuals exist, though clear and sharp data demonstrating this suggestion are remarkably hard to obtain. Too many alleged "proofs" turn out to be statistically deceptive, based on inadequate samples, or the result of highly subjective diagnoses. Justification for use of "proofs" of accident proneness may reflect the common law doctrines in which employers tried to justify accidents as being caused by the employee's action.

Statistical support for the existence of accident-prone individuals is elusive. Some safety professionals believe in accident proneness and feel it may be a passing phase in the individual and not a permanent characteristic. At most, it may be a problem encountered only by an insignificant minority. Realistically, objective analysis might reveal a supervisory deficiency or procedural weakness that can increase the risks of certain operations or interfere with the performance of individuals or groups of workers.

The same observation applies to psychological tests used as screening devices for new employees. So-called experts may make spectacular claims for the ability of these tests to predict accident proneness, but so far no test has proven itself to the general satisfaction of the safety profession. In the past, the work of psychologists like Dunbar and the Menningers aroused great interest among safety professionals. However, the best weapons in the practical, day-to-day fight against accidents tend to come from strong line management safety awareness and from the disciplines of engineering and behavioral psychology, such as human factors engineering, system safety, and risk management or assessment. (Refer to the discussions in Chapters 22, Motivation, and 23, Safety and Health Training.)

The field of industrial safety is one of progress and improvement, largely through the continued application of techniques and knowledge slowly and systematically acquired through the years. There appears to be no limit to the progress possible through the application of the universally accepted safety techniques of education, engineering, and enforcement.

Yet serious problems remain unsolved. A number of industries still have high accident rates. In far too many instances, management and labor either fail to work together or have different goals for the safety program. The efforts to form effective joint safety and health committees are helping labor and management meet common goals in safety and health.

The resources of the safety movement to tackle these problems are impressive: a growing body of knowledge, a corps of able safety and health professionals, a high level of prestige, and strong organizations for cooperation and exchange of information.

CURRENT ISSUES

Some problems of the safety movement are directly related to the movement's traditional strengths and weaknesses. A few of these problems are social and political, while others are essentially organizational.

Technology and Public Interest

There is no reason for the safety professional to be alarmed at the public's interest in product safety, a better environment, and general technological trends. An informed, educated public can be a powerful ally in safety work. Emphasis upon automation and more refined instrumentation will probably continue. New problems will arise, but they will generally be those that well-established methods of safety engineering can solve.

The use of new materials and techniques—particularly radioactive materials, automation, and lasers—is likely to present more serious difficulties to the safety professional. However, even here, the professional can draw on the movement's considerable experience to handle these problems.

Finally, safety professionals need to keep up with the rapid developments in communications and information technology as they affect the safety field. (See Chapter 9, Computers and Information Management.)

Political Problems

On the political side remains the age-old problem of industry-union-government relations. The key issue here is what regulatory role should government play in protecting worker health and safety and what aspects of national life should government regulate? This issue is likely to remain a source of controversy for years to come.

Organizational Problems

On the worldwide scale, a wide variety of organizations are attacking specific aspects of occupational safety and health problems. The National Safety Council is a strong, constructive, nonpolitical, nonprofit leader among these organizations. It has repeatedly sought and often achieved a cooperative division of labor between itself and other organizations in the safety and health field. One of the guiding principles of the Council has been that there is work enough and credit enough for all involved.

It remains to be seen whether the best organizational forms have been found for participation by all businesses in safety and health work. Safety and health professionals should be ready to consider new ideas and new solutions.

A Look to the Future

The greatest reasons for intensifying the safety effort are humane and moral. The worth of neighbors, friends, and family cannot be measured in dollars or coded into computer records. During the coming decade, the U.S. population will continue to grow, although more slowly than in the past. By 2000 it will be about 270 million people. The average life expectancy of these persons will increase due, in part, to health care advances.

The shift from extractive and manufacturing to service areas of employment will continue well into the next century. This fact will hopefully contribute to a decline in occupational death rates as there will be fewer people employed in high-risk industries.

The work force continues to undergo major changes as minorities and women move into more industries and management levels. Safety issues associated with the growing numbers of women workers are complicated, and many difficult choices regarding working conditions will have to be made.

As social changes continue to take place—including a high divorce rate, more single-parent families, and more two-income households—the effect on the structure and values of family life will be felt more strongly. Many perceive a reduced respect for either parental or social authority, a factor they believe creates considerable conflict and safety problems in the workplace.

One of the current trends is the evolution of a world market. The establishment of a European Union market in the early 1990s has had a major impact on world economics due to the EU regulations on products and goods. The complete restructuring of the economies of the former Soviet Union and Eastern Europe may have an even greater impact on the world markets during the 1990s and the early 21st century. As a result, there will be a great demand for standardizing accident record keeping, so that comparisons can be made between various segments of the global market. Labeling of hazardous material must also be standardized so that all personnel handling chemicals manufactured in one country and shipped to another will understand the warnings on the container.

Accident prevention information will become much more a global commodity than it has been. Although to some extent accident prevention information is shared with worldwide affiliates of large multinational corporations, the vast majority of information developed in one country rarely finds its way to others. There are excellent

behavioral techniques in use in Sweden today that are hardly known in the United States. As the world develops a global market, these national barriers will come down and information sharing will also become global.

The United States, Western Europe, and Japan are moving into a post-Industrial Revolution era or Information Age. Developing countries, however, are just coming out of the early Industrial Revolution as new technologies and manufacturing industries spread to these nations. This trend should create a strong demand for U.S. safety and health expertise in the developing countries.

What will the government's regulatory agencies be doing in the next decade? What kind of standards can be expected from U.S. OSHA, EPA, MSHA in the next few years? Probably the most far reaching will be the "generic standards" regulating exposure to chemicals. In 1989 OSHA promulgated its PEL (Permissible Exposure Limits) standard. Because of rules laid down in the 1970 OSHAct, OSHA was able to adopt at that time an existing federal standard (the 1968 Walsh-Healey Act) to establish permissible exposure levels for about 400 hazardous chemicals.

The next few years may see development of a generic exposure monitoring standard; however, a generic medical monitoring standard appears to be on indefinite hold in the legislative process. These standards are oriented to performance rather than to specification, that is, they will state an intended result and allow the employer the latitude to find ways to achieve the result. At the same time, OSHA has proposed a generic, performance-oriented standard requiring medical monitoring of personnel exposed to hazardous chemicals. This standard would require periodic medical evaluation by blood or urine analysis, as well as other diagnostic techniques. This concept has become controversial and has not progressed beyond the proposal status since many scientists have pointed out the low predictive power of generic testing.

The current high level of public concern in the U.S. for control of hazardous chemicals will probably translate into new legislation. Another area in which OSHA has become involved is setting a standard for protecting workers in hospitals and health care institutions from exposure to bloodborne pathogens. This standard was triggered by public and worker concern over the AIDS epidemic. Further in the future is a generic standard on exposure to biological agents, which is a highly specialized area. Though the pharmaceutical industry has been aware for many years of the hazards of biological agents and taken measures to protect its employees, methods of detection and protection are not widely known.

In addition, criminal prosecution of employers and safety and health professionals for negligence resulting in serious injury or accidental death of employees may become more common. Several recent state supreme court decisions in Illinois and Wisconsin are paving the way for continued activity by state district attorneys to press such criminal cases. As the safety profession becomes increasingly recognized, the public may hold some practitioners accountable for any omissions that may result in death or injury to workers in their organizations.

The days of the plant that operated incognito are gone forever. In the past, industry operators and owners felt the less anyone knew about their businesses, the better off they were. That philosophy became obsolete as the public and industry members alike realized the need to control accidents, injuries, and work-related illnesses. The trend toward openness and cooperation with the public, the government, and the media has spread to all industries. In fact, the leading companies will be those most successful in convincing the public and its local governments that they are good neighbors. Safety professionals will have to understand the legal and political ramifications of their companies' safety practices in light of the public's changing expectations.

These are issues of major importance and will need the expertise and guidance of everyone in the safety and health field. The future is, as it always has been, most uncertain. By working together, all those in the safety and health community can reduce some of that uncertainty by helping to make workplaces and off-the-job environments safer and healthier.

Opportunities for the Safety Professional

The last decade of the 20th century is an exciting time for the safety profession. Professionals in safety management, safety engineering, industrial hygiene, occupational medicine, and the new field of holistic medicine are discovering new and compelling reasons for cooperating more closely and for opening new areas of employment for the safety and health professional.

Other expanding employment opportunities for safety professionals lie in the safety departments of international (and some local) labor unions, on the staffs of a number of trade associations, and, of course, in government service. Safety consulting has expanded rapidly, both in individuals offering their talents on a contract basis, in safety and health consulting firms, and in consulting services offered by nonprofit associations and by a few industrial concerns.

Although advanced education courses and programs for safety professionals increased dramatically in the 1980s, the number could not keep pace with the expanding need. The result was a serious shortage of highly qualified safety professionals. At the same time, the population bulge in the prime working-age group created a surplus of less-skilled people in management. This situation had two effects on safety positions in companies and in government: increased competition for senior safety positions and increased demand for advanced safety training beyond that offered by technical colleges. In fact, one of the greatest growth opportunities for highly qualified safety professionals lies in teaching college safety and health courses.

These are the challenges of tomorrow—improving performance on the job and in the safety profession, coupled with obtaining the education necessary to compete and work effectively in the safety and health field.

SUMMARY

- According to the philosophy of accident prevention, the employer is primarily responsible for ensuring a safe, healthy work environment. Employees are held accountable for following prescribed safety standards and guidelines.
- During the Industrial Revolution, industry, government, and the public began to realize that productivity and safety were closely related. In 1913, the National Council for Industrial Safety, later renamed the National Safety Council, was founded. Other groups soon followed, and these groups began to keep statistics on accident and injury rates in industry.
- The safety movement began to identify health and safety issues and to develop methods to help protect workers on and off the job. Since the end of World War II, the safety movement has become an accepted and valued part of industry.
- Nations have started to develop international safety and health regulations due to the emergence of global markets, the rapid pace of industrialization, and advances in science and technology that put workers and consumers at risk.
- Over the past eight decades, safety workers have gradually become safety professionals. Today's safety and health problems require a highly trained, experienced, and dedicated team of professionals to find solutions.
- The National Safety Council is seeking to improve investigative methods and to establish an information exchange bank regarding work-related hazards. The Council also supports special research projects to study the effectiveness of safety measures and legislation.
- The safety movement faces serious challenges in the future, including new technologies, major changes in the work force, formidable environmental hazards, bloodborne diseases such as AIDS, and widespread drug and alcohol abuse.

GENERAL SAFETY BOOKS

In addition to this Manual, the following books cover the basics of occupational safety and health. Note: These resources serve as the basis for the questions on management in the Board of Certified Safety Professional examinations.

Browning RL. *The Loss Rate Concept in Safety Engineering.* New York: Marcel Dekker, 1980.

DeReamer R. *Modern Safety and Health Technology.* New York: John Wiley & Sons, 1981.

Ferry TS. *Modern Accident Investigation and Analysis,* 2nd ed. New York: John Wiley & Sons, 1988.

Firenze RJ. *The Process of Hazard Control.* Dubuque, IA: Kendall-Hunt, 1978.

Gilmore CL. *Accident Prevention and Loss Control.* New York: The American Management Association, 1970.

Grimaldi JV. *Safety Management,* 5th ed. Burr Ridge, IL: Richard D. Irwin, Inc., 1988.

Heinrich HW. *Industrial Accident Prevention,* 5th ed. New York: McGraw-Hill, 1980.

Petersen DC. *Analyzing Safety Performance.* Reprint. Goshen, NY: Aloray Publishers, 1984.

———. *Safety by Objectives.* Goshen, NY: Aloray Publishers, 1978.

———. *Techniques of Safety Management,* 3rd ed. Goshen, NY: Aloray Publishers, 1989.

Tarrents WE. *The Measurement of Safety Performance.* New York: Garland STPM Press, 1980.

REFERENCES

American Engineering Council. *Safety and Production.* New York: Harper & Brothers Publishers, 1928.

American National Standards Institute, 11 West 42nd Street, New York, NY 10036.

Andrews EW. "The Pioneers of 1912." *National Safety News* 66:24-25, 64-65, 1952.

Beyer DS. *Industrial Accident Prevention,* 3rd ed. Boston: Houghton Mifflin Co., 1928.

Campbell RW. "The National Safety Movement." Proceedings of the Second Safety Congress of the National Council for Industrial Safety, 188-192, 1913.

DeBlois LA. *Industrial Safety Organization for Executive and Engineer.* New York: McGraw-Hill, 1926.

Eastman C. *Work Accidents and the Law.* New York: Charities Publication Committee, 1910. Reprint. New York: Arno Press, 1969.

Felton JS. "History of Occupational Health and Safety." In LaDou J, (ed.). *Occupational Health and Safety.* 2nd edition. Itasca, IL: National Safety Council, 1994.

Heinrich HW. *Industrial Accident Prevention,* 5th ed. New York: McGraw-Hill, 1980.

Holbrook SH. *Let Them Live.* New York: Macmillan, 1939.

Menninger KA. *Man Against Himself.* New York: Harcourt, Brace and World, Inc., 1956.

Meyer RL. Series commemorating the Diamond Anniversary of the Council. *Safety and Health* (January through October 1987).

Mock HE. *Industrial Medicine and Surgery.* Philadelphia: WB Saunders, 1920.

National Safety Council, 1121 Spring Lake Drive, Itasca, IL 60143.

Accident Facts. Issued annually.

"Golden Anniversary Issue." National Safety News, 87:5, 1963.

Proceedings of the First Co-Operative Safety Congress, 1912.

Proceedings of the National Safety Congress. Issued annually from 1914-1925.

Proceedings of the Second Safety Congress of the National Council for Industrial Safety, 1913.

Safety and Health. Issued monthly.

Schaefer VG. Safety Supervision. New York: McGraw-Hill, 1941.

Schulzinger MS. "Accident Syndrome—A Clinical Approach." *Archives of Industrial Health,* 11:66-71; 1955.

Schwedtman FA. *Accident Prevention and Relief.* New York: National Association of Manufacturers in the United States of America, 1911.

Toynbee A. *The Industrial Revolution.* (First published in 1884.) Boston: Beacon Press, 1956.

U.S. Department of Labor, Bureau of Labor Statistics. *Occupational Injuries and Illnesses: Counts, Rates, and Characteristics, 1992.* Bulletin 2455. Washington, DC: U.S. GPO, April 1995.

REVIEW QUESTIONS

1. List the six reasons for preventing injuries and occupational illnesses as given in the text.
   a.
   b.
   c.
   d.
   e.
   f.
2. With the Industrial Revolution came innovations in the processes and organization of production known as the factory system. List four components of the factory system that emerged at that time.
   a.
   b.
   c.
   d.
3. What industry introduced the factory system in the United States?
   a. shoe manufacturing
   b. farm implement industry
   c. textile industry
   d. food processing industry
4. Define the fellow servant rule.
5. Define contributory negligence.
6. Define assumption of risk.
7. Which president's legislation set up the first workers' compensation laws covering only federal employees?
   a. Theodore Roosevelt
   b. Dwight Eisenhower
   c. Harry Truman
   d. Franklin Roosevelt
8. Which state passed the first effective workers' compensation act?
   a. New Jersey
   b. Wisconsin
   c. Washington
   d. New York

9. What industry was the first to realize that the actions of people were important in creating accident situations?
   a. textile
   b. railroad
   c. steel
   d. agriculture
10. When was the National Safety Council started?
   a. 1910
   b. 1905
   c. 1912
   d. 1915
11. What is the three "E" concept?
12. List the three trends in safety work and the safety profession that have emerged from the prevention and control measure concepts of the 1950s.
   a.
   b.
   c.
13. List the necessary elements for a joint safety and health activity to be successful in any workplace.
   a.
   b.
   c.
   d.
14. Regarding safety matters in the United States, which Act regulates companies with government contracts?
   a. Walsh-Healy Act
   b. Williams-Steiger Occupational Safety and Health Act
   c. Construction Safety Act
   d. The 1974 Health and Safety at Work Act
15. Which items were included in the national safety law brought about by the Williams-Steiger Occupational Safety and Health Act?
   a. one or more employees
   b. businesses affected by interstate commerce
   c. manufacturer
   d. businesses selling to the general public
16. List four factors that have spurred the drive for international standardization of health and safety regulations.
   a.
   b.
   c.
   d.
17. List five employer obligations relating to the safety and health of workers stated in the Framework Directive (89/391/EEC) of ISO 9000 and ISO 14000.
   a.
   b.
   c.
   d.
   e.
18. List three employee obligations relating to the safety and health of workers stated in the Framework Directive (89/391/EEC) of ISO 9000 and ISO 14000.
   a.
   b.
   c.
19. What were the purpose and significance of the death calendar used by the Russell Sage Foundation in Allegheny County, Pennsylvania?
20. What contributed to the lowering of the overall occupational death rate in the United States since World War II?
21. Explain why an independent union safety and health committee can be important.
22. What guidelines should both labor and management consider when establishing a joint safety and health committee at a workplace?
23. Discuss the major safety and health issues management must address.

2 Regulatory History

In the past three decades, Congress has enacted two major pieces of federal legislation impacting the field of occupational safety and health. These laws are the Occupational Safety and Health Act of 1970 (OSHAct) and the U.S. Mine Safety and Health Act of 1977. This chapter focuses only on these two Acts.

OCCUPATIONAL SAFETY AND HEALTH ACT

A new national policy was established on December 29, 1970, when President Richard M. Nixon signed into law the OSHAct (Public Law No. 91–596 found in 29 United States Code (USC) §§651–678). The Congress of the United States declared that the purpose of this piece of legislation was "to assure so far as possible every working man and woman in the Nation safe and healthful working conditions and to preserve our human resources."

The OSHAct took effect April 28, 1971. Coauthored by Senator Harrison A. Williams (Dem.-NJ) and the late Congressman William Steiger (Rep.-WI), the Act is sometimes referred to as the Williams-Steiger Act. It is regarded by many as landmark legislation because it goes beyond the present workplace and considers long-term health hazards in the working environment of the future.

The information provided in Part I of this chapter focuses on federal OSHA programs. State OSHA programs may differ from the federal program in certain areas, but are required to be equal to the federal requirements. However, unless specifically stated to the contrary, the recommendations in this chapter can be followed whether jurisdiction rests at the federal or state level.

LEGISLATIVE HISTORY

Historically, the enactment of safety and health laws has been left to the states. Prior to the 1960s only a few federal laws (such as the Walsh-Healey Public Contracts Act and the Longshoremen's and Harbor Workers' Compensation Act) directed any attention to occupational safety and health. Several pieces of legislation passed by the Congress during the 1960s, including the Service Contract Act of 1965, the National Foundation on Arts and Humanities Act, the Federal Metal and Nonmetallic Mine Safety Act, the Federal Coal Mine Safety and Health Act, and the Contract Workers and Safety Standards Act (Construction Safety Act), focused industry attention on occupational safety and health.

Each of these federal laws was applicable only to a limited number of employers. The laws were either directed at those who had obtained federal contracts, or they targeted a specific industry. Even collectively, all the federal safety legislation passed prior to 1970 was not applicable to most employers or employees. Up to that time, congressional action on occupational safety and health issues was, at best, sporadic, covering only specific sets of employers and employees. There was little attempt to establish the omnibus coverage that is a central feature of the OSHAct.

Proponents of a more significant federal role in occupational safety and health, mostly represented by organized labor, based their position primarily on the following:

1. With rare exceptions, the states failed to meet their obligation in regard to occupational safety and health. A few had reasonable or adequate safety and health legislation, but most states legislated safety and health only in specific industries. In general, states had inadequate safety and health standards, inadequate enforcement procedures, inadequate staff with respect to quality and quantity, and inadequate budgets.

2. In the late 1960s, approximately 14,300 employees were killed annually on or in connection with their job and more than 2.2 million employees suffered a disabling injury each year as a result of work-related accidents. The injury/death toll was considered by most to be unacceptably high.

3. The nation's work-injury rates in most industries were increasing throughout the 1960s. Because the trend was moving in the wrong direction, proponents of federal intervention felt that national legislation would help to reverse this trend.

The Act evolved amid stormy controversy in both houses of Congress as the legislators debated state versus federal roles, industry versus government control, and the like. Such issues were responsible for sharply drawn lines between political parties and between the business community and organized labor. After three years of political tug-of-war, numerous compromises were made to allow passage of the OSHAct by both houses of Congress.

ADMINISTRATION

Administration and enforcement of the OSHAct are vested primarily with the Secretary of Labor, Assistant Secretary of Labor for OSHA, and the Occupational Safety and Health Review Commission (OSHRC) as an appellate agency, which is discussed later. With respect to the enforcement process, the Secretary of Labor, through the Assistant Secretary, performs the investigation and prosecution aspects, and the OSHRC performs the administrative adjudication portion, with possible appeal through the courts.

Research and related functions and certain educational activities are vested in the Secretary of Health and Human Services. These responsibilities, for the most part, are carried out by the National Institute for Occupational Safety and Health (NIOSH) established within the Department of Health and Human Services (DHHS). Compiling injury and illness statistical data is handled by the Bureau of Labor Statistics (BLS), United States Department of Labor (DOL).

To assist the Secretary of Labor, the Act authorizes the appointment of an Assistant Secretary of Labor for Occupational Safety and Health. This position is filled by presidential appointment with the advice and consent of the Senate. The Assistant Secretary is the chief of the Occupational Safety and Health Administration (OSHA) established within the DOL. The Assistant Secretary acts on behalf of the Secretary of Labor. For the purposes of this chapter, OSHA is also synonymous with the term Secretary or Assistant Secretary of Labor.

The primary functions of the four major governmental units assigned to carry out the provisions of the Act are described in this section on administration.

Occupational Safety and Health Administration

The Occupational Safety and Health Administration came into existence officially on April 28, 1971, the date the OSHAct became law. This agency was created by the DOL to discharge the Department's responsibilities assigned by the Act.

Major Areas of Authority

The Act grants OSHA the authority, among other things, (1) to promulgate, modify, and revoke safety and health standards; (2) to conduct inspections and investigations and to issue citations, including proposed penalties; (3) to require employers to keep records of safety and health data; (4) to petition the courts to restrain imminent danger situations; and (5) to approve or reject state plans for programs under the Act.

The Act also authorizes OSHA (1) to provide training and education to employers and employees; (2) to consult with employers, employees, and organizations regarding prevention of injuries and illnesses; (3) to grant funds to the states for identification of program needs and for plan development, experiments, demonstrations, administration, and operation of programs; and (4) to develop and maintain a statistics program for occupational safety and health.

Major Duties Delegated

In establishing OSHA, the Secretary of Labor delegated to the Assistant Secretary for Occupational Safety and Health the authority and responsibility for safety and health programs and activities of the DOL, including responsibilities derived from:

1. Occupational Safety and Health Act of 1970
2. Walsh-Healey Public Contracts Act of 1936, as amended
3. Service Contract Act of 1965
4. Public Law 91–54 of 1969 (construction safety amendments)
5. Public Law 85–742 of 1958 (maritime safety amendments)
6. National Foundation on the Arts and Humanities Act of 1965
7. Longshoremen's and Harbor Worker's Compensation Act (Title 33, Chapter 18, §§901, 904, U.S.

Code; Act of March 4, 1927, Chapter 509, 44 Stat. 1424)

8. Federal safety program under Title 5 U.S. Code 7902.

Similarly, the Commissioner of the BLS was delegated the authority and given responsibility for developing and maintaining an effective program for collection, compilation, and analysis of occupational safety and health statistics; for providing grants to the states to assist in developing and administering programs in such statistics; and for coordinating functions with the Assistant Secretary for Occupational Safety and Health.

The Solicitor of Labor is assigned responsibility for providing legal advice and assistance to the Secretary and all officers of the Department in the administration of statutes and Executive Orders relating to occupational safety and health. In enforcing the Act's requirements, the Solicitor of Labor represents the Secretary in litigation before OSHRC and, subject to the control and direction of the Attorney General, before the federal courts.

To assist in carrying out its responsibilities, OSHA has established 10 regional offices in the cities of Boston, New York, Philadelphia, Atlanta, Chicago, Dallas, Kansas City, Denver, San Francisco, and Seattle. (See Directory of Federal Agencies at the end of this chapter.) The primary mission of the regional office chief, known as the Regional Administrator, is to supervise, coordinate, evaluate, and execute all programs of OSHA in the region. Assisting the Regional Administrator are Assistant Regional Administrators for (1) training, education, consultation, and federal agency programs; (2) technical support; and (3) state and federal operations. (Some functions are combined in certain regions.)

Area offices have been established within each region, each office headed by an Area Director. The mission of the Area Director is to carry out the compliance program of OSHA within designated geographic areas. The area office staff carries out its activities under the general supervision of the Area Director with guidance of the Regional Administrator, using policy instructions received from the national headquarters. The federal enforcement for implementing the enforcement portion of the OSHAct is carried out by the area offices in those states that do not have an approved state plan. In states with an approved plan, the area office monitors state activities. (See Federal-State Relationships, later in this chapter.)

Occupational Safety and Health Review Commission

The Occupational Safety and Health Review Commission is a quasi-judicial board of three members appointed by the president and confirmed by the Senate. OSHRC is an independent agency of the executive branch of the U.S. government and is not a part of the DOL. The principal function of the Commission is to adjudicate cases when an enforcement action taken by OSHA against an employer is contested by the employer, the employees, or their representatives.

OSHRC's actions are limited to contested cases. In such instances, OSHA first notifies the Commission of the contested cases. The Commission then hears all appeals on actions taken by OSHA concerning citations, proposed penalties, and abatement periods and determines the appropriateness of such actions. When necessary, the Commission may conduct its own investigation and may affirm, modify, or vacate OSHA's findings.

There are two levels of adjudication within the Commission: (1) the administrative law judge (ALJ), and (2) the three-member Commission. All cases not resolved in OSHA informal proceedings are heard and decided by one of the Commission's ALJs. The judge's decision can be changed by a majority vote of the Commission if one of the members, within 30 days of the judge's decision, directs that the decision be reviewed by the Commission members. The Commission is the final administrative authority to rule on a particular case, but its findings and orders can be subject to further review by the courts. (For further information, see Contested Cases later in this chapter.)

The headquarters of OSHRC is located at 1825 K Street, NW, Washington, DC 20006.

National Institute for Occupational Safety and Health

The National Institute for Occupational Safety and Health was established within the DHHS under the provisions of the OSHAct. Administratively, NIOSH is located in DHHS's Centers for Disease Control in Atlanta, GA. NIOSH is the principal federal agency engaged in research, education, and long-term training related to occupational safety and health.

The primary functions of NIOSH are (1) to develop and establish recommended occupational safety and health standards, (2) to conduct research experiments and demonstrations related to occupational safety and health, and (3) to develop educational programs to provide an adequate supply of qualified personnel to carry out the purposes of the OSHAct.

Research and Related Functions

Under the OSHAct, NIOSH has the responsibility for conducting research for new occupational safety and health standards. NIOSH develops criteria for establishing these standards and transmits the criteria to OSHA. OSHA is responsible for the final setting, promulgation, and enforcement of the standards.

The OSHAct also requires NIOSH to publish an annual listing of all known toxic substances and the concentrations at which such toxicity is known to occur. The inclusion of a substance on the list does not necessarily mean that it should be avoided. It does mean that the listed substance has a documented potential of being hazardous if misused and, therefore, care must be exercised to control the substance. Conversely, the absence of a substance from the list does not mean that it is nontoxic. Some hazardous substances may not qualify for the list simply because the dose that causes a toxic effect is not known.

Education and Training

NIOSH also has the responsibility to develop (1) education and training programs aimed at providing an adequate supply of qualified personnel to carry out the purpose of the Act and (2) informational programs on the importance and proper use of adequate safety and health equipment. The long-term approach to an adequate supply of training personnel in occupational safety and health is found in the colleges and universities and other institutions in the private sector. NIOSH encourages such institutions, by contracts and grants, to expand their curricula in occupational medicine, occupational health nursing, industrial hygiene, and occupational safety engineering.

Employer and Employee Services

Of principal interest to individual employers and employees are the technical services offered by NIOSH. The five main services are provided upon request to NIOSH's Division of Technical Services, 4676 Columbia Parkway, Cincinnati, OH 45226 (1–800–356–4674). These services are:

1. Hazard evaluation—Onsite evaluations of potentially toxic substances used or found on the job.
2. Technical information—Detailed technical information concerning health or safety conditions at workplaces, such as the possible hazards of working with specific solvents, and guidelines for use of protective equipment.
3. Accident prevention—Technical assistance for controlling on-the-job injuries, including the evaluation of special problems and recommendations for corrective action.
4. Industrial hygiene—Technical assistance in the areas of engineering and industrial hygiene, including the evaluation of special health-related problems in the workplace and recommendations for control measures.
5. Medical service—Assistance in solving occupational medical and nursing problems in the workplace, including assessment of existing medically related needs and development of recommended means for meeting such needs.

NIOSH and the Mine Safety and Health Administration test and approve personal sampler units for coal-mine dust and respiratory protective devices, including self-contained breathing apparatus, gas masks, supplied-air respirators, chemical-cartridge respirators, and dust, fume, and mist respirators.

NIOSH representatives, although not authorized to enforce the OSHAct, are authorized to make inspections and to question employers and employees in carrying out the duties assigned to the DHHS under the Act. NIOSH has both warrant and subpoena power, if necessary, to obtain the information needed for its investigations. It may also request access to employee records. However, it must obtain the consent of employees or use methods that maintain the employee's right to privacy concerning information in the records.

Injury and Illness Data

The responsibility for conducting statistical surveys and establishing methods used to acquire injury and illness data is placed in the BLS. Questions regarding record-keeping requirements and reporting procedures can be directed to any of the OSHA regional or area offices. (See the directory at the end of this chapter.)

Advisory Committees

The Act established a 12-member National Advisory Committee on Occupational Safety and Health (NACOSH) to advise, consult with, and make recommendations to the Secretaries of Labor and Health and Human Services with respect to the administration of the Act. Eight members are designated by the Secretary of Labor and four by the Secretary of Health and Human Services. Members include representatives from management, labor, occupational safety and health professions, and the public.

The Act also authorizes the appointment of 15-member advisory committees to assist OSHA in the development of standards. The Standards Advisory Committees on Construction Safety and Health and on Cutaneous Hazards are the two currently in place.

MAJOR PROVISIONS OF THE OSHACT

This section discusses the major regulations and provisions of OSHAct with which employers are expected to comply. The Act represents one of the most far-reaching efforts in U.S. history to provide safer, healthier, and cleaner conditions not only for employees but for communities living near office and manufacturing facilities and for consumers who use the products and services of these firms.

Coverage

Except for specific exclusions, the Act applies to every employer who has one or more employees and who is engaged in a business affecting interstate commerce. The law applies to all 50 states, the District of Columbia, Puerto Rico, and all U.S. possessions.

Specifically excluded from coverage are all federal, state, and local government employees. There are, however, special provisions in the Act for federal employees and potential coverage for state and local government employees. The Act requires each federal agency head to establish and maintain an occupational safety and health program consistent with the standards promulgated by the Secretary of Labor. Executive Orders setting requirements for federal programs have been issued over the years. OSHA regulations implementing the Executive Order are found in 29 *Code of Federal Regulations (CFR)* Part 1960 (1983).

Employees of states and political subdivisions of the states are excluded from the federal OSHAct. However, states with approved state plans are required to provide coverage for these public employees. Public employees in states without approved plans are not covered by the OSHAct in any manner.

However, two states—Connecticut and New York—have state plans covering only public employees. The District of Columbia and the states of Mississippi, New Hampshire, New Jersey, Rhode Island, and Wisconsin have laws specifically providing job safety and health protection to public employees.

The OSHAct also does not apply to those operations in which a federal agency (and state agencies acting under the Atomic Energy Act of 1954), other than the DOL, already has the authority to prescribe or enforce standards or regulations affecting occupational safety or health and is performing that function. An example of this exclusion is specific issues covered by Department of Transportation regulations in the railroad industry.

Also excluded from the OSHAct are operators and miners covered by the U.S. Mine Safety and Health Act of 1977. This Act applies to mines of all types: coal and noncoal, surface and underground.

In its yearly appropriations bills since 1977, Congress has placed restrictions on OSHA enforcement. Previously, for example, items exempt from inspection have included:

- farmers with ten or fewer employees on the day of inspection and the 12 months preceding the day of inspection
- any work activity in any recreational, hunting, fishing, or shooting area
- employers with 10 or fewer employees in selected Standard Industrial Classification industries with rates below the national average in lost workdays case rate (1989 rate was 3.9 per 100)

However, the exemptions do not apply to situations involving such issues as employee complaints, referrals, health hazards, accidents resulting in a fatality or inpatient hospitalization involving three or more employees, or discrimination complaints.

OSHA clarified its interpretation of coverage with respect to certain employees by issuing a regulation. (This policy regarding "Coverage of Employees Under the Williams-Steiger Occupational Safety and Health Act of 1970" is contained in Title 29 *CFR,* Chapter XVII, Part 1975.) OSHA has also stated that churches and religious organizations are not regarded as employers when performing ecumenical activities, nor are persons who, in their own residences, employ others to perform domestic household tasks. Further, any person engaged in agriculture who is a member of the farmer's immediate family is not regarded as an employee and hence is not covered by the Act.

Employer and Employee Duties

OSHA clearly delineates employer and employee responsibilities. Each employer covered by the Act:

1. has the general duty to furnish each employee with employment and places of employment free from recognized hazards causing or likely to cause death or serious physical harm (this is commonly known as the "general duty clause")
2. has the specific duty of complying with safety and health standards promulgated under the Act.

Each employee, in turn, has the duty to comply with the safety and health standards and with all rules, regulations, and orders that apply to employee actions and conduct on the job.

For employers, the general duty provision is used only where there are no specific standards applicable to a particular hazard involved. A hazard is "recognized" if it is a condition generally regarded as a hazard in the particular industry in which it occurs and is detectable (1) by means of human senses, or (2) by accepted tests known in the industry which can reveal its presence to the employer. An example of a "recognized hazard" in the second category is excessive concentration of a toxic substance in the work-area atmosphere, even though such concentration could be detected only through use of measuring devices. The general duty clause is applicable to a specific condition only if there is no OSHA standard addressing the condition.

During the course of an inspection, a compliance safety and health officer (CSHO) is concerned primarily with determining whether the employer is complying with the promulgated safety and health standards. However, the officer will also try to discover if the employer is complying with the general duty clause.

The law provides for sanctions against the employer in the form of citations and civil and criminal penalties if the employer fails to comply with the general and specific duties. However, there is no provision for government sanctions against an employee for failure to comply with the employee's duty. Although some may view this policy as unjust, it was not one of the controversial issues in the formative stages of the Act.

Both management and organized labor have long agreed that safety and health on the job is a management responsibility. The business community generally did not want the law structured to provide for government sanctions against an erring employee. This is because management can invoke its own measures against an employee who obstructs the employer's efforts to provide a safe workplace.

While the law expressly places upon each employee the obligation to comply with the Act's standards, final responsibility for compliance rests with the employer. As a result, employers should take all necessary actions to ensure that employees follow the promulgated standards. Employers should also establish within their safety system a means to detect when employees are not complying with applicable standards.

The duty of an employer to protect employees against health hazards in addition to safety hazards continues to gain emphasis. The growing awareness of the hazard of chemical exposure has caused OSHA to focus on the employer's obligation to monitor the work environment, to provide periodic medical examinations, and to make a range of protective measures available to guard against hazards of the work environment.

An emerging issue is the question of how far an employer must go to protect employees who are at increased risk of occupational injury or illness. The increased risk may come from lack of experience, language problems, special sensitivity to chemicals, or being a woman of child-bearing age. Medical screening is one method employers have begun to use to identify present and prospective employees who are at increased risk. However, the method became controversial when employees were denied employment, or assigned or reassigned, based on the results of the screening. Can such actions intended to protect the employee become discriminatory or invade privacy? In the early 1980s employees and other groups used the job-discrimination provisions of the Civil Rights Act to protect women workers in general, and pregnant women in particular, against termination and detrimental reassignment because of alleged hazardous working conditions. Although not all these issues have been settled, a 1991 U.S. Supreme Court decision struck down the use of fetal protection policies on the basis that they discriminate against women workers.

Employer Rights

An employer has the right to:

1. seek advice and off-site consultation as needed by writing, calling, or visiting the nearest OSHA office
2. request and receive proper identification of the OSHA CSHO prior to inspection
3. be advised by the CSHO of the reason for an inspection
4. have an opening and closing conference with the CSHO
5. file a Notice of Contest with the OSHA area director within 15 working days after receiving a citation notice and proposed penalty
6. apply to OSHA for a temporary variance from a standard if unable to comply because needed materials, equipment, or personnel are not available to make necessary changes within the required time
7. take an active role in developing safety and health standards through participation in OSHA Standards Advisory Committees, through nationally recognized standards-setting organizations, and through evidence and views presented in writing or at hearings
8. apply for, if a small business employer, long-term loans through the Small Business Administration (SBA) to help bring the establishment into compliance, either before or after an OSHA inspection
9. be assured of the confidentiality of any trade secrets observed by an OSHA compliance officer

Onsite Consultation

Congress has authorized, and OSHA now provides through a state agency or private contractors, free on-site consultation services for employers in every state. The onsite consultants help employers to identify hazardous conditions and to determine corrective measures.

The service is available upon employer request. Priority is given to businesses with fewer than 150 employees. These firms are generally less able to afford private-sector consultation. OSHA assigns higher priority to companies that use employees in highly hazardous jobs.

The consultative visit consists of an opening conference, a walk-through of the company's facility, a closing conference, and a written summary of findings. During the walk-through, the consultant tells the employer which OSHA standards apply to company operations and what they mean. The employer is informed of any apparent violations of those standards and, where possible, is given suggestions on how to reduce or eliminate the hazard.

Because employers, not employees, are subject to legal sanctions for violating OSHA standards, the employer determines to what extent employees or their representatives will participate in the visit. However, the consultant must be allowed to confer with individual employees during the walk-through in order to identify and judge the nature and extent of hazards. No citation or penalties are issued and OSHA is not notified of results except when an employer refuses to correct a serious hazard. Follow-up actions are taken to assure appropriate corrections are made.

Voluntary Protection Program

The purpose of OSHA's Voluntary Protection Program (VPP) is to emphasize the importance of, encourage the improvement of, and recognize excellence in employer-provided, site-specific occupational safety and health programs. These programs must not only meet, but exceed, the standards. When employers apply and are accepted, they are removed from the regular inspection list (except for valid formal employee complaints, fatalities, or catastrophic events).

Two voluntary programs have been established: Star and Merit. OSHA's Star Initiative, the agency's most demanding and prestigious voluntary protection program, is for worksites with outstanding workplace safety and health systems.

Safety and health programs can be based primarily on management initiatives and employee participation. Because of the special hazards in construction, firms in that industry are generally offered demonstration programs. Star participants are evaluated every three years.

What does it take to qualify for Star? First, a firm's accident record must meet stringent criteria. Star participants must have a three-year average injury incidence and lost-workday-case rate at or below the national average for their industry.

In addition, the worksite must have an effective, comprehensive safety and health program. Among the key elements OSHA considers are:

- Management commitment and accountability—the agency has found that top-flight safety and health programs have strong corporate commitment and accountability, including written, clearly defined assignment of responsibility for worker protection at every level of management.
- Hazard assessment—this means a comprehensive inventory of potential safety and health hazards and includes periodic review and updating, especially when processes are changed or new substances are introduced. Effective hazard assessment also includes an effective mechanism for inviting and responding to worker notices of possible hazards.
- Safety rules and enforcement—Star participants have their own policies and procedures that go beyond the protection specified by OSHAct standards.
- Employee training—In addition to formal training, many participants have regular safety meetings and toolbox meetings where safety procedures are reviewed.
- Self-evaluation—Comprehensive safety and health program audits ensure that the program continues to be an effective system for communicating to employees the company's commitment to safety and health and efficient procedures for responding to employee concerns and suggestions.

The Merit Program is a program for companies that have solid safety and health programs but need to make certain specific improvements to qualify for Star. As in Star, companies may use management initiative and employee participation. Merit participants are evaluated every year by OSHA.

When a firm applies for Star or Merit, an OSHA staffer is assigned to facilitate the application, answer any questions, and generally be the liaison between the applicant and the agency. In addition to analyzing written documentation supporting the application, OSHA will visit the workplace for an on-site review. The on-site review team will examine safety and health records, review the inspection history, conduct a walk-around check, and interview management and employees, if appropriate.

While this application procedure is and must be rigorous, most applicants report that it is not unduly burdensome, and that participation in the program makes the effort worthwhile.

OSHA Resource Materials

OSHA offers a variety of information services, publications, audiovisual aids, technical advice and speakers for industry and governmental agencies. The OSHA Training Institute, located in Des Plaines, Illinois, provides both basic beginning and advanced training courses in the areas of health, safety, and industrial hygiene. These courses are offered to federal and state compliance officers, state consultants, federal agency personnel and for interested parties from the private sector including employers, employees, and union representatives. Published materials are available from the area and regional offices or from the OSHA Publications Office, located at 200 Constitution Avenue, NW, Room N3101, Washington, DC 20210. These materials are normally limited to one free copy, with additional copies available for purchase.

Published materials available from OSHA include:

1. All About OSHA - OSHA 2056
2. Chemical Hazard Communication - OSHA 3084
3. Consultation Services for Employers - OSHA 3047
4. How to Prepare for Workplace Emergencies - OSHA 3088
5. Job Hazard Analysis - OSHA 3071
6. OSHA: Employee Workplace Rights - OSHA 3021
7. OSHA Handbook for Small Business - OSHA 2209
8. OSHA Inspections - OSHA 2098
9. Hazard Communications - GPO #929-022-00000-9
10. Recordkeeping Guidelines for Occupational Injuries and Illnesses - OBM #1220-0029
11. Employer Rights & Responsibilities Following an OSHA Inspection - OSHA 3000

Satellite training locations are available in many states. The nearest OSHA office can provide current information about locations.

Employee Rights

Although the employee has the legal duty to comply with all the standards and regulations issued under the OSHAct, many employee rights are also incorporated into the Act. Because these rights may affect labor relations as well as labor negotiations, employers should also be aware of the employee rights contained in the Act. These rights fall into three main areas related to (1) standards, (2) access to information, and (3) enforcement.

With respect to standards:

1. Employees may request OSHA to begin proceedings for adoption of a new standard or to amend or revoke an existing one.
2. Employees may submit written data or comments on proposed standards and may appear as an interested party at any hearing held by OSHA.
3. Employees may file written objections to a proposed federal standard and/or appeal the final decision of OSHA.
4. Employees must be informed when an employer applies for a variance of a promulgated standard.
5. Employees must be given the opportunity to participate in a variance hearing as an interested party and have the right to appeal OSHA's final decision.

With respect to access to information:

1. Employees have the right to information from the employer regarding employee protection and obligations under the Act and to review appropriate OSHA standards, rules, regulations, and requirements, which the employer should have available at the workplace.
2. Affected employees have a right to information from the employer regarding the toxic effects, conditions of exposure, and precautions for safe use of all hazardous materials in the establishment. The information can be provided through labeling or other means where such warnings are prescribed by a standard.
3. If employees are exposed to harmful materials in excess of levels set by the standards, the affected employees must be so informed by the employer, who must also tell them what corrective action is being taken.
4. If a CSHO determines that an alleged imminent danger exists, the officer must inform the affected employees of the danger and recommend that relief be sought by court action if the employer does not eliminate the danger.
5. Upon request, employees must be given access to records of their medical records and history of exposure to toxic materials or harmful physical agents that must be monitored or measured and recorded. Material Safety Data Sheets must be immediately available or accessible.
6. If a standard requires monitoring or measuring hazardous materials or harmful physical agents, employees must be given the opportunity to observe such monitoring or measuring.
7. Employees have the right of access to (1) the list of toxic materials published by NIOSH, (2) criteria developed by NIOSH describing the effects of toxic materials or harmful physical agents, and (3) industry-wide studies conducted by NIOSH

regarding the effects of chronic, low-level exposure to hazardous materials.
8. On written request to NIOSH, employees have the right to obtain the determination of whether a substance found or used in the establishment is harmful.
9. Upon request, the employees must be allowed to review the Log and Summary of Occupational Injuries (OSHA No. 200) at a reasonable time and in a reasonable manner.

With respect to enforcement:

1. Employees have the right to confer in private with the CSHO and to respond to questions from the CSHO during an inspection of an establishment.
2. An authorized employee representative must be given an opportunity to accompany the compliance officer during an inspection to aid in such inspection. (This is commonly known as the "walk-around" provision.) Also, an authorized employee has the right to participate in the opening and closing conferences during the inspection.
3. An employee has the right to make a written request to OSHA for a complaint inspection if the employee believes a violation of a standard threatens physical harm; the employee has the right to request that OSHA keep his or her identity confidential.
4. An employee who believes that a violation of the Act has occurred has the right to notify OSHA or a compliance officer in writing of the alleged violation, either before or during an inspection of the establishment.
5. If OSHA denies an employee's request for a special inspection, the agency must notify the employee in writing that the complaint was not valid and explain the reasons for this decision. The employee has the right to object to such a decision and may request a hearing by OSHA.
6. If a written complaint concerning an alleged violation is submitted to OSHA and the compliance officer responding to the complaint fails to cite the employer for the alleged violation, OSHA must furnish the employee or an authorized employee representative with a written statement explaining the reasons for its final disposition.
7. If OSHA cites an employer for a violation, employees have the right to review a copy of the citation, which must be posted by the employer at or near the place where the violation occurred. If "system-wide" agreements are made, all locations will be notified.
8. An employee has the right to appear as an interested party or to be called as a witness in a contested enforcement matter before OSHRC.
9. If OSHA arbitrarily or capriciously fails to seek relief to counteract an imminent danger and an employee

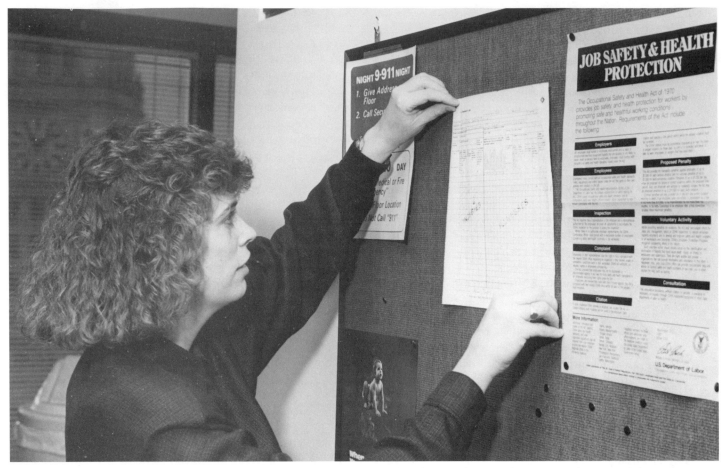

Figure 2-1. OSHA poster (OSHA 2203), "Safety and Health Protection on the Job," must be posted conspicuously at every plant, job site, or other facility. At left is the annual Log and Summary of Occupational Injuries and Illnesses, OSHA Form 200, that must be posted by February 1 of the following year and remain in place until March 1.

is injured as a result, that employee has the right to bring action against OSHA for relief as may be appropriate.

10. An employee has the right to file a complaint to OSHA within 30 days if the employee believes he or she has been discriminated against as a result of asserting employee rights under the Act.

11. An employee has the right to contest the abatement period fixed in the citation issued to the employer. This can be done by notifying the OSHA Area Director who issued the citation within 15 working days of the issuance of the citation.

The OSHA Poster

The OSHA poster (OSHA 2203, Figure 2-1) must be prominently displayed in a conspicuous place in the work environment where notices to employees are customarily posted. The poster informs employees of their rights and responsibilities under the Act.

Occupational Safety and Health Standards

The Act authorizes OSHA to promulgate, modify, or revoke occupational safety and health standards. The rules of procedure for promulgating, modifying, or revoking

standards are spelled out in Title 29 *CFR*, Chapter XVII, Part 1911. The current requirements are available at all OSHA area and regional offices and printed annually in the *Code of Federal Regulations*.

OSHA is responsible for promulgating legally enforceable standards that may require conditions or the adoption or use of practices, means, methods, or processes that are reasonably necessary and appropriate to protect employees on the job. It is the employers' responsibility to become familiar with the standards applicable to their firms and to make sure that employees have and use personal protective equipment required for safety. In addition, employers are responsible for complying with the Act's general duty clause.

To get the initial set of standards in place without undue delay, the Act authorized OSHA to promulgate any existing federal standard or any national consensus standard without regard to the usual rule-making procedures prior to April 28, 1973. The initial set of standards, Part 1910, appeared in the *Federal Register* of May 29, 1971. (Subscriptions to the *Federal Register* are obtained through the Government Printing Office, Washington, DC 20402.)

Standards contained in Part 1910 apply to general industry, while those contained in Part 1926 apply to construction. Standards that apply to ship repairing, shipbuilding, shipbreaking, and longshoring are contained in Parts 1915 through 1918, respectively. As new equipment, methods, and materials are developed, these standards are updated via modification. Agricultural standards are contained in Part 1928.

OSHA standards incorporate by reference certain other standards adopted by industry organizations. Standards incorporated by reference in whole or in part include those adopted by the following organizations:

- American Conference of Governmental Industrial Hygienists
- American National Standards Institute
- American Petroleum Institute
- American Society of Agricultural Engineers
- American Society of Mechanical Engineers
- American Society for Testing and Materials
- American Welding Society
- Compressed Gas Association
- Crane Manufacturers Association of America, Inc. The Fertilizer Institute
- Institute of Makers of Explosives
- National Electrical Manufacturers Association
- National Fire Protection Association
- National Plant Food Institute
- National Institute for Occupational Safety and Health
- Society of Automotive Engineers
- Underwriters Laboratories Inc.
- U.S. Department of Commerce
- U.S. Public Health Service

OSHA has the authority to promulgate emergency temporary standards in situations where employees are exposed to grave danger. Emergency temporary standards can take effect immediately upon publication in the *Federal Register*. They will remain in effect until superseded by a standard promulgated under procedures described in the Act. The law requires OSHA to develop a permanent standard no later than six months after publication of the emergency temporary standard. Any person adversely affected by any standard issued by OSHA has the right to challenge its validity by petitioning the United States Court of Appeals within 60 days after its promulgation.

Input from the Private Sector

Occupational safety and health standards promulgated by OSHA will never cover every conceivable hazardous condition that could exist in any workplace. Nevertheless, new standards and modification of existing standards are of significant interest to employers and employees alike. Industry organizations along with individuals and employee organizations need to express their views in two ways: (1) by responding to OSHA's advance notice of proposed rule making, which usually calls for information upon which to base proposed standards, and (2) by responding to OSHA's proposed standards, since it is within the private sector that most of the expertise and the technical competence lies. To do less means that industry and employees are willing to let the standards development process rest in the hands of OSHA.

Additional sources used by OSHA for the revision of existing occupational safety and health standards or the development of new ones are NIOSH criteria documents and standards advisory committees. These committees are appointed by the Secretary of Labor.

In order to promulgate, revise, or modify a standard, OSHA must first publish in the *Federal Register* a notice of any proposed rule that will adopt, modify, or revoke any standard and invite interested persons to submit their views on the proposed rule. The notice must include the terms of the new standard and provide an interval of at least 30 days (usually 60 days or more) from the date of publication for interested persons to respond. These persons may file objections to the rule and are entitled to a hearing on their objections if they request one to be held. However, they must specify the parts of the proposed rule to which they object and the grounds for such objection. If a hearing is requested, OSHA must hold one. Based on (1) the need for control of an exposure to an occupational injury or illness, and (2) the reasonableness, effectiveness, and feasibility of the control measures required, OSHA may either issue a rule promulgating an additional standard or modify or revoke an existing standard.

Variances from Standards

There will be some occasions when, for various reasons, standards cannot be met. In other cases, the protection already afforded by an employer to employees is equal to or superior to the protection that would be provided if the standard were strictly followed. The Act provides an avenue of relief from these situations by empowering OSHA to grant variances from the standards, providing that doing so would not degrade the purpose of the Act. The detailed "Rules of Practice for Variances, Limitations, Variations, Tolerances, and Exemptions" are codified in Title 29 *CFR*, Chapter XVII, Part 1905, 11(b).

OSHA can grant two types of variances—temporary and permanent. Temporary variances are generally concerned with compliance with new standards. Employers may apply for an order granting a temporary variance if they can establish that (1) they cannot comply with the applicable standard because they do not have the personnel, equipment, or time to construct or alter facilities; (2) they are taking all available steps to protect employees against exposure covered by the standard; and (3) their own programs will effect compliance with the standard as soon as possible.

Employer applications for an order for a temporary variance must contain at least the following:

1. name and address of the applicant
2. address(es) of the place(s) of employment involved
3. identification of the standard from which the applicant seeks a variance
4. representation by the applicant that he or she is unable to comply with the standard and a detailed statement of the reasons
5. statement of the steps the applicant has taken and will take, with dates, to protect employees against the hazard covered by the standard
6. statement of when the applicant expects to be able to comply with the standard and what steps have been taken, with dates, to come into compliance with the standard
7. certification that the employer has informed employees of the application and of their right to petition OSHA for a hearing. A description of how employees have been informed is to be included in the certification

Employers may also apply for a permanent variance from a standard. Variance orders can be granted if OSHA finds that employers have demonstrated, by a preponderance of evidence, that they will provide a place of employment as safe and healthful as the one that would exist if they complied with the standards.

Employer applications for a permanent variance order must contain at least the following:

1. name and address of the applicant
2. address(es) of the place(s) of employment involved
3. description of the countermeasures used or proposed to be used by the applicant
4. statement showing how such countermeasures would provide a place of employment that is as safe and healthful as that required by the standard for which the variance is sought
5. certification that the employer has informed employees of the application
6. any request for a hearing
7. description of how employees were informed of the application and of their right to petition for a hearing

An employer may request an interim order permitting either kind of variance until the formal application can be acted upon. Again, the request for an interim order must contain statements of fact or arguments why such interim order should be granted. If the request is denied, the applicant will be notified promptly and informed of the reasons for the decision. If the order is granted, all concerned parties will be informed and the terms of the order will be published in the *Federal Register*. In such cases, the employer must inform the affected employees about the interim order in the same manner used to inform them of the variance application.

Upon filing an employer's application for a variance, OSHA will publish a notice of such filing in the *Federal Register* and invite written data, views, and arguments regarding the application. Those affected by the petition may request a hearing. After review of all the facts, including those presented during the hearing, OSHA will publish its decision regarding the application in the *Federal Register.*

Beginning in the early 1980s, OSHA authorized its Regional Administrators to make "interpretations" of standards in a way that essentially became variances for individual employers. Such interpretations had no effect on other employers and reflected specific conditions at a particular workplace. OSHA has also been issuing "clarifications" of standards for employers asking for deviation from standards. While granting less than 10% of employers' requests for variances since enforcement began, OSHA has issued about eight times as many clarifications.

Record-Keeping Requirements

Most employers covered by the Act are required to maintain company records of all occupational injuries and illnesses. Regulations describing how to properly record and report injuries and illnesses are codified in Title 29 *CFR,* Chapter XVII, Part 1904. Such records consist of:

- log and summary of occupational injuries and illnesses, OSHA Form 200
- supplementary record of each occupational injury or illness, OSHA Form 101 (or state form)
- annual summary of the firm's total number of occupational injuries and illnesses—a proposed OSHA Form 300, if adopted in 1996, will be used in preparing the summary in the future. The annual summary must be posted by February 1 of the following year and remain posted until March 1 (Figure 2-1)

For details concerning recording and reporting occupational injuries and illnesses, see Chapter 9, Injury and Illness Record Keeping and Incidence Rates. Current OSHA forms for record keeping are available from all BLS regional offices and from OSHA area and regional offices. In states with an OSHA-approved plan, employers should check for any additional record-keeping requirements.

OSHA has made an effort to relieve small businesses from many burdensome record-keeping requirements. As a result, most employers who had no more than 10 employees at any time during the calendar year immediately preceding the current calendar year need not comply with the record-keeping requirements. However, any employer, regardless of size, can be notified in writing by OSHA or BLS that the organization has been selected to participate in a statistical survey of occupational injuries and illnesses. The employer will then be required to maintain a log and summary and to make reports for the time specified in the notice. Further, no employer is relieved of the obligation to report any fatalities or multiple-hospitalization accidents to the nearest OSHA area office.

In 1996, most employers with 10 or fewer employees who engaged in retail trade, finance, insurance, real estate, and services were also exempted from most of the record-keeping requirements. (Standard Industrial Classification 52–89, except 52–50, 70, 75, 76, 79, and 80.)

Reporting Requirements

Within eight hours after an accident occurs that is fatal to one or more employees or that results in the inpatient hospitalization of three or more employees, the employer must report the accident either orally or in writing to the nearest area director of OSHA. In states with approved state plans, the report must be made to the state agency that has enforcement responsibilities for occupational safety and health. If an oral report is made, it shall always be followed with a confirming letter written the same day. The report must relate the circumstances of the accident, the number of fatalities, and the extent of any injuries.

Workplace Inspection

Prior to the United States Supreme Court's decision on the controversial Barlow case—*Marshall v Barlow's Inc.,* 436 U.S. 307 (1978)—the DOL's compliance safety and health officers could enter, at any reasonable time and without delay, any establishment covered by the OSHAct to inspect the premises and all its facilities. (See Title 29 *CFR,* Chapter XVII, Part 1903.) However, since the Barlow decision, the OSHA compliance officer must obtain an inspection warrant and present it if the employer requires a warrant to permit an inspection. OSHA's entitlement to a warrant does not depend on demonstrating probable cause to believe that conditions on the premises violate the OSHA regulations. Rather, the agency merely has to show that reasonable legislative or administrative standards for conducting an inspection have been satisfied. An organization needs to determine, well in advance, if indeed an inspection warrant will be requested. As a general rule, it is not advisable for the employer to refuse entry to a CSHO who has no inspection warrant. Such action only delays an inspection and increases the officer's suspicion about working conditions. In many of these cases, OSHA can obtain an inspection warrant within 48 hours; however, generally a longer period in involved.

The OSHAct authorizes an employer representative as well as an authorized employee representative, if one is designated, to accompany the CSHO during the official inspection of the premises and all its facilities. Employee representatives also have the right to participate in both the opening and closing conferences.

Usually the authorized employee representative is the union steward or the chairman of the employee safety committee. Occasionally there may be no authorized employee representative, especially in nonunion establishments. In this instance, the CSHO will select employees at random and confer with them on matters of safety and health and work conditions.

An employer should not refuse to compensate employees for the time spent participating in an inspection tour and for related activities such as attending the opening and closing conferences.

Inspection Priorities

OSHA has established priorities for assignment of staff and resources. The priorities are as follows:

1. **Investigation of Imminent Dangers.** Allegations of an imminent danger situation will ordinarily trigger an inspection within 24 hours of notification.
2. **Catastrophic and Fatal.** Accidents will be investigated if they include any one of the following:

   - one or more fatalities
   - three or more employees hospitalized as inpatients
   - significant publicity and property damage
   - issuance of specific instructions for investigations in connection with a national office special program

3. **Investigations of Employee Complaints.** Highest priority is given those complaints that allege an imminent danger situation. Complaints reporting a "serious" situation are given high priority. If time and resources allow, the CSHO may attempt to inspect the entire workplace and not just the condition reported in the complaint, particularly if a high-hazard industry is involved.
4. **Programmed High-Hazard Inspections.** Industries are selected for inspection based on their death, injury, and illness incidence rates; employee exposure to toxic substances; and according to national and local inspection scheduling programs.
5. **Reinspections.** Establishments cited for alleged serious violations may be reinspected to determine whether the hazards have been abated, particularly if adequate abatement information is not provided by the employer to OSHA.

General Inspection Procedures

The primary responsibility of the CSHO, who is under the supervision of the OSHA Area Director, is to conduct an effective inspection to determine if employers and employees are in compliance with the requirements of the standards, rules, and regulations promulgated under the OSHAct. OSHA inspections are almost always conducted without prior notice.

To enter an establishment, the CSHO presents proper credentials to a guard, receptionist, or other person acting in such a capacity. Employers should always insist on seeing and checking the CSHO's credentials carefully before allowing the individual to enter their establishment for the purpose of an inspection (Figure 2-2). Anyone who tries to collect a penalty or promotes the sale of a product or service is not a CSHO.

The CSHO will usually ask to meet with an appropriate employer representative. It is recommended that employers furnish written instructions to security, the receptionist, and other affected personnel regarding the CSHO's right of entry and initial treatment, whom should be notified, and to whom and where the CSHO should be directed to avoid undue delay.

Opening Conference

The CSHO will conduct a joint opening conference with employer and employee representatives. Where it is not practical to hold a joint conference, separate conferences are to be held for employer representatives. If there is no employee representative, then a joint conference is not necessary. When separate conferences are held, a written summary of each conference should be made and the summary provided on request to employer and employee representatives.

Because the CSHO will want to talk with the firm's safety personnel, these employees should participate in the opening conference. The employer representative who accompanies the CSHO during the inspection should also participate in the opening conference.

At the opening conference, the CSHO will:

1. inform the employer that the purpose of the officer's visit is to investigate whether the establishment, procedures, operations, and equipment are in compliance with OSHAct requirements
2. give the employer copies of the Act, standards, regulations, and promotional materials, as necessary
3. outline in general terms:

   - scope of the inspection
   - records the officer wants to review
   - the officer's obligation to confer with employees
   - physical inspection of the workplace
   - closing conference

4. if applicable, furnish the employer with a copy of any complaint(s) that is the basis for the inspection.
5. answer questions from those attending the conference.

In the opening conference, the employer representative should find out which areas of the establishment the CSHO wishes to inspect. In some cases the inspection may include areas of the plant in which trade secrets are maintained. If this is the case, the employer representative should orally request confidential treatment of all information obtained from such areas. The employer should follow up with a trade-secret letter to the CSHO requesting the officer to keep information identified in the letter strictly confidential. The CSHO should not discuss any part of this information or provide copies to any person not authorized by law to receive the data without prior written consent of the employer.

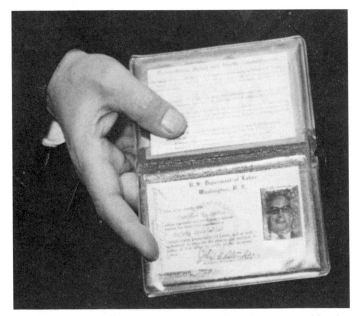

Figure 2-2. Bona fide OSHA compliance officers are equipped with official identification as shown here. The credentials are signed by the current or former Assistant Secretary of Labor for Occupational Safety and Health. If in doubt about the validity of the credentials, the employer should contact the nearest OSHA area office, determine whether or not the area office has scheduled an inspection at the establishment in question, and verify the serial number on the credentials.

During the course of the opening conference, the CSHO may ask to review company records. The CSHO is authorized to review only the records required to be maintained by the OSHAct, regulations, and standards. In general, these records include the "OSHA Injury and Illness Log and Summary" (OSHA Form 300) and the "Supplemental Record of Occupational Injuries and Illnesses" (OSHA Form 301). Such records should be made readily available to the officer. Prompt, thorough, and complete cooperation in the inspection will always make a good impression.

The CSHO may request information about the employer's current safety and health program in order to evaluate it. Naturally, a comprehensive safety and health program that shows evidence of effective accident prevention will be impressive to all concerned.

The CSHO will also ask the employer whether workers of another employer (for example, a maintenance or remodeling contractor) are working in or on the establishment. If so, the CSHO will give the authorized representative of those employees a reasonable opportunity to participate in the inspection of their work areas.

During the conference, the CSHO will meet with the employee's authorized representative and explain the person's rights. Generally, the representative will be an employee of the establishment inspected. However, the CSHO may judge that good cause has been shown to require a third party (such as an industrial hygienist or safety consultant) who is not an employee but is still an authorized employee representative to accompany the CSHO on the inspection to ensure an effective and thorough job. The final decision on this matter will rest with the CSHO.

The employer is not permitted to designate the employee representative. Employee representatives may change as the inspection process moves from department to department. The CSHO may refuse to allow any person to accompany him or her whose conduct interferes with a full, orderly inspection. If there is no authorized employee representative, the CSHO will consult with a reasonable number of employees concerning matters of safety and health in the workplace during the course of the inspection.

Inspection of Facilities

The CSHO will have the necessary instruments for checking items such as noise levels, certain air contaminants and toxic substances, electrical grounding, and the like. During the course of inspection, the CSHO will note and usually record any apparent violation of the standards, including its location, and any comments regarding the violation. The officer will do the same for any apparent violation of the general-duty clause. These notes will serve as a basis for the Area Director's citations or proposed penalties. For these reasons, the employer representative should find out what apparent violations the CSHO has detected during

the actual inspection of the facilities. The employer representative also should take the same notes as the CSHO during the inspection, including names of employees interviewed, so that the employer will have the same information as the CSHO.

It should be noted that the CSHO is only required to record apparent violations and is not required to present a solution or method of correcting, minimizing, or eliminating the violation. OSHA, however, will respond to requests for technical information regarding compliance with given standards. In such cases, the employer is urged to contact the regional or area office.

During the course of an inspection, the CSHO may receive a complaint from an employee regarding a condition alleged to be in violation of an applicable standard. The CSHO, even though the complaint is brought via an informal process, will normally inspect for the alleged violation.

In the course of a normal inspection, the CSHO may make some preliminary judgments regarding environmental conditions affecting occupational health. In such cases, the officer will generally use direct-reading instruments. Should this occur, and if proper instrumentation is available, it would be prudent for the employer to have qualified personnel at the establishment make duplicate tests in the same area at the same time under the same conditions. In addition, the employer representative should again take careful notes on the CSHO's methods as well as the results. If the inspection indicates a need for further investigation by an industrial hygienist, the CSHO will notify the Area Director, who may assign a qualified industrial hygienist to investigate further (referral). If a laboratory analysis is required, samples will be sent to OSHA's laboratory in Salt Lake City and the results will be reported back to the Area Director. Initial inspections may also be originated by industrial hygienists with subsequent referrals to a safety CSHO. Photos and videos may be taken by the CSHO to document apparent violations.

Closing Conference

Upon completion of the inspection, the CSHO will hold a joint closing conference with employee representatives and representatives of the employer. If a joint conference is not possible, separate conferences will be held. Again, the employer's safety personnel should be present at the closing conference. At this time the CSHO will advise the employer and employee representatives of all conditions and practices that may constitute an apparent safety or health violation. The officer should also indicate the applicable section or sections of the standards that may have been violated.

The CSHO will normally advise that citations may be issued for alleged violations and that penalties may be proposed for each violation. The authority for issuing citations and proposed penalties, however, rests with the Area Director or the director's representative.

The employer will also be informed that the citations will fix a reasonable time for abatement of the violations alleged. The CSHO will attempt to obtain from the employer a reasonable estimate of the time required to control or eliminate the alleged violation. The officer will take such estimates into consideration when recommending a time for abatement. Although the employer is not required to do so, it might be advantageous to give the officer copies of any correspondence or orders concerning equipment to achieve compliance. This act of good faith may help to establish a reasonable abatement period and may reduce the proposed penalty. The CSHO should also explain the appeal procedures with respect to any citation or any notice of a proposed penalty.

Informal Postinspection Conferences
Issues raised by inspections, citations, proposed penalties, or notice of intent to contest may be discussed at the request of an affected employer, employee, or employee representative at an informal conference held by the Area Director or his/her representative. Whenever the employer or employee representatives request an informal conference, both parties shall be afforded the opportunity to participate fully.

Follow-Up Inspections
Follow-up inspections will always be conducted for those situations involving imminent danger and may be conducted where citations have been issued for serious, repeated, or willful violations. Follow-up inspections will be ordered at the discretion of the Area Director.

The follow-up inspection should be limited to verifying compliance of the conditions alleged to be in violation. The follow-up inspection is conducted with all of the usual formality of the original inspection, including the opening and closing conferences and the walk-around rights of the employer and employee representative.

Violations
In addition to the general-duty clause, OSHAct occupational safety and health standards are used to determine alleged violations. There are five categories of violations: willful, serious, repeat, other than serious, and de minimis (very minor).

Willful Violations
The following definitions and procedures apply whenever the CSHO suspects that a willful violation may exist:

- A willful violation exists under the Act where the evidence shows either an intentional violation of the Act or plain indifference to its requirements.
- It is not necessary that the violation be committed with a bad purpose or an evil intent to be deemed "willful". It is sufficient that the violation was deliberate, voluntary, or intentional as distinguished from inadvertent, accidental, or ordinarily negligent.

- The determination of whether to issue a citation for a willful or repeated violation will frequently raise difficult issues of law and policy and will require the evaluation of complex factual situations. Accordingly, a citation for a willful violation shall not be issued without consultation with the Regional Administrator, who shall, as appropriate, discuss the matter with the Regional Solicitor. A repeat violation is a subsequent violation of the same or similar standard.

Serious Violation
A serious violation involves hazardous conditions that could cause death or serious physical harm to employees, and conditions that the employer knew, or should have known, existed.

OSHA's Field Operations Manual (Chapter IV, Violations) sets forth four steps for the CSHO to follow to determine whether a violation is serious or other-than-serious. Section 17(k) of the Act provides that a serious violation shall be deemed to exist in a place of employment if there is a substantial probability that death or serious physical harm could result from a condition which exists, or from one or more practices, means, methods, operations, or processes which have been adopted or are in use, in such place of employment unless the employer did not, and could not with the exercise of reasonable diligence, know of the presence of the violation.

The CSHO shall take four steps to make the determination that a violation is serious. The first three steps determine whether there is a substantial probability that death or serious physical harm could result from an accident or exposure relating to the violative condition. (The probability that an accident or illness will occur is not to be considered in determining whether a violation is serious.) The fourth step determines whether the employer knew or could have known of the violation.

Apparent violations of the general duty clause shall also be evaluated on the basis of these steps to ensure that they represent serious violations. The four elements the CSHO shall consider are as follows:

Step 1. The type of accident or health hazard exposure which the violated standard or the general duty clause is designed to prevent.
Step 2. The type of injury or illness which could reasonably be expected to result from the type of accident or health hazard exposure identified in Step 1.

- In making this determination, the CSHO shall consider all factors which would affect the severity of the injury or illness which could reasonably be predicted to result from an accident or health hazard exposure. (The CSHO shall not give consideration at this point to factors which relate to the probability that an injury or illness will occur.) The following are examples of a determination of the types of injuries which could reasonably be predicted to result from an accident:

- If an employee falls from the edge of an open-sided floor 30 ft. to the ground below, that employee could break bones, suffer a concussion, or experience other more serious injuries.
- If an employee trips on debris, that employee could experience abrasions or bruises, but it is only marginally predictable that the employee could suffer a substantial impairment of a bodily function.
- If an employee is exposed regularly and continually to beryllium at .004 mg/m^3, it is reasonable to predict that berylliosis or cancer could result.
- If an employee is exposed regularly and continually to acetic acid at 20 ppm, it is reasonable to predict that the illness which could result (irritation to nose, eyes, throat) would not involve serious physical harm.

Step 3. Whether the types of injury or illness identified in Step 2 could include death or a form of serious physical harm.

- Impairment of the body in which part of the body is made functionally useless or is substantially reduced in efficiency on or off the job. Such impairment may be permanent or temporary, chronic or acute. Injuries involving such impairment would usually require treatment by a medical doctor.

Examples include:

- amputation (loss of all or part of a bodily appendage which includes the loss of bone)
- concussion
- crushing (internal, even though skin surface may be intact)

Examples of illnesses which constitute serious physical harm include:

- cancer
- poisoning (resulting from the inhalation, ingestion, or skin absorption of a toxic substance which adversely affects a bodily system)
- lung diseases, such as asbestosis, silicosis, anthracosis
- hearing loss

Step 4. Whether the employer knew, or with the exercise of reasonable diligence, could have known of the presence of the hazardous condition.

The knowledge requirement is met if it is determined that the employer actually knew of the hazardous condition which constituted the apparent violation.

As a general rule, if the CSHO was able to discover a hazardous condition, it can be presumed that the employer could have discovered the same condition through the exercise of reasonable diligence.

Other-Than-Serious Violations

This type of violation shall be cited in situations where the accident or illness that would be most likely to result from a hazardous condition would probably not cause death or serious physical harm but would have a direct and immediate relationship to the safety and health of employees.

Repeated Violations

An employer may be cited for a repeated violation if that employer has been cited previously for a substantially similar condition and the citation has become a final order.

Identical Standard

Generally, similar conditions can be demonstrated by showing that in both situations the identical standard was violated.

Different Standards

In some circumstances, similar conditions can be demonstrated when different standards are violated.

De Minimis Violations

De minimis violations refer to conditions that represent no immediate or direct threat to safety or health. "De minimis" is short for the legal maxim, *De minimis non curat lex,* "The law does not concern itself with trifles." No written document is issued for such violations.

Citations

An investigation or inspection may reveal a condition that is alleged to be in violation of the standards or general-duty clause. In such instances, the employer may be issued a written citation that describes the specific nature of the alleged violation, cites the standard allegedly violated, and fixes a time for abatement. The employer must prominently post each citation, or copy thereof, at or near the place where the alleged violation occurred. All citations are issued by the Area Director or a designee and will be sent to the employer by certified mail.

A "Citation for Serious Violation" will be prepared to cover those violations which fall into the "serious" category. This type of violation must be assessed a monetary penalty.

A citation used for other-than-serious violations may or may not carry a monetary penalty. A citation may be issued to the employer for employee actions that violate the safety and health standards (either serious or other).

A verbal notice, in lieu of a citation, is issued for *de minimis* violations that have no direct relationship to safety and health.

If an inspection has been initiated in response to an employee complaint, the employee or authorized employee representative may request an informal review of any decision not to issue a citation. However, employees may not contest citations, amendments to citations, penalties, or lack of penalties. Employees may contest the time for abatement of a hazardous condition specified in a citation. They also may contest an employer's Petition for Modification of Abatement (PMA), which requests an extension of the abatement period. Employees must contest the PMA within 10 working days of its posting or within 10 working days after an authorized employee representative has received a copy.

Within 15 working days after the employer receives the citation, an employer may submit a written objection to the citation to OSHA. The OSHA Area Director then forwards the objection to OSHRC. Employees may request an informal conference with OSHA to discuss any issues raised by an inspection, citation, notice of proposed penalty, or employer's notice of intention to contest.

Petition for Modification of Abatement

Upon receiving a citation, the employer must correct the cited hazard by the prescribed date. However, factors beyond the employer's reasonable control may prevent the work from being completed on time. In such a situation, the employer who has made a good-faith effort to comply may file for a PMA date.

The written petition should specify (1) all steps the employer took to achieve compliance, (2) the additional time needed to complete the work, (3) reasons why additional time is needed, (4) all temporary steps being taken to safeguard employees against the cited hazard during the intervening period, (5) the fact that a copy of the PMA was posted prominently or at least near each place where a violation occurred, and (6) the employee representative (if there is one) who received a copy of the petition.

Penalties

Penalties may range to $7,000 for serious violations and to $70,000 for repeated or willful violations. Penalties are based on gravity of violation, good faith, size, and history of the employer. A "failure to abate" penalty may be a daily penalty for each day of failure to correct a violation past the stated abatement date.

Egregious Policy

If OSHA considers the apparent violations flagrant, the agency, instead of grouping similar violations, may propose a separate penalty for each instance or employee exposed.

Some of the factors used to determine the situation include:

- the number of worker fatalities, a worksite catastrophe, or a large number of similar injuries or illnesses

- a violation which results in high rates of injuries or illnesses
- the organization has an extreme history of workplace violations
- the employer seriously disregarded workplace safety and health responsibilities
- a large number of violations is found at the worksite.

States can adopt their own penalty structure, which may even exceed the one set up under OSHA.

As of April 1996, the egregious penalty policy for the general duty clause for each exposed employee is being litigated regarding its appropriateness.

CONTESTED CASES

An employer has the right to contest any OSHA action. The employer may contest one or more of the following: a citation, a proposed penalty, a notice of failure to correct a violation, or the time allotted for abatement of an alleged violation. OSHAct regulations that cover procedures for contesting cases are codified in Title 29 *CFR*, Chapter XX, Part 2200. On the other hand, an employee or authorized employee representative may contest only the time allotted for an abatement of an alleged violation.

Subsequent to initiating a formal contest of a citation, employers should request an informal conference with the Area Director or the Area Director's representative. Many times such informal sessions will resolve questions and issues, thus avoiding the formal contested case proceedings. The informal conference should occur within 15 working days of receipt of the violation and will not extend the Notice of Intent to Contest period. If the 15-day time period is exceeded, the citation will become a final order.

The informal conference will be attended by representatives from OSHA, the company, and the company's union representatives (if applicable). Conference members may be requested to obtain a more complete explanation of the violation cited by OSHA, to obtain an understanding of the specific standards cited in the violation, or to negotiate and enter into an informal settlement agreement. Other areas that may be discussed include the company's difficulty in meeting the abatement dates, employer defenses, problems concerning employee safety practices, and obtaining answers to any questions that the employer may bring to the table. The informal conference provides an opportunity for both sides to resolve the disputed citation and associated penalties in a relatively cordial atmosphere and is specifically intended to avoid the need for the more formal contest of the citation. The choice to contest the citation elevates the process to the Occupational Safety and Health Review Commission (OSHRC).

However, the informal conference may fail to resolve the dispute between OSHA and the employer. If the employer elects to contest the case, affected employees or the authorized employee representative are automatically deemed to be parties to the proceeding. In contesting an OSHA action, the employer must comply with the following rules that apply to a specific case:

1. The employer must notify the Area Office which initiated the action that the employer is to contest the case. This must be done within 15 working days after receiving OSHA's notice of proposed penalty; it should be sent by certified mail. If the employer does not contest within the required 15 working days, the citation and proposed assessment of penalties are deemed to be a final order of OSHRC and are not subject to review by any court or agency. As a result, the alleged violation must be corrected within the abatement period specified in the citation.

2. If any of the employees working at the site where the alleged violation exists are union members, a copy of the notice of contest must be served upon their union.

3. If employees who work on the site are not represented by a union, a copy of the notice of contest must either be posted at a place where the employees will see it or be served upon them personally.

4. The notice of contest must also list the names and addresses of those parties who have been personally served a notice, or, if such notice is posted, it must contain the address of the posted location.

5. In some cases the employees at the site of the alleged violation are not represented by a union and have not been personally served with a copy of the notice to contest. If so, posted copies must specifically advise the unrepresented employees that they may not be able to assert their status as parties to the case if they fail to properly identify themselves to the Commission or to the Hearing Examiner before the hearing begins or when it first opens.

6. There is no specific form for the notice of contest. However, such notice should clearly identify what is being contested—the citation, the proposed penalty, the notice of failure to correct a violation, or the time allowed for abatement—for each alleged violation or combination of alleged violations.

If the employer contests an alleged violation in good faith, and not solely for delay or variance of penalties, the abatement period does not begin until the OSHRC enters the final order. When a notice of contest is received by an Area Director from an employer, an employee, or an authorized employee representative, the Director will file with OSHRC the notice of contest and all contested citations, notice of proposed penalties, or notice of failure to abate.

Upon receiving the notice of contest from the Area Director, the Commission will assign the case a docket number. Ultimately, an Administrative Law Judge (ALJ) will be assigned to the case and will conduct a hearing at a location reasonably convenient to those concerned. At the hearing, OSHA presents its case and is subject to a cross-examination by other parties. The party contesting then presents its case and is also subject to a cross-examination by other parties. Affected employees or an authorized employee representative may participate in the hearings. The decision by the ALJ will be based only on what is in the record. Therefore, if statements go unchallenged, they will be assumed to be fact.

After the hearings are completed, the ALJ will submit the record and a report with decisions to OSHRC. If no Commissioner orders a review of an ALJ's decisions, they will stand as OSHRC's decision. If any Commissioner orders a review of the case, the Commission itself must render a decision to affirm, modify, or vacate the judge's decision. The Commission's orders become final 15 days after issuance, unless stayed by a court order.

Any person adversely affected or aggrieved by an order of the Commission may obtain a review of the order in the United States Court of Appeals. However, the person must seek a review within 60 days of the order's issuance.

SMALL BUSINESS LOANS

The Act enables small businesses to obtain economic assistance for health- and safety-related issues. It amends the Small Business Act to provide for financial assistance to small firms that must make changes to comply with the standards promulgated under the OSHAct or by a state under a state plan. Before approving any assistance, the Small Business Administration (SBA) must first determine that the small firm is likely to suffer substantial economic injury without financial help.

An employer can apply for a loan under one of two procedures: (1) before federal or state inspection in order to come into compliance, or (2) after federal or state inspection to correct alleged violations.

When an employer has not been inspected and requests a loan to bring the establishment into compliance prior to an inspection, the employer must submit to the SBA:

- a statement of the conditions to be corrected
- a reference to the OSHA standards that require the employer to make corrections
- a statement of the firm's financial condition showing that a loan is needed

The employer should submit this information to the nearest SBA field office along with any background material. The SBA will then refer the application to the appropriate OSHA Regional Office, Office of Technical Support. The OSHA Regional Office will review the application and advise SBA whether the employer is required to correct the described conditions in order to come into compliance and whether the proposed use of funds will accomplish the needed corrections. Direct contact with the applicant will be initiated by OSHA only after clearance with the SBA.

If the employer is making an application after an inspection to correct alleged violations, the procedure is the same as before inspection, except that the applicant also must furnish SBA a copy of the OSHA citation(s). SBA then refers the application to the OSHA Area Office that conducted the inspection. That office will notify SBA whether the proposed use of loan funds will adequately correct cited violations.

Forms for loan applications may be obtained from any SBA field office. In some instances, private lending institutions will be able to provide the form for SBA/bank participation loans.

FEDERAL-STATE RELATIONSHIPS

The OSHAct encourages the states to assume the fullest responsibility for administering and enforcing their own occupational safety and health laws. However, in order to assume this responsibility, such states must submit a state plan to OSHA for approval. If such a plan satisfies designated conditions and criteria, OSHA must approve the plan. The regulations pertaining to state plans for the development and enforcement of state standards are codified in Title 29 *CFR,* Chapter XVII, Part 1902. The states and possessions listed in Figure 2-3 have approved plans.

The basic criterion for approval of state plans is that the plan must be "at least as effective as" the federal program. It was not Congress's intent to require that state programs be a mirror image of the federal program. Congress believed rules for developing state plans should be flexible to allow consideration of local problems, conditions, and resources. The Act provides for funding up to half the costs of the implementation of the state program.

A state plan must include any occupational safety and health issue (industrial, occupational, or hazard group) for which a corresponding federal standard has been promulgated. A state plan cannot be less stringent, but it may include subjects not covered in the federal standards. However, state plans that do not include those issues covered by the federal program, in effect, surrender such issues to OSHA. For example, a state plan may cover all industry except construction. If such is the case, the state surrenders its jurisdiction for safety and health programs in construction operations to OSHA. It is then OSHA's obligation to enforce the federal standards for those operations not covered by the state plan.

Following approval of a state plan, OSHA will continue to exercise its enforcement authority until it determines on the basis of actual operations that the state plan is indeed being satisfactorily carried out. If the implementation of the state plan is satisfactory during the first three years after the plan's approval, then the relevant federal standards and OSHAct enforcement of such standards no longer apply to issues covered under the state plan. This means that for the interim period when dual jurisdiction exists, employers must comply with both state and federal standards.

While the state agencies administering the state plan are vitally concerned with its success, members of the state legislature do not always share their enthusiasm. The legislature must not only appropriate an adequate budget, but in many cases must pass legislation enabling the state agency to carry out all the functions incorporated in the state plan. At times the state agency responsible may fail to fully implement the state plan, and the state's performance falls short of being "at least as effective as" the federal program. In such instances, OSHA has the right and obligation to withdraw its approval of the state plan and once again assume full jurisdiction in that state.

MEDICAL ACCESS AND RIGHT-TO-KNOW

In the 1980s, two important standards came into effect: (1) final rules for access to exposure and medical records and (2) hazard communication. Both were designed to provide employees with information about the hazardous conditions to which they are, or have been, exposed, and to give them access to their own medical and exposure records and documented information about chemicals used in the workplace. Employee medical records should be retained by the company for 30 years from the termination date of the employee.

Medical Access Standard

Employers must maintain records on the exposure employees have had to dangerous substances during their working time. These records, together with an employee's medical records, must be available to the employee, or a designated representative, upon request. The standard embodies certain restrictions on the extent of the information that must be provided and on the procedure for providing it to designated representatives of employees.

Hazard Communication Standard

Over 20 states and local communities enacted statutes and ordinances, which in some instances do not go as far as the federal standard but in other cases go much further. The indication is that employers will face more stringent requirements to comply with both federal and state/local laws, where the two do not conflict.

Approximately one-fourth of the U.S. work force is exposed to one or more chemical hazards. Exposure to these hazards may cause or contribute to many serious health effects. Due to the seriousness of safety and health problems associated with these chemicals and the lack of information available to many employees, OSHA issued the Hazard Communication Standard (HCS).

Alaska Department of Labor
P. O. Box 21149
Juneau, ALASKA 99802–1149
(907) 465–2700

Industrial Commission of Arizona
800 W. Washington
Phoenix, ARIZONA 85007
(602) 542–4661

California Department of Industrial Relations
45 Fremont Street
San Francisco, CALIFORNIA 94105
(415) 972–8846

Connecticut Department of Labor
200 Folly Brook Boulevard
Wethersfield, CONNECTICUT 06109
(860) 566–4550

Hawaii Department of Labor and Industrial Relations
830 Punchbowl Street
Honolulu, HAWAII 96813
(808) 586–8842

Indiana Department of Labor
402 West Washington
Room W–195
Indianapolis, INDIANA 46204
(317) 232–2663

Iowa Division of Labor Services
1000 E. Grand Avenue
Des Moines, IOWA 50319
(515) 281–3447

Kentucky Labor Cabinet
U.S. Highway 127 South
Frankfort, KENTUCKY 40601
(502) 564–3070

Maryland Division of Labor and Industry
Department of Labor, Licensing and Regulation
501 St. Paul Place 4th floor
Baltimore, MARYLAND 21202–2272
(410) 333–4176

Michigan Department of Labor
Bureau of Safety and Regulation
7150 Harris Dr.
P.O. Box 30015
Lansing, MICHIGAN 48079
(517) 332–1814

Minnesota Department of Labor and Industry
443 Lafayette Road
St. Paul, MINNESOTA 55155
(612) 296–2342

Nevada Department of Industrial Relations
Division of Occupational Safety and Health
400 West King Street
Suite 400
Carson City, NEVADA 89710
(702) 687–3032

New Mexico Environmental Department
1190 St. Francis Drive N2200
Santa Fe, NEW MEXICO 87503-0968
(505) 827–2850

New York State Department of Labor
One Main Street
Brooklyn, NEW YORK 11201
(718) 797–7668

North Carolina Department of Labor
4 W. Edenton Street
Raleigh, NORTH CAROLINA 27601
(919) 733–7166

Accident Prevention Division
Oregon Department of Insurance and Finance
Labor and Industries Building
Salem, OREGON 97310
(503) 378–4271

Puerto Rico Department of Labor and Human Resources
Prudencio Rivera Martinez Building Ed.
505 Munoz Rivera Avenue
Hato Rey, PUERTO RICO 00918
(809) 754–5353

South Carolina Department of Labor
3600 Forest Drive
P.O. Box 11329
Columbia, SOUTH CAROLINA 29211-1329
(803) 734–9594

Tennessee Department of Labor
710 James Robertson Parkway
Nashville, TENNESSEE 37243
(615) 741–2582

Utah Occupational Safety and Health
160 E. 300 South
P.O. Box 146650
Salt Lake City, UTAH 84114-6650
(801) 530–6900

Vermont Department of Labor and Industry
Drawer 20
Montpelier, VERMONT 05620-3401
(802) 828–2765

Figure 2-3. States with approved plans.

Virgin Islands Department of Labor 2131 Hospital St. Christiansted St. Croix, VIRGIN ISLANDS 00820 (809) 773–1994	**Washington Department of Labor and Industries** 7273 Lyderson Way SW Room 334 - AX-31 Turnwater, WASHINGTON 98502 (360) 902–5800
Virginia Department of Labor and Industry Powers-Taylor Building 13 S. 13 St. Richmond, VIRGINIA 23219 (804) 786–5886	**Wyoming Department of Occupational Heath and Safety** 122 W. 25th St. Herszhler Building 2nd floor Cheyenne, WYOMING 82002 (307) 777–7786

Figure 2-3. (Concluded).

The Standard applies to all employers covered by OSHA in both manufacturing and nonmanufacturing sectors including construction. It also covers workers who may be exposed to hazardous materials under normal conditions or in a foreseeable emergency.

The basic purpose of the HCS is to establish uniform requirements to make sure that the hazards of all chemicals produced, imported, or used within the United States are evaluated. This hazard information must be transmitted to affected employers and employees.

This is accomplished through:

- hazard evaluation
- employee training
- container labeling
- Material Safety Data Sheets (MSDSs)
- a written hazard communication program

Written Hazard Communication Program

All employers using materials that may pose a hazard to employees must establish a written, comprehensive hazard communication program which includes provisions for container labeling, MSDS availability, and employee training. The program must also include a list of hazardous chemicals in each work area, how the employer will inform employees of the hazards of nonroutine tasks and unlabeled pipes, and how the employer will inform contractors in manufacturing facilities of the hazards to which their employees may be exposed. The program need not be lengthy or complicated, and must be available to employees.

Hazard Evaluation

The quality of a hazard communication program depends on the accuracy of the initial hazard assessment. The primary responsibility for hazard evaluation lies with the chemical manufacturer or importer. If a company uses a process that produces a chemical to which employees are exposed, the employer/owner is considered a "chemical manufacturer" and must evaluate the chemical's hazards. This is true for chemical intermediates or for decomposition products such as welding fumes.

A producer or user of a material that may be hazardous should develop a hazard evaluation, an inventory of chemicals used or produced. The inventory can be developed by reviewing purchase orders and performing a physical inventory of all containers of chemicals. After developing an inventory, the next step is to determine which substances are hazardous. The products in the inventory and their components should be compared to the following lists:

- 29 *CFR* 1910.1000–1047, Toxic and Hazardous Substances, OSHA
- "Threshold Limit Values for Chemical Substances in the Work Environment," American Conference of Governmental Industrial Hygienists (ACGIH), (last edition)
- National Toxicology Program (NTP), "Annual Report on Carcinogens" (latest edition)
- International Agency for Research on Cancer (IARC), "Monograph" (latest edition)

The next step is to determine whether any of the remaining chemicals on the inventories possess physical or health hazards. Consult the MSDS provided by the manufacturer, importer, or distributor, or discuss questions with a knowledgeable industrial hygienist or safety professional (call the National Safety Council).

Employee Training

An employer must provide training for employees exposed to hazardous chemicals. Training must be done when employees are first assigned to an operation and whenever a new hazard is introduced into the work area. Training must cover the following topics:

- existence and requirements of the Standard
- operations in the work area where hazardous chemicals are present and the hazards of these chemicals
- how the hazard communication program is implemented in the workplace, how to read and interpret information on labels and MSDSs, and how employees can obtain and use available hazard information

- measures employees can take to protect themselves from hazards
- specific procedures adopted to provide protection, such as work practices and the use of engineering controls or personal protective equipment

Material Safety Data Sheets

Chemical manufacturers and importers must develop an MSDS for each hazardous chemical they produce or import. You must obtain MSDSs for every hazardous chemical in your workplace. Copies must be readily available to employees. The HCS requires that specific information be on the MSDS:

- Section I—Manufacturer's name, address, phone number, and date sheet prepared
- Section II—Hazardous Ingredients/Identity Information—chemical identity of components, exposure limits (OSHA, PEL, ACGIH, TLV, and other recommended limits)
- Section III—Physical/Chemical Characteristics—boiling point, vapor pressure and density, specific gravity, melting point, etc., and the physical and chemical data which indicate the potential for vaporization
- Section IV—Fire and Explosion Hazard Data—flash point, flammable limits, extinguishable, media, unusual fire and explosion hazards, special firefighting procedures
- Section V—Reactivity Data—stability of product, potential for polymerization and decomposition, materials and conditions to avoid
- Section VI—Health Hazards—acute and chronic hazards, carcinogenicity, signs and symptoms of exposure, emergency and first-aid procedures
- Section VII—Precautions for Safe Handling and Use—procedures to be used for spills, waste disposal, handling, and storage
- Section VIII—Control Measures—personal protective equipment, ventilation, special worker or hygienic practices

The producer of the MSDS is responsible for the information on it and must ensure that all sheets are up to date.

Labeling

Chemical manufacturers, importers, and distributors must be sure that containers of hazardous chemicals leaving the workplace are labeled with:

- identity of the product
- written hazard warnings (in English)
- name and address of manufacturer or other responsible party

In the workplace, each container of hazardous chemicals must be labeled, tagged, or marked with:

- identity of the product (able to reference this name to a MSDS)
- hazard warnings (written in English), including target organ(s), if applicable
- graphic symbols. However, OSHA cites studies indicating that graphic symbols are not as quickly recognized as word statements. The warnings may be printed in other languages in addition to English.

There are several exemptions for in-plant labeling of containers. A sign or placard can be used for a number of stationary containers within a work area with similar contents. Operating procedures, process sheets, batch tickets, blend tickets, and similar written materials can be substituted for container labels if they contain the same information and are readily available in the work area to the employees.

Portable containers which are intended for immediate use by the employee who makes the transfer are also exempted. The employer is not required to label pipes or piping systems. However, the means the employer will use to train employees on contents of piping systems must be described in the written hazard communication program.

WHAT DOES IT ALL MEAN?

Congressional action in creating the OSHAct is only one step toward achieving the full purpose underlying the Act. Getting it to work with reasonable efficiency is the second and more difficult task. Achieving this purpose of providing a safe, healthy environment on and off the job will depend on the willingness and cooperation of all concerned—employees and organized labor as well as business and industry.

There is no doubt that the Act has given new visibility to the whole realm of occupational safety and health. Because many employee rights are incorporated into the OSHAct, it has given employees a significant role to play in occupational safety and health matters. It has moved the laggards from "little or no safety" to "some safety," but not to "optimum safety". It has raised occupational safety and health issues to a higher priority in business management. It has given new status and responsibility to professionals working in the occupational safety and health field. Management is now relying more heavily on these safety professionals for advice. And, the Act has bestowed a new status to nationally recognized organizations that develop industry standards.

The OSHAct has also given new impetus to the field of occupational health, a much more difficult discipline when compared with occupational safety. Much more needs to be done to determine what kinds of exposures are indeed hazardous to humans and under what conditions. Further, a great deal more needs to be done to determine what countermeasures are not only adequate but also reasonable and feasible to eliminate or minimize exposures to occupational health hazards. Far more research and data about occupational health will be required to achieve the best occupational safety and health programming.

The OSHAct has encouraged greater training for professionals in occupational safety and health. Several universities have developed new curricula and programs leading to various degrees in this field—and more are yet to come.

The OSHAct also gave new emphasis to the product safety discipline. Until the passage of the U.S. Consumer Product Safety Act, the OSHAct was the most significant piece of legislation affecting product safety ever passed by the Congress. Designers and manufacturers of equipment now used by industry have a moral (but not legal) obligation to design, deliver, and install such equipment in accordance with the applicable standards.

The OSHAct is not without limitations. Mere compliance with the requirements of the Act will not achieve optimum safety and health in terms of cost, benefits, and human values. All those concerned must recognize that occupational safety and health cannot be handed to the employer or to the employee by legislative enactment or administrative decree. At best, state or federal occupational safety and health standards are minimum standards and can cover only those areas that are enforceable—namely, control over physical conditions and environment.

As a matter of hard reality, enforcement standards simply do not adequately relate to the human in the human-machine-environment system. Important elements of a complete safety program, such as (1) establishment of work procedures to limit risk, (2) supervisory training, (3) job instruction training for employees, (4) job safety analysis, and (5) human factors engineering, by and large have not been included in the standards promulgated under the OSHAct. Neither do the standards address such issues as employee attitudes, morale, or teamwork.

For the most part, the occupational safety and health standards developed under the OSHAct are minimal criteria and represent a floor rather than a goal to achieve. Thus, to rely on mere compliance with these standards is to invite disaster since the residual risk after compliance often remains unacceptable. Effective accident prevention and control of occupational health hazards must go beyond the OSHAct.

Generally, a violation of a standard is only symptomatic of something wrong with the management safety system as a whole. Only complete occupational safety and health programming as described elsewhere in this Manual will achieve a level of risk acceptable to employers and employees. The real objective and the purpose of the OSHAct is improved occupational safety and health performance and not merely compliance with a set of standards.

ENVIRONMENTAL IMPACT

Managers responsible for occupational safety and health are increasingly affected by developments in environmental

law. In some cases, responsibility for compliance with environmental laws and regulations rests on the occupational safety and health manager or on a member of the same department. It is becoming more difficult to draw a clear line between safety and health in the plant and the safety and health of the surrounding community. Further, environmental laws and regulations contain provisions addressing issues of protection for anyone workers or the public—who comes into contact with dangerous waste products and other hazardous substances.

The most important pieces of legislation of concern to those responsible for occupational safety and health are the Resource Conservation and Recovery Act (RCRA); the Comprehensive Environmental Response, Compensation and Liability Act (CERCLA), commonly known as Superfund; the Clean Air Act; and the Clean Water Act. The 1986 amendments to Superfund, which set up procedures for reporting and controlling those hazardous substances with an environmental impact, are particularly important. These environmental laws impose requirements for the use of protective equipment and procedures, and for extensive reporting to federal, state, and local government agencies. They also affect the design and operation of many manufacturing processes and waste disposal systems.

As a result, occupational safety and health and environmental law compliance have become inextricably interrelated—and will continue to be so. (See Chapter 5, Environmental Management.) A memorandum of understanding to coordinate enforcement activities was signed by OSHA and the EPA, an accord OSHA described as a "precedent setting" agreement.

1. The agreement provides, among other measures, for EPA to assist OSHA with its special emphasis program for the petrochemical industry, and for OSHA to assist EPA in enforcement actions directed at lead pollution.
2. The memorandum is the second signed by the two agencies in recent months. In September, 1990, OSHA and EPA signed an agreement to set out procedures for joint inspection at some 30 hazardous waste incinerators.
3. The two agencies will work jointly on other programs affecting employers.

MINE SAFETY AND HEALTH ACT

On November 9, 1977, President Jimmy Carter signed into law the U.S. Mine Safety and Health Act of 1977, Public Law 95–164. The Act became effective March 9, 1978.

The U.S. Mine Safety and Health Act of 1977 (subsequently referred to as the Mine Act) is intended to ensure, so far as possible, safe and healthful working conditions for miners. It applies to operators of all types of mines, both coal and metal/nonmetal and both surface and underground. The Mine Act states that mine operators are responsible for preventing unsafe, unhealthful conditions or practices in mines that could endanger the lives and health of miners.

Mine operators are required to comply with the safety and health standards promulgated and enforced by the Mine Safety and Health Administration (MSHA), an agency within the DOL. Like OSHA, MSHA may issue citations and propose penalties for violations. Unlike the employees under OSHAct, miners (employees) are subject to government sanctions for violating safety standards relating to smoking in or near mines and mining machinery. Similarly, employers and other supervisory personnel may be held personally liable for civil penalties or may be prosecuted criminally for violations of Mine Act standards.

LEGISLATIVE HISTORY

Historically, the Bureau of Mines within the Department of the Interior administered mine safety and health laws. However, the bureau was eliminated at the end of 1995, and some of its functions were transferred to other sectors within the Department of the Interior. Before Congress passed the Mine Act, mine operators were governed by two separate laws, the Federal Coal Mine Safety and Health Act of 1969 and the Federal Metal and Nonmetallic Mine Safety Act of 1966. The U.S. Department of Labor, MSHA now administers mining safety laws.

Because the Bureau of Mines during its existence was also charged with promoting mine production, critics charged that this responsibility produced an inherent conflict of interest with respect to enforcement of safety and health laws. The establishment of the Mine Enforcement Safety Administration (MESA) in 1973 within the Interior Department failed to answer the criticism. Congress looked for alternative solutions, including the transfer of mine safety and health to the OSHAct. Finally, Congress resolved the issue by adopting the Mine Act, which repealed the Federal Coal Mine Safety and Health Act of 1969 and the Federal Metal and Nonmetallic Mine Safety Act of 1966.

ADMINISTRATION

The administration and enforcement of the U.S. Mine Safety and Health Act are vested primarily with the Secretary of Labor and the Mine Safety Health Review Commission. The agency that administers the investigation and prosecution aspects of the enforcement process is MSHA. The Mine Safety and Health Review Commission, an independent agency created by the Mine Act, reviews contested MSHA enforcement actions.

The Mine Act distinguishes between health research and safety research. Miner health research and standards development is the responsibility of NIOSH, in cooperation with MSHA.

Mine Safety and Health Administration

The Mine Safety and Health Administration, located within the DOL, administers and enforces the Mine Act.

MSHA is headed by an Assistant Secretary of Labor for Mine Safety and Health, who is appointed by the president with the advice and consent of the Senate. The Assistant Secretary acts on behalf of the Secretary of Labor. For the purposes of this chapter, MSHA is also synonymous with the term Secretary or Assistant Secretary of Labor.

MSHA is authorized to adopt procedural rules and regulations to carry out the provisions of the Mine Act. The agency also has the responsibility and authority to perform the following:

- promulgate, revoke, or modify safety and health standards
- conduct mine safety and health inspections
- issue citations and propose penalties for violations
- issue orders for miners to be withdrawn from all or part of the mine
- investigate mine accidents
- grant variances
- seek judicial enforcement of its orders.

Aiding the Assistant Secretary in carrying out the provisions of the Mine Act are, among others, (1) an Administrator for Coal Mine Safety and Health and (2) an Administrator for Metal and Nonmetal Mine Safety and Health. Each administrator is responsible for a Division of Safety and a Division of Health.

Mine Safety and Health Review Commission

The five-member Mine Safety and Health Review Commission serves as the administrative adjudication body. The Review Commission is completely independent from the DOL. The commission has the authority to assess all civil penalties provided in the Mine Act. It reviews contested citations, notices of proposed penalties, withdrawal orders, and employee discrimination complaints. Commission members are appointed by the president for six-year terms with the advice and consent of the Senate. The first commissioners took office for staggered terms of two, four, and six years.

The Commission appoints Administrative Law Judges (ALJs) to conduct hearings on behalf of the Commission. The decision of an ALJ becomes a final decision of the Commission 40 days after its issuance unless the Commission directs a review.

MAJOR PROVISIONS OF THE MINE ACT

This section describes the general scope of the regulations and guidelines that apply to the mining industry. Since passage of the Act, industry management has worked with union representatives and employees to upgrade the health and safety of workers in this hazardous occupation.

Coverage

The Mine Act covers all mines that affect commerce. The Act defines "mines" as all underground or surface areas from which minerals are extracted and all surface facilities used in preparing or processing the minerals. Structures, equipment, and facilities including roads, dams, impoundments, and tailing ponds used in connection with mining and milling activities are also included. The Mine Act provides that the Mine Safety and Health Administration develop and implement regulations, provide for state grants, and offer training at the National Mine Academy.

Because some facilities have operations that are under OSHA and others under MSHA, MSHA and OSHA have established an interagency agreement which, among other things, delineates certain areas of authority and provides for coordination between OSHA and MSHA in all areas of mutual interest. (This was published in 44 *CFR* 22827, April 17, 1979.) In case of jurisdictional disputes between OSHA and MSHA, the Secretary of Labor is authorized to assign enforcement responsibilities to one of the agencies.

Advisory Committees

The Act requires the Secretary of the Interior to appoint an Advisory Committee on Mine Safety Research. The Secretary of Health and Human Services is required to appoint an Advisory Committee on Mine Health Research. The Secretary of Labor may appoint other advisory committees as needed to aid in carrying out the provisions of the Act.

Miners' Rights

The Act affords miners (employees) a number of rights, including the following:

- Miners may request an inspection in writing if they believe a violation of a standard or an imminent danger situation exists in the mine. Similarly, written notification of alleged violations or imminent danger situations may be given to an inspector before or during an inspection.
- A representative designated by miners must be given the opportunity to accompany the inspector during the inspection process. Also, representatives have the right to participate in postinspection conferences held by the mine inspector on the premises.
- Miners are entitled to observe monitoring and to examine monitoring records when the standards require tracking exposure to toxic materials or harmful physical agents.
- Miners, including former miners, must be given access to medical and other records documenting their own exposures.
- Operators must notify miners if they are exposed to toxic substances in concentrations that exceed prescribed limits of exposure. Further, those miners must be informed of the corrective action being taken.
- Miners given new work assignments for medical reasons because of their exposure to hazardous substances must be paid at their regular rate if the related standard so provides.
- Miners who are not working because of a withdrawal order are entitled to be compensated subject to certain limits.
- Miners or their authorized representatives may contest the issuance, modification, or termination of any MSHA order or the time period set for abatement.
- Miners adversely affected or aggrieved by an order of the Review Commission may obtain judicial review.
- Miners may file a complaint with the Review Commission concerning their compensation for not working as the result of a withdrawal order issued by MSHA, or for acts of employer discrimination when notified by MSHA.
- Miners, through their authorized representative, may petition for a variance from mine safety standards.
- MSHA is required to send to the miners' authorized representative copies of proposed safety or health standards. In addition, the mine operator must post a copy of such standards on its office bulletin board.
- To keep miners informed, mine operators are required to post copies of orders, citations, notices, and decisions issued by MSHA or the Review Commission. Posting of these items shall be on the mine bulletin board (sec. 109a).
- Miners are entitled to receive training for their specific jobs and must be given refresher training annually. They are entitled to normal compensation while being trained. Miners who leave the operator's employ are entitled to copies of their training certificates.
- Operators may not discriminate against miners or representatives of miners for the exercise of miners' rights under this act.
- Miners suffering from black lung disease are entitled to extensive black lung benefits.

Duties

Mine operators are required to comply with the safety and health standards and other rules promulgated under the Act and are subject to sanctions for failing to comply. Similarly, every miner is required to comply with the safety and health standards promulgated under the Act. Any miner, whether management or shift worker, may be cited for smoking or carrying smoking materials, matches, or lighters in certain situations. For knowingly and willfully violating mandatory safety and health standards, any miner (management or lead-men) may be fined or sentenced to jail.

Miner Training

Mine operators are required to have a safety and health training program approved by MSHA which provides the following:

- At least 40 hours of instruction for new underground miners. The training must include the statutory rights of miners and their representatives under the Act, use of the self-rescue device and respiratory devices, hazard recognition, escapeways, walk-around training, emergency procedures, basic ventilation, basic roof control, electrical hazards, first aid, and the safety and health aspects of the task assignment.
- Twenty-four hours of instruction for new surface miners. The training must include all of the items for underground miners just listed, except escapeways, basic ventilation, and basic roof control, none of which are essential to surface mining.
- At least eight hours of annual refresher training for all miners.

The Mine Act requires that the training must be conducted during normal working hours and that the miners must be paid at their normal rate during the training period. Regulations concerning training and retraining of miners are codified at Title 30 *CFR,* Part 48. (See Chapter 16, Safety Training.)

MINE SAFETY AND HEALTH STANDARDS

The Mine Act authorizes MSHA to promulgate, modify, or revoke mine safety and health standards. To implement the initial set of standards without delay, the safety and health standards under the Coal Mine Safety and Health Act of 1969 were adopted under the Mine Act. These standards are codified in Title 30 *CFR,* Parts 70, 71, 74, 75, 77, and 90.

Similarly, the Mine Act adopted the mandatory standards that prevailed under the Metal and Nonmetallic Mine Safety Act of 1966. Later, many of the advisory standards were adopted as mandatory standards under the Mine Act. All of the metal/nonmetal standards are codified in Title 30 *CFR,* Parts 56, 57, and 58. Part 55 (Metal/Nonmetal) was eliminated on April 15, 1985, and recodified as Part 56. (A list of all of the standards promulgated under authority of the Mine Act is provided in the References, at the end of this chapter.)

If MSHA should determine that a standard is needed, it may propose a standard or seek assistance from an advisory committee. MSHA must publish the proposed standard in the *Federal Register,* and establish a time period of at least 30 days for public comment. MSHA may hold public hearings if objections are raised about a proposed standard. Upon adoption by MSHA, the standard must be published in the *Federal Register.* The new standard becomes effective upon publication or at a date specified.

Judicial Review

Any person adversely affected by any standard issued by MSHA has the right to challenge its validity by petitioning in the United States Court of Appeals within 60 days after promulgation of the standard. Although filing such a petition does not stay enforcement of the standard, the court may order a stay before conducting a hearing on the petition. Objections that were not raised during rule making will not be considered by the court, unless good cause is shown why an objection was not raised.

Input from the Private Sector

Mine safety and health standards promulgated by MSHA can never cover every conceivable hazardous condition that might exist in mines. Nevertheless, new standards and modification or revocation of existing standards are important to mine operators and miners alike. Mining operator organizations, miner organizations, and individuals should express their views during the rule-making process by responding to MSHA's proposed standards. Their participation is important because most expertise and technical competence lies in the private sector. To do less means that mining operators and miners are willing to let the standards-development process be controlled solely by MSHA.

Emergency Temporary Standards

MSHA has the authority to publish emergency temporary standards if it deems that immediate action must be taken to protect miners "exposed to grave danger" from toxic substances or physically harmful agents. The emergency temporary standard is effective immediately upon publication in the *Federal Register* and remains in effect until superseded by a permanent standard developed under normal rule-making procedures. MSHA is required to establish a permanent standard within nine months after publication of an emergency temporary standard.

Variances

When a petition for modification is filed by an operator or a representative of miners, MSHA may modify the application of any mandatory safety standard. The Act does not allow for variances of health standards.

A petition for modification may be granted under two conditions. First, MSHA must find that an alternative method of compliance will achieve the same measure of protection for miners as the standard would provide. Second, MSHA must determine that the standard in question will provide less safety to miners.

A petition for modification should be filed with the Assistant Secretary of Labor for Mine Safety and Health. If the mining operator submits a petition, a copy must be served on the miners' representative. Similarly, if the miners' representative petitions for modification, a copy must be served on the mine operator. The petition must include the name and address of the petitioner and the mailing address, identification, and name or number of the affected mine. It must also identify the standard, describe the desired modification, and state the basis for the request.

MSHA will publish a notice of the petition in the *Federal Register.* The notice will summarize information contained in the petition. Interested parties have 30 days to comment. MSHA then will conduct an investigation on the merits of the petition, and the appropriate Administrator will issue a proposed decision. The proposed decision becomes final 30 days after service, unless a hearing request is filed within that time.

Granting of Petition

If MSHA agrees, a petition can be granted if (1) the alternative method will achieve the same measure of protection for miners as compliance with the standard would provide or (2) application of the standard will result in a diminution of miners' safety.

A petition request is filed with the Director, Office of Standards, Regulations, and Variance. A copy of the request must be posted on the mine bulletin board. Until a petition is granted, there can be no discussion of mine operators requesting temporary relief from enforcement of a mandatory safety standard (30 *CFR* 44.16).

Accident, Injury, and Illness Reporting

For the purpose of reporting accidents, injuries, and illnesses under the Mine Act, the term "accident" includes:

- an unplanned inundation of a mine by a liquid or gas
- a fatality at a mine
- an injury to an individual at a mine that has a reasonable potential to result in the worker's death
- an injury that may result in death
- entrapment for more than 30 minutes
- an unplanned ignition or explosion of gas or dust
- an unplanned fire not extinguished within 30 minutes of its discovery
- an unplanned ignition or explosion of a blasting agent or an explosive
- an unplanned roof fall in active work areas where roof bolts are in use or a roof fall that impairs ventilation or impedes passage
- coal or rock outbursts that causes withdrawal of miners *or* that disrupts mining activity for more than one hour
- an unstable condition at an impoundment, refuse pile, or culm bank requiring emergency action or for which individuals must evacuate an area—or failure of an impoundment, refuse pile, or culm bank
- damage to hoisting equipment in a shaft or slope that endangers an individual or interferes with use of equipment for more than 30 minutes
- an event at the mine that causes a fatality or bodily injury to an individual who is not at the mine at the time of occurrence

"Occupational injury" means an injury that results in death, loss of consciousness, medical treatment, temporary assignment to other duties, transfer to another job, or inability to perform all duties on any day after the injury.

"Occupational illness" is an illness or disease that may have resulted from work at a mine or for which a compensation award is made.

All mine operators are required to immediately report accidents (as defined earlier) to the nearest MSHA district or subdistrict office. Similarly, operators must investigate and submit to MSHA, upon request, an investigation report on accidents and occupational injuries. The investigation report must include:

- date and hour of occurrence
- date the investigation began
- names of the individuals participating in the investigation
- description of the site
- explanation of the accident or injury
- name, occupation, and experience of any miner involved
- if appropriate, a sketch of the accident site, including dimensions
- description of actions taken to prevent a similar occurrence
- identification of the accident report submitted

All mine operators must submit to MSHA within 10 days of the incident a report of each accident, occupational injury, or illness on Form No. 7000–1. A separate form is to be prepared for each miner affected.

Accident investigation reports and the injury/illness reports filed by means of Form 7000–1 must be kept for five years at the mine office closest to the mine in which the accident, injury, or illness occurred.

Inspection and Investigation Procedures

Inspections of a mine are conducted by MSHA to determine if an imminent danger exists in the mine and if the mine operator is complying with the safety and health standards and with any citations, orders, or decisions issued. Mine inspectors from MSHA, called "authorized representatives" of the Secretary of Labor, or representatives of NIOSH have the right to enter any mine to inspect the site or to conduct an investigation. However, NIOSH representatives have no enforcement authority. The Mine Act's provision for conducting inspections without securing a search warrant has been held valid. MSHA inspectors, on certain occasions, can give advance notice of an impending inspection.

As in OSHA, MSHA inspectors cannot give employers advance warning of an inspection conducted to determine compliance. However, NIOSH may give advance notice of inspections carried out for research or other purposes.

Frequency

MSHA must inspect underground mines in their entirety at least four times a year. Surface mines are to be inspected at least two times a year. Normally, these inspections require multiple visits. MSHA must also conduct spot inspections based on the number of cubic feet of methane or other explosive gases liberated in mining operations during a 24-hour period. The Act authorizes MSHA to develop guidelines for additional inspections based on other criteria.

Miner Complaints

A miner's authorized representative, or any individual miner if there is no authorized representative, may request in writing an immediate inspection by MSHA if he or she has reasonable grounds to believe that a violation of a standard or an imminent danger situation exists. MSHA will normally conduct a special inspection soon after receiving the complaint. If MSHA determines that a violation does not exist, it must notify the complainant in writing. Similarly, before or during an inspection, the miners' authorized representative, or an individual miner if no representative exists, may notify the inspector in writing of any alleged violation or imminent danger situation believed to exist in the mine.

Health Hazard Evaluations

Upon written request of an operator or authorized representative of miners, NIOSH is authorized to enter a mine to determine whether any toxic substance, physical agent, or equipment found or used in the mine is potentially hazardous. A copy of the evaluation will be submitted to both the operator and the miners' representative.

The Inspection Procedure

A MSHA inspector will normally begin the inspection at the mine office. The officer will inform the mine operator of the reason for the inspection and request all needed records. The inspector's review of the records will likely focus on the preshift or on-shift examination record. Such records help the inspector determine where to concentrate attention during the inspection.

An operator's representative and a representative authorized by the miners must be given the opportunity to accompany the MSHA inspector during the inspection. Similarly, each must be given the opportunity to participate in the postinspection conference. One miner representative (who is an employee of the operator) must be paid the regular wage for the time spent accompanying the inspector.

Whenever the inspector observes a condition that appears to be a violation of the standards, he or she must issue a citation. If, in the opinion of the mine inspector, an imminent danger condition exists, then the inspector must issue a withdrawal order.

After completing the inspection, the inspector will hold a closing conference with the representatives of the mine operator and the miners to discuss all findings. The closing conference is mandatory. Occasionally, in the interests of those concerned, a separate closing conference

may be held with the mine operator and another with the miners' representative (see Program Policy Letter P94-III-1 (3/31/94), page 9, in 30 *CFR* 100.6 for additional information).

Withdrawal Orders

MSHA has the authority, under specified conditions, to order an operator to withdraw the miners from all or part of a mine. Miners idled by such an order are entitled to receive compensation at their regular rate of pay for specified periods of time. All miners in the affected area must be withdrawn except those necessary to eliminate the hazard, public officials whose duty requires their presence in the area, representatives of the miners qualified to make mine examinations, and consultants.

If an imminent danger is found to exist during an inspection, MSHA is required to order the withdrawal of all persons from the affected area, except those referred to in Section 104 of the Act, until the danger no longer exists. The order must describe the conditions or practices both causing and constituting the imminent danger and the area affected. The withdrawal order does not preclude issuance of a citation and proposed penalty.

Other situations for which MSHA may issue a withdrawal order include the following:

- If, during a follow-up inspection, MSHA finds that a mine operator has failed to abate a cited violation and there is no valid reason to extend the abatement period, MSHA must issue a withdrawal order until the violation is abated.
- If a mine operator fails to abate a respirable dust violation for which a citation has been issued and the abatement period has expired, MSHA must either extend the abatement period or issue a withdrawal order.
- If two violations constituting "unwarrantable failures" to comply with the standards are found during the same inspection, or if the second unwarrantable violation is found within 90 days of the first, a withdrawal order must be issued. An unwarrantable failure violation (second within 90 days) refers to a situation in which the operator knew or should have known that a violation existed and yet failed to take corrective action.
- Miners may be ordered withdrawn from a mine if they have not received the safety training required by the Act. Miners withdrawn for this reason are protected by the Mine Act from discharge or loss of pay.

Except for withdrawal orders issued for respirable dust violations and imminent danger situations, an operator or a miner may file a written request for a temporary stay of the order with the Review Commission. Also, they may request temporary relief from any modification or termination of a withdrawal order. The Review Commission may grant a stay of a withdrawal order provided granting such relief would not endanger the safety and health of miners.

Both operators and miners, or their representatives, may contest an imminent danger withdrawal order or any modification or termination of such an order. They must file with the Review Commission an application for review of the order within 30 days after receiving it or after receiving any modification or termination of such an order.

Citations

If a MSHA inspector or the inspector's supervisors believe that the mine operator is in violation of any standard, rule, order, or regulation promulgated under the Mine Act, they must issue a citation to the operator with "reasonable promptness." A citation may be issued immediately at the site of the alleged violation. In any case, the inspector must provide a citation for each alleged violation before leaving the mine property, unless mitigating circumstances exist.

Citations must be in writing, describe the nature of the violation, and include a reference to the provision of the Mine Act, standard, rule, regulation, or order allegedly violated. The citation, based on the inspector's opinion, will establish a reasonable time for the employer to correct the violation. Termination dates of citations are established after consultation with the mine operator and upon consideration of the hazard that is cited.

The Act requires the operator to post all citations on the mine's bulletin board. Copies are sent to the miners' representative, to the state agency charged with administering mine safety and health laws, and to those designated by the operator as having responsibility for safety and health in the mine.

Within 10 days after an operator receives a citation or an order for an alleged violation, the operator and/or miners' representative has the right to request a safety and health conference with MSHA management. The conferencing process is designed to allow parties an opportunity to present additional evidence or mitigating facts or circumstances concerning citations or orders issued by the inspector. The conferencing officer considers information from these sources and has the authority to modify, vacate, or affirm the citation.

This procedure enables mine operators and/or representatives of miners to resolve some issues prior to the penalty stage.

Penalties

MSHA must assess a civil penalty of not more than $50,000 for each violation of the Act. The agency may assess penalties up to $5,000 per day (maximum) for each day the operator fails to correct a cited violation. If an operator is convicted of willfully violating a standard, a federal judge may assess a fine of $25,000 and/or one year imprisonment. A penalty of up to $250 per occurrence may be assessed miners who willfully violate a standard that prohibits smoking or carrying smoking materials, matches, or lighters into or near a mine or mining equipment.

After the alleged violation has been corrected and after any safety and health conference has been conducted, MSHA will issue the proposed penalty. Nonsignificant and substantial (commonly termed "Non-S&S") violations that are abated in a timely fashion usually result in a minimal penalty. In determining the amount of the penalty for all violations, MSHA considers six criteria:

1. operator's history of previous violations
2. size of the operator's business
3. evidence of operator negligence
4. impact on the operator's ability to remain in business
5. gravity of the violation
6. demonstrated good faith to achieve rapid compliance after notification of the alleged violation

Most of the proposed penalty assessments are assigned a range of penalty points based on each of the above criteria. The total points are then converted into a dollar penalty. In addition, serious violations or those involving negligence are usually given a special penalty assessment.

CONTESTED CASES

Operators and miners (or miners' representatives) have 30 calendar days after receiving notice to contest a citation, a withdrawal order, or a proposed penalty. The notice of contest must be sent by registered or certified mail to the Assistant Secretary of Labor for Mine Safety and Health at the MSHA headquarters located at 4015 Wilson Boulevard, Arlington, VA. 22203. The miners' representative must also receive a copy of the notice.

If a mine operator fails to notify MSHA within the 30-day period and no notice is filed by any miner or miners' representative, the citation and/or the proposed penalty is deemed a final order of the Mine Safety and Health Review Commission and is not subject to review by any court or agency. However, it should be understood that the citation and the penalty have separate 30-day periods within which they may be contested. For instance, if a mine operator fails to contest the citation, he is not without options. When the proposed penalty is received at some later date, the mining operator has another 30 days to contest that penalty. In addition, if the penalty is contested, the citation may be reopened for negotiation at the same time.

A citation may be contested before the operator receives a notice of proposed penalty, even if the alleged violation has been abated. A notice of contest states what is being contested and the relief sought. A copy of the order or citation being contested must accompany the notice of contest.

Upon receiving the notice of contest, MSHA immediately notifies the Review Commission. Then a docket number and an ALJ are assigned to the case. The Review Commission will provide an opportunity for a hearing via the ALJ. The ALJ may hold an informal conference with all parties involved to clarify and settle the issues. If the issues are not resolved, then a formal hearing conducted by the ALJ will take place.

Mine operators, miners or representatives of miners, and applicants for employment may be parties to the Review Commission proceedings. Miners or their representatives may become parties by filing a written notice with the Executive Director of the Review Commission prior to the hearing.

The Review Commission's ALJs are authorized, among other things, to administer oaths, issue subpoenas, receive evidence, take depositions, conduct hearings, hold settlement conferences, and render decisions. The decision will include findings of facts, conclusions of law, and an order. A copy of the decision will be issued to each of the parties involved and to each of the Commissioners. Any person aggrieved by the decision of the ALJ may, within 30 days after an order or decision is issued, file a petition for a discretionary review by the Review Commission.

The Review Commission on its own motion and with the affirmative vote of two members may direct review of an ALJ's decision within 30 days of issuance only under two conditions: (1) when the decision may be contrary to law or to Commission policy or (2) when the petition raises a novel question of policy.

Any person adversely affected by a decision of the Review Commission, including MSHA, may appeal to the United States Court of Appeals within 30 days of issuance. The court may affirm, modify, or set aside the Commission's decision in whole or in part.

INTERNATIONAL REGULATIONS

As discussed in Chapter 1, Historical Perspectives, the rapid emergence of worldwide markets, the pace of technological and scientific change, and the sweeping social changes occurring around the world have all created a need for a uniform set of international safety and health standards. So far, Europe and the United States have led the way, although other nations are beginning to realize the importance of protecting the worker and the environment as they seek to industrialize their economies.

The goals of international standardization of safety and health regulations include:

- setting a common ground for market agreements and technological applications
- improving the quality and reliability of products and services
- protecting the user and/or the environment
- providing compatibility of goods and services

Some of the more important regulations include those developed by the European Union and the International Organization for Standardization.

European Union

The European Union (EU, formerly the EEC) has sought to create safety and health regulations that would afford the greatest protection for workers and consumers in member nations while providing uniform standards for all nations doing business either in Western Europe itself or with Western European nations. The countries involved are Austria, Belgium, Denmark, Finland, France, Germany, Greece, Ireland, Italy, Luxembourg, the Netherlands, Portugal, Spain, Sweden, and the United Kingdom. One standard, the Framework Directive (89/391/EEC) of June 12, 1989, encourages improvements in on-the-job safety and health of workers and describes the obligations of employers to ensure the safety and health of their employees. This standard has been adopted by EU countries and most nations doing business with these countries. (See Chapter 1 for a more detailed discussion of the Framework Directive; see also Chapter 4, International Legal and Legislative Framework, in the Environmental Management volume of the *Accident Prevention Manual*.)

ISO 9000 Series

The ISO 9000 series applies to companies who export goods and products to the EU countries. These standards describe a process for establishing quality management and quality assurance in companies. The basic goal of the 9000 series is to provide guidelines to help companies achieve consistency and uniformity of products or services throughout the entire chain of supply. (See Chapter 1, Historical Perspectives, and Chapter 12, Environmental Management.)

ISO 14000 Series

The ISO 14000 series is also related to establishing standards for quality assurance, aimed primarily at protecting the environment from the by-products of industrial processes. Companies can use their management systems developed under ISO 9000 regulations as a basis for establishing environmental management. The ISO 14000 is being phased in throughout the 1990s. The European Union has adopted the ISO standards as meeting the requirements of EU environmental standards. (See Chapter 1, Historical Perspectives, and Chapter 12, Environmental Management.)

Total Quality Management

Total quality management (TQM) principles, which emphasize continuous improvement, have been applied to safety and health issues on both the domestic and international fronts. TQM should not be thought of as a technical program, however; like any quality assurance program, it is a management system. As applied to health and safety procedures, TQM requires the following:

- commitment by top management to quality in all areas of organizational life, particularly health and safety
- training and education in safety matters at all levels of the organization
- continuous improvement programs implemented throughout the organization
- measure systems used to identify areas where improvements can be made
- communication fostered between management and employees, encouraging employee involvement and participation in quality improvement

MULTINATIONAL REGULATIONS

The current explosion in the growth of multinational businesses has raised a corresponding concern about regulating these "stateless" corporations. Because they act in more than one nation, regulations restricting these companies must be approved by all the countries involved. The conflict between encouraging economic development while protecting workers, consumers, and the environment does not appear to have an easy—or early—solution. Governments, nongovernmental organizations, unions, industries, and safety and environmental groups are all seeking to create uniform, enforceable regulations and standards that will help ensure the health and safety of employees, users, and the general environment. Some of the early efforts represent movement in the right direction but there is considerable room for improvement.

North American Free Trade Agreement

During the discussions that preceded passage of the North American Free Trade Agreement (NAFTA), there was considerable concern that the United States would be unable to hold its trading partners to the same high standards of environment protection as it does its domestic businesses. Amid increasing pressure from environmental groups, President Clinton added side agreements guaranteeing stringent environmental protection standards to NAFTA. In 1993, a U.S. District Court ruling attempted to require an environmental impact statement as part of the NAFTA deliberations.

Although the ruling was later overturned, it represented a major concern by nongovernmental groups about destruction of the environment resulting from unregulated international trade. The lack of a worldwide governing body with power to enforce environmental protection and worker health and safety laws is a major obstacle to guaranteeing compliance with international standards and guidelines.

General Agreement on Tariffs and Trade

Another major trade agreement in the world's economic system, the General Agreement on Tariffs and Trade (GATT) has also become the center of a controversy around its environmental implications. GATT was originally designed to eliminate trade barriers among nations and not to address environmental or health and safety issues. However, the hope among many groups was that the GATT organization would use its influence to persuade nations to act in a more ecologically sensitive manner. Unfortunately, the organization has often ruled that countries could conduct their business as they saw fit, even when doing so directly threatened wildlife or the environment. For example, if the United States wished to restrict tuna fishers from inadvertently killing dolphins, the United States would have to initiate its own agreements and treaty arrangements with the nations involved.

SUMMARY

- OSHA's primary responsibilities are (1) to promulgate, modify, and revoke safety and health standards; (2) to conduct inspections and investigations and to issue citations, including proposed penalties; (3) to require employers to keep records of safety and health data; (4) to petition the courts to restrain imminent danger situations; and (5) to approve or reject state plans for programs under the Act.
- The Occupational Safety and Health Review Commission (OSHRC) is a quasi-judicial, three-member board that hears cases when OSHA actions are contested by employers or employees.
- The primary functions of NIOSH are (1) to develop and establish recommended occupational safety and health standards, (2) to conduct research experiments and demonstrations, and (3) to develop educational programs to provide qualified safety and health personnel.
- With some exceptions, OSHAct applies to every employer in all 50 states and U.S. possessions who has one or more employees and who is engaged in a business affecting commerce. Under OSHA, the employer has a general and specific duty to provide safe, healthy work environments and comply with all applicable standards. Employees, in turn, must comply with all standards that apply to their situation and conduct on the job.
- Sanctions are in the form of citations for violating standards and civil and criminal penalties if the employer failed to comply with duties under the Act. OSHA can grant temporary and permanent variances from standards.
- Employers and employees are granted certain rights under OSHA regulations to seek advice and consultation with OSHA staff, participate in inspections, take an active role in developing safety and health standards, apply for financial assistance, and appeal or contest OSHA findings and decisions.
- OSHA requires employers to keep records of all occupational injuries and illnesses.

- General inspection procedures include an opening conference, walk-through of the establishment, documenting alleged violations, interviewing workers, and a closing conference. Employers or employees can request informal post-inspection conferences to discuss problems and solutions. Follow-up inspections are conducted to ensure correction of alleged violations.
- All citations except those for de minimis violations must be posted near the place where the violation occurred. The employer can object to or contest a citation and petition for a modification of the time for correcting the violation. If a citation or penalty is not contested, it takes immediate effect.
- The informal conference may be used before the citation is formally contested. If the parties do not agree with the results of the informal conference, the employer may formally contest the citation to the Occupational Safety and Health Review Commission (OSHRC).
- The OSHAct encourages states to assume the fullest responsibility for administering and enforcing their own occupational safety and health laws.
- Employers must inform employees of any hazardous materials used in the workplace and train them in methods of handling these materials and self-protection.
- The Resource Conservation and Recovery Act and the Comprehensive Environmental Response, Compensation, and Liability Act (Superfund) set up procedures for reporting and controlling hazardous substances that affect surrounding environments, mandated the use of protective equipment, and affected the design and operation of manufacturing processes and waste disposal.
- The Federal Mine Safety and Health Act of 1977 is intended to ensure safe, healthful working conditions for miners in all types of mines. Its primary duties are (1) to develop, revoke, or modify safety and health standards, (2) conduct inspections, (3) issue citations and propose penalties for violations, (4) order miners withdrawn from parts or all of mines, (5) grant standard variances, and (6) seek judicial enforcement of its orders. MSHA has the authority to propose new standards, emergency temporary standards, or variances.
- Miners (employees) have the right to request inspections, accompany inspectors during their plant tour, observe monitoring procedures, gain access to their own medical and employment records, be informed of their exposure to toxic substances, contest any alteration in MSHA orders, and receive training for their jobs. Mine operators must comply with safety and health standards and are subject to sanctions if they fail to do so.

- Within 10 days of an accident, mine operators must file a report on the incident with the nearest MSHA district or subdistrict office. They must follow up with a report on their investigation of the accident's causes.
- Mine inspections are usually carried out without prior warning to the operator. The inspector will conduct an opening conference and a walk-through of the premises, document all alleged violations, and hold a closing conference.
- Inspectors or their supervisors can issue citations for violations. Penalties are assessed for each violation and for each day the violation remains uncorrected. Operators, miners, or others affected have 30 days in which to contest the citations either at an informal conference or through formal procedures.
- International and multinational regulations seek to create standards and guidelines to ensure the health and safety of workers, consumers, and the environment wherever companies operate across national boundaries.

DIRECTORY OF FEDERAL AGENCIES

The Occupational Safety and Health Administration

National Headquarters:
OSHA
Occupational Safety and Health Administration,
U.S. Department of Labor, Department of Labor Building,
200 Constitution Avenue NW,
Washington, DC 20210; 202/523–8017.
NIOSH
National Institute for Occupational Safety and Health,
U.S. Department of Health and Human Services,
1600 Clifton Road NE,
Atlanta, GA 30333; 404/639–3061.
BLS
Bureau of Labor Statistics,
U.S. Department of Labor,
200 Constitution Avenue NW,
Washington, DC 20210; 202/523–7943.
OSHRC
Occupational Safety and Health Review Commission,
1825 K Street NW,
Washington, DC 20006; 202/634– 7943.

OSHA Regional Offices

Region I (Connecticut, Maine, Massachusetts, New Hampshire, Rhode Island, Vermont)
1st Floor, 133 Portland Street,
Boston, MA 02114; 617/565–7159.
Region II (New York, New Jersey, Puerto Rico, Virgin Islands)

(Room 670) 201 Varick Street,
New York, NY 10014; 212/944–3432.

    Region III (Delaware, District of Columbia, Maryland, Pennsylvania, Virginia, West Virginia)
Gateway Building, 3535 Market Street,
Philadelphia, PA 19104; 215/596–1201.

    Region IV (Alabama, Florida, Georgia, Kentucky, Mississippi, North Carolina, South Carolina, Tennessee)
1375 Peachtree Street NE,
Atlanta, GA 30367; 404/881–3573.

    Region V (Illinois, Indiana, Michigan, Minnesota, Ohio, Wisconsin)
J.C. Kluczynski Federal Building,
230 South Dearborn Street,
Chicago, IL 60604; 312/353–2220.

    Region VI (Arkansas, Louisiana, New Mexico, Oklahoma, Texas)
555 Griffin Square Building, Griffin & Young Streets,
Dallas, TX 75202; 214/767–4731.

    Region VII (Iowa, Kansas, Missouri, Nebraska)
Old Federal Office Building,
911 Walnut Street, Kansas City, MO 64106; 816/374–5861.

    Region VIII (Colorado, Montana, North Dakota, South Dakota, Utah, Wyoming)
Federal Building, 1961 Stout Street,
Denver, CO 80294; 303/844–3061.

    Region IX (Arizona, California, Hawaii, Nevada, Guam, American Samoa, Trust Territory of the Pacific Islands)
Federal Building, 450 Golden Gate Avenue,
San Francisco, CA 94102; 415/995–5672.

    Region X (Alaska, Idaho, Oregon, Washington)
Federal Office Building, 909 First Avenue,
Seattle, WA 98174; 206/442–5930.

The Mine Safety and Health Administration

    Mine Safety and Health Administration,
U.S. Department of Labor, Room 601,
4015 Wilson Boulevard,
Arlington, VA 22203; 202/235–1452

    National Institute for Occupational Safety and Health,
U.S. Department of Health and Human Services,
1600 Clifton Boulevard NE,
Atlanta, GA 30333; 404/329– 3061

    National Mine Safety and Health Academy,
P.O. Box 1166,
Beckley, WV 25801; 304/255–0451

    Mine Safety and Health Review Commission,
1730 K Street NW,
Washington, DC 20006; 202/653–5625

COMPILATIONS OF REGULATIONS AND LAWS

The safety and health professional and industrial hygienist should be familiar with three U.S. government publications:

- *Federal Register* (FR)
- *Code of Federal Regulations* (CFR)
- *United States Code* (USC).

    The first two are published by the Office of the *Federal Register*, National Archives and Records Service, General Services Administration. All three publications are available from the Superintendent of Documents, U.S. Government Printing Office, Washington, DC 20402. Every safety office should obtain them.

The *Federal Register*

The *Federal Register,* published daily Monday through Friday, provides a system for making publicly available regulations and legal notices issued by all federal agencies. In general, an agency will issue a regulation as a proposal in the *Federal Register* followed by a comment period, and then will finally promulgate or adopt the regulation in the publication. Reference to material published in the *Federal Register* is usually in the format A *FR* B, whereby A is the volume number, *FR* indicates *Federal Register,* and B is the page number. For example, 43 *FR* 58946, indicates volume 43, page 58946.

The *Code of Federal Regulations*

The *Code of Federal Regulations,* published annually in paperback volumes, is a compilation of the general and permanent rules and regulations that have been previously released in FR.

    The *CFR* is divided into 50 different titles, representing broad subject areas of federal regulations, for example: Title 29—"Labor" ; Title 40—"Protection of Environment" ; Title 49—"Transportation" ; etc. Each title is divided into chapters (usually bearing the name of the issuing agency), and then further divided into parts and subparts covering specific regulatory areas. Reference is usually in the format 40 *CFR* 250.XX, meaning Title 40 *CFR* Part 250 (Hazardous Waste Guidelines and Regulations), or 49 *CFR* 172.XX, (Hazardous Materials Table and Hazardous Materials Communications Regulations). The "XX" refers to the number of the specific regulatory paragraph.

    The *Code of Federal Regulations* is kept up to date by the individual issues of the *Federal Register*. These two publications must be used together to determine the latest version of any given rule or regulation.

The *United States Code*

Whereas the two previously described publications contain rules and regulations authorized by a law, the *United States Code* (U.S. Code) describes the actual law. The *U.S. Code* is the current official compilation by subject of the public, general, and permanent laws of the United States in force. It is prima facie the law. It is presumed to be the law. The presumption is rebuttable by production of prior unrepealed Acts of Congress at variance with the *U.S. Code*.–(Preface to the *United States Code.*) In other words, the *U.S. Code* is an arrangement by subject of the federal legislation of a public and permanent nature, from 1789 to date in force today; it does not include repealed and expired acts.

Editorial Selection of Statutes Included

Inclusion of legislation in the Code is under the supervision of a committee of the House of Representatives. Code sections included in one edition may be omitted from the next or may be changed from one title of the Code to another by editorial fiat. To avoid possible confusion, the date or supplement number of the *U.S. Code* edition cited should be given.

Editorial notes in the *U.S. Code* are printed with it but are not technically a part of it. This material, set out in fine print beneath the respective code sections, is meant to help the reader understand those sections.

Text of Code Sections

Sections are copied from the original enactment but often with changes of form, not of substance. Introductory words of the act, such as "Provided," or "That," at the beginning of the original statute section may be omitted; and the code title and section numbers are substituted in the body of the text for the official title and sections of the act from which they derived, when these are mentioned in the act. Other changes are also permitted. Statutory authority is cited in parentheses at the end of each code section or group of sections. Such authority may be Congressional acts or joint resolutions, presidential executive orders, or reorganization plans. The code is well indexed.

Tables of Contents to Code Titles

Each of the 50 titles into which the *U.S. Code* is divided is preceded by a table of contents, consisting of a table of chapters in that title, by number and caption. At the beginning of each chapter, there is a similar expanded table of contents.

REFERENCES

Bureau of National Affairs, Inc., 1231 25th Street NW, Washington, DC 20037. *Occupational Safety and Health Reporter.*

Commerce Clearing House, Inc., 4025 West Peterson Avenue, Chicago, IL 60646. Employment Safety and Health Guide.

La Dou J, ed. *Occupational Health and Safety,* 2nd ed. Itasca, IL: National Safety Council, 1994.

National Institute for Occupational Safety and Health, 5600 Fisher Lane, Rockville, MD 20857.

"The Advisor" (newsletter).

"Occupational Safety and Health Directory."

National Safety Council, 1121 Spring Lake Drive, Itasca, IL 60143.

Fundamentals of Industrial Hygiene, 4th ed., 1996

Safety and Health (magazine)

"OSHA Up-to-Date" (newsletter).

Price MO and Bitner H. *Effective Legal Research,* 4th ed. Boston, MA: Little, Brown, and Co., 1979.

Rothstein M. *Occupational Safety and Health Law,* 3rd ed. St. Paul, MN: West Publishing Co., 1990 (with yearly updates).

Superintendent of Documents, U.S. Government Printing Office, Washington, DC 20402.

Annual List of Toxic Substances.

Directory of Federal Agencies.

Federal Register.

Field Operations Manual.

Industrial Hygiene Technical Manual.

Occupational Safety and Health Act of 1970 (P.L. 91–596).

Occupational Safety and Health Regulations, Title 29 *Code of Federal Regulations (CFR).*

Part 11—Department of Labor, National Environmental Policy Act (NEPA) Compliance Procedures.

Part 1901—Procedures for State Agreements.

Part 1902—State Plans for the Development and Enforcement of State Standards.

Part 1903—Inspections, Citations and Proposed Penalties.

Part 1904—Recording and Reporting Occupational Injuries and Illnesses.

Part 1905—Rules of Practice for Variances, Limitations, Variations, Tolerances and Exemptions.

Part 1906—Administration Witnesses and Documents in Private Litigation.

Part 1907—Accreditation of Testing Laboratories.

Part 1908—Consultation Agreements.

Part 1910—Occupational Safety and Health Standards.

Part 1911—Rules of Procedure for Promulgating, Modifying, or Revoking Occupational Safety or Health Standards.

Part 1912—Advisory Committees on Standards.

Part 1912a—National Advisory Committee on Occupational Safety and Health.

Part 1913—Rules of Agency Practice and Procedure Concerning OSHA Access to Employee Medical Records.

Part 1915—Occupational Safety and Health Standards for Shipyard Employment.

Part 1917—Marine Terminals.

Part 1918—Safety and Health Regulations for Longshoring.

Part 1919—Gear Certification.

Part 1920—Procedure for Variations from Safety and Health Regulations under the Longshoremen's and Harbor Workers' Compensation Act.

Part 1921—Rules of Practice in Enforcement Proceedings under Section 41 of the Longshoremen's and Harbor Workers' Compensation Act.

Part 1922—Investigational Hearings under Section 41 of the Longshoremen's and Harbor Workers' Compensation Act.

Part 1924—Safety Standards Applicable to Workshops and Rehabilitation Facilities Assisted by Grants.

Part 1925—Safety and Health Standards for Federal Service Contracts.

Part 1926—Safety and Health Regulations for Construction.

Part 1928—Occupational Safety and Health Standards for Agriculture.

Part 1949—Office of Training and Education, Occupational Safety and Health Administration.

Part 1950—Development and Planning Grants for Occupational Safety and Health.

Part 1951—Grants for Implementing Approved State Plans.

Part 1952—Approved State Plans for Enforcement of State Standards.

Part 1953—Changes to State Plans for the Development and Enforcement of State Standards.

Part 1954—Procedures for the Evaluation and Monitoring of Approved State Plans.

Part 1955—Procedures for Withdrawal of Approval of State Plans.

Part 1956—State Plans for the Development and Enforcement of State Standards Applicable to State and Local Government Employees in States Without Approved Private Employee Plans.

Part 1960—Basic Program Elements for Federal Employee Occupational Safety and Health Programs and Related Matters.

Part 1975—Coverage of Employees under the Williams-Steiger Occupational Safety and Health Act of 1970.

Part 1977—Discrimination against Employees Exercising Rights under the Williams-Steiger Occupational Safety and Health Act of 1970.

Part 1990—Identification, Classification, and Regulation of Potential Occupational Carcinogens.

Part 2200—Review Commission Rules of Procedure.

Part 2201—Regulations Implementing the Freedom of Information Act.

Part 2202—Standards of Ethics and Conduct of Occupational Safety and Health Review Commission Employees.

Part 2203—Regulations Implementing the Government in the Sunshine Act.

Part 2204—Implementation of the Equal Access to Justice Act.

Part 2205—Enforcement of Nondiscrimination on the Basis of Handicap.

Mine Safety and Health Act of 1977 (PL 95–164).

Mine Safety and Health Regulations and Standards.

29 *CFR* Part 2700—Mine Safety and Health Review Commission, Rules of Procedure.

30 *CFR* Part 11—Respiratory Protective Devices; Tests for Permissibility; Fees.

Part 40—Representative of Miners.

Part 41—Notification of Legal Identity.

Part 43—Procedures for Processing Hazardous Condition Complaints.

Part 44—Rules of Practice for Petitions for Modification of Mandatory Safety Standards.

Part 45—Independent Contractors.

Part 46—State Grants for Advancement of Safety and Health in Coal and Other Mines.

Part 47—National Mine Health and Safety Academy.

Part 48—Training and Retraining of Miners.

Part 49—Mine Rescue Teams.

Part 50—Notification, Investigation, Reports and Records of Accidents, Injuries, Illnesses, Employment, and Coal Production in Mines.

Part 56--Safety and Health Standards: Surface Metal and Nonmetal Mines.

Part 57—Safety and Health Standards: Underground Metal and Nonmetal Mines.

Part 58—Health Standards for Metal and Nonmetal Mines.

Part 70—Mandatory Health Standards: Underground Coal Mines.

Part 71—Mandatory Health Standards: Surface Coal Mines and Surface Work Areas of Underground Coal Mines.

Part 74—Coal Mine Dust Personal Sampler Units.

Part 75—Mandatory Safety Standards: Underground Coal Mines.

Part 77—Mandatory Safety Standards: Surface Coal Mines and Surface Work Areas of Underground Coal Mines.

Part 90—Mandatory Health Standards: Coal Miners Who Have Evidence of the Development of Pneumoconiosis.

Part 100—Criteria and Procedures for Proposed Assessment of Civil Penalties.

42 *CFR* Part 37—Specifications for Medical Examinations of Underground Coal Miners.

Part 85—Requests for Health Hazard Evaluations.

Part 85a—NIOSH Policy on Workplace Investigations.

REVIEW QUESTIONS

1. What was the purpose of the Williams-Steiger Act (OSHAct)?
2. Prior to the passage of the OSHAct other pieces of national safety legislation had been passed in the United States; what was the main drawback of these pieces of legislation?
3. Labor was a strong proponent of having the federal government having a more significant role in occupational safety and health. What did they base their position on?
4. List the two organizations that are vested with the administration and enforcement of the OSHA Act.
   a.
   b.

5. Name the quasi-judicial board of three members appointed by the president and confirmed by the Senate to adjudicate cases.
6. What are the primary functions of NIOSH?
   a.
   b.
   c.
   d.
7. List the five main technical services provided by NIOSH.
   a.
   b.
   c.
   d.
   e.
8. The final responsibility for compliance rests with the _____.
9. What are the two voluntary safety programs established by OSHA for companies that have exceptional safety programs?
10. In what publication are possible federal safety standards published prior to their being brought up for formal consideration to enacted into law?
11. What two types of variances can OSHA grant?
   a.
   b.
12. Employers must report to OSHA within _____ _____ after an accident occurs that is fatal to one or more employees or that results in the inpatient hospitalization of three of more employees.
13. Are compliance safety and health officers required to present solutions or methods of correcting, minimizing, or eliminating a violation?
   a.
   b.
14. List the five categories of violations.

   a.
   b.
   c.
   d.
   e.
15. What four steps must a compliance safety and health officer take to determine that a violations is serious?
   a.
   b.
   c.
   d.
16. How may an employer contest a violation?
17. How long should employee medical records be retained by employers?
18. What is the basic purpose of the Hazard Communication Standard?
19. The quality of a hazard communication program depends upon what?
20. Who is responsible for the information on MSD's?

Short Answer

21. Who does the OSHAct cover, and who is excluded from its coverage?
22. What are the general duties of employers and employees under the OSHAct?
23. What is the purpose of the OSHA Voluntary Protection Program?
24. What concepts must a organization have present to qualify as an exemplary safety program?
25. What does the informal conference provide regarding a citation?
26. How does a state go about having a "state OSHA plan" approved?
27. How has the OSHAct given new visibility to occupational safety and health?

Part 2

Loss Control Information and Analysis

INTRODUCTION

During the twentieth century, safety and health issues have gradually changed from a focus on injury and disability management to a proactive program that emphasizes prevention and loss control. This change in perspective from post-event management toward aggressive prevention represents one of the major advances in safety and health management. Sophisticated tools have been developed to gather and process information so that appropriate and timely interventions can be made in workplace activities.

Part two begins with an overview of loss control programs. A new chapter on safety, health, and environmental auditing has been added. Increasingly, business and industry have come to view safety, health, and environmental (S/H/E) issues as inextricably linked and amenable to proactive management and control. The comprehensive audit has evolved into a sophisticated management tool that is part of a complete quality management program.

The tremendous variety and quantity of S/H/E data have produced a clear need for the development of information management techniques and systems. Chapter 9, Computers and Information Management, presents an extensive state-of-the-art review of the rapidly growing field of S/H/E information management. The rise of computer technology and its applications represents an important development for the safety and health professional.

3

Loss Control Programs

To be effective, a loss, or hazard, control program that identifies and controls or eliminates hazards must be monitored, planned, directed, and controlled; it cannot simply develop on its own. Management needs to establish program objectives and safety policies and to assign line management responsibility for the loss control program. Safety personnel and others must be able to perform specific steps to identify and control hazards. These steps are discussed later in the chapter. First, however, those charged with designing and participating in the loss control program need to understand the nature of hazards, their effects on the work process, and the basic causes of accidents and ways they can be controlled.

The following topics are covered in this chapter:

- loss control programs to combat workplace hazards
- factors leading to the cause of workplace accidents
- examining the principles and processes of loss control
- developing effective health and safety programs to protect workers
- basic considerations of the purchasing department
- the responsibility of safety and health committees in loss control programs
- the benefits of off-the-job safety programs

NEED FOR A BALANCED APPROACH

Before the concept of loss control was developed, accidents were regarded either as chance occurrences or acts of God—a view still held by some—or as an inherent consequence of production. Such approaches accept accidents as inevitable and, therefore, yield no information about causation and prevention. Control strategies are limited to mitigating the consequences of the occurrence.

In the early days of loss control, accident prevention activities focused on the human element. Findings indicated that a small proportion of workers accounted for a significant percentage of accidents. Control strategies were devised to reduce human error through training, education, motivation, communication, and other forms of behavior modification. During World War II, industrial psychology was aimed at matching employees to particular jobs. Personnel screening and selection were seen as the primary ways to prevent accidents. However, accident-proneness and other behavior models have a glaring weakness: while useful for understanding human behavior, they do not consider the interaction between the worker and the other parts of the system. (See the discussion in Chapters 22, Motivation, and 23, Safety and Health Training.)

The 1950s and 1960s saw the emphasis change to engineering and control programs aimed at machines and equipment. Further, the implementation of the Occupational Safety and Health Act of 1970 (OSHAct) placed emphasis on preventing accidents through control of the work environment and the elements of the workplace. This act, along with other legislation, specified standards and compliance rules and regulations.

In the 1980s and 1990s, there is the realization that even with all of the above, accidents still occur. In the 1980s concern with the commitment and culture of the organization were important issues. Clearly emphasis on any one area does not bring about meaningful changes. The balanced, personal, and human-interest approach stands to benefit the organization.

In the 1990s, the National Safety Council continues to work to increase awareness that an incident is a near-accident and that so-called accidents are not random events but rather preventable events. The Council changed its mission statement to eliminate the word "accident":

> The mission of the National Safety Council is to educate and influence society to adopt safety, health, and environmental policies, practices, and procedures that prevent and mitigate human suffering and economic losses arising from preventable causes.

With proper hazard identification and evaluation, management commitment and support, preventive and corrective procedures, monitoring, evaluation, and training, unwanted events can be prevented.

Management Oversight and Omission

Over the past few decades, many organizations seeking to reduce hazards have focused on system defects, which result from management oversight or omission, or on malfunction of the management system. A balanced approach to loss control looks at each component of the system and includes such weaknesses as inadequate training and education, improper assignment of responsibility, unsuitable equipment, or failure to fund hazard control programs. Because managers are responsible for the design, implementation, and maintenance of systems, management errors can result in system defects.

Examining Accident Causation

There are two basic approaches to examining how accidents are caused, after-the-fact and before-the-fact.

After-the-Fact

This approach relies on examining accidents after they have occurred to determine the cause and to develop corrective measures. Evaluation of past performance uses information derived from accident and inspection reports and insurance audits. This approach is too often used only after a serious accident has resulted in injury or damage,

or system ineffectiveness. Furthermore, accident frequency and severity rates do not answer the crucial questions, how, what, why, and when accidents occur.

Before-the-Fact

This method relies on inspecting and systematically identifying and evaluating the nature of undesired events in a system. One such method is the critical incident technique.

Critical Incident Technique

The critical incident technique measures safety performance and identifies practices or conditions that need to be corrected. This technique can identify the cause of an accident before the loss occurs. To obtain a representative sample of workers exposed to hazards, management selects persons from various departments of the plant. An interviewer questions a number of workers who have performed particular jobs within certain environments. They are asked to describe only those existing hazards and unsafe conditions they are aware of. These are called incidents (Figure 3-1). Management then classifies incidents into hazard categories, and identifies problem areas.

The investigative team can also analyze the management systems that should have prevented the occurrence of unsafe practices or the existence of unsafe conditions. The technique can lead to improvements in loss control program management.

The procedure needs to be repeated because the worker-equipment-environment system is not static. Repeating the technique with a new sample of workers can reveal new problem areas and measure the effectiveness of the accident prevention program.

Safety Sampling

Also called behavior or activity sampling, safety sampling is another technique that uses the expertise of those within the organization to inspect, identify, and evaluate hazards. This method relies on personnel—usually management or safety staff members—who are familiar with operations and well trained in recognizing unsafe practices. While making rounds of the plant or establishment, they record on a safety sampling sheet both the number and type of safety defects they observe. A code number can be used to designate specific unsafe conditions, such as hands in dies, failure to wear eye protection and protective clothing, failure to lock out source of power while working on machinery, crossing over belt conveyors, working under suspended loads, improper use of tools, or transporting unbanded steel.

Safety personnel or managers should make observations at different times of the day, on a planned or random basis in the actual work setting, and throughout the various parts of the plant. In a short time, they can easily convert

---

LIST OF TYPICAL INCIDENTS

An incident is any observable human activity sufficiently complete in itself to permit references and predictions to be made about the persons performing the act.

1. Adjusting and gaging (calipering) work while the machine is in operation
2. Cleaning a machine or removing a part while the machine is in motion
3. Using an air hose to remove metal chips from table or work (a brush or other tools should be used for this purpose, except on recessed jigs)
4. Using compressed air to blow dust or dirt off clothing or out of hair
5. Using excessive pressure on air hose
6. Operating machine tools (turning machines, knurling and grinding machines, drill presses, milling machines, boring machines) without proper eye protection (including side shields)
7. Not wearing safety glasses in a designated eye-hazard area.
8. Failing to use protective clothing or equipment (face shield, face mask, ear plugs, safety hat, cup goggles)
9. Failing to wear proper gloves or other hand protection when handling rough or sharp-edged material
10. Wearing gloves, ties, rings, long sleeves, or loose clothing around machine tools
11. Wearing gloves while grinding, polishing, or buffing
12. Handling hot objects with unprotected hands
13. No work rest or poorly adjusted work rest on grinder ($1/8$ in. maximum clearance)
14. Grinding without the glass eye shield in place
15. Making safety devices inoperative (removing guards, tampering with adjustment of guard, beating or cheating the guard, failing to report defects)
16. Using an ungrounded or uninsulated portable electric hand tool
17. Improperly designed safety guard, for example, a wide opening on a barrier guard which will allow the fingers to reach the cutting edge

---

Figure 3-1. This list represents a sample of work practices that, if not corrected, can result in an accident or loss.

observations to a simple report showing what specific unsafe conditions exist in which areas and what supervisors and foremen need help in enforcing good work practices. The information is unbiased and therefore irrefutable. What has been recorded is what has been observed.

ACCIDENTS AND LOSS CONTROL

This section focuses on the definition of hazards, their effects on the workplace, and efforts by management, safety professionals, and employees to control them. The more that management and workers realize that safety involves day-to-day teamwork and mutual support, the closer a company will come to achieving its safety goals.

Definition of Hazards

A workable definition of hazard is any existing or potential condition in the workplace that, by itself or by interacting with other variables, can result in deaths, injuries, property damage, and other losses. This definition carries with it two significant points.

• First, a condition does not have to exist at the moment to be classified as a hazard. When the total hazard situation is being evaluated, *potentially* hazardous conditions must be considered.
• Second, hazards may result not only from independent failure of workplace components but also from one workplace component acting upon or influencing another. For instance, if gasoline or another highly flammable substance comes in contact with sulfuric acid, the reaction created by the two substances produces both toxic vapors and sufficient heat for combustion.

Hazards are generally grouped into two broad categories: those dealing with safety and injuries and those dealing with health and illnesses. However, hazards that involve property and environmental damage must also be considered.

Effects of Hazards on the Work Process

In a well-balanced operation, workers, equipment, and materials are brought together in the work environment to produce a product or to perform a service. When operations go smoothly and time is used efficiently and effectively, production is at its highest.

When an accident interrupts an operation, it sets in motion a different chain of events and carries its own price tag. An accident is an unplanned, undesired event, not necessarily resulting in injury, but damaging to property and/or interrupting the activity in process. An accident increases the time needed to complete the job, reduces the efficiency and effectiveness of the operation, and raises production costs. If the accident results in injury, materials waste, equipment damage, or other property loss, there is a further increase in operational and hidden costs and a decrease in effectiveness. These cost factors are discussed in Chapter 7, Accident Investigation, Analysis, and Costs.

Controlling Hazards: A Team Effort

Traditionally, most managers have relied solely on their safety and operations people to locate, evaluate, and control hazardous situations. However, the more that is learned about hazard and loss control, the more evident it becomes that the job is too large for any individual or small group to do alone. Accident and hazard reduction requires a team effort by employees and management.

Here is how several departments and employee teams can work together.

- The engineering departments can design facilities to be free of uncontrolled hazards and provide technical hazard identification and analysis services to other departments. Their designs must comply with federal, state or provincial, and local laws and standards.
- Manufacturing departments can reduce hazards through efforts such as effective tool design, changes in processes, job hazard analysis and control, and coordinating and scheduling production.
- Quality control can test and inspect all materials and finished products. It can conduct studies to determine whether alternate design, materials, and methods of manufacture could improve the quality and safety of the product and the safety of the employees making the product.
- Purchasing departments can ensure that materials and equipment entering the workplace meet established safety and health standards, and that adequate protective devices are an integral part of equipment. They should disseminate information received from suppliers to line management and workers about safety and health hazards associated with workplace substances and materials.
- Maintenance can provide planned preventive maintenance on electrical systems, machinery, and other equipment to prevent abnormal deterioration, loss of service, or safety and health hazards.
- Industrial relations often administers programs directly related to health and safety.

Input also can come from the joint safety and health committee (discussed later in this chapter) and from quality circles and safety circles (see Chapter 25, Safety Awareness Programs).

Management Support

To coordinate the organizational and departmental efforts, a program of loss control is necessary as part of the management process (Windsor, 1979). Such a program provides hazard control with management tools such as programs, procedures, audits, and evaluations. Sometimes hazard control program teams neglect the basics in their rush to be competitive and innovative, to deal with complex employee relations issues and government involvement, and to address the technical aspects of the programs. A program of hazard control ensures that safety fundamentals also will be addressed. These basics include sound operating and design procedures, operator training, inspection and test programs, and communicating essential information about hazards and their control.

A loss control program establishes facility-wide safety and health standards and coordinates responsibilities among departments. For example, if one department makes a product and another distributes it, they share responsibility for hazard control. The producer knows the nature of the process, its apparent and suspected hazards, and how to control the hazards. The producer and the distributor are responsible for making sure this information does not end with the production department but is available to the purchaser or the next unit in the manufacturing process.

Coordination is also important when manufacturing responsibility is transferred from one department to another (as when a pilot program becomes a complete manufacturing unit). In addition, when a process is phased out, departments need to coordinate efforts to ensure that personnel who know the hazards are retained throughout the phase-out and that appropriate hazard control activities continue until the end.

Safety Management and Productivity Improvement

The process of identifying and eliminating or controlling hazards in the workplace is one way of making the best use of human, financial, technological, and physical resources. Optimizing these resources results in higher productivity. For purposes here, productivity is defined as producing more output with a given level of input resources.

Loss control, like productivity, quality, costs, and personal relations, is a strategic process. To be effective, it must be integrated into the day-to-day activities and management systems of the organization and must become institutionalized—an operating norm and a strategic part of the organization's culture.

There are other similarities between efforts aimed at hazard control and productivity improvement. To achieve both objectives, an organization must intelligently manage its financial and human resources and use the most appropriate technology. It must illustrate innovative, enlightened, and efficient use of its facilities, equipment, materials, and workforce, and have a trained, educated, and skilled workforce.

An accident interrupts the production process. It not only increases the time needed to complete a production task, it may also reduce the efficiency and effectiveness of the overall operation and increase production costs. Sometimes a succession of interruptions, or one long one, will prevent the production schedule or desired product quality from being met. Such conditions make it difficult to attract new business.

Production Accomplishment and Control

Control of an operation, by definition, means keeping the system on course and preventing problems from occurring. However, it also implies some allowance for variations within the system, provided they remain within controlled limits. Any production system has built-in control limits, both upper and lower. These limits provide direction and also any acceptable leeway for the system's operation.

There are many aspects to control, including control over the quality of products and services, personnel, capital, energy, materials, and the plant environment. Each of these factors interacts with the other factors to produce the desired effect.

Determining Accident Factors

In order to set realistic goals for its process, the organization should first determine the major factors likely to cause loss of control. It should then identify the location, importance, and potential effects. Control measures can then be instituted to help reduce risk and potential losses. Factors responsible for accident losses may be identified by either inspection or detailed hazard analyses. The control measures may be some type of process innovation or machine safeguarding, personal protective equipment, training, or administrative change. In addition to the control measures, monitoring systems should be used to continuously assess the effectiveness of these hazard-reducing controls.

Worker-Equipment-Environment System

Those professionals involved in establishing effective loss control programs must understand the interrelationships in the worker-equipment-environment system. (Chapter 13, Ergonomics Programs, in this volume examines the system in greater detail.) The present chapter explains the elements of the system (Figure 3-2). As shown in Figure 3-3, an accident can intervene between the system and the task to be accomplished.

Worker

In any worker-equipment-environment system, the worker performs three basic functions: (1) sensing, (2) information processing, and (3) controlling.

- As a sensor, the worker serves to monitor or gather information.
- As an information processor, the worker uses the information collected to make a decision about the relevance or appropriateness of various courses of action.
- The third function, control, flows from the first two. Once information is collected and processed, the

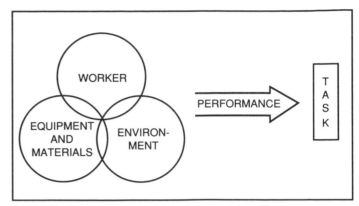

Figure 3-2. A system approach to hazard control recognizes the interaction between worker, equipment, materials, and environment in the performance of work.

worker keeps the situation within acceptable limits or takes the necessary action to bring the system back into an acceptable or safe range.

Evaluating an accident in the light of these three functions can pinpoint the causes. Did the error occur while the worker was gathering information as a sensor? Was the worker able to gather information accurately, for example, in adequate illumination without glare? Did the error occur as a result of faulty information processing and decision making? Did the error occur because an appropriate control option was not available or because the worker took inappropriate action?

In order for the system to move toward its production objectives, the employee must perform work effectively and avoid taking unnecessary risks. To do this, workers must be educated about the following (Firenze, 1978):

- necessary requirements of the task and the steps needed to accomplish it
- personal knowledge, skill, and limitations and how they relate to the task
- what will be gained if the worker attempts the task and succeeds
- what will result if the worker attempts the task and fails
- what will be lost if the worker makes no attempt to accomplish the task.

Equipment

Equipment (materials), the second component in the system, must be properly designed, maintained, and used. Hazard control can be affected by the shape, size, and thickness of tools; the weight of equipment; operator comfort; and the strength required to use or operate tools, equipment, and machinery. These variables influence the interaction between worker and equipment. Other equipment variables important in hazard recognition include speed of operation and mechanical hazards.

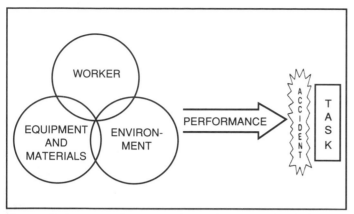

Figure 3-3. An accident causes the work system to break down. It intervenes between the worker, equipment, and environment and the task to be performed.

Environment
Special consideration must be given to environmental factors that might detract from the comfort, health, and safety of the worker. Emphasis should be placed on factors such as:

- layout: the worker should have sufficient room while performing the assigned task
- maintenance and housekeeping
- adequate illumination: poorly lit areas increase eyestrain and also the chance of making an accident-causing mistake
- temperature, humidity, noise, vibration, and control of emission of toxic materials.

Interpersonal relationships are another system factor that plays an important role in operational effectiveness. The task performed by one worker is related to tasks performed by others. Special consideration must be given to coordinating information, materials, and human effort.

ACCIDENT CAUSES AND THEIR CONTROL
Close examination of each accident shows that it can be attributed, directly or indirectly, to an oversight, omission, or malfunction of the management system regarding one or more of the following three items (refer also to discussion later in this chapter):

1. unsafe practices or procedures; either the worker or another person
2. situational factors, for example, facilities, tools, equipment, and materials
3. environmental factors, such as noise, vibration, temperature extremes, illumination.

If an adequate line management hazard control system is properly designed for the organization's workers,

equipment, and environment, then the likelihood of accidents occurring in the workplace is greatly reduced.

Unsafe Practices or Procedures
An unsafe practice is generally described as either a human action departing from prescribed hazard controls or job procedures or practices, or as an action causing a person unnecessary exposure to a hazard. Both workers and management can cause accidents by commission; for example, a worker may sharpen a wood gouge on a grinder without placing the tool on the grinder's rest. In this case, management should ask questions such as was the worker trained properly or pressured to rush a job, or were procedures enforced. Conversely, a supervisor contributes to the cause of an accident by omission when failing to have an oil spot on the floor wiped up.

An unsafe practice often is a deviation from the standard job procedures. Examples of such actions include:

- using equipment without authority
- operating equipment at an unsafe speed or other improper way
- removing safety devices, such as guards, or rendering them inoperative
- using defective tools.

Unsafe practices also can be a deviation from safety rules or regulations, instructions, or job safety analyses. Why the deviation occurred is the real issue. Some causes and their countermeasures are given in the following examples. When implementing a hazard control program, emphasis should be placed on the countermeasure. The following examples show how to do this.

- No known standard for safe job procedure exists. Countermeasure: Perform a job safety analysis (JSA) and develop a good procedure through job instruction training (JIT).
- The employee did not know the standard job procedure. Countermeasure: Train in the correct procedure.
- The employee knew, but did not follow, the standard job procedure. Countermeasure: Consider an employee performance evaluation. Test the validity of the procedure and motivation.
- The employee knew and followed the procedure. Countermeasure: Develop a safer job procedure.
- The procedure encouraged risk-taking, such as incentive pay for piecework. Countermeasure: Change the unsafe job design, procedure, or incentive program.
- The employee changed the approved job procedure or bypassed safety equipment. Countermeasure: Change the method or safeguards so safety measures cannot be bypassed.

- The employee did not follow the correct procedure because of work pressure or the supervisor's influence. Countermeasure: Counsel employee and supervisor; consider change in work procedures or job requirements.
- Individual characteristics, which may involve a disability, made the employee unable or unwilling to follow the correct procedure. Countermeasure: Counsel employee; consider a change in work procedures, workstation design, or job requirements; also consider in-depth training.

Many accidents are the result of someone deviating from the standard job procedures, doing something prohibited, or failing to do something that should be done. In other situations, however, the worker unfairly becomes the target for criticism when other factors actually caused the mishap. The following example illustrates this point. Suppose a newly hired worker, after receiving what was thought to be sufficient instruction on the use of a table saw guard, is required to make a particular cut that cannot be made with the guard in proper position. In this case, the required task causes the worker to remove the guard temporarily so the cut can be made. While removing the guard, the worker's hand slips off the wrench and is cut on the saw blade. Obviously the worker was instrumental in the accident situation, and consequently many people would view the procedure as unsafe. A closer analysis, however, reveals that the primary cause was the failure of training to identify equipment limitations.

In this instance, a failure in the management system contributed to the accident. First, a better guard should have been purchased. Second, the new worker should have been instructed more carefully in the use of the guard, including how to remove it when necessary. Most important, a contingency plan should have provided protection for times when the saw has to be used without adequate safeguarding.

An important first step in loss control is distinguishing between worker error and supervisory error, then addressing what caused the system to break down. Don't just look for a "scapegoat"; look for the cause(s). Human error is reduced when:

- supervisors and workers know the correct methods and procedures to accomplish given tasks
- workers demonstrate a skill proficiency before using the particular piece of equipment
- higher management and supervisors consider the relationship between worker performance and physical characteristics and fitness
- the entire organization gives top priority and continuous regard to potentially dangerous situations and the corrective action necessary to avoid accidents

- supervisors provide proper direction, training, and surveillance. The supervisor must be aware of the worker's skill level with each piece of equipment and process, and adjust the supervision of each worker accordingly. The supervisor shapes worker attitudes and actions by letting employees know that nothing less than safe work practices and the safest possible workplace will be accepted.

Situational Factors

Situational factors are another major cause of accidents. These factors are materials that make accidents likely, and unsafe operations, tools, equipment, and facilities. Examples are unguarded, poorly maintained, and defective equipment; ungrounded equipment that can cause shock; equipment without adequate warning signals; poorly arranged equipment, buildings, and layouts that create congestion hazards; and equipment located in positions that expose more people to a potential hazard.

Some causes of situational problems are:

- Defects in design—for example, a lightweight, unvented metal container for use with flammable materials or no guard on a power press
- Poor, substandard construction—for example, a ladder built with defective lumber or with a variation in the space between its rungs
- Improper storage of hazardous materials—for example, oxygen and acetylene cylinders stored in an unstable manner and ready to topple over with the slightest impact
- Inadequate planning, layout, and design—for example, a welding station located near combustible materials or placed where many workers without eye protection are exposed to the intense light of the welding arc.

An example of a situational problem occurred in a light industrial manufacturing plant where maintenance workers too often found themselves replacing a bearing on an expensive machine. Something had to be done to save downtime, labor, and the cost of the bearing. The industrial engineering and maintenance departments jointly devised the solution: a system that fed oil to the bearing at set intervals, keeping it well lubricated. It was no longer necessary to replace the bearing so often.

But the solution created new hazards. When oil was fed to the bearing, it dripped onto the floor in the aisle adjacent to the machine. Workers could slip on the oil spot and sustain serious injuries. Forklifts drove over the oil. With oil on the rubber wheels, the driver might not be able to stop the vehicle.

Had the maintenance and industrial engineering organizations been thinking of accident prevention, they could have avoided situational hazards by correcting their design. As an interim step, they might have collected the oil by placing a pan under the motor where the bearing was housed. They would then have time to install a tube that would return the oil to the system, thus saving oil while eliminating the hazard.

Environmental Factors

The third factor in accident causation is environmental, that is, the way in which the workplace directly or indirectly causes or contributes to accident situations. Environmental factors fall into four broad categories: human, chemical, biological, and ergonomic.

Human Factors

Noise, vibration, radiation, illumination, and temperature extremes are examples of factors having the capacity to influence or cause accidents and illnesses. Operations on a machine lathe, for example, may produce high noise levels that prevent workers from hearing other sounds and impair communication with others or may damage the workers' hearing over time. Thus, workers may be unable to warn one another of a hazard in time to avoid an accident.

Chemical Factors

Classified under this category are toxic gases, vapors, fumes, mists, smokes, and dusts. In addition to causing illnesses, these often impair a worker's skill, reactions, judgment, or concentration. A worker exposed to the narcotic effect of some solvent vapors, for example, may experience a loss of judgment and fail to follow safe procedures.

Biological Factors

Biological factors refer to those items capable of making a person ill through contact with bacteria, viruses, fungi, or parasites. For example, workers may suffer boils and inflammations caused by staphylococci and streptococci or experience groin itch caused by parasites.

Ergonomic Factors

See Chapter 13, Ergonomics Programs, and Chapter 19, Office Safety.

Sources of Situational and Environmental Hazards

Actions by purchasing agents; those responsible for tool, equipment, and machinery placement and for providing adequate machine guards; and those responsible for maintaining shop equipment, machinery, and tools may result in situational and environmental hazards.

Employee contributions to situational and environmental hazards include disregarding safety rules and regulations by (1) making safety devices inoperative, (2) using equipment and tools incorrectly, (3) using defective tools rather than obtaining serviceable ones, (4) failing to use engineering controls such as exhaust fans when required, and (5) using toxic substances in unventilated areas or without proper protection.

Purchasing agents can be instrumental in creating situational and environmental hazards if they disregard safety engineering recommendations. These agents may acquire tools, equipment, and machinery without adequate guards and other safety devices, especially if such items are selected with only cost in mind. Sometimes highly toxic and hazardous materials are purchased when less toxic and hazardous materials could be substituted. Other times purchasing agents fail to acquire from the vendor the necessary warning and control information that must be given to those in charge of the particular process. In many companies, however, the purchasing agent's choices are controlled by engineers, safety professionals, and government or consensus standards and other regulations. The safety and health professional, in conjunction with engineers, should provide the necessary criteria, specifications, etc. to assist the purchasing agents.

Those involved in layout, design, and placement of equipment and machinery also must consider adequate safeguarding and safety devices or equipment. Otherwise, they contribute to hazardous situations in the workplace. Examples are:

- placing equipment and machinery with reciprocating parts where workers can be crushed between the equipment and substantial objects
- installing electrical control switches on machinery where the operator will be exposed to the hazards of cutting tools or blades in order to start and stop the equipment
- installing equipment without providing for adequate lockout/tagout
- installing equipment and machinery guards that interfere with work operations
- locating high hazard workstations where they expose workers unnecessarily; for example, placing a welding station in the middle of a floor area instead of locating it in a corner or along a wall where better control over the welding arc is possible.

Those responsible for maintenance, both management and employees among others, sometimes cause hazards in the workplace. Examples are:

- improperly identifying high and low pressure steamlines, compressed air, and sanitary lines
- failing to detect or replace worn or damaged machine and equipment parts, such as abrasive wheels on power grinders
- failing to adjust and lubricate equipment and machinery on a scheduled basis

- failing to inspect and replace worn hoisting and lifting equipment
- failing to replace worn and frayed belts on equipment
- over-oiling motor bearings, resulting in oil being thrown onto the insulation of electrical wiring and onto the floor, and possible damage to the bearings
- failing to replace guards
- failing to lock out and tag unsafe equipment.

More details are included in this chapter under the heading Responsibility for the Hazard Control Program.

PRINCIPLES OF LOSS CONTROL

Loss control is the function directed toward recognizing, evaluating, and eliminating, or at least controlling, the destructive effects of occupational hazards. These hazards generally result from human errors and from the situational and environmental aspects of the workplace (Firenze, 1978). The primary function of a loss control system is to locate, assess, and set effective preventive and corrective measures for those elements detrimental to operational efficiency and effectiveness.

The process exists on three levels:

1. National—laws, regulations, exposure limits, codes, and standards of governmental, industrial, and trade bodies
2. Organizational—management of hazard control program, safety and health committees, task groups, teams, etc.
3. Component—worker-equipment-environment.

Loss control can be thought of as "looking for defects." First, there are fewer defects, or failures, than successes. Second, it is easier to agree on what constitutes failure than on what constitutes success. Failure is the inability of a system or a part of a system to perform as required under specified conditions for a specific length of time. The causes of failures often can be determined by answering a series of questions. What can fail? How can it fail? How frequently can it fail? What are the effects of failure? What is the importance of the effects? The manner in which a system, or portion of a system, can exhibit failure is commonly known as the mode of failure.

The opposite of failure is not necessarily total success. After all, totally error-free performance is an ideal state, not a reality. Rather, the opposite of failure is the minimum acceptable success. This is the condition in which operations are run with a minimum number of losses and interruptions, keeping efficiency and effectiveness of the operation within acceptable limits of control.

Management builds into each of its systems lower and upper limits of control. Each of these interfacing subsystems—maintenance, quality control, production control, personnel, purchasing, to name a few—is designed to move the system within acceptable limits toward its objective. This concept of keeping operations within acceptable limits gives substance and credibility to the process of loss control. In addition to familiarizing management with the full consequences of system defects, loss control can pinpoint hazards before failures occur. The anticipatory character of loss control increases productivity.

PROCESSES OF LOSS CONTROL

The processes of an effective loss control program should be directed toward evaluating, eliminating, and preventing workplace hazards. Management and safety officials can implement many preventive measures when designing a loss control program. An effective loss control program has six steps or processes:

1. hazard identification and evaluation
2. ranking hazards by risk
3. management decision making
4. establishing preventive and corrective measures
5. monitoring
6. evaluating program effectiveness.

Hazard Identification and Evaluation

The first step in a comprehensive loss control program is to identify and evaluate workplace hazards. These hazards are associated with machinery, equipment, tools, operations, materials, and the physical plant.

There are many ways to acquire information about workplace hazards. A good place to begin is with those who are familiar with plant operations and the hazards associated with them. (See Appendix 1, Sources of Help, for a description of many organizations that can be of help.) The critical incident technique (described earlier) is useful for obtaining information from workers and supervisors. Insurance company loss control representatives know those hazards most likely to cause damage, injuries, and fatalities. In addition to the National Safety Council (NSC), professional societies such as the American Society of Safety Engineers (ASSE), American Industrial Hygiene Association (AIHA), and the American Conference of Governmental Industrial Hygienists (ACGIH) have information about safety and health experience. Manufacturers of industrial equipment, tools, and machinery offer information about the hazards associated with their products, as do suppliers of materials and substances. Labor representatives and business agents can offer a perspective on hazards overlooked by others. Safety and health personnel in organizations doing similar work can be of inestimable value.

A second place to look would be old inspection reports, either internal (by a safety and health committee or company management and specialists) or external (by local, state or provincial, or federal enforcement agencies). The Occupational Safety and Health Administration (OSHA) can supply information describing violations uncovered in similar operations and outlining compliance regulations. (See Chapter 2, Regulatory History, and Appendix 1, Sources of Help, for descriptions of state agencies and private concerns that give onsite inspection and consultation services under OSHA and the National Institute for Occupational Safety and Health (NIOSH).)

Hazard information also can be obtained from accident reports. Information explaining how a particular injury, illness, or fatality occurred often will reveal hazards requiring control. Close review of accident reports filed in the past three to five years will identify the individuals and specific operations involved, the department or section where the accident occurred, the extent of supervision, and possibly the injured person's deficiencies in knowledge and skill.

OSHA incident rates also are useful. Although they are historical and reflect what has happened, not the current status of safety performance, they provide, from a large sample, data that reflect what actually has occurred in the workplace. Other valuable sources can be found in other chapters of this volume, in the *Occupational Safety & Health Data Sheets* of the NSC, and in the specifications for particular equipment and machines published by the American National Standards Institute (ANSI), Underwriters Laboratories Inc. (UL), American Society for Testing and Materials (ASTM), and the National Fire Protection Association (NFPA). Information about work activities, facilities, and equipment is distributed by the NIOSH.

Hazard analysis is another way to acquire meaningful hazard information and a thorough knowledge of the demands of a particular task. Analysis probes operational and management systems to uncover hazards that (1) may have been overlooked in the layout of the plant or the building and in the design of machinery, equipment, and processes; (2) may have developed after production started; or (3) may exist because original procedures and tasks were modified.

The greatest benefit of hazard analysis is that it forces those conducting the analysis to view each operation as part of a system. In doing so, they assess each step in the operation while keeping in mind the relationship between steps and the interaction between workers and equipment, materials, the environment, and other workers. Other benefits of hazard analysis include (1) identifying hazardous conditions and potential accidents; (2) providing information with which effective control measures can be established; (3) determining the level of knowledge and skill as well as the physical requirements workers need to execute specific shop tasks; and (4) discovering and

eliminating unsafe procedures, techniques, motions, positions, and actions.

The topic of hazard analysis—its underlying philosophy, the basic steps to be taken, and its ultimate use as a safety, health, and decision-making tool—will be treated in Chapter 6, Identifying Hazards.

Ranking Hazards by Risk (Severity, Probability, and Exposure)

The second step in the process of loss control is to rank hazards by risk. Such ranking takes into account the consequence (the severity), the probability, and the exposure index. The purpose of this second process is to address hazards according to the principle of "worst first." Ranking provides a consistent guide for corrective action, specifying which hazardous conditions warrant immediate action, which have secondary priority, and which can be addressed in the future (Figure 3-4).

Once safety personnel or others have ranked hazards according to their potential destructive consequences, the next step is to estimate the probability of the hazard resulting in an accident situation. Quantitative data for ranking hazard probability are desirable, but almost certainly they will not be available for each potential hazard being assessed. Whatever quantitative data exist should be part of the risk-rating formula used to estimate probability. Qualitative data—estimates based on experience—are a necessary supplement to quantitative data. Figure 3-5 shows how probability estimates should be made.

After estimating both consequence and probability, the next and final step is to estimate worker exposure to the hazard. The exposure classification scheme in Figure 3-6a is suggested for rating exposure.

Risk Assessment

When the hazards have been ranked according to all three criteria—consequence, probability, exposure—the next step is to assign a single risk number or risk assessment code (RAC).

Figure 3-4. Weighting criteria for hazard consequence

Severity: Consider the potential losses or destructive and disruptive consequences that are most likely to occur if the job task is performed improperly. Use the following point values for this activity:

1 Negligible—probably no injury or illness; no production loss; no lost workdays
2 Marginal—minor injury or illness; minor property damage
3 Critical—severe injury or occupational illness with lost time; major property damage; no permanent disability or fatality
4 Catastrophic—permanent disability; loss of life; loss of facility or major process

The higher numbers are assigned to the most severe consequences. This example uses a four-point scale.

Figure 3-5. Weighting criteria for evaluating probability of loss

Probability: Consider the probability of loss that occurs each time the job is performed. In this evaluation, the key question is How likely is it that things will go wrong when this job is performed? The probability is influenced by a number of factors, such as the hazards associated with the job, the difficulty of performing the job, and the complexity of the job. Use the following point values for this activity:

1 Low probability of loss occurrence
2 Moderate probability of loss occurrence
3 High probability of loss occurrence
The higher numbers are assigned to those jobs for which the likelihood of loss is greatest. This example uses a three-point scale.

Figure 3-6a. Weighting criteria for evaluating hazard exposure

Exposure: Consider the number of employees that perform the job, the number of times an individual employee performs the job, or both. This is an attempt to evaluate the frequency of exposure to the hazards of the job. Use the following point values for this activity:

1 A few employees perform the task up to a few times a day
2 A few employees perform the task frequently
3 Many employees perform the task frequently

The higher numbers are assigned to the jobs with the highest frequency of performance, whether by number of employees, by number of performances, or both. This example uses a three-point scale.

Total: After evaluating the job task according to these criteria, severity, exposure, and probability, add the point values. This produces a single number between 3 and 10. This combines the three evaluations into a single scale that establishes a ranking for job tasks, making it easier to select specific job tasks to analyze. Management can use these values to select particular jobs for immediate analysis (Figure 3-6b). The rating scale is a guideline and is not intended to be used as an absolute measurement system.

Figure 3-6b. Sample ranking system

Job Task	Severity 1–4 points	Exposure 1–3 points	Probability 1–3 points	Total
Conduct grinding mill area preshift inspection	3	2	2	7
Start complete grinding circuit	2	1	2	5
Start particle size monitor	4	2	3	9
Change steel rods	3	2	2	7
Routine rod mill shutdown	2	1	2	5
Charge rod mill with ore	2	2	2	6

Management Decision Making

The third step involves providing management with full and accurate information, including all possible alternatives, so managers can make intelligent, informed decisions concerning loss control. Such alternatives will include recommendations for training and education, better methods and procedures, equipment repair or replacement, environmental controls, and—in rare cases where modification is not enough—recommendations for redesign. Information must be presented to management in a way that clearly states the actions required to improve conditions. The person who reports hazard information must do so in a manner that promotes, rather than hinders, action.

After management's decision-makers receive hazard reports, they normally have three alternatives:

1. take no action
2. modify the workplace or its components
3. redesign the workplace or its components.

When management chooses to take no positive steps to correct hazards uncovered in the workplace, it usually is for one of three reasons:

1. It feels that it cannot take the required action. Immediate constraints—be they financial, crucial production schedules, or limitations of personnel—loom larger than the risks involved in taking no action.
2. It is presented with limited alternatives. For example, it may receive only the best and most costly solutions with no less-than-totally-successful alternatives to choose from.
3. It does not agree that a hazard exists. However, the situation can require additional consultation and study to resolve any problem.

When management chooses to modify the system, it does so with the idea its operation is generally acceptable but, with the reported deficiencies corrected, performance will be improved. Examples of modification alternatives are the acquisition of machine guards, personal protective equipment, or ground-fault circuit interrupters to prevent electrical shock; a change in training or education; a change in preventive maintenance; isolating hazardous materials and processes; replacing hazardous materials and processes with nonhazardous or at least less-hazardous ones; and purchasing new tools.

Although redesign is not a popular alternative, it sometimes is necessary. When redesign is selected, management must be aware of certain problems. Redesign usually involves substantial cash outlay and inconvenience. For example, assume that the air quality in a plant is found to be below acceptable standards. The only way to correct this situation is to completely redesign and install the

RECORD OF OCCUPATIONAL SAFETY AND HEALTH DEFICIENCIES

Location: Shipping

Pete Varga
Inspector

Deficiency No.	Date Recorded	Description of Hazardous Condition	Specific Location	Identification of Acceptable Standard	Hazard Rating		Corrective Action	Estimated Cost of Correction	Date Deficiency Corrected	Resources Used for Correction
					Conse-quence	Proba-bility				
S - 1	12/11/9-	Ungrounded Tools and Equipment	Throughout Shop	OSHA; Subpart S National Electrical Code, Article 250; 4S	I	A	Provide receptacles with the 3–prong outlet. Test each to make certain it is grounded. Make sure that all tools (other than double-insulated) have a grounding plug.	$5,000	1/9/9-	$4,900

Figure 3-7. One approach to recording and displaying hazard information for decision making. (Reprinted with permission from RJF Associates, Inc.)

plant's general ventilation system. The cost and inconvenience can be formidable.

Another problem is the fact that the new designs usually contain hazards of their own. For this reason, whenever redesign is offered as an alternative, those making the recommendation must establish and execute a plan to detect problems in design and the early stages of construction so hazards can be eliminated, reduced, or controlled.

One way to expedite decision making regarding actions for loss control is to present findings clearly so that management understands the nature of the hazards, their location, their importance, the necessary corrective actions, and the estimated cost. Figure 3-7 shows a record of occupational and safety health deficiencies, and illustrates one approach for recording and displaying hazard information for decision making. It indicates the hazard ranking, the specific location and nature of the hazard, and what costs are likely to be incurred. It also clearly states the recommended corrective action. At a glance, it shows if the corrective action has been taken and the final cost.

Establishing Preventive and Corrective Measures

After the safety team or others have identified and evaluated hazards and provided data for informed decisions, the next step involves implementing control measures.

Controls are of three kinds:

1. administrative (through personnel, management, monitoring, limiting worker exposure, measuring performance, training and education, housekeeping and maintenance, purchasing)
2. engineering (isolation of source, lockout procedures, design, process or procedural changes, monitoring and warning equipment, chemical or material substitution)

3. personal protective equipment (body protection, fall protection, etc.). See Chapter 7, Personal Protective Equipment, in the *Engineering & Technology* volume.

Before control installation takes place, it is essential that those involved in safety and health activities understand how hazards are controlled. Figure 3-8 illustrates the three major areas where hazardous conditions can be either eliminated or controlled.

- The first and perhaps best control alternative is to attack a hazard at its source. One method is to substitute a less harmful agent for the one causing the problem. For example, if a certain solvent is highly toxic and flammable, the first step is to determine whether the hazardous substance can be exchanged for one that is nontoxic and nonflammable, yet still capable of doing the job. If a nonhazardous substance meeting these criteria is not available, then a less toxic, less flammable substance can be substituted and additional safeguards employed.
- The second alternative is to control the hazard along its path. This is done by erecting a barricade between

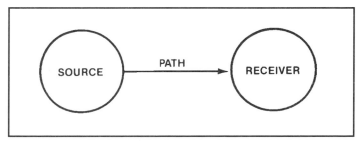

Figure 3-8. Three major areas where hazards can be controlled—the contaminant source, the path it travels, and the employee's work pattern and use of personal protective equipment.

the hazard and the worker. Examples of engineering controls are (1) machine guards, which prevent a worker's hands from making contact with the table saw blade; (2) protective curtains, which prevent eye contact with welding arc flashes; and (3) a local exhaust system, which removes toxic vapors from the breathing zone of the workers.

- The third alternative is to direct control efforts at the receiver, the worker. Removing the worker from exposure to the hazard can be accomplished by (1) employing automated or remote control options (for example, automatic feeding devices on planers, shapers); (2) providing a system of worker rotation or rescheduling some operations to times when there are few workers in the plant; or (3) providing personal protective equipment when all options have been exhausted, and the hazard cannot be corrected through substitution or engineering redesign.

Protective equipment may be selected for use in two instances: when there is no immediate way to control the hazard by more effective means, and when it is employed as a temporary measure while more effective solutions are being installed. There are, however, major shortcomings associated with the use of personal protective equipment:

- Nothing has been done to eliminate or reduce the hazard.
- If the protective equipment (such as gloves or an eye shield) fails for any reason, the worker is exposed to the full destructive effects of the hazard.
- The protective equipment may be cumbersome and interfere with the worker's ability to perform tasks, thus compounding the problem.

Chapter 7, Personal Protective Equipment, in the *Engineering & Technology* volume, discusses these subjects more fully.

Monitoring

The fifth step in the process of hazard control deals with monitoring activities to locate new hazards and assess the effectiveness of existing controls. Monitoring includes inspection, industrial hygiene testing, and medical surveillance. These subjects are covered in Chapter 6, Identifying Hazards.

Monitoring is necessary (1) to provide assurance that hazard controls are working properly, (2) to ensure that modifications have not so altered the workplace that current hazard controls can no longer function adequately, and (3) to discover new or previously undetected hazards.

Evaluating Program Effectiveness

The final process in hazard control is to evaluate the effectiveness of the safety and health program. Evaluation involves answering the following questions. What is being done to locate and control hazards in the plant? What benefits are being received, for example, reduction of injuries, workers' compensation cases, and damage losses? What impact are the benefits having on improving operational efficiency and effectiveness?

The evaluation team examines the program to see if it has accomplished its objectives (effectiveness evaluation) and whether they have been achieved in accordance with the program plan (administrative evaluation, including such factors as schedule and budget). Evaluation must be adapted to (1) the time, money, and kinds of equipment and personnel available for the evaluation; (2) the number and quality of data sources; (3) the particular operation; and (4) the needs of the evaluators.

Among the criteria management can use to determine the effectiveness of its safety and health program effort are the number and severity of injuries to workers compared with work hours; the cost of medical care; material damage costs; facility damage costs; equipment and tool damage or replacement costs; and the number of days lost from accidents.

An indicator of the effectiveness of a hazard control program is the experience rating given a company by the insurance carrier responsible for paying workers' compensation. Experience rating is a comparison of the actual losses of an individual (company) risk with the losses that would be expected from a risk of such size and classification. Experience rating determines whether the individual risk is better or worse than the average and to what extent the premium should be modified to reflect this variation. Experience modification is determined in accordance with the Experience Rating Plan (ERP) formula, which has been approved by the insurance commissioners in most states. Loss frequency is penalized more heavily than loss severity because it is assumed that the insured can control the small loss more easily than the less frequent, severe loss.

ORGANIZING AN OCCUPATIONAL SAFETY AND HEALTH PROGRAM

The purposes of a loss control program organization are to assist management in developing and operating a program designed to protect workers, to prevent and control accidents, and to increase effectiveness of operations. Figure 3-9 illustrates the major organizational components of a safety and loss control program.

Establishing Program Objectives

Critical to the design and organization of a safety and health program is the establishment of objectives and policy to guide the program's development. If the organization has a joint safety and health committee, it could be the body chosen to set the program objectives. It is assumed that those making recommendations to management would be employee representatives, supervisors, middle management, and safety and health professionals

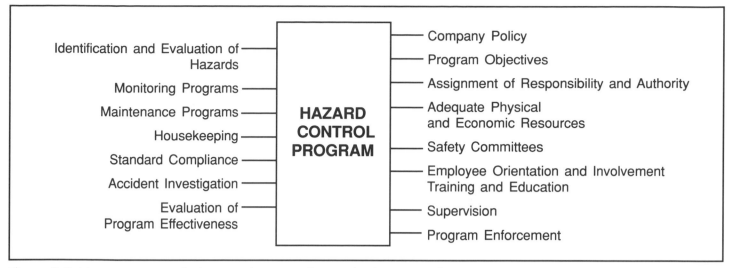

Figure 3-9. Major components of a loss control program. (Reprinted with permission from RJF Associates, Inc.)

(safety directors, managers, supervisors, and administrators; industrial hygiene technicians and professionals; and fire protection engineers).

Among the program objectives should be the following:

1. gaining and maintaining support for the program at all levels of the organization
2. motivating, educating, and training the program team to recognize and correct or report hazards located in the workplace
3. engineering hazard control into the design of machines, tools, and facilities
4. providing a program of inspection and maintenance for machinery, equipment, tools, and facilities
5. incorporating hazard control into training and educational techniques and methods
6. complying with established safety and health standards.

Establishing Organizational Policy

Once the objectives have been formulated, the second step is for management to adopt a formal policy. A written policy statement, signed by the chief executive officer/president of the organization, should be made available to all personnel. It should state the purpose of the hazard control program and require the active participation of all those involved in the program's operation. The policy statement also should reflect:

1. importance that management places on the health and well-being of employees
2. management's commitment to occupational safety and health
3. emphasis the company places on efficient operations, with a minimum of accidents and losses
4. intention to integrate loss control into all operations, including compliance with applicable standards

5. necessity for active leadership, direct participation, and enthusiastic support of the entire organization.

The policy of the American Telephone and Telegraph Company, for example, is probably the shortest statement, but it drives home the point that:

No job is so important and no service is so urgent—that we cannot take time to perform our work safely.

The National Safety Council's *Management Safety Policies,* 12304-0585, covers this subject more fully.

After management (in cooperation with labor) has established a safety policy, it should be publicized so that each employee becomes familiar with its content, particularly how it applies directly to him or her. Ways to publicize the statement include meetings, letters, pamphlets, and bulletin boards. The policy should also be posted in management offices to serve as a constant reminder of management's commitment and responsibility.

Responsibility for the Hazard Control Program

Responsibility for the safety program can be established at the following levels: board of directors, chief executive officers, managers, and administrators; department heads, supervisors, foremen, and employee representatives; purchasing agents; housekeeping and maintenance personnel; employees; safety personnel; staff medical personnel; and safety and health committees.

Management and Administration

Before any safety program gets underway, it must receive full support and commitment from top management and administration. The president, board members, directors, and other management personnel have primary responsibility for the safety program, which involves the continuing obligation to monitor the program's effectiveness. Management provides the motivation to get the program started and to oversee its operations. Management must initiate discussions with personnel during preplanning meetings and periodically review the performance of its safety program. Discussions should cover program progress, specific needs, and a review of company procedures and alternatives for handling emergencies in the event an accident occurs.

Specifically, responsibility at this level consists of setting objectives and policy and supporting safety personnel in their requests for necessary information, facilities, tools, and equipment to conduct an effective safety program and to establish a safe, healthy work environment. Management must realize that it is not fulfilling its organization's potential efficiency and effectiveness until it brings its operations at least into compliance with mandatory and voluntary regulatory safety and health standards.

Management and administration must delegate the necessary tasks at all organizational levels to ensure the safety program's success. Although management cannot delegate to others its responsibility for employee safety and health, it can assign responsibility for certain parts of the safety program. However, the authority to act must be delegated along with responsibility. While authority always starts with those in the highest administrative levels, it eventually must be passed down to middle and line management to achieve the desired results. If safety professionals, safety and health committees, department heads, supervisors, foremen, and employee representatives are to conduct a vigorous and thorough safety program, and if they are to accept and assert the authority delegated to them when circumstances warrant it, they must be fully confident that they have administrative support.

Management must understand that, although it can assert authority, it may encounter resistance unless it has enlisted employee support from the earliest stages of the program. If supervisors, employees, and their representatives are not aware of the reasons for and the benefits of an effective safety program, they may resist any changes in their methods of operation and instruction and may do as little as possible to assist the overall program effort.

Management must insist that safety and health information is an integral part of training, methods, materials, and operations. It must guarantee a system where loss control is considered an important part of equipment purchase and process design, operation, preventive maintenance, and layout and design. It must make sure that effective fire prevention and protection controls exist. Management also is responsible for informing subcontractors, at the time of negotiating the contract, of all applicable safety and health standards. Management must see that subcontractors comply fully with company and other safety regulations.

Management is required to safeguard employees' health by seeing to it that the work environment is adequately controlled. They must be aware that those operations producing airborne fumes, mists, smoke, vapors, gases, dust, noise, and vibration have the capacity to cause impaired health or discomfort among their workers. Management must be aware that occupational illnesses beginning in the workplace can take their toll later, even after a worker retires. To protect the future of its employees, management must maintain an effective industrial hygiene monitoring system.

Management should also provide meaningful criteria to measure the success of the safety program and to provide information upon which to base future decisions. It must decide the safety program goals for reduced accidents, injuries, and illnesses, and their associated losses.

There are many concrete ways management can show evidence of its commitment to safety: managing safety and health programs, attending safety and health meetings, periodic walk-through inspections, reviewing and acting upon accident reports, investigating accidents, reviewing safety records through conferences with department heads and joint employee-management committees, providing awards, and by setting a good example. Management must also audit the program to be certain that it was not "lost" in the management chain.

Department heads, supervisors, foremen, and employee representatives are in strategic positions within the organization to implement safety policies. Their leadership and influence should ensure that safety and health standards are enforced and upheld in each work area and that standards and enforcement are uniform throughout the workplace.

What are some responsibilities of department heads? They make certain that materials, equipment, and machines slated for distribution to their areas are hazard free or that adequate control measures have been provided. They make certain that equipment, tools, and machinery are being used as designed and are properly maintained. They keep abreast of accident and injury trends occurring in their areas and take proper corrective action to reverse these trends. They investigate all accidents occurring within their jurisdiction. They should see to it that all safety and health rules, regulations, and procedures are enforced in their departments. They require that a hazard analysis be conducted for certain operations, particularly those that they regard as dangerous, either from past accident history or from their perception of accident potential. They require that hazard recognition and control information be included in instruction, training, and demonstration sessions for both supervisors and employees. They actively participate in and support the safety and health committee and follow up on its recommendations.

Supervisors, foremen, and employee representatives carry great influence. With their support, top management can be assured of an effective safety and health program. Supervisors have a moral and professional responsibility to safeguard, educate, and train those who have been placed under their direction. Thus, they are generally responsible for creating a safe, healthy work setting and for integrating hazard recognition and control into all aspects of work activities. By their careful monitoring, they can prevent accidents.

For all practical purposes supervisors, foremen, and employee representatives are the eyes and ears of the workplace control system. On a day-to-day basis, they must be aware of what is happening in their respective areas, who is doing it, how various tasks are being performed, and under what conditions. As they monitor their areas, they must prevent accidents from occurring. If, despite their controls, they see danger, they must be prepared to intervene in the operation and take immediate corrective action. What are the chief safety and health responsibilities of supervisors? They train and educate workers in safe working methods and techniques. They should be certain that employees understand the properties and hazards of the materials they store, handle, or use. They make sure that employees observe necessary precautions, including proper guards and safe work practices. They furnish employees with the proper personal protective equipment, instruct them in its use, and make certain it is worn.

Supervisors demonstrate an active interest in and comply with loss control policy and safety and health regulations. They actively participate in and support the safety and health committees. They supervise and evaluate worker performance, with consideration given to safe behavior and work methods. They should first try to convince employees of the need for safe performance, and then, unpleasant though it may be, they should administer appropriate corrective action when health and safety rules are violated. Correcting employees requires tact and good judgment. Enforcement should be viewed as education rather than discipline. However, if a supervisor feels that a worker is deliberately disobeying rules or endangering his or her life and the lives of others, then prompt and firm action is called for. Laxity in the enforcement of safety rules undercuts the entire safety and health program and allows accidents to happen. Supervisors monitor their area on a daily basis for human, situational, and environmental factors capable of causing accidents. They should make sure that meticulous housekeeping practices are developed and used at all times. They should correct hazards detected in their monitoring or report such hazards to the persons who can take corrective action. They should investigate all accidents occurring within their areas to determine causes.

Foremen and employee representatives share much of the responsibility for safety and health with upper management. Their job is to inspect, detect, and correct. What are the specific responsibilities of foremen and representatives? They should encourage other workers to comply with the organization's safety and health regulations. They detect safety violations and hazardous machinery, tools, equipment, and other implements. They take corrective action when possible and report the hazard to the supervisor, along with the corrective action taken or still required. They may participate in accident investigations. They represent the interests of the workers on the safety and health committee.

Practical training aids for employee representatives, foremen, and supervisors do exist. The Council's *Supervisors' Safety Manual* and accompanying "Supervisors' Development Program" have been widely accepted by the industry. The Council's Joint Safety and Health Committee Training Course is also a good training tool and review for safety committees.

Housekeeping and Preventive Maintenance

These can be regarded as two sides of the same coin. No safety program can succeed if housekeeping and maintenance are not seen as important to the effort.

Good housekeeping reduces accidents, improves morale, and increases efficiency and effectiveness. Most people appreciate a clean and orderly workplace where they can accomplish their tasks without interference and interruption.

An industrial organization, by its very nature, contains tools that must be kept clean. In operations, workers use flammable substances and materials requiring special storage and removal. The processes generate dust, scrap metal filings and chips, waste liquids, scrap lumber, and countless other by-products that must be disposed.

Housekeeping is a continuous process involving both workers and custodial personnel. A good housekeeping program incorporates the housekeeping function into all processes, operations, and tasks performed in the workplace. The ultimate goal is for each worker to see housekeeping as part of their job performance and not as an extra task that someone else should do.

When the workplace is clean and orderly and housekeeping becomes a standard part of operations, less time and effort will be spent keeping it clean, making repairs, and replacing equipment, fixtures, and the like. When the worker can concentrate on required tasks without excess scrap material, tools, and equipment interfering with work, operations will be more efficient and the product of higher quality. Time will be used for work, not searching for tools, materials, or parts. When a plant is clean and orderly as well as safe, employee morale is heightened.

When everything has an assigned place, there is less chance that materials and tools will be taken from the plant or misplaced. In a few moments before quitting time, the supervisor can determine what is missing. Different colors of paint can be applied to tools to identify the department to which they belong. Tool racks or holders should be painted a contrasting color as a reminder to workers to return the tools to their proper places. When stored on a rack, the space directly behind each tool should be painted or outlined in color to call attention to a missing tool.

Savings and efficiency are increased when workers treat materials with the care they deserve by minimizing spillage and scrap, saving pieces of material for use in future projects, and returning even small parts to their storage area. When aisle and floor space is uncluttered, movement within the plant is safer, and workers can easily clean and maintain their machinery and equipment. When the plant has adequate work space and when oil, grease, water, and dust are removed from floors and machinery, workers are less likely to slip, trip, or fall, or inadvertently come into contact with dangerous parts of machinery.

Chances of fires also are minimized when a workplace is kept free from accumulations of combustible materials that can burn upon ignition or, in the case of certain material relationships, spontaneously ignite. Furthermore, an orderly plant permits easy exit by keeping exits and aisles leading to exits free from obstructions. A neat, well-organized workplace also makes it easier to locate and obtain fire emergency and extinguishing equipment.

Preventive maintenance is the orderly, uniform, continuous, and scheduled action to prevent breakdown and prolong the useful life of equipment and buildings. Preventive maintenance is a shared responsibility. Workers caring for tools and equipment accomplish specific maintenance tasks. Other maintenance duties, such as oiling, tightening guards, adjusting tool rests, and replacing wheels, are routinely performed by workers. Advantages to be gained from preventive maintenance include safer working conditions, decreased downtime of equipment because of breakdown, and increased life of the equipment.

Satisfactory production depends on having buildings, equipment, machinery, portable tools, safety devices, and the like in operating condition and maintained so that production activities will not be interrupted while repairs are being made or equipment replaced. Preventive maintenance prolongs the life of the equipment by ensuring its proper use. When tools are kept dressed or sharpened and in satisfactory condition, the right tool will be used for the job. When safe and properly maintained tools are issued, workers have an added incentive to give the tools better care. When repairs are made quickly, workers do not need to improvise—for example, using a strip of metal as a screwdriver or crowbar. Sound, efficient maintenance management anticipates machine and equipment deterioration and sets up overhaul procedures designed to correct defects as soon as they develop. Such a repair and overhaul system obviously requires close integration of maintenance with inspection.

Preventive maintenance has four main components:

1. scheduling and performing periodic maintenance functions
2. keeping records of service and repairs
3. repairing and replacing equipment and equipment parts
4. providing spare parts control.

Maintenance schedules can be set up on several different bases. Factors usually considered include:

- manufacturer's recommendations
- age of the machine
- number of hours per day the machine is used
- past experience
- machine changes with use.

The manufacturer's specifications contain standards that need to be maintained for safe and economical use of the machine. These specifications give maintenance personnel definite guidelines to follow. Examples of scheduled activities include lubricating each piece of equipment; replacing belts, pulleys, fans, and other parts; and checking and adjusting brakes.

Management should keep two types of maintenance records: The first is a maintenance service schedule for each piece of equipment. This schedule indicates the date the equipment was purchased or placed in operation, its cost (if known), where it is used, each part to be serviced, the kind of service required, the frequency of service, and the person assigned to the servicing. Each piece of equipment also requires a repair record, which includes an itemized list of parts replaced or repaired and the name of the person who did the work.

In addition to scheduled adjustments and replacements, maintenance personnel must repair malfunctioning or broken equipment in accordance with the manufacturer's specifications. Sometimes equipment must be sent back to the manufacturer or its service representative for repair. Maintenance personnel should be aware of their limitations and recognize that their experience and expertise are not sufficient to do all repairs. Those assigned repair responsibilities require special safety training. Many of the jobs to be performed include testing or working on equipment with guards and safety devices removed. Therefore, a statement of necessary precautions should accompany the repair directive. Maintenance personnel (along with others) have a responsibility to lock out and tag defective equipment, or simply tag if the equipment, such as a ladder, cannot be locked out. (See Chapter 6, Safeguarding, in the *Engineering & Technology* volume for the proper lockout/tagout procedure.)

The second maintenance record itemizes spare parts stock. Management should conduct routine surveys of spare parts requirements. To keep needed repair parts on hand, management should schedule review and reordering of stocked spare parts. When maintenance personnel keep purchasing agents informed of their anticipated stock needs, they help to reduce or eliminate downtime that occurs while workers wait for parts.

The difference between a mediocre maintenance program and a superior one is that while the former is aimed at maintaining facilities, the latter is designed to improve them. If conditions are good, a mediocre program will keep them that way but will not make them better; if conditions are not good, a mediocre program will not improve them. Preventive maintenance, on the other hand, is a program of mutual support that creates safe conditions, eliminates costly delays and breakdowns, and prolongs equipment life.

Employees

Employees make the safety and health program succeed. Well-trained and educated employees are the greatest deterrent to damage, injuries, and health problems in the plant or establishment.

What are the specific ways a safety program can be rooted in employee involvement and concern? Employees can observe safety and health rules and regulations and work according to standard procedures and practices. They can recognize and report to the foreman or supervisor hazardous conditions or unsafe work practices in the plant. They can develop and practice good habits of hygiene and housekeeping. They can use protective and safety equipment, tools, and machinery properly. They can report all injuries or hazardous exposure as soon as possible. Employees can help develop safe work procedures and make suggestions for improving work procedures. Management should encourage employees to participate on safety committees.

What shapes an employee's attitude toward safety? From the first day an employee goes to work, the employee starts to form attitudes about the job and organization. Substantial if subtle influence is exerted by the attitudes they observe in management, supervisors, and fellow workers. If these individuals regard safety as vital to an effective operation, a mark of skill and good sense, if they all participate actively and cooperatively in the safety program, if employees are recognized for having good safety records, then the employee will regard safety as something important, not as window dressing or an empty policy to which others pay lip service.

Introduction to the safety program should come on the employee's first day on the job. A three-pronged approach is suggested (Kane, 1979):

1. Coverage of general company policy and rules; discussion of various benefit programs, for example, hospitalization, pension plan, holidays, sick leave. The personnel department usually is responsible for giving this information immediately after the new employee is added to the payroll.
2. Discussion of general safety rules and the safety program. This part of the program should be the responsibility of a safety professional. The company's safety handbook should be given to the new employee, and the company's safety policy statement should be explained. The reasons behind the general safety rules should be explored with the employee, who is more likely to follow rules when they are understood.
3. Explanation of specific safety rules that apply to the new employee's department. At this point the supervisor's role overlaps with that of the safety professional. The supervisor can show how some hazards have been eliminated while others, which could not be designed out of the operation, are guarded against. This discussion provides an opportunity to talk about safe work practices and emergency procedures as well as to show how engineering controls and personal protective equipment can further reduce the effects of the hazard. The employee's responsibilities for safety and health also must be stressed. These responsibilities include reporting all accidents and incidents, checking equipment and tools before use, operating equipment only with proper authorization and prior instruction, and asking questions when the procedure is unclear.

Figure 3-10 lists the company safety rules developed by the Construction Advancement Foundation SAFE Committee and distributed to construction employees in Indiana.

Purchasing Agents

Those responsible for purchasing items for organizations are in a key position to reduce hazards associated with operations. The purchasing department has much latitude in selecting machinery, tools, equipment, and materials used in the organization. In maintaining standards of quality, efficiency, and price, the purchasing department must make certain that safety has received adequate attention in the design, manufacture, and shipping of items.

Depending upon the company organization, other departments—such as safety, engineering, quality control, maintenance, industrial hygiene, and medical—should indicate to the purchasing department what equipment and materials meet with their approval. The purchasing department is responsible for soliciting such guidance and direction. (Purchasing responsibilities are covered in depth later in this chapter under Purchasing.)

First, the purchasing agent must make certain all items comply with regulatory standards. A statement to this effect must be part of the purchase order. Purchasing agents also will be guided by (1) the standards of ANSI, Canadian Standards Association (CSA), and other standards and specifications groups; (2) products approved or listed by such agencies as UL and the NFPA; and (3) recommendations by such agencies as the NSC, insurance carriers or associations, the Factory Mutual System in its *Factory Mutual Handbook of Industrial Loss Prevention and Loss Prevention Data,* and trade or industrial organizations. (See Appendix 1, Sources of Help.)

COMPANY SAFETY RULES

All Employees Will Abide By The Following Rules:

1. Report unsafe conditions to your immediate supervisor.
2. Promptly report all injuries to your immediate supervisor.
3. Wear hard hats on the job site at all times.
4. Use eye and face protection where there is danger from flying objects or particles, such as when grinding, chipping, burning and welding, etc.
5. Dress properly. Wear appropriate work clothes, gloves, and shoes or boots. Loose clothing and jewelry should not be worn.
6. Never operate any machine unless all guards and safety devices are in place and in proper operating condition.
7. Keep all tools in safe working condition. Never use defective tools or equipment. Report any defective tools or equipment to immediate supervisor promptly.
8. Properly care for and be responsible for all personal protective equipment.

9. Be alert and keep out from under overhead loads.
10. Do not operate machinery if you are not authorized to do so.
11. Do not leave materials in aisles, walkways, stairways, roads or other points of egress.
12. Practice good housekeeping at all times.
13. Do not stand or sit on sides of moving equipment.
14. The use of, or being under the influence of, intoxicating beverages or illegal drugs while on the job is prohibited.
15. All posted safety rules must be obeyed and must not be removed except by management's authorization.
16. Comply at all times with all known federal, state and local safety laws as well as employer regulations and policies.
17. Horseplay causes accidents and will not be tolerated.

Violations of any of these rules may be cause for immediate disciplinary action.

Figure 3-10. Reprinted with permission from the Construction Advancement Foundation SAFE Committee.

Second, the purchasing agent must make certain tools, equipment, materials, machinery, and chemicals are purchased with adequate regard for safety. This requirement applies even to ordinary items such as boxes, cleaning rags, paint, and common hand tools. It is essential to compare safety features among the various brands when purchasing personal protective equipment and larger items, especially machines. Sometimes the cost of an adequately guarded machine seems out of proportion to that of an unguarded machine to which makeshift guards can be added. But experience has repeatedly proven that the best time to eliminate or minimize a hazard is in the design stage. Safeguards that are integral parts of a machine are the most efficient and durable.

Third, the purchasing agent must be cost conscious, realizing that every accident has both direct and indirect costs. The agent should understand that the organization cannot afford bargains that later result in accident losses and occupational disease.

Professionals in Loss Control

The roles of various professionals in the loss control program are described in National Safety Council's *Fundamentals of Industrial Hygiene,* 4th ed. The following discussion, therefore, describes briefly the responsibilities of the safety and health professional, the industrial hygienist, and staff medical personnel.

Safety and Health Professionals (Loss Control Specialists)

To ensure the effectiveness of the safety program, management usually places program administration in the hands of a safety director or manager of safety and health. The number of full-time safety professionals is increasing as the nature of their duties is better understood.

To administer a safety program effectively requires considerable training and many years of experience. A safety program has many facets: occupational health, product safety, machine design, plant layout, security, damage control, and fire prevention. Safety as a profession combines engineering, management, preventive medicine, industrial hygiene, and organizational psychology. It requires knowledge of system safety analysis, job safety analysis, job instruction training, human factors engineering, biomechanics, and product safety. The professional must have thorough knowledge of the organization's equipment, facilities, manufacturing process, and workers' compensation, and must be able to communicate and work with all types of people. Other desirable traits are the ability to see both management and employee viewpoints, and the skills of a good trainer (Figure 3-11).

The passage of the 1970 OSHAct requires that certain safety standards be met and maintained. Generally, organizations with moderate or high hazards and/or employing 500 or more persons need a full-time safety professional.

THE SCOPE OF THE PROFESSIONAL SAFETY POSITION

A safety and health professional brings together those elements of the various disciplines necessary to identify and evaluate the magnitude of the safety problem. He or she is concerned with all facets of the problem, personal and environmental, transient and permanent, to determine the causes of accidents or the existence of loss producing conditions, practices or materials.

Based upon the information collected and analyzed, he or she proposes alternate solutions, together with recommendations based upon his or her specialized knowledge and experience, to those who have ultimate decision-making responsibilities.

The functions of the position are described as they may be applied in principle to the safety and health professional in any activity.

In performing these functions the safety and health professional will draw upon specialized knowledge in both the physical and social sciences. He or she will apply the principles of measurement and analysis to evaluate safety performance. He or she will be required to have fundamental knowledge of statistics, mathematics, physics, chemistry, as well as the fundamentals of the engineering disciplines.

He or she will utilize knowledge in the fields of behavior, motivation, and communications. Knowledge of management principles as well as the theory of business and government organization will also be required. His or her specialized knowledge must include a thorough understanding of the causative factors contributing to accident occurrence as well as methods and procedures designed to control such events.

The safety and health professional of the future will need a unique and diversified type of education and training if he or she is to meet the challenges of the future. The population explosion, the problems of urban areas, future transportation systems, as well as the increasing complexities of man's every day life will create many problems and extend the safety and health professional's creativity to its maximum if he or she is to successfully provide the knowledge and leadership to conserve life, health, and property.

Functions of the Professional Safety Position

The major functions of the safety and health professional are contained within four basic areas. However, application of all or some of the functions listed below will depend upon the nature and scope of the existing accident problems, and the type of activity with which the safety and health professional is concerned.

The major areas are:

A. Identification and appraisal of accident and loss producing conditions and practices and evaluation of the severity of the accident problem.

B. Development of accident prevention and loss control methods, procedures, and programs.

C. Communication of accident and loss control information to those directly involved.

D. Measurement and evaluation of the effectiveness of the accident and loss control system and the modifications needed to achieve optimum results.

A. Identification and Appraisal of Accident and Loss Producing Conditions and Practices and Evaluation of the Severity of the Accident Problem

These functions involve:

1. The development of methods of identifying hazards and evaluating the loss producing potential of a given system, operation or process by:
   a. Advanced detailed studies of hazards of planned and proposed facilities, operations and products.
   b. Hazard analysis of existing facilities, operations and products.
2. The preparation and interpretation of analyses of the total economic loss resulting from the accident and losses under consideration.
3. The review of the entire system in detail to define likely modes of failure, including human error and their effects on the safety of the system.
   a. The identification of errors involving incomplete decision making, faulty judgment, administrative miscalculation and poor practices.
   b. The designation of potential weaknesses found in existing policies, directives, objectives, or practices.
4. The review of reports of injuries, property damage, occupational diseases or public liability accidents and the compilation, analysis, and interpretation of relevant causative factor information.
   a. The establishment of a classification system that will make it possible to identify significant causative factors and determine needs.
   b. The establishment of a system to ensure the completeness and validity of the reported information.
   c. The conduct of thorough investigation of those accidents where specialized knowledge and skill are required.
5. The provision of advice and counsel concerning compliance with applicable laws, codes, regulations, and standards.
6. The conduct of research studies of technical safety problems.
7. The determination of the need of surveys and appraisals by related specialists such as medical, health physicists, industrial hygienists, fire protection engineers, and psychologists to identify conditions affecting the health and safety of individuals.
8. The systematic study of the various elements of the environment to assure that tasks and exposures of the individual are within his psychological and physiological limitations and capacities.

Figure 3-11. (Continued on next page)

B. Development of Accident Prevention and Loss Control Methods, Procedures, and Programs

In carrying out this function, the safety and health professional:

1. Uses specialized knowledge of accident causation and control to prescribe an integrated accident and loss control system designed to:
   a. Eliminate causative factors associated with the accident problem, preferably before an accident occurs.
   b. Where it is not possible to eliminate the hazard, devise mechanisms to reduce the degree of hazard.
   c. Reduce the severity of the results of an accident by prescribing specialized equipment designed to reduce the severity of an injury should an accident occur.
2. Establishes methods to demonstrate the relationship of safety performance to the primary function of the entire operation or any of its components.
3. Develops policies, codes, safety standards, and procedures that become part of the operational policies of the organization.
4. Incorporates essential safety and health requirements in all purchasing and contracting specifications.
5. As a professional safety consultant for personnel engaged in planning, design, development, and installation of various parts of the system, advises and consults on the necessary modification to ensure consideration of all potential hazards.
6. Coordinates the results of job analysis to assist in proper selection and placement of personnel, whose capabilities and/or limitations are suited to the operation involved.
7. Consults concerning product safety, including the intended and potential uses of the product as well as its material and construction, through the establishment of general requirements for the application of safety principles throughout planning, design, development, fabrication and test of various products, to achieve maximum product safety.
8. Systematically reviews technological developments and equipment to keep up to date on the devices and techniques designed to eliminate or minimize hazards, and determine whether these developments and techniques have any applications to the activities with which the safety and health professional is concerned.

C. Communication of Accident and Loss Control Information to Those Directly Involved

In carrying out this function the safety and health professional:

1. Compiles, analyzes, and interprets accident statistical data and prepares reports designed to communicate this information to the appropriate personnel.
2. Communicates recommended controls, procedures, or programs designed to eliminate or minimize hazard potential, to the appropriate person or persons.
3. Through appropriate communication media, persuades those who have ultimate decision-making responsibilities to adopt and utilize those controls which the preponderance of evidence indicates are best suited to achieve the desired results.
4. Directs or assists in the development of specialized education and training materials and in the conduct of specialized training programs for those who have operational responsibility.
5. Provides advice and counsel on the type and channels of communications to insure the timely and efficient transmission of useable accident prevention information to those concerned.

D. Measurement and Evaluation of the Effectiveness of the Accident and Loss Control System and the Needed Modifications to Achieve Optimum Results

1. Establishes measurement techniques such as cost statistics, work sampling or other appropriate means, for obtaining periodic and systematic evaluation of the effectiveness of the control system.
2. Develops methods that will evaluate the costs of the control system in terms of the effectiveness of each part of the system and its contribution to accident and loss reduction.
3. Provides feedback information concerning the effectiveness of the control measures to those with ultimate responsibility, with the recommended adjustments or changes as indicated by the analyses.

Figure 3-11. (Concluded)

However, the nature of the operation may indicate the need for a full-time professional, regardless of the number of people employed.

The safety professional—whether called a safety engineer, safety director, hazard control specialist, loss control manager, a safety and health professional, or some other title—normally functions as a specialist on the management level. The loss control program should enjoy the same position as other established activities of the organization, such as sales, production, engineering, or research. Its budget reflects top management's commitment to the safety and health of its employees and includes the safety professional's salary, the salary of staff to help him or her, travel allowance, cost of safety equipment, cost of training and continuing education, and other related items. Safety professionals must define the needs of their program and, according to priorities, make short-range and long-range budget projections. With such projections in hand, they can present their needs to those with fiscal responsibility and stand a better chance of acquiring the resources to make their programs work. Figure 3-11 is the ASSE's description of the basic duties of the safety and health professional.

In general, the safety professional advises and guides management, supervisors, foremen, employees, and such departments as purchasing, engineering, and personnel on all matters pertaining to safety. Formulating, administering, monitoring, evaluating, and improving the accident prevention program are other responsibilities of the safety and health professional.

The safety professional usually investigates serious accidents personally or through the staff, reviews supervisors' accident reports, checks corrective actions taken to eliminate accident causes, makes necessary reports to management, and maintains the accident report system, including files. The safety and health professional recommends safety provisions in (1) plans and specifications for new building construction; (2) repair or remodeling of existing structures; (3) safety equipment; (4) designs of new equipment; and (5) processes, operations, and materials. The safety professional provides (or cooperates with the training supervisor to provide) safety training and education for employees. The safety staff also take continuing education courses and attend professional meetings like the National Safety Council Congress and Exposition (Figure 3-12).

Further, the safety professional makes certain that line managers know and understand their responsibility to comply with regulatory standards relating to safety and health and that standards, whether mandatory or recommended, are met; and accepts responsibility for preparing the necessary reports (for example, OSHA Forms 100, 102, and 200) for management. On jobs involving subcontractors, the safety professional informs line management of their responsibility to inform each subcontractor of relevant hazard control responsibilities; this is often spelled out in the contract. (See the discussion in the *Engineering & Technology* volume, Chapter 3, Construction of Facilities.) The safety professional also supervises disaster control, fire prevention, and firefighting activities when they are not responsibilities of other departments and stimulates and maintains employee interest in and commitment to safety.

Progressive management will give the safety professional the authority to order immediate changes on fast-moving and rapidly changing operations or in cases where delayed action could endanger lives (as in construction, demolition, or emergency work, fumigation, and some phases of manufacturing explosives, chemicals, or dangerous substances). Where the safety professional exercises this authority, it is done with discretion by accepting accountability to management for errors in judgment. However, the professional realizes that erring on the side of caution is more easily justified than taking unnecessary risks with worker health and safety.

Industrial Hygienists

The industrial hygienist is trained to anticipate, recognize, evaluate, and control health hazards—particularly chemical, physical, biological, and ergonomic agents—that exist in the workplace and have injurious effects on workers. Specialists work in fields such as toxicology, epidemiology, chemistry, ergonomics, acoustics, and ventilation engineering.

What are the specific responsibilities of the industrial hygienist in the hazard control program? The industrial hygienist (1) recognizes and identifies those chemical, physical, and biological agents that can adversely affect the physical and mental health and well-being of the worker; (2) measures and documents levels of environmental exposure to specific hazardous agents; (3) evaluates the significance of exposures and their relationship to occupationally and environmentally induced diseases; (4) establishes appropriate controls and monitors their effectiveness; and (5) recommends to management how to correct unhealthy (or potentially unhealthy) conditions. The roles of the industrial hygienists are detailed further in the Council's *Fundamentals of Industrial Hygiene.*

(Staff) Medical Personnel

Occupational health services vary greatly from one organization to another, depending on number of employees, nature of the operation, and commitment of the employer. One location may include a full-time physician, nurses, and technicians, with treatment rooms and a dispensary. Another may have only the required first aid kit and a person trained to render first aid and cardiopulmonary resuscitation (CPR).

The small operation has special need for persons thoroughly trained in first aid and CPR to take care of the employees on all shifts. Sometimes small facilities in the same locality share the services of a qualified physician on either a part- or full-time basis.

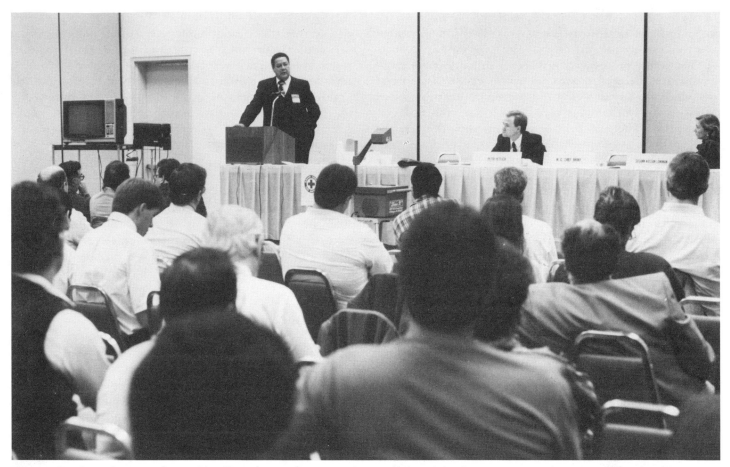

Figure 3-12. Each year, safety and health professionals continue their professional development at specialty sessions offered at the National Safety Council Congress and Exposition.

How much responsibility a part- or full-time nurse assumes generally depends on the availability of a licensed physician. It is not unusual for an occupational nurse to be solely and completely responsible for the occupational safety and health program with medical direction from a physician who rarely visits the workplace.

The American Association of Occupational Health Nurses define occupational health nursing as "the application of nursing principles in conserving the health of workers in all occupations. It involves prevention, recognition, and treatment of illness and injury, and requires special skills and knowledge in the fields of health, education and counseling, environmental health, rehabilitation, and human relations." In addition to being responsible for health care, the occupational nurse may on occasion monitor the workplace, perform industrial hygiene sampling, act as a consultant on sanitary standards, and be responsible for health education.

Chapter 10, Occupational Health Programs, deals specifically with the duties of the industrial physician, nurse, and first aid attendants. Regardless of size, health programs share common goals: to maintain the health of the workforce, to prevent or control diseases and accidents,

and to prevent or at least reduce disability and resulting lost time.

Staff medical personnel are a vital part of the total health and safety program. Their most obvious contribution is to provide emergency medical care for employees who are injured or become ill on the job. An outgrowth of this responsibility is to provide follow-up treatment of employees suffering from occupational disease or injuries. For some exposures, periodic examinations are required by OSHA regulations. A further outgrowth is medical personnel's promotion of health education and wellness programs for employees and their families. In addition, employee assistance programs (EAPs) are now commonly available for assisting employees with alcohol and drug issues along with other psychological problems.

Medical personnel foster a healthful environment by regularly scheduled workplace inspections. These inspections are necessary to keep medical personnel familiar with the materials and processes used and procedures performed by the employees under their care.

Another way that medical personnel are important to the health and safety organization is in the placement of employees within the legal constraints of the Americans with Disabilities Act. Give the medical staff an opportunity to post job offer evaluations that recommend which work accommodation is medically indicated. This evaluation objective should take into consideration the physical and mental findings of the employee so that an appropriate level of accommodation can be achieved.

PURCHASING

The safety department should have excellent liaison not only with the engineering department but also with the purchasing department, as discussed in the previous section. It should be the duty of the safety department to coordinate or develop written safety standards to guide the purchasing department. These standards should help to eliminate the hazards associated with a particular kind of equipment or material, for example, by substituting a safe material for a dangerous one, or safeguarding equipment for the protection of worker, machine, and product.

The purchasing agent is not closely concerned with educational and enforcement activities but is vitally concerned with many phases of the engineering activities. It is the agent's duty to select and purchase the various items of machinery, tools, equipment, and materials used in the organization. It is also the agent's responsibility—in part or to a considerable degree—to see that in the design, manufacture, and shipment of all these items, safety and health have received adequate attention.

In one facility, a lead hazard occurred when workers unloaded litharge (lead oxide) that had been shipped in 10-gallon paint pails with covers. Some of the litharge had leaked, coating the pails with a film of litharge on the outside. When the pails were moved, litharge was released into the air, creating a lead concentration 30 to 40 times the permissible limit, and contaminating the skin and clothing of workers.

Management and safety workers tried several solutions before finding the best approach. They eliminated the hazard by instructing the purchasing department to specify a rubber gasket under the pail lid as a part of the purchasing requirements. Thus the leakage, which created a serious health hazard, was easily controlled. The company also required that the shipper label the containers with information meeting the requirements of the Hazard Communication Standard.

Purchasing-Safety Liaison

Having a background knowledge of engineering specifications and codes and standards, the safety professional should be well prepared to advise the purchasing department.

What Purchasing Can Expect

The purchasing agent can reasonably expect that the safety professional will:

- give specific information about process and machine hazards that can be eliminated by change in design or by installing manufacturer-designed guarding

- supply similar information about other equipment, tools, and materials along with facts about injuries caused
- give specific information about health and fire hazards in the workplace
- provide information on federal and state safety requirements
- supply on request additional special information on accident experience with machines, equipment, or materials when such articles are about to be reordered
- request assistance in the investigation of accidents that may have been caused by faulty equipment or material
- request a list of equipment and materials requiring safety approval before going out for bids.

What the Safety Department Can Expect

Where there is effective liaison between the safety and purchasing departments, the safety professional can expect that the purchasing agent will:

- become familiar with the departmental and plant process hazards, especially in relation to machinery, equipment, and materials
- ask the safety department for information on hazards and accident costs, for federal and state safety requirements, and for lists of approved devices and appliances before making purchases
- become acquainted with the specific location and departmental use of machinery or equipment about to be ordered
- participate in accident investigations where injuries may have been caused through the failure of machinery, equipment, or materials.

Safety Considerations

The purchasing staff should always review applicable safety specifications and guidelines before purchasing supplies and equipment. For many articles, there is no need to consider safety. Some items, however, have a more important bearing upon safety than may be suspected.

The agent must exercise utmost caution when purchasing personal protective equipment, such as eye protection, respirators, and masks; equipment to move suspended loads, such as ropes and chains; equipment to move and store materials; and miscellaneous substances and fluids for cleaning and other purposes that might constitute or aggravate a fire or health hazard. The agent should specify adequate labeling that identifies contents and calls attention to hazards.

Investigation, however, may show that unsuspected hazards arise in the purchase of ordinary items, such as common hand tools, reflectors, tool racks, cleaning rags, paint for shop walls and machinery, and even filing cabinets. Among the factors purchasing agents need to consider include maximum load strength; long life without deterioration; reduction of sharp, rough, or pointed characteristics; less frequent need for adjustment; ease of maintenance; reduction of fatigue-causing characteristics; and minimal hazard to workers' health.

Here are a few examples of hazards created by purchased items that were considered safe. Goggles supplied to one group of workers were found to have imperfections in the lenses that caused eyestrain and headache, which led to fatigue and accidents. The toes of a laborer were crushed because his safety shoe had an inadequate metal cap and collapsed under a weight that would have been easily supported by a shoe meeting the ANSI Z41 standard. In another plant, workers were supplied with wooden carrying boxes, when a proper type of metal box could have eliminated the hazard of splinters and perhaps an infected hand. It is in the purchase of larger items, especially machines, that the more impressive examples of purchasing for safety are found. Today, machines of many types are manufactured and can be bought with adequate safeguards in place as integral parts of the machines. The enclosed motor drive is an outstanding example of engineering machine construction for safety.

When an order for equipment is about to be placed, the purchasing agent, if possible, should not consider any machine that has been only partly guarded by the manufacturer and needs to be fitted with makeshift safeguards. The agent should be in frequent consultation with the safety department before making any purchase where safety is a factor. The agent also should be particularly careful to see that every purchased machine complies fully with the safety regulations of the state in which it is to be operated.

Price Considerations

When considering plant purchases, the purchasing agent must struggle to reconcile quality, work efficiency, and safety with the price of an item. At times it may seem that the cost of a well-safeguarded machine is out of proportion to the cost of an unguarded machine, including the estimated expense of adding home-made safeguards. But the experience of many industrial organizations has proved again and again that the best time to safeguard a machine or process is in the design stage. Safeguards planned and built as integral parts of a machine are the most efficient and durable.

The purchasing agent, through accident information supplied by the safety department, should be familiar with the costs of specific accidents in the plant, especially those in which a purchased item has been found to contribute to an accident. As a result, the purchasing agent should be able to defend the decision to buy a slightly higher priced component where appropriate. Cooperation between the safety, engineering, manufacturing, and purchasing departments is absolutely necessary if accidents are to be eliminated.

These arguments should appeal to all executives responsible for the success of the industrial organization.

The executive who is already sold on safety is agreeable to expenditures reasonably justified in the interest of accident prevention. If an executive does not have this attitude, the purchasing agent should find ways to stimulate the executive's interest in the organizational safety program. In this undertaking, the purchasing agent can undoubtedly count upon the active cooperation of the safety professional.

In some instances, the purchase of machinery or equipment involves important engineering details. For such purchases, the company undoubtedly will have a system whereby engineers will first prepare definite specifications, perhaps including drawings. The safety professional then carefully checks these plans and specifications before the purchasing agent solicits bids and cost estimates. The purchasing agent will have the plans and specifications at hand when asking for prices. He or she will want to keep in close touch with the safety professional throughout the negotiations to use the latter's knowledge and experience in accident prevention.

After the purchase order has been made but before it is signed, there is one other important detail that should not be overlooked. This is a statement, in language that cannot possibly be misinterpreted, that the articles ordered must comply fully with the applicable federal and state safety laws and regulations of the locality in which they are to be used. This statement must be made a part of the purchase order.

Codes and Standards

In purchasing, the safety professional will a need to have a thorough knowledge of the facility's accident history, the costs involved in accidents, and the probable benefits of changes suggested. To fulfill this function in cooperation with the purchasing department, it is important that the safety professional be familiar with codes and standards. When a specific item of equipment is recommended, the safety professional should be able to state that it is a type approved by authoritative bodies and that it meets regulatory requirements.

Generally, the safety professional should consult with everyone concerned before setting up company standards to guide the purchasing department. There are many guidelines and standards that can be used as models. Accordingly, the safety professional (and all others concerned with setting company standards) should be familiar with the following:

- codes and standards approved by the ANSI and other standards and specifications groups—see Appendix 1, Sources of Help.
- codes and standards adopted or set by federal, state, and local governmental agencies, such as OSHA, the Bureau of Mines, and the National Bureau of Standards
- codes, standards, and lists of approved or tested devices published by agencies such as the NIOSH,

the Mining Safety and Health Administration, UL, and fire protection organizations. For fire protection, the standards and codes of the NFPA should be followed (see Sources of Help).

- safe practice recommendations of such agencies as the NSC, insurance carriers or their associations, and trade and industrial organizations.

Specifications

The engineering department, with the help of the safety department, should specify the necessary safeguarding to be built into a machine before it is purchased. Persons responsible for purchasing in an industrial plant are necessarily cost conscious. Consequently, the safety professional must become aware of accident costs associated with specific machines, materials, and processes. For instance, if the individual recommends spending several thousand dollars for a superior grade of tool, he or she must have evidence to justify the investment.

Because of highly competitive marketing, manufacturers of machine tools and processing equipment often list safety devices as accessories. It is important that the safety professional be familiar with regulatory-required auxiliary equipment and be able to justify its inclusion in the original order.

In some organizations, the safety professional is charged with checking all plans and specifications for machinery and other equipment. In many organizations, particularly where certain items, such as goggles or safety shoes, are to be reordered from time to time, various operating officials cooperate to prepare standard lists, and purchases are selected only from among the types and companies shown on these approved lists. In still other establishments, the responsibility for design, quality, safety, and other features rests with the employees who are to use the articles. In such cases, the purchasing agent is responsible only for price, date of delivery, and similar details.

In many companies where purchases are made in huge quantities and at a great investment of money, important duties are placed in three coordinate departments: (1) the engineering department, whose staff prepare plans and specifications for all machinery and equipment to be purchased; (2) the safety department, where the staff carefully check these plans and specifications for safety and carry out final inspections of articles purchased; and (3) the purchasing department, which still has latitude in making selections as well as in determining standards of quality, efficiency, and price.

Still another variable must be mentioned. Many companies have both a full-time purchasing agent and a full-time safety professional. However, in many other companies, especially smaller ones, these important duties are assumed by executives who devote part of their time to other activities. Nevertheless, the measures that should be taken to prevent accidents in the small plant are substantially the same as those taken in the large plant. The part the purchasing agent can play in the safety program is similar; the interest will be the same and the activities will vary only by degree. Success will lie in adopting as fully as possible all the suggestions that are presented here and in cooperating closely with others in the company to promote safety for all workers.

Specification of Shipping Methods

When the purchasing agent orders materials, it may be desirable to specify they be shipped in a particular manner. If safe and efficient shipping methods are worked out and then specified in the orders, the suppliers will be better able to deliver materials on time, in good condition, and in a shape or form that can be easily and safely handled by employees. The agent should specify that all hazardous materials must be labeled with Department of Transportation (DOT)-authorized shipping labels. Material Safety Data Sheets should accompany all chemicals or products that contain chemicals, as well as some substances like solid metal that are designed to be remelted or equipment like welding rods.

SAFETY AND HEALTH COMMITTEES

Safety and health committees can be invaluable to the loss control program by providing the active participation and cooperation of many key people in the organization. They also can be unproductive and ineffective. The difference between success and failure lies with the original purpose of the committee, its staffing and structure, and the support it receives while carrying out its responsibilities.

A safety and health committee is a group that aids and advises both management and employees on matters of safety and health pertaining to plant or company operations. In addition, it performs essential monitoring, educational, investigative, and evaluative tasks.

Committees may represent various constituencies or levels within the organization or may be management or workplace committees. The joint safety and health committee (discussed here) is responsible for:

- actively participating in safety and health instruction programs and evaluating the effectiveness of these programs
- regularly inspecting the facility to detect unsafe conditions and practices and hazardous materials and environmental factors
- planning improvements to existing safety and health rules, procedures, and regulations
- recommending suitable hazard elimination, reduction, or control measures
- periodically reviewing and updating existing work practices and hazard controls
- assessing the implications of changes in work tasks, operations, and processes
- field-testing personal protective equipment and making recommendations for its use or alteration based on the findings

- monitoring and evaluating the effectiveness of safety and health recommendations and improvements
- compiling and distributing safety and health and hazard communications to the employees
- immediately investigating any workplace accident
- studying and analyzing accident and injury data.

The OSHAct in Section 2(b)(13) clearly allows for the possibility of joint safety and health initiatives as a supplementary approach to accomplishing OSHA's objectives more effectively. Joint committees have considerable potential for reducing injuries and illnesses, thus leaving OSHA free to target enforcement according to the worst-first principle.

The joint committee concept stresses cooperation and a commitment to safety as a shared responsibility between management and workers. Employees can become actively involved in and make positive contributions to the company's safety and health program. The committee serves as a forum for discussing changes in regulations, programs, or processes, and potential new hazards. Employees can communicate problems to management openly and face to face, allowing information and suggestions to flow both ways. The knowledge and experience of many persons combine to accomplish the objectives of creating a safe workplace and reducing accidents. The approach can produce effective solutions to safety problems more easily. Because joint committees facilitate communication and cooperation, they usually raise employee morale as well.

Even though a joint committee represents both employees and management, the committee's analyses and recommendations—whether they pertain to policy or practice—should be reviewed and confirmed by experts when they relate to specialized areas (for example, electrical safety, exposure levels).

Labor-management cooperation was discussed in Chapter 1, Historical Perspectives. Committee organization and operation are covered in the Council publication "You Are the Safety and Health Committee."

OFF-THE-JOB SAFETY PROGRAMS

According to National Safety Council estimates, three out of four deaths and more than half of the injuries suffered by workers occur off the job. Annual production time lost due to off-the-job injuries averages 120,000,000 days compared with 75,000,000 days lost by workers injured in the workplace. Clearly, reducing the number and severity of off-the-job injuries should be a major concern for all industries.

There is a certain amount of confusion, however, as to what off-the-job (OTJ) safety really includes. Essentially, off-the-job safety is a term used by employers to designate the part of their safety program directed to employees when they are not at work.

The principal aim of off-the-job safety is to get employees to follow the same safe practices used on the job while pursuing outside activities. Therefore, off-the-job safety should not be a separate program but rather an extension of a company's on-the-job safety program. While companies have a legal responsibility to prevent injuries on the job, they have a moral responsibility to try to prevent injuries away from the job. The other reason for an off-the-job safety program is cost. Operating costs and production schedules are affected as much when employees are injured away from work as when they are injured on the job. (These costs are discussed in detail in Chapter 7, Accident Investigation, Analysis, and Costs.)

What Is Off-the-Job Safety?

Off-the-job safety is a logical extension of the occupational safety program. Accident/illness prevention at work is cost effective, while fulfilling an organization's moral and legal responsibilities. An effective off-the-job safety program meets these same needs: reduction of costly employee absences due to accidents, injuries, or deaths; and commitment to employee well-being. Preventing off-the-job incidents that could result in injury or illness can be accomplished by using methods proven successful for increasing safety awareness at work.

Responsible employers should also be concerned about the well-being of their employees' families. Many times injury to a family member will affect an employee's work performance. Family involvement in safety can be a key factor to reducing employee off-the-job injuries. This approach can be accomplished in many ways, but education and peer pressure are extremely important to the success of an OTJ safety program.

Complicating the company's efforts is the fact that current methods of gathering, recording, and measuring employee OTJ injury information vary substantially from company to company. Only when companies with OTJ safety programs treat their injury data uniformly can they compare their experience to other companies and help determine OTJ injury rates nationwide. While compiling OTJ injury data, companies should follow the practices recommended in ANSI Z16-.3-1994, *Recording and Measuring Employee Off-the-Job Injury Experience.*

Program Benefits

The aim of safety education, namely, changing the employee's behavior, is especially true of off-the-job safety. No asset is more important to a company than its employees. They should not only be protected during working hours but also be given every incentive to practice safety off the job. A company can realize three benefits from expanding its safety program to include off-the-job safety. The first is a reduction in lost production time and operating costs from both on-the-job and off-the-job injuries. Second, companies have found that efforts in off-the-job safety increase employees' interest in their on-the-job safety program. The third benefit, often overlooked, is that of better public relations.

Many benefits are derived by developing or revitalizing an off-the-job safety program. These include:

- fewer off-the-job accidents, injuries, and deaths
- fewer employee absences
- reduced operating costs
- safety awareness at home carries over to the workplace
- improved work efficiency and performance
- enhanced employee and employer relationships
- participation of employees and their families in community safety actions
- positive, viable demonstration of organization's commitment to employee-family well-being and social issues affecting the employee's family. As with any effective occupational safety program, management involvement and participation at all levels in off-the-job safety must be vocal, visible, and continuous.

Promoting Off-the-Job Safety

Techniques for promoting off-the-job safety are essentially the same principles and techniques used on the job. From a safety standpoint, operating power tools at home involves the same risks as operating the equipment at work; likewise, driving the family car is the same as driving a company vehicle.

The only difference between these two safety programs is that organizations do have to depend more on education and persuasion to get their message across. This is because once employees leave the office, plant, or job site, they tend to believe the risks they assume are a private matter and no longer fall under the organization's policies and guidelines.

As with any other program—whether it be attendance, quality control, or waste reduction—management support and guidance is essential. Once safety personnel show management the seriousness of the problem (through experience and cost records), there should be little difficulty in obtaining support.

Off-the-Job Safety Policy

The organization should communicate management commitment to employees and their families. For example, a written policy statement concerning off-the-job safety or reference to off-the-job safety should be in the occupational safety policy statement signed by the top organizational official.

Getting Started

Various methods and sources of information on off-the-job accidents are available to an organization. For example, management can keep records documenting the causes of employee absences due to accidents away from the workplace; health and accident insurance claims may record the accident cause on the form for payment; the

NSC's annual publication *Accident Facts* can be used to pinpoint the leading causes of accidental deaths and injuries nationally (these data help to determine the most common accidents occurring in a particular organization).

It is important to tailor an off-the-job safety program to the special needs of an organization. The location of the plant site, its environment, and the special interests of employee groups such as skiing, boating, mountain-climbing, cave exploring, hunting, camping, or flying are factors that can help in selection of topics and activities.

The types and number of injuries/deaths people suffer off the job, however, are not usually related to how dangerous the activities appear to be. Many injuries occur when people are doing ordinary, everyday activities that do not appear to be risky, such as lifting objects, working on the car, walking up and down stairs, taking a shower, and so on.

Select Program Details

A good topic breakdown provides a solid structural framework for the development of off-the-job safety programs. For example, content may be based on seasonal hazards; on home, traffic, and public accidents; on health risk assessment, such as exercise and fitness, stress management, alcohol and drugs, community right-to-know; or according to accident types and causes.

Timely, interesting, and practical topics will attract the attention of employees, create and maintain enthusiasm, and encourage active participation to develop patterns of safe behavior.

A seasonal emphasis outline might include:

- spring—good housekeeping, lawn mowers, garden tools, do-it-yourself activities, pruning/planting trees, bicycles/helmets
- summer—sunburn, swimming, camping, boating, hiking, field sports, fishing, insects, vacation hazards
- fall—hunting, home power tools, back-to-school hazards, home-heating equipment, yard cleanup, repair and storage of tools
- winter—winter sports, holiday safety, severe weather, overexertion, winter driving.

Topics can tie into programs of national scope or interest. These national programs provide radio, TV, newspaper, and other forms of publicity that help promote program content. National Child Passenger Safety Awareness Week in February, National Safe Boating Week in May, National Fire Prevention Week in October, and National Drunk and Drugged Driver Awareness Week in December are examples.

A program on home fire safety might include showing a video on the proper use of fire extinguishers or on general fire prevention or safety practices as well as handout literature on fire safety topics. Combined with this program could be a company discount for employee purchase of smoke alarms, fire extinguishers, and fire escape ladders for home, workshop, and auto use.

Employee, Family, and Community Involvement

The assistance of a special off-the-job safety group may be valuable. Membership in this group can include employees, family members of employees, local civic and school groups, or community people with an interest or role in safety in general. Organizations that use such a committee often find that its members contribute immeasurably to the success of the program by providing special information that represents their background and understanding of off-the-job safety. Another benefit of the group is that members share serious safety convictions with peers, friends, neighbors, and their families.

Every community has special interest groups and organizations already concerned with various phases of off-the-job safety that will lend their resources and personnel to assist in company activities. Among such groups are the safety council; Chamber of Commerce including the Jaycees; service clubs; Red Cross chapters; local newspapers; radio and TV stations; health, police, and fire departments; parent-teacher associations; rescue squads; and emergency service groups.

The local women's clubs can be an especially strong source of support for off-the-job safety. The clubs may adopt home, traffic, or recreation safety as a project for the year. In so doing, they ensure the participation of many homemakers and family members who are not exposed to a formal safety program.

A representative from safety and health disciplines, such as the ASSE, can present an off-the-job safety program to an organization's members or employees. He or she can obtain assistance from state safety organizations; medical, visiting nurse, and other associations; local utilities; poison control centers; state police; insurance companies; and public health groups.

A few of these groups maintain accident records, some participate in special programs, some publish bulletins on health and safety subjects, and some conduct courses in subjects related to off-the-job safety. Safe-behavior booklets and posters on seasonal activities are available from the NSC.

An organization can cooperate with the municipal recreation department in promoting swimming classes or courses in boating safety. Also, it can request that local police supervise an auto inspection clinic. Often insurance companies will provide the necessary equipment for testing drivers' physical qualifications and skills.

Organization personnel can, in turn, offer leadership and support for community activities. For instance, employees can help form local safety councils, assist schools and churches by making safety inspections upon request, or act as volunteer members of local fire departments.

Several small companies in a community may consider the possibility of pooling their resources and talents in a joint off-the-job safety program, as is sometimes done with disaster and rescue programs. Advance publicity ensures employee support by communicating objectives, plans, and activities. An informal letter from top management sent to employees' homes personalizes the organization's concern for the safety of not only the employees but their families as well.

The initial meeting for employees can be followed by other meetings that include family members. Many organizations hold picnics and other outings featuring various types of entertainment and safety exhibits as a means of reaching the families. An organization's newsletter, magazine, or paper is an effective way to carry the word to employees. Company bulletin boards often reinforce these messages. The community can be kept informed of plans and progress through spot announcements and stories on home, traffic, and recreational safety carried by the local radio and TV stations and the local newspapers. Preparation of such publicity can be financed by several companies together or by the local safety council.

Meetings of local organizations, such as church groups, women's and service clubs, PTA, and similar groups, are also good resources to educate the community about off-the-job safety programs and activities.

Programs that Worked

The following sections discuss promotional methods that have proven successful. They can easily be tailored to the needs of your organization.

Contest

Offering some incentive, as minimal as a savings bond or as generous as a college scholarship, can usually elicit good participation from employees' youngsters. The resulting posters, essays, calendars, or slogans can then become vehicles for promoting safety.

The two most widely used contest ideas involve essays (with a limit on the number of words) or posters (with restrictions on size and materials), such as, "What My Dad's/Mom's Safety Means to Me," "Vacation Safety," "Community Safety," and "The Importance of Off-the-Job Safety." Competitors should be divided into age groups—ages running from about 5–7 years old for the youngest, up to 16–18 years for the oldest. Each entrant should receive a token of participation, like a keychain, certificate, embroidered patch, or even a model of the organization's product. To ensure impartiality, judges may be from outside the community or the organization.

Traffic Safety

Offering the NSC's Defensive Driving Program, at the organization's expense and possibly with organization facilities and instructors, to all driving members of the employees' families is one of the most positive home, off-the-job, or community safety efforts an organization can make. It can improve the driving habits of those who drive, and can also promote car pooling as a safe answer to energy, traffic, and pollution problems.

Auto safety checks tie in well with vacation and holiday programs and work best on weekends. Have plenty of qualified inspectors available so participants will not be discouraged by long lines.

Company Picnic
Safety picnics should involve the whole family and can be as expensive as a completely catered affair or as simple as a family picnic. A safety theme can be included in drawings for door prizes, activities for all ages, and presentations of awards to or recognition of employees for safety achievements.

Family Night
Quite different from picnics, family-night gatherings are built around the presentation of some discussion or audio-visual presentation on safety, accompanied by refreshments. Sometimes the theme shifts from general safety to a program to make the family aware of and gain support for the safety efforts made at work. Sometimes an "open house" tour of the plant facilities can be combined with the family-night get-together.

Family first aid programs, including CPR training, can be held at business locations after business hours, or on weekends to involve the family. First aid kits available at a company discount purchase price could be offered for family use in the home, auto, or workshop.

Youth Activities
Sponsorship of activities like softball and football teams and bicycle rallies usually is not aimed directly at promoting safety. However, a poor safety record among young people taking part in the activity may hurt the sponsor's image. Therefore, financial sponsorship of such activities should be only part of an organization's participation. Employee leaders, who are knowledgeable in the activity and trained in safety and first aid (at the organization's expense), should be on hand to teach and help participants. Such actions can produce a strong, positive image of the organization in the community. Involvement of groups, such as the Boy and Girl Scouts, 4–H Clubs, Campfire, Inc., FFA, and FHA, will expand off-the-job safety efforts.

Recreational Programs
In any organization, many employees and their families are sports enthusiasts. Their pastimes may include hunting, boating, camping, swimming, fishing, skiing, to name a few. At the season openings of these activities, organizations may sponsor clinics featuring registered/competent instructors to check equipment and to provide instruction on improving skills. Sources of help include the local gun clubs, powerboat squadrons, the Coast

Guard Auxiliary, National Recreation and Park Association, the President's Council on Physical Fitness and Sports, the National Red Cross, YMCA, police and fire departments, and health organizations. All of these groups can provide ideas for safety activities.

Vacation-Holiday Program
Some organizations close down for regular summer vacation; others offer year-around vacation periods and three-day weekends. A good time to offer safe driving tips and safety literature to employees is at vacation times.

Other promotional methods include:

- Publicity distributed in-house that reinforces safety, both on-the-job and off-the-job (pamphlets, press releases, posters, and billboards are ideal vehicles for in-house promotions). The NSC has an excellent selection of this material, aimed at a variety of off-the-job safety topics. Employee-generated, original posters are also effective in personalizing safety efforts.
- Nearly every organization has some sort of in-house journal or newsletter that is either given to employees at work or sent to their homes. No other publication enjoys wider readership within an organization. As a result, it provides an excellent forum for safety education and safety program promotion; both employees and their families see it. Articles can be written on all aspects of a safety program, and the employee and family can be solicited for safety-related story ideas as well. The Council's bi-monthly newsletter, *Volunteers' Voice for Community Safety & Health,* gathers news for editors on a wide range of topics emphasizing round-the-clock safety and health
- NSC's periodical *Family Safety and Health* is an excellent way to promote off-the-job safety.

SUMMARY
Hazards are defined as any existing or potential condition in the workplace that, by itself or interacting with other variables, can result in deaths, injuries, property damage, and other losses. The two broad categories of hazards are (1) those dealing with safety and injuries and (2) those dealing with health and illnesses.
- The primary function of loss control is to help management locate, assess, and set effective preventive and corrective measures for hazards. Controlling hazards is a team effort between management and employees. Loss control programs set facility-wide safety and health standards and coordinate responsibility among departments. These programs must take into consideration the worker-equipment-environment system.

- Accidents can be attributed to the oversight, omission, or malfunction of the management system regarding work factors, human factors, and environment factors.
- Two basic approaches to examining accident causation are after-the-fact and before-the-fact investigations.
- An effective loss control program has six processes: hazard identification and evaluation, hazard ranking, management decision making, establishment of preventive and corrective measures, monitoring, and evaluating program effectiveness.
- To organize an effective safety and health program, management must establish program objectives, develop and publish an organizational policy, and delegate responsibility and authority for promoting the program.
- Management and administration, along with housekeeping and preventive maintenance, are a few of an organization's most potent weapons against hazards. A three-pronged approach to employee safety programs involves presenting general company policy and rules, discussing overall safety rules and the safety program, and explaining those safety rules that apply to each employee's department.
- Safety and health professionals, industrial hygienists, and staff medical personnel advise and guide organizations on safety matters, investigate accidents, maintain accident records, ensure line management's cooperation, and perform other duties as required to assist line management in safeguarding employees.
- Purchasing plays a key role in safety by ensuring that all items comply with government regulations and ordinances, that safety is engineered into items, and that safety is weighed into the purchasing price.
- The purpose of organizational safety and health committees is to aid and advise management and employees on safety and health, and perform monitoring, educational, investigative, and evaluative tasks in the safety program.
- Off-the-job safety programs educate employees to apply many of the same safety practices they use at work to activities away from the job.

REFERENCES

ANSI Z16-.3-1994, *Recording and Measuring Employee Off-the-Job Injury Experience.*

Board of Certified Safety Professionals of the Americas, 208 Burwash, Savoy, IL 61874. "Curricula Development and Examination Study Guidelines," Technical Report No. 1.

Boylston RB. "Managing Safety and Health Programs." Speech given before the Textile Section, National Safety Congress, October 1989.

Construction Advancement Foundation, Hammond, IN. Safety Manual.

Factory Mutual Engineering Corp. Loss Prevention Data. Norwood, MA 02062.

Dennis LE and ML Onion. Out in Front: Effective Supervision in the Workplace. Itasca, IL: National Safety Council, 1990.

Firenze RJ. *Guide to Occupational Safety and Health Management.* Dubuque, IA: Kendall/Hunt Publishing Co., 1973.

————. *The Process of Hazard Control.* Dubuque, IA: Kendall/Hunt Publishing Co., 1978.

————. *Safety and Health in Industrial/Vocational Education.* Cincinnati: National Institute for Occupational Safety and Health, 1981.

Johnson WG. *MORT Safety Assurance Systems.* New York: Marcel Dekker, Inc., 1980.

Kane A. Safety begins the first day on the job. *National Safety News* 53, January 1979.

Manuele FA. How effective is your hazard control program? *National Safety News* 53–58, February 1980.

National Association of Suggestion Systems, 230 North Michigan Avenue, Chicago, IL 60611. *Performance Magazine* (6 times a year).

"Suggestion Newsletter" (6 times a year).

National Safety Council, 1121 Spring Lake Drive, Itasca, IL 60143.

Accident Facts (published annually).

Management Safety Policies, Occupational Safety and Health Data Sheets, 12304-0585, 1995.

Supervisors' Safety Manual, 8th ed., 1992.

"You Are the Safety and Health Committee."

Peters GA. Systematic safety. *National Safety News* 83–90, September 1975.

Plog B., ed. *Fundamentals of Industrial Hygiene,* 4th ed. Itasca, IL: National Safety Council, 1996.

U.S. Department of Human Resources, National Institute for Occupational Safety and Health, Division of Technical Services, Cincinnati, OH 45226. Self-Evaluation of Occupational Safety and Health Programs, Publication 78-187, 1978.

U.S. Department of Labor, Occupational Safety and Health Administration. Organizing a Safety Committee, OSHA 2231, June 1975.

U.S. Department of Transportation, Office of Hazardous Materials, Washington, DC 20590.

"Newly Authorized Hazardous Materials Warning Labels." (Latest edition.) (Based on Title 49, *Code of Federal Regulations,* sections 173.402, —403, and —404; import or export shipments are covered in Title 14, CFR, section 103.13.)

Windsor DG. "Process Hazards Management," a speech given before the Chemical Section, National Safety Congress, October 17, 1979.

REVIEW QUESTIONS

1. Give three of the five benefits of hazard analysis?
   a.
   b.
   c.
2. What is the purpose of ranking hazards by risk?
3. Name the three major areas where hazardous conditions can be either eliminated or controlled and give an example of each.
   a.
   b.
   c.

Safety, Health, and Environmental Auditing

4

The events of recent years demonstrate that inadequate performance in the safety, health, and environmental (S/H/E) arenas can have severe consequences for both an organization and its management. A company responsible for environmental damage is at risk for fines, legal liabilities, and/or economic losses. Moreover, corporate managements continue to come under strong pressure from the public, customers, and regulators to create safe, healthy workplaces and to help preserve the environment. This chapter covers the following topics:

- the nature, purposes, and scope of S/H/E auditing
- the basic steps in conducting S/H/E audits
- major issues in designing and staffing audit programs
- future directions of S/H/E auditing and the emergence of auditing standards

Many organizations are devoting a sizable and increasing proportion of capital expenditures, operating costs, and managerial resources to programs for pollution control, occupational health and safety, product safety, and loss prevention, among others. In addition to using sophisticated environmental control equipment and other technology, companies are also formalizing many aspects of their safety, health, and environmental management systems. One such area of management that has received considerable attention since the mid-1970s is that of safety, health, and environmental (S/H/E) auditing.

This chapter discusses the development of S/H/E auditing, describes current practices, and identifies likely future trends.

DEVELOPMENT OF SAFETY, HEALTH, AND ENVIRONMENTAL AUDITING

In this chapter, the term "S/H/E auditing" is used to cover the full range of pollution control, occupational safety, process safety, industrial hygiene, occupational health and medicine, and product safety. Many companies have now established programs to monitor and audit the performance of safety, health, and environmental activities and have come to regard S/H/E auditing as a powerful management tool to help determine the compliance status and safety, health, and environmental performance of their operating facilities.

What Is Safety, Health, and Environmental Auditing?

Auditing, in its most common sense, is a methodical examination that involves analyses, tests, and confirmations of a facility's procedures and practices to verify whether they comply with legal requirements and internal policies and evaluate whether they conform with good safety, health, and environmental practices.

In this context, auditors base their judgments of compliance or deficiency on evidence gathered during the audit and documented in the auditor's working papers. In publicly held companies, "audit" has frequently been associated with the review of financial accounting statements by accountants. In addition, such terms as management audit, technical audit, operational audit, quality assurance audit, and energy audit are commonly used in many other corporate settings.

Purposes of Auditing

Motivations for developing a S/H/E audit program range from the desire to measure compliance with specific regulations, standards, or policies, to the goal of identifying potentially hazardous conditions for which standards may not exist. Thus, while auditing may appear to serve the universal need of evaluating and verifying safety, health, and environmental compliance, in practice auditing programs are designed to meet a broad range of objectives, depending on the needs of their various stakeholders.

Companies have established S/H/E auditing programs to:

- determine and document compliance status
- improve overall safety, health, and environmental performance at operating facilities
- assist facility management
- increase the overall level of safety, health, and environmental awareness
- accelerate the overall development of S/H/E management control systems
- improve the safety, health, and environmental risk management system
- protect the company from potential liabilities
- develop a basis for optimizing safety, health, and environmental resources
- assess facility management's ability to achieve S/H/E goals

Though these objectives can all be viewed as addressing compliance, they can also produce differences in program scope and focus. For example, some programs focus on determining present and past compliance over a specified time; others focus on determining compliance only at the time of the audit; still others focus on helping the facility manager achieve or maintain compliance.

The extent to which the goals and objectives of audit programs are documented and communicated varies from company to company. Some programs have a written corporate statement describing the audit program, while others have written position descriptions for the audit program manager and audit staff.

Audit Scope and Focus

Companies can define the scope and focus of an audit in organizational, geographical, locational, functional, and compliance contexts. *Organizational* boundaries address which of the company's operations (e.g., manufacturing,

R & D, distribution, and so on) are included in the audit program. They generally depend on the organizational structure, business unit reporting relationships, and corporate culture.

Geographical boundaries address how far or wide the program applies (state, province, regional, national, or international). Selection of geographical boundaries generally depends upon the location of facilities and offices, and the nature of products and services.

Locational boundaries address what "territory" is included in a specific audit. In many cases, the audit focus is mostly on the activities within the facility boundary, although some companies audit beyond this boundary. For example, an audit could also include off-site manufacturing or packaging activities, off-site waste disposal activities, local residences, or a nearby river or lake if there is a potential for environmental damage. Additionally, many companies include the activities of a tenant located on facility property in the scope of their audit.

A number of specific *functional* areas can be included in a S/H/E audit program. While most audit programs cover air and water pollution control, solid and hazardous waste management, employee safety, and industrial hygiene, many now also include occupational medicine, fire and loss prevention, process safety, and product safety. If all safety, health, and environmental disciplines are housed in the same organizational unit, the scope of the audit program is more likely to include many, if not all, of these subject areas.

Finally, *compliance* boundaries define the standards against which the facility is measured. These standards can include federal, provincial, regional, and local laws and regulations; corporate or division policies, procedures, standards, and guidelines; local facility operating procedures; or standards established by an outside group such as an industry or trade association.

CURRENT SAFETY, HEALTH, AND ENVIRONMENTAL AUDIT PRACTICES

An audit process built on a variety of experiences, such as safety audits and operations audits, and using verification techniques patterned after the financial audit, can be a powerful means of achieving the objectives of S/H/E auditing. A generalized description of such a S/H/E auditing process includes:

- reviewing the company's internal safety, health, and environmental management and control system, comprising:
  —*policies:* information, directives, guidelines, and standards concerning operations and performance
  —*procedures:* instructions to ensure that operations are carried out as planned
  —*controls:* checks and balances built into safety, health, and environmental operations, record keeping, and reporting

- developing a complete understanding of the internal control systems that are in place and recording this understanding in a flow chart or narrative form in the auditor's working papers
- evaluating the strengths and weaknesses of the internal control systems
- gathering audit evidence to meet the objectives of the audit program through inquiry, observation, and verification testing
- discussing the issues with management, presenting the exceptions found, and making judgments on their importance
- advising management on improvements needed in the safety, health, and environmental management systems
- creating a corrective action plan to address deficiencies/discoveries
- following up to maintain corrective actions

Safety, Health, and Environmental Audit Process

A number of basic activities are common among most audit programs. Some activities are undertaken before the on-site audit (preplanning), some during the audit field work (understanding systems, assessing controls, gathering and evaluating evidence), and others after the field audit has been completed (reporting the results and follow-up). Virtually all S/H/E audits involve gathering information, analyzing facts, making judgments about the status of the facility, and reporting the results to some level of management. A team approach is commonly used to conduct these activities. Even with these basic similarities, there can be a number of important differences. Figure 4-1 presents a simplified model identifying the key steps in the audit process. Most companies make some provision for including each of these steps in their audit process.

Preplanning

The S/H/E audit process begins with a number of activities before the actual on-site audit takes place. Initial arrangements relating to a facility audit include scheduling the visit, selecting the audit team, and gathering and reviewing background information.

Some companies audit all facilities on a repeat cycle (e.g., annually or every two years). In companies not auditing all facilities on a specific repeat cycle, the facilities that will be audited must be selected and scheduled. A list may be drawn up annually and modified throughout the year.

Companies select facilities for audit by a number of methods, typically on the basis of risk (i.e., perceived hazards, business importance, nature of operations, and so on). Initial arrangements relating to a facility audit include scheduling the visit, selecting the audit team, and gathering and reviewing background information. Initial notice of an upcoming audit (audit lead time) varies from company to

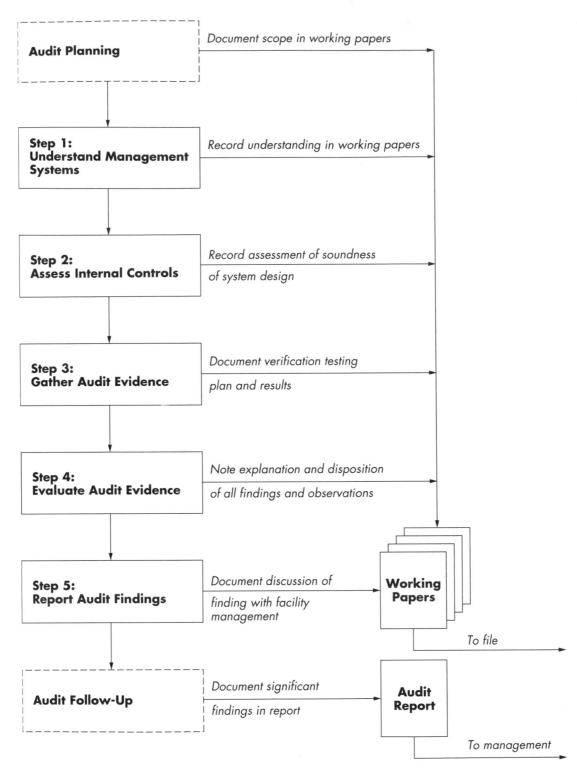

Figure 4-1. Key Steps in the Audit Process.

company and may be anywhere from one to six months. However, a few companies conduct "surprise" audits to obtain what they believe is a more accurate picture of facility operations.

Background information gathering generally begins well in advance of the audit and includes regulatory requirements, corporate policies, and facility information (organization, processes, layout). Some companies' audit teams visit the facility in advance of the audit to develop a basic understanding of facility processes and safety, health, and environmental management systems and to brief the facility staff on the objectives of the upcoming audit.

Field Work

Collection and review of advance information results in an audit plan that outlines the required audit field work steps, how each is to be accomplished, who will do them, and in what sequence.

Step 1: Understand Management Systems. Most on-site activities begin by developing a working understanding of how the facility manages activities that may affect safety, health, and environmental performance. This usually includes learning about facility processes, internal controls (both management and engineering), facility organization and responsibilities, compliance parameters and other applicable requirements, and any current or past problems. This step allows the team members to understand the actions taken within the organization that assist in regulating and directing its activities.

In developing this picture of internal management systems, auditors usually draw on information from multiple sources. These may include selected information provided by the facility in advance of the audit, staff discussions, and facility tours. Using questionnaires, discussions, and tours as background, auditors further investigate the more detailed aspects of management systems through in-depth interviews, guided discussions, and additional tours to specific sites. The data-gathering methods most used in Step 1 are inquiry and observation.

Step 2: Assess Internal Controls. After clearly understanding how various aspects of S/H/E compliance and performance are intended to be managed, auditors then evaluate the soundness of the facility's management systems to determine whether they are functioning as intended and will achieve the desired level of performance. For each of the areas or topics assigned, the auditor should ask himself or herself: "*If the facility is doing everything the way they say it happens, is that acceptable—will the facility be in compliance with applicable requirements, and is the company adequately protected?*"

In assessing the strengths and weaknesses of internal controls, auditors typically look for such indicators as clearly defined responsibilities, an adequate system of authorizations, capable personnel, documentation, and internal verification. It is far easier to identify significant weaknesses in internal controls than to determine adequacy. Each of these indicators usually requires significant judgment on the part of the auditor since there are no widely accepted standards an auditor can use as a guide to what is acceptable internal control. Thus, many auditors look to the audit program objectives, as well as to the company's basic safety, health, and environmental philosophy, for guidance about what is satisfactory internal control.

This step is especially important in that it determines, to a great extent, how the balance of the audit will be conducted. Where internal controls are judged to be sound, the auditor will spend time confirming the existence of the control systems and testing whether they function effectively on a constant basis. The auditor tends to rely on both the comprehensiveness and objective documentation of these internal controls when looking at the rest of the system. On the other hand, where internal controls are judged to be deficient or lacking, the auditor is not able to rely on their presence as an indication that the company is functioning well. In such cases, gathering evidence is limited to scrutinizing the outcomes of individual programs and processes or analyzing current compliance status against federal, state, and company standards.

Step 3: Gather Audit Evidence. Audit evidence forms the basis on which the team determines compliance with laws, regulations, corporate policies, and other standards. Evidence is gathered in a variety of ways, including record reviews, examination of available data, and interviews with facility personnel. Suspected weaknesses in the management system are confirmed in this step. Also, management systems that appear sound are tested to verify that they work as planned and are consistently effective.

Audit procedures (the means by which auditors collect audit evidence) fall into three broad categories:

1. Inquiry. The auditor asks questions both formally and informally. Audit questionnaires are common examples of formal inquiry.
2. Observation. The auditor collects evidence through what can be seen, heard, or touched. Because physical examination is often one of the most reliable sources of audit evidence, observation is a significant aspect of most S/H/E audits.
3. Verification testing. The auditor focuses on either the management system or physical equipment and performs systems tests. For example, retracing data would uncover errors in recording original data. Other common types of testing include verifying paper trails and equipment checks.

Programs vary regarding the amount and balance of inquiry, observation, and verification testing required. Some programs depend on inquiry as the primary means of gathering audit evidence. Inquiry is easy, provides rapid feedback, and does not require as many resources. Many of the more sophisticated S/H/E audit programs, however, require auditors to conduct a considerable amount of verification testing to determine whether management systems and equipment perform as they are supposed to. For each item to be audited, inquiry typically takes a matter of minutes, observation can take tens of minutes, and testing a matter of hours. Thus, the more items to be verified, the larger the resource commitment required. In most cases, more items could be verified (and more ways to verify each item are possible) than available audit resources allow. However, S/H/E audits usually serve as a check on the safety, health, and environmental management system rather than as a substitute for it. Therefore, most audit programs do not look at every situation, item, or document.

Step 4: Evaluate Audit Evidence. Once evidence gathering is complete, the audit findings and observations are evaluated. Audit evidence is reviewed in terms of program goals to determine both whether audit objectives are met and the significance of the audit findings.

Although auditors usually make preliminary evaluations of their observations throughout the audit, most audit teams devote a few hours at the end of the audit to jointly discuss, evaluate, analyze, and finalize these tentative audit findings.

Step 5: Report Audit Findings. The formal reporting process usually begins with an exit or close-out meeting between the audit team and facility personnel. Generally, any audit findings are informally communicated to facility staff as they are identified. However, during the exit meeting, the audit team formally communicates all observations and any findings noted during the audit. Any ambiguities about the findings are then clarified and their ultimate disposition discussed (e.g., for audit report, for local attention only).

Most companies prepare a written audit report. The purposes of the report are typically to provide management with information about compliance status; to initiate corrective action; and to document how the audit was conducted, what it covered, and what was found.

Some companies prepare a draft of the audit report on site. Most, however, prepare a draft audit report shortly after the on-site audit is completed. This draft usually undergoes review and comment before a final report is issued. Report reviewers often include the S/H/E affairs department, the legal department, facility management, and the audit team.

The content of the audit report varies considerably from company to company. Typically, audit reports contain a background or introduction section that describes the purpose and scope of the audit, outlines the audit approach, and identifies the audit team leader, team members, and other key audit participants. Most audit reports include sections on the facility's overall compliance with regulations, as well as compliance with the company's policies and procedures. Some audit reports identify all applicable facility operations; some include a detailed description of the facility and its history, or an impression of the facility management's ability to handle safety, health, and environmental crises. Still other audit reports contain recommendations for how to address the deficiencies identified.

The content of audit reports is strongly linked to the overall objective of the audit program and the needs of the report recipient. If the goal of the audit program is to provide assurance to management, the audit report often is limited to a factual description of the more significant findings and exceptions. On the other hand, if the goal of the audit program is to assist the facility manager, the audit report often is detailed enough to let the facility manager know precisely what was wrong and may include recommendations on how to improve the situation.

An effective reporting process communicates issues to appropriate persons within the company. Many companies have established a multiple or hierarchical reporting scheme. Under such a reporting process, the type of information and level of detail to be provided in an audit report depend on the problem identified and the individual who has to be notified. Some items may require reporting to corporate management and future follow-up; others may require only the attention of the facility manager.

Many companies' audit reports receive a relatively wide distribution. Typically, audit reports are distributed to the manager of the audited facility and the corporate S/H/E affairs department. Some companies also distribute the report to various levels of operating and line management, the legal department, and the members of the audit team. We have also seen a steady increase in the reporting of audit findings to senior management.

Although much has been written about protecting the confidentiality of the audit report, most well-established programs—especially those that conduct a large number of audits each year—do not routinely take steps to assert the confidentiality of the audit report. Instead, care is taken both in preparing the report to ensure that the report clearly, accurately, and appropriately reports the facts, and in ensuring that timely, appropriate follow-up occurs on all report items. The audit report is typically treated as internal management communication and distributed to those with a need to know. A few companies, however, establish their S/H/E audit program under the direction of counsel and treat the audit report and all associated documents as "attorney-client" privileged communication.

Follow-Up

Most companies have established formal procedures for responding to the audit report. The action planning process is initiated as audit findings are identified. It typically includes assigning responsibility for corrective action, determining potential solutions and preparing recommendations to correct any deficiencies noted in the audit report, and establishing timetables. Responses to the audit report are generally prepared by the facility manager and sent to line management and the audit program manager for review and, often, for approval. A few companies incorporate the action plan into the final report.

Typically, action plans are monitored by an individual with responsibility for follow-up—generally either operating management, S/H/E affairs, or, in a few cases, the auditors. In most instances, follow-up involves a written or oral inquiry about the status of the planned action. In companies where facilities are audited on a repeat basis within a specified time, the auditor or audit program manager is usually involved in action plan follow-up (typically reviewing the status of action plan implementation during the next audit). Where an audit team is unlikely to return to the facility for some time, operating management or S/H/E affairs usually assumes responsibility for the follow-up.

Basic Audit Tools

The S/H/E audit process is most commonly supported by some important tools: the audit protocol and the working papers. Although there is considerable latitude in current practice, most S/H/E audit programs use these devices in some form.

Audit Protocols

Names for the various documents that guide the auditor while conducting the audit include audit protocols, audit work programs, review programs, checklists, and audit guides. In this discussion, the term "audit protocol" will be used.

An audit protocol represents a plan of how the auditor is to accomplish the objectives of the audit. It lists the audit procedures that are to be performed to gain evidence about safety, health, and environmental practices. An audit protocol also provides the basis for assigning specific tasks to individual members of the audit team, for comparing what was accomplished with what was planned, and for summarizing and recording the work accomplished. A well-designed audit protocol can also be used to help train inexperienced auditors and reduce the amount of supervision required by the audit team leader. Many companies use the audit protocol to help build consistency into the audit, particularly where rotating audit teams are used.

The audit protocol itself is a listing of auditing procedures that are to be performed to gain evidence about safety, health, and environmental practices. Typically, a standard audit protocol is modified by the auditor to suit the circumstances of a particular situation. Modifications are made because of the nonapplicability of certain audit procedures (or regulations/policies), unique S/H/E hazards, or risks of potential materiality that may be present. The auditor may also modify the standard audit protocol according to the degree of reliance he or she places on the facility's internal controls systems.

Each procedure is typically annotated on the audit protocol with the specific working paper page reference. Thus, the completed protocol, combined with the working papers (described below), provides a record of the audit steps that were performed in carrying out the audit. The completed audit protocol also provides documentation of the rationale for any modifications. The completed audit protocol should also document the scope (or boundaries) of the S/H/E audit by identifying those aspects of the safety, health, and environmental areas under review, which were considered in carrying out the audit. The scope of the audit should also describe the period of time under review.

Working Papers

Working papers (the auditor's field notes) document the work performed, the techniques used, and the conclusions reached by the auditors. Working papers help the auditor achieve the audit objectives and provide reasonable assurance that an adequate audit was performed consistent with audit program goals and objectives. Working papers should include documentation of compliance or noncompliance.

Working papers are usually handwritten and include photocopies of documents selected by the auditor to help substantiate the findings of the audit. These papers are not a report that the auditor prepares from notes after the audit is complete; rather, they are the auditor's field notes to keep track of audit procedures undertaken, results achieved, and items requiring further information.

Most working papers include the following items:

- notes of audit planning
- completed copy of the audit protocol annotated with working paper references and auditor's initials
- results of compliance testing and evaluations of internal control
- descriptions of all functional tests (e.g., performance of pollution control equipment) and transactional tests (e.g., documentation of hazardous waste shipments) conducted during the audit
- documentation of all audit procedures and evidence obtained
- notes on any conferences held

Some working papers also contain a listing of the facility's materials and their uses, copies of regulatory permits for the area(s) under review, schedules of documentation and record keeping, and copies of previous audit findings and observations.

The working papers can be divided into two sections: the permanent or continuing file and the current file. The permanent file contains a record of items of continuing interest, while the current file contains items pertinent to the most recent audit. For example, the permanent file would contain:

- a listing of the facility's products and their uses
- copies of relevant regulatory permits and key regulatory correspondence
- internal controls questionnaires and flow chart
- schedules of documentation and record keeping
- copies of previous years' statements of findings and observations

The current file contains the auditor's evidence and final decision(s). It includes:

- notes of audit planning, including audit leader, team members and assignments, time budgets, rationale for any audit program modifications, and so on
- the completed copy of the audit protocol annotated with working paper references and auditor's initials
- the completed internal controls questionnaire, results of compliance testing, and evaluation of internal control
- schedules for compliance testing
- descriptions of all transactional and functional tests conducted during the audit, as well as an explanation of sampling plans employed
- documentation of all audit procedures and evidence obtained, including favorable as well as unfavorable evidence
- notes on any conferences or meetings held with facility management

To substantiate that the auditor has adhered to the proper protocols in conducting the audit, the working papers should provide sufficient detail of both the testing rationale and resulting evidence for each item in the audit protocol.

Designing Audit Programs

The basic objective of most S/H/E audit programs is to verify compliance with corporate, legal, and regulatory policies and rulings throughout the company. This objective can be best served under the following conditions:

- The S/H/E audit is conducted independently from those who are responsible for development or implementation of safety, health, and environmental management policies and programs.
- A mix of safety, health, and environmental engineering, operating, and auditing expertise is employed in the design and implementation of the audit process.
- Auditors serve as reviewers of the total system rather than as functional, technical experts while conducting a S/H/E audit. The company's professional safety, health, and environmental managers usually verify that the system is managerially and technically correct in design, operated by competent people, supported by management, and capable of operating properly, based on current performance data. However, the depth, detail, extensiveness, and variety of the validation is usually much greater in S/H/E auditing than in reviews or assessments conducted by those professional S/H/E managers responsible for ongoing administration of a safety, health, and environmental program.
- Definitions of important items are developed, clearly understood, and widely accepted. S/H/E auditors should not have to decide unilaterally what is important.

- Working papers document the planning and execution of the audit, describing, for each audit step: (1) what audit review and test activities were carried out; (2) what results were noted; and (3) what conclusions were reached.
- An understanding of the facility's internal management and control system is obtained and described in the working papers. The auditors then test whether the system is actually and consistently in use.
- All observations, findings (both compliance and noncompliance), and recommendations are documented with competent and sufficient evidence in written working papers.
- The evidence cited in the working papers provides a clear flow of logic from the testing rationale to the auditor's discoveries. Evidence is relevant, objective, free from bias, and persuasive.
- Findings are formally communicated to those with the organizational authority to review, evaluate, and, where appropriate, take corrective action.

Those charged with responsibility for a S/H/E audit program should keep in mind that in today's corporations, safety, health, and environmental considerations are likely to have an impact on many of the company's activities. In some matters, a company's safety, health, and environmental management will have primary responsibility for developing, implementing, and overseeing policies and programs. On the other hand, S/H/E performance is also likely to be influenced by a wide variety of other corporate activities. For example, business functions such as marketing, manufacturing, transportation, or finance often have wide influence on corporate performance within salient S/H/E areas such as pollution control, occupational health and safety, product safety, and loss prevention.

Corporate activities affecting human health and safety and the environment are regulated by several different federal, regional (state, province), and local agencies, and often have been viewed—especially by the regulators—as separate (and even unrelated) functions. As a result, there may be considerable variance in the corporation's managerial approach and stage of management system development within each safety, health, and environmental area. Companies must exercise care to establish effective S/H/E auditing approaches rather than attempting to use preconceived auditing systems across various safety, health, and environmental disciplines. An effective S/H/E audit program requires enough flexibility to review a wide variety of existing internal systems and at the same time provide meaningful feedback to those charged with developing and operating formal management systems.

Program Organization and Staffing

In deciding where to place the audit program within their organization, some companies emphasize that the audit function should be independent while others emphasize that other groups in the organization need to have easy access to the information gained from audits. Typically, companies place audit programs within a core corporate group, most commonly within the corporate safety, health, and environmental staff. However, in some companies the group is located in the internal audit department, regulatory affairs department, production or operations department, or legal department.

Companies choosing to place their S/H/E audit program in the internal audit department tend to view the program more as a corporate tool than as a safety, health, and environmental management tool. Production or operations departments are sometimes chosen to reflect or reinforce a company's philosophy that operations is responsible for S/H/E management. The legal department may be chosen because of sensitivity to potential legal issues involved in S/H/E auditing, e.g., potential disclosure of sensitive audit information.

Regardless of where the audit program is placed, the legal department is usually involved in the initial development of a S/H/E audit program and plays a central role in developing the audit reporting process. As the audit program becomes more established, the legal department's interaction with it may be limited to providing advice on regulatory interpretations and reviewing the audit reports. However, the legal department is usually more actively involved when an audit uncovers a significant noncompliance situation.

Companies staff their S/H/E audit programs in a variety of ways. Most established S/H/E audit teams include individuals with technical expertise, knowledge of S/H/E regulations, facility experience, and a strong knowledge of auditing procedures and techniques. Knowledge of safety, health, and environmental management systems and an understanding of similar companies' hazard control programs are also important staffing criteria. Audit teams usually include S/H/E specialists and may also include a facility manager, process or safety engineer, attorney, analytical chemist, internal auditor, industrial hygienist, or outside consultant.

Some companies have a full-time audit program manager; others staff their audit program only on a part-time basis. Of the companies that staff their programs on a part-time basis, some do not conduct enough audits to justify full-time staff while others want to vary participation in their program by rotating the audit team membership. Some rotate the membership of the audit team to involve a wider range of staff; other companies vary their team membership to get the specific expertise desired for a particular audit.

Using the same auditors for every audit, as opposed to rotating auditors, provides greater continuity from audit to audit, and generally greater confidence that the goals and objectives of the program are being met.

Typically, an audit program is funded in a manner consistent with the S/H/E and other corporate staff functions. In most companies, costs are absorbed as overhead and included in the budget of the organizational unit responsible for the program. However, in a few companies, audit costs are charged back directly to the audited facilities, especially when a company typically directly charges back the costs of many corporate staff activities. It is not unusual to charge travel and out-of-pocket expenses to an individual's assigned organizational unit when a company uses staff for the audit on a part-time or special-assignment basis.

FUTURE DIRECTIONS

What does the future hold? A number of important trends and further developments in audit program design, content, and coverage will occur over the next five years. These trends are described below.

Increased Growth

Over the next five years, the number of organizations with S/H/E audit programs will continue to grow. Although the existence of a S/H/E audit program is widespread among larger companies in the manufacturing and process industries, most of the new growth in audit programs will come from other segments of industry and from medium- and small-sized companies. This will be the case particularly as these organizations begin to appreciate that the up-front costs of auditing can significantly outweigh the eventual costs associated with noncompliance and cleanup.

Broader Scope

In the future, organizations with existing or new audit programs will expand the scope of their auditing efforts. The expansion will likely occur both geographically and functionally. Organizations that now conduct audits only of domestic facilities will broaden their scope to include overseas locations. In transporting auditing programs overseas, companies will have to address important considerations such as diverse local regulatory systems, travel and logistical requirements, language and cultural barriers, and different levels of company ownership and control. These considerations, if not fully examined and addressed, will impair the company's ability to audit the systems that have been developed to manage environmental risks.

Organizations whose audit programs focus only on environmental topics will expand them to include health and safety, and, in many instances, process safety and product safety issues. Given the many common issues shared by these topics, companies can realize economies in staff and resources if they place different audits under one central management. This expansion in scope, however, will not necessarily mean combining all topics into a single audit. Instead, the scope of the audits will be designed to match audit resources and objectives.

Increased Rigor and Depth of Review

Perhaps the most significant change in the years ahead will be the increased depth and rigor of review. More companies will shift the orientation of their programs from finding regulatory problems to verifying compliance and confirming that management systems are in place and functioning as they were designed to do.

- *Identify Problems.* Many companies begin with an assessment orientation, focusing first on identifying problems through inquiry and observation only. The problems identified may relate to regulatory compliance or may be in nonregulated areas.
- *Verify Compliance.* Effective environmental audit programs evolve beyond assessment to compliance verification. The focus here is not only on identifying compliance problems, but also on a systematic, rigorous verification of areas, both regulated and nonregulated, that appear to be functioning well and in compliance.
- *Confirm Functioning of Management Control Systems.* Because verification of compliance during the audit does not ensure that the facility will stay in compliance, a growing number of companies take a systems perspective. In this approach, the auditing effort extends beyond verification of compliance to include a review of the underlying programs, procedures, and systems that are in place to ensure ongoing compliance.

Increased Effectiveness of Field Resources

There will be a continuing shift in focus toward a more rigorous review and independent verification of the areas believed to be in compliance and operating smoothly. This is particularly true if there is a corresponding examination of the underlying facility-level environmental management systems that will require an increase in auditor effectiveness and perhaps an increase in the total amount of field resources allocated. It is anticipated that companies will find creative ways to accomplish this goal, despite the fact that budget and staffing constraints will most likely continue.

Increased Emphasis on Basic Skills

More and more companies are beginning to realize that there is a discrete set of basic skills required to be an effective auditor. These skills include:

- working knowledge of regulatory requirements applicable to the scope of the audit
- training and proficiency in basic auditing skills and techniques
- general familiarity with the type of facility operations being audited

- understanding of the S/H/E controls, procedures, and management systems of similar facilities

Over the next several years, increasing numbers of companies will work toward gaining a better balance among these four skills, with particular emphasis being placed on proficiency in basic auditing techniques.

Expanded Nature of Reporting

As top management continues to request assurance that facilities are operating well and are in compliance, the nature of reporting audit results will shift in the years to come from simple problem identification reports to performance reports, and ultimately to assurance reports.

- *Problem Reports.* Auditors list a number of problems identified at a facility and, perhaps, provide a list of positive findings as well.
- *Performance Reports.* Auditors summarize the overall results of a comprehensive audit not only by listing problems, but also by including a general defensible conclusion regarding the compliance status and overall S/H/E performance of a facility.
- *Assurance Reports.* Auditors provide top management with statements regarding overall compliance status, which is substantiated by systematic, organized audit documentation.

Emergence of Safety, Health, and Environmental Auditing Standards

As S/H/E auditing practitioners have developed a core group of common practices, they have begun to reach a consensus on principles and standards for S/H/E auditing. A U.S.-based example of this is the ASTM *Draft Standards Practice for Environmental Compliance Auditing*, 1994. The effort to define auditing standards has become more urgent with the advent of international standard-setting initiatives, including the European Union's Eco-Management and Audit Regulation and the International Organization for Standardization's environmental management standards (ISO 14000), described in Chapter 2, Regulatory History, of this volume.

Although practitioners continue to differ on audit approach and philosophy issues, they generally endorse the following standards:

- *Auditor Proficiency.* Audits are conducted by staff who have the qualifications, technical knowledge, training, and proficiency in the auditing discipline to perform their assigned auditing tasks. The organization managing auditing activities and the individual auditor take responsibility for ensuring staff proficiency. Audit team qualifications are suited to the objectives, scope, and complexities of the audit assignment.
- *Due Professional Care.* Due professional care is the application of diligence and skill in performing

audits. Exercising due professional care means achieving accuracy, consistency, and objectivity in the performance of audits; ensuring that the audit examination follows the procedures established for that audit; using good judgment in choosing tests and procedures; reporting audit findings in accordance with the audit's procedures; and developing conclusions and, if necessary, recommendations.

- *Independence.* Auditors are objective and independent of the audit site and/or activity to be audited, free of conflicts of interest in any specific situation, and not subject to internal or external pressures that could influence their findings. When complete independence is not feasible, the audit report should clearly communicate the limiting factors and the client and auditee's awareness of them.
- *Clear and Explicit Objectives.* The objectives of an audit are clearly established and fully communicated beforehand to the client and the auditee. The objectives of specific audits meet the needs of intended recipients of audit results and the provisions of accepted audit standards.
- *Systematic Plans and Procedures for Conducting Audits.* Audits are based on the use of systematic plans and procedures that provide uniform, consistent guidance in audit preparation, field work, and reporting. The audit is planned, resources allocated, and procedures selected and supervised to achieve explicit audit objectives.
- *Planned and Supervised Fieldwork.* Auditors and audit team leaders plan, implement, and supervise field work to foster efficiency, to ensure consistency in all parts of the audit and with the audit plan, and to achieve audit objectives. On site, auditors and team members collect the relevant and accurate information that is needed to meet audit objectives.
- *Thorough Review of Internal Controls.* Audits are conducted to review existing management systems and internal controls and to gather appropriate information for evaluating the reliability of internal controls in achieving environmental performance goals.
- *Audit Quality Control and Assurance.* Audits undergo quality checks to ensure accuracy and to encourage continuous improvement of audit management systems, procedures, and implementation. Quality control measures the extent to which an audit is conducted according to the objectives and scope of the audit, and to these standards.
- *Audit Documentation.* The auditor prepares documentation of ongoing activities during an audit in "working papers." Each subject reviewed in an audit should be documented sufficiently so that another auditor of similar skill could confirm the conclusions of the auditor without consulting other resources.
- *Clear and Appropriate Reporting.* For each audit, the auditor prepares a formal report that communicates information in accordance with the audit objectives. The report clearly communicates information and findings to the intended recipients in a timely manner and in sufficient detail and clarity to expedite corrective action.

These emerging standards draw on concepts from other types of auditing and on principles and practices that have emerged in S/H/E auditing since the 1970s. In the very near future, it is likely that S/H/E auditing standards will be much more sharply defined, opening the way for true auditor certification and independent audit verification.

SUMMARY

- Safety, health, and environmental (S/H/E) auditing is used to verify that a facility's procedures and practices comply with legal requirements and internal policies and conform to good safety, health, and environmental practices.
- The primary motivation for S/H/E audits ranges from the desire to measure compliance to the need to identify potentially hazardous conditions for which standards do not exist. Companies define the scope and focus of an audit in organizational, geographical, locational, functional, and compliance contexts.
- The S/H/E audit process involves preplanning, field work, and follow-up stages. Generally, a team approach is used to conduct the auditing activities of gathering background information, conducting the actual audit, and reporting the results. Basic audit tools include audit protocols and working papers.
- The role that auditing programs play in companies determines how much independence the programs are given and where they are placed in the organization. Companies may form audit teams composed of members from a variety of disciplines and departments; they may also appoint full-time or part-time managers.
- The number of companies with S/H/E audits is expected to increase. In addition, the audits will have increased depth and rigor of review, greater use of field resources, and increased emphasis on basic skills. Finally, S/H/E/ auditing standards that are emerging on both the domestic and international levels will provide guidelines for this increasingly important process.

REFERENCES

American Society for Testing and Materials (ASTM). *Practice for Environmental Site Assessments.* 2 Parts: *Phase 1* and *Transaction Analysis.* Philadelphia: ASTM, 1993.
Draft Standard Practice for Environmental Compliance Auditing, 1994.

Hall RM, Case DR. *All About Environmental Auditing.* Washington, DC: Federal Publications Inc., 1987.

McDaniel T, Shih J, Ardiente E. *Environmental Auditing for Continuous Improvement.* Pittsburgh, PA: AWMA Meeting, June 1993.

Priznar FJ. A guide to environmental auditing. *Scrap Processing and Recycling, J Inst of Scrap Recycling Ind* 50:101, March–April 1993.

Priznar FJ. Trends in environmental auditing. *Env Law Reporter*, Environmental Law Institute, 20 ELR 10179, Washington DC, 1990.

U.S. Environmental Protection Agency. *Policy Statement on Environmental Auditing.* Washington DC: 51 *Fed Reg* 25004, 1986.

REVIEW QUESTIONS

1. Briefly define safety, health, and environmental (S/H/E) auditing.
2. List six of the nine company objectives in establishing S/H/E auditing programs.
   a.
   b.
   c.
   d.
   e.
   f.
3. Name and briefly discuss the five criteria companies can use to define the scope and focus of an audit.
   a.
   b.
   c.
   d.
   e.
4. What are the seven key steps in the S/H/E auditing process?
   a.
   b.
   c.
   d.
   e.
   f.
   g.
5. List the two basic audit tools and explain why each one is so important.
   a.
   b.
6. Which department of a company is usually involved in the initial development of the audit program and plays a central role in developing the audit reporting process?
   a. Internal audit department
   b. Production/operations department
   c. Legal department
   d. Regulatory affairs department
7. In staffing their audit program team, companies should choose which combination of individuals?
   a. Facility manager, process or safety engineer, attorney, and an analytic chemist
   b. Process safety engineer, internal auditor, industrial hygienist, and an outside consultant
   c. Full-time audit program manager, attorney, analytic chemist, and an internal auditor
   d. Varies from company to company
8. As companies expand their auditing programs overseas, what types of considerations will they have to address?
   a. Diverse local regulatory systems
   b. Travel and logistical requirements
   c. Language and cultural barriers
   d. Different levels of company ownership and control
   e. All of the above
9. List five of the seven trends that will be seen in audit programs over the next five years.
   a.
   b.
   c.
   d.
   e.

5

Workers' Compensation

In a typical year, on-the-job injuries account for one fifth of the total accidental injuries from all causes in the United States. Injured workers, their families, their employers, and society as a whole suffer substantial economic losses as a result of these accidents. The statistics do not include the physical and mental suffering injured workers and their families undergo. Thus, when a worker dies, is disabled, or merely requires medical attention because of work-connected injury or disease, the economic consequences affect everyone. This chapter will discuss the following topics:

- the types of economic losses workers and their families may experience
- a historical review of workers' compensation laws and legislation
- six primary objectives underlying workers' compensation laws
- major coverage of workers' compensation and limitations on coverage
- basic benefits covering loss of income, medical payments, and rehabilitation
- how workers' compensation is administered
- issues in rehabilitation of injured workers
- four general classifications of disability
- how to manage a workers' compensation program

Effective loss control can prevent injuries and accidents and reduce their costs, thus benefiting workers, employers, and the entire economy. Many companies are continuing to show much greater interest in controlling the costs related to such incidents. This interest is sparked by the increasing importance that costs are being given in executive decision making. As a result, safety and health professionals have a great opportunity to influence management toward adopting more effective safety measures.

ECONOMIC LOSSES

Workers and their families may suffer two types of economic losses: (1) a loss of earnings and (2) additional expenses.

If a worker dies because of a work-related injury or sickness, the survivors lose the income the worker would have earned—less the amount spent on personal expenses—over the remainder of the individual's working career and retirement years. This loss can be substantial.

Total and permanent disability cause even greater earnings losses than death because the worker must be maintained even though unable to work. Permanent partial disability accounts for part of the economic losses due to disability, depending upon the proportion of annual earnings lost because the worker cannot function fully. An employee who is totally disabled temporarily loses any income for the time he or she is recuperating. Loss of even a month's earnings can be a serious financial problem for most workers. In addition to these earnings losses, the deceased or disabled worker often is unable to provide valuable household services that must now be foregone or be taken over by someone else at additional cost.

Although not all injured workers are disabled, nearly all will require some form of medical attention. In general, medical expenses usually amount to less than the total earnings loss; but for many workers, their medical expenses equal or exceed the income lost.

In addition to direct earnings losses, society also loses the taxes that injured employees would have paid and the products or services they would have produced. Some injured employees and their families become public assistance beneficiaries and must be supported by other members of society.

WORKERS' COMPENSATION IN THE UNITED STATES

In the early decades of industrialization in the United States, efforts to implement a system of compensation for industrial injuries lagged far behind the developments in Europe. However, toward the end of the 19th century, as work-related injuries and diseases and their consequences grew more severe and costly, the public and others began to demand radical change. The first tangible evidence of government interest in workers' compensation laws appeared in 1893 when legislators seized upon John Graham Brooks' account of the German system as a guide for their own efforts at reform. Their interest was further stimulated by the passage of the British Compensation Act of 1897.

Early Laws

In 1902, Maryland passed an act providing for a cooperative accident insurance fund. Although this represented the first legislation to embody the compensation principle in any degree, the scope of the act was restricted. Benefits, which were quite meager, applied only in cases of fatal accidents. Within three years, the courts declared the act unconstitutional. In 1908, a Massachusetts act authorized the establishment of private plans of compensation upon approval of the state board of conciliation and arbitration. This law had no practical significance, and proved to be a dead letter from the start.

By 1908, the United States still had no workers' compensation act. President Theodore Roosevelt, recognizing the injustice, urged passage of an act for federal employees in his message to Congress in January. He pointed out that the burden of an accident fell upon the helpless man, his wife, and children, and declared this state of affairs "an outrage." In 1908, Congress passed a compensation act covering certain federal employees. Though somewhat inadequate by some standards, it was the first real compensation act passed in the United States.

During the next few years, advocates of compensation continued to press for state laws. A law passed in Montana in 1909, applying to miners and laborers in coal mines, was declared unconstitutional. Nevertheless, many states appointed commissions to investigate the feasibility of compensation acts and to propose specific legislation. A significant number of laws resulted from these commission reports—all of which favored some form of workers' compensation legislation—which were combined with recommendations from various private organizations. However, widespread agreement on the need for compensation legislation did not end all conflict over reform. Special interest groups clashed over specific bills and over questions of coverage, waiting periods, and state versus commercial insurance.

In 1910, New York adopted a workers' compensation act of general application, whose coverage was compulsory for certain especially hazardous jobs and optional for others. None of the early state compensation acts expressly covered occupational diseases. Statutes that provided compensation for "injury" were frequently interpreted to include disability from disease. However, those acts that limited benefits to "injury by accident" expressly excluded occupational disease. Every state act except the one passed by Oregon required uncompensated waiting periods of one to two weeks before benefits were paid; several states provided retroactive payments after a prescribed period.

The 1911 Wisconsin workers' compensation act was the first law to remain effective, and was quickly followed by laws in Nevada, New Jersey, California, and Washington that same year. In 1916, the United States Supreme Court declared workers' compensation laws to be constitutional. Although 24 jurisdictions had enacted such legislation by 1925, workers' compensation was not provided in every state until Mississippi enacted its first law in 1948. Thus, the United States proceeded with a statewide workers' compensation system when other countries enacted nationwide workers' compensation. As a result, there are 50 different statewide workers' compensation acts in the United States today.

Compensation Legislation

All 50 states, the District of Columbia, Guam, and Puerto Rico have compensation acts. In addition, the Federal Employees' Compensation Act (FECA) covers all employees of the U.S. Government, while the Longshoremen's and Harbor Workers' Compensation Act covers maritime workers, other than seamen, and workers in certain other groups. The latter act provided compensation for workers in the "twilight zone" between ship and shore, because the U.S. Supreme Court had ruled they could not be covered under state compensation laws. (Each of the Canadian provinces and territories also has a compensation act or ordinance.)

Although economic changes and public policy have prompted increases in benefits and scope of the laws, the basic concepts have remained relatively unchanged. Employers and labor are both dissatisfied with certain aspects of workers' compensation. Labor attacks the system for inadequate benefits, coverage limitations, and exclusion of many injuries, illnesses, and disabilities that labor considers job related. Employers criticize the system for covering some injuries and diseases they do not consider job related and for its high cost relative to its apparent benefits. Thus, while the early advocates of workers' compensation conceived of it as a simple, efficient, equitable remedy to reduce litigation over industrial injuries, both labor and management have expressed doubt that their hopes can be realized.

OBJECTIVES OF WORKERS' COMPENSATION

A U.S. Chamber of Commerce publication, *Analysis of Workers' Compensation Laws,* cites these six basic objectives underlying workers' compensation laws:

- Provide adequate, equitable, prompt, and sure income and medical benefits to work-related accident victims, or income benefits to their dependents, regardless of fault
- Provide a single remedy and reduce court delays, costs, and workloads arising out of personal-injury litigation
- Relieve public and private charities of financial drains—incident to uncompensated industrial accidents
- Eliminate payment of fees to lawyers and witnesses as well as time-consuming trials and appeals
- Encourage maximum employer interest in safety and rehabilitation through an appropriate experience-rating mechanism
- Promote frank study of causes of accidents (rather than concealment of fault), reducing preventable accidents and human suffering.

Income Replacement

The first objective listed for workers' compensation is to replace the wages lost by workers who are disabled due to a job-related injury or illness. According to this objective, the replacement should be adequate, equitable, prompt, and sure.

To be adequate, the program should replace lost earnings (present and projected, including fringe benefits), less those expenses such as taxes and job-related transportation costs that would not continue. The worker, however, should share a proportion of the loss in order to provide incentives for rehabilitation and accident prevention. A two-thirds replacement ratio is found in most state statutes.

To be equitable, the program must treat all workers fairly. According to one concept of fairness, most workers should have the same proportion of their wages replaced. However, workers with a low wage may need to receive a high proportion of their lost wages in order to sustain themselves and their families. High-income workers who can afford to purchase private individual protection may have their weekly benefits limited to some reasonable maximum. However, if workers' compensation insurance is regarded primarily as a wage-replacement program, few people should be affected by this maximum.

The first objective also includes medical and vocational rehabilitation and return to productive employment. To achieve this goal, workers should receive quality medical care at no cost, care that will restore them as much as possible to their former physical condition. If complete restoration is impossible, workers should receive vocational rehabilitation that will enable them to maximize their earning capacity. Finally, the system should provide incentives for disabled workers and prospective employers to help employees return to productive employment as quickly as possible.

One of the objectives of workers' compensation is to allocate the costs of the program among employers and industries according to the degree to which they are responsible for the losses. Such an allocation is considered equitable because each employer and industry pays its fair share of the cost. In the long run, this allocation shifts resources from hazardous industries to safe industries and from unsafe employers within an industry to safe employers. Eventually, employers with the most unsafe operations will be driven out of the marketplace.

Critics argue that workers' compensation costs account for such a small part of overall operating expenses that they have little, if any, effect on a firm's resource allocation. As a result, unsafe employers would not need to resort to higher prices, and they would remain in the marketplace.

Accident Prevention and Reduction

Occupational accident prevention and reduction is the final commonly accepted objective of workers' compensation. Those who consider this objective to be important believe that the system should and can provide significant financial and other incentives for employers to introduce safety measures that will decrease the frequency and severity of accidents. More specifically, the pricing of workers' compensation should reward good safety practices and penalize dangerous operations. Employees also should have some incentive to follow safe work practices by sharing some of the losses. Injured workers should have the opportunity, and should be encouraged, to return to work as soon as they are physically able.

MAJOR CHARACTERISTICS

Compensation laws can be compulsory or elective. Under an elective law, the employer may accept or reject the act. However, if an employer rejects the act, it loses the three common-law defenses—assumption of risk, negligence of fellow employees, and contributory negligence. This means that in practice all the laws can be considered "compulsory." A compulsory law requires each employer to accept its provisions and provide for benefits specified.

Most jurisdictions require employers to obtain insurance or to prove financial ability to carry their own risk. Six states, two U.S. territories, and most provinces require employers to contribute to a monopolistic fund operated by the state or provincial agency. In some instances, employers may qualify as self-insurers. Thirteen states permit employers to purchase insurance either from a competitive state fund or from a private insurance company.

Covered Employment

Although most of the state workers' compensation laws apply to both private and public employment, none of the laws covers all forms of employment and occupation. For example, a few states restrict compulsory coverage to so-called hazardous occupations. Many laws exempt employers having fewer than a specified number of employees, usually less than three or four in any one location. Most of the laws also exclude workers in farming, domestic service, and casual employment. Many laws contain other exemptions, such as employment in charitable or religious institutions.

Federal workers are covered by FECA. Employees of the District of Columbia are covered by the District of Columbia Workers' Compensation Act, which went into effect in 1982. Its provisions closely follow those of FECA.

Two other major groups excluded from coverage by compensation laws are interstate railroad workers and maritime employees. Railroad workers whose duties involve any aspect of interstate commerce are covered by the Federal Employers' Liability Act (FELA). Maritime workers are subject to the Jones Act, which applies provisions of the FELA to seamen.

The Federal Employers' Liability Act is not a workers' compensation law. Instead, it gives an employee the right to charge an employer with negligence and prevents the employer from pleading the common law defenses that the worker is a fellow servant or assumes part of the risk; moreover, the act substitutes the principle of comparative negligence for the common-law concept of contributory negligence.

It is not known how many state and local employees are covered by workers' compensation or provided with such protection voluntarily by their employers. All states (as well as Puerto Rico, Guam, and the District of Columbia) provide some coverage of public employees but the extent of the benefits varies widely. Some laws specify no exclusions or exclude only such groups as elected or appointed officials. Others limit coverage to employees of specified political subdivisions or to employees engaged in hazardous occupations. In still others, the extent of coverage is left entirely up to the state or to the city or political subdivision employing government workers. Certain other groups, such as the self-employed, unpaid family members, volunteers, and trainees, generally are not protected by workers' compensation.

Limitations on Coverage

In view of the fact that some of the exemptions or exclusions in many state laws have persisted to this day, it may be helpful to review some of the reasons behind the original limitations. Nearly all state acts were prepared and enacted in the face of constitutional challenges and the outright opposition of certain business or government interests. Thus, each act was the result of political compromises.

Initially, workers' compensation was hailed as an innovation that would introduce greater certainty into the calculation and payment of benefits in contrast to the common law system. Under common law, workers could sue employers and, if successful, might be assured of an adequate payment; however, those who lost would be left with nothing but debts. To reduce this risk, the workers' compensation law specified the benefits that would be paid to all regardless of fault. Although the outcome of workers' compensation cases is far more certain than the ordinary suit where negligence must be shown, the law is not "automatically" applied.

In part, this remaining uncertainty arises from the wide variety of permanent partial disability cases that the schedules do not cover satisfactorily. Two factors usually prompt compensation litigation. One is uncertainty about whether an accident arose out of and in the course of employment; the other is the extent of disability. As workers' compensation comes to encompass more of the ailments to which the general population might be susceptible, it becomes difficult to separate impairments that are work related from those that are not. In addition, it requires an exercise of legal skills and medical judgment to assess the extent of disability in such difficult cases as occupational diseases, injuries to the soft tissue of the back, heart conditions, or situations where the only evidence before the commission may be a subjective complaint.

"Exclusive Remedy" for Work-Related Disabilities

Before workers' compensation laws were enacted in the states, an employee, in order to recover damages for a work-related injury, had to prove some degree of fault or negligence on the employer's part. Under what is now known as the "quid pro quo of workers' compensation law," employers accepted, or were required to accept, responsibility for injuries arising out of and in the course of employment without regard to fault. In exchange, employees gave up the right to sue employers for unlimited damages. These agreements are usually referred to in the state acts as "exclusive remedy" provisions, a term that is quite misleading.

In no state are workers' compensation benefits necessarily the only remedy available to an injured worker. Depending upon the wording of the applicable statute, workers may bring a negligence action against their employer, fellow workers, another contractor on the same job, or some other entity or individual who caused the compensable injury. For example, workers may sue the manufacturer of a piece of equipment which caused an injury. From the employer's viewpoint, the doctrine should be the "exclusive liability rule." As the employee sees the rule, it remains an "exclusive remedy" for obtaining "worker's compensation" from the employer. Neither liability nor remedy is perfectly exclusive.

Two concepts that are broadening the exclusive remedy provision are (1) the expansion of the dual capacity doctrine and (2) the intentional tort exception.

Under the first concept, an injured employee can sue an employer for an injury—even if it arose out of and in the course of employment—if the injury was caused by the employer's product or a service available to the public. (Example: The driver of a tire company delivery truck who is injured when a defective tire, made by the employer, causes the truck to have an accident. Or a hospital employee who, after an accident on the job, is injured as the result of negligent treatment by one of the hospital's medical staff.) In both cases the injury did not occur as a result of the employer-employee relationship but rather through a relationship more akin to that of a supplier or service provider and the public.

If an employer commits an intentional tort, i.e., either deliberately causes harm to an employee, is grossly negligent, or engages in reckless behavior that results in an injury, the employee has the right to sue the employer for damages. The rationale is that the exclusive remedy provision should not protect an employer against being sued for an injury resulting from a deliberate harmful action or from gross negligence and recklessness. In some states, failure of an employer to comply with OSHA standards may be evidence of deliberate harmful action by the employer.

Covered Injuries

Workers' compensation is presently intended to provide coverage only for certain work-related conditions, and not for all of an employee's health problems. Statutory definitions and tests have been adopted to distinguish between conditions that are compensable and those that are not. All jurisdictions, when drafting workers' compensation laws, relied to some extent on the English legal system (or on other statutes based upon the English model). Even though the statutory language of these laws is remarkably similar, there are variations in terminology and differences in interpretation; as a result, a condition considered compensable in one state may be held noncompensable in others.

The statutes usually limit compensation benefits to personal injury caused by accidents arising out of and in the course of employment. Although this restriction presents four distinct tests that must be met, in practice these tests are often considered in pairs: the "personal injury" and "by accident" requirements in one set, and the "arising out of" and "in the course of" requirements in the other.

If interpreted narrowly, personal injury would refer solely to bodily harm, such as a broken leg or a cut, while the "by accident" test would refer to the cause, such as a blow to the body or an episode of excessive or improper lifting. In practice, however, the distinctions are blurred.

The "by accident" concept is a carryover from English law. Early judicial interpretations of English law made it quite clear that the "by accident" requirement was intended to deny compensation to those who injured themselves intentionally. A number of U.S. jurisdictions, however, have applied the test in order to narrow the range of unintentional injuries that must be compensated.

One of the early casualties of the "by accident" requirement was occupational disease coverage. As the typical judicial holding was that "occupational disease" and "accidental injury" were mutually exclusive concepts, special legislation was required to provide coverage for workers suffering from work-related diseases.

Occupational Disease

Although workers' compensation laws initially had no specific provisions for occupational diseases, all states now recognize responsibility for them. Coverage extends to all diseases arising out of and in the course of employment. Most states do not provide compensation for a disease that is an "ordinary disease of life" or that is not "peculiar to or characteristic of" the employee's occupation. For example, in March 1996, the Virginia State Supreme Court ruled that carpel tunnel syndrome and "trigger finger" are not occupational diseases under the state's workers' compensation law. Then Virginia's Workers' Compensation Act was amended to define "occupational disease" in six parts, but generally as "a disease arising out of and in the course of employment, but not an ordinary disease of life to which the general public is exposed outside of employment." (Printing and Publishing Newsletter, July/August 1996, p. 3.)

Generally, compensation for occupational diseases is the same as for traumatic injuries, and medical care coverage is unlimited. A few states do not provide permanent partial disability benefits for certain diseases. Occupational diseases usually become evident during employment or soon after exposure. However, as with radiation disabilities, certain diseases may be latent for a long time. Most states have extended periods in which claims may be filed concerning latent, slowly developing occupational diseases.

Some states impose special restrictions regarding disability resulting from exposure to coal dust, asbestos, silica, cotton dust, or radiation. A number of states have established presumptions for police and firefighters who have heart attacks or respiratory conditions, but no attempt is made to chart them.

Hearing

The difficulty of distinguishing between occupational and nonoccupational hearing loss has led to enactment of special coverage provisions in many state statutes. They attempt to isolate the occupational component in the hearing loss and to compensate workers accordingly.

Black Lung Disease

One category of occupational ailment, black lung disease, is covered by a federal benefits program under the Federal Black Lung Act, part of the Coal Mine Health and Safety Act of 1969, as amended. The tremendous cost of compensation for disability arising from this and other occupational diseases has led to pressure to fund such coverage in whole, or in part, through federal programs.

Work-Related Impairment

The term "arising out of and in the course of employment," applied by almost every jurisdiction, is meant to clearly define the relationship between employment and an injury or disease for an employee to be eligible for workers' compensation. The phrase obviously lacks precision. Often it is quite difficult to determine whether a given set of facts can support an award of compensation.

The "course of the employment" aspect of this test refers primarily to the time frame of the injury. Virtually every jurisdiction holds that employees are considered within the course of employment—barring unusual circumstances or unreasonable conduct—from the moment they step onto the employer's premises at the start of the workday to the moment they leave at the day's end.

Although this test appears to be relatively simple to apply, it has often proved difficult. For example, what is meant by the term "premises"? Injuries that occur off premises but appear to deserve compensation lead plaintiffs to search for exceptions in the laws and encourage courts to modify the basic rules. In addition, many workers are not attached to particular premises. Even though an injury occurs off premises, as in travel to and from work, the employee may be compensated if a sufficient employment relationship can be established; perhaps the employer paid the worker for the time or expense of travel or provided a company vehicle for transportation. In these circumstances, the travel time to and from a worker's home may be included in the course of employment.

The "arising out of" segment of the test is intended to establish a causal relationship between the employment and the injury. For example, an employee cannot simply suffer a heart attack while at work and expect compensation. The person must show that the heart attack arose out of the employment. This means that at minimum (some states have more stringent rules) it must be shown that the stress and strain or exertion of the employment caused the heart attack and that it was not merely a spontaneous breakdown of the cardiovascular system.

The degree of employment relationship required varies from state to state and has been modified as workers' compensation law has evolved. Generally, it was felt that the hazard-causing injury must be peculiar to the particular employment or be increased by the employment before the injury could be said to "arise out of the employment."

Although it is difficult to determine what each jurisdiction will require to meet the "arising out of" test, two additional theories have been developed and followed. The first and more widespread is the "actual risk doctrine," which requires that the hazard resulting in injury be a risk of the particular employment regardless of whether it is a risk to which the general public is exposed. The second or "positional risk doctrine" could also be called the "but for" test. According to this theory, if the employment places the worker in a position where he or she is injured ("but for" the employment the injury would not have occurred), the injury is considered to "arise out of the employment."

BENEFITS

The three basic types of workers' compensation benefits are (1) loss of income, (2) medical payments, and (3) rehabilitation.

Income Replacement

Although 70% or more of recent workers' compensation cases are for temporary total disability, such cases account for only about 25% of cash benefits paid. At the same time, income benefits in the last few years to workers for permanent partial disabilities accounted for almost 66% of the total dollar amount.

Basic Features

In general, the cash benefits provided for temporary total disability, permanent total disability, permanent partial disability, and death are payable as a wage-related benefit— the weekly amount is computed as a percentage of the worker's wage. Although the benefit varies by state and by type of disability, it is commonly set at 66% to 100% of current wages. In some states, the statutory percentage varies according to the worker's marital status and the number of dependent children, especially for survivors' benefits.

For many beneficiaries, the benefit rate is limited to less than two-thirds of wages by another statutory provision—the maximum ceiling on the weekly benefits payable. Because of this ceiling, disabled workers whose wages are at or above the statewide average receive benefits below the statutory benefit rate in almost all states. However, for such individuals, benefits may exceed pre-injury take-home pay because they are tax-free.

Other restrictions on benefits set maximum time periods for receiving compensation or maximum dollar amounts that can be paid. Such limitations in permanent total disability and death cases may cut off benefits to workers or survivors even though their need for income continues. Only fourteen states limit the duration of total dollar benefits to widows and orphans.

To reduce administrative costs and to discourage workers from malingering, all states stipulate in their laws that benefits are payable only after a waiting period following the report of disability. This delay in payment, which ranges from three to seven days, applies only to cash indemnity payments and not to medical and hospital care. In all states, workers who remain disabled beyond the specified minimum waiting period receive payment retroactively for that time. In more than three-fourths of workers' compensation laws, the minimum period before retroactive payment of disability benefits begins is two weeks.

Benefits by Type of Disability

Income benefits vary depending on whether an employee's disability is temporary or permanent, partial, or fatal. Most compensation cases concern workers who incur temporary disability but recover completely. In a majority of the states, the percentage of the state's average weekly wage for temporary total disability is $100 or more. However, dollar ceilings on weekly income benefits mean that many disabled workers are not fully compensated for their earnings lost.

Benefits for permanent total disability are for those disabilities that prevent employees from performing any work in any well-known area or industry in the labor market and that are of indefinite duration. These benefits are similar to those paid for temporary total disability benefits. In a few states, the weekly payment for permanent disability benefits is less than for temporary disability. A small number of states limit the period of benefits for permanent disabilities, usually from 6 to 10 years.

Residual limitations on a worker's earning capacity after recovering from an injury (that is, after a permanent partial disability) are calculated on a relatively complex basis. Partial disabilities are divided into two categories: "scheduled" injuries, those listed in the law such as loss of specific bodily members; and "nonscheduled" injuries, those which are of a more general nature, such as back and head injuries.

Weekly benefits for scheduled injuries are calculated as a percentage of average weekly wages, usually the same as the benefit rate for permanent total disability. The maximum weekly benefit is for the most part the same as or lower than that for total disability. Nonscheduled injuries are paid at the same or similar rate but as a percentage of wage loss. This represents the difference between wages before injury and wages the worker is able to earn after injury.

Scheduled benefits are paid for fixed periods that vary according to the type and severity of the injury. For example, most state laws call for payments ranging from 200 to 300 weeks for loss of an arm and 20 to 40 weeks for loss of a great toe. The maximum benefits period for nonscheduled injuries for each state is either the same as or, more generally, less than the time limit set for permanent total disability. In the majority of states, compensation for permanent partial disability is paid in addition to benefits received during the healing period or while the worker is temporarily and totally disabled.

Death benefits are intended to furnish income replacement for families who depended on the earnings of an employee killed by a work-related incident or disease. As is true for other types of benefits, the amount of survivor benefits and the length of time they are paid vary considerably from state to state. If the survivor is a widow without dependents, benefits computed as a percentage of the deceased worker's wage often are less than those paid for permanent total disability. If the worker had dependent children, the benefits in many states will be augmented to help support them. In most states, the duration of these benefits is unlimited, although nine states have established limits that range from 7 to 20 years. In a number of states, payments to widows continue usually as long as the women do not remarry and to children until they are no longer dependent, usually to age 18. In many states, benefits to children continue to age 23 or 25 if they are in school. Benefits may be terminated earlier in four states that also limit total dollar benefits.

In addition to benefits for widows, widowers, and children, some states pay survivor benefits to dependent invalid spouses, parents, or siblings of the dead worker. Burial expenses are covered in all states.

Medical Benefits

For many years, medical care disbursements provided by workers' compensation have comprised about one-third of total outlay for benefits. Care includes first aid treatment, services of a physician, surgical and hospital services, nursing and drugs, supplies, and prosthetic devices. Some large employers, in addition to first aid facilities, employ staff physicians for workers. Most employers insure their medical care responsibility as they do the income benefits under workers' compensation.

Every state law requires the employer to provide benefits for medical care to the injured worker. In most jurisdictions, such treatment is provided without limit either through explicit statutory language or administrative interpretation. In the few states that limit the total medical care by specified maximum dollar amounts or maximum payment periods, management can decide to exceed the initial ceiling if circumstances warrant such an action. Also, if specified types of injuries or disease are denied cash benefits, medical care for these conditions also is denied.

An issue many employers face in providing medical benefits for workers' compensation is how to choose the physician who is to furnish care. Almost half of the states give the employer the right to designate the physician. In practice, the insurance company of the employer ordinarily will select the physician because the insurer is the one who handles the claim for benefits. When the doctor is chosen in this way, the medical care furnished may be more highly skilled and effective because of the selected physician's specialized experience. On the other hand, workers often feel that their own family physician will place more emphasis on their personal health and well-being. They believe other considerations, such as company interests or insurance costs, may influence a physician they do not select.

Two other sources of medical benefits for the disabled worker are Social Security and private disability programs provided as part of fringe benefit packages by larger employers, in particular. Any disabled worker whose disability is documented as lasting at least 12 months or results in death may receive (or survivors will receive) disability benefits if eligible under Social Security rules. The combination of Social Security benefits and workers' compensation payments cannot exceed 80% of the disabled worker's earnings prior to disability. These benefits are financed out of Social Security.

Even though qualifying for workers' compensation, disabled workers may not be able to meet Social Security's requirements for proving they cannot perform any kind of gainful work.

Private disability insurance programs usually are coordinated with workers' compensation plans. Frequently, a private disability program requires that employees file workers' compensation and Social Security disability claims before they can qualify for private carrier benefits.

Rehabilitation

Along with industrial safety, medical care, and cash compensation, rehabilitation of workers is recognized, at least theoretically, as one of the primary goals of the workers' compensation system. The most widespread benefits offered through workers' compensation laws to restore a worker to the fullest economic capacity are the special maintenance benefits authorized in more than half the states. These benefits usually are paid (sometimes in addition to the regular disability compensation) for various training, education, testing, and other services designed to speed the injured person's return to work.

Probably the main source of retraining and rehabilitation is the federal-state vocational program. The federal government, through the Federal Vocational Rehabilitation Act, is a major resource for funding such programs. Facilities operated by this program accept both individuals with work-related disabilities and others injured off the job. In all states, these institutions are directed by state vocational rehabilitation agencies. They provide medical care, counseling, training, and job placement. Unfortunately, not all workers' compensation cases referred for vocational rehabilitation can be accepted promptly, and many others are never brought to their attention. As a result, workers who may have been able to return to some type of job remain disabled.

One notable drawback preventing full use of available rehabilitation facilities is the complex, adversarial proceedings for determining a worker's right to benefits. Because the decision about whether benefits should be awarded for permanent partial or total disability (and how large the benefits should be) is based primarily on the worker's inability to work, the claim may offer a strong incentive for the person to put off rehabilitation. Further, in the many compromise settlements, the employer's (or insurer's) main goal is to pay an agreed amount of money and prevent any future liability for medical, vocational, or other needs arising from the injury. Such settlements also work against a full-fledged effort to restore the worker to full health and productivity.

ADMINISTRATION

The goal of workers' compensation is to provide for quick, simple, and inexpensive determination of all claims for benefits and to provide such medical care and rehabilitation services as are necessary to restore the injured worker to employment. Nearly all of the states have agencies to carry out these administrative responsibilities.

Objectives

An agency's responsibilities include close supervision over the processing of cases. The primary objective is to ensure that all parties comply with the law and to guarantee an injured worker's rights under the statute.

A key goal of the agency is to see that the injured worker gets the full benefit due. To do so, the agency must follow an injury case from the first report to the final closing. Some states not only check the accuracy of total payments but also require signed receipts for every compensation payment. Some require the employer to file a final receipt that itemizes the purpose of each part of the total benefits outlaid to facilitate a complete audit of individual payments.

Frequently, however, the legislation itself states that a workers' compensation agency must operate on the assumption that each injured worker is responsible for securing his or her rights and that the agency's primary function is to adjudicate contested claims. Even where the law does not favor this policy, lack of staff may force the agency into this restricted role.

Although many workers are unfamiliar with the provisions of their state's workers' compensation act, in only a few states does the agency administrator (as soon as possible after the injury is reported) advise the worker of his or her rights to benefits, medical and rehabilitation services, and assistance available at the commission's office. Too many states fail to insist that employers report accidents promptly, pay benefits on time, or submit final reports that list the amounts paid and how these amounts were computed. Although prompt reporting is usually required, some states impose no penalty if employers violate this statute.

Handling Cases

Workers' compensation claims may be either uncontested or contested. In uncontested cases, the two main methods followed are the direct payment system and the agreement system.

Under the direct payment system, the employer or insurer takes the initiative and begins paying compensation to the worker or dependents. The injured worker does not need to enter into an agreement and is not required to sign any papers before compensation starts. The laws prescribe the amount of the benefits. If the worker fails to receive this compensation, the administrative agency can investigate and correct any error. Jurisdictions whose laws provide the direct payment system include Arkansas, Michigan, Mississippi, New Hampshire, Wisconsin, and the District of Columbia; this feature is also provided for in the Longshoremen's and Harbor Workers' Compensation Act.

Under the agreement system, in effect in a majority of the states, the parties (that is, the employer or its insurer, and the worker) agree upon a settlement before payment is made. In some cases, the agreement must be approved by the administrative agency before payments start.

In contested cases, most workers' compensation laws provide for a hearing by a referee or hearing officer. The statutes also provide for either the worker or employer/insurer to appeal a decision of the referees or hearing officer to the commission or appeals board and from there to the courts. As the administrative agency usually has exclusive jurisdiction over the determination of the facts in a case, appeals to the courts usually are limited to questions of law. In some states, however, the court is permitted to consider issues of both fact and law in an appeal.

REHABILITATION

Most employees injured in work accidents return to their jobs after minor medical attention with little if any work-time lost. If the effects of the injury are temporary, the incident usually fades from memory. Even those who suffer days or weeks of disability and possibly endure substantial medical treatment may find the injury is not permanent. Although the loss of income and the medical expenses are distressing, eventually, when workers resume their jobs, they recover economically as well.

Unfortunately, a minority of those injured—as much as 10% of the total—experience injuries that disrupt their lives. Even when these workers receive effective medical care and eventually return to productive jobs, their lives are permanently changed by the event. Injuries for some are so severe that prolonged medical treatment and convalescence fail to restore them completely to full health and functioning. Residual disabilities prevent them from performing their former jobs. Only retraining and education, combined with special assistance, offer a prospect for future employment.

Some never return to work. If they do not die from their injuries, they live with such severe disabilities that they barely can manage for themselves. Often, the most that health services can do is to lighten the burden on those who must care for these persons. Treatment for workers whose livelihood is threatened by work-related impairments consists of medical rehabilitation and vocational rehabilitation.

Medical Rehabilitation

Each medical rehabilitation program, whether set up by an insurance company or workers' compensation agency, contains its own requirements for treatment, qualifications for eligibility, and definitions of service.

A disabled worker who requires medical rehabilitation receives whatever medical care is needed to treat the impairment and to restore lost function. The worker may report first to the plant nurse or physician for immediate attention. If the injury is serious, the person may then go to a hospital. Workers' compensation laws often obligate employers and insurers to pay the costs of rehabilitation medical care. Costs can be covered by having health service workers on salary, by contractual arrangement with health personnel, or by payment of hospital and doctor bills. The insurer may or may not have much influence in the selection or course of treatment.

For injuries associated with chronic disabilities, the insurer usually attempts to control the selection of the rehabilitation services, frequently by directing the worker to a particular specialist or facility with a particular expertise. Often the insurer pays for transportation to the specialist or facility as well as for rooms during treatment. Some insurance companies operate their own rehabilitation facilities, under individual or joint ownership, with medical personnel on salary, at least part time. When insurers contract to share rehabilitation programs or facilities, they may pay expenses case by case or through a rental agreement.

When informed of the potential need for rehabilitation, some agencies do little more than notify the worker and insurer that medical rehabilitation is worth considering. Other agencies conduct formal evaluations of the need for further medical care and recommend action. They seek to convince disabled workers of the wisdom of rehabilitation. When the workers agree, the insurers can be required to finance the care.

Vocational Rehabilitation

Vocational rehabilitation prepares the injured worker for a new occupation or for ways of continuing in an old one. Usually, vocational rehabilitation is assigned when medical treatment fails to restore the worker to the job held when the individual was injured. The worker's injury may be so severe or the work requirements such that even residual impairment can prevent the person from performing effectively. These workers need training to overcome or to compensate for their limitations. Many may even enter new occupations. In general, however, the more effective the medical rehabilitation, the less need for vocational rehabilitation.

The current definition of vocational rehabilitation makes a greater distinction between it and medical rehabilitation than is necessary. Although the difference in kinds of treatment offered seem clear enough—retraining as opposed to medical care—the two categories often overlap to a considerable degree. For example, in the public vocational rehabilitation programs in each state, services include medical diagnosis and evaluation, surgery, psychological support, the fitting of prostheses, and other health-related services along with education, vocational training, on-the-job training, and job placement.

The two programs also blend on an employee's medical records. Record keeping by workers' compensation insurers does not separate claimants who receive medical rehabilitation from those who receive vocational rehabilitation, although some distinguish between medical rehabilitation and acute medical care. In contrast, records kept by workers' compensation agencies usually separate vocational rehabilitation from other benefits.

Injured workers who need vocational rehabilitation are served by several means. An employer or insurer may channel the worker to whatever sources they think will provide satisfactory service. Some workers are referred to the public vocational rehabilitation program where services may be financed by taxes, although insurers may reimburse the public agency. Other insurers direct workers to private facilities where vocational training is conducted by technical schools or on the job. The cost of these services are always paid by insurers.

As with medical rehabilitation, some workers' compensation agencies support vocational rehabilitation and often will direct the worker into a program, if the insurer fails to do so. Several jurisdictions select candidates either in conjunction with screening for medical rehabilitation or separately. Workers who have serious injuries or permanent disabilities, or who are receiving extended compensation payments are reviewed by the agency for referral to the state's public vocational rehabilitation agency or to the insurer.

Some workers obtain vocational rehabilitation through their own efforts. If no one refers them, they may go directly to the public vocational rehabilitation office. Since 1920, the federal government and the states have cooperated financially (80% federal and 20% state funding) in supporting a vocational rehabilitation program that can be used by anyone with a vocational disability. Rehabilitation counselors, who usually determine a referral's acceptability, simply verify the disability without regard to its cause and explore ways of overcoming a worker's limitations. If the candidate shows relatively good prospects, the counselor designs a program to help restore the worker as fully as possible. For those who are unable to return to a paying job, the objective of vocational restoration may be to help them care for themselves and to free other members of the family to earn wages.

Once workers are established in a vocational rehabilitation program, they are assigned whatever resources the counselors think best fit their needs. Generally, the resources are not owned and operated by the vocational rehabilitation agency but are offered by private vendors or other public agencies. For example, a worker may be sent to a private rehabilitation center or school, enter a sheltered workshop such as those run by Goodwill Industries of America, or be enrolled in a public institution.

DEGREE OF DISABILITY

The question of determining the extent of disability is perhaps responsible for more litigation than any other single issue in workers' compensation. It requires not only correct application of legal principles but also evaluation of facts, subjective complaints and opinions, and attempts to predict the future.

As a general proposition (some jurisdictions use different terminology and slightly different classifications), disability can be classified in one of four categories: temporary total disability, temporary partial disability, permanent partial disability, and permanent total disability.

Temporary Total and Partial Disability

Temporary total disability occurs when an injured worker, temporarily incapable of gainful employment, has a good prospect of improving to the degree that he or she will be able to return to work either with no disability or with only a partial disability. Temporary partial disability is similar to temporary total in that it assumes the worker's physical condition has not stabilized and is expected to improve. The difference lies in the worker's current abilities. When temporarily partially disabled, the worker is capable of some employment, such as light duties or part-time work, but is expected to improve and regain much of his or her former capability.

The determination of temporary disability, either total or partial, is the least difficult. It requires merely evaluating the employee's present physical condition in light of the work opportunities available. In practice, evaluation of temporary disability is concerned only with the ability of the employee to return to work for the current or last employer. It is assumed that at some point an employee will be able to return to work for this employer. Given the difficulties involved in obtaining employment for workers still under medical care—and that any new employment probably will be temporary—most adjudicators have either expressly or in practice adopted the position that unless the worker can return to the last job held, or can be supplied with temporary light or part-time duties with this employer, he or she remains temporarily totally disabled. This is the case even though the worker might be able to perform another job whose duties are within the individual's temporary physical limitations.

Permanent Partial Disability

Permanent partial disability means the injured worker has attained maximum improvement without full recovery. The worker has benefited from medical and rehabilitative services as much as possible but still suffers a partial disability. Permanent total disability represents the same situation except that the disability is total.

Determining the extent of permanent partial disability depends on what the jurisdiction chooses to label "permanent partial disability." Three theories are used to establish guidelines for the payment of workers' compensation benefits for such disability: whole-person, wage-loss, and loss of wage-earning capacity theories. Their underlying philosophies differ somewhat, as do the factors to be considered in applying each one to a specific situation.

"Whole-Person" Theory

This theory is concerned solely with functional limitations. Here, the only considerations in assessing a case are whether the worker has in fact sustained a permanent physical impairment and, if so, to what extent it interferes with the person's usual functions and abilities. Age, occupation, educational background, and other factors are not considered.

"Wage Loss" Theory

The aim in applying this theory to determine what wages the worker would have been able to earn had the permanent impairment not occurred. If the worker's earnings dip below the estimated wage figure because of the impairment, he or she is paid compensation equal to some percentage of the difference between the wages that would have been earned and those actually earned. Here the degree of physical impairment is of little or no importance. The only concern is the actual wage loss incurred and whether it is due to the impairment.

"Loss of Wage-Earning Capacity" Theory

This theory requires a peek into the future. After the worker has reached maximum physical improvement, many factors (such as impairment, occupational history, age, sex, educational background) are considered in an effort to estimate, as a percentage, how much of the worker's potential earning capacity has been destroyed by a work-related impairment. The worker is awarded benefits on the basis of this computation. Benefits may be paid at the maximum weekly rate for a limited number of weeks or they may be based upon a percentage of the difference between wage-earning capacity before and after disability, to be paid up to a preestablished dollar or time limit.

Combinations

These three basic theories are capable of being used also in combination. For example, some states expressly or in practice use either the "wage loss" theory or the "loss of wage-earning capacity" theory but also provide a benefit floor determined by the worker's actual medical impairment. Thus, an employee who sustains a permanent impairment but no loss of wages or of wage-earning capacity would still receive some permanent disability benefits.

The use of schedules has relieved the tedious or controversial aspects of rating disabilities for a significant proportion of permanent partial disability cases. The typical schedule covers injuries to the eyes, ears, hands, arms, feet, and legs. It states that for 100% loss (or loss of use) of that body part, compensation at the claimant's weekly rate will be paid for a specified number of weeks. If loss or loss of use is less than total, the maximum number of weeks is reduced in proportion to the percentage of loss or loss of use. Only physical impairment is considered. The effect of the injury on wages or wage-earning capacity is ignored. If an injury is confined to a scheduled body part, the benefits provided by the schedule are exclusive, even though disability rating on a wage loss or loss of wage-earning capacity basis might result in greater benefits. Although this statement is true generally, some states provide additional benefits in the following circumstances: (1) if using one of the other theories results in higher benefits being paid, (2) if diminished wage-earning capacity continues after the scheduled amount is paid, (3) if the scheduled injury results in permanent total disability, or (4) if several scheduled injuries are sustained in the same accident.

Use of the schedule may also be avoided by showing that the effect of the scheduled injury, such as radiating pain, extends into other parts of the body. A few jurisdictions limit the use of schedules to amputation or 100% loss of use of a body part, as opposed to partial loss of use. Another group not only makes the schedule exclusive for permanent disability awards but requires that the weeks for which benefits are paid during the healing period be deducted from the number of weeks authorized by the schedule before an award is made for permanent partial disability.

Most U.S. states operate primarily on the "loss of earning capacity" theory. Even where statutory language seems to indicate clearly that only functional impairment is to be considered, the courts have managed to hold that loss of earning capacity is the real consideration. Even the use of schedules has been justified on an earning capacity basis as merely a legislative determination of presumed wage loss resulting from the impairment listed in the schedule.

Permanent Total Disability

Permanent total disability evaluation is, in most respects, merely an extension of the determination of permanent partial disability. In fact, it is a part of the same process, as the fact finder's only additional task is to determine whether the worker's wage-earning capacity is so destroyed that he or she is unable to compete in the job market.

Two aspects of the permanent total disability question warrant special attention. First, most states use certain presumptions that make the fact finder's job much easier. For example, it may be presumed that the loss of sight of both eyes or the loss of any two limbs will constitute permanent total disability. This relieves the fact finder of the difficult task of evaluating all the factors previously mentioned for other forms of disability. In some cases, these presumptions may be refuted by providing evidence that the worker has some wage-earning capacity, or the presumptions may be applied only for a limited period of time.

Second, the concept of permanent total disability must be defined. The injured employee need not be completely helpless nor unable to earn a single dollar at a job. The person's limitations need only preclude competing in the open job market and be such that no stable job market exists for a worker with this disability.

Insurance Incentives

The asserted insurance incentive of workers' compensation is based on the merit-rated pricing policy. "Merit rating" includes both experience rating and retrospective rating systems. All state funds use merit rating of some sort. Most states permit private insurers to rate employers on merit, although they are not required to do so. Under this procedure, the firm is charged a premium based on the dollar amount of claims for which it is liable. Consequently, a merit-rated firm has an incentive to reduce the amount of its claims through loss control measures. The strength of this incentive has been challenged. Only about one-fourth of insured firms, usually large ones, are eligible for merit rating.

The yearly accident record of firms with only a few employees is not a sufficiently reliable indication of their characteristic experience to be considered in establishing premium rates. On the other hand, merit-rated firms account for 85% of the dollar volume of premiums paid. In addition, self-insured firms, which pay approximately 14% of all benefits, are implicitly merit rated. If incentive effects are inherent in experience rating, they are not available to a large number of small firms and their employers.

Firms not eligible for merit rating are class rated. Under this procedure, all employers engaged in similar business operations within a state pay the same rate per $100 of payroll. These employers do not have a strong incentive to reduce the rates paid by their industry. The only accident prevention incentive generated for individual employers within an industry is that, as poor risks, they may be unable to obtain workers' compensation coverage. Other considerations include the relative quality of the safety and claims services provided by insurers, by management service organizations, and by employers themselves; by tax factors; and by the opportunity cost of paying an insurer a premium instead of paying losses and expenses as they occur.

Safety Incentives

The greatest contribution a safety and health professional can make to a firm's success is to work with managers throughout the organization in safeguarding employees from disabling occupational injury and disease. These injuries and diseases arising on the job hamper both the employee and those with whom he or she works in meeting various job-related objectives: workers miss target dates set for projects, their performance evaluations are disappointing, and often they must forgo opportunities for pay increases and promotions.

The safety and health professional who seeks to promote safety and health in the workplace should recognize the reasons that each worker, supervisor, and executive wants to succeed in his or her occupational efforts. No matter what these reasons may be, achieving career goals requires that individuals and their colleagues remain productive and efficient on the job. Thus, the perceptive safety and health professional can demonstrate to all employees that avoiding disabling injuries and diseases both on and off the job is in their own—not just their employer's—best interests. Doing one's job safely promotes personal and career success.

MANAGING A WORKERS' COMPENSATION PROGRAM

In the final analysis, the total costs of workers' compensation, in terms of both money and human suffering, are the organization's responsibility. Like every other item of cost, workers' compensation expenditures need to be managed carefully. Cost control should be a top management priority to ensure that dollars are spent as prescribed by the law. Injured employees are entitled to benefits, but businesses do not have to cover questionable or fraudulent claims.

To be successful within the legal limits requires that companies have a plan of action and treat each individual case fairly and consistently. Firms should focus on three goals for their workers' compensation program: (1) to prevent accidents, (2) to control costs, and (3) to respond to accidents promptly and efficiently.

When an organization's management style and philosophy demonstrate cooperation, commitment, teamwork, and communication between labor and management, the firm has set a tone for fairness when accident cases occur. Employees in such firms tend to feel more fulfilled both as workers and persons. Their confidence and self-discipline will be reflected in their work habits and productivity. As a result, such a working environment tends to prevent accidents and promote greater worker safety and productivity.

The organization's philosophy should also provide the basis for administering the workers' compensation program. The following are some points usually addressed in program management.

Hiring

A thorough interview and screening process can yield evidence of applicants' existing physical condition(s) and past medical conditions that may suggest future compensation liabilities. The value of preemployment physical examinations depends on the job, the employer's past experience, the physician's evaluation, and other factors. Although the objective is to hire the best people—without discrimination and in compliance with the Americans with Disabilities Act, 1990—companies must realize that at times they will inadvertently hire problem employees.

During the screening process, the interviewer should be trained to elicit complete answers from applicants and to probe more deeply when the responses are not satisfactory. The interviewer should make sure that all information is considered in the hiring decision.

First Report of an Accident

It is vitally important that workers and management report all accidents promptly. Employees must understand the value of the employer knowing when something has occurred so that immediate, efficient action can be taken.

Management should ensure that accident investigation reports are completed promptly and added to existing files. Employees should be accompanied if injured, ill, or if the person must be referred to outside medical services. This action demonstrates the company's concern for its employees.

There are advantages to having a centralized accident-reporting system. It can provide needed information quickly, help select appropriate medical care for the worker, and offer better monitoring and control of medical treatment.

Doctors and Medical Institutions

The company must choose doctors and medical institutions to provide care to injured employees. In most situations, the employee can see his or her personal physician; but in many cases, the organization provides at least the basic medical services. Even when outside or in-house special services may be needed, the organization may select the provider. Before selecting medical services, the employer should consult with other firms regarding their experience with specific physicians or institutions and seek second opinions on specific cases. Even though once in a while an employee may file a false claim, proper supervisory procedures and open communication about employee injuries and illnesses should eliminate most if not all fraudulent claims.

Rehabilitation

Getting a disabled worker back to work as soon as possible is a major goal of medical treatment and workers' compensation coverage. Adequate follow-up of those on leave is often necessary to achieve this objective. The added cost for rehabilitation is usually money well spent. The employer should find out how the prospective physician handles paperwork, bills, reports, patient care and treatment, and facilitates patient activities. Some physicians are likely to be more suitable to a company's needs than others.

Follow-Up

The employer should review ongoing bills, claims, and reports to make sure that services are being provided as specified in compensation acts. Management should maintain a file on each case until it is closed. Accident information arising from each incident must be fed back into the safety and health prevention program.

The employer should treat workers' compensation employees or agents fairly and consistently. If the company buys its insurance, it should not hesitate to seek bids from several different companies each year. The organization should keep in close touch with injured employees to see that follow-up and rehabilitation services are being properly administered.

An effective safety and health program has been shown to prevent accidents, lower workers' compensation claims, prevent some claims, and reduce the organization's overall costs. In addition, an analysis of high-risk hazards is a valuable tool in accident prevention. Typically, 20% of the accidents cause 80% of the costs. Hazard analysis and past accident experience can identify which hazards represent the highest risks and help an organization reduce, control, or eliminate these hazards. At the least, the company will be able to predict with some certainty the costs of accidents resulting from these hazards.

SUMMARY

- All 50 states and all U.S. territories and possessions have workers' compensation laws. Most compensation laws seek to replace wages lost by disabled workers, pay for medical and vocational rehabilitation, reduce litigation, encourage companies to ensure safe working conditions, and promote accident prevention and investigation.
- Although compensation laws are compulsory or elective, in practice all laws can be considered compulsory. Excluded workers include those in farming, domestic service, self-employment, some interstate commerce, and all federal and many state government employees (who are covered under government programs).
- Under an expansion of the dual capacity doctrine and the intentional tort exception, employers are liable for injuries to workers caused by the employer's product or service available to the public or caused by the employer's behavior.
- Workers' compensation covers only work-related injuries or illnesses, determined by the "personal injury" and "by accident" requirements on one hand and the "arising out of" and "in the course of" requirements on the other.
- Three basic types of workers' compensation benefits are (1) loss of income, (2) medical payments, and (3) rehabilitation expenses. All employers are required to provide medical benefits for employees to cover immediate and long-term care. Nearly all states have agencies to administer workers' compensation programs.
- Rehabilitation work may be medical or vocational or a combination of the two. Care is usually coordinated among the company, disabled worker, and various state agencies to select the physician(s), medical facilities, and vocational assistance that will provide the best rehabilitation services.
- Worker's disability is generally classified into four categories: temporary total disability, temporary partial disability, permanent partial disability, and permanent total disability. Various theories are used to help determine the degree of a worker's impairment and remaining capacity to find employment.

- Many companies are given insurance and safety incentives to reduce accident and illness rates. Merit-rating systems and occupational health and safety programs are effective ways to induce organizations to make safety and health concerns a top priority.
- A company's goals for its worker compensation program should be (1) to prevent accidents, (2) to control costs, and (3) to respond to accidents promptly and efficiently.

REFERENCES

English W. *Strategies for Effective Workers' Compensation Cost Control.* Des Plaines, IL: American Society of Safety Engineers, 1988.

LaDou J, ed. *Occupational Health and Safety,* 2nd ed. Itasca, IL: National Safety Council, 1994.

National Council on Compensation Insurance, 5 Marine View Plaza, Hoboken, NJ 07030. Rate and rating plan manuals.

National Safety Council, 1121 Spring Lake Drive, Itasca, IL 60143. *Accident Facts* (annually).

Printing and Publishing Newsletter, July/August, 1996.

U.S. Chamber of Commerce. *Analysis of Workers' Compensation Laws.* Washington, DC: U.S. Chamber of Commerce, 1995.

REVIEW QUESTIONS

1. Name two types of economic losses that workers and their families suffer when they are injured at work.
   a.
   b.
2. How can an effective loss control program benefit the entire economy?
3. The Federal Employees Compensation Act covers all employees of the U.S. government; which act covers the maritime workers?
4. Name the six basic objectives underlying workers' compensation laws.
   a.
   b.
   c.
   d.
   e.
   f.
5. What does a compulsory law require?
6. Under quid pro quo of workers' compensation law, what were employers required to accept?
7. What are the two concepts that are broadening the exclusive remedy provision?
   a.
   b.

8. What are the three types of workers' compensation benefits?

   a.

   b.

   c.

9. What is the minimum percentage of disability wages injured workers usually receive?

10. What two types of workers are excluded from workers' compensation?

    a.

    b.

11. Name the four categories of degree of worker disability?

    a.

    b.

    c.

    d.

12. What goals does a company hope to achieve through its workers' compensation program?

13. What is the goal of workers' compensation?

14. What federal act provides a major source of funding for the retraining and rehabilitation of workers?

15. What are two factors that prompt compensation litigation?

16. What is vocational rehabilitation, and when does it apply?

17. Which three theories are used to help determine the degree of a worker's impairment and remaining capacity to find employment?

    a.

    b.

    c.

18. When can an employee still sue the employer regardless of the exclusive remedy doctrine?

    a.

    b.

Identifying Hazards

6

A hazard is an unsafe condition or activity that, if left uncontrolled, can contribute to an accident. Before hazards can be controlled, they must be identified. This identification of hazards can be accomplished through a systematic hazard analysis program that includes job safety analysis, inspection, measurement and testing, and accident investigation. (See Chapter 7, Accident Investigation, Analysis, and Costs, for a detailed discussion of this topic.) Including all four functions means that analysis is performed before the operation begins, during the life cycle of the operation, and after indications that the system has broken down. This chapter covers the following topics:

- the philosophy and methods of hazard analysis
- the benefits and major components of job safety analyses
- planning and conducting inspections as a critical part of hazard identification
- methods of measurement and testing used to identify workplace hazards

HAZARD ANALYSIS

Hazard analysis is an analysis performed to identify and evaluate hazards for the purpose of their elimination or control. Data from hazard analysis can be regarded as a baseline for future monitoring activities. Before the workplace is inspected to ensure that environmental and physical factors fall within safe ranges, hazards inherent in the system must be discovered (i.e., hazard identification). Hazard analysis has proven to be an excellent tool to identify and evaluate hazards in the workplace.

Analyzing a problem or situation to obtain data for decision making is not new. Workers and their supervisors constantly make assessments—even if unconsciously—about their work to guide their actions. Written analyses carry the process one step further by providing the means to document hazard information.

Philosophy

Written analyses often serve as the basis for more thorough inspections. They can be used to communicate data about hazards and risk potential to those in command positions. They can also be used to educate those in the line and staff organizations who need to know the consequences of hazards within their operations and the purpose and logic behind established control measures. Management can request a formal, written analysis for each critical operation. Such analyses not only gather information for immediate use, but they also reap benefits over the long run. For instance, once important hazard data are committed to paper, they become part of the technical information base of the organization. These documents show the employer's concern for locating hazards and establishing corrective measures before an accident happens.

Traditionally, companies analyzed systems during the operational phase to uncover problems and failures that impaired system effectiveness. Applying hazard analytical techniques returned substantial dividends by reducing both accident and overall operational losses.

Hazard control specialists no longer concentrate solely on operations. They look at the conceptual and design stages of the systems for which they are responsible. They use analytical methods and techniques before the process or product is built to identify and judge the nature and effects of hazards associated with their systems.

This wide assessment has significantly altered the direction of hazard control efforts. When potential problems or failures can be located prior to the production or onstream process stage of a system's life cycle, specialists can cut costs and avoid damage, injuries, and death. Systems engineering was initially concerned with increasing effectiveness, not profits. Properly applied, however, it can point out profitable solutions to many of management's most perplexing operational problems.

What Is Hazard Analysis?

Hazard analysis is an orderly process used to acquire specific hazard and failure data pertinent to a given system. A popular adage holds that "most things work out right for the wrong reasons." By providing data for informed management decisions, hazard analysis helps things work out right for the right reasons. The method forces those conducting the analysis to ask the right questions and helps to answer them. By locating those hazards that are the most probable and/or have the severest consequences, hazard analyses provide information needed to establish effective control measures. Analytic techniques assist the investigator in deciding what facts to gather, determining probable causes and contributing factors, and arranging orderly, clear results.

What are some uses for hazard analysis?

- It can uncover hazards that have been overlooked in the original design, mock-up, or setup of a particular process, operation, or task.
- It can locate hazards that developed after a particular process, operation, or task was instituted.
- It can determine the essential factors in and requirements for specific job processes, operations, and tasks. It can indicate what qualifications are prerequisites to safe and productive work performance.
- It can indicate the need for modifying processes, practices, operations, and tasks.
- It can identify situational hazards in facilities, equipment, tools, materials, and operational events (for example, unsafe conditions).
- It can identify ergonomic situations through anthropometrics and work design (for example, worktable heights, chairs, reaching capabilities).
- It can identify work practices responsible for accident situations (for example, deviations from standard procedures).

- It can identify exposure factors that contribute to injury and illness (such as contact with hazardous substances, materials, or physical agents).
- It can identify physical factors that contribute to accident situations (noise, vibration, insufficient illumination, to name a few).
- It can determine appropriate monitoring methods and maintenance standards needed for safety.
- It can determine the possible results of failures/accidents, the persons or property exposed to loss, and the potential severity of injury or loss.

Formal Methods of Hazard Analysis

Formal hazard analytical methods can be divided into two broad categories: inductive and deductive.

Inductive Method

The inductive analytical method uses observable data to predict events and outcomes within a particular system. It postulates how the component parts of a system will contribute to the success or failure of the system as a whole. Inductive analysis considers a system's operation from the standpoint of its components, their failure in a specific operating condition, and the effect of that failure on the system.

The inductive method forms the basis for such analyses as failure mode and effect analysis (FMEA) and operations hazard analysis (OHA). In FMEA, the failure or malfunction of each component is considered, including the mode of failure. Management can trace throughout the system those effects of the hazard(s) that led to the failure and evaluate the ultimate impact on task performance. However, because only one failure is considered at a time, some possibilities may be overlooked. Figure 6-1 illustrates the FMEA format used at Aerojet Nuclear Co., Idaho Falls, Idaho. Figure 6-2 illustrates the OHA format used for industrial operations.

Once the inductive analysis is completed and the critical failures requiring further investigation are detected, then the fault tree analysis will facilitate an inspection (see deductive method below). The job safety analysis (JSA) section discussed later in this chapter also uses the inductive method for determining the safety risks and components of various jobs.

Deductive Method

If inductive analysis reveals what can happen, deductive analysis shows how. It postulates failure of the entire system and then identifies how the components could contribute to the failure.

Deductive methods use a combined-events analysis, often in the form of trees. The positive tree states the requirements for success (see Figure 6-3). Positive trees are less commonly used than fault trees because they can easily become a list of "shoulds" and sound moralizing.

Fault trees are reverse images of positive trees and show ways troubles can occur. The analyst selects an undesired event, then diagrams in tree form all the

FMEA Form

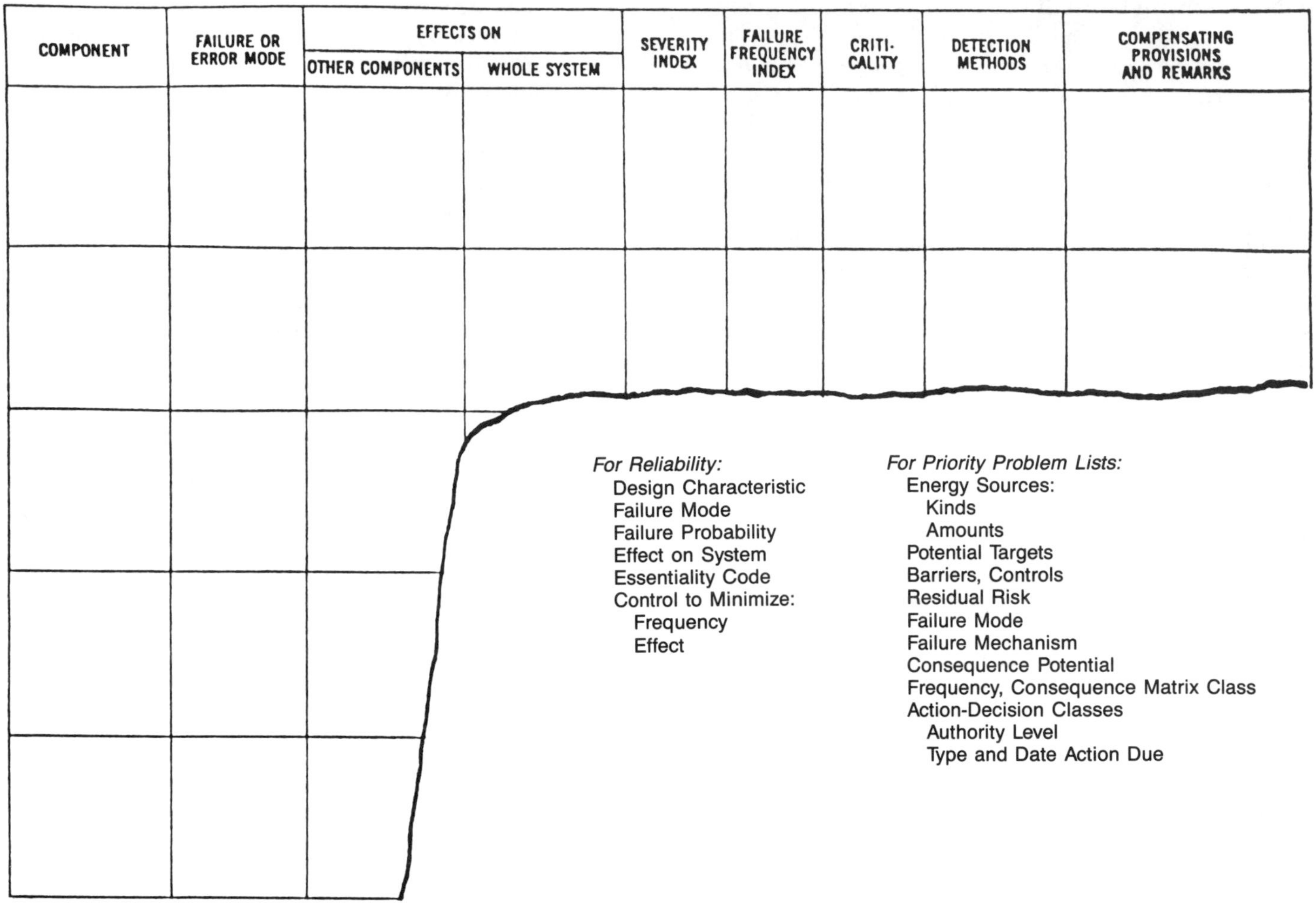

COMPONENT	FAILURE OR ERROR MODE	EFFECTS ON		SEVERITY INDEX	FAILURE FREQUENCY INDEX	CRITI-CALITY	DETECTION METHODS	COMPENSATING PROVISIONS AND REMARKS
		OTHER COMPONENTS	WHOLE SYSTEM					

For Reliability:
Design Characteristic
Failure Mode
Failure Probability
Effect on System
Essentiality Code
Control to Minimize:
Frequency
Effect

For Priority Problem Lists:
Energy Sources:
Kinds
Amounts
Potential Targets
Barriers, Controls
Residual Risk
Failure Mode
Failure Mechanism
Consequence Potential
Frequency, Consequence Matrix Class
Action-Decision Classes
Authority Level
Type and Date Action Due

Figure 6-1. Failure mode effect analysis form used by the Aerojet Nuclear Company (Johnson, 1980).

OPERATIONS HAZARD ANALYSIS									
Process	Operational Step	Task	Source of Potential Hazard	Triggering Event	Potential Effect on Equip., Material Environment	Personal Injury, Property Damage	RAC	Procedural Requirements	Safety & P.P.E.
Turning steel stock between centers on machine lathe.	Rough turning steel stock.	Select cutting tool and place in tool holder.	Improper tool used for rough cutting operation.	Starting lathe.	Tool jams in stock. Stock comes off centers. Uneven cut. Wasted stock.	Operator is hit in face with flying chips of steel.	2	A right-cut tool or roundhouse tool should be held in a straight tool holder. Lathe located to minimize exposure to other work stations.	Select proper tool for job. Operator to wear face protection while operating lathe.
		Place tool holder in the tool post and adjust cutting tool to proper location.	Tool holder extending too far from tool post.	Starting lathe.	Same as above. Breaking cutting tool.	Operator is hit in face with flying chips of steel from stock and broken cutting tool.	2	Tool post should be at end of T-slot. Face of tool must be on center and turned slightly away from headstock.	Operator to wear face protection while operating lathe.

Figure 6-2. This Operations Hazard Analysis (OHA) form is used for industrial operations. (Printed with permission from Indiana Labor & Management Council, Inc.)

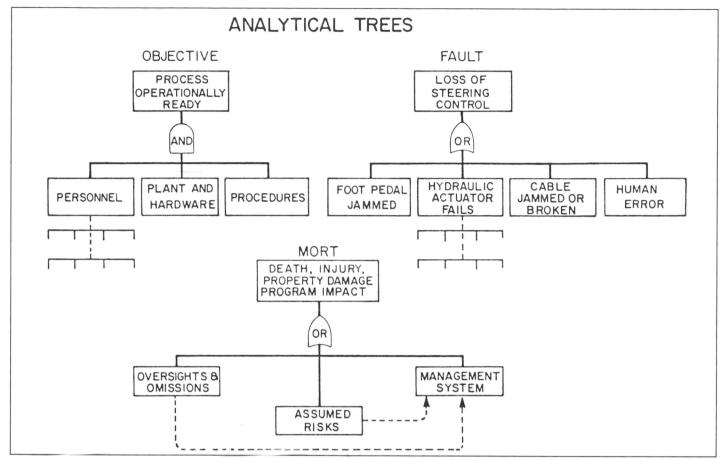

Figure 6-3. Analytical trees are nothing but "structured common sense." Trees are of two major types—the objective or positive trees, which emphasize how a job should properly be done, and the fault trees, which chart those things that can go wrong and produce a specific failure. A fault tree structured for one job can be generalized to cover a wide variety of jobs. The MORT diagram describes the ideal safety program in an orderly, logical manner; it is based on three branches: (1) a branch dealing with specific oversights and omissions at the work site, (2) a branch that deals with the management system that establishes policies and makes the entire system go, and (3) an assumed-risk branch visually recognizing that no activity is completely risk-free, and that risk-management functions must exist in any well-managed organization (Nertney, 1977). (Printed with permission from *Professional Safety*, February 1977.)

possible factors that can contribute to the event. The branches of the tree continue until they reach independent factors. The analyst can then determine probabilities for the independent factors occurring.

The fault tree requires a thorough analysis of a potential event and involves listing all known sources of failure. It is a graphic model of the various parallel and sequential combinations of system component faults that can result in a single, selected system fault. Figure 6-3 illustrates three types of analytical trees.

Analytical trees have three advantages:

1. They accomplish a thorough analysis without wordiness. Using known data, the analyst can identify the single and multiple causes capable of inducing the undesired event.
2. They make the analytical process visible, allowing for the rapid transfer of hazard data from person to person, group to group, with few possibilities for miscommunication during the transfer.
3. They can be used as investigative tools. By reasoning backwards from the accident (the undesired event), the investigator is able to reconstruct the system and pinpoint those elements responsible for the undesired event.

Cost Effectiveness

The cost-effectiveness method can be used as part of either the inductive or deductive approach. The cost of system changes made to increase safety is compared with the decreased costs of fewer serious failures or with the increased efficiency of the system. Cost effectiveness frequently is used to decide among several systems, each capable of performing the same task.

Choosing a Method

To decide what hazard analytical approach is best for a given situation, the hazard control specialist will want to answer five questions:

1. What is the quantity and quality of information desired?
2. What information already is available?
3. What is the cost of setting up and conducting analyses?
4. How much time is available before decisions must be made and action taken?
5. How many people are available to assist in the hazard analysis, and what are their qualifications?

Conducting a hazard analysis can be expensive. Before a hazard-analysis technique is chosen, it is important to determine what information is needed and how important it is.

It is beyond the scope of this manual to go into detail regarding other applications of system safety (see Johnson, 1980).

Who Should Participate in Hazard Analysis?

A hazard analysis, to be fully effective and reliable, should represent as many different viewpoints as possible. Each person familiar with a process or operation has acquired insights concerning problems, faults, and situations that can cause accidents. These insights need to be recorded, along with those of the person initiating the hazard analysis—usually the safety professional. Input from workers and employee representatives can be extremely valuable at this stage.

What Factors Need to Be Analyzed?

All machines, equipment, processes, operations, and tasks in any establishment or facility are good candidates for hazard analysis because they have the potential to cause accidents. Eventually, hazard analyses should be completed for all jobs, but the most potentially threatening should have immediate attention. In determining which processes, operations, and tasks receive priority, those making the decisions should take the following factors into consideration:

- *Frequency of Accidents.* Any operation or task with an associated history of repeated accidents is a good candidate for analysis, especially if different employees have the same kind of accident while performing the same operation or task.
- *Potential for Injury.* Some processes and operations can have a low accident frequency but a high potential for major injury (for example, tasks on a grinder conducted without a tool rest or tongue guard).
- *Severity of Injury.* A particular process, operation, or task can have a history of serious injuries and be a

worthy candidate for analysis, even if the frequency of such injuries is low.
- *New or Altered Equipment, Processes, and Operations.* As a general rule, whenever a new process, operation, or task is created or an old one altered (because of machinery, equipment, or other changes), the safety professional or supervisor should conduct a hazard analysis. For maximum benefit, the hazard analysis should be done while the process or operation is in the planning stages. No equipment should be put into regular operation until the safety professional has checked it for hazards, studied its operation, installed any necessary additional safeguards, and developed safety instructions or procedures. Adhering to such a procedure ensures that managers can train employees in hazard-controlled, safe operations and help to prevent serious injuries and exposures.
- *Excessive Material Waste or Damage to Equipment.* Processes or operations producing excessive material waste or damage to tools and equipment are candidates for hazard analysis. The same problems causing the waste or damage could also, given the right situation, cause injuries.

One of the first steps in hazard and accident analysis is performing job safety analyses. It can be specifically tailored to individual jobs or categories of jobs in the workplace. This next section describes the nature of job safety analyses and how they can be conducted.

JOB SAFETY ANALYSIS

Job safety analysis (JSA) is a procedure used to review job methods and uncover hazards that (1) may have been overlooked in the layout of the facility or building and in the design of the machinery, equipment, tools, workstations, and processes; (2) may have developed after production started; or (3) resulted from changes in work procedures or personnel.

A JSA can be written as shown in Figure 6-4. In the left column, the basic steps of the job are listed in the order in which they occur. The middle column describes all hazards, both those produced by the environment and those connected to the job procedure. The right column gives the safe procedures that should be followed to guard against the hazards and to prevent potential accidents. (See Figure 6-5 for a list of instructions printed on the back of JSA forms published by the National Safety Council.)

For convenience, both the JSA procedure and the written description are commonly referred to as JSA. Health hazards are also considered when making a JSA.

Benefits of JSA

The principal benefits of a JSA include:

- giving individual training in safe, efficient procedures
- making employee safety contracts

JOB SAFETY ANALYSIS	JOB TITLE (and number if applicable): Banding Pallets		DATE: 00/00/00	☒ NEW ☐ REVISED
INSTRUCTIONS ON REVERSE SIDE	TITLE OF PERSON WHO DOES JOB: Bander	SUPERVISOR: James Smith	ANALYSIS BY: James Smith	
COMPANY/ORGANIZATION: XYZ Company	PLANT/LOCATION: Chicago	DEPARTMENT: Packaging	REVIEWED BY: Sharon Martin	
REQUIRED AND/OR RECOMMENDED PERSONAL PROTECTIVE EQUIPMENT: Gloves - Eye Protection - Long Sleeves - Safety Shoes			APPROVED BY: Joe Bottom	

PAGE 1 OF 2 JSA NO. 105

SEQUENCE OF BASIC JOB STEPS	POTENTIAL HAZARDS	RECOMMENDED ACTION OR PROCEDURE
1. Position portable banding cart and place strapping guard on top of boxes.	1. Cart positioned too close to pallet (strike body & legs against cart or pallet, drop strapping gun on foot.)	1. Leave ample space between cart and pallet to feed strapping - have firm grip on strapping gun.
2. Withdraw strapping and bend end back about 3".	2. Sharp edges of strapping (cut hands, fingers & arms). Sharp corners on pallet (strike feet against corners).	2. Wear gloves, eye protection & long sleeves - keep firm grip on strapping - hold end between thumb & forefinger - watch where stepping.
3. Walk around load while holding strapping with one hand.	3. Projecting sharp corners on pallet (strike feet on corners).	3. Assure a clear path between pallet and cart - pull smoothly - avoid jerking strapping.
4. Pull and feed strap under pallet.	4. Splinters on pallet (punctures to hands and fingers) Sharp strap edges (cuts to hands, fingers, and arms).	4. Wear gloves - eye protection - long sleeves. Point strap in direction of bend - pull strap smoothly to avoid jerks.
5. Walk around load. Stoop down. Bend over, grab strap, pull up to machine, straighten out strap end.	5. Protruding corners of pallet, splinters (punctures to feet and ankles).	5. Assure a clear path - watch where walking - face direction in which walking.
6. Insert, position and tighten strap in gun.	6. Springy and sharp strapping (strike against with hands and fingers).	6. Keep firm grasp on strap and on gun - make sure clip is positioned properly.

Figure 6-4. This sample of a completed JSA shows how hazards and safe procedures are identified to help reduce the occurrence of accidents.

- instructing the new person on the job
- preparing for planned safety observations
- giving pre-job instruction on irregular jobs
- reviewing job procedures after accidents occur
- studying jobs for possible improvement in job methods.

A JSA can be done in three basic steps. However, before initiating this analysis, management must first carefully select the job to be analyzed.

Selecting the Job

A job is a sequence of separate steps or activities that together accomplish a work goal. Some jobs can be broadly defined by what is accomplished, for example, making paper, building a facility, and mining iron ore. On the other hand, a job can be narrowly defined in terms of a single action, such as turning a switch, tightening a screw, and pushing a button. Such broadly or narrowly defined jobs are unsuitable for JSA.

Jobs suitable for JSA are those assignments that a line supervisor may make. Operating a machine, tapping a furnace, and piling lumber are good subjects for job safety analyses because they are neither too broad nor too narrow.

Jobs should not be selected at random—those with the worst accident experience should be analyzed first if JSA is to yield the quickest results. In fact, some companies make such selections the focal point of their accident prevention program.

Selection of jobs to be analyzed and establishment of the order of analysis should be guided by the following factors:

- *Frequency of Accidents.* The greater the number of accidents associated with the job, the greater its priority claim for a JSA.
- *Rate of Disabling Injuries.* Every job that has had disabling injuries should be given a JSA, particularly if the injuries prove that prior preventive action was not successful.

INSTRUCTIONS FOR COMPLETING JOB SAFETY ANALYSIS FORM

Job Safety Analysis (JSA) is an important accident prevention tool that works by finding hazards and eliminating or minimizing them *before* the job is performed, and *before* they have a chance to become accidents. Use your JSA for job clarification and hazard awareness, as a guide in new employee training, for periodic contacts and for retraining of senior employees, as a refresher on jobs which run infrequently, as an accident investigation tool, and for informing employees of specific job hazards and protective measures.

Set priorities for doing JSA's: jobs that have a history of many accidents, jobs that have produced disabling injuries, jobs with high potential for disabling injury or death, and new jobs with no accident history.

Here's how to do each of the three parts of a Job Safety Analysis:

SEQUENCE OF BASIC JOB STEPS

Break the job down into steps. Each of the steps of a job should accomplish some major task. The task will consist of a *set* of movements. Look at the first *set* of movements used to perform a task, and then determine the next logical *set* of movements. For example, the job might be to move a box from a conveyor in the receiving area to a shelf in the storage area. How does that break down into job steps? Picking up the box from the conveyor and putting it on a handtruck is one logical set of movments, so it is one job step. Everything related to that one logical set of movements is part of that job step.

The next logical *set* of movements might be pushing the loaded handtruck to the storeroom. Removing the boxes from the truck and placing them on the shelf is another logical set of movements. And finally, returning the handtruck to the receiving area might be the final step in this type of job.

Be sure to list *all* the steps in a job. Some steps might not be done each time—checking the casters on a handtruck, for example. However, that task is a part of the job as a whole, and should be listed and analyzed.

POTENTIAL HAZARDS

Identify the hazards associated with each step. Examine each step to find and identify hazards— actions, conditions and possibilities that could lead to an accident.

It's not enough to look at the obvious hazards. It's also important to look at the entire environment and discover every conceivable hazard that might exist.

Be sure to list health hazards as well, even though the harmful effect may not be immediate. A good example is the harmful effect of inhaling a solvent or chemical dust over a long period of time.

It's important to list *all* hazards. Hazards contribute to accidents, injuries and occupational illnesses.

In order to do part three of a JSA effectively, you must identify potential and existing *hazards*. That's why it's important to distinguish between a hazard, an accident and an injury. Each of these terms has a specific meaning:

HAZARD—A potential danger. Oil on the floor is a *hazard*.

ACCIDENT—An unintended happening that may result in injury, loss or damage. Slipping on the oil is an *accident*.

INJURY—The *result* of an accident. A sprained wrist from the fall would be an injury.

Some people find it easier to identify possible accidents and illnesses and work back from them to the hazards. If you do that, you can list the accident and illness types in parentheses following the hazard. But be sure you focus on the *hazard* for developing recommended actions and safe work procedures.

RECOMMENDED ACTION OR PROCEDURE

Using the first two columns as a guide, decide what actions are necessary to eliminate or minimize the hazards that could lead to an accident, injury, or occupational illness.

Among the actions that can be taken are: 1) engineering the hazard out; 2) providing personal protective equipment; 3) job instruction training; 4) good housekeeping; and 5) good ergonomics (positioning the person in relation to the machine or other elements in the environment in such a way as to eliminate stresses and strains).

List recommended safe operating procedures on the form, and also list required or recommended personal protective equipment for each step of the job.

Be specific. Say *exactly* what needs to be done to correct the hazard, such as, "lift, using your leg muscles." Avoid general statements like, "be careful."

Give a recommended action or procedure for *every* hazard.

If the hazard is a serious one, it should be corrected immediately. The JSA should then be changed to reflect the new conditions.

Figure 6-5. Use these instructions for preparing a JSA.

- *Severity Potential.* Some jobs may not have a history of accidents but may have the potential for producing severe injury.
- *New Jobs.* Changes in equipment or in processes obviously have no history of accidents, but their accident potential may not be fully appreciated. A JSA of every new job should be made as soon as the job has been created. Analysis should not be delayed until accidents or near-misses occur.

After the job has been selected, the three basic steps in making a JSA are:

1. Break the job down into successive steps or activities and observe how these actions are performed.
2. Identify the hazards and potential accidents. This is the critical step because only an identified problem can be eliminated.
3. Develop safe job procedures to eliminate the hazards and prevent the potential accidents.

Breaking the Job Down into Steps

Before the search for hazards begins, a job should be broken down into a sequence of steps, each describing what is being done. Avoid the two common errors: (1) making the breakdown so detailed that an unnecessarily large number of steps results, or (2) making the job breakdown so general that basic steps are not recorded.

To do a job breakdown, select the right worker to observe—an experienced, capable, and cooperative person who is willing to share ideas. If the employee has never helped out on a job safety analysis, explain the purpose—to make a job safe by identifying and eliminating or controlling hazards—and show him or her a completed JSA. Reassure the employee that he or she was selected because of experience and capability.

Observe the employee perform the job and write down the basic steps. Consider videotaping the job as it is performed for later study. To determine the basic job steps, ask "What step starts the job?" Then, "What is the next basic step?" and so on.

Completely describe each step. Any possible deviation from the regular procedure should be recorded because it may be this irregular activity that leads to an accident.

To record the breakdown, number the job steps consecutively as illustrated in the first column of the JSA

JOB SAFETY ANALYSIS WORK SHEET
JOB: Using a Pressurized Water Fire Extinguisher

WHAT TO DO (Steps in sequence)	HOW TO DO IT (Instructions) (Reverse hands for left-handed operator.)	KEY POINTS (Items to be emphasized. Safety is always a key point.)
1. Remove extinguisher from wall bracket.	1. Left hand on bottom lip, fingers curled around lip, palm up. Right hand on carrying handle palm down, fingers around carrying handle only.	1. Check air pressure to make certain extinguisher is charged. Stand close to extinguisher, pull straight out. *Have firm grip, to prevent dropping on feet.* Lower, and as you do remove left hand from lip.
2. Carry to fire.	2. Carry in right hand, upright position.	2. Extinguisher should hang down alongside leg. (This makes it easy to carry and reduces possibility of strain.)
3. Remove pin.	3. Set extinguisher down in upright position. Place left hand on top of extinguisher, pull out pin with right hand.	3. Hold extinguisher steady with left hand. Do not exert pressure on discharge lever as you remove pin.
4. Squeeze discharge lever.	4. Place right hand over carrying handle with fingers curled around operating lever handle while grasping discharge hose near nozzle with left hand.	4. Have firm grip on handle to steady extinguisher.
5. Apply water stream to fire.	5. Direct water stream at base of fire.	5. Work from side to side or around fire. After extinguishing flames, play water on smouldering or glowing surfaces.
6. Return Extinguisher. Report Use.		

Figure 6-6. This JSA worksheet shows how to break down a job and analyze hazards and safe procedures.

training guide, illustrated in Figure 6-6. Each step tells what is done, not how.

The wording for each step should begin with an action word like "remove," "open," or "weld." The action is completed by naming the item to which the action applies, for example, "remove extinguisher," "carry to fire."

Check the breakdown with the person observed and obtain agreement of what is done and the order of the steps. Thank the employee for cooperating.

Identifying Hazards and Potential Accident Causes

Before filling in the next two columns of the JSA—Potential Accidents or Hazards and Recommended Safe Job Procedure—begin the search for hazards. The purpose is to identify all hazards—both those produced by the environment and those connected with the job procedure. Each step, and thus the entire job, must be made safer and more efficient. To do this, ask these questions about each step:

- Is there a danger of striking against, being struck by, or otherwise making harmful contact with an object?
- Can the employee be caught in, by, or between objects?
- Is there a potential for a slip or trip? Can the employee fall on the same level or to another?
- Can strain be caused by pushing, pulling, lifting, bending, or twisting?

- Is the environment hazardous to safety or health? For example, are there concentrations of toxic gas, vapor, mist, fume, dust, heat, or radiation? (See discussion in the National Safety Council's book *Fundamentals of Industrial Hygiene,* 4th edition.)

Close observation and knowledge of the particular job are required if the JSA is to be effective. The job observation should be repeated as often as necessary until all hazards and potential causes for accident or injury have been identified.

When inspecting a particular machine or operation, ask the question, "Can an accident occur here?" More specific questions include:

- Is it possible for a person to come in contact with any moving piece of machine equipment?
- Are rotating equipment, set screws, projecting keys, bolt heads, burrs, or other projections exposed where they can strike at or snag a worker's clothing?
- Is it possible to be drawn into the inrunning nip point between two moving parts, such as a belt and sheave, chain and sprocket, pressure rolls, rack and gear, or gear train?
- Do machines or equipment have reciprocating movement or any motion where workers can be caught on or between a moving part and a fixed object?
- Is it possible for a worker's hands or arms to make contact with moving parts at the point of operation

where milling, shaping, punching, shearing, bending, grinding, or other work is being done?

- Is it possible for material (including chips or dust) to be kicked back or ejected from the point of operation, injuring someone nearby?
- Are machine controls safeguarded to prevent unintended or inadvertent operation?
- Are machine controls located to provide immediate access in the event of an emergency?
- Do machines vibrate, move, or walk while in operation?
- Is it possible for parts to become loose during operation, injuring operators or others?
- Are guards positioned or adjusted to correspond with the permissible openings?
- Is it possible for workers to bypass the guard, thereby making it ineffective?
- Do machines, equipment, and appurtenances receive regular maintenance?
- Are machines placed so operators have sufficient room to safely work with no exposure to aisle traffic?
- Is there sufficient room for maintenance and repair?
- Is there sufficient room to accommodate incoming and finished work as well as scrap that may be generated?
- Are the materials handling methods adequate for the work in process and the tooling associated with it?
- If tools, jigs, and other work fixtures are required, are they stored conveniently, where they will not interfere with the work?
- Is the work area well illuminated with specific point-of-operation lighting where necessary?
- Is ventilation adequate, particularly for those operations that create dusts, mists, vapors, and gases?
- Is the operator using personal protective equipment?
- Is housekeeping satisfactory with no debris or tripping hazards or spills on the floor?
- Are there places where employees have access to machines (for example, the backside)?
- Are energy sources heat controlled for protection?
- Are energy sources controlled for maintenance?

All of these questions will be of most value if they are incorporated into an inspection form that can be filled out at regular intervals. Even though a question may not at first seem to apply to a specific operation, on closer scrutiny it may be found to apply. Using a checklist is a good way to make sure nothing is overlooked.

To complete column 2 of the JSA, the analyzer should list all potential causes for accidents or injury and all hazards yielded by a survey of the machine or operation. Record the type of potential cause of accidents and the agents involved. For example, to note that the employee might injure a foot by dropping a fire extinguisher, write down "struck by extinguisher."

Again, check with the observed employee after the hazards and potential causes of accidents have been recorded. The experienced employee will probably offer additional suggestions. You should also check with others experienced with the job. Through observation and discussion, the analyzer can develop a reliable list of hazards and potential accidents.

Developing Solutions

The final step in a JSA is to develop a recommended safe job procedure to prevent the occurrence of accidents. The principal solutions are as follows:

1. Find a new way to do the job.
2. Change the physical conditions that create the hazards.
3. Change the work procedure.
4. Reduce the frequency (particularly helpful in maintenance and material handling).

To find an entirely new way to do a job, determine the work goal of the job, and then analyze the various ways of reaching this goal to see which way is safest. Consider work-saving tools and equipment.

If a new way cannot be found, then ask this question about each hazard and potential accident cause listed: "What change in physical condition (such as change in tools, materials, equipment, layout, or location) will eliminate the hazard or prevent the accident?"

When a change is found, study it carefully to find other benefits (such as greater production or time saving) that will accrue. These benefits are good selling points and should be pointed out when proposing the change to higher management.

To investigate changes in the job procedure, ask the following questions about each hazard and potential accident cause listed: "What should the employee do—or not do—to eliminate this particular hazard or prevent this potential accident?" "How should it be done?" Because of his or her experience, in most cases the supervisor can answer these questions.

Answers must be specific and concrete if new procedures are to be any good. General precautions—"be alert," "use caution," or "be careful"—are useless. Answers should precisely state what to do and how to do it. This recommendation—"Make certain the wrench does not slip or cause loss of balance"—is incomplete. It does not tell how to prevent the wrench from slipping. Here, in contrast, is an example of a recommended safe procedure that tells both what and how: "Set wrench properly and securely. Test its grip by exerting a light pressure on it. Brace yourself against something immovable or take a solid stance with feet wide apart, before exerting full pressure. This prevents loss of balance if the wrench slips."

Some repair or service jobs have to be repeated often because a condition needs repeated correction. To reduce the need for such repetition, ask "What can be done to eliminate the cause of the condition that makes excessive repairs or service necessary?" If the cause cannot be eliminated, then ask "Can anything be done to minimize the effects of the condition?"

Machine parts, for example, may wear out quickly and require frequent replacement. Study of the problem may reveal excessive vibration is the culprit. After reducing or eliminating the vibration, the machine parts last longer and require less maintenance.

However, reducing the frequency of a job contributes to safety only in that it limits the exposure. Every effort should still be made to eliminate hazards and to prevent accidents through changing physical conditions or revising job procedures or both.

A job that has been redesigned may affect other jobs and even the entire work process. Therefore, the redesign should be discussed not only with the worker involved, but also co-workers, the supervisor, the facility engineer, and others who are concerned. In all cases, however, check or test the proposed changes by reobserving the job and discussing the changes with those who do the job. Their ideas about the hazards and proposed solutions can be of considerable value. They can judge the practicality of proposed changes and perhaps suggest improvements. Actually these discussions are more than just a way to check a JSA. They are safety contacts that promote awareness of job hazards and safe procedures.

Using JSA Effectively

The major benefits of a JSA come after its completion. However, benefits are also to be gained from the development work. While conducting a JSA, supervisors can learn more about the job they supervise. When employees are encouraged to participate in job safety analyses, their safety attitudes improve and their knowledge of safety increases. As a JSA is worked out, safer and better job procedures and safer working conditions are developed. But these important benefits are only a portion of the total benefits listed at the beginning of this discussion.

When a JSA is distributed, the supervisor's first responsibility is to explain its contents to employees and, if necessary, to give them further individual training. The entire JSA must be reviewed with the employees concerned so that they will know how the job is to be done—without accidents.

The JSA can furnish material for planned safety contacts. All steps of the JSA should be used for this purpose. The steps that present major hazards should be emphasized and reviewed again and again in subsequent safety contacts.

New employees on the job must be trained in the basic job steps, taught to recognize the hazards associated with each job step, and instructed in the necessary precautions. There is no better guide for this training than a well-prepared JSA. (See Chapter 23, Safety and Health Training, in this volume, for further discussion of job instruction training.)

Occasionally, the supervisor should observe employees as they perform the jobs for which analyses have been developed. The purpose of these observations is to determine whether or not the employees are doing the jobs in accordance with the safe job procedures. Before making such observations, the supervisor should prepare by reviewing the appropriate JSA to keep in mind the key points to observe.

Many jobs, such as certain repair or service jobs, are done infrequently or on an irregular basis. The employees who do them will benefit from pre-job instruction to remind them of the hazards and the necessary precautions. Using the JSA for the particular job, the supervisor should give this instruction at the time the job is assigned.

Whenever an accident occurs on a job covered by a JSA, the JSA should be reviewed to determine whether or not it needs revision. If the JSA is revised, all employees concerned with the job should be informed of the changes and instructed in any new procedures.

When an accident results from failure to follow JSA procedures, the facts should be discussed with all those who do the job. It should be made clear that the accident would not have occurred had the JSA procedures been followed.

All supervisors are concerned with improving job methods to increase safety and health, reduce costs, and step up production. The JSA is an excellent starting point for questioning the established way of doing a job. In addition, study of the JSA may well suggest ideas for improvement of job methods.

Once the hazards are known, the proper solutions can be developed. Some solutions may be physical changes that eliminate or control the hazard, such as placing a safeguard over exposed moving machine parts. Others may be job procedures that eliminate or minimize the hazard, for example, safe piling of materials. If these solutions do not completely or sufficiently control the hazard, personal protective equipment may be necessary to safely perform the job (see Chapter 7, Personal Protective Equipment, in the *Engineering & Technology* volume). A combination of these solutions may also provide a safe work environment.

The first stage of the hazard and accident analysis procedure is inspection. Management must know what problems and potential accident causes may be present in the workplace and what unsafe procedures or practices workers may be performing.

INSPECTION

Inspection should be conducted in an organization to locate and report existing and potential unsafe conditions or activities that, if left uncontrolled, have the capacity to cause accidents in the workplace.

Philosophy behind Inspection

Inspection can be viewed negatively or positively as either:

- fault-finding, with the emphasis on criticism
- fact-finding, with the emphasis on locating hazards that can adversely affect safety and health

The second viewpoint is more effective. This viewpoint depends on two factors: (1) yardsticks adequate for measuring a particular situation and (2) comparison of what is with what ought to be (Firenze, 1978). Failure to analyze inspection reports for causes of system defects ultimately means the failure of the monitoring function. Corrective action may fix the specific item but fail to fix the system.

Purpose of Inspection

The primary purpose of inspection is to detect potential hazards so they can be corrected before an accident occurs. Inspection can determine conditions that need to be corrected or improved to bring operations up to acceptable standards, both from safety and operational standpoints. Secondary purposes are to improve operations and thus to increase efficiency, effectiveness, and profitability.

Although management ultimately has the responsibility for inspecting the workplace, authority for carrying out the actual inspecting process extends throughout the organization. Obviously supervisors, foremen, and employees fulfill an inspection function, but so do departments as diverse as engineering, purchasing, quality control, human resources, maintenance, and medical.

Types of Inspection

Inspection can be classified as one of two types—continuous or interval inspection.

Continuous Inspection

This process is conducted by employees, supervisors, and maintenance personnel as part of their job responsibilities. Continuous inspection involves noting an apparently or potentially hazardous condition or unsafe procedure and either correcting it immediately or making a report to initiate corrective action. Continuous inspection of personal protective equipment is especially important.

This type of inspection is sometimes called informal because it does not conform to a set schedule, plan, or checklist. However, critics argue that continuous inspection is erratic and superficial, that it does not get into out-of-the-way places, and that it misses too much. The truth is that both continuous and interval inspections are necessary and complement one another.

As part of their job, supervisors make sure that tools, machines, and equipment are properly maintained and safe to use and that safety precautions are being observed. Toolroom employees regularly inspect all hand tools to be sure that they are in safe condition. Foremen are often responsible for continuously monitoring the workplace and seeing that equipment is safe and that employees are observing safe practices. When foremen or supervisors inspect machines at the beginning of a shift, a safety inspection must be part of the operation.

Continuous inspection is one ultimate goal of a good safety and health program. It means that each individual is vigilant, alert to any condition having accident potential, and willing to initiate corrective action.

However, in some instances, the supervisor's greatest advantage in continuous inspection—familiarity with the employees, equipment, machines, and environment—can also be a disadvantage. Just as an old newspaper left on a table in time becomes part of the decor, a hazard can become so familiar that no one notices it. The supervisor's blind spot is particularly likely to occur with housekeeping and unsafe practices. Poor housekeeping conditions may be overlooked because the deterioration is gradual and the effect is cumulative. A similar phenomenon may occur with unsafe practices, such as employees not following established production procedures.

In addition, no matter how conscientious the supervisors are, they cannot be completely objective. Inspections of their areas reflect personal and vested interests, knowledge and understanding of the production problems involved in the area, and concern for the employees. A planned periodic inspection of their areas by another supervisor can be used to audit their efforts. Furthermore, the supervisors who inspect another area may return to their own sections with renewed vision. Having looked at the trees day in and day out, they need occasionally to take the long view and see the forest.

Though this section is devoted primarily to discussing planned inspections, continuous inspections should be regarded as a cooperative, not a competitive, activity.

Interval Inspections

Planned inspections at specific intervals are what most people regard as "real" safety and health inspections. They are deliberate, thorough, and systematic procedures that permit examination of specific items or conditions. They follow an established procedure and use checklists for routine items. These inspections can be any one of three types: periodic, intermittent, and general.

Periodic inspection includes those inspections scheduled at regular intervals. They can target the entire facility, a specific area, a specific operation, or a specific type of equipment. Management can plan these inspections weekly, monthly, semiannually, annually, or at other suitable intervals. Items such as safety guard mountings, scaffolds, elevator wire ropes (cables), two-hand controls, fire extinguishers, and other items relied on for safety require frequent inspection. The more serious the potential for injury or damage might be, the more often the item should be inspected.

Periodic inspections can be of several different types:

- Inspections by the safety professional, industrial hygienist, and joint safety and health committees.
- Inspections for preventing accidents and damage or breakdowns (checking mechanical functioning, lubricating, and the like) performed by electricians,

mechanics, and maintenance personnel. Sometimes these persons are asked to serve as roving inspectors.
- Inspections by specially trained certified or licensed inspectors, often from outside the organization (for example, inspection of boilers, elevators, unfired pressure vessels, cranes, power presses, fire extinguishing equipment). These are often mandated inspections required by regulatory agencies, manufacturers, underwriters, or management.
- Inspections done by outside investigators or government inspectors to determine compliance with government regulations.

The advantage of periodic inspection is that it covers a specific area and allows detection of unsafe conditions in time to provide effective countermeasures. Measurement data collected at regular intervals indicate degenerative trends. The staff or safety committee periodically inspecting a certain area is familiar with operations and procedures and therefore quick to recognize deviations. A disadvantage of periodic inspection is that deviations from accepted practices are rarely discovered because employees are prepared for the inspectors.

Intermittent inspections are those made at irregular intervals. Sometimes the need for an inspection is indicated by accident tabulations and analysis. If a particular department or location shows an unusual number of accidents or if certain types of injuries occur with greater frequency, the supervisor or manager should call for an inspection. When construction or remodeling is going on within or around a facility, an unscheduled inspection may be needed to find and correct unsafe conditions before an accident occurs. The same is true when a department installs new equipment, institutes new processes, or modifies old ones.

Another form of intermittent inspection is that made by the industrial hygienist when a health hazard is suspected or present in the environment. This monitoring of the workplace is covered in detail in the Measurement and Testing section discussed later in this chapter. It usually involves:

- sampling the air for the presence of toxic vapors, gases, radiation, and particulates
- sampling physical stresses like noise, heat, and radiation
- testing materials for toxic properties
- testing ventilation and exhaust systems for proper operation.

Intermittent inspections may be initiated because of the following:

- increase in accident or illness rates in an area
- reports from employees in an area
- management directive
- reports of hazards from other departments, companies, manufacturers, or regulatory agencies

- random selection
- accident/severity potential
- reaction to an event (such as an accident, threat of sabotage, severe weather warning, and so on).

A general inspection is planned and covers places not inspected periodically. This includes those areas no one ever visits and where people rarely get hurt, such as parking lots, sidewalks, fencing, and similar outlying regions.

Many out-of-the-way hazards are located overhead, where they are difficult to spot. Overhead inspections frequently disclose the need for repairs to skylights, windows, cranes, roofs, and other installations affecting the safety of both the employees and the physical facility. Overhead devices can require adjustment, cleaning, oiling, and repairing.

Inspections of overhead areas are necessary to make certain all reasonable safeguards are provided and safe practices observed. Inspectors must verify that workers performing overhead jobs have suitable staging and fall protection equipment. They must also apply this safety directive to themselves during the inspection. They should look for loose tools, bolts, pipelines, shafting, pieces of lumber, windows, electrical fixtures, and other objects that can fall from building structures, cranes, roofs, and similar overhead locations.

Safety conditions change after dark, when the illumination is artificial. Therefore, when an organization has more than one shift, it is important to perform inspections at night to make sure illumination is adequate and the lighting system is well maintained. The safety professional should make this inspection, aided by a photometer and camera, where necessary. Even in organizations with no regular night shifts, some employees—maintenance personnel, firefighters, and night security guards—are required to work after dark. The safety professional occasionally needs to check on their night work conditions.

General inspections are usually required before reopening a facility after a long shutdown.

PLANNING FOR INSPECTION

An effective safety and health inspection program requires the following:

- sound knowledge of the facility
- knowledge of relevant standards, regulations, and codes
- systematic inspection steps
- a method of reporting, evaluating, and using the data.

An effective program begins with analysis and planning. If inspections are casual, shallow, and slipshod, the results will reflect the method. Before instituting an inspection program, these five questions should be answered:

- What items need to be inspected?
- What aspects of each item need to be examined?
- What conditions need to be inspected?
- How often must items be inspected?
- Who will conduct the inspection?

Hazard Control Inspection Inventory

To determine what factors affect the inspection, a hazard control inspection inventory can be conducted. Such an inventory is the foundation upon which a program of planned inspection is based. It resembles a planned preventive maintenance system and yields many of the same benefits.

Management should divide the entire facility—yards, buildings, equipment, machinery, vehicles—into areas of responsibility. These areas, once determined, should be listed in an orderly fashion. The analyst might develop a color-coded map or floor plan of the facility. Large areas or departments can be divided into smaller areas and assigned to each first-line supervisor and/or the hazard control department's inspector.

What Items Need to Be Inspected?

Once specific areas of responsibility have been determined, managers should develop an inventory of those items that may become unsafe or cause accidents. These would include:

- environmental factors (illumination, dust, fumes, gases, mists, vapors, noise, vibration, heat, radiation sources)
- hazardous supplies and materials (explosives, flammables, acids, caustics, toxic or nuclear materials or by-products)
- production and related equipment (mills, shapers, presses, borers, lathes)
- power source equipment (steam and gas engines, electrical motors)
- electrical equipment (switches, fuses, breakers, outlets, cables, extension and fixture cords, grounds, connectors, connections)
- hand tools (wrenches, screwdrivers, hammers, power tools)
- personal protective equipment (hard hats, safety glasses, safety shoes, respirators, hearing protection, gloves, etc.)
- personal service and first aid facilities (drinking fountains, wash basins, soap dispensers, safety showers, eyewash fountains, first aid supplies, stretchers)
- fire protection and emergency response equipment (alarms, water tanks, sprinklers, standpipes, extinguishers, hydrants, hoses, self-contained breathing apparatus, toxic cleanup, automatic valves, horns, phones, radios)
- walkways and roadways (ramps, docks, sidewalks, walkways, aisles, vehicle ways, escape routes)
- elevators, electric stairways, and manlifts (controls, wire ropes, safety devices)
- working surfaces (ladders, scaffolds, catwalks, platforms)
- material handling equipment (cranes, dollies, conveyors, hoists, forklifts, chains, ropes, slings)
- transportation equipment (automobiles, railroad cars, trucks, front-end loaders, helicopters, motorized carts and buggies)
- warning and signaling devices (sirens, crossing and blinker lights, klaxons, warning signs, exit signs)
- containers (scrap bins, disposal receptacles, carboys, barrels, drums, gas cylinders, solvent cans)
- storage facilities and areas both indoor and outdoor (bins, racks, lockers, cabinets, shelves, tanks, closets)
- structural openings (windows, doors, stairways, sumps, shafts, pits, floor openings)
- buildings and structures (floors, roofs, walls, fencing)
- miscellaneous—any items that do not fit in preceding categories. (List adapted from *Facility Inspection*, see References.)

There are many sources of information about items to be inspected, especially employees in an organization. For instance, maintenance employees know what problems can cause damage or shutdowns. The workers in the area are qualified to point out causes of injury, illness, damage, delays, or bottlenecks. Medical personnel in the organization can list problems causing job-related illnesses and injuries. Manufacturers' manuals often specify maintenance schedules and procedures and safe work methods.

Gathering the information about standards, regulations, and codes is a necessary first step in determining what items need to be inspected. A great deal of work already has been done for the safety inspector.

Building codes, building inspection books, and guides to building and facility maintenance also are useful references. Publications of the NFPA will help safety personnel ensure that fire hazards are being effectively controlled. In addition, insurance company surveys often contain checklists to help management determine the safety condition of their buildings. Research and reference material is contained in subsequent chapters of this manual. Other publications of the National Safety Council, such as *Accident Facts,* may prove useful as well.

State or provincial and federal governments also publish accident statistics. The *Federal Register* and the *Code of Federal Regulations* (*CFR*), Title 29, parts 1900–1950, give OSHA regulations. Figure 6-7 shows some special subjects addressed by the subparts of 29 *CFR* 1910 (General Industry) and the subparts of 29 *CFR* 1926 (Construction).

It is important to remember, however, that federal and state or provincial laws, codes, and regulations usually set up minimum requirements only. To comply with company policy and secure maximum safety, management must often exceed these requirements. Some OSHA

29 CFR 1910
OCCUPATIONAL SAFETY AND HEALTH STANDARDS

Subparts

C — Access to Employee Exposure and Medical Records
D — Walking-Working Surfaces
E — Means of Egress
F — Powered Platforms, Manlifts, and Vehicle-Mounted Work Platforms
G — Occupations Health and Environmental Control
H — Hazardous Materials
I — Personal Protective Equipment
J — General Environmental Controls
K — Medical and First Aid
L — Fire Protection

M — Compressed Gas and Compressed Air Equipment
N — Materials Handling and Storage
O — Machinery and Machine Guarding
P — Hand and Portable Powered Tools and Other Hand-Held Equipment
Q — Welding, Cutting, and Brazing
R — Special Industries
S — Electrical
T — Commercial Diving Operations
Z — Toxic and Hazardous Substances

29 CFR 1926
STANDARDS FOR THE CONSTRUCTION INDUSTRY

Subparts

C — General Safety and Health Provisions
D — Occupational Health and Environmental Controls
E — Personal Protective and Lifesaving Equipment
F — Fire Protection and Prevention
G — Signs, Signals, and Barricades
H — Materials Handling, Storage, Use and Disposal
I — Tools—Hand and Power
J — Welding and Cutting
K — Electrical
L — Scaffolding
M — Fall Protection
N — Cranes, Derricks, Hoists, Elevators, and Conveyors

O — Motor Vehicles, Mechanized Equipment, and Marine Operations
P — Excavations, Trenching, and Shoring
Q — Concrete, Concrete Forms, and Shoring
R — Steel Erection
S — Tunnels and Shafts, Caissons, Cofferdams, and Compressed Air
T — Demolition
U — Blasting and Use of Explosives
V — Power Transmission and Distribution
W — Rollover Protection Structures; Overhead Projection
X — Ladders and Stairways

Figure 6-7. This is a quick reference list of what subpart discusses certain major topics in 29 *CFR* 1910 and 29 *CFR* 1926.

publications indicate not only what standards are required but also what violations are most frequent. These same sources are helpful in the next step—determining critical factors to be inspected.

What Aspects of Each Item Need to Be Examined?

Particular attention should be paid to the parts of an item most likely to become a serious hazard to health and safety. These parts often develop problems because of stress, wear, impact, vibration, heat, corrosion, chemical reaction, and misuse. Such items as safety devices, guards, controls, work or wearpoint components, electrical and mechanical components, and fire hazards tend to become unsafe first. For a particular machine, critical parts would include the point of operation, moving parts, and accessories (flywheels, gears, shafts, pulleys, key ways, belts, couplings, sprockets, chains, controls, lighting, brakes, exhaust systems). Also to be checked are items related to feeding, oiling, adjusting, maintenance, grounding, attaching, work space, and location.

The most critical parts of an item are not always the most obvious. When the security of a heavy load depends on a cotter pin being in place, then that pin is a critical part.

What Conditions Need to Be Inspected?

The unsafe conditions for each part to be inspected should be described specifically and clearly. A checklist question that reads "Is ___ safe?" is meaningless because it does not define what makes an item unsafe. Inspectors should describe and not simply list unsafe conditions for each item. Usually, conditions to look for can be indicated by such words as "jagged, exposed, broken, frayed, leaking, rusted, corroded, missing, vibrating, loose," or "slipping." Sometimes exact figures are needed—for example, the maximum pressure in a boiler or the percent spread of a sling hook.

Checklists serve as reminders of what to look for and as records of what has been covered. They can be used to structure and guide inspections. They also allow on-the-spot recording of all findings and comments before they are forgotten. In case an inspection is interrupted, checklists provide a record of what has and what has not been inspected. Otherwise, inspectors may miss items or conditions they should examine or may be unsure, after inspecting an area, that they have covered everything. Good checklists also help in follow-up work to make sure hazards have been corrected or eliminated.

Many types of monitoring checklists are available, varying in length from thousands of items to only a few.

These checklists are useful in determining which standards or regulations apply to individual situations. Once the applicable standards are identified, the organization can tailor a checklist to its needs and uses and enter it into the computer system for action and follow-up.

The Centers for Disease Control of the U.S. Department of Health and Human Services (DHHS) have devised a suggested checklist for the safety evaluation of shop and laboratory areas. The worksheet is referenced to the OSHA "General Industry Standards."

Merely running through a checklist, however, does little to locate or correct problems. The checklist must be used as an aid to the inspection process, not as an end in itself. Of course, any hazard observed during inspection must be recorded, even though it is not part of the checklist.

The following 16 pages comprise a portfolio of various companies' inspection checklists (Figure 6-8). These are used with computer follow-up on inspection results, actions to be taken, and corrections made. Refer to the following portfolio of checklists.

The amount of detail included in the checklist will vary, depending upon the inspector's knowledge of the relevant standards and the nature of the inspection. An experienced inspector with thorough knowledge of the standards will need only a few clues as a reminder of items to be inspected. Checklists for infrequent inspection generally will be more detailed than daily or weekly ones.

The format of a checklist should include columns to indicate either compliance or action date. Space also should be provided to cite the specific violation, a way to correct it, and a recommendation that the condition receive more or less frequent attention. Whatever the format of the checklist, space should be provided for the inspector's signature and the inspection date.

Checklists can be prepared by the safety and health committee, by the safety director, or by a subcommittee that includes engineers, supervisors, employees, and maintenance personnel. The safety and health professional and the department supervisor should monitor checklist development and make sure all applicable standards are covered. In their final form the checklists should conform to the inspection route.

Choosing the inspection route means inspecting an area completely and thoroughly while avoiding:

- time-consuming backtracking and repetitions
- long walks between items
- unnecessary interruptions of the production process
- distraction of employees.

Often a closed-loop inspection will give good results. Sometimes it is valuable to follow the production path of the material being processed.

How Often Must Items Be Inspected?

The frequency of inspection is determined by four factors.

- What is the loss severity potential of the problem? The inspector should ask, "If the item or critical part should fail, what injury, damage, or work interruption would result?" The more severe the loss potential, the more often the item should be inspected. For instance, a frayed wire rope on an overhead crane block has the potential to cause a much greater loss than does a defective wheel on a wheelbarrow. Therefore, the rope needs to be inspected more frequently than the wheel.

- What is the potential for injury to employees? If the item or critical part should fail, how many employees would be endangered and how frequently? The greater the probability for injury to employees, the more often the item should be inspected. For example, a stairway continually used by many people needs to be inspected more frequently than one seldom used.

- How quickly can the item or part become unsafe? The answer to this question depends on the nature of the part and the conditions to which it is subjected. Equipment and tools used frequently can become damaged, defective, or worn more quickly than those rarely used. An item located in a particular spot can be exposed to greater damage than an identical item in a different location. The faster an item can become unsafe, the more frequently it should be inspected.

- What is the past history of failures? What were the results of these failures? Maintenance and production records and accident investigation reports can provide valuable information about how frequently items have failed and the results in terms of injuries, damage, delays, and shutdowns. The more often an item has failed in the past and the greater the consequences, the more it needs to be inspected.

The Occupational Safety and Health Administration, the Mine Safety and Health Administration (MSHA), the Federal Aviation Administration (FAA), the Nuclear Regulatory Commission (NRC), the Environmental Protection Agency (EPA), and other federal, state, provincial, and local regulatory agencies require periodic inspections. Consult the regulations and the agency responsible for enforcement for current information regarding inspection criteria and intervals. Frequency of inspections should be described in specific terms, for example, before every use, when serviced daily, monthly, quarterly, yearly.

Who Will Conduct the Inspection?

Answering the four previous questions—the items to be inspected, the aspects of each item to be inspected, the conditions to be inspected, and the frequency of inspections—will help to determine who is qualified to do the inspection. No individual or group should have exclusive responsibility for all inspections. Employees who perform

(Text continues on page 161.)

Loading Dock Safety Checklist

Date _____

Company Name _____

Plant Name _____ Plant Location _____

Dock examined/Door numbers _____

Company representative completing checklist:

Name _____ Title _____

A. Vehicles/Traffic Control

1. Do forklifts have the following safety equipment?
   - ☐ Seat belt ☐ Load backrest
   - ☐ Headlight ☐ Backup alarm
   - ☐ Horn ☐ Overhead guard
   - ☐ Tilt indicator
   - ☐ On-board fire extinguisher
   - ☐ Other _____

2. Are the following in use?	Yes	No
Driver candidate screening	☐	☐
Driver training/licensing	☐	☐
Periodic driver retraining	☐	☐
Vehicle maintenance records	☐	☐
Written vehicle safety rules	☐	☐

	Yes	No
3. Is the dock kept clear of loads of materials?	☐	☐
4. Are there convex mirrors at blind corners?	☐	☐
5. Is forklift cross traffic over dock levelers restricted?	☐	☐
6. Is pedestrian traffic restricted in the dock area?	☐	☐
7. Is there a clearly marked pedestrian walkway?	☐	☐
8. Are guardrails used to define the pedestrian walkway?	☐	☐

9. Comments/recommendations

_____
_____
_____
_____
_____

B. Vehicle restraining

1. If vehicle restraints are used:	Yes	No
a. Are all dock workers trained in the use of the restraints?	☐	☐
b. Are the restraints used consistently?	☐	☐
c. Are there warning signs and lights inside and out to tell when a trailer is secured and when it is not?	☐	☐
d. Are the outdoor signal lights clearly visible even in fog or bright sunlight?	☐	☐
e. Can the restraints secure trailers regardless of the height of their ICC bars?	☐	☐
f. Can the restraints secure all trailers with I-beam, round, or other common ICC bar shapes?	☐	☐
g. Does the restraint sound an alarm when a trailer cannot be secured because its ICC bar is missing or out of place?	☐	☐

Figure 6-8. The following 16 pages present a sample of inspection forms used in various organizations. Adapt these forms as necessary to your specific needs.

h. Are dock personnel specifically trained to watch for trailers with unusual rear-end assemblies (e.g. sloping steel back plates and hydraulic tailgates) that can cause a restraint to give a signal that the trailer is engaged even when it is not safely engaged? ☐ ☐

i. Are dock personnel specifically trained to observe the safe engagement of all unusual rear-end assemblies? ☐ ☐

j. Do the restraints receive regular planned maintenance? ☐ ☐

k. Do restraints need repairs or replacement? (List door numbers.)

l. Comments/recommendations

_____

_____

_____

	Yes	No
2. If wheel chocks are used:		
a. Are dock workers, rather than truckers, responsible for placing chocks?	☐	☐
b. Are all dock workers trained in proper chocking procedures?	☐	☐
c. Are chocks of suitable design and construction?	☐	☐
d. Are there two chocks for each position?	☐	☐
e. Are all trailers chocked on both sides?	☐	☐
f. Are chocks chained to the building?	☐	☐
g. Are warning signs in use?	☐	☐
h. Are driveways kept clear of ice and snow to help keep chocks from slipping?	☐	☐
i. Comments/recommendations		

_____

_____

C. Dock levelers

	Yes	No
1. Are the dock levelers working properly?	☐	☐

Figure 6-8. (Continued).

2. Are levelers long enough to provide a gentle grade into trailers of all heights? ☐ ☐

3. Is leveler width adequate when servicing wider trailers? ☐ ☐

4. Do platform width and configuration allow safe handling of end loads? ☐ ☐

5. Is leveler capacity adequate given typical load weights, lift truck speeds, ramp inclines and frequency of use? ☐ ☐

6. Do levelers have the following safety features?
   Working-range toe guards ☐ ☐
   Full-range toe guards ☐ ☐
   Ramp free-fall protection ☐ ☐
   Automatic recycling ☐ ☐

7. Do levelers receive regular planned maintenance? ☐ ☐

8. Do levelers need repair or replacement? (List door numbers.)

_____

_____

9. Comments/recommendations _____

_____

_____

D. Portable dock plates

	Yes	No
1. Is plate length adequate?	☐	☐
2. Is plate capacity adequate?	☐	☐
3. Are plates of suitable design and materials?	☐	☐
4. Do plates have curbed sides?	☐	☐
5. Do plates have suitable anchor stops?	☐	☐
6. Are plates moved by lift trucks rather than by hand?	☐	☐
7. Are plates stored away from traffic?	☐	☐
8. Are plates inspected regularly?	☐	☐
9. Do plates need repair or replacement? (List door numbers.)		

_____

_____

10. Comments/recommendations

_____

_____

E. Dock Doors

	Yes	No
1. Are doors large enough to admit all loads without obstruction?	☐	☐
2. Are door rails protected by bumper posts?	☐	☐
3. Do doors receive regular planned maintenance?	☐	☐

4. Which doors (if any) need repair or replacement?

_____

_____

5. Comments/recommendations

_____

_____

F. Traffic Doors

	Yes	No
1. Are doors wide enough to handle all loads and minimize damage.	☐	☐
2. Does door arrangement allow safe lift truck and pedestrian traffic?	☐	☐
3. Are visibility and lighting adequate on both sides of all doors?	☐	☐

4. Which doors (if any) need repair or replacement?

_____

_____

5. Comments/recommendations

_____

_____

G. Weather sealing

	Yes	No
1. Are the seals or shelters effective in excluding moisture and debris from the dock?	☐	☐
2. Are seals or shelters sized to provide an effective seal against all types of trailers?	☐	☐
3. Are seals or shelters designed so that they will not obstruct loading and unloading?	☐	☐

| 4. Are dock levelers weather sealed along the sides and back? | ☐ | ☐ |
| 5. In addition to seals or shelters, would an air curtain solve a problem? | ☐ | ☐ |

6. Do seals or shelters need repair or replacement? (List door numbers.)

_____

_____

7. Comments/recommendations

_____

_____

H. Trailer Lifting

1. How are low-bed trailers elevated for loading/unloading?

   ☐ Wheel risers ☐ Concrete ramps
   ☐ Trailer-mounted jacks
   ☐ Truck levelers

	Yes	No
2. Do lifting devices provide adequate stability?	☐	☐
3. Are trailers secured with vehicle restraints when elevated?	☐	☐

4. Do lifting devices need repair or replacement? (List door numbers.)

_____

5. Comments/recommendations

_____

_____

I. Other Considerations

	Yes	No
1. Dock lights		
a. Is lighting adequate inside trailers?	☐	☐
b. Is the lift mechanism properly shielded?	☐	☐
2. Scissors lifts		
a. Are all appropriate workers trained in safe operating procedures?	☐	☐
b. Is the lift mechanism properly shielded?	☐	☐
c. Are guardrails and chock ramps in place and in good repair?	☐	☐

Figure 6-8. (Continued).

3. Conveyors
   a. Are all appropriate workers trained in safe operating procedures? ☐ ☐
   b. Are necessary safeguards in place to protect against pinch points, jam-ups and runaway material? ☐ ☐
   c. Are crossovers provided? ☐ ☐
   d. Are emergency stop buttons in place and properly located? ☐ ☐

4. Strapping
   a. Are proper tools available for applying strapping? ☐ ☐
   b. Do workers cut strapping using only cutters equipped with a holddown device? ☐ ☐
   c. Do workers wear hand, foot and face protection when applying and cutting strapping? ☐ ☐
   d. Are all appropriate workers trained in safe strapping techniques? ☐ ☐

5. Manual handling
   a. Is the dock designed so as to minimize manual lifting and carrying? ☐ ☐
   b. Are dock workers trained in safe lifting and manual handling techniques? ☐ ☐

6. Miscellaneous   Yes No
   a. Are pallets regularly inspected? ☐ ☐
   b. Are dock bumpers in good repair? ☐ ☐
   c. Is the dock kept clean and free of clutter? ☐ ☐
   d. Are housekeeping inspections performed periodically? ☐ ☐
   e. Are anti-skid floor surfaces, mats or runners used where appropriate? ☐ ☐

   f. Are stairways or ladders provided for access to ground level from the dock? ☐ ☐
   g. Is the trailer landing strip in good condition? ☐ ☐
   h. Are dock approaches free of potholes or deteriorated pavement? ☐ ☐
   i. Are dock approaches and outdoor stairs kept clear of ice and snow? ☐ ☐
   j. Are dock positions marked with lines or lights for accurate trailer spotting? ☐ ☐
   k. Do all dock workers wear personal protective equipment as required by company policy? ☐ ☐
   l. Is safety training provided for all dock employees? ☐ ☐
   m. Are periodic safety refresher courses offered? ☐ ☐

J. General comments/recommendations

_____

_____

_____

_____

_____

For additional copies of this Loading Dock Safety Checklist, write to Rite-Hite Corporation, 9019 North Deerwood Drive, P.O. Box 23043, Milwaukee, WI 53223-0043.
For more information on loading dock safety, write for a complimentary copy of Rite-Hite's Dock Safety Guide.

This loading dock safety checklist is provided as a service by Rite-Hite Corporation, Milwaukee, Wis. It is intended as an aid to safety evaluation of loading dock equipment and operations. However, it is not intended as a complete guide to loading dock hazard identification. Therefore, Rite-Hite Corporation makes no guarantees as to nor assumes any liability for the sufficiency or completeness of this document. It may be necessary under particular circumstances to evaluate other dock equipment and procedures in addition to those included in the checklist. For information on U.S. loading dock safety requirements, consult OSHA Safety and Health Standards (29 CFR 1910). In other countries, consult the applicable national or provincial occupational health and safety codes.

Figure 6-8. (Continued).

OSHA CHECKLIST FOR HAZARD COMMUNICATION STANDARD

> ## *Does Your Company Meet The Requirements of This Standard?*
> The Occupational Safety & Health Administration (OSHA) requires certain manufacturers, distributors and employers to meet the requirements of the Hazard Communication Standard (29CFR 1910.1200). The following checklist has been developed to help you determine if your company is in compliance with the requirements of this standard.

	OSHA SECTION	YES	NO	ACTION TAKEN
A. HAZARD COMMUNICATION PROGRAM:				
1. The program is in writing.	1910.1200(e)(1)			
Our Written Program Provides The Following:				
2. Describes how hazards will be evaluated and described (employers may rely on the chemical manufacturer or importer).	1910.1200(d)(6)			
3. Tests all hazardous materials in the workplace (employers may rely on the chemical manufacturer or importers).	1910.1200(d)(1)			
4. Describes our labeling system.	1910.1200(f)			
5. Provides a list of hazardous chemicals referenced on MSDS for all hazardous materials used in the workplace (see Section B).	1910.1200(e)(1)			
6. Describes our employee education and training program.	1910.1200(h)			
7. Describes hazards of non-routine tasks.	1910.1200(e)(1)(ii)			
8. Describes how hazards of non-labeled pipes will be handled.	1910.1200(e)(1)(ii)			
9. Includes procedures for informing of on-site contractors of the hazardous substances in the workplace to which their employees may be exposed.	1910.1200(e)(1)(iii)			
10. Is available to employees, their designated representatives, assistant secretary of labor for OSHA, and the director of NIOSH.	1910.1200(e)(3)			
B. LIST OF HAZARDOUS MATERIALS IN THE WORKPLACE:				
Our List contains all hazardous chemicals, including, but not limited to:				
1. Raw materials.	1910.1200(e)(1)(i)			
2. Both isolated and nonisolated intermediates.	1910.1200(e)(1)(i)			
3. Final product.	1910.1200(e)(1)(i)			
4. Cleaning and maintenance chemicals.	1910.1200(e)(1)(i)			
5. Laboratory chemicals for which MSDS information has been received.	1910.1200(b) (ii),(iii)			
6. Waste products not regulated under RCRA, but which are hazardous under this standard.	1910.1200(e)(1)(i)			
7. Impurities and by products.	1910.1200(e)(1)(i)			
8. Waste treatment and products.	1910.1200(e)(1)(i)			
C. HAZARDOUS MATERIALS LABELING SYSTEM:				
1. All products containing hazardous materials leaving the workplace are labeled (applicable to chemical manufacturers, distributors and importers only).	1910.1200(f)(1)			
2. Stationary containers are labeled.	1910.1200(f)(4)			
3. Temporary containers used between workshifts or by different workers are labeled.	1910.1200(f)(6)			
4. A method has been established to insure that our labels are correct and up-to-date.	1910.1200(f)(4)(ii)			
D. CONTENTS OF HAZARDOUS MATERIAL LABEL:				
Our Labels Contain:				
1. A chemical name that coincides with name on MSDS.	1910.1200(f)(1)(i)			
2. The identity of hazards with words (in English), pictures or symbols.	1910.1200(f)(1)(ii)			
3. Hazards of immediate and direct consequences of mishandling are included.	1910.1200(f)(1)(ii)			
4. Information that does not conflict with DOT regulations.	1910.1200(f)(2) 49 CFR 172.101			
5. Other OSHA standards if material is already regulated.	1910.1200(f)(3)			
6. The name and address of a responsible party (or parties).	1910.1200(f)(1)(iii)			
E. IN-PLANT LABELING SYSTEM:				
1. Containers are labeled with the identity of hazardous chemicals and hazard warnings (unless hazard warning materials are used).	1910.1200(f)(4)(i)			
2. Hazard warning materials for hazardous chemicals in stationary process containers are readily accessible to the employee in the workplace.	1910.1200(f)(5)			

Figure 6-8. (Continued).

	OSHA SECTIONS	NO	ACTION TAKEN
3. The labels on incoming containers have not been removed or defaced unless immediately replaced with our own labels.	1910.1200(f)(7)		
4. The hazards in pipelines are identified, although they do not have to be labeled under this standard.	1910.1200(e)(1)(ii)		
5. Our labels are legible and in English.	1910.1200(f)(8)		
F. MATERIAL SAFETY DATA SHEETS:			
1. A MSDS is available for every hazardous chemical which an employer uses.	1910.1200(g)(1)		
2. Our MSDS are readily accessible to exposed employees in the work area throughout each work shift.	1910.1200(g)(8)		
G. PROCEDURES HAVE BEEN ESTABLISHED FOR:			
1. Updating our MSDS (or for receiving updated copies from our supplier).	1910.1200(g)(5)		
2. Taking appropriate action if a shipment is received without a MSDS.	1910.1200(g)(1)		
3. Getting new and updated MSDS to employees handling materials.	1910.1200(g)(8)		
4. Advising employees of any changes in MSDS.	1910.1200(h)		
5. Documentation of efforts to obtain MSDS from supplier (recommended practice but not required by this standard).	——		
H. HAZARDS OF NON-ROUTINE TASKS: Procedures have been established assessing the hazards of non-routine tasks as follows:			
1. All non-routine tasks involving the use or exposure to hazardous materials are identified.	1910.1200(e)(1)(ii)		
2. The hazards involved in the performance of non-routine tasks are described in writing.	1910.1200(e)(1)(ii)		
3. A MSDS is prepared or obtained for the hazardous materials involved in these non-routine tasks.	1910.1200(e)(1)(ii)		
4. A labeling system or written operating procedure has been established to identify the hazardous substances and their hazards involved in non-routine tasks.	1910.1200(e)(1)(ii)		
5. Special training has been established for the performance of non-routine tasks, including written operating procedures.	1910.1200(e)(1)(ii)		
I. EMPLOYEE EDUCATION & TRAINING: Procedures have been established to inform employees of:			
1. Covers all manufacturing, quality control, plant service, and R&D employees who may be exposed to hazardous materials.	1910.1200(b)(1)		
2. Requirements of the Hazard Communication Standard.	1910.1200(h)(1)(i)		
3. Operations where hazardous materials are present.	1910.1200(h)(1)(ii)		
4. Location and availability of the written hazard communication program including the hazardous chemical list and material safety data sheets.	1910.1200(h)(1)(iii)		
J. PROCEDURES FOR TRAINING EMPLOYEES INCLUDE:			
1. Information about physical and health hazards of chemicals in work area.	1910.1200(h)(2)(ii)		
2. Detecting the presence of hazardous materials - monitoring procedures, odors, visibility, etc.	1910.1200(h)(2)(i)		
3. Proper use and selection of personal protective equipment.	1910.1200(h)(2)(iii)		
4. Emergency procedures in the event of accidental exposure to hazardous materials, including emergency phone numbers and the location of eye washes and safety showers.	1910.1200(h)(2)(iii)		
5. How to determine hazards by reading a label.	1910.1200(h)(2)(iv)		
6. The location of MSDS and the procedure for reviewing them and/or obtaining a copy.	1910.1200(h)(1)(iii)		
7. How to obtain the correct MSDS for the hazardous substance used by the employee, such as use of the trade name as a key identifier.	1910.1200(h)(2)(iv)		
8. How the MSDS is updated or the procedure for obtaining updated copies from the chemical manufacturer, importer or distributor.	1910.1200(h)(2)(iv)		
9. The significance to the employee of each section of information on the MSDS, how to read it and what it means.	1910.1200(h)(2)(iv)		
10. The measures employees can take to protect themselves from chemical exposure. (Examples include eye washes, face shields, respirators, etc.).	1910.1200(h)(2)(iii)		
11. Training which is done prior to the handling of the hazardous chemical, including employees who may only temporarily do this work.	1910.1200(h)		
12. Updated training is considered when the employee has transferred jobs or departments.	1910.1200(h)		
13. Updated training is considered when significant changes in chemicals or operations have occurred.	1910.1200(h)		

LAB SAFETY SUPPLY CO. • **3430 Palmer Drive** • **Janesville, WI 53546** • **(608) 754-2345**

SRM-115

Figure 6-8. (Continued).

SELF-INSPECTION CHECKLISTS

General

	OK	ACTION NEEDED
1. Is the required OSHA workplace poster displayed in your place of business as required where all employees are likely to see it?	☐	☐
2. Are you aware of the requirement to report all workplace fatalities and any serious accidents (where 5 or more are hospitalized) to a federal or state OSHA office within 48 hours?	☐	☐
3. Are workplace injury and illness records being kept as required by OSHA?	☐	☐
4. Are you aware that the OSHA annual summary of workplace injuries and illnesses must be posted by February 1 and must remain posted until March 1?	☐	☐
5. Are you aware that employers with 10 or fewer employees are exempt from the OSHA recordkeeping requirements, unless they are part of an official BLS or state survey and have received specific instructions to keep records?	☐	☐
6. Have you demonstrated an active interest in safety and health matters by defining a policy for your business and communicating it to all employees?	☐	☐
7. Do you have a safety committee or group that allows participation of employees in safety and health activities?	☐	☐
8. Does the safety committee or group meet regularly and report, in writing, its activities?	☐	☐
9. Do you provide safety and health training for all employees requiring such training, and is it documented?	☐	☐
10. Is one person clearly in charge of safety and health activities?	☐	☐
11. Do all employees know what to do in emergencies?	☐	☐
12. Are emergency telephone numbers posted?	☐	☐
13. Do you have a procedure for handling employee complaints regarding safety and health?	☐	☐

Workplace

ELECTRICAL WIRING, FIXTURES AND CONTROLS

	OK	ACTION NEEDED
1. Are your workplace electricians familiar with the requirements of the National Electrical Code (NEC)?	☐	☐
2. Do you specify compliance with the NEC for all contract electrical work?	☐	☐
3. If you have electrical installations in hazardous dust or vapor areas, do they meet the NEC for hazardous locations?	☐	☐
4. Are all electrical cords strung so they do not hang on pipes, nails, hooks, etc?	☐	☐
5. Is all conduit, BX cable, etc., properly attached to all supports and tightly connected to junction and outlet boxes?	☐	☐
6. Is there no evidence of fraying on any electrical cords?	☐	☐
7. Are rubber cords kept free of grease, oil and chemicals?	☐	☐
8. Are metallic cable and conduit systems properly grounded?	☐	☐
9. Are portable electric tools and appliances grounded or double insulated?	☐	☐
10. Are all ground connections clean and tight?	☐	☐
11. Are fuses and circuit breakers the right type and size for the load on each circuit?	☐	☐
12. Are all fuses free of "jumping" with pennies or metal strips?	☐	☐
13. Do switches show evidence of overheating?	☐	☐
14. Are switches mounted in clean, tightly closed metal boxes?	☐	☐

Figure 6-8. (Continued).

	OK	ACTION NEEDED
15. Are all electrical switches marked to show their purpose?	☐	☐
16. Are motors clean and kept free of excessive grease and oil?	☐	☐
17. Are motors properly maintained and provided with adequate overcurrent protection?	☐	☐
18. Are bearings in good condition?	☐	☐
19. Are portable lights equipped with proper guards?	☐	☐
20. Are all lamps kept free of combustible material?	☐	☐
21. Is your electrical system checked periodically by someone competent in the NEC?	☐	☐

EXITS AND ACCESS

	OK	ACTION NEEDED
1. Are all exits visible and unobstructed?	☐	☐
2. Are all exits marked with a readily visible sign that is properly illuminated?	☐	☐
3. Are there sufficient exits to ensure prompt escape in case of emergency?	☐	☐
4. Are areas with limited occupancy posted and is access/egress controlled to persons specifically authorized to be in those areas?	☐	☐
5. Do you take special precautions to protect employees during construction and repair operations?	☐	☐

FIRE PROTECTION

	OK	NEEDED
1. Are portable fire extinguishers provided in adequate number and type?	☐	☐
2. Are fire extinguishers inspected monthly for general condition and operability and noted on the inspection tag?	☐	☐
3. Are fire extinguishers recharged regularly and properly noted on the inspection tag?	☐	☐
4. Are fire extinguishers mounted in readily accessible locations?	☐	☐

	OK	ACTION NEEDED
5. If you have interior standpipes and valves, are these inspected regularly?	☐	☐
6. If you have a fire alarm system, is it tested at least annually?	☐	☐
7. Are plant employees periodically instructed in the use of extinguishers and fire protection procedures?	☐	☐
8. If you have outside private fire hydrants, were they flushed within the last year and placed on a regular maintenance schedule?	☐	☐
9. Are fire doors and shutters in good operating condition?	☐	☐
Are they unobstructed and protected against obstruction?	☐	☐
10. Are fusible links in place?	☐	☐
11. Is your local fire department well acquainted with your plant, location and specific hazards?	☐	☐
12. Automatic Sprinklers:		
Are water control valves, air and water pressures checked weekly?	☐	☐
Are control valves locked open?	☐	☐
Is maintenance of the system assigned to responsible persons or a sprinkler contractor?	☐	☐
Are sprinkler heads protected by metal guards where exposed to mechanical damage?	☐	☐
Is proper minimum clearance maintained around sprinkler heads?	☐	☐

HOUSEKEEPING AND GENERAL WORK ENVIRONMENT

	OK	ACTION NEEDED
1. Is smoking permitted in designated "safe areas" only?	☐	☐
2. Are NO SMOKING signs prominently posted in areas containing combustibles and flammables?	☐	☐
3. Are covered metal waste cans used for oily and paint soaked waste?	☐	☐
Are they emptied at least daily?	☐	☐
4. Are paint spray booths, dip tanks, etc., and their exhaust ducts cleaned regularly?	☐	☐

Figure 6-8. (Continued).

	OK	ACTION NEEDED
5. Are stand mats, platforms or similar protection provided to protect employees from wet floors in wet processes?	☐	☐
6. Are waste receptacles provided, and are they emptied regularly?	☐	☐
7. Do your toilet facilities meet the requirements of applicable sanitary codes?	☐	☐
8. Are washing facilities provided?	☐	☐
9. Are all areas of your business adequately illuminated?	☐	☐
10. Are floor load capacities posted in second floors, lofts, storage areas, etc.?	☐	☐
11. Are floor openings provided with toe boards and railings or a floor hole cover?	☐	☐
12. Are stairways in good condition with standard railings provided for every flight having four or more risers?	☐	☐
13. Are portable wood ladders and metal ladders adequate for their purpose, in good condition and provided with secure footing?	☐	☐
14. If you have fixed ladders, are they adequate, and are they in good condition and equipped with side rails or cages or special safety climbing devices, if required?	☐	☐
15. For Loading Docks: Are dockplates kept in serviceable condition and secured to prevent slipping?	☐	☐
Do you have means to prevent car or truck movement when dockplates are in place?	☐	☐

MACHINES AND EQUIPMENT

	OK	ACTION NEEDED
1. Are all machines or operations that expose operators or other employees to rotating parts, pinch points, flying chips, particles or sparks adequately guarded?	☐	☐
2. Are mechanical power transmission belts and pinch points guarded?	☐	☐
3. Is exposed power shafting less than 7 feet from the floor guarded?	☐	☐
4. Are hand tools and other equipment regularly inspected for safe condition?	☐	☐

	OK	ACTION NEEDED
5. Is compressed air used for cleaning reduced to less than 30 psi?	☐	☐
6. Are power saws and similar equipment provided with safety guards?	☐	☐
7. Are grinding wheel tool rests set to within 1/8 inch or less of the wheel?	☐	☐
8. Is there any system for inspecting small hand tools for burred ends, cracked handles, etc.?	☐	☐
9. Are compressed gas cylinders examined regularly for obvious signs of defects, deep rusting or leakage?	☐	☐
10. Is care used in handling and storing cylinders and valves to prevent damage?	☐	☐
11. Are all air receivers periodically examined, including the safety valves?	☐	☐
12. Are safety valves tested regularly and frequently?	☐	☐
13. Is there sufficient clearance from stoves, furnaces, etc., for stock, woodwork, or other combustible materials?	☐	☐
14. Is there clearance of at least 4 feet in front of heating equipment involving open flames, such as gas radiant heaters, and fronts of firing doors of stoves, furnaces, etc.?	☐	☐
15. Are all oil and gas fired devices equipped with flame failure controls that will prevent flow of fuel if pilots or main burners are not working?	☐	☐
16. Is there at least a 2-inch clearance between chimney brickwork and all woodwork or other combustible materials?	☐	☐
17. For Welding or Flame Cutting Operations: Are only authorized, trained personnel permitted to use such equipment?	☐	☐
Have operators been given a copy of operating instructions and asked to follow them?	☐	☐
Are welding gas cylinders stored so they are not subjected to damage?	☐	☐
Are valve protection caps in place on all cylinders not connected for use?	☐	☐
Are all combustible materials near the operator covered with protective shields or otherwise protected?	☐	☐
Is a fire extinguisher provided at the welding site?	☐	☐
Do operators have the proper protective clothing and equipment?	☐	☐

Figure 6-8. (Continued).

Materials

	OK	ACTION NEEDED
1. Are approved safety cans or other acceptable containers used for handling and dispensing flammable liquids?	☐	☐
2. Are all flammable liquids that are kept inside buildings stored in proper storage containers or cabinets?	☐	☐
3. Do you meet OSHA standards for all spray painting or dip tank operations using combustible liquids?	☐	☐
4. Are oxidizing chemicals stored in areas separate from all organic material except shipping bags?	☐	☐
5. Do you have an enforced NO SMOKING rule in areas for storage and use of hazardous materials?	☐	☐
6. Are NO SMOKING signs posted where needed?	☐	☐
7. Is ventilation equipment provided for removal of air contaminants from operations such as production grinding, buffing, spray painting and/or vapor degreasing, and is it operating properly?	☐	☐
8. Are protective measures in effect for operations involved with X-rays or other radiation?	☐	☐
9. For Lift Truck Operations: Are only trained personnel allowed to operate forklift trucks?	☐	☐
Is overhead protection provided on high lift rider trucks?	☐	☐
10. For Toxic Materials: Are all materials used in your plant checked for toxic qualities?	☐	☐
Have appropriate control procedures such as ventilation systems, enclosed operations, safe handling practices, proper personal protective equipment (e.g., respirators, glasses or goggles, gloves, etc.) been instituted for toxic materials.	☐	☐

Employee Protection

	OK	ACTION NEEDED
1. Is there a hospital, clinic or infirmary for medical care near your business?	☐	☐
2. If medical and first-aid facilities are not nearby, do you have one or more employees trained in first aid?	☐	☐
3. Are your first-aid supplies adequate for the type of potential injuries in your workplace?	☐	☐
4. Are there quick water flush facilities available where employees are exposed to corrosive materials?	☐	☐
5. Are hard hats provided and worn where any danger of falling objects exists?	☐	☐
6. Are protective goggles or glasses provided and worn where there is any danger of flying particles or splashing of corrosive materials?	☐	☐
7. Are protective gloves, aprons, shields or other means provided for protection from sharp, hot or corrosive materials?	☐	☐
8. Are approved respirators provided for regular or emergency use where needed?	☐	☐
9. Is all protective equipment maintained in a sanitary condition and readily available for use?	☐	☐
10. Where special equipment is needed for electrical workers, is it available?	☐	☐
11. When lunches are eaten on the premises, are they eaten in areas where there is no exposure to toxic materials, and not in toilet facility areas?	☐	☐
12. Is protection against the effects of occupational noise exposure provided when the sound levels exceed those shown in Table G-16 of the OSHA noise standard?	☐	☐

Figure 6-8. (Continued).

Mechanical Inspection	
㉚ Koch Dry Ovens and Paint Booth Make Up Air Blowers	*Accr. No. 26023-61* *Ord. No. 759223* *Dept. No. 862*

Inspec's. Name: _____
Inspec. Date: _____

Inspect For: Security, Condition, Operation, Vibration, Belt Tension, Safety

To Be Inspected-
Jan-Mar-May-Jul-Sep-Nov

Dept. 862

Mach. No.	Bearings	Belts and Pulleys	Belt Guards	Lube & Lube Lines O.K.	Inspector's Comments
	Sub Assembly Paint System on Mezzanine - North to South				
7227					
7226					
7186					
7187					
7188					
7225					
7189					
	Paint Booth Make Up Air Bldg. - "V" Roof - East Side				
6790					
6791					
7219					
7221					
7220					
7218					
	Work in Process Paint System on Mezzanine - North to South				
6760					
6758					
6795					
6796					

Mach. No.	Bearings	Belts and Pulleys	Belt Guards	Lube & Lube Lines O.K.	Inspector's Comments
	Combine Finish Paint Syst. - Dry Ovens and Paint Booth Make Up Air Units Bldg. - "V1" South to North				
7251					
7252					
7259					
7260					
7261					
7262					
7297					
7298					
7300					
	Touch Up Paint Dry Oven (3-Ovens) Bldg. - "V2" North East Corner				
6390					

Figure 6-8. (Continued).

MECHANICAL INSPECTION (2) Inspector's Name _____ Inspection Date _____

EXHAUST BLOWERS & AIR CONDI. HEAT EXCHANGER & DOOR SEAL BLOWERS

3-Month Inspection
Feb-May-Aug-Nov

Acct. No. 26023-79
ORD. No. 759223
DEPT. NO. -862

Blower Machine Number	Check for Excessive Vibration	Chk. Belts and Pulleys for Ten. & Align. & Condi.	All Safety Shields are in Place and Secure	Blower Machine Number	Check for Excessive Vibration	Chk. Belts & Pulleys for Ten. & Align. & Condi.	All Safety Shields are in Place and Secure	Blower Machine Number	Check for Excessive Vibration	Same as Other	Same
"V" Bldg. Roof E. Side				W.I.P. Paint Dip				Wash Cool Down			
7426-Air Condi.Heat Exchanger				5765				7255			
7429-Air Condi.Heat Exchanger				6767				7256			
				6766				7257			
Sub.Assemb.Paint & Wash Ovens Mezzanine - North to South				6764				Combine Paint			
				W.I.P. Paint Spray				7268			
7228				6784				7269			
7178				6785				7270			
7179				6786				7271			
7181				6787				7272			
7180				6763				7263			
7183				W.I.P. Wash Dry Off				7264			
7182				6761				7265			
7185				6760				7266			
7224				6759				7267			
7223				6757				Combine Flash Off			
7187				W.I.P. Paint Dry				7292			
7190				6793				7293			
Flow Coat Booth				6794				Combine Paint Dry			
7192				6798				7295			
7193				6797				7296			
7194				W.I.P. Wash Booth				7299			
7195				6755				7301			
7196				6754				Combine Cool Down			
7197				6753				7303			
Spray Paint Booth				Comb. Finish Paint				7304			
7205				7241				7305			
7206				7242				7310			
7208				7243				7312			
7207				7249				7306			
7212				Wash Dry Off				7311			
7213				7250				Spray & Dry - N.E. Corner-"V2"			
7215				7253							
7214											

Figure 6-8. (Continued).

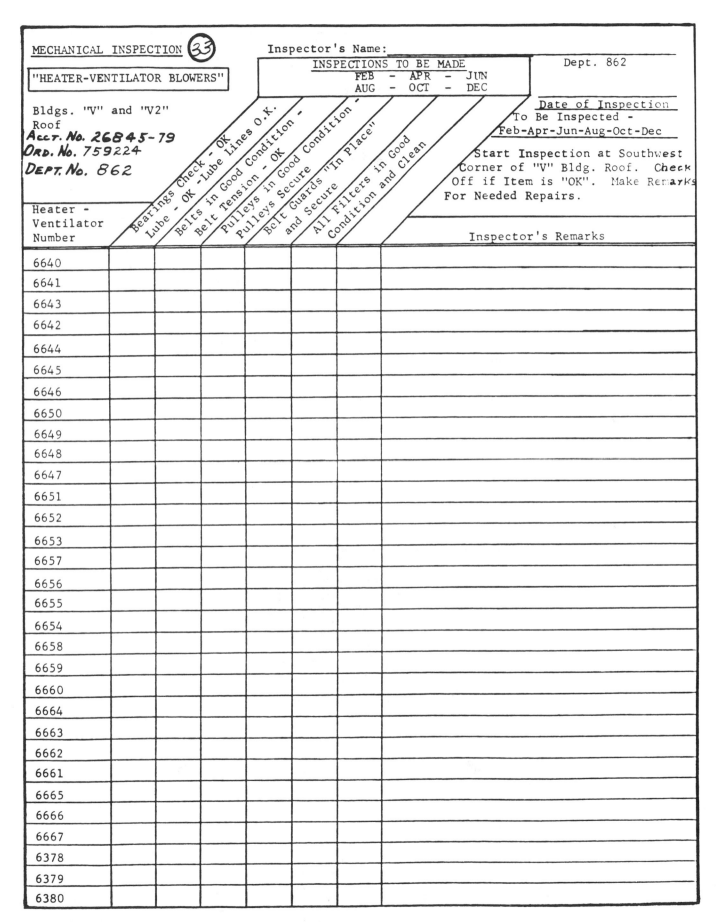

Heater - Ventilator Number	Bearings Check - OK Lube - OK - Lube Lines O.K.	Belts in Good Condition - Belt Tension - OK	Pulleys in Good Condition - Pulleys Secure	Belt Guards "In Place" and Secure	All Filters in Good Condition and Clean	Inspector's Remarks
6640						
6641						
6643						
6642						
6644						
6645						
6646						
6650						
6649						
6648						
6647						
6651						
6652						
6653						
6657						
6656						
6655						
6654						
6658						
6659						
6660						
6664						
6663						
6662						
6661						
6665						
6666						
6667						
6378						
6379						
6380						

Figure 6-8. (Continued).

PITTSBURGH ROLLS CORPORATION
CRANE INSPECTION REPORT

Crane No.................................Type.................................Capacity.................................

RUNWAY AND CONDUCTORS

Track Alignment.................................Spread.................................Fastenings.................................

Line Conductors.................................Conductor Supports.................................

TRUCKS AND MAIN COLLECTORS

Truck Wheels, Flat Spots?.................................Flanges.................................End Play.................................

Axle Bearings.................................Lubrication.................................

Truck Drive Bearings.................................Lubrication.................................Gears.................................

Gear Screws.................................Pinion.................................Key.................................Collectors.................................

GIRDERS AND DRIVE

Drive Shaft.................Couplings.................Bearings.................Lubrication.................

Foot Brake Shaft.................Couplings.................Bearings.................Lubrication.................

Bridge Brake Case.................Adjustment.................Lubrication.................

Walkway.................Railing.................Ladder.................

Bridge Motor Support.................Shaft Extension.................Couplings.................

Bridge Drive Gear Case.................Gears.................Lubrication.................

MOTORS

Location	Armature	Commutator	Brushes	Brush Holders	Bearings	Lubrication
Bridge						
Hoist						
Aux. Hoist						
Trolley						

CONTROLLERS

Location	Brushes	Brush Holders	Contacts	Wiring	Springs	Resistance
Bridge						
Hoist						
Aux. Hoist						
Trolley						

Trolley Wheels.................Trolley Wheel Bearings.................Lubrication.................

Trolley Gear Case.................Case Support.................Gears.................Lubrication.................

Hoist Gear Case.................Main Gear Train.................Comp. Train.................Lubrication.................

Mech. Brake.................Drift.................Elec. Brake.................Adjustment.................Lubrication.................

Drum.................Cable or Chain.................Cable Pin.................Limit Switch.................

Limit Switch Adjustment.................Hook.................Sheaves.................Lubrication.................

Trolley Conductors.................Trolley Collectors.................Cage Roof.................Door.................

Windows.................Foot Brake Treadle.................Cont. Levers.................Load Test.................

Bell or Signal.................

Inspected by.................Date.................

KEY WORDS—G=Good; F=Fair; W=Worn; A=Need Attention;
C=Need Cleaning; T=Too Tight

Figure 6-8. (Continued).

STATIONARY SCAFFOLD SAFETY CHECK LIST

PROJECT: _____

ADDRESS: _____

CONTRACTOR: _____

DATE OF INSPECTION: _____ INSPECTOR: _____

	Yes	No	Action/Comments
1. Are scaffold components and planking in safe condition for use and is plank graded for scaffold use?			
2. Is the frame spacing and sill size capable of carrying intended loadings?			
3. Have competent persons been in charge of erection?			
4. Are sills properly placed and adequate size?			
5. Have screw jacks been used to level and plumb scaffold instead of unstable objects such as concrete blocks, loose bricks, etc.?			
6. Are base plates and/or screw jacks in firm contact with sills and frame?			
7. Is scaffold level and plumb?			
8. Are all scaffold legs braced with braces properly attached?			
9. Is guard railing in place on all open sides and ends above 10' (4' in height if less than 45")?			
10. Has proper access been provided?			
11. Has overhead protection or wire screening been provided where necessary?			
12. Has scaffold been tied to structure at least every 30' in length and 26' in height?			
13. Have free standing towers been guyed or tied every 26' in height?			
14. Have brackets and accessories been properly placed: Brackets?			
Putlogs?			
Tube and Clamp?			
All nuts and bolts tightened?			
15. Is scaffold free of makeshift devices or ladders to increase height?			
16. Are working level platforms fully planked between guard rails?			
17. Does plank have minimum 12" overlap and extend 6" beyond supports?			
18. Are toeboards installed properly?			
19. Have hazardous conditions been provided for: Power lines?			
Wind loading?			
Possible washout of footings?			
Uplift and overturning moments due to placement of brackets, putlogs or other causes?			
20. HAVE PERSONNEL BEEN INSTRUCTED IN THE SAFE USE OF THE EQUIPMENT?			

Figure 6-8. (Continued).

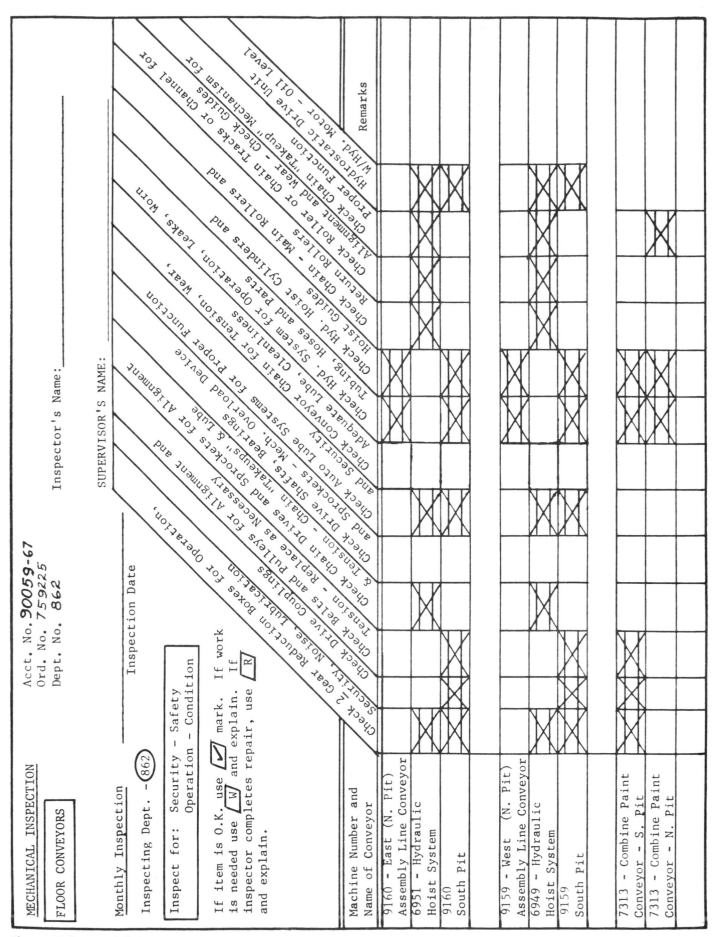

Figure 6-8. (Concluded).

DEPARTMENT *Maintenance*	UNIT *Workshop*	SUPERVISOR RESPONSIBLE *J. P. Smith*	APPROVED BY *Ralph T. Welles*	DATE *4/16/72*	PAGE NO. *1*
1. PROBLEMS	2. CRITICAL FACTORS	3. CONDITIONS TO OBSERVE	4. FREQUENCY	5. RESPONSIBILITY	
1. Overhead hoist	*Cables, chains, hooks, pulleys*	*Frayed or deformed cables, worn or broken hooks and chains, damaged pulleys*	*Daily—before each shift*	*Operators*	
2. Hydraulic pump	*High pressure hose*	*Leaks; broken or loose fittings*	*Daily*	*Shift leader*	
3. Power generator	*High voltage lines*	*Frayed or broken insulation*	*Weekly*	*Foreman*	
4. Fire extinguishers	*Contents, location, charge*	*Correct type, fully charged, properly located, corrosion, leaks*	*Monthly*	*Area safety inspector*	
5. General housekeeping	*Passageways, aisles, floors, grounds*	*Free of obstructions, clearly marked, free of refuse*	*Daily*	*Shift leader foreman*	

Figure 6-9. A hazard control inspection inventory should list the person responsible for each inspection. (Printed from *Principles and Practices of Occupational Safety and Health, Student Manual,* Booklet Three, U. S. Department of Labor, OSHA 2215.)

these inspections will benefit from training in hazard recognition. Some items will need to be inspected by more than one person. For example, while an area supervisor may inspect an overhead crane weekly and maintenance personnel inspect it monthly, the operator of the crane will inspect it before each use. When grinding wheels are received, they are inspected by the stockroom attendant, but they must be inspected again by the operator before each use.

As part of the hazard control inspection inventory, management should assign responsibility for each inspection. Figure 6-9 shows how the inventory can designate the proper person by title: area supervisor, operator, foreman, maintenance foreman, and so forth.

A suggested guide for planned inspections is as follows:

- Daily—area supervisor and maintenance personnel, who also can request suggestions from employees in their various workstations
- Weekly—department heads
- Monthly—supervisors, department heads, the safety department, and safety and health committees.

The safety department also may be actively involved in monthly, quarterly, semiannual, and annual inspections. Five qualifications of a good inspector are:

1. knowledge of the organization's accident experience
2. familiarity with accident potentials and with the standards that apply to his or her area
3. ability to make intelligent decisions for corrective action
4. diplomacy in handling personnel and situations
5. knowledge of the organization's operations—its workflow, systems, and products.

Safety Professionals

Clearly, the safety professional should spearhead the inspection activity. During both individual and group inspections, the professional can educate others in inspection techniques and hazard identification by using on-the-spot examples and firsthand contact. Supervisors, foremen, stewards, and safety and health committees can be shown what to look for when making inspections. The organization's fire protection representative or industrial hygienist usually works with the hazard control specialist in conducting inspections.

The number of safety professionals depends on the size of the company and the nature of its operation. Large companies with well-organized accident prevention programs usually employ a full-time staff. Sometimes large companies also have designated employees who spend part of their time on inspections.

In organizations where toxic and corrosive substances are present, the industrial hygienist will be part of the inspection team (see also the following section on Measurement and Testing). When an organization uses chemicals, the chief chemist will need to cooperate closely with the safety professional and fire protection representative in establishing inspection criteria. If the organization has no industrial hygienist, the safety professional needs to obtain training about the hazardous properties of substances, unstable properties of chemicals, and methods of control. An inspection conducted without this knowledge is incomplete and may miss potentially serious problems.

Company or Facility Management

Safety inspections should be considered part of the duties of company or facility management. By participating in inspections, management demonstrates its commitment to maintain a safe working environment. But the psychological effect of inspection by senior executives goes beyond merely showing an interest in safety. When employees know that management is coming to inspect their area, conditions that seemed "good enough" suddenly appear unsatisfactory and are quickly corrected.

First-line Supervisor or Foreman

Because supervisors and foremen spend practically all their time in the shop or facility, they are continually monitoring the workplace. At least once a day, supervisors need to check their areas to see that (1) employees are complying with safety regulations, (2) guards and warning signs are in place, (3) tools and machinery are in a safe condition, (4) aisles and passageways are clear and proper clearances maintained, and (5) material in process is properly stacked or stored. Although such a spot check does not take the place of more detailed inspections, it emphasizes the supervisor's commitment to maintaining safety in the area. A supervisor also should conduct regular formal inspections to make certain all hazards have been detected and safeguards are in use. Supervisors can perform such inspections weekly on their own and monthly as part of a safety and health committee.

Mechanical Engineer and Maintenance Superintendent

Either as individuals or as members of a committee, the mechanical engineer and the maintenance superintendent also need to conduct regular formal inspections. They can write necessary work orders on the spot for guards or for correcting faulty equipment.

Employees

As mentioned previously, employee participation in continuous inspection is one goal of an effective hazard control program. Before beginning the workday, the employee should inspect the workplace and any tools, equipment, and machinery that will be used. Any defects the employee is not authorized to correct should be reported immediately to the supervisor. Action resulting from this report must be reported to the employee to encourage further participation.

Maintenance Personnel

Maintenance employees can be of great help in locating and correcting hazards. As they work, they can conduct informal inspections and report hazards to the supervisor, who in turn should encourage the mechanics to offer suggestions.

Joint Safety and Health Committees

Joint safety and health committees (discussed in Chapter 3, Loss Control Programs, of this volume) conduct inspections as part of their function. They give equal consideration to accident, fire, and health exposures. By periodically visiting areas, members may notice changed conditions more readily than someone who is there every day. Another advantage provided by the committee is the members' various backgrounds, experience, and knowledge represented.

If the committee is large, the territory should be divided among teams of manageable size. Large groups going through the facility are unwieldy and distracting.

Other Inspection Teams

If there is no safety and health committee, a planned, formal inspection is still necessary. Management should assign an inspection team that includes the hazard control specialist, production manager, supervisor, employee representative, fire prevention specialist, and industrial hygienist. The important point is that inspections should be directed by a responsible executive who has the authority to ensure the work is carried out effectively.

Outside inspectors sometimes are needed to perform inspections. For example, insurance company safety engineers and local, state or provincial, and federal inspectors may lend their expertise to specific inspections.

Contractors' Inspection Services

For some technical systems, notably sprinkler systems, contracting companies furnish inspection services. Companies without either qualified safety professionals or a well-established maintenance program can avail themselves of such services.

For example, a sprinkler contractor may arrange with a customer for periodic inspection and tests of sprinkler equipment. The contractor and the client negotiate how often inspections are to be done. In some cases, the inspection will include other items, such as fire extinguishers, hoses, or fire doors. The contractor furnishes a comprehensive written report. The client can request that the contractor send copies of the report to the insurer.

The basic contract does not include maintenance work or materials required for alterations, repairs, or replacement. However, if the report indicates any maintenance needs, the client can have the contractor perform the work. Contract service does not relieve management of its primary responsibility for inspection and maintenance. Nevertheless, it does provide excellent inspection for small companies, buildings with mixed tenants, and companies with systems too complex for inspection by their own maintenance staffs.

CONDUCTING INSPECTIONS

Companies must not only conduct regular inspections but carry them out in a way that emphasizes the company's commitment to safety. Employees should feel part of the team rather than the target of management scrutiny. This section covers how to conduct inspections to maximize their potential for reducing accidents and motivating employees to support a company's safety goals.

Preparing to Inspect

Inspections should be scheduled for times when inspectors will have the best opportunity to see operations and work practices without much interruption. The inspection route should be planned in advance.

Before conducting an inspection, the inspector or inspection team should review all accidents that have occurred in the area. At this brief meeting, team members should discuss where they are going and what they will be looking for. During the inspection, before going into noisy areas, the team will need to discuss what they wish to accomplish in order to avoid arm waving, shouting, and other unsatisfactory methods of communication.

In addition to the regular checklist and accident reports, inspectors should have copies of the previous inspection report for that particular area. Reviewing this report makes it possible to check whether earlier recommendations have been followed and reported hazards corrected.

Those making inspections should wear the protective equipment required in the areas they enter: safety glasses and shoes, hard hats, acid-proof goggles, protective gloves, respirators, and so forth. If inspectors do not have or cannot get special protective equipment, they should not go into the area. They must be careful to "practice what they preach."

Inspectors also should be aware of any special hazards they encounter. For example, because welding crews and other maintenance crews move from place to place, they may be encountered anywhere in the facility. Inspectors should know what precautions are required where these crews are working.

Inspection Tools

Inspectors should have the proper tools ready before the inspection to make the process more efficient and to gather more precise data. Common tools include:

- clipboards
- inspection forms
- pens/pencils
- lockout/tagout supplies
- measuring tape/ruler
- flashlight.

Depending on the inspection area or type, the following equipment may also be useful:

- cameras
- tape recorder
- electrical testing equipment
- sampling devices (air, noise, light, temperature)
- sample containers
- calipers, micrometers, feeler gauges
- special personal protection equipment (see Measurement and Testing section)
- stopwatch.

Relationship of Inspector and Supervisor

Before inspecting a particular department or area, the inspector should contact the department head, supervisor, or other person in charge. This person may have important information for the inspection, particularly if conditions are temporarily altered because of construction, maintenance, equipment downtime, employee absence, and so forth.

If no rules prohibit it, the person in charge may want to accompany the inspector. The inspector can agree but should also emphasize that no tour guide is needed. The inspector must preserve independence and the opportunity to make uninfluenced observations.

If the supervisor of the area does not accompany the inspector, the supervisor should be consulted before the inspector leaves the area. The inspector should discuss each recommendation on particular hazards or unsafe conditions with the supervisor. Usually they can reach an agreement regarding the relative importance of each recommendation. Obviously, an inspector should not focus on numerous trivial items merely to make the report look complete. On the other hand, the inspector does not have the authority to overlook any condition that might cause an accident.

Even minor items that the supervisor can correct quickly should be reported. The inspector can note on the written report that the supervisor promises to correct the particular condition. This keeps the record clear and serves as a reminder to check the condition during the next inspection.

An inspector should not fail to report hazards merely because a supervisor regards such reporting as criticism. If a supervisor becomes defensive or resentful, the inspector can only repeat what the supervisor knows: the purpose of an inspection is fact-finding, not fault-finding. By retaining objectivity and refusing to let the issue of safety degenerate into a personality conflict, the inspector keeps matters on a proper professional footing and maintains a firm, friendly, and fair attitude.

Sometimes a supervisor will request the inspector's assistance in recommending new equipment, reassignment of space, or transfer of certain jobs from one department to another. When these suggestions deal with safety, the inspector will want to include them in notes and consider whether to make them part of the report. The inspector must be careful, however, not to promise either a supervisor or an employee more than actually can be delivered.

Relationship of Inspector and Employee

Unless company policy or departmental rules prohibit conversation with employees, the inspector can ask questions about operations, taking care, however, not to usurp the responsibility of the supervisor. If, for example, a member of a safety and health committee sees an employee who deviates from established safe work practices, it is better to ask the supervisor rather than the employee about the supposed infraction. The committee member may not fully understand the operation and may be incorrect in the assumption.

In another case, the employee may be performing a risky practice that has been sanctioned by those in authority. The employee could become defensive when questioned by the inspector. It is the supervisor's job to require compliance with company safe work procedures; it is the inspector's job to do the inspecting and reporting. If, however, the situation appears to present an immediate danger, the employee and management should be notified immediately.

Regulatory Compliance Inspection, Compliance Officer

Each regulatory agency has specific procedures and rules for inspecting an organization's facilities. The safety and health professional needs to know the specific procedures to be observed when an inspector calls. In general, an inspector will:

- call at regular business hours
- present credentials
- identify the purpose for the inspection, its length and scope
- request records and other documents as may be required
- perhaps consult with employees on the inspection
- perhaps request management to remove employees from an area of imminent danger and to take immediate action to correct the danger.

Chapter 3, Loss Control Programs, in this volume differentiated between deviations from accepted practices and workplace-induced human error. The inspection team needs to look for both. The inspector is not concerned with identifying the person who is responsible for the unsafe behavior (fault-finding). The goal is to identify the behavior (fact-finding) and see that it is corrected.

Unsafe behaviors will vary from one area to another. Among common items that might be noted are the following:

- using machinery or tools without authority
- operating motorized vehicles at unsafe speeds or in other violation of safe work practice
- removing guards or other safety devices or rendering them ineffective
- using defective tools or equipment or using tools or equipment incorrectly
- using hands or body instead of tools or push sticks
- overloading, crowding, or failing to balance materials or handling materials incorrectly, including improper lifting
- repairing or adjusting equipment that is in motion, under pressure, or electrically charged
- failing to use or maintain (or using improperly) personal protective equipment or safety devices
- creating unsafe, unsanitary, or unhealthy conditions by improper personal hygiene, using compressed air for cleaning clothes, poor housekeeping, or smoking in unauthorized areas
- standing or working under suspended loads, scaffolds, shafts, or open hatches.

Because the inspector's purpose is to locate unsafe practices, not pinpoint blame, the report should not specify any names. When the report states, "An employee in this area was observed," the supervisor has been advised of the need to enforce safe work practices. The inspector should not be seen as a police officer handing out tickets or, worse, as a snoop from "outside." Nor should information derived from inspections be used for disciplinary measures.

Sometimes it is necessary to closely observe workers on the job to understand their tasks. The inspector should explain to workers why he or she needs to observe the task and should always ask permission to watch. When employees understand that no one is trying to catch them in an error but rather that they have been chosen to demonstrate a task because of their exceptional skills, they probably will agree to being observed.

Recording Hazards

Inspectors should locate and describe each hazard found during inspection. A clear description of the hazard should be written down with questions and details recorded for later use. It is important to determine which hazards present the most serious threat and are most likely to occur. The hazard-ranking scheme described in Chapter 3 will simplify the job of classifying hazards.

Properly classifying hazards places them in the right perspective. This approach enables the inspector to briefly describe potential consequences and the probability of such consequences occurring. Inspection reports should enable management to quickly understand and evaluate problems, assign priorities, and make decisions.

On the other hand, the inspector should describe unsafe conditions or deviations from accepted practices in detail and identify machines and operations by their correct names. Also, the inspector must accurately name or number locations and identify in detail the specific hazards within them. Instead of noting "poor housekeeping," for example, the report should give the details: "Empty pallets left in aisles, slippery spots on the floor from oil leaks, a ladder lying across empty boxes, scrap piled on the floor around machines." Instead of noting "guard missing," the report should read, "Guard missing on shear blade of No. 3 machine, SW corner of Bldg. D."

Management must adopt some plan to note intermediate or permanent corrective measures. For example, if intermediate safety measures have been taken, the item could be circled. When permanent measures are taken, the item can be crossed out or marked with an X. Such a system identifies those items requiring further corrective action (Figure 6-10).

INSPECTION REPORT

Area Inspected _____ Building D _____ Date and Time of Inspection _____ —11:00 a.m. _____

Inspector and Title _____ Ron Baker, Hazard Control Specialist _____ Date of Report _____

Names of Those to Whom Report Is Sent: _____ Bob Firenze (Executive Director); Loren Hall (Department Head); file _____

No. of Items Carried Over from Previous Report _____ 3 _____   No. of Items Added to This Report _____ 4 _____   Total No. of Items on This Report _____ 7 _____

Item (asterisk indicates old item)	Hazard Classification		Hazard Description	Specific Location	Supervisor	Corrective Action Recommended	Corrective Action Taken
	Consequence	Probability					
*1	II	B	Guard missing on shear blade #2 machine. Work order issued to engineering for new guard 10/16/80. Wooden barrier guard in temporary use 10/23/79. Guard still missing.	S.W. corner, bay #1	Jay Rillo	Contact engineering to replace guard	Engineering says they will have guard by 11/24.
*2	IV	C	Window cracked. Work order issued for replacement 10/30/80.	South wall, bay #3	Joe Whitestone	Have maintenance replace window	Maintenance to replace all broken windows starting next week.
3	II	B	Oil and trash still accumulated under main motor. Was to be cleaned by 10/30/80.	Pump room	Tony Silva	Clean area; have supervisor talk to men	Cleaned out 11/21. Silva told men to keep area clear.
4	III	B	Mirror at pedestrian walk out of line	North end of machine shop	Tom Schroeder	Post temporary warning sign; call maintenance for adjustment	Sign posted 11/21—Butler has scheduled adjustment for 12/1
5	II	A	Three workers at cleaning tank not wearing eye protection	Electric shop	Hank Beine	Have supervisor give more training and education	Discussed with Beine—he held meeting on 11/25
6	I	A	Cable on jib crane badly frayed	Bay #3	Joe Whitestone	IMMEDIATE ACTION REQUIRED	Tagged crane out of Service Cable to be replaced 11/21
7	II	B	Guard rail damaged on stairway to second floor	Bay #1	Jay Rillo	Issue work order to carpenter shop to make replacement	Work issue order 11/21

Figure 6-10. This inspection report form simplifies procedures and emphasizes carryovers, new items, and responsibilities. The column at right is for noting corrective action taken later. (Printed with permission from RJF Associates, Inc. Bloomington, IN.)

If a committee is performing the inspection, one member assumes the task of keeping notes. Without such notes it is almost impossible to write a satisfactory inspection report.

Condemning Equipment

When a piece of equipment presents an imminent danger, the inspector should notify the supervisor immediately and see that the machine or equipment is shut down, tagged, and locked out to prevent its further use. See Chapter 6, Safeguarding, in the *Engineering & Technology* volume for lockout/tagout procedures.

When danger tags and locks are used, those persons authorized to condemn equipment must sign them. Only the inspector who places the tag should be permitted to remove it. This step should occur only when the inspector is satisfied that the hazardous condition has been corrected.

For more on lockouts, see Chapters 6, Safeguarding, and 10, Electrical Equipment, in the *Engineering & Technology* volume of this manual. No equipment or materials should be placed out of service without notifying the person in authority in the department affected.

Writing the Inspection Report

Every inspection must be documented in a clearly written report furnished by the inspector. Without a complete and accurate report, the inspection would be little more than an interesting sightseeing tour. Inspection reports are usually of three types:

1. Emergency—made without delay when a critical or catastrophic hazard is probable. Using the classification system described in Chapter 3, Loss Control Programs, this category would include any items marked IA or IIA.
2. Periodic—covers those unsatisfactory nonemergency conditions observed during the planned periodic inspection. This report should be made within 24 hours of the inspection. Periodic reports can be initial, follow-up, final, or a combination of all three.

3. Summary—lists all items of previous periodic reports for a given time.

The written report should include the name of the department or area inspected (giving the boundaries or location if needed), date and time of inspection, names and titles of those performing the inspection, date of the report, and the names of those to whom the report was made.

One way to make the report is to begin by copying items carried over from the last report that were not corrected. Each item is numbered consecutively. The item number can be followed by the hazard classification (IB, IIIC, and so on). Carryover items can be marked with an asterisk. The narrative should include the date the hazard was first detected. The inspector should describe each hazard and pinpoint its location. After the hazard is listed, the inspector should recommend corrective action and establish a definite abatement date. There should follow a space for noting corrective action taken later. In addition, a report should show what is right in a work area as well as what is wrong. When the report is from a committee, it should be reviewed by each member of the inspection team for accuracy, clarity, and thoroughness.

Generally, inspection reports are sent to the head of the department or area where the inspection was made. Copies are also given to executive management and the manager to whom the department head reports.

Follow-Up for Corrective Action

After the inspection report has been written and disseminated, the inspection process starts to return benefits. The information acquired and the recommendations made are valueless unless management takes corrective action. Information and recommendations provide the basis for establishing priorities and implementing programs that will reduce accidents, improve conditions, raise morale, and increase the efficiency and effectiveness of the operation.

Inspectors can list recommendations in the order in which the hazards were discovered or group them according to the individuals responsible for their correction. Recommendations are then sent to the proper member of management for approval. Where possible, management should set a definite time limit for correction for each recommendation and follow up to make sure corrective action has been taken.

Often the safety professional is authorized to make recommendations directly to the affected foreman, supervisor, or department if such recommendations do not require major capital outlays. Some organizations require that inspection reports be reviewed by the safety and health committee. This is particularly the case when recommendations apply to education and training and directly affect employees.

In making recommendations, inspectors should be guided by four rules (*Facility Inspection,* 1973):

1. Correct the cause whenever possible. Do not merely correct the result, leaving the problem intact. In other words, be sure the disease and not just the symptom is cured. If the inspector or supervisor does not have the authority to correct the real cause, they should bring it to the attention of the person who does.
2. Immediately correct everything possible. If the inspector has been granted the authority and opportunity to take direct corrective action, he or she should take it. Delays risk accidents.
3. Report conditions beyond one's authority and suggest solutions. Inform management of the condition, the potential consequences of hazards found, and solutions for correction. Even when nothing seems to come of a recommendation, it can pay unexpected dividends.

    For example, a company safety and health committee made a detailed proposal about guarding a particularly hazardous location, only to be told that the engineers had planned to move operations to another location. However, instead of feeling that it had wasted its time, the committee pointed out that the organization had serious communication problems, with the right hand not knowing what the left was doing. The committee recommended that effective management techniques be applied to the hazard control program.
4. Take intermediate action as needed. When permanent correction takes time, the hazard should not be ignored. Inspectors or supervisors should take any temporary measures they can, such as roping off the area, locking and tagging out equipment or machines, or posting warning signs. These measures may not be ideal, but they are better than doing nothing.

Some of the general categories into which recommendations might fall are setting up a better process, relocating a process, redesigning a tool or fixture, changing the operator's work pattern, providing personal protective equipment, and improving personnel training methods. Recommendations can also call for improvements in the preventive maintenance system and in housekeeping. Cleaning up debris and dirt may be considered the janitor's job, but preventing its accumulation is part of an effective hazard control program.

Management must realize that employees are keenly interested in the attention paid to correcting faulty conditions and hazardous procedures. Recommendations approved and supported by management should become part of the organization's philosophy and program. At regular intervals, supervisors should report progress in complying with the recommendations to the safety department, the company safety and health committee, or the person designated by management to receive such information. Inspectors should periodically check to see what progress toward corrective action is being made. Unsafe conditions left uncorrected indicate a breakdown in management communications and program application.

Sometimes management will have to decide among several courses of action. Often these decisions will be based on cost effectiveness. For example, it may be cost-effective as well as practical to substitute a less toxic material that works as well as the highly toxic substance presently in use. On the other hand, replacing a costly but hazardous machine may have to wait until funds can be designated. In this case, the immediate alternative may be to install machine guards. In all cases, action taken or proposed must be communicated to all persons involved.

MEASUREMENT AND TESTING

Two special sorts of inspection are conducted by the industrial hygienist and the medical staff. Testing for exposures to health hazards requires special equipment not always available to the hazard control specialist. In such cases, management can often obtain assistance from the state labor department industrial hygiene division or from the provincial department of labor and health. Another source of help can be the industrial hygienists employed by consulting firms and by insurance companies. Conducting physical examinations of employees exposed to occupational health hazards may require medical equipment that the organization's medical staff does not have. The following discussion is a summary of how to recognize, evaluate, and control health hazards in the workplace. Those interested in more details should consult the text *Fundamentals of Industrial Hygiene*, 4th edition (see References). This reference should help safety professionals understand their roles in this area of hazard control.

Kinds of Measurement and Testing

Occupational health surveillance monitors chemical, physical, biological, and ergonomic hazards. Four monitoring systems are used: personal, environmental, biological, and medical.

Personal Monitoring

One example of personal monitoring is measuring the airborne concentrations of contaminants. The measurement device is placed as closely as possible to the site at which the contaminant enters the human body. When the contaminant is noise, the device is placed close to the ear. When a toxic substance could be inhaled, the device is placed in the breathing zone.

Environmental Monitoring

Environmental monitoring measures contaminant concentrations in the workroom. The measurement device is placed in the general area adjacent to the worker's usual workstation or where it can sample the general room air.

Biological Monitoring

Biological monitoring measures changes in composition of body fluid, tissues, or expired air to detect the level of contaminant absorption. For example, blood or urine can be tested to help determine lead exposures. Similarly, the phenol in urine sometimes is measured to document benzene exposure.

Medical Monitoring

When medical personnel examine workers to see their physiological and psychological response to a contaminant, the process is termed medical monitoring. Medical monitoring can include health and work histories, physical examinations, x-rays, blood and urine tests, pulmonary function tests, and vision and hearing tests. The aim of such monitoring is to find evidence of exposure early enough to identify especially susceptible workers and to detect any damage before it becomes irreversible.

Biological and medical monitoring provide information after the exposure already has occurred. However, such programs also encompass arrangements to treat an identified health problem and to take corrective action to prevent further damage. To understand how industrial hygienists measure for health hazards, it is necessary to define some basic terms, to distinguish between acute and chronic effects, and to see how safe exposure levels are established.

Measuring for Toxicity

Measuring for toxicity involves several activities. These include determining a substance's capacity for injury or harm, inhalation hazards, influence of solubility, threshold limit values, and permissible exposure limits.

Toxicity

The toxicity of a material is not identical with its potential for being a health hazard. Toxicity is the capacity of a material to produce injury or harm. Hazard is the possibility that exposure to a material will cause injury or illness when a specific quantity is used under certain conditions. The key elements to be considered when evaluating a health hazard are:

- amount of material to which the employee is exposed
- total time of the exposure
- toxicity of the substance
- individual susceptibility.

Not all toxic materials are hazardous. The majority of toxic chemicals are safe when packaged in their original shipping containers or contained within a closed system. As long as toxic materials are adequately controlled, they can be safely used. For example, many solvents, if not properly handled, will cause irritation to eyes, mouth, and throat. Some also are intoxicating and can cause blistering of the skin and other forms of dermatitis. Prolonged exposure can cause more serious illness. But if workers use the solvents in a well-ventilated area and are given proper protective equipment to prevent the solvents from contacting skin, the substances can be used safely.

The toxic action of a substance can be divided into acute and chronic effects.

- Acute effects involve short-term (hours or days) high concentrations that can cause irritation, illness, or death. They can be the result of sudden and severe exposure, during which the substance is rapidly absorbed. Acute effects can be related to an accident, which disrupts ordinary processes and controls. For example, sudden exposure to very high concentrations of methane gas in a confined space can lead to loss of consciousness, coma, or death by asphyxiation.

- Chronic effects involve continued exposure to a toxic substance over an extended time (usually over 90 days). When the chemical is absorbed more rapidly than the body can eliminate it, the chemical begins to accumulate in the body. If the level of contaminant is relatively low, the effects, even if they are serious and irreversible, may go unnoticed for long periods because of the phenomenon known as the latency period. Latency refers to an extended time period, usually in years, between exposure and observed health effect.

Inhalation Hazards

Inhalation of harmful materials may irritate the upper respiratory tract and lung tissue, or the terminal passages of the lungs and the air sacs, depending upon the solubility of the material. Inhalation of biologically inert gases may dilute oxygen levels below the normal blood saturation value and disturb cellular processes. Other gases and vapors may prevent the blood from carrying oxygen to the tissues or interfere with its transfer from the blood to the tissue, producing chemical asphyxia.

Inhaled contaminants that adversely affect the lungs fall into three general categories: aerosols, toxic gases, and gases that produce systemic effects. Each is discussed later in the chapter in greater detail.

- Aerosols (particulates) are substances that, when deposited in the lungs, may produce either rapid local tissue damage, some slower tissue reactions, eventual disease, or only physical plugging.
- Toxic vapors and gases are hazards that produce adverse reactions in the tissue of the lungs themselves.
- Some toxic aerosols or gases do not affect the lung tissue locally but (1) are passed from the lungs into the bloodstream, where they are carried to other body organs, or (2) have adverse effects on the oxygen-carrying capacity of the blood cells themselves.

An example of the first type (aerosols) is particulates that can cause, over time, a variety of lung reactions, including production of scarring and progressive decrease in lung function.

An example of the second type (toxic gases) is hydrogen fluoride, a gas that directly affects lung tissue. It is a primary irritant of mucous membranes and causes chemical burns. Inhalation of this gas will cause pulmonary edema: after lung tissue is burned, the lungs fill with fluids that directly interfere with the gas-transfer function of the alveolar lining.

An example of the third type is carbon monoxide, a toxic gas that passes into the bloodstream without essentially harming the lung. The carbon monoxide passes through the alveolar walls into the blood, where it preferentially binds to the hemoglobin molecule so it cannot as easily accept oxygen, thus starving the body of oxygen. Cyanide gas prevents cell enzymes from using molecular oxygen; this state disrupts vital cell processes.

Sometimes several types of lung hazards occur simultaneously. In mining operations, for example, explosives release oxides of nitrogen into the air breathed by miners. These compounds impair the bronchial clearance mechanism, so that coal dust (of the particle sizes associated with the explosions) is not efficiently cleansed from the lungs.

Influence of Solubility

A compound that is very soluble—such as ammonia, formaldehyde, sulfuric acid, or hydrochloric acid—may pose less of a hazard. Although it is rapidly absorbed in the upper respiratory tract during the initial phases of exposure, it does not penetrate deeply into the lungs. Consequently, the nose and throat become very irritated, causing workers to leave the exposure area before they suffer serious harm. On the other hand, compounds insoluble in body fluids often cause considerably less throat irritation than do the soluble ones, but may penetrate deeply into the lungs. Thus, a serious hazard can be present without workers being immediately aware of it. With less irritation to the lungs, they have less warning that exposure is building up. Examples of such compounds (gases) are nitrogen dioxide and ozone. The immediate danger from these compounds in high concentrations is acute lung irritation or, possibly later, chemical pneumonia.

Numerous chemical compounds do not follow the general solubility rule. Such compounds are not very soluble in water and yet irritate the eyes and respiratory tract. They also can cause lung damage, even death in many situations. The supervisor must be sure that all hazardous compounds are identified and workers are properly protected.

Threshold Limit Values (TLVs)

Individual susceptibility to respiratory toxins is difficult to assess. Nevertheless, certain recommended limits have been established. A TLV refers to airborne concentrations of substances and represents an exposure level under which most people can work, day after day, without adverse effect. Because of wide variations in individual susceptibility, however, an occasional exposure of an individual at or even below the threshold limit may not prevent discomfort, aggravation of a preexisting condition, or occupational illness. The term TLV refers specifically to limits published by the American Conference of Governmental Industrial Hygienists (ACGIH). The TLV limits are reviewed and updated annually. The National Safety Council's *Fundamentals of Industrial Hygiene* explains this subject in detail. A brief overview follows. There are three categories of TLVs:

1. Time-weighted average (TLV-TWA) is the time-weighted average concentration for a normal eight-hour day or 40-hour week. It is believed that nearly all persons can be exposed day after day to airborne concentrations at these limits without adverse effect.

2. Short-term exposure limit (TLV-STEL) is the concentration to which persons can be exposed for a period of up to 15 minutes continuously without suffering:

   • irritation
   • chronic or irreversible tissue change
   • narcoses of sufficient degree to reduce reaction time, impair self-rescue, increase the likelihood of accidental injury, or materially reduce work efficiency, provided the daily TLV-TWA is not exceeded.

   No more than four 15-minute exposure periods per day are permitted with at least 60 minutes between exposure periods.

   A STEL is not a separate, independent exposure limit. Rather, it supplements the time-weighted average (TWA) limit in cases where workers suffer acute reactions to a substance whose toxic effects are primarily chronic. Short-term exposure limits are recommended only where toxic effects have been reported from high short-term exposures in either humans or animals.

3. Ceiling (TLV-C) is the concentration that should not be exceeded even for an instant.

Nearly one-fourth of the substances in the TLV list are followed by the designation "skin." This refers to potential exposure through skin absorption. This designation is intended to suggest appropriate measures to prevent absorption of substances through the skin.

Permissible Exposure Limits (PELs)

The first compilation of health and safety standards from the U.S. Department of Labor's OSHA appeared in 1970. Because it was derived from then-existing standards, the compilation adopted many of the TLVs established in 1968 by the American Conference of Governmental Industrial Hygienists. Thus threshold limit values—a registered trademark of the ACGIH—became, by federal standards, permissible exposure limits (PELs). These PELs represent the legal maximum level of contaminants in the workplace air.

The General Industry OSHA Standards currently list about 600 substances for which exposure limits have been established. These are included in subpart Z, "Toxic and Hazardous Substances," Sections 1910.1000 through 1910.1500. The PELs were updated in 1990, again including many of the newer, revised TLVs. However, recent legal controversy over PELs has reverted many of the 1990 levels back to earlier published standards. This area is a very dynamic facet of industrial hygiene. For current information on exposure limits, refer to current regulatory documentation.

U.S. OSHA action level is that point at which employers must initiate certain safety provisions: employee exposure measurement, employee training, and medical surveillance. A U.S. OSHA action level has not been defined for all employees. The action level for some OSHA-regulated chemicals like lead is usually set at about one-half the permissible exposure level (PEL).

Why is an action level set well below the PEL? Setting the action level at one-half the permissible exposure helps to protect employees from overexposure with a minimum burden to the employer. Where employee exposure measurements indicate that no employee is exposed to airborne concentrations of a substance in excess of the action level, employers in effect are exempted from having to initiate certain provisions in the standards.

The action level recognizes that air samples can only estimate the true TLV-TWA. For an extra margin of safety, companies set action levels lower than one-half the PEL. The adequacy of the TLVs to protect workers from illness is controversial, and a lower company action level may be desirable.

When to Measure?

The measurements done by the industrial hygienist can be divided into three phases:

1. problem definition phase
2. problem analysis phase
3. solution phase.

Problem Definition Phase

In many instances, inspectors take measurements to determine if there is a problem in the workplace. In addition, some OSHA regulations require measurement at certain specified intervals or any time there is a change in production, process, or control measures. Measurement often establishes that workers are not experiencing excessive exposure to hazardous materials. Such monitoring of the workplace helps to ensure a safe environment.

Monitoring, then, is frequently used to determine that employers are in compliance with OSHAct requirements, state or provincial regulations, commonly accepted standards, TLVs, PELs, and action levels. Newer health standards published by OSHA usually state:

Each employer who has a place of employment in which [toxic substance name] is released into the workplace air shall determine if there is any possibility that any employee may be exposed to airborne concentrations of [toxic substance name] above the permissible level. The initial determination shall be made each time there is a change in production, process, or control measures that may result in an increase in airborne concentrations of [toxic substance name].

When any hazardous substances are released into the workplace air, the employer must take the first step in the employee exposure monitoring program. For OSHA-regulated substances, there must be an actual exposure measurement to see if any employee has been exposed to concentrations in excess of the recommended levels. This step should be taken even if there is only a remote chance that employees have been exposed to a substance above recommended levels.

Where does sampling begin? Should the sample be taken at the worker's breathing zone? Out in the general air? At the machine or process that is emitting the toxic substance? Although OSHA requires sampling only in the worker's breathing zone, sampling at all three sites provides a clearer picture of the situation. Should the sample be taken for two minutes, two hours, or a whole day?

There are two major types of samples: grab and long-term samples. The grab sample is taken over so short a period of time that the atmospheric concentration is assumed to be constant throughout the sample. This usually will cover only part of an industrial cycle. A series of grab samples can be taken in an attempt to define the total exposure. However, doing so requires a sound knowledge of statistical sampling techniques.

The long-term sample is taken over a sufficiently extended period of time that the variations in exposure cycles are averaged. Usually one sample or a series of samples is taken to represent the employee's eight-hour average exposure. OSHA regulations usually require this type of sampling.

An adequate number of tests should be taken to define the TLV-TWA and to relate this level to recommended or regulatory exposure levels. But samples also must be taken to characterize the peak emissions during various portions of the process cycle.

If employee measurements indicate exposure at or above the action level, then OSHA requires that all employees so exposed be identified and their exposure measured. This step clearly determines the population at risk.

When exposure measurements are at or above the action level but not above the PEL or just below the TLV-TWA, the employer needs some statistically reliable means to ensure that exposure levels are not exceeding these values. Management should conduct periodic sampling of the affected area and order medical examinations to determine if any susceptible individuals are exhibiting effects at these exposures.

If employees are exposed above the PEL or TLV, then a more intensive monitoring program is necessary. Medical staff must examine workers exposed to these excessive levels to measure the effects on their health. Noninhalation exposures—such as skin absorption—also may occur. Therefore, accurate exposure evaluation may require breath, blood, and urine sampling.

The problem definition phase is an orderly progression. At each step of the process, employers can decide whether to proceed to the next higher step.

Problem Analysis Phase
Once the industrial hygienist has defined the problem in the first phase of the measurement process, he or she must determine its causes. Management can identify opportunities for improvements in the workplace, set objectives for solutions, and devise alternative solutions should the initial ones fail to solve the problem.

The following eight methods suggest some ways that exposure hazards can be controlled:

1. substitution of a less harmful material for a hazardous one
2. change or alteration of a process to minimize worker contact
3. isolation or enclosure of a process or work operation to reduce the number of persons exposed
4. wet methods to reduce generation of dust in operations
5. local exhaust at the point of generation
6. personal protective devices (see Chapter 7 in the *Engineering & Technology* volume)
7. good housekeeping, including cleanliness of the workplace, waste disposal, adequate washing, clean toilet and eating facilities, healthful drinking water, and control of insects and rodents
8. training and education—the OSHA hazard communication standard requires training for employees exposed to hazardous chemicals.

Solution Phase
Once the problem has been analyzed and a number of solutions proposed, the most effective, timely, and practical solution needs to be selected—one that provides optimum benefits with minimal risks. The details of the solution should be carefully worked out. In effect,

management needs to develop a blueprint describing what should to be done, how and by whom it should be done, and in what sequence the actions are to take place.

Once controls are installed, they must be checked periodically to be sure they are functioning properly. Follow-up monitoring and inspection will determine if the solution to a given hazardous exposure is controlling it within the specified limits. Thus, managers should regard the monitoring function as a circular, not horizontal process. If measurements at the solution phase reveal that controls are inadequate, the industrial hygienist must return to the first phase, that of defining the problem.

Who Will Do the Measuring?

Not every organization requires or can afford the services of a full-time industrial hygienist. Independent consultants can be hired to accomplish two major objectives:

1. Identify and evaluate potential health risks and accident hazards to workers in the occupational environment.
2. Design effective controls to protect the safety and health of workers.

Because any person can legally offer services as an industrial hygiene consultant, it is important that the consultant hired is a trained, experienced, and competent professional. A competent industrial hygiene consultant must have detailed knowledge of proper sampling equipment and analytic procedures and will probably hold the designation "certified industrial hygienist" (CIH).

Good sources of information and assistance regarding consultants are the American Industrial Hygiene Association and the American Society of Safety Engineers, the professional associations related to occupational and health safety, respectively (see the descriptive listing in Appendix 1, Sources of Help). Regional offices of the National Institute for Occupational Safety and Health (NIOSH) usually have lists of consultants in their area. Many insurance companies have loss prevention programs that employ industrial hygienists. The National Safety Council and its chapters having offices in major cities can offer assistance. The Council offers a full range of consulting services in safety and occupational health management (see Appendix 1, Sources of Help). For a state-by-state listing of governmental consulting service offices, see Sources of Help under U.S. Government Agencies.

ACCIDENT INVESTIGATION

A fourth function of monitoring in the total hazard control system is incident (accident) investigation, the subject of Chapter 7, Accident Investigation, Analysis, and Costs, in this volume. The following discussion demonstrates how accident investigation fits into the systems approach to hazard control.

Why Accidents Are Investigated

When viewed as an integral part of the total occupational safety and health program, accident investigation is especially important to determine direct causes, uncover contributing accident causes, prevent similar accidents from occurring, document facts, provide information on costs, and promote safety. Accident investigation concentrates on gathering all information about the factors leading to the accident.

Determine Direct Causes

Accident investigation determines the direct and contributing causes of accidents. At what points did the hazard control system break down? Were rules and regulations violated? Did defective machinery or factors in the work environment contribute to the accident? Poor machinery layout, for example, or the very design of a job process, operation, or task can contribute to an undesirable situation. Chapter 3, Loss Control Programs, outlined the three primary sources of accidents: human, situational, and environmental factors.

Uncover Contributing Accident Causes

Thorough accident investigation is very likely to uncover problems that indirectly contributed to the accident. Such information benefits accident-reduction efforts. For example, a worker slips on spilled oil and is injured. The oil spill is the direct cause of the accident, but a thorough investigation might reveal other contributing factors: poor housekeeping, failure to follow maintenance schedule, inadequate supervision, or faulty equipment (such as a lathe leaking oil).

Prevent Similar Accidents

Accident investigation identifies what actions and improvements will prevent similar accidents from occurring in the future.

Document Facts

Accident investigation documents the facts involved in an incident for use in any compensation and litigation that may arise. The report produced at the conclusion of an investigation becomes the permanent record of facts about an accident. It may become necessary to reconstruct an accident situation long after the occurrence. To do so, the details of the accident will have to be recorded properly, accurately, and thoroughly.

Provide Information on Costs

Accident investigation provides information on both direct and indirect costs of accidents. Chapter 7, Accident Investigation, Analysis, and Costs, gives details for estimating accident costs.

Promote Safety

Accident investigation yields psychological as well as material benefits. The investigation demonstrates the organization's interest in worker safety and health. It indicates management's sense of accountability for accident prevention and its commitment to a safe work environment. An investigation in which both labor and management participate promotes cooperation between these two groups.

Despite what many people believe, accident investigation is a fact-finding, not a fault-finding, process. When attempting to determine the cause of an accident, the novice investigator may be tempted to conclude that the person involved in the accident was at fault. But if human error is not the real cause, the hazard that produced the accident will go undiscovered and uncontrolled. Furthermore, the person falsely blamed for causing the accident will resent the unjustified accusation, as well as any disciplinary action. The worker will be less cooperative in the future and feel less respect for the organization's safety and health program. Investigators should always stress that the intent of accident investigation is to pinpoint causes of error and defects so similar accidents can be prevented.

Conducting an accident investigation is not simple. It can be difficult to look beyond the incident at hand to uncover causal factors, determine the true loss associated with the occurrence, and develop practical recommendations to prevent recurrence. A major weakness of many accident investigations is the failure to establish and consider all factors—human, situational, and environmental—that contributed to the accident. Reasons for this failure include:

- inexperienced or uninformed investigator
- reluctance of the investigator to accept full responsibility for the job
- narrow interpretation of environmental factors
- erroneous emphasis on a single cause
- judging the effect of the accident to be the cause
- arriving at conclusions before all factors are considered
- poor interviewing techniques
- delay in investigating accidents.

The trained investigator must be ready to acknowledge as contributing causes any and all factors that may have led, in any way, to the accident. What at first may appear to be a simple, uninvolved incident can, in fact, have numerous contributing factors that become more complex as analyses are completed. Immediate, on-the-scene accident investigation provides the most accurate and useful information.

When to Investigate Accidents

The longer the delay in examining the accident scene and interviewing the victim(s) and witnesses, the greater the possibility of obtaining erroneous or incomplete information. The accident scene changes, memories fade, and people discuss what happened with each other. Whether consciously or not, witnesses may alter their initial impressions to agree with someone else's observation or interpretation. Further, prompt accident investigation also expresses concern for the safety and well-being of employees.

As a general rule, all accidents, no matter how minor, are candidates for thorough investigation. Many accidents occurring in an organization are considered minor because their consequences are not serious. Such accidents, or incidents, are taken for granted and often do not receive the attention they demand. Management, safety and health committees, supervisors, and employees must be aware that serious accidents arise from the same hazards that produce minor incidents. Usually sheer luck determines whether a hazardous situation results in a minor incident or a serious accident.

In accident investigation, the investigator must give priority to the health and safety of affected personnel (including any victims). When possible, rescue and first aid procedures should be used that disturb the accident scene as little as possible. Measures to protect equipment should also preserve evidence. When the area is secure and victims have received medical attention, and appropriate notifications have been made, efforts can be concentrated on investigating the incident.

As with inspections, it is advisable to prepare investigation tools in advance. An investigation kit might include:

- camera and film
- tape recorder
- measuring devices
- sample containers
- interview/investigation forms
- flashlight
- barricade markers/tape
- warning tags and padlocks.

Having the necessary equipment ready will facilitate the investigation and certainly help to eliminate delays and other difficulties.

Who Should Conduct the Investigation?

Chapter 3, Loss Control Programs, discusses the question of who is to make the investigation: the supervisor or foreman, the safety and health professional, a special investigative committee, or a company safety and health committee. As a supplement to that discussion, the following section outlines the roles played by physicians and management in accident investigation. It also covers the responsibility of the safety professional in preventing further accidents from occurring during the investigation itself.

Physician

A physician's assistance is particularly important when human factors have been designated as direct or contributing causes of an accident. The physician can assess the nature and degree of injury and assist in determining the source and nature of the forces that inflicted the injury. The physician also can (1) determine what special biomedical studies, if any, are needed; (2) establish whether the injured person was physically and mentally fit at the time of the accident and whether the screening, selection, and preplacement process is adequate; (3) help judge the adequacy of safety and health protection procedures and equipment; and (4) help evaluate the effectiveness of the plans, procedures, equipment, training, and response of rescue, first aid, and emergency medical care personnel. The physician also can evaluate the effectiveness of measures aimed at early detection of medical conditions, mental changes, or emotional stress.

Management

Management and department heads should help investigate accidents resulting in lost workdays or major property damage. When management actively participates in accident investigation, it can evaluate the hazard control system and determine whether outside assistance is desired or required to upgrade existing structures and procedures. Management also must review accident reports in order to make informed decisions. When accident investigation reveals the need or desirability of specific corrective actions, management must determine whether the recommended action has been implemented.

Safety During the Investigation

In many cases the accident scene is a dangerous place. The accident may have damaged electrical equipment, weakened structural supports, and released radioactive or toxic materials.

The safety and health professional must be particularly alert to the hazards encountered by the investigating team and, when necessary, see that proper protective equipment is provided. Investigators need to be alerted to the hazards they may encounter and emergency procedures they should follow (see details in Chapter 15, Emergency Preparedness, in this volume).

What to Investigate

The accident investigation must answer many questions. Because of the infinite number of accident-producing situations, contributing factors, and causes, it is impossible to list all the questions that apply to all investigations. The following questions are generally applicable, however, and will be considered in most accident investigations (Firenze, 1978).

- What was the injured person doing at the time of the accident? Performing an assigned task? Maintenance? Assisting another worker?
- Was the injured employee working on an unauthorized task? Was the employee qualified to perform

the task and familiar with the process, equipment, and machinery?
- What were other workers doing at the time of the accident?
- Was the proper equipment being used for the task at hand (screwdriver instead of can opener to open a paint can, file instead of grinder to remove burr on a bolt after it was cut)?
- Was the injured person following approved procedures?
- Is the process, operation, or task new to the area?
- Was the injured person being supervised? What was the proximity and adequacy of supervision?
- Did the injured employee receive hazard recognition training prior to the accident?
- What was the location of the accident? What was the physical condition of the area when the accident occurred?
- What immediate or temporary actions could have prevented the accident or minimized its effect?
- What long-term or permanent action could have prevented the accident or minimized its effect?
- Had corrective action been recommended in the past but not adopted?

During the course of the investigation, the above questions should be answered to the satisfaction of the investigators. Other questions that come to mind as the investigation continues should be recorded.

Conducting Interviews

Interviewing accident or injury victims and witnesses can be a difficult assignment if not properly handled. The individual being interviewed often is fearful and reluctant to provide the interviewer with accurate facts about the accident. The accident victim may be hesitant to talk for any number of reasons. A witness may not want to provide information that might implicate friends, fellow workers, or the supervisor. To obtain the necessary facts during an interview, the interviewer must first eliminate or reduce an employee's fear and anxiety by establishing good rapport with the individual. The interviewer must create a feeling of trust and establish open communication before beginning the actual interview. Once good rapport has been developed, the interviewer can follow this five-step method.

1. Discuss the purpose of the investigation and the interview (fact-finding, not fault-finding).
2. Have the individual relate his or her version of the accident with minimal interruptions. If the individual being interviewed is the one who was injured, ask what was being done, where and how it was being done, and what happened. If practical, have the injured person or eyewitness explain the sequence of events that occurred at the time of the accident.

Being at the scene of the accident makes it easier to relate facts that might otherwise be difficult to explain.

3. Ask questions to clarify or fill in any gaps.
4. The interviewer should then repeat the facts of the accident to the injured person or eyewitness. Through this review process, there will be ample opportunity to correct any misunderstanding that may have occurred and clarify, if necessary, any of the details of the accident.
5. Discuss methods of preventing recurrence. Ask the individual for suggestions aimed at eliminating or reducing the impact of the hazards that caused the accident. By asking the individual for ideas and discussing them, the interviewer will show sincerity and place emphasis on the fact-finding purpose of the investigation, as it was explained at the beginning of the interview.

In some cases, contractual agreements may call for an employee representative to be present during any management interview, if the employee so requests.

Preparing the Accident Investigation Report
Chapter 7, Accident Investigation, Analysis, and Costs, and Chapter 8, Injury and Illness Record Keeping and Incidence Rates, outline specific ways to record and classify data: how to identify key facts about each injury and the accident that produced it, how to record facts in a form that facilitates analysis and reveals patterns and trends, how to estimate accident costs, and how to comply with regulatory record-keeping requirements.

An accident in any organization is of significant interest to employees, who will ask questions that reflect their concerns. Is there any potential danger to those in the immediate vicinity? What caused the accident? How many people were injured? How badly?

Those who investigate accidents should answer these questions truthfully and avoid covering up any facts. On the other hand, they must be certain they are authorized to release information, and they must be sure of their data.

Because the accident report is the product of the investigation, it should be prepared carefully and adequately to justify the conclusions reached. It must be issued soon after the accident. When a report is delayed too long, employees may feel left out of the process. If a final report must be postponed pending detailed technical analysis or evaluation, then management should issue an interim report providing basic details of the investigation.

Summaries of vital information on major injury, damage, and loss incidents should be distributed to department managers. Such summaries should include information on accident causes and recommended action for preventing similar incidents. Management should maintain incident and statistical report files as dictated by company policy.

Supervisors need to keep employees informed of significant accidents and preventive measures proposed or executed. Posting accident reports is one way to make information available.

Implementing Corrective Action
The preceding section on inspection emphasized that hazard control benefits accrue only after the inspection report is written and disseminated. Until corrective action is initiated, any recommendations—no matter how earnest, thorough, and relevant—remain "paper promises."

The same is true of accident investigation when it is used as a monitoring technique. Viewed from the perspective of hazard control, accident investigation serves as a monitoring function only when it provides the impetus for corrective action.

Whenever management and safety professionals review monthly accident reports, they exercise an essential auditing function. Management (including the chief executive officer) can demonstrate interest in safety by requiring prompt reporting of all serious or potentially serious incidents. They use accident reports to make decisions to prevent similar accidents from occurring, and they look for answers to certain key questions. Are all significant accidents being reported? Are all parts of the organization equally committed to the hazard control effort? Are there trends or patterns in accidents or injuries? What system breakdowns predominate? What supervisors require additional training? Are employees advised of the results of accident investigation and of preventive measures being instituted? What management deficiencies are indicated? Accident investigation as a monitoring function occurs after the hazard control system has already broken down. Although no amount of investigation can reverse the accident, accident investigation serves an important monitoring function. Past mistakes can be used to improve future operations. As George Santayana has written, "Those who cannot remember the past are condemned to repeat it."

SUMMARY
- Monitoring, a vital management tool, is a set of observation and data collection methods used to detect and measure deviations from plans and procedures in current operations. It involves hazard analysis, job safety analysis, inspection, measurement and testing, and accident investigation.
- Inductive or deductive hazard analysis is an orderly process used to acquire specific hazard and failure data pertinent to a given system. Factors such as frequency of accidents; potential for injury; severity of injury; new or altered equipment, processes, and operations; and excessive material waste or damage to equipment determine which tasks or processes are analyzed.

- Job safety analyses are the first step of hazard analysis. JSAs are used to identify and analyze potential hazards and causes of accidents within each job or within specific categories of jobs.
- The primary purpose of inspection is to detect potential hazards so they can be corrected before an accident occurs. Continuous inspections are a routine part of the job. Planned inspections are more formal procedures that may be periodic, intermittent, or general. Inspectors report all hazards—classifying and describing them carefully—and recommend corrective actions.
- Measurement and testing methods are used to monitor chemical, physical, biological, and ergonomic hazards. Four monitoring systems are used: personal, environmental, biological, and medical.
- Standards used to establish health and safety limits include threshold limit values (TLVs), time-weighted averages (TLV-TWA), short-term exposure limits (TLV-STEL), permissible exposure limits (PELs), and action levels. Measurements are divided into problem definition phase, problem analysis phase, and solution phase.
- Accident investigations are conducted to determine direct causes, uncover indirect accident causes, prevent similar accidents from occurring, document facts, provide information on costs, and promote safety measures and standards.
- Accidents must be investigated immediately to ensure accurate details and to preserve evidence. Investigators can be supervisors, safety professionals, special committees, or health and safety committees.
- Investigators must examine all human, situational, and environmental factors in determining accident causes, and apply the five-step method for interviewing accident victims and witnesses. An investigation report should be issued as soon as possible and contain recommendations for corrective action.

REFERENCES

Ferry T. *Modern Accident Investigation and Analysis,* 2nd ed. New York: John Wiley & Sons, 1988.

Firenze RJ. *The Process of Hazard Control.* Dubuque, IA: Kendall/Hunt Publishing Co., 1978.

Gowen LD. Using fault trees and event trees as oracles for testing safety-critical software systems. *Professional Safety,* 41:4 (April 1996), 41–44.

Johnson WG. *MORT Safety Assurance Systems.* New York: Marcel Dekker, Inc., 1980. (Also available through National Safety Council.)

National Safety Council, 1121 Spring Lake Drive, Itasca, IL 60143.

Supervisors' Safety Manual, 8th ed. Itasca: National Safety Council, 1992.

Plog B, ed. *Fundamentals of Industrial Hygiene,* 4th ed. Itasca, IL: National Safety Council, 1996.

REVIEW QUESTIONS

1. Define hazard analysis.
2. List the two formal methods of hazard analysis.
   a.
   b.
3. In determining which hazard analysis approach to use for a given situation, the hazard control specialist will need to answer which five questions?
   a.
   b.
   c.
   d.
   e.
4. In order to decide which processes, operations, and tasks receive priority, what are the five factors that need to be analyzed?
   a.
   b.
   c.
   d.
   e.
5. Define job safety analysis (JSA).
6. After a job has been selected to be analyzed, the three basic steps in conducting a JSA are:
   a.
   b.
   c.
7. Inspecting is the first stage of the hazard and accident analysis procedure. What is the general purpose of an inspection?
8. List and briefly explain the two types of inspections.
   a.
   b.
9. The toolroom employee examines all tools before sending them out to be used. What is this type of inspection called?
   a. General
   b. Intermittent
   c. Continuous
   d. Periodic
10. Gathering information about standards, regulations, and codes is the first step in determining what items need to be inspected. What sources of information are available to assist in this process?
11. List the four factors that determine the frequency of inspections.
    a.
    b.
    c.
    d.

12. Name and briefly define the four kinds of monitoring systems.
    a.
    b.
    c.
    d.
13. The "toxicity" of a material refers to:
    a. Its potential for being a health hazard
    b. Its capacity to produce injury or harm
    c. A standard measure of percent of particles
    d. All of the above

14. Define threshold limit values (TLVs).
15. Define permissible exposure limits (PELs).
16. List the six main outcomes of an accident investigation.
    a.
    b.
    c.
    d.
    e.
    f.
17. Why is it important to investigate accidents immediately?

7
Accident Investigation, Analysis, and Costs

T his chapter covers the investigation of incidents without injury as well as accidents that result in injuries. To increase awareness that an incident is a near-accident and that so-called accidents are not random events but rather preventable events, the term "incident" is used in this chapter wherever possible. The term incident is used in its broadest sense to include incidents that may lead to property damage, work injuries, or both. The following definitions are generally used in this chapter:

Accident. An unplanned, undesired event, not necessarily resulting in injury, but damaging to property and/or interrupting the activity in process.

Incident. An undesired event that may cause personal harm or other damage. In the United States, OSHA specifies that incidents of a certain severity be recorded.

With proper hazard identification and evaluation, management commitment and support, preventive and corrective procedures, monitoring, evaluation, and training, unwanted events can be prevented.

In some contexts it may be necessary to distinguish between incidents with no injuries and accidents resulting in injuries. To facilitate such record keeping, see the form titled Incident (No Injury Accident) Report Form (Figure 7-1).

The topics covered in this chapter include:

- basic types of incident investigation and analysis
- methods of conducting an investigation and analysis
- types of costs associated with incidents
- how to calculate incident-related costs
- the financial effects of off-the-job incidents

Successful incident prevention requires a minimum of six fundamental activities:

1. study of all working areas to detect and eliminate or control the physical or environmental hazards that contribute to incidents (see also Chapter 3, Loss Control Programs, and Chapter 6, Identifying Hazards)
2. study of all operating methods and practices and administrative controls
3. education, instruction, training, and enforcement of procedures to minimize the human factors that contribute to incidents (see also Chapter 22, Motivation, and Chapter 23, Safety and Health Training)
4. thorough investigation and causal analysis of every incident resulting in at least a lost-workday injury to determine contributing circumstances. Incidents not resulting in personal injury (so-called near-incidents or near-misses) are warnings and should also be

investigated thoroughly. This fourth activity, incident investigation and analysis, is a defense against any hazards overlooked in the first three activities, against hazards not immediately obvious, or against hazards resulting from circumstances difficult to foresee.
5. implementation of programs to change or control the hazardous conditions, procedures, and practices found in the preceding activities
6. program follow-up and evaluation to ensure that the programs achieve the desired control.

INCIDENT INVESTIGATION AND ANALYSIS

The ultimate purpose of incident investigation and analysis activities is to prevent future incidents. As such, the investigation or analysis must produce factual information leading to corrective actions that prevent or reduce the number of incidents. The more complete the information, the easier it will be for management to take effective corrective actions. For example, knowing that 40% of an organization's incidents involve ladders is not as useful as knowing that 80% of the organization's ladder incidents involve broken rungs. A good record-keeping system, as discussed in Chapter 8, Injury and Illness Record Keeping and Incidence Rates, is essential to incident investigation. The system allows the basic facts about an incident to be recorded quickly, efficiently, and uniformly.

All incidents should be investigated, regardless of severity of injury or amount of property damage. The extent of the investigation depends on the outcome or potential outcome of the incident. An incident involving only first aid or minor property damage is not investigated as thoroughly as one resulting in death or extensive property damage, that is, unless the potential outcome could have been disabling injury or death.

For purposes of incident prevention, investigations must be fact-finding, not fault-finding; otherwise, they can do more harm than good. This is not to say responsibility should not be fixed where personal failure has caused injury, nor that such persons should be excused from the consequences of their actions. It does mean the investigation itself should be concerned only with facts. The investigating individual, board, or committee must not be involved with any disciplinary actions resulting from the investigation.

Types of Investigation and Analysis

A variety of incident investigation and analysis techniques are available to the investigator, some of them more complicated than others. The choice of a particular method depends upon the purpose and orientation of the investigation. The Failure Mode and Effect approach discussed in Chapter 6, Identifying Hazards, may be useful when investigating situations involving large, complex, and interrelated machinery and procedures but may be of limited value for investigating incidents involving hand tools.

Incident (No-Injury Accident) Report Form
(This must be completed IMMEDIATELY after an accident when there is no injury.)

Exact location of incident: _____ Department:_____

Occurrence date: _____ Time: _____ Date reported:_____

Employee involved: _____ SS# _____ Employee ID:_____

Job title: _____ Employment date: _____ Time on present job:_____

1. Property damaged: _____

    Cost: $ _____ Length of downtime: _____

2. Unsafe condition at time of incident: (be specific) _____

    _____

    _____

3. Unsafe practice contributing to the incident: (be specific) _____

    _____

    _____

    _____

4. Witness(es) to incident: _____

5. Sequence of events: (detailed) _____

    _____

    _____

    _____

6. What can be done to prevent a recurrence of this incident?_____

    _____

Supervisor: _____ Department: _____ Date: _____
Immediately forward copies of this report to Department Management

Figure 7-1. This sample form can be used for reporting incidents that involve no injuries.

If management procedures and communications and their relationship to incidents are of great interest, the Management Oversight and Risk Tree analysis (MORT, see References at the end of Chapter 6, Identifying Hazards) could prove to be the best choice.

The incident investigation and analysis procedure outlined in this chapter follows the ANSI Z16.2 standard, *Information Management for Occupational Safety and Health*. Other similar techniques involve investigation within the framework of defects in man, machine, media, and management (the four Ms) or education, enforcement, and engineering (the three Es). For analysis purposes, these techniques involve classifying the data about a group of incidents into various categories. This approach has been referred to as the statistical method of analysis. Corrective actions are designed on the basis of most frequent patterns of occurrence.

Other techniques discussed in Chapter 6, Identifying Hazards, come under the systems approach to safety. Systems safety stresses a broader viewpoint that takes into account interrelationships between various events that could lead to an incident. As incidents rarely have one cause, the systems approach to safety can point to more than one place in a system where effective corrective actions can be introduced. This process allows the safety professional to choose the corrective actions best meeting the criteria for effectiveness, rapid installation, cost/benefit analysis, and the like. There are additional advantages to using systems safety techniques: management can implement them before incidents occur and can apply them to new procedures and operations.

Persons Conducting the Investigation

Depending on the nature of the incident and other conditions, the investigation is usually made by the supervisor. This person can be assisted by a fellow worker familiar with the process involved, the safety and health professional or inspector, the employee health professional, the joint safety and health committee, the general safety committee, or an engineer from the insurance company. If the incident involves unusual or special features, consultation with a state labor department or federal agency, a union representative, or outside expert may be warranted. If a contractor's personnel are involved in the incident, then a contractor's representative should also be involved in the investigation.

The supervisor or foreman should make an immediate report of every injury requiring medical treatment and other incidents he or she may be directed to investigate. The supervisor is on the scene and probably knows more about the incident than anyone else. It is up to this individual, in most cases, to put into effect whatever measures can be adopted to prevent similar incidents. The Accident Investigation Report in the next chapter illustrates one form that can be used to record the findings of an incident investigation (Figure 8-2).

The Safety and Health Professional. Ideally, the safety professional should be an adviser and guide to the supervisor on incident investigations and should verify the supervisor's finding and the adequacy of his or her investigation because sometimes a supervisor may attempt to cover up a supervisory error. The safety and health professional should conduct the investigation only in serious cases or when the supervisor is not adequately trained in incident investigation.

Special Investigative or Review Committee. In some companies, a special committee is set up to investigate and report on all serious incidents or to review the quality of incident investigations. To be acceptable to all involved, this committee should be composed of representatives of both management and workers. Thus, a report published by this committee would be more readily accepted not only by workers but also by management than would a report made solely by a safety and health professional. (See the discussion of safety committees in Chapter 3, Loss Control Programs.)

The Safety and Health Committee. In many organizations, especially those that are small or moderate sized, a number of safety activities, including incident investigation, are handled by a safety and health committee. Ordinarily, such investigation is conducted in a routine manner, but in important cases the head of the committee might call an extra meeting to initiate a special investigation.

Cases to Be Investigated

An incident causing death or serious injury obviously should be thoroughly investigated. Also, the near-miss incident that might have caused death or serious injury is equally important from the safety standpoint and should be investigated. Many companies investigate all OSHA recordable injuries.

Each investigation should be conducted as soon after the incident as possible. A delay of only a few hours may permit important evidence to be destroyed or removed, intentionally or unintentionally. Also, the investigator or committee should present the results of the inquiry as quickly as possible; this greatly increases their value in the safety education of employees and supervisors.

In addition, any epidemic of minor injuries demands study. A particle of emery in the eye or a scratch from handling sheet metal may seem to be a simple case; the immediate cause is obvious, and the loss of time is small. However, if such cases, or any others, occur frequently in the organization or in any one department, they need to be investigated to determine the underlying causes.

The chief value of such an investigation also lies in discovering any contributing causes. The energetic safety and health professional or manager appreciates this type of incident investigation because it can prove more valuable, though less spectacular, to safety efforts than an inquest following a fatal injury.

In any incident investigation, fairness and impartiality are absolutely essential. Otherwise, the value of this tool can be destroyed if employees suspect its purpose is to place blame or find a scapegoat. No one should be assigned to investigation work unless he or she has earned a reputation for fairness and is trained and experienced in gathering evidence. The person should clearly understand that incident investigations are conducted solely for the purpose of obtaining information to help prevent a recurrence of such incidents.

In the early years of the safety movement, incident prevention usually was a hit-or-miss activity. This approach has been replaced by more scientific techniques—see Chapter 6, Identifying Hazards.

In earlier years, a reduction in incident rates was prompted primarily by humanitarian appeal to management and workers. Now the most important methods are aimed at isolating and identifying incident causes in order to permit direct, positive, and corrective action to prevent their recurrence.

Like other phases of modern business management, incident prevention must be based on facts clearly identifying the problem. An approach to the incident prevention problem on this basis not only will result in more effective control over incidents but also will save the organization time, effort, and money. Incident analysis of individual cases identifies the facilities, locations, or departments in which injuries most frequently occur, and suggests necessary corrective actions to reduce incidents.

Sometimes an overall high rate is not identified with one or a few departments but instead represents a high frequency of incidents throughout the facility. Under such circumstances, it is even more important that an analysis of the incidents be made. A high incident rate may be hard to spot for several reasons. For example, similar incidents may occur frequently but at widely separated locations, so their high incidence is not apparent. Incidents may be more numerous in some machine operations than in others, or in certain procedures. Some unsafe practices that cause incidents may be committed repeatedly but at different times and in different places, so their importance as incident causes is not immediately recognized.

Analysis of the circumstances of incidents can produce these results:

1. Management can identify and locate the principal sources of incidents by determining, from actual experience, the methods, materials, machines, and tools most frequently involved in incidents, and the jobs most likely to produce injuries.
2. Investigations may disclose the nature and size of the incident problems in departments and among occupations.
3. Results will indicate the need for engineering revision by identifying the principal hazards associated with various types of equipment and materials.
4. The investigation can disclose inefficiencies in operating processes and procedures where, for example, poor layout contributes to incidents, or where outdated, physically overtaxing methods or procedures can be avoided.
5. An incident report will disclose the unsafe practices that need to be corrected by training employees or changing work methods.
6. The report also will enable supervisors to put their safety work efforts to the best use by giving them information about the principal hazards and unsafe practices in their departments.
7. Investigation results permit an objective evaluation of the progress of a safety program by noting in continuing analyses the effect of corrective actions, educational techniques, and other methods adopted to prevent injuries.

Minimum Data Required

The purpose of an incident investigation is twofold. First, to identify facts about each injury and the incident that produced it and to record those facts; and second, to determine a course of action to eliminate a recurrence. In addition, the investigation, instead of focusing solely on the injury and incident type, includes the entire sequence of events leading to the injury, as far back in time as the investigator feels is relevant. This expanded view of the incident sequence allows an employer to identify and implement a wider variety of corrective actions. Incident investigation records, individually and collectively, serve as guides to the areas, conditions, and circumstances to which incident prevention efforts can most profitably be directed.

An Accident Investigation Report form is used to help investigators gather, at a minimum, the basic information that should be recorded about each incident. Although many report forms are designed mainly for gathering data about injuries, they can also be used to investigate occupational illnesses that arise from a single exposure (e.g., a chemical burn caused by a splash of acid).

The Accident Investigation Report form, Figure 8-2 in the next chapter, shows a minimum data set that was developed to improve the quality of incident investigation and analysis. This minimum data set identifies the why of some of the incident characteristics as well as the who, what, when, where, and how. It acknowledges the existence of multiple causes of incidents by not restricting the investigator or analyst to choose a single causal act or condition. All questions on the form should be answered as completely and specifically as possible. If there is no answer for a question, or a question does not apply to the incident, the investigator should indicate this on the form. Supplementary information, such as photographs, drawings, or sketches, should be attached to the report. In a multiple-injury incident, investigators should complete a separate report form for each employee who is injured.

The minimum data set that investigators should record includes the following:

- If data will be used to compare one company with another, record data about *employer characteristics*. This includes the type of industry and the size of the company (number of full-time equivalent employees).
- Record *employee characteristics*. The victim's age and sex, the department and occupation in which he or she worked, and whether a full-time, part-time, or seasonal employee. Questions about the victim's experience are also asked. How long has the victim been with the company? How long in current occupation? How often had the employee repeated the activity engaged in when the incident occurred? Employee training records may be examined and assessed during this part of the investigation.
- Record the *characteristics of the injury*. Describe exactly the injury or injuries and the part or parts of the body affected by the incident. In the case of an occupational illness, provide the diagnosis and the body part or parts affected. On the form, check the highest degree of severity that applies to the injury.
- Prepare a *narrative description and accident sequence* that provides the exact location of the incident (attach any maps or diagrams to the report); a complete, specific breakdown of the sequence of events leading to the injury or near-miss; what objects or substances were involved in the incident; conditions such as temperature, light, noise, and weather pertaining to the incident; how the injury occurred and the specific object or substance that inflicted or was involved in the incident; whether any preventive measures had been in place; and what, if anything, happened after the injury occurred. The investigators should include only the facts obtained during the investigation; they should not record opinions or place blame.

In most incidents, the incident event itself and the injury event are different. For instance, an employee might be walking underneath a scaffolding when a hammer falls from the platform and strikes the employee on the head. The incident event (hammer falling from the scaffolding) is separate from the injury (hammer striking the employee's head). As a result, the form can draw out this distinction and record other events that led to the incident.

Preceding events can be something that happened that should not have occurred (the hammer falling off the scaffolding) or something that did not happen that should have occurred (no safeguards on the scaffolding platform). To determine whether a preceding event should be included in the incident report form, the investigator should ask whether the event set in motion a sequence of actions that resulted in the incident and subsequent injury.

- Record the *characteristics of the equipment* associated with the incident. These data should be incorporated into the narrative description in question 20 on the report form. Include the type, brand, size, and any distinguishing features of the equipment, its condition, and the specific part involved.
- Record the *characteristics of the task* being performed when the incident happened: the general task (such as repairing a conveyor) and the specific activity (such as using a wrench). The description should include the posture and location of the employee (for example, squatting under the conveyor) and whether the person was working alone or with others.
- Record the *time factors*. The investigation should record the time of day and whether it was the victim's first hour of the shift, second hour, or later. Also, what type of shift was it—day, swing, straight, rotating, etc.? Other information includes the phase of the employee's workday: performing work, rest period, meal time, overtime, entering or leaving plant.
- Record the *task and activity factors*. First, describe the general type of task the employee was performing at the time the injury occurred. Second, record the specific activity in which the employee was engaged. Indicate whether the injured worker was alone at the time or with a coworker or crew. Third, record the injured worker's posture in relation to the surroundings at the time of the incident (walking under a scaffolding, standing near a grinding wheel, kneeling beside a conveyor belt).
- Record *supervision information*. Indicate in the appropriate place on the form whether the employee was being supervised directly, indirectly, or not at all at the time of the incident. Also indicate situations in which supervision was not feasible at the time.
- Record the *causal factors*. Record the events and conditions that contributed to the incident. Be as specific and complete as possible.
- Describe the *corrective actions* taken immediately after the incident to prevent a recurrence, including interim or temporary actions.

Investigators can add other information to the form to comply with local or company requirements. These types of data might include:

- estimate or calculation of costs associated with the incident
- exposure data to be used in calculating incidence rates for injuries associated with certain activities
- management data for use in performance reviews
- information required for special studies (such as monitoring corrective action)
- information on accident patterns specific to a particular division, company, or industry.

The answers to the questions on this report form constitute the minimum information needed to proceed with an analysis. The nature of a company's operations or the interests of the analyst may suggest other questions to be answered in the investigation.

There are two types of analysis that can be done. First, the investigator can examine the individual incident to determine the corrective action or actions to prevent future occurrences of this specific sequence of events. Second, the investigator can do a statistical analysis to examine a group of similar occurrences for patterns lending themselves to corrective actions. Over time, this statistical analysis can show which corrective actions have been more effective than others.

Identifying Causal Factors and Selecting Corrective Actions

In any incident, many factors are at work that trigger a sequence of events resulting in an injury. The idea behind the corrective action selection procedure is to identify all the factors for which a corrective action is possible. Management then selects the ones likely to be most effective, most cost/beneficial, most acceptable, and so on, and implements those.

Figure 7-2 shows the Guide for Identifying Causal Factors and Corrective Action. This guide is used with the Accident Investigation Report in Chapter 8, Injury and Illness Record Keeping and Incidence Rates (Figure 8-2). The Guide contains four parts: equipment, environment, people, and management. Although these elements usually combine to produce products and profits, at times they result in incidents. The questions in these four sections help the investigator to analyze the contribution each factor made to the incident.

The structure of the Guide makes it easy to identify the causal factors. Questions are answered by placing an *X* in a circle or a box. An *X* in a circle means the item is a causal factor, while an *X* in a box indicates the item is not a causal factor.

The Comment column provides space to record the specific information about the incident being investigated. The Recommended Corrective Actions column has room to enter specific corrective actions for each causal factor. After listing all of the possible corrective actions identified on the Guide, each action must be evaluated for effectiveness, cost, feasibility, reliability, acceptance, effect on productivity, time required to implement, and any other factor deemed important, before deciding which ones to implement.

This systematic approach to selecting corrective actions ensures three basic steps. First, all major actions are considered; second, the analyst does not stop with familiar and favorite corrective actions; and third, each corrective action chosen for implementation is carefully thought out.

Classifying Incident Data

Although two incidents rarely happen in exactly the same way, incidents do follow general patterns. Because they must be grouped according to pattern for purposes of analysis, finding the patterns and common features of groups of cases is the statistical approach to incident analysis.

Setting Up Classifications

The ANSI Z16.2 standard, *Information Management for Occupational Safety and Health,* provides detailed classification categories for several important data elements: nature of illness or injury, part of body affected, source of injury or illness, event or exposure, secondary source of injury or illness, occupation, and industry. For other data elements, classifications must be set up for grouping the various data. For each basic fact, general classifications should be established to group similar data before the actual analysis work is done. For example, among the general classifications for hazardous conditions are the following:

- defects of agencies (which means undesired and unintended characteristics)
- dress or apparel hazards
- environmental hazards
- placement hazards
- inadequate safeguarding
- public hazards.

Within each one of these general classifications, more specific classifications are set up. Under defects of agencies, for example, are:

- composed of unsuitable materials
- dull
- improperly compounded, constructed, or assembled
- improperly designed
- rough
- sharp
- slippery
- worn, cracked, frayed, or broken
- other defects.

It is not always possible to establish classifications before the analysis is begun. In this case, management can develop classifications as it reviews reports and notice various incident situations.

For example, if an investigator is analyzing ladder incidents and finds that in a number of cases broken rungs caused the incidents, a specific category for "broken rungs" should be set up under the classification "defects of agencies."

It is recommended that general and specific classifications be used for most key facts. The point must be emphasized that for an analysis to be of maximum usefulness, classifications must be set up to encompass the situations pertinent to a particular organization.

Use of a Numerical Code

Regardless of the method eventually used to sort and tabulate the various key facts, the work will be made easier

GUIDE FOR IDENTIFYING CAUSAL FACTORS & CORRECTIVE ACTIONS

Case Number

Answer questions by placing an X in the "**Y**" circle or box for yes or in the "**N**" circle or box for no.

PART 1 EQUIPMENT

○ Y □ N **1.0 WAS A HAZARDOUS CONDITION[S] A CONTRIBUTING FACTOR?**
If yes, answer the following. If no, proceed to Part 2.

	Causal Factors	Comment	Possible Corrective Actions	Recommended Corrective Actions
○ Y □ N	1.1 Did any defect(s) in equipment/tool(s)/material contribute to hazardous condition(s)?		Review procedure for inspecting, reporting, maintaining, repairing, replacing, or recalling defective equipment/tool(s)/material used.	
□ Y ○ N	1.2 Was the hazardous condition(s) recognized? If yes, answer A and B. If no, proceed to 1.3.		Perform job safety analysis. Improve employee ability to recognize existing or potential hazardous conditions. Provide test equipment, as required, to detect hazard. Review any change or modification of equipment/tool(s)/material.	
□ Y ○ N	A. Was the hazardous conditions(s) reported?		Train employees in reporting procedures. Stress individual acceptance of responsibility.	
□ Y ○ N	B. Was employee(s) informed of the hazardous condition(s) and the job procedures for dealing with it as an interim measure?		Review job procedures for hazard avoidance. Review supervisory responsibility. Improve supervisor/employee communications. Take action to remove or minimize hazard.	
□ Y ○ N	1.3 Was there an equipment inspection procedure(s) to detect the hazardous condition(s)?		Develop and adopt procedures (for example, an inspection system) to detect hazardous conditions. Conduct test.	
□ Y ○ N	1.4 Did the existing equipment inspection procedure(s) detect the hazardous condition(s)?		Review procedures. Change frequency or comprehensiveness. Provide test equipment as required. Improve employee ability to detect defects and hazardous conditions. Change job procedures as required.	
□ Y ○ N	1.5 Was the correct equipment/tool(s)/material used?		Specify correct equipment/tool(s)/material in job procedures.	
□ Y ○ N	1.6 Was the correct equipment/tool(s)/material readily available?		Provide correct equipment/tool(s)/material. Review purchasing specifications and procedures. Anticipate future requirements.	

Figure 7-2. The Guide for Identifying Causal Factors and Corrective Actions assists the accident investigator to systematically consider four contributing accident factors: Equipment, Environment, People, and Management.

	Causal Factors	Comment	Possible Corrective Actions	Recommended Corrective Actions
☐ ◯ Y N	1.7 Did employee(s) know where to obtain equipment/tool(s)/material required for the job?		Review procedures for storage, access, delivery, or distribution. Review job procedures for obtaining equipment/tool(s)/material.	
◯ ☐ Y N	1.8 Was substitute equipment/tool(s)/material used in place of correct one?		Provide correct equipment/tool(s)/material. Warn against use of substitutes in job procedures and in job instruction.	
◯ ☐ Y N	1.9 Did the design of the equipment/tool(s) create operator stress or encourage operator error?		Review human factors engineering principles. Alter equipment/tool(s) to make it more compatible with human capability and limitations. Review purchasing procedures and specifications. Check out new equipment and job procedures involving new equipment before putting into service. Encourage employees to report potential hazardous conditions created by equipment design.	
◯ ☐ Y N	1.10 Did the general design or quality of the equipment/tool(s) contribute to a hazardous condition?		Review criteria in codes, standards, specifications, and regulations. Establish new criteria as required	
◯	1.11 List other causal factors in "Comment" column.			

PART 2 ENVIRONMENT

◯ ☐ Y N   **2.0 WAS THE LOCATION OF EQUIPMENT/MATERIALS/EMPLOYEE(S) A CONTRIBUTING FACTOR?**
If yes, answer the following. If no, proceed to Part 3.

	Causal Factors	Comment	Possible Corrective Actions	Recommended Corrective Actions
◯ ☐ Y N	2.1 Did the location/position of equipment/material/employee(s) contribute to a hazardous condition?		Perform job safety analysis. Review job procedures. Change the location, position, or layout of the equipment. Change position of employee(s). Provide guardrails, barricades, barriers, warning lights, signs, or signals.	
☐ ◯ Y N	2.2 Was the hazardous condition recognized? If yes, answer A and B. If no, proceed to 2.3.		Perform job safety analysis. Improve employee ability to recognize existing or potential hazardous conditions. Provide test equipment, as required, to detect hazard. Review any change or modification of equipment/tools/materials.	
☐ ◯ Y N	A. Was the hazardous condition reported?		Train employees in reporting procedures. Stress individual acceptance of responsibility.	

Figure 7-2. (Continued).

		Causal Factors	Comment	Possible Corrective Actions	Recommended Corrective Actions
☐ Y	○ N	B. Was employee(s) informed of the job procedures for dealing with the hazardous condition as an interim action?		Review job procedures for hazard avoidance. Review supervisory responsibility. Improve employee/supervisor communications. Take action to remove or minimize hazard.	
☐ Y	○ N	2.3 Was employee(s) supposed to be in the vicinity of the equipment/material?		Review job procedures and instruction. Provide guardrails, barricades, barriers, warning lights, signs, or signals.	
☐ Y	○ N	2.4 Was the hazardous condition created by the location/ position of equipment/ material visible to employee(s)?		Change lighting or layout to increase visibility of equipment. Provide guardrails, barricades, barriers, warning lights, signs or signals, floor stripes, etc.	
☐ Y	○ N	2.5 Was there sufficient workspace?		Review workspace requirements and modify as required.	
○ Y	☐ N	2.6 Were environmental conditions a contributing factor (for example, illumination, noise levels, air contaminant, temperature extremes, ventilation, vibration, radiation)?		Monitor, or periodically check, environmental conditions as required. Check results against acceptable levels. Initiate action for those found unacceptable.	
○		2.7 List other causal factors in "Comment" column.			

PART 3 PEOPLE

○ Y ☐ N **3.0 WAS THE JOB PROCEDURE(S) USED A CONTRIBUTING FACTOR?**
If yes, answer the following. If no, proceed to Part 3.6.

		Causal Factors	Comment	Possible Corrective Actions	Recommended Corrective Actions
☐ Y	○ N	3.1 Was there a written or known procedure (rules) for this job? If yes, answer A, B, and C. If no, proceed to 3.2.		Perform job safety analysis and develop safe job procedures.	
☐ Y	○ N	A. Did job procedures anticipate the factors that contributed to the accident?		Perform job safety analysis and change job procedures.	
☐ Y	○ N	B. Did employee(s) know the job procedure?		Improve job instruction. Train employees in correct job procedures.	
○ Y	☐ N	C. Did employee(s) deviate from the known job procedure?		Determine why. Encourage all employees to report problems with an established procedure to supervisor. Review job procedure and modify if necessary. Counsel or discipline employee. Provide closer supervision.	

Figure 7-2. (Continued).

	Causal Factors	Comment	Possible Corrective Actions	Recommended Corrective Actions
☐ ○ Y N	3.2 Was employee(s) mentally and physically capable of performing the job?		Review employee requirements for the job. Improve employee selection. Remove or transfer employees who are temporarily, either mentally or physically, incapable of performing the job.	
○ ☐ Y N	3.3 Were any tasks in the job procedure too difficult to perform (for example, excessive concentration or physical demands)?		Change job design and procedures.	
○ ☐ Y N	3.4 Is the job structured to encourage or require deviation from job procedures (for example, incentive, piecework, work pace)?		Change job design and procedures.	
○	3.5 List other causal factors in "Comment" column.			

○ ☐ Y N	**3.6 WAS LACK OF PERSONAL PROTECTIVE EQUIPMENT OR EMERGENCY EQUIPMENT A CONTRIBUTING FACTOR IN THE INJURY?** If yes, answer the following If no, proceed to Part 4. Note: The following causal factors relate to the *injury*.

	Causal Factors	Comment	Possible Corrective Actions	Recommended Corrective Actions
☐ ○ Y N	3.7 Was appropriate personal protective equipment (PPE) specified for the task or job? If yes, answer A, B, and C. If no, proceed to 3.8.		Review methods to specify PPE requirements.	
☐ ○ Y N	A. Was appropriate PPE available?		Provide appropriate PPE. Review purchasing and distribution procedures.	
☐ ○ Y N	B. Did employee(s) know that wearing specified PPE was required?		Review job procedures. Improve job instruction.	
☐ ○ Y N	C. Did employee(s) know how to use and maintain the PPE?		Improve job instruction.	
☐ ○ Y N	3.8 Was the PPE used properly when the injury occurred?		Determine why and take appropriate action. Implement procedures to monitor and enforce use of PPE.	
☐ ○ Y N	3.9 Was the PPE adequate?		Review PPE requirements. Check standards, specifications, and certification of the PPE.	

Figure 7-2. (Continued).

	Causal Factors	Comment	Possible Corrective Actions	Recommended Corrective Actions
☐ ○ ·Y N	3.10 Was emergency equipment specified for this job (for example. emergency showers, eyewash fountains)? If yes. answer the following. If no, proceed to Part 4.		Provide emergency equipment as required.	
☐ ○ Y N	A. Was emergency equipment readily available?		Install emergency equipment at appropriate locations.	
☐ ○ Y N	B. Was emergency equipment properly used?		Incorporate use of emergency equipment in job procedures.	
☐ ○ Y N	C. Did emergency equipment function properly?		Establish inspection/monitoring system for emergency equipment. Provide for immediate repair of defects.	
○	3.11 List other causal factors in "Comment" column.			

PART 4 MANAGEMENT

○ ☐ Y N	**4.0 WAS A MANAGEMENT SYSTEM DEFECT A CONTRIBUTING FACTOR?** If yes, answer the following. If no, STOP. Your causal factor identification exercise is complete.

	Causal Factors	Comment	Possible Corrective Actions	Recommended Corrective Actions
○ ☐ Y N	4.1 Was there a failure by supervision to detect, anticipate, or report a hazardous condition?		Improve supervisor capability in hazard recognition and reporting procedures.	
○ ☐ Y N	4.2 Was there a failure by supervision to detect or correct deviations from job procedure?		Review job safety analysis and job procedures. Increase supervisor monitoring. Correct deviations.	
☐ ○ Y N	4.3 Was there a supervisor/employee review of hazards and job procedures for tasks performed infrequently? (Not applicable to all accidents.)		Establish a procedure that requires a review of hazards and job procedures (preventive actions) for tasks performed infrequently.	
☐ ○ Y N	4.4 Was supervisor responsibility and accountability adequately defined and understood?		Define and communicate supervisor responsibility and accountability. Test for understandability and acceptance.	
☐ ○ Y N	4.5 Was supervisor adequately trained to fulfill assigned responsibility in accident prevention?		Train supervisors in accident prevention fundamentals.	
○ ☐ Y N	4.6 Was there a failure to initiate corrective action for a known hazardous condition that contributed to this accident?		Review management safety policy and level of risk acceptance. Establish priorities based on potential severity and probability of recurrence. Review procedure and responsibility to initiate and carry out corrective actions. Monitor progress.	
○	4.7 List other causal factors in "Comment" column.			

Figure 7-2. (Concluded).

if the classifications are assigned numerical codes. A numerical code simply refers to numbers assigned in sequence to a list of similar facts. Each data element should have its own number; but for different elements, the numbering series may be repeated.

The new ANSI Z16.2-1995 standard, *Information Management for Occupational Safety and Health,* recommends numerical codes for the nature of injury or illness, part of body affected, source of injury or illness, event or exposure, secondary source of injury or illness, occupation, and industry. Data coded according to this standard will be comparable to benchmarking data published by the Bureau of Labor Statistics and the National Safety Council.

With this method, the analyst reads each case only once, at which time he or she assigns code numbers to the different facts. Subsequent sorting of the various facts, whether by hand or using a computer, can be completed quickly by referring to the code numbers. After the analyst has reviewed the cases and assigned code numbers to the different key facts, he or she can easily and quickly sort or arrange the reports by any of the facts to reveal the principal data concerning the incidents.

Conducting the Analysis

Experience has proved that the most effective way to reduce incidents is to concentrate on the primary causes of one phase of the incident problem at a time rather than attempt to stop all incidents at once. The problem can be approached in different ways, any one of which should prove effective.

The analyst may group reports by occupation of the injured person. The individual would review each group of reports, then determine the most prevalent incident types, sources of injury, and agencies of incidents among different occupations. Such information is particularly helpful in planning employee training and in developing educational materials and programs.

Injury incidence rates computed by departments may reveal that injuries occur at sharply higher rates in some departments than in others. If this is the case, the analyst can examine incident reports in the high-rate departments to find the sources of the incidents and their causes. This method enables management to concentrate efforts on the locations where the most incidents occur.

If injury incidence rates reveal a high rate of occurrence throughout the organization, the analysis procedure usually starts with information about the injury, goes on to identify the injury-producing event, and then looks at the circumstances and causal factors. The same procedure can be followed to examine injuries occurring within a high-rate occupation or department.

The analyst can crosstabulate the injury data to show the relationship or interaction between the two categories. Table 7-A illustrates an analysis of the Nature of Injury versus the Part of Body. This crosstabulation, in addition to showing what types of personal protective equipment might be useful, also points out common injury patterns needing further investigation.

A crosstabulation can extend in several directions, producing the need for further, separate crosstabulations. For example, the analyst would use the categories in Table 7-A showing the highest frequency of injuries (Cut, laceration, puncture, or abrasion to fingers and Sprain or strain to the back) to construct a second (Table 7-B) and a third crosstabulation.

As shown in Table 7-B, by the second tabulation the number of cases in a category usually is small enough that the analyst can read individual incident reports to find common causal factors and to determine corrective actions. If the number of cases in a category is still too large, then additional crosstabulations, such as Location versus activity or Activity versus occupation, can reduce the number enough to allow study of individual incident reports.

Table 7-A. Nature of Injury versus Part of Body

Table 7-A. Nature of Injury versus Part of Body

Nature of Injury	Eyes	Head, Face, Neck	Back	Trunk	Arm	Hand, Wrist	Finger	Leg	Foot, Ankle	Toe	Internal, Other	Total
Amputation	0	0	0	0	0	0	1	0	0	0	0	1
Burn & scald (heat)	0	0	0	0	0	0	0	0	0	0	0	0
Burn (chemical)	2	0	0	0	0	0	0	0	0	0	0	2
Concussion	0	2	0	0	0	0	0	0	0	0	0	2
Crushing	0	0	0	0	0	1	0	0	0	2	0	3
Cut, laceration, puncture, abrasion	0	1	0	0	1	3	18	0	0	0	0	23
Fracture	0	0	0	0	0	2	5	0	1	0	0	8
Hernia	0	0	0	1	0	0	0	0	0	0	0	1
Bruise, contusion	0	2	0	1	0	3	2	3	0	1	0	12
Occupational illness	0	1	0	0	1	2	0	0	0	0	8	12
Sprain, strain	0	0	24	0	0	2	2	3	4	0	0	35
Other	1	0	0	0	0	0	0	0	0	0	5	6
Total	3	6	24	2	2	13	28	6	5	3	13	105

This crosstabulation shows how the nature of injury and the part of body interact. In this example, cuts most often affect the fingers and sprains and strains usually involve the back. Note that bruises and contusions affect several body parts.

Table 7-B. Source of Injury versus Type of Accident

| | Type of Accident | | | | | | | |
Source of Injury	Fall from elevation	Fall on same level	Struck against	Struck by	Caught in, under, or between	Rubbed or abraded	Bodily reaction	Overexertion
Machine	0	0	0	0	3	0	0	0
Conveyor, elev. hoist	0	0	0	0	0	0	0	0
Vehicle	0	0	0	0	0	0	0	0
Electrical apparatus	0	0	0	0	0	0	0	0
Hand tool	0	0	0	4	0	0	0	0
Chemical	0	0	0	0	0	0	0	0
Working surface, bench, etc.	0	0	0	0	0	0	0	0
Floor, walking surface	0	0	0	0	0	0	0	0
Bricks, rocks, stones	0	0	0	0	0	0	0	0
Box, barrel, container	0	0	0	0	0	0	0	0
Door, window, etc.	0	0	0	0	0	0	0	0
Ladder	0	0	0	0	0	0	0	0
Lumber, woodworking metals	0	0	0	0	0	0	0	0
Metal	0	0	0	9	0	0	0	0
Stairway, steps	0	0	0	0	0	0	0	0
Other	0	0	0	0	0	0	0	0
Unknown	0	0	0	0	0	0	0	0
None	0	0	0	0	0	0	0	0
Total	0	0	0	13	3	0	0	0

This crosstabluation shows Source of Injury versus Type of Accident for one of the most frequent injuries identified in Table 7-A. In this example, being struck by a metal object was the source of most finger cuts. Other causes were being struck by hand tools or becoming caught in machinery.

| | Type of Accident | | | | | | | |
	Contact with electrical current	Contact with temperature extremes	Radiations, caustics, toxic and noxious substances	Public transportation accident	Motor vehicle accident	Other	Unknown	Total
Machine	0	0	0	0	0	0	0	3
Conveyor, elev. hoist	0	0	0	0	0	0	0	0
Vehicle	0	0	0	0	0	0	0	0
Electrical apparatus	0	0	0	0	0	0	0	0
Hand tool	0	0	0	0	0	0	0	4
Chemical	0	0	0	0	0	0	0	0
Working surface, bench, etc.	0	0	0	0	0	0	0	0
Floor, walking surface	0	0	0	0	0	0	0	0
Bricks, rocks, stones	0	0	0	0	0	0	0	0
Box, barrel, container	0	0	0	0	0	0	0	0
Door, window, etc.	0	0	0	0	0	0	0	0
Ladder	0	0	0	0	0	0	0	0
Lumber, woodworking metals	0	0	0	0	0	0	0	0
Metal	0	0	0	0	0	0	0	9
Stairway, steps	0	0	0	0	0	0	0	0
Other	0	0	0	0	0	0	2	2
Unknown	0	0	0	0	0	0	0	0
None	0	0	0	0	0	0	0	0
Total	0	0	0	0	0	0	2	18

Choose a method of tabulating that is appropriate to the number of reports generated and the ways the data will be used. For analyzing a small number of reports (up to about 100), hand sorting and tallying is effective. This method's principal advantage is that the analyst is using original records and has all the information available for reference.

A personal computer or other data-processing equipment is best for large collections of cases. Computers are also useful for smaller data sets as well because they can sort and display cases very quickly. Their efficiency allows the investigator to concentrate on various incidents and to test alternative hypotheses easily.

Using the Analysis
Merely obtaining the information will not prevent recurrence of incidents. Management must correct the underlying and contributing conditions. Thorough analysis of

groups of incident investigation reports can point to corrective actions that might not be evident when studying an individual case. In particular, inadequate policies, procedures, or management systems often are apparent after looking at the forest rather than the trees.

The statistical evidence revealed in an analysis can provide the guidance to direct safety efforts along the most effective path. The analysis will provide objective support and justification for budget requests, training programs, or other management safety activities.

Involving Outside Employees and Contractors

Owners and employers must also recognize the importance of involving outside employees and contractors in incident investigation and analysis. (For a fuller discussion of this topic, see Chapter 21, Contractor and Nonemployee Safety, in this volume.) This involvement should begin before outside workers arrive on the job site. Employers should insist on the following basic safety arrangements, which can be written into any contract or stated in a written policy that is agreed upon by both parties.

- Employers should require the use of a system of permits for potentially hazardous activities. This policy ensures that workers have the skills to perform their jobs; understand the hazards of any equipment, chemicals, or processes they use; and know basic safety and emergency measures to follow.
- Employers should require outside employees and contractors to designate a responsible supervisor to coordinate safety activities on the job. He or she can ensure that workers know and comply with all company and regulatory safety standards that apply to their work. The supervisor should inform workers about new hazards or changing work conditions that may present additional risks. In addition, this person should be the one to report any incidents to the employer and assist in the investigation and analysis of injuries or illnesses that workers may experience.
- Employers should provide outside employees and contractors with safety guidelines that personnel must follow. All guidelines for work practices and procedures, use of personal protective equipment, and safe handling and disposal of materials and chemicals should be clearly stated in writing and communicated to all workers.

By stressing safety as part of the job requirements, employers can establish a spirit of cooperation between their own management and outside employees. This type of relationship tends to facilitate incident investigation and analysis, reducing downtime, helping to uncover the causes of incidents more quickly, finding the best solutions to prevent their recurrence, and minimizing costs associated with incidents.

ESTIMATING INCIDENT COSTS

This discussion concerns the elements of cost most likely to result from a work incident and presents a method whereby an organization can obtain an accurate estimate of the total costs of its work incidents. (This procedure for estimating costs was developed by Rollin H. Simonds, Ph.D., Professor, Michigan State University, under the direction of the Statistics Division, National Safety Council.)

Reliable cost information is one basis for making decisions upon which efficiency and profit depend. Even in so obviously desirable an activity as incident prevention, some proposed measures or alternatives must be evaluated on the basis of their potential effect on profits.

Although most executives want to make their company a safe place to work, they also have a responsibility to run their business profitably. Consequently, they may be reluctant to spend money for incident prevention unless they can see a prospect for saving at least as much as they spend. Without information on the cost of incidents, it is practically impossible to estimate the savings brought about by expenditures for incident prevention.

Annual reports stressing dollar savings are more meaningful to management than those using incidence rates. Facts about the costs of incidents also may be effectively used in securing the active cooperation of supervisors. Supervisors usually are cost conscious because they are expected to run their departments profitably. Monthly reports showing the cost of incidents or the savings resulting from good incident records can motivate supervisors to ensure safe operating procedures.

Definition of Work Incidents for Cost Analysis

Work incidents, for the purpose of cost analysis, are unintended occurrences arising in the work environment. These incidents fall into two general categories: (1) incidents resulting in work injuries or illnesses and (2) incidents causing property damage or interfering with production. The inclusion of the no-injury incidents makes "work incidents" roughly synonymous with the type of occurrences a safety department strives to prevent.

Method for Estimating

To be of maximum usefulness, cost figures should represent as accurately as possible the specific experience of the company. It is not useful to have a fixed ratio of uninsured to insured costs representing many different organizations in many different industries. Estimated costs of incidents in general do not take into account differences in hazards from one industry to another or the more important differences in safety performance from one company to another.

Because the distinctions between direct and indirect costs are difficult to maintain, they have been abandoned in favor of the more precise terms "insured" and "uninsured" costs. Using these data, a company can estimate its incident cost with reasonable accuracy.

Insured Costs

Every organization paying compensation insurance premiums recognizes such expense as part of the cost of incidents. In some cases, medical expenses, too, may be covered by insurance. These costs are definite and known. They comprise the insured element of the total incident cost.

In addition to these costs, many other costs arise in connection with incidents. Although the expense of damaged equipment is easily identified, others, such as wages paid to the injured employee for downtime on the day of the injury, are hidden. These items comprise the uninsured element of the total incident cost.

Uninsured Costs

While insured costs can be determined easily from accounting records, uninsured (frequently called "indirect") costs are more difficult to assess. The method described here is one way to calculate these added expenses associated with many incidents. The first step is to conduct a pilot study to ascertain approximate averages of uninsured costs for each of the following four classes of incidents:

1. Class 1—Cases involving lost workdays—days away from work or days of restricted work activity
2. Class 2—Medical treatment cases requiring the attention of a physician outside the facility
3. Class 3—Medical treatment cases requiring only first aid or local dispensary treatment and resulting in property damage of less than $100 or loss of less than eight hours in work time
4. Class 4—Incidents that either cause no injury or cause minor injury not requiring the attention of a physician, and that result in property damage of $100 or more, or loss of eight or more employee-hours (Figure 7-1).

Once average costs have been established for each incident class, they can be used as multipliers to obtain total uninsured costs in subsequent periods. These costs are then added to known insurance premium costs to determine the total cost of incidents.

Example of a Cost Estimate

An estimate of costs made by one company is given in the following example. First, a pilot study was conducted to obtain the average cost of each class of incident. Included in the study were 20 Class 1 incidents, 30 Class 2 incidents, 50 Class 3 incidents, and 20 Class 4 incidents. Costs were determined and averages developed as in Table 7-C.

During the entire year, the company had 34 Class 1 incidents, 148 Class 2 incidents, and 4,000 Class 3 incidents. No record was kept of the Class 4 incidents after the pilot study was completed. Instead, the ratio of the number of Class 4 to Class 1 incidents found in the pilot study was used. This ratio was shown to be about 1 to 1,

Table 7-C. Average Cost Determined by Pilot Study

Class of Accident	Number of Accidents Reported	Average Uninsured Cost
Class 1	20	$251.10
Class 2	30	80.80
Class 3	50	15.70
Class 4	20	507.10

and since there were 34 Class 1 incidents during the year, it was assumed there were about 34 Class 4 incidents. (A separate record could be kept of the number of Class 4 incidents.)

The average cost for each incident class was applied to these totals to secure the results shown in Table 7-D.

Because the final total is the sum of many estimates, it should not be implied that the total figure suggests absolute accuracy. The estimate should be rounded to three significant digits—in this case, to the nearest thousand dollars. As a result, in this instance, the analyst reported to the plant manager, "During the past year, incidents cost this company about $155,000 in compensation, medical expense, lost time, and property damage."

The average costs determined in this pilot study represent the actual experience of this particular organization. Until important changes take place in this company's safety program, in the kind of machinery used or persons employed, or in other aspects affecting costs, the same average costs can be used.

Adjusting for Inflation

The effects of inflation can quickly render obsolete the cost figures in a pilot study. To account for this effect, the cost factors should be adjusted for inflation each year. The wage-related cost elements can be multiplied by the change in the general level of wages in the company. Other cost elements can be brought up to date by multiplying them by the general inflation rate as measured by the change in the Consumer Price Index. Because these adjustments are only approximate, the pilot study should be repeated at least every five years to establish new benchmarks.

Table 7-D. Estimate of Yearly Accident Costs

Class of Accident	Number of Accidents	Average Cost per Accident (from pilot study)	Total Uninsured Cost
Class 1	34	$251.10	$ 8,537.40
Class 2	148	80.80	11,958.40
Class 3	4,000	15.70	62,800.00
Class 4	34	507.10	17,241.40
Total Uninsured Cost			$100,537.20
Insurance Premiums			54,400.00
Total Accident Cost for the Period			$154,937.20

Items of Uninsured Cost

Important to a pilot study is a careful investigation of each incident to determine all the costs arising out of it. The following items of uninsured or indirect costs are clearly the result of work incidents and are subject to reasonably reliable measurement. Less tangible losses, such as the effect of incidents on public relations, employee morale, or on wage rates necessary to secure and retain employees, are not included in this method of estimating costs but can be important factors in some cases.

Information on some of the items is derived from the Department Supervisor's Accident Cost Investigation Report (Figure 7-3). The items are discussed in the order in which they appear on the Investigator's Cost Data Sheet (Figure 7-4).

1. *Cost of Wages Paid for Time Lost by Workers Who were not Injured.* These are employees who stopped work to watch or assist after the incident or to talk about it, or who lost time because they needed the equipment damaged in the incident or because they needed the output or the aid of the injured worker.

2. *Nature and Cost of Damage to Material or Equipment.* The validity of property damage as a cost can scarcely be questioned. Occasionally, there is no property damage, but a substantial cost is incurred in reorganizing material or equipment. The charge should be confined either to the net cost of repairing or reorganizing material or equipment damaged or displaced or to the current worth of equipment, less salvage value, if damaged beyond repair. An estimate of property damage should have the approval of the cost accountant, particularly if the current worth of the damaged property differs from the depreciated value established by the accounting department.

3. *Cost of Wages Paid for Time Lost by the Injured Worker, other than Workers' Compensation Payments.* Payments made under workers' compensation laws for time lost after the waiting period are not included in this element of cost.

4. *Extra Cost of Overtime Work Necessitated by the Incident.* The charge against an incident for overtime work is the difference between normal wages and overtime wages for the time needed to make up lost production, and the cost of extra supervision, heat, light, cleaning, and other extra services.

5. *Cost of Wages Paid to Supervisors for Time Spent on Activities Concerning the Incident.* The most satisfactory way of estimating this cost is to charge the wages paid to the supervisor for the time spent away from normal activities as a result of the incident.

6. *Wage Cost Caused by Decreased Output of Injured Worker after Return to Work.* If the injured worker's previous wage payments are continued despite a 40% reduction in output, the incident should be charged with 40% of the worker's wages during the period of low output.

7. *Cost of Learning Period of New Worker.* If a replacement worker produces only half as much in the first two weeks as the injured worker would have produced in the same time for the same pay, then half of the new worker's wages for the two weeks' period should be considered part of the cost of the incident. A wage cost for time spent by supervisors or others in training the new worker also should be attributed to the incident.

8. *Uninsured Medical Cost Borne by the Company.* This cost is usually that of medical services provided at the plant dispensary. There is no great difficulty in estimating an average cost per visit for this medical attention. The question may be raised, however, whether this expense can be considered a variable cost. That is, would a reduction in incidents result in lower expenses for operating the dispensary?

9. *Cost of Time Spent by Management and Clerical Workers on Investigations or in the Processing of Compensation Application Forms.* Time spent by management or supervision (other than the foreman or supervisor covered in Item 5) and by clerical employees in investigating an incident, or settling claims arising from it, is chargeable to the incident.

10. *Miscellaneous Costs.* This category includes less typical costs, the validity of which must be clearly shown by the investigator on individual incident reports. Among such possible costs are public liability claims, equipment rental, losses due to cancelled contracts or lost orders if the incident causes an overall reduction in total sales, loss of company bonuses, cost of hiring new employees if this expense is significant, cost of above-normal spoilage by new employees, and demurrage. These cost factors and any others not suggested above would need to be well substantiated. Miscellaneous costs were found in less than 2% of several hundred cases reviewed in connection with this study.

Conducting a Pilot Study

The purpose of the pilot study is to develop average uninsured costs for different classes of incidents that can be applied to future incident totals. Therefore, it is desirable not to include the costs of deaths and permanent total disabilities. Such incidents occur so seldom that the costs should be calculated individually and not estimated on the basis of averages.

Management should build in some flexibility in the grouping of classes of cases. If an organization makes no distinction in the records between medical treatment cases requiring a physician's attention and those not requiring a physician's attention, the pilot study may combine Classes 2 and 3. (See Method for Estimating earlier in this chapter.)

DEPARTMENT SUPERVISOR'S ACCIDENT COST INVESTIGATION REPORT

Injury/Accident_____

Date_____Name of Injured_____Dept._____

TIME LOST

1. How much time did other employees lose by talking, watching, or helping at accident? Number of employees _____ x hours =

2. How much productive time was lost because of damaged equipment or loss of reduced output by injured worker?
   Estimate Hours =

3. How much time did injured employee lose for which he was paid on the day of the injury?
   Estimate Hours =

4. Will overtime be necessary?   Estimate Hours =

5. How much of the supervisors or other managements' time was lost as a result of this accident?
   Estimate Hours =

6. Were additional costs incurred due to hiring and training or replacement?
   Training Time Estimate Hours =

7. Describe the damage to material or equipment. _____

   _____

8. If machine and/or operations were idle, can loss of production be made up?
   Yes_____   No_____

9. Will overtime be necessary? Yes_____   No_____

10. Any demurrage or other cost involved? Yes_____   No_____

ADDITIONAL ACCIDENT COSTS

To compute the total costs of the accident, it is necessary to complete the following costs. Should the supervisor have access to this information it is advised he complete as much as possible. Safety Department will develop those costs not known by supervisor.

Figure 7-3. This cost form (8 × 11 in.) should be prepared by the department supervisor as soon after the accident as information becomes available on the amount of time lost by all persons and the extent of damage to product and equipment. It is sent to the safety department not later than the day after the accident.

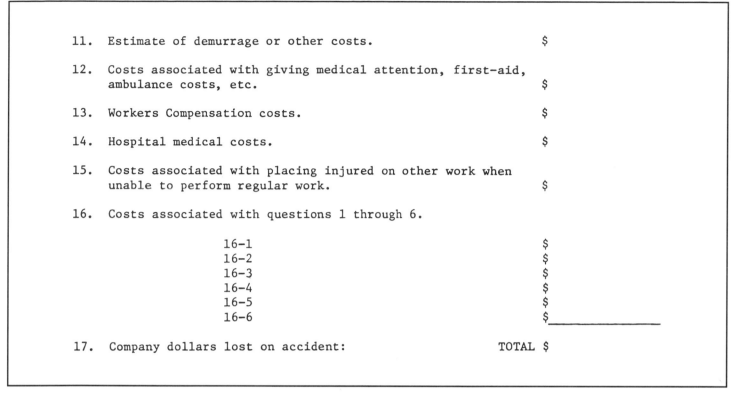

11. Estimate of demurrage or other costs. $

12. Costs associated with giving medical attention, first-aid, ambulance costs, etc. $

13. Workers Compensation costs. $

14. Hospital medical costs. $

15. Costs associated with placing injured on other work when unable to perform regular work. $

16. Costs associated with questions 1 through 6.

        16-1 $
        16-2 $
        16-3 $
        16-4 $
        16-5 $
        16-6 $_____

17. Company dollars lost on accident: TOTAL $

Figure 7-3. (Concluded).

The following discussion assumes the study of costs will be made with the injuries grouped in the recommended classes. The discussion covers Classes 1, 2, and 4. A different method must be applied to Class 3 injuries, and it will be discussed later.

Classes 1, 2, and 4

To analyze uninsured costs for incidents in Classes 1, 2, and 4, the supervisor in charge of the department where an incident occurs should secure for each incident the information indicated on the Department Supervisor's Accident Cost Investigation Report form (Figure 7-3). These data can be obtained during the supervisor's regular investigation of the incident. As soon as each report form is completed, it should be sent to the safety department.

Safety personnel then transfer information from the department supervisor's report to the Investigator's Cost Data Sheet (Figure 7-4). The safety department then assumes responsibility for securing the supplementary information from the accounting department, industrial relations department, and other departments where records on lost time and other necessary information are kept.

As an alternative, a member of the safety department could secure all information needed on the data sheet. In this case, the supervisor's report form is not used, and the supervisor is required only to report each incident in Classes 1, 2, or 4 to the safety department as soon as it occurs.

The investigator, before computing averages, should be certain that the pilot study has covered a sufficient number of cases of Classes 1, 2, and 4 to be representative.

This number will rarely be less than 20 cases. However, more cases should be studied if the costs of the cases in a particular class vary widely. Information should be secured on enough cases of each class so the average cost per case in each class is fully representative of past experience and will, by inference, be applicable to future experience.

Once a sufficient number of cases have been accumulated, the investigation of individual cases can be discontinued. For the data thus collected, separate averages should be calculated for the cases of each class. It is recognized that these costs are averages of the uninsured costs only.

Class 3 Injuries

These injuries are the common first aid cases in which no significant property damage results from the incident. They are the most difficult to analyze from the standpoint of cost. This is because the time lost is likely to occur repeatedly and only for short periods. Also, the injuries can occur so frequently as to place an undue burden on the supervisor and safety director if a complete report form and data sheet are required for each case.

The points of essential information needed are the average amount of working time lost per trip to the dispensary, the average dispensary cost per treatment, the average number of visits to the dispensary per case, and the average amount of supervisor's time required per case. The following method of developing averages for each of these items is recommended:

INVESTIGATOR'S COST DATA SHEET

Class 1_____
(Permanent partial or temporary
total disability)

Class 2 _____
(Temporary partial disability or
medical treatment case requiring
outside physician's care)

Class 3 _____
(Medical treatment case requiring
local dispensary care)

Class 4 _____
(No injury)

Name _____

Date of injury _____   Its nature _____

Department _____   Operation _____Hourly wage _____

  Hourly wage of supervisor $_____

  Average hourly wage of workers in department where injury occurred $_____

1. Wage cost of time lost by workers who were not injured, if paid by employer $_____

   a. Number of workers who lost time because they were talking, watching, helping ____

      Average amount of time lost per worker _____ hours _____ minutes

   b. Number of workers who lost time because they lacked equipment damaged in

      accident or because they needed output or aid of injured worker _____.

      Average amount of time lost per worker _____ hours _____minutes.

2. Nature of damage to material or equipment _____

_____

  Net cost to repair, replace, or put in order the above material or equipment $_____

3. Wage cost of time lost by injured worker while being paid by employer

   (other than workers compensation payments) $_____

   a. Time lost on day of injury for which worker was paid _____ hours _____ min.

   b. Number of subsequent days' absence for which worker was paid _____ days.

      (Other than workers' compensation payments) _____ hours per day.

   c. Number of additional trips for medical attention on employer's time on

      succeeding days after worker's return to work _____

      Average time per trip _____ hours. _____ min. Total trip time _____ hrs. _____ min.

   d. Additional lost time by employee, for which he was paid by company _____ hrs.

      _____ min.

(over)

Figure 7-4. This 8 × 11 in. form can be used to convert time losses into money losses. Initial time losses are obtained from the Department Supervisor's Accident Cost Report (Figure 7-3), and subsequent time losses are obtained from first aid and other departments as necessary. Wage rate information is obtained from the accounting department. The reverse side of the form (shown at the right) contains space for additional costs pertinent to the accident under study. See discussion under the heading, Conducting a Pilot Study.

4. If lost production was made up by overtime work, how much more did the work
    cost than if it had been done in regular hours? (cost items: wage
    rate difference, extra supervision, light, heat, cleaning for overtime.) $_____

5. Cost of supervisor's time required in connection with the accident $_____
    a. Supervisor's time shown on Dept. Supervisor's Report _____ hrs. _____ min.
    b. Additional supervisor's time required later _____ hrs. _____ min.

6. Wage cost due to decreased output of worker after injury if paid old rate $_____
    a. Total time on light work or at reduced output ____ days ____ hours per day
    b. Worker's average percentage of normal output during this period _____

7. If injured worker was replaced by new worker, wage cost of learning period $_____
    a. Time new worker's output was below normal for his own wage ____ days
        _____ hours per day. His average percentage of normal output during
        time _____ His hourly wage $_____.
    b. Time of supervisor or others for training _____ hrs. Cost per hour $_____

8. Medical cost to company (not covered by workers' compensation insurance) $_____

9. Cost of time spent by higher supervision on investigation, including local
    processing of worker's compensation application form. (No safety or pre-
    vention activities should be included.) $_____

10. Other costs are not covered above (e.g., public liability claims; cost of renting
    replacement equipment; loss of profit on contracts cancelled or orders lost if
    accident causes net reduction in total sales; loss of bonuses by company;
    cost of hiring new employee if the additional hiring expense is significant;
    cost of excessive spoilage by new employee; demurrage). $_____

Explain fully:
    Total uninsured cost ....$_____
Name of Company _____

Figure 7-4. (Concluded).

1. Secure an estimate of average working time lost per trip to the dispensary for first aid. Departmental time records should be consulted as they may show the amount of time each worker is absent from the job while receiving first aid. If so, a random sample of 50 to 100 records of persons known to have received first aid should be selected from different departments. The average time lost per dispensary visit is calculated by adding the absence time for all visits in the sample and dividing by the total number of visits.

2. If departmental records do not contain this information, it will be necessary to assign an investigator to observe a random sample of 50 or more persons visiting the dispensary. As before, to secure the average time, all the estimated time intervals of absence are added and then divided by the total number of persons observed.

3. Make an estimate of the average cost of providing medical attention for each visit by dividing the total cost of operating the dispensary for a year by the total number of treatments given during the year.

4. Calculate the average number of visits to the dispensary per case. This is done by dividing the number of treatments of Class 3 injuries in a representative period, perhaps a month or six weeks, by the number of Class 3 injuries reported during the same period of time.

5. Calculate the average amount of supervisor's time required per case. Where possible, this is accomplished by observing the activities of representative supervisors in connection with first aid cases.

Over time, a sufficient number of cases will have been studied to be representative both of the activities of supervisors in different departments and of different types of first aid cases. The average time spent by a supervisor is then computed by adding all the time intervals recorded and dividing by the number of cases.

In some instances, it may be impossible to make a time study of the supervisor's activities in connection with first aid cases. The only alternative is to secure from each supervisor an estimate of the time spent on the usual first aid case and then to add them and divide by the number of supervisors to find the average time. Determine the average value of this time by multiplying it by the average hourly wage of a supervisor.

The average total uninsured cost of a case in Class 3 is estimated from the data accumulated above as follows: the average amount of time lost for a trip to the dispensary (1, above) is multiplied by the facility's average wage rate, secured from the payroll department, to get the average cost per trip for the worker's time lost. To this figure is added the estimated cost of providing medical attention for a single visit (2). This figure is then multiplied by the average number of dispensary visits per medical treatment

case (3), and to this result is added the average value of supervisor's time required (4).

This method of recording costs is designed to provide estimates of the average uninsured cost per case for incidents causing localized property damage or, at most, a few injuries.

The method of cost investigation for incidents resulting in deaths, permanent total disabilities, or unusually extensive property damage is essentially the same as for others. However, the main difference is that every one is separately investigated and should be included in the final cost estimate as a separate item. In estimating the cost of a fire, for example, the investigator should bear in mind that the company's fire insurance will probably cover property damage in a major incident while for a less serious incident, fire damage would appear as an uninsured cost.

Development of Final Cost Estimate

Once the average for each class of case has been established, costs for any period in which a sufficiently large number of incidents has occurred to be representative can be estimated with considerable accuracy. This is calculated by multiplying the average uninsured cost per case for each of the four classes by the number of cases occurring in that class during the period.

If any deaths, permanent total disabilities, or unusually extensive property damage incidents have occurred, the investigator should add the specific uninsured costs of these to the estimated costs of the four classes of incidents.

To these uninsured cost totals the investigator then adds the cost of workers' compensation and insured medical expense. For self-insured companies, this figure will be the total amount paid out in settlement of claims plus all expenses of administering the insurance. For companies not carrying their own insurance, it will be the amount of their insurance premiums plus deductibles paid.

The method will have to be modified in accordance with the record-keeping systems of different companies. For example, most self-insurers will find it impossible to separate compensated medical expense from dispensary care. In that case, these items should be combined into one, and the dispensary cost omitted from the analysis of noncompensated costs on the data sheets. For an illustration of the development of a final cost estimate, see the example in Tables 7-C and 7-D.

As stated previously, presentation of incident costs or estimated cost saving is useful in securing management support. Because incident costs represent lost profit opportunities, incident dollars may be treated as lost profit dollars. The general method for achieving additional profits is through additional sales. It is possible then to estimate the sales necessary to recover profits lost due to incident expenses. Simply determine the net profit percentage to gross sales and divide incident costs by the net profit percentage. This will provide management with an estimate of the additional gross sales dollar volume necessary to replace profit lost due to incident expenses.

Another method of presenting incident costs to management is by equating lost incident dollars to lost purchasing opportunities for needed supplies or equipment.

OFF-THE-JOB DISABLING INJURY COST

The employer loses the same services whether the employee is injured off the job or on the job and incurs just about the same types of direct and indirect costs. Nevertheless, the costs of off-the-job (OTJ) incidents and illnesses are, at least in part, handled differently. This section explains the difference and presents sample calculations. (See Off-the-Job Safety Programs in Chapter 3, Loss Control Programs, and Off-the-Job Disabling Injuries in Chapter 8, Injury and Illness Record Keeping and Incidence Rates, in this volume.)

When incidents take place off the job, a major portion of the costs are borne by employers. Some of the costs are evident, such as insurance premiums and wages paid to absent employees. Some of the costs are hidden, such as training for new or transferred workers and medical staff time demanded for workers returning to work after an incident. For example, a new worker does not produce at the same level as an experienced worker; thus, the decreased productivity of the new worker indirectly increases the manufacturing overhead.

Other costs are more difficult to assess, although still very real. As incident rates in the community rise, so do insurance rates, taxes, and welfare contributions. Not all organizations are aware of the total costs that can result from off-the-job incidents and the impact they have on operations and profits. Enough experience has been accumulated, however, to develop a simplified plan for estimating such cost.

Categorizing OTJ Disabling Injury Cost

The cost of off-the-job disabling injuries (OTJ DI) to an organization falls into the following two categories: insured and uninsured. These are the same for on-the-job incidents that result in disabling injuries, described in the previous section.

Most uninsured costs are hidden. Aside from wage costs, most organizations do not keep records of uninsured costs. However, these costs are associated with all OTJ DI incidents and, therefore, affect profit margins.

- Insured-worker productivity, cost, product loss, and equipment damage.—Costs directly associated with the employee who sustained the OTJ DI injury are included in this expense subcategory.
- Noninjured-worker productivity cost, product loss, equipment damage, and administrative cost.—Costs incurred by personnel other than the employee who sustained the OTJ DI injury are included in this subcategory.

- Miscellaneous costs.—This subcategory includes loss of profit for cancelled contracts or orders and the costs of demurrage, telephone calls, transportation, or other miscellaneous expenses.

Estimating OTJ DI Costs

Some experts say the ratio of insured cost to uninsured cost is 3:2. In order to estimate a company's losses from employee OTJ DIs, management must first determine the insured cost. Next, using the 3:2 factor, it must escalate the insured cost to determine the total (insured and uninsured) employee costs. The insured cost for injuries to dependents of employees is then added to the total employee cost to ascertain total losses. The following examples illustrate calculation procedures.

Example 1

Company A is insured by an outside carrier. Twenty percent or $225,000 of its annual premium charge was required to pay for its previous calendar year OTJ DI incident experience. Of that total, $75,000 was required for 11 employee injuries and the remaining $150,000 was required for 22 employee-dependent injuries. These figures include the administrative fee paid by the company to the carrier.

The total cost for employee-dependent injuries is a conservative figure, since it does not include the administrative cost incurred by the company's insurance office staff to process claims. If this cost is known, it should be added to the employee-dependent injury expense category. When the $75,000 insured cost category is escalated to include the uninsured cost category, the total expense for employee injuries becomes $125,000.

Company A Estimated OTJ DI Costs

Insured cost for employee injuries =	$75,000
Uninsured cost for employee injuries =	50,000
Insured cost for employee-dependent injuries =	150,000
Total annual estimated OTJ DI cost =	$275,000

Example 2

Company B is insured by an outside carrier. Its carrier stated that $850,000 was paid for 138 employee injuries and $1,800,000 was paid for 279 employee-dependent injuries for its previous calendar year OTJ DI incident experience. Their administrative fee to the carrier was 6% of the total cost, resulting in a cost of approximately $900,000 for employee injuries and $1,900,000 for employee-dependent injuries. If the cost for insurance office staff claim-processing is known, it should be added to the total employee-dependent cost. Escalation of the insured cost category for employee injuries indicated a total of $1,500,000.

Company B Estimated OTJ DI Costs

Insured cost for employee injuries = $900,000

Uninsured cost for employee injuries = 600,000

Insured cost for
employee-dependent injuries = 1,900,000

Total annual estimated OTJ DI cost = $3,400,000

Example 3

Company C is self-insured. Insurance office records indicated that for the previous calendar year, OTJ DI incident experience for the amount of medical and health claims paid for 350 employee injuries was $2,280,000, and $4,700,000 was paid for 690 employee-dependent injuries. Insurance office staff administrative costs for claim-processing should be added to the employee-dependent cost category if that cost is known. Thus, $2,280,000 escalated to include the uninsured cost category for employee injuries resulted in a total cost of approximately $3,800,000.

Company C Estimated OTJ DI Costs

Insured cost for employee injuries = $2,280,000

Uninsured cost for employee injuries = 1,520,000

Insured cost for
employee-dependent injuries = 4,700,000

Total annual estimated OTJ DI cost = $8,500,000

Measuring the Effects of OTJ Safety Programs

Calculations of average costs are useful tools to support initiating or accelerating off-the-job safety awareness programs. The calculations also can be used to measure the effects of safety programs.

For example, 350 employees of Company C experienced OTJ injuries during the previous year for a total of approximately $3,800,000 and an average cost per incident of approximately $10,860. Based upon the OTJ DI loss experience, top management allocated $50,000 from the budget to initiate a safety awareness program. At the end of the year, the $10,860 average cost figure will be escalated to adjust for inflation and, in turn, the new figure will be used to calculate losses. For illustration purposes, it is assumed that the new average cost per incident figure is $11,500 and that employee injuries were reduced from 350 to 300. Calculations (300 × $11,500) indicate losses of $3,450,000, a savings of approximately $350,000 from last year's total. The estimated net return is $300,000, which is a 600% return on investment.

To further support justification for operating funds, safety personnel can add savings realized from reduced employee-dependent injuries to the employee savings total. This type of analysis provides a management tool to evaluate the impact of off-the-job disabling injuries on profit margins; thus, it can be used to gain management commitment to support operating budgets for safety awareness programs.

SUMMARY

- Successful incident prevention requires (1) study of all working areas to detect and control or eliminate hazards; (2) study of all operating procedures and administrative controls, (3) education, training, and discipline to minimize human factors; and (4) thorough incident investigation and analysis.

- The primary purpose of incident investigation and analysis is to prevent incidents by uncovering facts about each incident and determining how to prevent a recurrence. Several incident investigation and analysis techniques are available to management and can be adapted to the needs of each organization. Incident investigation generally should be made by the supervisor. The safety professional advises and guides the supervisor. In some companies, a special investigation or review committee reports on all incidents or reviews the investigation.

- All incidents should be investigated, particularly those involving deaths or serious injuries. Near-misses and a series of minor injuries also demand study. Employers should involve outside employees and contractors in safety on the job and incident investigation.

- Analysis of incident investigations can help management identify sources of incidents, pinpoint incident problems, reveal the need for engineering changes or correcting inefficiencies, disclose unsafe practices, enable supervisors to target safety efforts, and evaluate a safety program's effectiveness.

- Two incident analyses can be done: (1) to determine corrective actions and (2) to conduct a statistical analysis to uncover incident patterns so corrective actions can be devised.
 Statistical analyses can be used to classify various groups of data to identify key factors causing or contributing to incidents. These analyses also can be used to compare current incident rates with prior years' experience, with rates of other companies, or with rates for the industry as a whole.

- Most companies have a method for calculating direct (insured) or indirect (uninsured) costs associated with incidents. Pilot studies are used to determine the average uninsured costs for different classes of incidents that can be applied to future incident totals. Worker compensation and insured medical expenses are then added to these totals.

- Investigation, analysis, and cost estimates for off-the-job incidents are handled similarly to on-the-job incidents. These activities can show management the impact of off-the-job incidents on organizational operations and profits and help to gain support for worker education and safety training programs.

REFERENCES

Blankenship LM. *Nonoccupational Disabling Injury Cost Study*. K/DSA–457 Oak Ridge, TN: Martin Marietta Energy Systems, Inc., October 1981.

DeReamer R. *Modern Safety and Health Technology*. New York: John Wiley and Sons, Inc., 1981.

Grimaldi JV and Simonds RH. *Safety Management–Accident Cost and Control,* 5th ed. Burr Ridge, IL: Richard D. Irwin, Inc., 1988.

Johnson WG. *MORT Safety Assurance Systems*. New York: Marcel Dekker, Inc., 1980.

Kepner CH and Tregoe BB. *The New Rational Managers*. Princeton, NJ: Kepner-Tregoe, Inc., 1981.

National Safety Council, 1121 Spring Lake Drive, Itasca, IL 60143.

Accident Investigation, 2nd ed., 1995.

Occupational Safety and Health Data Sheet 601, *Off-the-Job Safety*, 12304–0601, 1986.

Information Management for Occupational Safety and Health, ANSI Z16.2–1995.

The Off-the-Job Safety Program Manual, 1994.

Nolter JC and Johnson RD. "How to conduct an accident investigation." In Slote L (Ed.). *Handbook of Occupational Safety and Health*. New York: John Wiley and Sons, 1987, pp. 99–123.

Tritsch S. "Accident investigation: How to ask why." *Safety & Health*, 1992;146(6):40–43.

Vincoli JW. *Basic Guide to Accident Investigation and Loss Control*. New York: VanNostrand Reinhold, 1994.

REVIEW QUESTIONS

1. Name the six (6) fundamental activities needed for a successful incident prevention program.
2. What is the primary purpose of an incident investigation?
3. What types(s) of incidents/accidents should be investigated?
4. Who is responsible for investigating an incident after it has occurred?
5. How soon after an incident has occurred should an investigation be started?
6. What can a company learn from incident and accident causation analysis?
7. List the eight (8) groups of minimum data that should be collected for each incident.
8. What are the three (3) basic steps in a systematic approach to selecting corrective actions?
9. What does the new ANSI 16.2-1995 recommend?
10. After the thorough analysis of groups of incident investigations, what corrective actions might be suggested that were not evident when studying an individual case?
11. What are the two (2) general categories of work incidents for the purpose of cost analysis?

Injury and Illness Record Keeping and Incidence Rates

8

I n this chapter, the terms "incident" and "injury" are restricted to occupational injuries and illnesses. In other chapters, "incident" is used in its broad meaning: unplanned, undesired events that interrupt the completion of an activity, and that may include property damage or injury.

The Williams-Steiger Occupational Safety and Health Act of 1970 (OSHAct) requires most U.S. employers to maintain specific records of work-related employee injuries and illnesses. Some employers are required to maintain injury and illness records under regulations issued by other federal agencies such as the Mine Safety and Health Administration (MSHA) and the Federal Railroad Administration (FRA). In addition to these records, many employers also are required to make reports to state compensation authorities. Similarly, insuring agencies may require reports for their records. Occupational injury and illness reports and records are now required of nearly every establishment by its management or the government.

Safety personnel are faced with two tasks—maintaining those records required by law and by their management, and maintaining records useful to an effective safety program. Unfortunately, the two are not always synonymous. A good record-keeping system requires more data than that contained in most standard federal and state forms.

This chapter covers the following topics:

- reasons for keeping incident records and how to establish an effective system
- which record-keeping regulations under the OSHAct companies must follow
- how to calculate incident rates for on-the-job injuries and illnesses
- how to calculate off-the-job incident rates.

Because the record-keeping requirements are subject to change, companies should contact an OSHA or Bureau of Labor Statistics (BLS) regional office for the latest information or consult 29 *CFR* 1904. Establishments subject to MSHA or FRA record-keeping requirements should consult the appropriate parts of the *Code of Federal Regulations* (30 *CFR* 50 and 49 *CFR* 225, respectively).

INCIDENT RECORDS

Records of incidents and injuries are essential to maintain efficient and successful safety programs, just as records of production, costs, sales, and profits and losses are essential to efficient and successful business operations. Records supply the information necessary to transform haphazard, costly, ineffective safety work into a planned safety program that controls both the conditions and the acts that contribute to incidents. Good record keeping is the foundation of a scientific approach to occupational safety.

Uses of Records
A good record-keeping system can help the safety professional in the following ways:

1. Provide safety personnel with the means for an objective evaluation of their incident problems and with a measurement of the overall progress and effectiveness of their safety program.
2. Identify high incident rate units, plants, or departments and problem areas so extra effort can be made in those areas.
3. Provide data for an analysis of incidents pointing to specific causes or circumstances, which can then be attacked by specific countermeasures.
4. Create interest in safety among supervisors or team leaders by furnishing them with information about their departments' incident experience.
5. Provide supervisors and safety committees with hard facts about their safety problems so their efforts can be concentrated.
6. Measure the effectiveness of individual countermeasures and determine if specific programs are doing the job they were designed to do.
7. Assist management in performance evaluation.

Record-Keeping Systems
The system presented in this section is a model that can be used to provide the basic items necessary for good record keeping. It is designed to dovetail with the present record-keeping requirements of the OSHAct and attempts to avoid a duplication of effort on the part of personnel responsible for keeping records and filing reports. Some of the forms presented in this section are also constructed with modern data-processing methods in mind. In general, a self-coding check-off form can save time for both the person who fills out the report and the person who is responsible for tabulating and processing the data.

A well-designed form takes into account the person who will fill it out and the way in which the forms will be processed. An incident report should accomplish three things: establish all causes contributing to the incident; reveal questions the investigator should ask to determine all environmental and human causes; and provide a means of accumulating incident data. Such a form is more likely to be filled out accurately and will present fewer problems for those who process and analyze the data. Care in the choice and design of forms will pay dividends in better, more reliable data. Centralized coding of data (e.g., in the safety office) has advantages as well. It increases reliability of the data and makes it possible to change the coding system without retraining as many people. A few coders who work with the system a great deal will have a better understanding of the codes and coding procedures than will supervisors who may code only a few reports each year.

The record-keeping system in this section is not the only way to keep records, but rather furnishes an example. The incident problems of individual establishments are unique and no one form or set of forms can provide every establishment with all the data for solving all of its individual problems. A system that does a good job of collecting the basic facts, however, makes it easier later on to pinpoint the data relating to a specific problem.

In addition to the system presented here, which is primarily a manual, paper-based system, there are a number of personal computer-based record-keeping systems available on the market today. Such systems greatly facilitate data analysis; some can be integrated into training management programs. (See also Chapter 9, Computers and Information Management, in this volume.)

The following sections deal with occupational injuries and illnesses. Property-damage incidents are covered in Chapter 7. Nonemployee incidents are covered in Chapter 21. Specific record-keeping requirements were explained in Chapter 2, Regulatory History.

Incident Reports and Injury Records

To be effective, preventive measures must be based on complete and unbiased knowledge of the causes of incidents. The primary purpose of an incident report is to obtain such information but not to fix blame. The completeness and accuracy of the entire incident record system depend upon information in the individual incident reports. Therefore, management must be sure the forms and their purpose are understood by those who must fill them out. Such employees should be given necessary training or instruction.

The First Aid Report

The collection of injury data generally begins in the first aid department. The first aid attendant or nurse fills out a report for each new case. Copies are sent to the safety department or safety committee, the worker's supervisor, and other departments as management may wish (Figure 8-1).

The first aid attendant or the nurse should know enough about incident analysis, record keeping, and investigation to be able to record the principal facts about each case. The company's occupational physician also should be informed of the basic rules for classifying cases. At times, his or her repeat treatment and opinion of the seriousness of an injury may be necessary to record the case accurately and to help determine if the specific injury could have been caused by this incident.

Incident Investigation Report

It is recommended that the supervisor make a detailed report about each incident, even when only a minor injury

Case No. ____________ Date ____________

First Aid Report

Name ____________________ Department ____________________

Occupation ____________________ Supervisor ____________________

Date of Occurrence __________ Time ______ a.m. p.m. Date Treated __________ Time ______ a.m. p.m.

Nature of Injury or Illness ________________________________

Treatment Given ________________________________

Sent: __ Back to Work __ To Doctor __ Home __ Hospital

Estimated lost workdays: ___ away from work ___ restricted work activity

Employee's description of Occurrence ________________________________

____

____

Report prepared by: ________________ Signed: ________________

Figure 8-1. This First Aid Report (4 × 6 in. or 10 × 16 cm) is prepared by the first aid attendant at the time an injured or ill person comes for treatment. A report serves as a record and permits quick tabulation of such data as department, occupation, and the key facts of the occurrence.

or no injury is the result. For purposes of OSHA summaries, only those reports that meet the minimum severity level need to be separated and tallied. Minor injuries occur in greater numbers than serious injuries, and records of these injuries can help to pinpoint problem areas. By working to alleviate these problems, workers and management can prevent many serious injuries. Minor injuries should not be regarded lightly, however. Complications may arise from these injuries, and their result can be quite serious.

Supervisors or team leaders should complete the supervisor's incident investigation report form as soon as possible after an incident occurs. They should send copies of these reports to the safety department and to other designated persons. Information concerning activities and conditions that preceded an occurrence is important to prevent future incidents. This information is particularly difficult to get unless it is obtained promptly after the incident occurs.

Generally, analyses of incidents are made only periodically, and often long after the incidents have occurred. Because at that time workers may find it impossible to recall the precise details of an incident, supervisors and others must record details accurately and completely at once or they may be lost forever.

All information may not be available at the time the incident report is being filled out. As a result, items such as total time lost and dollar amount can be added later. This should not, however, prevent the other items from being answered immediately after the incident occurs.

Three different supervisor's report forms are presented. They fulfill all of the information requirements of the present OSHA 101 form. The forms also include questions in addition to those contained in the present OSHA 101 form. These questions ask for additional basic data that should be known about each incident.

The first supervisor's report form (Figure 8-2) is an open-ended, narrative type of form. The next forms (Figures 8-3 and 8-4) are self-coding to allow keypunching of data items directly from the form without the extra step of recording this information for data-processing equipment. By using self-coding forms, data-processing equipment can easily be used to enter the information into the system and to produce a variety of summary reports (such as summaries by department and by type of incident). Detailed crosstabulations thus can be produced with little effort.

For definitions of the terms regarding severity of injury, what cases are recordable, and the like, please consult the OSHA sections of this chapter. For further clarification of OSHA definitions, see the Recording and Classifying Cases section of this chapter. Managers also may want to consult federal or state authorities (if their state has an approved plan in effect).

Employee Injury and Illness Record

After cases are closed, the first aid report and the supervisor's report are filed by source of injury (type of machine,

tool, material, etc.), type of event or exposure, or another factor that will facilitate use of the reports for incident prevention. Another form, therefore, must be used to record the injury experience of individual employees (Figure 8-5).

This form helps supervisors remember the experience of individual employees. Particularly in large establishments or plants where supervisors have many people working for them, they may not remember the total number of injuries—especially if the injuries are minor—suffered by individual employees. The employee injury card, therefore, fills a real need. It provides space to record such factors as date of injury, classification, costs, and lost workdays.

Because of the importance of the human factor in incidents, much can be learned about incident causes from studying employee injury records. If certain employees or job classifications have frequent injuries, a study of employee working habits, physical and mental abilities, training, job assignments, working environment, and the instructions and supervision given them may reveal as much as a study of incident locations, agencies, or other factors. An increase in the frequency of incidents by an employee bears further investigation.

Filing Reports

After injury reports have been used to compile monthly summaries, the incomplete reports can be kept in a temporary file for convenient reference as later information about the injuries becomes available. When injury reports are complete, they should be filed in a convenient system that permits quick access of information for special studies of incident conditions. For example, the reports can be filed by agency of the injury, by occupation of the injured person, by department, or by some similar item.

The employee injury card should be cross-referenced to the file location of the detailed incident report.

Periodic Reports

The forms discussed in the preceding paragraphs are prepared when the incidents occur; they are used to record the incidents and preserve information about contributing circumstances. Periodically, management should summarize this information and relate it to department or facility exposure in order to evaluate safety work and to identify the principal incident causes.

Monthly Summary of Injuries and Illnesses

Managers and supervisors should prepare a monthly summary of injuries and illnesses to reveal the current status of incident experience. This monthly summary of injury and illness cases (Figure 8-6) allows for tabulating monthly and cumulative totals and the computation of OSHA incidence rates. Space is also provided for yearly totals and rates. This form would be filled out on the basis of the individual report forms that were processed during the month or from OSHA form No. 200, Log and Summary of Occupational Injuries and Illnesses (Figure 8-7).

(Text continues on p. 212.)

ACCIDENT INVESTIGATION REPORT

Case Number

Company _____ **Address** _____

Department _____ **Location (if different from mailing address)** _____

1. Name of injured	**2. Social Security Number** / **3. Sex** ☐ M ☐ F / **4. Age** / **5. Date of accident**

6. Home address _____

7. Employee's usual occupation   **8. Occupation at time of accident**

9. Length of employment
☐ Less than 1 mo.   ☐ 1-5 mo.
☐ 6 mo. - 5 yr   ☐ More than 5 yr.

10. Time in occup. at time of accident
☐ Less than 1 mo.   ☐ 1-5 mo.
☐ 6 mo. - 5 yr   ☐ More than 5 yr.

11. Employment category
☐ Regular, full-time   ☐ Regular, part-time
☐ Temporary   ☐ Seasonal   ☐ Non-Employee

12. Case numbers and names of others injured in same accident
_____   _____
_____   _____

13. Nature of injury and part of body
_____

14. Name and address of physician
_____

15. Name and address of hospital
_____

16. Time of injury
A. _____ a.m.
            p.m.
B. Time within shift
C. Type of shift

17. Severity of injury
☐ Fatality
☐ Lost workdays—days away from work
☐ Lost workdays—days of restricted activity
☐ Medical treatment
☐ First aid
☐ Other, specify_____

18. Specific location of accident
_____

On employer's premises?   ☐ Yes   ☐ No

19. Phase of employee's workday at time of injury
☐ During rest period   ☐ Entering or leaving plant
☐ During meal period   ☐ Performing work duties
☐ Working overtime   ☐ Other _____

20. Describe how the accident occurred
_____
_____
_____

21. Accident sequence. Describe in reverse order of occurrence events preceding the injury and accident.
Starting with the injury and moving backward in time, reconstruct the sequence of events that led to the injury.

A. Injury event _____

B. Accident event _____

C. Preceding event #1 _____

D. Preceding event #2, 3, etc. _____
_____

Figure 8-2. This Accident Investigation Report (8 × 11 in. or 22 × 28 cm) captures a record of contributing circumstances to provide a basis for specific remedial action. Users should be trained to properly fill it out.

22. Task and activity at time of accident

General type of task

Specific activity

Employee was working:

☐ Alone   ☐ With crew or fellow worker   ☐ Other. specify

23. Posture of employee

24. Supervision at time of accident

☐ Directly supervised   ☐ Indirectly supervised

☐ Not supervised   ☐ Supervision not feasible

25. Causal factors. Events and conditions that contributed to the accident.
Include those identified by use of the Guide for Identifying Causal Factors and Corrective Actions.

26. Corrective actions. Those that have been, or will be, taken to prevent recurrence.
Include those identified by use of the Guide for Identifying Causal Factors and Corrective Actions.

Prepared by

Title

Department   **Date**

Developed by the National Safety Council
©1995 National Safety Council

Approved

Title   **Date**

Approved

Title   **Date**

Figure 8-2. (Concluded).

SERVICE NO. (NSC)

▶ 1-9 _____

CASE OR FILE NO.

▶ 10-15 _____

OSHA No. 101   NSC revision
(Meets OSHA requirements
when Instruction 1. has
been followed.)

SUPPLEMENTARY RECORD OF
OCCUPATIONAL INJURIES AND ILLNESSES

THIS REPORT IS

▶ 16, 1 ☐ First report   2 ☐ Revised report

EMPLOYER

1. **NAME** _____

2. **MAIL ADDRESS** _____

_____

3. **LOCATION,** if different from mail address _____

INJURED OR ILL EMPLOYEE

4. **NAME** _____

   SOCIAL SECURITY NO. _____

▶ **EMPLOYEE NO.** 17-26 _____

5. **HOME ADDRESS** _____

_____

▶ 6. **AGE**   27-28 _____

▶ 7. **SEX**   29, 1 ☐ Male   2 ☐ Female

▶ 8. **OCCUPATION** (specify) _____

   30-31, 01 ☐ Manager, official, proprietor
   02 ☐ Professional, technical
   03 ☐ Foreman, supervisor
   04 ☐ Sales worker
   05 ☐ Clerical worker
   06 ☐ Craftsman—construction
   07 ☐ Craftsman—other
   08 ☐ Machinist
   09 ☐ Mechanic
   10 ☐ Operative (production worker)
   11 ☐ Motor vehicle driver
   12 ☐ Laborer
   13 ☐ Service worker
   14 ☐ Agricultural worker
   15 ☐ Other
   16 ☐ Unknown

9. **DEPARTMENT** _____
   (Enter the name of department or division in which the injured person is regularly employed.)

CLASSIFICATION OF CASE

A. **INJURY OR ILLNESS** (see code on Log, OSHA No. 100)

▶ 32, 1 ☐ Injury (10)
   2 ☐ Occupational skin disease or disorder (21)
   3 ☐ Dust disease of the lungs (pneumoconioses) (22)
   4 ☐ Respiratory conditions due to toxic agents (23)
   5 ☐ Poisoning (systemic effects of toxic materials) (24)
   6 ☐ Disorder due to physical agents (other than toxic materials) (25)
   7 ☐ Disorder due to repeated trauma (26)
   8 ☐ All other occupational illnesses (29)

B. **EXTENT OF INJURY OR ILLNESS**

▶ 33, 1 ☐ Fatality
   2 ☐ Lost workday case
   3 ☐ Nonfatal case without lost workdays

▶ C. Number of workdays lost   34 36 _____

D. Permanently transferred or terminated

▶ 37, 1 ☐ Yes   2 ☐ No

THE ACCIDENT OR EXPOSURE TO OCCUPATIONAL ILLNESS

10. **PLACE OF ACCIDENT OR EXPOSURE** (mail address) _____

_____

11. **WHERE DID ACCIDENT OR EXPOSURE OCCUR?**

   a. On employer premises

▶ 38, 1 ☐ Yes   2 ☐ No   3 ☐ Unknown

   b. Place (specify) _____

▶ 39-40, 01 ☐ Office
   02 ☐ Plant, mill
   03 ☐ Shipping, receiving, warehouse
   04 ☐ Maintenance shop
   05 ☐ General or public area of employer premises (corridor, washroom, lunchroom, parking lot, etc.)
   06 ☐ Retail establishment (store, restaurant, gasoline station, etc.)
   07 ☐ Farm
   08 ☐ Motor vehicle accident
   09 ☐ Other
   10 ☐ Unknown

12. **WHAT WAS THE EMPLOYEE DOING WHEN INJURED?** (Be specific)

_____

_____

_____

   a. Task performed at time of accident

▶ 41-42, 01 ☐ Operating machine
   02 ☐ Operating hand tool (power or nonpower)
   03 ☐ Materials handling
   04 ☐ Maintenance & repair—machinery
   05 ☐ Maintenance & repair—building & equipment
   06 ☐ Motor vehicle driver, operator or passenger
   07 ☐ Office and sales tasks, except above
   08 ☐ Service tasks, except above
   09 ☐ Other
   10 ☐ Not performing task
   11 ☐ Unknown

   b. Activity at time of accident

▶ 43-44, 01 ☐ Climbing
   02 ☐ Driving
   03 ☐ Jumping
   04 ☐ Kneeling
   05 ☐ Lying down
   06 ☐ Lifting
   07 ☐ Reaching, stretching
   08 ☐ Riding
   09 ☐ Running
   10 ☐ Sitting
   11 ☐ Standing
   12 ☐ Walking
   13 ☐ Other
   14 ☐ Unknown

Figure 8-3. A self-coding supplementary record of occupational injuries and illnesses.

13. **HOW DID THE ACCIDENT OCCUR?** (Describe fully the events)

_____
_____
_____
_____

a. **AGENCY.** (Object or substance involved)

ACCIDENT AGENCY (1st column). The first object or substance involved in accident sequence.

INJURY AGENCY (2nd column). The agency inflicting the injury. See also section 15.

(Example: Worker fell from ladder and struck head on machine. Check "Ladder" under accident and check "Machine" under injury.)

ACCIDENT	INJURY	(Check one box in each column)

▶▶ 45-46, 01 ☐ 47-48, 01 ☐ Machine
02 ☐ 02 ☐ Conveyor, elevator, hoist
03 ☐ 03 ☐ Vehicle
04 ☐ 04 ☐ Electrical apparatus
05 ☐ 05 ☐ Hand tool
06 ☐ 06 ☐ Chemical
07 ☐ 07 ☐ Working surface, bench, table, etc.
08 ☐ 08 ☐ Floor, walking surface
09 ☐ 09 ☐ Bricks, rocks, stones
10 ☐ 10 ☐ Box, barrel, container (empty or full)
11 ☐ 11 ☐ Door, window, etc.
12 ☐ 12 ☐ Ladder
13 ☐ 13 ☐ Lumber, woodworking materials
14 ☐ 14 ☐ Metal
15 ☐ 15 ☐ Stairway, steps
16 ☐ 16 ☐ Other
17 ☐ 17 ☐ Unknown
18 ☐ 18 ☐ None

b. **ACCIDENT TYPE.** (First event in the accident sequence)

▶ 49-50, 01 ☐ Fall from elevation
02 ☐ Fall on same level
03 ☐ Struck against
04 ☐ Struck by
05 ☐ Caught in, under or between
06 ☐ Rubbed or abraded
07 ☐ Bodily reaction
08 ☐ Overexertion
09 ☐ Contact with electrical current
10 ☐ Contact with temperature extremes
11 ☐ Contact with radiations, caustics, toxic and noxious substances
12 ☐ Public transportation accident
13 ☐ Motor vehicle accident
14 ☐ Other
15 ☐ Unknown

This space may be used for additional information.

OCCUPATIONAL INJURY OR ILLNESS

14. **DESCRIBE THE INJURY OR ILLNESS** in detail and indicate the part of the body affected.

_____
_____
_____

a. **NATURE OF INJURY OR ILLNESS.** (Check most serious one)

▶ 51-52, 01 ☐ Amputation
02 ☐ Burn and scald (heat)
03 ☐ Burn (chemical)
04 ☐ Concussion
05 ☐ Crushing injury
06 ☐ Cut, laceration, puncture, abrasion
07 ☐ Fracture
08 ☐ Hernia
09 ☐ Bruise, contusion
10 ☐ Occupational illness
11 ☐ Sprain, strain
12 ☐ Other

b. **PART OF BODY.** (Check most serious one)

▶ 53-54, 01 ☐ Eyes
02 ☐ Head, face, neck
03 ☐ Back
04 ☐ Trunk (except back, internal)
05 ☐ Arm
06 ☐ Hand and wrist
07 ☐ Fingers
08 ☐ Leg
09 ☐ Feet and ankles
10 ☐ Toes
11 ☐ Internal and other

15. **NAME THE OBJECT OR SUBSTANCE WHICH DIRECTLY INJURED THE EMPLOYEE.** Also check one box in injury column under 13a.

_____
_____
_____

16. **DATE OF INJURY OR INITIAL DIAGNOSIS OF OCCUPATIONAL ILLNESS.**

a. **MONTH**

▶ 55-56, 01 ☐ Jan.   07 ☐ July
02 ☐ Feb.   08 ☐ Aug.
03 ☐ March   09 ☐ Sept.
04 ☐ April   10 ☐ Oct.
05 ☐ May   11 ☐ Nov.
06 ☐ June   12 ☐ Dec.

▶ b. **DATE OF MONTH**   57-58 _____

17. **DID EMPLOYEE DIE?**

▶ 59, 1 ☐ Yes   Date of Death _____
2 ☐ No

OTHER

18. **NAME AND ADDRESS OF PHYSICIAN** _____

_____

19. **IF HOSPITALIZED, NAME AND ADDRESS OF HOSPITAL** _____

_____

DATE OF REPORT _____

PREPARED BY _____

OFFICIAL POSITION _____

National Safety Council
Printed in U.S.A.

Figure 8-3. (Concluded)

ZURN / *a step ahead of tomorrow*

DIV. NAME: _____

SITE CODE: _____

OSHA FILE NO.: (Yr.-No.) _____

SUPERVISOR'S ACCIDENT INVESTIGATION REPORT

THIS FORM MUST BE COMPLETED AS SOON AFTER ACCIDENT AS POSSIBLE.

EMPLOYEE(S) _____ CLOCK NO. _____ ACCIDENT DATE ___/___/___

JOB TITLE/CRAFT _____ DEPT./SITE NAME _____

TIME / AGE GROUP	TYPE OF ACCIDENT	NATURE OF INJURY	PART OF BODY	UNSAFE CONDITION	UNSAFE ACT	UNSAFE PERSONAL FACTOR
UNDER 20	01 OVEREXERTION	26 BACK STRAIN-SPRAIN	51 HEAD	1 IMPROPERLY GUARDED	1 OPERATING WITHOUT AUTHORITY	1 IMPROPER ATTITUDE
20-29	02 FALL - SAME LEVEL	27 OTHER STRAIN-SPRAIN	52 EYE — 53 EAR	2 SAFETY DEVICES INOPERATIVE	2 WORKING ON MOVING OR DANGEROUS EQ	2 LACK OF KNOWLEDGE OR SKILL
30-39	03 FALL - DIFF LEVEL	29 DISLOCATION	54 FACE (OTHER)	3 DEFECTIVE	3 OPERATING AT UNSAFE SPEEDS	3 DEFECTIVE EYESIGHT
40-49	04 STRUCK AGAINST	30 FRACTURE	55 NECK	4 HAZARDOUS ARRANGEMENT	4 MAKING SAFETY DEVICES INOPERATIVE	4 DEFECTIVE HEARING
50-59	05 STRUCK BY	31 CONTUSION	56 SHOULDER	5 IMPROPER ILLUMINATION	5 TAKING UNSAFE POSITION OR POSTURE	5 FATIGUE
60 & OVER	06 CAUGHT BETWEEN	32 AMPUTATION	57 ARM — 58 ELBOW	6 IMPROPER VENTILATION	6 HANDLING MATERIALS INCORRECTLY - MANUAL	6 MUSCULAR WEAKNESS
___ A M	07 VEHICLE	33 OPEN WOUND	59 WRIST — 60 HAND	7 LACK OF SUITABLE PERSONAL PROT. EQ	7 USING DEFECTIVE TOOLS	7 HEART WEAKNESS
___ P M	10 ELECTRICAL	34 HEART ATTACK	61 FINGER	8 UNSAFE DRESS	8 USING HANDS INSTEAD OF TOOLS	8 EXISTING HERNIA
OVERTIME	11 TEMP EXTREME	35 VISION LOSS	62 BACK — 63 CHEST	9 HAZARDOUS DUST GASES OR FUMES	9 UNSAFE LOADING	9 INTOXICATED
DAY OF WEEK	12 REPETITIVE MOTION	36 BURN	64 LUNGS	10 UNCLASSIFIED INSUFFICIENT DATA	10 FAILURE TO USE PERSONAL PROT. EQ	10 UNCLASSIFIED INSUFFICIENT DATA
S M T W Th F Sa	13 RADIATION	40 ASPHYXIA	65 ABDOMEN	11 NO UNSAFE CONDITION	11 DISTRACTING, TEASING OR HORSEPLAY	11 NO UNSAFE PERSONAL FACTOR
TIME ON ___ YRS	14 INHALATION	41 HEARING LOSS	66 GROIN		12 NOT FOLLOWING RULES OR INSTRUCTIONS	
JOB ___ MOS	15 ABSORPTION	42 TENOSYNOVITIS	67 HIPS — 68 LEG		13 UNCLASSIFIED INSUFFICIENT DATA	
MALE	16 INGESTION	43 DERMATITIS	69 KNEE		14 NO UNSAFE ACT	
FEMALE	17 SLIP (NO FALL) TWIST	44 OCCUPATIONAL HEALTH	70 ANKLE			
OSHA — FATAL	18	45 FOREIGN BODY	71 FOOT — 72 TOES			
RECORDABLE — LOST TIME	19	46 MULTIPLE INJURY	73 BODY SYSTEM			
YES — DR VISIT ONLY	20	47 CUMULATIVE TRAUMA	74 MULTIPLE BODY PARTS			
NO — FIRST AID	24 OTHER	49 OTHER				

DESCRIBE IN DETAIL WHAT HAPPENED (Include name of machine, machine part or section, hand tool or material involved.) _____

_____

REGULAR JOB? YES ☐ NO ☐ WITNESS(ES) _____

REPORT DELAYED? YES ☐ NO ☐ IF YES, WHY? _____

DESCRIBE UNSAFE ACT OR FAILURE TO ACT BY EMPLOYEE(S) OR OTHERS: _____

_____

DESCRIBE ANY MECHANICAL, PHYSICAL OR ENVIRONMENTAL CONDITION THAT CONTRIBUTED TO ACCIDENT _____

_____

DESCRIBE ANY PERSONAL FACTORS THAT CONTRIBUTED TO ACCIDENT _____

WHAT **APPLICABLE** PERSONAL PROTECTIVE EQUIPMENT (PPE) **WAS BEING USED?** _____

_____ DID PPE FAIL? YES ☐ NO ☐

WHAT PERSONAL PROTECTIVE EQUIPMENT **SHOULD HAVE BEEN USED?** _____

_____ WAS PROPER PPE AVAILABLE? YES ☐ NO ☐

DESCRIBE ACTION(S) TAKEN TO PREVENT RECURRENCE _____

INDICATE DATE ACTION WAS OR WILL BE TAKEN: _____ BY WHOM: _____

___/___/___

DATE OF REPORT _____ NAME AND TITLE OF SUPERVISOR COMPLETING REPORT

SPECIFIC COMMENTS AND RECOMMENDATIONS BY DEPARTMENT HEAD _____

___/___/___

DATE _____ NAME OF DEPT. HEAD

DATE: _____ SAFETY SUPERVISOR: _____

This report should be completed and sent to Safety Dept. **no later than the next regularly scheduled workday** following the accident.

DIV./PLANT SAFETY — Form No. 80-PER, 4/89

CORP. SAFETY (FOR ALL LOST TIME INJURIES) — Form No. 80-PER, 4/89

DIV. SITE FIRST AID DEPT. — Form No. 80-PER, 4/89

EMPLOYEE FILE — Form No. 80-PER, 4/89

Figure 8-4. This Supervisor's Accident Investigation Report must be completed as soon after the accident as possible. The four-part form has distribution indicated on the bottom. (Courtesy Zurn Industries, Inc., Corp. Safety Dept.)

INJURY AND ILLNESS RECORD OF EMPLOYEE

Name _____ ID Number _____

Occupation _____ Department _____ Date Hired _____

Case Number	Injury or Illness	Date of Occurrence	Medical, Compensation, and Other Costs	Severity of Case			
				Fatal	Lost Workdays Involved?* Yes	No	First Aid Only

*Enter number of days lost in "yes" column or check mark in "no" column.

Figure 8-5. The Injury and Illness Record of Employee is a 4 × 6-in. (10 × 15-cm) card for recording injuries.

The monthly summary should be prepared as soon after the end of each month as the information becomes available but not later than the 20th of the following month. Because this report is primarily prepared to reveal the current status of incident experience, it is essential that this information be determined as soon as possible. If an incident report is still incomplete on the 20th of the following month because the employee has not returned to work, or if the classification of an injury is still in doubt at that time, an estimate of the outcome should be made by the company or consulting physician. The report must then be included with the completed cases in the monthly summary.

When definite information becomes available for estimated cases, any change in classification or OSHA lost workdays should be entered in the appropriate columns in the month of closing of the case. The adjustment is then included in the cumulative figures for the year through that month. This procedure provides reliable monthly data and an easy method of adjusting cumulative data.

Annual Report

Every establishment subject to the OSHAct is obliged to post its annual summary by February 1st for 30 days. The cumulative totals on OSHA form No. 200, Log and Summary of Occupational Injuries and Illnesses, serve as the annual report. Those establishments designated as part of the BLS annual survey also must report these figures to the requesting agency.

For management purposes, however, the annual report fulfills a more direct function. Managers and others prepare monthly summaries of injuries and illnesses primarily to show the trend of safety performance during the year. However, annual reports are prepared to compare data for the longer periods with the experience of previous years and with the experience of similar organizations and of the industry as a whole.

Monthly injury rates, especially in smaller companies, often show wide variations, making it difficult to evaluate safety performance correctly. For example, a small company may have only two or three injuries in a year. Thus, for the months in which these injuries occur, the rates will jump to extreme highs; in the other months, the rates will be zero. These variations will be smoothed out in annual totals, and the rate for the longer period will become more significant.

Annual reports should be prepared as soon after the close of the year as information becomes available, but again not later than the 20th of the following month. In some cases, an injury report may still be incomplete 20 days after the close of the year because the employee has

| | | | | OSHA Recordable Cases[1] | | | | | | | |
Period	Average Number of Employees	Number of Employee-Hours Worked	Costs (Compensation, other)	Deaths	Lost Workday Cases	Nonfatal Cases Without Lost Workdays	Total Cases	Total Lost Workdays	Total Case Incidence Rate[2]	Lost Workdays Incidence Rate[2]	First Aid Cases
Jan.											
Feb.											
2 mo.											
Mar.											
3 mo.											
Apr.											
4 mo.											
May											
5 mo.											
Jun.											
6 mo.											
Jul.											
7 mo.											
Aug.											
8 mo.											
Sep.											
9 mo.											
Oct.											
10 mo.											
Nov.											
11 mo.											
Dec.											
Year											

MONTHLY SUMMARY OF INJURIES AND ILLNESSES, 19___

Company _____ Plant _____ Department _____

[1]Refer to OSHA Log for numbers of cases: Deaths, columns 1+8; Lost Workday Cases, cols. 2+9; Nonfatal Cases Without Lost Workdays, cols. 6+13; Total Lost Workdays, cols. 4+5+11+12. Total Cases equals Deaths plus Lost Workday Cases plus Nonfatal Cases Without Lost Workdays.
[2]Incidence rate is number of "Total Cases" or "Total Lost Workdays" multiplied by 200,000 and divided by "Number of Employee-Hours Worked."

Figure 8-6. Results of the safety program can be gauged from data on this Monthly Summary of Injuries and Illnesses form. Rates computed for the month, year to date (cumulative), and year permit comparisons between time periods, and departments, facilities, and companies during the same period. Changes in the classification of injuries and other adjustments can be easily made.

not returned to work. In others, the classification of an injury may be in doubt at that time. In both these instances, an estimate of the outcome should be made by the company physician and the report included with the completed cases.

When annual records are closed, OSHA requires that the anticipated future lost workdays (both days away from work and days of restricted work activity) be recorded. Then, when a case becomes final, the records can be corrected to show the actual days lost.

If lost workdays that occurred in one year were included in the record for the next year, the latter year's experience would be biased and the record would fail to give a true picture of the severity of injuries.

Uses of Reports

Reports to Management
Management is increasingly interested in the incident experience of its company. Therefore, monthly and other periodic summary reports showing the results of the safety program should be furnished to the responsible executive. Such reports do not need to contain details or technical language. They can be supplemented by simple charts or

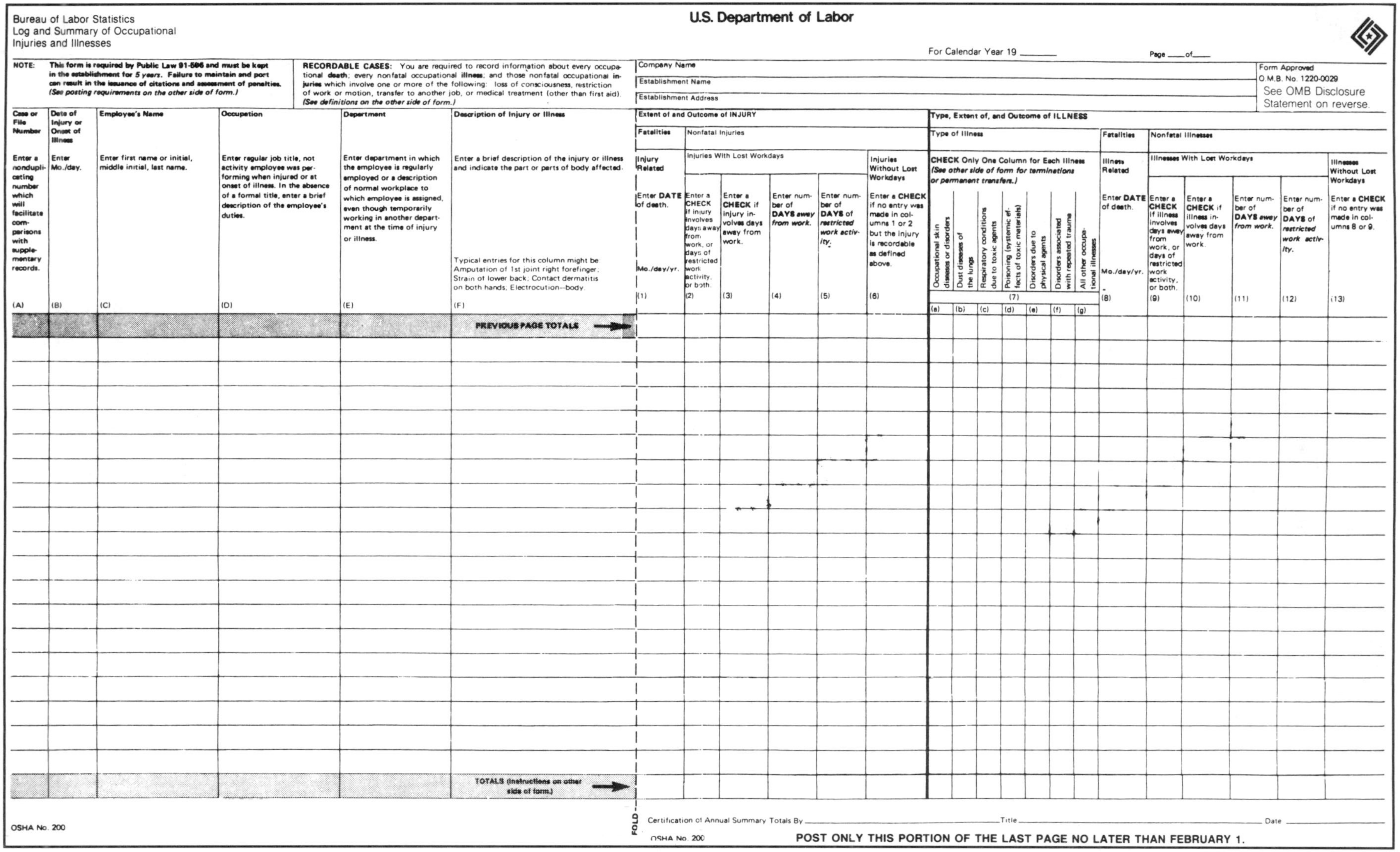

Bureau of Labor Statistics
Log and Summary of Occupational
Injuries and Illnesses

U.S. Department of Labor

For Calendar Year 19 ______ Page ___ of ___

NOTE: This form is required by Public Law 91-596 and must be kept in the establishment for 5 years. Failure to maintain and post can result in the issuance of citations and assessment of penalties. (See posting requirements on the other side of form.)

RECORDABLE CASES: You are required to record information about every occupational death; every nonfatal occupational illness; and those nonfatal occupational injuries which involve one or more of the following: loss of consciousness, restriction of work or motion, transfer to another job, or medical treatment (other than first aid). (See definitions on the other side of form.)

Company Name
Establishment Name
Establishment Address

Form Approved
O.M.B. No. 1220-0029
See OMB Disclosure Statement on reverse.

Case or File Number	Date of Injury or Onset of Illness	Employee's Name	Occupation	Department	Description of Injury or Illness
Enter a nonduplicating number which will facilitate comparisons with supplementary records.	Enter Mo./day.	Enter first name or initial, middle initial, last name.	Enter regular job title, not activity employee was performing when injured or at onset of illness. In the absence of a formal title, enter a brief description of the employee's duties.	Enter department in which the employee is regularly employed or a description of normal workplace to which employee is assigned, even though temporarily working in another department at the time of injury or illness.	Enter a brief description of the injury or illness and indicate the part or parts of body affected. Typical entries for this column might be: Amputation of 1st joint right forefinger; Strain of lower back; Contact dermatitis on both hands; Electrocution—body.
(A)	(B)	(C)	(D)	(E)	(F)
					PREVIOUS PAGE TOTALS →
					TOTALS (Instructions on other side of form.) →

Extent of and Outcome of INJURY

Fatalities	Nonfatal Injuries				
Injury Related	Injuries With Lost Workdays				Injuries Without Lost Workdays
Enter DATE of death. Mo./day/yr.	Enter a CHECK if injury involves days away from work, or days of restricted work activity, or both.	Enter a CHECK if injury involves days away from work.	Enter number of DAYS away from work.	Enter number of DAYS of restricted work activity.	Enter a CHECK if no entry was made in columns 1 or 2 but the injury is recordable as defined above.
(1)	(2)	(3)	(4)	(5)	(6)

Type, Extent of, and Outcome of ILLNESS

Type of Illness							Fatalities	Nonfatal Illnesses				
CHECK Only One Column for Each Illness (See other side of form for terminations or permanent transfers.)							Illness Related	Illnesses With Lost Workdays				Illnesses Without Lost Workdays
Occupational skin diseases or disorders	Dust diseases of the lungs	Respiratory conditions due to toxic agents	Poisoning (systemic effects of toxic materials)	Disorders due to physical agents	Disorders associated with repeated trauma	All other occupational illnesses	Enter DATE of death. Mo./day/yr.	Enter a CHECK if illness involves days away from work, or days of restricted work activity, or both.	Enter a CHECK if illness involves days away from work.	Enter number of DAYS away from work.	Enter number of DAYS of restricted work activity.	Enter a CHECK if no entry was made in columns 8 or 9.
(a)	(b)	(c)	(d)	(e)	(f)	(g)	(8)	(9)	(10)	(11)	(12)	(13)
(7)												

OSHA No. 200

FOLD

Certification of Annual Summary Totals By ______ Title ______ Date ______

OSHA No. 200

POST ONLY THIS PORTION OF THE LAST PAGE NO LATER THAN FEBRUARY 1.

Figure 8-7. This Log of Occupational Injuries and Illnesses, OSHA Form No. 200, is used to record injuries or illnesses that result in fatalities, or lost workdays, require medical treatment, involve loss of consciousness, or restrict work or motion. Complete instructions for using the log are on its reverse side (not shown here). The form measures 15 ¼ × 9 ¼ in. (34 × 56 cm). Because forms are subject to change, be sure to ask your OSHA Area Office for the latest forms.

graphs to compare current incident rates with those of the preceding period and the rates of other companies in the industry.

In a large company, departmental data help the executive visualize incident experience in various plant operations and provide a yardstick for better evaluation of progress made in the elimination of incidents. It can be particularly valuable to compare cost figures, if they can be obtained, for different periods.

Bulletins to Supervisors

Supervisors are primarily interested in their own department and workers. One of the most effective ways to create and maintain the interest of supervisors in incident prevention is to keep them informed about the incident records of their departments. Department injury rates based on sufficient amounts of exposure reflect the effectiveness of the supervisors' safety activities. Because interest increases with knowledge, management can send supervisors bulletins containing analyses of the principal causes of incidents in each department. These bulletins not only can maintain supervisors' interest at a high level but provide the supervisors with information to help them reduce injuries even further.

The agenda for employee safety meetings should emphasize information about outstanding injury and illness problems, frequent unsafe practices, hazardous types of equipment, and similar data disclosed by analysis of the incidents that have occurred in the department and plant.

Bulletin Board Publicity

Posting a variety of materials on bulletin boards is one of the best ways to maintain the interest of employees in safety. Incident records furnish many items, such as the following:

- no-injury records
- unusual incidents
- frequent causes of incidents
- charts showing reductions in incidents
- simple tables comparing departmental records
- standings in contests.

Reports to National Safety Council

The Council sponsors two recognition programs for its members requiring periodic reports of their occupational injury and illness experience. The Occupational Safety/Health Award Program, open to all employer members, is a noncompetitive evaluation of a company's annual experience. Awards are given if the records meet predetermined criteria.

The annual reports are tabulated to determine injury and illness incidence rates by industry. An annual "Work Injury and Illness Rates" pamphlet containing these incidence rates by industry is published and distributed to members. Each company can then compare its experience with the average experience of other companies from the same industry that participate in the Occupational Safety/Health Award Program. These rates (along with data from the BLS, state and national vital statistics authorities, state compensation authorities, and other federal, state, and private agencies) are shown in *Accident Facts*, the Council's annual statistical publication.

National Safety Council industrial employer members may also compete in more than 20 employee safety contests administered by the Council. Monthly bulletins are sent to contestants so that they can compare their experience with other competitors. Awards are made to the leaders of the individual contests at the end of each contest year.

The Concept of Bilevel Reporting

As mentioned earlier in this chapter, each company has different incident problems from those of organizations in other industries and even within the same industry. As a result, no individual form or set of forms can possibly include all of the information necessary to fully investigate the causes of all incidents. With this in mind, and because long forms are rarely welcomed or completed accurately, the concept of bilevel reporting has arisen.

The basic idea behind bilevel reporting is that further details about specific types of incidents are required in addition to the general information contained in the standard report form (such as the Supervisor's Incident Report Form). To obtain this additional data, a supplementary form, containing a few specific questions about the incident type under investigation, is prepared and made available. This supplementary form is then filled out and attached to the regular report, but only for those incidents the investigator needs to analyze in detail.

When a sufficient number of the bilevel forms have been collected, the supplementary form is discontinued and the results can be analyzed. This method requires only a minimum of time from the persons who fill out the forms in order to generate useful information. Several bilevel forms, each a different color for easy handling, can be used at any one time. They can be discontinued when their job is done and replaced by other supplementary forms, while the basic report form remains unchanged and in use.

OSHA RECORD-KEEPING REQUIREMENTS

The information in this section explains briefly the occupational injury and illness recording and reporting requirements of the OSHAct and Title 29 of the *Code of Federal Regulations*, Part 1904. It is taken from a June 1986 publication of the U.S. Department of Labor, BLS, titled "A Brief Guide to Record Keeping Requirements for Occupational Injuries and Illnesses."

Note: This document is under revision and should be approved in 1996 or 1997. Contact a regional or area OSHA office for the latest information.

In addition to the requirements of 29 *CFR*, 1904, many specific OSHA standards and regulations require maintenance and retention of records of medical surveillance, exposure monitoring, inspections, and other activities and incidents, and for the reporting of certain information to employees and to OSHA. These additional requirements are not covered in this section. (See Chapter 9, Occupational Medical Surveillance, in the *Engineering & Technology* volume.)

Employers Subject to the Record-Keeping Requirements

The record-keeping requirements of the OSHAct apply to private sector employers in all states, the District of Columbia, Puerto Rico, the Virgin Islands, American Samoa, Guam, and the Trust Territories of the Pacific Islands.

Employers with 11 or more employees (at any one time in the previous calendar year) in the following industries must keep OSHA records. The industries are identified by name and by the appropriate Standard Industrial Classification (SIC) code:

- agriculture, forestry, and fishing (SICs 01–02 and 07–09)
- oil and gas extraction (SICs 13 and 1477)
- construction (SICs 15–17)
- manufacturing (SICs 20–39)
- transportation and public utilities (SICs 41–42 and 44–49)
- wholesale trade (SICs 50–51)
- building materials and garden supplies (SIC 52)
- general merchandise and food stores (SICs 53 and 54)
- hotels and other lodging places (SIC 70)
- repair services (SICs 75 and 76)
- amusement and recreation services (SIC 79)
- health services (SIC 80).

If employers in any of the industries listed above have more than one establishment with combined employment of 11 or more employees, records must be kept for each establishment. Employers in other industries are normally exempt from OSHA record keeping, as are employers with no more than 10 full-time or part-time employees at any one time in the previous calendar year.

Even exempt employers, however, must comply with OSHA standards and display the OSHA poster. They must also report orally to OSHA within eight hours any incident that results in one or more fatalities or the inpatient hospitalization of three or more employees. Also, some state safety and health laws may require regularly exempt employers to keep injury and illness records, and some states have more stringent catastrophic reporting requirements. The following employers and individuals never have to keep OSHA injury and illness records:

- self-employed individuals
- partners with no employees
- employers of domestics in the employers' private residence for the purpose of housekeeping or child care, or both

- employers engaged in religious activities concerning the conduct of religious services or rites.

Employees engaged in such activities include clergy, choir members, organists and other musicians, ushers, and the like. However, records of injuries and illnesses occurring to employees while performing secular activities must be kept. Record keeping is also required for employees of private hospitals and certain commercial establishments owned or operated by religious organizations.

State and local government agencies are usually exempt from OSHA record keeping. However, in certain states, agencies of state and local governments are required to keep injury and illness records in accordance with state regulations. Employers subject to injury and illness record keeping requirements of other federal safety and health regulations are not exempt from OSHA record keeping. However, records used to comply with other federal record-keeping obligations may also be used to satisfy the OSHA record-keeping requirements. The forms used must be equivalent to the log and summary (OSHA No. 200) and the supplementary record (OSHA No. 101).

OSHA Record-Keeping Forms

Only two forms are used for OSHA record keeping: No. 200 and No. 101. Revisions to these forms have been proposed but as of this printing have not been approved. One form, the OSHA No. 200 (Figure 8-7), serves a dual purpose. It is used as the Log of Occupational Injuries and Illnesses, on which the occurrence and extent of cases are recorded during the year and as the Summary of Occupational Injuries and Illnesses, which is used to summarize the log at the end of the year to satisfy employer posting obligations. The other form, the Supplementary Record of Occupational Injuries and Illnesses, OSHA No. 101 (Figure 8-8), provides additional information on each of the cases that have been recorded on the log.

The OSHA Injury and Illness Log and Summary, OSHA No. 200

The log is used for recording and classifying occupational injuries and illnesses and for noting the extent of each case. The log consists of three parts: a descriptive section that identifies the employee and briefly describes the injury or illness, a section covering the extent of the injuries recorded, and a section on the type and extent of illnesses.

Usually, the OSHA No. 200 form is used by employers as their record of occupational injuries and illnesses. However, a private form equivalent to the log, such as a computer printout, may be used if it contains the same detail as the OSHA No. 200. It must also be as readable and comprehensible as the OSHA No. 200 to a person not familiar with the equivalent form. It is advisable that employers have private equivalents of the log form reviewed by BLS to ensure compliance with the regulations.

Bureau of Labor Statistics
Supplementary Record of
Occupational Injuries and Illnesses

U.S. Department of Labor

This form is required by Public Law 91-596 and must be kept in the establishment for *5 years.* Failure to maintain can result in the issuance of citations and assessment of penalties.

Case or File No.

Form Approved
O.M.B. No. 1220-0029

Employer

1. Name

2. Mail address *(No. and street, city or town, State, and zip code)*

3. Location, if different from mail address

Injured or Ill Employee

4. Name *(First, middle, and last)*

Social Security No.

5. Home address *(No. and street, city or town, State, and zip code)*

6. Age

7. Sex: *(Check one)*   Male ☐   Female ☐

8. Occupation *(Enter regular job title, not the specific activity he was performing at time of injury.)*

9. Department *(Enter name of department or division in which the injured person is regularly employed, even though he may have been temporarily working in another department at the time of injury.)*

The Accident or Exposure to Occupational Illness

If accident or exposure occurred on employer's premises, give address of plant or establishment in which it occurred. Do not indicate department or division within the plant or establishment. If accident occurred outside employer's premises at an identifiable address, give that address. If it occurred on a public highway or at any other place which cannot be identified by number and street, please provide place references locating the place of injury as accurately as possible.

10. Place of accident or exposure *(No. and street, city or town, State, and zip code)*

11. Was place of accident or exposure on employer's premises?   Yes ☐   No ☐

12. What was the employee doing when injured? *(Be specific. If he was using tools or equipment or handling material, name them and tell what he was doing with them.)*

13. How did the accident occur? *(Describe fully the events which resulted in the injury or occupational illness. Tell what happened and how it happened. Name any objects or substances involved and tell how they were involved. Give full details on all factors which led or contributed to the accident. Use separate sheet for additional space.)*

Occupational Injury or Occupational Illness

14. Describe the injury or illness in detail and indicate the part of body affected. *(E.g., amputation of right index finger at second joint; fracture of ribs; lead poisoning; dermatitis of left hand, etc.)*

15. Name the object or substance which directly injured the employee. *(For example, the machine or thing he struck against or which struck him; the vapor or poison he inhaled or swallowed; the chemical or radiation which irriatated his skin; or in cases of strains, hernias, etc., the thing he was lifting, pulling, etc.)*

16. Date of injury or initial diagnosis of occupational illness

17. Did employee die? *(Check one)*   Yes ☐   No ☐

Other

18. Name and address of physician

19. If hospitalized, name and address of hospital

Date of report

Prepared by

Official position

OSHA No. 101 (Feb. 1981)

Figure 8-8. The Supplementary Record of Occupational Injuries and Illnesses, OSHA Form No. 101, gives details of each recordable occupational injury or illness. Records must be available for examination by representatives of the U.S. Department of Labor and the Department of Health and Human Services and states accorded jurisdiction under the OSHAct. Records must be kept at least five years following the calendar year to which they relate.

The portion of the OSHA No. 200 on the righthand side is used to summarize injuries and illnesses in an establishment for the calendar year. Every nonexempt employer who is required to keep OSHA records must prepare an annual summary for each establishment based on the information contained in the log for each establishment. The summary is prepared by totaling the column entries on the log (or its equivalent) and signing and dating the certification portion of the form at the bottom of the page.

The Supplementary Record of Occupational Injuries and Illnesses, OSHA No. 101

For every injury or illness entered on the log, it is necessary to record additional information on the supplementary record, OSHA No. 101. However, the OSHA No. 101 is not the only form that can be used to satisfy this requirement. To eliminate duplicate recording, workers' compensation, insurance, or other reports (Figure 8-9) may be used as supplementary records if they contain all of the items on the OSHA No. 101. If they do not, the missing items must be added to the substitute or included on a separate attachment.

Completed supplementary records must be present in the establishment within six workdays after the employer has received information that an injury or illness has occurred.

Location, Retention, and Maintenance of Records

If an employer has more than one establishment, a separate set of records must be maintained for each one. The record-keeping regulations define an establishment as "a single physical location where business is conducted or where services or industrial operations are performed." Examples include a factory, mill, store, hotel, restaurant, movie theater, farm, ranch, sales office, warehouse, or central administrative office.

Under the regulations, the location of these records depends upon whether the employees are associated with a fixed establishment. The distinction between fixed and nonfixed establishments generally rests on the nature and duration of the operation and not on the type of structure in which the business is located. Generally, any operation at a given site for more than one year is considered a fixed establishment. Also, fixed establishments are generally places where clerical, administrative, or other business records are kept.

1. **Employees Associated with Fixed Establishments.** Records for these employees should be located as follows:

   a. Records for employees working at fixed locations, such as factories, stores, restaurants, warehouses, etc., should be kept at the work location.

   b. Records for employees who report to a fixed location but work elsewhere should be kept at the place where the employees report each day. These employees are generally engaged in activities such as agriculture, construction, transportation, etc.

   c. Records for employees whose payroll or personnel records are maintained at a fixed location, but who do not report or work at a single establishment, should be maintained at the base from which they are paid or the base of their firm's personnel operations. This category includes generally unsupervised employees such as traveling salespeople, technicians, or engineers.

2. **Employees Not Associated with Fixed Establishments.** Some employees are subject to common supervision, but do not report or work at a fixed establishment on a regular basis. These employees are engaged in physically dispersed activities that occur in construction, installation, repair, or service operations. Records for these employees should be located as follows:

   a. Records may be kept at the field office or mobile base of operations.

   b. Records may also be kept at an established central location. If the records are maintained centrally: (1) the address and telephone number of the place where records are kept must be available at the work site, and (2) someone must be available at the central location during normal business hours to provide information from the records.

Although the supplementary record and the annual summary must be located as outlined in the previous section, it is possible to prepare and maintain the log at an alternate location or by means of data-processing equipment, or both. Two requirements must be met: (1) sufficient information must be available at the alternate location to complete the log within six workdays after information is received that a recordable case occurred, and (2) a copy of the log updated to within 45 calendar days must be kept at all times in the establishment. This location exception applies only to the log and not to the other OSHA records. Also, it does not affect the employer's posting obligations.

The log and summary, OSHA No. 200, and the supplementary record, OSHA No. 101, must be retained in each establishment for five calendar years following the end of the year to which they relate. If an establishment changes ownership, the new employer must preserve the records for the remainder of the five-year period. However, the new employer is not responsible for updating the records of the former owner.

In addition to keeping the log on a calendar-year basis, employers are required to update this form to include newly discovered cases and to reflect changes that occur in recorded cases after the end of the calendar year. Maintaining or updating the log differs from simply retaining

FOR CARRIER USE ONLY: Insd. Location No.   Insd. Report No.

FORM 45: Employers First Report of Injury or Illness   PLEASE TYPE OR PRINT

Filing of this report does not affect your liability under the Workers' Compensation Act and is not incriminatory in any sense.

A | *45 | ILLINOIS UNEMPLOYMENT COMPENSATION NUMBER | | DATE OF REPORT | MONTH   DAY   YEAR | CASE OR FILE NUMBER |

B | EMPLOYER'S NAME | | | Is this a lost workday case? ☐Yes ☐No

C | DOING BUSINESS UNDER THE NAME OF | CITY, STATE | ZIP CODE

D | MAIL ADDRESS | CITY, STATE | ZIP CODE

E | EMPLOYER LOCATION IF DIFFERENT FROM MAIL ADDRESS

F | NATURE OF BUSINESS OR SERVICE | SIC CODE | TOTAL NUMBER OF EMPLOYEES AT THE LOCATION WHERE ILLNESS OR INJURY OCCURRED

G | NAME OF WORKERS' COMP. INSURANCE CARRIER | POLICY NUMBER | SELF INSURED | COUNTY WHERE INJURY OCCURRED
LIBERTY MUTUAL INSURANCE CO. | | YES ☐   ☐NO |

H | EMPLOYEE'S NAME   (LAST, FIRST, MIDDLE) | SOCIAL SECURITY NUMBER

I | HOME ADDRESS | CITY, STATE | ZIP CODE

J | MALE ☐ FEMALE ☐ | MARRIED ☐ SINGLE ☐ WIDOW(ER) ☐ DIVORCED ☐ | BIRTH DATE | MONTH   DAY   YEAR | NUMBER OF DEPENDENT CHILDREN UNDER 18 AT TIME OF INJURY OR ILLNESS

K | DATE AND TIME OF THE INJURY OR EXPOSURE | MONTH DAY YEAR ☐a.m. ☐p.m. | EMPLOYEE'S AVERAGE WEEKLY EARNINGS | LAST DAY EMPLOYEE WORKED | MONTH   DAY   YEAR

L | JOB TITLE OR OCCUPATION | DEPARTMENT NORMALLY ASSIGNED

M | ADDRESS OF LOCATION WHERE INJURY OR EXPOSURE OCCURRED | CITY, STATE | ZIP CODE

N | DID EMPLOYEE DIE AS A RESULT OF THE INJURY OR ILLNESS? YES ☐ NO ☐ | IF EMPLOYEE DIED AS A RESULT OF THE INJURY OR ILLNESS, GIVE DATE OF DEATH | MONTH   DAY   YEAR

O | WAS THE INJURY OR EXPOSURE ON THE EMPLOYER'S PREMISES? ☐YES ☐NO | DID THIS INCIDENT RESULT IN: ☐ OCCUPATIONAL INJURY ☐ OCCUPATIONAL DISEASE | Was Employee given Industrial Commission Handbook? YES ☐ NO ☐

P | NATURE OF THE INJURY

Q | PART OF THE BODY AFFECTED (BE SPECIFIC)

R | WHAT TASK WAS EMPLOYEE PERFORMING WHEN ILLNESS OR INJURY OCCURRED?

S | OBJECT OR SUBSTANCE RESPONSIBLE FOR INJURY OR ILLNESS (SOURCE)

T | HOW DID ACCIDENT OR ILLNESS OCCUR (TYPE)?

U | WHAT HAZARDOUS CONDITIONS, METHODS OR LACK OF PROTECTIVE DEVICES CONTRIBUTED?

V | WHAT UNSAFE ACT BY A PERSON CAUSED OR CONTRIBUTED TO THE INJURY OR ILLNESS?

W | HAVE MEDICAL SERVICES BEEN RENDERED TO THE EMPLOYEE? YES ☐ NO ☐ | IS OR HAS THE EMPLOYEE BEEN HOSPITALIZED? YES ☐ NO ☐

X | NAME AND ADDRESS OF PHYSICIAN | CITY, STATE | ZIP CODE

Y | NAME AND ADDRESS OF HOSPITAL | CITY, STATE | ZIP CODE

Z | REPORT PREPARED BY: (NAME—PRINT OR TYPE) | SIGNATURE | TITLE AND TELEPHONE NUMBER

ACCIDENT REPORTING DEPT., ILLINOIS INDUSTRIAL COMMISSION, 160 North LaSalle Street, Chicago, Illinois 60601
WITHOUT WRITTEN APPROVAL OF COMMISSION, THIS FORM MAY NOT BE REPRODUCED

12-CSF-1 R5   (Rev. 1-81)

Figure 8-9. Employers First Report of Injury or Illness is an equivalent to the OSHA No. 101 because it includes all of the same items. (Compare to Figure 8-8.)

records, discussed in the previous paragraph. Although all OSHA injury and illness records must be retained, only the log must be updated by the employer. If, during the five-year retention period, there is a change in the extent or outcome of an injury or illness that affects an entry on a previous year's log, then the first entry should be lined out and a corrected entry made on that log. Also, new entries should be made for previously unrecorded cases that are discovered or for cases that initially were not recorded but were found to be recordable after the end of the year in which the case occurred. The entire entry should be lined out for recorded cases that are later found nonrecordable. Log totals should also be modified to reflect these changes.

Recording and Classifying Cases

Guidelines presented here for determining whether a case must be recorded under the OSHA record-keeping requirements should not be confused with record-keeping requirements of various workers' compensation systems, internal industrial safety and health monitoring systems, the ANSI Z.16 standards for recording and measuring work injury and illness experience, and private insurance company rating systems. Reporting a case on the OSHA records should not affect record-keeping determinations under these or other systems. In addition, recording an injury or illness under the OSHA system does not necessarily imply that management was at fault, that the worker was at fault, that a violation of an OSHA standard has occurred, or that the injury or illness is compensable under workers' compensation or other systems.

Employees versus Other Workers on Site

Employers must maintain injury and illness records for their own employees at each of their establishments. However, they are not responsible for maintaining records for employees of other firms or for independent contractors, even though these individuals may be working temporarily in their establishment or on one of their job sites at the time an injury or illness exposure occurs. Therefore, before deciding whether a case is recordable an employment relationship needs to be determined.

Employee status generally exists for record-keeping purposes when the employer supervises not only the output, product, or result to be accomplished by the person's work, but also the details, means, methods, and processes by which the work is accomplished. This means the employer who supervises the worker's day-to-day activities is responsible for recording the worker's injuries and illnesses. Often, part-time employees are supervised by the organization they are working for and thus are considered employees for record-keeping purposes. Independent contractors are not considered employees; they are primarily subject to supervision by the using firm only for the results to be accomplished or end product to be delivered. Independent contractors keep their own injury and illness records.

Other factors that may be considered in determining employee status are (1) whom the worker considers to be his or her employer, (2) who pays the worker's wages, (3) who withholds the worker's Social Security taxes, (4) who hired the worker, and (5) who has the authority to terminate the worker's employment.

Method Used for Case Analysis

The decision-making process consists of five steps:

1. Determine whether a case occurred; that is, whether there was a death, illness, or injury.
2. Establish that the case was work-related; that it resulted from an event or exposure in the work environment.
3. Decide whether the case is an injury or an illness.
4. If the case is an illness, record it and check the appropriate illness category on the log.
5. If the case is an injury, decide if it is recordable based on a finding of medical treatment, loss of consciousness, restriction of work or motion, or transfer to another job.

Figure 8-10 presents this methodology in graphic form.

Determining Whether a Case Occurred

The first step in the decision-making process is to determine whether an injury or illness has occurred. Employers have nothing to record unless an employee has experienced a work-related injury or illness. In most instances, recognition of these injuries and illnesses is a fairly simple matter. However, some situations have troubled employers over the years. Two of these are:

1. Hospitalization for observation. In some instances, an employee visits or is sent to a hospital for a brief time for observation. However, the case is not recordable, provided no medical treatment was given or no illness was recognized. The determining factor is not that the employee went to the hospital, but whether the incident is recordable as a work-related illness or as an injury requiring medical treatment or involving loss of consciousness, restriction of work or motion, or transfer to another job.
2. Differentiating a new case from the recurrence of a previous injury or illness. Employers are required to make new entries on their OSHA forms for each new recordable injury or illness. However, new entries should not be made for the recurrence of symptoms from previous cases. It is sometimes difficult to decide whether a situation is a new case or a recurrence. The following guidelines address this problem:

   a. Injuries. The aggravation of a previous injury almost always results from some new incident involving the employee (such as a slip, trip, fall,

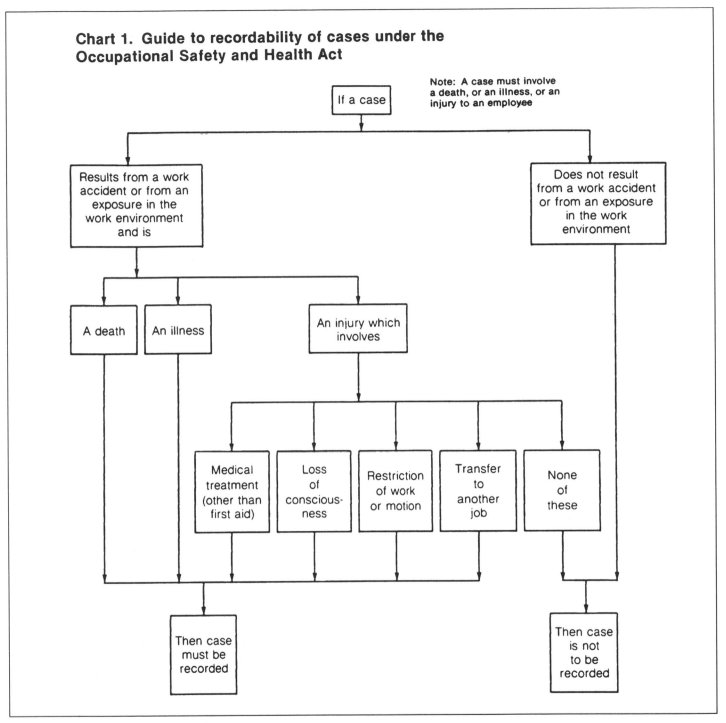

Chart 1. Guide to recordability of cases under the Occupational Safety and Health Act

Figure 8-10. Chart 1 of OSHA Guide.

sharp twist, etc.). Consequently, when work related, these new incidents should be recorded as new cases.

b. Illnesses. Generally, each occupational illness should be recorded with a separate entry on the OSHA No. 200. However, certain illnesses, such as silicosis, may have prolonged effects which recur over time. The recurrence of these symptoms should not be recorded as new cases on the OSHA forms. The recurrence of symptoms of previous illnesses may require adjustment of entries on the log for previously recorded illnesses to reflect possible changes in the extent or outcome of the particular case. Some occupational illnesses, such as certain dermatitis or respiratory conditions, may recur as the result of new exposures to sensitizing agents, and should be recorded as new cases.

Some of the most difficult cases are those involving the so-called cumulative trauma disorders

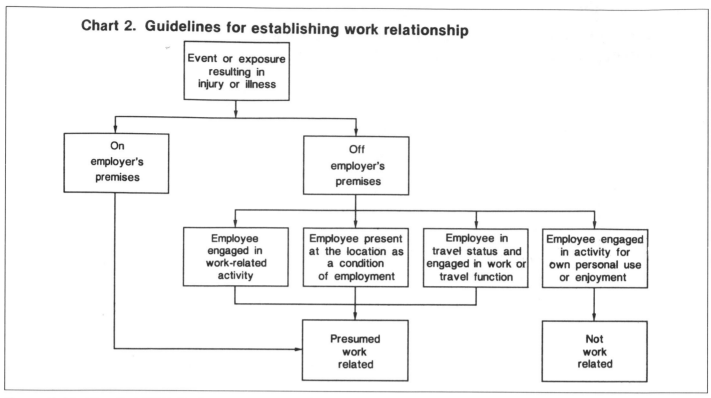

Figure 8-11. Chart 2 of OSHA Guide.

(CTDs). It can be very difficult to determine what is a "new" injury versus an exacerbation or worsening of a previously recorded injury.

Establishing Work Relationship

The OSHAct requires employers to record only those injuries and illnesses that are work related. Work relationship is established under the OSHA record-keeping system when the injury or illness results from an event or exposure in the work environment (Figure 8-11). The work environment is primarily composed of (1) the employer's premises and (2) other locations where employees are engaged in work-related activities or are present as a condition of their employment. When an employee is off the employer's premises, work relationship must be established; when on the premises, this relationship is presumed. The employer's premises encompass the total establishment, including not only the primary work facility but also such areas as company storage facilities. In addition to physical locations, equipment or materials used in the course of an employee's work are also considered part of the employee's work environment.

1. Injuries and illnesses resulting from events or exposures on the employer's premises. These conditions are generally considered work related. The employer's premises consist of the total establishment. They include the primary work facilities and other areas considered part of the employer's general work area.

The presumption of work relationship for activities on the employer's premises is rebuttable. Situations where the presumption would not apply include (1) when a worker is on the employer's premises as a member of the general public and not as an employee, and (2) when employees have symptoms that surface on the employer's premises but are actually the result of a nonwork-related event or exposure off the premises.

The following subjects warrant special mention:

a. Company restrooms, hallways, and cafeterias are all considered to be part of the employer's premises and constitute part of the work environment. Therefore, injuries occurring in these places are generally considered work related.

b. For OSHA record-keeping purposes, the definition of work premises excludes all employer controlled ball fields, tennis courts, golf courses, parks, swimming pools, gyms, and other similar recreational facilities. These areas are often apart from the workplace and used by employees on a voluntary basis for their own benefit, primarily during off-work hours. Therefore, injuries to employees in these recreational facilities are not recordable unless the employee was engaged in some work-related activity or was required by the employer to participate.

c. Company parking facilities are generally not considered part of the employer's premises for OSHA record-keeping purposes. Therefore, injuries to employees on these parking lots are not presumed to be work related and are not recordable unless the employee was engaged in some work-related activity.

2. Injuries and illnesses resulting from events or exposures off the employer's premises. When an employee is off the employer's premises and suffers an injury or an illness exposure, work relationship must be established; it is not presumed. Injuries and illness exposures off premises are considered work related if the employee is engaged in a work activity or if they occur in the work environment. The work environment in these instances includes locations where employees are engaged in job tasks or work-related activities, or places where employees are present due to the nature of their job or as a condition of their employment.

Employees who travel on company business shall be considered to be engaged in work-related activities all the time they spend in the interest of the company. This includes, but is not limited to, travel to and from customer contacts and entertaining or being entertained for the purpose of transacting, discussing, or promoting business, and so on. However, an injury/illness would not be recordable if it occurred during normal living activities (eating, sleeping, recreation); or if the employee deviates from a reasonably direct route of travel (side trip for vacation or other personal reasons). However, the employee would be considered in the course of employment once again when returning to the normal route of travel.

A traveling employee who checks into a hotel or motel establishes a "home away from home." Thereafter, the individual's activities are regarded in the same manner as for nontraveling employees. For example, if an employee on travel status is to report each day to a fixed work site, then injuries sustained when traveling to this work site would be considered off the job. The rationale is that an employee's normal commute from home to office would not be considered work-related. However, in some situations, employees in travel status report to or rotate among several different work sites after they establish their "home away from home" (such as a salesperson traveling to and from different customer contacts). In these situations, the injuries sustained when traveling to and from the sales locations would be considered job-related.

Traveling sales personnel may establish only one base of operations (home or company office). A salesperson with the home as an office is considered at work when in that office or when leaving the premises in the interest of the company. Figure 8-11 provides a guide for establishing the work relationship of cases.

Distinguishing between Injuries and Illnesses

Under the OSHAct, all work-related illnesses must be recorded, while injuries are recordable only when they require medical treatment (other than first aid) or involve loss of consciousness, restriction of work or motion, or transfer to another job. The distinction between injuries and illnesses, therefore, has significant record-keeping implications.

Whether a case involves an injury or illness is determined by the nature of the original event or exposure causing the case and not by the resulting condition of the affected employee. Injuries are caused by instantaneous events in the work environment. Cases resulting from anything other than instantaneous events are considered illnesses. This concept of illnesses includes acute illnesses that result even from relatively short exposure times.

Some conditions may be classified as either an injury or an illness (but not both), depending upon the nature of the event that produced the condition. For example, a loss of hearing resulting from an explosion (an instantaneous event) is classified as an injury; the same condition arising from exposure to industrial noise over time would be classified as an occupational illness.

Recording Occupational Illnesses

Employers are required to record the occurrence of all occupational illnesses, defined in the instructions of the log and summary as the following:

> any abnormal condition or disorder, other than one resulting from an occupational injury, caused by exposure to environmental factors associated with employment. It includes acute and chronic illnesses or diseases that may be caused by inhalation, absorption, ingestion, or direct contact.

The instructions also refer to recording illnesses which were "diagnosed or recognized." Illness exposures ultimately result in conditions of a chemical, physical, biological, or psychological nature.

Occupational illnesses must be diagnosed to be recordable. However, they do not necessarily have to be diagnosed by a physician or other medical personnel. Diagnosis may be by a physician, registered nurse, or a person who by training or experience is capable of making such a determination. Employers, employees, and others may be able to detect some illnesses such as skin diseases or disorders without the benefit of specialized medical training. However, cases more difficult to diagnose, such as silicosis or CTDs, would require evaluation by properly trained medical personnel. CTDs, for example, should be diagnosed from objective findings, not solely on the basis of complaints of pain. When and how to diagnose CTDs is a significant issue, as this category of illness or injury is a fast-growing occupational disorder.

In addition to recording the occurrence of occupational illnesses, employers are required to record each illness case in one of the seven categories on the front of the log. The back of the log form contains a listing of types of illnesses or disorders and gives examples for each illness category. These are only examples, however, and should not be considered as a complete list of types of illnesses under each category.

Recording and classifying occupational illnesses may be difficult for employers, especially the chronic and long-term latent illnesses. Many illnesses are not easily detected; and once detected, it is often difficult to determine whether an illness is work related. Also, employees may not report illnesses, either because the symptoms may not be readily apparent or because they do not think their illness is serious or work related. In addition, nonwork-related exposures to other medical conditions may be superimposed on the original presenting problem.

The following material is provided to assist in detecting occupational illnesses and in establishing their work relationship:

1. Detection and diagnosis of occupational illnesses. An occupational illness is defined in the instructions on the log as any work-related abnormal condition or disorder in workers' health (other than an occupational injury). Detection of these abnormal conditions or disorders—the first step in recording illnesses—is often difficult. When an occupational illness is suspected, employers may want to consider the following:

   a. A medical examination of the employee's physiological systems. For example:

      - head and neck
      - eyes, ears, nose, and throat
      - endocrine
      - genitourinary
      - musculoskeletal
      - neurological
      - respiratory
      - cardiovascular
      - gastrointestinal.

   b. Observation and evaluation of behavior related to emotional status, such as deterioration in job performance which cannot be explained

   c. Specific examination for health effects of suspected or possible disease agents by competent medical personnel

   d. Comparison of date of onset of symptoms with occupational history

   e. Evaluation of results of any past biological or medical monitoring (blood, urine, other sample analysis) and previous physical examinations

   f. Evaluation of laboratory tests: routine (complete blood count, blood chemistry profile, urinalysis) and specific tests for suspected disease agents (e.g., blood and urine tests for specific agents, chest or other x-rays, liver function tests, pulmonary function tests)

   g. Reviewing the literature, such as Material Safety Data Sheets and other reference documents, to ascertain whether the levels to which the workers were exposed could have produced the ill effects.

2. Determining whether the illness is occupationally related. The instructions on the back of the log define occupational illnesses as those "caused by environmental factors associated with employment." In some cases, such as contact dermatitis, the relationship between an illness and work-related exposure is easy to recognize. In other cases, such as CTDs, where the occupational cause may not be obvious, it may be difficult to determine accurately whether an employee's illness is occupational. In these situations, it may help employers to ask the following questions:

   a. Has an illness condition clearly been established?
   b. Does it appear that the illness resulted from, or was aggravated by, suspected agents, activities, or other conditions in the work environment?
   c. Are these suspected agents or activities present (or have they been present) in the work environment?
   d. Was the ill employee exposed to these agents or activities in the work environment?
   e. Was the exposure to a sufficient degree and/or duration to result in the illness condition?
   f. Was the illness attributable solely to a nonoccupational exposure?

Deciding if Work-Related Injuries are Recordable

Although the OSHAct requires that all work-related deaths and illnesses be recorded, the recording of nonfatal injuries is limited to certain specific types of cases: (1) those requiring medical treatment or involving loss of consciousness, (2) those involving restriction of work or motion, and (3) those requiring a transfer to another job. Minor injuries needing only first aid treatment are not recordable.

   1. *Medical Treatment.* It is important to understand the distinction between medical treatment and first aid treatment, because many work-related injuries are recordable only if medical treatment was given.

The regulations and the instructions on the back of the log and summary, OSHA No. 200, define medical treatment as any treatment, other than first aid treatment, administered to injured employees. Essentially, medical treatment involves the provision of medical or surgical care for injuries that are not minor through the application of procedures or systematic therapeutic measures.

The act also specifically states that work-related injuries involving only first aid treatment should not be recorded. First aid is commonly thought to mean emergency treatment of injuries before regular medical care is available. However, first aid treatment has a different meaning for OSHA record-keeping purposes. The regulations define first aid treatment as:

any one-time treatment, and any follow-up visit for the purpose of observation, of minor scratches, cuts, burns, splinters, and so forth, which do not ordinarily require medical care. Such one-time treatment, and follow-up visit for the purpose of observation, is considered first aid even though provided by a physician or registered professional personnel.

2. The distinction between medical treatment and first aid depends not only on the treatment provided but also on the severity of the injury being treated. First aid (1) is limited to one-time treatment and subsequent observation and (2) involves treatment of only minor injuries, not emergency treatment of serious injuries. Injuries are not minor if:

   a. they must be treated only by a physician or licensed medical personnel

   b. they impair bodily function (i.e., normal use of senses, limbs, etc.)

   c. they result in damage to the physical structure of a nonsuperficial nature (e.g., fractures)

   d. they involve complications requiring follow-up medical treatment.

Physicians or registered medical professionals, working under the standing orders of a physician, routinely treat minor injuries. Such treatment may constitute first aid. Also, some visits to a doctor do not involve treatment at all. For example, a visit to a doctor for an examination or other diagnostic procedure to determine whether the employee has an injury does not constitute medical treatment. Conversely, medical treatment can be provided to employees by lay persons, i.e., someone other than a physician or registered medical personnel.

The following classifications list certain procedures as either medical treatment or first aid treatment.

Medical treatment: The following are generally considered medical treatment. Work-related injuries for which this type of treatment was provided or should have been provided are almost always recordable:

- treatment of INFECTION
- application of ANTISEPTICS during second or subsequent visit to medical personnel
- treatment of SECOND OR THIRD DEGREE BURN(S)
- application of SUTURES (stitches)
- application of BUTTERFLY ADHESIVE DRESSING(S) or STERI STRIP(S) in lieu of sutures
- removal of FOREIGN BODIES EMBEDDED IN EYE
- removal of FOREIGN BODIES FROM WOUND; if procedure is COMPLICATED because of depth of embedment, size, or location
- use of PRESCRIPTION MEDICATIONS (except a single dose administered on first visit for minor injury or discomfort)
- use of hot or cold SOAKING THERAPY during second or subsequent visit to medical personnel

- application of hot or cold COMPRESS(ES) during second or subsequent visit to medical personnel
- CUTTING AWAY DEAD SKIN (surgical debridement)
- application of HEAT THERAPY during second or subsequent visit to medical personnel
- use of WHIRLPOOL BATH THERAPY during second or subsequent visit to medical personnel
- POSITIVE X–RAY DIAGNOSIS (fractures, broken bones, etc.)
- ADMISSION TO A HOSPITAL or equivalent medical facility FOR TREATMENT.

First aid treatment: The following are generally considered first aid treatment (e.g., one-time treatment and subsequent observation of minor injuries). These cases should not be recorded if the work-related injury does not involve loss of consciousness, restriction of work or motion, or transfer to another job:

- application of ANTISEPTICS during first visit to medical personnel
- treatment of FIRST DEGREE BURN(S)
- application of BANDAGE(S) during any visit to medical personnel
- use of ELASTIC BANDAGE(S) during first visit to medical personnel
- removal of FOREIGN BODIES NOT EMBEDDED IN EYE if only irrigation is required
- removal of FOREIGN BODIES FROM WOUND; if procedure is UNCOMPLICATED, and is, for example, by tweezers or other simple technique
- use of NONPRESCRIPTION MEDICATIONS and administration of single dose of PRESCRIPTION MEDICATION on first visit for minor injury or discomfort
- SOAKING THERAPY on initial visit to medical personnel or removal of bandages by SOAKING
- application of hot or cold COMPRESS(ES) during first visit to medical personnel
- application of OINTMENTS to abrasions to prevent drying or cracking
- application of HEAT THERAPY during first visit to medical personnel
- use of WHIRLPOOL BATH THERAPY during first visit to medical personnel
- NEGATIVE X–RAY DIAGNOSIS
- OBSERVATION of injury during visit to medical personnel.

3. Administration of TETANUS SHOT(S) or BOOSTER(S), is not considered medical treatment. However, these shots are often given in conjunction with more serious injuries; consequently, injuries requiring these shots may be recordable for other reasons.

1. *Loss of Consciousness.* If an employee loses consciousness as the result of a work-related injury, the

case must be recorded no matter what type of treatment was provided. The rationale behind this recording requirement is that loss of consciousness is generally associated with the more serious injuries.

2. *Restriction of Work or Motion.* Restricted work activity occurs when the employee, because of the impact of a job-related injury, is physically or mentally unable to perform all or any part of his or her normal assignment during all or any part of the workday or shift. The emphasis is on the employee's ability to perform normal job duties. Restriction of work or motion may result in either a lost-work-time injury or a nonlost-work-time injury, depending upon whether the restriction extended beyond the day of injury.

3. *Transfer to Another Job.* Injuries requiring transfer of the employee to another job are also considered serious enough to be recordable, regardless of the type of treatment provided. Transfers are seldom the sole criterion for recordability. This is because injury cases are almost always recordable on other grounds, primarily medical treatment or restriction of work or motion.

Categories for Evaluating the Extent of Recordable Cases

Once the employer decides that a recordable injury or illness has occurred, the case must be evaluated to determine its extent or outcome. There are three categories of recordable cases: fatalities, lost workday cases, and cases without lost workdays. Every recordable case must be placed in only one of these categories.

Fatalities

All work-related fatalities must be recorded, regardless of the time between the injury and the death, or the length of the illness.

Lost Workday Cases

Lost workday cases occur when the injured or ill employee experiences either days away from work, days of restricted work activity, or both. In these situations, the injured or ill employee is affected to such an extent that (1) days must be taken off from the job for medical treatment or recuperation, or (2) the employee is unable to perform his or her normal job duties over a normal work shift (which could be an extended hour shift of 8 to 12 hours), even though the employee may be able to continue working.

1. Lost workday cases involving days away from work refer to cases when the employee ordinarily would have been on the job but could not work because of the job-related injury or illness. The focus of these cases is on the employee's inability, because of injury or illness, to be present in the work environment during his or her normal work shift. If the

normal shift is 12 hours, then this extended shift is still equivalent to an eight-hour shift for purposes of counting lost and restricted days; i.e., a normal 12-hour shift does not count as 1.5 times as an eight-hour shift.

2. Lost workday cases involving days of restricted work activity are those cases in which, because of injury or illness, (1) the employee was temporarily assigned to another job, (2) the employee worked at a permanent job less than full time, or (3) the employee worked at his or her permanently assigned job but could not perform all the duties normally connected with it. Restricted work activity occurs when the employee, because of the job-related injury or illness, is physically or mentally unable to perform all or any part of the normal job duties over all or any part of the normal workday or shift. The emphasis is on the employee's inability to perform normal job duties over a normal work shift.

Injuries and illnesses are not considered lost workday cases unless they affect the employee beyond the day of injury or onset of illness. When counting the number of days away from work or days of restricted work activity, do not include the initial day of injury or onset of illness, or any days on which the employee would not have worked even if able to be on the job (holidays, vacations, etc.).

Cases not Resulting in Death or Lost Workdays

These cases consist of the relatively less serious injuries and illnesses that satisfy the criteria for recordability. However, the injuries and illnesses do not result in death or require the affected employee to have days away from work or days of restricted work activities beyond the date of injury or onset of illness.

Employer Reporting Obligations

This section focuses on the requirements of Section 8(c)(2) of the OSHAct and Title 29, Part 1904, of the *Code of Federal Regulations* for employers to make reports of occupational injuries and illnesses. It does not include the reporting requirements of other standards or regulations of OSHA or of any other state or federal agency.

The Annual Survey of Occupational Injuries and Illnesses

The survey is conducted on a sample basis. Firms required to submit reports of their injury and illness experience are contacted by OSHA or a participating state agency. *A firm not contacted need not file a report of its injury and illness experience.* Employers should note, however, that even if they are not selected to participate in the annual survey for a given year, they must still comply with the record-keeping requirements listed in the preceding sections as well as with those for reporting fatalities and multiple hospitalization cases provided in the next section.

Reporting Fatalities and Multiple Hospitalizations

According to 29 *CFR* 1904.8, all employers are required to report incidents resulting in one or more fatalities or in the inpatient hospitalization of three or more employees. (Some states have more stringent catastrophic reporting requirements.) The report is made orally to the office of the Area Director of OSHA nearest to the incident. However, a different reporting procedure applies if the state in which the incident occurred is administering an approved state plan under Section 18(b) of the OSHAct. Those 18(b) states designate a state agency to which the report must be made.

The report must be made within eight hours after the occurrence of each fatality or inpatient hospitalization of three or more employees. If death or injury occurs subsequent to the initial incident, it must be reported within 30 days of the incident. The report must contain the following information: the establishment name, the location and time of the incident, the number of fatalities or hospitalized persons, the name and phone number of a contact person, and a brief description of the incident.

Access to OSHA Records

To ensure the integrity of the OSHA record-keeping process, OSHA has established regulations regarding access to OSHA records. All OSHA records that are being kept by employers for the five-year retention period should be available for inspection and copying by authorized federal and state government officials. Employees, former employees, and their representatives are allowed access only to the log, OSHA No. 200.

Government officials with access to the OSHA records include the following: representatives of the Department of Labor, including OSHA safety and health compliance officers and BLS representatives; representatives of the Department of Health and Human Services while carrying out that department's research responsibilities; and representatives of states accorded jurisdiction for inspections or statistical compilations. "Representatives" may include Department of Labor officials inspecting a workplace or gathering information, officials of the Department of Health and Human Services, or contractors working for the agencies mentioned above, depending on the provisions of the contract under which they work.

Employee access to the log is limited to the records of the establishment in which the employee currently works or formerly worked. All current logs and those being maintained for the five-year retention period must be made available for inspection and copying by employees, former employees, and their representatives. An employee representative can be a member of a union representing the employee, or any person designated by the employee or former employee. Access to the log is to be provided in a reasonable manner and at a reasonable time. Employees can obtain redress through OSHA for any company's failure to comply with the access provisions of the regulations.

Penalties for Failure to Comply

Another procedure for ensuring the integrity of the OSHA record-keeping process is to establish and enforce penalties for failure to comply with record-keeping obligations. Employers committing record-keeping and/or reporting violations are subject to the same sanctions as employers violating other OSHA requirements such as safety and health standards and regulations.

INCIDENCE RATES

Safety performance is relative. Only when a company compares its injury experience with that of its entire industry, or with its own previous experience, can it obtain a meaningful evaluation of its safety accomplishments. To make such comparisons, a method of measurement is needed that will adjust for the effects of certain variables contributing to differences in injury experience. Injury totals alone cannot be used for two reasons.

First, a company with many employees may be expected to have more injuries than a company with few employees. Second, if the records of one company include all the injuries treated in the first aid room, while the records of a similar company include only injuries serious enough to cause lost time, obviously the first company's total will be larger than the second company's figure.

A standard procedure for keeping records, which provides for these variables, is included in the OSHA record-keeping requirements. First, this procedure uses incidence rates that relate injury and illness cases, and the resulting days lost, to the number of employee-hours worked; thus these rates automatically adjust for differences in the hours of exposure to injury. Second, this procedure specifies the kinds of injuries and illnesses that should be included in the rates. These standardized rates, which are easy to compute and to understand, have been generally accepted as procedure in industry, thus permitting the necessary and desired comparisons.

A chronological arrangement of these rates for a company will show whether its level of safety performance is improving or worsening. Within a company, the same sort of arrangement by departments not only will show the trend of safety performance for each department but will reveal to management other information to render safety work more efficient.

If it is found, for example, that the trend of incidence rates in a company is up, a review of the rate trends by department may reveal that this change is accounted for by the rates of just a few departments. With the sources of the highest company rates isolated, management can concentrate safety efforts at these points.

A comparison of current incidence rates with those of similar companies and with those of the industry as a whole serves a critical function. This step provides the safety professional with a more reliable evaluation of the company's safety performance than could be obtained merely by reviewing numbers of cases.

Formulas for Rates

Incidence rates are based on the exposure of 100 full-time workers using 200,000 employee-hours as the equivalent (100 employees working 40 hours per week for 50 weeks per year). An incidence rate can be computed for each category of cases or days lost depending on what number is put in the numerator of the formula. The denominator of the formula should be the total number of hours worked by all employees during the same time period as that covered by the number of cases in the numerator.

$$\text{Incidence Rate of Recordable Cases} =$$

$$\frac{\text{No. of injuries and illnesses} \times 200{,}000}{\substack{\text{Total hours worked by all employees} \\ \text{during period covered}}}$$

or

$$\frac{\text{No. of lost workdays} \times 200{,}000}{\substack{\text{Total hours worked by all employees} \\ \text{during period covered}}}$$

There are two other formulas that can be used to measure the average severity of the recorded cases:

$$\text{Average lost workdays} = \frac{\text{Total lost wordays}}{\text{Total lost workday cases}}$$

$$\substack{\text{Average days} \\ \text{away from work}} = \frac{\text{Total days away from work}}{\text{Total days-away-from-work cases}}$$

If these numbers are small, then it is known that the cases are relatively minor. If, however, the numbers are large, then the cases are of greater average severity and should receive serious attention. For example, to calculate the incidence rate for total recordable cases at the end of the year, one would simply multiply the number of recordable cases by 200,000 and divide that by the number of hours worked by all employees for the whole year.

The incidence rates may also be interpreted as the percentage of employees who will suffer the degree of injury for which the rate was calculated. That is, if the incidence rate of lost workday cases is 5.1 per 100 full-time workers, then about 5% of the establishment's employees incurred a lost-workday injury.

Significance of Changes

The incidence rate for lost workday cases is the best measure for comparing the occupational injury and illness experience among companies of various sizes. However, it is often not sufficient for determining the significance of month-to-month changes in the actual number of cases within a single company. For this comparison, all cases must usually be used, not merely the lost workday cases. This provides a more objective measure for determining the significance of month-to-month fluctuations in the number of cases, especially when a seemingly large increase or decrease from the monthly average appears.

Because the average number of cases per month is calculated from numbers of cases that are larger or smaller than the average itself, variation from the average is to be expected. The variation can either be random or caused; caused variation is significant and random variation is nonsignificant. Therefore, the significance of variations can easily be determined by distinguishing between those that are random or caused.

Manufacturing organizations are already faced with the task of determining the significance of variations in such factors as dimensions, weight, or performance of their products. To do this, they frequently employ a tool known as a quality control chart. This chart identifies and distinguishes between a random variation, which is said to be "in control," and caused variation, which is said to be "out of control." Being able to distinguish between the two types of variation permits management to concentrate safety efforts on out-of-control variations.

A similar control chart can be developed to evaluate the significance of changes in injury and illness experience. The first step in developing the chart is to calculate the average number of cases per month. Because this monthly average will fluctuate from year to year, several years' experience should be used to develop a stable average. Preferably, the average should be calculated using 60 months' experience. After the average number of cases per month has been determined, the upper and lower control limits (UCL and LCL, respectively) are calculated using the following equation:

$$\text{UCL and LCL} = n \pm 2\sqrt{n}$$

where n is the average number of cases per month.

Figure 8-12 shows a control chart developed for one company. Using 60 months' experience, it was determined

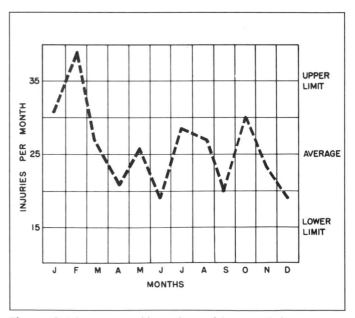

Figure 8-12. Upper and lower limits of this control chart were based on 60 months' experience, and show points at which the number of injuries can statistically be shown to have a contributing cause rather than by chance. Thus, action can be taken where it will be most effective.

that the company had an average of 25 cases per month. Substitution of 25 for *n* in the equation yielded the upper and lower control limits shown below:

$$25 \pm 2\sqrt{25} = 25 \pm 10 = 25 + 10 = 35 \text{ for the upper limit}$$

$$= 25 - 10 = 15 \text{ for the lower limit}$$

The control chart was then constructed and the company recorded the actual number of cases each month. Note that the actual monthly number of cases, with the exception of February, falls within the upper and lower control limits. These variations are random and do not represent significant changes from the monthly average of 25 cases. The variation for February is probably a caused variation, indicating that the experience for February is out of control and that corrective measures should be taken. Suppose, for example, an investigation revealed that an unguarded machine was the cause of the increase in cases for the month of February. It can be assumed that the installation of a mechanical guard corrected this cause and brought the injury and illness experience for March back in control.

The control chart can be a useful tool when it is used properly. As with any tool, it depends on the skill of the user to yield good results. Follow these rules when constructing a control chart:

1. Always use several years' experience when calculating the average number of cases per month to ensure that a stable average will be developed. Sixty months' experience is preferable, except as noted in (3), (4), and (5) following.
2. Count all cases when constructing a control chart. The use of lost workday cases alone may not provide enough data. A control chart cannot be constructed if the average number of cases per month is less than four because the lower limit will be negative. Since a facility cannot experience a negative number of cases per month, a chart based on an average of less than four would not yield meaningful results.
3. Calculate a new average each year—adding the latest year's cases and subtracting the oldest year's cases—to reflect any change in the monthly average number of cases.
4. Construct more than one chart if there are seasonal changes in employment. For example, if an operation employs 100 people for the first six months of each year and 400 people for the last six months, a separate chart must be constructed for each six-month period.
5. Construct a new control chart if external factors, such as a change in the level of employment, a change in the work environment, or a change in work hazards, bring about a permanent change in injury and illness experience. When calculating a new average in such a situation, do not use experience prior to the change.

There are important differences in the way the chart should be used for safety purposes as opposed to manufacturing. First, if the incident frequency falls below the lower control limit, it would be absurd to encourage more incidents so as to return the system to "control." Rather, one looks for changes in the system that may have accounted for the good performance. Corrective action is required only if the system goes out of control by exceeding the upper control limit. Second, while a "steady state" system (always in control) is desirable from a manufacturing point of view, steady *improvement* is expected from an occupational safety and health program. The long-term trend in monthly incident frequency (or incident rates) should be downward.

OFF-THE-JOB INJURIES

In recent years, off-the-job disabling injuries of employees have far exceeded on-the-job disabling injuries. Any unscheduled absence of employees can cause production slowdowns and delays, costly retraining and replacement, or costly overtime by remaining employees. As a result, many safety specialists are very concerned with off-the-job injuries their employees suffer. Moreover, activity to reduce off-the-job incidents should help to promote interest in this area of safety. (See also the Off-the-Job Safety Programs section in Chapter 3, Loss Control Programs, and the Off-the-Job Disabling Injury section in Chapter 7, Accident Investigation, Analysis, and Costs in this volume.)

The ANSI Z16.3 standard, as revised in 1989 and reaffirmed in 1995, provides a means for recording and measuring these off-the-job injuries. Definitions and rates used under the ANSI Z16.3 standard are compatible with those of the OSHA record-keeping requirements. Because the data on off-the-job injuries are not as easy to obtain, however, certain simplifications are introduced in ANSI Z16.3. Exposure (for use in rates per 200,000 employee-hours) is standardized at 312 employee-hours per employee per month (equal to $4\frac{1}{3}$ weeks less 40 hours per week at work and 56 hours per week for sleeping). Provision is also made in ANSI Z16.3 for recording home, public, and transportation injuries separately to allow for concentrated effort in problem areas.

SUMMARY

- Companies must maintain records of work-related employee injuries and illnesses as required by OSHA, state compensation authorities, and insurers. Good record-keeping systems provide data to evaluate incident problems and safety program effectiveness, identify high incident-rate areas, create interest in safety, enable the company to concentrate efforts on the more serious incident problems, and measure effectiveness of countermeasures against hazards and unsafe practices.

- The primary purpose of an incident report is to obtain accurate, objective information about the causes of incidents in order to prevent the incidents from reoccurring.
- Every establishment subject to the OSHAct is required to provide an annual summary of incident rates by February 1st. Bilevel reporting is used to gain further details about specific incidents in addition to the general information contained in standard report forms.
- OSHA record-keeping requirements apply to private sector employees in all states, District of Columbia, Puerto Rico, and other U.S. territories and possessions. Employers with 11 or more employees in specified industries must keep OSHA records in addition to any state or local requirements. Employers who are not in these specified industries or who have fewer than 11 employees are exempt from OSHA record-keeping regulations.
- However, all employers must report orally to OSHA the occurrence of one or more fatalities or the inpatient hospitalization of three or more employees resulting from a work-related incident within eight hours of the death or hospitalization.
- The location, retention, and maintenance of these records depends on the number of establishments and whether employees are associated with the fixed establishment. Employers are required to retain records for five years and to update them as new information about incidents and illnesses arises.
- All work-related illnesses must be recorded, while injuries are recorded only when they require medical treatment or involve a loss of consciousness, restriction of work or motion, or transfer to another job. Lost workdays cases are recorded when the employee cannot work because of an injury or illness, is temporarily reassigned, or has restrictions because of the injury or illness.
- All OSHA records should be available for inspection and copying by authorized federal and state governments. Employees, former employees, and their representatives are allowed access only to the log.
- OSHA incident rates help companies compare their safety performance with the performance of previous years or of the entire industry to evaluate their safety programs.
- Quality control charts help distinguish between random variations, which are in control, and caused variations, which are out of control. This distinction allows management to concentrate safety efforts on out-of-control variations.
- The ANSI Z16.3 standard provides a means for recording and measuring off-the-job injuries to help management develop safety programs for employees.

REFERENCES

American National Standards Institute, 11 West 42nd Street, New York, NY 10036.

American National Standard for Injury Statistics—Employee Off-the-Job Injury Experience—Recording and Measuring, ANSI Z16.3–1989, (R1995).

National Safety Council, 1121 Spring Lake Drive, Itasca, IL 60143.

Accident Facts (published annually).

Work Injury and Illness Rates (published annually).

Recht JL. Bilevel reporting. *Journal of Safety Research* 2(2): 51–54, 1970.

Tufte ER. *The Visual Display of Quantitative Information.* Cheshire, CT: Graphics Press, 1983.

U.S. Dept. of Labor, Bureau of Labor Statistics, Washington, DC 20212. *A Brief Guide to Recordkeeping Requirements for Occupational Injuries and Illnesses.* April 1986.

REVIEW QUESTIONS

1. Safety personnel must maintain records for what two reasons?
   a.
   b.
2. List five of the seven ways a good record-keeping system can help the safety professional.
   a.
   b.
   c.
   d.
   e.
3. The collection of injury data generally begins in which department?
   a. Human resources department
   b. Safety department
   c. First aid department
   d. Department where the accident occurred
4. Which nonsafety professional should make a detailed report of each incident?
   a. Human resources director
   b. Medical staff supervisor
   c. Legal department manager
   d. Injured worker's supervisor
5. How often should managers and supervisors prepare a summary of injuries and illnesses?
   a. After every 10 occurrences
   b. Weekly
   c. Monthly
   d. Quarterly
6. Which form serves as the annual summary report that is used by any company subject to the OSHAct?
7. What is the OSHA regulation for Form No. 200?

8. The OSHA record-keeping requirements are found in what publication?

9. What types of employers and individuals do not have to keep OSHA injury and illness records?

   a.

   b.

   c.

   d.

10. The distinction between injuries and illnesses has significant record-keeping implications. Under the OSHAct, all work-related illnesses must be recorded, while injuries are recorded only when:

    a.

    b.

    c.

11. All employers must report orally to OSHA the occurrence of one or more fatalities or the inpatient hospitalization of three or more employees resulting from a work-related incident within how many hours of the death or hospitalization?

12. What is the formula for calculating incidence rates of recordable cases?

13. Why are many safety specialists very concerned with off-the-job injuries their employees suffer?

14. The standard that provides a means for recording and measuring off-the-job injuries is:

    a. ANSI Z16.1

    b. ANSI Z16.2

    c. ANSI Z16.3

    d. ANSI Z16.4

9

Computers and Information Management

Advances in computer technology—from personal computers to local area networks and worldwide telecommunications—have led to the development of a wide array of computer systems for occupational safety and health functions. These systems can help ease the burden of regulatory compliance; maintain better records; analyze more information for decision making; and help prevent work-related deaths, injuries, illnesses, and toxic exposures. Computer systems have become smaller, easier to use, more affordable, and more accessible to more people than ever before. In addition, networking software has led to the rapid deployment of departmental systems that enable departments to communicate easily with one another in a business environment. (A list of computer common terms and their definitions is provided in Figure 9-1.)

To help managers understand the developments in this field and make informed decisions, this chapter covers the following topics:

- advantages of using computers to manage occupational health information systems (OHIS)
- the major components that make up the information system
- matching an organization's needs with the right type of system
- various types of software commercially available to manage information system functions
- steps in selecting or developing software for a particular company or department
- guidelines on working with software vendors
- implementing and evaluating an OHIS system
- current and future trends in OHIS software and hardware technology.

In 1970, information management in occupational safety and health became a critical function with the passage of the Occupational Health and Safety Act (OSHAct). Since then, environmental laws such as the Toxic Substances Control Act, the Resource Conservation and Recovery Act, and the Comprehensive Environmental Response, Compensation, and Liability (Superfund) Act (CERCLA) created an information management need that overwhelmed the resources of existing occupational safety and health programs. Right-to-know laws further added to the burden of proper compliance. All of these new regulations carried complex record-keeping and reporting requirements and stiff penalties for noncompliance. In today's regulatory environment, not only must workers be kept informed of hazardous substances being used in the workplace, but the community also has a right to the information.

As public awareness of workplace hazards grew, litigation involving occupational illnesses and deaths began to have a significant impact on organizations. Employers were asked to document their compliance in training, monitoring, and reporting procedures. Unfortunately, many employers found that their old paper records were inadequate and incomplete.

Basic Computer Terms

Bit A unit of data, a bit is the smallest piece of information that the computer can handle. Technically, a bit represents whether an electronic impulse is turned on or off.

Bug An error or irritation in a software program—usually not so severe that it causes a program to fail.

Byte This unit of measuring size is equivalent to eight bits, about the amount of information needed to store a single letter, number, or other character.

Central Processing Unit (CPU) A part of the insides of the computer that controls all of the work the computer does.

CD-ROM A high density storage system that uses a laser to write digital information. The medium is the same type of disc that is used in the music industry. The CD-ROM is the medium on which many software programs are now delivered, eliminating the need for software developers to provide 20 or more discs to load a new application of update.

CD-ROM Drive A device that is now often supplied with a PC computer that can read a CD-ROM disc.

Data Single pieces of information stored in a file. For example, in a file of accident records, the date of the accident would be one piece of data.

Database A collection of data that is organized for efficient storage, editing, and recall.

DOS Disk Operating System is the software that acts as the intermediary between the computer hardware and other software.

Dot-Matrix Printer An impact printer that uses an ink tape and a mechanical head to imprint on paper to produce hard copy.

File Collection of information stored on a disk. For example: a letter of correspondence might be saved as a single file.

Floppy Disk or Diskette A flexible, round electromagnetic device that stores computer programs and data. When used in a floppy drive (the part of the computer where the diskette is inserted) the information is read into the computer's memory for processing. Often the storage capacity of the disk exceeds the memory of the computer. The information on the diskette is permanent until erased or replaced.

Gigabyte (GB) This unit of measuring size equals 1000 Megs. It is used to describe the capacity of a hard disk drive, a tape drive or other storage device. For example: 1.0, 1.5, 2.0 GB.

Graphics A display of information in picture form. For example: a pie chart or a bar graph on-screen are graphic capabilities of a computer.

Graphical User Interface (GUI) A pictorial view of data displayed on a computer monitor. Certain computer environments offer graphical interfaces. Most can include the capability of producing any form of text, text size, drawings, photographs, and business graphs.

Hardware All of the physical components of the computer system. This includes the monitor, printer, keyboard, central processing unit, and everything else that you can touch; not floppy disks, however.

Hard Disk Resembling stacked floppy diskettes, this component is actual hardware that sits inside the computer. Hard disks, then, can hold much more information than a floppy. Information on the hard disk is read into the computer's memory for processing, and it is permanent until erased or replaced.

Hard Copy The paper copy of information, as printed out by the computer's printer.

Ink Jet Printer A printer that provides economical hard copy that is of higher quality than a dot matrix printer and is available to print in black and white or color.

Internet A network of computer system networks that provides access to remote sites containing a wide range of information subjects.

K This unit of measuring size is equal to 1,024 bytes. The term was used to express the memory capacity of the microcomputer. If a software program requires 256K to run, and your computer has only 128K, then your computer cannot run the software program unless you purchase additional memory-expansion hardware. Now, newer systems have larger memory needs (see megabytes).

Laser Printer A printer that produces very high quality output that approaches the print quality common to the magazine publishing industry.

Megabyte (MB or Meg) This unit of measuring size equals 1,048,576 bytes or 1,024 K. It is used to describe the capacity of a hard disk drive, a tape drive or other storage device. For example: 50, 100, 500 MB or Meg.

Memory The pan of the computer that stores programs and data currently being processed.

Modem The MOdulator-DEModulator or the hardware that allows computer-generated signals to be transmitted and received on telephone lines.

Monitor The hardware part that includes the screen.

Mouse A very popular form of pointing device.

Network A means of connecting one computer to another either in the same area or distant areas.

Operating System The software environment that governs the basic operation of a computer system.

Pointing Device A device that enables the computer user to select areas on a graphical computer screen. Most commonly a mouse, but may be a trackball, a light pen, or a hand-held stylus.

Printer The hardware device that produces hard copies.

Printer Port A connection often referred to as a parallel port of a computer to enable data to be transmitted to a printer.

Program A set of computer instructions on a disk.

RAM A term referring to random access memory. It represents the working space required for the CPU to perform its work. It is a space for temporary storage of data that is processed by the computer. Most GUI systems require a minimum of 8 MB to operate. Typically the more RAM that is installed in a computer, the better the performance will be.

Relational Database A collection of data that is organized for efficient storage, editing, and recall using a special indexing method that relates one stored record to another, thus simplifying retrieval of related data. Typical relational databases are accessible using query tools that facilitate information retrieval.

Figure 9-1. Commonly used computer terms.

Scanner A device that creates a digital image of a black and white or color document containing text, drawings, or photographs. The digital image can be stored for later use within a computer system.

Serial Port An input/output device that permits communication with a computer system. The term serial refers to the organization of data as it enters or leaves the computer system. Modems are typically attached to serial ports.

Software A series of programs that, together, allow the computer to function as a word processor, data base manager, or spreadsheet.

Windows A form of graphical user interface wherein work is performed within a number of individual windows that appear on a monitor screen. These windows can be moved, resized, and positioned in front of or behind others. Most graphical user interfaces use this means of managing and running application software.

World Wide Web A capability of the Internet that permits global information exchange using a hypertext management tool.

WYSIWYG A term that represents "what you see is what you get" and demonstrates one of the benefits of a graphical user interface. One example would be a print preview of a document produced by a word processor showing on the monitor what the user can expect to be printed.

Figure 9-1. (Concluded).

Even though during the past few years, many companies have reduced the size of their operations for various economic reasons, the need to manage occupational safety and health programs has increased. Unfortunately, budgets have not kept pace. Never has the need for efficiency been greater.

The need to comply with stiffer regulations, to produce better internal records management, and to control costs have led many organizations to purchase, develop, or enhance their computer systems to manage the growing workload and to disseminate more information.

ADVANTAGES OF COMPUTERIZED INFORMATION SYSTEMS

Computers provide several important advantages over manual information systems. Particularly in generating, manipulating, and retrieving data, the computer's speed, multitask efficiency, and storage capacity greatly outweigh the costs of starting up and upgrading the system.

Better Availability of Data

Any well-designed computer system (combining both computer software and hardware) can compile and store huge amounts of data and retrieve it easily for reporting and analysis. For example, an employer might want to know how many employees experienced back injuries in the workplace population. With a manual system, an employer would have to conduct a time-consuming search through medical files and/or injury files. With an automated system, however, this search can be completed in minutes with the help of specialized software developed for the task. Employees can perform even more complex inquiries without having to learn a programming language. Automated systems also can provide emergency information quickly, such as first aid and clean-up procedures for chemical spills.

Improved Decision Making

Similarly, the use of such systems can result in higher levels of quality assurance and professional decision making than manual methods can produce. Most companies undervalue this tangible benefit when generating their occupational health information system requirements. Management can improve the quality of its decision-making process by having immediate access to relevant information. Through the computer's powerful ability to organize and present data, managers and clinicians can analyze information from a variety of perspectives. The more quickly and easily data are transformed into information, the more effectively they can be used in analysis and in the development of more thoughtful decisions. In addition, standard forms and reports can be regularly generated to meet highly specific requirements.

Duplication Eliminated

Manual data management systems require that the same data be entered many times on paper forms and stored by many departments within an organization. For example, with a manual system, employees often had to keep duplicate records of occupational injuries in as many as four offices: safety, medical, human resources, and area supervision. In contrast, a computer system can keep one comprehensive file on each injury in a database that each department can use.

In this way, computers provide users with instant access to vast amounts of data organized in a variety of standardized and customized formats. Using relational databases, information can be entered once and automatically duplicated in designated data fields on command. The recent popularity of multiuser and networking computers eliminates the need to store data in more than one location. The technology allows authorized access to information that can be shared by members of different departments. This setup saves time, improves data access, and reduces overall operating costs.

Improved Communications and Quality of Data

With manual systems, important data can be lost in unfinished piles of paperwork. Keeping data current is very difficult, and manual statistics generated by one of the company's offices may not match those generated by another. For example, injury reports from one facility may not reach other facilities, decreasing the chance that all groups within the organization can benefit from the "lessons learned." A computerized occupational safety and health information system makes up-to-date data available to all authorized users. In addition, the advent of integrated fax or electronic mail (e-mail) systems can facilitate communication within and between facilities.

Standardized Data

Standardizing the data input and the way in which data are organized is particularly important in companies with many computer users. By making sure that the same data are input in the same way by everyone, a company improves its ability to analyze the data and increases its decision support capability. For instance, accident and incident rates can be analyzed for all locations once the common terminology and naming conventions have been adopted. This approach eliminates the need to analyze data from each location separately and then standardize the information for comparison purposes.

Improved Accuracy

As users enter data, state-of-the-art computer programs automatically check for out-of-range data, inconsistencies, and errors. The accuracy of data is also improved by checking entries for the completeness of records. For instance, a computer system can alert the occupational safety and health department that part of a worker's OSHA-required surveillance examination was neither scheduled nor conducted.

Improved Analytical Capabilities

With most manual systems, compilation and analysis of data from various records consumes professional time that could be better spent on program management activities. A good automated system performs analytical tasks quickly and formats the results for reports to management. Newer occupational safety and health systems directly link to office automation tools, such as spreadsheets, word processors, report writers, and presentation tools.

Analyzing data can reveal trends in employee health and give management time to act to prevent injuries and illnesses. For example, the ability to group employees by homogeneous exposure and correlate workplace exposure with medical findings leads to improved medical surveillance. Such analyses are not practical with manual record keeping but can be done routinely using an automated system.

Reduced Cost

Automated systems can save a company money above and beyond the costs of system development and operation. Increased employee productivity, decreased incident rates, and more effective program management can lead to better loss control. Prevention of even one or two serious injuries can pay for the entire up-front cost of a computer system.

Depending upon the extent of its use, computer systems also can provide a wide array of savings relating to reduced workplace absence, reduced lost time, and automation of time-consuming manual procedures. Every discipline within occupational safety and health can benefit from automated programs. For example, clerical work at one facility's occupational health clinic was reduced by 30% when the company switched to a computerized clinic management system. This result was not surprising, since at many of the employer's sites, data that had to be entered repeatedly into different paper records now could be entered only once into a computer database.

Computerized systems have had a significant impact on health care management. Data from thousands of work-related medical claims are now analyzed by managed care organizations to gain a greater understanding of individual practice patterns and to improve the quality of care occupational health managers purchase on behalf of their organizations.

PLANNING, DEVELOPING, AND IMPLEMENTING AN OHIS

Because of the high cost of development, and the complexity and rapid evolution of occupational safety and health requirements, many organizations choose to purchase commercially available software rather than develop internal software.

Five steps are key to implementing an automated occupational safety and health system successfully:

1. Develop a thorough understanding of the organization's current and evolving operations and identify realistic needs for data management. Consider reengineering work processes to eliminate activities that add little value to the management process.
2. Identify potential vendors, review software features and functionality, then select hardware to meet those needs.
3. Purchase existing software and/or hardware or develop a customized system based upon the identified needs.
4. Set realistic goals for start-up implementation and for training personnel.
5. Monitor implementation process, usage, and work processes to evaluate and improve these items.

Establishing a Comprehensive OHIS

Developing a comprehensive OHIS requires a clear understanding of the processes to be supported. A well-conceived set of user specifications will help ensure that the system will perform the desired functions. This holds true in setting expectations for off-the-shelf software programs or requirements for systems to be custom developed. If the system is to be custom built, either by an in-house department or by consultants, system developers should develop preliminary specification documents based on the detailed user specification statement. The following is a list of suggestions to consider:

- Identify the needs of each system user group and follow through with a precise set of needs. The overall requirements for a system are generated from these needs. For example, the company needs the system to support a hearing conservation program by generating letters to employees who have had a temporary hearing threshold shift (TTS). A requirement is to generate up to 50 letters per day within 20 working days of the audio testing. Additional precise requirements would add the following qualifier: The system needs to generate up to 50 letters per day that will be received by employees (at home) within 20 working days of the audio testing for a cost (including data entry, processing, and quality control) not to exceed $2.00 per employee tested.

- If a company is defining a system that will be used by several departments, management should generate user specifications and system requirements for each discipline. Requirements need to be developed by a task force composed of representatives of all affected departments in order to ensure that systems will support commonly agreed upon processes and specifications.

- Whether a company plans to buy or build, management must be sure to budget adequate time to evaluate the system to be implemented against the goals/requirements. The requirements need to be concise and complete enough so that the successes and the faults of the system are easily recognized and measured.

- If a company plans to build software, management needs to budget sufficient funds for future system modifications. Unfortunately, most systems are never finished. Finishing the last 10% of a system can often represent 50% of the entire systems cost, yet this last 10% frequently means the difference between success and failure. If the company purchases existing software, it should allow time to participate in user groups, if offered.

- Companies should develop or select systems that use an open architecture to help ensure OHIS hardware and software flexibility. In an *open* architecture system, the plans and specifications for the computer hardware are in the public domain, so any vendor can reproduce or improve the hardware. Closed systems, on the other hand, can restrict flexibility and increase the ultimate cost of the system. *Closed* architecture is proprietary to a single company and not accessible for outside vendors to use or enhance.

- Companies need to build in flexibility to permit change over time. In this regard, open architecture and the use of relational databases are invaluable. Organizational restructuring, reorganizing, downsizing, acquisitions, new product development and new business directions need to be part of the requirement discussions. Questions such as, "How will the system perform if the following scenario happens . . . ?" need to be part of the finished requirements. However, system flexibility has a price—it can make the software more complex and more costly to change over the long term.

Important Components of Data Management Software

Comprehensive occupational safety and health information system/integrated health data management systems share common components (Figure 9-2). Many can be integrated within a single, structured relational database; alternatively, several relational databases can share properly coded data elements. These common components include:

General Office Automation and Communications

One of the most important components of an OHIS database—perhaps the most important—is support of the office staff as they perform everyday tasks. Document preparation (word processing), spreadsheets, contact, time and personal information managers, and electronic mail have become vital parts of general office automation support. Without integration of these functions, the OHIS may not be well accepted. Most users already have these support programs on their desktop computers and they do not want to learn new systems, have two computers on the desk, or leave their desk to do their job. These technologies can enable occupational safety and health professionals to communicate in a more timely and complete manner.

Access to Personnel Data

Almost every OHIS will require information about the company (e.g., sites) and employees (e.g., name, mailing address, date of birth, gender, social security number, and other demographic information). Because of employee turnover or relocation, this information should be easy to update, indicating the last date information was changed. The most sophisticated systems now use data from company personnel (Human Resources) databases, either through periodic extracts or from real time linkages, where the actual data are stored in the personnel database.

Integration with Benefits Information

In an era of medical cost containment, linkage with benefits information can be valuable. Such information may include medically related absence and use of sick leave, health care use, workers' compensation, general health claims and costs, and detection of unusual occupational diseases that are being treated by the general medical community. There is growing interest in twenty-four-hour care where one health care vendor provides services for both work-related and non-work-related injuries or illnesses.

Documenting Workplace Conditions

Many systems manage safety, industrial hygiene, occupational health, physics, environmental, and other sampling, audit, and inspection data capable of describing and quantifying workplace conditions. Industrial hygiene sampling

IMPORTANT FEATURES OF OCCUPATIONAL SAFETY AND HEALTH INFORMATION SYSTEMS
GENERAL SAFETY & HEALTH

Office Automation
Integrate with Word Processor
Integrate with Spreadsheet
E-Mail
Facsimile Access
Remote Access
Drug and Alcohol Testing
Employee Assistance Program
Financial and Administrative Information
Disability Management
Total Quality Management (TQM)
Linkage to Other Systems and Databases
Predictive Equipment Inspection and Maintenance

SAFETY

Workplace Conditions
Work History and Job Tracking
Functional Job Requirements
Employee Job Capabilities, Work Restrictions and Accommodation
Regulatory Requirements
Injury Monitoring and Management
Safety Audits/Inspections
Document Accident/Incident Investigations
Protective Measures
Track Safety/Training Data
OSHA Reporting
EPA Reporting
Analysis and Reporting Capabilities

MEDICAL

Appointment Scheduling
Medical Record Keeping and Retrieval
Employee Health History
Health Promotion and Health Risk Appraisal
Injury Monitoring and Management
Exposure Related Medical Surveillance
Case Management
Analysis and Reporting Capabilities

INDUSTRIAL HYGIENE/ENVIRONMENTAL

Regulatory Requirements
Exposure Related Medical Surveillance
Chemical Inventory
OSHA Reporting
EPA Reporting
Toxicology
Exposure Monitoring
Protective Measures
Analysis and Reporting Capabilities

RISK MANAGEMENT

Integration with Personnel and Benefits Data
Document Accident/Incident Investigations
Loss Reporting
Analysis and Reporting Capabilities
Workers Compensation
Disability Management

Figure 9-2. Important features of occupational safety and health information systems.

results can be entered and used to produce exposure profiles of work processes within each workplace. Areas of concern can be readily identified and more frequently monitored. In integrated OHIS, workplace data can be correlated with employee health data to support occupational health surveillance programs. Also, if safety inspection findings, recommendations, target dates for completion, and completion dates are automated, a system can produce follow-up reports highlighting identified deficiencies that have not been corrected.

Maintaining Work History and Job Tracking

Current job titles, classifications, and prior job histories can become especially important in medical surveillance studies where test results or health outcomes are monitored for all workers in particular jobs or for those with selected exposures. Job histories can be particularly important when employees at the same locations are used as a control group to compare against workers exposed to hazardous substances.

Documenting Functional Job Requirements

Compliance with the Americans with Disabilities Act (ADA) requires health professionals to have more information than a current job title. Understanding functional job requirements—including physical, psychological, social, and environmental demands—becomes critical in medical assessment. Such descriptions should distinguish between essential and marginal job functions. This area can be managed with a relatively simple computer application. Specialized software is available to help develop functional job requirements and store descriptions for future recall.

Recording Job Capabilities, Work Restrictions, and Accommodation

Traditionally, health professionals have identified employee work restrictions in negative terms (e.g., "unable to lift more than 50 pounds," "legally blind," "sensitized to toluene diisocyanate (TDI)," or "may not operate hazardous equipment requiring use of upper extremities"). Job restrictions often have been systematized for use in medical or personnel information systems. The ADA challenges health professionals to state an individual's capabilities rather than limitations (e.g., "able to lift up to 50 pounds," "can be exposed to dust if uses a respirator," "can stand for two hours before requiring a 15-minute rest"). Contemporary systems also will include reasonable job accommodations that have been recommended and/or adopted.

Performing Appointment Scheduling

Computerized appointment scheduling is a valuable asset. The program can be set to issue reminder and follow-up notices to employees and facilitate tracking those who miss scheduled appointments. Such a system also can schedule future appointment dates for people requiring periodic medical evaluations, fit testing, or training. Similar functions can be used to alert industrial hygienists, safety personnel, and training managers about the need for workplace monitoring, corrective actions, and course work.

Maintaining Regulatory Requirements

Through simple automated text-retrieval systems, computers can help track current company policies and state or federal medical requirements, including the schedule and frequency of required examination procedures.

Performing Medical Record Keeping and Retrieval

Computerized systems can be used to collect the entire medical record of an employee or merely to summarize dates of past examinations and appointments, leaving actual medical information in hard copy or electronically imaged records. Many occupational health clinics are automating critically important information, and several employers have actually adopted paperless medical records.

The use of clearly defined responses (e.g., yes/no, present/absent) for employee health history and physical findings can significantly increase the value of these systems. However the development of these databases may be resisted by many clinicians who insist on using free text notes. The question of whether the electronic medical record is admissible as evidence in legal proceedings should be explored prior to converting from paper records. Procedures such as storing backup paper medical records signed by the health professional should also be considered, as this area is often covered by state laws.

Recording Employee Health History

Monitoring and protecting employee health is a key function of occupational safety and health systems. Medical personnel primarily manage and store records related to employee health. Preplacement examinations have replaced preemployment examinations. Preplacement examinations are used to determine from a medical standpoint whether an employee can perform the essential job functions. As part of this activity, along with periodic health surveillance, medical data are collected and recorded. Most medical histories are completed on paper; however, key historical information is often entered into computers for archiving and comparison over time.

Many employers have developed forms for preplacement and surveillance health histories that can be optically scanned to simplify data entry. Some medical departments use interactive computerized health histories. Interactive histories allow hundreds of questions to be included in a branching tree format, designed so that each employee only sees a fraction of the total available items. This requires the use of key branch points (e.g., gender, use of tobacco, or key symptoms for each organ system).

Automating health history information can help ensure appropriate medical follow-up of positive symptoms and

can even direct the physical examiner to do the most beneficial, cost-effective physical exam procedures. Automation also means that periodic examinations can be tailored in response to individual risk. A good example of surveillance is maintaining hearing conservation program records. Once noise monitoring, audiogram, and employee history data are entered and stored, these data can be used to produce monitoring schedules, threshold shift evaluations, notices to employees, and summary reports for program management. Similarly, potential chemical exposure in the workplace may require periodic biologic monitoring, laboratory tests, and medical examinations that must be documented for regulatory compliance.

Performing Injury Monitoring and Management

Many computer systems record the federal and state OSHA reports of first injury or illness, generating forms to ensure regulatory compliance. Tracking injuries by location and job can identify high-risk situations in need of follow-up and can monitor patterns of injury and illness.

Performing Health Promotion, Health Risk Appraisal, Risk Reduction, and Demand Management

Automating risk factor information such as blood pressure, cholesterol, weight, fitness, risky behavior, and smoking status can assist the development of health promotion programs. For example, individuals at high risk can be identified for targeted follow-up, health education mailings, and scheduled revisits. Computerized health risk appraisals range from inexpensive public domain software to highly sophisticated commercially available systems.

Aggregate reports can profile the health status of a population and even project health benefit expenses related to modified risk factors. Specialized programs include diet and nutrition analysis, fitness evaluation, body composition, and stress profiles. Systems can be used to track those at high risk and to monitor participation in risk-reduction, medical self-care and informed-consumer programs.

Performing Drug and Alcohol Testing

Random drug testing requires use of random-number generators to select individuals for testing on short notice. Specialized programs track individuals in a pool eligible for testing, urine collection procedures, laboratory test results, medical review of positive findings, referrals to substance abuse professionals, reporting of verified results to the employer, and treatment follow-up. The critical need for data security requires a high level of care in automating such information.

Supporting Employee Assistance Programs (EAPs)

Separate records are often kept by EAP counselors. Automated EAP systems or components of an OHIS with restricted access allow users to generate activity reports, track and follow-up individuals, and construct group report for evaluation purposes.

Documenting Medical Surveillance

An exposure-driven medical surveillance module permits users to track required examinations and make sure that every employee in a particular job or exposure group has been tested. This module can be linked to appointment scheduling systems and record pertinent test results.

Maintaining Chemical Inventory

Manufacturers and users of chemicals must track chemicals used to comply with the OSHA Hazard Communication Standard (HazCom). More than 50,000 different chemicals are used in the United States. Tracking the ingredients of mixtures is challenging, especially when formulations are proprietary. This information is often an integral part of occupational health information systems.

Performing Exposure Monitoring

Because industrial hygiene measurements generate a high volume of data, they have become increasingly automated. Potentially hazardous substances, appropriate measurement techniques, required periodicity of sampling, and actual measured exposures are included. Eight-hour, time-weighted average exposures can be easily calculated. By linking industrial hygiene data with job and personnel information, users can identify individuals who require medical surveillance. Flags can be set to identify any area exceeding allowable exposures, either by company action levels or external standards. Workplace exposures can be tracked by location or by groups of individuals. After facility surveys have been conducted, users can provide exception reports that point to identified hazards which have not yet been corrected (Figure 9-3).

Maintaining Toxicology Information

Under the OSHA HazCom, employers must compile and make available to employees copies of Material Safety Data Sheets (MSDSs). Automated information systems represent important advances that greatly help in monitoring environmental agents, including inventories of hazardous substances, toxicology data, and material safety information on hazardous substances within the workplace. These data include OSHA permissible exposure limits (PEL), action levels, and internally adopted standards, such as the American Conference of Governmental Industrial Hygienists (ACGIH) threshold limit values (TLVs). Software can be used to track ongoing events or problems requiring follow-up, such as hazard abatement plans and schedules. The agents managed include chemical, biological, and physical hazards. Inventories of agents can be maintained by department, area, process, job, or any combination of these.

Right-to-know and hazard communications requirements mandate that organizations identify substances in each workplace, train employees in their safe use, maintain material safety data sheets, and promptly respond to requests for information. Other regulations require that companies develop and maintain emergency response plans for handling unintentional releases of hazardous substances into the environment. Many companies have automated this information for substances they manufacture, and a number of vendors have compiled large numbers of MSDSs. Some organizations maintain centralized repositories of MSDSs that can be immediately faxed on demand to the worksite.

Compact disc read-only memory (CD-ROM) or telecommunication links to large databases allow users rapid access to MSDS and other toxicology information. These databases include CHEMINFO, Regulatory Information on Pesticides Products (RIPP), Transport of Dangerous Goods (TDG)/hazardous materials (49 *CFR*), Chemical Evaluations Search and Retrieval System (CESARS), the NIOSH Registry of Toxic Effects of Chemical Substances (RTECS), the Combined Health Information Database (CHID) of the National Institutes of Health, and the Superfund Amendments and Reauthorization Act (SARA) Title III.

Incident Management

Many organizations use automated systems to store, analyze, and report on incident data. Data for all types of incidents are entered, such as property damage, occupational injury, and near-misses. These data usually parallel the organization's first report and incident investigation forms. Most computer systems can produce (1) individual incident reports; (2) summary reports that categorize incidents by location, type, rate, severity, loss, and other factors; (3) OSHA Form 200 logs; and (4) analytical reports that pinpoint major causes and types of incidents within organizational subsets. As with financial systems, users can compare observations during a given time frame for a single site or among sites. Some companies use a 12-month rolling average to track, compare, and manage incident experience. For example, an organization might find that a disproportionate number of incidents are occurring in a particular operation. Management can then arrange for employees and supervisors in that operation to have additional training.

Incident Investigation

With the advent of powerful, low-cost computer systems and artificial intelligence capabilities, it is possible to uncover the underlying cause of many industrial incidents. Incidents routinely attributed to poor housekeeping, sloppy procedures, or indifference by workers and supervisors can be fully investigated and corrected once the root cause is identified. Because human error is so often responsible for incidents, management can target employee training programs accordingly. Employers can make many such systems portable by using lightweight notebook computers, which they can take directly to the

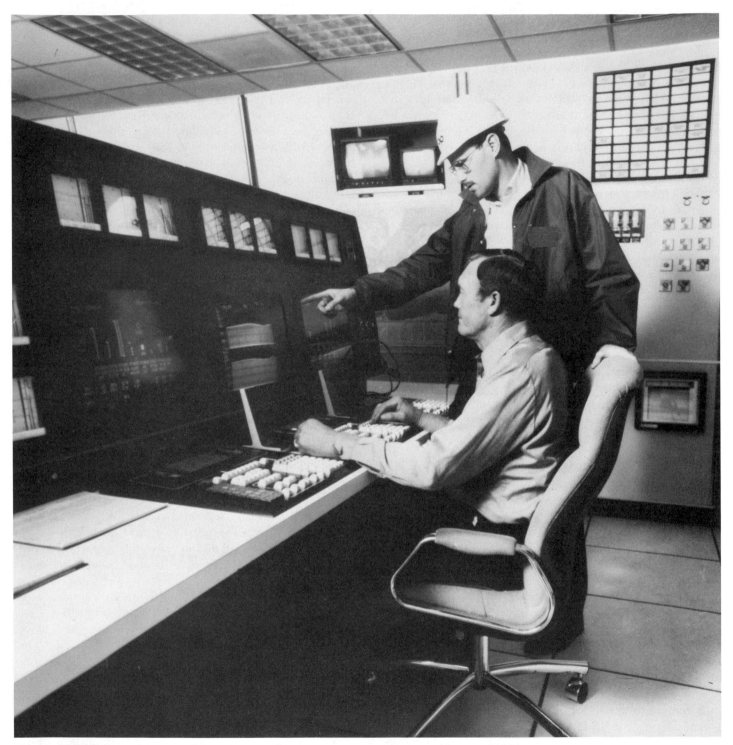

Figure 9-3. This computer-assisted program is used to monitor ongoing operations. (Courtesy UOP.)

scene of an explosion, fire, or spill. In this way, computer applications can reach far beyond the confines of physical facilities and provide benefits never before considered possible.

Protective Measures
Computer systems can maintain information on measures taken to protect employee safety and health through (1) engineering controls such as hoods for ventilation, (2) administrative controls for health hazards, and (3) the use and fit testing of personal protective equipment. These system functions allow an organization to record and store the information needed to facilitate regulatory compliance and to document the measures taken to protect employees. Access to such information at a later date is crucial, especially when an allegation surfaces that the employee

was harmed by an exposure to a hazardous chemical or physical agent.

Tracking Safety and Training Data

Information frequently tracked by computer includes personnel protective equipment, employee training (e.g., hazardous material handling, hazard communication), audits and inspections, fire prevention, and other safety information. Many companies have set up centralized incident reporting and tracking capabilities.

Computer systems are being used more often to deliver information to workers, supervisors, and managers. When that information is repetitive, highly complex, or technical, interactive computer-based training (CBT) is an ideal, cost-effective way to reach a large audience. For example, chemical handling safety, lockout/tagout, safe practices, heavy lifting, and emergency action planning are some of the training procedures being taught through computer systems.

Computer-assisted instruction can be used to record student responses, repeat material until mastery is obtained, and test for comprehension. Students set the pace with CBT and control how material will be presented—an ideal solution for teaching a diverse workplace population. To comply with recent regulatory initiatives, companies must maintain a computer record of the course, instructor, date of completion, and score.

Maintaining Predictive Maintenance Information

In some cases, incidents and injuries may be the result of faulty equipment maintenance. The implementation of a program based on predictive maintenance, inspection, and repair has been shown to reduce the number of equipment-related incidents. Such a program produces a set of reports designed to anticipate equipment failure, loss of productivity, or a regulatory violation.

Maintaining Financial and Administrative Information

Corporate occupational safety and health departments must accurately account for services charged back to operating organizations. Billing can be complicated when employers want a single consolidated invoice for multiple services provided by different elements of a hospital or clinic. Computer programs may need to incorporate many features of a general ledger sheet to accommodate these needs.

Performing Analysis and Reporting Functions

Although most OHIS still function largely as passive electronic storage cabinets, they must be able to present data in a variety of flexible formats. Increasingly, computer systems provide features such as error checking, determining if values are beyond acceptable ranges, and flagging situations requiring follow-up. Most OHIS products come with formatted standard reports. The program must allow users to generate customized reports easily by using ad hoc queries.

Implementing Total Quality Management (TQM)

Tracking performance indicators and quality outcomes measures has emerged from reengineering and continuous quality improvement movements within companies. In health care, outcome measures are being identified to monitor quality of care. As consumers of health care, employers are asking providers to implement total quality management (TQM) in their health care management. Thus, providers are beginning to use software systems to track variations in their performance, e.g., variation in selected services offered by the provider. Some provider systems are beginning to incorporate "decision support technology," which prompts the provider to make appropriate decisions. Examples include medical practice guidelines and protocols, such as those for diagnosing, treating, monitoring, and managing low-back pain and soft-tissue injuries.

Setting Up Linkage to Other Systems and Databases

Users must consider whether data from inexpensive, single-purpose software programs can be imported into larger integrated systems. Increasingly, off-site clinic vendors and employers are moving toward the integration of health data within the company's OHIS. Occupational health clinic users may need to obtain selected data from, and share these data with, other corporate databases for purposes of monitoring industrial hygiene, tracking, chemical inventories, evaluating safety, and managing personnel data.

SYSTEM SELECTION

When an organization considers buying or building a computer system, management should ask how the system will be used, who will use it, and what features are necessary to satisfy the users' immediate and future needs.

Matching Needs to Computer Systems

Prospective users should look at their current filing systems and work processes and ask, "Does this task need to be done and, if so, can it be done more efficiently with a computer system?" For instance, in the manual system for gathering data on workplace injuries, data collection may be inconsistent and sometimes incomplete. As a result, reports may also be incomplete, which would increase the time it takes to prepare an OSHA 200 report. Regardless of the user's area of expertise, there are a number of questions to answer before considering a computer system. Examples include:

- Where will the safety incident records come from? If there are branch plants or offices, will each branch forward its records to a central office for compilation or maintain them separately?
- Who will input the records? Will a clerical person be given the task or will the safety manager handle it?

- How many injury cases are recorded each year?
- What information must be maintained on each injury? Will workers' compensation claims data be included?
- What kinds of analysis does management want from a system? If the records are consistently maintained, what do they reveal about safety in the workplace?

If right-to-know compliance must be automated, management should answer the following questions:

- For how many chemicals must Material Safety Data Sheets (MSDSs) be maintained?
- Where is the chemical stored? In one facility or many? How will managers know if they have reached the reportable threshold?
- How is the chemical inventory maintained and where? Does the safety manager receive an MSDS with each new shipment of chemicals?
- How does management track which employees have been trained on which chemicals?
- Is there a need to prepare Tier reports or Form R reports for hazardous chemicals at the work site? How will they be done? Who will perform the calculations?

Documenting company needs will help management match their requirements with the features and functions a vendor can supply. Establishing overall goals will help in setting priorities. An organization may find that so many functions could benefit from automation that the task may appear overwhelming. Many occupational safety and health professionals have prioritized the needs and tackled one function at a time before proceeding to the next. In this way, automating processes becomes manageable and later implementations proceed smoothly with the benefit of early experience.

Budgetary considerations may also dictate the need to go slowly. Many off-the-shelf software products address one or more of an organization's needs, but may not interface or share data with other programs. In addition, although they may do an effective job for a given function, they may be impossible to integrate into an overall system.

Some vendors have solved the problem by developing complete OHIS systems that address the needs of occupational safety, health, training, worker's compensation, and industrial hygiene. Though these systems are often integrated, they may include modules that can be run independently. This gives the user the best of both worlds and allows for orderly implementation of new functions over time.

It is important to address the need for future integration when first designing or purchasing a system. This step will reduce the possible choices of future upgrades or changes to a more manageable list of vendor applications to review.

Off-the-Shelf and Customized Software

After developing a list of needs, the user group should have enough information to identify the computer and software system that best meet those needs. The following section provides several recommendations that may help potential system users make a wise decision.

Off-the-Shelf Software

Occupational safety and health software has been commercially available since the late 1970s. Two of the biggest advantages to choosing off-the-shelf software are cost and convenience. Because the cost of creating a software product is high, commercial sale allows those costs to be spread over many more users. An organization can benefit from a product that has been well tested, accepted by the profession, and made affordable. Several software vendors have established user-groups and advisory boards to guide their software development efforts. User group involvement is recommended. Most current programs can be installed by nonexperts with little or no training.

Traditionally, the major drawback to off-the-shelf packages has been their lack of flexibility, although over time flexibility has improved. It is rare to find a commercial package that can be customized to fit every situation. Nevertheless, newer technologies in relational database design allow software developers to set up data-driven tables that can be customized for each client.

Packaged software may force a company to change the way certain records are maintained or change some operating procedures. Sometimes a willingness to reexamine and modify work flow is beneficial. Despite a tendency to think that one's needs are different than those of others, experience among developers reveals otherwise. Keeping an open mind can reduce the need for an open pocketbook.

Although some software vendors offer their product in what is known as "source code" form to allow for modifications, the cost of these programs is generally higher. Once a client has modified a program, it is on its own. The developer may offer little assistance if modifications fail to perform as expected.

Customized Software

An alternative is to pursue custom-developed programs that fit an organization's exact needs. For a significantly higher cost in money and time, management will have a set of programs that mimic paper systems and that generate all the reports needed to exact specifications. Management must be prepared to wait patiently, however, while the system is designed, developed, tested, and implemented. Management involvement will be greater, too, because the software programmer will need highly detailed information about company needs and reporting requirements.

Ultimately, system selection will depend upon the functions desired, the working environment, a company's budget and a program's compatibility with other systems such as human resources or payroll.

Learning from Other Users

Companies can learn a great deal from the efforts of others, especially by studying how systems evolved and how expectations have risen as newer technologies emerged. Over the past few years, both needs and resources have changed considerably in response to user demands and industry/government developments.

First-Generation Mainframe Systems

The first generation of Occupational Health Information Systems came on-line in the 1970s. Because these systems were mainframe based, only the largest companies could afford the millions of dollars for the hardware and the even more millions for the software development. Some of these companies tried to recoup their investment by selling the software to other companies. This effort met with only limited success due to the lack of flexibility of the software to meet other companies' needs.

Most early systems were large data storehouses. Tons of data (and money) were put into these systems, but only limited provisions were made for getting the data out in usable form. This fatal flaw caused most systems to fail. Even extensive modifications could not fully meet the user's needs. Most of these systems had to be scrapped. Those that survived were rewritten to use relational database software in the mid 1980s.

All mainframes (and their systems) have a closed architecture, in which the specifications are protected by patents. In an open architecture system, any vendor can reproduce or improve the hardware. As a result of competitive pressure, open systems ultimately became a cheaper alternative to mainframes—a trend that will continue.

During this period, occupational safety and health departments began to realize they were "prisoners" of their data processing (DP) departments, now often called information systems (IS) or information technology (IT) departments. When DP implemented software for their internal OHIS clients, they applied the same methodology of back-charges used for other departments (i.e., accounting or engineering). This charge-back method turns DP into an internal profit center, allowing DP to build an empire by showing how much money they were making at the expense of the enterprise instead of answering the criticism of how much money they were costing the company. The result was that most occupational safety and health departments were charged for every page printed and for every medical record stored or accessed. Instead of having a system that encouraged use and thereby increased the efficiency of the department, the system discouraged use by punishing the user. Because of the technology involved, getting useful information from these systems was rarely successful.

Second-Generation Minicomputer Systems

The second generation of OHIS came on-line in the late 1970s through the 1980s. These systems, recognizing the limitations of proprietary, single-purpose mainframe systems, were based on minicomputers, which are cheaper than mainframes and easier to use but not as powerful. Although hospitals, mid-sized companies, and large clinics could all afford minicomputer hardware, few could afford the investment in generating the OHIS software.

Some minicomputer OHISs used early database technology, which could be tailored to meet different client needs. Unfortunately, the price of customization was high and minicomputer hardware could not take advantage of new developments in database and PC technology.

Minicomputer microprocessor manufacturers licensed their technology to others to create a variety of vendors offering similar systems. By increasing the number of vendors using the same processor, consumers were offered a wider selection of system features and functionality. At the same time, a currently popular multiuser operating system called UNIX was adapted to run on these microprocessors. UNIX has become the open system operating environment of choice for many mid-range multiuser computers. Several software developers built business software to operate in this environment, paving the way for departmental computing, thus bypassing an organization's reliance upon mainframe systems and DP department personnel involvement.

With the advent of UNIX, a number of developers created OHIS software to bring commercial applications to smaller organizations and departments within larger companies.

Third-Generation Microcomputers

Personal computers (PCs), such as the Apple II series, first appeared in the 1970s, and the IBM PC (and compatibles) appeared in the 1980s.

PCs were the first systems to be available with inexpensive connections to modems, scanners, medical testing equipment, CD-ROM, and so on. This fact has greatly enhanced the usability of these systems. The IBM-compatible PC represents an open architecture that has grown to dominate the desktop market. IBM's operating system was marketed under the trade name of PCDOS and was developed in collaboration with Microsoft. The latter company subsequently marketed their version of the same operating system called MSDOS to run in numerous manufacturers' IBM-compatible machines.

In the past decade, PCs and Macs have gained computer power at a rate never seen before in the industry. This increase of power, coupled with a decrease in price, has resulted in the sale of hundreds of millions of PCs and associated peripheral devices such as printers, modems, hard drives, network cards, and CD-ROM drives. Software technology has grown just as rapidly, creating a huge supply of software packages.

Occupational Health Information Systems have been implemented on both stand-alone and networked PCs. Most of these systems use relational databases with fourth-generation languages and graphical user interfaces (GUI). These developments have resulted in greater flexibility as well as cheaper systems. The recent popularity of client-server technology has increased system flexibility, and a number of developers are providing solutions that focus upon the need to share data and communicate between networks and other departments within organizations.

As the hardware has evolved from mainframe to mini-computers to microcomputers, the politics of software deployment in the business environment has also evolved. Systems have migrated from being data processing-controlled to user-controlled. This shift has caused turf wars between managers. This holds true for occupational safety and health professionals. The central data-processing structure within companies requires one to become more technically competent but also more politically savvy. The resolution of conflict in most companies will be the narrowing of focus and break-up of the DP empire that currently exists. The user will have increased options and responsibilities for systems, with DP providing technical support on request.

Information Systems Department Influence

Many factors influence an organization's choice and location of computer hardware for its information systems. These include currently available hardware, corporate standards for computer hardware, telecommunications, software, and the information requirements of system users. All system options and constraints should be scrutinized during the requirements study to create an economical, efficient, and integrated system. Rapid technology advances generally affect selection decisions.

Though there has been a trend to PCs, local area networks, and client-servers, some companies still use (1) a mainframe or minicomputer for all system functions, with terminals distributed to users and communication conducted over telephone lines; and (2) use of mini- or main-frame computers for regional database management, with terminals or PCs used for data input and communication lines among the larger machines for corporate-wide analyses. These systems require support from IS professionals, who may influence the selection of any computer system. In fact, the OHIS selection may require IS approval to ensure compatibility with other existing computer systems. If so, IS personnel should become familiar with OHIS needs as early as possible and participate in the selection process.

COMMERCIALLY AVAILABLE SOFTWARE

The following section describes groups of currently available occupational safety and health software products. A number of guides to software exist, such as *PC World*, *Computer Shopper*, *Byte*, *InfoWorld*, *LAN*, *PC Magazine*, *PC Week*, *Computer Language*, and *Dr. Dobb's Journal*. Some of the more well-known publishers of books that list software and software vendors and dealers include Bantam Computer Books, McGraw-Hill, Microsoft Press, Norton, and Osborne.

Integrated Health Data Management Systems

Large progressive corporations are turning to integrated health data management systems (IHDMS) for help in making complex decisions. Implementation of an IHDMS can help to prevent unnecessary duplication of effort, foster coordination among units, and provide a clearer picture of the company's health expenditures and priorities. These systems may include disability and general health benefits; occupational safety and health, wellness, and safety programs; and related databases (e.g., personnel) that cut across departmental lines for employees, dependents, and retirees.

Companies should develop opportunities to better target and evaluate their occupational safety and health programs. They should give special consideration to re-engineering critical processes to ensure that these processes contribute as much as they can to the organization's mission and support departmental goals. Team-building processes inherent in establishing and maintaining IHDMS should be employed in other aspects of corporate management. Most IHDM systems require mainframe or mini-computers to run programs and maintain records. However, some systems now operate on a PC platform, with work stations linked by PC networks. Full-scale IHDM systems can cost several million dollars to develop, with licensing fees usually exceeding $100,000.

Occupational Health Information Systems

Fewer than a dozen large-scale integrated systems are available that track personnel, safety, industrial hygiene, medical, toxicology/MSDS, Employee Assistance Program (EAP) and, occasionally, benefits information. During the past decade, a number of expensive integrated systems have dropped out of the marketplace. These were mostly mainframe systems using second-generation database management software or minicomputer applications. The few surviving systems use relational database technology. Many of the successful OHIS vendors have migrated their software applications to client-server technology to facilitate communication between the disciplines and departments involved. Many of these vendors have also developed graphical user interfaces to simplify data management and operate in the Microsoft Windows™ environment providing access to the office automation products available therein. Full-scale OHISs can also cost several million dollars to develop, with licensing fees often based upon company size. Fees exceeding $25,000 per site and $100,000 for enterprise systems are not uncommon for comprehensive systems. These integrated systems are often modular and contain most of the features described in the following categories.

Clinic Management

Though clinic management would be of minimal interest to occupational safety and health organization personnel, it may be valuable for a company to know what is available to the occupational medical providers who care for its employees. This category of commercial software addresses the more focused needs of occupational health clinics. Almost two dozen commercial software programs are available on PC platforms. Most include demographic and administrative information, appointment scheduling, injury/illness care and absence tracking, reporting capabilities, clinical results, accounting, and financial management. Some include alcohol and drug testing, hazard exposure, and other environmental information and ergonomics capabilities. Most vendors have fewer than 100 installed clients, so no software is dominant. Of interest is the continual entry of new software products into this competitive arena. Clinic management systems licensing fees are usually in the $5,000 to $25,000 range, with smaller annual maintenance/renewal fees.

Health Screening

A variety of smaller software programs focus on health screening. Examples include HIV testing, audiometric data storage and analysis, drug and alcohol screening, pulmonary function, and respirator use. These programs usually cost a few thousand dollars.

Compliance Records Management

More than a dozen software products now focus on injury reporting and disability management, a rapidly growing field. These products range from simple programs that complete OSHA 200 logs or provide absence duration guidelines, to full-scale disability case management systems that support case managers.

Ergonomics, Health Promotion, Fitness

Over a dozen software programs are available which perform health risk appraisal; most use algorithms promulgated by the Centers for Disease Control or the more recent Emory University Carter Center Update. A few have a more sophisticated risk computation science base and provide communication that helps motivate workers to adopt healthier lifestyles. Other health promotion software performs dietary nutrition analysis, fitness evaluation, and ergonomics assessment, focused on the spine and upper extremities.

Chemical and Environmental Programs

Dozens of programs have been developed for safety, industrial hygiene, toxicology, emergency response, and environmental medicine needs. Many provide rapid access to large chemical databases by use of CD-ROM or telecommunication linkages. Others enable employers to comply with federal regulations, including OSHA, SARA II, and TOSCA. Some are PC based, others require mainframe computers.

Decision Support, Epidemiology, Statistics

More than 50 low-cost epidemiological analysis software packages exist. Most of these build upon standard statistical analysis capabilities, but are tailored to the needs of epidemiologists.

Reference and Training

Software programs can track bibliographic information and medical references. They also provide health education information, drug information, computer-assisted instruction, and continuing education with CME credit.

National Safety Council

The National Safety Council develops and markets specialized software to assist in the management of hazardous materials. One example is the CAMEO data management system available from the National Safety Council in CAMEO DOS and Macintosh programs.

The CAMEO system can include databases on facility information and floor plans; Tier II Cards; a facility map out to 10 miles; vulnerability/risk screening and scenarios for each environmental health system (EHS); emergency contacts; fire hazard data sheets; and electrical isolation data sheets. ALOHA, the CAMEO Air Model, was used to develop the scenarios.

In the event of a fire, CAMEO would allow an incident commander quick access to the area's floor plan, where further data on chemicals and other hazards could be obtained. Several special databases or stacks were developed for this system. The locator stack consists of multi-layered floor plans of the reactor building, and even subdivides these into fire zones. The "Fire Hazards" stack contains key data on special hazards, such as flammable solids, firefighting suppressants, radiological concerns, protective clothing, and environmental safeguards. An "Isolation" stack provides information to turn off electrical current.

CAMEO allows emergency response personnel to access information much more rapidly than do traditional systems. With CAMEO programs in computers located at both the Emergency Operations Center and on an emergency response vehicle, personnel will be able to respond more effectively during an incident.

DEVELOPING A SYSTEM INTERNALLY

If an organization finally decides to develop its own system, managers must participate in the design and development process. Assistance from a systems analyst is essential. The intricacies of system development will depend on the size and complexity of the company's needs and the hardware and software environment in which it will be built. Naturally, a small PC-based system for management of one or more unrelated functions will require less effort and systems expertise than will a comprehensive occupational safety and health system.

Many organizations build and implement large systems in phases. This can work well if management prepares an overall plan for total system design that is followed during the phased development. Too often, groups have built system modules without considering total system requirements and necessary connections among modules, thus leading to project failure. Others have had budgets cut and have been unable to implement remaining hardware and software functions.

SOFTWARE SELECTION AND/OR DEVELOPMENT

The following recommendations are based on the authors' combined experience in occupational safety and health computer applications. This experience includes both consultations to, employment by, and management of corporations that were involved in implementing OHIS. (Figure 9-4 provides a summary of suggestions for choosing software.)

Define Needs and Uses

Companies must start by clearly defining their needs and specifications. Interdisciplinary task forces can be invaluable if they involve users and internal customers as well as local information system people. These groups can help management avoid making technically elegant decisions that fail to meet their needs. Companies may want to define their goals and identify priorities in decreasing order. The entire group should agree on the goals and priorities before proceeding.

Practical Suggestions for OHIS Software

- Define the organization's needs and intended uses.
- Explore the various available software packages and make a "buy/build" decision.
- Try out the software thoroughly and contact current users.
- Look carefully at the vendor's organization and the key people with whom the company will be dealing.
- Review the sample contract or agreement.
- Make sure there is enough flexibility to meet the company's needs.
- Look closely at group data as a valuable resource.
- Make sure the software meets regulatory compliance needs.
- Develop measured indicators and benchmarks.
- Select a system administrator.
- Clarify customer support, periodic updates, and annual fees.
- Cross-train two people to be liaisons with the vendor.

Figure 9-4. *Practical suggestions for evaluating, selecting, implementing, and/or developing OHIS software.*

Areas to address include:

- Program objectives and processes—e.g., regulatory compliance, patient scheduling, visit tracking, service billing, information storage and ready access, employee education, behavior change
- Users—e.g., degree of computer sophistication at all levels—end users, operators, systems managers
- Use—e.g., stand-alone task or part of integrated department or company-wide OHIS/IHDMS
- Feedback—specific information and types of reports that will be needed
- Resources available—personnel, computers, software, IS support, other users
- Size of operation and volume of data processing
- Speed and turn-around time—How quickly does the company need data entry reports? Could data entry/processing be done off-site?
- Group data needs—management reports, measures of success
- Total quality management (TQM) commitment.

Clear identification of needs can save time by helping to eliminate many products. An internal or external consultant can greatly assist management in choosing the right system. Successful systems are usually simple, user friendly (or at least tolerable), and flexible enough to adapt to changing needs.

Explore Options

Companies should evaluate their options in terms of how these options meet company needs. Most vendors are product oriented (i.e., they will try to sell a company what they have to offer, convincing management that they have just what the company needs). Managers should investigate the market thoroughly and be aware of the range of choices. None may suit a particular company's requirements exactly, hence the potential need for customization. Even if one product meets a company's purposes today, those needs are likely to change; and the software must adapt or be discarded. On the other hand, developing even the simplest program from scratch is filled with hazards (e.g., time, money, and the frustration of recreating the wheel). There is no magic solution that can satisfy all of a company's requirements.

If a company elects to build its own software program, management should check the literature and contact organizations that have had experience in this area. Management should always become familiar with commercially available alternatives and conduct a cost-benefit analysis before embarking on a major program development effort. They can enlist the participation of other departments and establish an integrated program as appropriate. Companies must be aware of the many pitfalls of developing a system, which are described in the next section.

Test Software

No amount of discussion or reading of marketing materials can replace making arrangements with a vendor to have a few hours of actual "hands on" work at the keyboard. Management can ask for a demonstration at an existing user site to see how the software operates. Run a number of sample employees, patients, exams, or industrial hygiene measurements through the system; put it through the paces. Every vendor uses examples that show the software in the best light. What about the typical person in the company's intended audience? Management should try to discover the limitations of the system being considered and talk with several experienced users of the product.

Investigate Vendor

Investigate the vendor carefully, asking the following questions. How long has the organization been in existence and what is its history? Is it stable? What is its financial condition? Will it be able to provide continued support after the next recession? How many clients does it have? How large is its programming staff? Are the annual renewal fees and number of current customers sufficient to support a programming staff? How long have the actual programmers worked with the vendor? What is their continuous quality improvement (CQI) process, if any? How does it work? How do they involve their customers in CQI? Do they have an active user group, or advisory group?

What are the backgrounds and qualifications of the principals and of those with whom the company will be dealing on a day-to-day basis? This is important because although many purchasers believe they are buying a product, in reality, they are buying a service and a relationship. Like most relationships, the true test comes after the purchase when problems arise and adjustments in the software or hardware need to be made. Talk to other clients who have similar uses for the system. Does the vendor come through and deliver as promised?

Review Sample Contract or Agreement

Does the vendor offer field tests, trial use of software, and a money-back guarantee if it does not work or if the company is not satisfied? Look at users' manuals and documentation; find out what kind of training, customer support, and software support is available. Is remote support provided, e.g., linking the company's computer directly to the vendor's modem? Will company personnel have access to the source code, possibly in an escrow account held by a third party, in case the vendor goes out of business?

Evaluate Flexibility

Perhaps an off-the-shelf product is all a company needs, but management may well want to customize the programs later. This step may be simple or complex, and could cost thousands of dollars. For example, how easy is it to add variables (e.g., new questions to a questionnaire) or to telecommunicate or to export data for external storage or processing? What is the organization's success rate in customizing its product for other clients? How close to estimated budget and time did it come? What about data analysis, storage, transfer, software limitations, import/export capabilities, compatibility?

Evaluate Group Data

Which group reports are available, and do they report what the company needs, e.g., can they report employees by location, payroll code, job class, or other variables? Will they provide the information needed to present to top management, e.g., on the health or illness of a defined population, or the potential cost savings or degree of improvement over time? What data-summarizing and query capability exists, and how complex is the query language? Many million-dollar systems have been developed or purchased without giving adequate consideration to analysis and reporting functions. How easy is report generation? Can a company design its own reports without needing to customize the software? Can data be exported to office automation tools such as word processors and spreadsheets? How easy is it to do so?

Evaluate Compliance Features

Does it cover current OSHA and other regulatory requirements? Does the vendor keep abreast of regulations?

WORKING WITH A SOFTWARE VENDOR

Once a company has decided to purchase or develop a software system, the next challenge is how to work more effectively with a vendor or developer to install and implement the system. The keys to success are planning and training.

Preplanning

Know Company Needs; Plan the Program Completely
The most important single task the company should complete is to think through its needs as clearly as possible at the beginning, then plan the system carefully. Planning the whole program not only helps management select the correct software package but also makes it easier to change, adjust or correct the software along the way. Careful planning can save a company considerable time and money. For example, if a manager is asked to justify the costs of a program or to show its effectiveness in the workplace, he or she is more likely to succeed if the right data are collected from the beginning.

Communicate Company Needs Clearly
Information system specialists can often help management think through program options and make good choices at

the beginning (e.g., data needed for later program evaluation). However, the company should oversee the specialists' contributions. Managers should beware of data processing departments motivated to extend or consolidate their empire. Also, it takes time to develop software, which means it is not unusual for users to forget what they requested. Managers should keep good notes and implement a formal sign-off process and change procedures for the software developed.

Build a Positive Relationship with Vendor Personnel

Company personnel should offer their questions, comments, and suggestions freely. An interested customer can bring out the best in a software team.

Clarify What the Company is Buying

The company should find out exactly what comes with the package and what will cost extra. What other services can the vendor provide? The price may never be lower than when you are negotiating the first dollar spent. What will cost extra? Management should get clarification in writing on such features as assistance in program design and evaluation, data analysis, graphics, and reporting.

Clarify Customer Support, Periodic Updates, and Annual Fees

Companies should be clear about which services are covered under the basic licensing/purchase fee and which features cost extra. Management should be willing to pay annual maintenance fees. In addition, the company needs to make certain that the vendor has a sufficient ongoing revenue stream to maintain the software and continuously improve it. Companies want to make sure that their vendors are in business in the future.

Develop Measured Indicators and Benchmarks

Working with a vendor to develop, customize, enhance, or maintain a system is challenging. Having agreed-upon criteria for success (or failure) and benchmarks for measuring how the effort is progressing (or not) helps to objectify the process.

Training

Implementing any new system is a joint effort between developers (internal or external) and users. Although many activities are associated with implementation, the most important one is user training.

Select a System Administrator

If a company has multiple users or sites using the software, it can arrange to have all communications go through one coordinator or systems administrator. This approach will simplify user support and assist the vendor in responding to the company's needs. Every system should have a manager or administrator who has ultimate responsibility for the day-to-day operation of the system, including (1) management of system security, (2) supervision of the data content, (3) problem solving, (4) coordination of changes to the system, (5) archiving of data, and (6) planning for future needs and applications. The system manager should know the application well and work closely with users to make sure the system meets user expectations.

Cross-Train Two People to be Liaisons with the Vendor

Companies should train employees in the system—developing in-house competence will be an asset in working with the vendor. But more than one person should be trained, so there is backup support in case of employee illness, absence, early retirement, or job transfer.

Train Company Users

The best computer system in the world will sit idle unless the system's users are trained and motivated. New computer systems change the way people work, and change can be quite threatening. Ask the vendor or a skilled training group for help. Training should address the transition from traditional ways of working to more computer-assisted methods. Doing double duty of front-end data entry while maintaining an old system leads to decreased productivity and disillusionment. To maximize the benefit of any system, establish a training program that adds new functions sequentially, reinforces learning, and embraces user input (Figure 9-5).

Establish Realistic and Achievable Goals

Training must establish realistic and achievable goals about the system's capabilities and requirements. Any changes required in record-keeping practices and potential transitional difficulties should be thoroughly explained and understood by the user. Implementation of new functions can be stressful; hence, emphasizing patience is crucial.

Provide On-the-Job Training to Supplement Formal Programs

Following formal training exercises, management should conduct on-the-job training as part of the work routine. If a graphical user interface is selected, workers must be trained in the fundamental uses of a pointing device or mouse and in window navigation. Occupational safety and health professionals can learn to master the use of terminals or PCs for data input and output in the context of their actual work.

Introduce Training Programs at the Right Time

Training should be timed to prevent any gap between training sessions and actual use of the system. When people learn something entirely new, they need to practice it immediately to reinforce the knowledge. The longer the gap between training and system use, the less people will retain. If lengthy delays occur, then a refresher session may be necessary.

Figure 9-5. An interactive training simulator instructs employees about operations, standard procedures, and emergency procedures. (Courtesy UOP.)

Periodically Reinforce User Learning

Success in using the system depends just as much on follow-up training as it does on initial training. Periodically, the system manager should determine if users' goals for the system are realistic and if they are using the system correctly. Error rates should be monitored and studied. If the rates are too high, management should identify the reasons and retrain the users.

EVALUATION AFTER IMPLEMENTATION

After an information system has been installed and users have been trained, management should evaluate the system regularly, e.g., once a year. Estimating an information system's value to an occupational safety and health program can be difficult for the same reasons that conducting a cost/benefit analysis is difficult. Problems can result from human, hardware, or software elements of the system. However, evaluation is an important follow-up in a setting of continuous improvement and total quality. The main factors to consider when evaluating a system include:

- completeness (functionality)
- reliability of system (hardware and software)
- user acceptance (ease of use and productivity)
- costs (overall operating costs)
- improved availability of information (reporting)
- new capabilities (future requirements)
- flexibility (adaptability).

If management detects problems in any but the last two items, they should take steps to correct the problem(s).

TECHNOLOGY UPDATE

Many companies are making PCs available to managers for word processing, spreadsheets, and electronic mail. They are part of an overall strategy of office automation. Having the right information made available to the right person at the right time improves decision making (Figure 9-6). The following section addresses some of the key technologies.

Relational Database Management Systems

Relational database management systems are an integral component of most current occupational safety and health systems. Database management systems provide tools for building files and reports, and they offer programming languages or commands for building complex programs (Figure 9-7). Most database management systems that are addressable by a common language called the structured query language (SQL) are most desirable. Newer software technologies permit applications to retrieve data from relational database systems provided by different vendors. Similarly, communications over local and wide area networks permit access to data that reside in computers in other locations at far distant sites.

Object Oriented Databases

A new generation of databases seems to come along every 5 to 10 years. The object oriented database is the currently available "next generation," but is not yet formal in any mainstream product. This new generation of database will allow elemental modules of OHIS programs to be shared universally. Object-oriented databases are critical to the development of information age OHIS and will finally take companies out of the cycle of generating new systems from scratch every 5 to 10 years.

Artificial Intelligence/Expert Systems and Fuzzy Logic

Computers follow the general method of solving problems of "If this is true, then do that . . . ." Artificial intelligence (AI), expert systems, and fuzzy logic techniques allow a computer to follow a general purpose approach in solving complicated problems and meets the need to learn by experience. Artificial intelligence has been confined to group data analysis under the supervision of a "human" professional and is not used in individual data analysis due to potential liability concerns.

There are two main types of systems—rule-based and self-learning. The former requires an expert to define the rule by which decisions are made and imbed them into the software program. The self-learning type periodically reviews data and automatically makes inferences that are used for decision support. Though artificial intelligence has been available for many years, it has been slow to be integrated in occupational health information systems.

Fourth-Generation and Higher Level Languages

It costs an estimated $100,000 per year to maintain "liveware" (salary, benefits, office space, computer software/hardware for a programmer). In order to generate more complex OHIS systems, the productivity of the programming staff must increase. With each generation of language, there is a 5- to 10-fold increase in productivity. Fourth-generation level programming languages write much of the code based upon simpler instructions. Many of today's PC-based database languages are fourth generation. A fifth generation is being developed, but will not be available for several years.

Operating Systems

The operating system of the computer enables the software to perform the tasks required to update, store, and record data. As users demand more user-friendly features, the migration from character-based systems to graphical user interfaces has accelerated. This means that most DOS and more mainframe applications will migrate to Windows and OS/2 type systems.

DOS

Microsoft's Disk Operating System (DOS) will continue to exist, but will only be used in bare-boned systems.

Windows, Windows 95, and Windows NT

Windows is evolving in several ways. It has added networking capabilities (Windows for Workgroups and NT) and has become 32 bit to enable larger programs to run faster. With Microsoft's new Windows 95 there is no longer a need to "run" Windows in a DOS environment. Windows 95 will be the operating system used for most PCs by the end of 1996. Windows NT will also evolve, first being used on network servers (NT is a string alternative to Novell 4.1), and later as the most popular desktop computer to run complicated multimedia desktop applications.

OS/2

IBM's more advanced and more powerful operating system for the PC was originally a joint project between IBM

**The Right Information at the Right Time:
The Right Decision**

Supervisor:
- Notifies others of how, when, and where injury occurred.
- Is notified of date employee can return to work and of any restrictions.

Employee Training:
- Schedules training programs directed at work-related injuries.
- Monitors attendance.
- Monitors effectiveness of preventive measures.

Risk Manager:
- Monitors incident reports, including work-related injuries.
- Manages need for preventive measures.

Physician:
- Evaluates, treats, and completes first report of injury.
- Notifies employer of date employee may return to work and of any work restriction.

Workers Compensation Manager:
- Is notified of work-related injury and circumstances.
- Manages claim.

Safety Manager:
- Is notified of work-related injury and circumstances.
- Performs investigation. Recommends preventive measures.

Figure 9-6. An office automation system that makes accurate information quickly available to the people who need it improves decision-making.

and Microsoft, but is now a sole IBM effort. OS/2 may become a niche product for users that make heavy use of mainframes and other IBM-based systems.

UNIX
Originally this was a minicomputer operating system predominantly used in research in university settings. It is powerful (can handle multiple processors), and widely available, but it is complicated. It is used in most organizations for sharing data among departments. However, the number of companies interested in a client-server system is growing. UNIX still plays a role as the operating system of choice for the "host" computer in many relational database applications within organizations. Novell's recent purchase of UNIX indicates there will be significant changes in how UNIX is marketed and how it is used in a client-server environment.

Networks
In the past, more speed and more capacity generally meant more money. Because of the rapid advances in PC technologies and declining prices, many companies have developed occupational safety and health systems relying solely on PCs. Some use local area networks (LANs) to link the PCs to one another either directly or through a shared mainframe computer. Others use a wide area network (WAN) to communicate at greater distances. Companies' reliance upon the mainframe has changed. In recent years mainframes have become repositories of data or data hosts, though they are used for very large applications in which massive amounts of data are processed. There are two types of networks that involve PCs: server-based and peer-to-peer.

Server-Based
This network is based around a centralized computer, which may be a PC or much larger computer. Data and software are obtained by the network "nodes" (desktop PCs) from this server. These networks tend to be complicated and usually require a dedicated technical staff to maintain. Examples of these server-based networks include Novell, LAN Manager, and Windows NT Server. This type of network provides higher levels of security and performance and supports more connections (up to 1,000 PCs connected together).

In the past decade, Novell has dominated this marketplace. Microsoft is challenging Novell's dominance with Windows NT Server. Novell will build upon its purchase of UNIX to compete aggressively with Microsoft's Windows NT Server. This will cause the marketplace to be volatile probably through 1996. Competition will also result in better security and "fault tolerance" (the ability of a network or computer to continue to function even though partially broken) for the large-scale user. With the addition of client servers, these network environments can handle the vast majority of mainframe applications.

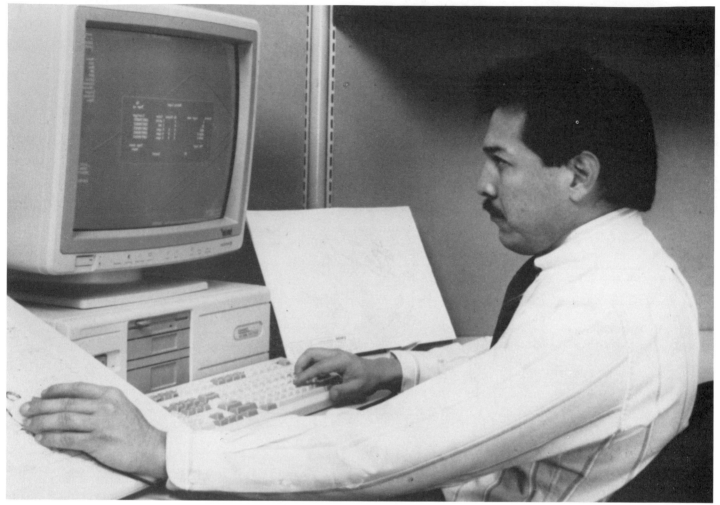

Figure 9-7. Cost- and time-effective safety management will increasingly require the use of computers. Shown here is an application that establishes current corrosion/erosion rates of piping, equipment, and tanks. (Courtesy United States Testing Company, Inc.).

Peer-to-Peer

A simple type of area network used by small companies or departments, peer-to-peer networks run on top of DOS and are typically used by 2 to 20 computers, although they can support up to 200. Both security and the speed of information flow between computers are limited compared to server-based networks.

Until recently LANtastic had this part of the marketplace to itself. Novell with "Novell Lite" and Microsoft with "Windows for Workgroups" have both attempted to penetrate this market. Their initial products have not been well received. For the next few years, LANtastic will still be the best choice. After that time only two vendors may still exist with products in the peer-to-peer area. It is too early to tell who those vendors will be.

Client Server

Computer networks often connect to large databases, which may be loaded into a host computer. This system is referred to as a client server. The client server approach is probably where most new OHIS are being developed. Information/data are requested from a server, then made available to the local PC client in the user's choice of application or database software (e.g., Paradox, Dbase, Foxpro). The server may be another PC, a PC with multiple processors, a minicomputer, or even a mainframe. In most cases, the local PC user will request the data needed in a standard relational database dialect, SQL. Client-server system architecture provides companies with greater performance and flexibility.

Improved Connectivity and Telecommunications

The cost of remote data access will probably continue to decrease as more fiber-optic communication networks come on-line. The Internet, a world wide network of networks, and the World Wide Web (discussed below) are accessible through commercial on-line services and local access providers.

Because of lower prices and faster transmission speeds, more users are accessing telecommunications networks and rapidly expanding their range. Security has lagged behind this development, but is an area that software vendors are addressing with programs such as Windows NT and planned upgrade versions of OS/2.

Many of the above technologies (e.g., client server, object-oriented database) will remove the difficulty and cost of these data connections. Future state-of-the-art systems that use fax boards, modems, electronic mail, cellular telephones, and other technology will be heavily interconnected and may not require wire cabling to make these connections. Similarly, as new models of medical and laboratory testing equipment (e.g., audiometers, tonometers, spirometers, gas chromatographs) are introduced, vendors and companies will have added the necessary hardware and software to allow connection to an OHIS.

Internet and the World Wide Web

The Internet is a network of networks. As mentioned above, a network is a group of computers that have been connected together so that they can communicate with each other. The Internet has been growing rapidly over the past few years until now an untold number of computer networks are linked together. Some of these networks are operated by government agencies, others by educational institutions, some by business entities, and even by individuals. What is fascinating about the Internet is that users can gain access to thousands of different systems regardless of their physical location. These systems contain, among other things, government archives, university databases, library catalogs, and local community resources covering millions of subjects. Not only can a company or individual user access such a vast array of information, they can access a myriad of data formats, such as text, pictures, video, and sound. The connectivity of the Internet is like a data highway that enables the user to communicate across large distances at nominal cost. This fact is one reason why the Internet is used so extensively for e-mail documents.

On a global level there is the World Wide Web (WWW)—a type of planet-wide Internet system. It is based on hypertext tools that allow users to connect around the world looking for information. The term "hypertext" refers to a text format linkage of highlighted words that provides a trail between documents. This technology allows users to move from one topic to another, searching for the desired subject matter. Several software tools have been developed to enable users to navigate the maze of data made available by users of the Web. These tools are called browsers and gophers. They can be purchased from Internet access suppliers or from software vendors.

Changes in the computer industry are occurring at such a rapid pace that it is difficult to predict the effect this technology will have on our business and personal lives. Efforts to make data exchange on the Internet safe to insure confidentiality may have a significant impact on how OHIS is used in the future. In addition, the economies that the WWW brings to industry are only beginning to be explored. The authors believe that much of the basic information occupational safety and health professionals rely upon will be distributed using technologies such as these.

Capacity and Storage

The capacity of a computer is measured by its memory and storage components. Memory is the work space of the computer, much like a manager's desk is his or her workspace area. Older computers may have as little as 640 kilobytes of work space memory (also known as RAM, random access memory) and newer PCs perhaps as much as 64 megabytes. Although the growth of graphical user interfaces has increased the need for computer power, technology appears to be keeping pace with the demand for speed. Similarly, document scanning and multimedia applications are also expanding the computer's uses in industry.

As the functions and features of an OHIS become more complicated and as users demand systems that are easier to use and compatible with different types of computer hardware, the software tends to grow bigger and operate more slowly. This holds true for operating systems as well as application programs. Today's software will continue to take up more computer memory and resources, but the evolution of processing power will more than make up for this drawback—and at prices lower than today's.

Hard disk drives, which are like large filing cabinets, are used for storing data in electronic files. Capacities now range from a meager 120 megabytes (about 6,000 pages) to 500 megabytes, to 1 gigabyte and 2 gigabyte sizes, to multiple-gigabyte drive clusters. The greater the storage, the more programs and data the computer can hold for rapid access.

Storage devices such as low-cost, high-capacity hard disk drives, CD-ROM, write once, read many (WORM), and optical disks have had a major impact on the availability of desktop information. Potential components of an OHIS, such as optically stored MSDSs and x-ray images, require vast amounts of data storage. Storage devices currently available are durable and can handle large amounts of data. For example, PCs now will come with 500 MB hard drives as standard. CD-ROMs are capable of storing 650 MB (like audio CDs), and are mostly read-only. Writing one time on a CD is now possible at nominal cost. Drawbacks are that data cannot be added later, and the CD-ROM drive is relatively slow. WORM drives are more expensive than CD-ROM, but faster. Material can be added on multiple occasions. The technology has been available for several years, but has not gained wide acceptance. WORM drives are a good choice for archiving data.

Optical disks holding from 21 MB to over a gigabyte of memory, have the same advantages and disadvantages as WORM, but users can write more than once on these disks. Optical disks are the best future technology if the price comes down. CD-ROM, because of wide appeal, probably has greatest current potential.

FUTURE TRENDS

The OHIS of the future will be a PC-based client-server system using a PC-based multiprocessor server. Each of the nodes (connected desktop systems) will use a graphics user interface (GUI) with a fourth-generation language and an object-oriented database as the underlying software. Much of the system software will be shared with other OHIS users, and system changes will be both inexpensive and quick to perform. Connections to other local company applications (other departments) and to keynote information sources will be transparent, that is, there will be no separate systems or barriers to users, eliminating data storage in more than one place. The system will store images of pictures and forms using CD-ROM or optical technology.

All measurement equipment will be directly connected to the system. Employees and clients with disabilities will gain improved access to the system. The performance of the hardware will far exceed anything available today and cost about the same as current systems. The software will be isolated from the hardware, allowing physical pieces of the system to be replaced, as better technologies become available. All of the above will result in a system that is more productive and flexible.

Users will be able to communicate with data repositories from the field using secure remote database access. As decision support technologies become more a part of daily work, the need to access data regardless of location will grow. This will provide a virtual work environment that allows data collection, validation, and record keeping to be performed at remote locations. The combination of faster computers, inexpensive storage, improved user interfaces, simpler software, graphical data output, and remote communications is revolutionizing industry. These developments will significantly improve the effectiveness of occupational safety and health professionals in fulfilling their day-to-day management responsibilities to the employee and company.

SUMMARY

- Computers are and will be used by most occupational safety and health professionals for a variety of activities. These tools help management make better decisions, manage incident data, assist in incident investigation, monitor workplace conditions and safety efforts, disseminate information on hazards and employee safety education, maintain files and employee health and medical records, and provide computer-assisted training programs.
- Computers also provide information on regulatory requirements to ensure compliance with various federal, state, and local ordinances. Operators can program computers to keep track of ongoing events or problems requiring follow-up and schedule physical examinations and other events required in an occupational safety and health system.
- Five steps are needed to develop and implement a successful computer system: (1) understand the organization's operations and needs, (2) identify and review appropriate software and hardware, (3) purchase or develop a system, (4) implement the system, and (5) evaluate the system. Management should tailor computer capabilities to its specific needs as closely as possible.
- Once the system is installed, user training should be conducted on the job. Operators should practice what they learn and receive follow-up training to ensure they are using the system to its best advantage.
- Every system should have a manager or administrator who is responsible for management of the security system, supervision of the data content, problem solving, coordination of changes to the system, archiving of data, and planning for future needs and applications. Periodically, the system should be evaluated for completeness, reliability, user acceptance, costs, improved availability of data, and new capabilities and flexibility.
- Factors influencing a company's choice of computer hardware include corporate standards, telecommunications systems, software, and information needs of users.
- The key features to look for in computer hardware are speed and capacity. Speed is determined by fast processing chips while capacity refers to computer memory and storage components.

REFERENCES

Best's Loss Control Engineering Manual. Oldwick, NJ: AM Best Co., published annually.

December J and Randall N. *The World Wide Web Unleashed,* 1st ed. Indianapolis, IN; SAMS, 1994.

Helander MG, ed. *Handbook of Human/Computer Interaction.* Amsterdam and New York: Elsevier, 1987.

LaDou J. ed. *Occupational Safety and Health,* 2nd ed. Itasca, IL: National Safety Council, 1994.

National Fire Protection Association, 1 Batterymarch Park, Quincy, MA 02269-9101. *Fire Protection Handbook,* 16th ed., 1986. *Protection of Electronic Computer/Data Processing Equipment,* NFPA 75. *Protection of Records,* NFPA 232.

Occupational Health and Safety Software Database. Available on CD. Hamilton, Ontario, Canada: Canadian Centre for Occupational Health and Safety, quarterly updates.

Peterson, Kent W. *Directory of Occupational Health & Safety Software,* 8th ed. Arlington Heights, IL: American College of Occupational and Environmental Medicine.

Ross DT and Schoman KE. "Structural Analysis for Requirements Definition." In Freeman P and Wasserman AI, eds. *Software Design Techniques,* 4th ed. Long Beach, CA: IEEE

Yenney, Sharon L. *Putting the Pieces Together: A Guide to the Implementation of Integrated Health Data Management Systems.* Washington, DC; Business Group on Health, 1992.

REVIEW QUESTIONS

1. List five of the eight advantages of computerized information systems.
   a.
   b.
   c.
   d.
   e.

2. Briefly describe the five key steps to implementing an automated occupational health information system (OHIS) successfully.
   a.
   b.
   c.
   d.
   e.

3. General office automation and communication, access to personnel data, documenting workplace conditions, maintaining regulatory requirements, performing drug and alcohol testing, managing incident data, and performing exposure monitoring are just a sampling of the many important components of:
   a. Data management software
   b. A supervisor's daily checklist
   c. The human resources department
   d. A safety professional's duties

4. When selecting a computer system, what are the three general questions management should ask?
   a.
   b.
   c.

5. With so many functions that can benefit from automation, the task of matching requirements with the features a vendor can supply may appear overwhelming. What should management do to streamline this process?

6. Name two advantages of choosing off-the-shelf software.
   a.
   b.

7. Customized software is an alternative to off-the-shelf software, although more costly in money and time. What is its advantage?

8. An organization decides to develop its own system. Who, at minimum, should participate in the design and development process?
   a. All anticipated end-users of the system
   b. Managers and a systems analyst
   c. A health care specialist
   d. All of the above

9. The successful implementation of a new system depends greatly on effective training. List four training considerations that affect implementation.
   a.
   b.
   c.
   d.

10. Once an information system is installed and in use, regular evaluations are essential. Identify the main factors to consider when evaluating a system.
    a.
    b.
    c.
    d.
    e.
    f.
    g.

Part 3

Safety/Health/ Environment Program Organization

INTRODUCTION

Part three presents detailed overviews of the fundamental elements of a successful safety/health/environment (S/H/E) program. While each workplace is unique, the basic elements of an effective S/H/E program are constant. Core programs in occupational health, industrial hygiene, and environmental management are described. Beyond these core programs are developments in ergonomics, employee assistance, emergency preparedness, and product safety. Many large and medium-sized businesses now find that it is cost-effective to have programs in both core and extended areas. In addition, transportation, office safety, laboratory, and contractor and nonemployee safety issues have become quite prominent as the U.S. economy has evolved from manufacturing to service industries. The descriptions given in these chapters provide an overview and guide for the initiation and development of a comprehensive S/H/E program.

10

Occupational Health Programs

The health and medical services that promote and support employee health in an organization are integral components of every occupational safety and health program. This chapter discusses the important functions and structure of occupational health programs; however, each organization must decide how to provide occupational health services to best meet the needs of its employees. The following topics are covered in this chapter:

- the type of employee health services and first aid provisions companies should provide
- the various occupational health professionals companies employ and their roles in employee health care
- the preplacement, on-the-job, and emergency medical services provided by companies
- the value of developing special programs for women and older workers and of establishing substance abuse programs for workers

Occupational health programs can range from comprehensive, with a large array of services, to programs with services that meet only mandatory legal requirements. One establishment may have a full-time staff of occupational health nurses, physicians, industrial hygienists, safety specialists, and technicians housed in a model occupational health unit while another may have only the basic first-aid kit with a trained person to provide first aid.

Ideally, occupational health programs, regardless of size, are composed of elements and services designed to promote and maintain the health of the workforce, prevent or control occupational and nonoccupational diseases and accidents, and reduce and prevent disability and resulting lost time. A good program should provide the following components:

- maintenance of a healthful work environment through establishing a comprehensive occupational health and safety program
- health examinations, including selective baseline and periodic surveillance for employees performing particular jobs as required by regulatory standards, fit-for-duty, return to work, and job transfer evaluations
- diagnosis and treatment services for occupational injuries and illnesses
- case management services
- immunization programs
- confidential health records kept separate from personnel records
- health promotion, education, and counseling
- open communication between the company's occupational health personnel and an employee's personal physician

Treatment of ill or injured persons should be provided within the legal standards of nursing and medical practice, using appropriate medical guidelines and standards for employee health care.

OCCUPATIONAL HEALTH PROGRAMS

Occupational health programs are concerned with all aspects of the employee's health and the employee's relationship with the workplace environment. The basic objectives of a good occupational health program are to:

- promote health and protect employees against health hazards in their work environment
- facilitate placement and ensure that individuals are assigned work that matches physical and mental capabilities and that they can perform with an acceptable degree of efficiency without endangering their own health and safety or that of their fellow employees
- promote adequate health care and rehabilitation of employees injured on the job
- monitor the work environment for hazards and abate them
- encourage workers to maintain their personal health

By applying occupational health principles to the workplace, management can place all employees, including those who are disabled, in jobs according to their abilities to perform the work (see Chapter 13, Workers with Disabilities, in the *Engineering & Technology* volume). This approach also promotes continuing health care and rehabilitation of workers who become ill or injured on the job. The achievement of these objectives benefits both employees and employers by improving health, morale, and productivity.

Industry experience has shown a strong relationship between accident prevention and occupational health. For example, some industrial chemicals, when improperly handled, represent serious hazards to health, property, and the environment. Depending on various environmental and workplace conditions, the vapor from a chemical can ignite, explode, or, if inhaled, cause symptoms ranging from dizziness to death. Dermatitis can be caused by the contact of a chemical with the worker's skin. (For details on the effects of specific chemicals, see the National Safety Council's *Fundamentals of Industrial Hygiene,* 4th edition.)

Safety and health professionals have demonstrated their ability to reduce the rates of accidental injuries by controlling many phases of the industrial environment through worker education and through improved management techniques (Boylston, 1990). These two elements must allow for the different physical and emotional characteristics of individual workers. Such characteristics account for variations in workers' job attitudes, productivity, safety and health practices, and personal health management.

Management often needs the services and skills of several professionals to attain the best results in its occupational health and safety programs (Felton, 1990; Rogers, 1994). These professionals include the following:

- Increasingly, physician specialists are trained in occupational and preventive medicine. They are assisted by specialists in orthopedic surgery, ophthalmology, radiology, surgery, dermatology, psychiatry, and other areas.
- Occupational health nurses use specialized knowledge, not only in basic nursing procedures and health maintenance, but also when dealing with legal, economic, and social issues; labor laws; and occupational health sciences (e.g., toxicology, safety, industrial hygiene). These nurses apply their knowledge to provide health care for workers. In many instances, nurses are the only full-time health care providers in a company.
- Industrial hygienists apply specialized knowledge in recognizing, evaluating, and controlling health hazards in the work environment.
- Safety specialists use specialized knowledge to design and implement strategies aimed at preventing and controlling workplace exposures that result in unnecessary injuries and deaths.
- Other occupational health professionals may include toxicologists, epidemiologists, counselors, and so on.

Working both individually and collectively, these specialists have helped to improve the occupational health and safety record of many industries. In some companies, these professionals are so well organized and effective in anticipating and correcting hazards that employees may actually be safer and healthier at work than in their own homes. In fact, one of the current challenges for occupational safety specialists is to educate workers in off-the-job safety at home, in recreational activities, and during travel so they will be able to return safely to work each day. If employees are injured off the job, they are as much a loss to the operation as if they were injured on the job.

However, safety professionals should be involved in these safety programs without being intrusive. Health care personnel can be influential in extending a safety program beyond the facility to include outside activities of employees and their families.

Some insurance carriers offer a consulting service to help organizations set up an occupational health program suitable for their needs. The consultants usually know which health care providers are available for this kind of service. The basic work, however, has to be done by the organization establishing such a program. Any justification for these services lies in their record of accomplishments and emphasis on an improved quality of life. Prevention is not only better than cure—it is easier and less expensive. In this regard, off-the-job safety programs and activities also can benefit a company of any size.

Employee Health Services

Occupational health services can be provided in any clean, private location in the company that is accessible to employees. However, the experience of many organizations indicates that the occupational health unit commands greater respect if management pays careful attention to providing a suitable and efficient location that has a pleasant, comfortable appearance and modern equipment. The location must have adequate space for multiple casualties during emergencies and suitable provisions for disabled employees (Guidotti, 1989).

The occupational health unit should be adequate for the size and needs of the workforce and should contain a minimum of four rooms or areas: a waiting room, a treatment room, a room for physical examinations or consultation, and a rest area. Rooms for special purposes can be added according to the needs and size of the workforce and company. The surgical treatment room should be large enough to treat more than one person at a time. Small dressing booths should be provided, which can give workers some degree of privacy.

First-Aid Provisions

Good administration of first aid is an important part of every safety program (Workers' Compensation Board, 1991). First-aid kits and supplies approved by the occupational health manager should be stored where they are readily accessible to trained personnel on each shift. Health care personnel should keep a careful record of each first-aid administration and send an incident investigation report to the injured worker's supervisor when first aid is administered.

In most companies today, the occupational health nurse manages the occupational health unit. This individual should routinely monitor first-aid provisions and train ancillary personnel in first-aid procedures and treatments. Although most companies do not employ a full-time medical doctor, they should maintain a good liaison with a local physician or physicians designated to handle certain injuries. The physician(s) should be invited to the company occasionally to tour the establishment and provide consultation when needed. The company-designated physician(s) should be familiar with the type of work done so he or she can evaluate injuries, illnesses, and other employee complaints.

In many small organizations and in field operations, it is neither practical nor efficient to maintain qualified professional health care personnel full time. In such cases, the best arrangement is to use trained first-aid attendants who follow procedures and treatments outlined by qualified health care professionals and who are available on an on-call or referral basis to treat serious injuries. If injured employees have their choice of treating physician, the employer should comply with all requests, if possible.

There are two kinds of first-aid treatments: emergency and prompt attention.

Emergency Treatment

Emergency care must be given for immediate, life-threatening conditions. First-aid staff provides care until proper medical treatment can be given (Hau, 1994). Proper first-aid measures reduce suffering and can lessen the severity of an injury or illness.

Prompt Attention

This type of first aid is used to treat minor injuries such as cuts, scratches, bruises, and burns. Ordinarily, the injured person would not seek medical attention for these injuries. By requiring that all employees immediately report for treatment when they are injured, regardless of the extent of the injury, the company can help reduce workers' risk of infection, disability, and missed diagnoses.

A first-aid program should include:

- properly trained and designated first-aid personnel on every shift
- instructions for calling an ambulance or rescue squad
- posted method for transporting ill or injured employees
- posted instructions for calling a physician and notifying the hospital that a patient is *en route*
- first-aid unit and supplies or first-aid kit approved by the health care professional
- first-aid manual with procedures
- list of reactions to chemicals and routes of exposure
- adequate first-aid record system and follow-up

The first-aid procedures manual should be developed by the occupational health nurse and physician or designate with input from the occupational health staff. All occupational health staff should review the manual. Health care staff should render emergency and first-aid treatment in accordance with appropriate guidelines or protocols and legal parameters of practice. In some situations, physician direction and practice may be required. The consulting physician should specify the type of medication, if any, to be used on injuries such as cuts and burns. The physician should specify the procedures to be followed when medication for temporary relief of nonoccupational ailments, such as for a toothache or headache, is administered. Procedures should also state that in areas where chemicals are stored, handled, or used, emergency flood showers and eyewash fountains (Figure 10-1) should be available and clearly identified. Material Safety Data Sheets (MSDS) should also be available.

Professional occupational health nurses practice under the state Nurse Practice Act. They must be cognizant of their legal scope of practice as defined by their specific state and the delegatory functions they can handle legally. Anyone responsible for first-aid treatment must understand clearly the limits of this work. Because improper treatment might adversely affect the employee and/or involve the company in serious legal problems, the first-aid attendant should be duly qualified and certified by the Mine Safety and Health Administration (MSHA), the National Safety Council, or the American Red Cross. These certificates must be renewed at specific intervals.

First-Aid Training

The National Safety Council Institute's first-aid textbooks and the MSHA manual of first-aid instruction are recommended for training employees. Many feel that accidents occur less frequently and are usually less severe among workers trained in first-aid procedures. It is, therefore, advisable that as many industrial workers as possible be given this training (Figure 10-2). The National Safety Council publishes posters and booklets that can be used for training employees. Other valuable sources are included in the reference list at the end of this chapter.

First-Aid Room

It is always advisable to set aside a conveniently located room for the sole purpose of administering first-aid treatment. The person administering first-aid should have a proper place to work. A first-aid room should be equipped with the following items:

- examining table
- cot for emergency cases, enclosed by movable curtain
- dustproof cabinet for supplies
- waste receptacle and biohazards-disposable containers
- small table
- chair with arms and one without arms
- magnifying light on a stand
- dispensers for soap, towels, cleansing tissues, and paper cups
- wheelchair
- stretcher
- blankets
- bulletin board to post all important telephone numbers for emergencies
- bed(s)
- appropriate medications in locked cabinet

Health care personnel often use oxygen in the treatment of many first-aid cases. Because of the danger of fire or explosion, smoking should be prohibited when oxygen is administered. Any type of resuscitating device should be used only by trained persons.

First-Aid Kits

Many types of first-aid kits are available to fill every need, depending on the type of accident that might occur (Figure 10-3). Commercial or cabinet first-aid kits, as well as unit kits, must meet regulatory requirements (Workers' Compensation Board, 1991). MSHA outlines specific first-aid items required at certain locations in mines and processing facilities.

Kits vary in size from the pocket model to almost a portable first-aid room. The size and contents depend on the intended use and the types of injuries to be treated. For

Figure 10-1. The emergency shower and eye wash should be well identified. *Left:* Harmful chemicals are quickly washed away and clothing fires doused immediately by the drenching action of the shower. *Right:* First-aid treatment to the eye must be prompt and consists of prolonged irrigation of the exposed eye with low-pressure water. (Photos reprinted with permission from Haws Drinking Faucet Co.)

example, personal kits contain essential articles only for the immediate treatment of injuries while departmental kits meet the needs of a group of workers. As a result, the quantity of material each kit contains depends on the size of the workforce. Trunk kits, which are the most complete, can be carried easily to an accident site or can be stored near working areas far from well-equipped emergency first-aid rooms. Although a trunk kit can include bulky items, such as a wash basin, blankets, splints, and stretchers, it can still be carried by only two persons.

Keeping first-aid kits strategically located throughout the company seems to work best when such kits are supervised properly (Figure 10-4). A single, trained individual should be responsible for maintaining each kit, treating serious injuries, and giving first-aid measures for minor cuts and scratches. This employee never provides more than immediate, temporary care. By combining a well-trained first-aid provider and well-maintained first-aid kits, companies can ensure that many minor injuries that might otherwise have been ignored, receive prompt, proper treatment. All miners, for example, receive first-aid training and annual refresher training.

In some organizations whose activities are widely scattered, employees may need to use first-aid kits and administer treatment themselves. Under these circumstances, the organization can control first-aid service by having the attendant in charge properly instructed in first aid and by seeing that the service as a whole has health care supervision.

In general, qualified health care personnel should supervise the maintenance and use of all first-aid kits and approve all materials provided. A member of the health services group should regularly inspect all first-aid materials and report on their content level and serviceability. Maintaining quantities of materials in the first-aid kit is easier if each kit contains a list showing the original contents and the minimum quantities needed. Health care workers should clearly label and date all bottles or other containers in the kit.

Recommended materials for first-aid kits are listed in the textbooks of the National Safety Council's First-aid Institute, the U.S. MSHA, and the American Red Cross. Suggestions are also available from the American Medical Association, the American Petroleum Institute, and from manufacturers of first-aid materials.

Stretchers

Company health services must be able to quickly transport a seriously injured person from the scene of an accident to

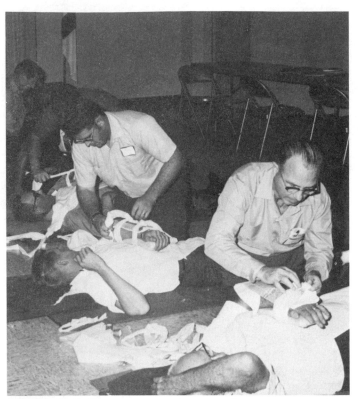

Figure 10-2. Demonstration and practice play an important role in first-aid training.

a first-aid room or a hospital (Workers' Compensation Board, 1991). Such promptness can help physicians determine the gravity of the injury and perhaps even increase the victim's chances for survival. The stretcher provides the most acceptable method of hand transportation for accident victims. It can be used as a temporary cot at the scene of an accident, during transit, and in the first-aid room or occupational health unit.

Although there are several types of stretchers, the commonly used army stretcher is satisfactory in most cases. However, when workers must hoist or lower the injured person out of an awkward place, it is better to use a lightweight stretcher to which the employee may be strapped and kept immobile.

Stretchers should be conveniently stored near places where employees are exposed to serious hazards. It is customary to keep stretchers, blankets, and splints in clearly marked, prominent cabinets. Stretchers should be clean and ready to use at all times. They should be protected against destructive vapors, gases, dust, or other substances, and against mechanical damage. If the stretcher is made of materials that will deteriorate, health care staff should test it periodically for durability and strength.

Any victim of a severe injury, such as a fall, head injury, spinal injury, or any injury resulting in unconsciousness, should be secured to and transported on a rigid spine board with proper restraint of head and neck as determined by a qualified person. Regardless of the transportation device, the employee should be secured to the device with straps or cravats.

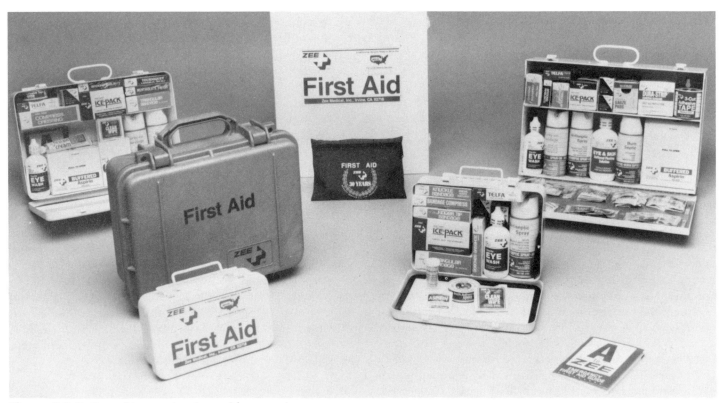

Figure 10-3. First-aid kits can be obtained for special purposes.

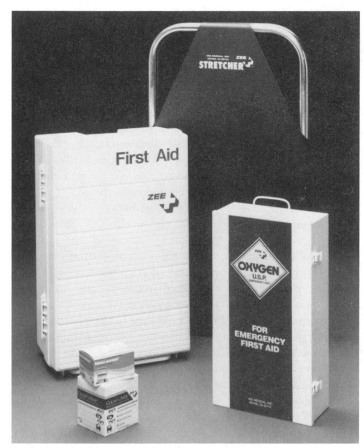

Figure 10-4. First-aid kits should be well-distributed throughout a facility. *Left:* They should be supervised by a trained person so employees do not try to "doctor" themselves. *Right:* Stretchers and oxygen kits should also be available in areas distant from the first-aid room. (Courtesy Zee Medical.)

OCCUPATIONAL HEALTH SERVICES

An effective occupational health service program should be planned by the occupational health staff in cooperation with management. The program must have the full support of top management if it is to be successful. Only with management support is it then possible to establish and maintain an adequate program, professional staff, and facilities for examinations, emergency cases, and record storage.

Over the past 20 years, occupational health services have evolved and expanded considerably. In addition, the roles of occupational health professionals who deliver health care have also changed. Today, companies use physicians more as consultants for treating serious illnesses and injuries and to provide limited direct care. Occupational health nursing practice, on the other hand, has greatly expanded. The occupational health nurse is typically the manger of the occupational health unit as well as the sole direct care provider of health services at the work setting.

Occupational Health Nurse

Occupational health nursing is a specialty practice within the nursing profession. The position requires a registered professional nurse who is licensed to practice in the state where he or she is employed. In addition, the occupational health nurse should know the basic principles of occupational health, health promotion and protection, occupational disease, related work processes and hazards, workers' compensation laws, insurance, health and safety regulations, sanitation, first aid, and record keeping (AAOHN, 1994). The occupational health nurse works with the other human and safety professionals to develop a comprehensive health program tailored to the needs of the company.

This health professional has a wide scope of practice directed toward improving, promoting, and protecting worker health. In the work setting, the occupational health nurse has major responsibility for the management and administration of the occupational health unit and for dispensing services. The occupational health nurse may be responsible for direct care services, preplacement, periodic, and other types of examinations and workplace monitoring surveillance programs. These activities are directed toward the identification of potential hazards harmful to workers. Using an interdisciplinary approach, the nurse works closely with the health care professionals to develop strategies to reduce employee risk.

A major component of the occupational health nurses' practice is the development, implementation, and evaluation of activities that support health promotion and protection strategies through primary, secondary, and tertiary prevention approaches. This process might include conducting employee health needs assessments; administering and interpreting health risk appraisals; designing and developing programs to provide positive lifestyle changes such as exercise, nutrition, and weight control; and initiating programs that support regulatory mandates such as hearing conservation and respiratory protection. The occupational health nurse works in collaboration with the physician and other health care professionals to provide for rehabilitation and return-to-work programs. This area encompasses the need to pay attention to physical and psychological needs of the worker and encourages active participation of management in the process.

Today more primary care services are being provided at the worksite. The occupational health nurse may give workers nonoccupational health-related care for minor health problems and offer chronic disease monitoring and follow-up using established guidelines and protocols. In addition, management services may be available to provide coordinated care for early intervention and reduction in worker disability and further injury.

The occupational health nurse is available to provide counseling to help employees clarify problems and to make informed decisions about their health needs. Referrals for employees in crisis situations are often handled by the occupational health nurse through employee assistance programs. Collaboration with community agencies enables the occupational health nurse to develop a network of resources to more efficiently and effectively develop health care services for employees and the company.

The occupational health nurse may play a major role in implementing many OSHA-mandated programs, such as hazard communication. For example, the nurse should have ready access to MSDS for chemicals at the facility and may participate in the overall management of the program and employee training.

The occupational health nurse must maintain a confidential, professional relationship with the employees, conforming to legal and ethical codes. The nurse must not divulge information contained in individual employee health records without the employee's written consent. The employee health files should be accessible only to health personnel.

Occupational Health Physicians

Physician services in industry depend on such considerations as the hazards in operations, the number of employees, and the type and extent of the occupational health program. The company can make arrangements for full-time, part-time, or consultant services.

Some large organizations have a full-time corporate medical director with part-time physicians serving their decentralized operations (Felton, 1990). Some physicians devote a scheduled number of hours, either daily or

weekly, to the medical service needs of a company and are available at other times for emergencies. Others arrange their service on an as-needed basis, for example, when job applicants require examinations, injured employees need medical care, or when other medical problems arise.

Management usually makes the on-call physician arrangement with a local physician. This service is most often used by companies (or establishments) with fewer than 500 employees, companies with a low incidence of accidental injuries, or companies with a minimum health service program for employees. The physician usually provides services on referral from the occupational health nurse (if there is one on site) when medical care is needed and for illnesses or injuries not requiring hospitalization. Otherwise, the company will make direct referrals to the physician's office. Frequently, the on-call physician has similar arrangements with a number of companies in the area—a particularly convenient system for small-plant or small-company clusters, such as in industrial parks. Also, it is not unusual to find physicians specializing in the care of industrial injuries and diseases and providing occupational medical services in some manufacturing centers or in associated free-standing or hospital-based occupational medicine clinics.

A physician who has a consulting service is not usually called in except to diagnose and treat serious injuries, illnesses, toxicological problems, and special kinds of injuries or disorders (such as eye injuries) that require the services of a specialist. State and local health departments often supply, without charge, medical, nursing, and engineering consultation. They also conduct industrial hygiene and radiological surveys for industries within their areas. Many insurance and private consulting companies offer this survey service to their clients.

Occupational physicians or those providing occupational health services should be familiar with all jobs, materials, and processes that are used in organizations (Levy & Wegman, 1995). An occasional inspection visit will help them keep abreast of current technologies, equipment, and processes. During these visits, the physician can develop an inventory of hazards to help in the diagnosis, treatment, and prevention of occupational health diseases. Physicians can also be involved in other company services that relate to the health of workers, such as food service, welfare service, safety programming, sanitation, and mental health.

Maintaining a true physician-patient relationship (with fairness to both employee and employer) is essential to the success of any occupational health program (Rosenstock, 1995). For example, employee patients deserve the same courtesy and professional honesty as do private patients. The first meeting of physician and employee usually occurs at the introductory physical examination. The examining physician—within the framework of professional discretion—should tell the worker the results of the examinations and, if necessary, refer the worker to a personal physician for further treatment.

The occupational physician should provide emergency medical care for employees who are injured or become ill on the job and require medical intervention. Necessary follow-up treatment for employees suffering from occupational diseases or injuries also should be arranged. The treatment of non-work-related employee injuries or diseases is usually the function of private medical practice. However, more primary care services are being provided on site by occupational health professionals including physicians, nurses and nurse practitioners, and occupational health nurses (Rogers et al., 1996).

Medical and surgical management in every case of industrial injury or disease should be aimed to restore disabled workers to their former earning power and occupation as completely and rapidly as possible. To help achieve these goals, the physician should promptly submit medical reports to those agencies that need to receive them. Furthermore, equitable administration of workers' compensation may depend on a physician's medical testimony regarding the injury and its possible consequences.

The occupational health nurse is also responsible for maintaining necessary records and reports. These records act as a guide to management and keep both the management and employees informed regarding the success of the health program. Records and reports are necessary to direct and evaluate preventive medical and safety engineering techniques, to chart progress in the reduction of accidents, and to meet the regulatory record-keeping requirements (see Chapter 2, Regulatory History, and Chapter 8, Injury and Illness Record Keeping and Incidence Rates).

Occupational Health Unit Location

The rooms housing the occupational health department should be easy for workers to find. If management places the department near the greatest number of employees, more workers are likely to report all minor injuries and have them treated. One potential location is near the work area entrance, which is also convenient for loading or unloading supply trucks and for bringing ambulances to the door, if necessary. Also, injured workers who are off duty but under treatment may come and go through a separate entrance. Management should try to put this department in a protected area so that a major company disaster does not destroy the first-aid or occupational health unit.

Physical Examination Program

Surveillance of workers' health status by qualified personnel is essential if an occupational health program is to provide maximum benefits for both the employee and employer. Therefore, the occupational health service should provide a regular physical examination program for all employees. The health care providers should discuss all significant findings with each worker. With the employee's written consent, a transcript of the data may be supplied to the worker's personal physician or to an insurance company. Courts, workers' compensation commissions, or health authorities may request this information by legal means, but it is preferable to obtain an employee's written consent to disclose his or her medical records. Occupational services personnel must strictly preserve the confidentiality of health examination records. Furthermore, the physician, nurse, or industrial hygienist should inform the employer of potentially harmful work-related exposures or hazards detected through their examination of workers and the environment (Rogers et al., 1996).

Health care providers determine the scope of a physical examination based on the type of operation involved—the nature of the industry; its inherent hazards; and the variations in jobs, physical and mental demands, and health exposures (McCunney, 1995). The values of different test procedures must be carefully weighed against their cost in time and dollars. Examinations probably will be different for different jobs; for example, the physical requirements for an ironworker who will be engaged in construction work are different from those of an office worker who will sit at a desk and do word-processing functions. However, certain basic examination criteria apply to each employee regardless of job assignment.

The various kinds of examinations may be classified as follows—preplacement, periodic, and special examinations such as job transfer, return-to-work, and termination. Aids to examinations such as x-rays, vision, or hearing procedures may be performed to obtain further data about an employee's health status.

Preplacement Examinations

The preplacement examination is done once an employee receives and accepts an offer of employment. This exam is used to determine and record the health condition of the prospective worker in order to match each employee to the right job. The applicant (or the personal physician with the applicant's approval) should be told of any conditions that need attention, because follow-up care may be necessary. The primary purpose of the preplacement examination program is to aid in selection and appropriate placement of all workers. The Americans with Disabilities Act requires employers to make reasonable accommodations for people with other disabilities. See Chapter 13, Workers with Disabilities, in the *Engineering & Technology* volume. Preplacement examination also provides baseline data for future evaluations.

It is not the health care provider's function to inform applicants whether or not they are to be employed. This is the prerogative and duty of management. Other factors besides health determine whether a candidate is suitable for employment in the company.

Periodic Examinations

Periodic examinations of all employees may be on a required or voluntary basis. These examinations are performed for employees for both preventive health monitoring and health surveillance purposes. Periodic health examinations also are conducted at intervals during employment to determine the worker's continued compatibility with the job assignment, and whether adverse health effects have occurred that may be attributable to the work or working conditions (Rogers, 1994). Management should institute a program for workers who are exposed to hazardous processes or materials or whose work involves responsibility for the safety of others, such as vehicle operators. In addition, process controls are usually established for substances such as lead, carbon tetrachloride, or many others that can cause occupational diseases. However, caution dictates that the occupational professional periodically examine workers to be certain that the engineering and hygiene controls are effectively safeguarding employees. Periodic examination also permits early detection of highly susceptible individuals and of practices or procedures that workers use to circumvent safety devices and policies.

How often workers are examined varies according to the quality of engineering controls (influenced by how rapidly the hazardous substance acts on the human body), the types of hazards, the findings produced on each examination, and in some cases the regulatory standards (Harris, 1994; McCunney, 1994). Thus, worker exposure to substances might require that employees receive examinations or laboratory tests on a weekly, monthly, quarterly, annual, or biannual basis.

In many cases, laboratory tests of blood or urine can serve as the major portion of a periodic examination program, with complete examinations being made less frequently. Deciding which type of special examination (laboratory, x-ray, etc.) is necessary for any exposure and interpreting the results require expert medical judgment (Jarvis, 1992). Appropriate regulatory standards, medical practice, and other sources offer additional guidance for performing tests.

Return to Work

Employees who experience on-the-job health problems often benefit from special examinations. For example, when there is a change in the employee's health status or in work conditions that may place the employee's health at risk, job transfer examinations are conducted to match the employee's capabilities with proposed new jobs.

Some organizations also find it worthwhile to require "return-to-work" examinations of employees who have been absent for more than a specified number of days as a result of an illness or injury. This type of examination is done to determine whether the health status of the employee might require a change in the worker's duties and to ensure proper placement to prevent further illness

and injury. The same disease can affect different people in widely different ways.

Return-to-work examinations also serve to evaluate workers after severe injuries or illness. Research has shown that the longer workers are away from their jobs, the less likely they are to return fully to productive employment. Rogers (1994) states that some employees' health problems may require them to have light duty, limited duty, or other work restrictions. This course of action should be determined and recommended by the occupational health professional after a thorough examination of the employee.

Light duty is an adaptation of the employee's original job to reduce the worker's tasks. Limited duty is defined as a new job that is appropriate to an injured worker's skills, interests, and capabilities. It is designed for individuals who cannot return to their original work area and is created for either temporary or permanent placement (Randolph & Dalton, 1989; Rogers, 1994).

Another process to rehabilitate workers is known as "work hardening." This is a progressive, individualized physical conditioning and training program designed to help injured workers return to the workplace as soon as possible. This goal is accomplished by gradually increasing physical and psychological requirements of the job, within the employee's current capabilities, until the employee can perform once again at acceptable work levels.

No matter what assessment for return to work is completed, occupational safety professionals must be sure to match returning workers with appropriate job demands, work conditions, and level of duties performed. The health professional should use job descriptions, job analyses, and discussions with employees in the same positions and the workers' supervisors to identify details and demands of each job.

When an employee returns to work following a serious injury, either occupational or nonoccupational, the health and safety professional should reevaluate the person's work capacities. Some work restrictions may be necessary. On the other hand, rehabilitation procedures may have actually improved a worker's ability to perform tasks or increased the range of jobs he or she can do.

Exit Examinations

Upon termination of employment, some organizations give employees a physical examination and document the findings. This procedure is particularly appropriate where operations exposed workers to health hazards such as lead, benzene (benzol), silica, asbestos dust, and excess noise levels, or where required by standards.

Laboratory Tests

Where workers will be exposed to toxic substances, appropriate laboratory tests may be indispensable. The use of general laboratory screening panels (Chem 24, and so on) is controversial. These tests are relatively nonspecific and can be affected by many unrelated problems, such as alcohol and viral infections.

Although not to be used routinely, x-ray tests may be appropriate in alerting a physician to a particular condition. However, x-rays are used much less frequently to detect pulmonary disease. Instead less potentially hazardous examinations, such as pulmonary function tests, are employed. These tests can detect loss of lung function before the disease reaches the stage where it can be seen on an x-ray. Furthermore, x-ray screening tests, as part of the preplacement examination, are of little if any significant value in predicting the onset of future physical conditions, such as back pain or injury. As a result, they should be avoided to reduce workers' unnecessary exposure to radiation and to save on cost.

Vision and Hearing Tests

Special devices have been developed for routine testing of several aspects of vision. Employees' near-sighted or far-sighted vision should be documented. If the company fails to match the visual requirements of a job with the visual abilities of employees, workers may become easily fatigued, inefficient, and frustrated. If the job involves working with colors or color-coded materials, health professionals should test workers' color vision. If the organization does not have vision-testing equipment available on site, management can contract with outside examiners to perform annual eye examinations and to fit prescription safety glasses, if needed.

Workers who will be exposed to high noise levels (generally 85 decibels (dBa) or greater for an eight-hour day) should be examined and tested for hearing acuity before job placement to determine prior hearing loss, if any. They should also be placed in a hearing conservation program (Royster, 1990). The audiometric examination is the accepted method of testing hearing acuity. Employees should be examined periodically thereafter to detect early hearing loss due to noise. Although protective hearing devices can reduce the amount and level of noise reaching the auditory nerve, hearing tests and records are also valuable aids to use when designing programs to prevent premature hearing loss.

Health History

A carefully taken personal health and occupational history may give as much if not more information about the worker's health status as the physical examination. The history also will indicate a worker's need for special tests and perhaps job-placement restriction.

The occupational health examiner must know the prospective employee's job title; a description of the job; duties, tasks, and related demands; potential work-related hazards; and any physical and/or emotional capacities required (Felton, 1990). The occupational health history is designed to determine previous work experiences and to assist the occupational health care provider in discovering preexisting conditions that may be worsened by conditions on the job (Rosenstock, 1995; Ginetti & Greig, 1981; LaDou, 1994; Stein & Franks, 1985).

A synthesis of data obtained from employees' personal and occupational histories and a review of systems can help health professionals make accurate diagnoses, prevent occupational disease and the aggravation of underlying medical conditions by workplace factors, identify potential workplace hazards, detect new associations between exposure and disease, and establish the basis for compensation of work-related disease (Kemerer & Raniere, 1990; Rosenstock & Cullen, 1986).

Emergency Medical Planning

The occupational health physician, nurse, and safety and health professional should confer with management to plan emergency procedures for handling large numbers of seriously injured employees in the event of a disaster, such as explosion, fire, or other catastrophe. Management should coordinate these plans with community plans for such events (see Chapter 15, Emergency Preparedness, in this volume).

Procedures should include the following:

- selection, training, and supervision of allied health and other personnel
- transportation and caring for the injured
- transfer of seriously injured personnel to hospitals
- coordination of these plans with the safety department, security, police, road patrols, fire departments, and other interested community groups

Employee Health Records

Health records contain an employee's health and medical history, examination and test results, medical opinions and diagnoses, descriptions of treatments and prescriptions, and employee medical complaints. Some exposure records may have to be maintained for 30 years, while medical records must be retained for the duration of employment plus 30 years.

Regulatory standards require employers to maintain accurate records of work-related deaths, injuries, and illnesses (see Chapter 8, Injury and Illness Record Keeping and Incidence Rates). The OSHA Access to Medical Records standard permits the worker, the worker's representative, and regulatory authorities to have access to employer-maintained health/medical and toxic exposure records. These specific conditions under which access is allowed apply to all employers in general industry, maritime, and construction whose employees are exposed to toxic substances or harmful physical agents. The records that interested parties are allowed to see include exposure records of other employees with past or present job duties or with working conditions related to those of the employee, records containing exposure information concerning the employee's working conditions, and MSDSs.

Occupational health records also provide data for use in job placement, in establishing health standards, in health maintenance programs, in treatment and rehabilitation, in workers' compensation cases, in epidemiologic studies, and in helping management with program evaluation and improvement. Such data are collected in the history interview, from the preplacement examination and any subsequent examinations, and from all visits the worker makes to the occupational health unit or first-aid room.

Health records also serve to establish a health status profile of each worker. The key to accurate diagnosis and treatment often lies in the adequacy and completeness of this profile; therefore, record maintenance is a professional responsibility. Further, to compile a complete history, health records should include absences caused by illness or off-the-job injury. Thus, record keeping of nonoccupational and occupational health incidents often reveals chronic or recurrent conditions. In such cases, early treatment (e.g., referral to family doctors) and preventive measures can reduce absenteeism and decrease accident rates.

Maintenance of health records, however, should not be so burdensome that the occupational health nurse (or first-aid attendant) becomes a file clerk. Recording forms and filing systems must be simple enough that they can be used and interpreted by a physician, nurse, or first-aid attendant. In addition, computerization of records may increase the efficiency of the record-keeping system.

Although the employer needs to know a worker's limitations for job placement purposes, all health/medical records are confidential. This means that only health care professionals have a legitimate or legal right to see employee health records. If a company does not have a resident medical director or nurse, its medical records are usually filed in the human resources (HR) department and must be kept strictly confidential by HR personnel.

Neck or Wrist Tags for Medic Alert
A universal symbol for emergency medical identification has been developed by the American Medical Association (Figure 10-5). The object of the symbol is to identify its wearer immediately as a person with a health-related condition requiring special attention. If the wearer is unconscious or otherwise unable to communicate, the symbol will indicate that vital medical facts are recorded on a health information card in the bearer's purse, wallet, or other location. The identification tag also gives a telephone number for obtaining more-detailed information.

These details should be known by all employees before they attempt to help an individual struck down by an accident or sudden illness (see Medic Alert Foundation and *Emergency Medical Identification* in the References).

Health Promotion and Wellness
Wellness has been defined as a way of life that promotes a state of health. The health of employees can directly

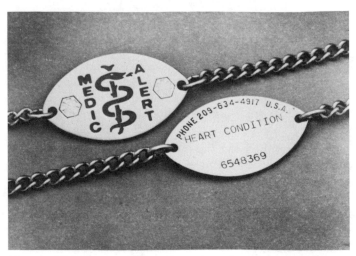

Figure 10-5. Persons with a physical condition for which emergency care may be needed should wear an identification tag. The tag provides a general indication of the problem and a phone number to call for more details if the tag wearer is unable to supply them. (Reprinted with permission from Medic Alert Foundation, Turlock, CA.)

affect the bottom line through decreased workers' compensation and health benefit costs. Maintaining and protecting good health is more effective and less costly than treatment and cure. The government publication *Healthy People 2000: National Health Promotion and Disease Prevention Objectives* presents a blueprint or plan of action to combat the leading preventable diseases and health-related problems. Priority areas identified include:

Health Promotion

- physical activity and fitness
- nutrition
- tobacco
- alcohol and other drugs
- family planning
- mental health and mental disorders
- violent and abusive behavior
- educational and community-based programs

Health Protection

- unintentional injuries
- occupational safety and health
- environmental health
- food and drug safety
- oral health

Prevention Services

- maternal and infant health
- heart disease and stroke
- cancer
- diabetes and chronic disabling conditions

- HIV infection
- sexually transmitted diseases
- immunization and infectious diseases
- clinical preventive services
- surveillance and data systems

To achieve these priorities will require a commitment on the part of society and its members to promote and adopt healthful lifestyles and behaviors conducive to optimal health, and by business and government to develop strategies, opportunities, and resources required to support the behavioral and environmental changes needed.

Health Promotion

Pender defines health promotion as "activities directed toward increasing the level of well-being and actualizing the health potential of individuals, families, communities, and society." Health promotion behaviors are described as continuing activities that are an integral part of one's lifestyle, such as physical exercise or eating nutritious foods directed at achieving the highest level of health.

Health Protection and Prevention Services

Prevention is best described as health-protecting behavior because primary emphasis is placed on guarding or defending an individual or group against specific illnesses or injuries. A major component of occupational health includes activities related to health promotion and protection for both individuals and groups of workers incorporating primary, secondary, and tertiary prevention strategies (Rogers, 1994).

Primary prevention takes place before disease or illness occurs. These types of programs serve to increase awareness and knowledge related to toxic and hazardous exposures and their effects and to promote lifestyle interventions such as smoking cessation, stress management, improved exercise and nutrition, use of seat belts, and proper use of prescription drugs.

Secondary prevention begins after a condition or disease is present. Intervention at this stage may lessen the complications and the disability resulting from disease or illness. Examples of this type of prevention are blood pressure screening, medical surveillance for occupational exposures, and preplacement, periodic, and termination physicals.

Tertiary prevention or rehabilitation begins when the disease or condition has stabilized and no further healing is expected. Examples are chronic illness monitoring, on-site therapy, modified duty, or assisting an employee with a work-related injury to reuse an injured hand. The employee may need to learn new skills or new ways to use old skills in order to return to work.

Wellness programs are designed to address and promote health and safety behaviors. According to the September/October 1994 NSC *Occupational Health Nursing Newsletter,* wellness programs should include the physical aspects of health issues and also should deal with emotional, spiritual, and social dimensions. O'Donnell (1994) indicates that health promotion and protection programs should be targeted at three levels, including awareness, lifestyle and behavioral changes, and supportive environments. Awareness programs are targeted at individuals and groups to increase their understanding of health and related risk factors. Use of pamphlets, health screening activities, or conferences are examples.

Lifestyle change programs are directed at helping individuals change their behavior, such as starting and maintaining exercises, eating nutritious foods, and enhancing communication and coping skills. Lifestyle change programs are most successful if they use a multistep process (i.e., introducing one or two programs/activities at a time); include a combination of educational and behavioral modification experiences; and are provided over a number of weeks or months. Supportive environments include families, friends, organizational/work cultures, coworkers, communities, and regulations that help to shape these environments (e.g., availability of jogging trails or on-site exercise classes). Fostering a supportive environment or changing an environment to encourage a concept of health will go a long way in improving employees' health behavior outcomes.

SPECIAL PROGRAMS

In today's work environment, companies must be aware of the special needs of some employee groups, such as women, employees with special health problems, and older workers. Government regulations and professional organizations have developed regulations and guidelines to help companies meet the needs of these groups so that these employees can remain productive members of the workforce.

Health Care for Women

In the last three decades the number of women in the workforce has more than doubled (NCHS, 1989). Many factors such as economics, women's changing roles, increased job opportunities, and gender antidiscrimination laws have influenced both job accessibility and availability in today's job market (Rogers, 1995).

Women comprise the largest group of workers in office, health care, and hospital environments and face various types of hazards, including exposure to chemical and physical agents, unsuitable ergonomic work conditions, and stress. Data suggest that 75% of the hospital labor force are women, and in several occupations women make up more than 90% of the total workforce. This is particularly true of jobs such as nursing (97%), dietary services (91%), and medical records (93%) (Zoloth & Stellman, 1987).

In addition, many workers involved in assembly-line production processes, such as in the electronics industry, are women. Besides exposures to chemical cleaning and degreasing substances, these workers are involved in processes that require rapid, repetitive motion of the fingers, hands, wrists, and arms, resulting in cumulative trauma disorders.

There are more than 18 million clerical workers in the United States, and 75% of them are women. Many jobs that women perform require long hours standing or sitting, sometimes in awkward positions. Sitting at a desk for long periods of time may cause blood pooling and edema in the lower extremities. Poor sitting posture or episodes of standing for long periods can create a postural load and may result in back pain, muscle stress, and general body fatigue. Jobs requiring repetitive tasks such as typewriting and word processing may result in or aggravate musculoskeletal conditions, including repetitive motion and carpel tunnel disorders.

The influence of work on reproduction is complicated because of the variety of work-related exposures that can occur. Many chemicals, including pharmaceuticals, anesthetic gases, heavy metals, pesticides, and organic solvents, can disrupt the menstrual cycle and affect the course of pregnancy or the development of the embryo or fetus (Harris, 1994). Of particular concern is that many women may not know when they first became pregnant and thereby remain at risk to potential exposure.

In the past, many companies used fetal protection policies to prevent women from working in areas where toxic chemicals that might harm a developing fetus were present. A 1991 Supreme Court decision struck down these policies as discriminatory to women, since they may also bar them from higher paying jobs. Critics of such policies have also argued that they did not deal with reproductive hazards posed to men by chemicals such as lead. The Court decision puts the responsibility on the employer to clean up the work environment and to provide information to the worker on the hazards that exist. Many experts state that as long as there is no negligence in this regard, the employer is not likely to be found liable for any injury to a fetus. Critics of the Supreme Court decision dispute this assertion.

Although reproductive risk is a complex subject, companies must inform women of the hazards they face on the job. Chemicals such as lead, mercury, some solvents, and therapeutic agents containing synthetic hormones may produce harmful effects on the fetus or lead to spontaneous abortion. Biological agents such as rubella (German measles) or physical hazards such as ionizing radiation may also produce such damage. Pregnancy may limit a woman's ability to do some strenuous physical work. For these reasons a woman should consult her doctor about any restrictions that should be placed on her work during pregnancy.

In line with the Pregnancy Discrimination Act and the Americans with Disabilities Act, an employer needs to make reasonable accommodations for women during pregnancy. Employers need to consult the latest regulatory interpretations of laws and court decisions in developing programs to deal with issues posed by workers who are pregnant or planning to conceive a child. For example, if employees desire medical advice regarding lead exposure and the ability to procreate a healthy child, the U.S. OSHA lead standard requires that employers offer medical examinations and consultations to both men and women on request.

Substance Abuse and Employee Health

Alcoholism and drug abuse continue to be among the leading illnesses in the workforce. Therefore, occupational health professionals and the employer must determine what they can do to help employees who have these illnesses. (See Chapter 14, Employee Assistance Programs.) Managers and supervisors should be alert to employees whose work and performance are deteriorating because of an alcohol problem. These employees should be referred for medical care as quickly as possible.

When alcohol and drugs are combined, they produce a variety of effects that severely impair workers. Concentrations of alcohol and drugs remain in the bloodstream much longer than most users realize, and the effects of this combination may appear unexpectedly. For example, they can make vehicle operators drowsy or impair their reflexes, which can increase the risk of accidents, injuries, and even deaths.

Dependence on alcohol or other drugs is a major cause of family strife, impaired job performance, increased insurance rates, occupational accidents, increased absenteeism, and rising crime rates. These illnesses know no boundaries, nor is there any "generation gap" among abusers—all ages, races, and socioeconomic groups are susceptible to their destructive effects.

To help detect and/or prevent substance abuse and other problems, every time an applicant or employee visits a physician or nurse, some health education and/or counseling should be given. Occupational health units are an ideal way to provide these services. Employers are paying an increasingly large part of the health care costs for employees and their dependents. The earlier health-related problems are detected, the less they are likely to cost. As a result, it makes economic sense for employers to maintain a sound occupational health department to provide proper care for employees and to offer timely referrals for further care, if needed. (See also Chapter 14, Employee Assistance Programs.)

Older Workers

It is clear that the American workforce is aging. By the year 2000, the median worker age is expected to be 39 years (Campeanelli, 1990) with nearly half (49%) of the workforce between the ages 35 and 54. Nearly all businesses have older workers. As a culture we have been conditioned to believe that aging means an inevitable decline in cognitive function and an increase in chronic disease, disengagement, nonproductivity, obsolescence, and retirement (Rude & Adams, 1993). However, recent research is showing that these common beliefs are erroneous. As Rude and Adams state, people age differently and at different rates; aging is not a fixed, mechanical experience; and new potential for growth, productivity, and creativity exists in the second half of life. As people age they become more individual, not more alike. Therefore, it is critical that the occupational health professional take a proactive approach in worker evaluations. Businesses intent on strengthening their future growth will value older workers and recognize their contributions. Industry and the public must shed the belief that older employees are past their prime and begin to realize the incredible resource they represent, if properly managed (Rude & Adams, 1993).

For these and other reasons, companies need to offer occupational health programs throughout employment that address both physical and social aspects of life. Although employees now have "work longevity," they also may exhibit specific age-related health problems that must be anticipated and dealt with appropriately. Examples of anticipated health problems include arthritis, hypertension, cardiovascular disease, hearing impairments, diabetes mellitus, and depression. Other lifestyle issues that may need attention are weight control, exercise, smoking, alcoholism, substance abuse, and stress (Hart & Moore, 1992).

The occupational health professional must teach employees how to be responsible for making good choices to protect and promote their health and for making sound health care decisions. Information given to employees during employment and special instructions given before retirement should include not only financial aspects but also how to use health care resources wisely, how to find information and resources to continue healthy lifestyles, and how to locate information about healthy adjustment to retirement, spouses, significant others, and family members. Programs targeted at screening for early disease detection and chronic disease monitoring are essential for health promotion of the aging workforce. In addition, several points in managing the aging workforce can be considered:

- understanding the normal aging process, its effects on health and impact on job assignments and team building
- addressing work-related stress that may accompany job changes and developing stress management and retraining programs to deal with this issue
- conducting aging programs for management and co-workers to prevent misunderstanding and foster effective use of older workers
- capitalizing on older workers' abilities and experience in job performance

The result will be an improved quality of life and work life, and reduced health and benefit costs.

SUMMARY

- Occupational health programs should provide a healthful environment, health examinations, diagnosis and treatment, health records, health education and counseling, and communication between company and occupational professionals. All health services provided by organizations must meet federal, state, and local codes.
- The principal goals of an occupational health program are to protect employee health, to match the right worker to the right job, to treat and rehabilitate injured employees, and to encourage personal health maintenance. Companies should also be involved in off-the-job safety and health programs.
- Management may work with other health professionals such as industrial hygienists and physical therapy specialists in delivering health care to employees. Experience has shown a strong relationship between occupational health and accident prevention.
- First-aid services, generally mandated by regulations, include properly trained personnel; a first-aid unit and supplies; company manual; list of reactions to chemicals; instructions for calling physicians, hospitals, and ambulances or rescue squads; methods for transporting ill or injured employees; and an adequate record system of injuries and treatment.
- First-aid treatment can be either emergency (before a physician can be summoned) or prompt attention (for minor injuries and illnesses). One person should be responsible in each work area for maintaining adequate supplies in good condition and for treating all worker injuries promptly.
- Occupational health nurses and physicians work together to provide occupational health services and health education to workers. Physicians and nurses must maintain a confidential relationship with employees and keep accurate, timely records.
- One of the most important functions of employee health programs is setting up a regular schedule of physical examinations of all workers. The basic purpose is to establish a baseline for worker health at the outset of employment, to detect or monitor any work-related illnesses or injuries, and to match the right worker with the right job.
- Physicians and nurses conduct preplacement, periodic, transfer, promotion, special, and termination examinations. The findings of these examinations are confidential and may be released only if a worker provides written consent. Workers have a right to review these records to determine how they may have been affected by hazards while on the job.
- Employee health records generally are retained for at least 30 years in industries where workers have been exposed to hazardous materials. Such records

contain the results of tests along with a complete health and occupational history, medical opinions and diagnoses, treatments, and employee complaints.

- Emergency medical planning is a responsibility of occupational health physicians, nurses, safety professionals, and management. These plans include procedures for training personnel in emergency measures, transporting and treating injured people, transferring the seriously injured to hospitals, and coordinating plans with other units and community departments and agencies.

- Most companies now have some type of substance abuse screening and referral or treatment programs in place. Abuse of alcohol and drugs greatly affects a worker's job performance, places other workers at greater risk, and costs industry several billion dollars a year in lost time, accidents, and lower productivity.

REFERENCES

American College of Surgeons, Committee on Trauma. *Emergency Care of the Sick and Injured.* Philadelphia: W.B. Saunders Company, 1982.

Boylston R. *Managing Safety and Health Programs.* New York: Van Nostrand, 1990.

Campanelli L. The aging workforce: Implications for organizations. *Occupational Medicine,* 5(4): 817–825, 1990.

Emergency Medical Identification. American Medical Association, 535 North Dearborn Street, Chicago, IL 60610.

Felton J. *Occupational Medical Management.* Boston: Little, Brown and Co, 1990.

Ginetti J. and Greig A. The occupational health history. *Nurse Practitioner,* December: 12–13, 1981.

Guide for Establishing an Occupational Health and Safety Service. Atlanta: American Association of Occupational Health Nurses, Inc., 1994.

Guidotti TL et al. *Occupational Health Services: A Practical Approach.* Chicago: American Medical Association, 1989.

Harris, RL and Cralley LJ. *Patty's Industrial Hygiene and Toxicology. Vol III, Part A. Theory and Rationale of Industrial Hygiene Practice: The Work Environment,* 3rd ed. New York: John Wiley & Sons, 1994.

Hart BG and Moore PV. The aging workforce: Challenges for the occupational health nurse. *AAOHN Journal,* 40(1): 36–40, 1992.

Hau ML. *While Help is on the Way* . . . Chicago: Health Products Marketing, 1994.

Jarvis C. *Physical Examination and Health Assessment.* Philadelphia: W.B. Saunders & Co, 1992.

Kemerer S and Raniere T. Cost effective job placement physical examinations. *AAOHN Journal,* 38(5): 236–242, 1990.

LaDou J, ed. *Occupational Health and Safety.* Chicago: National Safety Council, 1994.

Levy BS and Wegman DH. *Occupational Health: Recognizing and Preventing Work Related Disease.* Boston: Little, Brown & Co, 1995.

McCunney RJ. *A Practical Approach to Occupational and Environmental Medicine.* Boston: Little, Brown & Co, 1994.

Medic Alert Foundation, 2323 Colorado, Turlock, CA 95380.

Collins JG and Thornberry OT. *Health Characteristics of Workers by Occupation and Sex:* U.S., 1983–85. National Center for Health Statistics (1989). Advanced Data from Vital and Health Statistics, Pub No. 89–1250. Public Health Service, Hattsville, MD, 1989.

National Safety Council. *OSHA Up-To-Date,* Vol. 24, No. 11, 1995.

———. *Occupational Health Nursing Newsletter,* 1995.

O'Donnell MP and Harris HS. *Health Promotion in the Workplace.* Albany, NY: Delmar Publishers, 1994.

Rogers B, Randolph SA, and Mastroianni K. *Occupational Health Nursing Guidelines: Primary Clinical Conditions.* Boston: OEM Press, 1996.

Rogers B. Women in the workplace. In C Fogel and NF Woods, eds. *Women's Health Care.* London: Sage, 363–384, 1995.

Rogers B. *Occupational Health Nursing: Concepts and Practice.* Philadelphia: W.B. Saunders Company, 1994.

Rosenstock L and Cullen MR. *Textbook of Clinical Occupational and Environmental Medicine.* Philadelphia: W.B. Saunders Company, 1994.

Rosenstock L and Cullen M. *Clinical Occupational Medicine.* Philadelphia: W.B. Saunders Company, 1986.

Royster JD and Royster LH. *Hearing Conservation Programs: Practical Guidelines for Success.* Chelsea, MI: Lewis Publishers, Inc., 1990.

Rude J and Adams C. Doing business in an aging society. *The Business News,* August 23–September 5, 1993.

Stein C and Franks P. Patient and physician perspectives of work-related illness in family practice. *Journal of Family Practice,* 20(6): 561–565, 1985.

U.S. Department of Health and Human Services. Healthy People 2000: National Health Promotion and Disease Prevention Objectives. (DHHS Pub No 91–501212, 94–110), Washington, DC: U.S. Government Printing Office, 1991.

Wellness Councils of America. *Healthy, Wealthy and Wise: Fundamentals of Workplace Health Promotion.* Omaha, NE: WELCOA, 1993.

Williamson GC and Moore PV. Health care cost containment: A model for practice. *AAOHN Journal,* 35(11): 496–500, 1987.

Workers' Compensation Board of British Columbia. Dresser A, ed. *Industrial First Aid.* New York: Van Nostrand Reinhold, 1991.

Zoloth S and Stellman J. Hazards of healing. Occupational health and safety in hospitals. In Stromber AH, Larwood L, and Gutek BA. *Women at Work: An Annual Review.* Beverly Hills, CA: Sage Publications, 1987.

REVIEW QUESTIONS

1. List 5 of 8 components of a good occupational health program.
   a.
   b.
   c.
   d.
   e.
2. List the four (4) basic objectives of a good occupational health program.
   a.
   b.
   c.
   d.
3. What is dermatitis caused by?
4. What two methods of controlling the work environment can reduce the rate of accidental injuries?
   a.
   b.
5. The health service office (dispensary) should have a minimum of how many rooms?
   a. Two
   b. Three
   c. Four
   d. It doesn't matter
6. List and briefly define the two kinds of emergency treatments.
   a.
   b.
7. Why should all employees report for medical treatment immediately when they are injured?
8. List 5 of 8 elements that a first-aid program should include.
   a.
   b.
   c.
   d.
   e.
9. What federal agency outlines specific first aid requirements at certain locations in mines and processing plants?
10. What is the objective of medical and surgical management in cases of industrial injury or disease?
11. What is the primary purpose of preplacement examinations?
12. Explain the difference between light duty and limited duty.
13. What is the primary purpose of exit examinations?
14. Emergency medical planning procedures should include what four (4) items?
    a.
    b.
    c.
    d.
15. After an employee leaves a company, how long (maximum) must employee exposure records be kept?
    a. 30 years
    b. 20 years after the employee leaves the company
    c. 30 years after the employee leaves the company
    d. 20 years
16. List 5 of 8 ways in which the data from occupational health records can be used.
    a.
    b.
    c.
    d.
    e.
17. Define wellness.
18. Lifestyle change programs are most effective when they employ a(n) _____ process.
    a. aversive
    b. multistep
    c. baseline
    d. stochastic

11
Industrial Hygiene Program

Industrial hygiene, as defined by the American Industrial Hygiene Association (AIHA), is the science and art devoted to the recognition, evaluation, and control of environmental factors or stresses arising in or from the workplace which may cause sickness, impaired health and well-being, or significant discomfort and inefficiency among workers or among citizens of the community.

The AIHA defines industrial hygienists as occupational safety and health professionals concerned with the control of environmental stresses or occupational health hazards that arise as a result of or during the course of work. The industrial hygienist recognizes that environmental stresses may endanger life and health, accelerate the aging process, or cause significant discomfort. Working with management and medical, safety, and engineering personnel, the industrial hygienist can help to eliminate or safeguard against environmental conditions caused by chemical, physical, ergonomic, or biological stresses. This chapter covers the following topics:

- the nature, purpose, and scope of industrial hygiene
- the role of the industrial hygienist in recognizing and controlling environmental hazards
- four general categories of environmental conditions or stresses
- three definitive elements of industrial hygiene practices
- health and psychological problems associated with altered environments
- the inherent medical and physiology issues involved in shiftwork
- toxicity and hazards in the workplace.

Industrial hygiene includes the development of corrective measures to control health hazards by either reducing or eliminating the exposure. These control measures may include substituting safer materials for harmful or toxic ones, changing work processes to eliminate or minimize work exposure, installing exhaust ventilation systems, practicing good housekeeping (including appropriate waste disposal methods), and providing proper personal protective equipment.

The five factors necessary to assure an effective industrial hygiene program include (1) the anticipation and recognition of health hazards arising from work operations and processes; (2) evaluation and measurement of the magnitude of the hazard (based on past experience and study); (3) control of the hazard; (4) the commitment and support of management, since effective controls can be expensive and the need for these controls must be made clear within company management; and perhaps most importantly, (5) the industrial hygienist must be recognized by the workers within the facility as reliable, believable, and honest, so that recommendations for changes in work practices are diligently implemented and assurances that conditions are acceptable are indeed believed.

Most commonly, the industrial hygiene function is placed in the human resources department (traditionally with the medical function). Some companies combine the health and safety function, and others combine the environmental and industrial hygiene departments. Occasionally, the industrial hygiene function is found in the engineering department, or the security department.

For many small and mid-sized companies, a full-time industrial hygienist cannot be justified economically. In these cases, a consulting agreement with a qualified local consulting firm or individual consultant can be the most appropriate way for the company to obtain essential services. Ordinarily, the consultant reports to the safety director of the company. There are numerous qualified industrial hygienist consultants, and many companies are taking this approach, as described in the following section. Also, insurance companies and the consulting services offered by OSHA should not be overlooked. Insurance companies (usually the workers' compensation carrier) have a vested interest in improving the workplace environment, to reduce the potential for occupational illness.

INDUSTRIAL HYGIENIST ON THE OCCUPATIONAL HEALTH AND SAFETY TEAM

The industrial hygienist specializes in the recognition, evaluation, and control of health hazards. In most organizations, the industrial hygienist is principally responsible for chemical agent hazards, but will often be qualified to exercise responsibility for control of physical agents, such as noise and vibration, ionizing radiation, and nonionizing radiation. (In larger and more complex organizations, the control of radiation rests with the radiation specialist or the health physicist.) As the need for control of biological agents (especially for indoor air quality—such as prevention of legionnaires' disease or other communicable diseases) has become more widely recognized, many industrial hygienists have this responsibility as well. Depending on their expertise, industrial hygienists may also have responsibility for ergonomics. Industrial hygienists maintain lists of chemical, biological, and physical agents found within the facility, evaluate exposures to those agents, and institute controls to ensure that exposures to those agents are within tolerable limits.

In-House Industrial Hygiene Services

The industrial hygienist must have a good relationship with the engineering department, since many of the most effective controls on exposure will involve modifying the physical facility. The environmental staff—those responsible for control of emissions outside of the facility—should also be part of the industrial hygienist's major contacts. Controls to reduce exposures within the facility can lead to increased emissions; a unified approach to reducing exposures is essential.

Two of the most important departments for the industrial hygienist are the safety and medical departments.

Occupational Health and Safety Team

The chief goal of a facility's occupational health and safety program is to prevent occupational injury and illness by anticipating, recognizing, evaluating, and controlling occupational health and safety hazards. The medical, industrial hygiene, and safety programs may have distinct program goals, but all programs must interact with all components of the overall health and safety program. The occupational health and safety team consists of the industrial hygienist, the safety professional, the occupational health nurse, the occupational physician, the employees, senior and line management, and others depending on the size and character of the particular facility. All team members must act in unison to provide information and activities, supporting the other parts to achieve the overall goal of a safe and healthy work environment. Therefore, the separate functions must be administratively linked in order to effect a successful and effective program.

The first vital component of an effective health and safety program is the commitment of management. Serious commitment is demonstrated when management is visibly involved in the program and all personnel are in compliance with health and safety practices. Equally critical is the assignment of authority, as well as responsibility, to carry out the health and safety program. The health and safety function must be given the same level of importance and accountability as the production function.

PRACTICING INDUSTRIAL HYGIENE

The three definitive elements of industrial hygiene are recognizing, evaluating, and controlling occupational hazards. Recognizing health hazards is most important since it must take precedence before proper evaluation, or (if needed) control can take place. Upon recognizing a health hazard, the industrial hygienist should be able to identify the set of measures necessary for proper evaluation. When the evaluation is completed, the industrial hygienist—in consultation with other members of the occupational health team—can implement controls needed to reduce exposures to tolerable limits.

Recognizing Occupational Health Hazards

The first step in the process leading to the evaluation and control of harmful materials and processes is recognizing potential occupational health hazards. Such identification can be based on general knowledge of the characteristics of materials and processes, on clinical findings that disease or discomfort is present in the exposed population, on reports from others in scientific literature, in bulletins from trade associations or governmental agencies, in conversations with peers, or from reports by workers.

The principal tool of the industrial hygienist is observing the workplace. There is no real substitute for competent observation of:

- work practices used
- the extent of use of chemical physical agents
- apparent effectiveness of control measures.

A health hazard does not exist in isolation from the workplace—a chemical or physical agent is necessarily only a danger if workers can be exposed. The industrial hygienist must use all of his or her knowledge to evaluate the workplace. In addition, discussions with workers and managers are extremely helpful in evaluating the potential for exposure.

Environmental Factors or Stresses

The various environmental factors or stresses that can cause sickness, impaired health, or significant discomfort in workers can be classified as chemical, physical, biological, or ergonomic.

Chemical Factors

Chemical factors arise from excessive airborne concentrations of mists, vapors, gases, or solids in the form of dusts or fumes. In addition to the hazard of inhalation, some of these materials may act as skin irritants or may be toxic by absorption through the skin.

Physical Factors

Physical factors include excessive levels of ionizing and nonionizing radiations, noise, vibration, and extremes of temperature and pressure. See Chapter 20, Laboratory Safety, for a discussion of ionizing and nonionizing radiations.

Ergonomic Factors

Ergonomic factors include situations, conditions, or motions in the occupational environment that can result in accidents or illnesses, such as improperly designed tools, work areas, or work procedures, improper lifting or reaching, poor visual conditions, or repeated motions in an awkward position. Designing the tools and the job to fit the worker is of prime importance. Engineering and biomechanical principles must be applied to eliminate hazards of this kind.

Biological Factors

Biological factors are any living organism or its properties that can cause an adverse response in humans. They can be part of the total environment or associated with a particular occupation. Work-related illnesses due to biological agents have been widely reported, but in many workplaces their presence and resultant illness are not well-recognized. It is estimated that the population at risk for occupational biohazards may be several hundred million workers worldwide.

Exposure to many of the harmful stresses or hazards listed can produce an immediate response due to the intensity of the hazard, or the response can result from longer exposure at a lower intensity.

In certain occupations, depending on the duration and severity of exposure, the work environment can produce significant subjective responses or strain. The energies and agents responsible for these effects are called environmental stresses. An employee is most often exposed to an intricate interplay of many stresses, not to a single environmental stress.

Everyone is exposed on and off the job to a variety of chemical substances; most are not hazardous under ordinary circumstances, but they all have the potential to cause injury at some concentration. How a material is used is the major determinant of its hazard potential. Any substance contacting or entering the body is injurious at an excessive level of exposure and theoretically can be tolerated without effect at some lower exposure.

ALTERED ENVIRONMENTS

Traditionally, issues surrounding altered environments have focused on problems of exposure to physical hazards such as heat, cold, humidity, or altitude. Typically, the industrial hygiene staff has evaluated these environments with input from safety and medical personnel. Over the past 25 years, workers have found that they now are confronting new physical and psychological exposures and stresses that would have been unimaginable a few years ago. Two of the more common but complex altered environments are (1) the clean rooms found in many high-tech manufacturing settings, and (2) the problems associated with workplaces that operate on continuous 24-hour schedules. These two areas will be examined from a multidisciplinary perspective, as the solutions to the problems raised by clean rooms and shiftwork require information and expertise from medical, safety, hygiene, ergonomics, engineering, and human resources personnel.

Clean Room Environment

Every year, the electronics industry introduces new and more versatile products that perform a wide range of activities. At the heart of many of these new products is a silicon chip packed with integrated circuits. The manufacturing of these chips requires an unusual environment, known as a "clean room," that places significant new demands on worker safety, and ergonomics.

Clean rooms are designed to protect the chip from contamination by airborne particles. For example, normal ambient air can contain over 50,000 particles of greater than 0.5 micron (μm) per cubic foot of air (0.028 cubic meters). A typical clean room may have only 10,000 (0.5 μm) particles per cubic foot, with some production rooms having less than one particle per cubic foot. This level of air purity requires a series of engineering steps that can also create potential hazards for the workforce. For example, clean room ventilation hoods appear to be similar to standard hoods used in routine chemical or biological laboratories or operations which draw a steady stream of air away from the worker. In fact, these hoods typically push air away from the product and towards the worker, thereby potentially increasing employees' exposure to chemical vapors.

An additional problem that can increase worker exposure to air contaminants is the constant recycling of internal air with the clean room environment. The majority of air entering the clean room is not composed of external air but rather is recirculated from core areas. This desire to minimize outside air loading is based on the expense and difficulty of constantly cleaning, filtering, and controlling the temperature and humidity of external air. Usually, only 10% of the total clean room air is removed and replaced with outside air. Over the last few years, there have been considerable improvements in clean room ventilation, particularly with the development of vertical-laminar-flow (VLF) ventilation. With VLF ventilation, the air is constantly refiltered 9 to 10 times per minute. Unfortunately, the air and any chemical vapors are also rapidly redistributed into the entire work area. VLF ventilation is not restricted only to the semiconductor industry; in fact, a wide range of manufacturing and health care facilities use this type of ventilation. Health and safety professionals must be prepared to understand the mechanical engineering aspects of clean room ventilation so that they can successfully address worker questions and problems.

Health Problems

The ideal manufacturing environment in the clean room is similar to a semi-arid condition: hot (74°F), dry (35% relative humidity), and windy, due to the rapid number of air changes per hour. Although ideal for chip manufacturing, this environment is associated with a significant number of potential medical problems for workers, including dermatologic, allergic, and respiratory aliments. Chronic skin problems are one of the most frequent causes of medical transfer from the clean room. In general, these skin rashes have been described as "low humidity occupational dermatoses." The treatment of these rashes can range from application of simple moisturizers or lotions to mild hydrocortisone (steroid) creams. Difficult cases may require stronger steroid creams and, ultimately, the transfer of workers out of the clean room.

Upper respiratory problems, such as sinus, throat, and lung complaints, have also been reported in the clean room environment. Because these problems are usually attributed to the atmospheric and temperature conditions required by the manufacturing process, it is unlikely that anything short of medical transfer will be a satisfactory treatment. A similar, but less difficult, problem is associated with eye irritation. Usually, this problem can be treated with simple over-the-counter drops or lubricating fluids; however, some contact lens wearers may be forced to switch to glasses.

Another source of potential health problems is the constant use of protective garments, which are intended to protect the chips, not the workers. In street clothes, a person at a normal office workplace sheds 500,000 to 1 million particles per minute. This number increases by ten-fold when the worker is active and moving. Obviously, the clean room designer must prevent this level of contamination; therefore, a variety of suits have been designed that minimize particle shedding. These suits require special laundry and repair facilities. There have been reports in the literature of contact dermatitis in workers caused by laundry chemical residue left on clean room garments. Apparently, the seams of these protective garments can trap laundering agents and provide a potential exposure source.

An additional problem associated with protective garments is the development and aggravation of facial acne due to constant use of headgear. This problem has been attributed to the combination of relatively hot ambient air temperatures and the constant rubbing and irritation that can be produced by full facial garments. Similarly, health problems, primarily dermatologic, have been associated with the polyvinyl chloride gloves that are usually used in the clean room environment.

Overall, it is important to remember that the protective garments in the clean room environment are used to protect the product and not the individual worker. Therefore, it should not be surprising that the altered environment of the clean room can produce a wide variety of physical and psychological problems.

Psychological Problems

Given the extreme sterile conditions that are required in the clean room, it should not be surprising that workers report a myriad of psychological problems. Workers operate in a constant, unchanging environment that is completely isolated physically and emotionally from the outside. The use of protective equipment further accentuates the sense of isolation and dehumanization because workers cannot easily socialize and communicate with each other. To preserve the purity of the environment, workers are discouraged from frequently moving around the room, entering or exiting the area, and talking with fellow employees. Overall, these conditions make the clean room a relatively harsh workplace environment for people.

Interestingly, there have not been any large-scale psychological studies of workers in this environment. However, it would not be surprising to find somewhat increased rates of alienation, depression, or other mood disorders. Clearly, this environment presents both an opportunity and a challenge for specialists in occupational health, safety, hygiene, and human resources.

Shiftwork

Stimulated by the demands in World War II for increased production, the drive for continuous 24-hour work cycles has been steadily increasing across all industrial or service sectors. At least 25% of the U.S. labor force is engaged in some type of industrial or service activity that requires work around the clock. As the U.S. economy becomes ever more globalized, this trend will accelerate, particularly in communications and financial services sectors of the economy. Because humans have physiologically and psychologically evolved as "daytime" organisms, it is important to understand the ramifications of the move to

a 24-hour society. This type of altered workplace environment must be studied carefully by health and safety professionals to minimize the physiological and psychological disruptions that continuous operations can produce in humans.

Effects on Human Physiology

The human organism adapted over millions of years to an environment that accentuates activity during the daylight hours and sleep (protection) during the hours of darkness. Humans have a 24-hour activity cycle that is constantly reset by the rising and setting sun. This 24-hour cycle is known as "circadian" from the Latin *circa dies* or "about one day." Most physiological functions have cycles of high and low activity during a 24-hour period. Although some processes may have longer cycles, most chemical activities in the body follow this 24-hour period.

Interestingly, the actual human cycle is 25 ± 2 hours or 23 to 27 hours. In environments that are not ruled by artificial or mechanical time, most humans cycle about 27 hours. Apparently, like many other animals, humans are sensitive to the effects of bright lights, which reset their biological clocks back to 24 hours each day.

This constant resetting has important repercussions for the shiftworker, as most individuals prefer to add time to the day rather than to subtract time. This phenomena is frequently compared to east-west jet travel. As an individual flies westward, time is "added," while flying eastward "subtracts" time. Because there is a natural resetting in a slight eastward directions by one hour per day (to a 24-hour cycle) humans experience more biological disturbances when they fly eastward than when they go west. North-south movement does not affect the biological clock.

This circadian characteristic is critical in the workplace as the majority of shiftwork schedules were developed before the current understanding of the biological clock. Hence, most eight-hour shiftwork schedules rotate in an eastward or backwards direction such as day shift to night shift to evening shift as opposed to a clockwise rotational pattern. The counterclockwise pattern probably developed because it provides a slightly longer period of continuous time off. This time off is known as the "long change" and is very important to most industrial workers. A forward-rotating, eight-hour schedule has a shorter long change at the end of the full cycle rotation. Although the forward rotation pattern appears to make more medical sense, there are relatively little data that demonstrate a true physiological effect. The lack of convincing data may be due to the overwhelming effects of any eight-hour movement in either direction.

The human internal biological clock has only a limited ability to adapt to time shifts. Usually, shifts of only one to three hours per day are tolerated without some noticeable effect. However, the equivalent of one time zone shift that is produced by moving from one eight-hour shift to another (e.g., day or evening to night shift) overwhelms the body's ability to adapt, regardless of shift change directions. Despite the lack of published data demonstrating physiological improvement in forward rotation (west) versus backward (east) rotation, studies show that many shiftworkers prefer a forward rotational pattern.

Shift Rescheduling Projects

Worker surveys frequently reveal that most employees are dissatisfied with their shiftwork schedules. In response, companies have tried to modify, enhance, or totally redesign rotational schedules. Often these efforts are initiated by the workers themselves and are frequently based on experience gathered either from other nearby facilities or from newspaper articles reporting the latest research in circadian rhythms. To their frequent dismay, most workers and management discover that reevaluating and redesigning an operational shift schedule is neither quick nor easy.

Theoretically, the ideal shift schedule would be permanent shifts that did not rotate across the various hours of the 24-hour cycle. In fact, some facilities do have so-called permanent or dedicated shifts, for example, a worker is always assigned to the same shift—day, evening, or night—based on an eight-hour rotational pattern. Although this situation is sometimes successful, it also frequently fails as it is difficult to attract, train, and maintain a dedicated night crew. In addition, unless the night crew continues to maintain their same sleep/wake cycle on their days off, the physiological benefit of a permanent shift is quickly lost as soon as the workers revert back to a daytime schedule. It is critical to realize that the act of rapidly changing sleep-work-leisure patterns produces the physiological and psychological disruption associated with either shiftwork or long-distance east/west jet travel.

Despite the difficulties associated with shift schedule redesign, several guiding principles can be used:

- All schedules are a compromise between three driving forces—medical considerations, work preferences, and operational necessities.
- Schedule redesign is an interactive process among different groups that will never satisfy all parties.
- Worker preference surveys are important and should be conducted in a scientific, confidential manner.
- Detailed industrial hygiene and ergonomics analyses must be performed of all chemical and physical hazards if extended-hour, such as 10- to 12-hour, schedules are considered. This is because permissible exposure limits are time dependent. Any new schedule must be considered temporary and reevaluated after an appropriate trial period of at least three to six months.

Shift redesign projects can be a win-win proposition for both workers and management; however, both groups must recognize, define, and discuss certain so-called "boundary conditions" that may not be easily established.

Examples of boundary conditions include management's desire to control costs, the potential for workers to be overexposed to noise or chemical hazards due to extended-hour schedules, a decrease in time off, changes in contractually mandated work rules, and so on.

One important issue that requires in-depth evaluation is the decision to consider extended work schedules. The typical shift schedule is an eight-hour, four-crew configuration that rotates, either forward or backwards, through a 28-day calendar and averages approximately 42 hours of scheduled work per week. In this configuration, crews may spend five to seven "days" on a steady shift before rotation to the next scheduled time period. One variation of this pattern that is more common in Europe than in North America is the rapid rotation through the scheduled shifts during a workweek. For example, in a rapidly rotating eight-hour schedule, a crew could be scheduled for two day shifts, two night shifts, two evening shifts.

The purported advantages of rapid rotation are based on the slow adaptability of the biological clock. That is, supposedly it is easier for workers to keep a daytime internal clock rather than to adapt to constantly changing hours. Unfortunately, there is little or no evidence that the underlying premise of this schedule is correct. Given the uncertain benefit of one eight-hour schedule versus another, many workers feel that extended-hour schedules offer the best opportunity to improve their shiftwork experience.

Schedules of 10 and 12 hours are extremely popular in certain industrial sectors, such as refineries and chemical plants. These schedules maximize the amount of time off while still controlling salary costs. However, there are several disadvantages of these schedules:

- Some jobs are too physically or psychologically demanding to be done for 12 hours.
- Workers may have trouble staying alert and vigilant for extended periods.
- Routine overtime is difficult or impossible to schedule.
- Noise and chemical exposure standards must be modified to reflect the extended exposure period.
- Pay periods including holiday and overtime pay may have to be recalculated if costs are to be controlled.
- Older workers may not be able to handle extended hours.

Despite these disadvantages, none of which is insurmountable, extended work schedules do have strong support in the workplace because of the time-off issue. Instead of working 21 out of 28 days on a routine eight-hour schedule, a typical 12-hour pattern will produce an additional four days off. This time off is a powerful incentive because most shiftworkers want to spend more time at home and with their families. Before an operating facility considers using an extended-hour schedule it should conduct a careful review of health, safety, medical, and ergonomic issues.

Medical Issues

The medical effects of shiftwork have been well described in the occupational medicine and human factors literature. The National Safety Council's *Occupational Health & Safety* 2nd edition, presents some of the issues surrounding the evaluation of medical problems of shiftwork. Some workers, regardless of age, are unable to tolerate shiftwork and drop out of the rotation within two to three months. Other workers initially tolerate shiftwork well but slowly experience increasing problems after 10 to 20 years on shift. It appears that as workers age, particularly as they enter their 50s, their ability to tolerate shiftwork declines. Both groups show a cluster of problems that has been called shift maladaption syndrome (SMS).

Shift maladaption syndrome consists of a variety of medical and psychological complaints such as gastrointestinal problems (ulcers, reflux, chronic indigestion/constipation); sleep disorders (insomnia, fragmented and/or poor quality sleep); and psychological problems such as extreme mood swings and/or chronic irritability. There is no definitive test for SMS; instead, it is diagnosed by eliminating other alternative explanations and by removing the worker from the shiftwork for a trial period. Unfortunately, this latter strategy is fraught with potential legal and human resources issues because of workers' compensation and disability considerations. Nevertheless, certain medical problems, such as insulin dependent diabetes, seizure disorders, and bipolar disease generally disqualify an individual for shiftwork and should be carefully evaluated by the medical team.

Overall, the health and safety professional is confronted with a variety of new and rapidly evolving workplace environments. From "high-tech" clean rooms to problems of shiftwork, companies need a multidisciplinary team to respond to these challenges and to develop innovative solutions.

TOXICITY

A toxic effect is any reversible or irreversible noxious effect on the body—any chemically induced tumor or any mutagenic or teratogenic effect or death—as a result of contact with a substance via the respiratory tract, skin, eye, mouth, or any other route. Toxic effects are undesirable disturbances of physiological function caused by an overexposure to chemical or physical agents. Toxic effects can also arise as side effects in response to medication and vaccines. Toxicity is the capacity of a chemical to harm or injure a living organism by other than mechanical means. Toxicity entails a definite dimension (quantity or amount); the toxicity of a chemical depends on the degree of exposure.

The responsibility of the industrial hygienist is to define how much is too much and to prescribe precautionary measures and limitations so that normal, recommended use does not result in the absorption of too much of a particular material. From a toxicological viewpoint, the industrial hygienist must consider all types of exposure and the subsequent effects on the living organism.

Toxicity Versus Hazard

A distinction must be made between toxicity and hazard. Toxicologists generally consider toxicity as the ability of a substance to produce an unwanted effect when the chemical has reached a sufficient concentration at a certain site in the body; hazard is regarded as the probability that this concentration in the body will occur. Many factors contribute to determining the degree of hazard—route of entry, dosage, physiological state, environmental variables, and other factors. Assessing a hazard involves estimating the probability that a substance will cause harm. Toxicity, along with the chemical and physical properties of a substance, determines the level of hazard. Two liquids can possess the same degree of toxicity but present different degrees of hazard.

A chemical stimulus can be considered to have produced a toxic effect when it satisfies the following criteria:

- An observable or measurable physiological deviation has been produced in any organ or organ system. The change can be anatomic in character and may accelerate or inhibit a normal physiological process, or the deviation can be a specific biochemical change.
- The observed change can be duplicated from animal to animal even though the dose-effect relationships vary.
- The stimulus has changed normal physiological processes in such a way that a protective mechanism is impaired in its defense against other adverse stimuli.
- The effect does not occur without a stimulus or occurs so infrequently that it indicates a generalized or nonspecific response. When high degrees of susceptibility are noted, equally significant degrees of resistance should be apparent.
- The observation must be noted and must be reproducible by other investigators.
- The physiological change reduces the efficiency of an organ or function and impairs physiological reserve in such a way as to interfere with the ability to resist or adapt to other normal stimuli, either permanently or temporarily.

Entry into the Body

For an adverse effect to occur, the toxic substance must first reach the organ or bodily site where it can cause damage. Common routes of entry are inhalation, skin absorption, ingestion, and injection. Depending on the substance and its specific properties, however, entry and absorption can occur by more than one route, such as inhaling a solvent that can also penetrate the skin. Where absorption into the bloodstream occurs, a substance may elicit general effects, or, most likely, the critical injury will be localized in specific tissues or organs.

Inhalation

For industrial exposures to chemicals, the most important route of entry is usually inhalation. Nearly all materials that are airborne can be inhaled.

The respiratory system is composed of two main areas: the upper respiratory tract airways (the nose, throat, trachea, and major bronchial tubes leading to the lobes of the lungs) and the alveoli, where the actual transfer of gases across thin cell walls takes place. Only particles smaller than about 5 μm in diameter are likely to enter the alveolar sac.

The total amount of a toxic compound absorbed via the respiratory pathways depends on its concentration in the air, the duration of exposure, and the pulmonary ventilation volumes, which increase with higher work loads. If the toxic substance is present in the form of an aerosol, deposition and absorption occur in the respiratory tract. For more details, see Chapter 2, The Lungs, in *Fundamentals of Industrial Hygiene,* 4th edition.

Gases and vapors of low water solubility but high fat solubility pass through the alveolar lining into the bloodstream and are distributed to organ sites for which they have special affinity. During inhalation exposure at a uniform level, the absorption of the compound into the blood reaches an equilibrium with metabolism and elimination.

Skin Absorption

An important route of entry is absorption through either intact or abraded skin. Contact of a substance with skin results in four possible actions: the skin can act as an effective barrier, the substance can react with the skin and cause local irritation or tissue destruction, the substance can produce skin sensitization, or the substance can penetrate to the blood vessels under the skin and enter the bloodstream.

The cutaneous absorption rate of some organic compounds rises when temperature or perspiration increases. Therefore, absorption can be higher in warm climates or seasons. The absorption of liquid organic compounds may follow surface contamination of the skin or clothes; for other compounds, it may directly follow the vapor phase, in which case the rate of absorption is roughly proportional to the air concentration of the vapors. The process may be a combination of deposition of the substances on the skin surface followed by absorption through the skin.

Injection

A material can be injected into some part of the body. This can be done directly into the bloodstream, the peritoneal cavity, or the pleural cavity. The material can also be injected into the skin, muscle, or any other place a needle can be inserted. The effects produced vary with the location of administration. In industrial settings, injection is an infrequent route of worker chemical exposure.

Ingestion

Anything swallowed moves into the intestine and can be absorbed into the bloodstream and thereafter prove toxic. The problem of ingesting chemicals is not widespread in industry: most workers do not deliberately swallow materials they handle.

Workers can ingest toxic materials as a result of eating in contaminated work areas; contaminated fingers and hands can lead to accidental oral intake when a worker eats or smokes on the job. They can also ingest materials when contaminants deposited in the respiratory tract are carried out to the throat by the action of the ciliated lining of the respiratory tract. These contaminants are then swallowed and significant absorption of the material may occur through the gastrointestinal tract.

INDUSTRIAL HYGIENE CONSULTING SERVICES

The ideal circumstance for most companies is to have a full-time industrial hygienist on staff. However, because of economic factors, and the lack of fully qualified personnel, industrial hygienists can be employed in a consulting capacity. There are drawbacks to this approach. The consultant industrial hygienist is ordinarily called in only when problems arise. The person may not be as familiar with the facility and company personnel as a full-time industrial hygienist would be. The extent of this problem can be lessened by scheduling regular industrial hygiene consulting visits to discuss policy issues and inspect the facility during normal conditions. Some of these routine visits should coincide with visits of the medical consultant. Many companies with industrial hygiene consulting services also have medical consulting services.

Several sources are available to assist in the quest for a competent and qualified industrial hygiene consultant. First is the semiannual listing of industrial hygiene consultants published by the AIHA. The AIHA is happy to provide this list to inquirers. Presence of a name on the list does not guarantee competence, however. For an industrial hygiene consultant to be considered, the minimum requirement should be certification (i.e., designation as CIH) and familiarity with the industry or other occupational setting of interest. As with other professional services, personal recommendations from satisfied users of consulting services are often the best source for information—enabling a differentiation between otherwise apparently equivalently qualified consultants.

In addition to the qualifications of the person being considered as a consultant, resources of his or her firm can also be important. When quick turnaround of analytical results from air monitoring or other workplace monitoring is required, the availability of an in-house lab can be important. In any case, the laboratory considered for use should be accredited by the AIHA. (This accreditation should be required regardless of whether the analysis required is one for which specific accreditation is offered. The AIHA accreditation includes evaluation of such general areas as quality control and record keeping, in addition to performance on specific analytes.)

In some cases, trade associations have experience with consulting firms and can recommend those that are familiar with the industry of interest. Formal consulting agreements with the industrial hygiene consultant should include adequate time to discuss problems of mutual interest with the medical director and engineering staff.

CERTIFICATION AND LICENSURE

Certification of industrial hygienists began in 1960, when the AIHA and ACGIH established the American Board of Industrial Hygiene to set up certification requirements. Figure 11-1 shows the industrial hygiene code of ethics. Three classes of certification are currently recognized: Diplomates (permitted to use the designation Certified Industrial Hygienist—CIH), Industrial Hygienists in Training (IHIT), and Occupational Health and Safety Technologists (OHST). Certification is currently offered in comprehensive practice, and in the specialty area of chemical aspects—focusing on analytical chemistry. Beginning in 1993, specialized competence of diplomates in the two subspecialties of indoor air quality and hazardous materials management and remediation were formally recognized by the board. Before 1992, certification in the specialized fields of acoustical, air pollution engineering, radiological, and toxicological aspects of industrial hygiene was offered, and those certifications remain valid, although no new certificates will be offered. Certification is granted when education in the sciences equivalent to college graduation, successful performance in the field for a minimum of five years, recommendation by practicing industrial hygienists, and completion of two written (day-long) examinations are shown. The first of these examinations (the "core" examination) can be taken after one year of work in the field, and successful performance allows use of the IHIT designation. The second examination (covering comprehensive practice or a specialty) cannot be taken until the CIH candidate has completed a total of at least five years of acceptable experience. Maintaining certification requires documenting continued activity in certain defined professional activities and continuing education. The subspecialty recognitions are granted after successful completion of a written one-half-day examination. Diplomates are eligible for membership in the American Academy of Industrial Hygiene. The Board publishes a directory each year, which lists all Diplomates, IHITs, and OHSTs.

Although many industrial hygienists are becoming interested in seeking some form of government licensure, no federal licensure requirements were established by the end of 1996. However, the CIH designation is often required for industrial hygiene practice on projects or jobs supported by government funds. The CIH is increasingly recognized as a minimum requirement to assure that industrial hygiene work is being performed according to professional standards, especially when litigation is anticipated or feared. In 1993, Illinois enacted voluntary licensure requirements, and as of late 1996, six states enacted title protection that prohibits use of the CIH designation without board certification (AL, CA, CT, IN, NV, TN).

PROFESSIONAL ORGANIZATIONS

The American Industrial Hygiene Association (AIHA) is the predominant U.S. industrial hygiene organization. The AIHA was formed in 1939 by a small group of industrial hygienists, and has grown in the intervening years to more than 10,000 members. The only restrictions on membership are those of educational qualifications and practice of industrial hygiene—a full member must have an appropriate degree (usually in one of the physical or biological sciences), and have practiced industrial hygiene full-time for three years. The AIHA provides many services to the profession, including the accreditation of laboratories offering analytical services. There are local (regional) sections of the AIHA in all areas of the United States, and many foreign occupational health professionals are members, as well.

The American Conference of Governmental Industrial Hygienists (ACGIH) limits full membership to health and safety professionals practicing in government agencies and educational institutions, but offers affiliate membership to others in the health and safety field. Two services offered by the ACGIH are of particular value to the profession. One of these is the manual, *Industrial Ventilation—A Manual of Recommended Practice,* by the Industrial Ventilation Committee of ACGIH. This manual gives guidance in design of ventilation systems for the control of airborne health hazards, and is a standard reference source. The other is the annual publication of the TLV-BEI listings. The ACGIH, through its Threshold Limit Values Committee, publishes its listing of *Threshold Limit Values (TLVs) and Biological Exposure Indices (BEIs)* annually. The TLVs and BEIs are widely used as measures of allowable exposure limits; in fact, the 1968 TLVs were adopted into law as the Permissible Exposure Limits in the OSHAct. Both of these documents are discussed elsewhere in this chapter.

Many industrial hygienists are also members of organizations representing major and minor industries. The American Petroleum Institute, the Chemical Manufacturers Association, and the National Agricultural Chemical Association are examples of industry organizations with active committees on various aspects of occupational health on which industrial hygienists can serve.

Some industrial hygienists also maintain membership in organizations representing the disciplines of their original training. The American Chemical Society and the American Institute of Chemical Engineers are examples of societies claiming many industrial hygienists among their members. Industrial hygienists also belong to organizations with peripheral interests in industrial hygiene, or representing fields within which industrial hygienists have responsibilities. Such organizations as the Society for Epidemiological Research, the Air and Waste Management Association, the American Management Association, the American Society of Safety Engineers, the American Public Health Association, and the Genetic and Environmental Toxicology Association are examples. Finally, general

CODE OF ETHICS FOR THE PRACTICE OF INDUSTRIAL HYGIENE

Objective

These canons provide standards of ethical conduct for Industrial Hygienists as they practice their profession and exercise their primary mission: to protect the health and well-being of working people and the public from chemical, microbiological, and physical health hazards present at, or emanating from, the workplace.

Canons of Ethical Conduct

CANON 1

Industrial Hygienists shall practice their profession following recognized scientific principles with the realization that the lives, health, and well-being of people may depend upon their professional judgment and that they are obligated to protect the health and well-being of people.

INTERPRETIVE GUIDELINES

- Industrial Hygienists should base their professional opinions, judgments, interpretations of findings, and recommendations upon recognized scientific principles and practices which preserve and protect the health and well-being of people.
- Industrial Hygienists shall not distort, alter, or hide facts in rendering professional opinions or recommendations.
- Industrial Hygienists shall not knowingly make statements that misrepresent or omit facts.

CANON 2

Industrial Hygienists shall counsel affected parties factually regarding potential health risks and precautions necessary to avoid adverse health effects.

INTERPRETIVE GUIDELINES

- Industrial Hygienists should obtain information regarding potential health risks from reliable sources.
- Industrial Hygienists should review the pertinent, readily available information to factually inform the affected parties.
- Industrial Hygienists should initiate appropriate measures to see that the health risks are effectively communicated to the affected parties.
- Parties may include management, clients, employees, contractor employees, or others, dependent on circumstances at the time.

CANON 3

Industrial Hygienists shall keep confidential personal and business information obtained during the exercise of industrial hygiene activities, except when required by law or overriding health and safety considerations.

INTERPRETIVE GUIDELINES

- Industrial Hygienists should report and communicate information which is necessary to protect the health and safety of workers and the community.
- If their professional judgment is overruled under circumstances where the health and lives of people are endangered, Industrial Hygienists shall notify their employer, client, or other such authority, as may be appropriate.
- Industrial Hygienists should release confidential personal or business information only with the information owners express authorization, except when there is a duty to disclose information as required by law or regulation.

CANON 4

- Industrial Hygienists shall avoid circumstances where a compromise of professional judgment or conflict of interest may arise.

INTERPRETIVE GUIDELINES

- Industrial Hygienists should promptly disclose known or potential conflicts of interest to parties that may be affected.
- Industrial Hygienists shall not solicit or accept financial or other valuable consideration from any party, directly or indirectly, which is intended to influence professional judgment.
- Industrial Hygienists shall not offer any substantial gift, or other valuable consideration, in order to secure work.
- Industrial Hygienists should advise their clients or employer when they initially believe a project to improve industrial hygiene conditions will not be successful.
- Industrial Hygienists should not accept work that negatively impacts the ability to fulfill existing commitments.
- In the event that this Code of Ethics appears to conflict with another professional code to which Industrial Hygienists are bound, they will resolve the conflict in the manner that protects the health of affected parties.

CANON 5

Industrial Hygienists shall perform services only in the areas of their competence.

INTERPRETIVE GUIDELINES

- Industrial Hygienists should undertake to perform services only when qualified by education, training, or experience in the specific technical fields involved, unless sufficient assistance is provided by qualified associates, consultants, or employees.
- Industrial Hygienists shall obtain appropriate certifications, registrations, and/or licenses as required by federal, state, and/or local regulatory agencies prior to providing industrial hygiene services, where such credentials are required.

CANON 6

Industrial Hygienists shall act responsibly to uphold the integrity of the profession.

INTERPRETIVE GUIDELINES

- Industrial Hygienists shall avoid conduct or practice which is likely to discredit the profession or deceive the public.
- Industrial Hygienists shall not permit use of their name or firm name by any person or firm which they have reason to believe is engaging in fraudulent or dishonest industrial hygiene practices.
- Industrial Hygienists shall not use statements in advertising their expertise or services containing a material misrepresentation of fact or omitting a material fact necessary to keep statements from being misleading.
- Industrial Hygienists shall not knowingly permit their employees, employers, or others to misrepresent the individuals' professional background, expertise, or services which are misrepresentations of fact.
- Industrial Hygienists shall not misrepresent their professional education, experience, or credentials.

Figure 11-1. The joint Code of Ethics for the Practice of Industrial Hygiene endorsed by the AIHA, the ABIH, the AAIH, and the ACGIHG. (From *ACGIH Today!* 3(1), January 1995.)

scientific societies (such as the American Association for the Advancement of Science) also claim their share of industrial hygienists as members.

SUMMARY

- Industrial hygiene is the science and art devoted to the recognition, evaluation, and control of environmental factors or stresses arising in or from the workplace. These stresses may cause sickness, injury, or significant discomfort and inefficiency among workers or citizens in the community.
- The industrial hygienist works with management, medical, safety, and engineering personnel to help eliminate or guard against environmental pollution and other conditions that can endanger human health.
- The three definitive elements of industrial hygiene are recognizing, evaluating, and controlling occupational hazards. These include chemical, physical (excessive noise, radiation, etc.), ergonomic, and biologic hazards.
- Altered environments, such as clean rooms and shift work, also pose threats to human health. Hazards associated with clean rooms include arid environmental conditions, air pollution, health problems such as dermatitis, and psychological isolation.
- Shift work causes disruption in the human circadian rhythms, which causes physical and psychological problems for workers. Older workers frequently experience shift maladaption syndrome or SMS. Rescheduling shift work can alleviate some of these hazards.
- Toxic effects are undesirable disturbances of physiological function caused by an overexposure to chemical or physical agents. *Toxicity* is the capacity of a chemical to harm or injure a living organism by other than mechanical means. *Hazard* refers to the probability that a concentration of a chemical in the body will occur. Toxic substances can enter the body through inhalation, skin absorption, injection, or ingestion.
- Industrial hygienists must define how much exposure is too much and prescribe precautionary measures and limitations to protect workers' health. These professionals can be either full-time at a company or serve part-time as consultants.

REFERENCES

American Conference of Governmental Industrial Hygienists (ACGIH). *Industrial Ventilation: A Manual of Recommended Practice.* Cincinnati: ACGIH, published biannually, even-numbered years.

ACGIH. *Threshold Limit Values for Chemical Substances and Physical Agents and Biological Exposure Indices.* Cincinnati: ACGIH, issued annually.

LaDou J (ed.). *Occupational Health & Safety,* 2nd ed. Itasca, IL: National Safety Council, 1994.

Plog BA (ed.). *Fundamentals of Industrial Hygiene,* 4th ed. Itasca, IL: National Safety Council, 1996.

REVIEW QUESTIONS

1. Define industrial hygiene.
2. Who does the industrial hygienist work with to control environmental stresses or occupational health hazards in the workplace?
   a. Management
   b. Medical personnel
   c. Safety personnel
   d. Engineering personnel
   e. All of the above
3. List five control procedures that are used to reduce or eliminate exposure.
   a.
   b.
   c.
   d.
   e.
4. List five factors of an effective industrial hygiene program.
   a.
   b.
   c.
   d.
   e.
5. The industrial hygienist is responsible for monitoring what types of environmental factors or stresses that can cause sickness, impaired health, or significant discomfort?
   a.
   b.
   c.
   d.
6. Which department is the most important to the industrial hygienist?
   a. Engineering department
   b. Environmental department
   c. Safety department
   d. Medical department
   e. c and d
7. Who, at minimum, should be members of the occupational health and safety team?
   a.
   b.
   c.
   d.
   e.
   f.

8. What is the principle tool used by the industrial hygienist to recognize potential occupational health hazards?
   a. Communication with the environmental department
   b. Observations of the workplace
   c. Review of incident and accident records
   d. All of the above

9. List the broad types of health problems that are associated with "clean rooms"?
   a.
   b.
   c.
   d.

10. What percentage of the U.S. labor force is engaged in some type of industrial or service activity that requires 24-hour work cycles?
    a. 4%
    b. 12%
    c. 25%
    d. 30%

11. What are the medical and psychological ailments that stem from shift maladaptation syndrome?
    a.
    b.
    c.

12. Define toxicity.

13. Name the four ways that a toxic substance can enter the body.
    a.
    b.
    c.
    d.

12

Environmental Management

Management of environmental services has increasingly become a major focus for health and safety professionals. In many companies, there are separate environmental departments; however, as companies streamline and reengineer their professional support services staff, there has been a marked trend towards consolidation of the health, safety, and environmental staffs. Therefore, the traditional safety and health professional has increasingly found that knowledge of environmental affairs is critical. For companies involved in international business, the ISO 14000 standards potentially represent a revolution in corporate environmental management; therefore, the safety and health professional should become familiar with this new approach to environmental affairs. This chapter covers the following topics:

- a survey of the major issues that surround environmental management
- a brief synopsis of the major U.S. environmental regulations
- principles of proactive environmental management, waste minimization, and auditing
- a detailed discussion of the newly adopted International Standards Organization (ISO) 14000 series of standards covering environmental management
- a discussion of the global effects of environmental pollution and issues.

ENVIRONMENTAL REGULATIONS

The U.S. environmental regulatory system has always been characterized by both constant change and complexity. The mid-1990s have clearly illustrated this observation since major renewal legislation has still not been passed at the federal level. The history and legislative framework of both United States and international environmental regulations is presented in the National Safety Council's *Accident Prevention Manual for Business & Industry: Environmental Management*. The safety/health/environmental professional should always consult the most current source of legal/regulatory information since environmental laws and regulations are in a state of constant flux, particularly at the state and local levels. In the United States, the major environmental regulations cover the release, treatment, storage, and disposal of potentially hazardous materials into air, water (surface and ground water), and soil. Some of the most important laws are:

- **Comprehensive Environmental Response, Compensation, and Liability Act (CERCLA).** This law imposes liability on owners or operators of a facility from which there is a release of hazardous substances into the environment. Transfer of a property may not relieve an owner or operator of liability; in addition, the current owner may be held strictly liable. There are complex third party liability issues and provisions for cleanup cost recovery. The CERCLA or Superfund law has been extremely controversial

due to its perceived cost/benefit ratio, i.e., the costs of compliance and litigation have been staggering versus relatively modest cleanup of sites. As of this writing, CERCLA reauthorization is pending in Congress and the final outcome is uncertain since numerous changes have been proposed to the law. The last reauthorization of this law was the 1985 Superfund Amendments and Reauthorization Act (SARA) of 1986. These amendments covered emergency planning, community right-to-know, and toxic release reporting, among other notification requirements.

- **Resource Conservation and Recovery Act (RCRA).** This law regulates the generation, treatment, storage, and disposal of hazardous waste. RCRA requires an owner/operator of a facility to undertake corrective action to clean up a facility used for the treatment, storage, or disposal of hazardous waste. There is also a complex permit program with which the owner/operator must comply. RCRA is also undergoing revision, particularly the risk-based standards covering cleanup of what are termed solid waste management units (SWMUs). RCRA has increasingly become important since it broadly applies to any operating facility that is a generator of hazardous substances.

- **Underground Storage Tank Regulations (under RCRA and state control).** This law imposes operating, reporting, financial assurance, and potential cleanup obligations on persons and companies owning and operating underground petroleum storage tanks (USTs). The rules and regulations covering USTs are quite detailed and prescriptive. In addition, many of the states have their own regulations and enforcement policies. Underground petroleum storage tanks and petroleum releases have recently been the subject of important studies and recommended standards by the Lawrence Livermore Laboratory (California) and the American Society of Testing and Materials (ASTM). Both the Livermore and the ASTM work evaluated and proposed efficient and cost-effective approaches for the evaluation and risk-based cleanup of petroleum releases from USTs. Since USTs are so common and accidental releases are frequent, the safety/health/environmental professional should carefully review these two documents.

- **Clean Air Act (CAA) of 1990.** This act is a complex, multifaceted statute that is designed to regulate air emissions from stationary and mobile sources. The act is composed of eleven different "Titles" covering topics such as national ambient air quality standards (Title I), hazardous air pollutants (Title III), acid deposition control (Title IV), operating permits (Title V), and ozone protection (Title VI), among others.

- **Clean Water Act (CWA).** This act prohibits the discharges of pollutants from point sources and storm water into navigable waters of the United States without a permit. The act imposes liability on the person who is responsible for the operation and/or equipment that results in a discharge. The basic thrust of the act is to force compliance with both uniform technology-based effluent limitations, regardless of the quality of the receiving waters, and with more stringent limitations necessary to meet state established water quality standards.

- **Safe Drinking Water Act (SDWA).** This act imposes federal drinking water standards on virtually all public water systems. The act requires the establishment of drinking water standards for maximum contaminant levels (MCLs) for organic and inorganic chemicals, turbidity, coliform bacteria, and various measures of radioactivity.
- **Toxic Substances Control Act (TSCA).** This act governs the manufacture and use of chemical products. The importation and exportation are regulated under the act; in addition, there are specific regulations regarding the use, management, storage, and disposal of polychlorinated biphenyl (PCB) materials.
- **Oil Pollution Act of 1990.** This act imposes strict liability on responsible parties for removal costs and damages resulting from discharges of oil into navigable waters of the United States. An owner/operator of an onshore facility that resulted in a discharge would be considered a responsible party.

One by-product of these environmental regulations is the increasing concern that redevelopment of contaminated land is prohibitively expensive. The EPA has begun to recognize that the real or imagined fear of being caught in the spider's web of regulation, particularly the CERCLA process, has become a major barrier to redevelopment activities. In response to this problem, the concept of "brownfields" redevelopment has emerged. A brownfield site or property refers to contaminated land that can be redeveloped as a commercial or industrial site. The potential for residential usage is not eliminated; however, cleanup standards would be much stricter. The term "brownfields" is a word play off the concept of "greenfields," which typically refers to unused open space or farmland. Many companies have large portfolios of property that may be considered as contaminated due to historic activities. The opportunity for focused cleanup and reuse has attracted substantial interest across the United States. The brownfields movement represents one of the major reforms to dissatisfaction with the CERCLA (Superfund) process and promises to be a significant activity for environmental managers.

The increasing liability and cost of poor or fragmentary environmental management illustrates why most environmental professionals feel that an integrated approach to environmental affairs is required. The development of a proactive and comprehensive program for environmental management has moved from a discretionary activity to a virtual requirement if the company is going to prosper and thrive in the 21st century. The next section presents some of the approaches and opportunities that are available for development of proactive environmental management.

PROACTIVE ENVIRONMENTAL MANAGEMENT

With the advent of international, industry-specific, and country-specific standards for managing environmental impacts, the stage has been set for businesses to use these standards for their competitive advantage. Examples of recent standards include product recall requirements in Germany, recycling regulations in the Far East and Europe, and the general use of "green" marketing by a number of consumer product companies. For many multinational corporations, the question remains: How can competitive advantages be realized by proactively identifying and managing environmental, health, and safety (EH&S) issues?

A New Approach

The framework for corporate environmental management has traditionally been "command and control" through regulatory compliance. That began to change in the mid-1980s when the chemical industry enacted its Responsible Care program in Canada and subsequently in the United States. Responsible Care was unique in that it advocated voluntary efforts to address the nonregulated environmental aspects of doing business. Since then, similar proactive approaches have been adopted by other industry groups around the world. The benefits extend beyond actual compliance; they add value by improving a company's corporate image, enabling management of suppliers' risk and cultivating the industry-wide credibility needed for expansion beyond national boundaries.

The next major step toward the strategic management of EH&S factors will be the issuance in mid-to-late 1996 of the International Organization for Standardization (ISO) 14000 series of standards. These standards include ISO 14001, which encourages the developmental management systems (EMS). An EMS does not prescribe specific criteria; rather, it provides a process for identifying and managing the environmental aspects of producing, delivering, and using a company's products or services. It calls for setting goals, mission, and policy, defining responsibilities; and developing procedures for valuative and corrective action.

Although ISO 14000 standards are voluntary and do not require certification or registration, they may become a prerequisite of doing business with many governments and companies. Therefore, their adoption may become a *de facto* requirement for remaining competitive. Furthermore, the implementation of environmental management systems may also yield product and process improvements that become competitive advantages in their own right. A detailed discussion of ISO 14000 is presented later in this chapter.

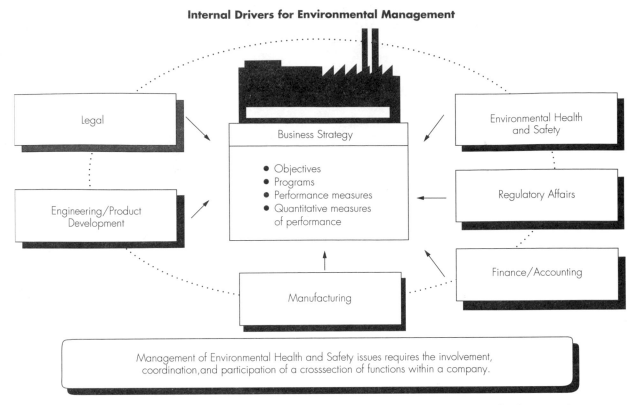

Figure 12-1. Environmental health and safety issues impact overall business operations. Consequently, they must be integrated into the strategic business planning process. (Courtesy Dames & Moore.)

Along with Responsible Care and the ISO 14000 series, a number of other standards act as drivers for companies to take action. These include national standards such as Europe's Eco-Audit Management Scheme (EMAS) and the British Standard BS-7750. While the focus is on environmental aspects, enterprising companies are developing management systems to address a variety of environmental, health, and safety issues. The challenges associated with developing these programs include:

- tracking strategic issues, such as emerging global trends, new product development, and competitors' EH&S initiatives
- elevating EH&S functions to contribute to process and business management decisions
- establishing appropriate responsibilities and accountabilities at all levels of facility and corporate management

Figure 12-1 illustrates the connections between many of these issues and an overall environmental management.

Strategies for Success

Competitive advantages can be realized through the alignment of critical EH&S factors with a company's *value chain:* the process by which products (or services) are conceived, produced, marketed, distributed, used, and recycled or disposed. Examples of implementing an EH&S program in the context of a value chain include:

- Research and Development: incorporate Design For Environment (DFE) or Design for Disassembly practices for new products or product modifications
- Procurement: identify those suppliers that follow DFE and ISO 14000 practices and encourage other suppliers to adopt them
- Production and Manufacturing: optimize processes to minimize the handling of hazardous materials and reduce consumption of energy and resources
- Marketing and Sales: educate customers and end-users as to the effective use and disposal/recycling of products, and identify "green" differentiators to gain market share or margin improvements
- Distribution: design for less packaging to reduce transport costs

The steps described above are examples of places in the organization where opportunities can support strategic business goals (Figure 12-2). The value chain approach aids managers in understanding how EH&S factors affect costs and anticipated revenues across organizational and product lines. It can also help filter out nonstrategic issues so that efforts are directed toward issues that affect goals for growth, diversification, margin improvement, or increased customer satisfaction. The ultimate aim is to move away from command and control and toward the proactive use of EH&S issues in support of broad corporate objectives. Many companies are using the ISO 14000 process as a way to facilitate this evolutionary change; therefore, a detailed

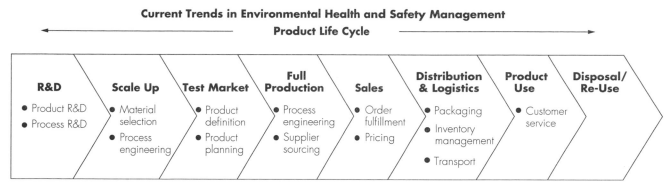

Figure 12-2. With the focus broadening beyond just compliance, environmental health and safety issues and strategy must be managed in the context of the Product Life Cycle. (Courtesy Dames & Moore.)

analysis of the ISO 14000 process is presented in the next section.

MANAGING FOR A HEALTHY ENVIRONMENT

Industry experts Christopher Hunt and Ellen Auster describe five distinct stages of environmental management program development. Their survey of corporations revealed departments that range from the "beginner" (stage one) through the "proactivist" (stage five). Small companies or ones that face minimal environmental risks do not require a "stage five, proactivist" environmental management program. Companies that use hazardous materials in their manufacturing processes or those with a variety of processes will, however, require a more active management approach. A program can be proactive for one company, but inadequate for another. A bakery may be able to operate successfully with a few guidelines and periodic reviews by a consultant. However, large manufacturers or chemical companies often require several hundred staff members who are environmental professionals.

Hunt and Auster describe "stage one, beginner" environmental programs as those of either older companies established before the environmental acts were passed or of firms, like banks or real estate developers, that do not deal directly with toxic materials, but that may encounter related risks from them. "Beginners" often ignore environmental concerns or deal with them by adding responsibility to an existing job description, such as plant manager or senior engineer. "Beginners" do not define corporate environmental policy, nor do they consider the potential effects of failing to have such a policy.

"Stage two, the firefighter" companies are described by Hunt and Auster as having a few people who spend time on environmental concerns or a small group of professionals who help individual installations respond to problems. Given this structure, environmental professionals have no option but to work at a crisis-intervention level. They have no opportunity to consider risks from serious problems that might occur in the future. Small or medium-sized firms that are at this level engage in environmentally hazardous activities and have not considered

the potential benefits of active environmental management.

Hunt and Auster describe "stage three, the concerned citizen" companies as those whose established environmental departments are either too understaffed or not high enough in the corporate structure to make significant changes. They are often staffed by technically competent professionals, like chemists, biologists, or geologists, whose backgrounds do not prepare them to offer business, legal, and public relations expert advice for environmental management. These companies, concerned about their environmental responsibilities, have funded programs that operate without serious senior management commitment to integrate them with the operating units.

At "stage four, the pragmatist" companies, environmental department staffs have sufficient expertise, funding, and authority. Staff members review all facilities and design better ways to limit toxic releases. They look to the future, evaluating potential risks and designing solutions whenever appropriate. They train key workers in environmental protection and write policy manuals for operating personnel. Formal reporting relationships are established. Although these departments are aggressive, their authority and program funding may be limited. At this stage of development, companies have not made environmental management a top-level corporate concern.

Environmental management is a top priority for companies that have reached "stage five, the proactivist." The environmental department is staffed with assertive leaders whose commitment to the environment goes beyond meeting regulations and planning for their firm's future environmental needs. These professionals are active in industry roundtables, sharing information and solutions with competitors to ensure that their performance does not result in polluted resources, adverse publicity, and stricter regulations. They participate in local, state, and federal policymaking meetings, aggressively pursuing solutions as they help policymakers determine how best to regulate their industry.

Not every organization needs to budget for a sophisticated, proactive environmental management department. The level of commitment appropriate for each company should be determined by its inherent environmental risks and its size.

Organizing an Environmental Management Program

Establishing an environmental compliance program can be an overwhelming proposition. However, any company whose business is regulated by U.S. EPA, the Occupational Safety and Health Administration (OSHA), Department of Transportation (DOT), other government agencies, or state or local authorities cannot afford to be without an environmental management program.

Corporate culture—the size of the company, diversity of its business units, and how it interacts with headquarters—will determine if the environmental management program will be small and basic, large and sophisticated, centralized or decentralized. Whether the company is large or small, however, senior management must communicate a strong commitment to responsible environmental practices through the ranks, down to the line worker and platform loader. Unfortunately, a sincere environmental management commitment that alters old practices often occurs only after a major disaster or stringent penalty. This does not have to be the case. Top management teams can learn from other companies' misfortunes.

Deciding where to begin can be the biggest obstacle to organizing an environmental management program. The ways to comply with governmental regulations can seem as endless as the thick tomes that transmit those regulations. William Friedman outlines five basic activities essential to any successful environmental compliance program (see References).

Prevent Common Violations

Mislabeling of chemicals and hazardous waste is probably the most commonly cited violation, and one that can most easily cause an environmental accident. It is also the easiest violation to avoid. Companies should attach proper labels to all containers. In the case of hazardous waste, use a label that meets the DOT's requirements.

Regulations require storage containers to be in good condition. Leaking containers are a common source of fines. Inspect tanks daily, and check drums and similar containers weekly. Make sure aisle space is adequate to give inspectors easy access to the facilities. When waste is stored in tanks, regulations require secondary containment to prevent leaks or spills.

Record Keeping

Environmental regulations have mandated that plants alter the way they handle and dispose of chemicals and hazardous wastes. They also mandate that company personnel create and maintain extensive records that document how the firm has met handling and disposal regulations. During an inspection by a regulatory agency, the mere existence of the records, their easy accessibility, and identifiability may be all an inspector needs.

It is also important to document all environmental management decisions. For example, whenever a facility has analyzed its waste streams and found them to be non-hazardous, records of the information supporting that determination should be maintained and made readily available to an inspector. Relying on a very sketchy explanation from a compliance manager (who may or may not be employed or present at the time of the inspection) can lead to an inspection report listing possible violations. Staff must then reconstruct the information and transmit it to the regulatory agency, which may still issue a violation. Accurate, available, and complete records save companies time, money, and legal headaches.

Create a Spill-Reporting Plan

Under numerous statutes and regulations (notably the U.S. federal Superfund law), companies must notify the U.S. National Response Center immediately after release of a reportable quantity of any hazardous substances or pollutants. As regulators decrease the time allowed between an incident and the reporting time, environmental authorities are strictly enforcing existing laws. In some cases, like air pollution discharges, the permitted time lag is only minutes.

An unintentional spill or release of a toxic material usually occurs during a plant emergency when personnel are literally putting out fires or responding to explosions or other emergencies. For that reason, every facility must have written instructions for employees to follow whenever a release or spill occurs. The reporting plan should detail the plant's procedures, telephone numbers of all agencies that should be notified, and who has the authority to report spills. Employees should be aware of these instructions and where they can be found at various locations. Spill-reporting plans not only permit companies to respond immediately to emergencies, but their existence is also evidence of a firm's environmental responsibility to regulatory agencies that might question the company's ability to respond to an emergency.

Set Realistic Limits and Schedules

Environmental managers should suggest realistic compliance schedules and discharge limitations to environmental regulators. Often in a spirit of cooperation, a manager will offer overly ambitious schedules and limits that ultimately cannot be met. Do not overcommit. Once limits are set in discharge permits and orders, it is difficult to have them changed. Before such limits are written into a company's legal documents, management should consult with government regulators to set realistic dates and limits.

When reporting a release that exceeds set limits, always include the reason and make an attempt to solve the problem. Companies have reported pollutant releases above limits, received no response from the regulatory agency and, over time, have assumed that the increased rate is acceptable. The company's own reports then establish its lack of compliance, which can prompt an enforcement action by the agency or serve as evidence in a citizen suit against the company. Whenever it appears that legal requirements cannot be met in the future, the compliance manager should request a change in limits or schedules, and detail the reasons why compliance is not possible.

Motivate Employee Action

Spills and releases are almost never caused by the person who writes a company's environmental compliance program. Mistakes that become environmental problems are usually made by the platform loader, line, or storage worker. If these employees are not educated or trained in environmental management practices, they often focus on doing their jobs quickly at the expense of environmentally important details and filling out "bothersome" reports.

Although training for all employees who handle environmentally sensitive substances is essential, it is all too often insufficient. Including adherence to environmental protection policies in employees' annual reviews and holding them responsible for cited violations brings home the importance of "little details" and "bothersome" reports. It is crucial, as well, to make employees aware that in this age of strict environmental enforcement, it is less costly for a company to adhere to regulations than to defend itself against a lawsuit.

Staff Skills and Backgrounds

The key player in any successful environmental management department is the manager. Whether his or her background is in engineering, science, or law is immaterial. The complex and diverse issues that the manager must deal with call for a person with top managerial skills, good internal and external networks, rapport with senior management, and the respect of others within the organization. Because of the diversity and sensitivity of day-to-day business, the manager should have a background in environmental issues, whether it be legal or scientific.

Most of the environmental management department staff will be scientists and engineers. Depending upon the industry, a company's environmental staff will include chemists, geologists, biologists, and so on. Because many corporate environmental issues are legal, companies also need a close association with an appropriate attorney. Without such a relationship, the company may overlook key environmental data or legal facts that can plague it far into the future. An attorney can take preventive legal measures that will avoid problems which could prove costly both financially and in terms of corporate image and public relations.

Some environmental management departments have an attorney on staff. In others, the corporate general counsel's office includes an attorney who is an environmental specialist. Smaller companies may rely entirely on outside counsel. Whichever approach is taken, it is essential to establish a firm relationship with an attorney skilled in environmental issues who will review environmental data and regulations and oversee all company transactions.

Counsel must have a close relationship with the environmental manager so as to respond to concerns and questions in a timely fashion.

A strong relationship must also be established between the environmental management department and the company's public relations department or outside public relations firm. A three-way partnership among the environmental manager, public relations executive, and attorney is the best assurance that the company's environmental efforts meet corporate guidelines, various regulations, and statutes and are reported to the public in the best light possible.

WASTE MINIMIZATION

Waste minimization can produce several benefits. It can save money by reducing waste treatment and disposal costs, raw materials purchases, and other operating costs. It can reduce potential environmental liabilities. It can protect public health and worker health and safety and protect the environment. Waste minimization can be accomplished by source reduction and recycling. Of the two approaches, source reduction is usually preferable to recycling because it has a lower total effect on the environment. The RCRA requires that generators of hazardous waste have a program in place to reduce the volume and toxicity of waste generated, to the extent that such a program is economically practical.

A waste minimization program must begin with an assessment, which requires a company to evaluate all solid waste streams to determine if they can be reduced or recycled. Next, the company must calculate the economic impact of reuse or recycling.

Technical evaluation of a waste stream should begin in the production process where the waste is generated. Factors such as changes in raw materials, production process, equipment, and operating conditions could change the characteristics of a solid waste to make it a better candidate for reuse or recycling.

Economic evaluation should consider the capital and operating costs associated with any onsite collection or treatment of the waste. In addition, it should consider disposal fees, transportation costs, raw material costs, and operating and maintenance costs. Economic projections should be as realistic as possible.

In choosing whether to reuse or recycle solid wastes, a company should consider the stability of the minimization technology. If reuse is the selected alternative, evaluate the effect minor changes in waste characterization will have on the reuse process. Waste is, by nature, inconsistent in composition. Therefore, if a company finds that minor fluctuations in waste composition will adversely affect the finished product or preclude the material from reuse, it should consider other alternatives.

In evaluating a solid waste for recycling, remember that recycling is a commodity-driven business, possible only if a market exists for the recycled product. An excellent example of a solid waste that is particularly suitable for recycling is the aluminum can. It takes less energy to recycle an aluminum can than it does to produce aluminum by conversion of bauxite ore. In this industry, manufacturers have done an excellent job of establishing facilities capable of recycling aluminum cans. Therefore, there is an excellent market for aluminum cans collected for recycling.

Glass is another material suitable for recycling. Once again, it is less expensive to produce glass through recycling than to manufacture new glass. However, the market for glass recycling is not as well developed. For example, a facility may find that at times a recycler is unable to accept its material because the recycled product cannot be sold. This does not preclude the use of recycling as a disposal alternative, however. If a facility selects recycling as a minimization technique, it must be prepared to store waste when the material cannot be recycled, or to develop disposal alternatives.

GROUNDWATER CONTAMINATION

One of the most serious and costly forms of pollution is groundwater contamination. When industrial pollution seeps into the groundwater and drinking water supplies, it can, at best, be costly to correct and, at worst, present long-term chronic health risks for people who live downstream from the facility.

The strata underneath the surface of the earth contain "horizons," some of which are water bearing. These water-bearing horizons vary in flow, characteristics, and water quality. The largest horizons are referred to as aquifers. These aquifers can range anywhere from continental to regional to local in size. Many are high-yield, drinking water–quality aquifers that serve as the primary drinking water source for a large portion of the population. In evaluating groundwater contamination, a firm should identify an area's geology and hydrology at least to a depth that penetrates the uppermost aquifer.

A company should classify site geology in accordance with the Unified Soil Classification System. This is accomplished by drilling bore holes and collecting soil samples at predetermined depth intervals. In good, cohesive soil conditions, a company can sometimes collect samples from the auger flites. However, when cohesive soil conditions do not exist and there is sidewall sloughing into the bore hole, samples from the auger flites will not allow proper soil classification. To properly classify the soils in noncohesive conditions, it is necessary to use a split spoon or sample tube device to collect samples at predetermined depths. Site hydrology includes the identification of groundwater flow paths, determination of groundwater flow directions (including horizontal and vertical flows), and determination of multihorizontal interconnects.

Horizontal groundwater flow direction is usually found by installing piezometers to determine the potentiometric surface of the groundwater. With these data, a team can identify the horizontal direction of flow by constructing a piezometric surface map.

The vertical flow component is determined by placing vertically nested piezometers in closely spaced, separate bore holes. The piezometers are screened at different depths to measure vertical variations in the hydraulic head. The data are used to construct flow nets, which are graphical representations of the vertical flow components.

After characterizing the site hydrology, an investigative team must determine temporal influences that might alter the piezometric surface. These influences might include location of offsite pumping wells, tidal variations, offsite and onsite land pattern changes, and seasonal variations in groundwater recharge.

Hydraulic interconnects between horizons can affect the horizontal and vertical flow path of groundwater contamination. Hydraulic interconnects are usually detected by installing a pumping well into the lower horizon, pumping at a predetermined rate, and monitoring the potentiometric surface of the upper horizon. Obviously, if there is a drawdown of the potentiometric surface at any of the piezometers during pumping of the lower horizon, hydraulic interconnect might exist.

Groundwater assessments require a great deal of experience in data collection and interpretation to produce meaningful results. If a facility is considering conducting a groundwater assessment, it is recommended that management seek the assistance of an experienced groundwater assessment firm.

ENVIRONMENTAL AUDITS

The EPA defines an environmental audit as "an investigation into the history and current status of a particular piece of property (site)." The purpose of an environmental audit is to:

1. Identify the presence and extent of environmental contamination or hazardous materials from current or previous site activities
2. Determine the level of compliance with current standards or regulations
3. Provide a general review of environmental risks associated with the site and its operations.

Companies that handle environmentally sensitive substances should conduct internal audits on a consistent, company-wide basis and request audits by outside consultants as needed. An audit is an important review of a site's environmental risks. Its minimum goal is to determine a company's level of compliance with regulations and standards.

Internal Audits

The main purpose of an internal audit is to ensure that a facility or department is safe and that the company is in compliance with environmental standards set by various regulations and statutes. Environmental management programs should institute consistent, comprehensive internal audits. They can be either "top-down" or self-audits. Each has its advantages and disadvantages.

In a "top-down" audit, staff from a centralized environmental management department review procedures, record keeping, and so on in operating departments or facilities throughout the organization. This type of review ensures professional auditing expertise. It also enables corporate staff to make sure of company-wide compliance with corporate environmental policy. Its disadvantage is a feeling of a "police action" that can create hostile relationships with operating staff who often feel that the environmental professionals are not familiar enough with their department or facility to conduct a comprehensive study.

Self-audits enable employees who work in the department or facility to assess problems in the operations with which they are familiar. However, they may be too close to the operations to be able to conduct an objective audit or to spot operational processes that could be improved to go beyond minimal environmental release standards. Operations staff may also lack incentive to identify problem areas, believing such problems will reflect badly upon themselves. Self-audits also do not help corporate staff become more aware of individual operations and areas that could be improved.

A company policy that combines self-audits with "top-down" audits (periodic department or facility self-assessments with follow-up audits by environmental professionals) can combine the advantages of the two approaches while minimizing their disadvantages. Environmental audits conducted on an internal basis, whether "top-down" or self-audit, should follow some predefined protocol. Records of this protocol should be retained in the company's records for review by environmental regulatory agencies and management.

As part of the procedure to develop the internal audit protocol, the company should develop a procedure for addressing deficiencies or violations detected. There is no point in developing self-examination information if the firm has no program to address remediation of the deficiencies. In fact, an internal audit can present a problem if the regulatory agency detects a deficiency or violation and determines, through a company's internal audit document, that it had prior knowledge of the fact but took no further action.

The primary purpose of any audit program is to translate the information gathered into new environmental programs or change existing programs to help the company resolve problems swiftly.

Consultant Audits

Environmental audits by outside consulting firms are conducted for a variety of reasons. Some companies use these audits as the first step in determining what needs to be done to comply with regulations. The main advantage to a consultant audit is that the company obtains an expert, unbiased opinion of the environmental condition of a facility. These audits are normally viewed by regulatory agencies to be more complete, with a higher degree of accuracy than "top-down" or self-audit protocols. They

are also used as a periodic review of a company's present overall performance on controlling air emissions, discharges to storm and sanitary sewers, groundwater, storage and handling of hazardous materials, and so on. Such a periodic audit can be integrated with an internal audit program.

An environmental consulting firm can be specifically contracted when there are serious concerns about compliance with environmental regulations and statutes. This can be a complete review of a company's compliance with all pertinent regulations, or it can focus on a specific section of the Clean Water Act, SARA, and so on. These audits often pay for themselves by reducing future compliance costs and avoiding costly penalties. An environmental consultant also can be engaged to do a site assessment before a real estate transaction.

CERCLA holds current and former property owners liable for environmental contamination of the property, no matter how long the hazard has existed or who caused it. Without an environmental audit before a real estate closing, an innocent buyer can own property that cannot be sold or that must be cleaned up at enormous cost. Many lending institutions have acquired environmentally contaminated property through foreclosure. Because of their potential liability, some of them are requiring an environmental audit before approving a loan.

United States federal regulations do not require an environmental audit to be part of every real estate transaction. Environmental audits of commercial, industrial, and some residential property may be required by state law. Whether mandated or not, an environmental audit will enable both the buyer and seller to be aware of the condition of the property.

When SARA was passed in 1986, it gave an "innocent purchaser' of contaminated property a narrow way of escaping liability. Liability may be avoided if the buyer can prove that an inquiry into previous ownership and uses of the property (an environmental audit) was made before its purchase. The completed audit form can be submitted as an exhibit in a court case, proving that the buyer did not know that any hazardous substances were released or disposed of at the site before the property was purchased.

Uses of Audits

Environmental audits serve many purposes. Site assessments before a real estate purchase help companies to avoid costly cleanup charges. When conducted to audit a company's compliance with regulations, they can prevent fines. If they include an assessment of environmental hazards to employees, audits promote good labor relations. When their results are reported to the community, they enhance a company's public image with citizens who may have been suspicious of the facility's activities.

In all cases, other than prepurchase site assessments, environmental audits serve as a "report card" showing where a facility has scored "A+" as well as areas that "need improvement." During one company's audit, a drum of trichloromethane was found set apart from other hazardous waste because it was going to be recycled. Only during the audit process was the fact that the drum had no secondary containment system discovered.

Whenever management plans an audit, the first step is a concise statement of company goals. Once management has determined what information it needs to gain from the audit, it can decide whether to conduct an internal audit or hire a consulting firm. Time and expertise are the deciding factors. Complex environmental regulations can require the knowledge of someone other than the environmental manager or department personnel. An extensive environmental audit also can take more time than busy internal staff have to give. (See also Chapter 4, Safety, Health, and Environmental Auditing.)

GLOBAL SOLUTIONS FOR GLOBAL PROBLEMS

No part of the globe is immune from chemicals that nations release into the ground and waterways or emit into the air. Water currents, like conveyor belts, deliver toxic chemicals to the Arctic Ocean. Carbon, sulfur, and other pollutants float on air currents from Eurasia to hover over the Arctic. The "imported" pollutants combine with "home-grown" nitrogen oxides from Alaskan oil fields to produce Arctic haze. The Arctic's annual mean level of photochemical smog rivals that of Los Angeles.

In the once-pristine Arctic, high levels of toxic substances have been discovered in seals and polar bears in recent years. these pollutants have made their way to the top of the food chain, where mercury is now found in mammalian milk in many parts of the Arctic. Sulfur dioxide, produced mainly by coal-fueled electrical utilities, and nitrogen oxide, the product of transportation sources and utilities, are chemically transformed in the atmosphere and transported as acid rain over national borders by prevailing winds.

Carbon dioxide, chlorofluorocarbons, methane, and nitrous oxide, produced in quantity by industrialized nations over the last 200 years, gather in the atmosphere and absorb the infrared waves emitted from the earth. This "greenhouse effect" may be causing global warming—the rise of the mean surface temperature of the earth by 0.6 C in the last 100 years. Increases of 2 to 5 C over the next 50 to 100 years are predicted. Those few degrees could turn arable land in the higher latitudes into deserts and could melt polar icecaps, causing a one-meter rise in sea levels. That increase would flood large population centers, leaving 50 million people homeless worldwide.

The last two decades have proven that we are all part of the world community. The effects of one country's pollution are felt around the globe. With this understanding, international policymakers have joined together to protect our earth and our atmosphere from environmentally hazardous practices.

The international community joined forces to meet the environmental challenge at the 1972 Stockholm Conference and has continued to make giant strides. The "Stockholm Declaration" articulated several principles of international environmental law. Principle 21, for example, holds that while countries have the right to develop, they have a responsibility not to damage the environment outside their borders. This includes the oceans and Antarctica, as well as other countries. Principle 21 now represents customary international law, law instituted by the agreements and practices of states. Its application to some worldwide environmental problems, however, is difficult. It is not easy to assess an individual country's responsibility regarding global warming.

Serious environmental accidents have spurred nations to cooperate more rapidly in developing international environmental protection laws. Following several major oil spills in the 1960s and 1970s, the International Maritime Organization reached rapid accords on oil-spill liability and regulations for oil discharges from ships. When scientific data about ozone layer damage from synthetic chemicals were released, the Montreal Protocol was negotiated and amended in record time. The Basel Convention's agreement was accelerated as soon as the hazardous waste and incinerator ash shipment to a Nigerian dump site came to light. Immediately after the Chernobyl nuclear accident, the International Atomic Energy Agency swiftly finished new treaties on responsibility for notification and assistance. A key feature codified a country's obligation to notify other nations if the risk of transboundary damage existed.

ISO 14000

On July 1, 1995, 39 participating nation members of the International Standards Organization's Technical Committee 207 (TC 207), meeting in Oslo, Norway, ushered in a new era of environmental management by approving as Draft International Standards (DISs), five documents of the ISO 14000 series on environmental management. After global voting, these become full-fledged ISO standards in the third quarter of 1996. They will dramatically change and broaden the ways in which environmental management is conducted and, potentially, regulated in the global economy.

Unlike environmental regulations, ISO 14000 is totally nondirective. It tells managers how to identify and manage environmental impacts, not *what* to manage. It assists organizations to (1) examine the complete range of environmental impacts, both adverse and beneficial, associated with their activities, products, and services, (2) logically manage those impacts which the organization can control and influence, and (3) balance both environmental and economic goals.

ISO 14000 is *not* a regulatory burden; it is a tool for systematically improving environmental management and, consequently, environmental performance. Its focus is on stimulating managers at all levels to consider the environment when making decisions and it seeks to maximize the role of the marketplace in encouraging improvements in environmental performance.

While ISO 14001 is a voluntary standard, there is nothing to prevent a country, state, or municipality from adopting it as a mandatory standard for performance improvement. ISO 14001 could, in fact, form the foundation for a model regulatory system unique in its reliance on business and market incentives to accomplish positive ends.

Historical Context

The initial era of environmental management in the United States began in 1970 with the first Earth Day, passage of the National Environmental Priorities Act, and the establishment of the U.S. Environmental Protection Agency (EPA). The period since has been marked by development of command and control approaches to regulating end of pipe emissions and hazardous waste. While the system has produced environmental gains, they have been achieved at high economic and social cost from litigation and unpopular regulation. Also, the focus of regulation on toxic releases to air, water, and land, has not addressed other, potentially significant global environmental impacts, such as natural resource depletion and nontoxic greenhouse gases, which present arguably greater potential for environmental damage.

ISO 14000 comes into existence in the context of Agenda 21, the United Nations Program of Action for Sustainable Development, which emphasizes the development of processes and procedures that achieve a balance between the needs of both present and future generations of life on the planet.

The Next Phase

ISO 14000 represents an optimistic second phase of environmental management, a beginning step in the necessary transformation of the Earth and its finite resources to a point where population, consumption, and the earth's carrying capacity are in balance. The DIS approved in Oslo comprise the fundamentals necessary to establish a new approach to environmental performance and performance verification. The following standards were approved:

- **ISO 14001—Environmental Management Systems Standard.** This is a straightforward, 19-page document describing the elements necessary to a certifiable Environmental Management System (EMS).
- **ISO 14004—Environmental Management Systems—General Guidelines on Principles, Systems, and Supporting Techniques.** This is an elaboration on ISO 14001, providing practical advice for implementing a new EMS or enhancing an existing EMS.
- **ISO 14010—Guidelines for Environmental Auditing—General Principles.** This is a uniform system for verifying environmental management and performance of all types.
- **ISO 14011—Guidelines for Environmental Auditing—Auditing of Environmental Management Systems.** This is an audit protocol for objectively

evaluating performance of an EMS, specifically distinguishable from an evaluation of environmental performance.
- **ISO 14012—Guidelines for Environmental Auditing—Qualification Criteria for Environmental Auditors.** This describes the skills, knowledge, experience, training, and personal attributes deemed necessary to conduct environmental audits.
- **ISO Guide 64—Guide for the Inclusion of Environmental Aspects in Product Standards.** This is intended for use by products standards writers to raise their awareness and understanding of the negative and positive effects of products on the environment.

Two guidance documents on environmental labeling, environmental performance evaluation, and life cycle assessment are being developed in the same international process and will be issued as DISs over the next several years: ISO 14001—The Environmental Management System Standard and ISO 14004—The Environmental Management System Guideline.

Certification

The most important document of the ISO 14000 series, ISO 14001, describes what needs to be done in order to establish either an ISO 14001 EMS in the absence of an environmental management program or to upgrade an existing environmental management program to the ISO 14001 level. ISO 14004, equally important, parallels ISO 14001 in structure and provides helpful guidance on how an organization might implement an EMS.

An organization may decide that commercial interests or social responsibility dictate certification of the EMS. Certification, under the standard, can be done by the organization itself (self certification) or it may be done by registrars or certifiers (third party certification) recognized by the American National Standards Institute (ANSI). Because rational approaches to environmental management are independent of market demands to certify, the decision to certify is made separately from the decision to adopt ISO 14001 as a framework for environmental management.

Meaning for Organizations

A potential end use of ISO 14001 in the United States is to establish a credible third party environmental audit system, paralleling the principles, standards, and credibility of financial auditing, and to file third party audit reports with environmental regulators, much as Form 10K is presently created and filed with the Securities and Exchange Commission. This approach would reduce regulatory intrusion on organizations and allow regulators to focus enforcement efforts on organizations not adopting the standard. A challenge for industry lies with doing what it can to help establish the public credibility and acceptability of ISO 14001 certification and the third party audit as substantive evidence of improved environmental performance.

ISO 14001 presents to U.S. industry and commerce an opportunity to be part of a new, constructive era of environmental-social responsibility that goes beyond regulatory compliance, is creative in its search for new ways of reducing environmental impacts, transparent in its operation, and powerful in its ability to focus attention on our collective environmental future. It also means a time for U.S. businesses to recognize the cooperative, collaborative approaches being taken by the U.S. EPA, various state environmental agencies, and many nongovernmental organizations, such as the Environmental Defense Fund, the World Wide Fund for Nature, and the International Network for Environmental Management. The opportunity to form productive alliances to benefit the environment is part of the power of ISO 14000.

Key EMS Elements

Some of the key elements of ISO 14001 that distinguish it from contemporary environmental programs include:

- **Requirements.** ISO 14001 requires that top management of the organization commit in its stated environmental policy to continual improvement of the EMS, prevention of pollution, and compliance with applicable environmental regulations. Beyond these three, there are no absolute requirements in the standard.
- **Environmental Aspects and Impacts.** These are activities, products, or services that can adversely or beneficially interact with (aspects) or change (impacts) the environment, which the organization can control or influence. This abridged definition of aspects and impacts suggests how ISO 14001 is markedly different from conventional approaches to environmental management that are rooted in regulatory compliance. ISO 14001 calls for recognition of the following as environmental aspects: emissions that are below permitted levels, emissions that are unregulated (such as CO_2), and consumption of energy and materials. Reference to adverse or beneficial impacts suggests balancing of adverse and beneficial impacts or a net beneficial impact test for new products and services. The concept of impacts that the organization can be expected to *influence* is a prelude to imposing environmental management standards on the organization's supply and distribution chain, its employees, and the communities in which it operates.
- **Best Practices.** While ISO 14001 encourages use of best practices, it also emphasizes evaluating the cost-effectiveness of best practices and does not require use of best practices that are not also cost-effective.
- **Performance Audits.** ISO 14001 requires periodic audits and continual improvement of the EMS, *leading* to improvements in environmental performance. It does not require environmental performance improvements, *per se.*

Implementation

Implementing an EMS calls for a variety of managerial skills. The objective is to embed an environmental management system so thoroughly into existing, free standing management systems that environmental management becomes an inherent function of those systems. Knowledge of quality management and risk management principles, in addition to environmental management, are helpful in establishing a number of the principles that are necessary to an EMS. Familiarity with the quality management cycle of plan, do, verify, and assess and the ability to deal confidently with the ambiguities and uncertainties characteristic of risk management are key to setting priorities for dealing with environmental problems. Quality Managers and Risk Managers reading ISO 14001 will recognize practices and tools borrowed from quality and risk management as well as environmental management techniques that can be applied in their domains.

Initial experience with ISO 14001 implementation indicates wide differences in current environmental management practices and approaches to implementation. This variance in implementation is one of the virtues of ISO 14001. It gives and encourages great latitude and interpretation in use of the standard, ultimately resulting in new and creative approaches to improving environmental performance.

Gap Analysis

Those organizations contemplating installing ISO 14001 as their environmental management system begin by conducting a gap analysis, i.e., a determination of where they are relative to the requirements of ISO 14001, and developing a strategy for migrating from the present position to ISO 14001 conformance.

Gap analysis involves reviewing the major requirements of ISO 14001 to determine the organization's level of conformance. Major requirements include:

- **Management Commitment.** Does top management show support for environmental management by assigning responsibility and authority for EMS implementation and providing resources, human and monetary, for running the EMS?
- **Environmental Policy.** As indicated previously, ISO 14001 calls for an environmental policy that contains organizational commitment to three nonnegotiable elements: (1) regulatory compliance, (2) pollution prevention, and (3) continual improvement of the EMS. Does existing policy reflect these elements?
- **Baseline Study (or Aspects Review).** ISO 14001 requires that an organization have a *procedure* for identifying "the environmental aspects of its activities, products, and services . . . to determine which . . . can have significant impacts on the environment." Since most organizations have *not* been managing the

broad, ISO 14001 definition of environmental aspects, it is unlikely that a procedure is in place or that a baseline study, known as an environmental aspects review, has been done.

- **Legal Requirements.** ISO 14001 requires an organization to have a procedure for accessing legal and other requirements to which it is subject. Generally, satisfaction of this requirement calls for establishment and maintenance of an inventory of environmental laws and regulations affecting the organization's activities, products, and services. Whether that is done by the organization itself or by an outside consultant or law firm is irrelevant so long as access is available. Other requirements include those industry standards or environmental principles, such as the Chemical Manufacturers Association Responsible Care program or the CERES Principles, to which the organization subscribes. If the organization has publicly subscribed to such standards or principles, the standards or principles are regarded by ISO 14001 as having the weight of regulation and become an element of certification.
- **Emergency Procedures.** ISO 14001 requires that an organization "establish and maintain procedures to identify potential for and response to accidents and emergency situations, and for preventing and mitigating the environmental impacts that may be associated with them." This requirement overlaps with the typical risk management practice of planning and preparing for emergency events.
- **Training.** ISO 14001 calls for the identification of training needs and ensuring that individuals having significant environmental responsibilities receive appropriate training. Specific training requirements are axiomatic to all management systems and specific to ISO 14001.

Environmental Aspects Review

The Environmental Aspects Review establishes a baseline of significant environmental aspects from which an environmental performance plan with objectives and targets can be developed. The aspects review process can involve bringing together a team of inside and outside experts knowledgeable about the organization's activities, facilities, and processes; environmental regulations; relevant geological and hydrogeological conditions; and surrounding plant, animal, and human populations to review available information on energy and material consumption, transportation, emissions to air, water, and ground, generation of solid wastes, distribution, sale, use, and disposal of products, environmental impacts of services, and environmental impacts of suppliers, employees, communities, and customers that the organization can control or influence.

Informative guidance on the conduct of an Environmental Aspects Review is contained in Annex A to ISO 14001. Annex A indicates that review of aspects should cover four key areas:

1. legislative and regulatory requirements
2. identification of significant environmental aspects
3. examination of existing environmental management practices and procedures
4. assessment of feedback from previous environmental incidents

The annex further says,

The process is intended to identify significant environmental aspects associated with activities, products and services. . . . Organizations do not have to evaluate each product, component, or raw material input. They may select categories of services, activities or products to identify those most likely to have a significant impact.

The standard and the annex give organizations latitude to determine what their environmental aspects are. Although significant environmental aspects could be identified through a program of site assessments, site assessments would likely be time consuming and an expensive way of developing information. Organizations are finding success with using a subjective evaluation done by persons knowledgeable about aspects of the organization's activities, products, and services and briefed on the objectives of Agenda 21, the Rio Declaration on the Environment and Sustainable Development.

Review of the Rio Declaration suggests the following broad categories of environmental aspects:

- water consumption
- water pollution
- energy consumption
- air pollution and ozone depletion
- waste disposal
- transportation
- toxic chemicals use
- product mass, use, recovery, and reuse
- biodiversity
- noise, odor, aesthetics, archeological, and architectural heritage
- sudden events and emergency preparedness

Other approaches to identifying significant environmental aspects are being developed for inclusion in an annex to the Environmental Performance Evaluation document. They include (paraphrased from Working Draft 4 of ISO 14031):

- Begin by identifying key activities of the organization (e.g., major manufacturing operations), then associating specific environmental aspects with those operations (e.g., emissions into the air of specific constituents), and then identify the type of impact related to that aspect (e.g., emissions that are potentially harmful to human health).

- Starting from the environmental perspective, identify key attributes in the organization's activities that may contribute to an environmental issue. Examples include raw material inputs, energy resources, discharges, emissions, or wastes. Examples of environmental issues may include global warming phenomena, water quality, or ambient air quality.
- Working from existing data on discharges and emissions, measure and assess these data in terms of quantity and hazards. This approach can be used to prioritize an organization's environmental aspects.
- Analyze environmental aspects (e.g., discharges or emissions) that are regulated and for which data have already been collected by the organization.
- Identify the significant environmental aspects of the organization's products, including manufacture, distribution, use, reuse, and disposal. These products can be traced back through the full range of the organization's activities.

With these as a starting point, organizations can identify their environmental aspects and determine which ones are significant. In organizations previously committed to regulatory compliance, ISO 14001 aspects reviews indicate that most of the significant environmental aspects are in the areas of energy and materials consumption. Approved toxic emissions have significantly declined due to the impact of waste minimization efforts.

The Environmental Aspects Review is critical to the success of the EMS. Good advice to organizations starting the process is to begin small and slowly expand the review process into all areas of the business.

Legal Considerations

Before the Environmental Aspects Review is conducted, the organization's legal counsel should be consulted on the issues of access and confidentiality of the information developed.

Logically, an organization installing an EMS begins by surveying and documenting its environmental exposures and using survey results to establish a baseline for prioritizing objectives and targets. The legal need to protect environmental information from disclosure was of great concern to the U.S. delegation to TC 207 in Oslo. European delegations, not sharing U.S. liability concerns, felt it appropriate to require a documented Environmental Aspects Review as a preliminary to implementing environmental management systems; reasoning that an organization should know where it *is* before deciding where it is *going*.

Consequently, ISO 14001 contains compromise language that stops short of *requiring* an aspects review:

> organization[s] shall establish and maintain a procedure to identify . . . [their] environmental aspects . . . in order to determine those which have or can have significant impacts on the environment.

In establishing and maintaining a *procedure*, documentation can be left to the organization. Although this semantic nuance allowed the U.S. delegation to vote in favor of the ISO 14001 DIS, a question remains for legal and public relations departments: Are attempts at privileging environmental evaluations a useful strategy in view of legal and market factors encouraging environmental transparency?

For U.S. firms, disclosure of environmental aspects should be considered within the context of requirements to report environmental problems and the consequences (i.e., positive or negative) to the organization. Legal considerations prompting voluntary disclosure include:

- state and federal prosecutorial guidelines favoring civil or criminal actions against companies that fail to fully disclose their environmental problems
- federal sentencing guidelines that mitigate penalties for companies that do disclose, while increasing penalties for those that do not
- Securities and Exchange Commission's Staff Accounting Bulletin No. 92, requiring publicly traded companies to disclose environmental liabilities in financial statements
- shareholder suits against directors and officers who withhold environmental information affecting investment decisions
- punitive damages and fines against companies that knowingly violate environmental regulations

Other competitive market conditions that would further prompt voluntary disclosure include:

- environmental management standards such as the European Union's Eco Management and Audit Scheme (EMAS) requiring a registry of environmental *effects* (aspects) to be created, validated, and publicly disclosed. This registry would facilitate an inference of similar environmental effects from similar operations in the United States.
- disclosure by one organization of an environmental aspect common to all organizations in a particular industry
- proliferation of "green" reports by organizations seeking to derive competitive advantage by highlighting their environmental stewardship over their operating facilities and products

Organizations should systematically evaluate their legal exposure and determine a logical strategy for conducting an Environmental Aspects Review. These evaluative steps can be taken:

- determine what environmental information is sensitive
- determine if disclosure will result in harmful consequences

- determine if disclosure has already been made or privilege violated; e.g., permit applications or information sharing beyond the scope of the privilege
- determine if information deemed to be sensitive, harmful, and confidential can be protected by attorney/client, attorney work product, or self-evaluative privilege

With the kind of universal knowledge likely to result from competition and the influence of EMAS, the standard of care required of managers in carrying out their environmental responsibilities will be increased. With this environmental transparency, the standard of care for managers is greatly increased; the possibilities for successful ignorance, willful or not, on the part of managers becomes remote; the credibility of nondisclosures becomes suspect.

Objectives and Targets

The ISO 14001 says, "The organization shall establish and maintain documented objectives and targets at each relevant function and level within the organization." It is intended that objectives will conform to the organization's environmental policy, coincide with the findings of the environmental aspects review, and be measurable.

Integration and Alignment

Integration of the EMS refers to the process of embedding the environmental program in other management systems. Alignment refers to specifically coordinating environmental planning on the same time cycle as strategic planning so that strategic business plans reflect environmental issues. In this way, the EMS becomes totally immersed in organizational activities and environmental performance gains come as a function of other management goals. For example, in Human Resources, identify all personnel who have environmentally related job functions and include in their performance appraisals a section that reviews environmental performance and encourages creative improvement of environmental performance. Review the existing training curriculum for opportunities to insert environmental awareness training.

Independent Audit Function

ISO 14001 says,

> The organization shall establish and maintain a program and procedures for periodic environmental management system audits . . . in order to determine whether . . . the environmental management system . . . conforms to planned arrangements for environmental management . . . [and] has been properly implemented and maintained.

In establishing the audit function, the organization can use its own internal auditors or call upon recognized third-party auditors that use either the ISO Environmental Auditing Guidelines or other recognized auditing principles. It is important to recognize that the required audit is of the Environmental Management System, not the organization's environmental performance.

Market Drivers

ISO 14001 relies upon the forces of the marketplace to impose environmental performance improvements. Behind voluntary ideals, however, looms a potential threat—more stringent national environmental management standards could become mandatory if voluntary adoption of ISO 14001 does not result in significant environmental performance improvement.

Market forces are evident in the competition to demonstrate "green" products and performance and to publicly disclose potentially positive aspects of environmental performance. In addition, the requirements by governmental agencies and other large consumers that supplier organizations become ISO 14000 certified will be a major market driver. Similar results have been seen with ISO 9000.

Competition

In many markets, "green" products and processes are thought to confer a competitive advantage. Evidence is found in the proliferation of environmental reports, "green glossies," advertisements extolling the environmental virtues of oil and paper companies, and product claims that hint at favoring sustainable material resources. Some consumer products companies have built their corporate and brand reputations on environmentally responsible products and commitment to the environment, e.g., the Body Shop and Patagonia.

Clearly, many organizations will see certification to ISO 14001 as an opportunity for competitive advantage. Although certification does not mean that environmental performance is superior, organizational commitment to continual improvement and to the rigors of self-examination required under ISO 14001 can stand for good faith efforts towards achieving superior environmental performance.

Disclosures

In markets where environmental disclosure can confer an advantage, environmental competitive advantage *and* performance standards will be established by those companies that have a policy of complete disclosure. For example, a paper company publicly acknowledging environmental problems with dioxin would set the disclosure standard for all other companies in the paper industry. It could be assumed by the investing and consuming public that other paper companies have similar or worse problems with dioxin. Hence, full disclosure of appropriate and accurate information can suggest management integrity.

SAB 92

In the United States, an effective disclosure tool has been the requirement by the U.S. Securities and Exchange Commission (SEC) that publicly traded companies disclose in annual and quarterly financial statements estimates of environmental liability. Initially, the compliance rate with this requirement, known as Staff Accounting Bulletin No. 92 (SAB 92), was low. The persistence of SEC Commissioners and staff members, however, has resulted in improved reporting coming from industry and an indication that companies are looking closely at their environmental liabilities. While the primary motivation of the SEC was sufficient environmental disclosure to allow investors to make informed investment decisions, the end result has been the invocation of the business school maxim, "what gets measured, gets managed." Consequently, this modest staff bulletin from the SEC, interpreting existing accounting principles, is having a great influence on management of environmental problems by companies.

EMAS

EMAS requires companies to establish a register of environmental effects, verified by a third party, and made available to the public. Individual companies having operations with significant environmental effects in Europe and similar operations in the United States are potentially disclosing the existence of environmental aspects in their U.S. operations. Alert interested parties, including environmental advocacy groups that monitor the environmental performance of industrial organizations, may seek to reconcile the validated existence of environmental problems in Europe with denials or nondisclosures of similar problems in the United States.

Influencing Suppliers

Organizations that have substantial market power are in a position to potentially require environmental performance improvements from their suppliers by only buying from ISO 14001 certified companies. A similar process has occurred with ISO 9000; i.e., only ISO 9000 certified suppliers are allowed to bid and compete as suppliers for certain major producers of consumer or industrial products.

Raising the environmental awareness of product specifications writers is the main thrust of the Environmental Aspects of Products Standard (ISO Guide 64). When spec writers consider the impact of product design requirements on material and energy consumption, they will often see opportunities for reducing environmental impacts by using alternatives. This opportunity is clearly present if a government or large organization adds these requirements to a large project bidder's specifications.

For example, as part of a bidding requirement, a purchaser of raw steel may specify that it be produced using the available energy input having the least impact on atmospheric carbon dioxide levels. Suppliers of steel for this project would have to evaluate and rank the carbon dioxide impact of thermal energy furnished by coal, oil, natural gas, oil, electricity, and the electric renewables, wind, direct solar, and geothermal. Suppliers would be contractually obligated to use the available energy option that their analysis indicates is most carbon dioxide

efficient even though that might add cost to the manufacturing process.

Many large consumers possessing economic market power will simply begin to require that their suppliers become certified to ISO 14001. Small- and medium-sized enterprises, however, appear largely unaware of the ISO 14000 documents. They will be hard hit when large customers require certification as a condition of doing business. The machine shop whose cash flow depends upon sales of a small part to a manufacturer, for example, stands to lose a great deal when that manufacturer will not buy any more parts until the shop is certified to ISO 14001. These businesses will quickly discover the advantage of ISO 14001 certification.

Mentoring

Many small enterprises will be helped by large organizations that agree to mentor them through the ISO 14001 certification process. The Merit Partnership of EPA's Region IX is actively providing incentives for large companies that agree to mentor their supply chain in ISO 14001 certification.

Environmental Performance Evaluation

The most critical need of improved environmental performance is the development of consistent methods for measuring impacts on the environment. Fundamental questions arise with respect to how to measure not only the present impact of human activities upon the environment but how we can begin to measure and move towards sustainable development that adequately serves the needs of today without depleting the resources required by future generations. To make progress on sustainable development, new approaches, Environmental Performance Indicators (EPIs), must be found. These are questions that are being addressed by the United Nations, the Club of Rome, the Wuppertal Institute, and, to a more limited extent, the Technical Committee 207 of the ISO (TC 207).

An imperative exists to account more thoroughly for the environmental costs associated with the production of goods and services and, simultaneously, consider the limits of sustainable economic growth and allocation of scarce resources. Although substantial progress is being made on measurement techniques, the concept of integrating sustainable development in economic policy is still in its formative stages.

Overview of ISO 14031, the Guideline Document

ISO 14031, Environmental Performance Evaluation (EPE), is the reference tool being developed by TC 207 to aid organizations in developing reliable, verifiable information to determine if the organization is meeting the environmental criteria set by management. ISO 14031 is scheduled to become a committee draft standard in November 1996; however, due to a built-in testing phase of eighteen months, it will be several years before ISO 14031 gets to Draft International Standard status.

Evaluation of environmental performance at the organizational level raises many collateral issues that impede the development of the EPE guideline. Many large organizations, particularly manufacturing organizations, have potentially significant environmental impacts and resist the establishment of standards that measure the impact of materials, energy, emissions, effluents, and wastes on the environment. Nevertheless, many organizations are currently evaluating their environmental performance before the development of an EPE guideline. Companies recognize that environmental trends exist that have clear implications for how business will be conducted in the 21st century. These organizations want to avail themselves of state-of-the-art methods for evaluating their impacts on the environment and also for the sustainability of their industry into the next century. In the future, estimates of environmental sustainability may be as important to businesses as forecasts of continued market demand are today.

Environmental impacts are *always* contextual. An organization cannot understand its environmental performance without an understanding of the circumstances surrounding that performance. For example, the environmental impact of water consumption depends upon the source of the water. Is it from a basin whose total draw down is in balance with precipitation into the basin or is it from an aquifer where draw down exceeds recharge? Consumption in the former circumstance has virtually neutral environmental impact; consumption in the latter would be of substantial environmental concern.

Environmental impact quantification, in part, involves considering the true costs of externalities. An externality is an accounting of indirect costs and can include direct and indirect environmental impacts (such as loss of enjoyment) as well as other economic impacts on employment and human welfare. For example, if permitted air pollutants produce respiratory disease, the costs of medical evaluation and treatment are paid for by society as a whole rather than by the emitter.

Underlying the costing of environmental externalities is an economic and social attitude toward the environment. How are organizations best motivated to make necessary social change? A simplistic answer is that change takes place automatically when environmental impact reduction strategies can be shown to coincide with and enhance economic gain; i.e., when it is in the best economic interest of the organization to change. Decision makers react positively to decision opportunities that increase revenues or reduce costs; hence the need to express environmental impacts in financial terms and relate their corrective actions to potential for revenue gain or cost reduction. When environmental impacts and corrective actions are expressed financially, the door opens to logical, benefit/cost-based decisions on reduction of environmental impacts and integration of environmental issues with strategic business plans.

Environmental managers and consultants work in terms of environmental performance indicators (EPIs) that are different for each contaminant. Concentrations of nitrous oxide, carbon dioxide, or various heavy metals, measured in parts per million/billion of volumes of air, water, or soil, are the EPIs of the environmental scientist. However, for business decision making, the language of environmental science is often insufficient for the following reasons: Scientific measurements of values of carbon dioxide or chlorofluorocarbons are not easily susceptible to full financial valuation. Scientific measurements describe the physical dimensions of contamination; however, business decision making requires an understanding of the uncertainty; i.e., what are the likely future economic consequences of environmental impacts. Scientific language is sometimes foreign to the executive decision maker, who thinks and acts in response to changes in revenues or costs, and needs to know the financial cost of environmental impacts.

A variety of analytic tools are used to measure environmental impacts, associated corrective/preventive actions, and the relationship of corrective actions to economic gain. These tools are used so that economically rational decisions can be made. The corrective action of choice becomes the one whose cost, after subtracting the value of measurable economic gain, achieves the largest impact improvement for monies spent. In many instances, the value of economic gain exceeds corrective action costs, producing a win for the environment and for the organization. The approach has several advantages:

- The financial approach is understood by corporate decision makers who might otherwise wrestle with environmental measurements presented in parts per million/billion.
- Linking corrective action costs to the potential future cost of a problem allows rational choices based on cost effectiveness between competing demands for limited corrective action funds.
- By trying to measure impacts (or externalities) rather than outputs, the true cost of an environmental problem is captured. Impacts consider not only the cost of reducing outputs but the social costs to environmental receptors.

Environmental Costing and Valuation Methods

Conversion of collected data into financial terms and quantifying uncertainty involves a variety of valuation methods, including:

- **Direct Cost Estimates.** An environmental problem is expressed in financial terms at the outset, for example, the cost to clean up a specific regulated contaminated site under Superfund.
- **Fines and Penalties.** The amount of fines and penalties is often stated in laws and regulations; what is unknown is the likelihood of penalty assessment and duration of fines. Fines and penalties are not environmental impacts. They are, however, costs to the organization

of noncompliance with regulatory requirements and relevant to corrective action strategies.

- **Third-Party Liabilities.** Estimating third-party liabilities adds a layer of complexity and uncertainty to the valuation process. These liabilities include impacts to people, ecosystems, and property not under the direct control of a company. Both the potential costs and uncertainties are often quite high.

- **Quantifications Requiring Subjective Valuations.** The full cost in damages to the environment is not captured in a wide array of circumstances, particularly where a level of emissions is permitted by law. These costs are transferred to society at large and become part of the price of economic well-being. Under ISO 14001, full costing and attribution of costs to responsible activities, products, or services will increasingly become a component of efforts to improve environmental performance. While the full amount of contaminating discharges can usually be easily measured, the values of impacts to human health or natural resources are not so easily determined.

  A common method for valuing impacts to natural resources is the contingent valuation method (CVM). This method relies on polling to determine preference or indifference for the value under study. Determining these values facilitates full-cost accounting and subsequent corrective action decisions.

- **Quantifications Requiring Value Establishment.** Value judgments provide an approximation of expected damage that could occur from contaminants that do not have an immediate, direct, and easily determinable environmental impact. For example, carbon dioxide emissions, and their long-term impact on global climate, is a potential environmental impact that could be quantified using the value judgment approach. This kind of valuation is most useful if conducted and established by an international body, such as ISO, so that there is a uniform approach for documenting aggregate impacts. With an established value scale, carbon dioxide as an emission by-product has a finite value that changes annually depending upon the global content of carbon dioxide in the air. The value of emissions and the value of increased forestation (representing ability to absorb greater amounts of carbon dioxide) would then be established. The importance lies not in the absolute accuracy of such projections but that some defensible estimate be developed so that organizations have a way of valuing their own emissions and the gains derived from corrective actions.

Life Cycle Assessment

As a means of relieving government of costly enforcement-based environmental regulations, some countries are establishing product-oriented incentives that are intended to yield environmental benefits. The basis for evaluating such incentives is the province of Life Cycle Assessment (LCA). ISO 14040, the Life Cycle Assessment Guideline, defines life cycle assessment as a

> systematic set of procedures for compiling and examining the inputs and outputs of materials and energy and the associated environmental impacts directly attributable to the functioning of a product or service system throughout its life cycle.

LCA is the subject of intense activity and considerable criticism within TC 207. There is a significant effort to develop a useful LCA standard within TC 207 because it is recognized that the European Union may develop and mandate a more stringent standard. The overall process is subject to criticism because LCA is a highly detailed cost evaluation that may be too expensive for use by small firms. In addition, the technical process of performing a LCA is undergoing constant modification. However, several examples of environmental actions that have been analyzed using LCA techniques have produced useful and important results.

France passed a fuel-tax exemption to promote consumer acceptance of bio-diesel fuel, i.e., diesel fuel produced from agricultural products such as soybeans. Burning bio-diesel fuel has been shown to reduce automotive tailpipe emissions. Subsequently, life cycle assessment established that the process of producing bio-diesel fuel resulted in other forms of pollution that exceed that of producing and burning petroleum diesel. In other words, substitution resulted in displacement of pollution from automotive tailpipes to bio-diesel production process wastes. Hence, the presumed substitution of environmentally preferable products must be carefully evaluated so that true environmental improvement is demonstrated.

The European Commission committee on waste management was given the task of developing regulations on packaging waste that would result in both flexibility in consideration of national priorities and environmental improvement throughout member countries. Existing regulations forcing a hierarchy of reduce, reuse, recycle resulted in costly investments in Germany and elsewhere that were losing money with no prospect of recovery.

Using life cycle assessment, it was determined that there was a trade off between recycling and disposal that depended on transportation distances. As distance to the recycling center increased, the environmental benefit of recycling decreased. Beyond certain distances, recycling actually increased pollution over that created by waste disposal nearer the source. As a result of the described study, the European Commission revised its package waste regulations for certain products to favor maximum recyclable content of packaging as opposed to mandatory and, possibly, counterproductive reuse.

Trends—Cause for Optimism

The management changes described in this chapter coalesce into discernible trends at various levels:

- **Capture of Relevant Environmental Costs.** Organizations everywhere, whether driven by social responsibility, market necessity, or regulatory agencies, are moving to include more of the environmental costs of their activities, products, and services in their management accounting systems. This could ultimately lead to a net benefit test for products and services: Do the environmental impact costs associated with the design, manufacture, marketing, delivery, use, and disposal of products or services result in social/economic benefits that exceed their environmental costs?
- **Inclusion of Beneficial Environmental Impacts.** Organizations are beginning to see value in creating and measuring *beneficial* environmental impacts—woodlands, wetlands, creature habitats, and other environmental assets. This could lead to an environmental balance sheet that offsets the adverse consequences of activities, products, and services with the beneficial consequences of other activities, products, or services. Advantages of identifying and quantifying beneficial environmental impacts could include their use as a tool for regulatory negotiations and, possibly, favorable tax treatment. Competitive pressure to produce environmental annual reports may, over time, increase managerial thinking and action toward neutral environmental impact. Environmental balance sheets might be formulated someday to evidence sustainable development practices.
- **Strategic Deployment of Environmental Improvement Activities.** Organizations, especially commercial and industrial entities, are increasingly aligning their environmental management planning with their overall strategic planning. For many corporations, this means expressing environmental impacts and corrective/preventive actions in financial terms and considering the potential that each corrective/preventive action has for producing an economic gain. Financial quantification of environmental impacts and ranking of corrective/preventive actions by cost effectiveness enables rational decisions on sequencing and scoping of environmental activities.

Environmental management has been thrust into a new, dynamic phase as a result of the promulgation of ISO 14001. While the results are not clear, some steps seem appropriate for managers to take now irrespective of a need or desire to adopt ISO 14001:

- Recognize that environmental management is a high organizational priority.
- Establish a dialogue with internal and external interested parties.
- Determine the regulatory requirements and environmental exposures associated with the organization's activities, products, and services.

- Develop management and employee commitment to the protection of the environment, with clear assignment of responsibility and accountability.
- Encourage environmental strategic planning throughout the product or process life cycle.
- Establish a disciplined management process for achieving targeted performance levels.
- Provide appropriate and sufficient resources, including training, to achieve targeted performance levels on an on-going basis.
- Assess environmental performance against appropriate policies, objectives and targets, and seek continual improvement where appropriate.
- Establish a management process to review and audit the Environmental Management System (EMS) and to identify opportunities for improvement of both the system and environmental performance.
- Coordinate EMSs with other systems (e.g., health and safety, quality, finance).

Overall, the opportunities for an integrated proactive environmental management program are substantial. While the management activities and time commitment necessary to achieve a fully developed environmental management program are significant, the potential payoffs for both the organization and the environment are equally large.

SUMMARY

- The safety/health/environmental professional must be familiar with the most current environmental regulations covering such areas as the release and recovery of hazardous substances, air, water, soil pollution, and toxic substances control.
- Environmental regulations in the Far East and Europe, such as ISO 14000, are helping companies take proactive steps in managing environmental, health, and safety issues. These regulations may be used to help establish international guidelines, develop certification criteria, and encourage voluntary compliance among manufacturers and suppliers.
- The ISO 14000 and similar regulations are providing international business with stronger guidelines and regulations for protecting the environment and safeguarding worker and consumer health and safety. These regulations also set standards for environmental protection.
- Industry experts have identified five stages of environmental management program development that can be found in many companies. These stages range from beginner to firefighter, concerned citizen, pragmatist, to proactivist. A company's stage often depends on its size, amount of hazardous or toxic material, and financial resources.
- For companies that have reached stage five, environmental management is a top priority. Top

- management must demonstrate a strong commitment to responsible environmental practices and communicate this priority throughout the company for the program to succeed. Program staff must include skilled professionals with backgrounds in the sciences, regulatory law, and environmental issues.
- Steps in the environmental management program include preventing common violations, maintaining accurate and thorough records, creating a spill-reporting plan, setting realistic limits and schedules for meeting environmental regulations, and training employees in program objectives and procedures. In addition, the company must establish a strong relationship with an attorney skilled in environmental law.
- Efforts to reduce and minimize waste can save companies money, reduce potential environmental liabilities, protect public health and worker health and safety, and protect the environment. Waste minimization can be accomplished through source reduction and recycling. Source reduction is generally more cost-effective and less of an impact on the environment than is reuse/recyling.
- Groundwater contamination is one of the more serious environmental issues of the past decade. Seepage of toxic and hazardous materials into aquifers has prompted local and federal agencies to force companies to take stronger measures in assessing and controlling this contamination.
- Government agencies or companies themselves can conduct environmental audits of facility operations to (1) identify contamination or hazardous materials from current or previous activities, (2) determine compliance with current standards or regulations, and (3) provide a general review of environmental risks associated with a site and its operations.
- Environmental audits can be internal (conducted by the company) or external, conducted by an outside firm. Internal "top-down" and self-audits do not always result in objective treatment of a company's environmental problems and risks. Audits performed by outside consultants provide objectivity and may pay for themselves in reduced fines and penalties.
- Environmental audits serve many purposes: avoiding costly cleanup charges, preventing fines, promoting good labor relations, enhancing a company's public relations with the community, and serving as a "report card" on a firm's environmental management functions.
- Complying with environmental regulations and standards can be a competitive advantage to companies who disclose their adherence. The public today is more sensitive to the responsibilities of industry to safeguard the environment.
- Organizations with substantial market power are in a position to require suppliers to improve their environmental record by buying only from certified companies.
- The Environmental Aspects Review establishes a baseline of significant environmental aspects to help companies develop an environmental performance plan with objectives and targets. Before conducting a review, a company should consult its legal council to protect sensitive company information from disclosure.
- ISO 14031, Environmental Performance Evaluation, is the reference tool being developed to help organizations determine if they are meeting their environmental protection goals. Companies must weigh the economic and marketing costs and benefits of achieving those goals.
- A variety of valuation methods can help companies estimate the costs of addressing environmental problems and assessing the fines and penalties involved in failing to comply with regulations and standards.
- Life Cycle Assessment provides a means of relieving government of costly enforcement-based environmental regulations by establishing product-oriented incentives.
- In the future, companies are likely to take a more proactive role in creating a better environment rather than focusing solely on preventing environmental damage.

REFERENCES

Friedman WJ. "Avoiding environmental liability in five simplified steps." *Chemical Processing.* (April 1988).

Hajost S. "The challenge to international law and institutions." *EPA Journal,* Vol. 16, No. 4 (July/August 1990).

Hunt CB and Auster ER. "Proactive environmental management: avoiding the toxic trip." *Sloan Management Review,* Vol. 31 (Winter 1990).

Keenan T. "Why is everyone talking about environmental audits?" *Industrial Safety & Hygiene News,* Vol. 24 (June 1990).

Kreiger GR, ed. *Accident Prevention Manual for Business & Industry: Environmental Management.* Itasca, IL: National Safety Council. 1995.

Leaf A. "Potential health effects of global climatic and environmental changes." *New England Journal of Medicine,* Vol. 321 (December 1989).

Main J. "Here comes the big new cleanup." *Fortune* (November 21, 1988).

Quarles J and Lewis WH, Jr. *The New Clean Air Act: A Guide to the Clean Air Programs as Amended in 1990.* Morgan, Lewis & Bockius, 1990.

Rhodes D. "Safety, environmental crimes and the tough new laws." *Professional Safety,* Vol. 35 (January 1990).

U.S. Department of Health and Human Services, Public Health Service, Centers for Disease Control and National Institute for Occupational Safety and Health. *Occupational Safety and Health Guidance Manual for Hazardous Waste Site Activities.* Washington, DC: U.S. Government Printing Office, 1985.

U.S. Environmental Protection Agency, Region 5, 230 S. Dearborn, Chicago, IL 60604.

An Introductory Guide to the Statuary Authorities of The United States Environmental Protection Agency; Our Air, Our Land, Our Water (April 1988).

EPA Property Searches for Buyers of Real Estate (Fact Sheet) February, 1991.

U.S. Occupational Safety and Health Administration, *CFR,* Title 29, Labor (for OSHA Hazardous Waste Training and Communications Regulations). Washington, DC: U.S. Government Printing Office, 1990.

———. *CFR,* Title 40, Protection of Environment, Washington, DC: U.S. Government Printing Office, 1990.

Young OR. "Saving the Arctic: challenge to eight nations." *EPA Journal,* Vol. 16, No. 4 (July/August 1990).

REVIEW QUESTIONS

1. What is the difference between ISO 14010 and ISO 14011?
2. What is a gap analysis when dealing with ISO 14001?
3. What are the four aspects that an Environmental Aspects Review should address?
   a.
   b.
   c.
   d.
4. What is the main purpose of the Eco Management and Audit Scheme?

13
Ergonomics Programs

Ergonomics, also called human factors engineering, studies the physical and behavioral interaction between people and their environments. This environment could be the workplace, the home, or even the car. This chapter covers:

- how to establish an ergonomics program that uses ergonomics principles to eliminate or reduce work-related stresses
- how to evaluate the workplace to identify potential problems or stresses created by tasks, procedures, or the environment
- how to evaluate and arrange machine displays and controls

The goal of ergonomics is to achieve a balance between the demands placed on people and their capabilities. When workers and their environments are mismatched, injury levels rise, production is inefficient, and other incidents occur that detract from organizational efficiency and worker welfare.

DEFINITION

Many different fields of study have contributed to industry's understanding of work-related stresses and solutions. These include:

- anatomy and physiology of the human body
- anthropometrics
- biomechanics
- psychology
- industrial design and engineering.

Ergonomists are people who have received training in these fields and who, using ergonomic principles, can develop practical solutions for workplace stresses. In approaching a problem, an organization may bring together specialists in these areas to analyze how people interact with machines, tools, work methods, and work spaces. Safety professionals, industrial hygienists, and occupational physicians and nurses may add their expertise to find solutions. Management and employees also play important roles in providing information, helping to evaluate work situations, and giving feedback on changes.

PURPOSE OF ERGONOMICS

The goal of ergonomics is to reduce the physical (and mental) stress associated with a given job; to increase the comfort, health, and safety of a work environment; to increase productivity; to reduce human errors associated with a task; and to improve the quality of work life. Ergonomics can also be described as making the most of the human/machine relationship to balance the capabilities of individual workers with the demands placed on them by the system.

Engineers and others in industry often regard the three elements of machine, raw materials, and end product as comprising a complete system. To an ergonomist,

however, a system is not complete until the human working within the system is considered. The success of the human/machine interaction often determines the success of the system. To make the most of this process requires identifying demands placed on the worker.

Such demands can be classified into three categories: physical demands, environmental demands, and mental demands. Due to the aging of the present U.S. workforce, the physical demands of a job often receive the greatest attention, although environmental and mental demands can be significant factors.

Physical demands are those placed on the musculoskeletal system of the body. Some examples are lifting, pushing, pulling, reaching, exerting force to perform a task, and the effort required to do the job, such as in using a tool or carrying totes/boxes, or pressing control buttons. Walking, standing, or sitting all day are also physical demands.

The environmental demands placed on a worker are a function of the physical environment in the workplace, such as vibration, temperature, humidity, noise, and lighting levels. There is also the psychosocial environment one works in, including such factors as the work organization, pace, shift schedule, need for overtime, and so on.

Mental demands include the information needed to perform a particular task, mental calculations or computations, any short-term memory demands, information processing, decision making, and the like.

These three types of demands are interrelated. For example, the length of time one can work at a high energy level on a hot, humid day is generally less than on a cooler, drier day. Likewise, the ability to process information and make decisions may be affected by the shift one is working and the level of physical fatigue.

All of these demands can affect a worker's performance on the job and should be carefully balanced with each person's capabilities. It is well known that a situation where the job demands are below a worker's capabilities can be as potentially harmful as one where the demands are too high. Workers may get bored when performing below their abilities. Boredom can lead to lapses in concentration and attention, resulting in product defects, quality problems, and even accidents. If the demands exceed workers' capabilities, the first consequence is usually fatigue. Physical fatigue can result in overstressing the muscles and aggravating disorders of the musculoskeletal system. In addition, a tired worker is more prone to have accidents or to produce poor quality products and is less likely to be efficient. Some other effects of poor ergonomic design are high training costs, unnecessary overtime costs, employee complaints, medical restrictions, absenteeism, and personnel turnover.

ESTABLISHING AN ERGONOMICS PROGRAM

Many industries have and continue to develop ergonomic programs in order to improve worker health, safety, and efficiency. The goal of an ergonomics program is to

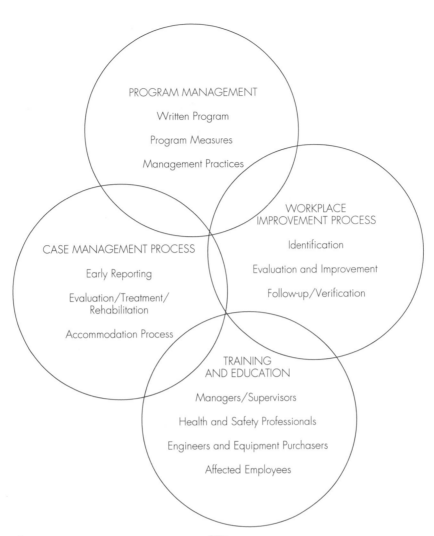

Figure 13-1. Components of an ergonomics program to manage CTDs.

reduce the human and monetary costs associated with inadequately designed workplaces, work processes, and work environments. To be successful, ergonomics should not be an "add-on" activity, but an inherent part of an organization's functions. An effective program combines both a proactive and a reactive approach and involves all affected personnel. The degree of involvement varies depending on each individual's roles and responsibilities within the organization.

Ergonomics should be an integral factor in the design (and execution) of all processes, jobs, or tasks. However, to make this discipline an inherent part of an organization's functions, it may be necessary first to implement a formal ergonomics program in the organization. In the 1990s the focus of many ergonomic programs has been on the prevention and management of musculoskeletal disorders (or cumulative trauma disorders, CTDs). The key components for such a program are given in Figure 13-1.

First, management commitment and support of ergonomic activities are vital to the success of any program. This commitment can be demonstrated by setting of goals for the program and providing the resources required to achieve those goals.

Second, case management is another vital component of an effective ergonomics program, especially if the major concerns revolve around musculoskeletal disorders. Timely identification of these disorders and effective treatment will help minimize many of the negative effects of injuries and illnesses. Safety professionals should work closely with health care providers to ensure that they have adequate information on any workplace or job characteristics that may influence the diagnosis and treatment of CTDs.

Third, training and education of all involved personnel is critical for the success of such a program. Such training should be appropriate to the role of each employee affected by the program.

The fourth, and most important component, is the workplace improvement process. When this process is incorporated into the design stage of new products and processes, it can be an important tool in proactively addressing ergonomic concerns. The basic approach to the workplace improvement process is similar to the traditional quality improvement cycle, as shown in Figure 13-2.

Typical issues addressed during ergonomic evaluations to identify potential problem jobs or tasks, include visual task interfaces (including illumination levels), thermal

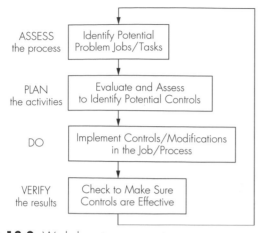

Figure 13-2. Workplace improvement process.

stress (heat/cold), physical work load, duration of work, vibration, and musculoskeletal disorders. The primary factors that are evaluated and assessed can generally be classified into the following categories:

- workplace characteristics and accessories
- physiological demands
- physical demands
- environmental demands
- design of displays, controls, and dials.

Based on these assessments, potential controls and modifications can be identified to improve the ergonomic impact of these factors on the worker. Each of these issues are discussed in turn in the sections below.

WORKPLACE CHARACTERISTICS AND ACCESSORIES

There are three types of workplaces: sitting workstations, standing workstations, and sit-stand workstations. Workplaces should be arranged so that the person using them can do so safely and effectively, without exceeding an acceptable level of energy consumption. For instance, a tool and die operator should be able to work in a comfortable position and be able to reach the materials needed easily.

The key to a good workplace setup is to design it so that it accommodates the greatest number of people (90% to 95% of the working population). For example, clearances should be designed to accommodate workers with the largest dimensions, while reaches should be designed with the smallest workers in mind. The human, or anthropometric, data required for such design decisions can be obtained from a number of sources, such as *Ergonomic Design for People at Work,* Vol. 1 (1986) by Eastman Kodak Company; *Anthropometric Source Book Vols. I–III* (1978) by NASA; or S. Pheasant's *Bodyspace, Anthropometry, Ergonomics and Design* (1986).

For most people, a comfortable working position is one in which the back is naturally curved, the head is held

erect, the shoulders are relaxed, the upper arms are close to the sides of the body, and the wrists are straight. This posture is often referred to as a "neutral posture." Operators should be able to work with their elbows bent at about 90°, and their feet should be well supported. The edges of the work surface should be smooth and well rounded, so that any contact stress is minimized (Figure 13-3). However, it is not possible or even desirable that workers hold these positions throughout a workday. Remaining in one position for long periods of time can place a static load on the muscle and rapidly build up fatigue. However, the postures the operators assume should be close to the neutral positions shown. Any significant deviations should be infrequent and for short durations only. Some of the factors that influence the postures assumed at work are discussed below.

Heights of Objects Handled

If the task to be performed requires little force, then the object being handled (work height) should be about elbow height. If the task to be performed involves precision work at close visual range, then the work height should be 3 in. to 4 in. (8–10 cm) *above* elbow level. If the operator has to exert downward force when performing a task at a standing workstation, then the work height should be a few inches (6–8 in. or 15–20 cm) *below* elbow height (Figure 13-4).

Elbow height will depend on whether the operator is sitting or standing while working. Further, it will vary with the physical dimensions of the operators. The workstation should preferably be individualized to the person using the area. If more than one person uses the workstation or if the height of the work varies, then an easily

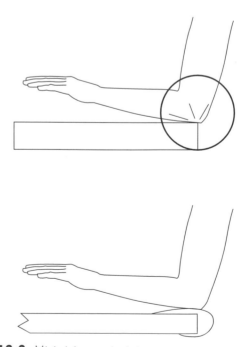

Figure 13-3. Minimizing contact stress.

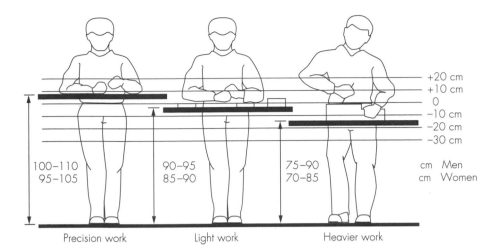

Figure 13-4. Recommended heights of benches for standing work. The reference line (+0) is the height of the elbows above the floor, which averages 105 cm for men and 98 cm for women.

adjustable workstation (e.g., pneumatic, crank driven, or electrically powered) is recommended. Either the height of the work surface itself should be adjustable, or the surface the operator is standing on (platform) or sitting on (chair) should be adjustable. If the height of the chair is increased, the operator should still have adequate foot support (i.e., provide a footrest or a foot ring).

Location of Work Material

Work material should be within the reach of the operator. The primary working area (i.e., the workspace together with the material and tools frequently used) should be within the elbow to wrist length of the operator. Secondary objects not often used could be within the shoulder to wrist length (i.e., the operator should be able to reach the items by stretching the arm, but without needing to lean forward). Infrequently used, light objects could be placed beyond this distance, but still within reach.

If space is at a premium, management should consider using the air space around a workstation. Some examples are hanging tools with an air or spring balancer or placing parts on a lazy-susan type shelf arrangement (Figure 13-5).

Figure 13-5. This reaming tool is supported above the work to relieve stress on the hands, arms, and shoulders.

Table 13-A. Evaluating Posture

Work Posture of Concern	Some Suggestions for Improvement
Looking down, up or sideways	• Locate commonly used displays in front of the operator at, or slightly below, eye level • Tilt work surfaces to allow easy viewing • Avoid glare on viewing surfaces (opt for matte finishes; use indirect lighting) • Arrange the workstation so that the primary work zone is in front of the worker
Raised elbows or shoulders	• Check the interaction between the work height and the operator's elbow height • Match work orientation/height with tool handle
Extended arm reaches (unsupported)	• Arrange the workstation so that the most frequently used objects are within the reach envelope
Rotation of the forearms	• If frequent, use a power tool • OR use an alternative tool (e.g., with a ratchet drive)
Wrist/hands that are bent (up/down, or left/right)	• Adjust work surface height to elbow level • Match the work surface height/orientation with the tool handle
Wide hand grasps	• Install fixtures/jigs to hold parts • Match tool handle size with hand size
Leaning forward or to the side, twisting	• Adjust the height of the work surface • Locate work materials so that the primary work zone is in front of the operator • Locate objects so that the operator has to take a step *towards* them, rather than twisting to reach them
Long periods of sitting/standing	• Vary tasks such that work positions are changed • Provide an easily adjustable chair (if sitting), or a sit/stand stool (if standing) • Install antifatigue mats or footrests/rails at standing workstations
Squatting or kneeling, or repetitive foot pedal work	• Provide cushioned surfaces to kneel on (e.g., mats, knee pads) • Modify foot pedals so that they are activated by releasing pressure, rather than applying pressure • Replace foot pedals with hand/finger-activated controls

Clearance and Accessibility

Although clearance and accessibility of work may be issues in a manufacturing situation, they are usually more critical for maintenance workers. Often equipment and work spaces are arranged to minimize the amount of space used. The engineer may consider the user when designing the machine or workspace, but the needs of the maintenance worker are rarely addressed. One often sees maintenance personnel crouched over a motor, or kneeling and reaching to get to the different parts of the equipment.

Access to components that need servicing should be adequate to accommodate both hands and still allow maintenance personnel to see what they are doing. If a tool has to be used, sufficient room should be provided to insert the tool *and* manipulate it with the required force. An opening of about 8 in. (20 cm) is recommended for one-handed tasks requiring force. If the equipment has to be adjusted, the operator should be able to see the displays or target while making the adjustments. Finally, the maintenance worker should be able to service the equipment without contacting hot surfaces, sharp edges, or electric currents. Note that any protective gear worn by the operator would not only restrict the range of motion of the operator, but may increase the size of the opening required to access the parts.

The position the operator assumes will clearly be influenced by the interaction among the physical characteristics of the workplace, the physical dimensions of the operators, and the range of motion of their joints—in other words, by the anthropometrics of the operators. The focus should be towards customizing the workstation so that the individual worker can assume a neutral posture while working. Some factors to look for when evaluating a workers' postures and suggestions to improve each situation are given in Table 13-A.

Design of Accessories

The design of the tools used greatly influences the posture of the operator. The joints affected are usually the wrist, hand, and arm. The design of the handle should be such that it enables the operator to work with the wrist in a neutral position (e.g., Figure 13-6). A quick review of the characteristics to be considered when choosing a tool are given in Table 13-B.

If workers wear gloves as they perform tasks, they should make sure the gloves are comfortable and fit well. An ill-fitting or stiff glove can reduce grip strength and interfere with manipulating the tool. A brief checklist covering the items discussed in this section is given in Table 13-C.

PHYSIOLOGICAL DEMANDS

The build up of fatigue can increase the risk of accidents and injuries and can distract operators from the task at hand. The physiological demands placed on an operator through the performance of a given task or job and through the thermal environment (temperature and humidity), influence the rate at which muscular fatigue develops. This section discusses two different situations that could lead to the build up of unacceptable levels of fatigue: performing tasks with a high-energy demand or doing static work. However, it is

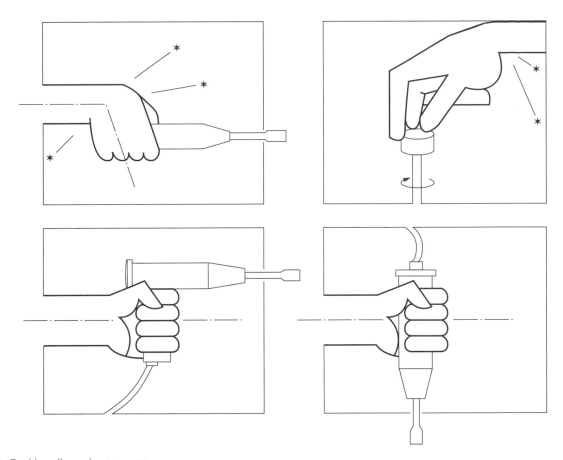

Figure 13-6. Tool handle and wrist position.

Energy Demands

The National Institute for Occupational Safety and Health (NIOSH) has developed some criteria for the rate of energy expenditure that the average worker can sustain over different periods of time. These levels are shown in Figure 13-7. Similar criteria have been adopted by both the American Conference of Governmental Industrial Hygienists (ACGIH), and the International Organization of Standards (ISO 8996). To determine the average energy expenditure for the day, it may be necessary to estimate the energy demand for each of the tasks performed during the day. The exact level of energy expended for a given task varies from individual to individual. However, an estimate for the energy demand can be calculated by using the NIOSH guidelines given in Table 13-D.

Example: An operator finishes a metal part with a grinder at a standing workstation.

Basal Metabolism 1.0 kCals/min
Standing 0.6 kCals/min
Work with both arms (heavy) 2.5
Total 4.1 kCals/min

important to note that given the same job and work environment, the rate of fatigue buildup is dependent on the muscular strength of each individual and on his or her level of fitness.

The rate of energy consumption for the whole shift is determined by using a time-weighted average. Thus, short peaks of high-energy demand may be balanced by periods of lower level activity in order to maintain the average demand for an eight-hour day. When estimating the energy demand for a whole day, remember to include the rest intervals (breaks and lunch).

For example, the worker's task is to finish parts using a grinder for six hours of an eight-hour day. He spends

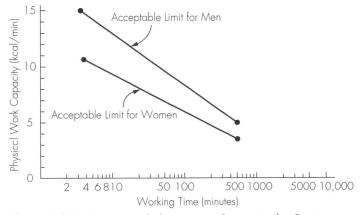

Figure 13-7. Recommended Maximum Capacities for Continuous Work. (Source: NIOSH Publ 81–122.)

Table 13-B. Tool Checklist

Tool Characteristics	Yes	No	Comments
Does the tool allow the operator to see the work interface?	❏	❏	
Does tool weigh less than 4 lbs (if held with one hand)?	❏	❏	
If weight is excessive, is tool balancer provided/used?	❏	❏	
Can continuous holding of the tool be avoided?	❏	❏	
Can the tool be operated without bending the wrist?	❏	❏	
Can the operator's fingertips close around the tool?	❏	❏	
Does the tool have a spring return?	❏	❏	
Can a pinch grip be avoided when operating tool?	❏	❏	
Can the tool be used with gloves?	❏	❏	
Can the tool be used with either hand?	❏	❏	

Trigger Characteristics	Yes	No	Comments
Can the controls/triggers be used without stretching the fingers	❏	❏	
Is the force required to activate the tool less than 316 L?	❏	❏	
Can the tool be run without continuous activation of the trigger?	❏	❏	
If it is not a precision tool, are single finger controls avoided?	❏	❏	

Handle Characteristics	Yes	No	Comments
Are sharp edges on the tool avoided?	❏	❏	
Does the tool handle extend beyond the palm area?	❏	❏	
Is the tool evenly balanced?	❏	❏	
Is there are flange/thumb stop at end of tool to prevent slipping?	❏	❏	
Is the tool comfortable to hold?	❏	❏	
Is the handle slip resistant?	❏	❏	
Is the handle made of/covered with compressible material?	❏	❏	
Is the finger clearance sufficient (>1" for one hand, >2" for two hands)?	❏	❏	
Is the tool handle at a comfortable temperature?	❏	❏	

Power Tools	Yes	No	Comments
Can impact tools/positive clutch tools be avoided?	❏	❏	
Is the force required to keep tool in position, minimal?	❏	❏	
Is the reaction torque at acceptable levels?	❏	❏	
Does the tool have an auto shut-off control?	❏	❏	
Is a battery powered design used to minimize trailing lines?	❏	❏	
Can the operator avoid handling the weight of the air or connection lines?	❏	❏	
Are tools hooked up with swivel type connections to the lines?	❏	❏	
Are the vibration levels appropriate?	❏	❏	
Are noise levels within acceptable levels?	❏	❏	
Is air exhaust directed away from the operator?	❏	❏	

General	Yes	No	Comments
Is the tool in good operating condition?	❏	❏	
Are debris and sparks directed away from the operator?	❏	❏	
Is the a guard for moving parts?	❏	❏	

about an hour of the day performing seated administrative tasks (e.g., tracking parts, updating information on a computer, logging activities) and has an hour total (half-hour lunch, two 15-minute breaks) of rest time. The average energy demand for the day would be:

Grinding 4.1 kCals/min *6*60
Administrative tasks (1.0 + 0.3 + 0.4)*1*60
Breaks(1.0 + 0.3 + 0.4)*1*60
Average rate for the day(24.6 + 1.7 + 1.7)
*60/8*60 = 3.5 k Cals/min

If the job's energy levels are above acceptable limits, the tasks should be modified to decrease the energy demand associated with them. Another option would be to decrease the proportion of time spent on tasks with a high energy demand.

Static Work

Static work represents another situation where an unacceptable build up of fatigue may occur. This type of work is performed by the muscles in any activity that involves holding a position for a certain period of time. Some

Table 13-C. Checklist for Workplace Characteristics

Use this checklist to evaluate the workstation and the work performed at it. If the answer to any of the questions is NO, consider ways to improve that element.

When Working, Can the Operator:	Yes	No
Keep the wrists straight?	❏	❏
Relax the shoulders?	❏	❏
Keep the back in its naturally curved position?	❏	❏
Keep the elbows at about 90°?	❏	❏
Keep their arms close to their sides?	❏	❏
Keep their neck straight and relaxed?	❏	❏
Work without twisting their body?	❏	❏
Easily reach all of the objects they work with?	❏	❏
Avoid contact with sharp-edged work surfaces?	❏	❏
Avoid keeping their body or arms or legs in one position for a long time?	❏	❏
Minimize unnecessary work motions?	❏	❏
Use a hammer or mallet rather than the hand to pound?	❏	❏
Shift position to help circulation?	❏	❏

Does the Workstation Have:

	Yes	No
Sufficient foot space to minimize any forward lean while standing?	❏	❏
The flexibility to allow changes in position (e.g., a sit-stand stool)?	❏	❏
Sufficient leg room to minimize forward lean while sitting?	❏	❏
A chair with good lower back support, if the operator sits while working?	❏	❏

examples of static work are holding things with the hands, bending or leaning forward for periods of time, standing without moving or sitting in the same position, working with the arms extended, and looking down or sideways or upwards for a sustained period.

To minimize the build up of fatigue, the first priority is to modify the workplace characteristics that contribute to the performance of static work. These include:

- installing a fixture or jig to hold the object
- redesigning the workbench or conveyor to minimize forward lean
- providing a sit-stand stool or an adjustable chair
- positioning frequently used objects closer to the operator
- placing displays, dials, and gauges frequently used in front of the operator.

In those situations where static work cannot be avoided, the work process may need to be evaluated to minimize fatigue build up. This can usually be achieved through the use of adequate rest and recovery intervals.

PHYSICAL DEMANDS

The physical/muscular demands placed on an operator usually arise from the task the person is performing. These demands can be the result of materials handling activities (lifting, pushing, pulling, or carrying) or of performing repetitive tasks. Demands that exceed the operator's capabilities may result in an injury or an incident, or impair the person's ability to perform the task well.

Materials Handling

Manual materials handling refers to the human activity necessary to move an object. This often means lifting and carrying but can also describe pushing and pulling. The main issues to consider when performing an ergonomic analysis of a materials handling task include:

- Is there any lifting activity?
- Is material pushed or pulled (even if it is a cart that is pushed or pulled)?
- Is the material carried from location to location?

General guidelines can be used to assess such jobs and identify possible improvements. Each of these issues is discussed in turn below.

Lifting

In 1981, NIOSH published guidelines and an equation for analyzing lifting tasks, with the goal of minimizing back

Table 13-D. Estimating Energy Cost of Work

A. Basal Metabolism		1.0 kCals/min

B. Body Position and Movement		kCals/min
Sitting		0.3
Standing		0.6
Walking		2.0–3.0
Walking uphill		add 0.8 per meter rise

C. Type of work:	Average (kCals/min)	Range (kCals/min)
Hand work		0.2 to 1.2
Light	0.4	
Heavy	0.9	
Work one arm		0.7 to 2.5
Light	1.0	
Heavy	1.8	
Work both arms		1.0 to 3.5
Light	1.5	
Heavy	2.5	
Work whole body		2.5 to 9.0
Light	3.5	
Moderate	5.0	
Heavy	7.0	
Very heavy	9.0	

Source: NIOSH Publication No. 86–113.

Table 13-E. Frequency Multiplier

Frequency = Average number of lifts per minute over a 15 minute period
Work Duration = The total amount of time, per shift,* spent lifting

	Work Durations					
	≤1 hour		>1 hour but ≤2 hours		>2 hours but ≤8 hours	
Frequency	V<30"	V≥30"	V<30"	V≥30"	V<30"	V≥30"
≤0.2	1.00	1.00	0.95	0.95	0.85	0.85
0.5	0.97	0.97	0.92	0.92	0.81	0.81
1	0.94	0.94	0.88	0.88	0.75	0.75
2	0.91	0.91	0.84	0.84	0.65	0.65
3	0.88	0.88	0.79	0.79	0.55	0.55
4	0.84	0.84	0.72	0.72	0.45	0.45
5	0.80	0.80	0.60	0.60	0.35	0.35
6	0.75	0.75	0.50	0.50	0.27	0.27
7	0.70	0.70	0.42	0.42	0.22	0.22
8	0.60	0.60	0.35	0.35	0.18	0.18
9	0.52	0.52	0.30	0.30	0.00	0.15
10	0.45	0.45	0.26	0.26	0.00	0.13
11	0.41	0.41	0.00	0.23	0.00	0.00
12	0.37	0.37	0.00	0.21	0.00	0.00
13	0.00	0.34	0.00	0.00	0.00	0.00
14	0.00	0.31	0.00	0.00	0.00	0.00
15	0.00	0.28	0.00	0.00	0.00	0.00
>15	0.00	0.00	0.00	0.00	0.00	0.00

*Note: The lifting equation only applies for work duration ≤8 hours.

Source: Applications Manual for the Revised NIOSH Lifting Equation (NIOSH Publication 1994–110).

injuries. These guidelines were revised and a new equation developed in 1991 (Revised NIOSH *Lifting Equation*) to incorporate updated information on the biomechanical, physiological, epidemiological, and psychophysical aspect of manual lifting. Although this equation has not been fully validated, it has been in use since its first publication.

Equation for Lifting

When using the NIOSH equation to analyze tasks, bear in mind that it applies strictly to two-handed lifting tasks and is based on the assumption that other manual handling activities (carrying, pushing/pulling) do not consume significant energy. Further, this equation does not apply if the operator is lifting in a constricted space, or lifting while sitting or kneeling or using tools (e.g., a shovel). The 1991 Revised NIOSH *Lifting Equation* assumes that the coefficient of static friction between the worker and the floor surface is at least 0.4, i.e., the worker can obtain a firm footing. The risk of low back injury is presumed to be the same regardless of whether the object is being lifted or lowered, unless the object is lowered by dropping it.

The revised NIOSH equation is used to calculate a recommended weight limit (RWL), which is the weight that a healthy population can handle (for a given set of conditions) without increasing their risk of developing lifting-related low back pain. The RWL is calculated for both the origin and the destination (the starting point of the lift and the end point of the lift). The lower of the two values obtained is defined as the *overall RWL for the task*.

The equation for calculating RWL in English units (FPS) is as follows:

$$RWL = 51*HM*VM*DM*AM*FM*CM \text{ lbs}$$

where HM = Horizontal Multiplier = $10/H$
and H is the horizontal distance of the hands from the midpoint between the ankles

VM = Vertical Multiplier = $1 - (0.0075 \times |V - 30|)$
and V is the distance of the hands above the floor

DM = Distance Multiplier = $0.82 + 1.8/D$
and D is the vertical travel distance—the absolute difference between the vertical heights at origin and destination of the lift (or lowering task)

AM = Angle Multiplier = $1 - 0.0032*A$
and A is the angle the object makes with the frontal (mid-sagittal) plane of the worker

FM = Frequency Multiplier obtained from Table 13–E

CM = Coupling Multiplier obtained from Table 13–F.

Table 13-F. Coupling Multiplier

| Coupling Type* | Coupling Multiplier | |
	V<30"	V≥30"
Good	1.00	1.00
Fair	0.95	1.00
Poor	0.90	0.90

*Note: If in doubt, choose the more stressful classification

Good coupling—is defined as those situations where the operators can wrap their hands around the handles of the object lifted

Fair coupling—is defined as those situations where the operators cannot comfortably wrap their hands around the handles or hand-hold cutouts or grip the object by flexing their hand about 90°.

Poor coupling—is defined as those situations where there are no handles or hand-hold cutouts, or the object is irregularly shaped, or nonrigid (e.g., bags that sag in the middle) or has sharp edges or loose parts or is otherwise hard to handle.

Source: Applications Manual for the Revised NIOSH Lifting Equation (NIOSH Publication 1994–110)

To improve the lifting situation, determine which multiplier or factor(s) has the greatest effect on the RWL (i.e., which multiplier is the smallest). Focus on these factors to increase the RWL.

Example. An operator is packing cases from a conveyor to a pallet right next to the conveyor. The operator stands in front of the pallet and twists 45° to reach to the conveyor. The boxes weigh 25 lbs each, arrive at a rate of two boxes per minute, and are placed in a single layer on the pallet. The sole task of the operator for the eight-hour shift is to pack these boxes. Because the boxes have no handles or hand-hold cutouts, the operators hold the boxes by grasping their hands around each one. The dimensions of

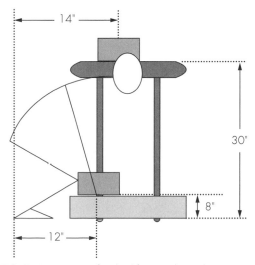

Figure 13-8. Parameters for the lifting task analysis.

Table 13-G. Values for the Multipliers for the 1991 NIOSH Lifting Equation

| Factor | Origin | | Destination | |
	Value	Multiplier	Value	Multiplier
H	14"	0.71	12"	0.86
V	30"	1.00	8"	0.84
D	22"	0.9	22"	0.90
A	45°	0.86	0°	1.00
F	2 lifts/min	0.65	2 lifts/min	0.65
C	Fair	1.00	Fair	0.95
RWL	**18.2**		**19.8 lb**	

the lift are given in Figure 13-8, while the values for each factor are given in Table 13-G.

The RWL is found to be below 25 lb (the weight of the box) at both the origin and the destination of the lift. In order to rectify this situation, one option would be to lower the weight of the boxes to 18 lb. If this is not possible, then each factor should be examined to determine which one has the greatest impact on increasing the allowable weight of the box. Clearly, the frequency of the lifts, F, has the lowest multiplier (0.65); therefore, slowing down the conveyor would be one solution. However, in many situations this may cause an unacceptable decrease in productivity. At the origin, the next lowest multipliers are H and A. At the destination, the lowest multipliers are V and H. By setting up the operator so that he or she does not have to twist to pick up the box from the conveyor and by bringing the boxes on the conveyor 1 in. closer to the operator, the RWL at the origin is found to increase to 25.5 lb. At the destination of the lift, raising the pallet to the same level as the conveyor (e.g., using a palletizer) increases the RWL to 27.5 lb (Table 13-H). Because the RWL at both the origin and destination are above the actual weight of the box (25 lb) the lifting task would now be acceptable, according to the 1991 Revised NIOSH *Lifting Equation*.

Pushing and Pulling

As a general rule of thumb, pushing is preferable to pulling, because it minimizes the risk of having the cart run over the operator or the operator's feet. Carts, if used, should be designed to be pushed, not pulled.

If material is manually placed on the cart or removed from the cart, it should be handled between elbow and waist height. Objects on the cart should not be stacked so high that the operator cannot see where he or she is going. Finally, cart handles should be approximately at elbow height. Castors should be well maintained and chosen to minimize the force required to move the cart. A well-designed cart would also include a brake system that would allow the operator to stop a moving cart, in an emergency. The recommended upper limits for the horizontal forces required to move a hand cart are given in Table 13-I (from Eastman Kodak Company, 1986).

Table 13-H. Values for the Multipliers, after the Task Is Modified

Factor	Origin Value	Origin Multiplier	Destination Value	Destination Multiplier
H	13"	0.77	12"	0.83
V	30"	1.00	30"	1.00
D	8"	1.00	8"	1.00
A	0°	1.00	0°	1.00
F	2 lifts/min	0.65	2 lifts/min	0.65
C	Fair	1.00	Fair	0.95
RWL	**25.5 lb**		**27.5 lb**	

Table 13-I. Recommended Upper Limits for Forces Required to Move a Hand Cart

Activity	Force to Start Motion	Force to Maintain Movement	Emergency Stops (within 3 feet)
Movement less than 10 ft	50 lbf	40 lbf	80 lbf
Movement over 200 ft	50 lbf	25 lbf	80 lbf

Source: *Ergonomic Design for People at Work, Vol, 2.* Eastman Kodak Company (1986), p. 380.

Carrying

Whenever possible, workers should avoid hand carrying objects, transporting them, instead, either on a cart or a conveyor. If workers have to carry objects, however, they should do so for short distances only. The maximum weight carried is a function of the distance to be covered, the arm and shoulder strength of the individual, the type of carry (one handed or two handed), the size of the object, and the design of the handles.

Repetitive Work

The main issue in repetitive work is to minimize the build up of fatigue in the muscles and to avoid overstressing the muscles. With this in mind, the key factors to be considered when evaluating repetitive work are:

- the posture or position of the joints when the task is performed
- the force exerted
- the repetition rate, or the amount of recovery time provided.

Posture

The position of the operator's body when performing different activities affects the force that is exerted (i.e., the stress on the muscles) and consequently the rate of build up of fatigue. For a work environment, the position that places the least stress on the muscles is generally accepted to be the one described in Figure 13-4 (see section Workplace Characteristics and Accessories). Use Table 13-B and the suggestions provided in Workplace Characteristics and Accessories to evaluate and improve an operator's posture.

Force

The greater the force exerted, the more the stress placed on the muscle. As of yet, there are no generalized or validated numerical limits on the force that can be exerted without overstressing the muscles used. However, some activities to look for and suggestions to address in each situation are given in Table 13-J.

Repetition and Recovery Time

Performing highly repetitive tasks could contribute to a build up of fatigue, especially if sufficient recovery time is not designed into the task. Although a job with a cycle time of less than 30 sec is often described as a repetitive task, there is no defined cut-off to classify a particular task as highly repetitive. Further, the exact amount of recovery time required varies from task to task. To identify high-repetition, low-recovery jobs, some key points to look for and suggestions to improve such conditions are given in Table 13-K.

Table 13-J. Evaluating for Force

Activity of Concern	Suggestions for Improvement
High gripping, pinching, pressing forces	• Provide cushioned grips • Optimize handle size for each person/task • Provide "return" springs whenever possible • Use well-fitting gloves
Concentrated pressure points	• Provide cushioned grips • Extend handle length • Investigate alternative handles • Pad sharp edges and corners
Using the body as a brace/vice (static loading)	• Use a fixture • Provide a tool balancer
Impact forces (i.e., hammering with the palm, or knee)	• Use a tool • Use padded, fingerless gloves • Consider redesign of task • Supplier interaction
High pushing or pulling forces	• Minimize the load on the cart • Caster maintenance • Shelves should have a low friction surface
High rotation, twisting, wringing force	• Use appropriate tool/power tool
Reaction torque from a power tool	• Use a reaction bar • Try an alternative tool

Table 13-K. Evaluating for Repetition and Recovery Time

Activity of Concern	Suggestions for Improvement
Identical or similar motions or patterns every few seconds	• Enlarge tasks • Share the muscle load • Use a power tool • Job rotation
Short task cycle times (> 2/min)	• Enlarge the task • Job rotation
Low task variety/long duration	• Enlarge the task • Job rotation
Piece rate work, machine-paced work	• Use accumulating conveyors
Static work	• Change positions frequently • Use tool balancers/holder • Use fixtures/vice

ENVIRONMENTAL DEMANDS

Environmental demands are those that arise from factors external to the operator. These include the effect of temperature, relative humidity, and the air movement in the work environment as well as vibration from machines and tools. These factors could impair the ability of a worker to perform his or her job and could contribute to the development of injuries or other incidents. (See Chapter 11, Industrial Hygiene in this volume.)

Effects of Heat

The human body is designed to work optimally at a temperature of 98.6 F ± 1.8 F. This is the core temperature, i.e., the temperature of the internal cavities/organs, not the skin temperature. If the core temperature rises, the operator may suffer from a heat disorder (heat stroke, heat exhaustion). The worker's ability to think and reason may also be impaired. The body has its own internal mechanisms that work to keep the body cool. As the heat and humidity levels of the work environment rise above the comfort zone, energy of the operator may be diverted towards keeping the body cool rather than to performing the task at hand. In general, the body tries to achieve a balance between the heat gained by the body and the heat lost to the surroundings. This balance is best described by the equation:

$$S = M \pm C \pm R - E$$

where S = heat stored

M = heat gained due to metabolic activity (muscular work)

C = heat gained or lost from the surroundings by convection

R = heat gained or lost from the surroundings by radiation

E = heat lost to the surroundings through the evaporation of sweat

Based on this equation, some methods to decrease the heat stored in the body would be:

• Decrease the metabolic demand, or the effort level of the task.
• Decrease the temperature of the surrounding air. If the whole area cannot be cooled, spot cooling may be an option. If the air temperature is less than 97F (skin temperature), increasing the velocity of air flow could decrease the heat lost through convection. Installing fans or blowers could help increase the air velocity. If, however, the air temperature is above 97°F, the operator may be gaining heat from the surrounding air, and increasing the velocity of the air flow may increase the rate of heat gain.
• Reduce the heat gained through radiation. Some options would be to place a reflective shield before the source of radiative heat (furnace, hot surface, and so on). Another option may be to wear reflective clothing.
• Increase the rate of sweat evaporation by decreasing the humidity level of the air, e.g., increase the air flow in an area.

Modifications in work practices could also help address this issue. Some factors to consider include:

• If the heat and humidity of the work environment are affected by the weather, try to reschedule heavy or difficult tasks for cooler days or cooler parts of the day. Avoid scheduling such activity during heat waves, unless it is an emergency.
• Provide a cool or air-conditioned area for rest and recovery times (e.g., break times).
• Provide a water fountain or other plentiful source of cool water near the work area.
• Make sure the operators pace themselves and take breaks whenever required. One method to determine the schedule of work in hot areas is to use the criteria developed by NIOSH (Publication No. 86–113) (Figure 13-9).

The wet-bulb globe temperature (WBGT) can be determined using a WBGT meter, and the metabolic demand can be estimated by using the method outlined in Table 13-D. For more detailed information, refer to NIOSH Publication No. 86–113: *Occupational Exposure to Hot Environments,* Revised Criteria 1986. If it is not possible to implement these modifications, or if the modifications do not completely resolve the heat-stress issue, the safety professional should investigate the use of protective clothing.

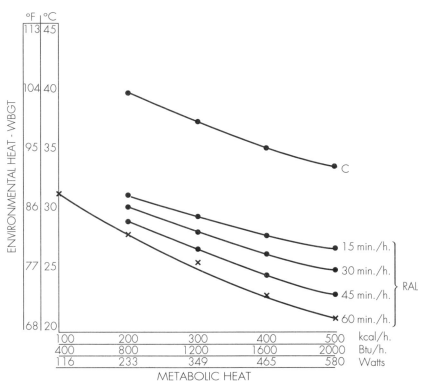

Figure 13-9. Recommended heat stress alert limits.

A number of devices are available in the market, based on either liquid cooling or air cooling. However, keep in mind that some suits may be restrictive or heavy and could increase the muscular effort or energy expenditure required to complete a task.

A safety professional should work with the operators to identify the garment that best suits the work and the worker. When working with the department or division to develop procedures for working safely in a hot environment, it is important to provide training for the operators on the key issues around such work.

Effects of Cold

The human body is less capable of coping with heat loss than with heat gain. In cold environments, therefore, workers should be adequately clothed to maintain their body temperature at about 98.6 F. Exposure to cold temperatures (air temperatures less than 61 F) can also reduce manual dexterity. If workers wear gloves to keep their hands warm, the gloves should fit well and should not significantly interfere with grip strength. Suggestions to minimize the effects of working in the cold include:

A. Work Practices

- Work in the sun as much as possible.
- Share the work load to avoid overheating and sweating.
- When possible, follow a work warm-up schedule.
- Use the buddy system to help detect early signs of frostbite. Avoid working alone. Periodically check under face protectors.
- Pay attention to shivering as a warning sign; workers should get out of the cold if this occurs.
- Wear several layers of clothing, and remove layers as the body warms up or as external temperatures rise.
- Clothing must be kept dry. Allow time for changing into dry clothes.
- Brush snow and moisture off clothes whenever possible.
- Evaporate perspiration by opening the neck, waist, arm, sleeves, and so on to provide fresh air circulation.

B. Protective Clothing

- When considering the insulation requirements of clothes, consider the work to be performed. In general, a person performing heavy work will not require as much insulation to be comfortable, because the body is generating heat.
- Insulated underwear and insulated overalls are usually sufficient to maintain comfort.
- Moisture control in clothes from both perspiration and precipitation is important. Waterproofed exterior clothing can eliminate absorption from snow, rain, fog, and other sources. Clothing should provide for venting of moisture from perspiration as well.
- A mask or scarf should be used during severe wind-chill conditions.
- Extendible fur-lined hoods can be used to protect the face.
- Outer garments that are wind resistant can be worn to reduce the wind chill.
- Footwear should be waterproof.
- Pants should be lapped over boot tops to prevent entry of snow.
- Air insole cushions and felt liners should be used with water-resistant boots.
- Heavy socks and additional dry socks should be available when heavy work is to be performed.
- Wear mittens instead of gloves when manual dexterity is not important. Use mitten/glove combinations.
- If using eye protection, use double-layered goggles with foam padding around the edges.
- Full-face type respirators must have separate respirator channels to prevent fogging and frosting of the face piece.

Vibration

Vibration is an external stressor, (i.e., has an external source) and as such is classified as an environmental factor. At particular frequencies, resonance can be set up in the human body, and this may lead to damage to the musculoskeletal system. There are two forms of vibration that cause concern: (1) whole-body vibration from driving trucks, forklifts, or off-road vehicles, and from the operation of large machinery; and (2) segmental vibration or hand-arm vibration from power tools.

Whole-body vibration is considered one of the risk factors for the development of low back pain, early degeneration of the lumbar spinal system, and herniated lumbar discs. Segmental vibration could contribute to the development of a number of disorders including vibration white-finger syndrome and carpal tunnel syndrome.

NIOSH has a published *Criteria for a Recommended Standard for Hand-Arm Vibration* (NIOSH Publication No. 89-106), but has not established exposure limits for such vibrations. However, the threshold limit values (TLVs)

booklet published by ACGIH (1995) does provide some guidelines, as do the ISO standards 5349–1986 and 2631–1. NIOSH recommends engineering and work practice (administrative) controls to minimize exposure to such vibration and to minimize the possibility of developing vibration-related disorders. Some suggestions offered by NIOSH include:

- selecting tools and machinery with the least amount of vibration
- modifying the vibration at the source
- using vibration-isolators for tools, machinery, and seats of transport vehicles
- providing protective clothing and equipment (e.g., antivibration gloves)
- using vibration-damping materials on the handles of tools and other contact surfaces
- providing adequate protection from the cold to minimize the possibility of vaso-constriction
- reducing the number of hours/shift of exposure to the vibration
- reducing the number of days/week the vibrating tools are used
- following a regular maintenance schedule for tools to ensure good lubrication and even wear
- decreasing the grip force required to reduce the coupling between the tool and the operator
- restricting the use of piecework and incentive pay.

MACHINE DISPLAYS AND CONTROLS

How workers activate machines and equipment and what information they receive through dials or gauges has a definite effect on work efficiency and worker safety. For example, if a worker must stop a machine in an emergency and pushes the wrong button, the result can be a serious accident.

The interaction between machine and operator must be considered when equipment is designed or purchased. This includes reaching and accessing parts of the equipment, setting correct working heights, properly locating controls and displays, and arranging convenient maintenance.

Most organizations use two major types of displays: visual and auditory. A display can be a dial gauge, instrument, or auditory alarm that conveys information to the operator about the status of equipment or processes. Information displayed in a confusing format can cause serious operator errors and possibly disastrous accidents.

Visual Displays

Visual displays are used for one of three purposes:

- Quantitative readings—to determine the exact quantity involved, such as a scale.
- Qualitative readings—to determine the state or condition at which the machine is functioning, which is

usually one of three conditions, such as *above, within,* or *below tolerance.*

- Dichotomous (check) readings—to check operations or to identify one or two levels, such as Off or On.

The purpose for which the display is to be read will dictate its best design. As a general principle, however, the simplest design is the best. The motion of the display should be compatible with (or in the same direction as) the motion of the machine and its control mechanism. A pointer that moves to the right, up, or clockwise to show an increase should have its corresponding control mechanism designed so that a rightward, upward, or clockwise movement of the control will increase the machine value and the corresponding display output value.

Multiple displays should be grouped according to their function or sequence of use. If dials are arranged in groups on a large control panel and must all be read at the same time, all the indicators should be pointing in the same direction when in the desired range. This method will reduce check-reading time and increase accuracy.

All displays should be labeled so the operator can quickly tell what the display means, what units are being used, and what is the critical range. If the equipment is located in a dimly lighted area, make sure all displays are adequately illuminated. Table 13-L lists types of information displayed and the best visual display for each.

Auditory Displays

Auditory displays should follow the principles outlined for visual displays. The question of whether an auditory or visual display should be used depends upon the situation. Table 13-M compares the relative advantages of auditory and visual displays. Other considerations for using auditory displays are found in the source materials at the end of this chapter.

To evaluate existing displays (visual or auditory), management should consider the following:

- Is the display intensity higher than the lowest threshold level for sight or hearing? (Each sense has its own threshold level; energy intensities below the level cannot be perceived.)
- Is the sense overloaded? What other demands are being made on sight or hearing at the same time?
- Is the display compatible with similar displays, controls, and machine movements?
- What environmental factors, if any, could mask the display?

Controls

A control is anything—a switch, lever, pedal, button, knob, or keyboard—used by an operator to put information into a system to affect its operations. Performance can be enhanced when the controls work as the operator expects them to, if they are designed to fit human dimensions, and if their operating characteristics are within the strength and precision capabilities of the operator. Figure 13-10 illustrates standard sizes for many common controls.

People expect things to behave in certain ways when they use operating controls. Most people in the United

Table 13-L. Recommended Visual Displays

Information Type	Preferred Display	Comments	Examples in Industry
Quantitative Reading	Digital readout or counter	Minimum reading time Minimum error potential	Number of units produced on a production machine
Qualitative Reading	Moving pointer or graph	Position easy to detect, trends apparent	Temperature changes in a work area
Check Reading	Moving pointer	Deviation from normal easily detected	Pressure gages on a utilities console
Adjustment	Moving pointer or digital readout	Direct relation between pointer movement and motion of control, accuracy	Calibration charts on test equipment
Status Reading	Lights	Color-coded, indication of status (e.g., "On")	Consoles in production lines
Operating Instructions	Annunciator lights	Engraved with action required, blinking for warnings	Manufacturing lines in major production systems

Six types of information to be gathered from displays are indicated in column 1. For each type the preferred display is indicated in column 2. Additional information is provided in column 3 about the reasons for choosing the preferred display or about characteristics of that display that make information transfer from display to operator more effective (e.g., color-coded lights). Some examples of industrial uses of these displays are given in column 4.

Source: Adapted from Grether and Baker, 1972.

Table 13-M. Visual versus Auditory Presentation of Signals

Use Visual Presentation If

- The person's job allows him or her to remain in one position
- The message does not call for immediate action
- The message is complex
- The message is long
- The message will be referred to later
- The auditory system of the person is overburdened
- The message deals with location in space
- The receiving location is too noisy

Use Auditory Presentation If

- The person's job requires him or her to move about continually
- The message calls for immediate action
- The message is simple
- The message is short
- The message will not be referred to later
- The visual system of the person is overburdened
- The message deals with events in time
- The receiving location is too bright or preservation of dark adaptation is necessary

The upper part summarizes conditions in the workplace, or conditions related to the information to be communicated, that make visual presentation the preferred method. The lower part provides a comparable list for auditory presentation of information. Visual presentation is preferred for complex messages in noisy environments where response time is not critical. Auditory presentation is preferred for simple messages in areas where people move around frequently and where response time must be rapid.

Source: Adapted from Deatherage, 1972.

States, for example, expect to turn on a light switch by flipping the switch up. A clockwise motion generally produces to an increase in volume, temperature, or other element. Conversely, people expect the reverse kinds of movements to turn a system off or to decrease a function or flow. Such responses are called "population stereotypes," and are common to nearly everyone in a given population. Table 13-N lists stereotypes for switches in the United States. Any control response that calls for a movement in a direction contrary to the established stereo-type is likely to produce errors. The designer is asking for trouble by requiring the operator to change a long-standing habit.

Safety also may be jeopardized if an operator misreads a poorly designed display and operates the wrong control or the right control in the wrong direction. Despite the fact that many accident reports classify these types of accidents as unsafe procedures or human error, in fact they are design errors. Retraining the operator probably will not prevent future incidents because, in an emergency, the operator will tend to fall back on the original stereotype behavior. The operator should never be required to manipulate controls in unnatural or unexpected ways.

Control Design Principles

Research has established the principles of compatibility, and coding in designing effective, safe controls.

Compatibility

As in the design of displays, control movement should be designed to be compatible with the display and machine movement. A lift truck, for example, with lift controls that move right or left to raise or lower the lift is bound to have a number of errors associated with its operation. The correct movement would be up and down.

Table 13-N. United States Stereotypes for Up and Down Switch Settings

UP	DOWN
On	Off
Start	Stop
High	Low
In	Out
Fast	Slow
Raise	Lower
Increase	Decrease
Open	Close
Engage	Disengage
Automatic	Manual
Forward	Reverse
Alternating	Direct
Positive	Negative

Some movement stereotypes for toggle switches used in equipment or production system control panels are given. The expected direction of activation of the switch (when mounted vertically) is given for 13 actions or conditions. These expectations are United States stereotypes and may vary in other countries.

Source: Adapted from Alexander D., 1976, Tennessee Eastman Company.

Path of C. motion	Control		Dimension (mm)	Force F [N] Moment M [Nm]		2 positions	>2 positions	Continuous adjustment	Precise adjustment	Quick adjustment	Large force application	Tactile feedback	Setting visible	Accidental actuation
Turning movement	Handwheel		D : 160 - 800 \n d : 30 - 40	D: 160 - 200 mm / 200 - 250 mm	M: 2 - 40 Nm / 4 - 60 Nm	◐	◐	●	●	◐	●	○	○	◐
	Crank	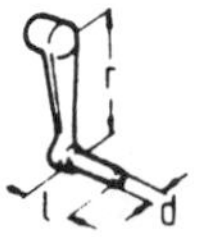	Hand (Finger) \n r : < 250 (< 100) \n l : 100 (30) \n d : 32 (16)	R: < 100 mm / 100 - 250 mm	M: 0,6 - 3 Nm / 5 - 14 Nm	◐	◐	●	◐	●	◐	◐	◔	○
	Rotary knob	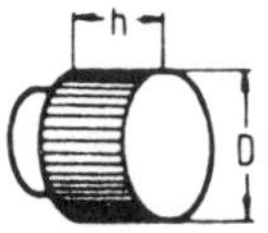	Hand (Finger) \n D: 25-100 (15-25) \n h : > 20 (>15)	D: 15 - 25 mm / 25 - 100 mm	M: 0,02 - 0,05 Nm / 0,3 - 0,7 Nm	◔	◔	●	●	◐	○	○	○	◐
Swiveling movement	Lever		d : 30 - 40 \n l : 100 - 120	10 - 200 N		●	●	◐	◔	●	●	◐	◐	○
	Joystick		s : 20 - 150 \n d : 10 - 20	5 - 50 N		●	●	●	●	○	◕	◐	◐	○
	Toggle switch		b : >10 \n l : >15	2 - 10 N		●	◐	○	○	●	○	●	●	○
	Rocker switch	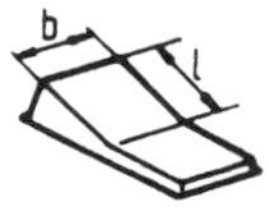	b : >10 \n l : >15	2 - 8 N		●	○	○	○	●	○	●	●	◔
Linear movement	Handle (Slide)	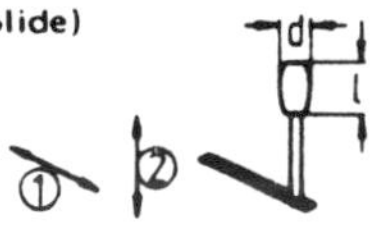	d : 30 - 40 \n l : 100 - 120	F_1 : 10 - 200 N \n F_2 : 7 - 140 N		●	●	●	◕	◐	●	◐	◐	○
	D-Handle		d : 30 - 40 \n b : 110 - 130	10 - 200 N		●	●	●	◐	○	◐	◕	◕	○
	Slide	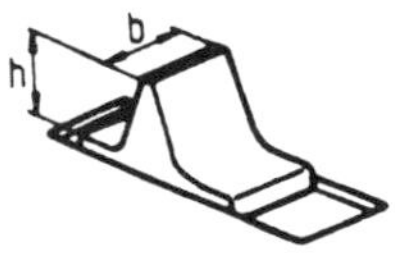	b : >10 \n h : >15	1 - 10 N \n (Thumb-finger grip)		●	◕	◕	◕	◐	◐	○	◕	◐
	Sensor key	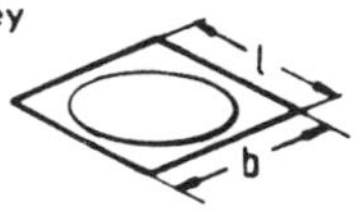	l : >14 \n b : >14			●	○	○	○	●	○	○	○	◐

b Recessed installation of the control

Figure 13-10. Hand- and foot-operated control devices and their operational characteristics and control functions.

Coding

Whenever possible, all controls should be coded in some way, such as by a distinguishing shape and texture, location, or color. A good coding system can reduce many errors, for example:

- Shape and texture to code controls is useful where illumination is low or where a control needs to be identified and operated through touch only.
- Location is another way to code controls. On forklift trucks, for example, all controls pertaining to mast operation can be grouped on the right side of the control panel. Location coding can also be achieved by using a minimum distance between controls.
- Color codes are useful for quick visual identification of various controls and for grouping controls for a particular operation.

All controls require some type of labeling to identify their function.

Arrangement of Controls

The arrangement of controls (and displays) should support the interaction between users and the controls or display components. Components of the system should be arranged with these considerations in mind:

- Convenience, accuracy, speed, or other criteria determine the location of each component.
- Components having related functions should be grouped.
- Controls and displays should be grouped according to their place in a critical set of operations. The important controls should be positioned in the most convenient location for rapid and easy use.
- Controls are frequently used in a sequence or pattern relationship. In these cases, the controls should be arranged to take advantage of such patterns by locating them close to each other.
- Less frequently used controls should be placed in more distant locations.

If there is conflict between any control arrangement considerations, compromises will have to be made. Generally, designers should first consider the frequency and sequence of use of a control. They should avoid control or display arrangements where workers must frequently transfer their entire body, eye, or hand from one place to another.

Control Evaluations

There are other questions to consider when designing controls:

- Where are the controls placed?
- Can they be reached easily?
- Are they spaced far enough apart?

- Are they labeled and coded?
- What type of control is used?
- Is it compatible with user needs?
- Do the controls themselves present a hazard?
- Which body limbs are involved?
- Are any of the muscles used to operate the controls being overloaded?
- Are similar control operations alike in design and function?
- How standardized are the controls?

SUMMARY

- The purpose of ergonomics is to reduce the physical (and mental) stress associated with a given job; to increase the comfort, health, and safety of a work environment; to increase productivity; to reduce human errors associated with a task; and to improve the quality of work life.
- The goal of an ergonomics program is to reduce the human and monetary costs associated with inadequately designed workplaces, work processes, and work environments.
- An effective ergonomics program combines both a proactive and reactive approach and involves all affected personnel.
- The key to a good workplace setup is to design it so that it accommodates the greatest number of people (90% to 95% of the working population).
- The build up of fatigue can increase the risk of accidents and injuries and can distract operators from the task at hand.
- Excessive physical/muscular demands resulting from materials handling activities or from performing repetitive tasks can contribute to an injury or incident or impair the worker's ability to perform the task well.
- Excessive temperature, relative humidity, air movement, or vibration from machines and tools can contribute to an injury or incident or impair the worker's ability to do the job.
- Controls to activate or deactivate machinery and information dials or gauges must be designed for worker safety and efficiency.

REFERENCES

American Conference of Governmental Industrial Hygienists. *Threshold Limit Values for Chemical Substances and Physical Agents, and Biological Exposure Indices (BEIs)*. Cincinnati, OH; ACGIH, 1995.

International Organization of Standards. Case Postale 56, Geneve 20, Switzerland CH-1211.

Ergonomics—Determination of Metabolic Heat Production. Geneva, Switzerland. ISO 8996 (1990).

Mechanical Vibration—Guidelines for the Measurement and the Assessment of Human Exposure to Hand-Transmitted

Vibration. Geneva, Switzerland: ISO 5349 (1986).

Evaluation of Human Exposure to Whole-Body Vibrations—Part 1: General Requirements. Geneva, Switzerland: ISO 2631-1 (1985).

Eastman Kodak Company. *Ergonomic Design for People at Work,* Vol. 1 and Vol. 2. New York: Van Nostrand Reinhold, 1986.

National Institute for Occupational Safety and Health. *Applications Manual for the Revised NIOSH Lifting Equation* (NIOSH Publication 1994–110). Cincinnati, OH: NIOSH, 1986.

Occupational Exposure to Hot Environments: Revised Criteria (NIOSH Publication 86–113). Cincinnati, OH: NIOSH, 1986.

Parsons, KC. *Human Thermal Environments.* Philadelphia, PA: Taylor and Francis, 1993.

Pheasant, S. *Bodyspace, Anthropometry, Ergonomics and Design.* Bristol, PA: Taylor and Francis, 1986.

Putz-Anderson, V. *Cumulative Trauma Disorders—A Manual for Musculoskeletal Diseases of the Upper Limbs.* New York: Taylor and Francis, 1988.

REVIEW QUESTIONS

1. Ergonomists have received training in which fields of study that have contributed to the industry's understanding of ergonomics problems and solutions?
   a. Anatomy and physiology of the human body
   b. Anthropometrics
   c. Biomechanics
   d. Psychology
   e. Industrial design and engineering
   f. All of the above
   g. Only a, c, and e
2. What are the five goals of ergonomics?
   a.
   b.
   c.
   d.
   e.
3. Name the three general categories of demands placed on the worker, and give examples of each.
   a.
   b.
   c.
4. List the four components of an ergonomics program designed to manage cumulative trauma disorders (CTDs).
   a.
   b.
   c.
   d.
5. What are the primary factors that are assessed during ergonomics evaluations?
   a.
   b.
   c.
   d.
   e.
6. Why can a situation where the job demands are below a worker's capabilities be just as harmful as one where the demands are too high?
7. List four factors that influence the postures assumed at work.
   a.
   b.
   c.
   d.
8. Describe a "neutral posture".
9. Above-average energy demand and static work result in an unacceptable build up of:
   a. Muscular strength
   b. Fatigue
   c. Stress
   d. All of the above
10. Guidelines and an equation for analyzing tasks to minimize back injuries were revised in 1991 by:
    a. The National Institute for Occupational Safety and Health
    b. The American Conference of Governmental Industrial Hygienists
    c. The International Organization of Standards

14 Employee Assistance Programs

All organizations seek to have a highly productive workforce with the lowest possible accident and lost time rates. However, within every workgroup are certain employees whose personal problems are so severe that they pose safety, productivity, and management challenges to the organization. These problems include alcoholism and drug abuse, marital and family problems, as well as significant psychological and emotional impairment. In addition, the problems of a worker's family often can be an equally severe drain on the worker's emotional stability. To help companies deal with these and other common employee problems, this chapter covers the following topics:

- how employee assistance plans (EAPs) work
- establishing an EAP tailored to a particular company's needs
- characteristics and benefits of internal EAPs
- external EAPs and their relationships with employers
- main products provided by EAPs to employees
- dealing with employee violence and harassment in the workplace

In most instances, when employees have serious problems, it is more responsible and cost effective for employers to offer them assistance than to use discipline or threats of termination to try to change their behavior. The goal of employee assistance programs (EAPs) is to enable employers to help their troubled employees (or their family members) resolve their personal problems as quickly and cost-effectively as possible and to return them to peak productivity.

The cost of ignoring troubled employees can be staggering.

> Estimates are that perhaps 17 percent of our workers . . . use alcohol and other drugs on the job. One consequence of this is that every single employee who does use alcohol or drugs on the job costs his or her employer between $4,000 and $5,000 per year above payroll. (EAPA, 1991, p. 8)

Much of this cost can be attributed to accidents, theft, and injuries on duty. Workers involved with alcohol or drugs are estimated to be caught up in work-related accidents 3 to 4 times more often than other workers. The Employee Assistance Society of North America (EASNA) estimates that up to 40% of industrial fatalities and 47% of industrial injuries can be linked to alcohol abuse. Worker problems need not be alcohol or drug related to pose safety and productivity risks. The worker who is depressed and exhausted after a night of family strife, or distracted because a child has run away from home will certainly have difficulty concentrating on the demands and risks of the job.

EAPs can reduce the costs associated with troubled employees and get workers back on jobs. A survey conducted by *American Management Magazine* in 1985 showed EAPs were responsible for declines of 33% in use of sickness benefits, 65% in work-related accidents, 30% in workers' compensation claims, and 74% in time spent on supervisor reprimands.

DEVELOPMENT OF EMPLOYEE ASSISTANCE PROGRAMS

The development of today's EAP began early in this century with occupational health programs designed to treat workers' illnesses. In the 1940s, companies such as DuPont, Eastman Kodak, and Consolidated Edison initiated "alcoholism programs" to address the needs of this particular troubled employee population. In 1962, the Kemper Group expanded its program to include persons with "other health and living problems" and their dependents. This approach gradually grew into what is called today the "broad-brush" approach to EAP, in which the EAP is charged with helping employees with any behavioral problem that may impair their productivity.

Today there are more than 17,000 EAPs in business and industry in the United States; 85% of the Fortune 500 corporations have an EAP. Although alcohol and drugs are now only two of the many problems EAPs address, these two issues remain central to any program's functioning. Substance abuse is directly correlated to safety problems in the workplace. Further, employee recovery from alcoholism or drug addiction is directly linked to cost savings and productivity improvement for the employer.

HOW EMPLOYEE ASSISTANCE PROGRAMS WORK
An EAP is

> a worksite-based program designed to assist in the identification and resolution of productivity problems associated with employees impaired by personal concerns including, but not limited to, health, marital, financial, alcohol, drug, legal, emotional, stress, or other personal concerns which may adversely affect employee job performance. (EAPA, 1991, pp. 3, 4)

The assistance is provided through referrals to outside counseling or other treatment services in the community, treatment services provided by the EAP itself, or by a wide variety of training and education programs available through the EAP. Services are available to both the employee and to members of the employee's immediate family.

Staffing and Philosophy of EAPs
EAPs employ trained professionals whose task is to help employees find the assistance they need. To accomplish this goal in the modern workplace, the EAP will always abide by certain principles and offer certain basic services.

The company will need to develop policies to establish the EAP and to encourage employee participation. Joint sponsorship or support of the EAP by both unions and management is a prerequisite for the launch of an EAP.

The principle of confidentiality is vital to any EAP's functioning. Only if employees feel that their employment is not threatened by participation in the EAP and that their problems will remain private will they willingly come forward to seek help. Policies regarding treatment for alcohol and drug abuse must establish what behavior is acceptable in the workplace and what are the consequences for violations.

EAPs are effective only if they attract voluntary participants, and if they also generate supervisory or team leader referrals of employees whose performance has noticeably declined. Therefore, all EAPs must include a supervisory training or team leader program about the EAP and its use. Parallel training for union stewards or coordinators will enable the unions to assist workers before they come to the attention of the supervisor.

Constructive confrontation is a basic supervisory tool that has been a foundation of EAP work since its inception. In constructive confrontation, the supervisor conveys to the troubled employee the adverse job consequences the employee may experience if his or her deteriorating job performance does not improve. The supervisor then refers the employee to the EAP. In this way, the supervisor becomes a key factor in helping the employee to face his or her personal issues.

Finally, the EAP will need a publicity vehicle to inform all employees about the services the EAP has to offer, the policies governing participation, and the details of its operation. The successful EAP will attract large numbers of volunteers, as well as a significant number of directed referrals from supervisors.

Procedures for Obtaining Assistance

Once the employee or dependent (known as a client) uses the EAP's services, either voluntarily or by supervisory referral, the following steps will almost always occur. A professional assesses the client's problem and recommends treatment. The assessment and recommendation are recorded by the professional in confidential EAP files. Treatment or other services will be rendered, either by the EAP itself, by referral to providers under the employer's health plan, or to community services that are either free or must be paid for by the client. If the client is an employee who has taken time off work for treatment, the EAP may assist with reintegration to the workforce as soon as practical.

EAP staff will then conduct follow-up sessions with the client to support his or her recovery and to provide additional services if they are needed. When the client's problem may impact safety in the workplace (such as a drug problem in an employee) or when the client was a supervisory referral, the EAP may monitor recovery to ensure that safety in the workplace is not compromised.

EAP and Community Treatment Resources

A primary responsibility of EAP staff is to identify and to evaluate treatment resources in the community. The staff must locate qualified treatment providers for any and all problems that clients may have and in locations accessible to the employee population. Thus, EAP staff may visit psychiatric hospitals and chemical dependency treatment programs, talk with therapists and counselors, identify credit counseling and legal services, and become familiar with self-help groups and community programs in the area.

The evaluation component of this task is essential. EAP staff must assess, on an on-going basis, the quality of treatment being given to the EAP's clients. If clients complain about a particular provider or the quality of treatment begins to decline, the EAP will need to locate new treatment providers who can better meet the needs of the EAP's clients.

Benefits of EAPs to Employers

Using an EAP provides many benefits.

- Morale improves as disruptive employees are helped.
- Managers spend less time working with troubled employees.
- Valued employees with personal problems remain with the company rather than resigning.
- Hiring and training costs are lower.
- Management spends less time and resources on discipline.
- Safety improves and liability declines.

Many companies have estimated savings of $5 to $7 for every $1 invested in their EAP.

SETTING UP AN EMPLOYEE ASSISTANCE PROGRAM

Regardless of the size of the company, or whether services will be provided by internal staff or by external contractors, the following steps are recommended to ensure the effectiveness of the EAP. First, an EAP advisory committee should be formed. This advisory committee should include representatives from management, human resources, the medical department, supervisory personnel, all labor unions, and the workforce (Figure 14-1). The success of the EAP generally depends on the support of all factions within the organization. The committee's functions include evaluating the need for an EAP (and for particular EAP services), planning and implementing the EAP, encouraging its use, setting goals, and evaluating the program and its progress toward those goals. When several small companies use the same EAP, each may place one or more members on a joint advisory committee that oversees all programs.

At the outset, companies must conduct a needs assess-

Management—Management must participate in all stages of assessment and definition to demonstrate support of the program. Open communication between employees and management is essential.

Labor representative—Participation by top level labor representatives is essential during the entire process. In order to achieve eventual acceptance of the program by employees, labor must be involved from initial information gathering to development planning to program implementation.

Employees—Because they are the users of EAP services, employees can give information on what is needed and how it can be accessed most easily.

Human resources/benefits coordinator—By monitoring use of current benefits, personnel staff can provide realistic information on the potential use of services and on the costs involved.

Health care—Advice from health care professionals is necessary for assessment and for selection of services.

Safety and health professional—The safety and health professional should be involved in the assessment of health and safety problems that may affect employee performance. If an EAP is adopted, the safety and health professional should perform follow-up evaluations.

Financial officer—Input from the organization's financial officer is necessary to define the feasibility and scope of the program.

Legal counsel—Legal advice is essential to assure that the EAP structure developed or contracted is fair, accurate, and legally defensible.

Public relations—The public relations department can contribute information on public opinion, from both inside and outside the organizational structure, and on methods of information distribution.

Figure 14-1. Depending on the size and structure of your organization, the EAP task force may include the members discussed here. If these positions do not exist, include persons who do handle pertinent information or seek outside advice.

ment to define the major employee problems within the workforce and the kind of EAP services required to address those problems. The needs assessment outlines the following:

- type of organization and industry
- number of worksites
- type of work/jobs
- size of workforce and demographics (sex, age, ethnicity, education, special needs)
- major employee problems (from benefits and workers' compensation data)
- risk management issues (from safety, medical, and insurance data)
- management and labor identification of problem issues
- regulatory requirements of government agencies
- resources available to EAP from the corporation

This needs assessment may be the joint product of the advisory committee, individuals from management assigned to complete the task, and/or outside consultants brought in to lend objectivity and industry perspective to the study.

Following the needs assessment, a plan should be developed that identifies any barriers to establishing an EAP that need to be overcome, the resources inside and outside the organization that will support the EAP, and the design and structure of the EAP that will best serve the corporation and its workers. For instance, the advisory

committee needs to examine the benefits plan to see if treatments that are likely to be recommended by EAP staff are adequately covered. They should also canvass providers in the community (therapists, hospitals, and so on) to determine what services are available or lacking. Finally, they should draft general policies governing the formation of an EAP and the rights of employees. The most important decision in the plan may be the choice of which EAP delivery system to use.

The major types of EAPs are:

- Internal EAP—services delivered by professionals employed by the organization
- External EAP—services delivered by a contracted vendor
- Union-based EAP—services delivered by trained union personnel to union members
- Consortium EAP—a group of smaller companies band together to jointly contract with an EAP
- Blended EAP—any combination of the above

If the needs assessment and planning have been carefully done, the selection of which EAP type to choose may be quite clear.

Because most EAPs are either internal or external, and because the characteristics of each differ in some significant ways, the two types are described separately in the following sections.

INTERNAL EMPLOYEE ASSISTANCE PROGRAMS

Because the internal EAP is staffed by employees of the corporation, the EAP can interface easily and cooperatively with other departments such as benefits, safety, medical, and all of the operations units. EAP staff are close to the workings of the organization and will have intimate knowledge of the special problems and concerns of employee groups. The placement of the EAP within the organization is important, as it is rare for managers ranking above the director of the EAP to seek help from the program. Yet assisting these individuals can have the greatest positive impact on company productivity. For this reason, many EAP directors/managers report directly to the president of the corporation. Alternatives include placing the EAP under the vice president for human relations, personnel, or risk management.

Size and Nature of Internal EAPs

Internal EAPs range in size from one professional to departments with 60 or more staff. The Employee Assistance Professionals Association recommends providing one counselor for each 2,500 employees, while the Employee Assistance Society of North America recommends a ratio of one counselor for every 6,000 employees. Professional staff should be appropriately licensed and certified as health professionals in their state of operation. Staff specialties may include psychologists, social workers, and licensed professional counselors. There is now a national credential called Certified Employee Assistance Professional (CEAP), which is given in recognition of the special set of skills required to offer comprehensive EAP services. In addition to general clinical interviewing skills, the ideal employee assistance professional will also be especially knowledgeable regarding alcohol and drug problems, possess marketing skills to help publicize the EAP within the company, and have the political sensitivity to evoke trust in both supervisors and union members.

The perception of confidentiality is especially important for the internal EAP, since embarrassment or suspicion on the part of employees about EAP participation can severely hamper its efforts. In addition, EAP offices need to be located at sites convenient enough to attract clients, yet separate enough from other employee areas to protect the privacy of those who seek EAP help. Policies and procedures defining the extent of EAP confidentiality, and guaranteeing that EAP participation will not adversely affect the employee's opportunity for advancement and promotion within the company, must be clearly written and widely distributed to all employee groups. The longer that EAP offices can be open during the week, the more accessible the EAP will be. However, this policy may present problems in terms of staffing for evenings and weekends, and may sometimes exceed the resources of the company.

The internal EAP has special advantages in its ability to publicize EAP services, train supervisors, and follow employees closely in the workplace after treatment. EAP staff have easy access to company newspapers, company bulletin boards, and other company media vehicles to promote EAP services. The staff can often participate in management meetings to identify new training opportunities and learn quickly about managers' current needs for new EAP "products." EAP can work in partnership with other departments to provide joint training for supervisors and managers on the best way to use EAP services. Because EAP offices are in or near the workplace, staff can work closely with employees as they complete treatment and return to their jobs.

Treatment and Recovery Services

Research shows that professional follow-up after behavioral treatment improves treatment success. Regular phone calls by EAP staff to troubled employees who are in treatment or have recently completed treatment will help workers integrate back into the workplace as they continue their recovery. If problems arise for these workers, a quick and confidential session with EAP staff can resolve issues or provide an opportunity to revise treatment strategy. Close follow-up and monitoring of recovering alcoholics and drug addicts after treatment can reduce the frequency of relapse and motivate employees to return to sobriety if a relapse does occur. In safety-sensitive industries, the EAP may be given the responsibility of notifying the medical staff to remove an employee from service if relapse occurs.

This type of intervention can be done confidentially (because medical staff, not EAP personnel, contact the supervisor), can strongly motivate the employee to resume recovery, and simultaneously protects the safety of the workplace. In the transportation industry (where the U.S. Department of Transportation encourages, and in some circumstances requires, monitoring of drug addicts) alcohol and drug recovery rates are frequently reported to be much higher than in industries where alcohol and drug monitoring by internal EAPs is less common.

Another strength of internal EAPs is their familiarity with the treatment needs of the employees and their knowledge of local community resources. EAP staff can easily visit the client who has been hospitalized for psychiatric problems or is in a local chemical dependency treatment center. This visit will help the client make a less stressful transition back to the workplace and will allow EAP staff to have first-hand knowledge of any changes in the quality of treatment or staff at the hospital. When many employees show an interest in a common problem or issue in the workplace (such as stress, domestic violence, grief issues, or parenting skills), the internal EAP staff can identify the need and quickly locate community resources to provide training on the issue in question. Because the EAP exists in the same community as its treatment providers, the program staff can establish a rapport with the therapists and hospitals that frequently work with the EAP's clients. This close working relationship may help improve the quality of treatment for clients and may enable the EAP to negotiate reduced fees to benefit both the clients and the company.

Cost Savings of EAPs to Employers

Because the mission of an EAP is to improve productivity and reduce employee costs, it is very important for the EAP to document its cost savings to the organization. The internal EAP can readily provide this documentation by working closely with workers' compensation, safety, benefits, and medical departments to track the employee costs of clients who come to the EAP. Internal EAPs routinely access benefits data to show that employees treated through EAP use fewer health benefits for themselves and their family members after completion of treatment. Safety and workers' compensation data can be assembled to show that EAP participants are injured less frequently at work and have fewer accidents. Lost-time data should show reductions in lost time after an employee enters the EAP. Similar EAP cost reductions could be shown for employee turnover, theft, arbitration, and legal costs, provided the internal EAP uses its position within the company to assemble this data.

EXTERNAL EMPLOYEE ASSISTANCE PROGRAMS

External EAPs can provide services to a wide variety of companies, from small privately held corporations to major Fortune 100 organizations. An external provider can focus its efforts on providing a specialty service while spreading the costs over a greater number of client organizations. The external provider can develop sophisticated information systems, hire personnel with greater expertise and experience in EAP, and develop other systems to better address issues such as quality assurance/quality improvement and provider network development.

Choosing an External EAP

The selection of an external EAP provider should include many considerations, keeping in mind that the external provider will be expected to offer all the services that an internal program would. Companies must take the time to choose an adequate external provider with whom they can build a quality EAP (Figure 14-2.).

The key elements to look for in a quality external EAP provider are:

- Does the EAP have the professional EAP expertise on staff? (e.g., Certified Employee Assistance Professionals [CEAPs], who have appropriate experience in all areas of the design, implementation, function, and maintenance of EAPs). Certification as an EAP

professional is based on commonly accepted standards of training and experience in the field. As a result, the presence of CEAP personnel on staff also confirms that the provider has the knowledge and ability to handle all basic EAP services and many other types of services.

- Does the EAP have an appropriate and adequate provider network? An appropriate provider network must include credentialed and trained EAP assessors (licensed clinicians) who are available to all employees. All assessors should be subject to a formal credentialing process by the organization that includes verification of personnel's education, licenses, professional liability insurance, and experience. The EAP network should also include adequate means to identify and access all other helping resources in the employer's service areas (i.e., those near all employer worksites).

- Does the EAP staff offer assessment, short-term counseling, and referral services one-on-one with a licensed clinician/assessor, or are these services performed over the telephone?

- How many sessions with a clinician/assessor are available for client assessment and short-term counseling? (Commonly, providers will offer one- to three-session, one- to six-session, or one- to eight-session modules at different prices.)

- Does the EAP follow up with the client and the treatment provider to make sure the client is receiving appropriate and adequate services?

- Does the EAP coordinate with the employer's insurer/managed care organization when the client must be referred for treatment that is paid for using health benefits?

- Does the EAP have an adequate system to maintain confidentiality of all program records?

- Is the EAP's system for quality assurance and quality improvement adequate and appropriate?

- Does the EAP have an adequate information system that is not leased? (Leasing opens another opportunity for breach of confidentiality.)

- Can the EAP produce program activity reports that will meet company information needs?

- Will the EAP provide an organization needs/resource analysis before implementing the EAP, technical assistance in the development of all program policies and procedures, supervisor procedural training, employee orientation, program promotion (posters, wallet cards, brochures, wellness seminars, newsletter articles, etc.), and management consultation regarding the supervisor referrals?

- Does the EAP offer 24-hour access to services, including crisis intervention, via toll-free phone lines?

- Does the EAP offer other desirable services that the purchaser may be interested in, such as critical

Professionals and facilities that provide services must meet at least the following criteria, but you may want to add others:

- Physicians should:
  a. Hold a degree in medicine (MD)
  b. Hold a license to practice in your state
  c. Have expertise in dealing with emotional and addiction problems, especially if they will be working on substance abuse cases
  d. Be recommended, if possible, by a physician or other reliable source who is familiar with the physician's services
  e. Be on staff at a hospital, with an appointment in his or her specialty.

- Psychologists should:
  a. Hold a state license or certification
  b. Hold a Doctor of Philosophy (PhD) in psychology
  c. Have two years supervised clinical practice.

- Alternate requirements for a psychologist are:
  a. Hold a Masters degree (MA) in psychology
  b. Have three years supervised clinical experience
  c. Have additional training in the treatment they will provide, such as chemical dependency or marital and family therapy.

- Social workers should:
  a. Hold a Masters degree (MA) in social work or a Masters of Social Work degree (MSW) from an accredited school of social work
  b. Have two years supervised clinical experience
  c. Be a Licensed Clinical Social Worker (LCSW) in states where social workers are licensed or an Accredited Clinical Social Worker (ACSW) in other states.

- Nurse practitioners should:
  a. Be licensed by the state as a Registered Nurse (RN)
  b. Comply with state credentials requirements for nurse practitioners
  c. Hold a Masters degree (MS), as recommended by the American Nurses' Association
  d. Have experience and expertise in dealing with the relevant area.

- Professional counselors should:
  a. Hold a Masters degree (MA) in counseling from an approved program at an accredited college or university
  b. Be a Licensed Professional Counselor (LPC) in states where counselors are licensed
  c. Have two years supervised clinical experience
  d. Have special training in the treatment they will provide, such as marital and family therapy.

- Facilities. Visit facilities under consideration and discuss with managers the following criteria:
  a. Proof of Joint Commission for the Accreditation of Hospitals accreditation (for in-patient treatment providers)
  b. Proof of state licensure
  c. Licenses or certifications of personnel
  d. Variety of resources
  e. Written objectives for program, services, and treatment, and for procedures followed to meet these objectives
  f. Individualized program, services, and treatment plans
  g. Assurance of confidentiality and compliance with state and federal confidentiality regulations

Figure 14-2. Professionals and facilities that provide services must meet at least these criteria.

h. Length of time in business (more than one year)
i. Marketing methods
j. Client references
k. Usage rate of other EAP clients
l. Methods and services consistent with organization's employees
m. Follow-up provided or available (with behavior change programs)
n. Location convenient to employees
o. Service times, length and number of sessions, and frequency convenient to employees
p. Minimum and maximum enrollment for classes
q. Flexiblity of service provision logistics
r. Safety precautions consistent with those of the organization
s. Waiver form for participants in the program
t. Evaluation of participants and method of sharing results with the organization
u. Evaluation of program, services, and treatment
v. Willingness to negotiate costs
w. Contract for program, services, and treatment.

Figure 14-2. (Concluded).

- incident debriefing, fitness-for-duty assessments, and violence in the workplace interventions?
- Does the EAP cover all personal problems or only selected issues? If some personal issues are excluded, how important are these to a company's work population?
- What additional personal issues and services are offered as optional coverage, and what are the costs?
- Does the EAP cover the immediate family/all household members under the program services?
- How long has the EAP been operating and how stable is its workforce? Having to change EAPs unexpectedly can be costly in many ways.
- For what length of time will the EAP guarantee its rates? What is included? Will the EAP offer any performance guarantees?
- What other companies with similar needs have used or are using the EAP? Be sure to request references from these companies.

Relationship with Employer

Although the relationship between the external EAP and the employer is obviously different from that of the internal EAP and the employer, the end results have been demonstrated to be quite similar. The same functions are performed by both the internal and the external programs, but a different approach must be taken by the external provider.

Before the implementation of the EAP, the external consultant must make a special effort to develop relationships with key people in departments such as benefits, safety, and medical, which are essential to the function of the program. This external relationship may seem to be a disadvantage at first glance; however, in fact, it may work

to the advantage of the external provider and, ultimately, to the EAP itself. Many times, because of internal politics, organizations are willing to give information more freely to outside consultants than they are to internal employees. As a result, in some cases, the external provider is able to wield greater authority as a paid consultant.

In addition, the perception of confidentiality is enhanced by the fact that the external EAP provider is a separate entity from the client organization. Because the people who provide the EAP services are not employees of the client organization, they are less likely to be pressured by management into divulging information about clients to anyone within the employer's organization. On the other hand, this distance between the external EAP and the client organization also creates a greater need for the EAP provider to perform an in-depth organizational analysis to define and clarify the structure, function and resources, and barriers to the successful operation of the EAP.

Both internal and external EAPs have demonstrated their ability to maintain a healthy, productive workforce. Each employer must decide which system is the better venue for its needs and situation.

PRODUCTS OF EMPLOYEE ASSISTANCE PROGRAMS

EAPs develop different services, or products, to meet the special needs of the organization(s) in which they function to improve productivity and safety. All EAPs offer the basic products of assessment of client problems, referrals for treatment, employee training, and record keeping. Among these basic products, different EAPs may offer different levels of service. For instance, one EAP may do assessments primarily through face-to-face sessions with the client, while another may do most assessments over the telephone. Some EAPs may offer only yearly supervisor training and some written material for each employee to explain the EAP, while others will train hundreds of employees every month to heighten awareness of EAP services.

Clearly, the number of staff and the budget of the EAP will determine the nature and time spent on these basic products. With the continued development of "broad-brush" EAPs and more sophisticated methods of meeting employers' needs, additional EAP products have become available. The EAP advisory committee will want to consider the following products to find those that best meet the needs of their organization(s).

Treatment Services

Some EAPs will provide clients with three to eight counseling sessions in their offices before making a decision whether to refer a client to a provider in the community. Thus, the EAP becomes a service provider itself before it refers the client. This service may include a "crisis-intervention" component in which the EAP responds quickly to workplace incidents by helping a troubled employee and/or any co-workers who have been affected by the troubled employee's behavior in the workplace. Most EAP counselors have been trained to treat clients by using short-term, problem-resolution focused therapy. Usually, these sessions are offered at no cost to the client.

The advantages of EAP-provided treatment include the fact that more clients will receive help for their problems (with referral, some clients inevitably fail to follow through and will not get help), and fewer providers will bill the company's insurance plan(s) for their services. Possible drawbacks to this product include client reluctance to trust an agent of the company to provide their treatment, and the possibility that the EAP counselor is not as skilled in treating the client's problem as an outside specialist might be.

Managed Care

Before the advent of managed care, when EAP counselors referred a client to a hospital or a therapist, they expected that the outside provider would decide on the nature of treatment to be delivered and when that treatment was complete. However, in the late 1980s, companies noticed that the cost of behavioral care (psychiatric, alcohol, and drug treatment) for their population of insured was rising at a double-digit rate each year (it was not unusual for a company's behavioral care cost per employee to double in three to five years). Many organizations targeted behavioral care cost as a problem area. As a result, various forms of managed care were developed to take control of the duration and type of behavioral care available to the client. Treatment providers discovered that under managed care, decisions about treatment must be shared with, or in some cases surrendered to, a third party—usually an insurer or some other financial entity. Since 1990, some

corporations have seen managed care reduce their behavioral care cost per employee by 10% to 50%, while other costs of doing business have steadily risen.

The debate about which outside entities should have the right to control behavioral treatment has been lengthy and sometimes bitter. Some organizations have resolved this conflict in a way that seems to please both providers and employees—they have given their EAP the responsibility of managing behavioral care. The EAP is well suited to manage decisions about the length and nature of behavioral care, because EAPs have always identified employee needs and found the best treatment for those needs.

Managing care involves approval or selection of the kind of treatment the client will need, approval or selection of the provider of the treatment, and authorization of the need to continue treatment after it has been initiated. If the provider and the EAP (in this example) disagree about treatment recommendations, a system of appeals ensures that the client's rights are protected. Ultimately, if the provider insists on treatment that has been determined to be unnecessary or inappropriate, the EAP may reduce payment or deny payment entirely for the unnecessary services. In fact, this rarely happens, because managed care by an EAP works best when the EAP and the provider regard each other as partners in treatment and work together to help the client.

Managing care often involves contracting with a network of preferred providers who have skills matching the needs of the company's employees, who will work cooperatively with the EAP, and who will agree to reduced fees to help the EAP reduce costs. An example of how an EAP can improve outcomes include such cases as employees who have often entered a psychiatric hospital for a drinking problem (when care was unmanaged). Under managed care, the EAP now directs such employees to enter a chemical dependency treatment center that EAP staff have determined has a high rate of success in treating alcoholism. This type of care is less expensive because it is not delivered in a full-service psychiatric hospital, and because the EAP has contracted for reduced rates. The employee stays sober after treatment for the first time because the treatment has been tailored to his or her needs. Or, an employee who has been in individual therapy for four years, but continues to have frequent injuries and absences at work, may be directed by the EAP to a psychiatrist for the first time. The psychiatrist determines that antidepressant medication will help the employee without interfering with his or her job. A year later the employee is stable and productive, and no longer needs either therapy or medication.

Treatment Follow-Up and Monitoring of Recovery
Although most EAPs offer some form of follow-up after treatment is complete, some EAPs offer more follow-up than a questionnaire or a phone call a few months after

services are rendered. Intensive follow-up is most often provided to chemically dependent clients. Research has shown that monitoring after treatment can significantly reduce relapse, especially in the drug-addicted client. For companies where safety is particularly critical, monitoring a drug addict after treatment may be very cost effective. Follow-up can consist of regularly scheduled calls or client visits to the EAP, regular reports to EAP staff from a professional aftercare provider, or random drug or alcohol testing coordinated by the EAP.

Some EAPs require all three monitoring tools to ensure recovery and detect relapse by an employee as soon as it occurs. Depending on the employer's policies, relapse by the employee may result in EAP referral for additional treatment, changes in job responsibilities, or termination. In any case, employee monitoring and follow-up will reduce relapse rates, and may eventually help the chronically relapsing employee either to change or to leave the organization.

Substance Abuse Professional Services and Other Regulatory Compliance

In 1988, the US Drug-Free Workplace Act was passed, requiring many organizations to provide information and services addressing drug abuse. EAPs are ideally suited to meet many of the Act's requirements. In 1991, the Omnibus Transportation Employee Testing Act required drug testing and EAP education for safety-sensitive employees in the transportation industry. This Act was revised in 1994 to include alcohol testing and provision of EAP information on alcohol abuse. The Department of Transportation (DOT) now requires substance abuse professional (SAP) assessment, referral, and monitoring recommendations for all employees who test positive on a DOT drug or alcohol test. The DOT lists Certified Employee Assistance Professionals as one of a select group of clinicians who are qualified for SAP work.

As a result, many corporations now expect this SAP service from their EAP. SAPs must adhere to a set of regulations issued by the DOT. They must ensure that all safety-sensitive employees with positive test results receive appropriate treatment and monitoring recommendations before they return to work in DOT-regulated positions. Thus many EAPs provide a set of substance abuse services required by law, allowing their organization to meet regulatory requirements without having to shop for these services from a variety of vendors.

Management of Workplace Threats and Violence

A growing safety and productivity problem for all companies is violence in the workplace. Employees are being stalked or threatened by fellow workers or nonemployees at an increasing rate. It is now known that employees threaten other employees with bodily harm much more frequently than is reported. And, of course, actual violence in the workplace—fights and/or attacks with a deadly weapon—is much more frequent than in the past. EAP managers and counselors are now being used as integral members of threat management teams, which are used by corporations to intervene whenever a threat or act of violence occurs. Because EAP staff deal routinely with troubled employees (some of whom may be potentially violent), the staff can help the organization develop sound threat management policies that will reduce the likelihood of violence and, should violence occur, help management deal effectively with the aftermath of such an incident.

The EAP may serve in a variety of ways to help deal with violence. EAP staff may be asked to assess or to arrange for an outside assessment of an employee's potential for violent behavior before he or she is allowed to return to the workplace. After returning, the employee who made threats or acted violently may be mandated to treatment monitored by EAP. Perhaps EAP's most important role is in working with the victim and co-workers of the victim after the incident. Victim and co-workers may require critical incident stress debriefing or therapy to deal with post-traumatic stress disorder (PTSD). The EAP can also provide training for the affected work group to help employees understand the dynamics of violence in the workplace and to help them return as quickly as possible to business as usual.

Critical Incident Stress Debriefing

Traumatic incidents in the workplace, such as employee violence, employee or customer death, fatal or near-fatal accidents, or natural disasters, can cause lasting symptoms or problems for many employees. After an incident, some employees will appear confused, anxious, depressed, or angry for an extended period of time. These are normal reactions to trauma and will occur in normal employees as well as in troubled personnel. All of these states will reduce productivity and safety, and may prompt employees who were more severely affected to resign.

In the past 10 years, a technique called critical incident stress debriefing (CISD) has been developed to reduce many of the aftereffects of traumatic incidents and to help identify those employees who will need mental health treatment. CISD can be done quickly by trained professionals with groups of 10 to 15 employees, usually in one or two sessions approximately two hours in length. Many times the CISD sessions are most effective if previously trained volunteers from the workforce assist the professional in the group meeting. These brief interventions have been shown to significantly help employees, and will reduce the number of PTSD cases that usually develop after an incident.

In many organizations, EAP staff are asked to coordinate the provision of CISD services, or they may actually assume the role of service provider for CISD sessions. The staff may be involved in organizing the training of volunteers to support the CISD effort. After an incident, EAP staff are responsible for ensuring that all employees who need CISD services are encouraged to participate. After each CISD session, the professional conducting the session can involve EAP staff in locating therapists or hospitals for individuals who may require ongoing mental health services.

Work/Family Services

Although some large corporations have separate work/family departments, in many organizations these services are offered by EAP. "Work/family" refers to all those issues in the lives of employees that occur outside the job, but impact the workplace. Issues include the need to arrange child care, to look after elderly parents, to support a relative who is ill, and to work in a way that supports a healthy lifestyle. An EAP may contract with a day care network to help working parents with children, or may coordinate day care services offered by the company. The EAP can direct employees to eldercare service agencies in their community or may contract with an eldercare service provider with special expertise in issues of the aging. Again, if the EAP can assist an employee with these tasks, the employee will be more focused on the job, more attentive to safety in the workplace, and more productive.

Employee Training

Training employees to be aware of and to use EAP services is one of staff's basic priorities. But some EAPs are beginning to widen their training programs beyond the basic services. For example, particular groups of employees may express a need to learn about stress management, parenting skills, couple's communication, smoking cessation, or dealing with difficult adolescents. If EAP staff can arrange for experts on these subjects to provide the requested training, or if the staff possess the expertise to offer the training themselves, the EAP can build employee awareness of its services and enhance the level of trust employee groups feel for the program.

Management Consultation and Organizational Development

Another basic service of EAP is management consultation, that is, helping managers deal with troubled employees before and after referral to EAP. In recent years, many EAPs have expanded this basic service to encompass consultation with management on any aspect of the company's operations that seems to be related to an increase in the number, or dysfunctionality, of troubled employees. This service can best be described as organizational development (OD) consultation (which may be the function of a separate department in some larger corporations).

Examples of OD work that may fall to EAP staff could include an analysis of why a certain work group has a sudden increase in carpal tunnel injuries, why a certain work location experiences a much higher rate of employee resignations, or why behavioral care costs in one particular division have risen much faster than they have in other areas of the company. In each case, the EAP's task is not to help a particular employee, but to diagnose a condition which may have caused a group of employees to lose productivity. If an accurate diagnosis is made, then the EAP can work with appropriate operations, safety, or personnel managers to resolve the problem. Companies may look to the EAP for this type of service because of its psychologically sophisticated staff and its access to information about troubled employees that is not available to other departments.

VIOLENCE AND HARASSMENT IN THE WORKPLACE

Although not a new phenomenon, violence and harassment in the workplace is an issue that is gaining recognition and drawing a response from the courts, government, and employers. The prevalence of workplace violence has been significant enough to draw a regulatory response on the federal and state levels. The U.S. Occupational Safety and Health Association (OSHA) recently issued voluntary guidelines addressing the problem of violence in health and social service workplaces, which it identified as a particularly high-risk work environment. OSHA will soon issue guidelines targeting the night retail industry, another particularly dangerous industry. The problem of violence in the workplace, however, is certainly not limited to these industries. No employer is immune from violence in the workplace.

The information provided in this section focuses on the legal duties employers have to combat the problem of workplace violence. The sources of these legal duties are varied, and even extend to the alleged perpetrator. This section covers the new OSHA guidelines on this topic and concludes with a brief discussion of ways employers can combat workplace violence.

Employers' Duties and Legal Obligations

Employers have legal duties to employees, customers, clients, and third parties in situations involving workplace violence. These legal obligations may stem from statutory requirements, contract, or the general duty clause of OSHAct.

OSHA Requirements

Federal law generally requires employers to provide their employees with a place of employment which is "free from recognized hazards that are causing or are likely to cause death or serious physical harm to . . . employees." (29 *U.S. Code* 654[a][1]) Encompassed within this general duty clause is an employer's obligation to do everything that is reasonably necessary to protect the life, safety, and health of employees, including the furnishing of safety devices and safeguards, and the adoption of practices, means, methods, operations, and processes reasonably adequate to create a safe and healthful workplace. These requirements are known as the general duty clause.

In what is likely to become the trend of the future, OSHA, on March 14, 1996, released voluntary guidelines to help health care and social service employers identify and prevent violence in workplace settings. Although these guidelines do not carry any regulatory weight, such guidelines will almost certainly come into play when OSHA is determining whether an employer violated the general duty clause. Although failure to implement the guidelines is not in itself a violation of the general duty clause, OSHA will not cite employers who have effectively implemented these guidelines.

OSHA decided to issue these voluntary guidelines for the health care and social service industry because the problem of violence in this industry was so acute. The OSHA guidelines for the health and social industry include four parts: (1) management commitment and employee involvement, (2) worksite analysis, (3) hazard prevention and control, and (4) safety and health training.

The first aspect of these guidelines emphasizes the establishment of teams or committees of management and employees to ensure zero tolerance for workplace violence. Worksite analysis involves identification of hazards, conditions, operations, and situations that could lead to violence. To ensure thorough identification, incidents of violence should be analyzed to establish trends. Jobs or locations with the greatest risk of violence should be pinpointed. High risk factors such as types of clients or patients should be identified, and the effectiveness of existing security measures should be analyzed. Hazard prevention and control includes: engineering controls and workplace adaptation (changing the physical facility to improve safety) and administrative and work practice controls (supervising the movement of clients and patients and requiring full disclosure of all assaults or threats to employees). Finally, all employees must be trained continually on the issues related to violence in the workplace.

Legal Precedents

Some duties and legal obligations of employers have been established by court cases, including:

- *Negligent Hiring.* The tort of negligent hiring is based on the principal that an employer has a duty to protect its employees and customers from injuries caused by employees whom the employer knows, or should know, pose a risk of harm to others. This duty is breached when an employer fails to exercise reasonable care to ensure that its employees and customers are free from risk of harm from unfit employees. Thus, an employer may be found to have been negligent in selecting an applicant for employment when, for example, the employer neglected to contact the applicant's former employers or to check references, when such investigation would have demonstrated that the applicant had a violent propensity or was otherwise unfit for the job.

The Colorado Supreme Court has expressly recognized the tort of negligent hiring and noted that at least 14 other states have also recognized the negligent hiring action (*Connes v. Molalla Transp. Sys., Inc.,* 831 P. 2d 1316, 1323, Colo. 1992).

- *Negligent Training.* Courts have also recognized a cause of action for an employer's negligent training of its employees that results in injury to a third person. For example, California courts have recognized that a medical university owes a duty to patients who are under the care of residents to see that the residents receive proper training and supervision. The court concluded that the university was negligent in discharging its contractual duties to educate and train the residents and to supervise their training (*County of Riverside v. Loma Linda Univ.,* 118 Cal. App. 3d 300, 317, 1981).

Courts may also recognize the theory of negligent supervision. The South Carolina Supreme Court has recognized that an employer may be liable for negligent supervision if an employee intentionally harms a third party while on the employer's property or while using the employer's equipment, if the employer knows or has reason to know that it has the ability to control the employee, and if the employer knows or should have known of the necessity and opportunity of exercising such control over the employee (*Degenhart v. Knights of Columbus,* 420 S.E.2d 495, 496, S.C. 1992).

- *Negligent Retention.* Where an employer does not realize an applicant's violent tendencies at the time of hiring despite following steps intended to prevent negligent hiring claims, an employer must still take steps to prevent violence in the workplace. The Court of Appeal of Minnesota has discussed the three theories of negligent hiring, negligent training/supervision, and negligent retention in a case where a former employee shot and killed a coworker a few days after the violent employee resigned from work (*Yunker v. Honeywell, Inc.,* 496 N.W.2d 419, 421, Minn. App. 1993).

- *Intentional and Negligent Infliction of Emotional Distress.* An employer's reckless or intentional failure to reasonably investigate applicants and employees may lead to a claim for intentional infliction of emotional distress. The tort of intentional infliction of emotional distress in the employment setting typically requires that: (1) the employer intended to inflict emotional distress or recklessly disregarded whether its acts would result in infliction of emotional distress, (2) the acts in fact caused severe emotional distress, and (3) the acts constituted an extraordinary transgression beyond the boundaries of socially tolerable conduct. Alternatively, where the employer owes a duty of care and negligently breaches that duty, causing employees or others to suffer emotional distress, the employer may be sued for negligent infliction of emotional distress.

Equal Employment Opportunity Laws

Where an employer fails to seriously consider an employee's concerns about threatened violence or harm by a coworker or other individual in the workplace based on impermissible grounds, such as race, sex, religion, national origin, pregnancy, age, marital status, workers' compensation claims, military service, sexual orientation, or disability, the employer may be held liable for violation of federal or state equal employment opportunity laws. As many incidents of workplace violence involve harassment or discrimination on the basis of sex or other protected category, employers also have to exercise caution to ensure that discrimination and harassment are not occurring or are being effectively remedied once they are discovered.

Contractual and Other Obligations

Employers may also have other employment contracts and agreements implied in law. Except under special circumstances, it is unlikely that employers expressly contract to provide employees with a workplace free from harm or violence by other employees or third parties. However, contractual obligations can arise in the absence of written or express agreements. In California, for example, the Supreme Court recognizes that an implied contract may be inferred from the language in a company's policy manual or employee handbook (*Foley v. Interactive Data Corp.*, 47 Cal. 3d 654, 1988).

Employers may have further duties to their employees based on the implied covenant of good faith and fair dealing. Based on the policy that neither party may act in bad faith to deprive the other of contractual benefits, this covenant is implied in every contract in California. The covenant requires the parties to be fair with one another in applying contract terms. Thus, it could be argued that, by entering into an employment relationship, an employer assumes, from the employment agreement, the duty to provide a safe working environment to enable employees to perform the bargained-for exchange, that is, the performance of services for the employer.

Finally, an employer has an obligation not to discriminate or retaliate against employees who express concerns regarding unsafe work conditions, such as threats of violence in the workplace. Many states have express statutory whistleblower laws that prevent such retaliation. In addition to such statutory claims, an employee who complains about unsafe working conditions or refuses to work in unsafe work environments and is terminated may be able to state a tort claim against the employer for wrongful termination in violation of public policy. In recent years, several states have allowed terminated employees to recover tort remedies, including punitive damages, from their employer notwithstanding the at-will nature of the employment relationship or the terms of an employment agreement where the termination violates fundamental principles of public policy.

Rights of the Alleged Perpetrator

Where the employer warns employees of an individual's violent tendencies, the employer could be found liable for defamation if the employer is under a mistaken belief that the alleged perpetrator is violent. Defamation occurs when a statement which is communicated to another individual is false, unprivileged, and the cause of injury (See W.H. Keaton, *et al., Prosser and Keaton on the Law of Torts*, 773–778, 5th ed. 1984). Regardless of whether the statement is a verbal (slander) or written (libel), the employer might be held liable for a false characterization that an employee is violent. To be sure that an employer is able to establish a good faith belief that the individual had violent tendencies, the employer should conduct a prompt investigation of the allegations of violence before warning other employees.

An employee who is accused of having violent tendencies and is terminated for such violent tendencies could file a wrongful discharge suit against the employer if the employee disputes his or her employer's characterization. An employer should take steps that minimize both the threat of violence and the risk of a wrongful discharge suit or grievance. Before a threat or actual act of violence even occurs, the employer should review its employee handbook and/or personnel policies to ensure that they specifically prohibit threats or acts of violence in the workplace. After a threat or actual act of violence allegedly occurs, an employer could consider suspending the alleged perpetrator while investigating the allegations to determine if the allegations have merit. This could serve both the goal of avoiding violence in the workplace and the objective of avoiding a wrongful discharge suit should the allegations of violence turn out to be hoax or otherwise unsubstantiated.

The protection of the alleged perpetrator's privacy rights arises from federal and state constitutions as well as court developed law. Although an employer has an obligation to investigate any threats of violence in the workplace in order to ensure the protection of other employees and customers, the employer must ensure it conducts its investigation legally and as unobtrusively as possible. In addition to a claim for violating the constitutional right to privacy, an employee could also file a tort claim for invasion of privacy, or other related privacy claims. To minimize the risk of such claims, the employer should verify the source of the information that the alleged perpetrator has engaged in violence so that it has a legitimate reason and a good faith belief for conducting an investigation that may intrude into the alleged perpetrator's private affairs.

Finally, the Americans with Disabilities Act (ADA), 42 U.S.C. section 12101, *et seq.* and related state statutes prohibit employers from discriminating against "qualified individuals" with physical or mental disabilities.

Combating Violence in the Workplace

An employer can call local law enforcement officials to pursue criminal sanctions against an individual who engages in workplace violence. Although local law enforcement will be familiar with the possible criminal laws on which to rely, employers should be aware of the basic categories of criminal laws that may be applied to a perpetrator of workplace violence. Virtually all states have some variation of the following criminal violations: laws prohibiting unlawful threats, threatening or annoying telephone calls, brandishing a weapon, assault and battery, and trespass by credible threat. Some states even have antistalking laws and hate crime laws.

Employers can also use the civil court system to respond to violence in the workplace. For example, a restraining order can be obtained on behalf of an employee who was a victim of workplace harassment or violence. Restraining orders serve two major purposes: (1) they prohibit specific conduct by the perpetrator, and (2) they establish a boundary around the victim that usually includes criminal prosecution if a violation occurs.

The best response to violence in the workplace involves a comprehensive, integrated plan that makes use of all available resources. This plan should include the following:

- Make preventing and controlling workplace violence a priority and form a management team to develop, review, and implement policies dealing with violence in the workplace.
- Develop emergency response procedures for incidents of workplace violence.
- Conduct an education program regarding early warning signs of potentially violent behavior and steps to be followed in responding to and investigating an incident of workplace violence.
- Establish clear internal and external lines of communication to effectively avert and respond to crisis situations.
- Prevent workplace violence through the use of proper prescreening, consistent enforcement of workplace rules, and the use of employee assistance programs and medical care resources.
- Increase physical security measures and develop cooperative relationships with local law enforcement authorities.
- Consider the use of the courts to prevent and redress incidents of workplace violence.

SUMMARY

- The goal of any EAP is to improve employee productivity by reducing lost time, reducing accidents and workplace injuries, and enabling employees to do their best on the tasks assigned to them in the workplace. EAP assistance is provided through referrals to outside counseling or other treatment services in the community, treatment services through the EAP, or a variety of training and education programs available through the EAP.
- To develop an EAP, a company establishes an EAP advisory committee, conducts a needs assessment, develops a plan for establishing the EAP, and drafts general policies governing the formation of an EAP and the rights of employees. The major types of EAPs are internal, external, union-based, consortium, and blended.
- Internal EAPs are staffed by employees of the corporation, which allows it to work easily with other departments of the company. Internal EAPs may also gain employee trust more easily, but workers may have doubts about confidentiality of their records.
- External EAPs often can provide a wider range of services than can internal EAPs. Companies must exercise care in choosing an external provider. In addition, workers may feel that an external EAP will protect the confidentiality of records.
- Products offered by EAPs include counseling sessions, managed care, treatment follow-up and monitoring, substance abuse programs, management of workplace violence and threats, critical incident stress debriefing, work/family services, employee training, and management consultant services.
- Violence and harassment in the workplace is receiving more attention from employers. Companies need to understand their legal obligations and liabilities, learn about the rights of the alleged perpetrator, and recognize how they can combat violence and harassment in their companies.

REFERENCES

Employee Assistance Professionals Association. *EAPs, Value and Impact.* Arlington, VA: 1991, pp. 2, 8.

_____. *EAPA Standards for Employee Assistance Programs, Part II: Professional Guidelines.* Arlington, VA: 1993, p. 19.

_____. *EAPs, Theory and Operation.* Arlington, VA: 1991, pp. 3, 4.

Employee Assistance Society of North America. *EASNA Standards for Accreditation of Employee Assistance Programs.* Berkely, MI: 1990, p. 4, 23.

National Institute on Drug Abuse. *National Household Survey on Drug Abuse.* Rockville, MD: U.S. Department of Health and Human Services, 1990, p. 8.

REVIEW QUESTIONS

1. An employee assistance program (EAP) is designed to assist in the identification and resolution of productivity problems associated with employees impaired by personal concerns including, but not

limited to:

a.

b.

c.

d.

e.

f.

g.

h.

2. How many more times do workers involved with alcohol or drugs become involved in work-related injuries than other workers?

a. 3-4 times more

b. 5-6 times more

c. 7-8 times more

d. 9-10 times more

3. How is EAP assistance provided?

4. What principle is vital to the functioning of any EAP in encouraging employees to willingly come forward to seek help, and why?

5. Once the employee (or client) contacts the EAP for services, either voluntarily or by a supervisor's referral, the following steps should occur:

a.

b.

c.

d.

e.

6. List five benefits provided by EAPs to employers.

a.

b.

c.

d.

e.

7. In setting up an EAP, an EAP advisory committee should be formed comprised of members from which group(s)?

a. High-level management

b. Supervisory management

c. All labor unions

d. Members of the work force

e. All of the above

8. What factors should be considered in a needs assessment to define the major types of employee problems and the kind of EAP services needed to address those problems?

9. List the five major types of EAPs.

10. What is considered a strength of an *internal* EAP?

a. EAP staff can interface easily and cooperatively with other departments

b. EAP staff will have more knowledge of the special problems and concerns of employee groups

c. EAP staff will become more familiar with the treatment needs of the employees

d. EAP staff will have more knowledge of local community resources

e. All of the above

11. Stemming from the OSHAct general duty clause, employers have legal duties to employees, customers, clients, and third parties in situations involving:

a. Family services

b. Workplace violence

c. Substance abuse treatment

d. a and c

12. List five of the nine EAP products that should be considered by the advisory committee to meet the needs of their organization.

a.

b.

c.

d.

e.

13. In monitoring a drug addict after treatment, what are three methods of follow-up?

a.

b.

c.

14. Which of the following is identified by OSHA as a particular high-risk environment in regards to violence?

a. Health and social services workplaces

b. Office environment

c. Industrial workplaces

d. Night retail industry

e. a and b

15

Emergency Preparedness

No industrial, commercial, mercantile, or public sector organization is immune from disaster. Emergencies can arise at any time and from many causes, but the potential loss is the same—injury and damage to people, the environment, and property. Advance planning for emergencies is the best way to minimize this potential loss. To help management cope with emergencies, this chapter covers the following topics:

- overview of the duties and priorities of management for emergency planning
- discussion of the types of emergency plans available
- the types of emergencies that companies may encounter
- what elements of emergency planning companies need to consider
- how companies can plan to protect their employees from personal attack
- the need to coordinate outside help during emergencies

MANAGEMENT OVERVIEW

A written comprehensive emergency management plan is intended to provide preplanned response to those unexpected or disastrous events that can strike any organization, such as a fire or natural disaster. Emergency planning is often assigned to the safety and health professional. In some instances this may be acceptable; however, ultimately the responsibility lies with the highest levels of management who best know a facility's resources, operations, and capabilities. The safety and health professional should act as the consultant, guiding line management through the process of identifying potential emergency events and developing primary and contingency plans to respond to them.

The safety of employees and the public must be the first concern in planning for an emergency. Preplanning should take into account the immediate short-term needs people are likely to have (i.e., injuries) as well as the long-term needs that may result from an emergency event.

Next, management should consider ways of protecting the property, operations, and the environment. In new facilities, this may mean locating certain buildings or operations well away from others or locating operational response teams in a more central area to allow for a quicker response. In general, all emergency plans must include salvage, overhaul, and possible decontamination operations.

The final steps in the plan should involve restoring business operations to normal. In emergencies, a facility's operations are likely to be damaged or shut down altogether. Management will need to decide how and when to resume operations in the face of such obstacles as temporary wiring, lack of heating, or the need for significant rebuilding. Moreover, the probability of environmental damage to the immediate and surrounding community will demand considerable attention from the organization's top management. Operations that could adversely affect the health of employees or community neighbors must meet the emergency planning requirements of regulatory agencies.

Regardless of the size or type of organization, management is responsible for developing and operating an emergency planning program designed to meet the eventualities. An effective plan requires the same good organization and administration as any business undertaking. No one emergency plan will do all things for all organizations. Each company must decide on a plan that fits its needs and budget.

Emergency plans involve organizing and training small groups of people to perform specialized services, such as evacuation, fire fighting, rescue, spill response, or first aid. These small, well-trained groups can serve as a nucleus that can be expanded to meet any kind of emergency. Even with outside help available, a self-help plan is the best assurance that losses will be kept to a minimum. Although planning should occur within each unit of the organization, the facility's emergency management staff should develop and implement a large-scale response scheme that encompasses all departments.

This comprehensive emergency management plan should meet all the different types of contingencies. Certain basic elements such as command functions, communications, and emergency staff personnel will be common to all emergency operations. Of course, the comprehensive plan should contain many procedures to deal with natural, technological, and nuclear hazards.

Organizations should also plan for an emergency where they may have to stand alone with no outside help. This is not a remote possibility in the event of a major disaster such as a tornado, hurricane, flood, or brushfire that may strike an entire area. At times like these, communities will either have to stand alone or join with others in a mutual-aid pact.

Self-help plans should include provisions for recall of off-duty personnel. Many of these call-back lists look good on paper, but one should try a practice run to see if they really work. Long holiday weekends can present real problems should an emergency occur during these times. Further, management should consider how to telephone the required personnel and how long it may take to secure 20 or more employees.

Before an organization initiates an emergency plan, it should identify and evaluate the potential disasters that might occur. The section Types of Emergencies later in this chapter discusses these in detail.

The next step is to assess and prioritize the potential harm to people, the environment, and property. The time of day and workshift patterns are other factors that should be considered in assessing the potential damage. Planning should take into consideration the impact of a catastrophe that might occur during weekends or holidays when no one or only a skeleton staff may be on hand.

To estimate potential damage to property, one should look not only at the general structures but their surroundings as well. For example, a building may be strong enough to resist an earthquake but a sudden 7–in. (18–cm) rainstorm might cause dangerous flooding that could short out all electricity. On the other hand, an exploding boiler located in an adjacent building would probably not harm the main facility.

Next, probable warning time should be considered. For example, although a flood may build up over a period of several days, a bomb scare may afford only a few minutes warning from a telephone call. Warning time should give management a chance to alert personnel and to mobilize the plan. It is desirable to have a number of different plans, depending upon the nature of the emergency and the actual (or estimated) time available.

Another factor is how much company operations must be changed to meet the emergency. For example, in anticipation of a heavy snowstorm, it may be necessary to send employees home early. Some equipment may be left on or idling instead of being shut down completely. Finally, companies will need to consider what power supplies and utilities may be needed, particularly those used for fire protection, lighting, ventilation, and communications.

A basic emergency preparedness plan will usually include a chain of command, an alarm system, medical treatment plans, a communications system, shutdown and evacuation procedures, and auxiliary power systems. Not every element discussed in this chapter will apply to every organization. Also, several of the functions may be combined and handled by one person, particularly in a smaller company. In general, however, the material in this chapter is directed to the more elaborate and expanded type of organization and planning process.

CHOOSING A PLAN

Once management has an idea of the company's risk to various hazards, has assessed the response capabilities, and has reviewed existing or previously written plans, it should decide on the type of plan that is best for the facility.

The type of facility and its associated hazards are factors that determine the complexity of the plan. However, other factors also must be considered, such as the availability of qualified personnel to write and maintain the plan, and the availability of funds to produce it. Nuclear power facilities, for example, are required to develop extremely sophisticated plans that cost utilities millions of dollars to develop, implement, and maintain. Other facilities, such as small manufacturing facilities where few chemicals are located in an area not prone to natural hazards, often require less complicated plans. Choosing the correct type of plan for a particular facility is very important, and there are a variety to choose from. The most common types used in industry include:

- action guides/checklists
- response plans
- emergency management plans
- mutual aid plans

Action guides or checklists are generally short, simple descriptions of basic procedures that must be followed, such as whom to call, necessary emergency information, and basic response functions. These guides and checklists are intended to be more of a reminder and should, therefore, not be used by an untrained individual. The materials are sometimes used in conjunction with other types of plans.

Response plans are usually highly detailed and tell all responsible individuals actions that must be taken to mitigate the problem at hand. Sometimes a response plan is written for each type of possible hazard at the site. For example, there might be a chemical spill response plan as well as a hurricane response plan, and so on. These plans usually deal with only the actions necessary to respond to an emergency and do not cover requirements for actions before a disaster (training, drills, and exercises) or after a disaster (i.e., recovery plans, business interruption plans, etc.).

An emergency management plan is the most comprehensive plan used in business. It usually states who does what and when before, during, and after a disaster. Often this type of plan will incorporate "implementing procedures" that state how something is to be done.

Finally, a mutual-aid plan can be developed with other nearby firms. Such a plan calls for firms to share resources and help one another during an emergency. This type of plan demands more coordination, but can be useful to small firms with limited resources or larger firms with high hazard potential.

Selecting the proper type of plan is important, but management should not lock itself into one of the mentioned plans. Instead, it should choose a type that answers the questions specific to its facility. For example, should the plan address only response activities during an emergency or focus on preparedness activities before and recovery activities after an emergency? Companies should also be sensitive to the cost and time associated with the development and maintenance of each type of plan.

TYPES OF EMERGENCIES

The first step in the emergency planning process begins with determining what types of hazards may affect the organization. Targeting specific hazards allows the company to create a comprehensive planning, organizing, and implementation program. Hazards posing a threat to the organization will vary according to its location, production/engineering processes, and work practices.

Many sources of hazard information can help management determine the likelihood of specific emergency events in a locality. These sources include historical knowledge and records of accidents, fire statistics, National Weather Service logs, U.S. Geological Survey studies, location of rail lines and airports, and local emergency management U.S. Hazard Vulnerability Community Analysis documents. The organization's management should agree upon a set of hazards that appear to have some chance of occurring. After this assessment, emergency response operations planning can begin, based on a realistic study of the potentially disastrous threats facing personnel, property, and environment. A discussion of types of emergencies follows.

Fire and Explosion

Except where fires result from large-scale explosions, warfare, civil strife, or hazardous chemicals, the fire emergency usually allows some time for marshaling firefighters and organizing an evacuation, if necessary. Many large fires originate as small blazes that, if caught early, could be controlled by trained personnel. Therefore, prompt action by a small, properly trained and equipped group can usually handle most situations. However, plans should include alerting extensive firefighting forces upon the first indication of any fire growing beyond the "small fire" stage. Plans should also include procedures for controlling chemical contamination that may mix with the runoff from water used to put out the fire. Also, firefighting and evacuation procedures should take into consideration toxic gases, smoke, and fumes that may be produced by the burning materials.

The main point is that small fires must be extinguished as soon as they start. The first five minutes are considered the most important. Good housekeeping, prompt action by trained people, proper equipment, and common sense precautions will prevent a small fire from becoming a disastrous blaze.

Specific information on fire extinguishing and control is in Chapter 11, Fire Protection, of the *Engineering & Technology* volume of this Manual.

Both small and large fires have the potential for producing significant environmental problems. These include:

- production of toxic gas and dust from combustion and decomposition
- thermal plumes that can carry the materials substantial distances
- disposal problem involving large quantities of contaminated water used in extinguishing the fire

Commonly used construction materials, furniture, carpets, and industrial materials can release significant quantities of toxic combustion products. The combustion of industrial and agricultural chemicals and materials may also be a source of hazardous gases, vapors, fumes, or dusts. The dispersion of these materials may be enhanced by the buoyancy effects of the hot gases caused by the fire.

Depending on weather conditions, these plumes may travel substantial distances and remain concentrated. Ash and dusts generated from some forest fires in the western part of the United States have been detected, for example, in the Midwest. To minimize these effects, those responsible for responding to fires must be aware of the potential hazardous airborne materials, be able to track their movement, and be prepared to measure their ground-level concentration. These activities may require access to or on-site collection of meteorological data and air sampling.

An additional environmental hazard in responding to fires is the disposal of the potentially contaminated wastewater produced in extinguishing a fire. For example, 100,000 gal of water can easily be used in fighting a simple residential fire. If a thousand gallons of a 10% concentrated pesticide is released in the course of the fire, the resultant concentration could be a 100 ppm solution. Introduction of such a liquid waste flow into a sewer system or surface waterway could have significant environmental ramifications. Therefore, diking, chemical neutralization, use of chemical absorbants, or a controlled burn may be considered necessary containment methods.

Floods

When a company or facility is located in a floodplain, it should have the protection of dikes of earth, concrete, or brick construction. The probable high-water mark can be obtained from the U.S. National Weather Service or the Army Corps of Engineers. The latter group also provides valuable assistance in planning floodwater control.

Floods—except flash floods caused by torrential cloudbursts or the bursting of a storage tank, dam, or water main—do not strike suddenly. Ordinarily, there is enough time to take protective measures when a flood seems imminent.

Another source of information on flooding in a particular area is the local emergency planning commission or the U.S. Geological Group. Many facilities are not in floodplains. However, the U.S. Geological Group has classified different types of floods that may hit a region at specified times, such as the flood that strikes every 100 years. These infrequent events should be included in a company's planning process, even if the chances of their occurring seem remote.

Companies must make careful preparations for their emergency responses to a flood situation. For example, a list of typical emergency equipment and material should include sandbags, battens for windows and doorways, boats, tarpaulins, fuel-driven generating equipment (such as gasoline-powered arc welding machines or motor-generator sets), standby pumping equipment, a supply of gasoline in safety containers to fuel this equipment, lubricating oil and grease, rope, life belts, portable battery-operated radio equipment, audio speakers, and so on.

In addition, companies who use tank cars should make some provision for moving these cars to higher ground and anchoring them. They should also move portable containers above the high-water mark, along with buoyant materials and chemicals soluble in water. Storage tanks under the probable high-water mark (including underground tanks) should be specially anchored to prevent floating. Workers can construct auxiliary dikes of sandbags or dirt around key areas.

Other precautions to take in anticipation of flood conditions include:

1. electrocution hazards and the need for proper grounding and electrical fault protection
2. bracing or storing important equipment, materials, and chemicals off the ground
3. availability of pumps and emergency power sources
4. protection against soil erosion, which may cause serious structural damage
5. provision and protection of potable water supplies

Hurricanes and Tornados

Areas most frequently exposed to the winds of destructive hurricane force are the Atlantic and Gulf coasts. However, some inland locations are not immune to this type of disaster.

The National Weather Service and other agencies have developed improved methods of detecting and tracking hurricanes; thus ample warning can be given for maximum protection of property and evacuation of personnel from threatened areas.

Organizations regularly exposed to this hazard have developed a system of tracking hurricanes on a map. At predetermined locations, a specified alert condition becomes effective, and each supervisor completes a checklist for that alert. As the hurricane progresses toward the facility through the 100–mile (160–km) circle, 50–mile (80–km) circle, and so on, the facility is shut down in an orderly manner.

Buildings constructed in areas where hurricanes occur should be built to withstand these destructive winds and tides. Basic preventive measures include equipping the facilities with storm shutters or battens that can be promptly attached, at least on the side from which the storm is expected to approach. If this is not done, the facilities could have windows shattered, the roof torn away, or other parts of the building destroyed. If the roof is lost or damaged, building contents can be drenched by the heavy rains that accompany the storm and by water from broken sprinkler pipes. Therefore, roofs should be securely anchored and tall structures (such as chimneys, water towers, and flagpoles) designed to withstand high winds.

As for tornado emergencies, although the central Mississippi Valley is considered the top tornado area of the United States, almost every state has experienced these destructive storms. They can inflict enormous damage quickly, although it is usually restricted to a small area.

It is not possible for the U.S. National Weather Service to give as much advance warning or to pinpoint the strike area as accurately as they can with hurricanes. Therefore, a company must be prepared to protect its personnel on short notice and to take quick action to protect and restore undamaged equipment and materials.

The National Weather Service radio bands can be monitored during storm conditions. In one midwestern city, several large companies have set up a cooperative warning network. A lookout is stationed atop the city's tallest building. Through a central network, not only the member companies but also the city's radio stations and civil defense are alerted in the event of an approaching tornado. Private weather consultants are also available.

Tornado and hurricane experience indicates that emergency plans should include the following steps:

1. establishing procedures for alerting and getting personnel to a safe place. If the building is not constructed to withstand these natural forces, emergency shelters should be located close to the work area. All personnel should be instructed in the procedure to follow, with and without advance warning.
2. assigning trained personnel to take care of power lines—dangling wires are a serious hazard
3. assigning trained people to remove wreckage to prevent injury to salvage and repair workers
4. scheduling regular meals and rest for the repair crews

Earthquakes

Many areas of the United States have the potential to experience damaging earthquakes. In particular, the Pacific Coast contains the San Andreas fault network and the Midwest contains the New Madrid fault system. Other major fault areas are located in North and South Carolina, Utah, and New England. No reliable earthquake warning system exists. This factor, coupled with the widespread damage that a quake can bring, makes this geological hazard a difficult phenomenon for planners to address.

"Earthquake-resistant construction" consists of building a structure so that it "floats above the bedrock and ballasting it as a ship is ballasted, by making lower stories heavy and upper stories light." Utility lines and water mains should be flexible and laid in trenches that are free of the building, rising in open shafts, and connected to fixtures by flexible joints. Lockers, cabinets, shelves, and so on, should be securely installed with seismic bracing and safety restraining strips on shelves containing bottles of chemicals.

The principal dangers of earthquakes come from structural failures of buildings, bridges, and other items; fires; flooding; broken utility lines; hazardous materials releases; water shortages and contamination; health problems; and damage to transportation facilities such as interstate highways, airport runways, navigable waterways, and railways. Water reservoirs or emergency water sources should be available as backup support for firefighting operations. Also, a system should be established for shutting down gas mains supplying companies. The method of turning off the gas supply depends on the size of the main and the pressure on the line that comes into the facility. Company officials should obtain expert advice when devising these plans.

See Figure 15-1 for recommended steps to take if an earthquake strikes. Be sure the emergency management plan provides for the actions recommended in Figure 15-1 if the facilities are in an earthquake risk area.

Civil Strife and Sabotage

Riot or civil strife is another item on the list of emergencies for which a company should plan in advance.

Civil Strife

An emergency involving civil strife raises the questions of the right to protect property versus individuals' legal right to assemble. A company should obtain from an elected legal authority in the community (the district attorney, for example) a statement explaining the company's rights in protecting its property and the company's legal responsibility for the safety of employees and other people—such as customers, supplier salespeople, and visitors—who may be on company property. A company's legal department can be helpful in determining such a position, but its opinion does not have the force of law.

Some of the problems involved include disruption of business when an office or facility is invaded by outsiders, protection against a mob intent on destroying company property, requests from neighboring companies for assistance during a riot, and the rights and responsibilities of armed company guards. Civil strife emergencies can be just as disastrous as any other type and should receive advance planning by manufacturing, mercantile, or commercial establishments.

Sabotage

Protection against sabotage is also an important consideration. The saboteur may be a highly trained professional or an amateur. He or she may be anyone—usually one of the least-suspected members of the organization. Because physical sabotage is frequently an inside job or requires the assistance—knowingly or unknowingly—of someone inside the facility, the principal measures of defense must be denying entry to suspicious persons. Evidence of sabotage should be reported to the U.S. FBI and, if military related, to the U.S. Department of Defense.

Work Accidents and Rumors

The "chain reaction" from a so-called "routine work accident" can result in an emergency situation. For example,

a break in a chemical line or accidental emission of toxic vapors may create an emergency.

The potential for work accidents to cause emergencies has increased due to the complexity of processes, the proliferation of industrial and agricultural chemicals, and the often close proximity of residential areas to industrial activities. For example, a small break in a chemical line could allow hazardous air contaminants to enter a facility's ventilation system creating a work-site emergency. A larger release of a chemical or toxic vapor may endanger a neighborhood. The toxic release may cause direct injuries or result in panic and create an emergency situation.

The potential for such events should be investigated and evaluated to protect individuals on site and in the surrounding community. The methods used in this evaluation may include the application of risk-assessment techniques. These techniques require management to identify hazards, calculate their probability of occurrence, and develop means to respond to these events. Various existing U.S. governmental regulations that require these types of analysis include:

- 40 *Code of Federal Regulations* (*CFR*), Parts 330 and 335 "Extremely Hazardous Substance List and Threshold Planning Quantities; Emergency Planning and Release Notification Requirements; Final Rule." Vol. 52, No. 77, Washington, DC, pp. 13378–13410.
- 29 *CFR* Part 1910.120 "Hazardous Waste Operations and Emergency Response."
- Hazard Communication Guidelines for Compliance. U.S. Department of Labor, Occupational Safety and Health Administration (OSHA), 29 *CFR* 1910.1200.
- Emergency Planning and Community Right-to-Know Act, Title III, Superfund Amendments and Reauthorization Act, *Public Law No. 99–499.*

Panic caused by rumors or lack of knowledge can also create an emergency. Plans for such situations should include establishing auxiliary areas in the building for medical treatment, a method of notifying employees of a current situation, a method of quickly taking a head count of workers, and sources of oxygen supplies available on short notice. A public relations coordinator should deal with the press, the public, and families.

Shutdowns

An emergency situation may occur following an unscheduled action, such as a disaster or strike; hence a fast shut-down procedure should be covered under an emergency plan. This plan should be based on a priority checklist, that is, all the tasks to be assigned and functions to be performed should be arranged in order of importance ahead of time so that if only a short notice is given, at least the most vital precautions are completed. This "crash procedure" is usually an adaptation of the routine procedure

ACTIONS TO TAKE IF AN EARTHQUAKE STRIKES

The greatest threat during an earthquake is from falling debris. Earthquakes are unpredictable and strike without warning. Therefore, it is important to know the appropriate steps to take when one occurs, and to be thoroughly familiar with these steps to be able to react quickly and safely.

IN THE OFFICE

During the earthquake	
Step	**Action**
1	**Remain inside** the building.
2	**Seek immediate shelter** under a heavy desk/table, or brace yourself inside a doorframe or against an inside wall.
	▪ Get at least 15 ft away from windows.
3	**Stay there.** If shaking causes the desk or table to move, be sure to move with it.
4	**Resist the urge to panic.** Organize your thoughts; mentally review the established psychological considerations for earthquake safety.
	▪ Don't be surprised if the electricity goes out, fire or elevator alarms begin ringing, or the sprinkler system is activated.
	▪ Expect to hear noise from broken glass, creaking walls, and falling objects.

Immediately after the earthquake	
Step	**Action**
1	**Remain in the same position** for several minutes after the earthquake in case of aftershocks.
2	**Do not attempt to evacuate** or leave your immediate area unless absolutely necessary or when instructed to do so by a proper authority.
3	**Check for injuries** and administer first aid. Recognize and assist co-workers who are suffering from shock or emotional distress.
4	**Implement your survival plan.** Establish a temporary shelter if rescue teams are expected to be delayed.
5	**Use stairway** when instructed to exit building.

AT HOME

During the earthquake	
Step	**Action**
1	**Remain inside** your house.
2	**Seek protection** from flying debris or fixtures. Brace yourself inside a doorframe or against inside walls. Seek cover beneath a table, desk, or bed.
3	**Stay in position** until the shaking stops.

After the earthquake	
Step	**Action**
1	**Remain calm.** Organize your thoughts by reviewing your home earthquake survival plan.
2	**Check for injuries** and administer first aid. Be prepared to respond to the psychological aftereffects generated by a major earthquake.
3	**Check water, gas and electric lines.** If you suspect damage, turn off the main valves and leave them off until advised by a utility company representative or other competent source.

Figure 15-1. In earthquake-prone areas, emergency preparedness training should include what to do if an earthquake occurs. (Courtesy Pacific Bell Safety Staff.)

AT HOME (CONTINUED)

	After the earthquake
Step	**Action**
4	**Do not use** candles, matches, or other open flames, or turn lights on/off, either during or after the tremor because of possible gas leaks.
5	**Turn on the radio** to receive emergency instructions. Reserve telephone usage for emergency calls only.
6	**Check your house for structural and internal damage.** Wear boots, if possible, to protect against shattered glass. • Chimneys are earthquake prone if not well enforced and should be approached with caution. Look for separation down the sides or for loose bricks. Unnoticed damage could result in a fire. • Check the interior of the house for dislodged items. The continued swaying motion from an earthquake will cause loose doors of medicine and kitchen cabinets/drawers to open and their contents to spill out. • Check closets and bulk storage areas (garage/basements) for items which may have toppled or collapsed during the earthquake spilling substances which could produce fumes or become potential fire hazards.

IN PUBLIC

	During the earthquake
Step	**Action**
	▪ **On the street**
1	**Enter the closest structure immediately**—do not look up. Enter a store, terminal, office building, etc., just get inside. NOTE: The greatest danger during an earthquake is falling debris.
2	**Remove yourself from windows** that may shatter.
3	**Brace against** an inside doorframe or against inside walls.
	▪ **In a stadium, amphitheater or church**
1	**Remain in your current location. Do not** rush to exits. The chaotic fleeing of large crowds diminishes the effectiveness of an evacuation procedure and frequently results in unnecessary injuries or deaths.
2	**Seek cover** under a bench or chair. If unavailable, crouch down, and cover your head with your arms to protect against falling debris.
3	**Keep away from overhead electric wires** or anything that might fall.
	▪ **In a vehicle**
1	**Stop the vehicle** if it is currently in motion. **Avoid** stopping either **on** or **under** a bridge or overpass.
2	**Remain inside vehicle** until the shaking stops.

	After the earthquake
Step	**Action**
1	**Remain calm.**
2	**Check for injuries** and administer first aid. Recognize and assist individuals who are suffering from shock or emotional distress.
3	**Await emergency evacuation instructions.**
4	**Watch for hazards** created by the earthquake when traveling to another location such as downed electrical wires, broken or undermined roadways, collapsed freeways, over-passes, or bridge structures.
5	**Stay away** from waterfronts or beach areas. Tsunamis may result as an aftereffect of the earthquake.
6	**Avoid sightseeing.** Emergency vehicles will need ready access to respond to emergency situations.

Figure 15-1. (Concluded).

used for scheduled shutdowns, such as for vacation or renovation. Naturally, the amount of warning time controls the speed of shutdown.

Whenever a facility or unit must be shut down, safeguards against fire, explosion, or chemical release take on added importance. The extent of these measures will vary with the size and purpose of the facility. It is important to organize a formal program for instructing personnel in emergency shutdown procedures. In particular, workers should do the following:

- remove lint, dirt, and rubbish from the area
- drain and clean dip and mixing tanks and other equipment where flammables have been used
- clean spray booths, ducts, and flammable liquid storage
- close gas, chemical, and fuel line valves; open switches on power circuits that may be out of service
- check serviceable condition of sprinkler systems, fire extinguishers, hydrants, alarms, and other protective apparatus
- anchor cranes

Prior to the closing, employees are alerted by special instructions to keep their workstations clean and fire-safe.

During the shutdown, continuous inspection of any maintenance or special operations, such as remodeling, must be maintained. Gas cutting and welding should be carefully supervised. Employees who remain on duty—the facility protection force, security staff, maintenance workers, supervisors, or executives—should be briefed in effective countermeasures in case a fire breaks out.

If there has not been sufficient notice to effect a normal shutdown, it may become necessary to allow personnel into the area to perform emergency functions. Emergency shutdown and spill containment procedures should be clearly established for high- and low-staffing situations and all personnel involved should be appropriately trained.

Company management should designate someone who can admit those personnel necessary to handle emergencies arising within the area. The security chief or the fire chief should arrange with local police and fire department officials for assistance if an emergency gets beyond local control. It is especially important that arrangements be completed to speed up admitting firefighters, security, or cleanup personnel and their equipment to the disaster site.

Some companies use facility protection service agencies to prevent loss from theft, fire, and accident hazards during shutdowns. Similar plans should be worked out with these people so that police and fire assistance can be obtained quickly when it is needed.

Wartime Emergency Management Planning

Emergency management planning (EMP) consists of the plans and preparations of business and industry managements to achieve a state of readiness during times of wartime disaster. One of the main differences between planning for peacetime emergencies and planning for wartime emergencies is that war may cripple an entire community. This difference makes it more important that emergency plans take into account the fact that the facility may have to stand alone. In war, outside sources of help—fire and police departments, hospitals and doctors, regular sources of supply for material and equipment—are not so readily available.

Even if a particular area is not attacked, facilities in that area may be requested to furnish transportation to evacuate the injured from damaged areas and to house and feed the evacuees. Facility emergency squads may also be required to assist other stricken buildings or areas. In a major wartime catastrophe, such as heavy bombing, there would probably not be enough hospital space available, making it necessary to keep the injured in temporary shelters for some time. In such cases, employees with the proper training might be required to administer sedatives and plasma and to treat injuries.

Hazardous Materials

As most organizations use a variety of chemical substances, management must be concerned about potential usage, handling, and disposal problems. Although there are many rules and procedures in place, the questions companies should ask include the following: What if a safeguard fails? What if the container cracks and substances leak out? In addition to normal hazards, can chemicals react with other substances to create further hazards to people and property? (See OSHA requirements in 29 *CFR* 1910.1200 Hazard Communication.)

Chemical hazards are discussed in the National Safety Council's *Fundamentals of Industrial Hygiene*, 4th ed. and *CFR* Title III, 1910.1200(g).

Radioactive Materials

Fires and other emergencies involving radioactive materials are becoming more common with the widespread, peaceful use of isotopes.

Radioactive Elements and Fire

Giraud (1973) (see References) makes the following observations. Radioactivity cannot by itself cause fires, nor can it be destroyed or modified by fire. However, a fire may change the state of a radioactive substance and render it more dangerous by converting it to a gas, aerosol, smoke, or ash and allowing it to contaminate a wide area.

Furthermore, fires can cause structural disruptions in stocks of fissionable materials and in the special equipment for their treatment or use. Such disruptions may, at worst, result in a nuclear chain reaction and initiate a critical nuclear accident.

Radioactive elements are found in various forms, depending on their uses. The human eye can detect no difference between an inactive element and the same element when rendered radioactive. Both appear equally harmless. A fundamental distinction must, however, be made between so-called "sealed" and "unsealed" sources.

- In the case of sealed sources, the radioactive substance is not accessible. The container has sufficient mechanical strength to prevent the substance from spreading during normal conditions of use. The capsule is made of stainless steel. The sources are of small dimension—approximately one centimeter.
- In unsealed sources, however, the radioactive substance is accessible. In normal conditions of use, there is no means of preventing it from spreading. Solid substances are kept in aluminum tubes, liquids are kept in flasks, and gases in glass ampules.

The fact that a substance is radioactive does not affect its general physical properties nor its behavior when heated to an abnormally high temperature—as, for instance, during a fire. The substance will, on contact with fire, undergo the normal transformations, depending on its initial form—i.e., solid, liquid, or gas. Melting, boiling, and sublimation can be expected, with the formation of combustion products corresponding to the chemical properties of the substance: slag, ash, powder, dust, mists, aerosols, fumes, or gases.

These combustion products are generally finer and less dense than the original substance, so they disperse more easily. Although the change in the physical state of the substance will not affect its radioactivity, the radiation hazard will be more difficult to control.

The protective containers currently in use have a widely varying resistance to fire. Therefore, the protection afforded to the contents will depend on the type of container used. In general, sealed sources are strongly fire resistant, and radioactive elements thus contained are well protected (Figure 15-2).

Unsealed sources, however, and solutions or gases in fragile containers easily fall victim to fire. What type of immediate action needs to be taken in the event of an accident with radioactive materials can be determined by the firefighting staff once the type of container is known. Actions to deal with the hazard will depend on the properties of the radioactive substance. (See OSHA regulations in 29 *CFR* 1910.1200 (f), Hazard Communication, 'Labels and Other Forms of Warning' and 29 *CFR* 1910.38, Employee Emergency Plans and Fire Preventing Plans.)

When, as the direct or indirect result of a fire, the protective container has been broken, the radiation hazards for rescue workers at the fire, or for personnel in the vicinity, are likely to be more serious than the danger of a conventional fire. Accordingly, the person in charge of the rescue work will sometimes be obliged to override the normal firefighting procedures to ensure proper confine-

Figure 15-2. "Radiation Yellow—III" label, which is affixed to each package of highly radioactive material. Different labels are required for different intensities and quantities of radioactive material. For details see Title 49—Transportation, *Code of Federal Regulations*, Part 172, Hazardous Materials Table and Hazardous Materials Communications Regulations.

ment of any radioactive elements released. If they are already affected by the fire, further hazards may arise.

The release of radioactive elements may result in contamination of surface areas. This may be caused by the spilling or splashing of radioactive substances or by the spreading of solid radioactive substances in paste, powder, or dust form. All possible precautions must be taken to prevent any further spread of the contamination. The means to be used, however, will differ with each case. In the first (spilling or splashing), absorbent materials should be used—such as powder, earth, sand, and so on. In the case of spreading, the substances should be slightly dampened with a spray of water—unless it is otherwise specified on the container. (See OSHA regulation 29 *CFR* 1910.1200 (g), Material Safety Data Sheets.)

Radioactive liquids can be contained by the methods normally used by the firefighting brigade. The contaminated area must be clearly marked and roped off to prevent the entry of unauthorized personnel (Figure 15-3).

Contamination of the atmosphere is caused by radioactive elements in the form of dust, aerosols, fumes, and gases. The spreading of such contamination is determined mainly by the prevailing weather conditions, and it is difficult to control. Such atmospheric contamination may lead to other toxic or corrosive hazards associated with the particular chemical. The most serious danger is that of

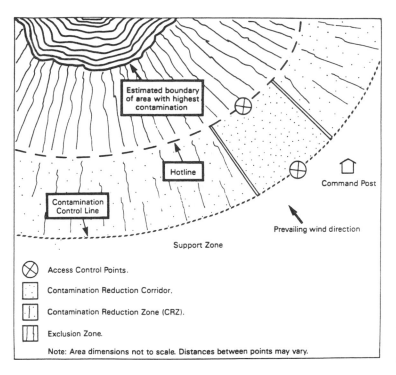

Estimated boundary of area with highest contamination

Hotline

Command Post

Contamination Control Line

Prevailing wind direction

Support Zone

⊗ Access Control Points.

Contamination Reduction Corridor.

Contamination Reduction Zone (CRZ).

Exclusion Zone.

Note: Area dimensions not to scale. Distances between points may vary.

Figure 15-3. This diagram shows site work zones for an emergency situation involving a radioactive hazard. Note that decontamination facilities are located in the Contamination Reduction Zone. (Courtesy NIOSH/OSHA/USCG/EPA, *Occupational Safety and Health Guidance Manual for Hazardous Waste Site Activities.*)

inhaling the substance when it is suspended in the air. Firefighters, accordingly, should wear self-contained breathing apparatus.

The danger of internal irradiation is always present whenever there is contamination by a source of penetrating radiation. It may also occur by accidental release of an alpha or beta emitter from its protective container, or by the destruction (even partial) of the protective container.

Weather-Related Emergencies

Throughout any year, unusually severe and unexpected weather events can require some changes in normal operations. For example, in North Dakota, the temperature occasionally may drop to –35 F (–30 C), yet most activities and travel are not normally affected. But, if the wind increases in strength or the temperature drops suddenly, people may need help as they travel or participate in other outside activities (Figure 15-4 provides a windchill chart). Employees could be alerted to the danger prior to leaving work and told when or how they are to be notified about whether the company will be open in the morning.

Management must plan for a variety of weather-related emergencies appropriate for the locale. In the event of extremely heavy snowfall, for example, what changes might be needed in operations? What should employees be told prior to leaving for home? An unusually heavy rainstorm may strand hundreds of customers in a store just a few minutes before closing. Are supervisors and clerks prepared to handle the situation? May they allow telephone calls in and out? How do they control the crowd? Hail or wind may start breaking glass windows while customers are shopping. What is the immediate action? Suppose that

adverse weather caused a power failure in a company or someone suddenly shut off all power and lights while crowds were shopping? The emergency lighting system may operate as intended, but employees, particularly key supervisors, must understand emergency plans and be prepared to act responsibly.

Other weather emergencies include droughts and extreme heat. These harsh conditions require attention to outdoor work schedules, water resources, and health effects. In addition, lightning is a major cause of weather-related deaths in the United States. Planning should include emergency procedures for halting work in areas vulnerable to lightning strikes or the loss of utilities when this particular danger is imminent.

PLAN-OF-ACTION CONSIDERATIONS

Once the risk assessment of potential emergencies has been completed, the next stage of the planning process should be preparing a plan of action. This plan should be supported by management and include input from both the public and private sectors. It is desirable to include union or labor representatives.

Generally, someone should be appointed emergency planning director or coordinator, perhaps with help from an advisory committee. Usually, because of their experience and training, the health and safety, medical, fire, and security departments will be involved. Of course, because production and maintenance will be affected, these departments must be consulted. Also, the legal staff needs to be aware of the plan. Finally, management should contact local law enforcement agencies and fire departments.

Estimated wind speed (mph)	Actual thermometer reading (°F)											
	50	40	30	20	10	0	–10	–20	–30	–40	–50	–60
	Equivalent temperature (°F)											
Calm	50	40	30	20	10	0	–10	–20	–30	–40	–50	–60
5	48	37	27	16	6	–5	–15	–26	–36	–47	–57	–68
10	40	28	16	4	–9	–24	–33	–46	–58	–70	–83	–95
15	36	22	9	–5	–18	–32	–45	–58	–72	–85	–99	–112
20	32	18	4	–10	–25	–39	–53	–67	–82	–96	–110	–124
25	30	16	0	–15	–29	–44	–59	–74	–88	–104	–118	–133
30	28	13	–2	–18	–33	–48	–63	–79	–94	–109	–125	–140
35	27	11	–4	–20	–35	–51	–67	–82	–98	–113	–129	–145
40	26	10	–6	–21	–37	–53	–69	–85	–100	–116	–132	–148
Wind speeds greater than 40 mph have little added effect.	Little danger for properly clothed person. Maximum danger of false sense of security.			Increasing danger Danger from freezing of exposed flesh.			Great danger					
Trenchfoot and immersion foot may occur at any point on this chart.												

Figure 15-4. Windchill factors. The human body senses "cold" as a result of both temperature and wind velocity. The numerical factor that combines the effects of these two is called "the windchill factor." Because of the extra clothing that people wear in cold weather, their physical size is greater than it is in warm weather. Be sure that equipment and controls are of adequate size and simplicity so that they can be run effectively and safely by persons wearing heavy clothing. (Courtesy Holmes Safety Association.)

Some managers may object to the cost and effort involved in giving immediate attention to emergency planning. However, this activity can be justified by weighing the cost of preparedness against the possibility of yearly losses from accidents, fires, floods, and other catastrophes.

Program Considerations

Once management has completed its advance planning and has evaluated the type of emergencies and their potential harm to people and property, the next step is to develop a working emergency management plan within the organization. In some cases, this step requires working with local agencies to protect a company's operations.

Advance planning is the key. Management must establish a written set of plans for action. The plans should be developed locally within the company (and corporate structure) and be in cooperation with other neighboring or similar organizations and with governmental agencies. It may not always be possible for every organization or agency to cooperate or participate fully, but through planned action each one can be aware of the resources and help available. In some instances when outside assistance is limited, a company may need to depend largely upon its own resources to deal with emergencies.

Often an emergency manual or handbook will be developed for the facility or organization. The following outline covers many of the items that might be included, but other items may be needed as dictated by the expected emergencies and the available resources:

1. company policy, purposes, authority, principal control measures, and emergency organization chart showing positions and functions
2. some description of the expected disasters with a risk statement
3. map of the facility, office, or store showing equipment, medical and first aid, fire control apparatus, shelters, command center, evacuation routes, and assembly areas
4. list (which may also be posted) of cooperating agencies and how to reach them
5. facility warning system
6. central communications center, including home contacts of employees
7. shutdown procedure, including security guard

8. how to handle visitors and customers
9. locally related and necessary items
10. list of equipment and resources that would be available and where they can be reached.

Some of these items will be discussed in more detail in the following pages. Management should rehearse the plan under realistic conditions to test the plan's effectiveness. For example, emergency lights may fail when needed, or the telephone service break down. These are real conditions that might occur in an actual disaster. Therefore, planning should include all possible as well as probable contingencies.

Chain of Command

Once management has decided to establish a disaster plan, it should appoint a director or coordinator and create an advisory committee representing the various departments. A basic guideline to follow in establishing the chain of command is (1) keep the chain as small as practical and (2) appoint personnel to crisis management positions based not on their title but rather on their ability to respond to a situation under extreme stress.

Experience has shown that the smaller the chain of command, the more efficient and effective its decisions and actions will be during a crisis situation. Normally, the chain of communication will pattern itself after the chain of command. Communications must be conducted in as efficient a manner as possible. Management should quickly provide accurate information or request the proper support. Long delays could affect the outcome of a given emergency situation.

It is normal to think that the ranking manager should assume the crisis manager's position. However, to do so may not be the proper action to take. Many people are excellent managers under normal conditions. They are able to give good direction if allowed time to think through the situation and weigh the pros and cons. But given a situation where extreme stress is introduced and quick, timely decisions are required, the same individual may not be able to perform. It is, therefore, imperative that the individuals selected to become a part of the crisis team be tested, through training exercises, to see if they can perform under emergency conditions. If the ranking manager is found not to be the proper choice for the crisis manager, he or she should be a consultant to whoever is placed in the crisis manager's position. This same philosophy applies to any of the positions established within the chain of command.

The emergency director should be a member of top management, whether it be a one-building or one-facility or a national organization. This individual will need to delegate authority and to speak for the organization. The head of the disaster-control organization must be cool and quick thinking and sufficiently healthy to withstand the arduous duties involved in an emergency. The emergency director's regular duties should be such that the greater part of the time will normally be spent at his or her own unit. However, the plan should always name an alternate director. The alternate should be a person who has authority and qualifications similar to the director and should be trained with the director.

The director (and alternate) should be the first to be trained. Management should maintain liaison with local emergency management authorities, if possible, to make sure that the plans are coordinated with those of the community and to keep the company informed on new developments. (See OSHA regulation 29 *CFR* 1910.156, Fire Brigades; 29 *CFR* 1910.1200, Hazard Communication; and EPA Regulations 40 *CFR* Part 311, OSHA 29 *CFR* 1910.120.)

The director of emergency management is responsible for:

- emergency operation center management
- communications
- firefighting
- security and law enforcement
- rescue operations
- emergency medical services
- transportation
- damage assessment
- mitigation and investigation
- public information and media briefings
- rumor control
- on-scene safety functions at the emergency site
- warning and evacuation of facility and community personnel
- utilities and engineering functions
- sheltering, feeding, and counseling functions
- morgue establishment and notification of survivors
- notification to the SARA Title III authorities in the event of a hazardous materials release

All of these functions are likely to be essential although some may be combined. The person (and alternates) responsible for each function should be selected with great care and trained by the director. These chiefs should be familiar with all parts of the plan and should have experience in the fields in which they are to serve.

Management should train assigned personnel to carry out their duties in accordance with the overall emergency plan. In small operations without regular security staff or firefighters, operating personnel will be trained to take care of these duties. Of course, the number of members on each of the teams depends on the circumstances of each facility. Each team captain should select personnel from the available volunteers, supervise their training, and procure their equipment. The strongest workers can be assigned to rescue squads because the work usually demands strenuous physical effort; those who are less strong could be useful in light salvage operations.

Because workers' whole-hearted cooperation is necessary to the successful operation of an emergency plan, shop stewards or other employee representatives should take part in the planning. They should understand that measures taken are to protect the lives and jobs of the workers as well as to protect property.

The organization should establish emergency reporting centers so that employees will know where to report should a disaster strike while the employees are away from work. Reporting centers give employees a feeling of security and continuity, and aid in taking a "roll call." To facilitate these arrangements, each employee should carry an identification card containing specific instructions on where to report, list of other reporting centers, basic employment record, and designation of the employee's next of kin in case they must be contacted or receive money due. The reporting center will keep a duplicate record for each employee assigned to the center. A one-facility company or a small company can consider using the home of a member of management, a supervisor, or an employee.

Training

One of the most important functions of the director and staff, on both the corporate and facility levels, is training. Training for each type of disaster is essential in developing a disaster-control plan and keeping it functioning. Employees must be taught to realize that an emergency plan is vital and real—it has no value if it remains simply an idea (Figures 15-5a, 5b and Figure 15-6). Training and rehearsals are time consuming, but they keep the program in good working order.

Simulated disaster drills—sometimes called paper drills—will help key people and employees respond to emergencies with greater confidence and effectiveness. There is no actual response required. Instead, key personnel operate under the direction of a drill coordinator who feeds information to them on a real-time basis and monitors the response.

Another type of drill is a full-scale dress rehearsal involving all personnel with simulated situations and injuries. Management must be careful to prevent injuries to participants and the public while making the full response as realistic as possible. Feedback from either type of drill is essential to improve emergency management plans.

Management should assure employees that the company is doing everything possible to prevent injury to them, that every employee is an essential and necessary part of the team, and that the disaster-control organization is ready for any emergency. Such assurance will go a long way toward preventing panic when a real disaster strikes.

IN CASE OF FIRE OR OTHER EMERGENCY

✔ **KEEP YOUR HEAD**—avoid panic and confusion.

✔ **KNOW THE LOCATION OF EXITS**—be sure you know the safest way out of the building no matter where you are.

✔ **KNOW THE LOCATION OF NEARBY FIRE EXTINGUISHERS**—learn the proper way to use all types of extinguishers.

✔ **KNOW HOW TO REPORT A FIRE OR OTHER EMERGENCY**—send in the alarm without delay; **notify the CHIEF OF EXIT DRILLS.**

✔ **FOLLOW EXIT INSTRUCTIONS**—stay at your work place until signaled or instructed to leave; complete all emergency duties assigned to you and be ready to march out rapidly according to plan.

✔ **WALK TO YOUR ASSIGNED EXIT**—maintain order and quiet; take each drill seriously—It may be "the real thing."

REMEMBER—IT IS PART OF YOUR
JOB TO PREVENT FIRES

Figure 15-5a. Sample emergency exit notice for general posting.

EMERGENCY EXIT INSTRUCTIONS
MACHINE SHOP—DAY SHIFT

Read Carefully

The following persons will be in command in any emergency, and their instructions must be followed:

CHIEF OF EXIT DRILL—H. C. Gordon, General Sup't.
MACHINE SHOP EXIT DRILL CAPTAIN—R. L. Jones, Foreman
MACHINE SHOP MONITORS—Dave Thomas and A. L. Smith

In event of FIRE in machine shop

✔ **NOTIFY THE GENERAL SUPERINTENDENT'S OFFICE**

✔ **PUT OUT THE FIRE, IF POSSIBLE**—If the fire cannot quickly be controlled, follow instructions given by Exit Drill Captain R. L. Jones or by the shop monitors. Leave by the exit door at the south end of the shop; if it is blocked by fire, use the door through the toolroom to the outside stairway.

In event of FIRE or EMERGENCY in other sections of building

The general alarm gong will ring for two 10-second periods as an "alert" signal. Continue work, but be on the alert for the "evacuation" signal, which will be a series of three short rings. At the evacuation signal:

✔ SHUT OFF ALL POWER TO MACHINES AND FANS

✔ TURN OFF GAS UNDER HEAT TREATING OVENS

✔ CLOSE WINDOWS AND CLEAR THE AISLES

✔ FORM A DOUBLE LINE IN THE CENTER AISLE AND FOLLOW MONITORS AND EXIT DRILL CAPTAIN TO EXIT—Walk rapidly, but do not run or crowd; do not talk, push, or cause confusion!

After leaving the building, do not interfere with the work of the plant fire brigade or the city fire department. Await instructions from the General Superintendent or your foreman.

Returning to the building

Return-to-work instructions will be given over the loudspeaker system or by telephone from the Superintendent's office.

Figure 15-5b. Sample individual instruction notice for general posting. Some facilities will include after-hours phone numbers and alternate communications protocols.

Should such an event occur, emergency forces will snap into action, workers will file quietly into their shelters or other designated areas, firefighters will be ready with hoses and equipment, and first-aid squads will stand by ready to aid the wounded. Such planning is further evidence of management's concern for employees.

Hazardous Materials/Spills Emergencies (HAZWOPER)

The federal OSHA regulation entitled Hazardous Waste Operations and Emergency Response: Final Rule became effective on March 6, 1990 (HAZWOPER). It addresses the many aspects of health and safety that are now legally required at hazardous materials sites; treatment, storage, and disposal facilities; and other hazardous materials emergency locations. The rule mandates certain requirements for monitoring instrumentation, site safety plans, respiratory and personal protective equipment, medical surveillance, engineering controls, work practices, training requirements, and other operational functions.

Generally, the *CFR* 1910.120 training standard addresses three categories of employees: hazardous

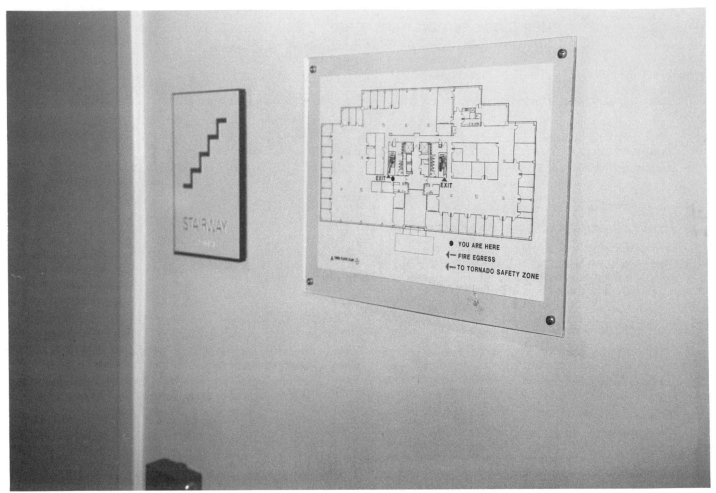

Figure 15-6. The clear identification of evacuation routes from all locations is an important aspect of emergency preparedness training.

materials site workers (paragraph [e]); treatment, storage, and disposal workers (paragraph [p]); and emergency responders to hazardous substance releases regardless of location. (See also OSHA regulation 29 *CFR* 1910.156, Fire Brigades.)

The HAZWOPER team consists of an organized group of employees, designated by the employer, who are expected to perform work to handle and control actual or potential leaks or spills of hazardous substances requiring possible close approach to the substance. The team members perform responses to releases or potential releases of hazardous substances for the purpose of control or stabilization of the incident. A HAZWOPER team is not a fire brigade and a typical fire brigade is not a HAZWOPER team. A HAZWOPER team, however, may be a separate component of a fire brigade or department.

Recommended minimum training requirements for the HAZWOPER team and appropriate courses are as follows:

Hazardous Materials Site Workers

This course is for general workers who are on site full time and probably have been exposed to a hazardous chemical or situation. They are required to have 40 hours

of classroom "hands-on" instruction away from the site followed by a minimum of three days actual field experience under a trained supervisor. All site workers must complete eight hours of refresher training each year, and the training should be documented.

Occasional Hazardous Materials Site Workers

This course accommodates inspectors, engineers, monitoring technicians, or others who are not likely to be exposed over permissible exposure limits. They are required to have 24 hours of off-site instruction followed by one day of actual field experience.

Hazardous Materials Site Supervisors

These site managers must have eight hours of specialized training after completing the 40-hour course. Supervisory training covers site safety plans, personal protective equipment selection, and health monitoring.

Treatment, Storage, and Disposal (TSD) Facility Workers

New employees in these work areas must have at least 24 hours of health and safety training. Current employees must demonstrate the equivalent amount of training from previous experience. All workers must complete eight hours of TSD refresher training each year.

Hazardous Materials Emergency Responders (Regardless of Location)

These workers are distributed into five categories. The training requirements apply to private facility employees as well as to community emergency services personnel.

First Responder Awareness

This is a four- to eight-hour course for responders who will only attempt to identify the involved hazardous material and then notify more qualified personnel.

First Responder Operations

An eight-hour training requirement exists, but 24 hours of instruction is highly recommended. This category covers responders who will identify materials, perform basic diking and containment operations, and initiate evacuation. This course requires training in personal protective equipment, decontamination, chemistry, and toxicology. However, employees are taught to respond in a defensive fashion; the course does not qualify them to attempt a patching or plugging operation.

Hazardous Materials Technician

This is a 24-hour course for workers who will perform the actual plugging, patching, or sealing of a container leaking a hazardous substance. The 24-hour First Responder Operations class is a prerequisite to this course.

Hazardous Materials Specialist

Additional training beyond the 48 hours required for the responder and technician levels must be taken in the specialty area of this skilled responder. No specific hours requirement exists, but this individual should be competent in the chemical, toxicological, and/or radiological behavior of the particular material involved.

Incident Commander

The incident commander must be a graduate of the 24-hour First Responder Operations course and have additional training (24 hours is recommended) in planning, decontamination, protective clothing, and command systems. The commander does not have to receive instruction in the technician or specialist categories. (See also the section Uniform Incident Command System later in this chapter.)

Command Headquarters

Although the command headquarters of a company will not withstand catastrophic disasters, management can plan for emergencies that may occur.

The disaster control organization should be coordinated from a well-equipped and well-protected control room and/or an alternate off-site command headquarters.

The headquarters should be equipped with telephones, sound-powered phones, public address system, maps of the facility, emergency lighting and electric power, sanitary facilities, a second exit, and two-way radios for communication both locally and with emergency management authorities if required.

Good communications are necessary for effective control and flexibility in a disaster situation. Communications include the (cellular) telephone, radio, messengers, and the facility's alarm system (discussed separately later in this chapter). The disaster plan should provide for adequate telephones in emergency headquarters to handle both incoming and outgoing calls. Panic and disintegration of the organization will develop quickly if these calls are not handled with dispatch. Operators should keep an accurate log of all incoming and outgoing messages.

Some means of communication independent of normal telephone service must be available during an emergency, such as the one provided by a battery-operated radio. Mobile cellular phones may or may not be operable during or following a disaster. The disaster plan must anticipate the possibility of losing normal telephone communications and electric power.

Uniform Incident Command System (ICS)

The intent of this incident command system (ICS) is to provide a comprehensive management structure that satisfies the requirements set forth in OSHA *CFR* 1910.120. As stated in the regulations:

> The ICS shall be established by those employers for the incidents that will be under their control and shall be interfaced with the other organizations or agencies who may respond to such an incident.

The command function within the ICS may be conducted in two general ways. Single command may be applied when there is no overlap of jurisdictional boundaries or when a single incident commander (IC) is designated by the agency with overall management responsibility for the incident. Unified command may be applied when the incident is within one jurisdictional boundary, but more than one agency shares management responsibility. Unified command is also used when the incident is multijurisdictional or when more than one individual designated by his or her jurisdictional agency shares overall management responsibility.

Every incident needs some sort of consolidated action plan. Written plans are usually required when resources from multiple agencies are used, when several jurisdictions are involved, or when changes of personnel or equipment are required. The action plan should cover all strategic goals, tactical objectives, and support activities required during the operation. In prolonged incidents it may be necessary to develop action plans covering specific operational periods. Command staff includes the following:

Incident Commander

The one function that will always be filled at every incident, regardless of size, is the incident commander. The IC has the responsibility of overall incident management. The incident commander's responsibilities include:

1. Assess the incident priorities.
   a. Safety: The IC must consider safety issues for all personnel at an incident. No industrial complex or form of property is worth the risk of even one life. Safety comes before all other considerations.
   b. Incident stabilization: The IC is responsible for determining the strategy that will minimize the impact that an incident may have on the surrounding area. The size and complexity of the command system developed and implemented by the IC should be directly proportional to the magnitude and complexity of the incident. The ICS structure must match the complexity of the incident, not the size. Situations that may appear hopeless must be managed and ultimately controlled.
2. Determine the incident's strategic goals and tactical objectives. The efforts of the resources available for handling any incident must be properly directed to minimize the damage. The clocks cannot be turned back. Damage that has already occurred cannot be alleviated, but further damage must be minimized. This is accomplished when the IC determines the broad strategic goals for the incident and then transforms these goals into obtainable, practical objectives.
3. Develop or approve and implement the incident action plan. The IC is the primary developer of the incident action plan. On most simple incidents, the action plan will be organized completely by the IC and may not need to be written down. In more complex incidents, the action plan will be a written document developed by a staff, headed by the IC. Action plans must be flexible and continually assessed. In the environmental business, conditions rarely remain constant. They are almost always dynamic.
4. Develop an incident command structure appropriate for the incident. The organizational structure is not based on the size or area of involvement; it depends on the complexity of the incident.
5. Assess resource needs and deploy as needed. The IC must continually evaluate and adjust the deployment of resources at all incidents. Initial assessment of the incident and the needed resources is only the first step. As soon as the IC determines the incident's strategic goals and tactical objectives and

then evaluates the resource needs to meet these goals and objectives, one of two actions will occur. Either the initial action plan will be successful or it will need to be revised. Additional resources may be needed, requiring reorganization. If the IC believes he or she has just enough resources for the required work, it is time to order additional help and/or other resources. Coming out exactly even means the IC is a gambler instead of a true manager.

Effective resource management requires that personnel safety be given the highest priority. Although everyone working at an incident must serve as his or her own safety officer, the ultimate responsibility for incident safety rests with the IC. As an incident escalates, the IC will need to assign a person as safety officer, with specific safety responsibilities. As stated in *CFR* 1910.120 Q3 (vii),

> The individual in charge of the ICS shall designate a safety official, who is knowledgeable in the operations being implemented at the incident site, with specific responsibility to identify and evaluate hazards and to provide direction with respect to the safety of operations for the incident at hand.
>
> (vii) When activities are judged by the safety official to be an IDLH condition and/or to involve an imminent danger condition, the safety official shall have the authority to alter, suspend, or terminate those activities. The safety official shall immediately inform the individual in charge of the ICS of any actions needed to be taken to correct these hazards at an incident site.

6. Coordinate overall site activities. Coordination is essential to effective incident management. Without it, resources will be wasted performing tasks that are not necessary to the overall success of the incident. The IC must constantly monitor the incident activities to ensure that the needed degree of coordination is present and that personnel are not working at cross duplication. The goal of the IC is to obtain the maximum productivity from all on-scene resources. Proper coordination will ensure that personnel and equipment are functioning within the action plan.

Liaison

The liaison individual(s) are the point of contact for assisting or coordination agencies. This function is assigned so the IC is not overloaded by questions from the number of assisting agencies that some incidents attract.

One of the most important responsibilities of the liaison individual is to coordinate the management of assisting or coordinating agencies. This is essential to avoid the duplication of efforts. It allows each agency to perform what it does best. Liaison management provides lines of authority, responsibility, and communication.

Operations

Operations is responsible for management of all tactical operations at the incident. Operations is implemented when the IC is faced with a complex incident having major demands in one or more of the remaining major functional areas. For example, the IC may be faced with a rapidly escalating incident with a significant need to evaluate strategy and to develop alternative tactical options. Faced with a major functional responsibility in addition to management of tactical operations, the IC may need to staff Operations to maintain an effective span of control.

The most common reason for staffing Operations is to relieve span-of-control problems for the IC. A complex incident, in which the IC needs assistance determining strategic goals and tactical objectives, may also require implementing Operations.

Planning

Planning is responsible for the collection, evaluation, dissemination, and use of information about the development of the incident and the status of resources. When faced with a complex or rapidly escalating incident, the IC may require assistance with the ICS Planning function. A wide range of factors may impact on incident operations. Planning must include an assessment of the present and projected situation. Proactive incident management is highly dependent on an accurate assessment of the incident's potential and prediction of likely outcomes. In addition to assessment of the situation status, there is a critical need to maintain information about resources committed to the incident and projected resource requirements.

Logistics

Logistics is responsible for providing facilities, services, and materials for the incident. As incidents grow in size, complexity, and duration, the logistical needs of the operating forces also increase. Even in a relatively simple incident, there are requirements for equipment, drinking water, and emergency medical care. When faced with a major incident the logistical requirements are significant. Long duration incidents of any type require provisions for feeding personnel, toilet facilities, refueling of vehicles/equipment, lodging, and a myriad of other service and support resources. Acquisition and the accurate distribution of material/equipment is a major functional responsibility for this position. This individual(s) works very closely with Planning personnel.

Finance

Finance is responsible for tracking all incident costs and evaluating the financial considerations of the incident. During large-scale incidents many significant purchases

and cash transactions are made. It is the responsibility of Finance personnel to insure that all disbursements are documented including, but not limited to, accurate invoicing for all services/materials used.

Personnel Requirements for Implementation

It shall be the ultimate responsibility of each Operations manager to insure that personnel assigned to the following tasks are competent and capable of fulfilling their assigned role. Figure 15-7 lists the elements that should be included in a basic incident command system.

Emergency Equipment

An emergency checklist should include equipment and material to be ordered as well as shutdown actions to be performed. For example, where it is not feasible to keep on hand the necessary emergency equipment and materials, one should maintain a resource or supply list and names of suppliers. The information should include type of equipment they can supply, after-hours information and phone numbers, and, if possible, prices. These sources must be outside of the immediate area.

Some of the items on the shutdown part of the checklist would include closing valves; protecting equipment that cannot be moved; closing and battening doors, windows, and ventilators to keep out looters as well as water; and plugging vents and breather pipes. The checklist should include a list of telephone numbers of supervisors and key employees to be notified.

Basic Incident Command System

Incident commander
Safety
Liaison
Operations
Sector Commanders
Supervisor
Foreman
Technicians
Planning
Assistant(s)
Subcontractors
Long-range logistical support
Logistics
Assistant(s)
Equipment and materials
Transportation
Rehabilitation
Food, water, and lodging
Finance
Accountant
Field clerks

Figure 15-7. This is an example of a basic incident command system.

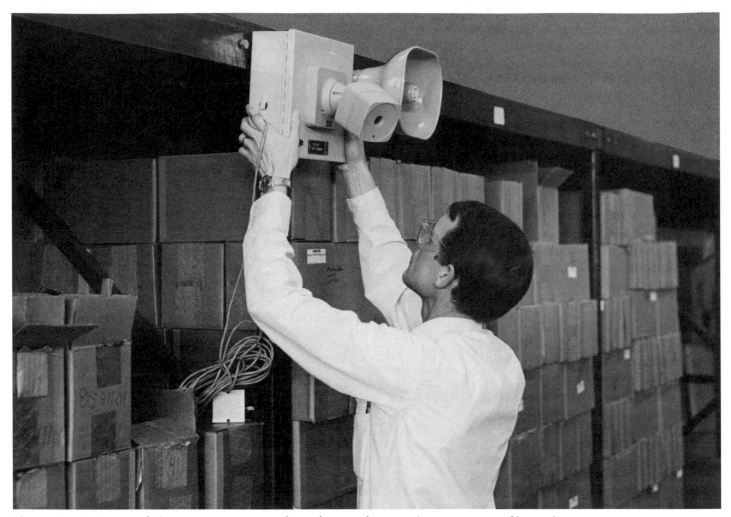

Figure 15-8. Spacing of alarm systems is important for notification of personnel in remote areas of large plants.

Other procedures must include shutting off electric and gas utility services at the main line before any water reaches them. Make sure hot equipment is cooled before water reaches it. Operators should coat all machine surfaces liberally with heavy grease, especially around openings to bearings. (This step applies even to machines that may not be under water because dampness can damage equipment.) Shut off all open flames so that any flammable liquid floating on the floodwaters will not be ignited.

If possible, keep a salvage crew at the site to continue preventive operations after the facility has been shut down. This crew can take further steps if the flood exceeds the estimated high-water level.

Alarm Systems

Most industrial operations have a special fire alarm system using existing signaling systems such as a facility whistle. However, to avoid confusing fire alarms with the regularly used signals, some facilities have special codes or other signaling devices for fires. This type of signal also may indicate the location of the fire, or separate signaling devices may be used for each building or working area within the company property.

The alarm system activating the emergency plan may or may not go through the communications center, but it should be touched off in the emergency headquarters office. All buildings should contain alarm systems (Figure 15-8).

In hospitals or other locations where both employees and nonemployees can hear an audible page system, a code name can be used to announce a fire and its location, such as "Doctor Red wanted in—". Employees must be trained to be alert for this subtle signal.

Electric alarms are preferred to mechanical ones, except in a large open shop area with only one alarm-summons station and one alarm-sending device, such as a manually operated gong. Manually operated alarms should supplement electric alarms. Closed-circuit systems of the type specified by NFPA standards are recommended (see References).

Companies in areas where municipal fire departments are available usually have a municipal alarm box close to

Figure 15-9. A command console for computer-based supervision and monitoring of large buildings or building complexes for loss-producing hazards including intrusion, fire, and other emergencies. Based on a central processing unit, the console includes computer display screen, high speed printer, and zoned annunciator panels.

the firm's entrance or in one of the buildings. Others may have auxiliary alarm box areas, connected to the municipal fire alarm system, at various points on the premises. Another system often used is a direct connection to the control dispatcher. This system may be touched off by a water alarm in the sprinkler system or be activated manually. If possible, connect the fire alarm system to the local firefighting alarm and make sure it has an independent power supply.

In some areas, private central station services are available and provide excellent protection. These central stations receive signals from facility fire alarm boxes, security staff, sprinkler head operations, and other hazard control points in the facility. Because these devices monitor facility security, they can relay information to fire or police departments without delay. The signal received at the fire department or assistance agency should pinpoint the site of the fire, or at least the building or area, so the fire can be found quickly (Figure 15-9). Automatic sending stations (thermostatic detectors) may be used but should not interfere with sending a manual alarm.

Regular checks should be made on the alarm system. Checks should include monthly inspections of all stations and a weekly test of the overall system to ensure it is in proper working condition. These weekly tests should be conducted at a prearranged time and under a variety of wind and weather conditions to determine if the signal can be heard in all parts of the facility at all times.

Fire Brigades

Fire prevention and fire protection must receive major attention in any emergency program because a fire can start from so many causes. Advance planning is important. (Note: U.S. organizations may elect to have fire brigades, in which case regulatory standards have to be followed.)

The company fire chief must be able to command people in addition to having special training in fire prevention and protection. A person who has had experience in city or volunteer fire department work, or a military service veteran with experience in firefighting may be a good choice. In a smaller company or facility, a master mechanic, maintenance department head, or other

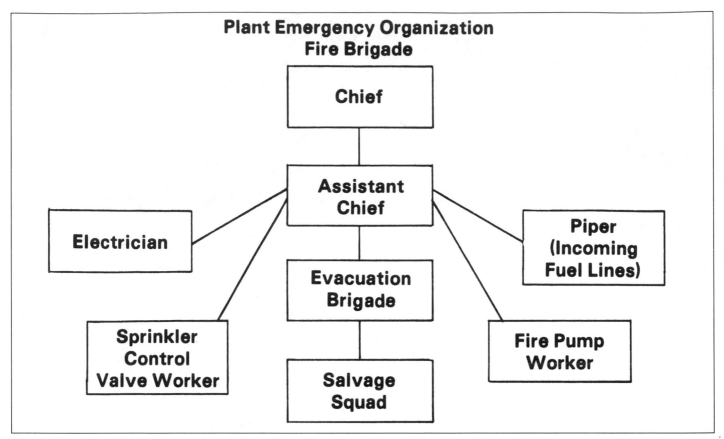

Figure 15-10. Because of the complexity of response needed to cope with an emergency, duties must be divided among facility emergency organization members to avoid confusion. Here is one facility's organizational chart. Current names, phone numbers, and those of alternates should be listed in the boxes. (Reprinted with permission from Textile Section Newsletter.)

employee with mechanical experience can be a good part-time fire chief.

Most U.S. states have a State Fire Service Training Program. As part of this service, there are now schools specifically oriented toward industrial fire protection. At the very least, all the fire brigade personnel or the guards should have advanced training. They are the ones most likely to use auxiliary firefighting equipment.

The fire chief may need one or more assistants who have complete knowledge of the facility and equipment, command the respect and obedience of those under them, and are qualified to perform the duties of the chief officer if necessary.

The size of the facility and the fire potential presented by the occupancy will determine the kind of firefighting organization (brigade) required. The majority of facilities may need only first-aid firefighters under the direction of departmental foremen or managers. In larger facilities, the fire brigade organization, directed by a full-time fire chief, is composed of full-time and emergency members. The full-time members maintain fire brigade equipment and are responsible for the permanent fire protection of the facility. Emergency members report for fire duty when the alarm is sounded.

Fire brigade apparatus should be selected only after a study has been made of the specific conditions to ensure the equipment will be adequate for any emergency. Management can obtain advice and assistance from the local fire department, the NFPA, State Fire Training Organization, or insurance companies.

The large facility fire brigade is usually organized into squads, each with specific duties. One company's organization chart is shown in Figure 15-10.

The evacuation squad evacuates all employees from the emergency area as quickly and orderly as possible, without injury. They search closed areas, such as washrooms, to determine that everyone has been evacuated.

The environmental monitoring squad identifies and watches environmental conditions. They determine hazardous conditions, make sure sites are not contaminated, and decide when an area can be re-entered safely.

Utility control squad members usually are maintenance personnel familiar with facility piping systems and the control of process gases, flammable liquids, and electricity. Sprinkler control squad members must understand the automatic sprinkler system—the direction of rotation of the valve they are to operate, the use of sprinkler stops, and the replacement of sprinkler heads, if this is not a maintenance department function.

Extinguisher Squad

Portable fire extinguishers are frequently operated by designated employees who work in the vicinity. However, as the size of the facility increases, it is advisable that special squads be selected, trained, and equipped for handling fire extinguishers (Figure 15-11).

Hose squad members are trained to operate fire hydrants and hoses. They should drill frequently with wet hose lines so they have the feel of a charged hose. After they become proficient in handling the hose, drills once or twice a year may be sufficient to maintain their preparedness (Figure 15-12).

The salvage squad is trained to protect as much stock and equipment as possible by controlling the directional flow of water and by covering stock with tarpaulins. Their training should include proper methods of throwing tarpaulins and using them to direct the flow of water. Squad members should also be familiar with the location of sawdust or other absorbent material and know how to use it in controlling water on floors.

The brigade-at-large or rescue team consists of maintenance personnel or specially trained workers. The main functions of this unit are to extricate casualties and eliminate hazards that may endanger other workers involved in controlling the emergency. Members respond to all alarms

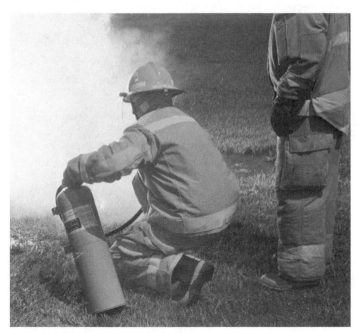

Figure 15-11. Fire extinguisher training is very important. Underwriters Laboratories Inc. considers any fire extinguisher in the hand of a trained operator to have 2 times more firefighting capacity than when used by a novice. Here the fire chief instructs a worker in the use of a pressurized water fire extinguisher. (Courtesy Ansul Fire School.)

with a utility truck containing such rescue equipment as ropes, chains, block and tackle, ladders, cutting torches, saws, axes, and jacks. The amount of equipment, of course, will depend upon the size of the facility and the hazards involved, but every effort should be made to anticipate possible problems.

Under the direction of the brigade officer, personnel in this unit also control utilities, ventilating fans, and blowers. They close fire doors, windows, and other openings in division walls, and open windows and doors leading to fire exits. If escape equipment is the swinging section type, rescue personnel should be the first to operate the escapes and to secure the steps to the ground.

Depending upon the size and inherent hazards of the facility, it may be necessary to train squads in handling and erecting ladders, using foam lines, recharging foam generators, specialized rescue techniques, and chemical spill response procedures.

Fire Pump Team

There should be at least two competent people for pump duty in the main pump room. Management must assign one trained person to each pump located elsewhere.

Firefighter Training

The brigade should go through complete drills, preferably every week. Later, the leaders can hold less elaborate drills less often. Drills should be unannounced and thorough in every respect, closely approximating fire conditions (Figure 15-12).

No matter how thoroughly the industrial fire brigade is trained, management should make sure they cooperate closely with adjoining or nearby facilities and the public fire department. It would be a good idea to do mutual training with the local fire department so that each side knows the expectations of the other. This approach builds a better relationship in the short run and helps to unite the two groups as a single firefighting team for the long term. As stated before, the brigade chief, safety and health professional, or designated employee should immediately call the municipal fire department whenever a fire is reported at the facility.

In fire emergencies, security staff should open the yard gates and be ready to direct fire apparatus to the blaze. If facilities have railroad tracks, rail crossings that may be needed in an emergency should be kept free of all cars, hand trucks, and the like.

The brigade chief is in command at a fire until the officer in charge of the public fire department takes over. The brigade chief then serves as an advisor on processes and special hazards.

The fire station itself should be centrally located but not exposed to possible fires. It should be built of fire-resistant material or located in a sprinklered part of the facility and protected with portable extinguishers.

Figure 15-12. Under direction of an instructor (white helmet), an emergency response drill team approaches a 300-gallon oil tank filled with burning fuel oil. Note the shower spray above the lead firefighters.

Larger facilities may require mobile units, such as light hand-drawn trucks outfitted for special hazards.

Facility Protection and Security

Security is management's responsibility. Government agencies, consultants, and insurance representatives can provide assistance and advice in establishing a policy. Because the basic problems of security are protection of property and control of persons, an organization does not have to establish a new department to handle this function. It could simply build its emergency security force around present security and administrative personnel.

Personnel need training to maintain order, handle crowds, cope with the threat of panic, and prevent on-site looting. They map emergency routes to shelters, both inside and outside the grounds.

Fires have often been caused by security guards either smoking while on duty or overlooking fire hazards. Heavy losses can result from their failure to discover fires promptly, from shutting off sprinklers before determining if fires have been extinguished, or from failing to know the proper sprinkler valves to close after a fire is extinguished.

As a supplement to automatic alarm and signal systems, a firm can employ an intelligent, well-trained, and physically fit security guard to prevent or minimize fire loss. Because fires that start in an idle facility produce more damage, the security guard becomes an important part of a facility's fire prevention and detection organization.

Security functions include protecting the premises against pilferage, burglary, vandalism, and espionage. The time and route of inspection rounds should be scheduled irregularly, to avoid creating any detectable pattern.

The first inspection round immediately after the facility closes is the most important. Evidence shows that most fires are likely to start just after employees have left; they are caused either from machines or processes left running unattended or from carelessly discarded smoking materials.

The security guard should have enough time on the rounds to make a thorough inspection of the premises. The route should require no more than 40 minutes and should include all hazardous areas. The guard should be provided with an approved flashlight or other illuminating device; where practical, lights along the security route should be left on.

The guard can look for violations of smoking rules, improper storage of chemicals or flammable materials; and leaking oil, gasoline, gas, or other flammable materials, and should immediately report unsatisfactory conditions to the management. Security guards should be physically capable of turning on a fire alarm, dealing immediately with small fires, and with such matters as shutting off gas and closing fire doors. If the guard ever fails to report on schedule at the end of a regular patrol, the incident should be investigated immediately.

Additional protection and security measures call for closing off certain windows and other openings in facilities that are not vital to operations and limiting the number of entrances and exits. (All measures must be consistent with good fire prevention practices.) Companies might also install protective wire mesh over windows along public thoroughfares and use floodlights in critical areas of facilities at night.

Emergency Medical Services (EMS)

First aid and medical services should be headed by the company doctor, if available, as discussed in Chapter 10, Occupational Health Programs, in this volume. In the organization of the medical phase of the emergency plan, those responsible must select and train personnel; decide what measures, equipment, and supplies are needed; and establish first aid stations and a treatment center.

All employees should be encouraged to enroll in a cardiopulmonary resuscitation (CPR) first aid course. People assigned to first aid and medical units should pass standard and advanced courses in these areas. The National Safety Council First Aid Institute offers training in CPR and first aid. Local chapters of the American National Red Cross also provide training programs. Because more than 140,000 Americans die every year from injuries, and one in three suffers a nonfatal injury, the Council believes that every person at some time will encounter an emergency requiring first aid. The Council trains and certifies instructors who in turn teach classes nationwide. The courses meet OSHA requirements for workplace first-aid and CPR training.

In addition, any time an emergency sprinkler system or some type of shower facility is installed in a chemical work area, management should have a light or alarm installed in case of an emergency. In many locations, especially cold climates, people have been known to suffer severe cases of hypothermia when they had a chemical accident and drenched themselves to wash the chemicals off. They had no way to summon help and suffered from exposure to the cold. In addition, supervisors should keep copies of Material Safety Data Sheets (MSDS) on hand to send with an injured worker when the person is taken to the hospital or medical facility. This information will help emergency and in-house physicians treat the employee. (See also ANSI Z358.1, *Emergency Eye Wash and Shower Equipment*.)

If a major disaster occurs, there may not be enough trained doctors and nurses available to treat victims. In such an eventuality, care beyond the first-aid level will have to be provided by emergency medical technicians (EMTs), paramedics, and first responders who have received additional training. Such a medical team can be developed by recruiting volunteers, preferably with some medical or paramedic experience. In addition to first-aid training, more advanced instruction by regular medical personnel or local hospitals should be provided. Companies should keep in mind when establishing an EMS program that while most of the medical aid will be given in a central medical station, some may be administered on the job site, possibly under hazardous conditions.

Plans should be made for representatives of the medical team to check all personnel at the disaster scene for trauma. The team can provide a written clearance for victims to leave the facility when they are able to do so. Representatives of the investigation team may interview personnel before they leave to be sure that all eyewitness information is recorded.

Welfare and medical services include devising steps to prevent epidemics and to inspect food and sanitation facilities. In the planning stages for these activities, management should designate certain company trucks as ambulances and make sure they are stocked with the necessary equipment. Two-way radio communication for such ambulance service is essential. Management should also make provisions to have nonperishable food and water rations on hand. There should be close coordination of facility EMS measures with the local civil defense, health, and medical services.

The chemical service responsibility of the EMS unit requires that they provide gas masks and other protective equipment to workers if toxic or corrosive chemical gases are released. The units must also have trained personnel, equipment, and supplies for chemical defense and decontamination. The company should plan a priority sequence for decontamination—that is, water supply first, power facility second, machinery areas third, warehouse areas fourth, and so on. Medical team personnel will also be responsible for any radiological monitoring required after a nuclear incident. The mere knowledge that such monitoring equipment is available can be a morale builder for workers. After a disaster, no one should be permitted to drink water until it has been examined for contaminants.

Warden Service and Evacuation

The warden service is responsible for maintaining employee control during emergencies, including (1) guiding employees to safe areas; (2) directing employees, including physically disabled workers, away from hazardous areas; and (3) averting panic. In smaller organizations, the wardens may take charge of the shelters. In some cases, the warden service may be responsible for seeing that process and equipment shutdown is carried out smoothly.

This type of service was devised primarily for areas of high population densities, such as commercial structures, factories, and residential areas. For facilities with several hundred or several thousand employees, the use of warden teams can be vital during emergencies. In facilities with only a few personnel, the machine operators will have to be trained in shutdown details.

Management, in checking and providing for safe exits, emergency lighting, and evacuation drills, should refer to the NFPA's standards and to appropriate regulatory standards. Smooth, safe functioning of an evacuation plan requires that those in charge have a thorough knowledge of all operations and employees; the number, type, and location of available exits; alternate exits; and location of hazards, as well as a knowledge of warning and evacuation facilities. The subject of building exits is covered in Chapter 19, Office Safety; in Chapter 2, Buildings and Facility Layout, in the *Engineering & Technology* volume; and in the Life Safety Code, NFPA 101. (See OSHA regulations 29 *CFR* 1910.157 (a)(2) and 29 *CFR* 1910.38.)

Most facilities have a rigid rule that only especially appointed people on the fire brigade shall go to the vicinity of the fire. Everyone else should proceed on signal to a safe location or assembly area designated by the evacuation plan.

Transportation

When transportation facilities are disrupted or traffic is temporarily restricted, many employees may find it impossible to get to work. The organization may need to provide transportation with trucks and cars. Advance planning for car pools and pick-up stops will greatly facilitate such a procedure.

The transportation responsibility includes arranging for ambulance service, transporting employees to and from work, and moving emergency service crews and supplies as needed. Any planning for adequate transportation service and traffic control requires cooperation with the public police department, emergency management planning authorities, and possibly the military.

An emergency transportation unit might consist of a group of regularly assigned drivers. Station wagons with the seats removed and company trucks can be used when it becomes necessary to handle stretcher cases in evacuating any injured. The unit will transport auxiliary firefighters, first aid teams, and salvage and rescue workers to the scene of the disaster at the earliest possible instant. The unit can also deliver needed equipment and material from outside suppliers.

When developing emergency transportation plans, the organization must anticipate sources of motor fuel.

PROTECTION FROM PERSONAL ATTACK

Protecting the employee from muggings, rapes, and robberies is a concern of management, especially when such attacks occur on company property. Even if these attacks happen elsewhere, they can affect workers' performance on the job.

The Crime Problem

Safety and health personnel are not professional crime fighters. However, the welfare of the workforce on the job is their responsibility, which may include the security of grounds, buildings, or facility. For example, a woman may be attacked while waiting in her car for her husband to get off work, or a man may be mugged in the parking lot on payday, or some part of the office or facility may be burglarized. In each case the safety of employees or family members is jeopardized.

How should management cope with security and crime problems within the confines of the facility or building? How does a safety and health professional include this area of concern with other duties? Finally, how are any associated costs budgeted?

There are no pat answers to these questions. Each must be answered in terms of a particular industry, location, and workforce. All should be answered, however, with an attitude of common sense and practicality that addresses itself to the welfare of the employee and the organization.

Building and Premises Survey

Management should survey the grounds and buildings of each facility. What does the building or facility look like at night? Is it well lit with floodlights? Does an armed guard control electronic gates to the facility or yard? Or is it set back from the street in an open, landscaped industrial park? Whatever the physical setup, management should study it for exposed areas such as unguarded access to the parking area from a busy street or highway, or viaducts or catwalks that allow easy entrance to a restricted area.

Although not all mugging/burglary crimes take place in dark, secluded areas, many of them happen in those locations simply because discovery is less likely. Thus a security survey of an industrial complex should focus on finding any secluded or remote areas that may place employees at risk. These areas include tunnels between buildings, street underpasses, poorly lighted stairwells, dock areas, and freight elevators. They all offer hiding places for unauthorized personnel who may attack employees. Don't overlook washrooms and locker-room facilities. Burglars have been known to hide in toilet stalls until after hours and then shop through the facility and offices for valuables.

Implementation of the Survey

Once management has drawn up a list of security problems, planners should establish priorities and decide how to implement them. For example, installing lights and a fence around the parking lot may deter a car thief from the area or prevent an assailant from hiding in the back seat of an unlocked car. In addition, the fence and gate may slow up employees leaving the area and make shift-change traffic more organized and safe. Adequate lighting in a stairwell could prevent attacks and reduce the risk of employees slipping or tripping on the stairs.

Often the facility or building survey reveals security gaps that can be closed by the maintenance department. For example, they can repair or board up broken windows and remove trash barrels or other containers an intruder might use to reach and squeeze through a window or trap door. For small organizations or industrial operations that may be housed in one building, lights around the periphery of the structure are a relatively low-cost method of deterring intruders. Decals can be displayed prominently in a window announcing on-premises security alarms. (Many establishments display such a decal whether or not an alarm system is actually installed.)

Professional Assistance and Community Relations

The experience of local law enforcement agencies can be a major source of help in resolving security needs. In addition, they can offer advice regarding any local laws or codes that require the company's compliance.

Some police departments, particularly in suburban or rural areas, patrol industrial sites throughout the evening hours. Find out what the local police force does. At the same time management should investigate types of crime in the area and recommend special procedures to prevent such crimes from happening on site. For instance, an adjacent race car track may bring large crowds near the company's warehouse. Burglars may use a race as a diversion to break into the complex without being noticed.

Even the unpredictability of weather should be considered in a security program. Do heavy rains, snow, or fog render some security procedures inoperable? If so, the procedures may have to be changed. The police and fire department may tell the organization of special procedures they are required to take in the event of unusual weather conditions that might affect some of the company's security procedures. For example, the safety and health professional should find out if bus routes or traffic must be rerouted behind the facility or building if the viaduct floods or the surrounding area catches fire.

Based on discussions with the law enforcement agencies, management may decide to work with additional security professionals. These can include surveillance companies, insurance firms, burglar alarm and detector manufacturers, and lighting companies. Naturally, any security equipment management selects should be based on the firm's actual needs.

Just as multifacility industrial complexes share core medical facilities, it may be possible for an individual company or public agency to participate in cooperative security plans. Several organizations can share expenses for additional lighting, fences, and security patrols. However, this step should be taken only with the approval of the organization's insurance carriers.

Personnel Protection

Once the facility is secure under the supervision of security staff and/or detector systems, the safety of employees is also greatly improved. For instance, not all unauthorized persons questioned by a guard or detected by a surveillance system intend to steal property. Some may wish to harm an employee, perhaps even selecting someone at random.

A well-organized employee identification system is basic to an organization's security. All too often people assume that a person sauntering through an area, be it restricted or not, has a legitimate reason for being there. By challenging the stranger's presence and asking for a company identification card or verification from another employee, workers and managers can help to maintain security. At first, employees may object to "police" overtones of "identify or else." However, selling security to them on the basis of, "It's for a company's own protection," can win them over. Even the strongest resisters can be persuaded if management supports this approach with examples of problems resulting from little or no security measures.

Management can require workers to wear photo-identification cards clipped to a shirt pocket or collar as a condition of employment. In addition, the card can be color-coded to identify first aid personnel, fire brigade personnel, maintenance workers, and others selected for specific tasks during an emergency. This system works especially well when the organization cooperates with other firms or groups. Different colors help people easily identify members of each group. Finally, asking the employees themselves about improving security can often elicit ideas that fit in well with specific operations. The approach "we want you to be safe on the job" is a more positive way of selling security to employees than the "no-admittance-beyond-this-point" or "don't-do-this" approach.

Female Employees

Rape is a hideous crime. It can occur at any place, at any time. It is a crime in which the victim is left with serious emotional scars and/or physical injuries.

Security measures to prevent rape should include adequate lighting, visible presence of security staff, and a guard escort service, if necessary. Female employees should be informed that, with the exception of "date rape," rape victims are usually victims by chance, not necessarily because they know or recognize the assaulter.

Tell female employees about self-protection during a training session set up just for the subject of security. Contact local police or sheriff's departments to come in and do programs on rape prevention. Encourage comments from the women, particularly suggestions on ways they will feel more secure. Distribute the National Safety Council's *Just Another Statistic* (see References) or similar literature that may be available from the local law enforcement agency. Some programs distribute whistles for the women to take with them. A number of cities report a reduction in rapes after establishing whistle campaigns. Although stopping rape is one of the reasons for using the whistle, it can be used by men and women in other emergency situations. For example, the whistle is a good signal to use when a person sees something suspicious happening or is personally threatened and wants to summon help.

Sometimes to enhance the safety of a workforce, management can institute additional security measures for the in-transit employee. In-transit security is a sensible procedure to prevent muggings and robberies as well as rapes. Cabs and minibus rides from the facility gate to public transportation are standard procedures in many work situations. Guard escort from the facility gate to the employee's car is provided in some high-crime areas. For the most part, these efforts have been established primarily to protect employees who work evening shifts. Similar protection is afforded office workers by having them work in groups or by requiring the presence of a supervisor or other management staff during overtime hours.

Mutual Assistance

Organizations can cooperate with surrounding organizations and facilities, local law enforcement authorities, and transportation systems to develop in-transit security. The cost of operating a minibus service from several establishments to the local bus or train can be shared by the participating parties. Or, a similar agreement can be made with a local taxi company. Some companies located in high-crime areas have borne the expense of taxi rides to employees' homes in an effort to ensure the safety of their workforce.

OUTSIDE HELP

The chance of survival and recovery from a disaster or major accident is greater when organizations pool knowledge, equipment, and personnel with their neighbors. Therefore, emergency plans should include a provision for exchanging aid with other organizations in the community.

These plans should be drawn up ahead of time for both groups. Everyone needs to know beforehand what equipment and services will be available and how it can be called in. Plans should include provisions for calling another organization after hours should a problem develop outside normal working times.

Mutual Aid Plans

A number of industrial communities are organized to assist their members in the event of an emergency or disaster. These organizations include manufacturing facilities, large offices, stores, hotels, utility companies, chemical facilities, law enforcement organizations, hospitals, newspapers, and radio and television stations. They operate independently of or as supplements to any individual emergency management teams.

It is usually not possible to have adequate supplies available for a really large disaster such as a major earthquake. The best defense is to have adequate supplies in other areas committed for standby use, with communication channels and a plan for rapid transportation of supplies to the stricken area. Adequate emergency medical supplies and firefighting equipment are especially important. Companies should develop a plan for rapid and accurate communication, as discussed earlier in this chapter.

Mutual aid plans with neighboring companies and community agencies should include establishing an organizational structure and communication system; standardizing an identification system, procedures, and equipment (such as fire hose couplings); formulating a list of available equipment; stockpiling medical supplies; sharing facilities in an emergency; and cooperating in test exercises and training.

One item often overlooked in most mutual-aid contracts is who will pay any costs, such as repairs to equipment or workers' compensation for injured employees. Most often, workers' compensation is covered by the person who is considered the full-time employer.

Frequently these "cooperatives" establish a task force composed of personnel from each member company. Their training is supplemented by detailed written instructions. Each facility marks bulldozers, floodlights, and tools for emergency use. Training on a community basis might include instruction by members of the public fire department to facility fire brigade members. In addition, members of a construction or wrecking company may provide actual training to show the salvage and rescue teams how to handle heavy weights and to work safely among debris.

Contracting for Disaster Services

Some companies contract for disaster service. The service may include an annual retainer, plus additional pay for the actual hours worked. For example, a wrecking company can be engaged to supply the workers and equipment necessary to clear debris created by a disaster. Contracting for such a service removes the burden from companies of providing trained personnel and maintaining idle emergency equipment that could easily be damaged in the very disaster for which it was designed.

One must remember that sometimes in a major disaster, some of the contractors a company has on standby may go instead to the highest bidding firm. In such cases, management must have contingency plans.

Municipal Fire and Police Departments

Firefighters from the station most likely to respond to an alarm should be fully acquainted with all fire hazards in the facility. The organization may encourage such cooperation by inviting local fire officials to inspect the area prior to an emergency. The officials can become familiar with the location, construction, and arrangement of all buildings, as well as all special hazards, such as toxic, corrosive, or flammable gases, liquids, and materials. Because public fire department rescue equipment can supplement facility rescue units, they can make sure that equipment is compatible with the municipal equipment.

Therefore, the local fire department can formulate an efficient plan of attack before an emergency occurs. Such a procedure is far better than waiting for an incident to happen, then running the risk that local firefighting units will misunderstand the situation and initiate an improper response.

The public police force can help to quell large-scale disturbances and to assist in evacuating the facility should a major disaster occur. Planning for this outside help should include making arrangements for traffic control, particularly where a parking lot empties immediately onto a public highway.

Industry and Medical Agencies

Details of the following services are given in the Appendix 1, Sources of Help.

In 1970, the Chemical Manufacturers Association (CMA) created the Chemical Transportation Emergency Center (CHEMTREC). (See Appendix 1, Sources of Help.)

The Toxicology Information On-Line Network (TOXLINE) has been designed to provide current and prompt information on the toxicity of substances. It is intended to be used by health professionals and other scientists involved in antipollution, safety, drug, health, and other disciplines. (See Appendix 1.)

A number of other emergency and specialized information sources are listed in the Appendix 1, Sources of Help.

Governmental and Community Agencies

During a community-wide disaster, a large number of governmental and private agencies are available to assist industries. These include the Federal Emergency Management Agency (FEMA), the U.S. Army Corps of Engineers, the Salvation Army, the American Red Cross, the U.S. Public Health Service, the National Weather Service, and National Oceanic and Atmospheric Administration (NOAA). To be effective in coping with a disaster, the efforts of all of these groups must be coordinated and directed toward a common end. Therefore, each organization should have an updated listing of all cooperating agencies; the administrator's name, address, and telephone number; and the task assignment of the agency. If possible, safety and health professionals and emergency planning directors should meet periodically with these administrators to discuss mutual problems and disaster control techniques.

The emergency planning director should become thoroughly familiar with the authority, organization, and emergency procedures that are established by law and which will become effective upon declaration of a civil emergency.

It is in the organization's best interests to be active in the county Local Emergency Planning Committees (LEPCs). In this way, they can cooperate with other organizations from other industries and groups in the community. The main purpose of an LEPC is handling hazardous materials problems along with natural disasters and other crises.

In wartime, the federal, state, and local governments are responsible for relief measures. The American Red Cross has offered to assist the government in providing food, clothing, and temporary shelter on a mass-scale basis during the emergency period immediately following enemy attack. In many communities, local emergency management officials have requested American Red Cross chapters to assume all or part of this responsibility, acting under emergency management authorities.

In natural disasters, the American Red Cross is responsible for assisting families and individuals to meet disaster-caused needs that cannot be met through their own resources. The relief operations of local American Red Cross chapters are coordinated with the activities of the local, state, and federal governments. The resources of the national organization are available to supplement chapter assistance.

SUMMARY

- Advanced emergency management planning is the best way to minimize potential loss from natural or human-caused disasters and accidents. Emergency planning must provide for the safety of employees and the public, protect property and the environment, and establish methods to restore operations to normal as soon as possible.
- Basic emergency management planning usually includes establishing a chain of command, an alarm system, medical treatment plans, communication system, shutdown and evacuation procedures, and auxiliary power systems.

- Organizations can use a variety of government and private sources to find out what types of hazards are most likely to occur in their area. Once the initial background work has been done, the next step is to develop a working emergency management plan.
- The chain of command established for emergency plans should be kept as short as practical and should be staffed with employees selected for their ability to respond well in high-pressure situations.
- Emergency plans generally call for establishing a command headquarters or center to coordinate communications and disaster response. An incident command system can be established to manage a company's response to an emergency.
- Most operations have special alarm systems for fire and security. Many organizations establish local fire and hazardous materials brigades known as hazardous materials response teams (HAZWOPER), as defined by OSHA and various statutes. Other services that companies often organize for emergencies include emergency medical services, warden service, and emergency transportation units.
- One major responsibility is protecting employees from personal attack—muggings, rapes, and robberies. To help ensure worker safety, the organization can make a security survey of buildings and grounds, implement steps to correct a hazard or breaches of security, enlist the advice and aid of local law enforcement agencies, and train workers in the basics of personal protection.
- In some cases, an organization may need to request outside help to survive and recover from a disaster or major accident. Mutual aid plans with other firms and organizations in the community can pool the knowledge, resources, equipment, and personnel of many organizations. These cooperatives may establish a task force composed of members from each organization to organize and administer emergency responses.

REFERENCES

American Insurance Services Group, Engineering and Safety Service, 85 John Street, New York, NY 10038.
Fire Hazards and Safeguards for Metalworking Industries, Technical Survey No. 2.
Fire Safeguarding Warehouses, Technical Survey No. 1.
American National Standards Institute, 11 West 42nd Street, New York, NY 10036.
Emergency Eye Wash and Shower Equipment, ANSI Z358.1-1990.
The Conference Board, 845 Third Avenue, New York, NY 10022.
Studies in Business Policy, No. 55, "Protecting Personnel in Wartime."
Environmental Protection Agency, OS–120. "Chemicals in Your Community."

Factory Mutual Research Organization, 1151 Boston Providence Turnpike, Norwood, MA 02062.
Handbook of Industrial Loss Prevention.
Loss Prevention Data.
Federal Emergency Management Agency, 500 C Street SW, Washington, DC 20472. (Available through Superintendent of Documents, U.S. Government Printing Office, Washington, DC 20402.)
Attack Environment Manual, June 1987.
Emergency Planning and Community Right-to-Know Act, Title III, Superfund Amendments and Reauthorization Act, Public Law No. 99–499.
In Time of Emergency, H–14, Oct. 1985.
Mass Casualty Planning: A Model for In-Hospital Disaster Response, Aug. 1986.
Giraud R. Radioactive elements. *National Safety News*, June 1973. Adapted from author's article in *Revue Technique du Feu, Entreprise moderne d'edition*, 4 rue Cambon, 75 Paris 1, France.
International Association of Fire Chiefs. 4025 Fair Ridge Dr., Fairfax, VA 22033. *Nuclear Hazard Management for the Fire Service*, 1975; "Disaster Planning Guidelines for Fire Chiefs."
Kerr JW. Preplanning for a nuclear incident. *Fire Command!* April, 1977.
National Fire Protection Association, 1 Batterymarch Park, Quincy, MA 02269–9101.
Explosion Prevention Systems, NFPA 69.
Facilities Handling Radioactive Materials, NFPA 801.
Fire Protection Handbook, 17th ed.
Guard Service in Fire Loss Prevention, NFPA 601.
Health Care Facilities, NFPA 99.
Industrial Fire Hazards Handbook, SPP 57A.
Installation of Signalling Systems, NFPA 13.
Life Safety Code, NFPA 101, 1991 edition.
Professional Competence of Responders to Hazardous Materials Incidents, NFPA 472.
National Petroleum Council, 1625 K Street NW, Washington, DC 20006.
Disaster Planning for the Oil and Gas Industries.
Security Principles for the Petroleum and Gas Industries.
National Safety Council, 1121 Spring Lake Drive, Itasca, IL 60143
Fire Prevention and Control on Construction Sites, Occupational Safety and Health Data Sheet 12304–0491, 1988.
Fire Prevention in Stores, Occupational Safety and Health Data Sheet 12304–0549, 1990.
First Aid Institute.
Fundamentals of Industrial Hygiene, 4th ed.
Hazardous Materials, Booklet.
Just Another Statistic, Booklet.
Security from personal attack. *National Safety News*, July 1973.
(See other appropriate topics treated in both volumes of this Manual, especially those pertaining to organization, training, medical and nursing services, fire extinguishment and control.)

Noll GG, Hildebrand MS, Yvorra JG. *Hazardous Materials: Managing the Incident.* Fire Protection Publications, Oklahoma State University, Stillwater, OK 74078.

OSHA. 29 *CFR* 1910.38, "Employee Emergency Plans and Fire Preventing Plans."

——. 29 *CFR* 1910.120, "Final Rule for Hazardous Waste Operations and Emergency Response."

——. 29 *CFR* 1910.156, "Fire Brigades."

——. 29 *CFR* 1910. Section 3088, "How to Prepare for Workplace Emergencies."

——. 29 *CFR* 1910.1200, "Hazard Communication Guidelines for Compliance."

——. 29 *CFR* 1910.1200 (f), "Labels and Other Forms of Warning."

——. 40 *CFR* 311, 330, 335, "Extremely Hazardous Substance List and Threshold Planning Quantities; Emergency Planning and Release Notification Requirements; Final Rule."

——. 49 *CFR* Part 172, "Hazardous Materials Table and Hazardous Materials Communications Regulations."

Pacific Telesis Group. *Earthquake Survival Guide.* San Francisco: 1987.

Underwriters Laboratories Inc., 333 Pfingsten Road, Northbrook, IL 60062.

Classification of Fire-Resistance Record-Protection Equipment.

Gas Shutoff Valves—Earthquake.

U.S. Department of Health & Human Services. *Occupational Safety and Health Guidance Manual for Hazardous Waste Site Activities.*

REVIEW QUESTIONS

1. What should be the first *concern* in planning for an emergency?
   a. Protecting the property
   b. Ensuring the safety of employees and the public
   c. Restoring business operations to normal
   d. Protecting the environment
2. The first step in the emergency planning process begins with determining what types of hazards/emergencies may affect the organization. Name eight of the eleven types of emergencies.
   a.
   b.
   c.
   d.
   e.
   f.
   g.
   h.
3. Many sources of hazard/emergency information can help management determine the likelihood of specific emergency events in a locality. List five of the seven sources.
   a.
   b.
   c.
   d.
   e.
4. The second step in the planning process should be:
   a. Developing a checklist of emergency equipment and shutdown actions
   b. Designating HAZMAT team members
   c. Testing all alarm systems
   d. Preparing a specific emergency plan of action
5. What are six of the twelve considerations that should be addressed within the specific plan of action?
   a.
   b.
   c.
   d.
   e.
   f.
6. Why is disaster training one of the most important functions of the director and staff, on both the corporate and facility levels?
7. Which OSHA regulation addresses aspects of health and safety that are now legally required at hazardous materials sites; treatment, storage, and disposal facilities; and other hazardous materials emergency locations?
8. What is the function of the HAZMAT team?
9. What can an organization do to protect employees from personal attack?
   a. Make a security survey of the grounds and buildings of each facility
   b. Implement steps to correct a hazard or breaches of security
   c. Enlist the advice and aid of local law enforcement agencies
   d. Train workers on the basics of personal protection
   e. All of the above

16
Product
Safety
Management

Although no one can completely prevent product misuse by customers, the producer or manufacturer can do a great deal to minimize or defend against product liability claims. The principal way to achieve this goal is to sell only reasonably safe products and, when necessary, to include instructions for their proper use and to warn about known misuse and foreseeable misuse. The key to achieving a reasonably safe and reliable product and, at the same time, reducing a company's product liability exposure is to incorporate product safety into all aspects of the design, manufacturing, and marketing of the product. This is done by establishing and auditing an effective product safety management (PSM) program. This chapter describes how a company can use effective product safety management procedures to protect its customers and itself. The following topics are discussed:

- establishing and coordinating a PSM program
- conducting a PSM program audit
- conducting PSM audits of each department in a company
- auditing the quality assurance and testing department
- conducting audits of functional activities within a company

A PSM program must include a comprehensive process of evaluating user injury risks from all product-related sources. Therefore, many people from various parts of an organization need to be involved in this risk management process. From the start, management must perform ongoing risk evaluations and determine what corrective actions are necessary if existing controls prove inadequate.

To establish and audit a PSM program, management must begin by (1) becoming committed to the fact that it will be necessary to spend money and resources to save money in the long term; (2) selecting a program coordinator; and (3) selecting a program auditor. The same person typically serves as both program coordinator and auditor, depending on the size, needs, and structure of the individual company.

ESTABLISHING AND COORDINATING THE PROGRAM

Regardless of a company's size, establishing and coordinating a satisfactory PSM program requires a comprehensive systems analysis of all operations and production stages, from design through manufacturing, quality assurance, and shipping. To keep all these functions organized and coordinated, someone must oversee the program, either alone or with a committee's assistance.

Program Coordinator

Because the success of the PSM program requires the cooperation and coordination of all departments, the program head or committee chairman should be carefully chosen. The individual must be able to exert stringent control over all phases of product development, from initial product design through product sale and distribution. The extent of the individual's role will depend, of course, on the size of the company. In a small to medium-sized company, management will select either a PSM manager or an auditor on the basis of the person's broad experience in areas of safety technology.

It is not important whether the position is full time or part time, and, if part time, what the coordinator's primary function is within the organization. What is important is the PSM coordinator's level of authority. Can the coordinator take needed action without having to go through several supervisory levels? Does the coordinator have easy access to top management? Is the coordinator permitted to implement most program plans or suggestions?

If the organization has a PSM committee to coordinate program activities, the head of the committee must have a similar level of authority. Likewise, committee members should have the authority to speak for their respective departments.

Responsibilities

A program coordinator must have sufficient authority to take action and to carry out the following responsibilities:

- function as a staff member for corporate management
- assist in setting general PSM program policy
- recommend special action regarding

  - product recall
  - field modification
  - product redesign
  - special analyses

- participate with other appropriate personnel in the review of product literature and warnings
- conduct and review complaint, incident, or accident analyses
- coordinate appropriate PSM program documentation
- ensure an adequate flow of verbal and written communications
- develop sources of product safety and liability prevention data for use by operating personnel
- maintain a liaison with business, professional, and governmental organizations on matters pertinent to product safety and liability prevention
- conduct PSM program audits, where appropriate

Ground Rules

The following ground rules must also be clearly defined by management if the PSM program coordinator is to be effective:

- The purpose of the PSM program coordinator must be clearly defined.

- The authority and responsibility of the PSM program coordinator must be clearly specified by top management and understood by the PSM coordinator.

As previously noted, a PSM program involves most of the departments in a company and requires the coordination of many disciplines. As a result, the PSM program coordinator must ensure a thorough and systematic approach when implementing the program.

Committee Coordination

Whenever a PSM program committee is used, a company should keep the group's size within manageable limits—generally, no more than five or six members. In a large corporation, management may find it better to appoint a small corporate committee or individual corporate coordinator and assign a separate program coordinator for each corporate division or department.

The committee is especially important when the company is launching a new product safety effort within the organization. The role of the committee is to recommend policies and guidelines, offer advice to top management, and audit product safety performance.

Committee membership varies according to the organizational structure of a particular company. However, key members of the committee will almost always be selected from the following areas:

- product safety
- design or engineering
- manufacturing
- purchasing
- quality assurance
- service or installation
- risk management/insurance
- legal

In addition, sales, marketing, advertising, personnel, public relations, facility safety, and representatives should be designated. These individuals can serve as resources to the PSM program committee when their expertise is required.

CONDUCTING A PSM PROGRAM AUDIT

The purpose of a PSM program is to develop a means to perform (1) an evaluation of the product during design and manufacture, distribution, sale, and consumer use and (2) to control any accident and hazard potential through good product safety management techniques. The major goal of the program audit is to reduce or eliminate the causes of product liability exposure. Some of these causes are

- unsafe product designs (failure to review product design safety or failure to use state-of-the-art technology or materials)

- inadequate manufacturing and quality assurance procedures
- inadequate preparation and review of consumer warnings and instructions
- misleading representation of products or services

Preliminary Procedures

The auditor should undertake the following preliminary activities before beginning the formal audit. They will make the job easier to perform and enhance its effectiveness.

- Explain the purpose of the audit to management.
- Arrange mutually agreeable audit dates and review with involved members of management.
- Define how the audit's operational plan (schedule) will involve members of management.
- Review appropriate PSM program-related documentation and procedures.
- Review appropriate product-related procedures.
- Review appropriate product-related technical and standards information.
- Acquire all necessary product-related printed instructional and precautionary materials.

By carrying out these preliminary procedures, the program auditor usually can obtain the complete cooperation of involved management in each major department and of safety personnel.

Management's Commitment

Wholehearted commitment and support of the PSM program by all levels of management gets the program moving and keeps it rolling. As is the case for an occupational safety and health program, management must have a written policy outlining its responsibilities to provide support, to set basic objectives, and to establish priorities for a product safety and liability prevention program within the organization. Some evidence of management's commitment to the program might include the following:

- posted letter or bulletin
- formal statement of management policy on the subject
- use of a PSM program coordinator and committee
- use of a PSM program auditor

Top management should tell all key people within the organization that they have an important role to play in the program and that they must commit enough time and effort to make the program successful. In short, management should clearly communicate to all employees by word and deed that controlling product losses and liability is a major company objective.

Ideally, the chief executive officer issues a written policy stating the company's commitment to product safety and liability prevention. It should be distributed to management, all company departments, and every employee. Management should always provide some tangible evidence that it is truly committed to product safety and liability prevention and that employees have been advised of this fact.

In addition, the auditor must work with and cultivate the respect and cooperation of supervisory personnel. Supervisors and middle-management personnel are essential to the PSM program. They have direct contact with the employees who make the products, and, it is hoped, follow the PSM program guidelines for product safety.

Role of the Safety and Health Professional

The role played by the safety and health professional in a PSM program will vary according to the size of the company. However, in all cases, the safety and health professional's role is vital in implementing a successful program. In a small to medium-sized company, the safety and health professional, because of broad experience in areas of safety and health technology, may be selected by management to be a PSM program coordinator or auditor. However, as the company product line increases, management may designate the head of the engineering or design department for this assignment, appointing the safety and health professional as an assistant. Nevertheless, some companies do hire a full-time product safety director.

Regardless of the specific role the safety and health professional plays within the company's PSM program, the auditor should evaluate the person's performance to determine if he or she is, in fact, adequately contributing to the company's overall PSM program. Following are some of the contributions that safety and health professionals can make based on their knowledge of facility operations, experience in safety training, and understanding of accident investigation techniques:

- Evaluate and offer comments on the company's PSM program.
- Evaluate and comment on the product-safety-related training sessions developed under the PSM program.
- Assist those who will be performing product accident investigations.

In addition, members of the safety and health department can provide product safety surveillance in production areas to help prevent accidents. These individuals should be advised that all product-safety-related complaints or problems they uncover must be discussed with the facility engineer and the production department and formally documented. Copies of the reports must be sent to the PSM program coordinator. Although it is important to document complaints and problems in the interest of safer products, documentation can be a two-edged sword. Writings that point out problems with a product or memos critical of quality or product safety, particularly when taken out of context, can easily become a document that can be used effectively against a manufacturer at a product liability trial. In such situations, managers should be aware of the issue, and the company should have an active document retention program.

Because of its past experience in developing and implementing employee safety programs, the safety and health department may be aware not only of potential product hazards but also of ways in which customers can misuse those products. Therefore, a knowledgeable safety and health department representative should serve as a consultant to the team performing design reviews, hazard analyses, and product safety audits.

DEPARTMENTAL AUDITS

Although virtually all departments are involved in the PSM program, certain departments—primarily engineering, manufacturing, service, legal, purchasing, human resources, and risk management/insurance—play more significant roles. For the purposes of this chapter, only these departments will be discussed.

Engineering or Design Department Functions

The primary function of the engineering or design department should be to design and manufacture salable, reliable products that can be used with reasonable safety. It is more practical and cost-effective for the manufacturer to build reliability and safety into the product than to suffer the consequences of catastrophic product liability losses.

Products should be reasonably safe during (1) normal use; (2) normal service, maintenance, and adjustment; (3) foreseeable uses for which the product is not intended; and (4) reasonably foreseeable misuses. Courts have stated that a manufacturing company is responsible for ensuring that its products are safe for any reasonably foreseeable use or misuse to which the customer might put them.

Because the engineering or design department is such a critical area within PSM program activity, the program auditor should check the following:

- Does the organization evaluate product hazards prior to production?
- Is anyone within the organization formally assigned this responsibility?
- Is there a formal written or prepared design review procedure, even if it is not referred to as such?

At certain predetermined points in the design and manufacture of all complex products, tests or inspections of the products are made to determine compliance with manufacturing requirements. This is because problems detected at predetermined points in the process can be corrected more quickly and economically than can those detected after manufacture.

Safety Audit

A company cannot make a case that it adequately protected the user of a product from hazards that it did not know existed. For that reason, safety audits are of major importance in the product design and development stages. For example, in one simple audit safety process, the hazards associated with the product are identified and classified as to seriousness of the hazard—i.e., catastrophic, critical, occasional, remote, or improbable. Following these steps, the company can develop a hazard risk index that allows the design team to assign priorities for hazard remediation. It is during these audits that employees can present design alternatives, have them reviewed for feasibility, and identify hazards that cannot be eliminated so that product warning labels can be developed.

Product safety audits can be done in-house or by an outside consultant or both. Using someone outside the organization helps validate the process and reduces the number of hazards that potential litigants can claim were reasonably foreseeable. This validation can be of great value in the event of product liability litigation.

Formal Design Review

At predetermined points in the design process, the design should be checked for compliance with requirements in order to achieve the best product design possible. The appropriate check for the product engineering design process is called the formal design review (FDR).

The FDR is a scheduled systematic review and evaluation of the product design. The personnel carrying out the review are not directly associated with the product's development. However, as a group, they are knowledgeable about and responsible for all elements of the product throughout its life cycle, including design, manufacture, packaging and labeling, transportation, installation, use and maintenance, and final disposal.

Each designer and reviewer should ask the question "Would I feel safe using the product or having family members use it?" Unless the answer is an unqualified yes, each reservation or problem should be recorded and addressed by the designer and manufacturer.

Codes and Standards

The PSM program auditor must determine whether products conform to all applicable safety standards (including state, provincial, or federal codes and regulations; testing or inspection laboratory requirements; industry standards; technical society standards; and machine safeguarding standards). These regulations and standards, in most cases, should be considered as minimum requirements for product safety and reliability. In some cases, where no formal standards apply, reviewers should determine whether in-house design standards are being used. If so, what criteria are applied to judge the adequacy of the standard? Before each survey, the auditor should become familiar with all standards applicable to the company's product design process.

Human Factors

The PSM program auditor must determine if human factors have been considered in product designs. (See Chapter 13, Ergonomics Programs, in this volume.) Some of these human factors include the physical, educational, and mental limitations of the people who will use the products. The program auditor also should determine whether the company has weighed the possibility that customers might use the product in ways other than it was designed to be used but that might be considered reasonable.

Critical Parts Evaluation

Critical parts or components are defined as those "whose failure could cause serious bodily injury, property damage, business interruption, or serious degradation of product performance." Individual organizations, however, may have different criteria for defining a critical part, such as "one that is unusually expensive, difficult to acquire, or requires lengthy order lead time." When analyzing critical parts, the program auditor must be certain that everyone is using the appropriate definition.

The auditor determines whether products are being analyzed for critical parts or components. If so, are the critical items receiving any special attention? Have they been field-tested and designed to outlast the product itself? If this is impractical, is a special effort made to warn customers and users of possible hazards of critical-part failure? Has any effort been made to instruct customers and users in inspection techniques to detect impending failures? Have maintenance procedures been outlined for critical parts or components on the product itself and in operating and maintenance instructions? Has careful consideration been given to the expected life of the product?

All parts or components must have a life expectancy compatible with the life expectancy of other parts and of the total product. A part or component whose life expectancy is shorter or incompatible with that of the rest of the product should be considered "critical." This is because it may become the source of a product liability lawsuit, an expensive product recall, or a field-modification program.

Packaging, Handling, and Shipping

The PSM program auditor determines if product packaging will help prevent deterioration, corrosion, or damage of the product. Packaging requirements must cover conditions affecting the product at the manufacturing site, during transit, and under normal storage at the customer's location. Packaging methods should be established for each individual product. Product packages should be marked to indicate special requirements—for example, DO NOT STORE IN HOT AREAS, THIS END UP, USE NO HOOKS. Special containers and transportation vehicles should be used where necessary to prevent damage.

The program auditor also must carefully evaluate shipping procedures. A product shipped without proper instructions or with hidden damage may cause an accident when used. A company should select its packaging materials, cartons, and carriers not simply for economic reasons, but also to deliver a product in perfect condition to the distributor or customer.

The product, as shipped, must agree in every respect with the purchase order accompanying it. All appropriate labels, manuals, warnings, and descriptive materials must be included and should be checked at the time of shipping.

Warning Labels

It is not enough to design and make a satisfactory product; an organization must also label its products correctly and warn potential consumers and users of any dangers involved. The legal duty of a manufacturer to label products and warn of potential danger is increasingly cited as the basis for product liability lawsuits.

The program auditor must carefully examine the product instructions and labeling to be sure they conform to pertinent regulations and recent court decisions affecting a company's field of operations. The basic rule, as stated in *Restatement of the Law, Torts,* Second Series (see References), is that

> a manufacturer or supplier must exercise reasonable care to inform its consumers of a product's dangerous condition or the facts which make it likely to be dangerous if he knows or has reason to know that the product is likely to be dangerous for the use for which it is supplied and has no reason to believe that those for whose use the product is supplied will realize its dangerous condition.

Generally, a manufacturer has no duty to warn its consumers of danger if the danger is well known. However, duty to warn and the adequacy of warning is a question for a jury. Their interpretation of "duty" and "adequacy" will vary, depending on the circumstances of product use and any injury or damage customers or users sustain.

In general, the courts have tended to equate insufficient warning with no warning at all. Therefore, a company must not only give clear instructions on how to use the product but also must provide specific warnings about any dangers or possible misuses that could result in injury.

The courts have consistently held that a manufacturer has a duty to warn customers about any reasonably foreseeable use of a product beyond the purposes for which it was designed. For a warning to be considered adequate, it must advise the user of the following:

- hazards involved in the product's use
- how to avoid these hazards
- possible consequences of failing to heed these warnings

All warnings should be designed to comply with the requirements of ANSI Z535.4 1991, the ANSI standard for product safety signs and labels.

Instruction manuals accompanying the product should repeat hazard warnings and show how the hazards can be reduced or avoided. In addition, the manuals and/or videos should instruct consumers on how to inspect the product upon receipt and how to assemble, install, and inspect it periodically. If the manuals and videos include troubleshooting hints, the dangers involved should be thoroughly explained—for example, "NO USE OF UNAUTHORIZED PARTS," "DO NOT REMOVE BACK PANEL," or "HIGH-VOLTAGE HAZARDS." The manual/video needs to describe the preventive maintenance program required to maintain the product in safe working order.

Videos can be excellent instructional tools either by themselves or packaged with manuals, especially for more complex products. There is about a 25% illiteracy rate in the United States. As a result, if a company relies solely on printed materials, it will miss about a significant portion of its intended audience. Customers who do not understand how to assemble or operate a complex product will be more prone to make mistakes and may be a source of product liability claims. Companies who choose to produce an instructional video should keep the following points in mind:

- Show *only* the safe, proper way to assemble or use the product.
- Review all warnings and cautions, several times where appropriate.
- Be sure to present the product's characteristics and capabilities without exaggeration.
- List the product's limitations and the hazards of misuse.
- Have the video and script reviewed by appropriate legal and technical staff, including the safety professional.

Finally, the program auditor must evaluate sales brochures, product advertising, and all aspects of marketing and selling. These elements may need to be reviewed by the engineering or design department to ensure that the product's capabilities are accurately depicted and that only safe operating and maintenance procedures are shown. In addition, the auditor should determine whether all warning labels, hazards, and instructions developed by the engineering or design department have been reviewed by the company's legal counsel. This step will ensure that product users are receiving adequate instructions for use, warnings about potential product hazards, and instructions for proper maintenance.

Manufacturing Department

After a reasonably safe and reliable product has been designed, the manufacturing department must turn the design specifications into a finished product. If this is not done under proper controls and supervision, manufacturing errors could result in an unsafe and possibly unreliable product reaching customers.

The manufacturing department can contribute to a company's overall PSM program in many ways. The most important are listed below. By evaluating each of them (as it applies to a particular situation), the program auditor can effectively measure the manufacturing department's contribution.

- Motivate manufacturing employees by letting each person know that he or she is making a vital contribution to a quality product.
- Instill pride in employees in their work and in the company's product by
  - using up-to-date manufacturing equipment
  - keeping the facility's manufacturing capacity within bounds
  - providing adequate work space for each task
  - keeping the workplace clean and well lit
  - implementing good equipment maintenance practices
- Provide standardized on-the-job training procedures.
- Implement zero defects or error-free performance or other error-elimination programs.
- Design a program or procedure to identify and eliminate all production trouble spots. This should be accomplished in cooperation with the inspection and testing personnel of the quality assurance department.
- Prevent unauthorized deviations from design specifications and work procedures.
- Manufacture products in accordance with documents procedures.
- Participate in safety audits on new product designs. This is often accomplished by serving on the company's PSM committee, if one exists.

Sometimes product specifications are not realistic for the machines and equipment available in the shop. When this is the case, the production department must advise the engineering department immediately of the problem. Production workers should not try to maintain cost or production schedules by deviating from the specifications without first consulting the engineering department. If deviations are necessary or unavoidable, they should be made only with the approval of the engineering or design department.

Record Keeping
Manufacturing records should be kept for the life of all products and particularly for their critical parts or components. These records are vital to have on file in the event of product recalls or field-modification programs and for successful defense in product liability suits. Records on critical components should be complete enough to identify the batch, lot, and supplier of the raw materials and the finished products in which they were used.

Discontinued and New Products
Two other important areas of manufacturing department risk exposure are the existence of discontinued products

(products on the market but no longer being manufactured) and the development of new products or product lines. The PSM program auditor must determine if discontinued products or the manufacture of new products present additional hazards to workers, management, or consumers.

Service Department
Most companies provide service contracts for their customers, either through the dealer or distributor or through service subcontractors. These contracts can significantly increase a company's product liability exposure because through them the company has extended its exposure beyond the controlled environment of its manufacturing facility.

The PSM program auditor must carefully evaluate the company's service department controls. The auditor will determine the effectiveness not only of these controls and their contribution to the company's PSM program activities but also of the company's use of service department feedback.

In many companies (depending upon the type of product), service department personnel are required to maintain close contact with customers. Consequently, more than any other employees, they are familiar with customer needs and characteristics. They are most likely to hear customer complaints, reactions, and compliments regarding the company's products. They also see product misuses and usually know something about incidents and accidents that occur. In fact, service department personnel are often the first to hear of product accidents. Service department personnel should be trained to report any previously unreported new health complaints in accordance with the requirements of the Toxic Substances Control Act (TSCA, Section 8C).

Legal Department
The legal department in any organization should play a significant role in both the prevention and defense of product liability cases. Too often, companies use their legal department staff only when product safety problems or product liability litigation is imminent. Firms often fail to recognize the importance of involving legal personnel in the day-to-day aspects of product safety and liability prevention.

Legal personnel should act as advisors to the PSM coordinator. The legal department should be assigned the following responsibilities to maximize its contribution to the company's PSM program activities.

Review for Potential Liability
Legal personnel should review all product-related literature for potential liability.

Coordination
Legal personnel should coordinate all product investigations and product claim defenses, including working with the insurance carrier, retaining local counsel, and retaining statements of negotiations.

Marketing Department

Unfortunately, after a "reasonably" safe product has been designed and manufactured, product claims can still be incurred because of the way in which the company presented the product to customers and users. Customers often must rely on the company's sales personnel; advertising and sales brochures; and operating, service, and maintenance instructions for their knowledge of the product's capabilities and hazards.

If the customer is led to believe that the product has capabilities it does not have or if the instructions do not adequately warn of the product's hazards, an injured customer may have legal cause for action where none existed before. Therefore, the PSM program auditor must review the company's advertising and sales materials and evaluate whether

- They are clear and accurate.
- They overstate the product's capabilities.
- They encourage the customer to believe the product has uses for which it was not designed or intended.
- The product can safely do what the company's advertising and sales materials claim.
- Only safe operating procedures are depicted.
- The company's product is illustrated only with safety devices in place.
- Advertising and sales materials have been reviewed and approved by the engineering or design and legal departments regarding their accuracy and potential liability risks.

Warranties and disclaimers developed by the company must also be reviewed by the program auditor to determine whether they are

- reasonable and practical for the uses intended
- included with each of the company's products
- clear and concise
- prominently displayed and easily recognizable by the customer or product user
- thoroughly reviewed by the company's legal department or legal counsel

The program auditor must verify that the marketing department has retained (for the life of the product) all sales and distribution records that can identify purchasers. These records are essential if product recall or field-modification programs are to succeed. In addition, these records should indicate, whenever possible, how the company's products will be used, particularly when the company is selling to subcontractors or assemblers.

The company also must ensure that its sales personnel and dealers know how to describe the capabilities of the products they are selling or distributing without incurring undesired implied or expressed warranties. For example, sales personnel must never exaggerate the capabilities of the company's products or claim uses for which they were never designed.

Purchasing Department

Some companies are not large enough to have a formal purchasing department. However, every company has someone responsible for acquiring the raw materials and components needed to manufacture products.

The purchasing activity carries significant product liability potential for the company. Consequently, the PSM program auditor must carefully examine the company's activities to determine whether it is performing acceptably in this area.

The primary responsibilities of the purchasing department include the following:

- becoming familiar with all material specifications set by the engineering, design, and manufacturing departments
- always purchasing quality raw materials, parts, and components that meet the specifications set by the various departments
- evaluating, in conjunction with the company's quality assurance department, the capabilities and reliability of suppliers, using vendor-rating systems (compile a list of approved suppliers).

Human Resources Department

Employee job qualifications and work attitudes have a considerable influence on the quality of a company's products. Employees who lack proper job skills or who are unmotivated increase the company's chances of producing defective products.

The company's human resources department must work to (1) select, train, and place new and transferred employees and (2) continually upgrade the morale and performance goals of present employees and others who are involved in critical product safety matters within the company. It is also essential that the personnel department ensures that product safety duties and responsibilities are included in job descriptions and that their performance is considered in personnel reviews. This important policy is often overlooked by many companies.

Insurance Department

The purpose of risk management is to protect the company's assets from loss of any kind. Often this is understood to mean simply the purchase of insurance. However, effective risk management involves every aspect of a company's operations. Risk management professionals often possess the necessary expertise to serve as PSM coordinators or auditors.

Similarly, companies often believe that the only function of the insurance specialist or department is to buy insurance and to report claims to the insurance company. In reality, because of the insurance department's experience in handling product claims, it usually possesses valuable knowledge of the factors that often give rise to product liability claims. As a result, management should use the insurance department as an information clearinghouse in conjunction with the PSM program coordinator.

The insurance department usually reports liability claims to the insurance carrier and coordinates any accident investigations with the carrier. Occasionally, the department will coordinate the preparation of a legal defense (if one is necessary) between the legal department and the insurance carrier. Depending upon the company, the legal department may perform these functions instead.

Public Relations Department

A public relations department, like the safety and health department, generally plays a smaller role in a company's PSM program activities than many of the other departments. Nevertheless, the program auditor should not overlook it in an overall evaluation of the PSM program.

Product accidents, product recalls, or product field-modification programs may be reported by the news media and require the company to prepare press releases on the issue. If these statements are positively written, they can help to prevent unfavorable publicity. In fact, correct handling of publicity on product accidents, product recall, or product field-modification programs can be used by the public relations department to show the public the company's diligent efforts to design, manufacture, and sell a safe, reliable product. The public relations department should also distribute product information to the media when the company has incorporated safety advances into its products.

QUALITY ASSURANCE AND TESTING DEPARTMENT

The term *quality assurance* refers to the actions taken by management to ensure that manufactured, assembled, and fabricated products conform to design or engineering requirements. At one time, the terms *quality assurance* and *statistical quality assurance* were considered synonymous, because statistical techniques were considered to be the major tools of quality assurance. Over the years, the definition of quality assurance has broadened to include any actions that ensure product conformity to design and other requirements and the achievement of customer acceptance and satisfaction.

Before discussing the evaluation of a company's quality assurance activities, it must be pointed out that quality assurance policy, department organization, function, and responsibility vary according to a company's size, management policy and organization, type of product, facility location, number of facilities, economic resources, and other variables. When evaluating a company's quality assurance department, it is important for the program auditor to determine the following:

- Is the quality assurance program adequate to help attain the company's quality objectives? An incomplete quality assurance function does not necessarily mean that the system is inadequate. Necessary

functions can be added later. Others may not even be required because of the type of products, manufacturing processes, or size of a company.
- Does the quality assurance program function as planned? To answer this question, the program auditor must go beyond the quality assurance manual and review the actual implementation of the system.

In evaluating this function, the PSM program auditor must ask basic questions: How? Why? When? Where? What? Who? The auditor should then follow through with "Explain how it works," "Let's see examples," and "Show me the records." Asking these questions may reveal that the company has instituted a quality assurance program, but it does not prove that the system has actually been installed throughout the facility or that the system is functioning as planned.

The PSM program auditor must see tangible evidence that the quality assurance program has been implemented and is functioning. To get a total picture of the program, the auditor should review all paperwork, procedures, instructions, and records.

The program auditor should spend some time with the quality assurance shop personnel, if possible. These individuals can expound on the pros and cons of the program, thus providing information about the program's strengths and weaknesses that might not otherwise be available.

The basic quality assurance program functions that must be evaluated by the PSM program auditor are as follows:

- organization and manuals
- engineering and product design coordination
- evaluation and control of suppliers or vendors
- evaluation of manufacturing (in-process and final assembly) quality
- evaluation of special process control
- evaluation of the measuring equipment calibration system
- sample inspection evaluation
- evaluation of nonconforming material procedures
- evaluation of the material status and storage system
- evaluation of the error analysis and corrective action system
- evaluation of the record retention system.

One of the most important requirements for effective operation of a quality assurance program is the company's positive interest and concern. This begins with top management. In most instances, top management must delegate quality assurance responsibilities to a coordinator and clearly define the assignment of responsibility and authority throughout all levels of management, supervision, and operation.

Manuals

Managers should develop a quality assurance manual, in which the form and content vary according to a company's requirements. The manual is usually divided into two sections: policy and procedures.

Policy

- states company quality assurance policy and objectives
- establishes organizational responsibilities
- establishes systems for implementing quality assurance policy.

Procedures

- state operational responsibilities
- give detailed operating instructions.

The policy section of the manual outlines general goals, while the procedures section contains the daily, detailed operations of the quality assurance department.

Engineering and Product Design Coordination

The quality assurance department, by looking at operations from a different viewpoint, can assist the engineering or design department in its research, development, design, and specifications functions. Quality assurance normally is not involved in engineering or design or research unless it has specific data to share. Quality assurance performs such functions as inspection, testing, recording, maintenance, equipment calibration, and data accumulation.

Evaluation and Control of Suppliers

It is just as important to control the quality of purchased materials and services as it is to establish and enforce such controls for internal functions. The degree and extent of quality assurance established for a supplier will depend on the complexity and quantity of the supplier's products and its quality history. The primary criteria in selecting suppliers are their ability to maintain adequate quality and to establish good lines of communication with the company.

Manufacturing Quality (In-process and Final Assembly)

Quality assurance of items produced in-facility is accomplished by planning, inspection, process control, and equipment calibration.

Manufacturing Planning

Planning of production ensures that product and manufacturing instructions will be uniform throughout the process. Planning normally is a joint effort of the manufacturing department and quality assurance department. Manufacturing personnel initiate the internal paperwork, including all data necessary to produce the product. Quality assurance personnel review the paperwork to confirm that the data adhere to established parameters and include quality requirements.

Manufacturing Work Instructions

All work affecting product quality should be supported by documented, step-by-step instructions appropriate to each task, the working conditions, and employee skill levels. Instructions also must include quantitative and qualitative means for determining that each operation has been satisfactorily completed. The amount of detail depends on the skill levels of the workers and on the complexity of each task.

Inspection Instructions

Instructions to inspection personnel should give them specific guidelines to ensure that products conform to design specifications.

Manufacturing Inspection

Product inspection usually includes one or more of the following general methods:

- visual inspection
- dimensional inspection
- hardness testing
- functional testing
- nondestructive testing (NDT)
- chemical-metallurgical testing

Product inspection can be conducted in-process rather than after completion of all operations. If the inspection is conducted in-process, it should not be done until all previous inspections have been completed and the product has been certified as acceptable up to that point.

Records

Records should be kept of the manufacturing operations, including incoming materials, assemblies or processes, and final inspection results.

Special Process Control

Special processes used by manufacturing to change the physical, mechanical, chemical, or dimensional characteristics of product production include the following:

- heat treating
- plating
- fusion welding
- stamping and forming
- batch mixing
- chemical mixing
- adhesive bonding

Because special processes greatly influence the quality of the completed product, they must be carefully yet economically controlled. A 100% inspection process is often infeasible, because it might require destructive testing to determine product conformance to specifications, or similar tests. The processes, therefore, should have self-monitoring controls to ensure that all processed items will be of the same desired quality.

Calibration of Measuring Equipment

Measuring and process control equipment and instruments used to ensure that manufactured products and processes conform to specified requirements must be calibrated periodically. Any equipment employed either directly or indirectly to measure, control, or record manufacturing processes should be calibrated against certified standards so that it can be adjusted, replaced, or repaired before it impairs product quality.

Sample Inspection Evaluation

A sample inspection is often used by quality assurance to determine the quality of a lot without inspecting all items. This technique can be very beneficial if used correctly.

Even when a 100% inspection is performed under the most favorable conditions, it is only 85 to 90% effective because of human and other error. A sampling inspection is not necessarily 100% effective, but it can approach that level. The sampling disadvantage, which is far outweighed by its advantages, is that occasionally the sample-gathering procedure for a lot does not give a true picture of the lot quality: good lots can be rejected or bad lots can be accepted. The goal, however, is to make the acceptance of good lots far more likely than the acceptance of bad lots.

Nonconforming-Material Procedures

The term *nonconforming material* is usually applied to products that are rejected because they do not meet established requirements. Raw materials, parts, components, subassemblies, and assemblies can be classified as nonconforming material whenever they fail to meet specifications at any point in the manufacturing process. The company should have a system to control nonconforming material that includes the following:

- identification
- segregation
- disposition
- reinspection
- customer notification
- supplier reporting
- records

Material Status and Storage

The company must have a way to identify whether a product (or lot) either has not been inspected, has been inspected and approved, or has been inspected and rejected. All raw materials, components, subassemblies, assemblies, and end products should be labeled to note their process, inspection, and test status.

Material storage refers to the temporary holding of raw materials and recently purchased or in-process parts and assemblies. Stores personnel must have a control system that documents at all times the status and condition of stored raw materials, components, and products. The program auditor should evaluate the effectiveness of controls used by the company to make sure the stock on hand is the stock ordered and that the system used to identify the materials in the storeroom is adequate.

Error Analysis and Corrective Action System

The error analysis and corrective action system is a follow-up to the nonconforming-material system. The objective of this system is to use the discrepancies reported in the nonconforming-material system to do the following:

- Analyze manufacturing errors.
- Request corrective actions from those responsible for the errors, evaluate responses, and determine the effectiveness of the actions taken.
- Analyze manufacturing rejection rates and report to management the dollar losses due to scrappage, rework or repair, and reinspection costs.

Record-Keeping System

Good records are the backbone of a quality assurance program because they document the history of a product. These records indicate that the company's quality assurance system is functioning as planned, substantiate that a product was inspected, and often contain the actual inspection findings. Records can be vital in the defense of a product in a lawsuit.

Good quality assurance records also allow a company to trace any product from start to finish. These documents should include the following:

- lists of raw materials from which products are produced
- inspection results for purchased and manufactured products
- inspection results from each inspection station
- special-processes control data
- calibration data
- sample inspection data
- nonconforming-material data
- error analysis and corrective action data
- shipping data

Records should be stored in metal cabinets in a low-hazard area and kept for the life of the product.

FUNCTIONAL ACTIVITY AUDITS

These final sections cover two interrelated activities: record-keeping and field-information systems. These important activities are referred to as functions to differentiate them from the departmental categories previously used (design and manufacture, quality assurance, insurance, and others). The term *function* implies that these activities, to be properly accomplished (from a PSM program standpoint), must be performed by several or possibly all of the departments within a company rather than just one department.

Record Keeping

A company must not only manufacture and market safe, reliable products; it also must be able to prove in court the safety of those products. Complete, accurate records can be convincing evidence in a court of law. Consequently, a company should retain records that document all phases of its manufacturing, distributing, and importing activities—from the procurement of raw materials and components through production and testing to the marketing and distribution of the finished products. However, as indicated earlier, documents can become problems in the wrong hands and if taken out of context. Records that demonstrate noncompliance with established procedures in the manufacturing process or that simply point out problems with a product can contribute to a ruling against a company in a product liability suit.

In addition to their usefulness in court, comprehensive PSM program records enable a company to identify and locate products that its data collection and analysis system indicates may have reached the customer in a defective condition. Should the company need to implement a product recall or field-modification program, these PSM program records can play a key role in the success of such efforts.

Both federal regulations and internal PSM program requirements dictate the form and content of a company's record-keeping program, the types and quantity of records retained, and how long the company should retain them.

Regulations

A company should be aware of regulations affecting record-keeping requirements. Some of the more important governmental regulations to consider when developing a PSM record-keeping system both in the United States and Canada areas follows:

- *United States*
  Consumer Product Safety Act (Public Law No. 92–573)
  Federal Hazardous Substances Act (15 USC 1261)
  Federal Food, Drug and Cosmetic Act (21 USC 321)
  Poison Prevention Packaging Act (Public Law No. 91–601)
  Occupational Safety and Health Act (Public Law No. 91–596)
  Child Protection and Toy Act (Public Law No. 91–113)
  Magnuson-Moss Warranty—Federal Trade Commission Improvement Act (Public Law No. 93–637)
  ANSI Z535.1, .3, .4–1991
- *Canada*
  Consumer and Corporate Affairs Canada Act
  Consumer Packaging and Labelling Act
  Hazardous Products Act
  Industrial Design Act
  National Trademark and True Labelling Act

Patent Act
Seals Act
- *International*
  ISO 3864(1984); Safety Colours and Safety Signs 89/392/EEC; Machinery Safety Directive, Amex 1, Section 1.7.2, Warning of Residual Risks
  prEN 500 99–1; Indicating, Marking, and Actuating Principles. Part 1—Visual, Audible, and Tactile Signals

Internal PSM Program Requirements

A company must maintain all pertinent information related to the design, manufacture, marketing, testing, and sale of its products. Of particular value are records documenting design decisions, since faulty design is an almost universal claim in product liability lawsuits. A company can mount an effective defense if it can introduce evidence proving that a certain design or material was chosen with the safety of the consumer in mind. Also, records documenting why certain production techniques were chosen, such as having a part forged rather than cast, should be retained. Records can describe various product tests and the reasons for any design or manufacturing modifications and improvements. When a product hazard cannot be eliminated, a company's records should show conclusively why it cannot.

An important means of product liability protection is a complete, up-to-date set of records that can establish the care taken by a company to manufacture and market safe, reliable products. The PSM program auditor must bear in mind that a company's size, product lines, and organizational structure will determine the type and number of records that should be kept.

Period of Retention

It can be difficult for a company to determine what records should be retained and for how long. Product liability claims can be brought years after the manufacture or sale of an item has ceased.

Generally speaking, unless a company can accurately define the life of its product, it should be prepared to retain all PSM records in perpetuity. If this creates record storage or maintenance problems, a company can use a microfilm or computerized record-keeping system. The PSM program coordinator must ensure that key personnel fully understand why records are being maintained, what they contain, and for how long they should be kept.

Reasons for Keeping Records

Several reasons exist for keeping PSM records:

- to comply with regulations covering the design, manufacture, and sale of the company's products
- to demonstrate management's commitment to market a quality product
- to avoid wasting time and money redoing what has already been done
- to establish how much care is needed to produce and sell a safe, reliable product

- to enable the company to trace a product or customer
- to establish a sound data base for items such as insurance costs, sources of supply, and product recall or field-modification expense requirements.

Field-Information System

Because a company must receive information from the field about product performance, the PSM program auditor should evaluate the effectiveness of the company's field-information system for each product line and its distribution system. The program auditor must thoroughly analyze (1) the information system's ability to identify and trace a product from raw material form through final sales and distribution stages; (2) the product's ability to acquire and use field data (complaints, incidents, and accidents); and (3) its ability to use these two factors to implement product field-modification or recall actions, where appropriate.

Data Collection and Analysis of Complaints, Incidents, and Accidents

Every company should have a reporting system to acquire and evaluate product information from the ultimate testing laboratory—the customer. This information comes from a variety of sources both within and outside the organization (service personnel, salespeople, repairers, distributors, and retailers). A reasonably detailed report used with a data collection and analysis system can help the company accurately evaluate each complaint, incident, and accident and assess any problem as it develops.

For maximum effectiveness, data should be directed to one individual in the company, the PSM program coordinator. The program auditor must be able to verify, through company records, the results of any field data analysis and what corrective action, if any, was taken.

Diversification among manufacturers makes it impossible to create one data collection and analysis system applicable to all companies. However, every data reporting system must provide the answers to the following questions:

- Who is the customer?
- What type of product is involved?
- What is the problem?
- How is the product being used?

Analysis of Causes

The complaint report should provide answers to questions such as these:

- Is the product being used in the manner intended?
- How is it mounted or installed?
- What environmental conditions has the product been subjected to?
- Has the product been altered or modified by the customer?

When it is determined that a potential product hazard exists, a company should promptly instigate field investigations to determine its cause (design, manufacturing, or quality assurance). The cause should then be analyzed to discover if it is part of a developing trend or if various product problems are unrelated. This information should be given to the PSM program coordinator, who will decide whether to recommend further action.

The type of action taken might include changes in design, manufacturing, quality assurance, or advertising procedures, or the implementation of a product field-modification or recall program. If a company takes no action, it is actually saying that the data collection and analysis system has no value in the manufacture of safe, reliable products. Examination of TSCA, section 8c, files is also a good source of potential audit findings.

The best way to evaluate a company's data collection and analysis system is for the auditor to request a review of the company's complaint, loss, field data, and similar files. If the company does not have such files, or if they are poorly maintained, it indicates that the firm does not make use of this information to help it manufacture safe products. Examination of TSCA, Section 8c, files is also a good source of potential audit findings.

Product Recall or Field Modification

Any manufacturer may be faced with the need to institute a product recall or field-modification program. This action also can involve those who supply component parts, materials, and services for products. Therefore, the PSM program auditor should determine whether a company is capable of implementing a successful product recall or field-modification program should the firm's data collection and analysis system indicate the need for one.

Whenever a company finds that substantial performance or safety defects exist in some or all of its shipped products, it must recall or field-modify these products. In addition, recall or modification should be considered when the company is aware of or has itself developed an improvement or changed its product (for example, added a guard or a fail-safe). If the improvement is developed after the sale of earlier product models, a company should advise its customers of the product improvement, the hazard it eliminates, and how they can incorporate it into their earlier product models.

Without an adequate tracing system established during product planning and manufacture, it may be difficult to identify and locate specific lots, batches, quantities, or units to be recalled or modified in the field. Every company's record-keeping system should be extensive enough to enable management to trace a product from design through production to delivery of the product to the customer.

Should a product recall or field-modification program become necessary, a company can minimize the quantity that must be recalled or modified and the cost to do so if it can pinpoint the exact number of defective products, parts, or components that need to be changed.

Field-Modification Program Plan

A product recall or field-modification program plan will be different for each company. The plan must be defined by the PSM program auditor after evaluating the firm's record-keeping program and product traceability systems capability.

Based on information acquired from complaints, incidents, and accident reports made by customers, distributors, or dealers, and from regulatory agencies, a company must be able to determine immediately if a substantial product hazard exists. Techniques for this step include on-site investigation of the complaint, incident, or accident; hazard and failure analyses of the unit involved in the complaint, incident, or accident by the company or an independent laboratory; analyses of other units of the same product batch; and careful evaluation of in-house tests or other records. If a company finds that a substantial product hazard does exist, it should implement an appropriate product recall or field-modification plan.

Independent Testing

As an integral part of the PSM program, a company should conduct adequate product testing. This can be done either in-house or through affiliation with an independent testing laboratory. In certain instances, companies maintaining bona fide laboratory or adequate testing facilities should be encouraged to certify on their own authority and reputation that their products comply with existing standards. In those instances where no external standards exist, products should comply with appropriate company standards developed by safety and engineering personnel.

PSM PROGRAM AUDIT REPORT

The PSM program auditor will generate a sizable body of information during evaluation. This information must be assembled, organized, and presented to management in a clear and concise format.

For management's benefit during review, the audit report should begin with a summary of findings. This summary section clearly states the program auditor's opinion of the company's overall PSM program and any recommendations to correct deficiencies detected during the audit. If recommendations are made, they must be supported by data in the body of the report and should be practical and straightforward, using the company's already existing systems and procedures where possible. The report should indicate each recommendation's importance and the possible consequences if the company fails to comply. The program auditor should then estimate a realistic time frame for compliance with the recommendations.

It must be clear to management how the program auditor developed his or her overall opinion about the company's PSM program. If management doubts the auditor's conclusions, the value and impact of any efforts to assess and improve an existing PSM program or control system will be seriously impaired.

SUMMARY

- The major goal of a product safety management (PSM) program audit is to reduce or eliminate the causes of product liability exposure. The auditor must have the support and cooperation of all management personnel.
- The safety and health professional can evaluate the program's content, train personnel, and investigate accidents; maintain safety surveillance in production areas; and identify potential hazards related to the manufacture and use of products.
- The auditor should review the engineering, manufacturing, purchasing, service, legal, risk management/insurance, and marketing departments. Companies have a legal duty to warn customers of any potential hazards or risks related to using or misusing the product.
- The PSM auditor must determine if the quality assurance program is adequate to carry out the company's quality objectives. The auditor evaluates quality assurance manuals, policies, procedures, and other materials and processes.
- Functional activity audits cover record-keeping and field-information systems. Records must be retained to satisfy government regulations; to prove the company's commitment to manufacture safe, reliable products and to protect the consumer; and to trace customers and product lots should a product recall or field-modification program become necessary.
- Field-information systems are essential to respond to consumer complaints or suggestions regarding product performance.
- The PSM audit report to management should begin with a summary of findings, contain adequate data to support all recommendations, and provide realistic, measurable implementation steps to correct deficiencies or to improve current procedures.

REFERENCES

American National Standards Institute, 11 West 42nd Street, New York, NY 10036.

"Catalog of American National Standards" (issued annually).

Product Safety Signs and Labels, ANSI Z535.4–1991.

Baldwin S, et al. *The Preparation of a Product Liability Case,* 2nd ed. New York: Little, Brown and Company, 1992.

Burditt M, ed. *Product Safety Management and Engineering,* 2nd ed. Des Plaines, IL: American Society of Safety Engineers.

Canavan M. *Product Liability for Supervisors and Managers.* Reston, VA: Reston Publishing Company, 1981.

Commerce Clearing House, Inc., 2700 Lake Cook Road, Riverwoods, IL 60615, *Products Liability Reports.*

Eads G and Reuter P. *Designing Safer Products—Corporate Responses to Product Liability Law and Regulation.* Santa Monica, CA: Rand Institute for Civil Justice, 1983.

Hammer W. *Product Safety Management and Engineering.* 2nd ed. Englewood Cliffs, NJ: Prentice-Hall, Inc., 1993.

Kolb J and Ross S. *Product Safety and Liability.* New York: McGraw-Hill Book Company, 1980.

Leach G. *Product Safety Checklist.* Aurora, IL: Stephens-Adamson, Inc., 1979.

National Bureau of Standards, Washington, DC 20234, *An Index of U.S. Voluntary Engineering Standards,* Spec. Pub. No. 329, Standards Information Service.

National Safety Council, 1121 Spring Lake Drive, Itasca, IL 60143.

Occupational Safety and Health Data Sheets (listing available).

Product Safety Management Guidelines, 2nd ed., 1996.

Product Safety Management Training Course, Establishing and Auditing an Effective Product Safety Management Program.

"Product Safety Up-To-Date" (published bimonthly).

Product Liability Prevention Technical Committee. *Product Recall Planning Guide.* Milwaukee, WI: American Society for Quality Control, 1981.

Pyzdek T. *What Every Engineer Should Know About Quality.* WEESKA Series. New York: Marcel Dekker, Inc., 1990.

Restatement of the Law, Torts, Second Series. Philadelphia, PA: American Law Institute, 1965.

Schaden R and Heldman V. *Product Design Liability.* New York: Practicing Law Institute, 1982.

Seiden RM. *Product Safety Engineering for Managers: A Practical Handbook and Guide.* Englewood Cliffs, NJ: Prentice Hall, Inc., 1984.

Thorpe JF and Middendorf. *What Every Engineer Should Know About Product Liability.* New York: Marcel Dekker, Inc., 1979.

Underwriters Laboratories Inc., 333 Pfingsten Road, Northbrook, IL 60062, *Standards for Safety Catalog.*

Weinstein AS. *Product Liability and the Reasonably Safe Product.* New York: John Wiley and Sons, 1978.

REVIEW QUESTIONS

1. The key to achieving a reasonably safe and reliable product while reducing a company's product liability exposure is to incorporate product safety into all aspects of the:
   a. Designing of the product
   b. Manufacturing of the product
   c. Marketing of the product
   d. All of the above
   e. Only a and b
2. The three basic steps in establishing and coordinating a product safety management (PSM) program are:
   a.
   b.
   c.
3. Although all departments should be involved in PSM program audits, which departments' audits are most significant?
   a.
   b.
   c.
   d.
   e.
   f.
   g.
4. Define quality assurance.
5. Functional activity audits should be performed by all of the departments within a company and should encompass which four activities?
   a.
   b.
   c.
   d.
6. Complete and accurate record keeping can be beneficial in what two ways?
   a.
   b.
7. When developing a PSM record-keeping system, which of the following regulations do companies need to be aware of?
   a. United States regulations
   b. Regulations in Canada
   c. International regulations
   d. All of the above
8. List the information that must be included in a PSM program audit report.
   a.
   b.
   c.
   d.
9. List the four questions to which a PSM program should provide answers.
10. Who has the responsibility for assuring the safety and reliability of a product?

17

Retail/Service/ Warehouse Facilities

The service industry is the fastest growing industry in the United States. Due to price competition, service companies must continuously strive to improve their operational effectiveness without compromising safety issues. Because many jobs in the service industry are physically demanding and many positions are filled by inexperienced employees, the industry is at risk for a high incidence of work-related injuries.

This chapter provides an overview of the safety and health programs that should be established for employees in retail/service/warehouse facilities and of the special safety-related issues confronting employers in these facilities. For more detailed discussions of each topic, see the specific chapters referenced in each section. For discussion of issues related to the safety of outside contractors, vendors, clients, and customers, see Chapter 21, Contractor and Nonemployee Safety. The topics covered in this chapter include:

- management recognition of safety, health, and environment programs
- OSHA regulations relevant to service and retail industries
- specific safety, health, and environmental programs that should be established

Loss control should be incorporated into the organization's total quality management programs (TQM). Effective management of safety, health, and environmental risks will not only reduce injuries and accidents but substantially increase the quality of services and ensure better product management.

IMPORTANCE OF SAFETY, HEALTH, AND ENVIRONMENT PROGRAMS

Service and retail facilities have a high exposure to risk in terms of both customer and employee injuries. According to the National Safety Council, the direct cost of accidents is only a small part of the total expenditures (insurable cost), while indirect costs of accidents can range from 5 to 20 times the direct costs. In the long run, no organization can efficiently conduct business if it has a high volume of accident losses. The service industry, like other industries, must pass along the cost of such losses to its customers.

To stay competitive, each company must establish a comprehensive loss control plan that includes:

- creation of a safety and health culture within the company
- clear safety policies and written procedures and safety manuals
- identification of responsibility and authority regarding safety issues
- safety committees
- safety and health training, auditing, and inspections

- emergency preparedness plan
- accident investigation and analysis

Policy Statement

Each company should have a clear statement of policy regarding its position on safety, health, and environmental matters. This policy should be signed by the president and general manager and posted for all employees to review. It should identify those responsible for safety matters and the steps employees should take if they have questions and/or concerns on any environmental or safety matter.

Employee Manual

Each company should develop an employee manual that defines not only the level of safety expected from employees but also gives specific instructions. An example would be a statement on safe practices for forklift operators, which might read as follows:

> Only trained and certified forklift drivers will drive any forklift on company property.

The policy should also specify safe work practices and dos and don'ts regarding use of forklifts at the company. Maintenance and emergency procedures on forklift safety should also be included.

Safety Committee

The company should establish a safety committee to help define safe working conditions, to hear concerns of employees, to monitor operating conditions, and to serve as a communications liaison between management and employees. These committees should be given real authority and functions and not merely be regarded as token gestures toward creating safer workplaces.

Safety Audits

A critical part of total quality management safety and plant environmental control is safety auditing. A formal procedure should be established to systematically and comprehensively audit each operating task within a company. The auditors should have expertise in the type of operation being audited (e.g., warehousing, trucking, retailing). Often, establishing a team composed of an outside auditor, management auditor, and a facility manager works best. (See Chapter 4, Safety, Health, and Environmental Auditing, in this volume.)

Training

Each company should conduct a general safety orientation for all employees, along with task-specific safety training. Workers in high-risk jobs (e.g., chemical handling, heavy lifting, or forklift driving) need specialized safety training. Training manuals should be used to ensure consistency across the company and within particular jobs. In some cases, the best training is done by employees who know their jobs well and use a well-written manual to teach others about safety policies and safe work practices. The company must document all training. (For more details, see Chapter 23, Safety and Health Training.)

Investigation and Action

A vital part of any quality control program is to learn from mistakes and to modify practices to reduce future exposures. Thus, companies should thoroughly investigate all accidents, recommend changes, and take corrective actions. Employee involvement in these investigations can reduce the chance of recurrence and strengthen corporate commitment to safety. (For more details on this topic, see Chapter 7, Accident Investigation, Analysis, and Costs.)

Corporate Safety Culture

Effective implementation of the steps described above will greatly enhance the safety culture of a company and reduce work-related losses. In creating a safety culture, the service industry, like other industries, also must be concerned about government regulations that pertain to health and safety both within the workplace and in the community where the company operates. These regulations include Occupational Safety and Health Administration (OSHA) standards, state workers' compensation laws, Environmental Protection Agency (EPA) regulations, Department of Transportation (DOT) regulations, and general liability issues.

OSHA REGULATIONS

Several OSHA regulations address specific safety issues within the service industry. The major relevant OSHA regulations include:

- 29 *CFR* 1910—Occupational Safety and Health Standards
- Subpart C—General Safety & Health Provisions
- Subpart D—Walking-Working Surfaces
- Subpart E—Means of Egress
- Subpart F—Powered Platforms, Manlifts, and Vehicle-Mounted Work Platforms
- Subpart G—Occupational Health and Environmental Control
- Subpart H—Hazardous Materials
- Subpart I—Personal Protective Equipment
- Subpart J—General Environmental Controls
- Subpart K—Medical and First Aid
- Subpart N—Materials Handling and Storage
- Subpart O—Machinery and Machine Guarding
- Subpart P—Hand and Portable Powered Tools and Other Hand-Held Equipment
- Subpart R—Special Industries
- Subpart S—Electrical

Companies should comply with these regulations and make sure that their policies, written procedures, and safety notices conform to these requirements.

General Duty Clause

The overall OSHA standard requires each employer to provide employees with a safe place of employment free from recognized hazards that are causing or are likely to cause death or serious physical harm to the employees. The General Duty Clause is frequently used by OSHA inspectors as a reference for citing those activities that are felt to represent an important hazard but are not specified in the detailed regulations. In order for OSHA to apply the General Duty Clause, it must be established that there is a hazard to which employees were exposed. A serious hazard must be proven to exist and OSHA must be able to document that employees were exposed to the hazard. If an employer fails to provide a safe place of employment, OSHA may issue a citation under the General Duty Clause.

This clause could be used by OSHA to cite a company in the service industry for the following:

- Cumulative trauma disorders such as carpal tunnel and tendinitis are cited under the General Duty Clause due to the lack of a comprehensive, formally approved set of OSHA standards on ergonomics.
- Injuries and hazards relating to manual lifting could be cited under the General Duty Clause if no specific standard exists for that particular task.
- A violation of the General Duty Clause relating to a serious training deficiency may be present if the training is not addressed under another general or specific training requirement.

In summary, employers have a duty to take whatever actions are necessary to protect workers from hazards. This duty extends beyond the hazards addressed by existing standards and includes hazards which are associated with recognized employee exposure and may or may not be included in current OSHA safety and health standards. Failure to adhere to this duty may result in OSHA citations and/or fines under the General Duty Clause.

Posting Requirements

Each employer must post a notice, furnished by OSHA, informing employees of the safety requirements and protection mandated by OSHA. The notice must be posted in each store or work area in a conspicuous space where other postings are kept for employees to read. The notice can be obtained by contacting the nearest Department of Labor.

Reporting and Record Keeping

Employers are required to prepare and maintain two forms for OSHA record keeping. One form is the OSHA 200 log, which serves as the log and summary of occupational injuries and illnesses. All occurrences are recorded, and the summary of occupational injuries and illnesses is used at the end of each year. A second form, the OSHA 101 form, provides detailed information on each of the cases recorded on the OSHA 200 log. (See also Chapter 2,

Regulatory History, and Chapter 8, Injury and Illness Record Keeping and Incidence Rates.)

Means of Egress

The general requirements for all means of egress from a building include regulations that provide for:

- exits sufficient to permit the prompt escape of occupants in case of fire or other emergency
- a building constructed, arranged, equipped, maintained, and operated to avoid undue danger to the lives and safety of its occupants
- exits of kinds, numbers, locations, and capacities appropriate to the individual building or structure
- exits arranged and maintained to provide free and unobstructed egress from all parts of the building or structure at all times when it is occupied
- clearly visible exits that are indicated in such a conspicuous manner that every occupant will readily know the direction of escape from any point
- adequate and reliable illumination for all exits
- fire alarm facilities in every building or structure of such size, arrangement, or occupancy that a fire may not itself provide adequate warning to occupants
- at least two means of egress remote from each other in every building or structure

Compliance with the above conditions does not eliminate or reduce the need for other provisions for the safety of people using a structure under normal occupancy conditions.

Maintenance and workmanship of exits are addressed by three requirements:

1. Doors, ramps, passages, signs, and all other components of the exit shall be of substantial and reliable construction.
2. All exits shall be continuously maintained free of all obstructions or impediments.
3. Any device or alarm installed to restrict the improper use of an exit shall be so designed and installed that it cannot, even in cases of failure, impede or prevent emergency use of such exit. This prohibits the use of a lock and hasp on exit doors and sets requirements for push-bar alarm systems.

Companies that handle large quantities of materials and have limited storage space for products and merchandise may be more likely to violate these standards. Management and supervisors should inspect means of egress on a regular basis to ensure they remain accessible at all times. Management needs to pay particular attention to proper maintenance of an exit's components and hardware and to enforce rules prohibiting blocking exit ways with stored products and merchandise, equipment, mop buckets, hand trucks, debris, or other obstacles.

Hazard Communication

The hazard communication standard (HAZCOM) is based on the concept that employees have a need and a right to know the identities and hazards of the chemicals to which they are exposed in the workplace (29 *CFR* 1910, 1915, 1917, 1918, 1926, and 1928). Employees also have the right to know what protective measures are available to prevent exposure to these chemicals.

The main requirements of the hazard communication standard are:

- written hazard communication program
- labels and other forms of warning on containers of chemicals
- Material Safety Data Sheets (MSDSs) collected and available for review by employees
- information and training provided for potentially exposed employees

Medical Services and First Aid

The 29 *CFR,* Subpart K, 1910.151-.153, standard also requires that employers provide access to and information about medical services and first aid. The standard states the following:

a. The employer shall ensure the ready availability of medical personnel for advice and consultation on matters of facility health.
b. In the absence of an infirmary, clinic, or hospital in near proximity to the workplace, which is used for the treatment of all injured employees, a person shall be adequately trained to render first-aid. First-aid supplies approved by the consulting physician shall be readily available.
c. Where the eyes or body of any person may be exposed to injurious corrosive materials, suitable facilities for quick drenching or flushing of the eyes and body shall be provided within the work area for immediate emergency use.

The words "near proximity" have been interpreted to mean within 5 to 10 minutes. If proper medical facilities are not within a 5- to 10-minute drive or a rescue squad would not arrive within 5 to 10 minutes, someone at the worksite should be adequately trained in first-aid procedures.

Another requirement of this OSHA subpart is to make emergency service telephone numbers available to workers so they can obtain help quickly. Posting emergency phone numbers by company telephones is sufficient and emphasized in areas without "911" capability. All employees should be trained in emergency response procedures to minimize the damage or severity of any injury or illness.

Walking and Working Surfaces

The OSHA standard (29 *CFR,* Subpart D, 1910.21-.32) also requires development of a comprehensive program to identify and address the facility physical hazards. The following elements are components of the program:

a. Housekeeping: All places of employment, passageways, storerooms, and service rooms should be kept clean, orderly, and in a sanitary condition.

b. Floor Loading Protection: Floors in buildings used for mercantile, business, industrial, or storage purposes, should be posted to show maximum safe floor loads.

c. Stairway Railings and Guards: Every flight of stairs having four or more risers should have a standard railing on all open sides. Handrails should be provided on at least one side of closed stairways, preferably on the descending right side.

d. Fixed Stairways: Fixed stairways should have a minimum width of 22 in. (56 cm). Fixed stairways should be provided for access from one elevation to another where operations necessitate regular travel between levels.

e. Dockboards (Bridgeplates): Dockboards should be strong enough to carry the load imposed on them.

f. Toeboards: Railings protecting floor openings, platforms, scaffolds, etc., should be equipped with toeboards whenever persons can pass beneath the open side, when there is moving machinery, or when there is equipment from which falling material could cause a hazard.

g. Portable Ladders (Stepladders and straight ladders): Stepladders should be equipped with a metal spreader or locking device of sufficient size and strength to securely hold the front and back sections in open position.

h. Fixed Ladders: All rungs should have a minimum diameter of three-fourths of an inch, if metal, or one and one-eighth inches, if wood. They should be a minimum of 16 in. (41 cm) wide and should be spaced uniformly no more than 12 in. (30 cm) apart.

i. Aisles and Passageways: Where material-handling equipment is used, sufficient safe clearances should be allowed for aisles, at loading docks, through doorways, and wherever turns or passage must be made.

j. Floors, General Condition: All floor surfaces should be kept clean, dry, and free from protruding nails, splinters, loose boards, holes, or projections. Where wet processes are used, drainage must be maintained.

k. Open-Sided Floors: Every open-sided floor, platform, or runway four feet or more above adjacent floor or ground level should be guarded by a standard railing with toeboard on all open sides, except where there is entrance to a ramp, stairway, or fixed ladder.

l. Railings: A standard railing should consist of top rail, intermediate rail, and posts, and shall have a vertical height of 42 in. (107 cm) from upper surface of top rail to floor, platform, etc.

Lockout/Tagout

This OSHA standard (29 *CFR* 1910.147) helps safeguard employees from hazardous energy while they are performing servicing or maintenance on machines and equipment. According to the standard employers must do the following:

• Develop a control program to prevent the accidental start-up of machinery or equipment being repaired or serviced.

• Use locks when equipment can be locked out.

• Ensure that new equipment and overhauled equipment can accommodate locks.

• When lockout procedures cannot be employed, tagout procedures should be in place. Tagout procedures require that all switches, valves, levers, and so on be tagged to instruct all parties not to open or operate such controls until the tag is removed.

• Identify and implement specific procedures (generally in writing) for the control of hazardous energy, including preparation for shutdown, equipment isolation, lockout/tagout application, release of stored energy, and verification of isolation.

• Institute procedures for release of lockout/tagout, including machine inspection, notification, and safe positioning of employees and removal of the lockout/tagout device.

• Obtain standardized locks and tags that identity the employee using them. The locks and tags should be of sufficient quality and durability to ensure their effectiveness.

• Conduct inspections of energy control procedures at least annually.

• Train employees in the specific energy control procedures and provide refresher training as part of the annual inspection of control procedures. Document all training.

• Adopt procedures to ensure safety when equipment must be tested during servicing, when outside contractors are working at the site, when a multiple lockout is needed for a crew servicing equipment, and when shifts or personnel change.

The following machines and equipment are covered by the OSHA lockout/tagout standard: dough mixers, meat saws, meat wrappers, meat grinders, balers, and compactors. Also included are carpentry and maintenance shop equipment such as radial saws, table saws, lathes, drilling machines, bench grinders and wire buffers, and similar equipment that could cause injury to a person during maintenance or servicing of such equipment if it could be energized by another employee or by stored energy. All persons using such equipment should be trained in their safe use and such training documented. (See also Chapter 6, Safeguarding, in the *Engineering & Technology* volume.)

Electrical Installations and Equipment

According to OSHA citations, the electrical standards most frequently violated in the service industry are:

a. General electrical standard (29 *CFR* Subpart S, 1910.301–.399). This standard states that companies must maintain sufficient access and working space around electrical panels, service entrances, circuit breakers, etc., to permit ready and safe operation and maintenance. A good rule of thumb is to keep all storage and material at least three feet away from all major electrical equipment.

The standard also requires that electrical equipment operating at 50 volts or more (most appliances and equipment) shall be guarded against accidental contact. This means that all covers, boxes, and cabinets provided with most equipment must be properly installed and in place at all times. Proper warning signs are required to be in place where exposed live parts are available, such as during maintenance and repair.

b. Flexible cords and cables. This standard is cited for improper use of extension cords. Long runs of extension cords can be both fire and tripping hazards.

c. Cabinets, boxes, and fittings. This standard offers guidelines for physically protecting electrical installations.

d. Examination, installation, and use of equipment. This section requires that electrical equipment be examined and found free of recognized hazards prior to use.

e. Identification of disconnecting means and circuits. This section requires that each means of disconnecting equipment and appliances (circuit breakers, etc.) be legibly marked to indicate its purpose.

Machine Guarding

A company must be sure to guard any machine part function or process that may cause injury. Where the operation of a machine or accidental contact with it can injure the operator or others in the vicinity, the hazard must be either controlled or eliminated. Management can conduct routine audits to ensure that such guards are in place and being properly used by trained employees. Power tools should never be loaned to contractors, clients, or untrained employees.

Powered Equipment

The OSHA 29 *CFR*, Subpart P, 1910.241–.247 standard

requires that all companies using powered industrial trucks establish a safety program that includes the following:

- General rules for safe operation are developed and implemented.
- Only trained and authorized operators are permitted to operate powered industrial trucks.
- Employers must provide training for operators in safe work practices and equipment operation.

GENERAL LIABILITY ISSUES

Each state has its own workers' compensation and general liability laws. These laws have considerable variability and must be consulted for specific guidelines because a business is responsible for compliance to state rules and regulations. Companies in the service industry must follow the laws of the state(s) in which they operate, particularly in regard to matching the right worker with the right job. Not surprisingly, management commitment to compliance has increased in the past few years because the industry realizes that in a low-margin industry, controlling safety, health, and environmental costs is critical.

Hiring and Placement

The warehouse/distribution sector is a demanding part of the service/retail industry that requires employees to be physically fit. Consistent with the Americans with Disabilities Act (ADA), once a company makes a tentative job offer to an employee, a drug screening and physical examination are conducted. The physical examination also includes a review of the individual's past medical history.

Recently, many warehousing environments have begun to introduce a physical capacity screening process, which measures a person's strength and endurance. The screening helps an employer determine whether an applicant is physically capable of performing a job that requires a significant amount of manual labor. The results of the screening are matched against the specific demands of the job. When the applicant's capabilities meet or exceed the job demands, the person is considered qualified physically to do the job. Otherwise, the employer must determine if a reasonable accommodation can be made so that the employee can perform the essential functions of the job. (See also Chapter 13, Workers with Disabilities, in the *Engineering & Technology* volume.) If the employer's efforts meet the specifications set by the ADA and the individual still cannot perform the work, the employer is justified in turning down the applicant.

Service/retail/warehouse organizations should also conduct initial job orientations to ensure that the new employee is made aware of company safety and health regulations and procedures. Most orientation programs cover the following topics:

- employee responsibilities and discipline
- safety rules
- accident/incident reporting
- general fire protection

- emergency procedures
- security alarms and inventory controls

(See also the sections on New Employee Training and Orientation and Job Instruction Training in Chapter 23, Safety and Health Training, in this volume.)

Loss Control

Many jobs in the distribution industry involve substantial amounts of materials handling. This type of work can result in a wide variety of injuries and illnesses as defined by OSHA. To help reduce the high frequency and severity rates of occupational injuries and illnesses, many service industry companies use annual Risk Control Action Plans. These action plans outline specific efforts necessary to obtain their Risk Control Objectives. After a thorough analysis of individual risks in company operations, the company uses the plan to establish measurable individual risk-control efforts and numerical goals that will be used in performance reviews for all management personnel.

The service industry has put considerable effort into providing safety training and assigning operational accountability to management. All supervisory personnel up through middle management are given training in risk-control management which includes:

- cost and efficiency consequences of accidents
- modified duty (light duty)
- accident analysis
- planned inspection
- workers' compensation laws
- OSHA regulations
- ergonomics
- fire protection and prevention

One area of special emphasis has been the active training of supervisory personnel, safety committees, and general employees in proper accident investigation techniques. The attachment is an example of an outline Accident Investigation Workshop from the National Safety Council.

Because of the number of customers who shop at retail establishments, these stores are especially susceptible to customer accidents. For this reason, retail companies should analyze in detail the potential causes of customer accidents (falls, slipping, cuts, and so on) then take measures to reduce the risks. This process is particularly critical, since a substantial number of lawsuits can be expected from these types of accidents.

On the other hand, for the distribution side of the industry, the greatest risk is posed by accidents involving outside vendors or drivers who routinely visit the distribution centers to monitor or deliver the product being stored in that facility. To reduce the risk of this type of loss, many companies restrict outside vendors and drivers to certain areas, such as the side or back of a building or special delivery entrance way. In addition, companies can require written review and compliance with in-house policies and procedures.

Because property losses are an important negative cost, many companies train their employees during orientation about the importance of eliminating property damage. The warehousing industry has a high risk of different types of property damage because of the volume of power equipment used to move material and products in the warehouse. For examples of accident report forms that can be used to report and analyze the causes of employee and customer losses, see Figures 17-1 and 17-2. (See also Chapter 7, Accident Investigation, Analysis, and Costs.)

Audit and Inspection Program

As discussed in Chapter 4, Safety, Health, and Environmental Auditing, audits and inspections are an important part of a company's health and safety program. The service industry has developed audit and inspection programs to ensure that loss-control initiatives are implemented and followed. Many companies are beginning to use both external and internal auditors to conduct a comprehensive evaluation of loss-control initiatives. Typically, these are annual audits that look at the entire company's loss-control issues. For example, an external audit could include the following topics:

- management direction and compliance with corporate policies
- compliance with federal and state regulations
- management accountability and compliance with corporate safety policies
- hiring and placement
- physical conditions
- accident analysis
- incident follow-up
- ergonomic control
- fleet operations
- business interruption

Internal audits can be as comprehensive but are more commonly conducted by supervisors or employees on a daily or weekly basis. These audits are more focused and include operational safety issues such as:

- materials handling
- body positioning
- environment
- equipment
- racking
- employee exposure potentials
- fire and life safety
- motorized equipment
- chemical management

See Figure 17-3 for a sample internal audit form.

Materials Handling

Materials handling is by far the area which has the greatest risk for injuries and property damage in both the retail and the distribution industries. This fact is due to the volume

LIBERTY PERSONNEL: PLEASE FORWARD TO DIVISION STATISTICAL

Employee Injury Data Entry Report - Distribution Center

DIVISION _____ LOCATION CODE ___ ___ ___ ___ ___ ___

DIVISION ADDRESS _____ STATE_____ ZIP _____

THIS IS AN EMPLOYEE INJURY REPORT BASED UPON THE SUPERVISOR'S INTERPRETATION OF INITIAL ALLEGATIONS OF FACTS AS RELAYED BY OTHERS.

WHO? INJURED EMPLOYEE'S NAME _____ MALE _____ FEMALE _____

AGE _____ LENGTH OF EMPLOYMENT: YEARS _____ MONTHS _____ FULL TIME _____ PART TIME _____

WHEN? DATE OF INJURY ___ ___ ___ TIME OF ACCIDENT ___:___ ☐ AM ☐ PM HOURS WORKED BEFORE INJURY _____
MONTH   DAY   YEAR   SHIFT _____

WHAT REPORTEDLY HAPPENED? BRIEF FACTUAL DESCRIPTION _____

WHERE DID THE ACCIDENT OCCUR
(CHECK ONE)

OUTSIDE

- ☐ 70 Parking Lot
- ☐ 75 Personnel Entrance/Sidewalk
- ☐ 97 Trailer
- ☐ 69 Yard
- ☐ 10 Transportation-While Driving
- ☐ 95 Not Otherwise Classified

INSIDE

- ☐ 68 Battery Recharge
- ☐ 96 Bindery Area
- ☐ 11 Building Maintenance
- ☐ 87 Cake Production
- ☐ 94 Camera Area
- ☐ 22 Dairy
- ☐ 88 Dough Production
- ☐ 01 Dry Grocery
- ☐ 05 Freezer
- ☐ 09 Garage
- ☐ 02 General Merchandise

OCCUPATION   CODE ___
(See back of this form for Occupation Code Numbers)

- ☐ 78 In-Plant Vehicle Shop
- ☐ 71 Lunchroom
- ☐ 82 Mailroom
- ☐ 21 Meat
- ☐ 12 Office
- ☐ 67 Packaging
- ☐ 83 Print Shop
- ☐ 04 Produce

- ☐ 06 Railcar Receiving
- ☐ 13 Recoup/Salvage
- ☐ 14 Repack Department
- ☐ 90 Sanitation
- ☐ 74 Transportation at Store or Backhaul
- ☐ 07 Truck Receiving Dock
- ☐ 08 Truck Shipping Dock
- ☐ 95 Not Otherwise Classified

TYPE OF ACCIDENT *(CHECK ONE)*

FALL OR SLIP

- ☐ BS Fall - From Racking
- ☐ A3 Fall - From Elevation (Other Than Rack)
- ☐ AA Fall - Same Level
- ☐ A2 Slip or Trip (Not Fall)
- ☐ A7 Falls Not Otherwise Classified

MATERIAL HANDLING *(Complete Backside Of Pink Copy If Coded For Order Selector as Occupation)*

- ☐ HD Carrying, Holding
- ☐ HB Lifting/Lowering
- ☐ HC Pulling, Pushing
- ☐ WY Repetitive Motion
- ☐ H3 Twisting
- ☐ H4 Reaching
- ☐ JC Not Otherwise Classified

STRUCK BY OR AGAINST

- ☐ RP Powered Object
- ☐ RS Stationary Object
- ☐ RR Falling, Flying Object
- ☐ RM Not Otherwise Classified

MISCELLANEOUS

- ☐ NL Caught In Between or Under
- ☐ Y1 Contact with Sharp Object
- ☐ YD Electrical Contact/Shock
- ☐ XN Explosion or Flash Fire/Back
- ☐ XS Friction
- ☐ X1 Hot Liquid, Steam
- ☐ XZ Hot Surfaces, Objects
- ☐ X9 Misconduct
- ☐ GD Motor Vehicle Accident
- ☐ YA Not Otherwise Classified

TYPE OF INJURY/ILLNESS *(CHECK ONE)*

- ☐ 300 Abrasion
- ☐ 100 Amputation
- ☐ 110 Asphyxiation/Choking/Drowning
- ☐ 160 Bruise/Contusion
- ☐ 120 Burns (Other Than Chemical)
- ☐ 130 Chemical Burns
- ☐ 140 Concussion
- ☐ 165 Crushing Injury
- ☐ 170 Cut/Lacerations
- ☐ 190 Dislocation
- ☐ 200 Electric Shock
- ☐ 235 Foreign Object/Body (Eyes)
- ☐ 210 Fracture
- ☐ 220 Frostbite or Freezing
- ☐ 230 Hearing Loss
- ☐ 420 Heart Attack
- ☐ 240 Heat Exhaustion
- ☐ 250 Hernia
- ☐ 350 Infection
- ☐ 150 Infectious Disease
- ☐ 260 Inflamation (Irritation of Muscles, Joints)
- ☐ 145 Loss of Consciousness
- ☐ 400 Multiple Injuries
- ☐ 175 Puncture
- ☐ 180 Rash/Dermatitis, Skin Inflamation
- ☐ 311 Spasms/Muscle Cramps/Sprains/Strains
- ☐ 994 Not Otherwise Classified - Illness
- ☐ 995 Not Otherwise Classified - Injury

BODY PART *(CHECK ONE)*

- ☐ 100 Head (Multiple)
- ☐ 150 Scalp
- ☐ 140 Face
- ☐ 120 Ear
- ☐ 146 Nose
- ☐ 130 Eye
- ☐ 144 Mouth/Tongue/Teeth

- ☐ 201 Neck/Throat

- ☐ 300 Upper Extremities (Multiple)
- ☐ 310 Arm
- ☐ 313 Elbow
- ☐ 320 Wrist
- ☐ 330 Hand (Multiple)
- ☐ 350 Thumb
- ☐ 340 Fingers

- ☐ 400 Trunk (Multiple)
- ☐ 410 Abdomen/Stomach
- ☐ 412 Pelvis
- ☐ 420 Back
- ☐ 430 Chest
- ☐ 440 Hips
- ☐ 441 Buttocks
- ☐ 411 Groin
- ☐ 450 Shoulder

- ☐ 500 Lower Extremities (Multiple)
- ☐ 510 Leg (Multiple)
- ☐ 560 Thigh
- ☐ 513 Knee
- ☐ 550 Calf
- ☐ 520 Ankle
- ☐ 530 Foot
- ☐ 540 Toes

- ☐ 700 Multiple Body Parts - Exterior

- ☐ 860 Internal Body (Multiple)

(Body diagram labels: 100, 201, 400, 300, 300, 400, 500, 500)

POSSIBLE CONTRIBUTING FACTORS (AGENCY)
(CHECK ONLY ONE)

- ☐ 0200 Animal/Insect
- ☐ 0620 Bag
- ☐ 5602 Bike
- ☐ 0600 Box(es)
- ☐ 2246 Box Cutter/Utility Knife
- ☐ 0700 Building Structure/Walls, Pillars
- ☐ 5920 Building Heating/Refrigeration Equipment
- ☐ 5600 Car/Auto
- ☐ 5610 Cart
- ☐ 0900 Chemical
- ☐ 5620 Compactor/Bailer
- ☐ 1300 Conveyor
- ☐ 0761 Dock Leveler, Dock

- ☐ 0762 Dock Plate
- ☐ 0780 Dolly Crank
- ☐ 0771 Door - Dock
- ☐ 0770 Door - Personnel
- ☐ 0773 Door - Trailer
- ☐ 2620 Elevator
- ☐ 4110 Fastener, Nail, Screw
- ☐ 1710 Fire/Flame/Smoke
- ☐ 1951 Floor (Uneven/Holes)
- ☐ 1952 Floor (Slippery Surface)
- ☐ 1960 Floor Sweeper
- ☐ 8000 Foreign Object
- ☐ 5640 Fork Lift - Stand Up
- ☐ 5641 Fork Lift - Sit Down
- ☐ 2000 Glass

- ☐ 2200 Hand Tools - Non-Powered
- ☐ 2300 Hand Tools - Powered
- ☐ 2600 Hoisting Equipment
- ☐ 6250 Ice/Snow
- ☐ 2245 Knife
- ☐ 2800 Ladder
- ☐ 2900 Liquids
- ☐ 3000 Machinery
- ☐ 2240 Meat Slicer
- ☐ 5930 Noise
- ☐ 7700 Other Person
- ☐ 7300 Pallet/Slips
- ☐ 7320 Pallet Jack - Non-Powered
- ☐ 7310 Pallet Jack - Powered
- ☐ 4510 Paper/Cardboard

- ☐ 5940 Product Spill
- ☐ 5650 Rack
- ☐ 9700 Rope
- ☐ 5826 Scaffold
- ☐ 5603 Scooter
- ☐ 5840 Stairs/Steps
- ☐ 0601 Tote/Tub/Pan/Tray
- ☐ 5625 Trailer
- ☐ 6450 Truck/Tractor/Shuttle
- ☐ 2550 Welding Equipment
- ☐ 4030 Wire/Banding
- ☐ 5700 Wood
- ☐ 6700 Not Otherwise Classified

Date Report Completed _____ Completed By_____ Title _____

PS 1133 (Rev. 8/95)   **WHITE COPY TO INSURANCE CARRIER**   **YELLOW COPY** *(If Completed)* **TO HOME OFFICE RISK MANAGEMENT**   **PINK COPY - KEEP FOR YOUR RECORDS**

Figure 17-1. This sample form can be used to report and analyze employee injuries.

Customer Store Incident Report

Store #: _____   Manager Completing This Report: _____

Date: _____   Time: _____

Name of Injured: _____ Age: _____

If a Minor, Parent's Name: _____

Address _____

Phone Number (Home)_____ (Work) _____

Personal Injury _____ And/or Property Damage _____

Describe Fully How Incident Occurred_____

_____

_____

_____

_____

_____

Nature of Injury/Damage (Describe Fully)_____

_____

_____

_____

_____

Witnesses (Name, Address and Phone Number) _____

_____

Figure 17-2. A report such as this sample form should be completed whenever a customer is involved in an accident that involves personal injury or property damage.

Safety/Sanitation Checklist

Completed By: _____   Date: _____

Location: _____   Time: _____

Employees Observed: _____ _____

_____ _____

_____ _____

Directions: For each item, check entire department/location. If satisfactory, make a checkmark in the "Safe" column. If not, make a checkmark in the "Unsafe" column and explain why in the "Comments" column. Leave blank if item is not applicable.

SAFETY

	SAFE	UNSAFE	COMMENTS
Manual Materials Handling			
Loads handled as close to the body as possible.	_____	_____	_____
Pivots feet and avoids twisting of trunk.	_____	_____	_____
Unnecessary awkward postures avoided.	_____	_____	_____
"Z-pick" selecting pattern followed.	_____	_____	_____
Avoids walking on inappropriate surfaces (i.e., pallets)	_____	_____	_____
Places feet properly when climbing up or down racks.	_____	_____	_____
Watches for obstacles when climbing up or down steps.	_____	_____	_____
Uses hook to select products when necessary.	_____	_____	_____
Body Positioning			
Avoids putting hands or feet under pallets.	_____	_____	_____
Avoids walking under forklift's raised head.	_____	_____	_____
Works in a deliberate motion when in confined space.	_____	_____	_____
Environment			
All exit doors accessible for emergency exit.	_____	_____	_____
No plastic in slots.	_____	_____	_____
All aisles/floors free from physical debris or spills.	_____	_____	_____
When debris/spill is detected, it is removed/guarded.	_____	_____	_____
Items slotted according to ergonomic specifications.	_____	_____	_____
First-aid kit well stocked.	_____	_____	_____
Product stacked to prevent sliding/falling/collapse.	_____	_____	_____
Smoking rules observed.	_____	_____	_____
Sprinkler heads have 24" clearance above stock.	_____	_____	_____
Equipment			
Material-handling equipment in safe operating condition.	_____	_____	_____
Faulty equipment not used and referred for repair.	_____	_____	_____
Eyewash station in good order.	_____	_____	_____
Racking			
Rubber steps installed on racks as needed.	_____	_____	_____
Rubber pads installed on sharp edges of racks.	_____	_____	_____
All racks are structurally sound.	_____	_____	_____

Additional Comments:

_____

_____

_____

Figure 17-3. This sample checklist can be used for internal audits.

Safety/Sanitation Checklist

SANITATION

Warehouse Inspection	SATISFACTORY	UNSATISFACTORY	COMMENTS
No open product indicating pilferage.	_____	_____	_____
No damaged product in slots.	_____	_____	_____
No cigarette butts found.	_____	_____	_____
No product stored directly on the floor.	_____	_____	_____
No cleaning or processing chemicals near food.	_____	_____	_____
Dead, crawling, or flying insects encountered.	_____	_____	_____
If rodent evidence is observed, are rodent traps set?			
Open bait.	_____	_____	_____
Floor is clean.	_____	_____	_____
No objectionable odor.	_____	_____	_____
Stock is properly rotated in all departments.	_____	_____	_____
No damaged product in overstock and/or picking slots.	_____	_____	_____
White lines and curbs are clean.	_____	_____	_____
Areas under cart rack have been cleaned.	_____	_____	_____
All empty pallets have been removed from the area.	_____	_____	_____

Docks (Inside)

	SATISFACTORY	UNSATISFACTORY	COMMENTS
Floor and plates clean.	_____	_____	_____
Cracks in wall openings.	_____	_____	_____
Doors in good repair, closed, and locked.	_____	_____	_____
Lights operating.	_____	_____	_____
Blacklight available and working in receiving office.	_____	_____	_____
Receiving crews monitor receipts for insect/rodent signs.	_____	_____	_____
Plastic wrap removed at time of receiving.	_____	_____	_____

Additional Comments:

_____

_____

_____

Figure 17-3. (Concluded).

of product that has to be moved by manual handling or by motorized equipment such as forklifts and power pallet jacks. The industry is developing comprehensive training programs to address these risks. The training starts with new employee selection and orientation and continues with job-specific training regarding proper lifting and proper use of motorized equipment. To help control the risk of injuries that occur while lifting material or product, the industry offers many types of training programs in proper lifting techniques. (See also Chapters 14, Materials Handling and Storage, and 17, Powered Industrial Trucks, in the *Engineering & Technology* volume.)

To help control injuries involving forklift and pallet jack operations, the distribution industry has implemented a certification program requiring forklift and pallet jack operators to have initial training and annual refresher training (Figure 17–4). For some businesses, the certification process also includes medical testing to ensure continued fitness to work. However, these medical programs are controversial because their ability to predict employees' physical performance is weak. In addition, these programs may conflict with ADA regulations, i.e., reasonable accommodation of a medical condition.

Fleet Safety Programs

The high costs, both direct and indirect, of vehicle accidents can affect companies in several ways. The company usually has to pay out more than it recovers from any insurance policy, must pay higher insurance premiums, and experiences a negative impact on profit levels. The distribution industry in particular has a high risk of vehicle accidents because of the volume of product being shipped.

To address this accident potential, companies should develop a fleet safety program which includes:

1. driver selection (application review)

   - past employers
   - driver's license

Forklift Training Outline

I. Safety
    A. Daily safety check
    B. Personnel protection
    C. Starting and stopping
    D. Travel
    E. Loading
    F. Accidents
    G. Parking
II. Forklift Responsibilities
    A. Letdowns
    B. Incoming freight
    C. Prevents
    D. As you work
    E. Pallet handling and sorting
    F. Shift-end
III. Operational Procedures
    A. Common practices
    B. Grocery specific practices
    C. Perishable specific practices
    D. Freezer specific practices
IV. Telxon Procedures
    A. General
    B. Putaways
    C. Transfers
    D. Date rotation
V. Reserve practices
    A. Receiving
    B. Putaways
    C. Single selection bays
    D. Drive-in bays
VI. Housekeeping
    A. Common practices
    B. Grocery specific practices
    C. Perishable specific practices
    D. Freezer specific practices
VII. Productivity
    A. Service
    B. Time management
    C. Commitment to continuous improvement

Figure 17-4. Topics that should be covered in a forklift training program.

- driving experience
- prior accidents
- prior traffic violations

2. motor vehicle reports (MVRs)

- MVRs obtained on all applicants prior to hiring
- MVRs obtained annually on existing drivers
- objective standards set and used for the analysis of MVRs, both preplacement and annually
- new drivers' preplacement physical examinations that incorporate DOT-required parameters

3. driving test

- determination of applicant's ability to drive (DOT 391.31)

- road tests with a preplanned route of ten miles or more requiring a minimum of 30 minutes

4. documented orientation and training

- steps to take in the event of an accident (company-specific forms and checklists should be available in all company vehicles)
- state and federal regulations involved in fleet operations
- expected safe driving practices
- proper care and operation of the vehicle
- defensive driver training classes
- inspection responsibilities (pre-trip and post-trip)
- safe cab entrance and exit procedures
- use of seat belts
- procedures for ceasing operations
- emergency response for tank rupture or fuel leak
- written fleet safety manual reviewed with and provided to new drivers

5. follow-up training

- annual safety-related training for all drivers
- corrective action, including training, given after a chargeable accident

6. accident reporting

- established procedures for all drivers to follow at the scene of an accident in which they may be involved
- investigation of all serious accidents and a detailed analysis completed by the safety manager and the transportation manager at the scene of the accident
- review of all accidents by accident review committee to determine the underlying and immediate causes of the accident

7. vehicle inspections

- daily documentation of all pre- and/or post-trip inspection reports

This comprehensive fleet safety program can help companies prevent serious injuries to workers, nonemployee pedestrians, and nonemployee drivers and reduce costs associated with accidents. (See also Chapter 18, Transportation Safety Programs.)

Ergonomics

Ergonomics is the study of people at work and the use of various methods to create a "good fit" between the employee and the job. The goal of a good ergonomics program is to prevent and minimize accidents and illnesses due to repetitive physical and psychological stresses on the body. (See also Chapter 13, Ergonomics Programs.)

The service industry, especially the distribution and warehousing section, has considerable potential for ergonomics-related problems. Many companies are beginning to develop a variety of ergonomics programs to safeguard worker health and to contribute to a more efficient operation. For the program to succeed, it must have strong support from senior management. In addition, an in-house ergonomics team should be established and include supervisory personnel and personnel from engineering, human resources, operations, purchasing, health and safety, and appropriate employees. Ergonomic programs need to be site-specific and include at least four basic activities:

1. *Identify the Problems.* Management must know which jobs require the most study and which type of injury patterns occur most frequently on those jobs.
2. *Study the Physical Demands of the High-Priority Jobs.* The company can develop checklists to help identify specific activities that give rise to particular injury patterns.
3. *Formulate a Written Action Plan.* The plan should include a description of the action to be taken and a realistic completion date.
4. *Maintain the Effort.* This goal can be accomplished through ongoing periodic reviews of job demands and through obtaining management and employee comments regarding emerging problems, changes in processes, or other ergonomic issues that need to be addressed.

Emergency Preparedness

The two essential steps in preparing for and successfully managing emergencies are creating a plan and practicing the procedures. Many retail, service, and warehouse facilities have substantial inventories. Because of the potential impact of wholesale losses due to sudden, unexpected problems, it is important for companies to be adequately prepared and to practice for a variety of emergencies. Fire, earthquake, or other emergency drills are an important way to protect employees and property, to better manage crisis situations, and to reduce the overall costs of such disasters. During or after practice drills, employees often raise critical questions that could help save lives and preserve property in the event of a real emergency. The company should document all such exercises and their results.

An emergency management plan should be established and used to prepare for and react to possible disasters. The plan allows authorized personnel to find the right information quickly and to respond to the disaster. The plan can also serve as a training document to acquaint new personnel with the company's procedures for coping with such emergencies.

The geographic location of a facility often determines the types of emergencies that will be included in the plan. It is important that a cross-section of employees conduct a detailed analysis of the company's needs so that an appropriate plan is developed for all potential emergencies.

Contingency plans should be developed for the following potential emergencies:

- security for facilities and inventory
- fires in the workplace or on the grounds
- chemical release spills
- natural disasters
- riots/strikes
- bomb threats
- power failures
- product recalls/tampering
- violence in the workplace
- natural disasters, such as tornadoes, earthquakes, hurricanes, floods, fires, and so on.

(See also Chapter 15, Emergency Preparedness, for more information on developing a plan and conducting drills.)

The emergency preparedness plan should also address post-emergency recovery procedures that facilitate return to normal business operations as soon as possible. Although no recovery plan can cover all aspects of a situation, it should provide the following minimum recovery procedures:

- computer backups to allow for the continuance of operations
- guide for successful recovery of operations after an emergency
- definition of responsibilities of those involved in the recovery
- procedures to ensure continued review and update of the emergency recovery plan

Violence and Crime in the Workplace

In the past few years, violence and crime in the workplace have risen sharply. According to a June 1996 study of violence in the workplace, retail trade and service industries account for more than half of workplace homicides and 85% of nonfatal workplace assaults (NIOSH, p. iv). As noted in the draft document, *Guidelines for Workplace Violence Prevention Programs for Night Retail Establishments* (OSHA, June 1996), employers can be cited for violating the General Duty Clause of the OSHAct if there is a recognized hazard of workplace violence in their establishments and they do nothing to prevent or abate it. (See also Chapters 14, Employee Assistance Programs, and Chapter 15, Emergency Preparedness, for additional discussion of this topic.)

Service and retail employers should recognize the need for procedures to prevent or appropriately respond to violence in the workplace. There are three major types of exposure to workplace violence: violence in the course of a crime, violence by a current or former customer, and violence that is employment related. Workers in the retail and service environment are at risk for all three types of workplace violence.

In the distribution industry the greatest risk is from employment-related violence. This type of violence may be caused by a current or former employee, supervisor, or manager who may be a spouse, a friend, or an acquaintance of an employee. The primary target of employment-related violence is a co-worker, supervisor, or manager. In committing the assault, the individual generally is seeking revenge to settle a personal score or for what he or she perceives as unfair treatment. This could include an unsatisfactory review, some type of disciplinary action, loss of pay or benefits, demotion, or termination.

Studies have shown that many cases of employment-related violence can also involve domestic or romantic disputes. For these reasons, most distribution industries are beginning to formulate a plan and provide specific training to address this issue. Such training includes how to handle personnel issues before they escalate, and what to do when an incident occurs.

Risk Factors

The NIOSH report notes those factors that may increase a worker's risk for a workplace assault:

- contact with the public
- exchange of money
- delivery of passengers, goods, or services
- having a mobile workplace, such as a taxicab or police cruiser
- working with unstable or volatile persons in health care, social services, or criminal justice settings
- working alone or in small numbers
- working late at night or during early morning hours
- working in high-crime areas
- guarding valuable property or possessions
- working in community-based settings (NIOSH, p. 14)

Obviously, employees who have the greatest exposure to violence committed in the course of a crime are those who have face-to-face contact and exchange money with the public. They often work alone or in small numbers, and work late at night. All companies—particularly those whose employees have these risk factors—should develop formal plans and training programs for employees to reduce employee risk from violent assault. These plans should include using methods of crime deterrence whenever possible.

Prevention Strategies

According to "Armed Robbers and Their Crimes," a report based on interviews with convicted armed robbers, robbers are generally not deterred by surveillance cameras, the threat of arrest, police patrols, or bullet-resistant barriers. Therefore, prevention plans that rely on only these strategies may not prevent robberies. The NIOSH report suggests a variety of environmental designs, administrative controls, and behavioral strategies for a comprehensive approach to reducing workplace violence:

- environmental designs

  - using time-controlled drop safes
  - carrying small amounts of cash
  - posting signs and printing notices that limited cash is available
  - implementing cashless transactions using automatic teller account cards or debit cards, where possible
  - physical separation of workers from customers, clients, and the general public
  - making high-risk areas visible to more people (e.g., by clearing window areas)
  - increasing external lighting
  - controlling the number of entrances and exits, escape routes, access to work areas, and hiding places
  - use of numerous security devices (closed-circuit cameras, alarms, two-way mirrors, card-key access systems, panic-bar doors locked from the outside only, trouble lights, geographic locating devices for mobile workplaces)
  - body armor

- administrative controls

  - staffing plans and work practices that prohibit unsupervised movement
  - increasing the number of staff on duty
  - using security guards or receptionists to screen persons entering
  - controlling access to work areas
  - establishing policies and procedures for reporting, and assessing threats to employees
  - training employees on recognition of potential for violence, methods for defusing violent situations, use of security devices and protective equipment, and procedures obtaining medical care and post-trauma care

- behavioral strategies

  - training employees in nonviolent response and conflict resolution
  - establishing hazard recognition and prevention strategies

(See also NIOSH, *Violence in the Workplace: Risk Factors and Prevention Strategies.*)

Theft Prevention

Companies that have direct contact with customers who visit the place of business to purchase a product may be at risk for both product losses and potential injuries due to shoplifting. To reduce this risk, these establishments must implement a theft prevention program that includes a comprehensive policy which all employees are trained to follow consistently. An organization that does not have such a policy or does not consistently follow that policy will significantly increase their liability. The policy should include:

- techniques used by shoplifters
- how to detect a potential shoplifter
- methods of surveillance
- how to apprehend a shoplifter

Employers should emphasize in the policy statement what specific activities are to be followed in case of a robbery. It is important that all employees understand that their main goal should be to avoid violence and to prevent anyone from being injured. The important message which must be understood is "Don't do anything that would jeopardize personal safety. Don't be a hero."

SUMMARY

- Good safety management in the service/retail industry is based on a strong corporate commitment to safety and its investment in time to develop specific guidelines and policies for training employees. This investment pays off through reduced injury rates, reduced workers' compensation problems, reduced regulatory fines, better quality service, and better profit margins.
- Companies can use the principles of total quality management (TQM) to reduce their risk of losses and to enhance productivity. The TQM cycle includes developing a clear policy statement and specific documents, employee training and disaster preparedness, management and employee accountability for action and safety records, internal and external safety audits, and the establishment of safety committees or other oversight groups.
- OSHA regulations that particularly apply to the service industry include the General Duty Clause, posting requirements, reporting and record keeping, means of egress, hazard communication, medical services and first aid, walking and working surfaces, lockout/tagout, electrical installations and equipment, machine guarding, and powered equipment.
- Because the service industry is physically demanding, companies must make sure that employees are physically fit to do their work. They should also provide employees with proper training in safe work practices and procedures.
- Companies in retail and distribution can manage their risks and losses by training employees in proper accident reporting and analysis procedures, introducing motor fleet safety programs, conducting safety audits, training workers in materials handling, introducing ergonomics to the workplace, planning for emergencies and disasters, and ensuring greater security against crime and violence on the job.

REFERENCES

Rosemary Erickson. "Armed robbers and their crimes." Athena Research Corp., 2508 11th Avenue W, Seattle, WA 98119, 1996.

NIOSH. *Violence in the Workplace: Risk Factors and Prevention Strategies.* Current Intelligence Bulletin 57. DHHS NIOSH Publication No. 96-100. NIOSH: U.S. Department of Health and Human Services, June 1996.

REVIEW QUESTIONS

1. The indirect costs of accidents are _____ times the direct costs.
   a. 2
   b. 5 to 20
   c. 2 to 4
   d. 10 to 50
2. List five of the eight components of a sound risk management program.
   a.
   b.
   c.
   d.
   e.
3. An employee manual should not only define the required level of safety, but also _____.
4. Briefly define the general duty clause.
5. What is the difference between the OSHA 200 and the OSHA 101 logs?
6. List five of the nine OSHA requirements for all means of egress.
   a.
   b.
   c.
   d.
   e.
7. Briefly define HAZCOM.
8. List the four main requirements of HAZCOM.
   a.
   b.
   c.
   d.
9. Briefly define tagout procedure.
10. Which of the following machines/equipments are not covered by tagout standards?
    a. Balers
    b. Meat wrappers
    c. Ventilation hoods
    d. Wire buffers
11. Posting emergency phone numbers by company is a sufficient action under OSHA regulation.
    a. True
    b. False
12. All storage and materials should be kept at least

_____ feet from all major electrical equipment.

   a. 3

   b. 5

   c. 10

   d. 15

13. During physical capacity screening, an applicant is considered qualified when their capabilities _____ the job demands.

14. List four of the six topics that an orientation program should cover.

   a.

   b.

   c.

   d.

15. For large corporations, an effective audit program should be conducted _____.

   a. Daily

   b. Weekly

   c. Monthly

   d. Yearly

16. _____ is the area which has the greatest risk for injuries.

17. Briefly define ergonomics.

18. Ergonomics programs should contain which four basic activities?

   a.

   b.

   c.

   d.

19. The leading cause of death in the workplace is _____.

   a. Violence

   b. Chemical exposure

   c. Asbestos exposure

   d. Accidents

20. What are the two major types of exposures to workplace violence?

18
Transportation Safety Programs

Transportation safety programs cover a range of vehicles from company cars to corporate jets, each type with its own safety issues and problems. The fleet equipment discussed in this chapter includes trucks, passenger cars, buses, and motorcycles. (Powered industrial trucks and hand trucks are covered in the *Engineering & Technology* volume in Chapters 14 and 17, Materials Handling and Storage, and Powered Industrial Trucks, respectively. Haulage and off-road equipment are discussed in Chapter 18, Haulage and Off-Road Equipment, in the *Engineering & Technology* volume.) Because so many companies now have corporate airplanes, this chapter also includes a section on safety issues and programs for these vehicles. This chapter covers the following topics:

- cost of vehicle collisions
- basic elements of a vehicle safety program
- safe practices and procedures in the repair and maintenance shop
- U.S. Department of Transportation drug and alcohol testing requirements
- elements of a company air transportation safety program

Many readers of this manual will have overall responsibility for the safety program of their entire company. However, most of them will not consider themselves the fleet safety manager. In fact, many do not even consider the fact that their company has a fleet of vehicles. They often overlook such vehicles as those used by maintenance workers, security personnel, company management, and so on. Yet for many industries, these types of vehicles represent their firm's greatest risk of work-related injuries and deaths and civil liability losses. This chapter cannot address the entire range of vehicles that a company or industry uses—only the highlights of a total fleet safety program. For more details, see the National Safety Council's *Motor Fleet Safety Manual*, 4th edition.

Safe vehicle operation is not the result of chance, but training, skill, planning, and action. Unfortunately, many companies fail to pay enough attention to the safe operation of motor vehicles. The reason for this lapse may be the difficulties of organizing an adequate safety program and of providing good driver and fleet supervision.

The majority of all motor vehicle collisions are caused by driver error or poor operating practices—both factors that can be controlled. Only a small percentage are due to mechanical failure of vehicles or to improper maintenance of equipment. As a result, an organization's vehicle collision prevention efforts should focus primarily on two principal collision factors—driver error and vehicle failure.

Companies can control driver error by implementing a program of driver selection, training, and supervision; vehicle failure can be reduced by a systematic preventive maintenance program. As experience has shown, the unsupervised fleet usually has higher collision costs than the supervised one.

COST OF VEHICLE COLLISIONS

The total cost of a vehicle collision almost always exceeds the amount recovered from the insurance company. Collision control in a large motor vehicle fleet is critical because increased insurance premiums and indirect costs reduce profits. Insurance premiums can fall or rise with the collision frequency and costs.

As discussed in Chapter 7, Accident Investigation, Analysis, and Costs, property damage is only one of the costs resulting from a collision. There also are indirect costs which, as with most work-related accidents, may be several times the direct costs. Typical indirect costs include the following:

- salary paid and loss of service of employees injured in a collision
- added workers' compensation costs resulting from a disabling injury
- vehicle's commercial value while it is out of service, the cost of the replacement vehicle, or rental costs
- cost of supervisory time spent in investigating, reporting, and cleaning up after the collision
- poor customer and public relations resulting from a company vehicle having been involved in a collision
- cost of replacing or retraining an injured employee
- time lost by coworkers while discussing the nature of the collision and the extent of the victims' injuries

Besides appealing to the company's profit motive, the safety and health professional can make the most telling argument for controlling collisions by pointing out the company's moral obligation to act reasonably and responsibly toward its employees and the public. The employer, through the authority to hire, supervise, discipline, and discharge employees, exercises considerable control over their driving performance. Management can require employees to enroll in driver safety education programs and provide proper supervision on the job.

VEHICLE SAFETY PROGRAM

A vehicle safety program should provide for the following:

- a written safety policy, developed, supported, and enforced by management
- a person designated to create and administer the safety program and to advise management
- a driver safety program, including driver selection procedures, driver training, and safety-motivating activities. Proper supervision and implementation are mandatory for success.
- an efficient system for collision investigation, reporting, and analysis; determination and application of

appropriate corrective action; and follow-up procedures to help prevent future collisions
- a vehicle preventive maintenance program

Responsibility

The first requirement of an effective driver safety program is that all employees from the president down to the individual vehicle operator accept responsibility for safe operation of company vehicles. Management must define standards of acceptable driver and vehicle performance and establish criteria and procedures to evaluate and correct job performance to meet these standards. Companies must make sure all employees understand that collision prevention is a requirement of continued employment.

In larger companies, a fleet safety professional is usually responsible for supervising the program for the safe operation of all automotive equipment. However, in a small industrial concern, this function may be assigned part-time to a supervisor or manager. Some of the duties of the person include the following:

- Advise management on collision prevention and safety matters.
- Develop and promote safety activities and work-injury prevention measures throughout the fleet.
- Study and recommend fleet safety programs regarding equipment and facilities, personnel selection and training, and other phases of fleet operation.
- Evaluate driver performance and skills requirements.
- Conduct or arrange for effective safety training and procure or prepare and disseminate safety education material.
- Review collisions to determine their causes and recommend corrective action to management.
- Compile and distribute statistics on collision-cause analysis and experience; identify problem persons, operations, and locations.
- Maintain individual driver safety records and administer the safe-driver award incentive program.

Driver Safety Program

A driver safety program should include the following five basic collision prevention procedures.

- Initiate a driver training program.
- Develop standards to determine ways collisions can be prevented.
- Require immediate reporting of every collision.
- Recommend performance goals to management; compute and publish the fleet collision record.
- Establish competency and skills levels, set objectives, and maintain a collision record for each driver.

The planning and administration of a safety program for motor fleets are described in greater detail in the National Safety Council's *Motor Fleet Safety Manual,* 4th edition (see References).

A motor vehicle collision can be defined as any incident in which the vehicle comes in contact with another vehicle, person, object, or animal, in a way that results in death, any degree of personal injury, or any extent of property damage, regardless of where the incident took place or who was responsible.

This definition includes even minor collisions such as fender scratches. All vehicle collisions, major and minor, are important to the person responsible for vehicle safety, whose major concern is eliminating poor driving habits or careless attitudes. Even minor collisions, not just spectacular ones, provide clues that can help to prevent recurrences.

Collision Reporting Procedures

Drivers should be required to make out a complete collision report on a standard Accident Report form for every incident in which they are involved. The National Safety Council Form Vehicle 1 is a good example of an accident report form (Figure 18-1). (See also the National Safety Council's *Motor Fleet Safety Manual,* 4th edition, for illustrations of other helpful forms.) Make sure drivers know how to fill out accident report forms accurately and thoroughly. Drivers should complete the report at the scene of the collision, if possible, and then send the form to their supervisor no later than their next assigned shift.

Companies can supply drivers with an accident report packet to be carried in their glove compartments. This packet (Figure 18-2) should provide for a memorandum report of the collision (such as National Safety Council Form Vehicle 2), an accident report checklist, pencils, plain paper, Courtesy Cards, and telephone numbers of insurance representatives. Drivers will then have what they need to record all pertinent information at the scene of a collision.

Specially printed Courtesy Cards can help the driver quickly get names and addresses of key witnesses at the collision site (Figure 18-3). Drivers should be told it is important to identify as many witnesses to the collision as possible.

Failure to report a collision, no matter how slight, or falsification of data on an accident report should be cause for disciplinary action against the driver. Management can ensure timely collision reporting by requiring a documented vehicle inspection before and after each trip.

In the case of serious collisions, especially those resulting in a fatality or severe personal injury, a representative of the company—the supervisor, safety director, or claim agent—should conduct an investigation of the collision at the scene as quickly as possible. The purpose of such an investigation is to verify the driver's information on the accident report form and to obtain other data that might prove valuable for collision prevention work or for defense against liability claims.

National Safety Council
GREEN CROSS FOR SAFETY

MOTOR TRANSPORTATION
DRIVER'S ACCIDENT REPORT
READ CAREFULLY – FILL OUT COMPLETELY

FORM VEHICLE 1

(For office use)

File No. _____
☐ Preventable
☐ Not Preventable
☐ Reportable
☐ Not reportable
_____
_____

COMPANY _____

DIVISION _____ ADDRESS _____

ACCIDENT INFORMATION

DATE OF ACCIDENT _____, 19 _____

DAY OF WEEK _____

TIME _____ A.M. P.M.

M O V I N G
☐ Another com'l vehicle
☐ Passenger car
☐ Pedestrian
☐ _____

F I X E D
☐ Building or fixture
☐ Parked vehicle
☐ _____

T Y P E
☐ Head On   ☐ Rear End   ☐ Front End
☐ Sideswipe   ☐ Other (Describe) _____
☐ Right Angle   _____

☐ Non Collision (Describe)
_____
_____

LOCATION

PLACE WHERE ACCIDENT OCCURED:

ADDRESS OR STREET ON WHICH ACCIDENT OCCURED: _____

☐ AT INTERSECTION WITH: _____

CITY/TOWN, STATE _____, _____

☐ NOT AT INTERSECTION: _____ FEET N S E W ☐☐☐☐ OF _____

Nearest Intersecting Street/Road; House Number, or Landmark: Bridge, Milemarker, etc.

POLICE

PRESENT? ☐ YES ☐ NO

Name of Force _____
☐ Local ☐ County ☐ State

Officers Name _____

Badge No. _____

Report No. _____

Tickets/Arrests   Driver   1 2 ☐ ☐

Other _____

DRIVER / PASSENGER / PEDESTRIAN INFORMATION

DRIVER VEHICLE ONE

Driver's Name:	Company I.D. Number:
Driver's Address:	City/State:
Phone No.:	Driver's License Number:
Vehicle Number(s):	Run: / Route:
Date Of Birth: / Employment Date As Driver:	Hours Since Last 8 Hours Off:
Parts Of Vehicle Damaged:	

Was Street Lighted?   ☐ Yes ☐ No
Was Vehicle Lighted? ☐ Yes ☐ No

DRIVER VEHICLE TWO

Driver's Name:	Phone No.:
Driver's Address:	City/State:
Driver's Occupation:	Driver's License Number:
Age:	Insurance Co.:
Make Of Vehicle: / Year: / Type:	Vehicle License No.:
Registered Owner:	
Owner's Address:	
Parts Of Vehicle Damaged:	
Others In Vehicle:	

Was Vehicle Lighted? ☐ Yes ☐ No

PASSENGER / PEDESTRIAN

	Name	Address, City, Zip		Home Phone
PASSENGER ☐ PEDESTRIAN ☐ NO. 1	Passenger In Vehicle # / Date Of Birth / Sex	What Was Pedestrian/Passenger Doing:		
PASSENGER ☐ PEDESTRIAN ☐ NO. 2	Name / Passenger In Vehicle # / Date Of Birth / Sex	Address, City, Zip / What Was Pedestrian/Passenger Doing:		Home Phone

I N J U R I E S		INJURED: YES / NO	DESCRIBE INJURIES	INJURED TAKEN TO/BY
	DRIVER VEHICLE 1			
	DRIVER VEHICLE 2			
	PASSENGER VEH ___			
	PASSENGER VEH ___			
	PEDESTRIAN 1			
	PEDESTRIAN 2			
	OTHER			

Figure 18-1. Driver's Accident Report.

VEHICLES/PEDESTRIANS/PASSENGERS

(CHECK ONE OR MORE FOR EACH DRIVER)—YOU ARE No. 1	PREVIOUS TO ACCIDENT		WHEN FIRST IN DANGER		AT IMPACT	
	No. 1	No. 2	No. 1	No. 2	No. 1	No. 2
Going straight ahead	☐	☐	☐	☐	☐	☐
Slowing	☐	☐	☐	☐	☐	☐
Stopped in traffic	☐	☐	☐	☐	☐	☐
Park or stopped in zone	☐	☐	☐	☐	☐	☐
Backing	☐	☐	☐	☐	☐	☐
Starting	☐	☐	☐	☐	☐	☐
Passing	☐	☐	☐	☐	☐	☐
Being passed	☐	☐	☐	☐	☐	☐
Changing lanes	☐	☐	☐	☐	☐	☐
Turning left	☐	☐	☐	☐	☐	☐
Turning right	☐	☐	☐	☐	☐	☐
Entering zone/Pulling to curb	☐	☐	☐	☐	☐	☐
Leaving zone/Pulling from curb	☐	☐	☐	☐	☐	☐
Other (explain)	☐	☐	☐	☐	☐	☐
YOUR SPEED	____ MPH		____ MPH		____ MPH	
SPEED OF OTHER VEHICLE	____ MPH		____ MPH		____ MPH	
DISTANCE YOUR VEHICLE FROM OTHER VEHICLE	____ FEET		____ FEET			
DID YOU SOUND HORN? ☐ YES ☐ NO HOW FAR AWAY? ____ FEET						
DID YOU APPLY BRAKES? ☐ YES ☐ NO HOW FAR AWAY? ____ FEET						
AFTER IMPACT—VEHICLE MOVED _____ FEET			AFTER IMPACT—OTHER VEHICLE MOVED _____ FEET			

GIVEN CONDITIONS, WHAT WAS SAFE SPEED FOR:

VEH. 1 _____ MPH VEH. 2 _____ MPH

VEHICLE

1	2	
☐	☐	Did not have right-of-way
☐	☐	Following too closely
☐	☐	Failure to signal intentions
☐	☐	Speed too fast for conditions
☐	☐	Disregarded traffic signs or signals
☐	☐	Improper passing
☐	☐	Improper turning
☐	☐	Improper backing
☐	☐	Improper traffic lane
☐	☐	Improper parking
☐	☐	No improper driving
☐	☐	Defective brakes
☐	☐	Defective steering
☐	☐	Defective lights
☐	☐	Defective tires
☐	☐	No defects
☐		_____
	☐	_____
		(Specify other)

PEDESTRIAN

- ☐ Walking with traffic
- ☐ Walking against traffic
- ☐ Coming from behind parked vehicle
- ☐ Crossing at intersection
- ☐ Crossing not at intersection
- ☐ Alighting from a vehicle
- ☐ Working in roadway
- ☐ Playing in roadway
- ☐ _____
  (Specify other)

PASSENGER

- ☐ Boarding vehicle
- ☐ Alighting from vehicle
- ☐ Caught in doors
- ☐ Seated
- ☐ In motion in vehicle
- ☐ Other (describe)

_____
_____
_____
_____

ENVIRONMENTAL CONDITIONS (CHECK ALL THAT APPLY)

WEATHER (check one)
- ☐ CLEAR
- ☐ CLOUDY
- ☐ RAINING
- ☐ SNOWING
- ☐ FOGGY
- ☐ OTHER

SURFACE (check one)
- ☐ DRY
- ☐ WET
- ☐ ICY
- ☐ SNOWY
- ☐ OTHER _____

TRAFFIC CONTROL (check one)
- ☐ STOP SIGN
- ☐ YIELD SIGN
- ☐ TRAFFIC SIGNAL
- ☐ FLAGMAN
- ☐ NO CONTROL
- ☐ OTHER

LIGHT (check one)
- ☐ DAWN
- ☐ DAY
- ☐ DUSK
- ☐ DARK-NO LIGHTS
- ☐ ARTIFICIAL LIGHT
- ☐ OTHER

ROADWAY No. of Lanes
- ☐ DIVIDED ____
- ☐ UNDIVIDED ____
- ☐ ASPHALT ____
- ☐ CONCRETE ____
- ☐ GRAVEL ____
- ☐ OTHER ____

ALIGNMENT (check one)
- ☐ STRAIGHT
- ☐ CURVE
- ☐ BRIDGE
- ☐ INTERSECTION
- ☐ RAMP
- ☐ RAILROAD
- ☐ OVERPASS
- ☐ UNDERPASS
- ☐ LEVEL
- ☐ UPHILL
- ☐ DOWNHILL

WITNESSES

NAME	ADDRESS	PHONE

INDICATE ON THIS DIAGRAM WHAT HAPPENED

Use one of these outlines to sketch the scene of your accident, writing in street or highway names or numbers.

1. Number each vehicle and show direction of travel by arrow:
2. Use solid line to show path before accident; dotted line after accident
3. Show pedestrian by: ———○
4. Show railroad by: ++++++++++++
5. Show distance and direction to landmarks; identify landmarks by name or number.
6. Indicate north by arrow, as: ⬆

INDICATE NORTH BY ARROW

ACCIDENT DESCRIPTION

DRIVER'S ACCOUNT OF ACCIDENT _____

_____
_____
_____
_____
_____
_____
_____
_____
_____

This report is accurate to the best of my knowledge   DRIVER(S) _____   DATE _____

Figure 18-1. (Concluded).

National Safety Council
Form Vehicle 2
Stock No. 229.32

National Safety Council

ACCIDENT REPORT PACKET

KEEP THIS ENVELOPE IN YOUR VEHICLE

OPEN ONLY IN CASE OF ACCIDENT

FIRST —

Stop Immediately **and determine damage. Avoid obstructing traffic if possible.**

Place **emergency flags or flares.**

Aid **the injured and see to it that they receive medical attention as soon as possible.**

Report **accident to local police and your company.**

SECOND —

Get **witnesses to sign courtesy cards.**

Record **information on reverse side of this envelope at scene of accident.**

IMPORTANT —

Make no statement to anyone except:

A. An officer of the law.

B. Your company's representative.

C. Your insurance company's representative. Make no settlements. Do not argue about the accident.

If the accident involves an unattended vehicle or fixed object, take reasonable steps to locate and notify the owner. If the owner cannot be found, leave a notice in a conspicuous place on the vehicle or object, listing your name and address, the company name, and a brief description of the accident. Whenever possible get a witness signed statement.

TURN ENVELOPE — FILL IN INFORMATION

ACCIDENT MEMORANDUM

Date of Accident.................................., 19...... Day of Week.................................. Hour........... a.m. p.m.

☐ CITY ☐ SUBURBAN ☐ RURAL — PLACE WHERE ACCIDENT OCCURRED
County.............................. City, town or township..............................

If accident was outside city limits indicate distance from nearest town. Use two distances and two directions if necessary. ....miles.......... north-south / ....miles.......... east-west of ☐ limits of / ☐ center of .... City or Town

ROAD ON WHICH ACCIDENT OCCURRED.............................. Give name of street or highway number (U.S. or State)

☐ AT ITS INTERSECTION WITH.............................. Name of intersecting street or highway number
OR
☐ NOT AT INTERSECTION (Check and complete one) ....feet.......... north-south / ....feet.......... east-west of.............................. Show nearest intersecting street or highway, house number, curve, bridge, rail crossing, alley, driveway, culvert, milepost, underpass, numbered telephone pole, or other identifying landmark. Show exact distance, using two directions and two distances if necessary.

OTHER DRIVER'S NAME
ADDRESS
CITY STATE DRIVER'S LICENSE No.
OTHER VEHICLE OWNER'S NAME
ADDRESS
CITY STATE VEHICLE LICENSE No.
TYPE VEHICLE MAKE YEAR No.
DAMAGE TO OTHER VEHICLE AND/OR PROPERTY

INJURED PERSONS	Age	Sex	Injuries
Name			
Address			
Name			
Address			
Name			
Address			

POLICE Name Badge No.

BE SURE WITNESS CARDS ARE COLLECTED *PLACE THEM IN THIS FOLDER*

INDICATE ON THIS DIAGRAM WHAT HAPPENED

DRIVER'S SIGNATURE

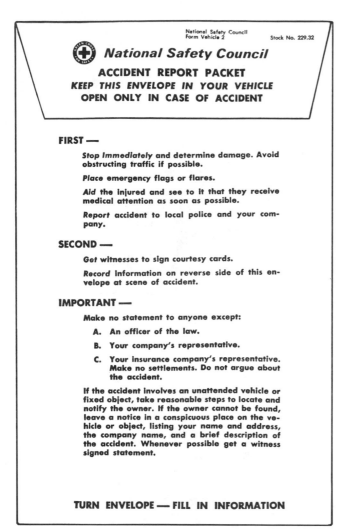

Figure 18-2. This Accident Report Packet is available as NSC Form Vehicle 2. At left is the front side of an envelope that has almost everything a person needs to complete an accurate report and obtain the identification of witnesses.

Interviewing Drivers

As soon as possible after a collision, the investigator should interview the driver involved to determine whether the collision could have been prevented. If this is determined to be the case, based on the interview and written reports, the collision should be classified as preventable. The fleet professional must then explain to the driver what actions contributed to the collision and make sure the driver understands what to do to prevent similar collisions in the future. For example, if the driver's lack of skill contributed to the collision, the safety investigator should recommend remedial driver training.

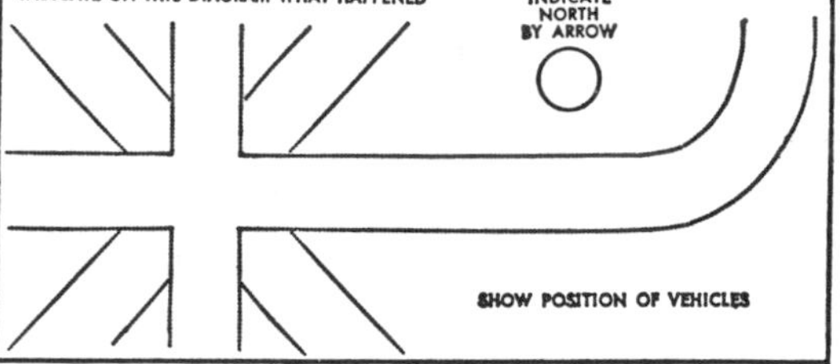

In some cases, the driver may disagree with the decision of the supervisor. For such instances, many firms have established an internal accident review committee to referee contested decisions. Should the driver wish to appeal the employer's determination that a collision was preventable, the National Safety Council maintains an Accident Review Committee that will render an unbiased opinion about the case.

If, on the other hand, the investigator believes the driver did everything possible to prevent the collision, it should be classified as nonpreventable. Responsibility for preventing collisions involves more than carefully observing traffic rules and regulations. Drivers must practice

WILL YOU KINDLY FAVOR THE DRIVER BY FILLING IN THIS CARD?

ACCIDENT AT ______________________ (show street no. or intersection)

DATE ______ Time ______ ☐ A.M. ☐ P.M.

DID YOU SEE THE ACCIDENT HAPPEN? ☐ YES ☐ NO
DID YOU SEE ANYONE HURT? ☐ YES ☐ NO
WERE YOU RIDING IN A VEHICLE INVOLVED? ☐ YES ☐ NO

IN YOUR OPINION WHO WAS RESPONSIBLE?
☐ OUR DRIVER ☐ OTHER DRIVER ☐ PASSENGER ☐ PEDESTRIAN

NAME ______
ADDRESS ______
CITY & STATE ______ TELEPHONE ______

National Safety Council THANK YOU
Rep. 20M 86003 Stock No. 229.38
Form Vehicle 12

Figure 18-3. When this Courtesy Card is distributed, filled in, and collected, it will help determine who saw the accident, in case witnesses are needed later.

"defensive driving" to prevent collisions, that is, they must assume other drivers will do the unexpected and will fail to observe normal traffic laws and safe driving practices.

Driver Record Cards

Management should maintain a record card or computerized record for each employee who drives a company vehicle. This record not only furnishes a history of all collisions a driver has been involved in but also provides information needed for safe driver awards or other forms of safety recognition (Figure 18-4). The date of each collision and the collision category, preventable or nonpreventable, should be entered on this record. The investigator should review the records regularly to determine trends and recurring incidents.

When a driver becomes a collision repeater, management should make every effort to rehabilitate the person through counseling, retraining, closer supervision, or reassignment. When all efforts fail to reduce preventable collisions, the driver should be discharged or assigned to nondriving duties in the best interests of both the firm and the employee.

Fleet Collision Frequency

A useful collision control tool is monthly or quarterly computing of the fleet's collision frequency rate per 1,000,000 vehicle miles driven. Vehicle miles should be computed from odometer readings of all vehicles and not left to rough guesses based on route mileages unless the operations of the fleet are stable from day to day. The standard formula for figuring a fleet accident rate is:

$$\text{Fleet accident frequency rate} = \frac{\text{No. of accidents} \times 1{,}000{,}000}{\text{millions of miles driven}}$$

By keeping a monthly record of frequency rates, the investigator can:

- analyze changes in group safety performance
- compare records from several years to find seasonal or other trends
- compare the fleet's performance with similar fleets

The fleet safety program can be planned based on an analysis of the collision frequency rate, collision causes, and trends. Accident frequency rates also are of interest to

National Safety Council
MOTOR TRANSPORTATION
AWARD AND ACCIDENT RECORD

JAN	FEB	MAR	APR	MAY	JUN	JUL	AUG	SEP	OCT	NOV	DEC
			X								

Name Smith, (Last) Bill (First) E. (Middle)

Company Highway Express

Location Midwest Division

Address 123 (Number) Home St., (Street) Any Town

Badge Number 123 Date Employed Feb. (Month) 1 (Day) 19-- (Year) Age 29 (At Emp.)

SAFE DRIVER AWARD RECORD

Earned 2 Year N. S. C. Award during period from 3-2-19-- to 3-2-19-- with Company Overnight Freight, Inc. Certified by F. J. O'Connell, Supt.

DRIVING TESTS

Date	Score	Remarks
2-4-19--	90	Slow reaction time

Award earned	Certificate Number	Date Award earned	Preventable Accidents: File No	Preventable Accidents: Date	Non-Driving Time: From	Non-Driving Time: To
3	468950	3/2/19--	1016	4/10.19--	12/1/19--	1/1/19--
4	501850	3/2/19--				
5	600100	4/2/19--				

REMARKS: 12-1-19-- took on month leave of absence.

Form Vehicle 6 PRINTED IN U.S.A. STOCK No. 229.36

Figure 18-4. An Award and Accident Record should be kept as part of each driver's personnel record. This form is useful in administering a company award plan and in counseling accident repeaters.

management because they can prove whether the safety program is effective.

The National Fleet Safety Contest, conducted annually by the National Safety Council, issues a bulletin giving the accident frequency rates of all participating fleets, thus providing a means of comparison among similar companies.

Selection of Drivers

For some jobs that require driving, such as sales representative or technician, other qualifications need to be considered besides a safe driving record. However, if the interviewer does not regard driving as an important part of the job, the prospective employee's competence as a driver may not be investigated completely. As a result, when a person is to be hired for a job in which driving is a regular or even occasional function, the company should make every effort to select an individual who can be expected to drive safely. In making that selection, the following factors should be considered.

Experience

Individuals who have a record of frequent vehicle collisions should not be assigned to drive company vehicles. An individual's safety record should be investigated (1) in a personal interview, (2) by consulting with former employers, and (3) by checking the state motor vehicle department or license bureau for accident reports and the National Driver Register service either for out-of-state or for two or more in-state license violations. A number of private services can provide motor vehicle record information electronically via computer within 24 hours.

Personal Traits

Researchers have reported a close relationship between one's ability to drive safely and such personal traits as dependability, good judgment, courtesy, pleasant personality, and the ability to get along with other people. Conversely, those who tend to be antisocial, argumentative, and impulsive often prove to be problem drivers. Likewise, dissatisfied, timid, cocky, troublesome, or otherwise temperamental or unstable individuals often are not consistently good drivers.

Selection Standards

In determining standards of driver selection, management should make a careful analysis of the particular job and of pertinent driver qualifications. The next step is to decide what qualifications the applicant must have to perform the job effectively. Management must be able to justify each qualification imposed. It is helpful to study the employees who are performing the job in an average or better-than-average manner. Their skills and abilities should serve as the basis for what is expected of new employees. In addition, in the United States interstate and hazardous materials carriers must comply with the driver qualification

regulations of the U.S. Department of Transportation (DOT), which include drug and alcohol testing.

Canadian provinces are encouraged to adopt the National Safety Code. Canadian carriers are required to comply with the provincial regulations adopted from the National Safety Code.

Management should always rank a candidate's safe driving skills as critically important on the job. Otherwise, collision losses might completely offset any advantages gained through a driver's other special abilities (i.e., sales abilities, and so on).

Information-Gathering Techniques

After essential job tasks and employee qualifications have been determined, management's next step is to develop methods for gathering and evaluating data about each applicant. Standard employment procedures include:

- application forms
- personal references
- interviews
- psychological tests
- driving tests
- motor vehicle records
- physical examination

Illustrations of the forms and more descriptive detail can be found in the National Safety Council's *Motor Fleet Safety Manual,* 4th edition (see References).

The application form is used to record details of an individual's past employment and other personal data to help the interviewer determine the best candidate for the job. Because of the Right to Privacy Act, no personal information can be requested on the written application or during the interview that does not have a direct bearing on a position's occupational requirements.

Questions on the form should cover the basic qualifications for the job and should be arranged in logical sequence. Specifically, questions about driving experience should include mileage, number of years applicants have been drivers, and types of vehicles they operated; seasons of the year and geographical areas in which candidates operated vehicles; preventable and nonpreventable collisions they experienced; number of convictions received for traffic violations; number and type of driver's licenses held; and safe driver awards they received. The Office of Motor Carriers requires such specific information on the application form; other regulatory agencies may have other requirements.

Personal References

The applicant should furnish the names and addresses of previous employers in the space provided on the application form; the interviewer should check these references. The interviewer can also look up an applicant's record with the appropriate state motor vehicle department.

Organizations should consult their legal advisor to guard against any liability resulting from personnel investigation activities or any other aspects of the hiring process.

A properly conducted interview is intended to reveal additional facts about the applicant's employment experience, knowledge of traffic regulations, attitude, personality, appearance, family life, and general background in order to help the interviewer make a good decision. The interviewer should conduct the session in private and make sure the applicant is seated and put completely at ease. The interviewer should keep in mind at all times the inventory of basic job qualifications. A written checklist of these can serve as a guide. After the interview, the applicant can be evaluated for each of the qualifications listed.

Driving Tests

Each applicant should undergo an actual driving test or a written examination on traffic regulations as part of the selection procedure. Some firms use a written examination to evaluate driving ability. After an individual is hired, the examination can again be used to monitor his or her developing skills when undergoing an extensive driver training course.

There are two types of driving tests: driving range (off the road) and in traffic (on the road). The requirements of a good driving test in traffic can be listed as follows:

- The test should be designed to sample a number of typical driving situations.
- The test should include typical maneuvers to test a driver's ability.
- The test should use a standard scoring procedure and predetermined test route so that all drivers are evaluated equally.
- The examiner should have an objective checklist concerning the driver's performance (to reduce subjective judgment) and determine if driving errors may be corrected by proper training.

Driver Performance Measurement Test

Michigan State University has developed the Driver Performance Measurement Test. This test can help assess safe driving ability during selection by pointing out the precise unsafe habits likely to result in collisions. Information can be obtained from Michigan State University, Lifelong Education Highway Traffic Safety Program, telephone number 517/355-3270.

The Michigan State University (MSU) system requires the preparation of a thoroughly calibrated test course on public roads. As many as 50 traffic situations may be charted and correct/incorrect driving procedures determined for each. After the course is validated by MSU, nearby companies can use it to test their drivers, once individual company observers have been trained and certified into the MSU system.

Although the driving range test course obviously requires considerable organization and test time, it does give a standardized measure of driving performance.

Some companies use a truck "rodeo" type of driving course, where maneuvering skill can be measured. Skills may be important to measure in a city driver who, for instance, may be backing into tight alleys and loading docks.

In some trucking operations, companies may test job applicants' skills in nondriving tasks, including required paperwork and reports. Experienced tank truck driver applicants may be asked to hook up hoses or load trucks. A driver claiming experience with hauling doubles (trailers) may be required to hook up a set of trailers. The applicant's driving test performance will indicate if driver training is needed and if weaknesses can be corrected during the training period.

Tests for interstate and hazardous materials drivers must meet DOT federal requirements. The DOT has developed regulations covering driver licensing, drug testing, training, physical examinations, and other qualifications. Depending on the type of motor vehicle operated, organizations must comply with all or some of these requirements. As mentioned earlier, Canadian operations need to check the Provincial Regulations adopted from the National Safety Code.

Medical Examination

All applicants should be examined by a physician who is familiar with the various governmental requirements before being hired. For U.S. firms engaged in interstate commerce or hazardous materials transport, medical examination of all new drivers is mandatory; and drivers must be reexamined at least every two years.

Drivers should also meet certain psychophysical, vision, depth-perception, and hearing qualifications. Drivers should be advised of substandard results and the best way to compensate for them.

Acceptance Interview

Management should contact all applicants who pass the various requirements and tell them whether they have been hired or placed on a waiting list. The interviewer should reinforce applicants' enthusiasm for the job. Applicants should feel that although company employment standards are rigorous, they have attributes the company sincerely wants. When applicants are accepted, the interviewer and others should welcome them into the company as valued new employees. Their new boss or supervisor should give further orientation about the company, what training they will receive, when and where to report for work, and an explanation of wages and working conditions.

Legal and Social Restrictions

The employer must abide by certain state and federal hiring requirements. Laws forbid discrimination against any job applicant on the basis of characteristics such as sex, creed, race, ethnic background, disability, and so on. These are not unreasonable restrictions and certainly should not force the employer to hire the unqualified. No applicant should be hired for a position for which he or she is not qualified. Such an individual runs the risk of being involved in collisions, incurring possible injuries, and creating an unfavorable work record.

Driver Training

Companies should strive to provide individual training by a skilled instructor to all employees assigned to drive company motor vehicles. Various types of training can be offered.

- basic—for new employees
- remedial—for drivers who get into trouble
- refresher—for periodic updating of all drivers
- special—for operators of specialized equipment

The company must plan a course to fit each job, assemble appropriate training materials, and provide training facilities. The driver training course should cover the following points.

State and Federal Driving Rules

Most state and provincial motor vehicle departments publish the rules and regulations for those seeking driver's licenses. The training course should cover the salient points found in such booklets. Employees subject to U.S. licensing requirements and U.S. DOT examinations should attend appropriate training courses. The same is true of employees of Canadian operations as well as those of other countries.

Company Driving Rules

Each company has rules governing the use of company vehicles: for instance, how the vehicles may be obtained, who may drive them, where they may be operated, where parked and under what conditions, speed to be observed, and so on. Instructors should cover these rules thoroughly in the training course and give drivers written guidelines to take with them on the job.

What to Do in Case of a Collision

This topic should cover how to prepare company accident report forms, how individual driver records affect the employee, what to do in case of a vehicle accident away from the facility, mandatory reporting requirements, and so on.

Defensive Driving

This concept embraces all the common-sense rules of safe and courteous driving. Defensive driving instruction seeks to impart to prospective drivers a strong sense of responsibility not only for their safety, but for the safety of others

who are less skilled and who have had less training and practice (Figure 18-5). The company's safe driver award or incentive plan should be explained, along with policies for disciplinary action and criteria for determining preventable collisions.

Safe Driving Program

Most fleet professionals regard it as part of their job to provide effective safety motivation so drivers will use more of their driving skills more of the time. To provide this motivation directly or indirectly, the safety program should include:

- a required detailed report of every collision
- driver interviews after each collision to determine whether it could have been prevented
- a record of each driver's safety performance
- continuous safety instruction and reminders—through the use of all media: company newsletters, bulletins, booklets, posters, bulletin board displays, meetings, and direct conversation
- safe driving performance recognition

A cautionary word about safe driver awards and prizes. Although these awards and cash or merchandise prizes for safe driving are strong motivators, company management must understand that a recognition program supports a safety program, and is not a substitute for it.

Safety Devices

Some motor vehicle collisions are due to lack of safety devices and/or inadequate vehicle maintenance. Therefore,

Figure 18-5. Training sessions in the National Safety Council "Defensive Driving Course" use audiovisual materials to aid retention of good driving tips.

one of the fundamental requirements for safe operation is that all vehicles be equipped with the necessary safety devices. Tires, brakes, and steering mechanisms, headlights, taillights, horn, and safety restraints must be maintained in first-class condition and used when required (Figures 18-6a and 18-6b). Additional safety devices include the following:

- directional signals
- windshield wipers
- windshield defroster
- fire extinguisher
- power steering/brakes
- low air-pressure warning system
- rock guards over the drive tires/mud guards on other tires
- adequate outside mirrors
- backup lights/required warning lights

DRIVER'S VEHICLE INSPECTION REPORT

AS REQUIRED BY THE D.O.T. FEDERAL MOTOR CARRIER SAFETY REGULATIONS

CARRIER: ____________________

DATE __________ TIME: __________ A.M. __________ P.M.

Check any defective item and give details under "Remarks"

TRACTOR NO. ____________________

☐ Air Compressor	☐ Front Axle	☐ Springs
☐ Battery	☐ Fuel Tanks	☐ Steering
☐ Body	☐ Heater	☐ Tachograph
☐ Brake Accessories	☐ Horn	☐ Tires
☐ Brakes	☐ Lights	☐ Transmission
☐ Clutch	☐ Loses Water	☐ Wheels
☐ Defroster	☐ Mirrors	☐ Windows
☐ Door Handles	☐ Oil Pressure	☐ Windshield Wiper
☐ Drive Line	☐ Radiator	
☐ Engine	☐ Rear End	☐ OTHER
☐ Fifth Wheel	☐ Safety Equipment	

TRAILER(S) NO.(S) ____________________

☐ Brake Connections	☐ Roof
☐ Brakes	☐ Springs
☐ Coupling Chains	☐ Tarpaulin
☐ Coupling (King) Pin	☐ Tires
☐ Doors	☐ Wheels
☐ Hitch	
☐ Landing Gear	☐ OTHER
☐ Lights	

☐ CONDITION OF THE ABOVE VEHICLES IS SATISFACTORY

REMARKS: ____________________

Driver's Signature ____________________

☐ Above defects corrected ☐ Above defects need not be corrected for safe operation of vehicle

Mechanic's Signature ____________________

Driver reviewing repairs ____________________ (Signature)

Figure 18-6a. Although maintenance personnel are responsible for giving the driver a vehicle that is in top mechanical condition, the driver must inspect the vehicle at the start of each day. Using a checklist ensures that no point is forgotten. (Courtesy of JJ Keller & Associates, Inc., Neenah, WI. Used with permission.)

- audible backup signal for trucks/power equipment
- slip-resistant surfacing on fenders, floors, and steps
- safety belts/restraint/protection devices/air bags
- tires/chains (when required)
- automatic sander (buses)
- antijackknife device (where required)
- reflective markings

In addition, the following devices are recommended for dump trucks:

- light or indicator to show when the body is raised
- "CAUTION" sign on the rear of packer-loader trucks
- cab protector or canopy
- built-in body prop

Certain hazardous materials and cargo require placarding and special precautions when they are loaded and unloaded. The U.S. Department of Transportation (DOT), Environmental Protection Agency (EPA), and/or other regulatory requirements must be followed where applicable.

Injuries incurred from accidents or collisions during the loading and unloading of materials, such as lumber, pipe, equipment, and supplies, are especially numerous but can be avoided if these precautions are followed:

- The bulk and weight capacity of the truck should be observed.
- Loads that may shift should be blocked or lashed. Tiedowns (ropes, chain, boomers) should be tightened on the right side or top of the load.
- If material extends beyond the end of the tailgate, a red flag (or, at night, a red lamp) should be fastened to the end of the material. No material should extend over the sides.
- Before loading or unloading a truck, the brakes must be securely set and the wheels blocked to protect the workers both on the truck and on the ground (Figure 18-7).
- To reduce the danger to the driver from falling material while the truck is being loaded or unloaded, the truck should be spotted so the load does not swing over the cab or seat.
- If a truck cannot be so located and does not have a protective canopy over the cab, the driver should dismount and stand clear of the truck.
- A truck should not be moved until all workers are either off the truck or properly seated on seats provided and are protected from injury if the load should shift during transit.
- To avoid falling when unloading a flatbed truck, employees should keep away from the sides of the truck, especially when shoes, floors, and loads are wet or muddy.
- Workers must be alert for pinch points when loads are being pulled, hauled, or lifted.

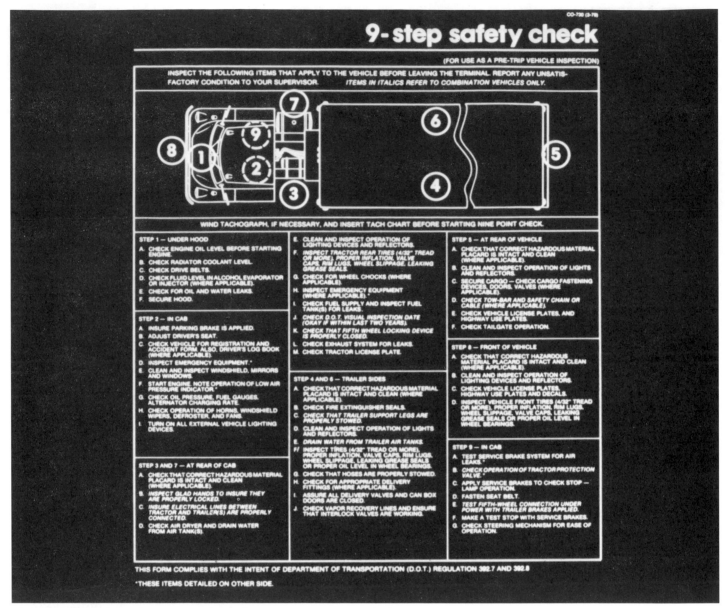

Figure 18-6b. This truck inspection chart will ensure no items are overlooked.

- All safe practices for materials handling, such as using mechanical handling equipment, getting sufficient help, and so on, should be observed (Figure 18-8; see also Chapter 14, Materials Handling and Storage, and Chapter 18, Haulage and Off-Road Equipment, in the *Engineering & Technology* volume).
- Specialized training depending on the classification of the cargo (for example, flammable, corrosive, radioactive) should be provided, as appropriate.
- When loading and unloading detached trailers with a lift truck, be sure the wheels are properly blocked/checked. It is important to place the chock properly when trailers are being loaded and unloaded at the dock. Preferred location of chocks is at the rear set of wheels. If a trailer is not properly blocked, the vehicle may move because of an incline or be set in

motion by the loading or unloading operation (Figure 18-7).
- Under a heavy forklift load, landing gears have collapsed from the weight because of dolly metal rust or fatigue, defective struts, or some other cause. If inspection indicates this possibility, the trailer nose can be supported by screw or hydraulic jacks—one on each side of the nose—to provide additional support for the landing gear assembly (Figure 18-8).

Preventive Maintenance

A well-managed motorized equipment program, whether highway or off the road, includes a properly designed and implemented preventive maintenance (PM) program. Such a program, based on either the mileage or the operating hours of the equipment (as recommended by the

Figure 18-7. This 8 in. (20 cm) high heavy-duty cast aluminum chock provides secure chocking for trucks or trailers. (Courtesy Worden Safety Products, Inc.)

manufacturer or company procedures), determines when the employer should change the oil, rotate or replace tires, and undertake minor and major routine maintenance. The objectives of such a program are to:

- prevent collisions and delays
- minimize the number of vehicles down for repair
- stabilize the work load of the maintenance department
- save money by preventing excessive wear and breakdown of equipment and unscheduled downtime

The PM program should cover all mechanical factors relating to safe operation of motorized equipment, such as brakes, headlights, rear and stop lights, turn signals, tires, wheel rims, coupling devices, emergency equipment, windshield wipers, muffler and exhaust systems, steering mechanisms, glass fixtures, horns, and rearview mirrors. The manufacturer of the equipment and lubricant supplier can help with maintenance specifications.

Drivers and operators can play an important role in a PM program if they are properly instructed and motivated. Because they are most familiar with the vehicle and how it normally operates, they are usually the first to notice when minor or major mechanical defects develop. If possible, each driver or operator should be assigned a specific vehicle and be given responsibility for reporting defects. This encourages drivers and operators to take better care of their vehicles.

Drivers regulated by U.S. DOT must perform a pre-trip and post-trip vehicle inspection. However, all drivers should inspect their vehicles thoroughly before beginning the workday. Any major or safety-related defects must be reported and corrected before that vehicle is used. Minor items not affecting the vehicle's operation can be corrected at the next scheduled PM.

The company should provide vehicle condition report forms to all drivers and operators for the inspection, and

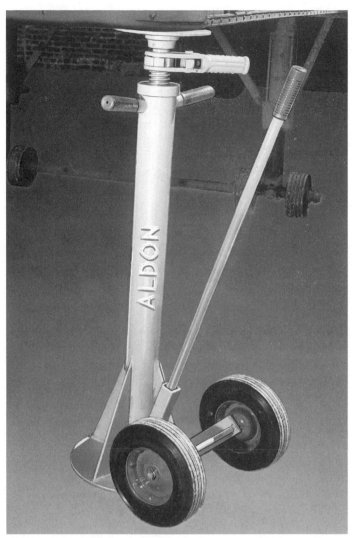

Figure 18-8. Properly placed trailer stabilizing jacks will help prevent trailer tipping or falling. (Courtesy The Aldon Company.)

a copy of the form should be kept in the vehicle cab. The form should cover the following items:

- Brakes should apply evenly to all wheels to prevent a vehicle from swerving when coming to a stop. This even pressure also gives maximum braking effectiveness.
- Headlights should function and be properly aimed to avoid blinding other motorists and to provide maximum lighting of the road ahead and to each side. The dimming switch and upper and lower beams should work properly at all times.
- Connecting cables on a combination vehicle should be fastened securely enough to be unaffected by vehicle vibration. All other cables, such as brake and electric, should be free of defects.
- Make sure all stop lights, turn lights, rear lights, warning lights, and side-marker lights function properly.

- Tires should be inflated to recommended pressure, and regularly checked for adequate tread and any cuts, breaks, or other defects. Dual tires should be well matched.
- Windshield wipers must wipe clean and not streak or skip any of the glass, which can obscure or blur a driver's view of the road.
- Steering wheels should be free from excessive play. Front wheels should be properly aligned.
- All window glass should be free from cracks, discoloration, dirt, or unauthorized stickers that might obscure vision.
- Horns should respond to a light touch and be easy to locate.
- Rearview mirrors should give the driver a clear view. Portions of outside rearview mirrors can be conventional and convex to provide maximum sight advantage.
- Any stalling or lugging problems of the engine should be investigated and corrected immediately.
- All instruments must work properly.
- Exhaust systems should be checked to protect the driver against carbon monoxide gas leaks. The exhaust manifold, pipe connections, and muffler should be inspected periodically.

Emergency equipment in every vehicle must include a fire extinguisher, essential tools for road repairs, spare bulbs, flares, reflectors, flags, and such other equipment deemed necessary in case of fire, collision, or road breakdown. These items should be checked regularly to ensure they are in good working order and are readily available. Interstate and hazardous materials vehicles must be equipped with emergency items required by the appropriate regulatory agency.

Many governmental agencies require periodic safety tests and inspections for all vehicles. The maintenance facility must know the applicable inspection standards, and the company should see to it that all vehicles meet these requirements.

REPAIR SHOP SAFE PRACTICES

The vehicle maintenance facility supervisors must constantly evaluate the work habits of automotive repair employees, determine their safety attitudes, and cooperate with management to create a safe work environment. An effective safety program can help to ensure that all employees engage in safe work practices and that all federal and state or provincial regulations are followed.

Servicing and Maintaining Equipment

Workers may suffer a wide range of injuries while servicing and maintaining trucks. The equipment often requires mechanical aids for handling heavy parts.

Serious injuries are likely to occur when equipment undergoing repair shifts or moves unexpectedly. Workers should set equipment brakes and block the wheels. If work must be done under a raised vehicle, the vehicle must be propped up in such a way that it cannot come down should someone or something inadvertently strike the hoist or jack control levers or pedals and release the load.

Workers often use a jack to raise equipment and then leave it in place as an unstable support. Because serious injuries may occur when a vehicle falls off a jack, it is important that workers set the jack on a firm foundation and make sure it is exactly perpendicular to the load. When the vehicle has been raised to the desired height, it should be supported by stanchions, blocking, or other secure support (Figure 18-8).

Employees should never work near the engine fan or other exposed moving parts until the engine has been stopped. If they must run the engine to inspect or check on moving parts, they should keep a safe distance away and not attempt to make an adjustment. Workers should wear close-fitting unframed clothing, safety shoes, and goggles, but remove jewelry, especially rings, while repairing equipment.

Employees risk burn injuries when servicing vehicle engines that have been running. To open a radiator, the employee should use a heavy work glove, bleed off any steam, and then remove the cap. Gasoline or alcohol spilled on hot engines can cause a serious fire. Suitable funnels and safety containers should be used when transferring liquids.

Tire Operations

Workers face a particularly serious hazard when inflating vehicle (i.e., truck) tires should the locking ring blow off at high pressure (Figure 18-9). Using a tire safety rack will greatly reduce the potential for injury. Workers need to check all tires to make sure the valve has been removed and the tire fully deflated before disassembling it. When removing a tire from a dual wheel, fully deflate both tires before removing the one to be repaired. Workers must inflate tires in steel cages that will restrain flying objects should a blowout occur. A locking ring must be seated properly and must not be yanked free by being twisted. Always replace a defective locking ring or rim with a new one and keep ring and rim seats clean. Carefully separate parts, rims, and rings for various types of wheels to eliminate a possible mismatch.

Workers should use inflators they can preset, and that have locking attachments so hands or arms are not in the danger area, even if the tire is in a cage. Blowouts can occur because of overinflation, improper placement of the tire on the rim or wheel (which may pinch or chafe the tire or tube), or improper mounting of lock-rings or rims. Only employees trained in tire repair and thoroughly familiar with the hazards and safe methods involved in handling tire equipment should inspect, install, repair, and replace tires and rims.

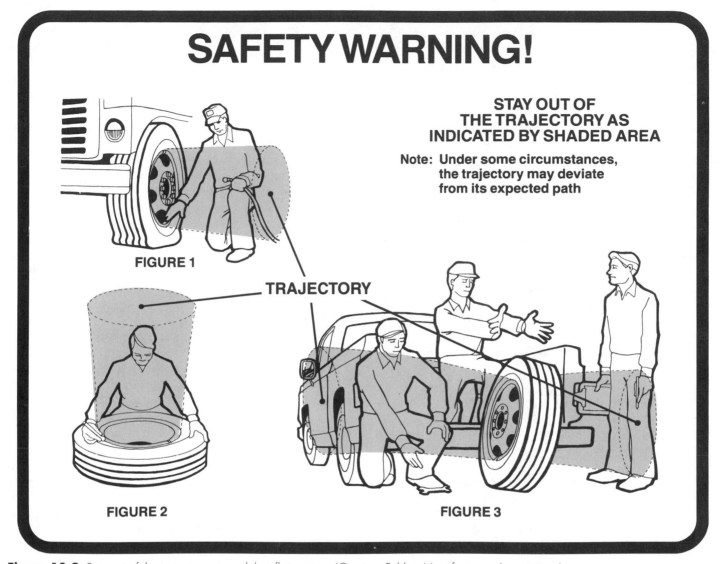

Figure 18-9. Stay out of the trajectory area while inflating tires. (Courtesy Rubber Manufacturers Association.)

Other potentials for injury include strains or hernias resulting from lifting heavy tire assemblies. Companies should provide mechanical lifting and moving devices so workers are not required to lift heavy tires.

Casing compounds used for filling tire cuts and rubber cement and flammable solvents used for patching inner tubes should be kept in safety cans. Routinely inspect electric heating elements used for vulcanizing or branding tires and replace any defective wiring.

Where power-driven rasps or scrapers are used for casings or inner tubes, the operators must wear eye protection and an approved particulate filtering respirator. Attaching a local exhaust or vacuum system to these machines will keep the fine rubber dust out of the workroom air.

Fire Protection

Because of fire hazards, heavy-duty trucks should be equipped with Type B-C fire extinguishers listed by the Underwriters Laboratories for use on burning oil, gasoline, grease, and electrical equipment. Place the extinguisher in a convenient location in the cab or mount it on the vehicle, and train drivers to operate it. Some regulatory agencies require a monthly inspection of fire extinguishers.

The maintenance facility should also have an adequate number of fire extinguishers of the ABC type. Supervisors must show all employees how to use the equipment.

If workers perform cutting, burning, or welding tasks near fuel oil tanks, they should have an extinguisher on hand. A wet tarpaulin may be used to cover fuel, oil tanks, or combustible materials to protect them against sparks and excessive heat. Do not perform such work on a fuel tank or other fuel container until it has been drained and thoroughly purged of vapors. Hydrocarbon vapor meters, properly calibrated and adjusted, will determine vapor levels. Use of the "hot work" permit system is recommended.

Fueling requires certain precautions to avoid fires. Workers must shut off the engine and extinguish all smoking materials. Use safety containers and a grounded fuel hose for fueling. When the tank is being filled, the metal spout of the hose should firmly contact the tank to ensure bonding and to dissipate static charges strong enough to ignite fuel vapors and cause an explosion or fire.

Fires occur in shops each year because gasoline and similar flammable solvents are used for cleaning parts. As a result, use only safe, nontoxic and nonflammable cleaning liquids. Parts cleaning equipment should be provided with a self-closing lid in the event a fire occurs. The risk of a fire in a shop can be reduced further by good housekeeping, especially the proper disposal of oil-soaked combustible waste and similar materials in covered metal containers.

Fire prevention procedures apply as well to all motorized equipment, such as power cranes, shovels, bulldozers, and trucks. Details are covered in the *Engineering & Technology* volume, Chapter 17, Powered Industrial Trucks, and Chapter 18, Haulage and Off-Road Equipment.

Lubrication and Service Operations

In lubrication operations, make sure floors are kept free of grease and oil to prevent slips and falls. Spills should be wiped up immediately or covered with an oil-absorbent compound.

Advise workers to keep their hands away from sharp or rough edges and to obtain immediate first-aid treatment for all cuts and scratches. Also, train workers not to put their hands or face in front of the grease gun nozzle when the handle is pulled. Instances have been reported in which quantities of grease have been forced under the skin of workers by high-pressure grease guns. Tops of grease cylinders should be securely fastened into place; otherwise, covers may blow off and seriously injure anyone who is nearby. Inspect all equipment weekly and make repairs when needed.

Workers should be warned of the danger of inhaling sprayed or atomized oils. They should stand clear of the lubricant spray, which settles quickly, and must not direct the spray at other employees. When draining used oil from machinery, many state and federal regulations require that it be collected for recycling or properly disposed of by an approved waste handling company. Drained antifreeze and other drained lubricants such as car oils must be considered a hazardous waste and should be treated accordingly.

Wash Rack Operations

When washing vehicles, a person can slip and fall on wet floors, suffer cuts or abrasions from the sharp or rough edges of the vehicle, or experience burns from careless use of hot water or steam. The concrete floor of the wash rack should be rough troweled to produce a slip-resistant surface. While washing vehicles, employees should wear safety toe rubber boots, preferably with slip-resistant soles and heels, and a rubber coat or apron and gloves.

Workers should never point the high-velocity streams of hot or cold water at another person because serious injuries can result. They should direct the hose, particularly when washing under the vehicle, in such a way as to avoid being struck by a backlashing stream of water and dirt.

Supervisors should provide workers with product safety information covering the washing solutions used. Where a hot-water or steam hose is used, employees must wear personal protective equipment to avoid skin contact and to prevent burns. Heavy-duty gloves and face shields should be used when necessary. A portable fan can blow steam away to enable the operator to see the work. Supervisors should schedule a periodic cleanup of the entire wash rack and associated equipment.

Battery Charging

Although most battery-operated vehicles now are recharged by means of on-board chargers and plug-in cords, some batteries still require recharging outside the vehicle. Other circumstances occasionally call for battery removal and charging, such as battery servicing, replacement, or emergency jump-starts. The principal hazards of battery charging operations are acid burns, back strains from lifting, electric shocks, slips, falls, and explosions from buildup of hydrogen gas. Management should train workers to observe proper safety procedures whenever they are servicing or charging a battery. They should not charge a frozen battery—most of the charge will go to heat the battery and not to energize it.

Employees should wear safety apparel suitable for battery work. This includes splash-proof eye and face protection and acid-proof gloves, aprons, and boots with slip-resistant soles. Workers should wear rubber boots and aprons when filling batteries. Goggles and face shields should be used when working around batteries to prevent acid burns to the eyes and face.

Flooring should be constructed of wood slats and kept in good condition to prevent slips and falls and to protect against electric shocks from batteries being charged. Install fire doors between battery-charging rooms and adjacent areas where flammable liquids are handled and stored.

Always follow the manufacturer's recommendations about charging rates for various sized batteries to prevent rapid generation of hydrogen. Potentially explosive quantities of oxygen and hydrogen accumulate in battery cells. This is particularly true if the battery is defective or if a heavy charge has been or is being applied. The lower the water level in the battery, the larger the cavity in which gas can accumulate. Care should be taken to prevent electric arcing while batteries are being charged, tested, or handled. Make sure tools and loose metal (and even lifting hoist chains) cannot fall on batteries and cause a short circuit, which, in turn, can inflict serious burns on workers or cause an explosion.

When manual lifting is necessary, tell employees to request sufficient help to prevent strains, sprains, or hernias. Use hand carts for transporting batteries.

Workers must handle acid carboys with special care to prevent breakage and possible injury caused by splashing acid. Acid carboys should never be moved without their protecting boxes nor stored in excessively warm locations or in the direct rays of the sun. Instead, use carboy tilters. A safety shower and an eyewash fountain are required by certain regulatory agencies wherever acids or caustics are used (see Chapter 10, Occupational Health Programs).

First Aid for Chemical Burns
Many batteries contain an acid electrolyte although some, such as the nickel-iron battery, contain an alkali solution. Whether it be acidic or alkaline, if electrolyte gets on a person's skin or eyes, it must be immediately washed off with large quantities of running water for 15 minutes or more. Supervisors and workers should understand that because neutralizing agents are mishandled so often, these agents can do more harm than good. Use them only if first-aid directions are printed clearly on labels or on company or facility directions. Get medical aid for the injured employee at once.

First Aid for Burns of the Eye
Flush the eye thoroughly with large amounts of clean, cool water for 15 minutes or more. Place a sterile dressing over the eye to immobilize the lid and get medical aid at once. Check with the company physician for emergency help.

Gasoline Safety
The handling and storing of gasoline should comply with the provisions of the *National Fire Protection Association Flammable and Combustible Liquids Code,* NFPA 30 and NFPA 30A, which is discussed in Chapter 11, Fire Protection, of the *Engineering & Technology* volume. Never use gasoline for any cleaning tasks. Solvents with higher flash points are equally effective and much safer. Even when higher flash-point solvents are used, if carburetors or gas-line parts are cleaned, the solution should be changed regularly. The admixture of even small quantities of gasoline will lower the solvent's flash point and increase the danger of fire and explosion.

Workers can remove grease, oil, and dirt from metal parts with nonflammable solutions or with high–flash-point solvents in special degreasing tanks with adequate ventilating facilities and fire protection. Advise workers not to use gasoline to remove oil and grease from garage floors. Nonflammable commercial cleaning compounds are available for such tasks.

Workers should never use gasoline to remove oil and grease from their hands. Strong commercial soaps will effectively remove greasy dirt from the skin without risk of injury. In addition, protective creams and ointments, if applied before starting work, will protect an employee's skin from dirt and grease. Likewise, workers should never use gasoline to clean work clothes.

Gasoline spills should be cleaned up immediately; many absorbent substances can be used to soak up spills. Dirt, sand, or other material should be used to prevent gasoline from entering sewers, drains, and so on. If quantities of gasoline get into the sewage system, notify the fire department and state EPA at once so the hazard can be handled properly. Because gasoline vapor is heavier than air, it collects in low spots, such as basements, elevator pits, and sumps. Ventilate these places until gasoline vapors are dissipated to a safe level. Do not dispose of spilled gasoline by flushing it down a sewer drain, sink, toilet, and the like. Instead, collect spilled gasoline into an approved container and dispose of it through a waste-handling company, or other approved means.

Other Safe Practices
Management should also make sure that workers understand and use safe practices for other types of equipment that pose safety and health hazards. Workers should receive training in these practices during orientation as well as on the job whenever equipment is changed or replaced, jobs are rotated, employees return after leaves of absence, and refresher training is needed. More details on mechanical and chemical safety can be found in the *Engineering & Technology* volume and in *Fundamentals of Industrial Hygiene,* 4th edition, respectively.

Using Jacks and Chain Hoists
Vehicles jacked up or hung on chain hoists should always be blocked with stanchions or pyramid jacks (make sure these items are carefully inspected before use). The best jack for maintenance garage work is the hydraulic-over-air type—if one system fails, the other will operate. Do not use ordinary pedestal jacks, especially the type supplied for passenger cars, as the vehicle can be tipped or jarred off the jacks and cause injury.

When someone is working under a blocked-up vehicle, other employees should not do any work on the car that might knock it off the blocks. Employees who work under vehicles should be safeguarded from danger when their legs protrude into passageways. Erect barricades around them or make sure the worker's entire body is under the vehicle. Advise employees to wear suitable eye protection such as safety glasses, goggles, or plastic eye shields to avoid injury from dirt and metal chips falling into their eyes. When necessary, fog-resistant eye protection should be used.

Removing Exhaust Gases
Local exhaust and ventilating equipment can prevent accumulation of vehicle exhaust gases within a repair shop.

Repairing Radiators

Where radiators are boiled out or tested for leaks, the operator should have both chemical goggles and a face shield of clear plastic. The entire face should be protected.

Cleaning Spark Plugs

All mechanics using blast-type spark plug cleaners should wear goggles or face shields.

Replacing Brakes

Workers should not use air pressure when cleaning around brake drums and backing plates while replacing brakes. The pressure can send asbestos, metal particles, or dust from the brake area into the air and cause respiratory problems. Instead, workers can clean brakes with HEPA vacuums, chemical wash solutions, or steam cleaners. Train workers to wear approved respiratory protective equipment until brake components are free of dust. Often the work area may not be free of dust, and respiratory equipment may still be needed.

Controlling Traffic

Supervisors should regulate movement of vehicles inside maintenance facilities. Traffic lanes and parking spaces should be painted on the floors, with arrows and signs indicating the direction of traffic flow. Vehicles with air brakes should not be moved until sufficient air pressure has been built up. Move vehicles using low gear and low speeds inside shop areas, especially up and down ramps.

Drivers must stop their vehicles at entrances or exits and make sure the way is clear or sound their horn before passing through. Signs requiring this procedure should be posted in conspicuous places. Install mirrors at blind corners.

Other potential sources of injury include jumping across open inspection pits, falling off ladders, materials handling, and using hand tools improperly. Prevention of all work injuries requires proper selection and training of employees, careful supervision of their work habits, review of all injury causes, and compliance with safety programs.

Training Repair Shop Personnel

Supervisors and managers should train apprentices and new employees to do each job in the most efficient manner. Make sure job instruction includes the safety rules and regulations pertaining to each job and the reasons for such rules. New mechanics must be thoroughly indoctrinated concerning the company's policy toward safety and the safety program.

The new employee, having been indoctrinated and trained to work safely, must be motivated to observe accepted safe practices while performing job tasks. Many motivational methods are available to accomplish this goal, including safety supervision; safety contests; and meetings, posters, safety bulletins, and pamphlets (see Chapter 22, Motivation, and Chapter 23, Safety and Health Training).

U.S. DOT REQUIRED DRUG AND ALCOHOL TESTING

The Omnibus Transportation Employee Testing Act of 1991 requires alcohol and drug testing of safety-sensitive employees in the motor carrier, railroad, aviation, and mass transit industries. The Department of Transportation published rules mandating antidrug programs and alcohol misuse prevention programs in February 1994. These rules also expand and supplement the U.S. Drug Free Workplace Act of 1988 that mandated drug testing of aviation, interstate motor carrier, railroad, pipeline, and commercial marine employees. The February 1994 rules generally required implementation beginning on January 1, 1995, for employers with 50 or more employees in safety-sensitive jobs, and January 1, 1996, for employers with fewer than 50 employees in such jobs.

The DOT rules include a drug and alcohol testing rule (49 *CFR,* Part 40) that establishes procedures for urine drug testing and breath alcohol testing. The urine drug-testing procedures rule was issued in December 1989 and governs drug-testing programs mandated by the Federal Railroad Administration (FRA), Federal Transit Administration (FTA), Research and Special Programs Administration (RSPA), and the United States Coast Guard. The 1994 amendments to Part 40 add breath alcohol testing procedures and additional urine specimen collection procedures that provide for split urine specimens. The DOT rules cover safety-sensitive employees in commercial transportation as defined by each DOT agency.

Drug Testing and DOT Requirements

Transportation employers, even self-employed individuals who must conduct urine drug-testing programs under regulations issued by the various agencies of the Department of Transportation, must ensure that all testing is conducted in accordance with 49 *CFR,* Part 40. Under the DOT rules, employers are required to test employees for five drugs: marijuana, cocaine, opiates, amphetamines, and phencyclidine (PCP). The rule requires five types of drug tests: preemployment, postaccident, reasonable cause, return-to-duty, and random or unannounced testing. Random testing must be given to a certain portion of the employer's safety-sensitive pool. The random testing works like a lottery, and the percentages of employees to be tested are prescribed by the DOT.

Employers should know that they may expand the testing for other drugs only with DOT agency approval. At present, however, there is no DOT approval to test beyond the panel of five drugs because the Department of Health and Human Services (DHHS) does not have any approved testing protocols and positive thresholds to expand testing. Because testing must occur at a

DHHS-certified laboratory, the uniform standards crucial to the accuracy and integrity of the testing process are not now in place for any additional drugs.

A urine drug testing custody and control form must be filled at the time of a DOT urine collection and accompany all specimens to the laboratory. This form must conform to the requirements of 49 *CFR*, Part 40. Employees shall be provided a wrapped or prepackaged specimen bottle or collection container in which to urinate. The specimen bottle must be sealed with a tamper-proof seal, signed, and dated by the collector.

Collection site personnel must be trained to carry out the required collection procedures or, if they are licensed medical professionals or technicians, they must have instructions for conducting the required collection procedures. Generally, supervisors of employees being tested should not collect the specimen. Collection site personnel must be the same gender as the donor when a collection is conducted under direct observation. When a collection is conducted in a public restroom or other facility that does not afford the donor complete privacy, a medical professional or technician of either gender may collect the specimen; however, if a nonmedical collector is used, the collector must be the same gender as the donor.

Written instructions and/or procedures concerning specimen collection must be provided to collection site personnel, employer representative, and employees. Any specimen, if used, must be transported along with associated paperwork from collection site to a DHHS-certified laboratory in a shipping container which can be sealed with tamper-proof tape and initialed to prevent undetected tampering. The collection site shall arrange to ship the specimen to the laboratory in shipping containers that minimize the likelihood of transport damage and are sealed with tamper-proof tape. The collection site person shall sign and enter the date of shipment on the tape.

Employers subject to 49 *CFR*, Part 40, must use laboratories certified under the Department of Health and Human Services "Mandatory Guidelines for Federal Workplace Drug Testing Programs" 53 *CFR*, 11970, and subsequent amendments. All DOT tests must be performed in a certified laboratory to meet DOT requirements. The laboratories are required to have internal quality control procedures to monitor each step of the drug testing process. At least 10% of all specimens tested in the laboratories must be quality control specimens introduced into the testing process by the lab's quality assurance section. Additionally, employers are required to submit three blind quality control specimens to the laboratory for every 100 employee specimens sent for testing. These quality control specimens are called blind performance tests because they are not known to the laboratory. Blind quality control specimens can be either blanks (negatives) or spikes (positives).

If a false positive performance test error is determined to be a technical or analytical procedural error, the employer will, in addition to notifying the DOT agency concerned, instruct the laboratory to submit to this DOT agency all quality control data from the batch of specimens which included the false positive specimen. Depending on the information provided, DHHS will take appropriate action on the laboratory's certification.

A Medical Review Officer (MRO) shall review all laboratory confirmed positive results. The MRO shall be a licensed physician; have a knowledge of substance abuse disorders; and may be on the employer's staff or be a private physician under contract to the employer. The MRO should not be an employee of the laboratory conducting the drug testing unless there is a clear separation of functions to preclude any conflict of interest. The MRO will review positive results reported by the laboratory, examine alternative medical explanations for any positive test results, and may include a personal interview or physical examination of the employee.

A positive test is not positive until the MRO confirms the test results. The MRO does not discuss the results with the employer or report any findings until his/her verification is completed. Disclosure of any medical information provided by the employee to the MRO as part of the test-verification process is prohibited except under the following circumstances:

- A DOT regulation requires such disclosure.
- The MRO believes the information could result in the employee being found medically unqualified under a DOT agency rule.
- The information indicates that continued performance by the employee could pose a significant safety risk.

Alcohol Testing and DOT Requirements

Because alcohol is a legal substance, the DOT rules define specific prohibited alcohol-related conduct. Performance of safety-sensitive functions is prohibited under the following conditions:

- When an employee has an alcohol concentration of 0.04 or greater as indicated by an alcohol breath test.
- When the employee is using alcohol.
- Within four hours, and eight hours for aviation crew members, after an employee has used alcohol.

In addition, employees are prohibited from refusing to take an alcohol test and from using alcohol within eight hours after a collision or until tested.

Testing procedures that ensure accuracy, reliability, and confidentiality of test results are outlined in 49 *CFR*, Part 40. These procedures include training and proficiency requirements of breath alcohol technicians, quality assurance plans for the breath-testing devices, requirements for a suitable test location, and protection of employee test records.

The DOT requires alcohol testing to be performed for preemployment, postcollision, reasonable suspicion, return-to-duty and follow-up, and random tests. Random alcohol testing must be conducted just before, during, or just after an employee's performance of a safety-sensitive duty. The employee is randomly selected for testing. As with drug testing, the testing dates and times are unannounced and are given with random frequency throughout the year, similar to a lottery. The random rate is set for each industry each year.

The DOT rules require breath testing using evidential breath-testing devices (EBTs) approved by the National Highway Traffic Safety Administration (NHTSA). Two breath tests are required to determine if a person has prohibited alcohol concentration. A screening test is conducted first. Any result less than 0.02 alcohol concentration is considered a negative test. If the alcohol concentration is 0.02 or greater, a second or confirmation test must be conducted. The employee and the individual conducting the breath test (called a breath alcohol technician) complete the DOT-prescribed form to ensure that the results are properly recorded. The confirmation test, if required, must be conducted using an EBT that prints out the results, date and time, a sequential test number, and the name and serial number of the EBT to ensure the reliability of the results. The confirmation test result determines any actions taken by the DOT and/or the company.

Employees who violate the alcohol misuse rules will be referred to a substance abuse professional for evaluation. Any treatment or rehabilitation should be provided in accordance with the employer's policy or labor/management agreements. The employer is not required under these rules to provide rehabilitation, pay for treatment, or to reinstate the employee in his/her safety-sensitive position. Any employer who does decide to return an employee to a safety-sensitive duty must ensure that the employee has been evaluated by a substance abuse professional, has complied with any recommended treatment plan, has taken a return-to-duty alcohol test with a result less than 0.02, and is subject to unannounced follow-up alcohol tests.

Setting up a Testing Program

There are many service companies located across the United States who provide both drug and alcohol testing services. Employers should interview several companies, comparing the pricing and services offered to determine the best provider for their company's needs. In addition to testing, DOT regulations require companies to keep records, offer employee training, and provide employee information on drug and alcohol testing and results.

AIR TRANSPORTATION SAFETY

Today's air transportation industry has demonstrated that it is the safest means of moving passengers and freight.

The National Safety Council's 1994 edition of *Accident Facts* notes that the commercial aviation industry accident rate was 0.0002 per million passenger miles flown while the commuter airlines experienced a rate of 0.008. Compared to the motor vehicle collision rate of 1.83, the rail travel rate of 0.2, and water transportation rate of 0.2, air transportation has a far safer record (Figure 18-10).

Many major airlines and aviation corporations realize the importance of safety and have developed competent, professional corporate departments devoted to aviation safety. These flight safety departments are responsible for creating policies, procedures, and programs that ensure a safe environment for aircrews and provide the world with the safest form of transportation.

Flight Safety Department

Flight safety departments provide expert safety personnel who interact with federal and international flight operation regulatory agencies such as the Federal Aviation Administration (FAA) and the National Transportation Safety Board (NTSB). These departments are responsible for investigating incidents and accidents throughout the corporate system and identifying the causative factors and measures to prevent a reoccurrence.

It is imperative that the aviation industry maintain and expand safety programs that shift investigative activities from cause-analysis investigation to prevention-analysis investigation. Flight safety departments maximize profit for the corporation and present a positive message to customers, employees, and governmental regulators by taking strong action to prevent incidents and accidents. A safe aviation operation will reduce aircraft accidents, avoid fines, and save money through the control of losses and the management of risk.

The major activities and responsibilities described above must be assumed by all aviation corporations' flight safety departments. In most cases, regulatory mandates also include compliance, surveillance, or reporting.

An airline safety department should act as an independent department reporting to the highest level of management. This reporting structure, promoted by the FAA, enables the department to take an unbiased approach when investigating incidents/accidents. The structure also enables the flight safety department to use multidepartmental resources to conduct an accident-prevention analysis, unbiased by any particular departmental affiliation. As a result, the various safety reports and recommendations of a safety department investigation are more likely to be accepted as a true reflection of the corporation's safety concerns and issues.

Data Collection and Analysis

The collection of data from various sources will help flight safety departments research, document, and analyze pertinent summary information relating to safety-of-flight operational events. These data are analyzed, statistically

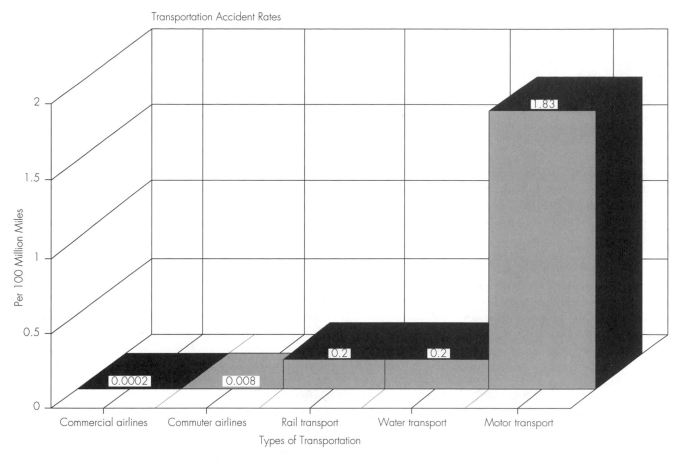

Figure 18-10. Comparative accident rates for various modes of transportation.

and graphically, to identify developing trends that may either contribute to incidents/accidents or help to prevent them.

Data management provides functional managers with information on operational irregularities from various points of view. This activity requires a corporate reporting system that will allow the following:

- integration of all available recording sources
- creation or streamlining of the initial data entry
- providing a comprehensive database capable of accurate, efficient, and timely reporting of all safety events

Creating corporate-wide reporting systems will reduce inefficiencies between various departments throughout the organization. Greater cost savings can be achieved when management at all levels takes responsibility for collecting and handling safety-related data to be entered into the system. The departmental manager will then be responsible for ensuring adequate control or elimination of identified hazards.

Trend analysis is used to provide documentation for industry issues currently under scrutiny, to support operational procedures, and to develop new policies arising from internal concerns and from communication with other aviation corporations. For example, using trend information, various aviation safety departments were able to determine that portable electronic devices created an electromagnetic interference that interrupted navigational and communications equipment.

Information Dissemination

Database trends must be analyzed, interpreted, and used to develop corrective actions. The resulting information then is applied to policy and procedures decisions by various constituents of the aviation industry, with special emphasis in the following departments:

- maintenance and engineering
- flight and flight services
- consumer relations
- insurance

In addition, trend information provides accurate and timely reports to senior management and may be used in documenting the company's procedural compliance and incident recurrence issues.

Safety Education

Companies should establish and implement their own internal safety education programs to help ensure compliance with regulations and federal mandates. Examples of typical programs available off the shelf, although not specific to the airline industry, are those dealing with accident preparedness and bloodborne pathogens.

Airline flight safety departments should also provide employees with safety education outside the company or in conjunction with other interested groups. Programs developed in conjunction with the FAA and the NTSB provide a better understanding of an air carrier's operational requirements. These programs also foster the investigator's awareness of various areas of expertise available in the airline industry as well as the airlines' culture.

Safety departments produce and offer training videos for fire fighting and emergency response personnel to give employees an understanding of an aircraft's specific procedures and to tailor response activities to various aircraft. By providing safety personnel to give periodic speeches and offering audiovisual presentations to these emergency response organizations, the company also promotes a better understanding of its operations and enhances teamwork and goodwill.

Airline safety departments can assist in the training of airport rescue fire-fighting personnel. The departments provide informational briefing cards and videos specific to a certain type of aircraft. This information is especially helpful at fire stations in which personnel are not familiar with the aircraft equipment, such as in rural areas and developing countries. If necessary, the information should be provided in various languages.

Flight safety personnel also work in conjunction with individuals who assist customers or employees in the event of an accident or incident. The safety department should oversee the organization of airport disaster exercises to test compliance with regulations in emergency response. These exercises assist airport stations and corporate headquarters personnel in refining their response plans. Exercises dealing with emergency response should be conducted on a regular basis to develop a well-designed plan.

Accident/Incident Preparedness

Developing effective damage control plans can produce savings through the management of property and monetary losses. However, the airline flight safety department must develop a comprehensive plan of operation *before* a crisis arises. The plan should improve station personnel's preparedness and designate specific responsibilities for responding to a minor or major accident or incident. The degree of preparedness and awareness of response activities must be the same regardless of the severity of the accident. Employees' actions must become automatic, without needing a reference or having to ask basic questions such as "Where do we go?" or "What should we do?"

Aircraft and Systems Engineering

Safety engineering and system safety techniques are critical elements in the identification of potential loss or damage to equipment and facilities. Safety department personnel should also have the educational background and ability to perform engineering system failure analyses of the aircraft, maintenance and flight procedures, and any new technologies introduced into the workplace. System engineering techniques may be used to detect hidden failures in aircraft equipment before purchasing. These techniques may be used to conduct a hazard analysis to detect deficiencies in equipment following its placement into service, and to provide recommended corrective action.

Cabin Safety

Airline flight safety departments are responsible for the implementation of flight, flight service, and field service policies and procedures for on-board cabin equipment and cabin safety issues. These responsibilities also include the design of cabin safety cards to meet federal regulations. In the past several years, public interest and regulatory and congressional action have focused attention on many areas never before considered. Due to the increased requirements for cabin safety improvements, the aviation industry is aware of the conditions that may injure or adversely affect passengers and employees while in flight.

Safety Audits

Various types of safety audits should be conducted to identify noncompliant activities, emphasizing maintenance, cabin safety, flight, and flight training. The safety department should also perform audits of corporations that provide materials and services to the airline. The audits must evaluate the equipment, staffing, work operations, and final products to ensure that operations are in compliance with corporate as well as federal regulations. Flight safety department personnel should also inspect airfields at which the company intends to begin service. These new stations should be inspected for items such as crash and fire rescue teams, runway capabilities, air traffic control systems, and previous occurrences at that station.

Aircraft Accident/Incident Investigation

A key component of the flight safety department's charge is the responsibility for investigating all real and potential aircraft incidents and accidents. The department should be the team leader for the company's accident investigation team unless the investigation is delegated to local management. The safety department, however, will review the results of all investigations. All aircraft accidents must be thoroughly investigated to determine the underlying causes so that corrective action can be recommended and implemented. All records pertaining to an accident or incident are retained by the safety department, which also acts as the coordinator between the airline and the NTSB.

Investigation follow-up is also a major responsibility of the safety department. The flight safety community is switching from a philosophy of placing blame for accidents and incidents to a philosophy of prevention. A key reason for this new approach is the extensive efforts arising from every incident investigation to prevent a recurrence. Safety departments have taken a major step forward concerning this issue by increasing the involvement of responsible departments in the investigation and recommendation process and by documenting investigations through reports.

Because the airline's safety department is the corporate team leader for all major accidents, it provides and maintains an investigation kit. The kit should contain various items used to protect the investigators and aid in the investigation, such as bloodborne pathogen protective clothing and investigative tools such as cameras, notebooks, and satellite telephones. (See also Chapter 7, Accident Investigation, Analysis, and Costs.)

National Transportation Safety Board

The National Transportation Safety Board (NTSB) is charged with the investigation of aircraft accidents, which may range from a private aircraft failing to land safely at a private airport to the loss of an aircraft with many passengers on board.

In the event of a major or highly publicized incident or accident, the NTSB will send a "Go Team" from Washington, DC, to begin an immediate investigation. The Go Team will assemble and arrive at the site as soon as possible, usually within six hours. The closest NTSB field office representative will take charge of the site (after local fire fighters and emergency response personnel have declared it safe) until the Go Team arrives. The Go Team is composed of specialists in the areas of powerplants, systems, structures, operations, air traffic control, weather, survival factors, and human performance. An investigator in charge will be appointed to coordinate all aspects of the investigation. Typically, a board member or NTSB public affairs member will accompany the Go Team to communicate with the news media.

The flight safety department for the company involved in the incident will begin to assemble personnel and equipment to assist in the investigation and to provide technical assistance to the NTSB investigation. Among the department's responsibilities is to notify the NTSB or other agencies immediately of an aircraft accident or incident. Most of the interaction between the airlines and the NTSB originates from mandatory reportable items.

The following highlight is a listing of terms used by the NTSB (and generally accepted worldwide) during the investigative phase.

- *Aircraft Accident.* An occurrence associated with the operation of an aircraft, with the intention of flight, in which any person suffers death or serious injury, or in which the aircraft receives substantial damage.
- *Aircraft Incident.* An occurrence, other than an accident, which affects or could affect the safety of operations associated with the operation of an aircraft.

- *Fatal Injury.* An injury which results in death within 30 days of the accident.
- *Serious Injury.* Any injury which is required under Title 49 *CFR,* Part 830, to be reported to the NTSB. These injuries include those which:
    - require hospitalization for more than 48 hours
    - cause the fracture of any bone (except fingers, toes, or nose)
    - involve any internal organ
    - cause second- or third-degree burns or any burns affecting more than 5% of the body surface
- *Substantial Damage.* Damage or failure that adversely affects the structural strength, performance, or flight characteristics of the aircraft and that would normally require major repair or replacement of the affected component. Single-engine failure or failure limited to that engine, damage to landing gear, wheels, tires, flaps, brakes, or wingtips are not considered "substantial damage."

Legal and Insurance Issues

The flight safety staff also works with the corporate insurance department, hull insurance carrier, and corporate legal department by providing paralegal services in claim defense preparation, plaintiff interrogatory requests, service in technical analysis and support, and expert witness testimony.

Industry Liaison

Liaison with other carrier flight safety departments and the free exchange of timely safety information are critically important to improve and maintain safety within the airline industry. Lessons learned from aircraft accidents and incidents experienced by other airlines may lead to discoveries that help prevent the same incidents from happening in another airline. In addition, flight safety liaisons with trade associations can actively lobby for legislation concerning airline safety activities.

SUMMARY

- Most motor vehicle collisions are caused by using improper driving procedures; only a small percentage are the result of mechanical failure. Companies can control driver error by introducing a program of driver selection, training, and supervision, while vehicle failure can be reduced by implementing a preventive maintenance program.
- The total cost of vehicle accidents almost always exceeds the amount recovered from the insurance company and includes direct and indirect expenses of collisions. The costs of vehicle collision prevention programs are more than justified when compared with potential losses related to collisions.
- A vehicle safety program should include a written safety policy, a designated safety program manager, efficient accident investigation and reporting systems, and a preventive vehicle maintenance program.

The safety and health professional or fleet manager is responsible for supervising the program and reporting on safety issues to top management.

- A driver safety program should include a training program, collision prevention measures, reporting procedures, driver performance goals and incident reports, and a method for establishing competency and skills levels and collision/safety records for each driver.

- A motor vehicle collision can be defined as any incident in which the vehicle comes in contact with another vehicle, person, object, or animal with resulting injury or property damage. All collisions must be reported, and reports provide information that may help to prevent similar accidents in the future.

- Collision reporting procedures should enable all employees to follow the same steps and use the same forms to document details and to record names and addresses of witnesses. An investigator should interview drivers immediately after a collision to help the company calculate collision frequency rates and identify unsafe drivers.

- Companies must follow careful procedures when hiring employees who will be driving vehicles as part of their jobs. Applicants can be given driving tests to evaluate their driving skills and provide either basic, remedial, refresher, or special training, as required.

- Vehicles should have proper safety devices in good working order to help prevent collisions. Workers must take precautions to avoid being injured when loading or unloading trucks or when handling hazardous or toxic materials.

- A preventive maintenance program can be based either on mileage or operating hours of the equipment. Drivers should report any malfunctions or problems they encounter when driving their vehicles.

- Management and workers must adhere to proper procedures for working on vehicles to maintain safe practices in a maintenance facility. Repair shop workers should receive special training in safe work practices, and new employees work under close supervision.

- The Department of Transportation mandates drug and alcohol testing of all employees who work in safety-sensitive jobs, such as operating motor vehicles. Employers must establish testing programs in compliance with DOT rules and regulations.

- The primary goal of a flight safety department is to enhance aviation safety by preventing accidents and incidents through comprehensive safety programs and investigation. Safety education can be used to inform workers and others of an organization's operations and provide other airlines with information to develop their own plans.

- To reduce aviation accidents and incidents, an organization must have a well-developed and practiced response plan that can be implemented immediately. The primary purpose of collision and incident investigation is to provide the data necessary to prevent similar occurrences.

REFERENCES

Air Transportation Association, *Air Transport 1995—The Annual Report of the U.S. Scheduled Airline Industry.* Air Transportation Association, 1301 Pennsylvania Avenue, NW, Washington, DC, pg. 1.

Alcohol and Drug Problems Association of North America (formerly North American Association of Alcoholism Programs), 444 North Capitol Street NW, Suite 706, Washington, DC 20001.

Alliance of American Insurers, 1501 Woodfield Road, Suite 400 W, Schaumburg, IL 60173. *Code of the Road.*

American Automobile Association, 1000 AAA Drive, Heathrow, FL 32746. "Driver Training Equipment" (catalog).

American National Standards Institute, 11 West 42nd Street, New York, NY 10036.

Low Lift and High Lift Trucks, ANSI/ASME B56.1–1983 (R1993).

Powered Industrial Trucks, ANSI/NFPA 505 (R1995).

Recording and Measuring Employee Off-the-Job Injury Experience, ANSI Z16.3–1989.

American Society of Safety Engineers, 1800 East Oakton, Des Plaines, IL 60016.

Dictionary of Terms Used in the Safety Profession.

Photographic Techniques for Accident Investigation.

Profitable Risk Control.

Safety Law—A Legal Reference for the Safety Professional.

American Trucking Associations, Inc., 2200 Mill Road, Alexandria, VA 22314.

ATA Hazardous Materials Tariff, 1983.

Bulletin Advisory Service (3 vols.), 1983.

Effective Truck Terminal Planning and Operations, 1980.

Fundamentals of Transporting Hazardous Materials, 1982.

Fundamentals of Transporting Hazardous Waste, 1980.

National Truck Driving Championship, 1986.

Associated General Contractors of America, Inc., 1957 E Street NW, Washington, DC 20006. *Manual of Accident Prevention in Construction.*

Association of American Railroads. *Rules Governing the Loading, Blocking and Bracing of Freight in Closed Trailers and Containers for TOFC/COFC Service, 1992.* Washington, DC: Association of American Railroads.

Products in Closed Trailers and Containers for TOFC/COFC Service, 1990.

Association of Casualty and Surety Companies, 110 William Street, New York, NY 10038.

Guide Book, Commercial Vehicle Drivers

Truck and Bus Drivers Rule Book

Baker JS. *Traffic Accident Investigation Manual,* 9th ed. Chicago: Traffic Institute, 1986.

Boeing Company. *Company Flight Safety.* Seattle, WA: Boeing Company, February 1993, p. 1.

Implementation Guidelines for the FAA Anti-Drug Program, Federal Aviation Administration Office of Aviation Medicine.

Current WF. *Does Drug Testing Work?* The Institute for a Drug-Free Workplace, 1992.

Procedures for Transportation Workplace Drug and Alcohol Testing Programs, Department of Transportation 49 *CFR,* Part 40, 1994.

Heavy Construction Contractors Association, P.O. Box 505, Merrifield, VA 22116. (General.)

Kirk-Othmer Encyclopedia of Chemical Technology, 4th ed. New York: Wiley Interscience, 1991.

The National Committee for Motor Fleet Supervisor Training. *Motor Fleet Safety Supervision, Principles and Practices.* The National Committee, A 364, Engineering Bldg., Michigan State University, East Lansing, MI 48824.

National Fire Protection Association, 1 Batterymarch Park, Quincy, MA 02269.
Flammable and Combustible Liquids Code, NFPA 30, 1990.
Fire Prevention Code, NFPA 1, 1992.
Fire Protection Handbook, 17th ed., 1991.
Hazardous Chemicals Data, NFPA 49, 1994.
Life Safety Code Handbook, 1994.
National Electrical Code Handbook, 1993.

National Safety Council, 1121 Spring Lake Drive, Itasca, IL 60143.
Accident Facts. (Published annually.)
Aviation Ground Operation Safety Handbook, 4th ed.
Defensive Driving Program Materials.
First Aid Program Materials.
Fleet Accident Rates—Manual. (Published annually.)
National Fleet Safety Contest.
Motor Fleet Safety Manual, 4th ed.
Public Employee Safety and Health Management.
"Street and Highway Maintenance"
"Vehicular Equipment Maintenance"
Safe Driver Award Program.
Standards for School Buses and Operations Manual.
Supervisors' Safety Manual, current edition.

New York University Center for Safety Education, Washington Square, New York, NY 10003. Publications list.
Pilot Judgment Training and Evaluation, Volumes I–III (DOT/FAA/CT–82–56), Bunnell, FL: Embry-Riddle Aeronautical University, June 1982.

National Private Truck Council of America. *Driver Training Manual.* Washington, DC: Private Truck Council of America, 1981.

U.S. Department of Defense, Department of the Army, The Pentagon, Washington, DC 20310.
Driver Selection and Training, TM 21–300.
Drivers' Manual. TM 21–305.
General Safety Requirements. EM 385–1–1, U.S. Army Corps of Engineers.
"Methods of Teaching."
Motor Transportation, Operation, FM 25–10.

U.S. Department of the Interior, Bureau of Mines, 2401 E Street NW, Washington, DC 20241.
Minerals Yearbook.
Also various handbooks, miners' circulars, and other publications.

U.S. Department of Labor, 200 Constitution Avenue NW, Washington, DC 20210.
Occupational Safety and Health Standards, CFR, 1910 and 1926.
OSHA Compliance Guide, Volume 3.
OSHA Compliance Operations Manual.
OSHA Recordkeeping Requirements.
OSHA Job Hazard Analysis.

U.S. Department of Transportation, 400 Seventh Street SW, Washington, DC 20590.
Federal Motor Carrier Safety Regulations.
Hazardous Materials Emergency Response Guidebook.
Manual on Uniform Traffic Control Devices for Streets and Highways. (Also identified as American National Standard D6.1).
Model Curriculum for Training Tractor-Trailer Drivers.
Title 49, *Code of Federal Regulations,* Parts 390–397, *Motor Carrier Safety Regulations.*

U.S. Government Printing Office, North Capitol and H Streets NW, Washington, DC 20401.
Code of Federal Regulations:
Title 29—"Labor."
Title 40—"Protection of the Environment."
Title 49—"Transportation."

REVIEW QUESTIONS

1. An organization's vehicle collision prevention efforts should focus primarily on what two factors?
   a.
   b.

2. Which of the following statements is true when looking at the cost of vehicle collisions?
   a. The total cost of a vehicle collision is usually more than the amount recovered from the insurance company.
   b. The total cost of a vehicle collision is usually less than the amount recovered from the insurance company.
   c. The total cost of a vehicle collision is usually equal to the amount recovered from the insurance company.

3. List the five basic elements that a vehicle safety program should provide.
   a.
   b.
   c.
   d.
   e.

4. What factors should be considered when first screening an individual who is applying for a job that requires driving?

   a.

   b.

5. List the four objectives of a preventive maintenance program.

   a.

   b.

   c.

   d.

6. Repair shop workers should receive training in safe practices whenever:

   a. Equipment is changed or replaced

   b. Jobs are rotated

   c. Employees return after leaves of absence

   d. Refresher training is needed

   e. All of the above

7. Which regulation has established procedures for urine drug testing and breath alcohol testing of all employees who work in safety-sensitive jobs?

8. Name the two federal and international flight operation regulatory agencies that are responsible for investigating and preventing incidents and accidents in the aviation industry?

   a.

   b.

9. Why should companies implement their own internal safety education programs?

10. List the methods used by safety department personnel to help educate emergency response organizations, and explain what they accomplish.

   a.

   b.

   c.

19

Office Safety

Many large organizations today, such as insurance, governmental, and financial companies, consist almost entirely of office workers. Accidental injuries can be just as painful, severe, and expensive when they occur to office workers as when they happen to production workers. Through an effective office safety and health program, incidents and accidents can be managed. This chapter covers the following topics:

- statistics on types and rates of office accidents
- how to control office hazards through an effective safety and health program that covers the work environment, equipment, and procedures
- how to set up computer workstations that meet the ergonomics needs of the operators
- programs that can be established to promote and maintain interest in office accident prevention.

It is true that the risk of an occupation-related accident or injury to office workers is lower than the risk to employees involved in manufacturing or transportation. However, office risks often go unrecognized and unmanaged, and some could eventually lead to serious injuries and property loss.

A company safety program cannot be fully effective if there is only partial participation by employees and management. A safety program that is not vigorously pursued in company offices probably will not be vigorously pursued in the factory, shop, or plant. If office workers are exempt from safety and health policies, then production workers may feel that following rules to avoid hazards is an unnecessary burden, and, perhaps, an unfair exercise of authority by management. Exempt office workers may not understand the importance of safety and health policies and procedures and may scoff at production-oriented safety and health activities.

The safety and health professional who expects to sell safety to management must get management involved in a total safety program. The emphasis must be on preventing office accidents as well as production accidents.

SERIOUSNESS OF OFFICE INJURIES

One reason office safety and health programs are not more widespread is that many people believe office injuries are minor. This is a serious mistake. One aerospace firm, for example, paid out $102,000 over eight years at one facility just for injuries incurred by people falling out of chairs. Approximately 25,000 people worked at the facility, and more than half were office workers. Of the 14 chair accidents, the two worst cost the company a total of $97,000. Not only are medical and wage replacement benefits expensive for disabling accidents, but there also are hidden costs such as the loss of productivity. (See Chapter 7, Accident Investigation, Analysis, and Costs.)

Studies made by the State of California Department of Industrial Relations and the Equitable Life Assurance

Society of the United States (see References) show that this is not an isolated example (Figure 19-1). The California State Department of Industrial Relations analyzed reports filed by more than 3,000 California employers on disabling injuries to employees. Those employers together employed more than one million office workers. "An office worker" was defined in the California study as "a person primarily engaged in performing clerical, administrative, or professional tasks indoors in an office at the employer's place of business." The definition did not include salespersons, claims adjusters, social workers, medical and teaching personnel (other than clerical or administrative), and certain stock, order, and inventory clerks.

When the results of the study were extrapolated nationwide, investigators found that on-the-job office accidents amounted each year to about 40,000 disabling injuries at a direct cost (indemnity benefits and medical expenses) of about $100 million. That figure does not include any indirect costs for employers, workers, or the nation.

Of the accidental fatalities occurring to office workers, approximately half are due to work-related automobile accidents. That amount can be significantly reduced by using a defensive driving program (DDC) such as that developed by the National Safety Council.

The California study did not include employees of the federal government, maritime workers, and railroad workers in interstate commerce. An Equitable Life Assurance Society study, on the other hand, did include salespeople, claims adjusters, medical personnel, and supply and warehouse personnel. The insurance company study included approximately 8,000 employees of one company working in one building, about 5% of whom were maintenance personnel. (See Chapter 8, Injury and Illness Record Keeping and Incidence Rates). The average number of days charged per disabling injury was 5.9. See Tables 19-A and 19-B for details of the studies.

Who Gets Injured?
The California and the Equitable Life studies pointed to whom most injuries occurred, and how they occurred.

New Surroundings
The studies showed the importance of teaching office employees when beginning to work in a new environment to look for hazards and to correct them. Studies revealed a substantial increase in the number of injuries in the first year after a company moved into a new office building. The change upsets established routines and presents unknown hazards. Even going to and from work becomes more hazardous as employees explore new driving and transportation routes.

New or Young Employees
The California study found that new and younger employees do not have a higher accident rate than the longer

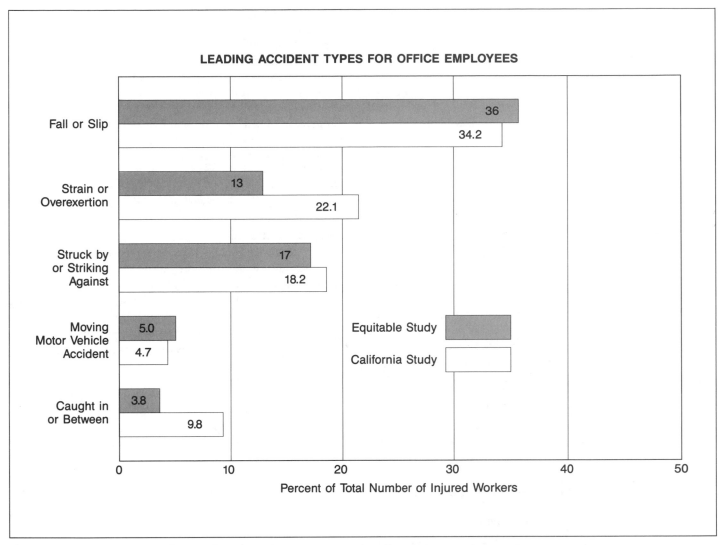

Figure 19-1. A comparison of two studies of accidents occurring in the office worker population.

employed and older workers (Table 19-B). Office employees who had been on the job for one and two months had an injury percentage rate of 2.0% and 3.9%, respectively. Those who had been employed for three to five years had the highest accident percentage rate (17.2%). The study showed that only 3% of the injuries occurred in the 18-to-19-years bracket.

Gender of Employee

In the California study, 70% of the disabling injuries occurred to women, who comprised 68% of the office labor force. For office occupations, the estimated disabling work injury rate for women was also very close to the injury rate for men. The rate was approximately 7.8 disabling injuries per 1,000 women employed in office work compared with 7.1 for the same number of men.

The injury statistics compiled from the Equitable study were similar to the California study. The rate of injury accidents per thousand male employees found in the Equitable study was about the same as it was for female employees. However, the rate of total days lost from disabling injuries was two and one-half to three times higher for men than for women.

Types of Disabling Accidents

Falls are the most common office accident and account for the most disabling injuries, according to both surveys (Figure 19-1). They cause from two to two and one-half times as high a disabling injury rate among office as among nonoffice employees. Falls were the most severe office accident and were responsible for 55% of the total days lost because of injuries.

Most chair falls occurred when a person was sitting down, getting up, or moving about on a chair. A few were caused by people leaning back and tilting their chairs in the office or cafeteria, or putting their feet up on the desk. Although stairs would seem to be more hazardous than chairs, people recognize the stair hazard and are more cautious. Furthermore, people are not as often exposed to the stairs as they are to chairs.

Another major accident category is falls occurring on the same level. That includes slipping on wet or slippery floors and tripping over equipment, cords, or litter left on the floor. Good housekeeping procedures can reduce accidents in this category.

A final category was falls from elevations, caused by standing on chairs or other office furniture, and by falls from ladders, loading docks, or other miscellaneous elevations. Those falls (not including stairs) accounted for approximately 2% of the disabling injuries in the California study. Falls on stairs accounted for almost 5%.

Overexertion

Almost three-fourths of the strain or exertion mishaps occurred while employees were trying to move objects—carrying or otherwise moving office machines, supplies, file drawers and trays, office furniture, heavy books, or other loads. Often, the employees were moving office equipment or furniture without authorization from their supervisors. A significant number resulted when the employee made a sudden or awkward movement and did not involve any outside object. Reaching, stretching, twisting, bending down, straightening up, and cumulative trauma were often associated with these injuries.

Objects Striking or Struck by Workers

Objects striking office workers accounted for about 11% of the injuries in both studies. Most of these injuries were sustained when the employee was struck by a falling

object—file cabinets that became overbalanced when two or more drawers were open at the same time, file drawers that fell out when pulled too far, office machines and other objects that employees dropped on their feet when attempting a move, or typewriters that fell from a folding pedestal or rolling stand. In addition, a number of employees were struck by doors being opened from the other side. Office supplies or other material and equipment sliding from shelves or cabinet tops caused a few injuries in this classification.

Striking against objects caused approximately 7% of the office injuries in both studies. Two out of three of these injuries were the result of bumping into doors, desks, file cabinets, open drawers, and even other people (Figure 19-2) while walking. Hitting open desk drawers or the desk itself while seated at a desk, or striking open file drawers while bending down or straightening up caused most of the rest of these injuries. Other incidents included workers bumping against sharp objects such as office machines, spindle files, staples, and pins. This category also comprised infected cuts incurred when employees handled paper, file drawers, and supplies.

Caught In or Between

The final major classification was accidents where the worker was caught in or between machinery or equipment. Mostly, this was getting caught in a drawer, door, or window. However, a number of employees got caught in duplicating machines, copying machines, addressing

Table 19-A. Disabling Accidents from Falls in Offices

Equitable Life—8,000 employees (an eight-year study)		Disabling Accidents	Days Lost
In hallways and work areas, caused by running, slipping, tripping over wires, desk drawers, file cabinet drawers, etc.		53	553
From chairs		21	120
Stairs		16	117
Escalators or elevators		8	55
	Total	98	845

California Survey—1,000,000 employees (a one-year study)	Disabling Accident Totals
Falls (all categories)	4,360
Falls or slips on stairs or steps	752
Falls from other elevations	370
Falls on the same level	3,238

Table 19-B. Percentage of Disabling Work Injuries* to Office Workers by Occupation and Length of Service

Occupation	Total Disabling Work Injuries	Length of Service					
		1 mo.	2 mo.	4–6 mo.	3–5 yr.	6–10 yr.	11–20 yr.
Clerical and kindred	12,858	2.1	4.3	4.0	17.1	15.7	9.2
Professional, technical, and kindred	1,418	1.5	2.5	7.5	17.7	15.7	12.1
Managers and administrators	2,000	1.4	2.4	3.4	18.1	16.0	18.9
Totals	16,276	2.0	3.9	8.2	17.2	15.7	11.0

*These figures are true only for the California Study.

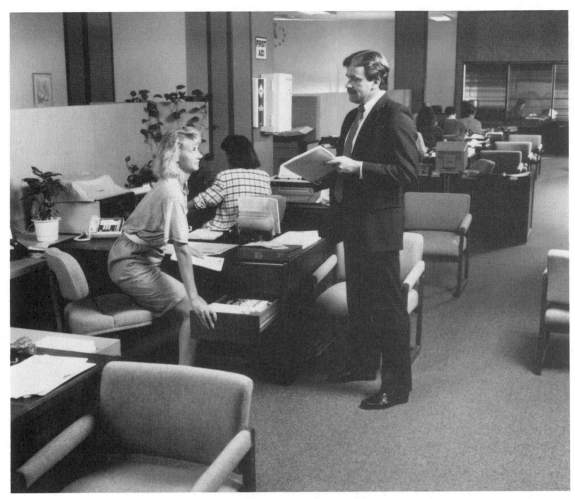

Figure 19-2. Desk and file drawers should always be closed when not in use. Labels on drawers help relieve stress and strains. (Courtesy Zee Medical.)

machines, and fans. Several injured their fingers under the knife edge of a cutter.

Miscellaneous office accidents included foreign substances in the eye, spilled hot coffee or other hot liquid, burns from fire, insect bites, electric shocks, and paper cuts.

CONTROLLING OFFICE HAZARDS

Office accidents can be controlled by eliminating hazards or, when they cannot be eliminated, by reducing exposure to them. Management can eliminate or reduce hazards most easily when the office is in the planning stage, when equipment is purchased, or when new office procedures are set up.

Layout and Ambience

Offices should be laid out for efficiency, convenience, and safety (Figure 19-3). The principles of workflow apply to offices as well as to factories.

Stairways and exits (including access and discharge) should comply with NFPA 101, *Life Safety Code;* floor and wall openings should comply with ANSI *Construction*

Safety Requirements for Temporary Floor and Wall Openings, Flat Roofs, Stairs, Railings, and Toeboards, A10.18. Handrails, not less than 30 in. or more than 34 in. (0.8 and 0.9 m) above the upper surface of the tread are specified for one side of stairs up to 44 in. (1.1 m) wide, and both sides for stairs wider than 44 in. Stairs wider than 88 in. (2.2 m) require an intermediate (center) rail.

Exits, particularly stairways, should be checked frequently to be sure that they are unobstructed and well illuminated. Exit hallways or paths should have emergency lighting. Exit doors, if locked, should not require the use of a key for operation from inside the building.

Doors are another frequent source of accidents in offices. Glass doors should have some conspicuous design, either painted or decal, about 4 ft. above the floor and centered on the door so that people will not walk into it (see Chapter 21, Contractor and Nonemployee Safety, for details). Safety glass complying with ANSI standard Z97.1 should be installed rather than plate glass. Sometimes local codes specify the type that must be installed.

Frosted safety glass windows in doors provide visual clues to prevent accidents while preserving privacy. Solid

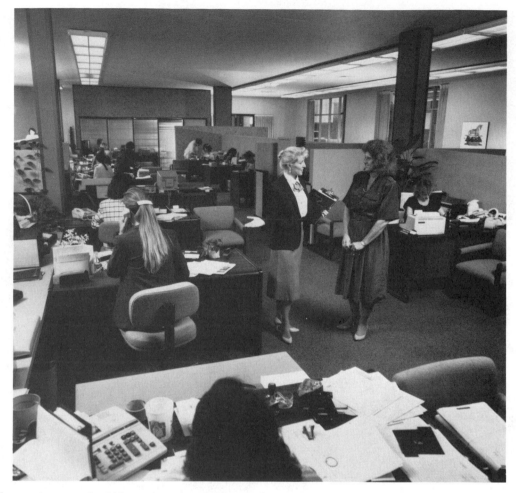

Figure 19-3. Office workstations should be arranged so that required equipment is easily accessible. Provide wheeled carts for transport of light loads, such as computer printouts or books. (Courtesy Zee Medical.)

doors present a hazard because they can be approached from both sides at the same time and one person can be struck when the door opens. Employees should be warned of this hazard and instructed (1) to approach a solid door in the proper manner, that is, away from the path of the opening door; (2) to reach for the door knob so that, if the door is suddenly opened from the other side, the hand receives the force of the impact rather than the face; and (3) to open the door slowly if it opens outward.

Another hazard is the door that opens directly onto a passageway. If the door opens directly into the path of oncoming traffic, somebody might bump into the edge of the door. If doors that open onto hallways cannot be recessed, they should be protected with short-angled deflector rails or U-shaped guardrails that protrude about 18 in. (46 cm) into the passageway. As an alternative, the area they swing over can be marked as suggested in the following paragraph. Another procedure is to place storage lockers or benches along the wall near the door to provide the safety of a recessed door.

Some offices having tile floors will have white or yellow stripes or tape on the floors to mark traffic flows or to guide people away from a rapidly opening door. The floor in front of a swinging door also can be marked or painted as a warning or a warning sign can be posted. As a final precaution, it is good practice to have the door hinges on the upstream side of the traffic; that is, on the right-hand side as one faces the door from the hallway. Doors are covered in the *Life Safety Code,* NFPA 101.

Lighting

Adequate light, ventilation, and other employee services have an important influence on employee morale. Business growth often requires installing more desks and other equipment than originally planned. Overcrowding is undesirable both for appearance and the physiological effect on employees, especially if it overtaxes ventilation facilities. Smaller offices can be made to appear larger and less crowded if walls, woodwork, and furniture are the same color.

Illumination levels recommended by the Illuminating Engineering Society for an office are listed in Table 19-C. Also, see Chapter 2, Buildings and Facility Layout, in the *Engineering & Technology* volume of this Manual.

Table 19-C. Levels of Illumination for Offices

	Recommended Illumination* (Footcandles)
Cartography, designing, detailed drafting	200
Accounting, auditing, tabulating, bookkeeping, business machine operation, reading poor reproductions, rough layout drafting	150
Regular office work, reading good reproductions, reading or transcribing handwriting in hard pencil or on poor paper, active filing, index references, mail sorting	100
Reading or transcribing handwriting in ink or medium pencil on good quality paper, intermittent filing	70
Reading high-contrast or well-printed material, tasks and areas not involving critical or prolonged seeing such as conferring, interviewing, inactive files, and washrooms	30
Corridors, elevators, escalators, stairways	20 (or not less than 1/5 level in adjacent areas)

*Minimum on task at any time.

Some accidents can be attributed to poor lighting. However, many other factors associated with poor illumination are contributing causes of office accidents. Some of these are direct glare, reflected glare from the work, and harsh shadows, all of which hamper sight.

Excessive visual fatigue can be an accident-causing element. Accidents also can be prompted by delayed eye adaptation when moving from bright surroundings into dark areas and vice versa. Some accidents attributed to an individual's carelessness can actually be traced to difficulty in seeing.

Office design can facilitate good lighting. If offices depend primarily on daylight, for example, employees engaged in visual tasks should be located near windows. North light is preferred by drafters and artists. However, employees generally should not face windows, unshielded lamps, or other sources of glare. Indirect lighting can produce high levels of illumination without glare. Furthermore, walls and other surfaces should avoid annoying reflections. Ceiling, walls, and floor act as secondary large-area light sources and, if finished with the recommended reflecting paints or wall coverings, will increase light and reduce shadows. Finally, modern office lighting must be designed to accommodate CRT/VDT users.

One of the best guides for reducing glare and reflections in the computer environment is "Solving the Problem of VDT Reflections" by Mark Rea in the October 1991 issue of *Progressive Architecture*.

Ventilation

For large interior spaces, forced ventilation is needed if the space is to be used as an office area. All mechanical ventilation and comfort conditioning systems require careful planning and installation by qualified specialists. Private offices installed around the outer walls of a large office space should not cut off light and ventilation to other employees. If fans are used in an office, they should be guarded, secured, and installed where they cannot fall.

The designs of many office buildings seal in office air—odors, smoke, office solvents and chemicals, molds, fungus, and other contaminants. In these buildings, it is particularly important to design and install adequate ventilation systems. Maintenance should regularly inspect the systems, change filters, and update them if problems appear.

Electrical

Management must protect workers against electrical equipment hazards in an office. In some cases, the hazard can be avoided completely, such as not using electric-key switches. In other cases, hazards can be reduced by using UL-listed equipment, ground-fault circuit interrupters (GFCI), sufficient well-located receptacles, and arranging cords and outlets to avoid tripping hazards.

Employees should not use poorly maintained, unsafe, or poor-quality, non-UL-listed electrical equipment such as coffee makers, radios, and lamps. Such appliances can create fire and shock hazards.

The company should make sure that a sufficient number of outlets are installed to reduce the need for extension cords. Those that are necessary should be clipped to the backs of desks or taped down. If cords cannot be dropped from overhead and must cross the floor, cover them with rubber channels designed for this purpose. Cords should not rest on steam pipes or other hot or sharp metallic surfaces.

Outlets should accommodate three-wire grounded plugs to help prevent electric shock to operators. Floor outlets should be located, if possible, so that they are not tripping hazards and cannot be accidentally kicked or used as a foot rest. A floor outlet protruding above floor level is frequently shielded by a desk or some other piece of furniture. However, when the desk is moved, the outlet can become an immediate tripping hazard unless it is appropriately covered. Such floor design is not common any longer; underfloor or cellular floor raceways are usually used in new construction.

The *National Electrical Code,* ANSI/NFPA 70, requires GFCI in restroom areas. In all areas, wall receptacles should be so designed and installed that no current-carrying parts will be exposed, and outlet plates should be kept tight to eliminate the possibility of shock or collision injury.

Cords for electrically operated office machines, fans, lamps, and other equipment should be properly installed and frequently inspected for any defects that could cause shocks or burns. Switches should be provided, either in the equipment or in the cords, so workers do not have to remove the plugs to shut off the power. (See Chapter 10, Electrical Equipment, in the *Engineering & Technology* volume.)

Office electrical service should be designed to accommodate changes in equipment and technology. Electrical, coaxial, computer, and telephone cables are easily serviced and moved when run through modular channels dropped from the ceiling.

Installation or repair of any electrical equipment should be done by qualified workers using only approved materials. Because defective wiring may constitute both shock and fire hazards, management and workers must follow all recommendations of the *National Electrical Code,* ANSI/NFPA 70.

Equipment

Workers should never place an office machine on the edge of a table or desk. Maintenance staff or manufacturer representatives should secure machines that tend to creep during operation. They can either affix the machines directly to the desk or table or place them on a nonslip pad. In particular, typewriters on folding pedestals should be fastened to the pedestal.

Workers should place heavy equipment and files against walls or columns; files also can be placed against railings. File cabinets should be bolted together or fastened to the floor or wall so workers cannot tip them over.

Floors

Improper floor surfaces are one of the major causes of office accidents. They should be as durable and maintenance free as possible. Management should select floor finishes for slip-resistant qualities. Well-maintained carpet provides good protection against slips and falls. Maintenance staff must repair defective tiles, boards, or carpet immediately. They should also replace or repair worn or warped mats under office chairs and rubber or plastic floor mats with curled edges or tears. These conditions create tripping hazards.

Highly polished and extremely hard but unwaxed surfaces such as marble, terrazzo, and steel plates represent slipping hazards. Slip-resistant floor wax can increase the coefficient of friction and can reduce their slipping hazard. Maintenance staff, however, must not apply wax so thickly that a smeary coating results. They should not use an oil mop on a waxed floor because it creates a soft, smeary coating that could become a slipping hazard (see Chapter 21, Contractor and Nonemployee Safety).

Special slip-resistant protection should be used on stairways and at lobby or elevator entrances. Be sure these especially hazardous areas are always maintained in the best possible condition. Floor mats and runners often provide a better, more slip-resistant walking surface. Their use is discussed in the National Safety Council's Industrial Data Sheet 12304-0595, *Floor Mats and Runners.* A well-planned routine maintenance program is needed to keep entrance, cafeteria, and vending area floors dry. Refer to Chapter 21, Contractor and Nonemployee Safety.

Outdoor Areas

Parking lots and sidewalks represent major problem areas. Many slip and fall accidents occur in the company parking lot. To reduce the hazards, maintenance workers must keep the lot clean, remove debris, fill potholes, and correct uneven surfaces. In colder climates, effective snow and ice removal controls should be used during the winter months. Chapter 21 covers the subject in greater detail.

Aisles and Stairs

The suggested minimum width for aisles is four feet. All aisles and passageways must meet the width requirements of NFPA 101, *Life Safety Code.* Keep passages through the work area unobstructed, and place wastebaskets where people will not trip over them. Install telephone cables and electrical outlets so their wires do not create a tripping hazard in passageways. These and other obstructions, such as low tables and office equipment, should be placed against walls or partitions, under desks or in corners. Avoid building stepoffs from one level to another in an office; if one exists it should be well marked and guarded with a railing.

File drawers should not open into aisles, particularly narrow ones, unless extra space is provided. Pencil sharpeners and typewriter carriages must not jut out into aisles.

Storage

Materials stored in offices sometimes cause problems. In general, materials should be stored only in areas specifically set aside for the purpose. Where possible, the storage area should be located so general traffic patterns do not have to be crossed to reach the stored items. Workers should not store or leave anything on the floor in a passageway where it could become a tripping hazard.

Train workers to stack materials in stable piles that will not fall over. They should put the heaviest and largest pieces on the bottom of the pile. When materials are stored on shelves, the heavy objects should be on the lower shelves. Workers should not stack objects on windowsills if there is a danger the objects may break the window or fall through it.

Supervisors and managers should plan storage areas to make items easily accessible. Appropriate stepladders should be provided where necessary. Office falls can occur when workers stand on chairs, counters, or shelves to reach inconveniently stored items. Rolling ladders are discussed in the following section, Safe Office Equipment.

Companies should prohibit smoking in mailing, shipping, print shops, or receiving rooms. They should also ban smoking in other areas where large quantities of loose paper and other combustible material may be stored and in areas where flammable fluids are used, such as duplicating rooms or artists' supply areas.

Workers should store flammable and combustible fluids and similar materials in safety cans, preferably in locked and identified cabinets. Only minor quantities should be left in the office, and bulk storage should be in properly constructed fireproof vaults. (See Chapter 12, Flammable and Combustible Liquids, in the *Engineering & Technology* volume of this Manual.)

Safe Office Equipment

Good quality office furniture not only contributes to the safety of the office but also enhances its appearance. This, in turn, improves the attitudes of both employees and visitors.

Chairs, especially, should be comfortable and sturdily built with a wide enough base to prevent easy tipping. Five-legged chairs are more stable and discourage employees from tilting back on their chairs. (See Chapter 13, Ergonomics Programs, in this volume.) The casters on swivel chairs should be on at least a 20-in. (0.5 m) diameter base, but a 22-in. base is preferred. The casters should be securely fixed to the base of the chair and well constructed. Loose or broken casters are a frequent cause of chair falls. About 20% of the chair falls in the California study were due to chair defects.

Companies should purchase chairs with easy-to-adjust seat heights and back supports. Show employees how to properly adjust their chairs. The correct fit will make the employee more comfortable and help to reduce acute and chronic back strain—enabling office workers to work more safely and productively.

Even if good quality desks and file cabinets are purchased, it is possible that occasionally one will have a sharp burr or corner on it. Supervisors or maintenance staff should inspect office furniture when it is received and remove such burrs or corners immediately. Drawers on desks and file cabinets should have safety stops to prevent workers from pulling them out of the drawer slot.

Other safety tips can help to prevent accidents and injuries. Purchase office machines, such as rotary files, copying machines, paper cutters, and paper shredders, with well-designed guards. Glass tops on desks and tables can crack and cause safety hazards. Durable, synthetic surfaces are safer. Make sure workers have enough noncombustible wastebaskets. When smoking is permitted provide safety-type ashtrays that are large and stable enough to safely contain smoking materials.

Office fans should have substantial bases and convenient attachments for moving and carrying. They must be well guarded, front and back, with mesh to prevent workers' fingers from getting inside the guard. Many cut fingers result when people try to move fans by grasping the guard or try to catch falling fans. Train workers not to handle fans until they shut off the power and the blades stop turning.

Rolling ladders and stands used for reaching high storage should have brakes that operate automatically when weight is applied to them. All stepladders should have nonskid feet.

Computers

If computers are installed in a building with overhead sprinklers, keep sprinkler protection in service, but get advice on necessary protection against both fire and water damage to computer hardware. Actually, water damage is not to be feared as much as previously thought. Most new computers are less susceptible because of their solid-state circuitry. In addition, tests have shown that water does not harm magnetic tape. Most of the damage suffered by computers in a fire results from the heat. One of the best ways to prevent a damaging fire is to keep combustible materials such as paper, tapes, and cards at an absolute minimum in the room with the computer. When safeguarding such an investment, call in a fire protection adviser as well as a computer installation expert (see Chapter 9, Computers and Information Management).

Chemical Products

Organizations often underestimate the number and types of hazards represented by office chemicals. The safety and health professional should assess all chemical products used in copying and duplicating machines and in print shops, and should assess all adhesives and cleaning materials. Workers must be informed of any dangers and instructed in the safe use of hazardous chemical products.

If possible, substitute nontoxic and nonflammable solvents for those used in printing and duplicating or other operations. (Details are given in *Fundamentals of Industrial Hygiene,* 4th edition, part of this *Occupational Safety and Health Series.*) If chlorinated bleaches are purchased for cleaning purposes, make sure that they will not be mixed with strongly acidic or easily oxidized materials. Purchase a good grade of slip-resistant floor wax.

Purchasing Equipment

The company safety and health professional should work with the purchasing agent in buying office furniture and equipment. Both should be aware that although advertisers sometimes stress the safety features of office equipment, the machines may be delivered without these important safeguards. Mechanical hazards of heavy office equipment can be determined by careful, expert inspection before purchase. These hazards can almost always be eliminated or minimized, although sometimes at substantial expense.

The safety and health professional should also inform the purchasing department of precautions to be taken in connection with chemicals, dyes, inks, and other supply items. Particular attention should be paid to toxic, irritant, or flammable properties. Where hazards are unavoidable, manufacturers, suppliers, or the safety, health, and environmental department should supply labels and specific instructions for careful handling or issue instructions when workers receive the material.

The purchasing department should gather all pertinent information from the manufacturer on equipment design and electrical and space requirements, and should try to determine the composition of proprietary compounds. They can forward this information to the safety and health professional (or safety and health department) for an opinion concerning inherent safety hazards before purchasing new equipment or supplies.

Office Machines

Machines that have external moving parts that could be hazardous should have enforced safety procedures and constant training and retraining of operators as necessary. Tell employees that if any office machine gives a shock, appears defective, sparks, or smokes, they should turn it off, pull the plug, and inform the supervisor.

Some office machines may be noisy, especially the telex, computer printers, and printing equipment. This noise is usually more of an annoyance than a health hazard, but may need to be evaluated by a professional for possible adverse effects. Various covers are available to dampen machine noise.

Printing Services

Larger Offset Presses

Only qualified operators should operate presses. Check the operation of offset presses. Is the operator putting his or her fingers on the blanket while the press is in motion? One offset press department had seven finger-injury accidents in the first two weeks of operation, all caused by press operators who put their fingers in the running press to remove dirt or other particles from the plate. Presses should conform to all guarding regulations imposed by local, state or provincial, and federal agencies.

Make sure the area around the presses is free from clutter and well lighted. The flooring should be resilient, or rubber mats should be provided to minimize operator fatigue and to prevent slipping.

Loose clothing and long hair are hazardous around these machines. Use a safe, nontoxic substance to clean the

presses; office supervisors and press operators should understand the fire and possible health hazards involved and follow all instructions for safe use and storage. Cleaning materials should be disposed of in a safe, acceptable manner.

Gathering and Stitching Machines

Supervisors should make sure guards are installed on open sprockets and collector chain drives of gathering and wire stitching (or stapling) machines to protect employees from hand and body injury. The operating arm on the end of the gathering machine should be guarded.

Install hinged drop-guards to cover any exposed operating mechanism that creates nipping hazards under the machine and along the working area where operators fill the pockets. Nonskid material should cover the floors and work platform at this area. Supervisors should train operators to open signatures in the middle and place them on the saddle or rod between the hooks on the moving chain. If the hook is not put on the rod or chain correctly, workers must shut off the machine before attempting to straighten the hook out. Operators should also shut down the machine when threading stitcher heads, making any adjustments, or removing jams.

Folding Machines

Here are points to be stressed for safe operation of folding machines:

1. Before jammed paper is pulled from the machine, shut the motor off to avoid getting hands in the feed rollers.
2. Finger clearance at the folding knife should be checked before pulling out paper, putting tape on rollers, or adjusting plates and roller pressure.
3. Workers should walk down the steps of folder feeder platforms facing forward, never backward.
4. On large-sized folders, all steps and platforms should be protected by railings.

Defective staples protruding from reports or booklets should be removed to avoid cuts from them while books are being jogged, trimmed, or wrapped. Workers should be trained to cup their hands over the work when removing defective wire staples. Employees engaged in this operation should wear eye or face protection, and passers-by should be protected against flying staples by screens or by isolation of this work.

Basic Office Safety Procedures

Because the major category of office accidents is slips and falls, employees should never run in offices. Also, a number of office accidents can be prevented if everyone walking in passageways would keep to the right. Convex mirrors should be placed at corners and other blind intersections. Collisions at a door, as discussed earlier in this chapter, can be prevented if people stand away from the path of its swing when they go to open it.

People carrying material must be able to see over and around it when walking. They should not carry stacks of materials on stairs, but use the elevator instead. If one is not available, the person should make more trips, if necessary. People should not have both arms loaded when using stairs; one hand should be free to use the handrail.

When using stairs outside at night or in a dimly lit area, workers should go single file, keep to the right, and always hold the handrail. People should not crowd or push on stairways. Falls on stairs often occur when the person is talking, laughing, and turning to friends while going downstairs. Other safety rules for stairs include: do not congregate on stairs or landings, and do not stand near doors at the head or foot of stairways.

Good housekeeping is essential. Employees should not be permitted to litter in their work areas and should wipe up all spilled liquids immediately. Pieces of paper, paper clips, rubber bands, pencils, and other loose objects must be kept off the floor.

Broken glass should be swept up at once. Do not allow employees to discard loose broken glass in a waste container. It should be wrapped in heavy paper and marked Broken Glass for Disposal. Glass that has shattered into fine pieces can be picked up with damp paper towels.

All tripping hazards, such as defective floors, rugs, or floor mats, should be reported to the maintenance department and immediately repaired. Many falls could be prevented if employees wore supportive footwear with nonslip soles; high heels should be discouraged.

Chair Falls

Some habits can lead to chair falls. Supervisors should instruct employees not to scoot across the floor while sitting on a chair or lean sideways from the chair to pick up objects on the floor. Discourage workers from leaning back in the chair and placing their feet on the desk. It is possible to fall over backwards in this position.

People should properly seat themselves in their chairs. They should form the habit of placing a hand behind them to make sure the chair is in place. Sitting down on the edge of the seat rather than in the center, or backing up too far without looking, or kicking the chair out from under can result in a sudden fall to the floor. Standing on a chair with castors to reach an overhead object is particularly dangerous and must be forbidden.

Filing cabinets, as discussed earlier in this chapter, are a major cause of injuries. These include bumped heads from getting up too quickly under open drawers, mashed fingers from improperly closing drawers, and hand injuries and strains from moving the cabinets.

Some precautions are necessary against these accidents:

- People should never close file drawers with their feet or any other part of their body. They should use the drawer handle to close the cabinet, making sure their fingers are not curled over the edge when the drawer closes. File drawers should be closed immediately after use. File cabinets should be designed to allow only one drawer to be opened at a time.
- Employees should open only one file drawer in the cabinet at a time to prevent the cabinet from toppling over. As previously stated, where possible, have the file cabinets bolted together or otherwise secured to a stationary object to safeguard against this chance of human failure.
- Do not open a file drawer if someone else is close by or underneath and could be injured by the drawer. Do not leave open drawers unattended—not even for a minute. Whoever opens a file drawer should warn others working in the area so they do not turn around or straighten up quickly and bump into or trip over an open drawer.
- No one should ever climb on open file drawers.
- Small stools used in filing areas are tripping hazards when left in passageways. They should be stored where they cannot cause falls.
- Filing personnel should wear rubber finger guards to prevent cut fingers from metal fasteners or paper edges.

Office personnel should never move desks or files; they should be moved by maintenance workers, preferably using special dollies or trucks. In general, furniture should not be rearranged without authorization from office management. When desks or cabinets are moved, workers should consider whether they will obstruct floor space or aisles before making the move. If a telephone terminal box on the floor or electrical outlet box is exposed after moving furniture, the box should be marked with a tripping hazard sign until it is removed. Maintenance staff must remove the outlet and, if it is needed, relocate it; this step is far cheaper than the medical costs of a fall.

Do not run electric cords under rugs; they sometimes come out because of traffic movement and form tripping hazards. They also are fire hazards. New outlets should be installed to eliminate the necessity for extension cords.

Materials Storage

There are a number of precautions to be taken when storing materials. Neat storage makes it easier to find and recover materials without dropping or knocking over other items. Supervisors must keep employees from stacking boxes, papers, and other heavy objects on file cabinets, desks, and window ledges, or from placing these materials carelessly on shelves where they could tumble down. If heavy objects fall toward a window, the glass might break and cause an accident.

Instruct workers not to place card index files, dictionaries, or other heavy objects on top of file cabinets and other high furniture. Caution workers not to throw loose razor blades, thumbtacks, and other sharp objects into their drawers but to store them in small boxes. Tools with blades and points should have the cutting or sharp end stuck in foamed-polystyrene blocks.

Lifting

Occasionally it is necessary for office personnel to lift heavy objects, such as files, books, boxes, and computer tapes. For these times, make sure office workers are trained in proper lifting techniques.

Other Hazards

Some additional precautions follow: (1) never use a spindle (spike) file in the office; (2) never store pencils in a glass on the desk with points outward; (3) never leave a knife or scissors on a desk with the point toward the user or hand sharp-pointed objects to anyone point first; (4) equip paper cutters with guards that afford maximum protection (bar guards or single-rod barriers found in some cutters are not considered full protection); and (5) do not leave glass objects on the edges of desks or tables where they can easily be pushed off. Make sure that office machinery is operated only by authorized persons.

Some offices have an employee lounge or eating area with a hot plate for brewing coffee or a microwave oven for warming lunches. In these areas, spilled beverages can be a burning and a slipping hazard. (See the discussion under Food Service in Chapter 8, Industrial Sanitation and Personnel Facilities, in the *Engineering & Technology* volume of this Manual.)

If employees travel on company business, a safe driving program should be part of the company's safety program.

Supervisors should encourage employees to report all broken chairs, missing casters, stuck drawers, cracked glass, and other hazards for correction. Management should establish a policy for immediate correction of these defects and set up a formal program requiring quarterly office safety inspections.

Fire Protection

Fire Hazards

To prevent spontaneous combustion fires, store all solvent-soaked or oily rags used for cleaning duplicating equipment in a metal safety container. Management should prohibit smoking within 10 ft (3 m) of where flammable solvents are used in duplicating or any other office operation. Workers should be trained in handling solvents to prevent eye injuries from splashes and wear proper protective equipment.

In recognition of the health hazards caused by smoking, many organizations do not allow smoking at any place in the facility. Other safety rules regarding smoking include (1) never allow smoking on elevators and (2) do not throw matches or cigarettes into wastebaskets; the contents usually are highly combustible. Supervisors or department heads should establish procedures so cleaning and maintenance personnel do not collect possible smoldering combustible material from ashtrays and throw it into combustible containers, such as cardboard boxes or cloth bags.

Some waste containers made from plastic or other flame-resistant material may actually be combustible if subjected to fire or intense heat. Such fires can generate dangerous toxic gases and dense smoke that can easily endanger a whole office. To control for this hazard, use only metal or fire-safe tested materials designed to contain fire.

Fire Extinguishers

Portable fire extinguishers in a fully charged, operable condition must be kept in their designated places at all times when not in use (Figure 19-4). (See Chapter 11, Fire Protection, in the *Engineering & Technology* volume of this Manual, for the correct type of extinguishers for specific office hazard areas.)

Employees in general should know what to do in case of fire. Supervisors must train workers to operate extinguishers and fire hoses, if provided, and show them how to react in case of fire or other emergency. (Panic and confusion can be as dangerous as flame and smoke.)

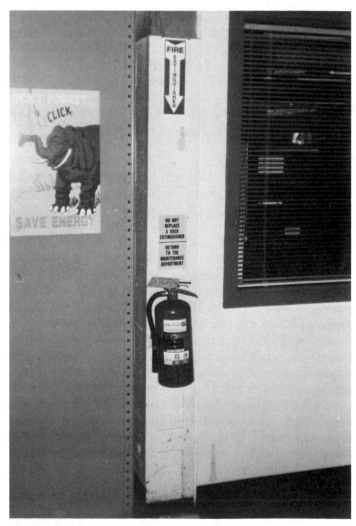

Figure 19-4. Note that this fire extinguisher is placed in an easily accessible location, has nothing stored in front of it, has a sign indicating its location and other instructions regarding replacement, and carries an inspection tag. (Courtesy Signode.)

When a fire is discovered, the employee should do three things: (1) turn on the alarm (no matter how small the fire is), (2) alert fellow workers, and (3) use the proper firefighting equipment, but only if the employee has been trained to do so, and always has a safe path of escape while fighting the fire.

Office employees should receive annual fire and other emergency training. The training should include the use of portable fire extinguishers, procedures for reporting emergencies, and location of escape routes and shelters.

Emergency Plan

Every office should have a written emergency plan. Supervisors should be appointed in every area to safely guide people out of the building. Every department should be assigned a specific route, and an alternate, in case that exit is blocked (See Chapter 15, Emergency Preparedness, and Figure 15-6). A fire drill may save lives in an emergency situation in the future.

When the alarm sounds, supervisors should direct the show, but every employee must play a part. The group should move calmly along, without hurrying or pushing, and wait on a different floor or outside the building for the signal to return. In a real emergency, the officials in charge would authorize return to the building. (See Chapter 15, Emergency Preparedness.)

ERGONOMICS IN THE OFFICE

Ergonomics is the science of optimizing a system of designing for the capabilities and limitations of the human interacting with it. When the system under consideration is the modern office, the primary components of the "system" are usually a computer, the table or desk on which it rests, a telephone, and a chair (since most offices are seated workstations). When designing a seated workstation, the factors to be considered are: (1) visual demands, (2) reaches required, and (3) the muscular strength exerted to perform the task. The following section will discuss ways of optimizing these factors. Because each individual working in an office environment is different, layout suggestions will be discussed in terms of anthropometric dimensions, and not absolute measurements. (See also Workplace Characteristics and Accessories in Chapter 13, Ergonomics Programs.)

Visual Demands

In order to optimize the visual interface, the computer screen should be placed at or slightly below seated eye height and at a comfortable distance for reading. This may necessitate raising or lowering the monitor. If the monitor is placed on top of the CPU it can often be lowered by taking it off the CPU and placing the monitor on the table top. Then, if necessary, it can be raised a couple of inches by using risers or monitor holders.

The goal is to have the operator's head positioned comfortably on his or her neck without needing to hold the head back, or to look down or to the side to see the screen. This goal has implications for bifocal wearers whose eyeglass prescription may require that they look at the screen through the bottom half of their lenses. They may be forced to tilt the head back to read a screen that is placed at eye level.

There are a number of options to accommodate such situations. For example, operators can use "computer glasses," a full-frame prescription for the distance they prefer to sit from the screen. These glasses also give a larger field of vision through which to see, so there is less head movement required to view all parts of the screen. An option for those who need or strongly prefer to use bifocals while at the computer is to lower the monitor so that they can see it comfortably through the bottom half of their bifocals without having to tip their head back.

The monitor should be at a distance that suits the visual acuity of the operator. The operator should be able to read the information on the screen without leaning forward. The monitor should be far enough from the operator's head so that he or she does not have to move it to read the whole screen. The operator should be able to scan the screen simply by moving the eyes. The optimal distance is usually 18 in. to 30 in. from the operator. Finally, the monitor should be placed directly in front of the operator, to minimize twisting of the neck or trunk.

Glare can sometimes be a concern, either from a window in the office or from overhead fluorescents. Glare from the windows can be eliminated by placing the monitor at right angles to the window, or by covering the window. Glare from overhead lights can be minimized by reorienting the monitor perpendicular to the light source and by slightly tilting the screen downwards. If these techniques don't work, antiglare screens can be effective. The contrast and brightness should be adjusted to optimize for clear visibility and minimal reflections from the surroundings. If it is a color monitor, a color scheme that emphasizes contrast while minimizing reflections is often based on a pale background with dark lettering. A final option would be to reconfigure the lighting in the work area.

Hardcopy material ideally should be located at the same distance from the operator's eyes as the screen, and at the same height. This position would minimize eye movement from copy to screen and minimize the head/neck motion that is often necessary when referring to copy that may be placed flat on the desk. Another option would be to place the copy between the keyboard and the monitor.

Studies have shown that the rate of blinking when reading a screen is decreased, thus increasing the chances of drying out the eyes. Supervisors or trainers should remind computer operators to blink frequently and to keep their eyes hydrated, especially if they wear contact lenses. Computer workers who need glasses should be reminded to get their eyes checked regularly to make sure their prescriptions remain current. Inadequately corrected eyesight may promote poor posture and eye strain. It is also a good idea to periodically relax the eyes by closing them for a few seconds, doing eye exercises, or refocusing them on a distant object.

Reach

The objects most frequently used by computer operators are the keyboard and the mouse. In order to minimize "reaches," the keyboard and mouse should be placed so that operators can use both of them without stretching out the arms. Workers should be able to work with the shoulders relaxed and the upper arms close to their body. Other objects that may be often reached for should be placed close to the operator (e.g., the telephone, writing materials). An often overlooked culprit in shoulder problems is the coffee cup. If computer workers are frequently stretching out their arms to pick up a coffee cup and put it back down, they may build up fatigue in the shoulder muscles.

Muscle Exertion

The main muscles that are used when typing on the keyboard are those of the hand and fingers. Because the strength of the fingers and hand are greatest when the wrist is straight (not bent side to side *or* up and down), the keyboard and the mouse should be placed so that operators can use them with their wrists straight. As keyboards are usually placed on a flat work surface, the operator would have to bend the elbows in order to flex and extend the wrist. The muscles used to bend the elbow (biceps) are strongest at about 90 degrees of flexion. Therefore, the keyboard should be placed at approximate elbow height. To achieve this support the surface for the keyboard and mouse may need to be raised or lowered. Another option would be to raise or lower the chair. Using a wrist rest in conjunction with the keyboard and mouse could also help minimize wrist extension. If the operator types by moving the whole arm across the keyboard, there may not be significant radial or ulnar deviations. If however, such deviations are placing an unacceptable stress on the operator's wrists, a curved or split keyboard could help reduce such stress.

In general, using a keyboard is not a high-force task; however, some studies have shown that the level of force exerted while typing may be associated with the development of wrist related musculo-skeletal disorders. A light touch on the keyboard would therefore be recommended.

Back/leg concerns are related primarily to the chair and the space under the work surface. There should be sufficient room for the operator's legs, so that the operator isn't required to lean forward to reach the keyboard or read the monitor. Operators should be able to sit all the way back in their chairs without their knees pressing up against the edge of the chair. If the seat pan length is too long, a lumbar pillow can help to cut down on the seat length as well as provide lumbar support. The operator's weight should be evenly supported by the seat pan, along the length of the thighs. The feet should be well supported while sitting. If the operator's feet cannot be placed flat on the floor when sitting back in the chair, a foot rest may be required.

Another factor to be considered is the stress placed on the back while sitting. Some studies have shown that sitting increases the intradiscal pressure and so possibly the risk of back injury. The natural curves of the back should be maintained while sitting, so that the pressures on the discs remain even from front to back. This means that the chair should have a convex curve that matches the "small" of the back. If the chair does not have such a curve, a back support pillow (similar to those used for long distance driving) can help.

General Considerations

Avoid resting the elbows, forearms, or wrists on sharp edges, since this could place pressure on the ulnar nerve and possibly on the muscles and ligaments in this region. Such "contact stress" can be reduced by rounding and/or cushioning edges on which the elbow, forearm or wrists are supported.

Balancing the telephone between the shoulder and the ear could require an office worker to bend the neck to keep the phone in place. If office workers spend a significant portion of their work hours communicating via telephone, they may assume this position frequently. In order to enable the operator to work with the neck and shoulders relaxed, a shoulder rest may be necessary. If the operator's hands need to be free for using the keyboard and mouse while on the telephone, a headset or speakerphone may provide a better option.

Due to the multifunctional capacities of computers, office workers could spend many hours without leaving their workstation. "Micro-breaks" (a break of 30 to 45 seconds every 30 to 45 minutes) have been found to minimize the buildup of fatigue in the muscles.

To summarize, the computer workstation should be laid out so that operators can work with their:

- neck at a comfortable angle for viewing the monitor
- back firmly against the chair backrest
- elbows close to their sides and bent at about 90 degrees
- wrists straight and not resting against any sharp edges
- legs evenly supported by the seat pan
- feet flat on the floor or on a footrest.

See Table 19-D for a computer workstation checklist. Any modifications to a computer workstation should be discussed with the workstation operator and a qualified person with training in ergonomics. Items in the checklist may be interrelated. For example, if a monitor is lowered but not tipped back, greater discomfort could result.

The focus of any modification of a work area should be to customize the workstation to fit the individual(s) using it. People come in all different sizes, the furniture they are

Table 19-D. Computer Workstation Checklist*

Monitors	Yes	No
1. Is the top of the monitor at least 15° below horizontal eye level?	❏	❏
2. Is the top of the monitor farther from the eyes than the bottom of the monitor?	❏	❏
3. Is the screen free of glare and reflections?	❏	❏
4. For noncolor-sensitive work, can the screen be set with dark letters on a white background?	❏	❏
5. Is the front of the screen at least 25 in. (63.5 cm) from the user's eyes?	❏	❏
6. Can the font size of the software be increased if it is not large enough to allow for at least a 25-inch viewing distance?	❏	❏
7. Is the screen free from perceptible flicker?	❏	❏
8. Are the screen contrast and brightness set for maximum clarity?	❏	❏
9. Do the characters on the screen appear sharp and well defined?	❏	❏
10. Is the field of view of the operator free from bright light sources that are causing discomfort?	❏	❏
11. Is a copy holder available so hardcopy (if frequently referenced) does not have to be laid flat on the work surface?	❏	❏
Seating		
1. Does the seat height allow the operator to rest his/her feet flat on the floor or a footrest without requiring a thigh–torso angle of less than 90°?	❏	❏
2. Can the operator sit in full contact with the backrest without the back of the legs contacting the front of the seat pan?	❏	❏
3. Is the seat wide enough to support the user's thighs?	❏	❏
4. Does the chair have a lumbar support?	❏	❏
5. Are adjustable armrests available, if requested?	❏	❏
6. Are the casters appropriate for the flooring surface?	❏	❏
Keyboard and Mouse Support Surfaces		
1. Does the height of the keyboard or mouse support surface allow the operator to work without bending the wrist backward?	❏	❏
2. Does the keyboard or mouse support surface allow for elbow angles between 75° and 135°?	❏	❏
3. Are the keyboard support surface and other surfaces free of sharp edges that are likely to contact the operator's wrists or forearms?	❏	❏
4. Can the elbow be close to the body while keying or mousing?	❏	❏
5. Are the shoulders relaxed while keying or using a mouse or other input device?	❏	❏
6. Are the wrists held off the wrist surface of the palm rest while keying?	❏	❏
Administrative		
1. Is the operator able to intersperse noncomputer work (e.g., filing, copying) with computer work?	❏	❏
2. Can the operator take "micro-breaks" to stand up, stretch, or focus the eyes at a far or intermediate distance?	❏	❏
3. Is consideration given to reducing stress in the workplace?	❏	❏
4. If a phone is used at the same time as computer work is done, is a headset provided?	❏	❏

* This checklist should be used in consultation with a qualified person who knows the ways in which items are interrelated.

given varies, and the components of their computer system may be different as well. What works for one individual with a given workstation may not work for another. Suggestions for possible modifications to make a computer workstation more comfortable include the following:

- Monitoring holders can position the monitor at the correct height for comfortable viewing.
- "Computer glasses" can help eliminate neck bending for bifocal or trifocal wearers and allow a larger field of corrected vision.
- Reorienting the monitor in the work area can help eliminate glare from windows or overhead fluorescent lights.

- Glare screens can be fitted on monitors if glare is still a problem. Parabolic louvers are retrofits for overhead fluorescents which can eliminate glare. Usually however, the whole room would need to be retrofitted.
- A copyholder can position hardcopy at the same distance and plane from the eyes as the monitor.
- Adjusting the height of the chair or table can help achieve the correct height in relation to the keyboard.
- A computer table or under table adapter, such as an articulating keyboard arm, can also help put the keyboard at the correct height for keying.
- A wrist or forearm rest can help minimize wrist

extension and eliminate sharp edges against soft tissue.

- Maintaining the natural curves of the back while sitting is very important. If the design of the chair does not support this, lumbar pillows may help.
- A footrest can provide support for the legs if they are not fully supported on the floor.
- To the extent possible, noncomputer work should be interspersed with computer work.
- "Micro" breaks to stand up, to stretch, and to refocus the eyes on something in the distance can help minimize the buildup of fatigue.

SAFETY ORGANIZATION IN THE OFFICE

The supervisor is, of course, the key person in the office safety program. However, even the hardest working supervisor will have difficulty maintaining full-time interest in safety all alone. On the other hand, the office safety committee can help to maintain interest in the accident prevention program, but it cannot substitute for good management.

Safety and Health Training

Safety training has a tendency to be overlooked in an office environment. All office workers should be provided with safety training that focuses on accident prevention, fire prevention, fire emergency response, and medical emergency response. Depending on the specific nature of the office environment, hazard communications training may be required. If an office environment contains an art department, print shop, or duplicating center, hazard communications training should be given to all employees.

To develop proper safety behavior, managers must provide safety instructions for all new office employees. The personnel or industrial relations department can supply an accident prevention brochure or a set of printed rules. They also should arrange for all explanations of procedures as quickly as possible during the employee's early workdays. (See also Chapter 22, Motivation, and Chapter 23, Safety and Health Training.)

Unfamiliar surroundings, new equipment, or altered work tasks increase the likelihood of accidents, even among veteran employees. Therefore, these people also should be trained when beginning a new job and given specific instructions for each piece of equipment. No one should ever be permitted to use a machine unless fully instructed in its operation and shown the location of fire equipment, how to use it, and how to summon medical aid.

Managers should have safety instruction in safe office operation because they are likely to be as unaware of accident hazards as the employees. However, prevention of accidents requires the dedicated vigilance of the supervisor throughout every working day. If he or she fails to carry out this function, the number of accidents due to unsafe practices by employees will continue undiminished.

Routine safety and health training should be part of company policy, and employees should receive continuing information on work-related hazards and safe practices. Training meetings are recommended on such topics as slip and fall prevention, proper lifting, fire safety, emergency procedures, office chemical safety, and off-the-job safety. Safe attitudes and behavior are not merely put on when an employee enters the office and taken off when the individual walks out the front door (see Chapter 23, Safety and Health Training, for more detailed discussion).

All office employees who must enter production areas where safety hats, eye protection, and hearing protection are necessary must be provided with these items and shall be required to wear them. Every employee who visits the facility should have a card of the general safety rules that apply to the facility and should be familiar with them. The same requirement should be enforced for all visitors. Safety rules should apply to everyone if the program is to be successful.

Another critical aspect of training focuses on ergonomic and stress-related areas. Supervisors need to educate workers to recognize common physical complaints that could be caused by the ergonomic design of the workstation or how they use equipment at the workstation. Common physical complaints include eyestrain; back pain; neck pain; shoulder and arm symptoms; wrist, hand and finger symptoms; leg and foot symptoms; headache; and fatigue. Training materials should emphasize how employees can change their work habits or adapt equipment and furniture to reduce these complaints.

Office workers tend to work under a high level of stress. One definition of stress is an employee's physical and emotional reaction to change. A definite focus of training should be stress reduction and occupational wellness. Workers can learn how to recognize stressors in the work environment, detect physical symptoms of stress overload, and use various methods to alleviate stress. Stress reduction programs can include progressive relaxation, positive imagery, values clarification, proper nutrition, and exercise.

Many other programs can be developed that focus on occupational wellness. Chapter 23 of this volume contains additional information on safety and health training.

Finally, a company safety program cannot succeed unless it has the wholehearted backing of its top management. Supervisors must know that their accident prevention performance is watched and that good performance is appreciated. The Council's *Supervisors' Safety Manual* contains information on this topic that can guide both top management and the safety and health professional.

Safety and Health Committee

In planning an office safety program, management should ensure that worker representation on the safety and health committee reflects the composition of the workforce in company departments or divisions. The organization of the office safety committee can be the same as that of the company and joint safety committees discussed in Chapter 3, Loss Control Programs.

The office should be on the inspection itinerary of the company's safety and health professional. The office supervisor should accompany the safety inspector on every inspection, along with an office safety committee member. (See Chapter 6, Identifying Hazards, for inspection procedures.)

The committee's responsibilities include helping all supervisors maintain safe working conditions in office areas. The committee reports directly to the safety director or whoever is in charge of the program. Along with the department head, it can make periodic inspections of the office to look for accident or fire hazards. The committee makes recommendations, many of them based on suggestions from supervisors and other employees. It also can help to prepare and revise company safety rules.

Often the committee is in charge of office-wide communication, training, and incentive programs. These are designed to maintain peak interest in safety, and use such means as posters, bulletins, and contests.

Accident Records

If a safety program is to succeed, the company needs to keep accurate accident records. Not only do accident investigations and analysis of records spotlight problems that must be corrected, but the records show if the company is making progress in accident prevention.

Office employees, like facility workers, should report every accident, no matter how minor the injury. The reports should be detailed and made as soon as possible following the accident or near-accident. Unsafe conditions or procedures indicated in the reports should be corrected as soon as possible, because near-miss accidents are warnings of worse accidents to come.

Records are the concrete foundation of the safety structure. They tell the "who, what, when, why, and how" of accidents in the office—and help supervisors and employees to prevent repeat performances. Accurate records also provide guidelines on which company insurance rates are based.

The average office will not have enough major injuries to warrant extensive investigation and analysis. However, it is important that records be kept to pinpoint problems and prevent future accidents. If, for instance, a large number of falls are injuring workers, supervisors can double-check possible hazards and devote special attention to the problem in meetings and other communications. Standard report forms are available. (See Chapter 8, Injury and Illness Record Keeping and Incidence Rates, and Chapter 7, Accident Investigation, Analysis, and Costs.)

SUMMARY

- Because the risks of serious injury are as great to office workers as they are to production workers, a company's safety policy must include office workers in its program. Office workers and managers must be informed of the hazards and safe work procedures that apply to their jobs.

- Researchers have found that new surroundings and length of service increase the chances of worker accidents and injuries. The most common categories of major injuries are falls, strains or other injuries related to overexertion, workers either being struck by objects or striking them, and workers being caught in or between machinery or equipment.

- Office accidents can be controlled by eliminating hazards or reducing exposure to them. Offices should be laid out for efficiency, convenience, and safety.

- Adequate lighting, proper office design, and proper ventilation can reduce or eliminate eyestrain and visual fatigue and ensure that contaminants do not accumulate in office areas. Management must protect office workers against hazards from electrical and moveable equipment and from hazardous flooring.

- Supervisors and managers should plan storage areas for safety and to make items easily accessible. Workers should be trained in how to stack materials in stable rows or piles and to keep passageways free.

- Workers must also be trained in the safe use and storage of chemicals used in copying, printing, and duplicating machines and for cleaning purposes. Office machines must be properly guarded and designed with built-in safeguards.

- The safety and health professional should work with management to ensure that office furniture and workstations are designed according to good ergonomic principles. Following these principles can reduce or eliminate worker fatigue and injury.

- The supervisor and safety and health professional should ensure that workers receive safety training and know safety procedures for their jobs. Employees should understand how to prevent fires in office work areas and what to do during a fire drill or outbreak of fire. Every company should draw up written emergency plans in the event of a natural or human-caused disaster.

- The company should establish a formal safety program for office workers and establish training sessions for all employees. The company needs to keep accurate records on all accidents and injuries and to help pinpoint major problems so that management can design effective solutions.

REFERENCES

American National Standards Institute, 11 West 42nd Street, New York, NY 10036.

Safety Performance Specifications and Methods of Test for Glazing Materials Used in Buildings, ANSI Z97.1–1984. (R1994)

Practice for Office Lighting, ANSI/IES, RP1–1992.

Construction Safety Requirements for Temporary Floor and Wall Openings, Flat Roofs, Stairs, Railings, and Toeboards, ANSI A10.18-1983.

National Fire Protection Association, Batterymarch Park, Quincy, MA 02269.

Life Safety Code, NFPA, 101, 1994.

National Electrical Code, NFPA 70, 1993.

National Safety Council, 1121 Spring Lake Drive, Itasca, IL 60143.

Fundamentals of Industrial Hygiene, 4th ed., 1996.

Motor Fleet Safety Manual, 4th ed., 1995.

Starting an Office Safety Program, 1990.

Supervisors' Safety Manual, 8th ed., 1992.

Rea M. Solving the problem of VDT reflections, *Progressive Architecture* (October 1991).

Scott D. *Sitting on the Job: How to Survive the Stresses of Sitting Down to Work—A Practical Handbook.* Boston: Houghton-Mifflin, 1989.

State of California, Department of Industrial Relations, Division of Labor Statistics and Research, 525 Golden Gate Avenue, San Francisco, CA 94102.

Disabling Work Injuries to Office Employees, 1963 and 1978 editions.

"Work Injuries and Illness in California," annually.

Statistical Abstracts of the United States, U.S. Department of Commerce, latest edition.

REVIEW QUESTIONS

1. What is the most common and most severe type of office accident?
   a. Objects falling on worker
   b. Overexertion by worker
   c. Worker falling
   d. Worker caught in or between machinery
2. List five of the nine different elements of an office layout that can be hazardous to workers.
   a.
   b.
   c.
   d.
   e.
3. In addition to contributing to the safety of an office, which of the following improves the attitudes of both employees and visitors?
   a. Eliminate the overcrowding of desks and equipment
   b. Adequate light and ventilation
   c. Good quality office furniture
   d. All of the above
   e. Only a and c
4. Improper floor surfaces are one of the major causes of office accidents. What should be done to make floors safer?
   a.
   b.
   c.
   d.
   e.
5. Define ergonomics.
6. What three basic factors should be considered when designing a seated workstation?
   a.
   b.
   c.
7. List four safety guidelines that help prevent visual and/or muscular fatigue when looking at the computer screen.
   a.
   b.
   c.
   d.
8. Which of the following tells the "who, what, when, why, and how" of accidents in the office?
   a. The safety and health professional
   b. Accident records
   c. The office management
   d. All office workers

20

Laboratory Safety

Over the past few decades, the versatility and adaptability of lasers and ionizing radiation equipment has led to a rapid increase in their use for diagnostic, monitoring, and precision toolwork applications. Yet these sophisticated tools also pose special hazards to workers. The hazards are associated not only with the high-energy nature of lasers and ionizing radiation but with the high level of skill and training needed to operate and maintain such equipment safely. Industry safety associations and state and federal governments have created standards and regulations to help ensure that high-energy equipment meets rigorous guidelines and that workers are protected. However, companies still need to be vigilant in their efforts to comply with these standards and regulations and to identify the hazards unique to their workplace and to their workers.

This chapter discusses the following topics:

- the nature of lasers and the hazards associated with these sophisticated tools
- safety regulations and recommended safety programs to protect workers from laser hazards
- the nature and hazards of ionizing radiation
- the basic elements of a carefully planned and implemented radiation safety program
- the nature and hazards of nonionizing radiation and recommended safety measures

LASER SAFETY

The term "laser" is an acronym for light amplification by stimulated emission of radiation. Laser radiation is a form of electromagnetic radiation characterized by its wavelength. The wavelengths typically associated with laser radiation range from 180 nanometers to 1 millimeter. This range of wavelengths encompasses several regions of the electromagnetic spectrum: ultraviolet, visible, and infrared. Although laser radiation is often referred to as "light," many lasers produce radiation in the ultraviolet and infrared regions, which cannot be seen by the human eye.

Principles of Lasers

As a specific type of radiation, lasers are versatile and beneficial when used properly but potentially hazardous when used unwisely. Lasers produce a beam that is coherent and directional. Lasers are also monochromatic, meaning the output of energy of a laser is a single wavelength.

Probably the most important quality of laser radiation is coherence, that is, all the photons act in unison, like different voices singing the same note together. Coherent radiation is a very efficient way of delivering energy because there is no destructive interference. A coherent beam produces a concentrated and powerful effect when striking a surface. It is also part of the reason why a 1 milliwatt helium-neon (HeNe) laser appears brighter than the sun.

Lasers are also directional. Directional beams keep their shape for long distances and will retain these properties even after reflecting off a mirror-like surface. This directionality allows the laser to be guided but can also present a hazardous condition at a distance from the laser itself. Finally, lasers are monochromatic, meaning the output energy of a laser is of a single wavelength.

Although lasers may take various forms, they are all constructed with an optical cavity and a source of excitation. An optical cavity contains the active medium—a solid, liquid, or gas—and two mirrors. The medium is excited by high-voltage electricity or other power source until it begins to emit radiation in all directions. Mirrors placed in parallel at each end of the medium reflect the laser radiation between them until standing electromagnetic waves form in the cavity. One type of mirror is totally reflective and the other is partially reflective to allow the release or emission of the laser radiation. Although the active medium will generally dictate the type of laser wavelength emitted, the length of the optical cavity plays a substantial role as well. Figure 20–1 shows a schematic diagram of a simple laser system.

Laser Types

Industry uses a variety of active mediums for laser sources, either liquid, gas, or solid. Liquid lasers, such as dye lasers, are seldom used in industry because many have hazardous or toxic properties. Gas lasers are common, with the more common of these being helium-neon (HeNe) and carbon dioxide (CO_2), along with the newcomer excimer lasers. Gas lasers are not terribly efficient at converting energy to laser radiation because of the space between the molecules in the optical cavity. The powers of gas lasers are typically determined by the optical cavity volume.

HeNe lasers are usually sealed-cavity, low-power lasers that emit a visible (red) beam. HeNe lasers are used to help align parts of precision equipment. They are also used as bar code readers in inventory control systems. Most of the low-powered gas lasers are designed to emit a continuous wave or beam of laser radiation.

High-power gas lasers are mostly CO_2. The emission from a CO_2 laser is invisible (infrared). These laser systems can be scaled up to deliver massive amounts of energy by using a flowthrough design. One application of these large lasers is their use in processing metals. The newest gas laser to be widely used is the excimer. This laser emits wavelengths in the ultraviolet (UV) region of the spectrum. It uses acid gases in the active medium, presenting an additional hazard to the laser user. Applications for this laser are constantly being developed.

Solid or solid-state lasers are typically used in industries that cut, weld, and mark. One of the most popular of these lasers is neodymium:YAG (Nd:YAG). YAG is an acronym for yttrium, aluminum, and garnet. These lasers use an optical energy source, whereas most lasers use electricity or another laser. The optical energy is provided by

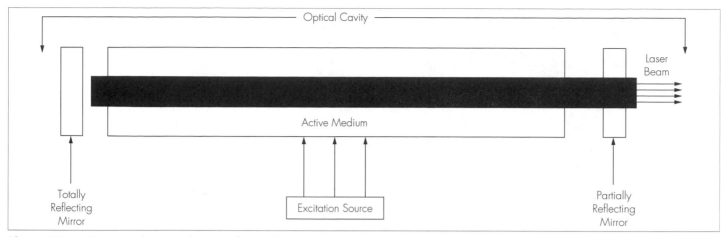

Figure 20-1. This is a schematic diagram of a simple laser system.

flash lamps, usually high-pressure xenon arc lamps. The visible and ultraviolet energy from the lamps is absorbed and converted into near infrared radiation. Though these lasers can be used to emit a continuous beam, most Nd:YAG lasers are pulsed.

Semiconductor lasers are becoming more popular because of their size and their efficient conversion of electrical energy into laser radiation. Most of the semiconductor lasers are made of a gallium-arsenide (GaAs) compound. These lasers have wavelengths from about 640 vm to about 950 vm.

HAZARDS OF LASER RADIATION

Hazards associated with laser equipment can be divided into beam hazards and nonbeam hazards. Beam hazards, from direct or scattered laser radiation, affect primarily two organs of the human anatomy: the eyes and the skin.

The skin is the less susceptible of the organs, because it can repair itself when injured. The main hazard to the skin is thermal burns produced from the extreme heat of a laser beam. Some lasers emitting UV laser radiation add a photochemical effect, which produces the equivalent of a sunburn on exposed skin. Workers burned by lasers should seek immediate medical attention.

Unlike the skin, eyes have less ability to repair themselves and are more vulnerable to injury. The eyes have three basic structures that can be affected by laser radiation: cornea, lens, and retina. The cornea, which is the outer covering of the eye, helps focus visible light radiation used in normal vision and absorbs the energy of the deep UV and the far infrared laser radiation. The hazard for the cornea is both thermal and photochemical laser burns.

The lens, which provides fine focusing ability, is protected from the deep UV and far infrared by the cornea. Near UV and near-infrared laser radiation passes through the cornea and deposits much of their energy in the lens. This can cause the lens to lose elasticity and may result in cataracts.

The retina is the most sensitive structure in the eye. Visible and near-infrared light passes through the cornea and lens to the retina. The light receptors contained in this structure focus the laser beam and can intensify the thermal and photochemical damage to the retina. Lasers can also damage the retina through thermoacoustic effects. This occurs when the highly focused beam forms a steam bubble near the retina. When the bubble pops, it sends out shock waves that damage retinal tissue and blood vessels. The degree of permanent damage caused by thermoacoustic injury is determined by what part of the retina was struck and whether blood reached nerve tissue. In general, brief pulses of laser light lasting only millionths of a second can produce more damage to the eye than can longer pulses, because shorter bursts retain more heat.

Other factors that influence the degree of eye injury that workers can experience from laser light include pupil size, distance from the source of energy to the retina, energy and wavelength of the laser, and pigmentation of the target area. Because people are highly dependent on their vision, the effects of laser radiation on the eyes are the primary concern of health and safety professionals.

Nonbeam hazards associated with lasers are many and potentially more dangerous than the beam hazards. Lasers generally use high-voltage electrical power as a source of excitation for the active medium. As a result, electric shock is by far the greatest hazard associated with lasers. In addition, because of the inefficiency of many lasers, the power supply provides much more power than the laser can convert, and the remaining energy is given off as heat. This heat must be safely dissipated to minimize risk of personal injury, equipment damage, and property damage.

Chemical hazards are associated with the use of some lasers. Some of the potential chemical hazards include asphyxiant gases such as carbon dioxide. Acid gases such as hydrogen chloride and potentially toxic by-products from the laser action on a surface are also chemical hazards.

Other nonbeam hazards requiring evaluation in the laser workplace are noise, dust, hot surfaces, cryogenic gases, explosion of high-pressure flash lamps, toxic laser dyes, electrocution or shock, and fire. High-powered lasers also have the potential to ignite combustible materials. Workers should keep flammable materials out of the areas where Class 4 and some focused Class 3 lasers are used all the time.

Hazard Classification

Laser products are classified by the manufacturers according to the requirements of Title 21 of the *Code of Federal Regulations* Part 1040.10 (21 *CFR* 1040.10), the U.S. federal performance standards. The classification scheme, which designates hazards as Class 1–4, is based on the hazard level the laser presents (Table 20–A).

These classifications are based on the normal operation of the laser product. Maintenance and service procedures may actually expose the individual to levels of laser radiation far above those during normal operation. The level of exposure to these individuals is not indicated by the classification label. Before work begins, the company should conduct an evaluation of the hazards that workers will encounter under actual working conditions.

Hazard Controls

Companies can use engineering and administrative controls and personal protective equipment to control hazards associated with lasers (see also Chapter 7, Personal Protective Equipment, in the *Engineering & Technology* volume). If a company correctly implements these measures, it can reduce employees' exposure to hazards.

Engineering Controls

In general, engineering controls, such as protective housings, interlocked enclosures, ventilation systems, beam enclosures, and so on, should be the primary method used to control exposure to laser hazards. Ideally, these controls require no action by the individual worker to be effective.

Interlocked laser beam enclosures can range from plastic panels on a framework to metal boxes with doors large enough to accommodate a person or vehicles. Laser beams can also be routed over or below walkways by the use of mirrors or elevated enclosures or tunnels. Interlocked laser beams either shut off the electrical power or drop a shutter into the beam, cutting it off. Interlocks can also be operated remotely, and are used for laser systems that cover large areas. For Class 4 lasers, viewing portals, viewing screens, and optical instruments must be connected to interlocks or reduce beam intensity to acceptable levels. Status lights, loudspeakers, and buzzers can be installed to protect employees working inside and outside the area of laser operations.

One simple engineering control is a key-in-lock system, which prevents unauthorized people from using the laser. This method is especially effective when the lasers are easily accessible by many people. Once an operator is

Table 20-A. Laser Classification Scheme

Class 1:	Levels of laser radiation are not considered to be hazardous.
Class 2a:	Levels of laser radiation are not considered to be hazardous if viewed for any period of time less than or equal to 1000 seconds but are considered to be a chronic viewing hazard for any period of time greater than 1000 seconds.
Class 2:	Levels of laser radiation are considered to be a chronic viewing hazard.
Class 3a:	Levels of laser radiation are considered to be, depending upon the irradiance, either an acute intrabeam viewing hazard or chronic viewing hazard, and an acute viewing hazard if viewed directly with optical instruments.
Class 3b:	Levels of laser radiation are considered to be an acute hazard to the skin and eyes from direct radiation.
Class 4:	Levels of laser radiation are considered to be an acute hazard to the skin and eyes from direct and scattered radiation.

finished using the laser, he or she should shut off the device and remove the key, preventing further use of the machine.

Although engineering controls are effective when lasers are installed, they are less useful when workers set up the device or carry out maintenance work. As a result, management should make sure that workers use special caution during these times, including the use of administrative controls and warning signs. For example, nearly 37.2% of laser accidents occur during the process of aligning the beam. Workers should use special protective equipment, such as laser alignment goggles, so that the laser beam will not enter the eyes. ANSI Z136.1 specifies the types of special warning signs that workers should use during setup. Finally, workers and management should remove all unnecessary reflecting surfaces, such as chrome chairs, mirrors, or highly polished floors or furnishings. Even painted walls and stipple-polished tools can serve as highly reflective surfaces for gas laser beams.

The personal protective equipment used for lasers, such as safety eyewear, must meet the provisions of the latest edition of ANSI Z87.1. Workers should select laser eyewear that is tailored to the specific laser wavelength used and severity of exposure. This eyewear can be color-coded to indicate the specific use.

Administrative Controls

Administrative controls are used during setup and maintenance and require. The Laser Institute of America's LIA/ANSI standard Z136.1–1993 specifies calculating the nominal hazard zone (NHZ), where people could be exposed to laser beam levels above the MPEs, and blocking people from entering this zone by barrier signs, flashing lights, and warning signs. Specifications for these items can be found in the ANSI Z535 series of standards, which specify colors, warning words, standard symbols, and warning sign layout. For example, the word CAUTION is used on signs and labels for Class 2 and 2a. The laser warning symbol is the distinctive sunburst surrounded by a triangle (Figure 20-2). The word CAUTION is used for Class 2 and 3a lasers, but other Class 3a, all of Class 4 lasers are marked with the word DANGER. Existing laser-safety signs and labels are grandfathered in Z136.1–1993. Companies are also required to provide medical surveillance and training programs if they use Class 3b and 4 lasers.

Because administrative controls require the individual to take some action, they should not be used in place of engineering controls. Instead, they should serve as backup in the event the engineering controls fail. In cases where engineering controls are determined ineffective or impractical, administrative controls can assume a primary role. Examples of administrative laser safety controls are adherence to written standard operating procedures (such as lockout/tagout), observing warning signs and labels, establishing a nominal hazard zone (NHZ) for a laser product, training the laser users and maintenance personnel, and so on.

Personal Protective Equipment

Personal protective equipment (PPE) is the third type of control measure. These devices are designed to minimize a worker's accidental exposure to a hazard and should not be used as a primary means of protection. The most recognizable laser PPE is the laser goggles or glasses, which provide a defined amount of eye protection at specific

Figure 20-2. The internationally recognized sunburst laser warning icon: ANSI Z136.1–1993 makes limited use of a red symbol for DANGER signs. Symbols and narrative text now appear in black, and colors are used to indicate degree of hazard (yellow and black for CAUTION, orange and black for WARNING, and red, black, and white for DANGER). The long tail symbolizes how laser beams retain their power over long distances.

wavelengths. However, companies must make sure that the laser PPE is specifically chosen for each laser. Eye protection is determined by reducing the exposure to less than the maximum permissible exposure (MPE). The required protection value is called the optical density (OD). The OD calculation can be found in American National Standards Institute (ANSI) standard Z136.1. The ODs for specific wavelengths must be provided on the laser glasses to help management select the proper eye protection. (See also Chapter 7, Personal Protective Equipment, "Face and Eye Protection," in the *Engineering and Technology* volume.)

SAFETY REGULATIONS

The main U.S. standard for laser safety is the Laser Institute of America's LIA/ANSI Z136.1 *American National Standard for Safe Use of Lasers,* revised in 1993. The Center for Devices and Radiological Health (CDRH) of the Food and Drug Administration, 21 *CFR* Part 1040, has promulgated regulations for safety in the use of commercial laser devices. Lasers are also covered by OSHA's General Duty Clause (Occupational Health and Safety Act of 1970, Section 5(a)(1)). This clause allows OSHA the flexibility to use industry or national standards to determine if the workplace is free of recognized hazards. Additional requirements discuss posting, labeling, and exposure limits. Although these limits are in the OSHA regulations, OSHA developed Technical Publication 8.1.7 to assist inspectors in the evaluation of laser hazards and programs. Lasers also are covered specifically by OSHA in the construction standards part of the OSHA regulations (29 *CFR* 1926.54—Nonionizing Radiation). The requirements in this section begin with "Only qualified and trained employees shall . . . operate laser equipment."

The LIA/ANSI and CDRH standards focus on the hazard control class type of regulation, which requires classifying the severity of the hazard posed by a laser device. While the two main measures of severity are wavelength and power or energy output, other measures include pulse duration and size or spread of reflection of the beam. The LIA/ANSI and CDRH standards set accessible emission limits (AELs) for each class of laser. Safety precautions become more stringent as the output of the laser increases from Class 1 to Class 4.

- Class 1 lasers require no special safety precautions.
- Class 2 and 2a warrant limited precautions such as wearing safety goggles.
- Class 3a lasers present moderate risks only if the worker stares at the light source or view through a telescope.
- Class 3b require more stringent safety precautions, such as shielding and personal protective equipment.
- Class 4 lasers require rigorous safety precautions.

The precautions listed in LIA/ANSI Z136.1–1993 are summarized in the chapter on nonionizing radiation in the

National Safety Council's *Fundamentals of Industrial Hygiene,* 4th edition.

The LIA/ANSI also specifies maximum permissible exposures (MPEs) for lasers, which are equivalent to OSHA PELs set for chemicals or the MPEs set for radio-frequency/microwave radiation set by ANSI IEEE C95.1–1991. The AELs used to define safety limits for laser classes are equal to the MPEs for maximum allowable viewing time stated in the standard for visible and IR-A lasers multiplied by the size of the pupil opening.

The LIA/ANSI Z136.1–1993 offers guidelines on exposure times for employees working with lasers, depending on the wavelength of the radiation. The averaging time for exposures to visible lasers is 0.25 s, and 10 s for IR exposures. The standard also gives protocols for evaluating the risks of exposure to pulsed lasers, although the protocols are complicated and difficult to apply. Managers and other employees working with pulsed lasers should read paragraph 8.2.2 of the LIA/ANSI standard carefully.

Other regulatory concerns with lasers are addressed by OSHA in its construction safety standard 29 *CFR* 1926.54 (use of lasers for alignment) and by the ANSI standard Z136.1–1993 (general industry use of lasers). The OSHA construction standard specifies that workers who could be exposed to direct or reflected light above 5 mW must be given laser eye protection. In addition, only mechanical or electronic devices may be used to guide alignment of lasers. The standard also set exposure limits for workers. OSHA 29 *CFR* 1910.132 covers standards for the use of eye and face PPE. Use of lasers in fiber-optic communications is covered in Z136.2 and Z136.3, issued by the LIA/ANSI Z136 committee.

The Food and Drug Administration (FDA) publishes regulations that contain the performance standards that laser manufacturers must meet to sell lasers in the United States. The FDA's Center for Devices and Radiological Health (CDRH) regulates electronic products, including lasers, through the use of performance standards. The laser product standard is found in 21 *CFR,* Part 1040—Performance Standards for Light-Emitting Products. These standards provide the requirements for the construction, classification, and labeling of laser products.

Laser manufacturers and importers must certify that their laser products comply with the performance standards according to the requirements in 21 *CFR* 1000 to 1010. The FDA does not certify equipment compliance against performance standards. They will agree or disagree with the information submitted in the Product Report.

RECOMMENDATIONS FOR SAFETY PROGRAMS

Laser safety programs being developed by companies should be based on laser hazard classifications and/or hazard evaluations. Although no two laser safety programs will be identical, an effective laser safety program should contain the seven elements discussed below.

Written Program

The program document should describe the overall purpose of the business' laser safety program, with responsibilities and authorities assigned. The document should reference the ANSI Z136.1 or similar standard as the basis for the company's laser safety practices.

Leadership Support

As part of the leadership support for safety programs in general, an influential and respected top manager should sign the program into effect. This individual might be the level production manager or the champion of the company's safety committee.

Laser Safety Officer

The appointment of a laser safety officer (LSO) is critical as it will be his or her duty to manage the laser safety program. Training for the LSO is very important. Because hazard recognition and control methods form the basis of a successful laser safety program, the LSO should be viewed as the local resource for knowledge and information about laser hazards.

Inventory Control

A laser inventory is an important tool for the LSO because it provides information on laser hazard classes that the laser safety program must cover. As the inventory changes, so must the laser safety program. The inventory also serves as a list for periodic inspections and hazard evaluations. In some states with radiation control programs for lasers, the inventory helps the LSO meet the registration requirements (mostly Class 3b and 4).

Medical Surveillance

Medical surveillance is the practice of evaluating employees to be sure they have the physical capabilities needed to perform a task. In the case of lasers, the practice deals specifically with the evaluation of the eye or skin for laser damage. The examination (generally done by an ophthalmologist) helps identify any preexisting conditions and the extent of injury from exposures. The ANSI standard recommends that users of Class 3b and 4 lasers receive examinations before starting laser work and following any suspected laser injury. No periodic examinations are recommended.

Safety Training

The type and content of laser training depend on the particular hazards. Important ideas to communicate are the recognition of hazards and methods of protection from laser hazards. Service and maintenance personnel should be trained to protect themselves against the highest hazard level of accessible laser radiation, since many laser products have lower hazard classifications based on enclosure of the beam. Repair and maintenance technicians should also be trained in high-voltage safety and cardiopulmonary resuscitation. Sometimes these workers must remove or circumvent safety interlocks to perform a maintenance or service task. With safety interlocks defeated, the accessible laser radiation can be far in excess of the laser hazard indicated by the classification. Companies should prepare specific procedures and checklists to help maintain proper protection of the employees during and after such service.

Inspections

Several types of inspections are part of a laser safety program. These include an installation, postservice, and annual inspection. The inspections focus on issues such as engineering controls (enclosures, interlocks, and so on) and administrative controls (labels and signs, procedures, and so on). Personal protection equipment (PPE) use and applicability is a part of these inspections. The users should demonstrate the proper selection, care, and maintenance of their safety equipment.

In addition to laser-emitting machines, workers also face hazards from radioactive materials and machines that produce radiation. These types of hazards require their own safety controls and PPE to protect workers from biologic injury and illness.

IONIZING RADIATION SAFETY

Radioactive materials and radiation-producing machines are commonly found in the workplace. Some of the common uses of radiation in the workplace are thickness gauges, nondestructive testing such as radiography, static elimination, and smoke detection. Providing a safe and healthy workplace that uses radiation sources requires an understanding of the hazards and the regulations that apply. Compliance with applicable regulations is difficult without knowing where to start. Let's begin with some basics of radiation.

Principles of Radiation

In this section, the term "radiation" means a photon or charged particle that has enough energy to liberate an electron from an atom. This electron liberation is called "ionization." Ionizing radiation comes in several types, the most common of which are alpha (α) particles, beta (β) particles, gamma (γ) rays, and x-rays. The first three types of radiation are emitted from the nucleus of a radioactive atom. Radioactive materials are present in our natural environment, or they may be artificially produced. X-rays are produced when high-speed electrons are decelerated (called *bremsstrahlung*). Some of the electrons' kinetic energy is converted into x-rays. Because x-rays are produced by high voltage, they can be turned off. Due to this fact, commercial industry uses x-rays in nondestructive testing. X-rays can be produced by machines designed to emit this type of radiation or as a by-product of the machine's primary process.

Radioactive materials that emit the other three types of radiation (α, β, and γ) cannot be turned off like x-rays. The emitted radiation is produced from the transformation (commonly called decay) of a radioactive atom.

Radioactive materials decay with a specific half-life. This half-life is the length of time it takes for half of the original radioactive atoms to decay to another nuclear configuration. The duration of a half-life ranges from fractions of a second to billions of years. Each radioactive material has a unique half-life. The properties of radioactive materials (nuclear configurations, half-life, radiation and their energies) are found in references such as the *Chart of the Nuclides*.

Radiation Hazards

As radiation passes through a substance (such as air, water, tissue, or bone), some energy is deposited that results in the substance being ionized. The deposited or absorbed energy can produce damage, generally assumed to be proportional to the energy absorbed (dose). At high dose levels, this relationship can be shown. At low dose levels, the exact relationship of energy and damage is still being debated. As a conservative approach, the dose-response relationship is assumed to be constant (linear) regardless of dose or dose rate.

Radioactive contamination is different from radiation. Contamination is the presence of radioactive material, which emits alpha, beta, and gamma rays, where it is not wanted. When unwanted radioactive materials are in the body, the body is considered internally contaminated. There will always be some radioactivity in the body because of the natural radioactivity in our environment. When the radioactive contamination is on the skin of the body or surface of an object, it is called skin or surface contamination. There are three levels of contamination.

Alpha Radiation

Alpha radiation is not an external hazard. Alpha particles on the skin deposit most of their energy in the dead layer of epidermal tissue that covers the body, resulting in little damage or harm. But once inside the body there is no dead layer of cells to protect the living tissues, and alpha particles then deposit all of their energy within a few cell layers. Alpha radiation potentially can cause greater damage to the cells in the immediate vicinity of the source than do other types of radiation.

Beta Radiation

Beta radiation is an external and internal hazard. A highly energetic beta particle can penetrate about one centimeter into the skin. This radiation can deposit energy in the base layers of the skin, but not the internal organs. The lens of the eye is particularly vulnerable to beta radiation. A beta particle deposited in the lens potentially can cause cataracts. Though beta particles are not as energetic as alpha particles, the potential damage area for beta radiation extends beyond the immediate vicinity to adjacent tissues, including nearby internal organs.

Gamma and X-Ray Radiation

Gamma and x-ray radiation are external and internal hazards. The properties of these two types of radiation are identical. Gamma and x-ray radiation can pass completely through the human body, including the internal organs and bones. Because of their ability to penetrate tissue, gamma and x-rays are whole-body radiation hazards.

Radiation Protection Methods

The guiding principle or goal in radiation safety is to keep radiation exposures a̲s l̲ow a̲s r̲easonably a̲chievable, a principle known as ALARA. Exposure is the potential of a radiation field to deposit energy in an individual. There are a few proven methods that, when used, can lower an individual's exposure to a radiation hazard. These radiation protection methods include engineering and administrative controls as well as personal protective equipment.

The basic radiation protection methods use the concepts of time, distance, shielding, and quantity.

- Time: Reducing the time the individual is exposed to a radiation field reduces the individual's radiation exposure.
- Distance: Moving farther from the source of radiation reduces the radiation field by the square of the distance.
- Shielding: Placing radiation-absorbing materials between the source and the individual will reduce the radiation exposure.
- Quantity: Limiting the amount of radioactive material will limit the potential for external and internal exposure.

Radiation Standards

Under the authority of the Atomic Energy Act of 1954, radiation standards provide dose limits to ionizing radiation. This limiting dose is measured in the units of rem, Roentgen Equivalent Man. The occupational limit in the United States for whole-body radiation dose is 5 rem per year. A more common unit of measure of dose is the millirem, or 0.001 rem. There are limits for the extremities, the eyes, and the skin. The latest revision of the occupational radiation standards includes the summation of internal and external radiation doses. There is a limit for a fetus during pregnancy of a declared pregnant woman separate from the mother's occupational limit. The general public limit is 100 millirem.

The basic standards for protection against ionizing radiation, such as those for the extremities and skin or for pregnant women, are written into the Nuclear Regulatory Commission (NRC) regulations. The protection standards are published in Title 10 of the *Code of Federal Regulations* Part 20 (10 *CFR* 20). Other notable parts of Title 10 cover instructions to workers (10 *CFR* 19), licensing of radioactive materials (10 *CFR* 30–40), radioactive waste (10 *CFR* 61), transportation (10 *CFR* 71) and fees (10 *CFR* 170). Although the NRC sets radiation protection standards and rules for radioactive materials, they do not govern x-rays, because x-rays are not produced by radioactive materials,

or x-ray machine design or use. The Food and Drug Administration provide x-ray machine manufacturers with performance standards for x-ray machine design. X-ray machine performance standards are found in Title 21 of the *Code of Federal Regulations* Part 1020.

The NRC has authority over the use of radioactive materials and grants licenses for the privilege of using these materials. There are four basic types of licenses used in industry (not nuclear power). These licenses are a broad license, a specific license, a general license, and a special nuclear materials license. Depending on which license a company has or needs, different requirements apply. NRC has also granted exemptions for certain types of products or uses of radioactive materials.

The NRC has agreements with most states that these states will regulate radiation protection as strictly or more strictly than the NRC. In addition to the radioactive materials usage, most states (nonagreement states included) regulate the use of x-rays as this type of radiation is not covered by the NRC. Some states now have regulations that also cover a hazard known as the use of n̲aturally o̲ccurring r̲adioactive m̲aterials (NORM).

RADIATION SAFETY PROGRAM

Companies that use radiation sources need to have a radiation safety program to protect workers and to comply with safety standards and regulations. The basic objective of any sound radiation control program is to reduce unnecessary exposure to ionizing radiation. A radiation safety program should use some or all of the elements discussed below. The exact mix of these elements will depend on the size and number of the radiation sources and the exposure potentials during their use.

Radiation Safety Officer

A radiation safety officer (RSO) should administer the radiation safety program. RSOs should be well trained and educated to meet the radiation needs of the program as well as the managerial requirements. The RSO should provide advice and assistance to the users and the radiation safety committee, if a committee is needed. The RSO is the individual authorized by a company to officially communicate with the regulatory agencies, both at the state and federal level. This person should have the authority to act on behalf of the company or business.

The RSO is also the individual specifically named on a license to receive communications from the regulatory agencies. A good RSO will help the users comply with applicable regulations and be the cornerstone of a sound Radiation Protection Program.

Exposure Control

Limiting worker exposure to radiation is an essential element of a radiation program, whether the radiation comes from radioactive sources or x-ray producing equipment. Reducing exposure supports the philosophy of ALARA, which is now a part of the regulations. The control of radiation exposure includes minimizing both the external and the internal sources of radiation exposure (dose). Methods of controlling radiation exposures (doses) include posting signs marking Radiation Areas, restricting access to these areas or to radiation-emitting equipment, and training the users in safe procedures and practices. Workers can also be separated from the radiation hazard by distance or shielding with appropriate materials to reduce the radiation received to below the maximum permissible dose.

Contamination Control

Loose radioactive material is the primary source of internal contamination. There are four basic paths that radioactive materials use to enter the body: ingestion, inhalation, injection, and absorption. The potential for ingestion and inhalation is higher than the potential for injection or absorption in most cases. Preventing radioactive materials from gaining access to these paths helps prevent internal contamination and exposure. Control methods and practices to protect individuals from internal radioactive contamination can include wearing gloves; instituting a no smoking, eating, or drinking policy in work areas; respiratory protection; and adequate ventilation.

To develop and implement appropriate controls for radiation exposure or contamination, a company must first determine what kinds of radiation hazards are present in the workplace.

Radiation Detection Instrumentation

Workers and management must use instruments to detect radiation or contamination levels because radiation cannot be detected with the five senses. These instruments take advantage of the ionizing ability of radiation in order to detect it. Passive instruments are able to determine the dose that an individual has received (dosimetry). There are also active instruments for determining radiation field strengths and contamination levels.

There are two common types of portable survey equipment: gas-filled detectors and scintillation detectors. Gas-filled detectors are the most popular and versatile. Within this category there are two types of meters, an ionization chamber and a Geiger–Mueller (G–M) detector. Each type of meter has strengths and weaknesses. G–M detectors respond much faster and can be designed to measure radiation or contamination levels. Ionization chambers measure radiation dose independent of the radiation energy. Both instruments are useful and belong in a sound radiation program.

Scintillation detectors are generally more sensitive than the gas-filled detectors. Scintillation detectors are normally designed to measure a single radiation type, such as alpha particle detection. An alpha detector is designed to measure surface contamination levels. If the radioactive materials program uses alpha-emitting materials, an alpha

detector helps locate contamination so it can be removed. However, the information obtained from such survey meters is only as good as the calibration. Survey instrument calibrations should be based on standards set by the National Institute of Standards and Technology (NIST), formerly the National Bureau of Standards (NBS). The survey instruments should be calibrated to within 10% of the dose rate standard. These calibrations should be performed at least annually, but can be required as often as once a quarter. Recalibration helps ensure the accuracy and consistency of radiation survey measurements.

Dosimetry is used to measure the dose received by individual workers. Commercial dosimetry processors use either film or thermoluminescent dosimeters as the radiation detector. These dosimetry processors have tested their dosimeters' accuracy through the National Voluntary Laboratory Accreditation Program (NVLAP). This program tests the dosimetry in eight radiation categories. Any dosimetry processor used by a radiation safety program should be accredited through the NVLAP process.

Bioassay involves determining the extent of a worker's internal contamination by radioactive materials. This process can be accomplished by direct measurement or analyzing the individual's urine or feces. The method of bioassay will depend greatly on the type of material suspected in the contamination. The bioassay process should be performed only by well-qualified experts who have the equipment and protocols to provide accurate results.

Inventory Control

This aspect of radiation safety is governed by the radioactive materials license and the conditions of acceptable use. Knowing what and where radioactive materials are used and stored helps companies design an effective radiation survey program. Companies are required to keep the amount of radioactive materials they use below the license limits in order to comply with regulations.

Safety Training and Qualifications

Safety training programs and course content will depend on a company's radiation machines or uses of radioactive materials. As a minimum, users should be trained in radiation protection methods and regulatory requirements for their operations. Users should also be trained in the operations they will perform before starting work. The qualifications of principal users should be specified by a knowledgeable radiation safety officer or a radiation safety committee. Principal users will manage the use of the radiation source and employees who work with that source.

Transportation of Radioactive Materials

Transporting radioactive materials either as sources or as waste on a public road is controlled by the requirements of the Department of Transportation (DOT). The DOT specifies the packaging, marking, and labeling of these and other hazardous materials offered for transport. The specifics for radioactive materials are found in Title 49 of the *Code of Federal Regulations* Part 400 through 478. NRC uses 10 *CFR* 71 to determine acceptability for transport packaging.

Radioactive Waste Disposal

Since the enactment of the Low Level Waste Policy Act (LLWPA) in 1980, the disposal of radioactive waste and materials has become an expensive but necessary part of doing business. Waste reduction methods, perceived as costly 10 years ago, are routinely used to reduce current volumes and disposal charges. Waste minimization and effective contamination control will help to reduce the amount and thereby the cost of waste disposal. There has been a dramatic reduction in the amount of radioactive waste generated since the passage of the LLWPA.

NONIONIZING RADIATION

In terms of overall workplace exposures to the electromagnetic spectrum, nonionizing radiation sources have become extremely common. Typically, the electromagnetic spectrum is divided into four general parts: (1) subradiofrequency—0–3kHz, (2) radiofrequency/microwave (RF/MW)—3kHz–300GHz, (3) optical radiation—300GHz–3E15 Hz, and (4) ionizing radiation.

Electromagnetic Radiation

Electric and magnetic fields are present whenever electricity is generated, transmitted, or used. Electric fields represent the forces or strength that electric charges exert on charges at a distance, whereas magnetic fields are produced by the motion of the charge. These fields vary as a function of the frequency, or the number of times that the field oscillates per second. Power systems produce an oscillating electromagnetic field (EMF) of 50–60 times per second, or 50–60 hertz (Hz). At frequencies between 3–3000 Hz, the electric and magnetic fields are essentially independent of each other.

Hazards of Exposure

EMFs have become significant workplace issues. Many people are concerned that constant low-level exposures to EMFs, for example, video display terminals (VDTs), may be associated with health effects damage to reproductive organs or the development of cancer. Although there is no hard evidence that these effects have occurred, some studies have shown that low-level EMFs can induce nausea, metallic taste in the mouth, and dizziness. Researchers suggest that exposure to low-level fields may cause problems in employees who work unusual shifts and lose track of normal day-night cycles. Some evidence indicates that this effect is caused by the interference of EMFs with people's natural rise and fall in body temperature and in hormones.

Other hazards of EMFs include interference with electronic medical implants, metal prostheses, and demagnetizing of credit cards, computer disks, and so on. People with electronic implants, such as pacemakers, must be cautious around EMFs to avoid resetting the devices and causing dangerous irregularities in heartbeat. In addition, even weak magnetic fields can magnetize some metal objects, producing potential hazards.

Despite over a decade of active and detailed research, issues over EMF exposures are still unsettled. For a detailed discussion of the complexities and controversies associated with EMF exposures, see Chapter 11, Nonionizing Radiation, in *Fundamentals of Industrial Hygiene*, 4th edition, and Chapter 11, Public Health Issues, in the *Accident Prevention Manual for Business & Industry: Environmental Management*.

Control of EMF Hazards

Electric fields are relatively easy to stop by shielding the source and transmission power lines. Even a shield such as chicken wire can block a 60 Hz electric field. Shielding can be metal or magnetic alloys to form a barrier, blocking the electric field. The key is that the mesh openings should be no more than ¼ of the wavelength in size.

Other controls used include waveguides and coaxial cables. Waveguides are open metallic conduits with flanged ends that can be fastened to other waveguides, sources, or receivers. Copper tape can be used to patch leak points in enclosures or in waveguides.

Distance is also an effective control, as the potentially harmful effects of electric fields tend to diminish sharply as one moves farther away. For example, many occupational safety and health guides recommend that computer operators sit at least 2 ft (0.6 m) away from their video terminals to minimize EMF exposure.

Standards

Health regulations applying to electric fields tend to fall into two categories: exposure criteria, which describe how much a person can be exposed to, and emission criteria, describing how much be can released into or present in the environment near a source. The International Commission for Non-ionizing Radiation Protection (ICNIRP) has issued general public and occupational exposure criteria based on induced current flow considerations. The American Conference of Governmental Industrial Hygienists (ACGIH) also established threshold limit values (TLVs). A number of states have established emission criteria for transmission power lines by stating the maximum fields that can exist along the edges of the right of way occupied by a power transmission line.

The demand for guidelines to protect workers on the job has resulted in several ideas, two of which deserve mention. The first is prudent avoidance, which relies on reducing magnetic fields when and where possible. The second, 2 mG, represents an averaged exposure level used to define high average exposure scenarios in epidemiological studies. However, it does not define the field strengths that actually might cause harmful effects. Much work remains to be done in this area of worker protection and accident prevention.

In addition, the Institute of Electrical and Electronic Engineers' C95.1–1991 standard, *American National Standard Safety Levels with Respect to Human Exposure to Radio Frequency Electromagnetic Fields* offers exposure and control criteria. Although this standard does not distinguish between occupational and nonoccupational exposures, OSHA can enforce it. The regulations applying to radio-frequency and microwave exposure in OSHA's *Code of Federal Regulations* are obsolete.

SUMMARY

- Lasers produce high-energy beams that are coherent, directional, and monochromatic. Lasers can use either liquid, gas, or solid as the active medium, with the most common being gas lasers. This type of equipment can produce a massive energy beam used for heavy industrial work.
- Hazards associated with laser equipment are classified as Class 1–4 and involve beam and nonbeam hazards. Beam hazards affect primarily the skin and eyes, subjecting these organs to varying degrees of thermal and photochemical burns. Nonbeam hazards include electric shock, heat, toxic byproducts, noise, dust, and explosion of high-pressure flash lamps.
- Companies must identify the unique hazards present in their workplaces and use engineering and administrative controls and personal protective equipment to safeguard workers. Safety goggles and other protective eyewear must be selected for the specific type of laser used.
- The FDA and OSHA have generated specific guidelines and regulations for the use of lasers that require manufacturers and employers to comply with performance standards and safe practices.
- Companies must develop laser safety programs that include such elements as written policies, top management support, appointment of laser safety officers, inventory control, medical surveillance, safety training, and inspections of hazard controls and PPE.
- Ionizing radiation generally includes alpha, beta, gamma, and x-ray radiation. Unlike x-rays, the other types of radiation cannot be turned off and present ongoing hazards to employees in the workplace. Alpha is an external radiation hazard; beta is an external and internal hazard; and gamma and x-ray radiation are whole-body radiation hazards.
- The basic radiation protection methods are limiting workers' time of exposure, keeping the source of radiation at a distance, shielding workers, and limiting the amount of radioactive material used. The Atomic Energy Act, OSHA, and Nuclear Regulatory Commission have established written standards and regulations for the use of radiation and for worker safety.
- An effective radiation safety program includes the following elements: appointment of a radiation safety officer, exposure and contamination control, use of radiation detection instrumentation, inventory control, safety training and qualifications, guidelines for transporting radioactive materials, and guidelines for complying with radioactive waste disposal.
- Nonionizing radiation presents some hazards, but the far-reaching effects of these hazards on human health are not well understood. This type of radiation can be controlled largely by shielding.

REFERENCES

American Conference of Governmental Industrial Hygienists. *Threshold Limit Values* (latest ed.). Cincinnati: OH: American Conference of Governmental Industrial Hygienists.

———. *Documentation of the Threshold Limit Values and Biological Exposure Indices,* 6th ed., Vol III, PA–45–PA–70. Cincinnati, OH: American Conference of Governmental Industrial Hygienists, 1993.

Anderson LE. Biological effects of extremely low-frequency electromagnetic fields: In vivo studies. *AIHAJ* 54:186–196, 1993.

Bonneville Power Administration. *Electrical and Biological Effects of Transmission Lines—A Review,* DOE/BP–945. Portland, OR: Bonneville Power Administration, 1993.

Breysse PN, Gray R. Video display terminal exposure assessment. *Appl Occup Environ Hygiene* 9:671–677, 1994.

Department of Energy and Public Policy, Carnegie Mellon University. *Part 2: What Can We Conclude from Measurements of Power Frequency Magnetic Fields?* Pittsburgh: Carnegie Mellon University, 1993.

Laser Institute of America. *American National Standard for Safe Use of Lasers,* Z136.1–1993. Orlando, FL: Laser Institute of America, 1993.

———. *Safe Use of Optical Fiber Communication Systems Using Laser Diode and LED Sources.* Z136.2-198. Orlando, FL: Laser Institute of America, 1993.

———. *Safe Use of Lasers in Health Care Facilities,* Z136.3–1988. Orlando, FL: Laser Institute of America, 1993.

Sliney D, Wolbarsht M. *Safety with Lasers and Other Optical Sources—A Comprehensive Handbook.* New York: Plenum, 1985.

Swedish Board for Technical Accreditation. *Test Methods for Visual Display Units,* MPR 1990:8 1990–12–31. Northridge, CA: Standards Sales Group, 1990.

———. *User's Handbook for Evaluating Visual Display Units,* MPR 1990:10 1990–12–31. Northridge, CA: Standards Sales Group, 1990.

Varanelli AG. Electrical hazards associated with lasers. *Laser Appl* 7:62–64, 1995.

REVIEW QUESTIONS

1. The term "laser" is an acronym for what words?
2. Lasers use which of the following as the active medium?
   a. Liquid
   b. Gas
   c. Solid
   d. All of the above
3. What two organs of the human anatomy are primarily affected by beam hazards, specifically thermal and photochemical burns?
   a.
   b.
4. Why are nonbeam hazards more dangerous than beam hazards?
5. Why should engineering controls, such as protective housings, interlocked enclosures, and ventilation systems, be the primary method used by companies to control exposure to laser hazards?
6. Which of the following have generated specific guidelines and regulations for the use of lasers that require manufacturers and employers to comply with performance standards and safe practices?
   a. Laser Institute of America
   b. Center for Devices and Radiological Health
   c. Occupational Safety and Health Administration
   d. Food and Drug Administration
   e. All of the above
   f. Only a, b, and c
7. List the seven elements that should be included in laser safety programs developed by companies.
   a.
   b.
   c.
   d.
   e.
   f.
   g.
8. What are the four basic radiation protection methods?
   a.
   b.
   c.
   d.
9. The role of a radiation safety officer (RSO) includes:
   a. Performing medical evaluations of employees' eyes or skin for laser damage
   b. Managing the laser safety program
   c. Communicating with regulatory agencies on behalf of the company
   d. All of the above

21

Contractor and Nonemployee Safety

Good management is evident in the conduct of a business's routine operations and in areas that directly affect relations among managers, employees, outside contractors and their employees, and the firm's customers. Two of these areas, contractor and nonemployee safety, also affect company operations, employees, and the products or services they provide or sell. When accidents occur, operations are interrupted no matter who is involved. With the increase in consumer product safety legislation (see Chapter 16, Product Safety Management) and greater cost of liability claims, management must pay close attention to these areas of potential loss. To help management establish effective safety policies and work procedures with contractors, this chapter covers the following topics:

- safety issues and guidelines in hiring contractor workers
- how to manage contractor safety at all stages of the contractor-employer relationship
- basic safety responsibilities of contractors toward employees
- safety and legal issues in nonoccupational incidents and injuries
- work and other areas in companies that pose special liability issues
- loss control efforts regarding escalators, elevators, stairway, fire, transportation, and attractive nuisances

A company must ensure that its loss control plans include preventing contractor and nonemployee incidents or accidents. These unwanted events can be minor or catastrophic. A firm can be liable for damages or for injuries sustained by someone, such as a customer or temporary office worker, from the minute that person enters the property (including the parking lot). In addition, a firm can be liable for the actions of its employees when they are sent off company property on business should their actions result in damage or personal injury to others.

SAFETY AND CONTRACTOR OVERSIGHT

Today, many companies hire outside contractors to perform a variety of tasks, such as maintenance and repair, clerical duties, construction, and computer installation and training. Based on the "general duty" clause of OSHAct and a 1993 regulation regarding contractor safety performances, employers have a responsibility for the safety of outside employees who either work on the employer's premises or who are sent off site on the employer's business. In 1993, OSHA published a rule emphasizing the qualifications of contractors on pre-bid reviews and the employer's responsibility to obtain information about the contractor's safety record. Although the rule, "Process Safety Management of Highly Hazardous Chemicals, Explosives, and Blasting" (29 *CFR* Part 1910.119), applies

to places where hazardous chemicals are used, subparagraph (h)(2)(i) states:

> The owner (employer) when selecting a contractor, shall obtain and evaluate information regarding the contractor employer's safety performance and programs.

More employers are accepting responsibility not only for the safety performance of their own workers but for the improved safety performance of outside contractors they hire. A comprehensive, systematic safety program can produce the following immediate and long-term benefits:

- reduced injuries and liability risks
- reduced potential regulatory action
- reduced potential for damage to the employer's facility and contractor's equipment
- increased productivity and lower overall costs

Employers with the best safety records understand the key factors that influence contractor safety and how to select contractors who have good safety records. These employers also develop and implement safety programs for outside workers as part of their overall company safety policy.

Factors Influencing Contractor Safety

Studies of West Coast companies who routinely hired contractors and subcontractors revealed that employers can strongly influence the safety performance of these outside workers (CII, 1991). The studies highlighted five key factors in companies with excellent safety records:

1. *Strong Employer Management.* Employers clearly define the jobs contracted workers will do and keep tight control of schedules, responsibilities, training, and problem solving.
2. *Effective Coordination of Job Tasks.* All in-house and outside employees know their roles and responsibilities and to whom they report. Management makes an effort to foster a spirit of teamwork and cooperation among all employees.
3. *Employer Emphasis on Safety.* Employers emphasize safety in their daily communications, reflecting top management's commitment to safe work practices and to safety in general.
4. *Strong Interpersonal Skills of Supervisory Personnel.* Recognition of the individuality of workers and respect for their experience, ideas, and feelings enhances their adherence to safety practices and procedures.
5. *Safe Work Environment in the Employer's Facility.* If outside employees see that safety is already a top priority in the company, they will take safety more

seriously. Employers should make sure that their workplaces conform to all current safety regulations and that written materials, posters, MSDS, and other safety-related materials are posted in a prominent location.

Guidelines for Selecting Safe Contractors

One way to help manage the risk of hiring outside workers is to select those contractors who have good safety records. The preamble to OSHA's new process safety management ruling (29 *CFR* Part 1910.119) explains that

> an employer should be fully informed about a contract employer's safety performance. Therefore, [OSHA] is requiring an evaluation of a contract employer's safety performance . . . and safety programs . . . . The final rule . . . does not require that employers refrain from using contractors with less than perfect safety records. However, the employer does have the duty to evaluate the contract employer's safety record . . .

Employers have three sources of objective information to evaluate the safety performance of outside contractors: contractor safety practices and policies, experience modification rates (EMR) for workers' compensation insurance, and OSHA incidence rates for recordable injuries and illnesses.

Safety Practices and Policies

In general, companies that hold their management and workers accountable for safety have the best safety records. Employers can interview contractor management to determine their attitudes and commitment toward health and safety and examine their safety training materials, on-the-job training, hazard recognition program, substance abuse program, and other safety-related matters. Some of the key items to look for include written policies stating management's commitment to safety, thoroughness and accuracy of accident and illness data, frequency of safety inspections and safety meetings, and work practices and disciplinary measures that stress safe methods of working and protecting employees.

Experience Modification Rates (EMR) for Workers' Compensation Insurance

The insurance industry uses EMR for workers' compensation insurance as a means of determining equitable premiums to charge companies for workers' compensation insurance. These rating systems consider the average accident losses for a given firm's type of work and amount of payroll and predict the dollar amount of expected losses due to work-related injuries and illnesses that the employer will pay over a set period of time. In general, the lower a firm's rates, the better its safety record is likely to be.

Incidence Rates

Employers can also check a contractor's safety record by asking to see their accident and illness records. OSHA requires employers to record and report accident information on Injuries and Illness Summary Form 200. These records must be retained for five years. It is easy to calculate a contractor's or subcontractor's incidence rate by multiplying the number of incidents by the standard base number of 200,000 and dividing the total by the number of hours employees worked. The lower the incidence rate, the fewer accidents and illnesses the company sustained.

These three sources of information can help employers make good choices among the range of outside firms that are available. Employers should request from prospective contractors the EMRs and OSHA incidence rates for the three most recent years. Although the EMR is a more objective measure of a company's safety record than is the OSHA incidence rate, both will indicate past safety performance.

CONTRACTOR EMPLOYEES' SAFETY

Research shows that concerned employers can do a great deal to improve the on-the-job safety performance of outside workers. The Stanford University Department of Civil Engineering sent questionnaires to owners and contractors to determine what measures they took to ensure safe working conditions (*The Business Roundtable,* 1989). The responses, arranged in order of importance, were as follows:

1. Outside workers were required to obtain permits certifying their skill and knowledge to perform potentially hazardous activities.
2. Employers required the contractor or subcontractor to designate a supervisor to be responsible for safety coordination on the job site.
3. Employers provided outside employees with company safety guidelines that had to be followed.
4. Employer-contractor safety meetings were regularly held.
5. Employers conducted regular safety audits of outside employees' work.
6. All accidents were reported to employer management immediately.
7. All accidents involving outside employees were investigated promptly.
8. Employers stressed safety as part of the pre-bid process.
9. Employers maintained statistics of outside workers' accidents.
10. Employers conducted periodic safety inspections.
11. Employers set goals for safety on the job.
12. Safety was considered as a major factor in prequalifying bids.
13. Where appropriate, employers set up a safety

department or function specifically to monitor outside worker safety.

14. Safety guidelines were established in the body of the contract.
15. Employer orientation sessions informed outside workers of safety hazards on the job and how to control, avoid, or eliminate them.

Although none of the employers carried out all 15 measures, those with the best safety records implemented at least eight to ten. However, several key activities, described in the following sections, were common to all employers.

Pre-Award Meetings

Before the contract or bid is awarded to outside firms, the employer should clearly specify the scope of the work to be done and emphasize the importance of safety. This information is best communicated in face-to-face meetings with contractor management. These meetings are particularly important when the work to be done is hazardous or will be carried out in a hazardous area.

Safety Requirements in Work Contract

Once the contract is awarded, the employer should specify in detail in the contract how safety will be managed and monitored and what bearing safety requirements will have on costs. The contract language should go beyond the usual standard statements (e.g., "the contractor must comply with all federal, state, and local safety regulations"). Employers should require a detailed safety plan that includes such information as safety inspections, enforcement, staffing, permits required, testing for substance abuse, basic safety training, and accident reporting and investigation. The contract should also specify regular reporting of accident and illness statistics to the employer.

Safety Orientation

Before work begins, the employer should meet with contractor management and safety staff to conduct a safety orientation and review session. The employer should address and document the following items:

- Work site safety requirements, including providing safety manuals and standards related to the proposed work.
- Detailed outline of the safety responsibilities of the employer, outside management, and outside workers.
- Any special hazards that exist at the work site, including a review of hazardous materials and relevant MSDSs.
- Training requirements, including orientation, for outside personnel.
- Schedule to review safety auditing, performance, and training programs after the first few days and weeks on the job.

Safety Record Keeping

Employers should require contractors to turn in monthly and, if appropriate, annual reports of incidence rates. Employers will need to total the number of accidents and contractor employee hours worked in order to analyze these data. To show that they are serious about the safety of outside workers, employers can establish a monitoring program such as the following:

- Regular (weekly) safety progress reviews between employer and outside management.
- Regular safety progress reports that include such items as new contractor employees, documentation of training given, summary of injuries and lost workdays, safety meetings conducted, and number of attendees.
- Reports of safety inspections by contractor management and a summary of steps taken to correct safety problems.
- Safety inspections by employers and a review of the findings.
- Investigation of any incident or accident that occurs at the facility. Employers should review the results of accident investigation reports with contractor management. They should also keep detailed files for third-party litigation situations.
- Chemical screening programs administered by contractor management that are at least as strict as the employer's.
- Annual evaluation by the employer of the safety performance of each contractor. The results of these evaluations should be placed in the employer's file.
- Careful auditing of each contractor firm for compliance with all safety obligations and adherence to various safety programs as specified in the contract and in discussions with the employer. Nonsafety-related activities are not included in this audit.

Besides conducting their own safety efforts, employers must also insist that contractor management actively participate in creating a safe work environment and ensure employee health and safety on the job.

CONTRACTOR SAFETY RESPONSIBILITIES

When requesting bids from contractors or subcontractors, employers should request that each firm submit a written safety program. The extent of the program will depend on the safety requirements of the particular job. The following items are some of the safety provisions that should be included.

1. The contractor should have a written safety program document containing:

   - management policy statement
   - safety goals and objectives and a method of measuring the program's effectiveness
   - safety responsibilities for managers, supervisors, safety representatives, and employees

- written procedures for safety activities such as new employee safety orientation, training, enforcement, inspections, personal protective equipment, medical services, and substance abuse program
- written operating procedures, with methods on how to obtain permits

2. The safety document should contain hazard communication procedures that conform to OSHA Hazard Communication standard (29 *CFR* 1926.59):

   - written hazard communication program and designated administrators
   - MSDSs available for all materials used on the job
   - all containers properly labeled and showing hazard warning
   - documented employee hazard training program

3. The contractor should have a safety handbook for employees, covering responsibilities of management and employees and important safety procedures, warning signs and symbols, and so on.
4. Contractor management should provide orientation for all new employees, covering safety responsibilities, safety rules, first-aid facilities, accident reporting, safety meetings, PPE use, and reporting of unsafe conditions or actions.
5. All safety enforcement procedures should spell out the consequences of safety violations and the disciplinary actions that will be taken.
6. Job safety inspections should be conducted by contractor or subcontractor managers, safety personnel, and supervisors. Specific areas for inspection include facility operations, building and ground conditions, housekeeping programs, electrical circuits and equipment, lighting, heating and ventilation, machinery, chemicals, fire prevention, maintenance, and personal protective equipment.
7. Accident investigation procedures should cover how to report all OSHA recordable injuries and illnesses to the employer in a timely manner, how to investigate the causes of the accident, and how to prevent a recurrence.
8. Contractors should issue work permits, such as safe work permit, hot work permit, excavation permit, confined space entry permit, crane permit, lockout/tagout, and others.
9. Contractor management must provide specified personal protective equipment for hazardous conditions and employee training in their proper use and care. The employer should have the right to verify that the contractor is providing the right equipment and that contractor employees are actually using this equipment.
10. First aid and medical services must be provided to meet OSHA requirements. However, small contractor firms may elect to use the employer's services. If so, this decision should be stated in the contract.
11. Contractors' substance abuse programs should be as comprehensive as the employer's.
12. Emergency response guidelines must be prepared in coordination with the employer's safety personnel.
13. Periodic (monthly) safety reports should be submitted to the employer, including a record of new contractor employees, documentation of training, summary of injuries and illnesses, and safety meetings conducted and number of employees who attend.

Safety on the job is the shared responsibility of the employer and contractor management and employees. A strong partnership between the two organizations can reduce or eliminate risks, injuries, and illnesses; help control health and insurance costs; and improve employee productivity and morale.

The second major area of risk for many companies involves safety issues related to nonemployees, such as customers, visitors, and others who enter the building or work site of a company. The next sections describe the liability of companies and steps they can take to reduce that liability.

NONEMPLOYEE ACCIDENT PREVENTION

In today's business environment, more people are involved in serving and selling to the public than are engaged in making the products that are sold. Service businesses include specialty shops, department stores, discount stores, shopping center complexes, restaurants, fast-food service operations, hotels and motels, automotive service and dealerships, hardware and building supply stores, amusement parks, banks and office buildings, and franchise businesses. Accidents involving the patrons, guests, or visitors of these businesses make up the category of nonemployee injuries.

Service and sales operations must constantly monitor patron activities for potential problems, even though they cannot control customers' personal activities or habits. For example, inebriated guests or patrons represent a serious risk when they improperly dispose of smoking materials or their instability causes them to slip, fall, or suffer other injuries. Children can also be involved in accidents in the business environment. Whenever possible, design the original facility for predictable use and misuse.

LEGAL SIDE OF NONOCCUPATIONAL INJURIES

Customer or product claim cases can end up in a courtroom. The legal terminology and aspects of the law that deal with accident claims include the following definitions and explanations. These are not intended to be comprehensive but to provide a quick review of some principal considerations.

Tort

A tort is a private or civil wrong resulting in an injury—a violation of a right not arising out of a contract. It may be either (1) a direct invasion of some legal right of the individual, (2) the infraction of some public duty by which special damage accrues to the individual, or (3) the violation of some private obligation by which damage accrues to the individual. Torts result from negligence, accidents, trespass, assault, battery, seduction, deceit, conspiracy, malicious persecution, and many other wrongs or injuries.

Negligence

Negligence is the failure to exercise that degree of care which an ordinarily careful and prudent person ("reasonable individual") would exercise under similar circumstances. To establish a proper claim of negligence, however, there must be (1) a legal duty to use care, (2) a breach of that duty, and (3) injury or damage.

Degree of Care

The degree of legally required attention, caution, concern, diligence, discretion, prudence, or watchfulness depends upon the circumstances. For example, a high degree of care is demanded from people who invite others onto their premises by written, verbal, or implied invitation. All sales and service enterprises must exercise a high degree of care for the safety of their patrons. As long as a business is open, it assumes a responsibility for the well-being of its customers.

Invitee

An invitee is one whose presence on the premises is upon the invitation of another, such as a patron at a sports stadium or a person who visits an exhibition hall, even though no admission is charged.

Licensees

Licensees are neither "invitees" nor "trespassers." They have not been specifically invited to enter upon the property but they have a reasonable excuse (by permission or by operation of the law) for being there. These could be vendors, delivery personnel, people visiting executives, or purchasing agents for business purposes, and the like. Policemen and firefighters who enter property in the course of their duties have sometimes been held by the courts to be invitees (patrons) and sometimes licensees (nonpatrons).

Limitations of Recovery

American businesses are not automatically insurers of their clients or patrons. All states apply some formula to limit the ability of insured persons to recover monies from a company through a civil lawsuit. These legal doctrines fall into two broad categories. The oldest, called contributory negligence, bars any financial recovery by an injured party if the victim contributed to the original accident in any way.

A newer, and perhaps more humane approach, is found in the legal doctrine called comparative negligence. This doctrine, adopted by most states, requires a court to limit the recovery of an injured party based upon how much their own action contributed to the original incident. Both doctrines illustrate the need for thorough and truthful investigation of every incident. It can strengthen a company's ability to reduce the overall cost of doing business and help to prevent future lawsuits.

Assumption of Risk

Claimants cannot collect damages when they were aware of peril or danger yet were willing to proceed with their original intention and undertook the action. "That to which a person assents is not regarded by law as an injury." For example, a skier who falls while descending a slope legally assumes the normal risks associated with this sport, unless some special negligence in the design or maintenance of the slope and its environment is a contributing factor. If so, the skier assumes no risk when the owner or operator is negligent. However, an injury caused by a mechanical defect in a chair lift or tow rope could prove costly for the company running the operation or for the owners of the ski resort.

Hold Harmless

A clause in a contract agreeing that one party will assume all liabilities, losses, or expenses involved is a hold-harmless agreement. For example, a department store may have a hold harmless agreement with a manufacturer who supplies a particular type of merchandise. If a consumer claims an injury was caused by that product, the manufacturer will reimburse the store, should it be held liable, and will pay for legal and other expenses incurred by the store in defending itself. Even though the consumer purchased the item from the department store, a hold-harmless agreement may relieve the store of the financial effect of a tort.

Attractive Nuisance

This item refers to liability associated with a dangerous condition that is generally a threat to children. It excuses trespassing and penalizes an organization for failure to keep children away from the hazard or for failure to protect or eliminate a hazard that may reasonably be expected to attract children to the premises. A swimming pool is often considered an attractive nuisance because children are drawn to it, regardless of the protective features.

Burden of Proof

The burden of proof means that the injured party must prove injury or damage and its causal relation to the event or item that created the accident. A defendant is not liable if he or she is without fault. Proof must be established by facts, not opinion, suspicion, rumor, hearsay, gossip, or emotional reaction. Proof is a conclusion drawn from the evidence.

Honest and sincere witnesses often report widely different impressions about the same events. Thus, it is essential to assemble and preserve evidence quickly. Signed statements taken shortly after the accident or an all-important photograph can often make a big difference in a trial. In liability claims, the burden of proof rests upon the plaintiff (claimant).

This chapter cannot cover all possible nonemployee accidents. Instead it describes the most common ways that nonemployees can suffer an injury. Accident prevention techniques are broad and varied. It will be up to each safety and health professional to examine the company's operations in light of these guidelines. The following sections review some major problem areas and discuss typical accidents that occur in modern business facilities.

PROBLEM AREAS

When starting a nonemployee loss control process, the safety and health professional should look over the records of past nonemployee injuries to identify accident trends. Next, choose general trend areas, such as stairs, doors, those due to poor lighting, faulty interior design, and poor traffic flow. Change facilities and/or operations so that both employees and nonemployees are given a safe business and shopping environment.

Glazing

Modern office building, store, and plant design often uses glass extensively in doors, show windows, panels, and enclosures. Such areas can result in a confusing pattern, especially to the first-time visitor or patron.

Unmarked glass panels and doors can be the cause of severe injuries and cuts should employees or patrons walk or fall through them. Companies must follow recommended practices and standards to correct or eliminate this hazard. The safety and health professional or manager should understand regulatory standards on this issue.

Tempered glass panels, required for many uses, are given a special heat treatment during production so that, if broken, the pieces of glass will fall as small, blunt beads instead of sheet glass plates.

The improper use of glazing materials is a continuing source of legal problems for companies dealing with the public. A complete check of all glazed items is important.

Some plastic glazing offers advantages similar to glass but without the risk of damage or injury. Unfortunately, some plastics have higher impact resistance and tend to scratch or be more easily abraded than glass. As a result, workers may find it harder to break plastic windows or doors than glass if they need to escape from an area.

A few precautions will allow companies to enjoy the beauty and functionality of glass while lessening the chances for injury:

- Make glass doors more visible to adults and children by placing decals or pressure-sensitive tape at their respective eye levels. Sand-blasted or etched designs serve the same purpose.
- Decals or pressure tapes will also prevent glass panels from appearing to be doorways. A tall, attractive plant, placed in the center of the panel, will let people know the panel is there.
- Installing safety bars reduces the size of the open glass areas and lessens the chance of glass breakage. The bars should be at the door-handle level on sliding doors and should be on both sides of a swinging door.
- Keep doorways and areas that are close to glass panels free of tripping hazards; common hazards include scatter rugs and toys. Indeed, this is a good rule for all occupied areas.

Parking Lots

Most shopping or business centers have parking lots for patron use, usually self-service. Bicycles, mopeds, and motorcycles as well as automobiles, vans, and trucks require parking facilities. If these self-service parking lots are well designed and attractive, they can almost eliminate the risk of damage to any vehicle in the lot.

Parking lot problems depend to some extent on how the lots are used. Self-service parking lots for industrial plant employees, for instance, usually differ from public lots at shopping centers, theaters, stadiums, stores, schools, restaurants, or motels. Usually, public lots have a more steady flow of traffic, more children present, and more obstacles such as shopping carts obstructing traffic ways. More importantly, the lots are used by persons with varied levels of driving skill who may be less familiar with the lot's layout. Some public parking lots at places such as theaters, sports arenas, and schools have the combined hazards of extreme traffic fluctuation and great variations in driver skill and alertness.

An easy-to-use layout, adequate signs, and conspicuous markings help to make a parking lot safe and attractive. These steps represent the first way to reduce hazards. Of several other safety measures, none is more important than enclosing the parking lot with a curb or fence so that cars cannot enter or leave traffic unexpectedly. This step should significantly reduce accidents in the area. Parking lots should have separate, well-marked entrances and exits, placed so that they favor right-hand turns. Single-lane entrances should be at least 15 ft (4.5 m) wide; exits should be at least 10 ft (3 m) wide. Where entrance and exit must be combined, the double-lane drive should have at least 26 ft (8 m) of usable width. When the double-lane combination is necessary, it should have median curbs or strips to positively control the flow of traffic.

In general, entrances and exits should be:

- at least 50 ft (15 m) from intersections
- away from heavily traveled highways or streets
- well marked and well illuminated
- as few in number as possible.

Pedestrian traffic must be considered in parking lots. The most common injury occurs from hazards such as potholes, changes in elevation, and cracks in the walking surfaces due to uneven settling of the ground, wear of heavily loaded vehicles, and effects of cold weather. Maintenance staff should conduct periodic inspections to identify and rectify these hazards and to maintain a safe walking surface for patrons. Stairways and ramps should be constructed in accordance with ANSI standard A1264.1–1989, *Safety Requirements for Floor and Wall Openings and (Nonresidential) Stair and Railing Systems.* They should be well marked, well illuminated, and guarded with handrails. The needs of the physically disabled must also be considered in the design of the lot and in any walkway to the building. See Chapter 13, Workers with Disabilities, in the *Engineering & Technology* volume.

Parking aisles should be perpendicular to the buildings, so that pedestrians walk down the aisles rather than between parked cars. Where possible, marked lanes, islands, or raised sidewalks should be provided between parking rows, particularly where pedestrian traffic is especially heavy. These walks should be wide enough so that car bumpers overhanging them do not restrict pedestrians from walking to or from their cars. Companies should minimize the use of curbs as they present a tripping and falling hazard.

Angle parking has both advantages and disadvantages. Although fewer cars may fit in the lot, angle parking is easier for customers and does not require as much room for sharp turns.

The area allowed per car in parking lots varies from 200 to more than 300 ft² (18.5 to 28 m²) if aisles are included. (See Chapter 2, Buildings and Facility Layout, in the *Engineering & Technology* volume.)

Parking Stalls

The design of parking stalls depends on (1) the size and shape of the lot, (2) the traffic pattern in the lot, and (3) the type of driver and pedestrian who use the lot. (See Figure 13-13a and 13-13b in Chapter 13, Workers with Disabilities, in the *Engineering & Technology* volume for stall dimensions for the physically disabled.)

Bumpers on stalls have these advantages:

- They prevent drivers from driving forward through facing stalls and proceeding in the wrong direction in one-way aisles.
- They encourage drivers to pull forward against the bumper, thereby preventing rear overhang that might reduce aisle width.
- They break up the huge expanse of an open lot that may tempt drivers to cut across aisles and endanger pedestrians and other drivers.
- They block cars from accidentally rolling down inclines or running through walkway areas.

Stall bumpers can cause some problems, however:

- They may require maintenance.
- They may interfere with drainage or snow removal.
- They may cause pedestrians to trip and fall. (Painting them a bright yellow or other distinctive contrasting color may help to reduce such incidents.)
- They may reduce the flexibility of traffic flow.

Signs and Lighting

Traffic signs in parking lots should conform to recommended standards. They should be like those used as street and highway signs for easy recognition.

For example, stop or yield signs should be installed at main crosswalks for pedestrians, wherever exits cross public sidewalks, and wherever exits enter main thoroughfares.

A well-lit parking lot reduces accidents and discourages crime. The amount of light recommended for parking at night usually ranges from 0.5 to 1.0 foot candles (decalux) per sq ft at a height of 36 in. (0.9 m). Lights are usually mounted 30 to 35 ft (9 to 10 m) high on poles whose bases are protected against impact by cars.

Design

Built-in bumps on lot surfaces have been used to discourage speeding. However, because they can also damage cars, their use seems questionable. An alternative device is to keep straight lanes short and to provide sharp curves that force drivers to reduce their speed.

Parking lot operators should control unauthorized use of company lots. Local police or organizational security personnel should check parking areas after hours. If lots cannot be fenced or entrances cannot be protected against unauthorized entry, prominent warning signs stating the lot is guarded will help to minimize the risks of unauthorized use and vandalism.

Those responsible for supervising lots should cooperate closely with local traffic authorities. It is also important to cooperate with police, sheriff, and fire departments. Fire trucks will need access to hydrants in a lot and to buildings served by the lot. Zoning boards are also interested in problems connected with parking lots because of the impact on traffic flow and congestion.

Shopping Carts

Many companies want to keep shopping carts out of parking lots. To achieve this goal, companies should have customers drive their cars to a loading point where an attendant puts their goods into the car or wheels the cart to the car, unloads it, and brings it back.

If shopping carts are permitted in the parking lot, they should be collected frequently and temporarily stored in a designated space. Do not allow carts to accumulate in stalls in heavily trafficked aisles. Cart-collecting areas should be well marked, well illuminated and, preferably, separated from traffic by barriers in the pavement. Encourage customers to leave carts in those areas by installing signs and access lanes. To keep carts from rolling into traffic lanes, install low curbs or a shallow depression. People who supervise the lot should make sure that children do not play with the empty carts or use them as scooters and racers.

In inclement or cold weather, shopping carts should be stored indoors. The buildup of ice and snow on carts that are stored outdoors will melt when the carts are brought inside and can cause a slipping hazard on the floor.

In addition to shopping carts, stores should also provide motorized carts to help transport disabled customers. This provision will help stores comply with regulations in the Americans with Disabilities Act (ADA).

Other Vehicles

The use of bicycles, mopeds, and motorcycles as commuting vehicles is fairly commonplace. More people will use these vehicles if (1) the routes are safer and (2) secure parking is available. Current parking facilities are not generally adequate for these vehicles. Like most motorists, operators of bicycles, mopeds, or motorcycles prefer to park near their destination. In planning for such parking, two major problems arise: (1) convenience and (2) security.

Bicycle storage is fairly easy, but the problem of security is more difficult. Twelve to 15 bicycles can be stored in the space it takes to park one car. A lesser number of mopeds or motorcycles will also fit in the same area.

Three basic types of bicycle parking facilities are in current use. The common bicycle rack comes in many styles and shapes and requires chaining or clamping the bike to the rack. Unless the bicycle frame and both wheels are locked, the owner can lose part of the bike. Second, riders can use a hitching post with a chain secured to lock the bike. Finally, they can store bicycles in a key-operated locker much like baggage lockers. These are totally enclosed, protected from thieves and the weather—an advantage where weather protection is a concern, especially for all-day parking. Moped owners and motorcyclists face most of the same convenience, security, and weather problems as bicyclists.

Entrances to the Building

Entrances to department stores and office buildings must be adequate in number and size to meet all building codes. Revolving doors should have governors that limit their speed to 12 turns per minute. Maintenance should replace all worn weatherstripping and keep sidewalks and driveways cleared and in good repair to prevent tripping and falling hazards. Entrance lighting should be a minimum of 10 foot candles.

Stores that serve the public also must be accessible to the disabled under the ADA. Some significant issues include width of floor aisles, accessibility of entrances and exits, bathroom access, and assistance in reaching shelves of different heights.

Types of Ramps

To provide passage from the sidewalk to the customer's car, it is customary in shopping centers to construct a ramp. Indented ramps are the best type, ones that actually cut into the sidewalk with an easy slope. Rails on each side can prevent people from falling into one of the recessed sides if they are sufficiently deep to be a hazard. Do not paint the ramp because paint seals the concrete surface and can create a slipping hazard.

Entrances to buildings are of considerable importance from a safety viewpoint. Often entrance and exit doors are automatic and are activated when anyone steps onto a carpet. With increased use of automatic entry/exit doors to provide customers with easy access to retail locations, these stores also have increased exposure to customer injury claims caused by malfunctioning automatic doors. Companies must provide periodic maintenance and adjustment to ensure that the doors operate safely and are free of mechanical problems. If the vestibules are glass-enclosed, decals should be applied to alert people against walking into glass panels, as discussed earlier.

Walking Surfaces

Slips and falls are the most likely sources of nonemployee injuries. These injuries can occur almost anywhere at any time. Few surfaces can be ignored, including everything from asphalt roads; concrete walks; wooden, tiled, or rug-covered floors; and special surfaces on stairs and conveyances (moving sidewalks, escalators, elevators), to bridges and catwalks.

The natural properties of any surface can change substantially when people track in mud, snow, dirt, and moisture. Moisture-absorbent mats, runners, or rugs can reduce such hazards. Floor maintenance requires special attention to eliminate the hazard of torn or curled-up floor coverings. The company should implement administrative controls to ensure that these hazards are quickly removed. Floors, stairs, and other walking surfaces should be kept nonslippery, clear, and in good repair.

Another method that can provide a safer walking surface is to use slip-resistant flooring products to help prevent falls. Most slip-resistant materials can be installed in the same manner as any other vinyl tile product. However, the slip-resistant products still require periodic maintenance to preserve their appearance and to ensure that the product remains hazard free.

Slip meters have been developed by testing agencies and insurance companies to measure the slipperiness of floors. One type of instrument (Figure 21-1), mounted on three leather "feet," is pulled across the floor by a motorized winch. The dial on top measures the intensity of the pull required to start moving it. This is converted into a "slip index," or the coefficient of friction (0.5 to 0.6 is

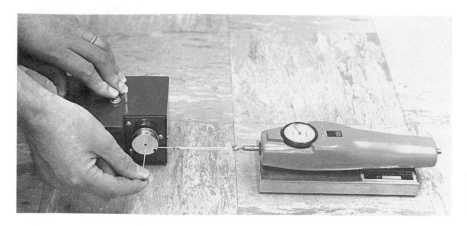

Figure 21-1. Slip meters, ranging from the motorized type (shown here) to a simple spring scale and heavy block pulled by hand, can be used to gauge the slipperiness of floors. (Courtesy Liberty Mutual Insurance Company.)

ideal). Although important, this "coefficient of friction" is not foolproof. Floor slipperiness may change and usually increase because of moisture, oil, grease, foreign or waste materials, and incorrect cleaning or waxing.

Falls on Floors

Falls on floors occur in various ways and from various causes. A person may slip and thus lose traction, or trip over an open drawer, box in the aisle, or other object. The primary mechanical causes of falls on floors are unobserved, misplaced, or poorly designed movable equipment, fixtures, or displays; poor housekeeping; and defective equipment. The condition of a person's shoes or type of footwear soles and heels are likely to be major contributing factors in slips and falls.

Inadequate illumination also can cause falls. Light values at floor level should be uniform with no glare or shadows. Also, there should be no violent contrasts in light levels between floor areas, such as from bright sunlight outside the entrance to a dimly lit lounge or restaurant.

Other factors causing falls include a patron's age, illness, emotional disturbances, fatigue, lack of familiarity with the environment, and poor vision. Because a company cannot readily control many of these factors, it should make sure that the walking surface is as safe as possible. For example, management should not place mirrors and other distracting decorations in areas visible from steps or from approaches to steps or escalators.

Types of Floor Surfaces

A wide variety of floor surfaces are available. In office buildings, hotels, mercantile stores, and similar establishments, the use of masonry (terrazzo, cement, or quarry tile) floors is common at entrances, in lobbies, on stairways, and sometimes throughout the ground floor and in upper floor corridors. Decorative materials such as terrazzo, marble, and ceramic tile are most often applied to interiors while concrete and granite are generally considered more practical for exterior use (Tables 21-A and 21-B).

In other public areas of these buildings, the base floor, usually of concrete or wood, is generally surfaced with one or more of the popular resilient floor-covering materials. Management generally uses carpeting in limited areas of various stores and hotels. Elsewhere, asphalt, linoleum, rubber, or plastic in either sheet or tile form, will usually be found. Obviously, safety, initial cost, durability, and maintenance costs are some factors governing the choice of floor covering. In every case, follow the recommendations of the floor material manufacturer. Companies should not improvise. Many accidents happen when managers attempt to improve the original appearance of a floor material at the expense of safety.

Most flooring materials, whether wood, masonry, or the resilient types, are reasonably slip resistant in their original, untreated condition. However, some of the masonry materials are exceptions to the rule. A highly polished marble, terrazzo, or ceramic tile, used for ornamental effect, can be slippery even when dry. When moisture is present, its slipperiness will be greatly increased. Improper surface-treating preparations and improper cleaning materials and methods also increase a floor's slipperiness. Unless a slip-resistant material is added to the compound during construction, the only preparation that should be used on such floors is a penetrating, slip-resistant sealer.

Floor Coverings and Mats

Reduce the possibility of slips and falls by using good carpeting, bound edging, and flush floor-level mats and runners (Figure 21-2). Over a given period of time, carpet threads tend to become loose, creating a tripping and falling hazard. This condition is more of a risk in wall-to-wall carpeting. Maintenance staff should inspect carpeting regularly for potential hazards. When loose threads are found, have the carpet restretched or replaced to eliminate the hazard.

Whenever possible, provide a contrasting color on carpeted areas that meet and continue on treads and risers (or treads only) of stairways. If material such as an extruded metal runner is used to provide self-cleaning removal of snow, ice, or mud at entryways, it should be flush and not present a tripping hazard. Maintenance staff

Table 21-A. Physical Properties of Floor Finishes

Types of Finish	Resistance to			Quality of			
	Abrasion	Impact	Indentation	Slipperiness	Warmth	Quietness	Ease of Cleaning
Portland cement concrete *in situ*	VG-P	G-P	VG	G-F	P	P	F
Portland cement concrete precast	VG-G	G-F	VG	G-F	P	P	F
High-alumina cement concrete *in situ*	VG-P	G-P	VG	G-F	P	P	F
Magnesite	G-F	G-F	G	F	F	F	G
Latex-cement	G-F	G-F	F	G	F	F	G-F
Resin emulsion cement	G-F	G-F	F	G	F	F	G-F
Bitumen emulsion cement	G-F	G-F	F-P	G	F	F	F
Pitch mastic	G-F	G-F	F-P	G-F	F	F	G
Wood block (hardwood)	VG-F	VG-F	F-P	G-F	F	F	G
Mastic asphalt	VG-F	VG-F	VG-F	VG	G	G	G-F
Wood block (softwood)	F-P	F-P	F	VG	G	G	G-P
Metal tiles	VG	VG	VG	F	P	P	G-F
Clay tiles and bricks	VG-G	VG-F	VG	G-F	P	P	VG
Epoxy resin compositions	VG	VG	VG	VG	F	F	VG

Code: VG—Very Good; G—Good; F—Fair; P—Poor; VP—Very Poor.

Figure 21-2. A heavyweight rubber or plastic mat with a nubby finish or raised design and beveled edges tends to lie flat and stay in place. Rotating the mat distributes the wear and minimizes "bald" spots in high-traffic locations.

must pay careful attention to rubber mats, rug runners, and the like to prevent them from becoming tripping hazards. Often times their edges become rumpled, corners and ends are torn or do not lie flat, and excess wear causes tears. Management should replace mats and runners at the first sign of wear or other unsafe conditions.

Eliminate slips and falls on throw, oriental, and area rugs by using skid-resistant rug pads. When using rugs over carpet, attach a skid-resistant underlay to the rugs. These underlay pads can be purchased from carpet dealers. Floor mats, runners, and carpeting are used wherever water, oil, food, waste, and other material on the floor might make it slippery.

A company should establish clear procedures for placing, cleaning, removing, and storing mats. Those who put mats in place during inclement weather should have specific instructions about when and where mats should be put down and removed. If workers do not place the mats promptly and close enough to the door, the entranceways may become slippery; and patrons and others track in water and dirt beyond the entranceway, creating a hazard and maintenance problem in other locations. Maintenance staff should follow definite procedures for inspecting and checking the condition of mats and for maintaining them in a safe condition. Consider a dry mop in the entry way to help keep floors dry and free of slipping hazards.

Table 21-B. Guide to Floor Materials and Surfacings

Floor types*	Characteristics	Use of Abrasives	Dressing Materials
Asphalt tile	Composed of blended asphaltic and/or resinous thermoplastic binders, asbestos fibers, and/or other inert filler materials and pigments.	Abrasive materials of various types may be used to reduce slipperiness of floors. Colloidal silica** can be incorporated in wax and synthetic resin floor coatings.	**Wax or wax-base products**—For most purposes, wax has several advantages. This is especially true of Carnauba wax, an ingredient generally used in so-called wax products. This wax, a Brazilian palm tree product, dries in place with a very hard and glossy finish, but with a characteristically slippery surface. Because of its many good qualities, it is widely used as a base for floor surface preparations, both in paste and emulsion forms. Other waxes, notably petroleum wax and beeswax, have their place in floor dressing formulas; they are softer and less slippery than Carnauba, but are still slippery to a degree depending on the formulation.
Linoleum	Cork dust, wood flour, or both, held together by binders consisting of linseed oil or resins and gum. Pigments are added for color.		
Rubber	Vulcanized, natural, synthetic, or combination rubber compound cured to a sufficient density to prevent creeping under heavy foot traffic.	Slip-resistant except when wet.	
Vinyls	Composed of inert, nonflammable, non-toxic resins compounded with other filler and stabilizing ingredients.	Adhesive fabric with ingrained abrasives can be used. They are patterned in strips, tiles, and cleats.	
Terrazzo	Consists of marble or granite chips mixed with a cement matrix.	Silicon carbide or aluminum oxide can be included in mix when floor is laid. Also an abrasive-reinforced plastic coating can be painted on.	**Slip-resistant sealant** will typically improve slip-resistant quality if renewed periodically.
Concrete	Made of portland cement mixed with sand, gravel, and water and then poured.		
Mastic	Like asphalt tile in composition but is heated on the job and troweled onto the floor to form a seamless flooring. Such floors are often used over concrete to give a new durable, resilient surface.	(Same as asphalt tile)	**Synthetic resins**—These preparations, known as "synthetics," "resins," or "polishes," are intended to supply the desirable characteristics of wax without producing the same degree of surface slipperiness. They include soaps, oils, resins, gums, and other ingredients, compounded in various ways to produce the desired result.
Wood	May be either soft or hard, in a variety of thicknesses and designs.	Metallic particles and artificial abrasives in varnish or paint give good nonslip qualities to various floors.	
Cork tile	Made of molded and compressed ground cork bark with natural resins of the cork to bind the mass together when heat cured under pressure.	(Same as asphalt tile)	**Other materials**—Paint products (paint, enamel, shellac, varnish, plastic) are semipermanent finishes used principally on wood and concrete floors. They do not materially increase the slipperiness of the base.
Steel	Iron containing carbon in any amount up to about 1.7 percent as an alloying constituent, and malleable when used under suitable conditions.	Surface can be touched up with an arc welding electrode so the shape of raised places on the surface resembles angle worms. Also an abrasive reinforced plastic coating can be painted on to any desired thickness, dries hard as cement, and has a sandpaper like finish. If a temporary nonskid surface is needed, two uses of mats can be employed: (1) flexible rubber mats made from old automobile tires; (2) rubber or vinyl runners.	
Clay and quarry tile	Kiln-dried clay products are similar to bricks and are extensively used in areas requiring wet cleaning.	Typically resistant to abrasives.	May be treated by etching. May be formulated as nonslip by adding carborundum or aluminum oxide when mixing the clay before kilning.

*Floors and stairways should be designed to have slip-resistant surfaces insofar as possible; adhesive carborundum strips may be used on stair treads or ramps and at critical concrete areas. Etching with mild hydrochloric (muriatic) acid solution will lessen slip problems.

**Colloidal silica is an opalescent, aqueous solution containing 30 percent amorphous silicon dioxide and a small amount of alkali as a stabilizer.

Stair rails, treads, and surfaces should be designed per standards, codes, and regulations, and be kept in good repair and checked frequently for defects. Make sure nothing is stored on the stairways and landings that could contribute to falls.(See also Chapter 19, Office Safety.)

Merchandise Displays

Workers must construct displays so they are stable to prevent articles from falling and injuring or tripping the customers. Repair and smooth sharp or broken displays and counters so they do not cut or catch passersby.

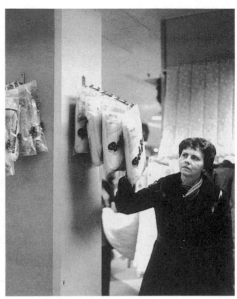

Figure 21-3. Point-of-purchase display hangers must be located in such a way that the human eye cannot contact them (such as shown at left where both the upper and lower shelves project beyond the hangers). Shelf hangers must be safeguarded if they project into the aisle (center). Locating projections above eye level is another safe method (right).

Marketing people in some stores (like supermarkets) try to stack merchandise high on a display table to catch customers' attention. However, if these displays are stacked too high, they become a hazard for the shopper. A customer reaching for an item can set off a cascade of boxes, cans, bottles, or other items that can strike or injure the person or create a tripping hazard.

When a product is stacked too high, the customer may need to step onto a lower shelf to reach it. Staff should stack shelves evenly by layers, placing heavy items on the lower shelves and lighter items on the top shelves (Figure 21-3, left).

In many stores, workers may card and hang nonfood accessories such as hardware items, notions, and kitchen equipment on pegboard panels. These panels or sections should be adequately recessed to accommodate the extended J-hooks (with minimum J-radius of 1 m). Remember, because shoppers will be bending over to reach lower items, the hooks above should not extend so far as to be a potential risk to a person's eyes or face. Specially safeguarded extenders can eliminate this problem (Figure 21-3, center).

Merchandise with sharp or cutting edges should not be in open displays unless the edges are covered or otherwise protected. Protective sleeves or plastic coatings or covers serve to guard the edges from customer handling. Likewise, all electrical and mechanical display elements should have adequate protective features.

Construct display platforms using color(s) or lighting that contrasts with the floor or carpet, and do not allow them to obstruct aisles. Round or clip all corners of platforms. Displays and mannequins should be at least 6 in.

(15 cm) off the floor so that customers will not tip them over accidentally. Make sure the display or the mannequin is fastened to its base.

Floor displays should be at least 3 ft (0.9 m) high so they are visible without becoming a tripping hazard. They should not be at the ends of aisles where shopping carts can dislodge them.

If a company hangs its displays from the ceiling, it should be sure that ceilings are structurally sound and that all code requirements have been met. In addition, any duplex electrical receptacles should be located 18 in. from the ceiling on certain columns with another outlet 12 to 18 in. from the floor.

Housekeeping

In mercantile establishments, it is estimated that fixtures, displays, and other portable equipment are involved in over 40% of customer falls. It is, therefore, essential that management provide safe equipment and that the loss control process particularly emphasize safe placement and use of that equipment. Also, workers should remove dress racks and stock trucks from the sales area and return them to the stock room as soon as the items have been emptied.

All electrical wiring and extension cords hooked to store machines, displays, special decorations, and the like should not lie exposed on the floor. They can present a serious tripping hazard. Where necessary, install wires or cords in low-profile channels.

Poor housekeeping accounts for one third of all customer falls. Each employee needs to realize that it is part of his or her responsibility to maintain good housekeeping in the work area and to promptly report unsafe floor conditions, such as tears in carpets and holes in the floors. Train workers to wipe up or barricade spills immediately or as soon as possible until the hazard can be removed. A special warning sign can be used.

ESCALATORS, ELEVATORS, AND STAIRWAYS

Many business establishments move people between floors by escalators, elevators, stairways, and ramps. See also the discussion in Chapter 15, Hoisting and Conveying Equipment, in the *Engineering & Technology* volume.

Escalators

Escalators can be operated from a low of 70 feet per minute (fpm) to a speed of 125 fpm (0.36 to 0.64 m/s). The average recommended speed is 90 fpm (0.46 m/s). A 4-ft (1.2-m) wide escalator can move 4,000 to 8,000 passengers per hour. If escalators are operated faster or slower than people anticipate, passengers can be injured, especially children or elderly patrons.

Accidents may occur when passengers are entering or leaving escalators or while they are riding them. Common problems that can result in such accidents are:

- Unsafe floor conditions and poor housekeeping at landings
- Sales counters, mannequins, display bases, and similar units that hamper the movement of passengers
- Lights or spotlights facing passengers as they step on or off at landings
- Mirrors near escalator landings causing passengers to misjudge their step and stumble
- Merchandise signs or displays distracting riders, causing them to bump into one another or to fall
- Failure of passengers to step on the center of a step tread, causing them to fall, possibly against others
- Overshoes, particularly the thin plastic type, sneakers, or other objects catching in the comb-plate when pressed against a riser or catching at the side of the moving steps
- A passenger "riding" a hand on the handrail beyond the combplate and back into the handrail return. This can cause an injury if exposed fingers run into the handrail guard.
- An Emergency Stop button, unprotected against accidental contact. This hazard can cause an incident if a passenger (or object) mistakenly presses it and halts the escalator unexpectedly.
- A package, stroller, or other conveyance placed on the escalator and jamming between the balustrades or slipping from the grasp of the person trying to hold it
- Parts of the body becoming involved with moving parts of the escalator resulting in falls, lacerations, and amputations of the toes, fingers, or other parts of the body

- A passenger walking or running up or down a moving escalator
- Children sitting on escalator steps
- A passenger who fails to hold onto a moving handrail and stumbles or falls

Escalator Standards and Regulations

When installing or modifying escalators, check local and state ordinances. Also refer to National Safety Council Data Sheet 12304–0516, *Escalators* (available through the NSC library). The American National Standard (American Society of Mechanical Engineers) A17.1, *Safety Code for Elevators and Escalators,* is known as the *Elevator Code.* Some of its important points are:

- Hand and finger guards are to be protected at the point where the handrail enters the balustrade.
- Prominent caution signs should be displayed (Figure 21-4).
- Comb-plates with broken teeth should be replaced immediately.
- Strollers, carts, and the like must be prohibited on escalators.

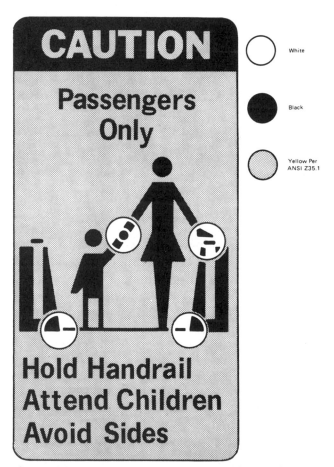

Figure 21-4. Caution sign recommended for escalators by ANSI/ASME standard A17.1–1993. Sign should be 4 in. (102 mm) wide by 7 in. (197 mm) high.

The *Elevator Code* also states that balustrades must have handrails that move in the same direction and substantially at the same speed as the steps. The company should also follow city, state or provincial, and local code requirements.

Inspections

Examine all escalators from landing to landing every day. This includes riding them before the store opens to discover any defects.

Once a week, inspect the escalators:

- Replace any broken treads or fingers in step treads and comb-plates.
- Examine handrails for damage.
- Check balustrades for loose or missing screws and for damaged or misaligned trim.

Semiannually, inspect:

- step chain switches
- governor
- top and bottom oil pans
- skirt clearances
- step treads and risers
- steps
- machine brake
- skirt switch
- handrail brushes.

Start and Stop Controls

Escalators should have an EMERGENCY STOP button or switch located at the top and the bottom landing. These must stop but not start the escalator. By placing a STOP button at the base of a newel, with either recessed design or a cover, the company can prevent a person or object from pressing the button by accident. All employees should be trained in the location and use of emergency switches.

Elevators

Business establishments with elevators face other accident problems. One of the most common is a customer being struck by the door. This is particularly common in self-operating elevators.

Inspections

A logbook containing the following information should be kept:

- day, month, year, and time of inspection
- observations by mechanics or inspectors
- all breakdowns, including causes and corrective action(s)

Entries should be initialed and dated.

A company must comply with all regulatory standards and should use the ANSI/ASME standard A17.1, the Elevator Code referred to under Escalators. At least every three years, all elevators should have a balance test and a contract load test. Make any required adjustments.

Once a year, hand-test the following controls:

- governors
- governor cable grip jaws
- gripping jaws of car and counterweight
- releasing carrier
- cutoff switches
- tail rope and trip rod drums
- safety rails

Spot-check automatic elevators at the start of each day for level floor stops, brakes, and other mechanical operations. Test the alarm bell to be sure it rings and that its signal registers in the maintenance department. See also Chapter 15, Hoisting and Conveying Equipment, in the *Engineering & Technology* volume for more details.

Stairways

Because stairways are so important in emergency exits from buildings, they are covered in detail in the National Fire Protection Association's (NFPA's) publication *Life Safety Code,* NFPA 101, Chapter 5, Means of Egress. That chapter also includes ramps, exit passageways, smokeproof towers, outside stairs, fire escape stairs and ladders, illumination, exit marking, and escalators and moving walks. Only inside stairways are discussed in the following section.

The typical stair accident occurs when someone slips while descending a stairway. It has been found that most fall victims tend to look at stairway treads far less than do other people, thus they are more prone to accidents. See the U.S. Department of Commerce study *Guidelines to Stair Safety* (listed in References at the end of this chapter).

The following list represents major areas of concern, along with their corrective actions, that companies need to address.

- Because stairways are not level walking surfaces, they require different walking habits. Companies can minimize their use by posting signs directing people to the escalators and elevators. Also, never chain or otherwise lock exits to and from stairs so they cannot be used in an emergency.
- Design handrails so that employees and patrons can grasp them firmly and slide their hands along the rail without encountering obstructions. When designing handrails, keep in mind that a simple round profile permits the fingers to curl around the rail, in effect "locking" the hand in place. Two inches is the recommended maximum diameter for a typical handrail.
- Highlight treads and handrails to help riders quickly and easily distinguish them from the riser and wall

surfaces. Adequate lighting is essential; NFPA specifies a minimum of one foot-candle (decalux), measured at the floor. Arrange lights so they do not create glare surfaces or temporarily blind stair users and so the failure of one unit will not leave any area in darkness. If one side of the stairwell is open to adjoining space, it is a good idea to close off that view to prevent distractions that may cause people to fall.

- The edge of the step or tread (nosing) should be easy to see. If possible, use contrasting carpeting to distinguish the approach to the steps from the stairs themselves. If not, special nosing may be installed to provide definition between the steps. Nosings must be securely fastened. Uncarpeted stairs should be edgemarked.
- A continuous handrail must be provided; see details in Chapter 19, Office Safety, in the section Aisles and Stairs. The railing preferably should be of a lighter color, because people seem to be more inclined to use a lighter-colored railing than a dark one that looks dirty and greasy. The railing should be kept clean. The handrail should extend to the top and bottom of the staircase so that it may be grasped before stepping on the first step or leaving the last step of the flight. Be sure that the railing does not extend into a passageway and become a hazard.
- Stair treads need to be stable and provide good traction. Outdoor stairs need extra slip resistance and should have adequate water runoffs that lead away from other walking surfaces.
- Worn or defective treads or other parts should be replaced immediately.
- Stairs should be kept clear of obstructions. Make sure sharp handrail and guardrail ends are removed or covered to prevent injury. Clearly mark and protect glass areas adjacent to stair landings or at either end of the stairway to prevent people from walking through them. Fixtures should not project into the stairway.

PROTECTION AGAINST FIRE, EXPLOSION, AND SMOKE

If fire breaks out or an explosion occurs, a quick, orderly evacuation of the premises will protect all persons including visitors. Base all evacuation plans on the premise that most visitors will be on the property for the first time. They need to be directed to exits by numerous special direction signs, even though exit locations might be well-known to regular employees.

Businesses that generally attract large numbers of customers, guests, or patrons should provide well-marked exits, emergency lighting, and enough direction signs to guide people to safe exits. In a fire, dense, penetrating smoke can be as deadly as the heat or flame and can quickly sear people's lungs. Therefore, it is essential to evacuate the fire area immediately.

Enclosed stairwells provide the best fire escape routes. Doors to such stairwells must never be obstructed, locked, or propped open. A public address system, staffed by a qualified and trained person, can be used to direct the building's evacuation and issue life-saving instructions.

Usually the early detection of a fire and the use of a good evacuation plan can prevent panic and personal injuries. More details are given later in this chapter and in Chapter 15, Emergency Preparedness.

Every business should consider installing a fire-detection and suppression system for maximum safety. Management should make sure that the fire plan is updated regularly and that fire drills are held at least semi-annually. These steps are necessary so that employees are familiar with the plan and where to meet outside the building should a fire break out.

Fire Detection
Properly engineered fire detection systems are sound investments, but the best installation is useless if no one responds to the alarm. Systems should have a direct connection to the local fire department or security service.

The Four Stages of Fire
Fire is a chemical combustion process created by the rapid combination of fuel, oxygen, and heat. A full discussion is found in Chapter 11, Fire Protection, in the *Engineering & Technology* volume. Most fires develop in four distinct stages: incipient, smoldering, flame, and heat. Detectors are available for each stage.

- Incipient stage. No visible smoke, flame, or significant heat develops but a large number of combustion particles are generated over time. These particles, created by chemical decomposition, have weight and mass, but are too small to be visible to the human eye. They behave according to gas laws and quickly rise to the ceiling. Ionization detectors respond to these particles.
- Smoldering stage. As the incipient stage continues, the combustion particles increase until they become visible—a condition called "smoke." No flame or significant heat has developed. Photoelectric detectors "see" visible smoke.
- Flame stage. As the fire condition develops further, ignition occurs and flames start. The level of visible smoke decreases and the heat level increases. Infrared energy is given off that can be picked up by infrared detectors.
- Heat stage. At this point, large amounts of heat, flame, smoke, and toxic gases are produced. This stage develops very quickly, usually in seconds. Thermal detectors respond to heat energy.

Burning Plastics
Some fuels such as plastic waste receptacles can produce considerable toxic smoke when they burn. Therefore, the

use of nontoxic and noncombustible materials (such as metal cans) is preferable. Managers and supervisors should be aware that other materials, such as building finishes, carpentry, furniture, and office fixtures, also can give off toxic smoke. For example, polyvinyl chloride (PVC) in a single foot of one-inch sized PVC rigid nonmetallic conduit involved in a fire:

- can produce a sufficiently heavy, dense smoke to obscure 3,500 ft^3 (100 m^3) of space
- can generate enough hydrogen chloride to provide a lethal concentration of HCl in approximately 1,650 ft^3 (45 m^3) of space.

Engineering and Control Procedures

The best fire-detection system is only as good as its weakest component. Management should utilize the services of a fire protection consultant in engineering the system and establishing the control procedures. Here are four steps to consider:

1. Select the proper detector(s) for the hazard areas. For example, a computer area may involve ionization or combination detectors. A warehouse may have infrared and ionization detectors. In low-risk areas, thermal detectors or combinations of detectors may be used.
2. Determine the spacing and locations of detectors to provide the earliest possible warning.
3. Select the best control system arrangement to provide fast identification of the exact source of alarm initiation.
4. Ensure notification of responsible authorities who can immediately respond to the alarm and can take appropriate action. Every detection system must have an alarm signal transmitted to a constantly supervised point. If this cannot be ensured on the premises, the signal must be transmitted to a central station, fire department, or other reporting source.

Sprinkler systems can also be viewed as detection systems as well as extinguishing systems. Routine maintenance must be provided for these systems on a regular basis.

Response Methods

Early warning systems are as important during hours of occupancy as they are when the premises are closed.

When detection pinpoints a trouble spot, immediate response by a responsible trained company representative is all important. Shaving seconds and minutes can mean the difference between lives lost or saved and between fire confined or allowed to spread out of control.

Some systems that companies use include the following.

Twenty-four-hour Supervisory Service

If an installation has 24-hour, seven-days-a-week supervision at some point in the building, or building complex,

then management should hook alarm, trouble, and zone signals into this location.

Less Than 24-hour Supervisory Service

For periods when an installation does not have responsible personnel to respond to the alarm, a backup system should be provided. The NFPA advises connecting to a central station supervisor's service or other service.

Such systems should include a way to initiate fire and trouble signals to the central station transmitter equipment. In the case of a trouble signal, a representative can be dispatched immediately to investigate the trouble and to notify proper representatives of the property under surveillance.

Central Station Monitoring Unavailable

In areas where no central station supervision is available, the local fire or police department may accept installation of a remote fire alarm panel at their headquarters or firehouse.

Central Station or Telephone Leased-line Tie-in Unavailable

If none of the foregoing possibilities exists, then management should consider using qualified and licensed telephone answering services. Also, automatic dialing units connected to responsible officials are another alternative.

High-Rise Building Fire and Evacuation Controls

Just what is a high-rise building? These are four basic criteria to designate a high-rise structure. First, the size of the building makes personnel evacuation impossible or impractical. Second, part or most of the building is beyond the reach of fire department aerial equipment. Third, any fire within the building must be attacked from within because of building height. Fourth, the building has the potential for "stack effect."

High-rise megastructures all share one characteristic—they are intended to house people. As buildings go higher up, the population density per square foot of ground area increases, posing a whole new set of problems concerning the health, safety, and welfare of building occupants. Actually, each building is a sealed life-support system. While engineering approaches have improved heating and cooling systems, they are often extremely wasteful of energy. Because these buildings are usually airtight, occupants are at greater risk of death or injury from smoke and toxic combustion by-products.

The ladders on most fire department aerial equipment are limited to approximately 85 ft (26 m). This means that a building higher than about eight to ten stories cannot be served by this equipment. Because of the height problem and increased floor areas, fires must be fought from within the building. The principal firefighting equipment includes automatic sprinklers, hose standpipes, and portable extinguishers, as well as hose lines from the building exterior.

When a building must be evacuated, management should have an orderly plan in place. An evacuation checklist is given in Figure 21-5. Several sources are available to help companies devise a plan. The U.S. Occupational Safety and Health Standards Subpart E, Means of Egress (Title 29, *Code of Federal Regulations (CFR)*, Chapter XVII, Part 1910, Subpart E; Revised as of July 1, 1979; Amended by 45 FR 60203, September 12, 1980) mandates in Section 1910.38:

(a) The emergency action plan shall be in writing. (The only exception to this regulation is for locations with ten or fewer employees.)

(a) (5) Training (ii) The employer shall review the plan with each employee covered by the plan at the following times:

(A) Initially when the plan is developed,

(B) Whenever the employee's responsibilities or designated actions under the plan change, and

(C) Whenever the plan is changed.

Although OSHA is oriented to the employees, obviously such a plan should include delegation of evacuation leaders and a series of steps to evacuate both employees and nonemployees in an emergency.

Stack Effect

Every building has its own peculiarities for creation of a "stack effect," or spread of fire, smoke, and toxic fumes from one area to another. Among the factors are structure configuration, height, number and size of openings, wind velocities, temperature extremes, number and location of mail chute openings, and elevator shafts, all of which create varying air flows that tend to accelerate and intensify an interior fire. Unprotected air-conditioning systems are an open invitation to catastrophe. If there is no automatic smoke and heat detection, no automatic fan shutdown, and no automatic fire dampers, smoke and toxic fumes can be quickly drawn into the exhaust or return air duct system and promptly distributed to all other floors and areas of the building served by the air-conditioning system.

The NFPA recognizes this potential. Its Standard No. 90A, *Installation of Air Conditioning and Ventilating Systems,* states that in systems of over 15,000 cfm (7 m³/s) capacity, smoke detectors should be installed in the main supply duct downstream of the filters. These detectors automatically shut down fans and close smoke dampers to stop the recirculation of the smoke, or they may incorporate automatic exhaust.

In planning evacuation, assume that children, elderly, and physically disabled will be involved. In addition, some people panic in a fire situation. The quantity and size of staircases will undoubtedly prohibit complete evacuation. Tests indicate that with an occupant load of 240 persons per floor, total evacuation of an 11-story building can take up to 6 minutes, while an 18-story building can take up to 7 minutes. Exits are just not designed to handle all occupants simultaneously.

Most codes do not consider elevators to be an exit component and prohibit their use during fire emergencies. But codes generally also require that one or more elevators be designated and equipped for firefighters. Key operations can transfer automatic elevators to manual operation and bring the elevators to the street floor for use by the fire service. The elevators should be situated so the fire department can find and use them easily.

Many elevators use capacitance-type call buttons, which may bring them to a stop on the fire-involved floor. Then they cannot move because smoke interrupts the light beam and keeps the doors open. Other possibilities include the inadvertent arrival of the elevator at the fire-involved floor by a passenger who does not know there is a fire, or a person who pushes the call button and then, in panic, uses the staircase for exit. With problems of this magnitude, it can be assumed that complete evacuation is impossible.

Crowd and Panic Control

Any commercial establishment may be faced with an unruly crowd because of an emergency, panic, or even a planned demonstration. Self-interest dictates that management preplan for such events to protect the facility and its employees, along with patrons and bystanders. Different measures are required depending on whether the panic occurs in the building or outside of it and who is involved. In shopping areas, companies should have directional signs displayed at many areas in the building. Exit signs are especially important. But these alone do not reduce the higher risk potential involved when a great number of people are gathered, such as at sports events, entertainments, schools, and other places.

In almost every emergency, some panic is likely. Employees need periodic drills and practice in handling emergency situations with customers, some of whom may be confused and others who may be physically disabled. Management cannot overlook the threat of an unusual occurrence (such as a riot, bombing, and the like) and must make plans to handle such a possibility.

Although high-rise buildings pose new and special problems, other public places such as theaters and amusement and recreation facilities must also provide well-planned emergency procedures. Panic and the press of frantic, hysterical people have caused considerable loss of life in many emergencies. Often, such losses could have been prevented through the strict observance of building and fire codes to eliminate hazards and "death traps" caused by improper design and lack of firefighting and disaster-control measures.

Civil strife and sabotage are covered in Chapter 15, Emergency Preparedness.

Demonstrations

The following procedures concern controlling public demonstrations at a store, but they can easily be adapted to the needs of any establishment.

Evacuation Preparedness Checklist

All questions in this checklist should be answered with "yes," "no," "NA" (not applicable), or "U" (undetermined). For all answers that are not "yes," or "NA," the persons responsible for the specific areas in question that need correction should be noted.

Floor Diagrams
❑ Are floor plans prominently posted on each floor?
❑ Is each plan legible?
❑ Does the plan indicate every emergency exit on the floor?
❑ Does a person looking at the plan see an "X" indicating "you are here"?
❑ Are room number identifications for the floor as well as compass directions given?
❑ Are directions to stairwells clearly indicated?
❑ Are local and familiar terms used on the diagram to define directions to emergency exit stairwells? For example, are particular areas identified, such as mail room, cafeteria, personnel department, wash rooms?

Exit Paths to Stairwells
❑ If color coding of pillars and doors, or stripes and markings on floors are used, are they properly explained?
❑ Is additional clarification needed?
❑ Are paths to exits relatively straight and clear of all obstructions?
❑ Are proper instructions posted at changes of direction en route to an emergency exit?
❑ Are overpressure systems and venting systems operative?

Elevators
❑ Are signs prominently posted at and on elevators warning of the possible dangers in using elevators during fire and emergency evacuation situations?
❑ Do these signs indicate the direction of emergency exit stairwells which are available for use?

Elderly and Physically Disabled
❑ Are there elderly or physically disabled persons who will need assistance during a fire and emergency evacuation of premises?
❑ What provision is made for their removal during an emergency?
❑ Who will assist? How will the disabled be moved?

Emergency Exit Doors
❑ Are all emergency exits properly identified?
❑ Are exit door location signs adequately and reliably illuminated?
❑ Do exit doors open easily and swing in the proper direction (open out)?
❑ Are any exit doors blocked, chained, locked, partially blocked, obstructed by cabinets, coat racks, umbrella stands, packages, etc.?
   Note: Blockage must be removed immediately and subsequently prohibited.
❑ Are all exit doors self-closing?
❑ Are there complete closures of each door?
❑ Are all exit doors kept closed, or are they occasionally propped open for convenience or to allow for ventilation?
   Note: This practice must be prohibited.

Emergency Stairwells
❑ Are stair treads and risers in good condition?
❑ Are stairwells free of mops, pails, brooms, rags, packages, barrels, or any other obstructing material?
❑ Are all stairwells equipped with proper handrails?
❑ Does each emergency stairwell go directly to the ground floor exit level without interruption?
❑ Does the stairwell terminate at some interim point in the building?
❑ If so, are there clear directions at that point which show the way to completion of exit?
❑ Is there provision for directing occupants to refuge areas out of and away from the building when they reach the ground floor?
❑ Are directions provided where evacuees can congregate for a "head count" during and after the evacuation has been completed?
❑ Is there adequate lighting in the stairwell?
❑ Are any bulbs and/or fixtures broken or missing?
❑ Where? Describe locations?
❑ Are exits properly identified?
❑ Are they illuminated for day, night, and power-loss situations?
❑ Are any confusing nonexits clearly marked for what they are?
❑ Are floor numbers displayed prominently on both sides of exit doors?

Emergency Lighting
❑ In the event of an electrical power failure or interruption of service in the building, is automatic or manually operated emergency lighting available?
❑ If not, what will be used?
❑ Where are stand-by lights kept?
❑ Who controls them?
❑ How would they be made available during an emergency?
❑ Is there an emergency generator in the building?
❑ Is it operable?
❑ Is it secured against sabotage?
❑ Is a "fail-safe" type of emergency lighting system available for the exit stairwells that will function automatically in event of total power failure?
❑ How long can it provide light?
❑ Is the emergency lighting tested on a regular monthly basis with results recorded? Who maintains such record?

Communications
❑ How should occupants of the building be notified that an emergency evacuation is necessary?
❑ Is one or more communication system available to each floor? (P.A. system, Muzak, stand-pipe phones, battery-operated "pagers," etc.)
❑ If messengers must be used, have they been properly instructed?
❑ Is the communications system(s) in good working condition?
❑ Under what emergency conditions is it used and who operates it?
❑ Can announcements be prerecorded by someone with a calm but authoritative voice?
❑ Is the communications system protected from sabotage?
❑ Do all occupants know how to contact building control to report a dangerous situation?
❑ Is the building's emergency communications system tested monthly? By whom and to what extent?

Figure 21-5. Reprinted from National Safety Council Occupational Safety and Health Data Sheet 656, *Evacuation System for High-Rise Buildings* (available from NSC Library).

Demonstration Outside of Store Building. Advise employees to call security and/or management. Security should telephone police, advise them of the situation, and follow their instructions.

Arrange for two or more key personnel to assume previously assigned positions at all store entrances and other key points; they should know security's telephone number to relay information and to receive instructions. They should never leave their assigned posts unless relieved or advised accordingly. Caution them to remain calm and not to interfere with the entrance or exit of customers or employees.

Those employed in portable, high-valued merchandise departments (diamonds, furs, etc.) should arrange to have such merchandise placed in an assigned secure area. Those employees working in departments selling firearms, knives, axes, straight razors, bows and arrows, and even meat cutlery, should have them removed from the selling floor to a secure area. Proceed with "business as usual" in all other departments.

Because rumors can create panic or problems among employees and customers present, all employees should be advised of two or three emergency interior telephone numbers to verify information and squelch rumors.

Demonstration Moving into Store. (People carrying signs, groups linking arms across aisles and taunting employees, fights between individuals.)

1. Advise all employees to avoid any comments, antagonism, or physical contact with marchers, to answer all queries courteously, and above all to keep calm.
2. In areas where demonstrations are taking place, have employees stop selling, lock their registers, remain in their areas keeping as calm as possible, and await further instructions from their supervisors.
3. In areas where "business as usual" is being maintained, arrange for frequent cash pickups.
4. Key personnel and employees should take their assigned places, as discussed above.
5. In the event such a demonstration turns into group looting or group "hit and run" stealing, employees should not attempt to make any apprehensions. Security personnel will follow previous orders for such conditions, as advised by management. Remember, personal safety is more important than property protection.

Self-Service Operations

The best-known self-service operation is found in gasoline stations. Here, customers put the gasoline in their vehicles without employee assistance. The employee merely sees that customers observe the safety rules and follow the prescribed procedure, usually posted on the pump housing or on a nearby sign. All states require that customers turn off their engines and refrain from smoking or using open flames.

Evacuation of the Physically Disabled

The evacuation of physically disabled persons from hotels, stores, and other facilities is an added problem for both management and the safety professional. Communication, especially, can be a problem. Special written instructions can be given to people with a hearing impairment; verbal instructions can be given to the visually impaired.

More details on how to help the physically disabled are presented in Chapter 13, Workers with Disabilities, in the *Engineering & Technology* volume.

TRANSPORTATION

Some businesses by their very nature have special accident control problems. Much of their loss-control effort must be directed at protecting the nonemployee from harm.

All types of transportation from commercial airlines, railroads, marine, and buses to local transit, taxi, and school bus operations must be vitally concerned with preventing passenger injuries. Not only does this involve maintenance and vehicle or unit operation but also the safety of areas around vehicles, such as terminals, stations, school-bus loading areas, and the like.

Courtesy Cars

When a company provides transportation for customer courtesy and convenience (such as a hotel or motel courtesy car or bus), it should follow the same precautions used in commercial operations. These include providing the safest vehicle, maintaining it in proper working condition (meeting all regulatory requirements), and operating it with a professional driver who is trained and skilled in all facets of the vehicle's operation.

Company-Owned Vehicles

Another source of damage and injury claims arises out of the operation of company motor vehicles by employees. Because the odds are that one of every five drivers will have a collision in any given year, liability from such accidents is a constant threat and often a real dollar drain for insurance protection and claim settlement. Management must be aware of the liability resulting from an employee's use of his or her own car on company business.

PROTECTION OF ATTRACTIVE NUISANCES

Every business can suffer losses caused by the public's curiosity about their equipment or operations. Some examples follow.

Any unattended vehicle or machine left in an operable condition is attractive to the young (and the so-called "young in heart"). "Tamper-proof" locks discourage the unauthorized use of vehicles or machines and, if necessary, security officers should be employed to watch over the equipment.

Frequently, partially finished road repairs, construction, or storms and adverse weather can present serious hazards for motor vehicles. Companies should barricade

Figure 21-6. Basic lifesaving equipment must be available at every pool. Shown here at each lifesaving platform are ring buoys with the throwing rope attached. (Courtesy National Spa and Pool Institute.)

such hazards and warn motorists away from them. Refer to the U.S. Department of Transportation's *Manual on Uniform Traffic Control Devices for Streets and Highways.*

Many contractors or builders provide special, safe observation facilities for "sidewalk superintendents," that is, members of the public who like to watch construction projects. Local authorities and insurance engineers should be consulted for regulations and control measures.

One of the more common attractive nuisances is a swimming pool, particularly one where the public can easily enter the swimming area. Many hotels, motels, and public areas provide swimming pools for their patrons (Figure 21-6). Pool accidents usually result from inadequate protective barriers around pools, absence of lifeguards or qualified adult supervision, disregard for the rules of good pool conduct, and the failure to teach youngsters drowning-prevention techniques.

Pool precautions include:

- Screens, fences, or other enclosures to control admittance. A tamper-proof lock and pool alarm may provide additional protection.
- Accurately marked pool depths in feet or meters, on both pool deck and poolside.
- No diving boards unless the pool has been constructed and staffed for diving.
- Adequate safety measures to prevent diving accidents. The depth of water in a pool is extremely difficult for the eye to measure. This presents a hazard for persons diving. Every year numerous injuries occur when people unknowingly dive into shallow water. To prevent such accidents, the following is recommended.

- In shallow water areas, post warning signs such as "DANGER," "NO DIVING ALLOWED," "SHALLOW WATER."

- Designate specific diving areas where the water is sufficiently deep.
- Use either pool markings or plastic or wood floats strung on rope to differentiate between the diving and no-diving areas.

- Do not store pool cleaning agents near flammable or reactive chemicals. Containers of pool cleaners that are not empty should not be thrown in the trash.
- Make sure that after chemical treatment, pool water has the proper pH balance before allowing swimmers to use the pool.
- Basic lifesaving equipment, which should include a lightweight, strong pole with blunt ends at least 12 ft (3.7 m) long, or a ring buoy with long throwing rope.
- A competent pool manager and a lifeguard on duty whenever the pool is in use.
- A telephone nearby, such as in the bathhouse or changing room. Emergency telephone numbers should be on hand—the nearest available physician, ambulance service, hospital, police, and the fire and/or rescue unit.
- Decks around the pool kept clear of debris and no breakable bottles allowed in the area. Make sure all cups and dishes used at poolside are unbreakable. Provide litter baskets and replace defective matting.
- No games near the pool that could injure anyone.
- Electrical equipment conforming to local regulations and the latest National Electrical Code requirements. Any electrical appliance used near the pool must be protected by a ground-fault circuit interrupter (GFCI).
- No swimming allowed in the pool during a thunderstorm.
- All pool appliances and equipment maintained properly. Periodic safety checks should be made.
- Sensible pool rules established, posted, and enforced.

SUMMARY

- Employers who accept responsibility for the safety of outside contract workers can have a significant influence on the safety programs of those contractors. To help ensure a safe workplace and to protect contractor workers' health and safety, employers can establish their own safety programs, select safe contractors, and insist on written, implemented safety programs developed by contractors.
- Companies must be familiar with their legal duties and liabilities regarding nonemployee accidents and injuries. Two of the most common sources of nonemployee accidents are (1) unmarked glass doors, panels, and windows and (2) parking lots.
- Other areas that commonly present hazards and must be safeguarded include building entrances, walking surfaces, merchandise displays, escalators, elevators, and stairways. Companies should implement safety measures to reduce or eliminate tripping and falling hazards and to post warning signs and other safety markings where nonemployee patrons can easily see them.
- Every business should install adequate fire detection equipment to sound the alarm at any of the four stages of a fire. Fire detection and fire response systems must be adequate for each building's requirement and should be tied into company and municipal fire-fighting departments.
- Management and employees must be trained in and practice evacuation procedures to handle crowds and control panic among nonemployees. Company employees should be able to conduct patrons and visitors, including those physically disabled, quickly and efficiently to well-marked exits.
- Companies should also develop procedures to handle public demonstrations on their property. Employees should know how to communicate with security staff, summon police, handle demonstrators who enter the building, and guard their own safety.
- Courtesy cars and company-owned vehicles can be sources of nonemployee accidents and injuries. Companies must make sure their courtesy vehicles are in good repair and operated by a skilled driver.
- Firms must follow state and local ordinances and guidelines for preventing the public from gaining access to attractive nuisances and being injured or killed as a result.

REFERENCES

American National Standards Institute, 11 West 42nd Street, New York, NY 10036.
Elevators and Escalators, ANSI/ASME A17.1–1993.
Safety Requirements for Workplace Floor and Wall Openings, Stairs, and Railing Systems, ANSI A1264.1–1995.

The Business Roundtable. *Improving Construction Safety Performance: A Construction Industry Cost Effectiveness Project Report.* New York: The Business Roundtable, Report A-3, 1982, 1989.

Chemical Manufacturers Association. *Improving Owner and Contractor Safety Performance.* Washington, DC: American Petroleum Institute, 1991.

Construction Industry Institute (CII). *Managing Subcontractor Safety.* Austin, TX: University of Texas, 1991.

Guidelines to Stair Safety, Department of Commerce, U.S. Government Printing Office, Washington, DC 20402.

International Conference of Building Officials, 5360 South Workman Mill Road, Whittier, CA 90601. Uniform Building Code, section 3303(j).

Matwes GJ and Matwes H. *Loss Control: A Safety Guide-book for Trades and Services.* New York: Van Nostrand Reinhold Co., 1973.

National Fire Protection Association, 1 Batterymarch Park, Quincy, MA 02269.

Installation of Air Conditioning and Ventilating Systems, (1989) NFPA 90A.

Life Safety Code, (1990) NFPA 101.

National Electrical Code, (1993) NFPA 70.

National Safety Council, 1121 Spring Lake Drive, Itasca, IL 60143.

Dennis L and Onion M. *Out in Front: Effective Supervision in the Workplace.* Chicago: National Safety Council, 1990.

Safety and Health Reprints.

Occupational Safety and Health Data Sheets.

Falls on Floors, 12304-0495, 1991.

Fire Prevention in Stores, 12304–0549, 1990.

Floor Mats and Runners, 12304–0595, 1986.

Sidewalk Sheds, 12304–0368, 1990.

National Spa and Pool Institute, 2111 Eisenhower Avenue, Alexandria, VA 22314.

Superintendent of Documents, U.S. Government Printing Office, Washington, DC 20402. Commercial Practices, Title 16, *Code of Federal Regulations,* Chapter II, July 22, 1996, Consumer Product Safety Commission.

U.S. Department of Transportation. *Manual on Uniform Traffic Control Devices for Streets and Highways,* D6.1.

REVIEW QUESTIONS

1. How does "Process Safety Management of Highly Hazardous Chemicals, Explosives and Blasting" (29 *CFR* Part 1910.119) apply to contractor and nonemployee safety?

2. Studies of west coast companies who routinely hired contractors and subcontractors highlighted five key factors in companies with excellent safety records. Cite three.
   a.
   b.
   c.

3. Regarding (29 *CFR* Part 1910.119), the final rule does/does not require that employers refrain from using contractors with less then perfect safety records.

4. Why do insurance companies use EMR (experience modification rates) when considering workman compensation insurance?

5. What is indicative of lower rates?

6. When should the Safety Orientation take place between the employer and contractor management?

7. Identify three of the five recommended items to be addressed and documented by the employer.
   a.
   b.
   c.

8. What is an "attractive nuisance"? Describe one.

9. What is the most important feature of a parking lot?

10. How often should escalators be examined?

11. Where should emergency stop buttons be located for an escalator?

12. What type of information should an elevator logbook contain?
    a.
    b.
    c.

13. How often should all elevators have a balance test and contract load test?

14. What are the four stages of a fire?

15. Identify two of the criteria that designate a high-rise building.
    a.
    b.

16. Briefly discuss the disadvantages of an unprotected air-conditioning system.

17. Your place of employment is suddenly the target of an outside demonstration. Your first two duties are:
    a.
    b.

18. The demonstration has moved inside. What precautions should be taken?
    a.
    b.
    c.
    d.
    e.

19. What are two of the most common sources of nonemployee accidents on the premises?
    a.
    b.

20. Before allowing swimmers into the pool after a chemical treatment, what should be done?

Part 4

Program Implementation and Maintenance

INTRODUCTION

After problems are identified and programs developed in response to these problems, it is necessary to implement and maintain proposed solutions. Part four describes some motivational, training, and awareness techniques developed to enhance and maintain core and extended safety, health, and environment programs. The importance of these techniques is presented and described in the first three sections of this manual.

22

Motivation

The industrial safety and health professional assists line management in achieving maximum production by preventing or mitigating work-related fatal or injury accidents. The occupational environment is composed of interacting components, primarily the worker, materials, and equipment. A comprehensive safety program addresses all aspects of the work environment and recognizes that safe behavior-management stands at the center of the program. Consequently, a major responsibility of line management, with the safety and health professional's assistance, is to motivate workers to follow safe practices and procedures. This chapter is designed to promote understanding of human behavior in the work environment and covers the following topics:

- the importance of gaining employee commitment and involvement in safety programs
- key psychological factors in safety issues
- methods used to motivate workers
- worker attitudes and behavior
- motivational techniques that work
- motivation and learning.

Motivating workers is an important facet of an effective safety and health program. Motivation involves moving people to action that supports or achieves desired goals. In occupational safety and health, motivation increases the awareness, interest, and willingness of employees to act in ways that increase their personal safety and health, and that of co-workers, and that support an organization's stated goals and objectives. The ultimate success of a motivational model in changing employee attitudes and behavior depends on visible management leadership. In addition, the motivational techniques used should support the mainline safety and health management system, not take its place. Similarly, evaluation of the work of these techniques should be measured in terms of how well they achieve their support roles, such as maintaining employees' interest in their own safety, rather than by their effect on injury rates.

Many companies are moving away from traditional approaches to managing employee safety and health. The traditional approaches exhibited such characteristics as top/down communication, minimal employee participation, and a dependence on discipline to influence safety behavior. The challenges of motivating employees, changing attitudes, and controlling behavior continue to resist uniform, simple solutions.

ACHIEVEMENT OF COMMITMENT AND INVOLVEMENT

Achievement of permanent performance improvement is a product of employee and management commitment and involvement. Employee acceptance must occur and management leadership must be visible.

When behaviors are repeated and reinforced, the foundation has been laid for personal acceptance to occur. In other words, employees are willing to take the responsibility for paying attention to the behavior or process changes necessary to achieve specified safety and health objectives until they become part of their habitual action patterns.

A change in work habits is not attained overnight. Rather, it takes time to accomplish and must have continuous management support. Employees must perceive that management at all levels is behind the objectives of the safety and health program and supports the methods chosen to achieve them. If at all possible, the techniques used to motivate employees should blend with an organization's culture. They should not be viewed as special or outside the management mainstream, but rather as part and parcel of the motivational methods that are used to achieve high levels of production and quality.

When this mesh is present, the likelihood that safety and health needs will be met within the normal business function is high and, therefore, the potential for permanent improvement is increased. When motivation and behavior change techniques do not fit in with a company's management style or are looked upon as nonmainstream activities, even though they receive considerable initial attention, they are unlikely to become permanent.

PSYCHOLOGICAL FACTORS IN SAFETY

The full role and responsibilities of the industrial psychologist are too broad to be covered in this chapter. However, those psychological factors that most directly influence safety program success—individual differences, motivation, emotions, stress, attitudes, behaviors, and learning processes—will be described.

Individual Differences

Individual differences among employees are an ongoing challenge to managers. Although differences are often obvious and constant, some factors are common to all people, a fact that managers and safety personnel need to understand when dealing with work groups.

Motivation

When someone has an internal drive to acquire something, it can be said he or she is motivated to acquire it. It is important to realize that someone can be equally motivated to not want something. For instance to use a guard to protect one's fingers from a saw is, perhaps, indicative of motivation for safe practices. However, the desire to ignore a safety device because it might decrease production is also a motivator. Safety personnel and others need to consider the concept of conflicting motivators in any attempt to understand human actions.

Emotions

Although emotions can be constructive at times, they also can be counter-productive—working to the detriment of both the individual and the safety program. Emotions such as anger, fear, and excitement can interfere with the thought processes, resulting in behavior that conflicts with a rational approach to work tasks.

Stress

Stress, a physical, chemical, or emotional factor that causes bodily or mental tension, may be a factor in disease causation. Stress is always present and can be both good and bad. How employees choose to react to the stress on or off the job is very important, impacting decisions to use safe or unsafe behaviors.

Attitudes and Behaviors

Industry has recognized the effect employee attitudes can have on production, plant morale, turnover, absenteeism, plant safety, and other work-related issues. As a result, management has spent much time and money trying to determine workers' attitudes. Although workers' attitudes toward safety and health procedures and policies can make a difference between an effective and ineffective safety program, accurately determining an individual's attitude can be difficult. Attitudes are internal and very difficult to measure. Accordingly, it is necessary to observe and measure employee behaviors. Changing behaviors can help to modify attitudes.

Learning Processes

Finally, management should be concerned with how people learn on the job. One cannot understand motivation, attitudes, emotions, or even individual differences without some consideration of the learning process involved in establishing each worker's perspective and personality.

Much of the success of a safety program depends on its acceptance by those to whom it is directed. The safety and health professional and line management must effectively explore and find factors common to the group that can be used to promote safe conduct on the part of all employees. The basic question is, "What factors associated with human behavior can be used to increase the effectiveness of safety programs?"

Several of these psychological factors are discussed below.

INDIVIDUAL DIFFERENCES

When a chemist analyzes a chemical compound, its exact nature and composition can be accurately specified. When this compound is used, its behavior can be predicted. The action and reaction of any one sample are the same for all other samples of the compound.

When studying human behavior, however, the psychologist is not dealing with the same degree of certainty. Other than what is communicated by the individual, the psychologist does not know the psychological make-up or history of the person.

In addition, the behavior of one person is not the same as the behavior of another person. Because behavior can be influenced by the individual's attitude, heredity, and environmental variables, Person A in the sample is not equivalent to Person B in the same way one cubic centimeter of distilled water equals all other cubic centimeters of distilled water.

The known fact that people differ has been referred to as the "personal equation" or, more commonly, "individual differences." The personal equation presents many challenges for both the safety and health professional and manager. Yet within the framework of individual differences, certain general patterns common to a group do exist.

For example, human behavior is typically motivated by some belief, need, or drive. Regardless of what the individual does, there is usually some purpose driving the behavior. For example, the purpose may be to resolve or reduce some basic tension. The degree and nature of this tension, or dissonance, will depend in part on the individual's attitude, values, or perception of situational variables (e.g., job expectations, work conditions, physical ability). Whether or not an employee works safely depends on: (1) the present situation—is the employee rushed, stressed, fatigued, or in poor health? (2) past experiences—were accidents avoided in the past, or what amount of training does the employee have? and (3) workplace and methods design—were the job procedures and work setting designed to promote safe and healthy behaviors?

Average-Person Fallacy

Although each person differs from every other person, the important fact is how one deals with individual differences. Figure 22-1 presents the distribution of scores in a normal curve. Many human characteristics (for example, anthropometric characteristics or IQ) are assumed to be distributed according to the normal distribution (often referred to as the "normal curve"). Note that in such a distribution, half the persons are below the average line (called the mean) and half the persons are above it.

Often managers and safety and health professionals realize individual differences exist, but it is physically impossible to deal with each person separately. They may erroneously attempt the next best thing—appeal to the "average person."

Unfortunately, inept use of the "average" as the main characteristic of a population has distorted statistical data and produced a misconception called the average person. People, in reality, are highly unlikely to be average. For example, in body dimensions, which can be measured more accurately than emotional, behavioral, or intellectual characteristics, less than 4% of a test group had three common average dimensions. Less than 1% were average in five or more dimensions at the same time. Therefore, when an appeal is aimed at the average, it misses much of the population.

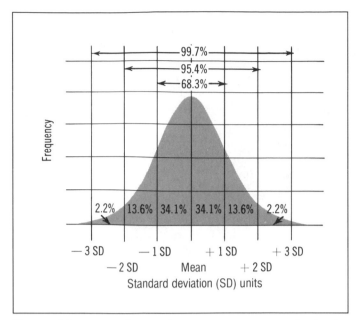

Figure 22-1. Many human characteristics are assumed to be distributed according to the normal curve shown here. Because the normal curve has a known shape it is possible to state the percentage that lies between +1 and −1, standard deviation (SD), or between any other two points expressed in SD units. Thus, it is easy to describe a normal distribution of data using simple statistics—the mean (average) and the standard deviation (SD). For example, 68% of all the measured values lie within ±1 standard deviation about the mean; 95% are within ±2 SD, and so on. Given the mean and the standard deviation, therefore, the complete distribution in a normal distribution can be determined.

A better approach is to use percentiles. When designing the system, instead of setting the standards to fit the average, design it to fit all but the upper 5% and the lower 5%. The system will then fit 90% of the population.

The use of percentiles (for example, between the 5th and 95th percentile), rather than averages, has long been the technique for relating individual test scores to a group as a whole. This system can also serve as a guiding rule in the design of human-machine relations.

Fitting the Person and the Job

Very often when replacing or relocating personnel, managers strive to find the individual possessing most of the characteristics deemed necessary. Job specifications and employee requirements are given in terms of a minimum, but very seldom are psychological factors expressed within a minimum and maximum (see Chapter 13, Ergonomics Programs, in this volume).

Often the ideal candidate is pictured as being the most intelligent, most loyal, most productive, and so on. In other words, management tends to look for the perfect person rather than for the right one. For example, a deaf person properly trained for a task in which appropriate noise levels cannot be controlled would experience fewer

problems than a hearing person properly trained for the same task, because of the issue of wearing protective hearing devices.

Another example is the person who comes to work under any condition, whether ill or not. This person may be as undesirable as the one who uses every excuse to take off. The worker with a high fever, bad cold, or sore back is a potential hazard to all concerned, especially if on heavy medication.

Physical Characteristics of the Individual

Reaction time, psychomotor skills (for example, manual dexterity), and visual abilities seem to have at least some bearing on safe performance. Whether they are directly responsible for accidents is not clear or constant, but it appears that a certain minimum degree of physical competence is required for successful, accident-free performance. Some types of jobs demand superior physical abilities while others do not.

Individual Characteristics and the Job

Human resources people in industry screen candidates on the basis of the specific characteristics required to perform a particular job. In many cases, both physical characteristics (such as size, visual acuity, and steadiness) and personality characteristics are important. As a result, many individuals are not considered for a job because they do not have all the necessary qualifications.

The same principle applies in other areas of matching the right worker to the right job. When designing equipment, for example, human engineering experts consider physical limitations as well as other human characteristics in an effort to make human-machine interactions as nearly perfect as possible. (See Chapter 13, Ergonomics Programs, for a discussion of human factors engineering in relation to occupational safety.)

Where hazards cannot be eliminated, safeguarding provides protection against the potential failure of people to use equipment correctly. Both safe design and safeguarding minimize the effect of individual differences on accident frequency and severity.

Methods of Measuring Characteristics

Regardless of the technique used to screen, place, and motivate employees, management needs a method to measure safety program effectiveness. Techniques used to obtain feedback range from the across-company accident rates to the within-company approach of safety sampling, or critical incident technique (see also Chapter 8, Injury and Illness Record Keeping and Incidence Rates, in this volume).

Measuring techniques can be assessed by determining reliability and validity. Reliability refers to consistency of measurement. The reliability of a given measurement, such as a test or instrument, can be estimated in various ways. In general, however, the concept of reliability refers to how stable measurements remain. This is the case whether stability is assessed (1) across time or between settings, (2) using the same or a different group of individuals, or (3) for internal consistency or consistency between alternate forms of the same test or instrument.

Underlying the concept of reliability is measurement error. All gauges used to assess performance will have, in differing degrees, some inherent measurement error. Measurement error refers to the estimated fluctuations likely to occur in individual or group performance because of factors beyond anyone's control.

Validity refers to how well the test or instrument measures what it is supposed to measure. The validity of a test or instrument can be assessed in various ways: (1) the relevance or plausibility of items (or the overall instrument), with regard to the given behavior (face validity); (2) whether all aspects within a given concept are adequately covered (content validity); (3) how well the test or instrument measures a specific construct or trait (construct validity); and (4) a demonstrable relationship between performance, as measured by the given test or instrument, and some other related behavior (such as IQ and school aptitude) (Anastasi, 1988).

A measurement can be reliable without being valid, but a valid measurement also must be reliable. To illustrate that a reliable measure is sometimes not valid, consider a yardstick. A yardstick is very reliable—it gives a consistent measurement every time it is used. It is also valid for measuring the length of a table. But if a yardstick is used for weighing the table, it is no longer a valid measure.

In addition to reliability and validity, a measuring technique must be practical. A technique can possess high validity but be so cumbersome and intricate that it can only be used in special situations by highly skilled technicians. In spite of its statistical value, such a technique is almost worthless in most workplaces.

Two sampling techniques used for evaluating potential accident-producing behavior are the critical incident technique, described in Chapter 3, Loss Control Programs, and behavior sampling.

- The critical incident technique involves the following process. A random sample of employees is interviewed to collect accident information concerning near-misses, difficulties in operations, and conditions that could have resulted in death, injury, or property loss. Those participating are asked to describe any incidents coming to their attention. This technique can be useful in investigating worker-equipment relationships in past or existing systems, evaluating modifications to existing systems, or developing new systems.
- The behavior sampling or activity sampling technique involves the observation of worker behaviors at random intervals and the classification of these behaviors according to whether they are safe or unsafe (place the worker at risk). Calculations are

then made to determine either the percentage of time the workers are involved in at-risk practices, the percentage of workers involved in at-risk practices during the observation period, or the percentage of unsafe versus safe behavior observed. Using this technique, management can apply various components of a safety program (such as safety lectures, posters, brief safety talks, safety inspections, motion picture films, supervisory training) and immediately note their influence on workers' unsafe behavior.

Research studies have shown that feedback can be introduced into this process with clearly positive effects on worker behavior and accident rates. Feedback charts are prepared that show the percent of safe behavior observed during each sample. These charts are posted in the workplace and serve as positive reinforcement for safe behavior (see Analyzing and Changing Behavior later in this chapter).

When screening employees or potential employees in regard to their accident potential, human resources personnel must exercise caution. To date, no systematic screening procedures have been developed that meet both reliability and validity criteria and at the same time are adequate for use in all industries or even in a specific industry. While in theory such a screening procedure is possible, the state of scientific knowledge in occupational safety research is too limited to support the development of such screening measures. Therefore, the reduction in human-error potential by means of a good ergonomics program (Chapter 13, Ergonomics Programs) is still the best line of defense.

FACTORS IN MOTIVATION

Kanfer (1990) indicates that motivation includes three variables:

- **Direction of Behavior** in terms of the performance of actions that accomplish defined objectives
- **Intensity of Action** in terms of the amount of personal attention and thought given to the performance of goal-oriented actions
- **Persistence of Effort** that desired performance lasts over time

Direction of behavior requires two supporting actions. First, the behaviors to be achieved (or output to be obtained) must be specified. Second, employees must clearly understand how to achieve the desired objective(s).

Safety and health objectives can be defined very narrowly in terms of specific behaviors, or broadly in terms of process improvements. Once objectives have been set, employees must attain the knowledge necessary and learn the skills to achieve them. Unless these two elements are present, motivational efforts are futile.

Intensity of action means that employees integrate safety and health objectives into their job assignments with the same degree of mental and emotional effort that they expend for other work objectives. It also implies that employees may have to be willing to spend extra time to incorporate safety and health practices into their routine work patterns and, if necessary, accept the potentiality that additional work may be required as some new behaviors are acquired.

To ensure that employees are seriously involved, the reinforcement of appropriate behaviors and/or performance levels should be as strong as possible. Also, communication of performance results to employees is critical to the change process to demonstrate achievement, emphasize its importance, and identify areas still needing improvement.

Persistence of effort relates directly to the nurturing and maturing of the attitudes or action tendencies that support improved safety and health performance throughout the organization. For this continuity of purpose to occur, both employees and management must be committed. Employees must be willing to modify personal behavior in accordance with company safety goals and objectives. Management must be visibly committed and active in its support for employee safety and health.

Specification of Safety Objectives

Conard (1983) and Cohen (1987) have suggested several broad classes of behaviors that have worksite hazard control implications. The majority focus on prevention, but some deal with the mitigation of unwanted effects during and after incident occurrence.

In the area of prevention, critical behaviors include:

- proper use and operation of equipment to maximize safe performance
- adherence to work procedures that maximize safe performance
- avoidance of actions that increase risk of injury or illness
- recognition of physical and process-related hazards
- observance of good housekeeping, maintenance, and personal hygiene practices.

With respect to injury/illness/incident mitigation, critical behaviors include:

- proper and consistent use of personal protective equipment and other controls
- recognition of illness-related symptoms
- proper response(s) to emergency situations.

In each of these classes, the specific behaviors to be identified and practiced or avoided vary from company to company. Company operating policies, rules, and procedures normally address general safety practices. OSHA and related laws specify selected areas of compliance. The company's way of doing things, bolstered by group norms and motivation, also tends to influence safety priorities.

Often, however, the information from these sources is not sufficiently specific to cover the complete array of tasks that must be performed at the most basic operational levels. Furthermore, policies, rules, and regulatory standards seldom, by themselves, carry the motivational understanding of workplace safety needs and improvement opportunities that can come from injury/illness/incident investigations as well as through periodic inspections and audits. These procedures frequently pinpoint safety deficiencies in a concrete way and with an immediacy that directly influences employees to improve their performance.

To supplement investigations, inspections, and audits and add considerably to their potential for stimulating safe behavior, employee involvement in the identification of safety needs is also advisable. Job safety analysis, which allows employees to participate in the identification of hazards and means for their elimination, lays an important foundation for behavior change. This type of participation should be reinforced and continued through an open communication system that welcomes safety discussions on the part of employees, and offers a means for action in response to suggestions and corrective action when problems or hazards are reported. In this regard, the use of employee safety program perception and attitude assessments can play a significant role both in specifying process safety priorities and in enhancing employee morale. The role of surveys as a motivational stimulus to improved performance is detailed later in this chapter.

Once safety improvement objectives are specified, employees must learn how to attain or achieve them. Training or retraining of employees is often a necessary step at this juncture of the improvement process and lays the groundwork for immediate positive results. With the proper thrust and scope, training in hazard identification and other skills develops an employee's capacity to make knowledgeable contributions to improved job safety performance.

Reinforcement of Desired Behaviors

Once safety objectives are defined and the behaviors necessary to attain them are known, they must become part of the job performance action pattern. To achieve this, employees must pay attention to job safety requirements and put what they have learned into practice. This means that undesirable, habitual, and comfortable behaviors must be eliminated or altered, while new behaviors are substituted.

If employees are involved in specifying safety and health improvement objectives and receive adequate training, the process of behavior change and reinforcement has already begun. Reinforcement in this context refers to anything that encourages the repetition of a behavior.

Two important characteristics influence the effectiveness of reinforcement. First, it is generally agreed that positive reinforcement—such as personal recognition or performance awards, which focus on increasing the occurrence of desired behaviors—is more efficient in achieving higher levels of safety performance than forms of disciplinary action focused on eliminating unwanted behaviors. Second, the closer in time a reinforcement is associated with a behavior, the stronger its effect.

Both of these principles flow from learning theories that were developed largely on the basis of animal behavior studies. Principles derived from these studies have been used as a basis for explaining how learning occurs in humans. These studies have also led to the development of a variety of "need" and "drive" theories about why people behave as they do. Based on these theories, it can be observed that timely, positive reinforcement of specific behaviors and action patterns, and general feedback about performance that has produced the desired result(s) are considered essential features of the learning and motivation process. (See the later section on Management Theories of Motivation.)

Group feedback helps to create a positive atmosphere for behavior change. When a work group accepts common safety improvement objectives, its members tend to reinforce one another's behaviors. To make this happen, feedback at the group level should be maintained. Whether or not positive change is achieved, feedback provides momentum for future accomplishments.

Group feedback requires objective evaluation or measurement of progress. The establishment of a measurement system begins with the definition of the specified safety and health objectives so that they can be observed. Observation can be direct when specific behaviors are involved and can take the form of counting the occurrences of unsafe behaviors. To observe progress toward achieving broadly defined process improvement goals, surveys or other performance indicators may be required. These examine such factors as the current level of employee involvement in safety and health activities and the effectiveness of safety and health management system processes.

Whatever measurements are used, feedback at the group level requires the collection of baseline data and a commitment to continue the measurement process. Feedback maintains individual interest in and attention to the desired safety objectives. It also helps to make the newly elicited safety action patterns part of a work group's job performance norm.

One of the best ways of motivating employees is by a joint labor/management safety committee. Such a committee involving the workers and management for safety and health endorsed by all levels of the company can have extremely positive results.

Peer pressure, positive involvement, and recognition combined with utilizing the expertise from every company level for the improvement of safety and health can be a strong motivational tool for the safety and health professional.

Complexity of Motivation

Motivation is perhaps one of the most complex issues in the field of human behavior—people have many needs, all continually competing with and influencing behavior. Hersey and Blanchard (1988) called motives the "whys" of behavior. Motives and needs create activity and direct the behavior of people.

Psychologists and researchers cannot give clear-cut, concise answers to all the questions that might be asked about other people's motivation. Rather they attempt to describe some basic concepts such as hierarchical motivation, affiliation motivation, and achievement motivation. Understanding these concepts can help managers, safety and health professionals, and others learn about what can motivate workers.

The hierarchy of needs was developed by Abraham Maslow. Hierarchical motivation simply means some needs have a higher priority than others in people's lives (Figure 22-2). Once an individual has satisfied the physiological need to sustain life (food, clothing, shelter) the other needs can become dominant. Once physiological needs are satisfied, safety and security needs take precedence over social needs (recognition, affection, social approval). When the safety needs are satisfied, then the social needs become more pressing. For example, a line manager may feel that acceptance and recognition by workers is a major need to be satisfied. Upper management must realize that managers who care less about enforcing an unpopular safety rule and more about the affection and recognition of the workers are not performing their job adequately. Employee safety has the highest priority.

This hierarchy of needs is not static. The strength of certain needs will change depending on the individual's level of satisfaction. As an individual satisfies a need, that

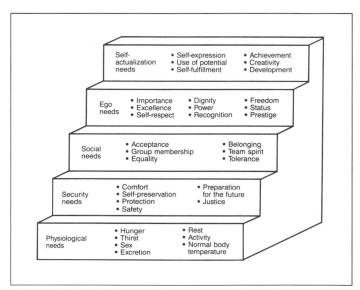

Figure 22-2. Abraham Maslow investigated people's various needs and arranged them into a hierarchy. Note that safety needs are near the base, and that more intellectually satisfying needs are near the top.

need is no longer as strong a motivator as an unsatisfied need. By satisfying related needs, the individual advances to the succeeding hierarchy level.

A second area of interest to safety and health professionals is the difference between affiliation motivation and achievement motivation. Affiliation-motivated people need to be accepted by others and to feel that they belong. They will be motivated by needs which gain them acceptance by the group or retain membership in the group. Achievement-motivated individuals are more concerned about the outcome, or results of a task. Achievement-motivated people are not extreme risk takers. They will attempt to provide solutions to problems rather than leaving the outcome to fate or chance.

People may not be achievement motivated because this requires the development of a need. A person may have a high need for achievement, but if the individual is not challenged the need might not be aroused. For example, the individual may fail to understand how to control a significant safety problem. By presenting the problem as a challenge, it increases the likelihood that the person will want to solve the problem. Thus, to develop achievement-motivated workers, management must create a need that provides clear objectives, a reasonable probability of success, and measurable feedback.

The expectation of success or fear of failure are two important categories of achievement motivation. The employee whose goal is work safety is success oriented because his or her motivation is the intrinsic value of the goal. The person who fears the potential negative consequences of not working safely, such as a serious injury, wants to avoid failure rather than pursue the goal for the sake of achievement.

Safety and health professionals and other individuals should never assume that a worker's satisfied need will remain satisfied forever and thus they do not need to continue to give recognition, or affection, or social approval. For example, situational change at work or home can create motivational changes within the needs of the individual requiring continued feedback.

People strive to satisfy their needs but do not always attain them. Also, individuals will seek to satisfy their needs in different ways at different times. For instance, the need for acceptance may be satisfied through social approval one time while at another time it may be satisfied through affection from family members. Thus, in dealing with people, one must recognize that what was effective yesterday may not work today, although the person's behavior is to some degree similar.

Being aware that people continually seek to satisfy their needs makes it somewhat easier for the safety and health professional or anyone else to deal effectively with people in the work situation. The safety and health professional must develop the ability—and help managers to develop the same ability—to help people attain the type of satisfaction they require.

Safety and health professionals and managers must spend time together and with workers so they can come to know one another and the workers very well. The safety and health professional must clearly understand what the workers want from their job. Is it good working conditions, promotion and growth, interesting work, job security, or good wages? They can then determine with a fair amount of accuracy what a worker's behavior at that moment means in terms of need satisfaction. This awareness is based on the little cues in the individual's behavior that the safety and health professional and manager can learn over time to recognize, provided they have made the effort to get to know workers.

Careful observation is required to obtain this type of knowledge. Safety and health professionals should advise line management that any supervisory training program includes effective worker observation methods.

Job Satisfaction

In the interest of further understanding motivation in the work environment, many studies have been conducted to determine what constitutes job satisfaction. Generally, these studies assess the job elements workers claim contribute to their job satisfaction (or dissatisfaction). The results of these investigations suggest that satisfaction of psychosocial needs rather than physiological needs may be the major motivational aspect of job satisfaction.

Table 22-A presents the results of various surveys of job satisfaction conducted over the years. The numbers represent rankings of the factors considered in each study. Because different language and alternatives were used in each survey, the factors have been paraphrased to represent the various elements.

The results of these types of surveys suggest that high pay is not in itself a primary job motivator. Although workers expect a just and equitable income, they appear to expect only what others would be paid for comparable work. Although workers might feel dissatisfied if underpaid, higher pay alone does not guarantee job satisfaction.

On the other hand, steady work or job security does appear to be a primary job motivator. Workers want the security of knowing if they perform their jobs well, they will have a job in the future. Job security, as a component of job satisfaction, can explain the willingness of a worker to remain in a lower paying, stable job instead of accepting a higher paying, less-stable job.

Other factors involved in job satisfaction include type of work, opportunities for advancement, and good working companions. Note that all of these factors are related to the psychosocial needs of feeling important and belonging to a peer group. Likewise, comfortable working conditions (rated high by a number of employees) are probably associated with a desire to be humanely treated.

The results of these surveys on job satisfaction are important when considering the safety program within the context of personnel policy. In as much as the safety program

Table 22-A. Summary of Different Surveys on Job Satisfaction in Order of Importance of Significant Factors

	Women Factory Workers	Union Workers	Nonunion Workers	Men	Women	Employees of Five Factories
Steady work	1	1	1	1	3	1
Type of work				3	1	3
Opportunity for advancement	5	4	4	2	2	4
Good working companions	4			4	5	
High pay	6	2½	2	5½	8	2
Good boss	3	5½	5	5½	4	6
Comfortable working conditions	2	2½	3	8½	6	7
Benefits		5½	6	8½	9	5
Opportunity to learn a job	8					
Good hours	9	7½	7	7	7	
Opportunity to use one's ideas	7	7½	8			
Easy work	10					

is designed to ensure the well-being of the employee, it helps to maintain the employee's continued ability to do the work (which, in turn, promotes job security).

Likewise, the safety program represents management's interest in the co-workers and working conditions of the employee. All of these aspects of safety programming should be anticipated and incorporated into the approach that is taken with both supervisory staff and employees. Honest and sincere positioning of the safety program to enhance employee welfare makes practical sense in light of current knowledge about job satisfaction.

Management Theories of Motivation

There are many theories in psychological literature addressing human motivation. Within this literature, specific theories have evolved with special reference to management as it exists in industrial organizations. Although only two such theories are presented here, other equally cogent points of view, such as Blake and Mouton's attitude-based Managerial Grid III or Hersey and Blanchard's behavior-based Situational Leadership, could be discussed. (For additional information see the Suggested Reading section at the end of the chapter.)

Because theories of human motivation lack sufficient data to support all their tenets, they might best be viewed as philosophies of management. They are important to the safety and health professional because, if accepted, they can influence the direction in which management seeks to develop and implement a safety program.

Theory X and Theory Y

In an attempt to analyze how management regards human motivation, McGregor (1985) discovered two basic ways in which managers view workers. The view a company accepts determines the management practices that are adopted to run the company.

Theory X, according to McGregor, assumes the worker is essentially uninterested and unmotivated to work. In order to change this attitude, management must motivate workers through various external motivators—rewards (bonuses, prizes, etc.) or punishments (docked pay, fewer privileges, etc.). In effect, the worker is motivated to work by virtue of the external reward and punishments offered. Thus, under Theory X policy, management uses control and direction as the means of worker motivation.

Theory Y, according to McGregor, assumes the worker has the potential to be interested and motivated to work. In fact, work is assumed to be as natural and desirable as other forms of human activity, such as sleep and recreation. Under such circumstances, management is confronted with the role of organizing work so the worker's job coincides with the goals and objectives of the organization. Thus, a Theory Y manager views the task as constructively using the worker's self-control and self-direction as the instruments through which work is accomplished.

By emphasizing responsibility and goal orientation, management capitalizes on the motivation already present within the worker. If conflicts occur between the worker's goals and management's goals, they are resolved through mutual exploration and discussion. It is always assumed, under Theory Y policy, that the worker's inherent motivation is essential to accomplish the organization's goals.

Both Theory X and Theory Y proponents exist, and management systems successfully operating on the basis of each of these theories can be found throughout American industry. The important point is for safety and health professionals to remember that safety programs can work under whichever system is operating within an organization. While the technique of implementation may differ, Theory X and Theory Y approaches to human motivation can both be used to motivate workers to adopt safe behavior.

Job-Enrichment Theory

Another analysis of human motivation in occupational environments has been developed by Herzberg (1966). Although quite comparable to the Theory X and Theory Y, Herzberg is explicit in both detail and philosophy. His concept of job enrichment, in many ways an extension of Theory Y, has been a major force in the development of management and leadership strategies.

The classic approach to motivation concerns itself with changing the environment in which a person works. This includes the circumstances surrounding the individual on the job (good or poor lighting, an agreeable or offensive manager), and the incentives given in exchange for work (money, a pat on the back, a writeup in the company bulletin, and so on).

Herzberg believes worker concerns about environment are important but are not sufficient alone for effective motivation. People are motivated best by using the work itself to satisfy their desires and needs.

Herzberg holds there is no conflict between the classic (environmental) approach to motivation and his approach to motivation through work itself. He regards both as important. The classic approach is called hygiene whereas Herzberg's approach is called "job enrichment." Figure 22-3 presents a contrast of the classic hygiene approach and the job-enrichment approach to motivation.

The hygiene approach may be understood by the following analogy—a person is provided with pure drinking water and waste disposal; both are necessary to keep this person healthy, but neither makes him any healthier. By extension, good rapport with others may enhance an individual's job satisfaction, but job satisfaction alone will not necessarily motivate the individual to develop safe work habits.

Further, treating a person better does not enrich the job, although the individual can become unhappy if not treated well. Again, a salary increase can keep an employee from becoming dissatisfied for a time, but sooner or later another increase will be required to boost motivation again. Nor does worker protection enrich a job. Even in hazardous industries such as coal mining or

bridge building, the worker may regard protection like gas testing and life lines as a part of the job. In such cases, the protection will not be a motivating force; on the other hand, the worker will be very unhappy if no protective effort were made.

Herzberg's idea that work itself can be a motivator represents an important behavioral science breakthrough. Traditionally, work has been regarded as an unpleasant necessity but it has not been thought of as a potential motivator. Although automation is helping to phase out the unstimulating aspect of many jobs, a job should provide an opportunity for personal satisfaction or growth. When it does, it becomes a powerful motivating force.

People, Herzberg further theorized, must be given the opportunity to do work they think is meaningful. Merely complimenting an individual who is doing a routine job or saying that the worker is doing something meaningful accomplishes little. The worker often does not feel this is true. Job rotation is not the answer, either; it does not enrich a job—it only enlarges a worker's responsibilities and tasks.

Herzberg also observed that new technology can make workers feel obsolete. As Herzberg noted, "Resurrection is more difficult than giving birth." Obsolescence must, therefore, be eliminated by continued retraining—not just an occasional session. Jobs should be updated, and people doing the jobs must keep abreast of the latest equipment and techniques.

Even though a company provides the hygiene factors, it also must provide a task that has challenge, meaning, and significance. An unchallenged individual who does not quit but stays on is usually resigned to the work and suffers from poor morale. That, says Herzberg, is the price a company pays for not motivating people.

In summary, following are seven principles of job enrichment.

- Organize the job to give each worker a complete and natural unit of work.
- Provide new and more difficult tasks to each worker.
- Allow the worker to perform specialized tasks in order to provide a unique contribution.
- Increase the authority of the worker.
- Eliminate unneeded controls on the worker while maintaining accountability.
- Require increased accountability of the worker.
- Provide direct feedback through periodic reports to the worker.

ATTITUDE, BEHAVIOR, AND CHANGE

The link between attitude and behavior on one hand and motivation on the other is still a matter of debate. Although many social and industrial psychologists believe that motivation is primarily about changing or molding worker attitudes and behaviors, others believe that there

Hygiene Approach (Classic)	Job-Enrichment Approach
Company policies and administration	Achievement
Supervision	Recognition
Working conditions	Work itself
Interpersonal relations	Responsibility
Money, status security	Professional growth

Figure 22-3. Contrast of the hygiene approach to motivation with the job-enrichment approach.

is still no clear evidence indicating that attitudes predict future behavior. This section discusses some of the current theories on how people's attitudes are formed, how they affect behavior, and the problems encountered when management attempts to influence or change worker attitudes and behavior.

Attitude

Many theories on attitude formation and change suggest that attitudes consist of three components: feeling, knowing, and acting. Such a theoretical approach distinguishes, yet integrates, such concepts as knowledge, emotions, and behavior or behavioral tendencies. Other theories regard attitudes as the expression of one's values towards something or someone. Most of these theories agree, however, that attitudes (at least in part) reveal a person's tendencies to evaluate objects, persons, or situations favorably or unfavorably.

Katz and Stotland (1959) identified the following three attitude components:

- *Affective (Feeling) Components.* Affective, or feeling, components are the positive or negative emotions underlying attitudes. Although some attitudes can be quite irrational, they involve little more than feelings. People may not like doing something a safe way, but cannot explain why.
- *Cognitive (Knowing) Components.* Attitudes can differ due to the extent of knowledge involved in the attitudes. For example, some people work safely because they have worked through the problem and have decided that working safely is the best way. Others may not work safely because their attitude towards safety is based on incorrect information or knowledge. In fact, misinformation can be a source for many attitudes.
- *Action Components.* As stated earlier, attitudes may bear little relationship to behavior. An individual may express a very strong attitude about working safely, and yet the person fails to wear the personal protection equipment provided for the job. In such a case, the attitude may lack a significant action component.

A person's response or attitude is dependent, in part, on that individual's previous experiences. For example, if someone sees a person on the street who resembles a friend, the immediate response may be a smile, friendly gestures, and warm voice quality. When that person is perceived to be a stranger, there is an immediate change to another facial expression, gesture, voice quality, and so on. Such reactions can be triggered by many factors—a certain look from another person, a manner of speaking, a mustache or the lack of one, hair color, or style of handshake. All individuals have certain feelings about other people or situations and tend to act according to their attitudes, which have been formed over the years.

Some of these attitudes are latent, or hidden. The responses based on latent attitudes can lie dormant within the individual until triggered by some event. Given the appropriate stimulus, the attitude surfaces and the behavior exhibited is consistent with the individual's feelings. A certain word, for example, elicits different responses in different people. The words "union," "management," "labor," and even "safety" carry with them certain connotations that touch off different attitudinal reactions in individuals, depending on the kinds of experiences they have had with the subject.

Because attitudes play such an important role in everyday relationships, the safety and health professional should understand how they are developed, their effects on individuals, and what can be done to change them.

Determination of Attitudes

Direct personal experience is thought to be one way attitudes are established. The three components of attitude—feeling, knowing, and action—can be influenced through personal experiences. In particular, experiences individuals have had in the past, especially those involving strong emotions, determine attitudes. Many people have had experiences that they learned to associate with fear, sorrow, pain, or happiness. All of these emotions tend to make people act the same way when a similar experience occurs.

For example, if a worker has been fired several times for what the individual believes are superficial reasons, the worker's attitude toward management or the manager may be fearful and hostile. There might have been sound reasons for the dismissals or layoffs, but to admit this would be a threat to the person's pride; thus, the individual puts the blame elsewhere. Because of the effects these dismissals have had on the employee and the family, such as causing financial stress, the employee may become very angry. Management now is the enemy. The person's present hostile, fearful attitude is set and will be difficult for the next manager to change without concerted effort on both parts.

Attitude Change

Numerous factors are associated with attitude change. Commitment and responsibility are core concepts in the formation and changing of attitudes. Simply imitating a particular view will have little impact on an individual's attitude. If the person truly wants to develop a new attitude or change an existing attitude, he or she must learn it in a specific situation where the attitude is tied to action. Attitudes that are acquired at only a verbal level cannot be expected to influence behavior.

For example, suppose the safety and health professional wants to bring about a change in the safety attitude of an individual or group. He or she must design the change to include an activity that requires the participants to link the new attitude to the activity. Without providing individuals with the means or methods to experience change, the safety professional cannot expect simple slogans to change workers' attitudes.

In summarizing the large body of psychological literature on attitude formation and change, McGuire (1968) suggested three components of the process (Table 22-B). These include (1) types of attitude-change situations (column 1), (2) variables associated with the communication process (column 2), and (3) behaviors associated with attitude change (column 3). The main points of the process are as follows:

- *Relevant situational factors* in changing attitudes include:
  - suggestive situations (where the desired attitude is repeatedly presented)
  - conformity (where social or peer pressure is used to elicit the desired response)
  - group discussion, persuasive messages, intensive indoctrination

  Each situation is associated with varying degrees of attitude change. Also, the implications of each depend on the type of communication variables involved.
- *Source variables* refer to characteristics associated with the person or represented organization presenting the message. The effect of the message on subsequent attitudes can vary depending on:
  - the credibility, attractiveness, and power of the message
  - the order in which specific issues are presented
  - the differences between the sender and receiver of the message.
- *Channel factors* consist of ways the message is presented, for example, direct personal experience or mass media.
- *Receiver factors* include whether the audience is actively involved in the process and how much the audience can be influenced.
- *Destination factors* refer to the degree of post-communication message decay across time, and time latency associated with delayed action.3

According to McGuire (1968), the receiver (or audience) must proceed through five steps for attitude change to occur. These are attention, comprehension, yielding, retention, and action. According to this model, each step depends on the preceding one.

To add to the complexity of this model, communication variables, especially in daily situations, can affect one another and interfere with the message. Also, the sequential-step process may have more intuitive than objective support. Nevertheless, this model should alert safety and health professionals and others to the fact that attitude change is not a simple process.

Organizational Development

Behavioral scientists, concerned about the amount and rate of change in a technological society, have studied the impact of such change on the industrial environment. These researchers have evolved an approach, called "organizational development," which attempts to assess a corporation's ability to adjust to such conditions as rapid growth, new technology, increasing diversity, and management system problems. Organizational development is designed to assess and improve the attitude and structure of organizations so they can better adapt to a change.

Effective organizational development requires a strong organizational culture. The development of a strong culture will aid a company's success by providing the employees the means to identify and act on the values of the company (Deal and Kennedy, 1982). The organization's culture provides workers with informal rules on how they are to act. It is through the company's culture that the organization can either reinforce the safety efforts of the workers or create barriers to the development of an effective safety program.

Generally, an organizational development approach gathers information from employees and management to determine the climate and capacity of the organization to adapt its objectives to the new technological environment. Based on such feedback, an attempt is made to develop organic systems to replace mechanical systems within the organization. Organic systems are characterized by a preoccupation with people as they work together; mechanical systems are characterized by a preoccupation with the structure of a company. Figure 22-4 presents a summary of the differences between mechanical and organic systems.

To implement the ideas represented by organizational development, management can use a number of procedures to effect changes in the organization. Since the changes are typically people-oriented, management

Table 22-B. Factors Associated with Persuasive Communication

Attitude Change: Situation Types	Communication Process	Attitude Change: Behavioral Step
1. Suggestive	1. Source	1. Attention
2. Conformity	2. Message factors	2. Comprehension
3. Group discussion	3. Channel factors	3. Yielding
4. Persuasive messages	4. Receiver factors	4. Retention
5. Intensive doctrination	5. Destination factors	5. Action

Source: McGuire (1968).

Mechanical Systems	Organic Systems
Emphasis upon individual performance	Emphasis on relationships in group
Chain of command concepts	Confidence and trust among everyone
Adherence to delegated responsibility	Adherence to shared responsibility
Division of labor and management	Participation in multimember teams
Centralized management control	De-centralized sharing of control
Resolution of conflict through grievance procedures	Resolution of conflict through problem solving

Figure 22-4. Organizational development seeks to implement organic systems in place of mechanical systems.

should recognize the need to motivate employees to accept the changes within both management and work groups. Typically, people can be educated and trained to learn how to work with the new system. These techniques have merit in implementing change only so long as top management agrees to the changes and provides incentives within the organization for their adoption.

Ultimately, organizational development seems to provide the necessary means for preparing an organization to plan for the future. Included in such an effort should be the recognition of how changes in an organization will affect the current safety program. By involving all levels of managers and workers in developing and managing safety programs, it is likely that the programs will be able to meet the future needs of modern organizations.

Behaviors Versus Attitudes

Attitudes refer to internal predispositions to behavior; as such, they are difficult to observe and measure. Behavior refers to observable actions, which can be measured. This distinction is vitally important because measurement of behaviors lays the groundwork for effective safety management.

Although attitudes are difficult to change directly, changing behavior is not as difficult. A good example of this is people's attitudes toward using seat belts. As the behavior of wearing seat belts has changed over the years, so have attitudes toward wearing them.

Behavioral Management

More attention and recognition have been given to the importance of behavioral management in safety programs. Behavioral management refers to the systematic identification, measurement, and control of safety-related behaviors. This section briefly reviews this approach.

Developing and planning for behavior change requires that the safety and health professional or manager has the ability to understand the situational variables that influence the behavior of individuals. Why would a person have a strong desire to work safely, yet continually perform in an unsafe manner? Why are there discrepancies between the expectations and results of the safety efforts in the work environment?

Often, when there is a discrepancy between results and expectations, managers may believe that there is a need for additional employee education or training. Ferdinand Fournies (1987) asked over 4,000 managers—"Why don't subordinates do what they are supposed to do?" The responses included the following:

- They don't know what they are supposed to do.
- They don't know how to do it.
- They don't know why they should do it.
- There are obstacles beyond their control.
- They don't think it will work.
- They think their way is better.
- They're not motivated—poor attitude.
- They're personally incapable of doing it.
- There's not enough time for them to do it.
- They are working on the wrong priority items.
- They think they are doing it.
- There's poor management.
- They have personal problems.

Robert Mager and Peter Pipe (1984) developed a useful procedure that provides the safety and health professional and manager with the means to systematically and accurately analyze performance discrepancies and create change. Mager and Pipe's first step in analyzing performance is to identify the nature of the discrepancy by asking, "What is the issue?" The safety and health professional must accurately describe the safety discrepancy. Examples would include driving a forklift with the forks elevated above a specified level, not wearing personal protective equipment, or conducting maintenance without locking out the equipment. The second step is to ask the question, "Is it important?" If the answer is "yes," the process continues to the third step and question, "Is it a skill deficiency?"

If a person drives the forklift with the fork too high because of a lack of skill, the safety and health professional may need to provide formal training if the individual has not had prior experience on a forklift. A second factor could be due to the fact that the person has not driven the forklift very often and requires additional practice. The third factor could be that the person has driven the forklift often, but requires appropriate feedback.

If the person's forklift driving behavior is not due to a skill deficiency—"They could do it if they wanted to"—additional knowledge from education and training may not change the individual's performance. In these types of situations, the safety and health professional must determine if:

- the appropriate performance will result in punishment
- there is no reward for the appropriate performance
- the appropriate performance may not matter to the individual
- there are obstacles to attaining the appropriate performance.

In these types of situations it is important to remove the punishment, arrange positive consequences, or remove the obstacles. Does the plant layout require the forklift driver to constantly keep raising and lowering the forks to clear congested areas, thus reinforcing the driver's decision to drive with the forks elevated? Can the work area be redesigned?

In some situations, such as not wearing personal protective equipment, the safety and health professional may not need to determine if the discrepancy is related to a skill deficiency. By analyzing the situation, the safety and health professional or manager may be able to design a simpler way to do the job that does not require the use of the protective equipment. If a simpler way is identified, the job should be changed or on-the-job training should be provided.

In addition to redesigning the job, poor or nonperformance may be due to an individual's potential. The individual may be unable to learn the job, lacks the physical or mental potential, or is overqualified for the job. If the safety and health professional or manager accurately determines that the person lacks the potential, a decision must be made to either transfer the person to a different job or terminate the relationship. It is important to realize that Mager and Pipe believe that transfer or termination options are used more than they should be. This could be due in part to the decision maker's lack of understanding regarding human performance and how to make improvements or an indication of the decision maker's inability to identify other options.

Because an individual's behavior can be the result of competing factors—created by either internal needs or situational variables—the safety and health professional or manager may identify a variety of solutions. By understanding the factors that can influence performance discrepancies, the best possible solution can be selected and implemented.

Evaluating critical behaviors in the work place can assist in changing safety-related behaviors. The basic steps of the process are:

- *Identify Critical Behaviors.* This means to write, in observable terms, what employees should do to properly perform their jobs. The safety and health professional can list a few critical behaviors or a complete inventory, depending on the scope and results desired.
- *Conduct Measurement through Observation.* Trained observers watch the workplace to determine if the listed behaviors are performed safely or unsafely. The total number of observed behaviors is divided into the number of safe behaviors to obtain a percentage figure for safe behaviors.
- *Give Performance Feedback.* The percentage figure for safe behaviors is shown on a graph displayed in the workplace. At regular intervals, behaviors are again observed and the new safe behavior figures added to the graph. Studies show this critical feedback will improve safety behaviors. Praise and recognition from managers or peer pressure can be effective ways to encourage and reinforce safe behaviors.

Structural Change

In the previous discussion of attitude formation and attitude change, no direct relationship between attitude change and behavioral change was assumed. In practice, however, line management and safety and health personnel are primarily interested in ways of changing employees' attitudes only if their behavior also changes; for instance, they not only change their attitude about wearing personal protective equipment, but they also change their behavior by actually using it consistently. While the research literature indicates that many techniques—training and counseling, for example—can change attitudes, there is much less support for a direct relationship between attitude changes and subsequent behavior changes (Ajzen and Fishbein, 1977).

In effect, line management and the safety and health professional must consider techniques other than simple attitude change when considering employee behavior. Evidence suggests it is possible to change behavior by changing the structure of the work environment. Examples of structural changes involve changing situational variables, including changing job contents, modifying the physical arrangements of work, changing worker interaction patterns, and rearranging work procedures.

In each case, it is not necessary to expend the time and effort to change attitudes before changing behavior directly. Rather, the introduction of the structural change can modify behavior and with it, possibly change employee attitudes.

Although the structural change model has not been applied extensively in the past in relationship to occupational safety, it warrants consideration by the safety and health professional. Human behavior can be changed by altering the very circumstances under which the individual works. Unsafe practices, for example, cannot occur when the structure of the work and the work groups prevents them from happening.

MOTIVATION MODELS

Among the many models that can be used to motivate employees to improve safety and health performance, two have had considerable acceptance. One, the organization behavior management (OBM) model, is tied directly to the use of reinforcement and feedback to modify behavior. The other, the total quality management (TQM) model, is based on attitude adjustment methods used to achieve quality improvement goals in industry.

As shown in Table 22-C, both management approaches are responsive to the motivational variables and supporting actions that have been described. The models vary, however, on the safety emphasis used to motivate employees. As a result, they differ in terms of their efficiency in accomplishing the requirements of the three motivational variables.

The theoretical orientation of the models differs with regard to the specification of safety and health objectives. As Table 22-C shows, OBM emphasizes external behavior change, while TQM emphasizes internal attitudinal changes as a prerequisite to behavior change. Accordingly, OBM objectives can probably be communicated in a more direct and simple fashion to employees, while TQM principles may be more complex.

Training provided to employees through OBM and TQM differs as well. OBM training is specific in its coverage of critical behavior to be changed, while TQM education includes behavioral as well as attitudinal skills development, such as team building and problem solving.

Reinforcement and feedback are provided by both OBM and TQM models, but the observational techniques used in OBM are more formal and specific to the objectives (i.e., specific behavior changes) than are those used in TQM. Also, the reinforcement/feedback schedules used in OBM are more structured than those found in TQM, thus facilitating behavior changes more rapidly.

Employee commitment to improved safety performance on a permanent basis as selected in both attitude and behavior changes appears more directly attained by TQM than OBM, because of TQM's emphasis on process or root cause improvements. Process changes produced by TQM flow from employee involvement and empowerment as individuals and teams. Employees are much more likely to support change if the objectives to be achieved and the methods used to achieve them are based on their own recommendations rather than imposed by management.

OBM's emphasis on critical behaviors or special cause improvements includes some degree of employee participation in the selection of behaviors to be changed and, possibly, the incentives to be used, but control is maintained through a relatively rigid observation and

reinforcement system. As a result, unless the system is continued, the permanence of the behavior change remains questionable.

With regard to management commitment, TQM requires a definite change from the philosophy that employees should be "managed" to conform to existing systems to a viewpoint that processes and systems can be improved or even completely revamped, and that employees are in the best position to know how those processes work and should work. It also calls for the creation of a corporate culture that emphasizes flexibility and responsiveness to employee needs and trust between labor and management.

The OBM approach does not demand a shift from the traditional management philosophy that seeks to manage employee behavior. It does, however, call for the acceptance of a new management style that focuses on a different and relatively exacting observation system that is not widely used and that requires a relatively substantial amount of time and resources to maintain.

SELECTED MOTIVATIONAL TECHNIQUES

No matter how intrinsically effective a motivational model is in changing employee attitudes and behavior, its ultimate success depends on visible management leadership. This is a prerequisite for any program, whether it is focused on production, product or service quality, or employee safety and health. It is the consistent finding across all successful safety programs, despite their differences. As Cohen and Cleveland (1983), in their study of outstanding National Safety Council member companies, conclude: "No safety program was quite like any other. However, all had one major thing in common: safety in each instance was a real priority in corporate policy and action."

Besides this essential ingredient, there are many other productive motivational techniques that can be used to promote employee safety and health or otherwise influence employees to take "self-protective" action against workplace hazards. These have been reviewed by a number of authors, including Cohen, Smith and Anger (1979), Cohen (1987) and Peters (1991).

It must be emphasized that the role of motivational efforts is to support the mainline safety and health management system, not take its place. Safety and health programs are more likely to succeed when the programs are

Table 22-C. Comparison of OBM vs. TQM Approaches to Employee Motivation

Motivational Variable	Supporting Action	Safety Emphasis OBM Model	Safety Emphasis TQM Model
Direction of Behavior	Specify Objectives	Behavior	Attitude/Behavior
	Provide Training	Behavior Training	Process Education
Intensity of Action	Give Reinforcement	Behavior Occurrence	Process Improvement
	Maintain Feedback	Behavior Data	Operating Indicators
Persistence of Effort	Commit Employee	Behavior Change	Continuous Improvement
	Commit Management	Style Change	Cultural Change

energized through slogans, performance recognition and discipline in combination with safe and healthful workplace conditions; well-designed tools, equipment and workplace layout; effective maintenance; and appropriate training and supervision.

Similarly, evaluation of the worth of motivation techniques should not be measured by reduction in injuries, property damage, or workers' compensation costs. Rather, their effectiveness should be judged in terms of how well they achieve their support roles, such as maintaining employees' interest in their own safety, communicating management's interest in employee safety and health, generating employee involvement in safety activities, increasing morale, and reminding employees to take special precautions.

Cohen et al. (1979) point out that these approaches "seek to establish a generalized tendency to act in safe and healthful ways through appropriate attitude change, increased knowledge or heightened awareness." This discussion briefly highlights three of them.

Communication

Communications of various kinds are used to enhance the general effectiveness of any motivational effort. The communication process can be summed up thus: *Who* says *what*, in *what way*, to *whom,* and with *what effect*? Accordingly, communication programs usually involve a source, message, media, target, and objectives.

Communications vary in terms of their coverage and impact. Safety posters, banners, and other mass media are high in coverage and most effective in increasing general awareness about safety and health issues and in presenting on-the-spot directions or safety reminders. They can also be a useful vehicle for making employees aware of management's general interest in their welfare. As is the case with mass media in general, their impact is lessened when used alone because they provide no opportunity for interaction, further information, or response to questions. However, the impact of their message can be increased when combined with opportunities for person-to-person or two-way communication, either through group discussions or individual contacts. Though low in coverage value, these methods can be high in impact and lead to changes in behavior.

Credibility of source is very important in safety and health communications. Bauer (1965) describes two types of credibility:

- problem solving—based on competence and trust
- compliance—based on likability and prestige

In this model the problem-solving source is more likely to prompt lasting behavior change than the compliance-based source. Cohen et al. (1979) note that, in the workplace, supervisor competence (i.e., knowledge of the task and the hazards involved, ability to set a good example) is an important factor in enabling employees to regard supervisors as credible sources of safety and health information.

With regard to communication content, the use of fear or scare tactics has been a topic of research and controversy for years. This strategy attempts to change safety and health attitudes about the risks involved in hazardous behaviors by instilling fear in a target audience and then reducing that fear by providing methods to prevent the danger or lower the risk. Workplace examples include personal protective equipment use campaigns, while non-workplace examples include anti-smoking campaigns and seat belt use programs. The main argument against using scare tactics is that receivers may block out or suppress the message and that it is not long-lasting in its effect.

Based on his review of the research literature, Peters (1991) recommends the following when using fear messages:

- The message should attempt to evoke a high (versus low) level of fear.
- The suggested preventive actions should be relatively detailed, specific, and presented immediately after the fear response is evoked.
- The preventive actions should be presented in a block, rather than interspersed with information designed to elicit fear.
- The suggested preventive actions must be perceived by the target audience to be effective in preventing danger.
- The source of the communication should have high credibility. Most research suggest the direct supervisor as being the best source for fear messages.
- Person-to-person, two-way communication is more effective than written, one-way communication.
- Increased effectiveness can be achieved by gaining the support of significant others, such as co-workers, supervisors, and family.
- Finally, safety and health communications should consider the target group(s) at which messages are aimed. For example, research has shown that fear messages are more effective with new employees than with seasoned employees who can use their experiences to discount the message. Additionally, fear messages have been found to be especially effective in influencing employees who are not under direct supervision and are expected to comply with safety regulations on their own.

As an aid to both defining targets and establishing objectives, employee surveys are recommended to assess current levels of safety and health knowledge, attitudes toward safety management programs and practices, and compliance with rules and procedures. Such measurements assist in pinpointing education and persuasion priorities and set a baseline for later evaluations of the effectiveness of communication efforts.

Cohen (1987) provides these general guidelines for communications:

- Statistics or risk data have the greatest impact if they are specific to a particular workplace or location.
- Persons in the target group should be involved in planning communication content.
- Pilot testing of communication programs is advisable.

Finally, Planek (1969) points out that because of the unknowns involved in the attitude change/behavior change process, evaluations of the success of communications programs should be guided by the following principle: *Observed success in conveying knowledge or changing attitude is not an indicator of behavior change unless further observation of behavior is made.*

Awards/Incentives/Recognition

The use of incentives, awards, and recognition to motivate employees to perform safely is an accepted feature of both OBM and TQM models. In the OBM model, use of incentives to reinforce employee behavior is critical to program success. In TQM, rewards, promotions, and other incentives are used to recognize individuals for contributions to process improvement. Also, at the group, team, or company level, special days or other functions are used to celebrate achievement.

Broadly speaking, the use of incentives of any type—including salary increases and promotions—can be viewed as having a positive influence on employee attitudes and behavior. When safety and health are made part of the decision to reward employee performance, these factors take on added significance as important job-related requirements.

Cohen et al. (1979) comment that the use of incentive and award programs is controversial. Given the wide use of incentives and awards of some kind in all phases of social enterprise, one may wonder why this controversy exists. The problems with such programs tend to arise when they are misused or abused. At the organization level, incentive or award programs for employees can be substituted for legitimate safety and health management programs. As has been indicated, motivational techniques support mainline programs but are never a substitute for them. At the employee level, abuse can result in the failure to report an injury or incident for fear that either an individual or work group will not receive an award.

Many companies use incentives and awards as part of their safety and health program on a continuing basis. Certainly, the OBM literature is filled with success stories supporting the use of incentives, and there is a great amount of anecdotal evidence citing the successful use of a wide variety of reward programs.

There is also some evidence presented by Deci (1975) that distinguishes between the effects of rewards that are perceived as "controlling," as can happen in the OBM approach to the reinforcement of safe behaviors, and those that are viewed as "informational," which is the emphasis in the TQM model. Studies that explore these differences (Kanfer, 1990) have found that rewards that recognize personal competence, as occur in TQM, are stronger than those that simply provide positive performance feedback, as is the case in OBM. One explanation for this finding is that employees perceive informational rewards, which recognize achievement and personal competence, to be under their own control rather than in the hands of another person who gives or withholds rewards based on the performance being observed. Accordingly, the focus for control of informational rewards is within the employee or intrinsic, as opposed to being outside the employee or extrinsic, as is the case for controlling rewards.

Awards that recognize achievement at the company or corporate level are also worthy of discussion. In the arena of quality alone there are two national awards of note, the President's Award for Quality and Productivity Improvement and the Malcolm Baldridge National Quality Award, both of which have attracted the attention of top U.S. corporations. Similarly, the National Safety Council provides awards for outstanding records in occupational safety and health.

The purpose of these awards is also to recognize achievement and reinforce the actions that lead to successful corporate performance at all organizational levels. When these awards are applauded throughout recipient companies, they can be a powerful stimulus for continuous improvement. Indeed, the Council's Occupational Safety/Health Award Program draws applications from more than 6,000 member companies annually.

Here again, however, problems of misuse can occur insidiously in organizations that receive national awards for excellence in safety and health. When management uses attainment of an annual award as the sole criterion of program achievement, two forms of misuse are possible. Management may have a self-satisfied attitude that causes it to ignore or be otherwise unresponsive to safety and health system and performance inadequacies that have not caused lost workdays, injuries, or deaths. Or, it may adopt a "win at all costs" posture that generates counterproductive stress and can lead to abuses in reporting or in the proper handling of employee injuries, and illnesses. In either of these cases, a national organizational award loses its effectiveness and becomes a hindrance to continuous safety and health improvement.

In summary, the appropriate use of awards, incentives, and recognitions can play an important helping role for organizations that use them wisely. They can, as Cohen et al. (1979) conclude, "add interest to an established hazard control program, which could enhance self-protective actions on the part of the work force."

Employee Surveys

Employee surveys in the form of questionnaires and/or interviews are becoming a widely used means of uncovering safety and health management systems strengths and weaknesses. They play a critical role in TQM approaches to employee motivation as a source of information on safety and health needs as well as a measure of performance improvement.

Surveys also stimulate employee participation in safety and health. As Planek (1993) points out, "It makes sense to include employees when designing program plans intended to increase their safety and health. When gathered in an objective and representative fashion, candid employee input can provide unique insights about the current status of safety and health and assist the selection of improvement priorities." DeJoy (1986) points out other positive effects of employee participation through surveys, such as heightened employee awareness and interest in safety, and the perception that management is not only interested in employee safety and health but also that it is open to employee input on the subject.

At a minimum, a basic employee survey should consider the following five factors:

- *Management Leadership.* Clearly communicates a safety and health vision supported in words and action.
- *Supervision Involvement.* Reinforces and communicates management's vision through open employee interactions, example, training, control, and recognition.
- *Employee Responsibility.* Personally supports company safety/health objectives and adopts the attitudes and action necessary to achieve them.
- *Safety Support Activities.* Generate employee cooperation and action to achieve safety and health objectives, such as a management/nonmanagement safety committee.
- *Safety Support Climate.* Sustains employees' perception that management is completely committed to employee safety and health as a top priority.

Like the other techniques that are discussed in this chapter, the benefits that can be derived from employee surveys depend on how they are used. For example, if survey results are acted on by management, they can greatly enhance employee morale and provide a strong indication that management is serious about employee safety and health. This image can be further enhanced if results are communicated to employees, at least in summary form. If, in contrast, management fails to follow up on findings, the survey activity can prove detrimental to morale and damaging to management's image with regard to employee safety and health.

This situation is particularly critical in organizations whose management style has been traditionally nonparticipative. An employee survey can be an effective way to initiate increased employee participation and to communicate this change in management style to employees.

However, if there is no intention to change the prevailing management style, the use of a survey will be seen simply as window dressing.

The primary reason for conducting an employee survey is to obtain as accurate a picture as possible of how the safety and health management system is operating. Whether or not an organization has open communications, it is frequently difficult to get an objective and balanced view of employee reactions to safety and health programs and activities. Communication barriers, whether personal on the part of individual employees or organizational in nature, can produce unbalanced impressions about program strengths as well as weaknesses.

For this reason, it is necessary to ensure that all employees are given an equal opportunity to participate in a survey. This may be done by including all employees in the activity or by selecting a random sample of employees that truly represents the workplace population. If the former approach is taken, the activity takes on added morale-building value, because all employees are given an opportunity to express their opinion. However, economic considerations may preclude this possibility, particularly in large organizations. In such cases, a random sample of respondents is appropriate if the proper selection techniques are used.

It takes time, effort, and resources to conduct an employee survey. Accordingly, management must be clear in how it intends to use survey findings and resolved to do whatever it takes to obtain the most objective and, therefore, most accurate and reliable results possible.

LEARNING PROCESSES

It is important for the safety and health professional to consider learning and the principles that affect it because training is a major consideration in safety programs. Often, to change behavior, trainers must substitute new learning for old habits.

Motivational Requirements

Work-based educational systems must place emphasis on stimulating the individual to want to learn. Materials must be designed to apply to practical situations in workers' lives, and trainers have to develop the workers' interest in the materials. To train a worker to use a computer, for example, the trainer might have the person create a simple letter or copy and store files related to the department's work. Trainers try to motivate their students and encourage their learning process. With the changing demographics of the U.S. workforce, effective employee education and training programs that can result in knowledgeable, safe, and productive workers is now more important than ever before.

The safety and health professional must recognize that if workers are going to learn safe procedures, they must want to do so. It is not wise to assume that, because management sees the value of safe procedures, the workers also will. Management may be motivated to start a safety program because it will lower insurance costs, reduce the amount of waste, and increase the number of units produced. Workers cannot be expected to want the program for the same reasons. To sell a safety program to the employees, one must know what will motivate them. Here the safety and health professional can capitalize on the needs discussed previously, and probably can find many more that are consistent with the aims of the safety program. Everything that will motivate the worker to learn the right procedures should be used. (Chapter 25, Safety Awareness Programs, discusses this in more detail).

To help motivate workers toward safe procedures, the safety and health professional should show the whole picture and emphasize how the workers will be affected:

- Point out the risks of improper work habits, and their costs to both the company and the individual.
- Show how improper work habits affect the product/service provided.
- Show the long-term impact of improper work habits on the employee, the company, and the product.
- Seek the workers' aid in addressing these problems in a functional manner for the benefit of all.

Principles of Learning

Some consideration of basic learning principles is valid whether the learning is to be done in a college classroom or in a work area. When training procedures are based on these principles, learning is more efficient and thorough.

Reinforcement

Through experimentation, psychologists have found that positive reinforcement can often facilitate learning. In practice, reinforcement can, for example, make work more efficient and more productive. When a worker's pay increases because he or she produces more units, the person has received a reward for learning. This type of reinforcement can have negative aspects as well—the worker who figures out a hazardous shortcut to produce more units may actually be rewarded for an unsafe practice.

Employees must be recognized and rewarded only for safe work methods. Higher productivity based on safe work methods satisfies the dual need for achievement and recognition. This fact in itself not only reinforces employee learning, but also influences other employees who see or hear about it. A supervisor who recognizes the same needs in other workers can reinforce this learning through praise of greater productivity and telling workers how much improvement has been shown. Thus, the publicity given to a safety record, a bonus, a promotion, or anything else that satisfies individual needs would serve as a reward to reinforce whatever learning has brought about the desired behavior.

Reinforcement tends to be more effective if it is given soon after the desired behavior occurs. Praise, for example, should be given at the time or within a reasonably short time after the desired behavior happens, or if delayed, associated verbally with the desired behavior. If reinforcement is delayed, the safety and health professional or supervisor should let the individual know why he or she is receiving the reward.

The principle of positive reinforcement also applies to employee participation in a safety program whether through suggestion systems, safety and health committees, discussion groups, or training sessions. In all these areas an individual gains personal worth if his or her opinions are requested and positively reinforced.

A safety program that gathers the ideas of all, either individually or by representation, satisfies the need people have for being "in the know." In this way, the safety and health professional can create a positive atmosphere for the program and instill in employees a sense of obligation and responsibility for its success. Research has demonstrated that when employees feel they helped to create a program, there is more chance for its success.

Research on the effectiveness of disciplinary actions suggests that it can have mixed consequences. Discipline generally is thought to be less effective than reinforcement, perhaps because discipline provides indirect cues or information (what not to do), as opposed to positive reinforcement that provides direct information about what employees should do (Church, 1963).

Recognizing an individual's correct performance will lead to a more positive attitudinal response on the worker's part than will any discipline. A supervisor's praise for a task well done may influence an employee's attitude toward other parts of the job, including training procedures, safety devices, and the safety program. The proper use of recognition thus can lead to worker acceptance of efficient and safe work methods.

Knowledge of Results

Another aspect of recognition is knowledge of results. People like and need to know how they are doing on the job. When workers know that a new procedure is helping to increase their production, they have received reinforcement for their learning and effort.

Knowledge that a job was done correctly and safely is rewarding to employees. However, the contrary is also true. If a job was not done correctly, then employees must know the correct rules and procedures that are a part of job performance.

One phenomenon that often occurs in the learning process is known as a plateau. At this point, a worker's progress levels off for some time before he or she experiences another increase in learning. Often some individuals become discouraged during this time, and their learning may decline if it is not supported. The trainer needs to understand this phenomenon and communicate to the individual that such a leveling off is normal and that further progress will come later.

Practice

The safety and health professional is interested in developing safe working habits in employees that will become almost automatic on the job. As a result, merely putting a worker through training sessions is not enough. Despite any apparent mastery of safety methods, the next time the employee performs the task, he or she may make one or more mistakes. To be sure habit patterns are firmly entrenched, workers must practice what they learn. Job instruction training (JIT) programs include one of the important aspects of training, follow-up by the manager. This helps to ensure mastery of safe working methods. Within a reasonable time, depending on the job complexity, follow-up will no longer be needed.

In addition to the learning principles covered so far, whole versus part learning has been a challenge to industrial trainers for many years. Whether trainers should teach a work procedure as a whole or break it down and teach it part by part has pros and cons. Both methods have advantages and disadvantages depending on job complexity, the trainee, the kind of job breakdown used, and the level of support provided by the manager. A combination of the two approaches, using the whole method, but with sufficient flexibility to emphasize meaningful parts of the task wherever necessary, may be most effective.

Meaningfulness

Studies in verbal learning have demonstrated how important it is that the material used be meaningful to learners. Meaningfulness to management and workers is understanding the value of producing a marketable product or service (which contributes to job satisfaction, job retention, profit, and the like) and the negative consequences of failing to engage in appropriate actions, leading to higher production costs, higher product rejection rates, accidents, and their associated losses.

Meaningfulness is essential in safety training because the worker needs to understand why a certain procedure is better than another. Adequate explanation, how a given movement or change in position can eliminate hazards with no decrease in production, gives meaningfulness to the procedure. With this understanding, workers are more likely to be motivated to learn the safe procedure. Without it, they will be inclined to use their own methods until they learn, perhaps by an accident, how inadequate unapproved work procedures are. The safety and health professional and line supervisor should reinforce the advantage of workers' understanding the reason for having protective clothing, safety devices, safety meetings, and discussions, as well as the need for full and complete accident reports.

Selective Learning

Out of each day's many experiences, people for a myriad of reasons select those they will retain. This selection process is closely related to motivation and for that reason motivational aspects of a job training program need to be considered. Safety trainers must be sure the workers retain the most important facts. Relating subject matter to individual needs will help to ensure the proper selection and retention.

Frequency

People do best those things they practice most. This principle is certainly important to the safety and health professional, for it emphasizes the necessity of frequently applying safety practices in training programs and on the job. Giving the worker a copy of safety rules and regulations in hopes these will be learned and used is not enough. Instructors must devise ways to bring these rules and regulations to workers' attention frequently and regularly.

Trainers must insist not only on frequent practice, but on the trainee's following the correct method. Daily use of safe methods will create safety habits in employees that will become routine. The safety and health professional must make sure the work method practiced is the safe one.

Recall

Closely related to the principle of frequency is that of recall. What is learned last usually can be most easily recalled. As has been indicated, handing a worker a set of safety rules and regulations does not ensure learning. Those who received printed instructions or a few training demonstrations a long time ago may not be able to recall what the rules and regulations are. Safety and health professionals must devise ways for workers to come into constant contact with these regulations through activities such as contests, reviews of safety regulations, and committee work.

Primacy

The law of primacy must be taken into consideration in two aspects of the safety program. First, the worker's initial contact with safety procedures should be positive and of major importance. If this initial contact is of a negative nature—such as being tossed a book of rules and told "Make sure you learn them"—the worker is left with the impression that safety is unimportant. From the first day of employment, individual safety rules as well as the whole safety program are of great and equal importance.

Secondly, in training programs, developing habit patterns using safe methods should assume primary importance. The manager must be certain the worker does not learn how to work any other way than by the safe method. It is more difficult to establish good patterns if one has first learned the poor ones.

Intensity

Those things made most vivid to the worker will be retained the longest. Safety programs already use this principle in safety publicity with eye-catching posters, slogans, and the like.

Part of the safety and health professional's responsibility is to help management enhance the worker's interest in the program. In this way, the training material will be remembered for a long time. To some degree, the trainer must position the safety program in the same way advertisers position their commercials to catch the public's attention.

Transfer of Training

Another principle of learning that trainers should remember is that all new learning occurs within the context of previous learning experiences. Current learning (or performance) can be influenced by previous learning, a phenomenon known as transfer of training. Positive transfer of training occurs when the previous learning experience helps current learning or enhances current performance. Negative transfer of training occurs when previous learning makes the current learning experience more difficult or in some way inhibits current performance.

Generally, learning to make identical responses to new stimuli results in positive transfer. For example, learning to drive in a new (or different) automobile is facilitated by the fact that identical responses (such as accelerating or steering) are required in the new automobile, similar to those in the old automobile. If the new stimuli are similar to previous cars (for example, the location of the controls, their texture, and their direction of movement), positive transfer should be high. In such situations, positive transfer increases with higher level of similarities between the two learning situations.

Learning to make new responses to what appear to be identical stimuli, but are actually different, results in negative transfer. For example, if the controls of the cars appear identical but in each car the "fifth" gear is in the opposite position, negative transfer can be expected for a time. If the fifth gear position is located at the upper right position on the floor shifter in one car, the person will experience some difficulty learning to drive a car in which the fifth gear position appears on the lower right. Or, if the Off position of one toggle switch is the same as the On position of another, the potential for accidents is serious.

Such negative transfer can cause errors, delayed reactions, and generally inefficient performance on the new task. In task situations, the amount of negative transfer increases the more apparently similar the two learning situations seem to be.

With an understanding of transfer phenomena, it is possible to maximize, within the industrial environment, opportunities for positive transfer and minimize those for negative transfer. This requires careful planning of machine purchases and work procedures to make certain the new tasks required of a worker make use of (and do not conflict with) previous learning experiences.

Short-Term and Long-Term Memory

This discussion of learning would not be complete without some consideration of the memory process, which occurs as learning takes place. Learning and forgetting are not mutually exclusive of each other. As one learns, one also forgets some of what was previously learned. Curves of forgetting indicate that most learning is lost immediately after it has taken place. Depending on the complexity of the job, the amount of learning lost will vary after each day's training session. There should be less forgetfulness or initial mistakes in successive days of training, and more need for systematic refresher sessions to recall what was learned.

To minimize the process of forgetting, trainers must consider the various principles of learning and the motivational aspects associated with them. This needs to be kept in mind from the initial stages of employment through the everyday work situations in the company.

SUMMARY

- Attainment of permanent performance improvement is a product of employee and management commitment and involvement. Employees must be willing to take responsibility for paying attention to the behavior changes needed to achieve safety and health objectives.
- The psychological factors that influence safety behavior include individual differences, motivation, emotions, stress, attitudes, behaviors, and learning processes.
- Individual differences among employees present an ongoing problem in industrial settings. Yet employers often make the mistake of gearing their safety efforts toward a nonexistent "average person" instead of matching each person to a specific job.
- Motivation includes direction of behavior, intensity of action, and persistence of effort. To affect employee motivation, employers must establish specific safety objectives, reinforce desired behaviors, understand the needs theory of Maslow, recognize the motivational power of job satisfaction, understand the X and Y theories of management, and be familiar with job enrichment theory.
- Attitudes are comprised of three elements: feeling, knowing, and action. Attitudes are formed in a variety of ways but tend to be persistent and difficult to change. Commitment and responsibility are the key factors in forming and changing attitudes.
- Attitudes are important to safety in a company, but safety-related behaviors are critical. Behavioral management's "action"-directed focus involves the identification, measurement, and control of employees' safe work-related behaviors. Steps involved in changing safety-related behaviors can include (1) providing training, (2) arranging practice, (3) arranging positive consequences that reinforce correct performance, (4) removing obstacles that prevent satisfactory performance, or (5) finding a safer way to do the job.

- Two motivational models used in many companies are organization behavior management and total quality management. Selected motivational techniques that have proven successful include communication, awards/incentives/recognitions, and employee surveys.
- By understanding the principles of how people learn, management can establish effective employee development and training programs that can influence, change, and reinforce the concept of safe work-related behavior. Management must understand and apply learning principles such as reinforcement, feedback, practice, frequency, transfer of training, and supporting short-term and long-term memory.

REFERENCES

"Accident Causation Theory: The Emerging Pluralism," *Occupational Hazards* (October 1983):45,10.

Ajzen I and Fishbein M. "Attitude-Behavior Relations: A Theoretical Analysis and Review of Empirical Research." *Psychological Bulletin* (1977)84: 888-918.

Anastasi A. *Psychological Testing*, 6th ed. Toronto, Canada: Macmillan, 1988.

Bauer RA. Communication as transaction. *The Obstinate Audience*, DE Payne (ed.) Ann Arbor, MI: Foundation for Research on Human Behavior, 1965.

Bolton R. *Social Style/Management Style*. New York: American Management Associations, 1984.

Braun WD, PhD. *Welcome Stress. It Can Help You Be Your Best*, 1st ed. Minneapolis: CompCare Publications, 1983.

Chhakar JS and Wallin A. "Improving Safety Through Applied Behavior Analysis." *Journal of Safety Research* (1984) 15:141-151.

Church RM. "The Varied Effects of Punishment on Behavior." *Psychological Review* (1963) 70:369-402.

Cohen A. Protective behaviors in the workplace. *Taking Care: Understanding and Encouraging Self-protective Behavior*, N Weinstein (ed.). Cambridge, England: Cambridge University Press, 1987.

Cohen A, Cleveland B. "Safety program practices in record-holding plants." *Professional Safety* 28(3):26-33, 1983.

Cohen A, Smith M, Anger W. "Self-protective measures against workplace hazards." *Journal of Safety Research* 11:121-131, 1979.

Conard RJ. *Employee Work Practices*. NIOSH Contact Report 81-2905. Cincinnati, OH: National Institute for Occupational Safety and Health, 1983.

Cooper CL and Smith MJ. *Job Stress and Blue Collar Work*. Chichester, England: John Wiley & Sons, 1985.

Deal T and Kennedy A. *Corporate Cultures: The Rites and Rituals of Corporate Life*. Reading, MA: Addison-Wesley, 1982.

DeJoy D. "A behavioral-diagnostic model for self-protective behavior in the workplace." *Professional Safety* 31(12):26-30, 1986.

Dipboye RL and Howell WC. *Essentials of Industrial and Organizational Psychology*, Rev. ed. Homewood, IL: The Dorsey Press, 1982.

Fournies F. *Coaching for Improved Work Performance*. Blue Ridge Summit, PA: Liberty House, 1987.

Glasser W, MD. *Control Theory*. New York: Harper & Row, 1984.

Hersey P and Blanchard K. *Management of Organizational Behavior: Utilizing Human Resources*, 5th ed. Englewood Cliffs, NJ: Prentice Hall, 1988.

Herzberg F. *Work and the Nature of Man*. Cleveland: World Publishing Co., 1966.

Hulse SH, Deese J, Egeth H. *The Psychology of Learning*, 5th rev. ed. New York: McGraw-Hill Book Co., 1980.

Kanfer R. Motivation theory and industrial and organizational psychology. *Handbook of Industrial & Organizational Psychology*, MD Dunnette, LM Hough (eds.). Palo Alto, CA: Consulting Psychologists Press, 1990.

Katz D and Stotland E. "A Preliminary Statement to a Theory of Attitude Structure and Change." In S Kock (ed.), *Psychology: A Study of a Science*, Vol 3, 423-475. New York: McGraw-Hill, 1959.

Mager R and Pipe P. *Analyzing Performance Problems Or You Really Oughta Wanna*, 2nd ed. Belmont, CA: David S. Lake Publishers, 1984.

Marks LN, RN. "OHN's Can Provide Listening Ear Dealing with Employee's Problems." *Occupational Health and Safety*, (May 1986) 55,5.

Maslow AH. *Motivation and Personality*, 2nd ed. New York: Harper & Row, 1970.

McGregor D. *The Human Side of Enterprise*. New York: McGraw-Hill Book Co., 1985.

McGuire WJ. "The Nature of Attitudes and Attitude Change." In Lindzey G and Aronson E (eds.), *The Handbook of Social Psychology*, 2nd ed., Vol. 3, Reading, MA: Addison-Wesley, 1968.

Miller NE and Dollard J. *Social Learning and Imitation*. New Haven, CT: Yale University Press, 1962.

"Open Your Ears and Learn to Listen." *Safety and Health Magazine*, (July 1990) 44–47.

Peters R. "Strategies for encouraging self-protective employee behavior." *Journal of Safety Research* 22:53-70, 1991.

Planek TW, Fearn K. "Reevaluating occupational safety priorities." *Professional Safety* 10:16-21, 1993.

Rader CM, PhD. and Haber PB, PsyD. "A Psychological Profile of Industrial Injured Workers." *Occupational Health Nursing* (November 1984) 32,11.

ReVelle JB and Boulton L. "Worker Attitudes and Perceptions of Safety." *Professional Safety*, (December 1981) 26,12.

Weisinger H, PhD. *Dr. Weisinger's Anger Work-Out Book*. New York: Quil, 1985.

"What Does It Take To Change Unsafe Behavior?" *Safety and Health*, (July 1990) 60-63.

Suggested Readings

Baldwin BA. "Professional Burnout: Occupational Hazard and Preventable Problem." *Pace* (January/February, 1981) 8,1:20ff.

Bennis WG and Nannus B. *Leaders: The Strategies for Taking Charge*. New York: Harper & Row, 1986.

Blake RR and Mouton JS. *The Managerial Grid III*, 3rd ed. Houston, TX: Gulf Publishing, 1984.

Deal T and Kennedy A. *Corporate Cultures: The Rites and Rituals of Corporate Life*. Reading, MA: Addison-Wesley, 1982.

Dennis LE and Onion, ME. *Out in Front: Effective Supervision in the Workplace*. Chicago: National Safety Council, 1990.

Drucker PF. "New templates for today's organizations." *Harvard Business Review* Jan.-Feb. 1974.

Fournies F. *Coaching for Improved Work Performance*. Blue Ridge Summit, PA: Liberty House, 1987.

Hersey P and Blanchard, K. *Management of Organizational Behavior: Utilizing Human Resources*, 5th ed. Englewood Cliffs, NJ: Prentice Hall, 1988.

Herzberg F. "One More Time: How Do You Motivate Employees?" *Harvard Business Review* Jan.-Feb., 1968.

Hinricks JR. "Psychology of Men at Work." *Annual Review of Psychology* (1970) 21.

Lazarus RS. "Little Hassles Can Be Hazardous to Health." *Psychology Today* (July, 1981) 15: 48-62.

Machlowitz M. *Workaholics*. New York: Addison-Wesley Publishing Company, Inc., 1980.

Mager R and Pipe P. *Analyzing Performance Problems Or You Really Oughta Wanna*. 2nd ed. Belmont, CA: David S. Lake Publishers, 1984.

Patrick P. *Health Care Worker Burnout: What Is It, What To Do About It*. New York: Inquiry Books, 1980.

Zimbardo PG. *Psychology and Life*, 12th ed. Glenview, IL: Scott, Foresman and Company, 1988.

REVIEW QUESTIONS

1. Define motivation.
2. List five of the seven psychological factors that directly influence the success of a safety program.
   a.
   b.
   c.
   d.
   e.
3. What two characteristics influence the effectiveness of reinforcement of desired behaviors?
   a.
   b.
4. Whether or not an employee works safely depends on what three factors?
   a.
   b.
   c.
5. What was Abraham Maslow's contribution to the study of motivation?
6. According to various types of studies and surveys, satisfaction of which of the following is the major motivational aspect of job satisfaction?
   a. Physiological needs
   b. High salary
   c. Psychosocial needs
   d. All of the above
7. McGregor discovered two basic ways in which managers view workers. Give a brief description of each.
   a.
   b.
8. What did Herzberg theorize to be the most effective concept of motivation in the workplace?
   a. Job enrichment
   b. Environment approach
   c. Hygiene approach
   d. New technology
9. What are the three components of attitude?
   a.
   b.
   c.
10. On what is a person's response or attitude usually dependent?
11. List four of the seven principles of job enrichment.
    a.
    b.
    c.
    d.
12. Briefly explain the difference between behaviors and attitudes.
13. List seven of the ten principles of learning.
    a.
    b.
    c.
    d.
    e.
    f.
    g.

23
Safety
and Health
Training

When a safety or health problem is caused by lack of appropriate attitude, knowledge, or skills, one solution to meet that problem is training. Training is primarily focused on behavior or performance change; in a work setting, the goal of effective training is learning that leads to improved on-the-job performance. When safety and health are the primary issues, the goal is that workers learn how to keep themselves and others safe and healthy. This chapter covers the following topics:

- training as part of a safety program
- benefits of training
- when training is the right solution
- adult learning needs
- principles for selecting or developing a safety and health training program
- how to deliver a safety training program
- types of available training methods.

A key element in every successful organization, in any successful accident prevention program, and in any occupational safety and health program is effective job orientation and safety and health training. Workers must have proper training to do their jobs safely and efficiently.

The responsibility for, and the implementation of, employee training rests with management. Management must recognize that for the organization to achieve its objectives, employees need to perform at a certain level. In order to achieve a satisfactory level of performance, management needs to establish worker training policies, which in turn, will lead to training programs for the workers. To help achieve the organization's objectives, those responsible for safety and health, whether safety and health professionals or line managers, must:

- be trained in the proper methods of leadership and supervision
- know how to detect and eliminate or control hazards, investigate accidents, and handle emergency situations
- understand how to train employees in the safest and most efficient methods of job performance
- know and support all company health and safety policies and procedures.

The safety and health professional must also ensure that all training provided meets all regulatory training obligations and complements/supports the business goals of the company (see Roles of the Safety Professional in Training later in this chapter).

TRAINING AS PART OF THE SAFETY AND HEALTH PROGRAM

This section focuses on the definition of training, its benefits, when to choose training as a solution, how the safety professional supports training objectives, and the characteristics of effective safety and health training programs. Effective training programs ensure employees know how to do their jobs using safe methods.

Definition and Goals of Training

Training, generally speaking, is one specific solution to meet a safety or health need caused by lack of appropriate behavioral skills, related knowledge, and/or attitudes. Training provides the "how-to-do" of a subject and usually the "what." It only provides the "why" to the extent that people need to know it in order to complete the task. (See also Chapter 22, Motivation.)

Training is primarily focused on behavior change. In education, the focus is on *information* about something that may or may not be used on the job; in training, the focus is also on *how* to do something properly and how to apply the new information and skills on the job.

Benefits of Training

The benefits of safety and health training include:

- reinforcement of the operational goals of the organization
- improved performance
- fewer incidents/accidents
- reduced costs.

Training and Nontraining Solutions

Sometimes a training program is not needed in order to impact on-the-job performance. Millions, perhaps billions, of dollars are wasted every year because people concluded that training will solve a problem that needs a different solution. In many situations, the solution may be a nontraining solution, or a combination of training and nontraining solutions. Nontraining solutions are actions taken to resolve safety and health problems that do not require training. For example:

- A high-humidity work area may have a sign showing the safety goggles that may be worn when working in the area.
- A supervisor may offer a bonus to the employee with the best overall safety record.
- The engineering department may redesign the work flow of a manufacturing site to decrease the amount of injuries caused by people walking near heavy machinery.

It is important to analyze the problems and assess training needs before making decisions about solutions. Management should consider all training and nontraining alternatives to safety and health challenges. For example, a certain task is performed infrequently and workers often fail to remember all the steps (or perhaps the task is too complex to perform from memory). In a situation like this, nontraining solutions such as those listed below can be as effective as a training program.

- Job aids such as:

  task procedures
  Material Safety Data Sheets (sections on protection information, special precautions, spill/leak procedures)
  flowcharts
  checklists
  diagrams
  troubleshooting guides
  decision tables

- reference manuals
- help desks or hotlines
- reward systems
- improved physical work environments
- improved work processes.

Training and nontraining problems or needs, and their related solutions, generally can be categorized as: (1) selection and assignment, (2) information and practice, (3) environment, and (4) motivation/incentive. These four major categories of training and nontraining problems and solutions are defined as follows.

1. Selection and assignment refers to considerations and processes used to hire people and assign them specific responsibilities and on-the-job tasks. Improving the hiring process helps ensure the right worker is assigned to the task.
2. Information and practice is the act of providing all necessary information and developing the skills needed to use safe work practices. This area includes clearly communicating tasks, goals, and instructions; and providing feedback, training, job aids, guided practice, and experience. This is the only area which can sometimes be improved through a training solution, since effective training provides many opportunities for skills practice.
3. Environment refers to all day-to-day influences on performance. These include, but are not limited to, equipment, floor plan, access to tools, working relationships with others, and working relationships with the safety and health professionals of the organization. For example, in work areas where people should wear safety goggles but may forget, a sign can be posted showing the use of the proper safety goggles. Or, if walkways are located near areas where heavy machinery is used, the engineering department may redesign the workflow of the site to decrease the potential for injuries.
4. Motivation/incentive refers to any link between safety and rewards for safe behavior, such as bonuses, pay increases, contest prizes, and so on. For example, a supervisor may offer a bonus to the employee with the best overall safety record. In safety training, in particular, the reduced risk of injury and death is a potential motivator. For

detailed discussion of motivation, see Chapter 22, Motivation.

Training Roles of the Safety Professional

A safety professional may play many roles in fulfilling a responsibility for training. The safety professional may conduct the training, coordinate its development with internal or external instructional designers, purchase existing materials, contract with a trainer, or coordinate the entire training function. No matter what the role, it is important to have a clear idea of the training goals, based on the needs of the company or facility and the workers.

Structured Training

If the solution to a specific, on-the-job problem is indeed training, then it should be planned, designed, developed, and delivered in a structured manner with clear evaluation of the training and its effectiveness in solving the problem and impacting behavior. Such training is called performance based.

Performance-based training is a learning experience (training) that is implemented to solve a specific, on-the-job problem or to encourage a specific behavioral change. Performance-based training can be measured or evaluated by analyzing a worker's performance. This training is directly related to the job the worker is expected to perform, linked to the organization's safety and health goals, measured and evaluated by the worker's observable performance, and created using a proven systematic approach that has five phases. The five phases are as follows:

1. Assessment: In this phase, the safety professional consults others and conducts research to determine whether training is the right solution for the problem and what knowledge, skills, and attitudes people need to develop.
2. Design: In this phase, the safety professional or contracted trainer identifies instructional goals, chooses training media, and most importantly, the sequence in which workers will learn new information and skills. This work is based on needs assessment results, an audience analysis, and specific performance-based objectives.
3. Materials acquisition or development: In this phase, the safety professional or contracted trainer buys or creates training materials, or does some combination of both.
4. Delivery: In this phase, the safety professional or contracted trainer delivers the training.
5. Evaluation: In this phase, the safety professional gets feedback from workers, supervisors, and others involved in the performance-based training. Training is then improved based on this feedback.

There are four major types of performance-based training:

- Instructor-led training: This training is presented in a classroom-like setting and everyone follows the pace set by the instructor (also commonly called facilitators or trainers). This type of training generally includes participant materials.
- Self-paced training: This is usually completed on an individual basis, by working through a textbook and/or workbook. Self-paced training may also be completed through computer-assisted instruction.
- Computer-based training or computer-assisted instruction: This is a version of self-paced training in which the learner is guided through the course by a computer. The computer provides the course content, guides learning activities, and administers tests. Computer-based training can also provide the safety professional with a quick list of who has completed the training, and some programs can provide tests and scores.
- Structured on-the-job training: This is similar to instructor-led training, except that it is conducted at the learner's actual place of work. The instructor is usually a trained supervisor who acts as a coach or guide. This type of training may or may not include participant materials.

Adult Learning Needs

A major factor in the success of training is the extent to which adult learning needs are taken into consideration. Safety and health issues can be technical and difficult to learn. Also, people may see safety and health training as a disruption in their busy work days.

To keep the participants motivated and involved in safety and health training, the participants' learning needs must be considered. To accomplish this, adult learning principles must be used in the design and delivery of training.

There are many adult learning principles. For example, one principle is based on the belief that each person learns best through one of three senses:

- hearing (11% of humans are auditory learners)
- sight (83% of humans are visual learners)
- touch or activity (69% of humans are kinesthetic learners).

Training following this model would be tailored to each learner's best method of learning.

More generally applicable are the four needs common to all adult learners. This information is adapted from Clay Carr, a widely-respected training professional and author of the book, *Smart Training: The Manager's Guide to Training for Improved Performance.* These same principles are generally accepted by most adult learning experts.

- Adults need to know why they are learning a particular topic or skill because they need to apply learning to immediate, real-life challenges.

- Adults have experience that they apply to all new learning.
- Adults need to be in control of their learning.
- Adults want to learn things that will make them more effective and successful.

These principles should be applied to any safety and health training developed or contracted. Each will be discussed in more detail.

Need to Know Why

Adults need to know why they are learning a particular topic or skill because they need to apply learning to immediate, real-life challenges. Therefore, they will learn best when:

- They see clear demonstrations of how the training directly applies to their jobs.
- They have opportunities to apply the new information or skill to solve problems during the training.
- They have opportunities to think about how the new information or skill can be used.

An example of meeting this need is a supervisor demonstrating how to safely operate machinery, then providing feedback while watching the workers actually operate the machinery.

Need to Apply Experience

Adults have experience that they apply to all new learning. Therefore, they will learn best when:

- They have opportunities to share their experiences during training. They can make use of comparison and contrast, of relationships, and association of ideas.
- They have opportunities to think about how the new information or skill relates to their past, present, or future experiences.
- Their experience, and opinions formed from their experience, is taken seriously by the facilitator.

To meet this need, for example, during a training on personal protective equipment (PPE), provide time for participants to talk about their own PPE practices at their facility. They can also share their observations on the practices of their co-workers.

Need to Be in Control

Adults need to be in control of their learning. Therefore, they will learn best when:

- They choose, or at least influence, the training they receive.
- They are in a flexible learning environment.
- They are actively involved in the learning process rather than passive receivers of information. (The facilitator is a guide for the learning experience.)

- They have opportunities to voice concerns and see them addressed. They take part in activities such as group discussions, role plays, simulations, etc.

An example of meeting this need is fluorescent light factory workers who are involved in a safety and health committee. Their responsibilities include assessing workers' needs for training, identifying appropriate training solutions, and evaluating the training once it has been delivered.

Need for Success

Adults want to learn things that will make them more effective and successful. Therefore, they will learn best when:

- They know why they are taking the training and how it will benefit them.
- They can ask and answer the questions, "What will this do for me? How will it help me get ahead? What hazards will it help me control?"

An example of meeting this need is a safety management techniques course that includes action planning for future personal and professional development.

Literacy Levels

Many workers read, write, use math skills, comprehend information, and use problem-solving abilities at a level called "low literacy" or "mid-literacy." Low literacy generally refers to workers whose skills and/or abilities are at or below a fourth grade level. Mid-literacy generally refers to workers whose skills and/or abilities are between a fifth through eighth grade level.

Low or mid-literacy may result in workers having trouble understanding large amounts of information that is presented in complex textbooks, manuals, and lectures. This must be taken into consideration when designing and delivering safety and health training programs. For instance, training can be modified to be written at a fourth or fifth grade reading level or to rely on less printed material. Use of symbols and illustrations in safety and health training and promotion can minimize the risk of the message being misunderstood.

DEVELOPING THE TRAINING PROGRAM

The safety professional should work with an internal or external instructional designer to develop any training program. The development process should include a needs assessment, written performance objectives, selection of materials to fit the objectives and workers, outline of content, timing, testing, and evaluation. The training program should meet the same criteria whether it is developed internally or selected from external sources.

Assessment

In order to create an effective safety training program, the safety and health professional needs to assess current worker performance, comparing it to desired worker performance to identify training needs and goals, and to conduct learner analysis. A needs assessment is important to an organization because it helps to:

- distinguish between training and nontraining needs
- understand the problem or need before designing a solution
- save time and money by ensuring that solutions effectively address the problems they are intended to solve
- identify factors that will impact the training before its development.

Need(s) assessment is the process of determining the "who, what, when, where, and why" of the training need once it is determined that a safety or health problem can be solved through training. Identifying the specifics of the training need(s) is the first phase of developing effective training.

The safety and health professional tries to determine at a general level *who* may need training, *what* they need to be trained to do, *when* they need to be trained, *where* they need to be trained, and *why* they need the training. This may be done by answering:

- Who does the job?
- What do they do in their jobs?
- When do they do the job?
- Where do they do the job?
- Why do they do the job?
- How do they do the job?

Primarily, a needs assessment is conducted to determine the specifics of the problem or need that must be addressed through the required training. This needs assessment may also show that other nontraining solutions are necessary to completely address the issue. When a regulatory agency is not involved, the safety and health professional has the responsibility to determine if training is needed at all before determining a complete picture of the (perceived) training need. This is one of the main reasons it is important for safety and health professionals to have an understanding of nontraining solutions. In the initial stages of an assessment, always remember to consider the many solutions that could potentially solve the problem.

In the needs assessment phase, the safety and health professional should identify a training goal(s)—a statement that describes how the training will satisfy the safety and health need or solve the problem, as well as state the general purpose of the training. Training goals are used to provide direction for further analyses—learner, job, and setting analyses—and to develop learning objectives. Training goals are also valuable tools for the safety and health professional to use in communicating the purpose of training to others. When writing a training goal, it should clearly:

- describe how the training will satisfy the safety or health problem
- communicate the purpose of the training.

While the training goals are being identified, the safety and health professional should also identify the specific characteristics of the learners (workers) who will be involved in the training. Learner analysis is also referred to as the process of identifying the characteristics of the learners. The results of learner analysis impact the training design. These characteristics include general demographics and the learner's preferences, attitudes, knowledge, skills, and previous experiences. The results of a learner analysis include:

- specific characteristics that will impact the design and delivery of the training (also called an audience description)
- recommendations for how the design of the training will accommodate the specific characteristics of learners.

The steps to completing a learner analysis include:

1. Identify general characteristics of learners who will be attending the training, such as job titles/position and age.
2. Identify characteristics of the learners' preferences regarding training.
3. Identify characteristics of the learners' attitudes or feelings regarding training.
4. Identify characteristics of the learners' knowledge or skills, especially as related to content of the training.
5. Identify ways of adapting to these characteristics in training and delivery.

Two other types of analyses conducted in the needs assessment phase include job analysis and setting analysis. Job analysis is the process used to determine the procedures, decisions, knowledge, and skills required for a worker to perform a job function. Setting analysis is the process of identifying specific characteristics of the training environment (where training will occur) that will impact training. Ideally, it is best to obtain setting information prior to design of the training. However, this is not always possible. The training course should be flexible enough to be delivered in a variety of settings.

Objectives

Writing performance objectives and evaluating learning progress are two of the most important steps for developing effective training programs. Performance objectives provide the safety and health professional with a structure or framework for developing training. Objectives are also important guides in the:

- selection and development of course content
- selection and development of learning activities

- measurement of the learner's performance.

Objectives are important to the workers being trained because they:

- provide a target for performance (or behavior)
- help learners identify their focus
- inform learners how they will be evaluated.

Objectives are especially critical in the safety and health arena because there is too much risk (potential injury, illness, or loss of life) to workers to allow a "hit or miss" approach. Additionally, a vast amount of time and money is potentially wasted if training objectives are not determined in advance.

Unlike goals, objectives are measurable and observable. Objectives differ from learning activities in that they describe the results, not how to get results in the classroom. The four parts of an effective objective (sometimes referred to as the ABCD method of objective writing) are:

- audience (learners): always identify the audience
- behavior: identify what learners must do in order to demonstrate mastery
- condition: identify what learners will be given or not given in order to do the behavior
- degree: specifies how well the audience members (learners) must perform the behavior.

Generally, if not specified in the objective, most people will assume that 100% is the standard performance. Performance can be measured in terms of quantity, quality, or both. The current philosophy behind loss control supports measuring quality and quantity.

By detailing training objectives, the safety and health professional can develop programs that meet an organization's established safety and health goals. The objectives should provide specific information about what learners will be able to do as a result of the training.

Materials

The materials of an effective safety and health training program should provide (1) an introduction, (2) presentation of the information, (3) practice and feedback, and (4) summary, evaluation, review, or transition. The introduction can be an overview, a rationale, a warm-up activity, questions, or a story or analogy. The information should be presented in a manner that helps the workers organize and remember the important facts. The following types of materials can be used to achieve this goal:

- text
- graphics
- examples
- job aids
- checklists
- graphs
- tables

- data
- reports
- relevant articles
- glossary
- table of contents.

Practice and feedback should provide the workers with the opportunity to try out the new knowledge and skills and to receive information about their performance. Materials that help guide the process of practice and feedback include:

- activity directions
- activity time allowance
- activity worksheets
- question and answer worksheets
- business scenarios
- performance checklists
- sample solutions, expert answers.

The summary, evaluation, review, or transition phase should stress important training points, answer workers' questions, and make a connection to the next training topic or the job. Supporting materials for this phase include text, graphics, question and answer sheets, listing of next steps, and goal sheets for implementing new knowledge and skills on the job.

Organization and Timing

The organization and timing of the content of the safety and health training program is achieved by outlining the content. To outline, the training developer selects an objective and identifies any actions the worker must take to achieve the objective and related topics the worker must know to accomplish the actions. The training developer then sequences the content in the appropriate order (perhaps with input from the safety and health professional or management) and estimates the amount of time needed for each point. The time available may be predetermined. If more time is needed, the trainer can either request it or narrow the scope of the objectives.

Testing and Evaluation

In their commitment to improving the knowledge, skills, and attitudes of workers, safety and health professionals must also be willing to measure whether workers have achieved the training objectives. Performance testing is one means the safety and health professional can use to measure whether learners have met the learning objective. Performance tests can only be created after performance objectives have been developed. There are three main types of performance tests: pretests, review tests, and post-tests.

Pretest

A pretest measures how well learners can perform objectives prior to training. Pretests are generally optional and are often provided as a means for learners to identify how much they already know before starting the training. Results are also used by trainers to customize the program to meet audience needs. In other situations, pretest (combined with post-test) results are used to assess the learning progress made by an individual or class.

Review Test

A review test measures how well the learners can perform the objectives while training is in progress. Learners take review tests during training to determine when they are ready to move on to the next objective (topic or task). Review tests often take the form of activities or exercises. Participants' progress is informally assessed by the trainer.

Post-Test

A post-test measures how well the learner can perform the objectives after training. Post-tests often test on the same information included in the pretest.

When designing a training program, safety and health professionals will need to decide whether to use, and when to use pretests, review tests, and post-tests to measure learners' performance. In safety and health training, tests are highly recommended when:

- the training involves a certification or qualification process
- the organizational culture supports its use
- the risks of not mastering the objectives may include injury, death, or significant financial loss.
- the effectiveness of training might be questioned
- qualitative and quantitative data are needed pertaining to training and/or safety and health issues.

DELIVERING TRAINING

Many companies regard orientation as an excellent opportunity to begin training workers in safety policies and safe work practices. In fact, regulatory agencies state that an employer not only should provide such training but is obligated to do so. Safety and health training can be delivered through an orientation program and through written policies and procedures manuals.

New Employee Training and Orientation

Safety and health training begins when a new worker is hired. The person is usually open to ideas and to information about the way the company operates. From the first day, the new employee formulates opinions about the management, supervisors, workers, and the organization. Personnel managers may say they have never hired a worker with a bad attitude. This may not be entirely true, but many employees' negative attitudes are developed after they have been on the job.

Timeliness of instruction is a key issue in the orientation program. For example, although statistical data differ, it is generally agreed that new employees are significantly more prone to work-related accidents. This fact is attributed to the inexperience of new workers, their unfamiliarity with procedures and facilities, and their zealousness to do the work. There are also a significant number of workers injured between the four- to six-year range. This is generally attributed either to a change in work duties (new or transferred employee) or worker complacency. These issues can be covered in an ongoing training program and help to reduce the number of accidents.

The following subjects are suggested as part of the orientation program:

- company orientation: history and goals
- policy statements
- benefit packages
- organized labor agreements (if applicable)
- safety and health policy statement (if separate)
- acceptable dress code (as required)
- personnel introductions
- housekeeping standards
- communications about hazards
- personal protective equipment
- emergency response procedures: fire, spill, etc.
- accident reporting procedures
- near-miss accident reporting
- accident investigation (supervisors)
- lockout/tagout procedures
- machine guarding
- electrical safety awareness
- ladder use and storage (if applicable)
- confined space entry (if applicable)
- medical facility support
- first aid/CPR
- hand tool safety
- ergonomic principles
- eyewash and shower locations
- fire prevention and protection
- access to exposure and medical records.

These subjects, among many others, cover essential information that is far too important to overlook or leave to casual learning. A formal program should be developed not only to provide the worker with this information, but to create a strong link between employees and the safety and health philosophy of the organization.

Many of these subjects are left to the manager, supervisor, or team leader to cover with employees when they arrive at the worksite. This person must have time to cover these points with the employees. Unfortunately, the importance of going over these subjects with the employees is not often recognized until an accident occurs.

Managers, supervisors, or team leaders should be familiar with current information presented to new workers to avoid the age-old problem of contradicting the training material. They should also become involved in the development of training programs to keep the information current and practical in relation to work demands. When they start disregarding or contradicting the training program, the entire program—and the company's image—loses credibility with the workers.

The manager, supervisor, or team leader should reinforce the training program content by demonstrating how it will apply to the worker's specific job assignment. For example, the new employee may have been trained how to read the warning labels on chemical containers. The manager, supervisor, or team leader can then enhance this information by conducting a tour of the department, pointing out all hazardous materials, and identifying the protection provided through personal protective equipment and control measures.

Regulatory Obligations

All regulatory agencies publish information about the employer's responsibility to provide training to their employees. There may still be difficulties in the final interpretation, and the best solution for these problems is to communicate with the agency involved, if necessary, and the organization's legal counsel. Table 23–A lists the references published by the Occupational Safety and Health Administration (OSHA) that reflect training requirements. Table 23–B identifies the requirements for training as defined by the Mine Safety and Health Administration (MSHA). Remember that these regulatory obligations represent only the minimum requirements and are rarely comprehensive enough to provide a truly safe and healthful environment for employees.

Manuals

The use of written policies and procedures manuals has been a traditional means of providing information to new employees. The company should exercise care in preparing the manuals to ensure that the information is easy to understand, complete, and that all rules are enforced. Policies and procedures manuals should cover items such as first aid, personal protective equipment, electrical safety, and housekeeping, just to name a few. These manuals can be used as key references by the participants during training.

TRAINING METHODS

Trainers can select one of several different training methods when preparing a program. Each method has strengths and weaknesses. The technique selected should be determined by the objectives to be met, type of student participation, time allocated, facility being used, and equipment available. Everyone learns at different speeds, and through different methods. Trainers must have the training and teaching skills to address these elements of human behavior. The following are the most common types of training techniques used in industry:

1. On-the-job training (OJT)
   a. job instruction training (JIT)
   b. coaching
2. Group methods
   a. conference
   b. brainstorming

(Text continues on page 533.)

Table 23-A. Index to OSHA Training Requirements

Hazard	Part	Subpart	Section
Blasting or Explosives	1910.109	H	(d)(3)(i)(iii)
	1926.901	U	(c)
	1926.902	U	(i)
	1915.10	B	(a) thru (b)
	1916.10	B	(a) thru (b)
	1917.10	B	(a) thru (b)
Carcinogens			
4-Nitrobiphenyl	1910.1003	Z	(e)(5)(i) thru (ii)
alpha-Naphthylamine	1910.1004	Z	(e)(5)(i) thru (ii)
4, 4'-Methylene bis (2-chloroaniline)	1910.1005	Z	(e)(5)(i) thru (ii)
Methyl chloromethyl ether	1910.1006	Z	(e)(5)(i) thru (ii)
3, 3'-Dichlorobenzidine (and its salts)	1910.1007	Z	(e)(5)(i) thru (ii)
bis-Chloromethyl ether	1910.1008	Z	(e)(5)(i) thru (ii)
beta-Naphthylamine	1910.1009	Z	(e)(5)(i) thru (ii)
Benzidine	1910.1010	Z	(e)(5)(i) thru (ii)
4-Aminodiphenyl	1910.1011	Z	(e)(5)(i) thru (ii)
Ethyleneimine	1910.1012	Z	(e)(5)(i) thru (ii)
beta-Propiolactone	1910.1013	Z	(e)(5)(i) thru (ii)
2-Acetylaminofluorene	1910.1014	Z	(e)(5)(i) thru (ii)
4-Dimethylamino-azobenzene	1910.1015	Z	(e)(5)(i) thru (ii)
N-Nitrosodimethylamine	1910.1016	Z	(e)(5)(i) thru (ii)
Vinyl chloride	1910.1017	Z	(j)(1)(i) thru (ix)
Cranes and Derricks	1910.179	N	(m)(3)(ix)
	1910.180	N	(h)(3)(xii)
Decompression or Compression	1926.803	S	(a)(2)
	1926.803	S	(b)(10)(xii)
	1926.803	S	(e)(1)
Employee Responsibility	1910.109	H	(g)(3)(iii)(a)
	1926.609	U	(a)
Equipment Operations	1910.217	O	(f)(2)
	1926.20	C	(b)(4)
	1926.53	D	(b)
	1926.54	D	(a)
	1910.252	Q	(c)(6)
Fire Protection	1916.32	D	(e)
	1917.32	D	(b)
	1926.150	F	(a)(5)
	1926.155	F	(e)
	1926.351	J	(d)(1) thru (5)
	1926.901	U	(c)
Forging	1910.218	O	(a)(2)(i) thru (iv)
Gases, Fuel, Toxic Material, Explosives	1910.109	H	(d)(3)(i) and (iii)
	1910.111	H	(b)(13)(ii)
	1910.266	R	(c)(5)(i) thru (xi)
	1910.106	H	(b)(5)(vi)(v)(3)
	1916.35	D	(d)(1) thru (6)
	1926.21	C	(a) and (b)(2) thru (6)
	1926.350	J	(d)(1) thru (6)
General	1926.21	C	(a)
Hazardous Material	1915.57	F	(d)
	1916.57	F	(d)
	1917.57	F	(d)
Medical and First Aid	1910.94	G	(d)(9)(i) and (vi)
	1910.151	K	(a) and (b)
	1915.58	K	(a)
	1917.58	F	(a)
	1926.50	D	(c)
Personal Protective Equipment	1910.94	G	(d)(11)(v)
	1910.134	I	(a)(3)
	1910.134	I	(b)(1), (2) and (3)
	1910.134	I	(e)(2), (3) and (5)
	1910.134	I	(e)(5)(i)
	1910.161	K	(a)(2)
	1915.82	I	(a)(4)
	1915.82	I	(b)(4)
	1916.57	F	(f)
	1916.58	F	(a)
	1916.82	I	(a)(4)
	1916.82	I	(b)(4)
	1917.57	F	(f)
	1918.102	J	(a)(4)
	1926.21	C	(b)(2) thru (6)
	1926.103	E	(c)(1)
	1926.800	S	(e)(xii)
Pulpwood Logging	1910.266	R	(c)(5)(i) thru (xi)
	1910.266	R	(c)(6)(i) thru (xxi)
	1910.266	R	(c)(7)
	1910.266	R	(e)(2)(i) and (ii)
	1910.266	R	(e)(9)
	1910.266	R	(e)(1)(iii) thru (vii)
Powder-Actuated Tools	1915.75	H	(b)(1) thru (6)
	1916.75	H	(b)(1) thru (6)
Power Press	1910.217	O	(e)(3)
Power Trucks, Motor Vehicles, or Agricultural Tractors	1910.109	H	(d)(3)(iii)
	1910.109	H	(g)(3)(iii)(a)
	1910.178	N	(1)
	1910.266	R	(e)(9)
	1910.266	R	(e)(6)(viii)
	1928.51	C	(d)
Radioactive Material	1916.37	D	(b)
Signs—Danger, Warning, Instruction	1910.96	G	(f)(3)(viii)
	1910.145	J	(c)(1)(ii)
	1910.145	J	(c)(2)(ii)
	1910.145	J	(c)(3)
	1910.264	R	(d)(1)(v)
Tunnels and Shafts	1926.800	S	(e)(xiii)
Welding	1910.252	Q	(b)(1)(iii)
	1910.252	Q	(c)(1)(iii)
	1915.35	D	(d)(1) thru (6)
	1915.36	D	(d)(1) thru (4)
	1916.35	D	(d)(1) thru (6)
	1917.35	D	(d)(1) thru (6)

Table 23-B. OSHA and MSHA Training Requirements

The continued importance of training is evidenced by the requirements of both the Occupational Safety and health Administration (OSHA) and the Mine Safety and Health Administration (MSHA).

OSHA REQUIREMENTS

Listed next are the major parts of the OSHA regulations (Title 29—Labor, *Code of Federal Regulations*) covering training requirements. (Table 23–A gives a convenient index of the types of hazards and the parts of the regulations requiring training to protect against the hazards.)

- Part 1910, Safety and Health Training Requirements for General Industry
- Part 1915-18, Safety and Health Training Requirements for Maritime Employment
- Part 1926, Safety and Health Training Requirements for Construction
- Part 1928, Occupational Safety and Health Requirements for Agriculture

MSHA REGULATIONS

The following is a summary of the training requirements under the MSHA Regulations, 30 *CFR*, PART 48—TRAINING AND RETRAINING OF MINERS (See §48.2 and §48.22 for definitions of terms used in Part 48)

Subpart A—Training and Retraining of Underground Miners

§48.1 Scope

The provisions of this subpart A set forth the mandatory requirements for submitting and obtaining approval of programs for training and retraining miners working in underground mines. Requirements regarding compensation for training and retraining are also included.

§48.3 Training Plans; Time of Submission; Where Filed; Information Required; Time for Approval; Method for Disapproval; Commencement of Training; Approval of Instructors

(a) Each operator of an underground mine shall have an MSHA approved plan containing programs for training new miners, training newly-employed experienced miners, training miners for new tasks, annual refresher training, and hazard training for miners.

(b) The training plan shall be filed with the District Manager for the area in which the mine is located.

(d) The operator shall furnish to the representative of the miners a copy of the training plan two weeks prior to its submission to the District Manager. Where a miners' representative is not designated, a copy of the plan shall be posted on the mine bulletin board 2 weeks prior to its submission to the District Manager. Written comments received by the operator from miners or their representatives shall be submitted to the District Manager. Miners or their representatives may submit written comments directly to the District Manager.

(e) All training required by the training plan submitted to and approved by the District Manager as required by this sub

part A shall be subject to evaluation by the District Manager to determine the effectiveness of the training programs. If it is deemed necessary, the District Manager may require changes in, or additions to, programs. Upon request from the District Manager the operator shall make available for evaluation the instructional materials, handouts, visual aids and other teaching accessories used or to be used in the training programs. Upon request from the District Manager the operator shall provide information concerning the schedules of upcoming training.

(f) The operator shall make a copy of the MSHA approved training plan available at the mine site for MSHA inspection and for examination by the miners and their representatives.

(g) Except as provided in §48.7 (New task training of miners) and §48.11 (Hazard training) of this subpart A, all courses shall be conducted by MSHA approved instructors.

(h) Instructors shall be approved by the District Manager.

(i) Instructors may have their approval revoked by MSHA for good cause which may include not teaching a course at least once every 24 months. A decision by the District Manager to revoke an instructor's approval may be appealed by the instructor. . . . Such an appeal shall be submitted to the Administrator within 5 days of notification of the District Manager's decision. Upon revocation of an instructor's approval, the District Manager shall immediately notify operators who use the instructor for training.

(j) The District Manager for the area in which the mine is located shall notify the operator and the miners' representative, in writing, within 60 days from the date on which the training plan is filed, of the approval or status of the approval of the training programs.

(k) Except as provided under §48.8(c) (Annual refresher training of miners) of this subpart A, the operator shall commence training of miners within 60 days after approval of the training plan, or approved programs of the training plan.

(l) The operator shall notify the District Manager of the area in which the mine is located, and the miners' representative of any changes or modifications the operator proposes to make in the approved training plan. The operator shall obtain the approval of the District Manager for such changes or modifications.

(m) In the event the District Manager disapproves a training plan or a proposed modification of a training plan or requires changes in a training plan or modification, the District Manager shall notify the operator and the miners' representative in writing.

(n) The operator shall post on the mine bulletin board, and provide to the miners' representative, a copy of all MSHA revisions and decisions which concern the training plan at the mine and which are issued by the District Manager.

§48.4 Cooperative Training Program

(a) An operator of a mine may conduct his own training programs, or may participate in training programs conducted by MSHA, or may participate in MSHA approved training programs conducted by State or other Federal agencies, or associations of mine operators, miners' representatives, other mine operators, private associations, or educational institutions.

(b) Each program and course of instruction shall be given by instructors who have been approved by MSHA to instruct

Table 23-B. (Continued)

in the courses which are given, and such courses and the training programs shall be adapted to the mining operations and practices existing at the mine and shall be approved by the District Manager for the area in which the mine is located.

§48.5 Training of New Miners; Minimum Courses of Instruction; Hours of Instruction

(a) Each new miner shall receive no less than 40 hours of training as prescribed in this section before such miner is assigned to work duties. Such training shall be conducted in conditions which as closely as practicable duplicate actual underground conditions, and approximately 8 hours of training shall be given at the minesite.

(b) The training program for new miners shall include the following courses:

   (1) Instruction in the statutory rights of miners and their representatives under the Act; authority and responsibility of supervisors.

   (2) Self-rescue and respiratory devices.

   (3) Entering and leaving the mine; transportation; communications.

   (4) Introduction to the work environment.

   (5) Mine map; escapeways; emergency evacuation; barricading.

   (6) Roof or ground control and ventilation plans.

   (7) Health.

   (8) Cleanup; rock dusting.

   (9) Hazard recognition.

  (10) Electrical hazards.

  (11) First aid.

  (12) Mine gases.

  (13) Health and safety aspects of the tasks to which the new miner will be assigned.

  (14) Such other courses as may be required by the District Manager based on circumstances and conditions at the mine.

(c) Methods, including oral, written, or practical demonstration, to determine successful completion of the training shall be included in the training plan.

(d) Upon proof by an operator that a newly employed miner has received the courses and hours of instruction set forth in paragraphs (a) and (b) of this section within 12 months preceding initial employment at a mine, such miner need not repeat the training, but the operator shall give and the miner shall receive and complete the instruction and program of training set forth in paragraph (b) of §48.6 (Training of newly employed experienced miners), and §48.7 (New task training of miners), if applicable, before commencing work.

§48.6 Training of Newly Employed Experienced Miners; Minimum Courses of Instruction

(a) A newly employed experienced miner shall receive and complete training in the program of instruction prescribed in this section before such miner is assigned to work duties.

(b) The training program for newly employed experienced miners shall include the following:

   (1) Introduction to work environment.

   (2) Mandatory health and safety standards.

   (3) Authority and responsibility of supervisors and miners' representatives.

   (4) Entering and leaving the mine; transportation; communications.

   (5) Mine map; escapeways; emergency evacuation; barricading.

   (6) Roof or ground control and ventilation plans.

   (7) Hazard recognition.

   (8) Self-rescue and respiratory devices.

   (9) Such other courses as may be required by the District Manager based on circumstances and conditions at the mine.

§48.7 Training of Miners Assigned to a Task in Which They Have Had No Previous Experience; Minimum Courses of Instruction

(a) Miners assigned to new work tasks as mobile equipment operators, drilling machine operators, haulage and conveyor systems operators, roof and ground control machine operators, and those in blasting operations shall not perform new work tasks in these categories until training prescribed in this paragraph and paragraph (b) of this section has been completed. This training shall not be required for miners who have been trained and who have demonstrated safe operating procedures for such new work tasks within 12 months preceding assignment. This training shall also not be required for miners who have performed the new work tasks and who have demonstrated safe operating procedures for such new work tasks within 12 months preceding assignment. The training program shall include the following:

   (1) Health and safety aspects and safe operating procedures for work tasks, equipment, and machinery.

   (2) (i) Supervised practice during nonproduction.

      (ii) Supervised operation during production.

   (3) New or modified machines and equipment.

   (4) Such other courses as may be required by the District Manager based on circumstances and conditions at the mine.

(b) Miners under paragraph (a) of this section shall not operate the equipment or machine or engage in blasting operations without direction and immediate supervision until such miners have demonstrated safe operating procedures for the equipment or machine or blasting operation to the operator or the operator's agent.

(c) Miners assigned a new task not covered in paragraph (a) of this section shall be instructed in the safety and health aspects and safe work procedures of the task, prior to performing such task.

(d) Any person who controls or directs haulage operations at a mine shall receive and complete training courses in safe haulage procedures related to the haulage system, ventilation system, firefighting procedures, and emergency evacuation procedures in effect at the mine before assignment to such duties.

(e) All training and supervised practice and operation required by this section shall be given by a qualified trainer, or a supervisor experienced in the assigned tasks, or other person experienced in the assigned tasks.

§48.8 Annual Refresher Training of Miners; Minimum Courses of Instruction; Hours of Instruction

(a) Each miner shall receive a minimum of 8 hours of annual refresher training as prescribed in this section.

(b) The annual refresher training program for all miners shall include the following courses of instruction:

   (1) Mandatory health and safety standards.

   (2) Transportation controls and communication systems.

Table 23-B. (Continued)

    (3) Barricading.

    (4) Roof or ground control and ventilation plans.

    (5) First aid.

    (6) Electrical hazards.

    (7) Prevention of accidents.

    (8) Self-rescue and respiratory devices.

    (9) Explosives.

  (10) Mine gases.

  (11) Health.

  (12) Such other courses as may be required by the District Manager based on circumstances and conditions at the mine.

(d) Where annual refresher training is conducted periodically, such sessions shall not be less than 30 minutes of actual instruction time and the miners shall be notified that the session is part of annual refresher training.

§48.9 Records of Training

(a) Upon a miner's completion of each MSHA approved training program, the operator shall record and certify on MSHA form 5000-23 that the miner has received the specified training. A copy of the training certificate shall be given to the miner at the completion of the training. The training certificates for each miner shall be available at the minesite for inspection by MSHA and for examination by the miners, the miners' representative, and State inspection agencies. When a miner leaves the operator's employ, the miner shall be entitled to a copy of his training certificates.

(b) False certification that training was given shall be punishable under section 110 (a) and (f) of the Act.

(c) Copies of training certificates for currently employed miners shall be kept at the minesite for 2 years, or for 60 days after termination of employment.

§48.10 Compensation for Training

(a) Training shall be conducted during normal working hours; miners attending such training shall receive the rate of pay as provided in §48.2(d) (Definition of normal working hours) of this subpart A.

(b) If such training shall be given at a location other than the normal place of work, miners shall be compensated for the additional cost, such as mileage, meals, and lodging, they may incur in attending such training sessions.

§48.11 Hazard Training

(a) Operators shall provide to those miners, as defined in §48.2(a)(2) (Definition of miner) of this subpart A, a training program before such miners commence their work duties. This training program shall include the following instruction, which is applicable to the duties of such miners:

    (1) Hazard recognition and avoidance;

    (2) Emergency and evacuation procedures;

    (3) Health and safety standards, safety rules, and safe working procedures;

    (4) Use of self-rescue and respiratory devices

    (5) Such other instruction as may be required by the District Manager based on circumstances and conditions at the mine.

(b) Miners shall receive the instruction required by this section at least once every 12 months.

(c) The training program required by this section shall be submitted with the training plan required by §48.3(a) (Training plans: Submission and approval) of this subpart A and

shall include a statement on the methods of instruction to be used.

(d) The operator shall maintain and make available for inspection certificates that miners have received the hazard training required by this section.

(e) Miners subject to hazard training shall be accompanied at all times while underground by an experienced miner, as defined in §48.2(b) (Definition of miner) of this subpart A.

§48.12 Appeals procedures

The operator, miner, and miners' representative shall have the right of appeal from a decision of the District Manager.

Subpart B—Training and Retraining of Miners Working at Surface Mines and Surface Areas of Underground Mines

The provisions of this subpart B set forth the mandatory requirements for submitting and obtaining approval of programs for training and retraining miners working at surface mines and surface areas of underground mines. Requirements regarding compensation for training and retraining are also included.

§48.23 Training Plans; Time of Submission; Where Filed; Information Required; Time for Approval; Method for Disapproval; Commencement of Training; Approval of Instructors

(a) Each operator of a mine shall have an MSHA approved plan containing programs for training new miners, training newly-employed experienced miners, training miners for new tasks, annual refresher training, and hazard training for miners.

    (2) Within 60 days after the operator submits the plan for approval, unless extended by MSHA, the operator shall have an approved plan for the mine.

    (3) In the case of a new mine which is to be opened or a mine which is to be reopened or reactivated after the effective date of this subpart B, the operator shall have an approved plan prior to opening the new mine, or reopening or reactivating the mine unless the mine is reopened or reactivated periodically using portable equipment and mobile teams of miners as a normal method of operation by the operator. The operator to be so excepted shall maintain an approved plan for training covering all mine locations which are operated with portable equipment and mobile teams of miners.

(b) The training plan shall be filed with the District Manager for the area in which the mine is located.

(d) The operator shall furnish to the representative of the miners a copy of the training plan 2 weeks prior to its submission to the District Manager. Where a miners' representative is not designated, a copy of the plan shall be posted on the mine bulletin board 2 weeks prior to its submission to the District Manager. Written comments received by the operator from miners or their representatives shall be submitted to the District Manager. Miners or their representatives may submit written comments directly to the District Manager.

(e) All training required by the training plan submitted to and approved by the District Manager as required by this subpart B shall be subject to evaluation by the District Manager to determine the effectiveness of the training programs. If it is deemed necessary, the District Manager may require changes in, or additions to, programs. Upon

Table 23-B. (Continued)

request from the District Manager the operator shall make available for evaluation the instructional materials, handouts, visual aids, and other teaching accessories used or to be used in the training programs. Upon request from the District Manager the operator shall provide information concerning schedules of upcoming training.

(f) The operator shall make a copy of the MSHA approved training plan available at the mine site for MSHA inspection and examination by the miners and their representatives.

(g) Except as provided in §48.27 (New task training of miners) and §48.31 (Hazard training) of this subpart B, all courses shall be conducted by MSHA approved instructors.

(h) Instructors shall be approved by the District Manager

(i) Instructors may have their approval revoked by MSHA for good cause which may include not teaching a course at least once every 24 months. A decision by the District Manager to revoke an instructor's approval may be appealed by the instructor. Such an appeal shall be submitted to the Administrator within 5 days of notification of the District Manager's decision. Upon revocation of an instructor's approval, the District Manager shall immediately notify operators who use the instructor for training.

(j) The District Manager for the area in which the mine is located shall notify the operator and the miners' representative, in writing, within 60 days from the date on which the training plan is filed, of the approval or status of the approval of the training programs.

(k) Except as provided under §48.28(c) (Annual refresher training of miners) of this subpart B, the operator shall commence training of miners within 60 days after approval of the training plan, or approved programs of the training plan.

(l) The operator shall notify the District Manager of the area in which the mine is located and the miners' representative of any changes or modifications which the operator proposes to make in the approved training plan. The operator shall obtain the approval of the District Manager for such changes or modifications.

(m) In the event the District Manager disapproves a training plan or a proposed modification of a training plan or requires changes in a training plan or modification, the District Manager shall notify the operator and the miners' representative in writing.

(n) The operator shall post on the mine bulletin board, and provide to the miners' representative, a copy of all MSHA revisions and decisions which concern the training plan at the mine and which are issued by the District Manager.

§48.24 Cooperative Training Program

(a) An operator of a mine may conduct his own training programs, or may participate in training programs conducted by MSHA, or may participate in MSHA approved training programs conducted by State or other Federal agencies, or associations of mine operators, miners' representatives, other mine operators, private associations, or educational institutions.

(b) Each program and course of instruction shall be given by instructors who have been approved by MSHA to instruct in the courses which are given, and such courses and the training programs shall be adapted to the mining operations and practices existing at the mine and shall be

approved by the District Manager for the area in which the mine is located.

§48.25 Training of New Miners; Minimum Courses of Instruction; Hours of Instruction

(a) Each new miner shall receive no less than 24 hours of training as prescribed in this section. Except as otherwise provided in this paragraph, new miners shall receive this training before they are assigned to work duties. At the discretion of the District Manager, new miners may receive a portion of this training after assignment to work duties: Provided, That no less than 8 hours of training shall in all cases be given to new miners before they are assigned to work duties. The following courses shall be included in the 8 hours of training: Introduction to work environment, hazard recognition, and health and safety aspects of the tasks to which the new miners will be assigned. Following the completion of this preassignment training, new miners shall then receive the remainder of the required 24 hours of training, or up to 16 hours, within 60 days. Operators shall indicate in the training plans submitted for approval whether they want to train new miners after assignment to duties and for how many hours. In determining whether new miners may be given this training after they are assigned duties, the District Manager shall consider such factors as the mine safety record, rate of employee turnover and mine size. Miners who have not received the full 24 hours of new miner training shall be required to work under the close supervision of an experienced miner.

(b) The training program for new miners shall include the following courses:

(1) Instruction in the statutory rights of miners and their representatives under the Act; authority and responsibility of supervisors.

(2) Self-rescue and respiratory devices.

(3) Transportation controls and communication systems.

(4) Introduction to work environment.

(5) Escape and emergency evacuation plans; firewarning and firefighting.

(6) Ground control; working in areas of highwalls, water hazards, pits and spoil banks; illumination and night work.

(7) Health.

(8) Hazard recognition.

(9) Electrical hazards.

(10) First aid.

(11) Explosives.

(12) Health and safety aspects of the tasks to which the new miner will be assigned.

(13) Such other courses as may be required by the District Manager based on circumstances and conditions at the mine.

(c) Methods, including oral, written or practical demonstration, to determine successful completion of the training shall be included in the training plan.

(d) Upon proof by an operator that a newly employed miner has received the courses and hours of instruction set forth in paragraphs (a) and (b) of this section within 12 months preceding initial employment at a mine, such miner need not repeat the training, but the operator shall give and the miner shall receive and complete the instruction and program of training set forth in paragraph (b) of §48.26 (Training of newly employed experienced miners) and

Table 23-B. (Continued)

§48.27 (New task training of miners), if applicable, before commencing work.

§48.26 Training of Newly Employed Experienced Miners; Minimum Courses of Instruction

(a) A newly employed experienced miner shall receive and complete training in the program of instruction prescribed in this section before such miner is assigned to work duties.

(b) The training program for newly employed experienced miners shall include the following:
  (1) Introduction to work environment.
  (2) Mandatory health and safety standards.
  (3) Authority and responsibility of supervisors and miners' representatives.
  (4) Transportation controls and communication systems.
  (5) Escape and emergency evacuation plans; firewarning and firefighting.
  (6) Ground controls; working in areas of highwalls, water hazards, pits, and spoil banks; illumination and night work.
  (7) Hazard recognition.
  (8) Such other courses as may be required by the District Manager based on circumstances and conditions at the mine.

§48.27 Training of Miners Assigned to a Task in Which They Have Had No Previous Experience; Minimum Courses of Instruction

(a) Miners assigned to new work tasks as mobile equipment operators, drilling machine operators, haulage and conveyor systems operators, ground control machine operators, and those in blasting operations shall not perform new work tasks in these categories until training prescribed in this paragraph and paragraph (b) of this section has been completed. This training shall not be required for miners who have been trained and who have demonstrated safe operating procedures for such new work tasks within 12 months preceding assignment. This training shall also not be required for miners who have performed the new work tasks and who have demonstrated safe operating procedures for such new work tasks within 12 months preceding assignment. The training program shall include the following:
  (1) Health and safety aspects and safe operating procedures for work tasks, equipment, or machinery.
  (2) (i) Supervised practice during nonproduction.
      (ii) Supervised operation during production.
  (3) New or modified machines and equipment.
  (4) Such other courses as may be required by the District Manager based on circumstances and conditions at the mine.

(b) Miners under paragraph (a) of this section shall not operate the equipment or machine or engage in blasting operations without direction and immediate supervision until such miners have demonstrated safe operating procedures for the equipment or machine or blasting operation to the operator or the operator's agent.

(c) Miners assigned a new task not covered in paragraph (a) of this section shall be instructed in the safety and health aspects and safe work procedures of the task, prior to performing such task.

(d) All training and supervised practice and operation required by this section shall be given by a qualified trainer, or a supervisor experienced in the assigned tasks, or other person experienced in the assigned tasks.

§48.28 Annual Refresher Training of Miners; Minimum Courses of Instruction; Hours of Instruction

(a) Each miner shall receive a minimum of 8 hours of annual refresher training as prescribed in this section.

(b) The annual refresher training program for all miners shall include the following courses of instruction:
  (1) Mandatory health and safety standards. The course shall include mandatory health and safety standard requirements which are related to the miner's tasks.
  (2) Transportation controls and communication systems.
  (3) Escape and emergency evacuation plans; firewarning and firefighting.
  (4) Ground control; working in areas of highwalls, water hazards, pits, and spoil banks; illumination and night work.
  (5) First aid.
  (6) Electrical hazards.
  (7) Prevention of accidents.
  (8) Health.
  (10) Self-rescue and respiratory devices.
  (11) Such other courses as may be required by the District Manager based on circumstances and conditions at the mine.

(d) Where annual refresher training is conducted periodically, such sessions shall not be less than 30 minutes of actual instruction time and the miners shall be notified that the session is part of annual refresher training.

§48.29 Records of Training

(a) Upon a miner's completion of each MSHA approved training program, the operator shall record and certify on MSHA form 5000-23 that the miner has received the specified training. A copy of the training certificate shall be given to the miner at the completion of the training. The training certificates for each miner shall be available at the mine site for inspection by MSHA and for examination by the miners, the miners' representative and State inspection agencies. When a miner leaves the operator's employ, the miner shall be entitled to a copy of his training certificates.

(b) False certification that training was given shall be punishable under section 110 (a) and (f) of the Act.

(c) Copies of training certificates for currently employed miners shall be kept at the mine site for 2 years, or for 60 days after termination of employment.

§48.30 Compensation for Training

(a) Training shall be conducted during normal working hours; miners attending such training shall receive the rate of pay as provided in §48.22(d)

(b) If such training shall be given at a location other than the normal place of work, miners shall be compensated for the additional costs, such a mileage, meals, and lodging, they may incur in attending such training sessions.

§48.31 Hazard Training

(a) Operators shall provide to those miners, as defined in §48.22(a) (2) (Definition of miner) of this subpart B, a training program before such miners commence their work duties. This training program shall include the following instruction, which is applicable to the duties of such miners:
  (1) Hazard recognition and avoidance;
  (2) Emergency and evacuation procedures;
  (3) Health and safety standards, safety rules and safe working procedures;

Table 23-B. (Continued)

(4) Self-rescue and respiratory devices; and,

(5) Such other instruction as may be required by the District Manager based on circumstances and conditions at the mine.

(b) Miners shall receive the instruction required by this section at least once every 12 months.

(c) The training program required by this section shall be submitted with the training plan required by §48.23(a) (Training plans: Submission and approval) of this subpart B and shall include a statement on the methods of instruction to be used.

(d) In accordance with §48.29 (Records of training) of this subpart B, the operator shall maintain and make available for inspection, certificates that miners have received the instruction required by this section.

§48.32 Appeals Procedures

The operator, miner, and miners' representative shall have the right of appeal from a decision of the District Manager.

---

   c. case study

   d. incident process

   e. facilitated discussion

   f. role playing

   g. lecture

   h. question and answer

   i. simulation

3. Individual methods

   a. drill

   b. demonstration

   c. testing

   d. video-based training

   e. computer-assisted training (interactive)

   f. reading

   g. independent study

   h. seminars and short courses.

These methods are discussed in detail in the following sections.

On-the-Job Training

On-the-job training (OJT) is widely used because it allows the worker to produce during the training period. However, three considerations must be addressed when using OTJ training:

1. The trainer must possess proper training skills.

2. A training program should be developed to ensure that all workers are trained in the same way to perform their tasks in the safest and most productive manner.

3. Adequate time must be allotted to the trainer and trainee to be sure the subject is well covered and thoroughly understood.

Job Instruction Training

A variation of OJT is known as job instruction training (JIT), also referred to as the four-point method. Instruction is broken down into four simple steps:

1. preparation

2. presentation

3. performance

4. follow-up.

This method of OJT has been highly successful. Workers are taught each job skill from a formal schedule of training. The program is adjusted to each student so that workers learn one task before beginning the next.

In all training programs, the selection of the trainer is critical to program success. The one-on-one relationship between trainer and trainee allows for better consistency and communication. In the four-point method, this trainer-trainee relationship works in the following ways:

Step 1. Preparation. During the preparation stage the trainer puts the workers at ease. He or she explains the job and determines what the worker currently knows about the subject. This stage also includes the preparation of the proper learning/working environment.

Step 2. Presentation. In the second step, the trainer demonstrates the work process. The student watches the performance and asks questions. The trainer should present the steps in sequence and stress all key points.

Step 3. Performance. In the next step, the worker performs the task under close supervision (Figure 23–1). The trainer should identify any discrepancies in the work performance and note good performance. The worker should explain the steps being performed. This ensures that the worker not only can perform the task but understands how and why the task is done. This stage continues until the trainer is satisfied with the worker's competence at the job.

Step 4. Follow-up. The final step is the follow-up of performance. The trainer and/or the supervisor must monitor the worker's performance to be sure the job is being performed as instructed and to answer any questions the worker may have.

Of the various OJT methods, JIT is probably the most flexible and direct. Through practicing, the trainee is expected to develop and apply the learned skills in a typical work environment while under the guidance of a trained worker/trainer. The trainer must know the job thoroughly, be a safe worker, and have the patience, skill, and desire to train.

The advantages of training in the JIT method are:

1. The worker can be more easily motivated because the training/guidance is personal.

Figure 23-1. This supervisor is explaining an enclosed operation to one of his workers. He is instructing him in the correct and safe operation of the machine and will make frequent follow-ups on his progress until certain of his proficiency.

2. The trainer can identify and correct deficiencies as they occur.
3. Results of the training are immediately evaluated because the worker is performing the actual job on actual equipment. The work performed can be judged against reasonable standards.
4. Training is practical, realistic, and demonstrated under actual conditions. Workers can easily ask questions.

Timing is an important element in this type of training. The trainee can receive help when needed, and the trainer can provide feedback as the training progresses. The trainer can also determine when the trainee is ready to move on to new levels of training.

It may be helpful to prepare a chart of tasks or subjects for which workers must receive training. This chart will make it easier to keep track of workers' training progress and the levels of competence they achieve.

Of the other methods of OJT that can be used, the most common is known as "coaching" or the "buddy system." This system is considered effective in some situations, but the following challenges are associated with it:

1. Trainers may be selected for their availability rather than for their training skills. A trainer who lacks basic teaching skills can undermine the entire orientation of a new worker.
2. Each trainer may have his or her own way of performing the tasks being taught. This lack of continuity can make it difficult to control hazards in the workplace and lead to many accidents.
3. Key elements of orientation can be overlooked in the training program and may not be realized until an incident or accident occurs.
4. Poor techniques or bad habits can be spread from one worker to another. Short cuts or safety violations can be demonstrated to new workers as the "way we do it."
5. Safety performance may not be emphasized during the training. Job performance should never be separated from safety standards in any training provided to workers.

Group Methods

Group techniques encourage participation from a selected audience. These methods allow trainees to share ideas, evaluate information, and become actively involved in the

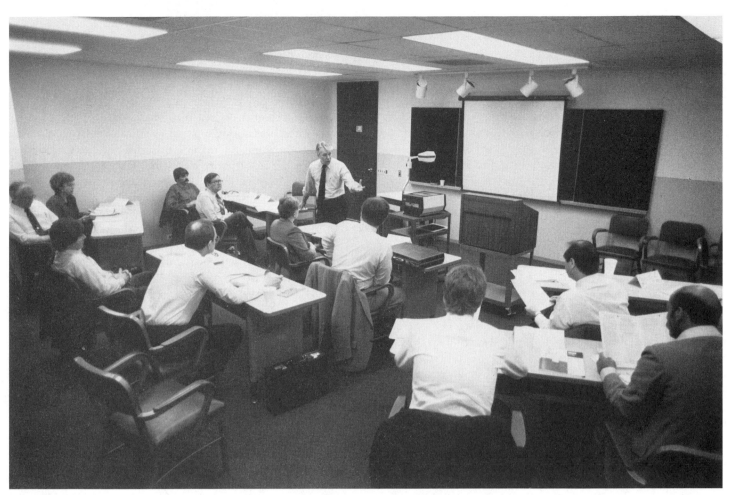

Figure 23-2. In a participation class, the number of people in the conference group should be small enough so open discussion can occur.

planning and implementation of company policy. Several types of group training are used, but all require skilled facilitators to be successful.

Conference or Meeting Method

The conference method of training is widely used in business and industry for education-sharing purposes. The reason this method is so readily accepted is because of the knowledge each participant brings to the group. In this process, sometimes called meetings, the trainer/facilitator controls the flow of the session as the participants share their knowledge and experience. The skill of the facilitator can mean the success or failure of these sessions. Facilitators must use various techniques to draw information and opinions from members. The number of people involved should be limited to allow open discussion from all participants. The opinions of each member should be recorded, and a summary of the group conclusions provided to those who were involved, as well as to those who should be kept abreast of the information (Figure 23–2).

The conference technique is also a valuable method of problem solving. There are several situations where a safety and health professional can use the knowledge and expertise of members of the organization to address safety and health issues. It is important that at the beginning of the conference, members identify their goals and expectations for the session. For example, if a conference has been called to identify possible solutions to a safety and health concern, the group must understand that they are to make recommendations only, not to establish policy or procedures. By defining the actual role of the conference at the start, the groups can avoid misconceptions and misdirection. When a group is asked to make recommendations, they should be kept informed of the results of those recommendations.

If management fails to establish these ground rules and to provide follow-up information, the members may feel their efforts are ineffective. On the other hand, proper control and guidance of a conference can ensure its success and make it a gratifying experience for the participants.

Brainstorming

Brainstorming is a technique of group interaction that encourages each participant to present ideas on a specific issue. The method is normally used to find new, innovative approaches to issues. There are four ground rules:

1. Ideas presented are not criticized.
2. Free-wheeling creative thinking and building on ideas are positively reinforced.
3. As many ideas as possible should be presented quickly.
4. Combining several ideas or improving suggestions is encouraged.

A recorder should be selected to write down all the ideas presented. The moderator must control the flow of suggestions, cut off negative comments, and solicit ideas from each member.

Brainstorming allows ideas to be developed quickly, encourages creative thinking, and involves everyone in the process. The group can go beyond old stereotypes or the "way it's always done."

Case Study

Case studies are written descriptions of business decisions or problems that learners will use as a basis for demonstrating predetermined skills and knowledge. There are two distinct advantages to building case studies: first, case studies provide an opportunity for the learners to use skills and knowledge acquired during the course. Second, case studies can serve as an evaluation tool for trainers to measure the degree of proficiency attained during a course or module. Here are other key benefits:

- Students, during a case study, begin to internalize the critical principles being taught and retain the information for longer periods of time.
- Case studies emphasize practical or critical thinking skills.
- The student's perspective is broadened through an interaction with others.
- Case studies encourage reflection, application, and analysis.
- Case studies reinforce the value of discussion and interaction with others.

Planning, thinking, and adhering to the instructional objectives are paramount in designing an effective case. The key is to start at the end and work backwards toward the beginning. Ask the following questions:

- What questions must be answered?
- What skills or knowledge should be exercised?
- What specific performance objectives should be measured?
- What learning objectives are to be addressed by the case study?

A case study may involve an actual situation or be fictitious. The goal of the activity is to develop group members' insight and problem-solving skills. The study normally is presented by defining what happened in a particular incident and the events leading up to the incident. The group is then given the task of determining the actual causes or problems, the significance of each element, and/or the acceptable solutions.

Incident Process

This is a type of case study in which the group works with a written account of any incident. The group is allowed to ask questions about facts, clues, and details. The trainer provides the answers to the questions, and the group must assemble the information, determine what has happened, and arrive at a decision. The trainer must guide the group to prevent arguments and to prevent one or two members from dominating the discussion. This is a useful training method that encourages employee participation in the accident prevention program. Situations can be real cases in the company or developed from potential hazards that exist.

The moderator must be capable of controlling group process and progress and of preventing the group from missing the true or root causes. For example, an employee was struck in the eye by a foreign body and an investigation revealed the employee was not wearing safety glasses while operating a bench grinder. The group must seek the root cause of the accident and not settle for the common conclusion "the worker failed to follow procedures" or "the worker failed to wear eye protection." They must specify why the supervisor or management failed to enforce the proper procedures (assuming there was an established procedure). Why did management allow this lax supervision?

Facilitated Discussion

Facilitated discussion or dialogue is the management of discussion about the course content so that the learning objectives are met, the discussion flows logically from topic to topic, and the applications to the learners' jobs are made clear. Facilitated discussion requires that the trainer/facilitator have the skills to accept all ideas and contributions as valid, show how they relate to the course objectives, and manage the time element and the flow of information to meet the course objectives.

The benefits of facilitated discussion include the following:

- ensures that the learners are involved and challenged
- builds a bond between the trainer/facilitator that encourages the free exchange of ideas and information.

Role Playing

This training method is effective for evaluating human relations issues. Members attempt to identify the ways people behave under various conditions. Although this technique is not an effective method of problem solving, it can uncover issues not previously considered. This method is particularly helpful in identifying and changing personnel issues such as poor morale or negative attitudes.

Lecture

With this method, a single person can impart information to a large group in a relatively short time. This method is normally used to communicate facts, give motivational speeches, or summarize events for workers. There is little time or opportunity for interaction by the attendees. Follow-up for these sessions must be well planned in advance to be successful.

Question and Answer (Q and A) Sessions

Normally Q and A sessions follow training periods after the trainer has summarized the material presented. Workers can use this method to clarify individual concerns or facts. However, workers will often need time to prepare and organize their thoughts before they can ask questions. In situations where they must absorb a large amount of material, allow them time to reflect on or apply the knowledge and to formulate questions. The trainer can plan to have a follow-up session or allow workers to present their questions personally. Question-and-answer sessions are helpful in clarifying issues of policy or changes in schedules or events.

Simulation

When actual materials or machines cannot be used, trainers can use a simulation device. This method is used effectively in aircraft pilot training, railroad engineer training, and other applications. Various methods are employed in management training programs as well, such as the "in-basket technique," "war games," and others. One simulation demonstrates the loss of eyesight to workers to encourage them to wear safety glasses on the job. The only limit to this technique is the trainer's creativity.

Simulation is most effective when the workers can participate. Careful planning and attention to detail are required. The initial costs of these sessions can be high because of the equipment required and time involved in conducting training.

Individual Methods

Drill

Using the elements of practice and repetition, this method of instruction is valuable for developing worker skill in fundamental tasks and for performing under pressure.

Workers required to perform in crisis situations should be trained under conditions that resemble the crisis as closely as possible. For example, when instructing workers in cardiopulmonary resuscitation (CPR), the trainer must try to instill the tension that workers will experience when they attempt to resuscitate a real victim.

Demonstration

As discussed in the section on JIT, the method of demonstration allows the trainer to perform the actual task and then have the worker repeat the performance. Trainers must be sure the job is performed exactly as required to prevent workers from developing poor habits and performance standards (and supervisors must see that employees follow the designated procedure). If the conditions used in demonstrations are not similar to the actual workstation or equipment on the job, this method will yield few useful results.

Testing

This technique is normally used to determine if workers understand the necessary information and can apply the knowledge when required. Developing good tests is a skill that requires constant review to ensure that training objectives are being met. Poor tests can reduce workers' morale and undermine training objectives.

Video-Based Training

An increasing number of training programs are designed as videotape presentations. Tapes are available on nearly any subject. This method of instruction is effective if properly applied. The use of videotapes does not eliminate the need for professional instruction, but can enhance a classroom presentation. Videos are available from the National Safety Council and numerous private companies. Trainers should screen the tapes to make sure they fulfill the needs of the training program.

Production companies can produce a training video on budgets ranging from hundreds of dollars to millions of dollars. Before selecting a production company—or, for that matter, deciding to make the video for the organization—the initial step is to determine how the video will fit into the overall training program. The major factors are the course design, the purpose of the course, and how the objectives are met. Some distinct advantages to using a video in a training class:

- Video offers the learner an opportunity to see examples of tasks and processes being performed correctly.
- The cost of producing training videos has gone up while the difficulty in producing them has gone down. Newer formats, such as HI-8 and S-VHS, offer lighter cameras and greater ease of use.

If the video portion is the core of the training program, a "higher-end" type of production should be used. This will usually include hiring a production company (independent producer), professional scriptwriter, director, and editor. Plan on post-production costs (editing, graphics, animation, sound design, and music) to consume two-fifths of the overall budget.

If the video portion of the training program, however, is supplemental or ancillary, a safety and health professional can tape the material and hire a professional editor to assemble the footage.

Computer-Assisted Training

Interactive computer programs are being developed for many areas of employee training. They allow the workers to receive information by reading or watching a video presentation and then respond to situations and questions. If the correct response is entered, the computer will advance the program to the next section; if the wrong response is entered, the program will repeat the information and retest the workers. The system is valuable for several reasons:

1. Workers can work at their own pace.
2. Records can be automatically kept of all training. The amount and type of records maintained can be modified to meet the company requirements.
3. Correct answers are required before a worker can proceed to the next lesson, or remediation methods are built into the program.
4. Workers receive training as time is available in their schedule, rather than having to meet training schedules.

With computer-assisted training programs, instructions can guide workers step by step through a curriculum designed to meet the goals of the individual, the company, and/or any regulatory obligations. The company can keep records that include the amount of time each worker spends in training, the type of material or information presented, and the success of the training. This program works extremely well for organizations with small workforces or those that cannot remove large groups of workers from their jobs at any one time. (See also Chapter 24, Media.)

Reading

Companies should provide employees with written safety materials such as monthly newsletters and supervision and safety magazines. In addition, organizations may establish a library where employees can research information on subjects such as work procedures, safety, leadership, health care, family or home safety, and other subjects of concern. It is important, however, that management does not assume everyone has the ability to read and comprehend all of the written material provided. Companies cannot replace instruction or training programs simply by handing an employee a training manual. Written material is meant to supplement or to serve as a reference for training.

Independent Study

Home-study courses or correspondence courses are used by many companies. They can help workers to advance within the organization or to improve their knowledge of their jobs and industries. A major advantage of this method is that the worker does not lose any time from work and can complete the course at his or her own pace. Another advantage is the low cost of home-study

programs. Normally they are centered around a textbook assignment, followed by self-tests using multiple choice, true/false, fill-in, or essay questions. Several courses also provide laboratory or performance materials such as television, radio, or computer repair programs that work on actual equipment. Some home-study programs come with videotaped presentations for workers to view.

The National Safety Council offers home-study programs for supervisor training—"Supervising for Safety" and "Protecting Workers' Lives."

Seminars and Short Courses

Seminars, short courses, and workshops, for safety and leadership information and skills are offered by many colleges and universities as well as by insurance companies and private organizations. The Safety Training Institute of the National Safety Council offers on-site instruction for workers, supervisors, and management. Seminars, short courses, and workshops range from one-hour sessions to several days.

SUMMARY

- Training is focused mainly on behavior change, showing workers how to do something properly and to apply their knowledge on the job. In some cases, however, nontraining solutions are more appropriate.
- Training and nontraining problems and solutions are categorized as selection and assignment, information and practice, environment, and motivation/incentive.
- Adult learners have special needs and requirements that trainers must recognize for the programs to be effective. Performance-based instruction generally works well with adult learners.
- To design an effective training program, the safety professional must assess workers' needs, analyze learners' characteristics, develop specific objectives, develop materials and schedules, and design testing and evaluation methods.
- Training begins with new employee orientation. The use of written policies and procedures manuals is one way of meeting new employee training needs and of conforming to regulatory standards.
- Training methods include on-the-job training, job instruction, group methods, and individual methods.

REFERENCES

"A Bibliography of Programs and Presentation Devices." C Hendershot, 4114 Ridgewood Drive, Bay City, MI 48706.

"A Catalog of Ideas for Action Oriented Training." Didactic Systems, Inc., PO Box 4, Cranford, NJ 07016.

Carr C. *Smart Training: The Manager's Guide to Training for Improved Performance.* New York: McGraw-Hill, Inc., 1992.

Dennis L. and Onion M. *Out in Front: Effective Supervision in the Workplace.* Itasca, IL: National Safety Council, 1990.

"The In-Basket Method." Bureau of Industrial Relations, Department of Training Materials for Industry, The University of Michigan, Graduate School of Business Administration, Ann Arbor, MI 48104.

"Library of Programmed Instruction Courses." E I du Pont de Nemours and Company, Inc., Education and Applied Technology Division, Wilmington, DE 20017.

Mager RF. *Preparing Instructional Objectives.* 2nd rev. ed. Belmont, CA: David S. Lake Publishers, 1984.

Mager RF and Beach KM Jr. *Developing Vocational Instruction.* Belmont, CA: David S. Lake Publishers, 1984.

Mager RF and Pipe P. *Analyzing Performance Problems of 'You Really Oughta Wanna'* 2nd ed. Belmont, CA: David S. Lake Publishers, 1984.

National Society for Programmed Instruction, PO Box 137, Cardinal Station, Washington, DC 20017.

"Simulation Series for Business and Industry." Science Research Associated, Inc., Department of Management Services, 155 North Wacker Drive, Chicago, IL 60606.

Supervisors Safety Manual. 8th ed. Itasca, IL: National Safety Council, 1993.

"Training Requirements of OSHA Standards," OSHA No. 2254. U.S. Department of Labor. Available from U.S. Government Printing Office, Washington, DC 20402, or local OSHA regional office.

Zoll AA. "The In-Basket Kit." Addison-Wesley Publishing Company, Reading, MA 01867.

REVIEW QUESTIONS

1. List four benefits of safety and health training.
   a.
   b.
   c.
   d.
2. Identify seven of the twelve nontraining solutions that can be as effective as a training program.
   a.
   b.
   c.
   d.
   e.
   f.
   g.
3. Define performance-based training.
4. What are the five phases of the systematic approach of performance-based training?
   a.
   b.
   c.
   d.
   e.
5. Adults learn best through which of the following senses?
   a. Hearing
   b. Sight
   c. Touch
6. Describe the four adult learning principles that should be applied to all safety and health training.
   a.
   b.
   c.
   d.
7. List the three most common types of training methods used in industry.
   a.
   b.
   c.
8. Group training encompasses which of the following techniques?
   a. Brainstorming
   b. Case study
   c. Role playing
   d. Simulation
   e. All of the above
   f. Only b and c
9. Define computer-assisted training.

24

Media

A medium is a channel of communication. It can be anything that carries information between a source and a receiver, such as television, films, diagrams, printed materials, computers, and trainers. The purpose of media is to facilitate communication, and the purpose of this chapter is to facilitate effective use of media, whether a particular type of medium is being used for safety and health training, general instruction, or presentations in meetings. The term "trainer" is used most frequently in this chapter, but the information presented applies to speakers in all types of communication situations. This chapter discusses the following topics:

- identifying a company's training and communication needs
- selecting and designing the right media for a company's operations and workforce
- determining the best computer graphics, visuals, films, and video to use
- deciding when and where to use computer-based training materials and supplemental materials
- preparing trainers to present employee training programs
- evaluating the effectiveness of a training program.

Media used in for all purposes take advantage of the two senses people use most when learning—sight and hearing (Figure 24-1). Learners retain 83% of what they experience through sight, and 11% through hearing. Only 1.0% is learned through taste; 1.5% through touch; and 3.5% through smell. Further, extensive research shows that people retain 50% of what they both see and hear, but only 20% of what they learn through hearing alone and only 30% of what they learn through sight alone.

The full range of instructional or training media includes books and workbooks, chalkboards, marker pen boards, tablets, bulletin boards, display cases, flannel boards, posters, flip charts, slides and other projected transparencies, videos, computer disks and tape, films, recordings, models, and fully operating equipment. Media equipment (hardware) includes projectors, simulators, tape and cassette players for sound and video, cameras for stills and video, monitors, television equipment, and equipment for computer-interactive training, such as personal computers and CD-ROM players (Figure 24-2). Live demonstrations and experiments also are classified as audio and visual media events.

DEFINING TRAINING AND COMMUNICATION NEEDS

Safety and health professionals who want to make full use of the power of media must first understand employee training needs and how media can help to support training objectives. Regulatory agencies and legislative bodies are mandating safety and health training in far more detail than ever before. For an in-depth discussion of safety and

Actual Direct Experience

Simulations: "Hands On" Devices

Demonstrations: "Show and Tell"

Field Trips: Familiarization

Exhibits: Displays

Live Television

Motion Pictures

Still Pictures: Photos and Slides

Auditory Aids

Graphics: Charts and Graphs

Words: Spoken and Written

Scale of Sensory Experience

Concrete

Abstract

Figure 24-1. The more concrete the medium of communication, the more effective it is.

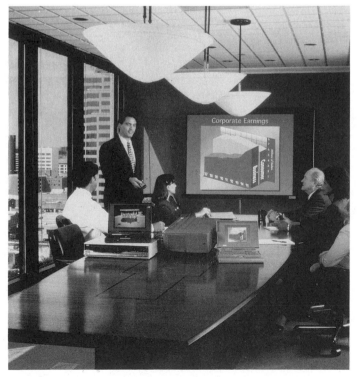

Figure 24-2. This portable multimedia projector allows the trainer to conduct presentations virtually anywhere and to enhance learning through effective visuals. (Courtesy 3M Visual Systems Division.)

health training, see Chapter 23, Safety and Health Training, in this volume. In general, employers no longer rely on class rosters as sufficient evidence to prove that workers received adequate training. Instead, they must train workers in specific competencies and document that such training has been effective. For example, new regulations or changes in existing rules will require additional training time, even for veteran employees (see the Lockout/Tagout section in Chapter 6, Safeguarding, in the *Engineering & Technology* volume).

Accident Evaluation

Accident investigations are beneficial in determining areas of weaknesses or failures in training. Although identifying high accident areas and analyzing possible training deficiencies are reactive measures, such an approach can be proactive if it helps management to prevent similar incidents. Each form filled out after an accident should indicate whether training for management and workers is effective in preventing recurrence (see Figure 8-4, Supervisor's Accident Investigation Report, in Chapter 8, Injury and Illness Record Keeping and Incidence Rates). By compiling and studying those forms, management can pinpoint areas where the training program should be reviewed and improved or where a new training program may be needed.

Personnel Changes

A training need exists whenever personnel changes occur. Situations can range from new employee training to refresher training for workers returning from vacations, health-related leaves, or even shift changes.

New Employees

New employees need training in their required job skills and safety practices. When workers are cross-trained on several jobs and/or machines so they can fill in for other employees when needed, they require even more safety training because they are learning a variety of procedures.

Veteran Employees

Employers should also provide refresher courses; training for reassigned employees; and training on new equipment, processes, and procedures to ensure that workers follow safe practices at all times.

Promoted or Transferred Personnel

One of the best ways to help employers embrace a safety orientation is to make sure that personnel who have been promoted are thoroughly trained in their new safety responsibilities. A supervisor who has come off the line, for example, needs training in his or her new role of taking charge of a work group. Someone who has been transferred does not automatically know where first-aid equipment is kept at the new location, what to do in case of an emergency, and what the housekeeping practices should be at the new workstation. This is true even though the worker may be familiar in general terms with company policies.

Meeting Presentations

Effective communication of safety and health issues to higher levels of management is an important aspect of the safety and health professional's job. Delivery of facts using the appropriate media can help clarify the message and improve the relationship between the safety and health professional and management.

SELECTING AND DESIGNING TRAINING MEDIA

This section introduces the factors involved in selecting and designing media for training, including advantages and limitations related to various media forms.

Establishing Objectives

For both management and employees to get the most out of training, managers must establish clear objectives early in the development of training programs. Once trainers know what they need to accomplish, they can begin to concentrate on the details of how to achieve the objectives, including which medium is best suited for the situation.

Cultural Sensitivity

In addition, trainers should ask the following questions about program materials:

- If more than one language is spoken in an office or factory, has the company provided translations for key materials?
- Is the translation not only accurate but sensitive to cultural differences that may exist in the workforce?
- Do the visual messages, as well as the text, reflect this sensitivity?

Type of Message to be Communicated

To be successful, training should improve and/or increase workers' performance, skills, or knowledge, as well as correct deficiencies. Effective training principles call for behavioral objectives, that is, specific changes in workers' behavior that can be measured by performance-based testing. Behavioral objectives have several components:

- audience: describes learner characteristics
- behavior: describes what learners must do to demonstrate mastery
- condition: describes what learners will and will not be given to perform behavior
- degree: describes how well learners must perform (standard performance)

A good example of a training objective might be the following: "After completing the program, the worker will respond to a simulated hazardous materials emergency by obtaining the proper MSDS and handling the emergency according to the MSDS directions with less than 10% error."

In contrast, the following is not a good training objective: "After viewing this training tape, workers will feel better motivated to follow recommended procedures for wearing appropriate eye protection when working with lasers." "Feel," "believe," "support," and similar words are not "performance" or "behavior" oriented and how well the audience must perform the behavior is not indicated. Teaching, reinforcement, and practice are necessary if workers are to retain and assimilate information. Carefully chosen media help employees to understand the information they receive, shorten the learning curve, reinforce concepts and skills, and help to create interest in the topic.

Audience Characteristics and Setting

A training presentation for a group of 200, in an auditorium setting, will require a different approach than a hands-on, "how-to" session given by a supervisor to a small group of employees. The larger, more formal session can be handled best by a lecture format, with appropriate media used for reinforcement of key points such as, slides, video, or film. The factory floor discussion, which may take place in a small conference or training room, probably can benefit from a discussion format. A flip chart, chalkboard, or large pad on which the supervisor can list ideas the group contributes are appropriate media for this setting.

Training materials should be designed with the target audience in mind. Language levels should match the audience's comprehension or literacy level. For example, if an instructor uses too many technical terms without explanation, the talk will be difficult for a general audience to understand. On the other hand, too simple a presentation is not appropriate for a group of technical engineers. (See also Literacy Levels in Chapter 23, Safety and Health Training.)

Role of Trainer

Each type of media training material available has an appropriate role to play in training. Advantages and limitations of many of the more widely used media tools are given in Table 24-A.

One factor sometimes neglected in the selection of media is the role a trainer expects to play in a particular training session and how he or she perceives training responsibilities. Good trainers continually ask themselves these questions: "Is what I want to do appropriate for the training objective in this particular session? Does the medium I am considering enhance my training role, as I see it?" (Table 24-B.)

Trainer as Expert

Trainers can choose the "expert" role, communicating information in a more formal presentation. The trainer answers questions from the audience but allows little opportunity for dialogue. Such a choice may be appropriate when trainers need to convey information to a large audience, for example, explaining new or revised regulations or policies to the workforce. Appropriate media

Table 24-A. Major Features and Limitations of Various Media

Type and Popular Size	Audience Size	Limitations	Strong Points	Comments
MOTION PICTURES 16mm	M/L	Camera and projector expensive; require trained operator, except for self-threading models. Film not easily changed or updated.	Effective for training and motivating. Uniform professional message. Optical sound unerasable. Sharper image than 8 mm for given projection size. Single-frame, stop-motion projectors are available.	Silent version less costly, but less effective. Many companies prefer video.
SLIDES 2 × 2 in. (35mm, 126, or 127) 2¼ × 2¼ in. (120 film) 3¼ × 4 in. (superslides) (theater projector)	S/L	Slides may get out of sequence, reversed, etc. Cardboard mounts not durable	Effective for training and motivating. Less of a "canned" show since slides may be rearranged. Slides can be made and processed quickly. Color inexpensive.	Taped message or reading script easily added or changed. Remote control and multiple projection possible. Many companies prefer video.
FILMSTRIPS 35mm sound	S/L	Strips and records not easily updated. Message might not be effective or suitably paced for user's needs.	Effective for training and motivating. Message uniform. Sounds easily added to tape or disk.	Silent strips with scripts less expensive, but still effective. Seldom used any more.
OVERHEAD PROJECTORS 10 x 10 in. 7 x 7 in.	S/L	Transparencies positioned by hand. Projector close to screen; it or user may block view unless screen is raised or set at an angle. Ready-made material not widely available.	Effective for training. User can write on transparency while facing audience. No need to darken room. Transparencies easily made and filed. Presentation informal and flexible.	Color transparencies or overlays easily made.
OPAQUE PROJECTORS 10 x 10 in. max.	S/M	Projectors require manual operation. Material in books may be difficult to store or ship. Room must be darkened. Copy may be too small.	Effective for training. No transparencies required; small objects, printed material, drawings, and photographs used "as is."	Copies or originals can be hinged or put on rolls to maintain sequence. Seldom used any more.
CLOSED-CIRCUIT TELEVISION, VIDEO-TAPE CASSETTE	S/M	Initial investment expensive. Requires adequate lighting. In color or black and white. Copies must be made one at a time unless duped by lab.	Instant replay. Excellent for training situation where trainee must "see himself or herself in action." Has relatively low operating cost. Can be shown in lighter room.	Small number of people can view screen. TV is a culturally natural transmission medium.
COMPUTER-BASED INTERACTIVE TRAINING	S	High initial expense. May be difficult to copy. Requires that trainees have access not only to computers, but to videodiscs, videotapes, and so on.	Low operating costs. High retention of material presented, user paced. Can automatically record and verify training and test scores for individual employees. Presentation and message uniform, does not require presence of instructor. Effective if a machine or process must be shown in detail or simulated. Best method for simulating hazardous conditions or those where a mistake would have serious consequences in the "real world."	Computer-based training works best if delivered through computers that trainees use regularly.

Table 24-A. (Continued)

Type and Popular Size	Audience Size	Limitations	Strong Points	Comments
FLANNEL, HOOK AND LOOP, MAGNETIC 12 × 36 in. to 48 × 24 in.	S/M	Presentation requires advance preparation. Few ready-made presentations available. Flannel board material may fall off if not applied correctly or if board too nearly vertical.	Effective for training. Message easily changed, yet can be filed and reused. Permits informal presentation with desirable audience contact. Dramatic, "slap-on" effect builds interest.	Boards suitable for heavier displays; cost slightly higher than cards or pads.
FLIP CHARTS, CARDS, AND PAPER SHEETS 38 × 48 in. 28 × 36 in. 18 × 24 in.	S	Limited to small groups. Good lighting necessary. Speaker must print legibly.	Effective for training, informing, or discussion. Good audience contact. Material easily prepared; can be added during talk and saved. Low-cost pads easily obtainable. Permits reference to other sheets both during the discussion and later for writing of minutes.	Ready-made letters, color, sketches, cut-outs easily added. Colored paper effective. Paper sheets used in place of chalkboards, no erasing.
CHALKBOARDS (portable and wall mounted) 36 x 48 in., large for wall mounted	S	Board must be erased before reuse and recall not possible. Good lighting necessary. Ordinary chalk marks hard to see. Dust from chalk and erasers annoying.	Effective for training or discussing a limited number of points. Presentation informal. Portable chalkboards also useful for holding charts or displays.	Colored or fluorescent chalk adds life to talk. Magnetic boards available.
POSTERS AND BANNERS 8 1/2 × 11 1/2 in., 17 × 23 in., and larger	S/L	Only one or two ideas can be presented at a time; considerable time needed for changing.	Effective for motivating; support training. Specific messages can be posted at points of hazard or to meet timely situations. Ample posters available.	Homemade posters supplement general posters.
WORKING MODELS, EXHIBITS, AND DEMONSTRATIONS	S/L	May require special training to use. Live action is subject to errors.	Action can closely simulate actual conditions. Permit group participation.	

would include projected visuals, such as slides and overheads, and a take-home handout.

Trainer as Mentor

At times, trainers can assume the role of a mentor who teaches his or her skills to other workers. Supervisors or team leaders who have come up from assembly-line work or who have had experience on specific equipment often are called upon to pass on their skills and expertise to new employees or to promoted or transferred workers. Hands-on film demonstrations, either those commercially available or company-produced videos, would be appropriate media. If the organization uses interactive training, a CD-ROM module—if one is available on the subject—is another choice. For example, such a module might show a worker how to use a full-face mask respirator: the respirator parts and their functions; how to put on and take off the respirator; and how to test, maintain, and store the equipment.

Trainer as Facilitator

A third role the trainer can play is that of a resource person or guide who helps to match individual student needs with appropriate materials. Such a role assumes the trainer is flexible and has a broad knowledge of available teaching aids, including various media. Trainers who function successfully in this role work to create a training environment in which students are not afraid to ask questions or to admit their need for extra help.

Student workbooks or worksheets, which can be completed at the employee's own pace, can serve as the basis for individualized discussions. The trainer uses the material to help employees understand their shortcomings and to suggest remedies. Management also can provide

Table 24-B. Some Safety Uses for Videotapes

Videotape is a versatile medium. Here are a few ways it is currently being used to put across the industrial safety message:

- **Security surveillance**—Mounted cameras observe distant gates, loading docks, and payroll departments.

- **Job review**—Used in conjunction with Job Safety Analysis, Task Safety Analysis, and Step Safety Analysis for observation and review of both good and bad procedures.

- **Management training and development procedures**—First train the trainer, then have him train employees down the line. Playbacks assure that procedures are practiced correctly by the instructor before others are taught to do as he does.

- **Incident investigation**—When brought to the accident scene, a recording is made of the physical set-up, personnel present, time, weather, lighting, and other conditions pertinent to the accident.

- **Motivation and enforcement**—Showing the employee how his performance can lead to an accident is an effective way of gaining his cooperation. OSHAct violations can be spotted and corrected. Before-and-after scenes can prove dramatic.

- **Training in sophisticated equipment, complex procedures**—Videotapes can show and repeat processes a step at a time, permitting interruptions for questions.

- **Informing distant audiences of a procedure or announcement affecting all units**—Duplicates of a tape can be made and mailed anywhere so that all personnel are informed simultaneously.

audiotapes and videotapes that employees can take home, enabling students to go over the material until they understand it. Interactive training using CD-ROM disks also may be an option if the organization has such equipment. Workers can practice privately without feeling pressured to keep up with faster learners in the class.

Cost Factors

To determine whether the cost of a training medium is justified, management must consider what the expenditure will buy, how quickly the material will be outdated, how it will be used, and how many students will view it. Numerous options should be considered: selecting commercial training materials, modifying materials or having them custom-designed, designing and producing materials in-house.

A bottom-line cost for a single medium never tells the whole story; management must always rely on personal judgment to determine if the return on investment (ROI) is appropriate for the organization and its needs. For example, interactive computer simulation of an airplane cockpit may seem expensive. However, by allowing a pilot to practice correct responses to emergency conditions, such simulation may prevent an airplane crash, with its potential loss of life and property damage. Because the up-front expense of interactive training is in design and production, costs per student for training are affordable if many students will use the material. One major company estimates that interactive training technology is saving $100 million in training costs.

Even in small companies, a training department can easily justify purchasing a computer graphics package to produce attractive slides, overhead transparencies, and posters if the company conducts worker training frequently. (See also the section on Computer Graphics later in this chapter.) The training department can make the case that effective training will help cut down on employee turnover as well as accidents, thus lowering company costs. Inexpensive teaching aids that also can be used for in-the-field training or mobile classrooms (e.g., portable chalkboards, flip charts, audiotape players, and even VCRs) make it possible for organizations with limited resources to present information with the added impact of sight and sound reinforcement. On the high end of the cost spectrum, custom-produced videos, commissioned by a trade association or other industry group, may give users access to the latest in technology and training materials that they would be unable to afford on their own.

Commercial versus In-House Production

Today's workers are far more experienced and sophisticated as viewers than employees were in the pretelevision era. According to the American Academy of Pediatrics, by the time the average person reaches age 70, he or she will have spent approximately seven years watching television. Data from A.C. Nielsen, a market research organization, indicate that adolescents 12 to 17 years of age watch more than 23 hours of television a week. As a result, viewers expect that in-house-produced media will equal commercial productions in quality.

Companies with in-house production organizations can consult with trainers about their needs for slides, transparencies, posters, and videos. The media department can provide cost estimates, show trainers what has already been done for other company departments, and—should a custom-designed presentation be required—work with the trainers from conception to execution. If trainers want a second or third opinion (and they should always get more than one bid on any major film or video project), they can include their in-house organization as one possible alternative. They also should look outside the company for others, such as independent production houses or organizations that specialize in custom training materials.

State-of-the-art technology and equipment have become so complex that the average training department's personnel cannot match the skills and expertise of media professionals. The trainer is the expert on what should be communicated and on the required training methods. Media professionals are the experts on how to manipulate the media to get the best effects and quality. (See the discussion on video later in this chapter for an easy-to-use guide on how to select the appropriate level of video quality.)

The wide range of commercial products, however, should not exclude the use of home-made materials. Trainers can easily make inexpensive flip charts, paper and pencil charts, chalkboard presentations, instant photos, and audiotape interviews. As described in the section on

Here Is How to Use It
There is a correct and an incorrect way to pick up a heavy object. The correct way is to keep the back straight, the knees bent and spread, and the load close to the body. The incorrect way is to reach way over and lift. (Twisting the back complicates the bad effect.)

To demonstrate this effectively, a special model can be used. If used with its block spine locked, the model simulates lifting with strong leg muscles; the ribbon on the spine remains limp, indicating very little tension of the back muscles. It demonstrates that the back cannot be kept straight without bending the knees.

To demonstrate improper lifting, the model is used as is shown in the sketch. The legs are bent only slightly (or held straight). One hand lifts the handle just ahead of the fulcrum at the hips in order to lift the weight in the hands. The back arches under the strain and visibly pulls each block apart.

The model can also be made to show proper foot placement (see drawing at left).

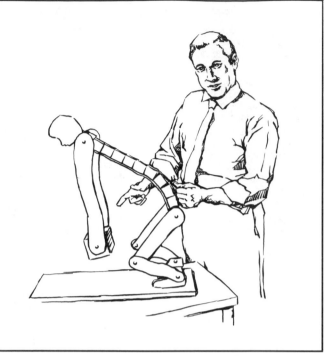

Figure 24-3a. A model is used to demonstrate lifting techniques.

computer graphics, some software packages are fairly easy to use and will produce acceptable media materials.

The choice between professionally developed or home-made materials always depends on a number of factors that the trainer must weigh according to their relative importance for his or her organization. These factors include the number of showings, how many people will be delivering the training, size and composition of the audience(s), degree of customizing (is this aid for general use, or is the information for a particular department or facility only?), and the importance of the message.

Use of Complementary Training Tools

In addition to materials that stress sight and hearing, instructors also can use training tools that engage the sense of touch. These tools help reinforce learning by allowing workers to practice what they have just learned or to experience a real-world application of the lesson. (See Chapter 23, Safety and Health Training, for a thorough discussion of live presentations, including lectures, discussion, demonstrations, and performances.)

For example, exhibits and models make effective three-dimensional displays for use in training. Cardiopulmonary resuscitation (CPR) training, for instance, requires a mannequin on which students can practice. A cut-away model showing layers of muscles or a skeleton showing vertebrae can be effective in demonstrating how injuries can occur and what body parts are affected (Figures 24-3a and 24-3b).

Trainers can use exhibits to demonstrate the safe operation of a machine or process. Exhibits can also feature examples of first-aid equipment, protective clothing,

rescue equipment, respiratory protective equipment, and fire protection appliances.

Companies at times may organize demonstrations of fire prevention and suppression with the assistance of the local fire department (Figures 24-4a and 24-4b). Trainers also can arrange demonstrations of good and bad lighting; impacts on safety hats, safety shoes, and eye protection devices; and first aid treatment and transport of injured workers.

Organizing the Media

Designing effective media is easy to do if trainers remember the basic purpose of media: to communicate a message. After trainers have selected a training objective, considered how to enhance their training sessions with the media, they must organize the materials. Every presentation begins with an idea or concept and builds from the basics. Here are several ways of organizing a session:

- Start with a question or problem, and move toward an answer or solution.
- Present a problem, fill in the details, and end with a solution (a good way of using case histories effectively).
- Describe "what if?" possibilities; after discussion, reach a workable solution.
- Do a time sequence, developing visuals to illustrate chronological order.
- Present information, using a near-to-far or far-to-near sequence.
- List "for" and "against" points in support of, or opposed to, a main idea.
- Explain the features and benefits of an idea or problem.

Construction Specifications

The spine is made with 9 blocks, each 2 in. square and 1⅝ in. high. Each is drilled at the center to accommodate a standard screen door spring. The T-shaped head and shoulder piece is about 8 in. long, 2 in. thick, and supports the arms on shoulders about 5 in. apart. The hip block is 2 in. square, with sloping sides so the legs will spread open in front. Add a ⅜-in. spacer between the hip block and each leg.

Assemble the body by using wood screws to attach ends of spring to the head and hip pieces. Tack a piece of 2-in. wide belting to the *front* of the spine blocks to hold them in alignment.

The arms and legs can be shaped from ¼-in. plywood.

A block, approximately 5 in. square and 3½ in. high, represents the lifting weight. Elastic tape or multiple rubber bands are stretched from the shoulder to the hip, along the *back* of the spine blocks to represent the spine muscles.

A metal handle can be secured to the lower end of the hip block, as shown in the drawing.

The model can be mounted on a 1×12-in. board, 18 to 24 in. long.

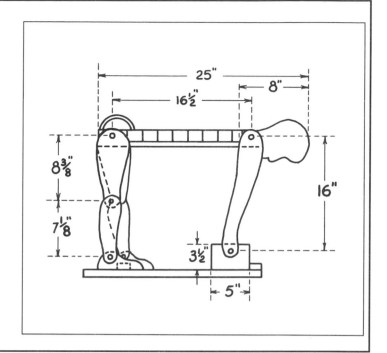

Figure 24-3b. Instructions for constructing the model.

Figure 24-4a. Railroad safety officers are receiving hands-on training in firefighting. (Courtesy of Association of American Railroads.)

Figure 24-4b. As part of hazardous material response training, these students extinguish a liquified petroleum gas fire at the AAR Transportation Test Center. (Courtesy Association of American Railroads.)

To check the organization, develop a simple outline of the media to be used in the session. Work through, in advance, key training points that media can help to explain. Media should clarify or dramatize an important point.

Trainers should ask themselves the following questions. Does the outline flow logically? Does it match the order in which the trainer presents ideas? Although trainers will want to build from simple thoughts to more complex concepts, they should not forget to put at least one visual early in the presentation to communicate clearly what topics will be explained. Such a visual (called an advance organizer) focuses attention on succeeding material, because it lets the audience know what is coming.

Chart Formats

Media work best in training when trainers use easily understood charts and graphs that provide accurate information. For projected visuals, such as overhead transparencies and 35mm slides, trainers have a choice of formats.

For most projected images, use the horizontal format (called "landscape" by many computer graphics software packages). Like a landscape, such visuals are wider than they are long. Because fewer lines fit on a horizontal visual than on a vertical one, trainers are forced to limit their content, resulting in a clearer visual.

Long lists, forms, and detailed flow charts often project better when designed for a vertical format (called "portrait" in computer graphics). Like a portrait, such visuals are longer than they are wide. One disadvantage of portrait format is that viewers' heads sometimes get in the way, making it harder for most of the audience to see text at the bottom of the screen.

Text Charts

Text charts are one of the most commonly used media. They are easy to create, either by keying copy into the computer, using a typewriter, hand-lettering, or using press-on letters available at art stores or office supply stores.

Text charts are made in four styles:

- **Titles.** Title charts are good for introducing presentations. For maximum effectiveness, include no more than eight lines. As a general rule, use the top one-third of the chart for titles and the center of the chart for the message or announcement of the topic.
- **Lists.** Lists are a good way of helping people organize their thoughts about a central idea. They can show sequential steps in a process, such as lockout/tagout training for a particular machine, or can present ideas that then serve as a springboard for discussion. In a simple list, each idea is emphasized equally. Items in the list can be highlighted by putting a bullet or icon in front of each one.

  A good way to stimulate discussion on each bulleted idea is to present them as single thoughts. The trainer can make the first visual using only the first line of the list; the second visual, with two; the third visual, with three, and so on, building the list as the talk continues. If the trainer is using projected visuals, another easy way to control what the audience sees, bit by bit, is to create a single visual with three or four ideas on the list. Then place a piece of paper under the visual to block out all but the area viewers should see. Pull the paper down, revealing the next idea on the list, when ready to begin discussion on that topic. This technique is especially good for ideas that follow a logical sequence.
- **Organization Charts.** An organization chart is a quick way to illustrate lines of authority and corresponding responsibilities. Use an organization chart, for instance, to indicate safety committees for each shift. Many available computer graphics software programs automatically do the drawing, once the trainer has keyed in the names and titles of the staff and indicated the reporting relationships.
- **Tables.** Tables are good for hand-outs, but difficult to follow on projected visuals unless they are extremely simple. Resist the temptation to put too much information on a table. Viewers cannot absorb details that quickly. A variation of a table is a column

chart, with related data displayed in two or three columns. "Days Without Reportable Accidents" could be the overall title, with June, July, and August listed as column headers, and number of days under each month. Column charts are also useful for listing ideas or problems, with potential benefits in one column and drawbacks in another.

Font and Format Selection

As a rule, the clearer the posters, slides, and overheads are, the more memorable they will be. Thus, avoid selecting fonts that are too complicated, placed unattractively on a page or on the screen, or too small to be read comfortably.

Choosing Font Styles

Font is a term for a whole family of letters with a particular design (also called "type"). Some fonts have serif lines or curves that project from the ends of letters. Times Roman, Bookman, and Bodoni are examples of serif fonts. Serif fonts may be easier to read when used for the body of text charts because they are widely found in newspapers and magazines.

Fonts without these lines are called sans serif. Helvetica, News Gothic, and Arial are three sans serif fonts. They look more modern, but some studies show that multiple lines of sans serif font may be harder to read, especially for people more accustomed to serif fonts. Most designers recommend restricting use of sans serif fonts to headlines and outlines rather than to the body text.

For big, bold lettering, Helvetica and Gothic Bold (both sans serif) have long proved popular. They make attractive headlines and display type to accompany body text set in Times Roman or Bookman (both of which are serif fonts).

Script, which looks like handwriting, is a third style of font, but is the hardest of all to read. If another, less common font is desired, trainers should consult with a graphic designer or compositor (typesetter) to make sure fonts and graphic images do not clash, distracting from the importance of the safety or health message being conveyed.

Font Size

Any visual is ineffective when lettering is too small to be read comfortably. People who wear prescription lenses, especially bifocals, and older workers may need to have larger type. Design lettering so it can be read easily from the back row of the audience; the message will have a much better chance of getting across.

In general, use 16 point or larger font for projected visuals to make them easy to read. Remember, larger is better. Keep at least a 2-point difference between text and headline, or text and subheads, to make a visual distinction between the two. For example, if the text font is 14 point, make a headline in 16 point font. For projected visuals to be read comfortably, the designer should be able

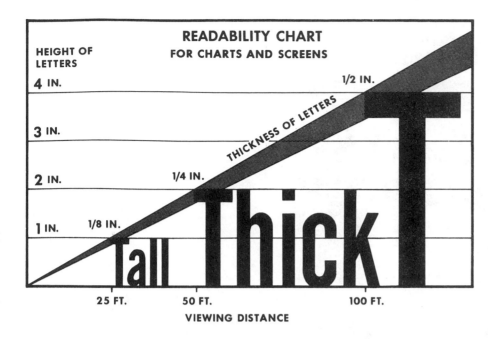

Figure 24-5a. This chart shows what heights and thicknesses letters and symbols should be for easy viewing of projected and nonprojected visuals.

to read the smallest line of type on the visuals at arm's length without a magnifying glass.

Use Figures 24-5a, 24-5b, and 24-6 as guidelines for recommended heights and thicknesses of lines and symbols. Block letters show up better than handwritten copy. To be easily read at a distance of 50 ft (15 m), printed or projected letters should be at least 2 in. (5 cm) high and ¼ in. (8 mm) thick.

For use with overhead projectors, material prepared with characters at least ¼ in. high, with spacing of at least ¼ in. between lines, will give a letter height of 2 in. on a screen 6 ft (1.8 m) wide. Text should be clearly visible at a distance of six times the screen width.

Aligning Text

Text set on the page is aligned in straight or ragged (uneven) margins. Text said to be "flush left" has a straight left margin and a ragged, or uneven, right margin. Text set "flush right" has a straight right margin and a ragged left margin. When text is "full justified," it has both left and right straight margins.

Text charts are generally easier to read, and easier to produce, if designers set the text flush left. This is because viewers are accustomed to looking first at the left-hand side of a page. For clarity, double-space between lines on a text chart and keep lines short. Try to put no more than eight lines on a single text chart.

Capitalization

Studies show that viewers find it easier to read text set in upper and lower case than text set in all capital letters, or "all caps." All caps works for short headlines, but not for body text.

If trainers want to emphasize certain words, boldface (dark) fonts work better in headlines. Inside a paragraph, however, italic font works better than either boldface or

all capital letters. In this modern computer era, underlining is rarely used as a means of emphasizing words or sentences.

Plenty of white space in visuals makes them more effective. Double-space text for high readability. For projected visuals, use no more than six lines of text in a title chart, six words per line.

Computer graphics and desktop publishing software programs make it easy to create text charts and legends for graphs by letting users change and control the font. Traditionally, font size is measured vertically, in points, and line lengths are measured in units called picas. As a rough rule, 12 points = 1 pica, and 6 picas = 1 inch (2.54 cm). The conversion from picas to inches is slightly different on a type gauge (a special, inexpensive ruler available in most art supply stores) because the computer software rounds off the conversion.

Computer instruction manuals and "how to use" books about popular software programs available at most computer stores are good sources of helpful information. They often suggest appropriate font sizes for varying usage. Some even specify a particular size as the default, unless the user deliberately overrides the program. Usually, the designer can preview how a chart will look and make any adjustments before printing out the final product. Font specification sheets show different styles and sizes of fonts. If trainers consider working with an outside media service bureau to have overheads, transparencies, or even posters made, the bureau's designers can give a quick overview of what will look best for the particular purposes.

Because each font has its own personality, the number of characters per line may vary even when the fonts are the same size. Some fonts have a relatively high character count per line, and look smaller than others. Commercial printers and service bureaus can help trainers choose a font that will give visuals the clearest, cleanest look.

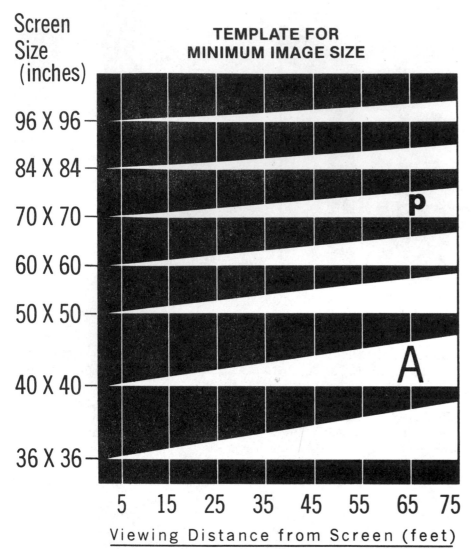

Screen Size (inches)

TEMPLATE FOR MINIMUM IMAGE SIZE

96 X 96
84 X 84
70 X 70
60 X 60
50 X 50
40 X 40
36 X 36

5 15 25 35 45 55 65 75

Viewing Distance from Screen (feet)

Figure 24-5b. The correct type size for overhead projection can be scaled by placing the transparency over this template. For example, lettering for a transparency to be projected on a 70- × 70-in. screen must be at least as large as the "p" ($^1/_4$ in. high) if it is to be viewed from a distance of 65 ft. For this same viewing distance, but with a 40- × 40-in. screen, a letter of minimum size "A" ($^1/_2$ in. high) should be used when making the transparency.

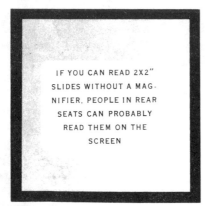

IF YOU CAN READ 2X2"
SLIDES WITHOUT A MAG-
NIFIER, PEOPLE IN REAR
SEATS CAN PROBABLY
READ THEM ON THE
SCREEN

Figure 24-6. Rule-of-thumb for lettering to be shown on a 2- × 2-in slide. (Printed with permission from Eastman Kodak Company.)

Graphs

Graphs are a visual summary of data. To make them easier to understand, always put a legend or short title alongside the graph to tell viewers what the information is about.

Graphs work most effectively when they are matched to the information presented. Although there is no rule about which type of graph to use, trainers should find the following suggestions helpful:

Circle Graphs (Pie Charts)

A circle graph, so-called because of its circular form and pie-shaped wedges, is useful to show the relationship of each part to the whole. For example, a circle graph can show that out of 1,000 reportable accidents, 250 of them took place at only one facility. This would clearly demonstrate the need to improve safety at this facility. Circle graphs also help to illustrate good safety records. Of those 1,000 reportable accidents, for example, perhaps only 16 took place in the shipping department.

Remember, use a circle graph to show data from only one time period. Do not compare two years' accident records within a single circle. Instead, use two separate circles, one for each year.

Creating a circle graph is fairly easy. Most computer graphics software packages will make a graph automatically if a user merely keys in the data. If drawing the circle graph by hand, remember that the total circle contains 360 degrees. The angle needed for each component of the graph must correspond to the correct percentage of the circle. In the case above, the 250 accidents represent $\frac{1}{4}$ of the 1,000 accidents. Consequently, the circle "wedge" drawn must represent $\frac{1}{4}$ of the 360 degrees of the circle, or a 90-degree angle.

$$250/1,000 = 25\%/100\% \times 360 \text{ degrees} = 90 \text{ degrees}$$

All the components in a circle graph should add up to 100% of the total amount represented, just as the angles for each of the wedges should total the 360 degrees.

To break down data further, create a chart within a chart. For example, a wedge is only part of the whole circle, but can become its own series of data. Pull the wedge to one side of the visual for extra emphasis, and turn the information contained in that wedge into a column chart.

Assume that of those 250 hypothetical accidents, 50 occurred in the warehouse, 175 happened in the factory, and 25 took place in the facility office. For the column chart, which is basically a vertically stacked circle, 250 becomes the 100% total. Consequently:

$$50/250 \times 100\% = 20\% \text{ warehouse accidents}$$
$$175/250 \times 100\% = \text{about } 70\% \text{ factory}$$
$$25/250 \times 100\% = 10\% \text{ facility office}$$

A computer graphics package can calculate the percentages and draw the column automatically. The column

will look more balanced if the largest section (with the biggest percentage) is placed on the bottom. Emphasize that section's importance by giving it a dark color or heavy pattern. Link the circle and the column for an effective visual (Figure 24-7).

Because most people look at the right-hand side of a graph first, place the most important "wedge" of the circle at 3 o'clock on an imaginary clock face. Moving counterclockwise (backwards) around the clock, each wedge following the most important one should be a slightly lighter color. If the chart is in black and white, rather than color, give the most important segments black or dark gray colors. Make the least important elements light gray or white.

Avoid putting too many wedges in a circle, so as not to confuse the audience; six or seven wedges is about the limit an average viewer can comprehend. When there are too many wedges, it is hard to compare their sizes.

Computer graphics packages allow trainers to make proportional circle graphs, or three-dimensional ones. Simply because they can, however, does not mean they should. Two circles on a single visual, equal in size but representing different amounts of data, are often difficult for viewers to interpret. For example, suppose that Facility A has only half as many accidents as Facility B. One circle could represent the number of accidents at Facility A, and a circle twice as large could represent the number of accidents at Facility B (Figure 24-8).

Although the proportional circles shown in Figure 24-8 are statistically accurate, they may mislead the

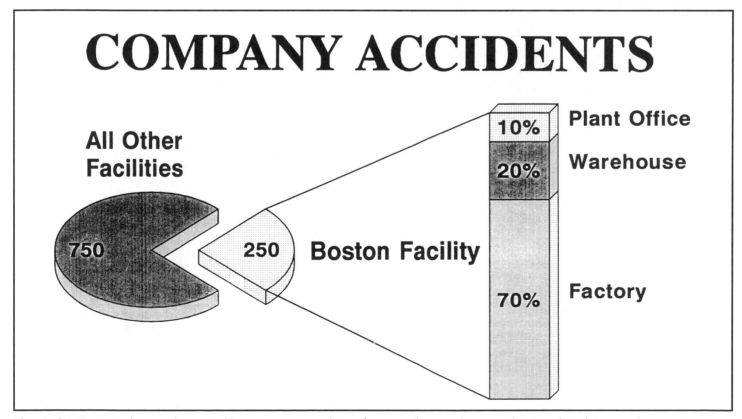

Figure 24-7. A pie chart can be created by a computer graphics software package to show accident (or other) data at a glance. (Courtesy Karen Zmrhal.)

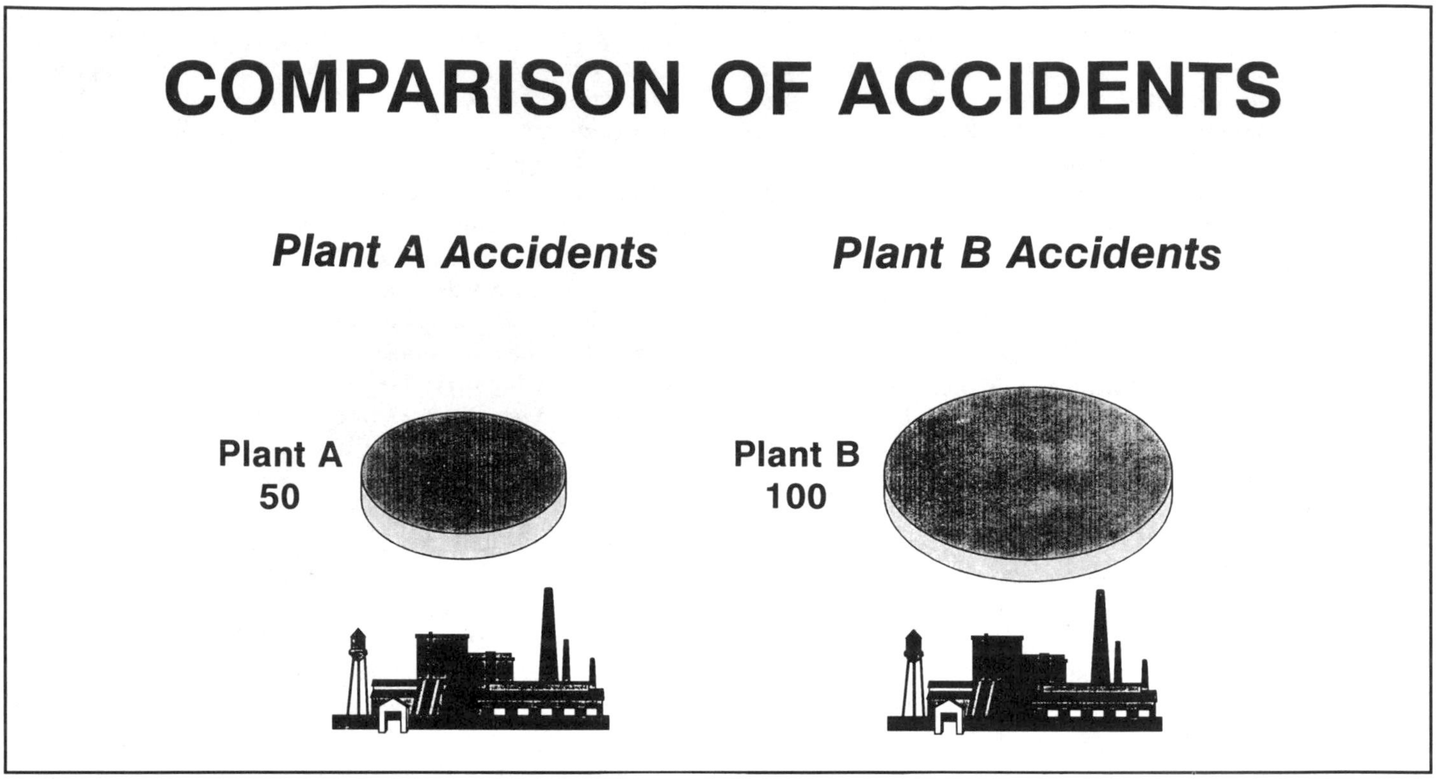

COMPARISON OF ACCIDENTS

Plant A Accidents

Plant B Accidents

**Plant A
50**

**Plant B
100**

Figure 24-8. Two pies of different size may be clearer than one pie with several slices. In this computer-graphics-generated chart, the area of the pie on the left is 50% of the area of the pie on the right. (Courtesy Karen Zmrhal.)*

audience. Most people find it hard to judge circular proportions quickly, because they tend to think in terms of line lengths.

Three-dimensional circles (which show the "depth" of the graph) also can be misleading. If the top of the circle is made lighter to show perspective, that section may appear to be too large to represent its data correctly.

Bar Charts

Bar charts are one of the simplest ways to display data to an audience. They are used to illustrate comparisons of volume over time. For example, a bar chart can show the dollar amounts a company has paid in Workers' Compensation over the past five years. Designers can set the bars horizontally or vertically, provided there are only 10 to 12 bars on a single visual. If bar labels are long, a horizontal bar chart is a good choice. When the data permit, arrange the bars in sequential order, either low to high, or high to low. Ranking lets viewers compare data more easily.

Computer graphics software packages allow users to choose among several designs, such as overlapping bars, stacked bars, or paired bars. When choosing an overlap option, start the series with the smaller values first, so that the shorter bars in the overlapping groups come before larger ones.

Line Charts

Line charts are plots of points that trace changes in data over time. For example, "How many accidents did the

company have in the machine shop from 1982 to 1991?" is a good question to answer with a line chart. The years, which are the independent variable, go on the X, or horizontal, axis, and the number of accidents per year go on the Y, or vertical, axis. Plot the points, and draw a line connecting them or use a computer graphics package to draw the line chart.

Area Charts

Area charts are a good way to show volume, especially to emphasize changes in data. These charts use lines and patterns to separate different kinds of information. Trainers can use area charts as an alternative to a line chart if they are working with a single series of data. For example, to show dramatically how accidents have declined at a factory, an area chart can make the information easy to remember. Designers can combine data series by constructing an area chart with several layers: e.g., the total number of accidents and the number of accidents in the office, the factory, and the warehouse.

To make a multiple stacked chart, put the darkest pattern on the bottom, and work from dark to light, with the lightest pattern on the top of the stack. Either shade the chart by hand, or use a computer graphics software package to create the chart from data values keyed in.

Drawings and Sketches

A simple drawing, done by hand or with a computer program, can often help make a process or procedure clearer.

Keep the illustration as simple as possible, and always include a brief title or explanation next to the drawing.

For consistency within a single presentation, try to keep elements of the visuals, if they are repeated, in approximately the same place for each of the slides or transparencies. If a company logo is used on each visual, always place it in the same location on each slide or transparency. Do the same on all drawings for titles or explanations.

Color

The skillful use of color will add greatly to the impact of visuals. On charts (including those that are all text) a too-light background makes information difficult to read. Overheads with lettering on a clear background project well; so do transparencies and slides that use "reverse lettering," or white text on a darker background.

Remember that approximately 10% of the population have difficulty distinguishing red-green color. If lettering or graph markings are in either color, such persons will see them only as varying shades of gray. The impact of the message will be lost.

If using a computer graphics program to generate text charts and graphs—whether print materials, overhead transparencies, or slides—control the impulse to try every color combination the program offers. Experts suggest using no more than four colors in a projected visual.

Experiment with color schemes and test them for visibility and legibility. A good color concept on paper may not look good on screen, especially for those viewers sitting in the back of the room. Make sure lettering and background offer sufficient contrast; green lettering on a gray background, for instance, blurs at a distance if the values of each color are too similar.

COMPUTER GRAPHICS

Most computer graphics software packages provide templates, that is, predesigned styles that specify type size, justification, and color schemes. Users need only to key in their data. The automatic layout features ensure consistent, attractive results. Software packages usually include a predefined color palette, used when the program displays charts on the computer monitor or sends chart output to a film recorder to create slides. If the trainer does not like the standard color palettes, he or she can modify them to create a customized color combination.

Some software can read data directly from existing spreadsheets and worksheets, including text, titles, labels, and graph settings, thus eliminating time-consuming double entry of data.

Low-end computer graphics software and high-end word processing software have document-formatting capabilities. After the user has created text, he or she can adjust the type style and type size, and print the output on a dot-matrix or laser printer. The computer will produce sheets that can be run through a plain paper copier to make black-on-white overhead transparencies.

High-end computer graphics software has several options for output. If the company has a film recorder, the trainer can make slides instantly. Print the output and make overhead transparencies on laser or ink-jet printers. Some graphics software offers options that allow sequence presentation over a monitor. Although the monitor limits the size of the audience, hardware can be purchased or leased to project the sequence on a screen. The computer output is diverted from the monitor to the window-like device that is set on or built into an overhead projector.

Trainers can also send their computer images to a service bureau, which will turn them into 35mm slides, color overhead transparencies (mounted or unmounted), printed colored handouts, or even colored posters. Either mail the files on computer disk to the service bureau or transmit them directly over the telephone lines by using a modem and special communications software.

Choosing a Computer Graphics Package

A number of factors must be considered before deciding if a computer graphics package is cost- and time-efficient. Here are suggested questions to ask:

- Have the trainers analyzed their training schedule? How many presentations or displays do they or their departments expect to put together within the next 12 months or the next budget cycle? What percentage of them could benefit from better visuals?
- What are the minimum computer hardware requirements for the graphics package? Today's sophisticated software packages require a computer with a hard disk drive. Even if the company has a laser printer, users may not be able to print a full page of high-quality graphics for complex charts if the printer has less than 1.5 megabytes of memory or less than 300 dpi (dots per inch).
- Do users, or staff in the organization who will make the visuals, have unrestricted access to a computer and printer that meet those requirements, or will they have to schedule computer time? How far ahead will they have to schedule time to get the visuals made? How long, realistically, will it take to make them?
- Who within the organization actually will use the software: the trainer, a supervisor, a secretary? How enthusiastic are those people about taking on a new task? How proficient (or willing to learn) are they already with computers? Can they read the manuals, or are they intimidated by the documentation? Can their work schedules be adjusted to allow for the new time demands? Will they be asked to prepare visuals for other departments, now that the company has the graphics package?

- What training is available on how to use the software, and what will it cost? Reading a manual or "how to use" text will not be enough to teach the necessary skills. Is training given in a lecture format only, or is hands-on training possible? Some computer graphics packages are more complicated to learn than others and require considerable time and repetition before a user is proficient. Users who make visuals infrequently may have to relearn parts of the procedure.
- Where does the user have to go for training? Is the vendor willing to train on site if enough workers from an organization use the software? Some community colleges offer training in popular packages. If so, can employees attend? Can management arrange an in-district tuition rate or discount?
- Can users' work schedules be rearranged to permit them to attend a training session? Are there fees for the training? Can more than one person from a company attend training sessions at a reduced rate? After the training session, what telephone support is available, and from whom? Is there a charge? Will the company have to pay long-distance tolls to talk to the software manufacturer's technical department?
- Will the vendor or software manufacturer supply names and phone numbers of others within an industry who are using the software for similar purposes? If so, users should call them and ask for their opinion of the software. Would they be willing to buy the package again? Why or why not? If they hedge, ask for details; then ask the vendor or manufacturer more specific questions.
- If service bureaus exist in a firm's area, users should ask two or three of them to recommend software that will produce the type of visual output that could be processed by the bureaus. Make sure the bureaus are already using the software for customers and ask for references that the company can contact.

When using computer graphics software to create projected visuals, trainers need to consider how they will present the information to an audience. They may have to trade off top-of-the-line (but costly) quality for budget considerations. For example, high-resolution projected visuals look wonderful on screen. However, unless the company has an in-house graphics department willing and able to produce them, management almost certainly will have to order the visuals from an outside service bureau.

Using a Service Bureau

Virtually every major city has at least one service bureau that specializes in producing slides, overheads, prints, and posters from the computer files created with a computer graphics software package. Companies not close to a major city can send their files directly to a bureau by modem over the telephone line. They also can mail the computer disk or send the disk via overnight air express. Turnaround time is usually 24 to 48 hours, plus transit time to get the slides or transparencies back.

Costs for this professional service may be a significant factor. If a firm is training a small group of people in a half-hour meeting that will not be repeated, trainers can almost certainly get by without expensive media aids. They can use the word-processing or computer graphics program to create overheads, even if they project them in black and white, with shaded patterns to indicate varying data. On the other hand, if users are showing safety data at a stockholders' meeting, giving a presentation to a top executive, or speaking at a conference, they will want the best media possible, regardless of cost.

Look closely at the services such a bureau can provide before hiring them. Ask for names of clients and companies, and check references. One question to ask: does the bureau deliver what it promises on time? No matter how professional a bureau's output may look, if users are depending on slides or transparencies for an important presentation, they cannot afford to have their visuals arrive late. Some bureaus will produce a sample slide or transparency from a company's computer files for a nominal fee, giving users a chance to evaluate the bureau's work.

PROJECTED VISUALS

Projected visuals include slides, transparencies for overhead projectors, objects used in opaque projectors, films, television, and video.

This chapter does not cover slides and films in great detail, because excellent information is available from the manufacturers of films and projectors, camera stores, libraries, schools, and publications. The National Safety Council's *Safety and Health* magazine, for instance, carries detailed articles on producing slide shows, using slides, and related topics, and data sheets are available through the National Safety Council library (see References).

Overhead Transparencies

Overhead projection offers a number of advantages to trainers (Figure 24-9):

- Visuals can be shown while room lights are left on, so audience can take notes.
- Trainers can project to any size audience.
- Trainers can face the audience at all times.
- Trainers can use a pen or pencil as a pointer to indicate an item on the transparency. The shadow of the pointer shows up on the screen, helping to focus audience attention.
- Overhead projectors are easy to operate and usually available.

Overhead transparencies are easy to make when trainers use a computer graphics software package (Figures 24-7

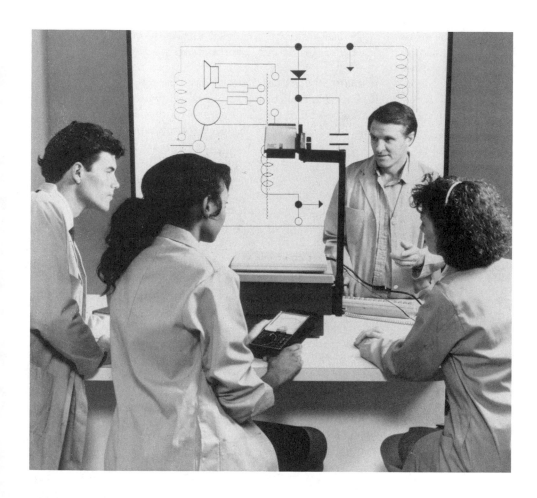

Figure 24-9. This computerized overhead projection system assists the trainer in an explanation of electrical circuitry. (Courtesy 3M Visual Systems Division.)

and 24-8). However, if they can type, draw, use transfer lettering (available at art stores or office supply stores) or clip art, (available in purchased books or on computer disk), they can create overheads easily and inexpensively.

Trainers can type, draw, or letter their message with dark ink on plain paper, send it through a plain paper copier in their office or at an office supply store, and get a black-on-clear transparency. They can also make a reverse transparency, using white lettering on dark background. Using a service bureau or outside provider, trainers can take a 35mm slide or photo and have color transparencies made professionally.

Special Effects

Trainers can highlight certain words or phrases on their overheads by spot-coloring them ahead of your presentation with permanent color pens. Or, if they're using the same visuals for different audiences, they can use washable color pens and write directly on the transparencies while showing them. Afterwards, they can wipe off the markings.

Overlays help explain step-by-step concepts and build the visual's story in a meaningful way. Two or more imaged transparencies are used in sequence over each other. Overlay visuals are hinged to the frame on one side

with tape to allow the base visual to be presented first. Then the overlay is flapped over it to complete the message.

A technique good for title visuals is an overlay. Make two transparencies from the same data; one, in a positive color; the other, in a reverse image. Frame the two transparencies together, but slightly offset, so the net effect is that of shadowed lettering.

Framing Transparencies

Because frames help to block out distracting light from the screen, add rigidity for handling and storage, and provide a convenient border for writing notes, most trainers do frame transparencies, even though they can use them unframed. Frames also provide a space to number and date visuals so trainers can keep them in order.

Traditional cardboard frames, preframed films, and flip-frame transparency protectors (clear plastic sleeves) are available. The latter two are lighter in weight than conventional cardboard and take less room in storage.

Projection Panels

One option for presenting overhead transparencies is for trainers to use a computerized projection panel to display the output of their computers. Such panels are available with color or monochrome displays. By connecting the computer into an overhead projector that shows the computer images "live," the panel lets trainers make changes in their data and immediately show those changes on the screen. The technique is often used to dramatize "what if" data or spreadsheet figures.

Changing data can be used in combination with overhead transparencies for even more dynamic impact. The trainer can set up two projectors and two screens. On one set, transparencies are used for "static " information, such as safety goals or workdays lost to accidents. Combine the second projector and a production panel to display the "what if" information. "What's the dollar impact if we reduced the number of accidents by 5%?" by 10%?" and so on. As the speaker manipulates the computer spreadsheet to show the changing figures, the audience sees the changes being made, because they're projected on the second screen.

The two-projector/projection panel technique also works well as a discussion springboard for problems. Display the problem on the first screen, and show the range of solutions on the second. As students talk about each alternative, adjust the computer data to show the impact of each possibility, and project the display on the second screen.

Slides

Slide shows, whether developed by computer graphics, photography, or a mix of the two technologies, are another technique for presenting materials. If sound is added, programs can be recorded and sound synchronized to slides on a cassette recorder, which can automatically advance the slides with the narration.

If cassette recorders are equipped with a stop button on the visual synchronizing mechanisms, the speaker can stop the show at appropriate points. He or she can pose questions, get a class response either in writing or verbally, and can start the show again, explaining and discussing correct answers.

Slides offer certain advantages:

- easily updated
- inexpensive
- easy to make
- produced by photography, computer graphics software packages, or a combination of techniques
- commercially available
- targeted to audience or training needs, since specific slides can be substituted in a series
- can be coupled with additional slide and/or film projectors for multimedia impact
- audio tape for automatic advance available.

Slides, however, have some potential disadvantages:

- easily get out of order
- can stick in the holding tray

- can easily be projected upside down or backwards
- if poorly made, can be distracting.

Media Preproduction

Creating a storyboard is one effective way of visualizing and assembling a slide show. On index cards, or on paper on which the trainer has drawn frames about 2 x 3 in., he or she can sketch in what the first visual should be. When the trainer has made a quick sketch of what audiences will see (just as the camera will see it), he or she should down a brief description of the scene on the same card.

The trainer should arrange the cards in order on the storyboard. An 80-slide show needs somewhere between 70 and 80 cards mounted on the storyboard, running in order from top left to bottom right. Edit the cards until they are in the right the sequence, then number them. A well-thought-out storyboard will help a trainer fit images together in a sequence that supports and enhances the presentation.

If the trainer is shooting (or commissioning) slides that some day might be transferred to film or videotape, there are several technical points to keep in mind as the photography is planned. For instance, viewers will not remember slight visual differences in a series of projected slides. Film and video cameras, however, pick them up. The following suggestions come from the Association for Multi-Image International, Inc.

Aspect Ratio

The ratio of width to height of a projected picture image (W:H) is called the aspect ratio. The standard in slide production is 3:2, but the aspect ratio of film and videotape is 1.33:1. If a trainer wants the full image of a slide transferred to film and video, he or she will have a dark band at the top and bottom of the image unless the critical information is kept within the film and video aspects. Many camera reticles (the lines seen as one looks through the camera to focus it) are marked with these areas. The trainer should be sure that everything that will appear on the screen falls within the "lines" as the photograph is snapped.

Type and Character Sizes

Slides have a higher resolution (the clarity or fineness of detail visible on-screen) than 16mm film or videotape. Consequently, if the trainer is planning a transfer to film or video, he or she should make the type and character sizes larger than normal on the slides. Be especially sure that complicated typefaces such as serif or extended italic type are larger because of the lack of definition of video.

Color Choices

Trainers should limit the graphics colors they select for their slides if the slides may eventually be transferred to video. Video has a reduced contrast range, causing poor reproduction of dark, intense colors or of highly-saturated colors, especially reds.

Differences in Brightness

If the brightness of any one slide varies greatly from the average of the entire show, color-correct when slides are transferred by having the lab use neutral density filters.

Trainers should not underestimate the time, money, and effort it takes to produce a good slide show. They will need to compare the time and costs associated with purchased slide packages, computer graphics software, in-house visual production unit (if one exists), or an outside production company before deciding which route to go.

Safety Considerations

Photographers—whether the trainers themselves, their company professionals, or outside photographers from a production company—should observe these safety precautions when on location.

Safe Background

It's not enough simply to make sure an object or person who's the subject of a photo is in complete compliance with all pertinent safety regulations. Persons in the background who might also show up in the picture also need to be observing safety precautions and wearing all the personal protective equipment required for the job they're doing.

Note: A safety violation or evidence of poor housekeeping showing up in a photo that's supposed to be promoting safety and health is likely to attract a lot more attention than the main safety message.

Safety Equipment

Photographers, their helpers, and everyone else who goes into an area where personal protective equipment is required for the purpose of taking safety photos must wear all the equipment specified, even though it might appear as if they personally face little risk of injury.

It's essential that persons promoting safety follow all the rules for using personal protective equipment, as well as all safety rules for that facility or work site.

Safe Practices

Photographers, crew members, and visitors must observe all safety rules, including no-smoking rules and rules that prohibit certain types of electrical equipment in areas where sparks, radio frequency (RF), or electromagnetic interference (EMI) could be hazards. Rules must be observed, both for the protection of the visitors, and to demonstrate respect for safety principles.

Lighting Equipment

Photo floodlights and other special equipment used on a photo shoot should be in good working condition and use wiring, plugs, and outlets that are automatically grounded. Electrical cables must be strung in a way so they will not cause a fall or some other accident. When appropriate, a facility electrician or safety personnel should be on hand to help a photographer avoid mistakes or accidents. All the equipment used on a safety assignment should be UL-listed or approved by another equivalent agency.

Fall Protection

Photographers should use sturdy ladders and work platforms to climb on when going after elevated photo angles. They should not climb on machinery or tables. Fall-arrest harnesses and lifelines ought to be used when working from heights.

Opaque Projectors

Opaque projectors are useful for showing printed material and even three-dimensional objects. Printed material up to 10 in. (25 cm) square or an object up to 2 1/2 in. (6.5 cm) thick is placed in the machine, and the image is projected onto the screen by means of a powerful light and a mirror. The room must be darkened for effective viewing.

With an opaque projector, material that cannot conveniently be transferred for overhead projectors or photographed for slide projectors can be quickly shown. An example of such material would be a safety catalog or a safety poster in full color. Of course, to be suitable for use in an opaque projector, printed material must have type large and clear enough to be easily read by the audience when the material is projected.

Another use of the opaque projector is to project a picture, map, or other shape onto a pad or chalkboard so it can be traced. The size of the projection can be varied to the exact size required, for a neat, professional-looking drawing that can also be used as a flip chart.

Films

Although video is now generally preferred, films are excellent tools for training and motivating workers. Because films cost more compared to other media, trainers should plan their use carefully. The need for a quality product and the equipment and technology required to produce a film are reasons enough to use a company's in-house media department or go to an outside production company to make training films.

In addition, a wide selection and variety of 16mm safety films are available from insurance companies, local safety organizations, commercial film libraries, trade associations, industrial producers, and the National Safety Council. Many films can be rented or purchased, or previewed before a purchase decision is made. Because design and production costs (the largest cost components of films) have been absorbed by the producer and spread out over many renters or buyers, such films may not be personalized to each company's situation. However, they are almost always more economical than those a company's own media department can produce.

Video

Technical training by video has become extremely common, because videos offer a number of advantages:

- Job site brought to the classroom. A major advantage of video is its ability to show work-site situations. Students easily identify with on-screen visual representation of conditions or problems.
- Familiar format. Almost everyone is used to receiving information from a television screen, and seven out of 10 U.S. households have video cassette recorders (VCRs).
- Instant playback. If a trainer desires, he or she can replay a sequence, pointing out specific techniques or incidents. Students viewing the video can see material over again until they feel they have mastered the content.
- Availability of easy-to-use equipment. It is much simpler to pop a prerecorded video cassette into television play-back equipment than it is to locate a 16mm projector and screen, thread the projector properly, and show the film. Virtually no special training is needed to run the video cassette player. Some companies allow workers to check out safety videos for home viewing and self-study, much as commercial chains rent prerecorded tapes.

Showing Prerecorded Tapes

Safety videos on a wide variety of subjects are available, either as single topics or as part of a series. Many videos, including a number of those offered by the National Safety Council, come with additional learning material designed for reinforcement: textbooks, instructor's guides, transparencies, and program guides.

Working With a Video Production Company

Because people are so familiar with television programs and videos, they are highly conscious of production quality. Unless a company's staff are experienced professionals, the firm should consider using an outside training organization, the company's own media production department, or an outside video production company to achieve the visual quality and impact needed for effective training.

Before trainers commission a custom-designed training video, they will need to answer the following questions:

- *Benefit/Impact.* How will the project save the company money, improve productivity, safety, or efficiency? Can the trainer provide "hard data," with potential benefits?
- *Background.* Why does the trainer believe that a video is necessary? Can he or she describe the situation or problem that needs to be communicated? What solutions has the company already tried? What solutions, other than video, are they considering?

- *Objectives.* Can the trainer describe the session objectives in behavioral terms and list them in order of priority? Should the video primarily train (impart specific skills); inform (give background, ideas, facts); or motivate (difficult to measure effectiveness or results)? What should the audience be able to do—or stop doing—after they see the video?
- *Audience Demographics.* What are workers' ages, years with the company, gender, education, etc.?
- *Audience Interest/Need.* Why should the audience be interested in the situation shown in the tape or in solving the work-related problem? How will they benefit from the video?
- *Audience Knowledge/Experience.* What does the audience know about the situation or problem? What is their past experience or involvement? (Before video production is started, the producer should poll the target audience for their input.)
- *Audience Attitudes/Prejudices.* What attitudes do the audience members already have about the situation or problem? What motives might the audience have for paying attention to, or ignoring, the video?
- *Company Goals and Strategies.* How does the intended video support the company goals and strategies?
- *Generic Value.* Can the proposed video be used in more than one facility? Can it be written generically and still meet company objectives? Can alternate scenes be shot to make the program generic? Would other organizations be interested in the program?
- *Use.* How will the program be used? What will the viewing environment be? Will the video be used as a "stand-alone" program or as part of a course? Will any support materials (leader's guide, student study guides, handouts, booklets, etc.) be needed to go along with the program? Who will develop them?
- *Evaluation.* Will the video be evaluated formally, with specific questions, or informally through comments? Who will do the evaluation, and what is the evaluation strategy? Will the video be previewed with a sample from the target audience before production is finalized?

Choosing a Production Company

If management decides to produce a custom-designed video, trainers will want to get several bids before selecting one firm. The organization's media production company, if it exists, should be one of the candidates.

As a basis for bid comparisons and to help manage client expectations, one major company has developed a program classification matrix describing four levels of video production. These levels range from a "talking head" video where someone simply faces the camera and delivers the message (Class I) to a comprehensive, top-of-the-line program comparable to broadcast television (Class IV), using professional talent, studio sets, extensive editing, and custom music.

Use the program classification matrix as a tool in selecting the right video production. Choose blocks across the rows that represent trainer expectations, total the columns, add the answers, divide by 10, and round up or down to the nearest whole number.

Trainers may want an outside consultant to help them evaluate the bids, check references, and evaluate past productions of any company they are considering. Remember that production costs do not include distribution. Get bids on duplicating and distributing the video from several vendors.

If a company has a culturally diverse workforce, trainers should consider using a production firm specializing in translations to do voice-overs, or narration, for the video. For maximum effectiveness, use a competent, experienced organization that is sensitive to nuances of language and cultural differences.

Shooting In-House Videos

Photographing short onsite videos with a camcorder is not difficult to do and may be an option when budget restraints have priority over production quality. Often, camera stores run free or low-cost classes in "how-to" techniques. Local television studios may offer cable access classes in video production; high schools and community colleges frequently have training classes.

Before investing time or money in making videos, be sure, that the video can be played on company-owned equipment, or that management can acquire appropriate, cost-efficient equipment.

Trainers who decide to shoot their own videos should make them short and simple. They may want to have a colleague videotape their training sessions before giving presentations to students in order to review, analyze, and improve their performance.

For onsite shooting, plan a script carefully, using storyboard techniques previously discussed in this chapter. Be sure to obtain appropriate releases from those who appear in the video.

COMPUTER-BASED TRAINING

Computer-based training, also called computer-assisted training, is a viable, high-tech media option that a number of companies have used with good results.

Benefits and Drawbacks

The benefits of this media include:

- More efficient learner-centered training—Students proceed at their own learning rates.
- More timely training—A training module can be given to a student whenever he or she needs it.
- An increased student-to-instructor ratio—One instructor can monitor more students who are using computers than is possible in a conventional classroom setting, if enough computer workstations are available in one location.

Because concepts and "how-to" training are standardized, computer-based training can be more effective. All students receive the same information, regardless of their location or their teacher. Interactive computer-based training provides standardized feedback, giving the same responses each time a student answers a question. Also, because students have the opportunity to practice until they master the skills addressed in the training, they become more proficient (Figure 24-10).

If individualized training is desired, interactive computer-based training can provide pretests and posttests for students, tracking and recording their performance as they complete various segments of the program. Some interactive computer-based training programs can be customized to adjust the work within a module, based on a student's performance.

Drawbacks to computer-based training include the cost and time necessary to develop or purchase high-quality training, based on specific training needs. Generic computer-based training packages are difficult to match with an individual company's objectives. Organizations that wish to develop their own programs (or work with an outside organization that will do so) must be prepared to invest substantial time and money. Programs often need to be tested and refined several times during development.

A second drawback to computer-based training lies in its standardization. Not all people learn in the same way or think alike. If computer-based training is used by itself, without an instructor to ask questions, students may learn to respond with the "correct" answers but may not understand those answers or why they are "correct." When faced with a situation on the job that is slightly different from the computer version, they may not know how to handle the problem.

Finally, because computer-based training is impersonal, keeping students "on target" and motivated can be difficult. Well-designed training materials can help, but the student's attention span may be shorter than would be the case in a more conventional training class that promotes interaction between participants and the trainer.

Assessing Needs

Deciding whether or not computer-based training is appropriate for an organization is easier if managers use a systematic approach. They will need to evaluate their training requirements carefully. Are they training for specific skills needed to operate certain equipment? If so, hands-on time with the equipment is necessary. If the problem is shared by many companies across a variety of industries, computer-based training software packages may have already been developed.

Programs in widespread use in business, such as word processing and spreadsheets, often come with self-paced on-screen tutorials. Training on how to use such software may be available on audiotape or video. On the other hand, if managers want to show new employees how to

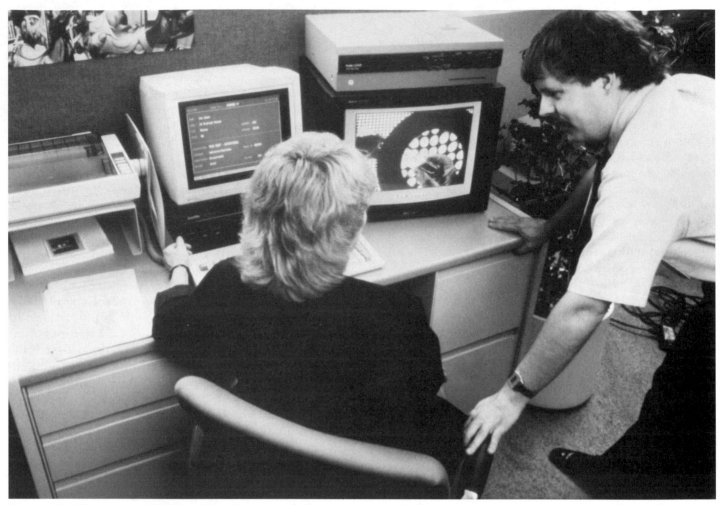

Figure 24-10. Interactive, computer-based training is designed to allow the student to work independently, correct mistakes, and reinforce correct responses. (Courtesy Eastman Kodak Company.)

change machine tool set-ups by reprogramming a computer-controlled machine, the company will almost certainly have to develop its own computer-based training, or find a different vehicle for training.

If company training programs must meet regulations that require documented employee training, trainers may find that commercial packages already have been developed on these topics for the industry. Before investing in any packages, and the necessary hardware to run them, however, trainers should ask if vendors will release names of companies that have used the computer-based training successfully, or that can provide evidence that the training, in fact, does the job it is designed to do.

A substantial investment in computer-based training also may be hard to justify if new procedures, new equipment, or new regulations make the program outdated soon after its production and distribution.

Learning Formats for Computers

Computer learning is achieved by a question-and-answer, or challenge-and-response, format. These formats require the student to participate actively in the training. Two types of interactive materials are often stressed: conceptualization and simulation.

Conceptualization

Primarily used for training that depends on theory or other abstract ideas, conceptualization is a good choice when employees must learn "soft skills," such as human relations or management training. This type of training helps to communicate psychological or philosophical viewpoints.

Simulation

The use of simulations that duplicate the look and feel of real experiences has proven to be a valuable learning tool. The most sophisticated simulators are probably those used by the aerospace industry to train pilots. To the extent possible, the simulated cockpit replicates visual, auditory, and tactile experiences of real flight. The primary advantage is to place the pilot in emergency situations under controlled conditions so that safety procedures can be practiced. Trainers can evaluate both the procedures and the pilot's performance.

Simulated electrical panels are used widely for electrical training. The electronic components for a machine or systems are installed on a board, so that trainers can create problems requiring students to practice troubleshooting skills in a classroom setting.

Fire training institutes and some fire departments use simulations to create various firefighting and rescue situations. Trainees then practice the recommended techniques under controlled conditions.

The National Aeronautics and Space Administration (NASA) uses special water-filled tanks and weighted suits for astronauts to simulate working in a weightless environment. Some mining companies have replicated underground conditions in surface training labs to enable miners to practice machine operations under controlled conditions.

Computer-Managed Training

Trainers can choose to use computer-managed training with computer-based training. Under computer-managed training, computers monitor students on their learning time, attendance, and participation; the training materials students have used; tests taken; and the final results achieved. The computer can keep track of how a student performs and suggest additional materials or extra drill, if the performance needs improvement. If some students complete the modules ahead of schedule and are ready for a higher level of training, the computer can generate a list of names for the instructor's review.

Interactive Training

Interactive training, a technology that uses a CD-ROM (laser-operated) based technology in combination with a computer video display terminal screen, allows a person to be trained systematically by simply touching the screen to choose from various options. The user does not have to know anything about computers, nor key in the answers.

The system combines sight, sound, and kinesthetics in a single presentation, thus reinforcing learning. Because training is one on one and requires touch-screen action by the worker before the module continues, employees stay motivated and involved in learning.

Interactive technology offers a number of advantages, including:

- *Reduced Learning Time.* In some studies, learning time can be reduced up to 50%. Immediate feedback provides constant reinforcement of concepts and content.
- *Reduced Delivery Costs per Student.* In interactive training, the largest costs cover design and production, not duplicating or distributing the film. Delivery cost per student decreases as more students use the same program. In traditional training systems, which depend heavily on teachers, delivery cost remains constant or rises as numbers of students increase. Even for a custom interactive program, the cost-per-student breakeven point occurs with a relatively small number of students. For large organizations with many students who need to learn the same material, savings can increase dramatically.

- *Reduced Risk.* Interactive systems can allow students to explore potentially dangerous subjects without compromising safety. For example, a course on basic electronics and maintenance lets students accidentally "touch" the wrong parts without risking electrocution, yet "see" the consequences on screen.

Because interactive systems do not allow students to move on to new material until they have demonstrated their mastery of fundamentals, trainees gain a strong foundation on which to build further skills. Yet, because training is self-paced, students can repeat and review materials until they are confident they have mastered the information.

SUPPLEMENTAL MATERIALS

Supplemental materials can enhance and support a company's media training tools. Whether supplemental materials are stand-alone safety reminders or designed as part of a comprehensive presentation package, they can increase safety awareness and reinforce training messages.

Posters

The purpose of posters is to create safety awareness and educate or reinforce. A number of commercially available safety posters on various topics are offered by a variety of organizations, including the National Safety Council. State-of-the-art procedures, single-topic reminders, and compliance/regulatory warnings are good visuals.

Displayed at various locations throughout an organization, posters can help employees avoid hazards and can be a useful part of a company's comprehensive safety program. For more detail, see the discussion of posters and displays in Chapter 25, Safety Awareness Programs, in this volume.

If a trainer or someone in the department is making posters, keep them simple and short for maximum impact. Each poster should convey only one main idea, with no more than two or three additional details for reinforcement. Eye-catching visuals can be used; for example, a drawing of wiggling toes sticking out of a cast that is decorated with signatures of fellow workers, and the message, "Wearing Safety Shoes Might Have Prevented This," makes the point graphically.

Ideally, posters should be large: 15 x 20 in. (38 x 51 cm) is a minimum size, although 22 x 28 in. (56 x 71 cm) is better. If posters are too small, people will pass by without noticing them. Vivid colors such as golds, yellows, hot orange, or brilliant blues grab attention immediately. For even more emphasis, highlight with a splash of contrasting color.

Type should be as large as possible—38 to 72 points, or letters at least one inch high for headlines. Trainers can use markers, transfer letters, cut paper, or India ink for lettering.

Pictures

A good safety photograph also attracts attention. Photos can show the safe way to perform an activity, such as lifting a package, putting on a respirator, or using a computer keyboard correctly to minimize the risk of carpal tunnel syndrome.

Like posters, photographs should convey one main idea to be most effective. Pictures of people are always more interesting than those of equipment alone; close-ups are more attention getting than long, "establishing" shots. If the photographer is shooting a series of photos, however, he or she may want a long "establishing" shot to set the location and orient the viewer, plus some medium shots and close-ups for variety. Someone skilled at focusing may want to use a tighter close-up by including only the important details, such as focusing on a worker's hand to illustrate proper tool use.

To ensure top-quality photos, match the film to the environment. Daylight film works best with daylight or flash, while tungsten film works best with tungsten light. If the photographer has to shoot with the "wrong" film, he or she should use the correct color-compensating filter. Color-correction filters can also help maintain the correct color balance with fluorescent lighting, found in many offices or factories. Fluorescent lighting tends to produce greenish tones. If subjects wear bright, strong colors, the green cast will not be as noticeable.

When shooting outside or in a well-lighted room, film with an ASA (ISO) reading of 400 can take advantage of available light. The higher the ASA (ISO) number, however (check the box), the grainier the photograph when enlarged or projected.

To improve the chances of getting successful photographs, overshoot; that is, take substantially more pictures than needed. Film is the cheapest commodity; the time of the people involved is far more costly. Taking four or five photos of a single set-up to get one usable picture makes economic sense. Bracket exposures; adjust the camera meter and settings for what will be the ideal exposure. After snapping the picture, take at least two more, one f-stop under and one f-stop over, the recommended setting. If trainers want to brush up on photography, a camera store, an equipment manufacturer, a high school, or community college will often offer free or low-cost classes on simple techniques.

On the other hand, if the project calls for a number of photographs, particularly in a factory or large office setting, or if the upcoming event is a major one, it is almost always cost-efficient to use a professional photographer. Media professionals have the lighting equipment and know-how to make a shoot run smoothly. Clear the way with the supervisor and with whatever authorities might be involved, including the facility's manager, and any governmental body that might have jurisdiction. Get releases from all persons photographed (including employees and managers) before including them in pictures. The company's legal department can draw up the correct form. Make sure the release is open-ended and nonspecific enough so that if the company wants to recycle the slide into another show, it will not be necessary to acquire a second permission signature.

Handouts

Handouts represent another supplemental way of reinforcing training concepts or of testing specific skills taught in the presentation. Employee safety booklets, guidebooks, manuals, compilations of case studies, and "hard copy" of overhead transparencies and slides are good learning aids. Many are available commercially from various sources, including trade associations and the National Safety Council.

Service bureaus can turn overheads or slides into hard-copy handouts. However, the projected visuals will be summaries, with few words and lines on each overhead or slide. Handouts give the speaker a chance to expand on the information shown on screen, including more details or cases.

When trainers plan to use such handouts as part of a comprehensive program that includes other visual media, they should make sure not only to preview visuals, but to check the printed handouts carefully. Both media should give the same message, not confuse students.

Occupational Safety and Health Data Sheets

Occupational safety and health data sheets, such as those available from the National Safety Council library, can reinforce safety concepts by providing supervisors and workers with facts, diagrams, photos, and analyses in one convenient handout. Because such data sheets are periodically revised, they can help trainers, supervisors, and others keep up to date on the latest regulations, compliance issues, or correct factory procedures.

Audio Cassettes

Used as stand-alone training or as components in a comprehensive package, audio cassettes can promote safety effectively. Purchased cassettes—some available with response booklets and self-tests—are valuable for individualized training or refresher training. Some are available commercially with training in a choice of languages. Because students can replay audio cassettes, they can go over material until they have mastered it. Cassettes also can be part of classroom presentations.

If trainers make tapes themselves, they should use a hand-held microphone to filter out ambient room noise. Sixty-minute tapes (even if the recorded material is not that long) work better in a recorder than 90-minute tapes. Trainers can do brief on-the-job interviews that may prove useful later for bulletins, newsletters, and future meetings. They also can dictate a running commentary while photographing safe or unsafe operating conditions, new processes, or equipment that require subsequent study. They can record minutes in safety meetings or valuable discussions in training sessions.

In addition, trainers can tape safe, efficient job procedures, provided the trainer allows sufficient time for performing each task. The trainee then listens to the tape and follows the training while performing the work.

PRESENTATION OF TRAINING MATERIALS

Most modern meeting rooms and classrooms are well designed for media presentation. Occasionally, trainers may find themselves leading a training session in a less-than-ideal environment. The tips in the following sections will help in any situation.

Tips for Presentations

- Make sure the room to be used is, in fact, available. Know in advance what training activities are planned and where the participants can go if forced to switch rooms at the last minute. Will the planned equipment work in an alternative location? Can three-pronged adapters or extension cords be obtained fast, if necessary? Is a chalkboard on wheels available? Is there chalk and an eraser? Will the number of tables and chairs needed fit in the substitute room? Will someone move additional tables and/or chairs there, if needed?

- Collect everything trainers could possibly need for the presentation: if they are using a chalkboard, extra chalk and erasers; if they are using an audiotape cassette player, extra batteries as well as an AC adapter cord. Trainers should have essential items, such as spare projector bulbs and fuses, and make sure trainers or the staff know how to replace bulbs and fuses.

- Be sure the person responsible knows how to operate all the equipment to be used. Have trainers coach and rehearse their talk with a substitute in case the first helper is unable to attend the meeting.

- Make an agenda of the meeting with approximate times listed for each part of the presentation. Trainers should give the agenda to their helpers well ahead of the session. Have an extra copy of the agenda on hand should a last-minute substitute helper become necessary.

- Staff should know when the speaker wants the lights on and off, or adjusted to a particular level. Determine who will adjust the lighting controls.

- Check the equipment a day ahead of time to be sure it is in good working order. If equipment is shared with other users, trainers should make sure they have signed up for the required equipment well in advance. Double-check the day before the presentation to make sure no one has preempted the reservation. Does all equipment "match?"—i.e., is the VCR player the correct size for the videotape? Will the slide tray fit the projector? Don't take anything for granted; assume that if something can go wrong, it will, and prepare accordingly.

- Make sure that prerecorded media (i.e., videotapes, films, and audiotapes) are of acceptable quality for presentation. Preview everything well in advance, if possible, in case alternatives are needed. Time any prerecorded media in advance to be sure the presentation will run smoothly and on schedule.

- Also preview projected visuals made in-house. If possible, use the same room, equipment, and lighting chosen for the presentation. Slides that look good on a slide sorter may be under- or overexposed when seen on the actual screen to be used at the meeting. Leave slides or transparencies on screen, and sit in the furthermost chair. Can the visuals be read comfortably? Are they sharp and in focus?

- Before the presentation begins, preset the brightness and audio controls on projectors, audio equipment, and video monitors to avoid disrupting the program later to make adjustments.

- Make sure all the slides to be used in a presentation are in order and good condition. When the slides are in proper sequence, draw a line diagonally across the top of the pack. In this way, a slide that happens to be turned accidentally on its side or moved out of sequence will be noticeable at a glance.

- Use trip guards to tape down electric cords, so they won't pose a fall hazard (duct tape will work if trip guards are not available).

- For effective communication, presenters should face the audience or keep their faces turned in profile while discussing slide shows, movies, and videos.

- If a loudspeaker is used, place it near the screen and slightly elevated for realistic-sounding audio reproduction. If a clip-on microphone is used, test it ahead of time to be sure it actually works and that the volume is set at the correct level for the meeting room conditions.

- Make sure microphones are placed at convenient locations if audience participation is planned so that everyone can hear what is said.

Seating

Check the presentation room for safety features and for seating arrangement. Aisle space should be adequate, and exits should be available for emergency use. The seating arrangement should give every member of the audience an unobstructed view of the visual. White matte screens do better at an angle of 40 degrees or less, while beaded screens are not as effective for viewing at an angle. To view a projected visual, the most desirable seating area is within a 30-degree angle from the projectors for a matte screen and within a 20-degree angle for a beaded screen (Figures 24-11a and 24-11b).

The recommended minimum viewing distance is at least twice the screen width; the maximum should be no more than six times the screen width. A 4-foot-wide (1.2 m) image, therefore, could be viewed at a maximum distance of about 24 ft (6.3 m); a 6-foot-wide (1.83 m) image

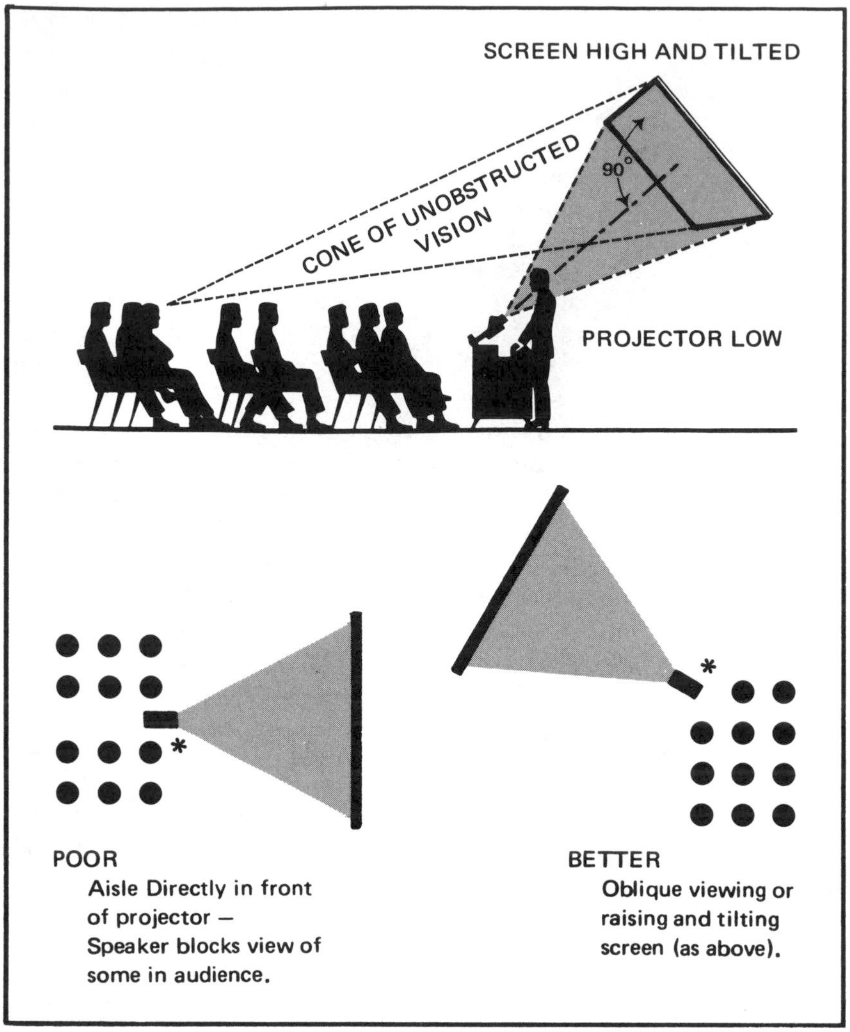

Figure 24-11a. To obtain optimum viewing area, especially when instructing large groups, tilt the overhead screen (*top*) or place it toward the corner of the room (*lower right*). Both placements reduce the possibility of the instructor and equipment blocking the vision of some viewers, as might be the case when the projector is located in front of the group (*lower left*).

could be viewed comfortably at a maximum distance of about 48 ft (14.6 m).

For meeting room or auditorium seating, allow 21 to 24 in. (53 to 61 cm) width for each chair and 36 in. (91 cm) for each row of chairs, or about 5 to 6 sq ft (0.46 to 0.56m²) per viewer. In classrooms or conference rooms, allow twice as much space.

Various seating arrangements work better for certain kinds of meetings. If a conference table is not available and the group consists of five persons or less, set up an overhead projector on a desk and project visuals onto a portable screen or against light-colored walls (see also Figure 24-2). Meetings with six to 12 people work well with a single center table. For groups of 20 and under, use a U-shaped table to encourage participation and discussion.

Tables arranged in a herringbone pattern are good when the presentation is mostly a lecture format, and the speaker wants to focus attention on talk and visuals. The tables provide work space, yet the participants can still see each other and interact. A more formal auditorium setting, or an amphitheater with elevated levels, probably provides little or no audience participation. In such settings, the speaker almost certainly needs a microphone and a screen and visuals large enough so the audience can see them clearly.

In some extremely wide rooms, the seats cannot be arranged so the audience sits within the recommended viewing angle. For these situations, project slides or overheads simultaneously on two or three screens, separated by at least 25 ft (7 m).

Trainer/Speaker Aids

To be most effective in a training session, trainers should take full advantage of attention-getting devices, such as pointers, easels, charts, teleprompters, props, and aids.

Pointers

A pointer, usually made of wood or aluminum, is a necessary item when a speaker is using a chalkboard or charts on an

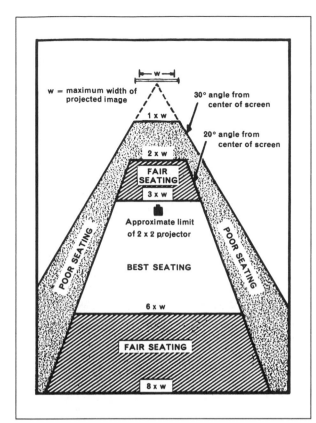

Figure 24-11b. The shading indicates both good and poor seating areas when ordinary (beaded) projection screens are used. A lenticular screen would widen the angle, which is measured from the center-line. The best viewing area is within 20 degrees of the projection axis for a beaded screen, and within 30 degrees for a matte screen.

easel. However, for overhead transparencies projected on a screen, a speaker can use a pencil or pen, laid on the transparency itself, to call attention to a word, phrase, or portion of a chart. Using a pointer allows the speaker to face the audience, while at the same time referring to the visual material. A speaker should not fidget with the pointer, however, since that distracts from the presentation.

If visibility is a problem, use a pointer with a fluorescent tip. In a dark or semi-dark room, a battery-operated or 110-volt flashlight-type pointer or a laser pointer will project a spot of light or a bright arrow onto a screen from a considerable distance. (CAUTION: Make sure no one in the audience looks directly at the source of the laser light to prevent injury to sight.) If the speaker must stand at some distance from the chart or screen, a high visibility electric pointer is useful, even in a fully lighted room. A telescoping, pocket-sized pointer is handy for trainers/speakers who must carry a pointer with them.

Easels
Portable aluminum easels are helpful for holding flip charts and paper pads. Be sure the easel is sturdy enough to hold materials and will not collapse during a speech. Some easels fold for use on a table.

To be effective easels should be used close to the audience. However, the impact of the display is often lost when audiences are large or are more than 20 to 30 ft (6 to 9 m) away from the board.

Charts
Well-chosen and well-prepared charts and diagrams allow the audience to "see" what the speaker is saying. Charts tell a story faster and more clearly than can an oral presentation alone. Trainers who use charts (rather than projected visuals) should practice showing them before their talks to make sure the charts appear in the correct sequence and that the text is readable. Trainers who fumble with the charts or switch them haphazardly distract from their message, losing any impact the information may have had. If trainers discuss material other than topics illustrated by the charts, they should cover the charts temporarily to focus audience attention on what they are saying.

Teleprompters
Teleprompters or other cuing devices can be used for dramatic or formal presentations in which trainers/speakers are required to follow a prepared script. Teleprompters require technical help to set up. Trainers must know how to use the equipment so their presentations will appear natural.

Boards and Pads
Chalkboards, marker boards, and paper pads lend themselves to small group training and provide inexpensive visuals. Each is an excellent tool for stimulating discussion. Participants can comment on points the trainer previously listed, and the trainer can quickly record student responses or suggestions.

Although chalkboards come in several colors, light green is considered standard. A dustless chalk should be used, preferably a strong, bright color for maximum visibility. Fluorescent chalks are particularly effective.

Marker boards have replaced many chalkboards. Colored felt-tip markers are used to write or draw on the polished white surface. Marks are erased with an ordinary cloth with little or no dust. One advantage to these boards is that they serve a dual purpose as a projection screen and marker board for small audiences.

Large paper pads, available at art stores or office supply stores, are easy to use. Clamped on a lightweight or folding easel, such pads are portable and always ready for use. A speaker can keep training material written on the pad or discard it after the session. Feedback or questions from students, recorded and saved on the paper, can serve as a nucleus of ideas for a future training session.

In brainstorming sessions, where the leader's objective is to stimulate quickly as many ideas as possible without detailed analysis or criticism, pads are of real value. As each chart or page is filled, it can be tacked to a bulletin board or clipped to a wire running along one side of the conference room. Group members then have a continuous record of what has been discussed. Latecomers can use these sheets to catch on to the discussion, and the speaker has a good record of the meeting.

Inexpensive rolls of white paper can be used for charts; occasionally newsprint "ends" are available at small cost from a local publisher. Even brown wrapping paper will do if bright-colored contrasting markers are used to write down the ideas. The speaker can write and read material while unrolling the chart like a scroll.

Flip charts, similar to paper pads, but usually prepared in advance, often are used in more formal meetings. Blank pages can be provided for on-the-spot notes. Portable units are easy to make and transport to different locations.

Magnetic boards also provide an inexpensive way to display materials. Trainers can make a magnetic board of either a spray-painted sheet metal plate or a steel-backed chalkboard. Magnetic boards are also available commercially. Small objects or cutouts mounted on small magnets or on magnetic tape can be placed on the board and moved at will.

Trainers often use this type of visual for training operators of vehicles such as forklift trucks. The mobility of the objects—toy vehicles or cutouts of trucks, together with the cutouts of aisles and loads—enables the instructor to give a realistic demonstration of safe practices.

Organizing Props and Presentation Aids

As a first step in preparing for a training session or speech, go through the talk and the media materials well ahead of time. Make a duplicate copy of the outline and in the margin mark items needed for the talk. Make a separate list of those items to ensure that nothing is left out.

Use a simple notepad or a preruled accounting notebook page with multiple columns to keep track of all requirements. Head the columns: "Have," "Need to reserve," "Need to buy," "Need to rent." Down the left side, list all required media props and presentation aids in the order in which they will be used during the talk. Don't forget to include items like chalk, erasers, markers, extra pencils or pens, etc. Next, check the appropriate column for that particular presentation aid. In this way, a speaker can easily see what has to be acquired.

Revise the list at least once before the presentation is scheduled, updating the status of each item. Use the far right column to indicate who will be responsible for seeing that the item is on hand. However, even if some tasks are delegated, the trainer is still responsible for double-checking that everything is ready and in the proper place well before the presentation.

Preparation and Rehearsals

Successful trainers and speakers never give presentations without preparing well in advance. The smooth, seamless talk that looks so effortless was undoubtedly rehearsed ahead of time, often more than once. By practicing their presentations, trainers/speakers are able to stay within their time limits and to deliver a more effective talk. When a trainer or speaker is confident in what he or she is doing and handles accompanying media well, audiences are free to concentrate on the message being delivered.

If a presentation includes several speakers, it is a good idea for all participants to show their materials to one another in advance. This is especially important if company executives or important visitors will be in the audience. This approach will help avoid duplicating information or media. Individuals may wish to revise their delivery or practice certain skills, depending upon suggestions they receive from other trainers/speakers.

EVALUATION

Evaluation helps trainers get the maximum benefit from the time and effort they have spent in creating and using media materials. They need to know how they are doing and what areas require improvement.

Effective evaluation involves more than handing out forms asking audiences how they liked a speaker or presentation. A well-designed evaluation strategy can yield a great deal of useful information. Such a strategy, planned in advance, should consider:

- What will be evaluated? At times, trainers want their audiences to evaluate media quality. Other times, they will ask for feedback on how well the media materials were used. A third question to consider is, "Did the materials and training session influence worker motivation or behavior?"
- Who will do the evaluation? The trainer? Audiences? The supervisor? Executives? Outside consultants? Establish who will evaluate the sessions.
- When will the evaluation be done? Trainers can lose thoughtful comments and feedback if evaluating becomes a routine fill-out-the-form exercise after each training session. Instead, consider polling individuals in the audience to keep interest alive.
- How will the evaluation be implemented? Will formal rating sheets be used? Will the audience fill out 3 x 5 index cards and drop them in a suggestion box? Do trainers want to be ranked "by the numbers" or are they looking for descriptive comments?

No one strategy or form is "right" for a particular organization or situation. By working together with others in the facility, trainers should be able to devise an effective evaluation strategy for their organization's needs. Also, they should evaluate the strategy itself, at least once a year, to see how it can be improved.

Creating an Evaluation Climate

If trainers want honest comments, they need to create a climate that encourages feedback and respect for each other's ideas. If trainers become defensive when criticized, or if workers believe those who make suggestions will be penalized, trainers will not get the thoughtful appraisal they need.

One useful approach which often works involves asking employees who say they don't like something to tell the trainer how they would make it better. By forcing them to focus on the positives, the trainer not only gets suggestions that might work, but reinforces the team feeling of "we're in this together."

Just as trainers do in a brainstorming session, they should accept all comments without allowing negative discussion on any ideas presented (see Chapter 23, Safety and Health Training). The purpose is to reinforce creative thinking and involvement, weaning the group from the "we've always done it this way" limitations.

Evaluating Media Materials

Criteria trainers can use as a starting point to judge the quality of media materials include the following:

Appearance

Were diagrams and graphs appropriate for the information being displayed? Were easy-to-read fonts chosen? Was lettering large enough to be seen easily by the audience, no matter where a viewer was sitting or whether he or she wore glasses? If more than one color or pattern was used, were they distinct enough so viewers could distinguish information easily?

Were the visuals focused properly? Were they appropriate for the message? Was the information they contained up to date? If "how-to" skills were shown, was the training paced slowly enough so that viewers could understand and follow the teaching?

Audio

If sound was part of the media presentation, was it clear and distinct? Was the volume level comfortable, but adequate? Was the sound consistent throughout the session? Was the sound appropriate for the visuals? Was the level of language appropriate for the audience and message? If technical terms were used, were they defined immediately?

For live presentations, were microphones free of annoying electronic feedback or distortion? Were there enough microphones, correctly placed, for all who spoke?

If a soundtrack was used in a second language, was the message accurately translated? Did the translation take into account any cultural differences or sensitivities?

Props

If chalkboards, flannel boards, hook and loop boards, or easels were used, were they placed so the audience could see them? Could each member of the audience see the material they contained easily?

Evaluating Media Use

Did the media help make the presentation message clear? Did they reinforce what was being taught? Were the materials well-suited to the presentation format, whether it was individual or small group training, computer-based training or simulation, discussion, lectures, or a presentation by a panel? Were the materials integrated into the presentation appropriately, that is, at appropriate places, and for an appropriate length of time?

Had the trainer rehearsed enough with the materials to use them well, and with confidence? Was the room comfortable? Was the seating arrangement satisfactory? Was lighting appropriate for the materials used? If projected visuals were used (overheads, slides, films, videos), did the person responsible for the projection handle the equipment properly and efficiently?

Evaluating Audience Reaction

Trainers can use the following questions as a checklist to evaluate audience reaction, adding their own questions to tailor to checklist for each presentation.

- Was the objective of the presentation motivational? If so, did the session stimulate or reinforce motives? Did the materials make a direct link to the core desire or identified need?
- Was the objective of the presentation skills improvement? If so, did the materials provide illustrations and examples of application? Did the presentation reinforce the message by also using learning activities?
- Was the objective of the presentation informational? If so, did the materials present facts and supporting material in a clear, logical structure? Was the pacing of the total presentation appropriate, so that the audience could acquire new knowledge?

SUMMARY

- Because people learn and retain material better when they use both sight and hearing, media have become indispensable tools in occupational safety, health, and environmental training.
- The selection of media depends on the role the trainer must play, audience size, cost of the materials, whether in-house or outside personnel will make the materials, and whether the materials justify the time and cost.
- Management must establish measurable training objectives tailored to a specific audience and message.
- Media should be designed to reinforce a message. Graphics used in media presentations include circle graphs, bar charts, line charts, and area charts. Companies can use computer graphics software to design their own materials. These visuals should be easy to read from all angles of a meeting room and from the back of the seating area.

- Projected visuals, films, videos, and computer-based training are three of the most commonly used media in training. A company can make its own materials or hire a service bureau or production company to create them.
- The most common computer-based systems include computer-managed training and interactive training, which uses conceptualizations and simulations to teach workers. Simulations allow students to practice high-risk procedures and safety techniques without incurring injury.
- Supplemental materials can enhance and support media and make a trainer's/speaker's presentation more effective and memorable.
- To create effective media presentations, trainers should do careful preplanning to be sure they have confirmed or checked such items as the meeting area, all supplies and materials, equipment, media materials, seating arrangements, and all safety concerns.
- The final step is evaluation of the training session. Trainers need to establish a climate that encourages open, honest feedback from participants. Media materials should be evaluated on the basis of their quality and their use.

REFERENCES

Periodicals

Audio-Visual Communications. Media Horizons, 50 West 23rd Street, New York, NY 10010.

Fast Forward. Association of Audio-Visual Technicians, Box 9716, Denver, CO 80209.

In Plant Video/AV Communicator. (bi-monthly, February through December), PTN Publishing Corp., 210 Crossways Park Drive, Woodbury, NY 11797.

Technical Photography. PTN Publishing Corp., 2210 Crossways Park Drive, Woodbury, NY 11797.

Books

Bruccoli M, ed. *Audio Visual Market Place*. New York: R. R. Bowker Co. (annual).

Bunyan, JA. *Why Video Works: New Applications for Management*. White Plains, NY: Knowledge Industry Publications Inc., 1988.

Carlberg S. *Corporate Video Survival*. New York: Macmillan Publishing Co., Inc., 1991.

Cole M and Odenwald S. *Desktop Presentations*. New York: AMACOM, Division of American Management Association, 1990.

Hansell KJ. *The Teleconferencing Manager's Guide*. White Plains, NY: Knowledge Industry Publications Inc., 1989.

Kerlow IV and J Rosebush. *Computer Graphics for Designers & Artists*, 2nd ed and rev. New York: Van Nostrand Reinhold Co., 1993.

National Safety Council, 1121 Spring Lake Drive, Itasca, IL 60143.
Occupational Safety and Health Data Sheets (available through NSC library).
Nonprojected Visual Aids, 12304–0564, 1993.
Photography for Safety, 12304–0619, 1991.
Posters, Bulletin Boards, and Safety Displays, 12304–0616, 1991.
Projected Still Pictures, 12304–0574, 1989.

Wiese M. *Film & Video Budgets*. rev. ed. Studio City, CA: Michael Wiese Productions, 1990.

Associations and Government Agencies

American Society for Training and Development, 1630 Duke Street, Alexandria, VA 22313.

Association of Audio-Visual Technicians, 2378 South Broadway, Denver, CO 80210.

Association for Multi-Image International, Inc., 8019 North Himes Avenue, Suite 401, Tampa, FL 33614.

International Association of Business Communicators, 1 Hallidie Plaza, Suite 600, San Francisco, CA 94102.

International Television Association, 6311 North O'Connor Road, LB 511, Irving, TX 75039.

National Safety Council, Safety Training Institute, 1121 Spring Lake Drive, Itasca, IL 60143.

U.S. Office of Education, Bureau of Adult and Vocational Education, 400 Maryland Avenue SW, Washington, DC 20036.

Women in Communications, Inc., National Headquarters, 2101 Wilson Boulevard, Suite 417, Arlington, VA 22201.

(Other data books and pamphlets are available from distributors and manufacturers of cameras, films, computer graphics software, and other visual media equipment. Trade journals in the fields of computers, visual media, photography, education, sales management, advertising, and training also contain excellent information.)

REVIEW QUESTIONS

1. Define medium and its purpose.
2. To be effective, what should training accomplish?
3. Identify three situations where media can help support training objectives.
   a.
   b.
   c.
4. What are some of the roles a trainer uses?
5. What does a trainer do as a facilitator?
6. How does management justify the cost of training?
7. Name four of the five factors to consider in deciding between commecial versus in-house training materials.
   a.
   b.
   c.
   d.

8. What is the basic purpose of supporting media?
9. What factor must be considered in designing color charts or graphs?
10. How many colors should one use in a visual?
11. What are some disadvantages to using slides in a presentation?
12. What personal protective equipment should a photographer use when working from heights?

25 Safety Awareness Programs

This chapter deals primarily with promoting and maintaining managers', supervisors', and employees' interest in safety. The chapter covers promotions (what an organization does to promote safe behavior and practices within its workplace) and campaigns (what it does to publicize safe behavior and practices to workers' families and areas served by the organization).

To promote and maintain interest in safety, the safety and health professional must inform and train key executives and managers about the safety program's objectives and how to achieve them. However, management must also demonstrate its commitment and interest and actively support a solid safety program. Only with top management leadership, support, and commitment will activities to promote employees' interest be successful. A well-promoted program involves both management and employees in safety and health activities. (See Meetings of Executives, later in this chapter.)

If top management seems uninterested in such issues, the safety professional should not assume that executives are indifferent or opposed to safety. Many times they may simply be unaware of the importance and basic benefits of an organized safety program. This chapter will discuss the following topics:

- the importance of promoting and maintaining interest in safety awareness
- objectives and benefits of safety awareness
- factors to consider when initiating a safety awareness program
- the role of safety and health professionals and line managers in promoting safety awareness
- the function of safety committees and observers in safety programs
- involving workers in quality and safety circles
- types of meetings and planning programs in safety promotion
- the uses of contests, awards, and competitions in promoting safety
- posters, displays, and other promotional methods for promoting safety
- the benefits of good public relations
- how organizations can make safety news
- planning and producing a safety publication.

REASONS FOR MAINTAINING INTEREST

Why is it necessary to maintain interest in safety (1) if the workplace has been designed for safety, (2) when work procedures have been made as safe as possible, and (3) after supervisors have trained their crews thoroughly and continue to enforce safe work procedures? The answer is simple: Even with these optimum work conditions, accident prevention basically depends upon the desire of people to work safely.

All hazardous conditions, unsafe practices, and loss-control problems cannot be anticipated. Employees frequently must use their own imagination, common sense, and self-discipline to protect themselves; they must think beyond the immediate work procedures to act safely in potentially hazardous situations when on their own.

Indications of Need for a Program

Various yardsticks indicate supervisor and employee attitudes toward accident prevention. These measures also indicate problems that the safety professional needs to address by creating a program to stimulate and maintain employee interest in safety.

- An increased rate of injuries, accidents, and near-accidents may signal that a program is needed. If such an increase cannot be explained through engineering methods, training, or supervision, then perhaps employees are forgetting or ignoring work rules, failing to stay alert, or taking chances. A program to develop and maintain their interest in safety will help to reverse this trend.
- If housekeeping is deteriorating, protective equipment is not being used, and guards are not being replaced, it is time to improve supervisory interest in safety awareness and to promote renewed and increased management interest in safety.
- Incomplete or missing accident reports also indicate decreasing supervisory interest and, perhaps, even a failure of employees to report minor accidents and injuries. Motivation to ensure better reporting is then in order.

Program Objectives and Benefits

A well-planned program, although it cannot be expected to do everything, can help workers become more committed to safety. For example, the program can:

1. Help to develop safe work habits and attitudes—but it cannot compensate for unsafe conditions and unsafe procedures.
2. Focus attention on specific causes of accidents, although by itself it cannot eliminate them.
3. Supplement safety training, yet it cannot be considered a substitute for a good training program.
4. Give employees a chance to participate in accident prevention activities, such as suggesting safety improvements in job procedures.
5. Provide a channel for communication and cooperation between workers and management, because accident prevention is certainly a common meeting ground.
6. Improve employee, customer, and community relations, because it is evidence of management's commitment to accident prevention (Figure 25-1).

The major objective of safety awareness is to maintain interest in safety to involve management and employees in preventing accidents. Usually, though, it is difficult to determine the degree of success such a program achieves.

Figure 25-1. A sticker like this one (actual size shown here) can be placed on surfaces to remind everyone of a company's continuous efforts to work safely.

This is because companies with these activities also have sound basic safety programs: working conditions are safe, employees are well trained and safety minded, and supervision is heavily involved in promoting safety measures.

However, one company that already had a good basic program attributed a reduction in its work injury rate to stepped-up efforts to maintain interest in safety. It was based on an idea submitted by an employee: Each month candy bars were distributed to injury-free employees. Wrapped with some of the candy bars were slips that could be traded for free pairs of safety shoes. Another company gave each employee who had an injury during the month a package of gum with the slogan "Something to chew on" along with a friendly safety message and wishes for an injury-free future.

One poultry-processing company was able to boost its safety program by giving away "big ticket" items. Prizes ranged from appliances up to an all-expense paid trip to Hawaii for eligible employees and their guests. An expenditure of $40,000 netted the organization a 29% reduction in lost workday cases and a reduction of $450,000 in workers' compensation costs.

SELECTION OF PROGRAM ACTIVITIES

Modern advertising and marketing techniques have much in common with those used to "sell" safety. Just as steady and imaginative sales promotion is needed to sell most products and services, safety also requires constant, skillful promotion. This approach makes the basic elements of accident prevention easier for workers to understand and accept.

Many safety and health professionals make the common mistake of instituting promotional activities with no preliminary planning or objectives in mind. They may use films, videos, or posters solely because they are available at low cost. These managers are doing their employers and co-workers a disservice. The same is true of safety and health professionals who spend a disproportionate amount of time on committee work, contests, or "homemade"

visual aids because of their bosses' or their own personal interests. This haphazard approach cannot be used to develop an entire safety program. Such cosmetic activities are only part of safety in the workplace. They offer no substitute for a sound, well-planned program based on managerial involvement, responsibility, and accountability.

Basis for the Program

For a program to be effective in maintaining interest in safety, it must be based on employer and worker needs. Management should select activities not simply because they will be popular, but because they will yield results. To develop suitable activities and promotional materials, management must know the needs of supervisors and employees.

To find out what employees really thought of their safety program and just how interested they were, one company inaugurated a safety inventory plan. Each year after the regular stock inventory had been taken, safety inventory cards were distributed to all employees—salaried workers as well as hourly.

The cards were distributed by supervisors who asked employees to take stock of their jobs and environment with regard to safety. The following year's safety program was planned on the basis of the questionnaire returns, which ran better than 90%. Many suggestions for improving the safety program were received and subsequently put into practice.

Factors to Consider

When planning safety awareness activities, several major factors should be considered.

Company Policy and Experience

If a company ordinarily uses activities such as committees, mass meetings, award programs, and contests in areas other than safety, then these should be used for the safety program. It is generally unwise to spend much time on activities that are unknown to company policy and experience, unless it is believed that a new "sales pitch" is justified.

On the other hand, if supervisors and employees are overinvolved in committee work and activities such as sales promotion, quality control, and tool damage programs, similar activities for safety may be lost or considered burdensome. Other approaches might prove more effective. However, if management believes safety is important, effective measures will be taken regardless of the burden.

Once a promotional program is under way, it should not impair other aspects of the accident prevention program or other company activities. For example, safety meetings should not take much more time than meetings for quality control, sales, or industrial relations.

Budget and Facilities

Budget considerations always affect plans for a safety promotion program. At first the program will require

extra effort, time, and money. The safety and health professional can justify this expenditure, however, as an investment that will produce direct and indirect benefits from both a financial and employee-relations standpoint. If the program is to be successful, management must allocate sufficient funds to carry it out.

In selecting program activities, line management and the safety and health professional should consider the available facilities. Films, for instance, require not only a projector and screen, but also a darkened room free of background noise. In many companies, safety-oriented facilities, publications, or services may be obtained through the industrial or public relations departments. Public relations also may be a valuable source in helping to plan promotional activities. The Safety and Health Department or the Customer Service Department can provide information as well.

Types of Operations

The nature and organization of company operations affect the choice of activities and materials for maintaining interest in safety. When operations are widely scattered and diversified, as in the construction, railroad, marine, motor transport, sales, and air transport industries, the job of selecting and disseminating safety information becomes more complicated.

In decentralized operations, the safety and health professional must help management choose materials, such as publications, videos, and films, that can be used easily in the field. Local supervisors must be able to conduct meetings, present material, and handle posters.

The safety and health professional must also consider the needs of employees involved in widely different kinds of work at far-flung locations. For example, a poster program may be used as one means for maintaining interest in safety at each outlying location. If so, the supervisor can designate a trustworthy employee to receive posters and take care of their distribution and posting. This individual can also be sure that poster boards and display cases are kept clean, attractive, and free of extraneous paper.

Another method is to equip a trailer with permanent displays and transport the company safety story to remote locations. Rear-screen projectors (35mm) and videotapes are also well suited for use in scattered locations and in the field. A relatively new and inexpensive method of communication is the video camera. Videos can be used to communicate a variety of short subjects to distant work sites. This method will ensure that the same message is communicated in the same way to all employees. In addition, video cassette players are now available with a continuous play mode. After a tape is shown, the VCR automatically rewinds the tape and replays it as many times as desired.

For many organizations, the types of educational materials used for different groups of employees may vary considerably. For instance, movies may be suitable for the day shift when many employees can be taken off the job, but would not be appropriate for a small night shift when no one can leave the job site even for a short time.

However, night crews and maintenance employees should not be overlooked. Their work is as vital as that of other employees to the overall accident prevention effort. Programs for them will have to be planned to fit their specific requirements and time constraints.

Types of Employees

The types, backgrounds, and educational levels of employees must be considered when choosing safety promotion activities. For example, migrant workers frequently do not receive sufficient job training. Material for these employees should present basic safe practices for their jobs in brief, easily understood form and in the appropriate languages when their English skills are still poor. Management can use local colleges or universities as a resource to assist in translating material into other languages. Be aware that different dialects, even in the same language, may result in the safety message being misunderstood. A similar approach should be used with temporary workers or those assigned from union halls. Employees who have difficulty understanding or reading English need visual material. Material in Spanish is available from the Inter-American Safety Council (see Appendix 1, Sources of Help).

Basic Human Interest

If employees seem bored or uninterested in safety activities, an extra push is needed to involve them. One approach is to base promotional activities on employee interests, such as bowling or fishing, as an incentive.

The wise choice of these activities depends on an understanding of basic human needs and emotions, such

as those listed in Figure 25-2. The safety activities suggested for employees should have general appeal.

Other Considerations

The tasteful use of employee human interest materials featuring children, animals, and cartoon figures along with activities and contests fostering good-natured competition play a big part in most companies' safety promotion programs. These elements can be as effective in safety promotion as they are in sales and marketing. Their wise use should not compromise company policy or offend anyone.

In one instance, an Ohio company capitalized on the interest of many people in wagering. Any employee in a carpool who forgot to buckle his or her safety belt had to pay for lunch or dinner. The idea proved so popular the company decided to expand the program by using posters and announcements.

Ideas for maintaining interest often use humor. The "light touch" is essential and should be good natured. Ridicule should never be used; it only arouses resentment (Figure 25-3).

A positive, constructive approach is generally better than a negative approach. However, the latter sometimes is preferable if it is more dramatic. A picture showing the consequences of an accident, such as a fall on a slippery floor, may have a greater impact than one depicting the safe practice that could have prevented the accident.

Variety is Essential

Often a simple change, such as a different type of contest, a bulletin board redesign, or revising the format of safety meetings, can renew workers' interest. The activity itself may not be more effective, but its new form stimulates thought and discussion. Although safe work practices should become routine, their presentation should not.

BASIC HUMAN INTERESTS AND CORRESPONDING ACTIVITIES

Basic Interest Factors	Ways to Use These Factors
Fear of painful injury, death, loss of income, family hardship, group disapproval or ridicule, supervisory criticism.	**Visual material:** emotional or shocker posters, dramatic films, pictures and reports of serious injuries on bulletin boards, in company papers.
Pride in safe workmanship, in good records, both individual and group.	**Recognition** for individual and group achievement; trophies, personal awards, letters of appreciation.
Recognition: desire for approval of others in group and family, for praise from supervisors.	**Publicity:** photos and stories in company and community papers, on bulletin boards.
Participation: desire to be "one of the gang," "to get in the act."	**Group and individual activities:** safety committees, suggestion plans, safety stunts, campaigns.
Competition: desire to win over others, such as shown in sports.	**Contests** with attractive awards.
Financial gain through increased departmental or company profits.	**Monetary awards** through suggestion systems, profit-sharing plans, promotions, increased responsibility.

Figure 25-2. Ways to put six basic human interest factors to effective use in promoting safety. Basic needs and desires that motivate people are shown on the left. The right column lists direct appeals safety promotional programs can make.

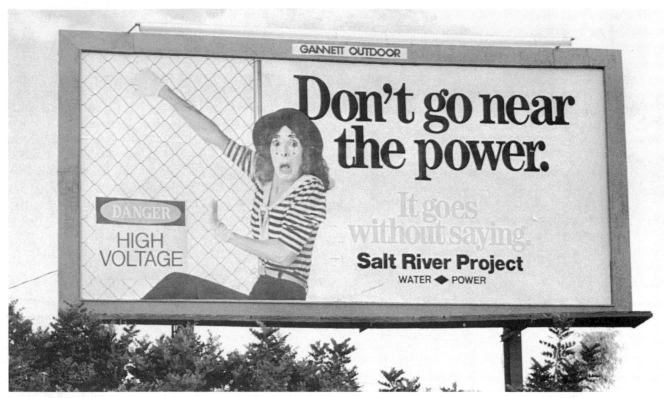

Figure 25-3. The return on investment for public safety advertising is high, according to the Salt River Project. This billboard is one of 22 in the area that carry a new message each month and reach an estimated 76% of adult Valley residents. The effort costs between $50,000 and $70,000 and has been in effect 10 years. The advertising contributes to halving the insurance premium in five years. (Reprinted with permission from Salt River Project, Arizona.)

Activities that require employee participation generate more interest than do those that involve only seeing and listening. Many companies have worked with the National Safety Council to produce films, videos, and movies in their plants. These firms report an upsurge in safety interest when some employees are asked to act in a video or film emphasizing safety procedures or practices that may have seemed routine.

In some instances, management urges employees to submit suggestions for equipment guards or to help in the selection of personal protective equipment. Companies find that these workers are more inclined to use the guards and the personal equipment than they would be if they had not been asked for their opinions. For the same reasons, including employees in the process of drafting safety rules encourages compliance on the part of those workers who participate in the project. Companies have also found that workers who serve on a safety and health committee tend to develop increased awareness of safety responsibilities.

STAFF FUNCTIONS

Creating and maintaining employee interest in a line management responsibility falls to the safety and health professional. This individual is responsible for assisting line management in planning the safety program. Line managers and supervisors are responsible for planning, organizing, and executing the program.

Role of the Safety and Health Professional

Safety and health professionals coordinate the program, supply the ideas and inspiration, and enlist the support of management, supervision, and employees. Programs should be designed to involve both management and employees. Employees like to receive awards; managers like the public-relations aspects of presenting them.

Safety and health professionals may work with local safety councils, chapters of the American Society of Safety Engineers, and other civic or technical groups interested in accident prevention. They can gain much by attending the National Safety Congress and Exposition and regional safety conferences. Here they learn what other companies are doing and how those ideas can be translated into practical activities for their own organizations. After participating in roundtable discussions, listening to speakers, and meeting people with similar interests, safety and health professionals often return to their jobs with renewed enthusiasm.

Because they frequently are called upon to address groups, safety and health professionals should be able to present their ideas clearly, effectively, and convincingly. Polishing public speaking skills will also help when dealing with people on a one-to-one basis. Safety professionals should know when and how to use visual aids.

Showmanship tactics, however, should be used only with considerable discretion. If they backfire, they can hurt the credibility of the safety and health professional, the company, and the safety program in workers' eyes.

Safety and health professionals should help educate line management in two areas. First, managers should learn how to keep working conditions as safe as possible. Second, they must know how to motivate workers to follow safe procedures consistently, as a part of good job performance.

A vast amount of program material useful to the safety professional is available through National Safety Council publications. Safety and health professionals also are invited to submit interesting data, difficult problems, or "gimmicks" of any nature to the Council for help in solving safety problems and to share ideas and solutions with others. The Council has information on every phase of safety, gleaned from the experience of members in various industries. One company's solution to a problem may prove invaluable to other firms.

Role of the Line Manager

The line manager is the key person in any program designed to create and maintain interest in safety. This is because the manager is responsible for translating management's policies into action and for promoting safety activities directly among the employees. How well this responsibility is met will determine to a large extent how favorably employees will view safety activities.

Ranking managers must learn that under existing laws they are directly accountable to their organization and society for their employees' safety. Management, with the safety professional's assistance, must see that supervisors receive adequate safety training.

The supervisor's attitude toward safety is a direct reflection of upper management's attitude towards safety and is a key factor in the success not only of specific promotional activities but also of the entire safety program. Supervisors look to upper management and employees look to the supervisor for leadership. Line managers, including supervisors, who are sincere and enthusiastic about accident prevention can usually maintain employee interest. Conversely, if line managers give only lip service to the program or ridicule any part of it, their attitude can cancel any good that the safety and health professional may be able to accomplish.

Many supervisors are reluctant to change their modes of operation or to accept new safety engineering ideas, much less to enthusiastically support contests, safety stunts, committee projects, and other activities used to promote and maintain interest in safety. It is line management's task to sell these supervisors on the benefits of accident prevention. They must be convinced that promotional activities are not "frills" but important projects

that can help to prevent injuries. Line management's wholehearted cooperation is essential to the success of the entire program. It can be pointed out that a successful program makes management's job easier and less time consuming.

One way line managers can support safety is by setting a good example and educating their workers in safe practices. They must follow safety rules and procedures, and wear safety glasses and other personal protective equipment whenever they are required. Teaching safety is also an important function of supervisors. To be successful in this area, they cannot depend upon safety posters, a few warning signs, or even general rules to do their teaching for them. Supervisors themselves must first be trained to teach if they are to be competent in this area. (See Chapter 23, Safety and Health Training, for details on training courses and techniques.)

When enforcing safety and health rules, line managers should not suddenly adopt a "get tough" approach after being lax. They should be consistently firm and fair all along. Otherwise, workers who have the impression the supervisor cannot recognize unsafe conditions and practices or simply does not care will not take the new tougher line very seriously.

Supervisors should be encouraged to take every opportunity to exchange ideas on accident prevention with workers, to commend them for their efforts to do the job safely, and to invite them to submit safety suggestions. They can also be most effective in relaying personal reminders on safety to employees. Such reminders are particularly appropriate in the transportation and utility industries, where crews are on their own from terminal to terminal.

Line managers, including supervisors, should request and receive help from the safety department. Help may also come through correspondence, educational materials for distribution, and personal visits from safety personnel. Supervisors should also receive adequate recognition for independent and original activity.

SAFETY COMMITTEES AND OBSERVERS

Various types of safety and health committees have many different functions. (For further details, see the Council publication, *You Are the Safety and Health Committee*.) However, the basic function of every safety committee is to create and maintain interest in safety and health and thereby help to reduce accidents.

Some organizations prefer other types of employee participation to formal safety committees. This is because they feel safety and health committees require a disproportionate amount of administrative time, generally tend to pass the buck, may stir up more trouble than they are worth, and may be a scapegoat for supervisors who want to unload their responsibilities.

The answer to these objections is not to abolish the committees but rather to reexamine their duties, responsibilities, and methods of operation. Such analysis often can lead to constructive changes enabling a committee to fulfill its original objective—that of stimulating and maintaining interest in safety.

Committee membership should be rotated periodically. This ensures a fresh perspective and also increases the number of employees who are trained to look at operations with safety in mind.

Involving employees in safety inspections, either alone, as observers, or as part of a formal safety and health committee has the same basic objective: to get more employees actively involved and interested in the safety and health program. Planning, organizing, publicizing, and following definite procedures will streamline the work of both committees and observers and help ensure effective results.

QUALITY CIRCLES AND SAFETY CIRCLES

Hazard recognition and control is one area in which employee involvement produces substantial results. The following discussion is taken from the Indiana Labor and Management Council's booklet *Worker Involvement in Hazard Control* (see References), used with permission.

There are two reasons for the popularity of involving employees in hazard recognition and control. The first is management's desire to use all available resources to increase productivity and quality in the face of growing competition. The second is management's understanding that employees want to accept new challenges and to participate in activities that affect their work life.

Popular forms of worker involvement programs are quality and safety circles. A quality circle is a group of employees, performing similar work or sometimes varied work, who meet weekly. In the meetings, they learn about and apply basic techniques to identify problems within their area(s), analyze them, and recommend solutions to management. In some instances, circle members discover hazards during their analysis of other plant problems and make excellent recommendations for their control.

Safety circles are a type of quality circle used by some companies to reduce the number of accidents and injuries by keeping safety and all its important features foremost in the minds of the employees. This implies a change in the employee's role from passive to active, while management's role becomes less negative (fewer don'ts) and more positive (more do's).

In many situations, safety circles are established on a plant-wide and departmental level. Safety circle meetings are held monthly for approximately one-half hour to one hour. Each circle meeting is usually preceded by a presentation explaining the successes and problems that the safety circle team experienced during the preceding month. The circle reviews a breakdown of all injuries, including first-aid cases, as a means of measuring safety progress and as a means of pinpointing trouble areas. Usually each member of the safety circle is assigned a specific

responsibility to review. Companies employing the safety circle concept report an improvement in their accident and injury experience.

MEETINGS

Safety and health meetings may be conducted for managers, supervisors, employees, or other groups. In every case, the purpose of the meeting is to stimulate and maintain interest and commitment in safety and health issues. If meetings fail to achieve this, their format or content should be changed to make them effective or they should be discontinued and a new approach taken.

A formal agenda for all attendees should be distributed prior to the safety meeting. An agenda helps to focus attention on specific issues, and avoids wasting time on nonrelevant topics.

Types of Meetings

The following types of safety and health meetings commonly arouse and maintain interest in accident prevention (see also Chapter 23, Safety and Health Training).

1. Meetings of operating executives and supervisors to formulate policies, initiate and maintain a safety and health program, or plan special activities.
2. Mass meetings of all employees, sometimes including families, or even the entire community to serve special purposes such as launching a major new program or contest and "selling" safety to everyone affected by accidents and injuries.
3. Departmental meetings to discuss special problems, plan campaigns, or analyze accidents.
4. Small group meetings to plan the day's or week's work so that it can be done safely, to discuss specific accidents, or to review safety instructions.

Meetings of Executives

When a safety and health program is inaugurated, it is especially important that the chief executive officer (CEO) of the company or top plant manager call a meeting to announce the general accident prevention plans and policies to all line managers and other operating executives. Under some labor contracts this must be, or should be, a joint union-employee announcement. If these persons meet at regular intervals to discuss operating problems, this announcement can be made at a regular gathering. Otherwise, the CEO or manager should call a special meeting. Periodic safety inspections by executives help to lend credence to management's interest in safety.

After this first meeting, the group may hold sessions periodically (usually monthly) to evaluate and coordinate the safety program, to check on the progress being made in accident prevention, to appraise proposed activities, to set policies, and to make decisions. It is also desirable for this group to review and/or investigate all fatalities, multiple amputations, and other serious injuries.

Departmental meetings serve many safety and health purposes. They may be used to discuss the company safety program so that employees will better understand its policies and procedures, to provide information about accident causes and accident types, or, in a purely inspirational manner, to create an awareness of hazards and a desire to prevent accidents. Departmental meetings also can be used to review and/or investigate all injuries involving lost workdays or restricted work.

Many departmental safety meetings are held monthly and most are conducted by the supervisor. The safety department usually assists in planning and provides materials, such as visual aids.

The program for a departmental meeting may include:

1. Reports on injuries, illnesses, or other concerns in the department since the last meeting, a safety inspection in the department, and the department's standing in a contest. (The total time spent on reports should be limited so that this part of the meeting does not become tiresome.)
2. Discussion by the supervisor of specific safety practices or unsafe conditions that need to be improved.
3. Talk, demonstration, or audiovisual presentation on an appropriate accident-prevention subject. The speaker may be the supervisor, a department employee, the company safety and health professional, an outside expert, or a company executive.

Departmental meetings give the supervisor an opportunity to point out the dangers of particular unsafe practices or conditions. By condemning those practices, the supervisor sets a good example and lets workers know they are to follow the same rules and procedures. Most workers welcome an opportunity to share their safety ideas in these meetings.

At the conclusion of departmental meetings, the supervisor may prepare written reports for the plant (or company) safety committees and managers.

Supervisors also can hold small group meetings with people doing similar work at or near the workplace. The supervisor may discuss the causes of a recent accident workers have witnessed or learned about. Employees should be encouraged to join in the discussion with the goal of reaching some conclusions about how the accident might have been prevented.

The supervisor may present a problem that has developed because of new work or new equipment. Again, all should participate and offer their views.

At times the supervisor may present a film or chart talk on a subject related to the work of the group members. Other audiovisuals such as models or exhibits may be used. Safety devices or pieces of equipment or material may be brought in and discussed.

"Production huddles" are instruction sessions about a specific job that include safety information. Such meetings are particularly useful with maintenance crews when an unusual job is about to start. The plans for doing the job safely and efficiently are discussed and a procedure is agreed upon. Public utility line crews who use this type of meeting call it a tailboard or tailgate conference. Before starting a job, the crew gathers around the truck and discusses the job, laying out the tools and materials they will need and agreeing upon each person's tasks.

A particular advantage of small group meetings is that they provide excellent opportunities for presenting all types of information, including safety information, directly to employees. They also stimulate an exchange of ideas that can benefit the accident prevention program. To be successful, each safety and health meeting must have a tangible message, imaginative presentation, opportunity for audience participation, and a conclusion that spurs action toward an attainable goal.

Mass Meetings

Mass meetings are held for special purposes, such as launching a contest or an award program, presenting awards, introducing new equipment, explaining a change in company policy, or celebrating an exceptionally fine safety record with an event such as "safety day."

In companies with plants in different cities, a top executive may call a meeting of employees during a plant visit. The talk may cover safety as well as other subjects. One company president makes an annual round of plants with the safety director and speaks at a safety rally of all employees at each plant.

Under certain conditions, particularly in smaller communities, large meetings can be held in a local theater or public hall. These meetings require fairly elaborate arrangements and considerable publicity to ensure good attendance on employees' part.

A mass meeting in a public hall using the "family safety night" theme allows not only employees to attend but also their families and friends. In addition to a presentation or speech targeted at the program theme, management should provide for some other entertainment. Often good talent can be found right in the plant or shop.

Mass meetings afford an excellent opportunity to use an outside speaker who can talk with authority and in a crowd-pleasing manner on general accident-prevention work. Such speakers can be found in nearby plants, insurance companies, city administrations, automobile clubs, or community safety councils.

If management wants to include movies, videocassettes, or slide shows relating to accident prevention at the meeting, they can find a suitable selection from those available through the National Safety Council or regional film service organizations, whose locations may be obtained from the Council.

Planning Programs

Making the safety meeting interesting is of the utmost importance. Speakers should not complain about or scold workers on their job performance. Talks should be brief and start and end on time. The subject matter should be studied in advance to make sure it is pertinent and is not a repeat of other recent talks.

Large occasional meetings need even more preparation and timing than do small meetings. Speakers, including company executives, should review their remarks with the meeting planner to ensure they will add to the desired purpose. Films, videos, and other visuals should also be reviewed in advance.

Persons responsible for employee meetings should analyze them periodically to determine whether they are accomplishing their purpose. It is all too easy for meetings to degenerate into dull routine. Only continual effort and planning will prevent this trend.

A plan of action to develop a successful safety and health meeting includes these points:

- Prepare in advance. The preliminary arrangement determines the results. Do not conduct a meeting without preparation.
- Select a major topic. Make it timely and practical— one that the group can discuss.
- Obtain facts and figures. Be sure they are correct and complete. Prepare a visual, such as a simple chart or table, whenever possible.
- Map the presentation. Decide on the best way to present the subject of the meeting. Try to anticipate the group's reaction and questions. Outline results you hope to accomplish.
- Set a timetable. Allow adequate time, but set a reasonable limit.
- Have an agenda distributed in advance.
- Be sincere. Managers' sincerity and interest in employees' welfare must be unmistakable.
- Introduce the topic. Tell in simple terms what the meeting is all about. Use a punch line or other short, to-the-point lead-in.
- Present facts, arouse interest. State pertinent facts in an interesting manner.
- Promote group discussion. Ask questions that cannot be answered "Yes" or "No." Prompt members of the group to think individually and collectively. Let them talk.
- Agree on some action. Try for group agreement on methods of correction and improvement. Write these down.
- Summarize the meeting. Review briefly what has been discussed and decided. Follow up in the various departments in writing, if possible.

CONTESTS

Management and supervisors should always keep in mind that contests (such as housekeeping contests, interdepartment contests, etc.) are not substitutes for management interest, safe procedures, and "built-in" safety. Nevertheless, although a successful accident prevention program depends on good management, good training, and efficient operations, some special effort may be needed to maintain worker interest in safety. Moreover, the interest value of contests has direct bonus values in good publicity and improved employee morale.

Therefore, a competition should be held only after management has taken the basic steps in a safety program: developed a policy statement, adopted a recordkeeping system, safeguarded equipment, and installed a first-aid department. Such substantial demonstrations of management's interest, sincerity, and responsibility greatly help win the active participation of supervisors and workers in a contest. However, a contest serves only as part of a safety program, not as a substitute for one.

Purpose and Principles

Safety contests are operated purely for their motivational value. A contest that does not motivate workers to be more safety minded is worthless. In most contests, the competing groups are departments of the same plant or divisions of the same company. This type of contest is useful when investigations of accident injuries reveal that most of them are caused by employees' unsafe practices. If accidents are the result instead of unsafe conditions or a combination of unsafe practices and unsafe conditions, the contest should be reconsidered. They might well focus negative attention on one of a company's most conscientious, safety-minded employees injured through no fault of his or her own.

Generally, contests are based on accident experience and are operated over a stated period. A prize is offered to the group attaining the best record according to the contest rules. Contests have been important almost from the time of the first safety programs. Over the years, companies have developed a fairly well-established set of operating principles:

1. A contest should be planned and conducted by a committee representing all competing groups.
2. Competing groups should be natural units, not arbitrary divisions.
3. Methods of grading must be simple and easily understood.
4. The grading system must be fair to all groups.
5. Awards must be worth winning and generate interest among employees.
6. Good publicity and enthusiasm are important.

Contests may run for various periods—from a few months to a year. Those who recommend longer periods believe that if workers are kept on their toes for a longer time, safe working practices are more likely to become a habit. Some safety professionals, however, believe that greater interest can be aroused and maintained during a short period and therefore prefer short and more frequent competitions. Management should devise contests of different lengths to see which is most effective.

A safety contest stock certificate idea was developed by the Maxwell House Division of Kraft General Foods and ran for one year. For each week of the contest that a department worked without a disabling injury, employees received a stock certificate worth 50 cents. Dividends were paid on this stock at the rate of 10 cents for the first 1,000 consecutive safe hours worked; 25 cents for the first 10,000 consecutive safe hours worked; 50 cents for the first 50,000 consecutive safe hours worked; and one dollar for the first 100,000 consecutive safe hours worked. This meant that if a department worked 100,000 consecutive workhours without a disabling injury it received a total of $1.84 in dividends for each share of stock held.

There were penalties, however. If a disabling injury occurred in a department, it was penalized one month's stock earnings. This meant that during that particular month the department could not be awarded any stock or dividends. It also meant that if an accident should occur in the latter few days of the month the department would lose all dividends and stock certificates previously issued for that month.

Injury Rate Contests

In a contest based on injury rates, the measure of safety performance is the Occupational Safety and Health Administration (OSHA) incidence rate (described in detail in Chapter 8, Injury and Illness Record Keeping and Incidence Rates).

Injury rate contests carry with them an inherent risk of possible abuse and must be closely monitored to ensure honesty. All injuries must be reported. These contests tend to put peer pressure on employees not to report their injuries, to bring administrative pressure to bear on those doing the recording not to count certain injuries, and to tend to make management focus attention on the contest rather than on the safety program. If these conditions arise, the entire safety program is discredited.

Contests should not be based upon injury severity, because severity data cannot be determined promptly. Also, the classification is frequently a matter of chance and contributes little to devising accident prevention measures. Nor is using a combination of frequency and severity classifications helpful, for the same reasons. Finally, contests should not be based on reducing the number of reported first aid cases; people will simply stop reporting such injuries.

In-Plant Contests

To be effective, a contest must run for a specific length of time. The contest will be more successful if the campaign is launched with advance publicity and fanfare. Often the president or other high official makes the original announcement, presents awards, and otherwise lends prestige to the contest.

Competition, if properly organized, can do much to develop teamwork in safety. Some employees who may give little thought to their own work habits can be influenced to cooperate with their fellow workers if they know that their actions can either discredit their department or "team" or result in a payoff for safe work.

Council Award Programs

The National Safety Council firmly believes in the value of award programs for maintaining interest in employee safety and health. The award programs are available to employer members, and some include participation of nonmembers through cosponsor agreements with trade associations.

The Occupational Safety/Health Award Program is a noncompetitive recognition program. Awards can be earned by organizations achieving a perfect safety record—no deaths or cases involving days away from work. Organizations with nonperfect records may also earn recognition if they achieve the criteria for a significant reduction of their incidence rates. There are four levels of awards:

- Award of Honor
- Award of Merit
- Award of Commendation
- President's Citation.

The Occupational Safety/Health Contest is a competitive award program. Participants are able to compete in one or more of 20 different industry contests. Each industry contest is subdivided into divisions according to the products being produced. Whenever possible, organizations of like size are grouped together.

Awards are given based on rank within each division. The site having the lowest incidence rate for cases involving days away from work is ranked first. First, Second, and Third Place awards can be earned.

The Council designated the Safe Worker Award Program to reward and encourage the accident-free performance of individual employees. Each employee is awarded a Safe Worker lapel pin or other incentive for each 12-month period during which no cases of days away from work are incurred. The employee continues to earn an incentive item for each additional year of safe work achieved, striving for the pinnacle of 45 years.

The National Fleet Safety Contest is a competitive award program comparing fleets from the same industries, having similar vehicles, or of the same size. The program is open to all National Safety Council members and members of cosponsoring associations whose fleets operate in the United States and Canada.

Data from contestants are collected on a monthly basis from January 1–December 31 of each year. This data includes the cumulative number of miles driven, the cumulative number of accidents experienced, and the number of vehicles in the fleet. From this data, an accident rate is developed and used as an initial means of comparison.

Outlined by the National Fleet Safety Contest Rules, the number of fleets within the following awards are presented: First Place plaques, Second Place plates, or Third Place plates. Two noncompetitive awards, Perfect Record and Certificate of Achievement, can also be earned by participating fleets.

The National Safety Council Safe Driver Award is the recognized trademark of professional drivers designed to prevent accidents and promote safety in the professional driving industry. Member organizations of the National Safety Council enrolled in the Motor Fleet Safety Service, or the School Transportation Safety Service, are eligible to participate in this program.

Companies certify drivers who have driven a continual number of years without a preventable accident. Through this certification, the driver receives a wallet-sized certificate denoting the number of years of safe driving service. Other incentive items, such as lapel pins and patches, may be purchased to signify this accomplishment.

The Safe Driver Award rules spell out what is expected of professional drivers in the way of safety performance—driving a motor vehicle without having a preventable accident. The award rules constitute a yardstick by which drivers can measure their own performance and by which supervisors can measure the performance of individual drivers.

The 'XYZ Company' Contest

The "XYZ Company" is engaged in steel fabrication and erection. Rules for one of its safety contests are given in Figure 25-4, but some explanation is needed.

Rule 1

Although contests can run for any period of time, the company chose a one-year period. This was done for two reasons. First, the period allows development of safe working habits—the useful objective of the contest. Second, because some of the departments were relatively small, the one-year period eliminates random (chance) factors from influencing an individual department's experience record.

Rule 2a

Hazards in steel fabrication and erection vary greatly; therefore, fabricating units compete in one division and erection units in another. Operations are similar enough to have a common basis for determining standings—see Rule 4.

RULES FOR THE 'XYZ COMPANY' SAFETY CONTEST

Rule 1. The contest shall begin January 1, and end December 31, 19 . . . .

Rule 2. The contest shall consist of two divisions.
    a. Fabricating units shall participate in Division I.
    b. Field erection departments under the direction of a superintendent shall participate as separate units in Division II, which shall be divided on the basis of size into two groups: Group A shall consist of the five erection units working the largest number of man-hours, and Group B shall consist of all other units. Units shall be tentatively grouped by size during the first three months, and the final classification shall be made on the basis of total man-hours worked at the end of four months. No further changes in size groups will be made after April 30, 19 . . . .

Rule 3. Recognition awards shall be:
    a. Trophies to the winners in Division I and Groups A and B of Division II.
    b. Engraved certificates to plants and erection units ranking second and third in Division I and in Groups A and B of Division II.

Rule 4. a. The winners shall be the contestants having the lowest weighted incidence rates.
    b. In the event that two or more contestants in any classification have had no chargeable injuries during the contest period, the winner shall be the contestant who has worked the largest number of man-hours since the last chargeable injury.

Rule 5. All injuries and illnesses resulting in death or days away from work shall be counted, as defined by OSHA record-keeping requirements.

Rule 6. A sum of $50.00 shall be presented to the units that have had no lost workday cases during the first six months of the contest or have reduced their incidence rates for the first six months 50 per cent in comparison with the rate for the preceding six months' period. The award shall be divided into prizes of $12, 10, 8, 6, 4, 2, and eight $1 prizes and raffled to employees. No employee may win more than one prize.

Rule 7. Standings shall be compiled monthly and published in a bulletin that will be distributed to all managers, superintendents, and foremen.

Rule 8. All questions pertaining to the definitions of injuries and rules shall be referred to the Contest Committee, whose decisions shall be final.

Rule 9. Awards shall be presented at an appropriate ceremony to be announced at the end of the contest.

Figure 25-4. A typical set of rules for a safety contest. See text for a point-by-point discussion.

Other companies may find that hazards differ sharply from one plant or operation to another, and the similar units may be too few to group. Such plants or operations may compete on an equitable basis in several ways.

- The participant achieving the largest percentage reduction in its frequency rate in comparison with a base period, such as the previous year, may be the winner. Because each unit competes only against its past record, this method provides a fair basis for comparison of rates.
- The workers' compensation insurance rates for different types of units in the same state have been used to establish a handicap factor to compensate for differences in hazards. If the rates per $100 payroll are $3.00 for Plant A, $2.00 for Plant B, and $1.50 for Plant C, factors for the units are 3, 2, and 1.5, respectively. The frequency rate of each plant is adjusted by dividing the rate by its factor. Plants are ranked from the lowest to the highest on the basis of the adjusted rates.
- The national average rates for different types of units may be used similarly for determining standings. If Plant A achieved a frequency rate of 10.0 and the national average for units in this industry was 12.0, Plant A's rate is 0.83 of the national average. Plant B's rate in comparison with its national average rate is 0.75. Therefore, Plant B would rank nearer the top than Plant A. Average rates for most industries are published annually in the National Safety Council's Work Injury and Illness Rates.

Rule 2b

Separation of large and small units is essential. Generally, a small unit finds it easier to go through an entire contest period without a disabling case than does a larger unit. If the number of units is sufficient, management can set up three size classifications. Unequal size groups can compete if they include units that can attain no-injury records. For this reason, the erection departments of the company in this example were divided into Group A—the five largest departments—and Group B—the remaining units. Five contestants in a group are usually considered minimum.

Rule 3

Awards should be specified. "XYZ Company" follows the general practice of giving first, second, and third place awards.

Rule 4a

The frequency rate is most often used as the basis for determining standings because it can be computed promptly and easily.

Rule 4b

There should be a satisfactory method of determining the winner between two or more units having perfect records

to give the smallest contestant a fair chance. For example, selecting the winner on the basis of the largest number of employee-hours worked since the last lost time case is fair to all units, regardless of size.

Rule 5

A standard method of counting cases is essential to avoid controversies and maintain confidence in standards. Use the incidence rate described in Chapter 8, Injury and Illness Record Keeping and Incidence Rates.

First aid and other minor injuries should be excluded for purposes of the contest. If they are counted, workers may fail to obtain treatment so that cases will not be recorded. The result could be an increase in infections.

Rule 6

Various "special rules" may be used in a contest that runs six or more months to maintain employee interest.

Rule 7

Frequent contest bulletins keep everyone informed about standings. Supervisors can post bulletins and discuss standings at safety meetings. Safety personnel can announce contest results in plant publications and in other ways to build and maintain interest.

Rule 8

Questions about rule interpretation will arise and must be settled fairly. This is an important function of the contest committee, which, in turn, may refer decisions about disputable cases to outside judges. A committee of the American National Standards Institute has been set up to interpret standard definitions.

Rule 9

See the section Meaningful Awards, later in this chapter, for award ideas.

It is important that competing plants or groups be informed of the latest findings. Figure 25-5 shows a typical contest bulletin.

Interdepartmental Contests

Interdepartmental competitions get "close to home" and place responsibility for a good showing on supervisors. Because workers have a greater personal interest in the standing of their own department than in the record of the entire plant or other unit, this type of competition has proved popular for creating interest in safety among both supervisors and workers.

An interplant contest plan often may be adapted to an interdepartmental competition. A company operating a number of similar facilities may take advantage of workers' interest in their departmental records by conducting a competition among the same departments in various organizations. Public utilities may conduct a contest among districts, and other nonmanufacturing companies may follow a similar plan. Because hazards in similar operating

STANDINGS IN THE 'XYZ COMPANY' SAFETY CONTEST
JANUARY-JUNE

Oakland leads at the halfway mark!

Tulsa moved into second place.

Chicago slipped from second to third place in June.

Corbin, in last place, had no lost workday cases during June. Good work!

Tulsa still leads for the President's Award for largest improvement over the previous year's record.

Plant	Rank	January-June Incidence rates*	% Increase + or decrease – over last year
Oakland	1	2.8	–30%
Tulsa	2	2.9	–40%
Chicago	3	3.5	+ 5%
Cincinnati	4	4.4	–20%
Corbin	5	5.0	+22%

*Incidence rate is number of cases resulting in death or days away form work per 200,000 employee-hours.

Tips on How to Win, No. 6

One-sixth of our lost workday cases occur in the use of cranes and hoists. Have foremen hold a safety meeting on safe practices in hitching loads and other crane operations for shop workers. We'll furnish a film. Use posters on the subject. Require safe methods. See the enclosed bulletin for suggestions on how to solve this major accident problem.

Figure 25-5. A simple monthly contest bulletin gives essential information about current standings and also includes a suggestion for improvement in "Tips on How to Win."

units are about the same, the basis of standings may be the frequency rate. Departments or other units may be grouped according to size.

Most departmental contests, however, are conducted among dissimilar departments in one plant; the departments often differ greatly in size. The difficulties due to variations in hazards and numbers of employees from one department to another may be overcome using the same techniques discussed above regarding differences between plants.

Always measuring the numbers of accidents of this year versus last year can have unwanted results. Too much emphasis on negative measurements can result in underreporting because participants do not want to look bad or lose the contests.

Another approach can be considered; that is, measuring the success of a safety program by using positive numbers. Here are a few examples:

1. Count the number of safety and health suggestions submitted by a department and count the number of positive ideas and corrections made because of these suggestions. This effort could be tied into production and quality suggestion programs.
2. Give credit for accident investigation reports that are submitted complete, on time, and with positive recommendations for preventing an incident from happening again (avoid accepting reports recommending "I told Joe to be more careful next time!").

3. Encourage employees to report all accidents, including near-accidents, first-aid cases, property damage without injuries, and accidents without injuries.
4. Give credit for safety meetings that are planned, outlined, and run by supervisors and managers.
5. Use safety and health inspections in a system that counts the number of corrections that were made because of these inspections.

Intergroup Competitions

Intergroup contests are particularly suitable for units that employ fewer than 400 people and have small departments in which hazards vary sharply. Employees are divided into teams of from 20 to 50 workers. To equalize factors of size and difference in hazards, each team has a proportionate number of employees from the most and least hazardous occupations. Each group is led by a captain, whose principal duty is to contact members of the team and create interest in winning. Team members may be members of contest committees.

Interest is often promoted by naming teams after prominent baseball, football, or other outstanding sports teams, and the entire competition may be named after a league or other sports organization. Workers can draw their team names from a hat and show their membership by wearing colored pin-on buttons. Supervisors can place contest signs in competing departments.

Using colored buttons to identify workers helps supervisors or others to keep score more accurately. Because members of different teams often work together, the buttons make it easy to ensure that every case involving a competitor is charged against that group's record.

Intraplant or Intradepartmental Contests

With set standards of performance, an intraplant or intradepartmental safety contest helps to emphasize the supervisor's and employees' responsibilities for avoiding accidents. These standards often include no-injury records for varying periods, achievement of lower injury rates compared with a previous period, and improvement over the average injury rates of similar units or of the industry. In this type of contest, units of an organization do not compete with one another; rather, each unit attempts to match or surpass the established standards.

Personalized Contests

Contest that focus on individual skills and achievements are very effective motivators.

Safe Driver Award Program

This award program is the recognized trademark of the professional driving industry for those who have proven their skills as accident-free drivers. Over 7 million drivers have earned this prestigious award since 1930. Employees of these eligible organizations must be certified by their company in order to receive awards designating the number of years of safe driving service.

Safe Worker Award Program

Recognize your safe workers with distinctive incentive items that encourage continued safe performance. Recognition items indicating one or more years of accident-free performance are part of the program. All full-time employees of Council member organizations and members of Council Chapters are eligible to participate in this program. Registration for the program is available through the Council's Safety and Health Motivation Programs department.

Some plants single out for special recognition employees who have safe records. Managers give certificates to those who have worked one, two, five, and ten years without an injury. Holders of certificates have found them useful in obtaining promotion and even in seeking employment with other organizations.

Periodic raffles of merchandise or cash, or the use of a new car for three or four months, regardless of the injury record of a department or plant, also have been used successfully to encourage and acknowledge the efforts of safety-conscious workers. Employees who are involved in disabling accidents (or drivers who have had a "preventable accident") during the period become ineligible for drawings.

Various sweepstakes plans have proved popular and effective in maintaining interest in a good record from month to month. One such plan was operated successfully by a branch plant of a well-known paper company. Here's how it worked.

On the payday prior to the beginning of each month all hourly employees received cards with serial numbers at the pay office window. The workers wrote their names and departments on the cards, tore off the stubs and put them in a box, and retained their portions of the cards. Names of employees in departments having no disabling injuries during the month then were drawn for prizes ranging from $5 to $25. The company contributed a total of $75 per month.

Since eligibility for the drawing depended on a perfect departmental record, each worker had to be careful about all job-related actions. Workers frequently corrected others for unsafe practices that might spoil chances for prizes.

If no accidents occurred during a three-month period, supervisors participated in a drawing for their own prizes ranging from $5 to $15. This feature proved helpful in enlisting their cooperation.

Some companies give awards like first aid kits or trading stamps to those in every department who have worked during a given period without an accident. This approach is effective only if going a month, for instance, without an accident is unusual.

Overcoming Difficulties

Although there are several difficulties in operating contests, they can be overcome rather easily. One of the most common difficulties is that some departments are inherently more hazardous than others. To overcome this problem, management can establish handicaps based on annual rates set by insurance companies or on average accident frequency for the different kinds of work.

Another method is to base standings on improvement over past records. Thus, a department with a past average accident rate of 20 and a current rate of 15 would be acknowledged as achieving a 25% improvement. They would win over a department with a past frequency of 15 and a current frequency of 12—only a 20% improvement. Usually, in both methods an average of rates over three to five years is used as the base.

Another problem is that a department may have so many accidents at the beginning of the contest period that it is out of the running and loses interest. Having shorter contest periods helps to overcome this difficulty. Another remedy is to have different awards for different achievements. An award for the department having the longest run of injury-free workhours is one example.

Dividing responsibility for the cause of an accident may become a problem. For example, the unsafe practices of one supervisor's worker (or an unsafe condition in this supervisor's department) can result in an accident involving an employee from another department. This would affect another supervisor's record. These situations must be anticipated and the contest planned to deal with them fairly.

Noninjury-Rate Contests

The use of noninjury-rate contests is gaining favor among safety professionals to motivate program performance. They are used to target problem areas and are not subject to the same potential misuse as are injury-rate contests.

For example, an employee contest can be based on the safe worker. The contest uses either a monetary award or a status award, such as a drawing for a television set or a private parking space. All employee names are included in the competition. During the contest period each worker spotted performing an unsafe practice or not wearing required protective equipment has his or her name removed from the contest. To maintain credibility, those experiencing recordable injuries would also be eliminated.

Supervisory contests, such as a "Supervisor's Safety Club," generally are based on how well supervisors achieve program objectives. These objectives range from the percentage of a supervisor's employees receiving safety talks to the number of self-inspections conducted and deficiencies promptly corrected. Here again, to maintain credibility the recordable injury of an employee would remove the supervisor from the contest. Awards range from steak dinners and sports events to a Supervisor's Safety Club jacket. Managers overseeing several supervisors or departments would have similar competitions and similar awards.

Contests at the plant level generally cover multiplant performance in a corporation. In addition to how well corporate safety program objectives are met, injury experience should be included in the contest criteria. Such contests usually have a corporate trophy to be displayed at the winning plant rather than providing any monetary consideration.

Other noninjury-rate contests include safety slogan, poster, housekeeping, and community contests. No matter what the contest, the objective is to get the maximum number of people talking, thinking, and participating in safety.

Slogan, Limerick, and Poster Contests

Safety slogan contests vary. One can be for the best safety slogan submitted by an employee. Another may ask the employees or their spouses if they can repeat the "slogan of the week" or the message on a certain safety poster.

Company magazines may conduct contests to "finish the limerick" or "write a rhyme" or write "twenty-five words on the best way to be safe." Often these are open to both employees and family members.

The value of homemade posters is in their special application to a particular industry or company. If the posters are the result of an employee contest, their motivational value will be increased, possibly exceeding that of the industry-made variety of posters. An important ingredient of such a contest is to get employees and their families participating in the planning and judging stages as well. Frequently, employee poster ideas—aside from the quality of the artwork—are so good that companies even submit the winning contest entries to the National Safety Council for possible conversion into printed safety posters.

Departments often conduct housekeeping contests. This type of competition is aimed at fundamental causes of accidents and usually tries to eliminate unsafe practices and conditions. Housekeeping contest plans differ from one company to another. The following plan has been used successfully by a metals firm.

Once a week a committee of three management representatives inspects each department and reports unsafe conditions to the superintendent. A copy of the report is furnished to the works manager, and another is kept for the use of later inspection committees. A demerit for each unsatisfactory condition is charged to the department. If the condition is not rectified within one week, an additional demerit is added.

At the end of the month the demerits are totaled, and departments are rated on the basis of the proportion of demerits to the total number of employees in the department. If Department A employed 175 people and had 25 demerits, its rating would be 85.7. This figure is obtained by dividing 25 (number of demerits) by 175 (number of employees), multiplying by 100, and subtracting the product from 100. Standings are posted monthly on the bulletin boards in each department.

Awards are made at a mass meeting of the employees and guests. Names of employees in winning departments are placed in a box from which is drawn the name of the winner for the month. The winner's picture is posted on a special bulletin board, and a short talk on safety by a representative of management is broadcast throughout the plant. (A general rule prohibits an employee from winning more than one award during the contest.)

The name of the winning department is inscribed on a plaque each month. The head of the winning department receives the plaque from the previous month's winner at the mass meeting. At the end of the year, the department that has won the plaque the greatest number of times receives it permanently.

Community and Family Contests

Many companies have stimulated interest by sponsoring safety essay or poster contests for children of employees, local school children, or young art students. The publicity before and after such contests, plus the interest generated by the posters themselves and the judging, not only stimulates the interest of employees, but also promotes the company's community and public relations.

More than one company has launched a safety poster or essay contest for the children of employees. Management knows full well that employees will give their children considerable help and that there will be much favorable discussion about the contest in locker rooms, lunchrooms, and car pools.

To promote greater safety at railroad crossings, Texas railroads sponsored a contest at the Texas Press

Association Convention. The journalist who made the closest guess of the total number of grade crossings in Texas won a prize. Because of the contest, safety at grade crossings received considerable attention in the press.

Miscellaneous Contests

There is an endless number of different types of contest possibilities. Often they can be combined with injury reduction contests. Contests can be held for attending safety meetings, for wearing safety shoes, for reporting unsafe conditions or unsafe practices, or for off-the-job or public safety activities of individuals, departments, or branch plants.

Safety contests are popular with both management and employees. However, the safety and health professional should always determine before recommending one whether the contest will take valuable time and effort away from providing safer equipment or better training for supervisors and employees.

Contests and Other Publicity

Management should publicize all stages of a contest as fully as possible. It can announce the contest through placards and stories in employee newsletters. Special signs, banners, or posters should publish standings at frequent intervals. Supervisors can hand out bulletins to employees urging them to keep the record perfect. Trade journals like Safety and Health and National Safety Council newsletters are other publicity outlets. In a smaller community, outstanding safety performance by a well-known company deserves—and usually gets—excellent publicity in the local newspaper and on the radio station.

The publicity value of a successful contest is considerable, although difficult to estimate. It should be commensurate with the significance of the occasion.

The presentation of an award to an employee who has gone 25 years without a lost time injury has human interest value for both internal and external communications resources. Recognition of an exceptionally fine "no-injury" record by a plant or company, or presentation of a National Safety Council award, would also call for widespread publicity.

Some companies purchase radio or television time to announce the results of a contest. One company had large campaign-type buttons made, and photos of children wearing them appeared in the local press.

Publicity (including photographs) can be sent to the local newspaper. The information should be prepared as a press release, indicating the nature of the contest or award. See Communications by the Safety Office later in this chapter for details of preparing a news release.

Meaningful Awards

An award serves several purposes: an inducement, a good-will builder, a continuing reminder, and a publicity tool. To serve these purposes, however, an award must be meaningful to those participating in the contest.

Employees sense when management is giving awards only for "sales" or publicity purposes and is providing little or no safety effort. If a number of awards are given, the sheer volume can detract from the value of the program.

The value of awards lies in their appeal to basic human interest factors, such as pride, need for recognition, urge to compete, and desire for financial gain (Figure 25-2). Some organizations present monetary awards as a bonus for making an extra safety effort. The distribution of U.S. Savings Bonds for safe records gets away from the direct monetary nature of a financial award. Other organizations, however, prefer gift or plaque awards appealing to the other basic human interest factors.

The originality or cleverness of an award or of its presentation is an important factor. Select awards that are worthy of good publicity, photograph well, and provoke employee conversation. Refreshments for all employees in a department after the completion of a set number of injury-free hours probably will create more favorable comment than would the presentation of a fancy plaque to the department supervisor. Drawing for a small cash prize or a grab bag prize would attract more interest than a routine presentation of the same award. An award to an employee's spouse for completing a home safety checklist, for writing a safety slogan, or for contributing to the company paper also generates more interest than would the same award given on the job.

- One Council member reported an award idea that received an unusual amount of publicity, both within the plant and locally. A local automobile agency loaned the company a new car that an injury-free employee, whose name was drawn from a hat, drove for a week. The employee also had a special "reserved for John Doe" parking space in the company lot. The only expense for the employer was a few dollars to cover special insurance. Even a special parking space awarded on a rotating basis can be effective.
- Another way to gain interest is to let the employees participate in selecting the award, planning its presentation, and helping with publicity. Frequently, they will suggest a humorous or novel award or publicity approach that may attract more interest than one planned by management.
- Payment of bonuses as awards for good safety records evokes considerable differences of opinion. Some managements and safety people feel that this approach is unwarranted, because all employees are paid to work safely. Others believe it can enhance an already successful program.
- Some companies raffle household or sports merchandise. Interest in safety among supervisors and workers often is developed by personal awards like wallets, knives, or key cases, often suitably inscribed.

• Many employees value attractive pins or engraved cards commending them for years of employment without an accident. One company places a safety record sticker on the employee's hard hat. Others provide special badges, pins, or shoulder patches to recognize safety achievement or service on a safety committee.

In addition to contest awards, recognition should be given to those who have saved lives, served on safety committees, submitted valuable safety suggestions, or made other significant contributions to accident prevention.

Award Presentations

To make an award presentation something special, one company rented an auditorium and invited civic and labor leaders, company executives, other dignitaries, and employees' families to a large celebration party. Another approach is for the president of the company, or some other high official, to present the awards at a general meeting, a picnic, or a dinner (or breakfast) that may even include entertainment. The reason for inviting VIPs is not only to add prestige to the presentation, but also to promote their interest and commitment to safety.

Such presentations require planning and must be in keeping with the importance of the occasion. The location chosen for the event should be appropriate and comfortable, not noisy or crowded. An award to an individual might be made in an executive office; groups might receive awards in a conference room, private dining room, or company lounge or cafeteria (during a nonrush period).

Inform participants about the agenda. Familiarize those who make the presentation with the significance of the award, the achievement it recognizes, and the background of the individual(s) who earned it. Ask the press to cover the event. Managers can post press photographs after the actual presentation to take advantage of desirable backgrounds, such as the plant or company name, some prominent trademark, or another eye-catching effect. Feature the award itself prominently in the picture. (See the sections following Campaigns later in this chapter and the National Safety Council Data Sheet 619, *Photography for the Safety Professional,* available through the Council library, for other ideas.)

For a group award—such as a company, plant, or department completing an injury-free year—management can provide free refreshments for a specified time ranging from one coffee break (per shift) to a full 24-hour period. Another, more elaborate award was given by a company employing about 300. At the end of a year in which there had been no disabling injuries, the president took the group to a major league baseball game. The following year, after the company maintained its no-injury record, the employees, plus their families, were invited to an all-day picnic and cruise.

Such activities help build better employee relations, as well as promote more interest in safety on the job and within their communities.

POSTERS

Posters can reach a large audience with brief, simple messages designed to accomplish one or more missions: to convey information, to change attitudes, and/or to change behavior. They are used to communicate with people who are going about their normal activities. As a result, the audience must notice, read, and take in the message in a very short time—often in a matter of seconds (Figure 25-6).

Safety posters are one of the most visible evidences of accident prevention work. Because of this, some companies have mistakenly assumed that posters alone can accomplish safety instruction and have neglected such essentials as real management support, guarding, and job instruction. In fact, hit-or-miss use of posters in a plant where no other safety work is done is likely to have a negative influence, making employees feel that the company is not sincere. (See the National Safety Council Data Sheet 616, *Posters, Bulletin Boards, and Safety Displays.*)

Purposes of Posters

Posters properly used have great value in a safety program through their influence on attitudes and behavior. One has only to see the efforts that commercial advertisers make to acquire space in business areas or near factory gates to appreciate the value of posters inside the workplace.

When selecting posters, keep their specific purposes in mind. They are used to:

1. remind employees of common human traits that cause accidents
2. impress people with the value of working safely
3. suggest behavior patterns that help prevent accidents
4. inspire a friendly interest in the company's safety efforts
5. foster the attitude that accidents are mistakes and safety is a mark of skill
6. remind employees of specific hazards.

Posters also help to support special campaigns, for instance, using guards, wearing eye protection, maintaining good housekeeping, offering safety suggestions, or driving carefully. They promote traffic, home, and even pedestrian safety by encouraging safe habits both on and off the job.

Effectiveness of Posters

A number of studies have underscored the effectiveness of posters for training and motivating.

Figure 25-6. Posters convey safety messages forcefully and often humorously and help emphasize a safety message in a memorable way.

- One study, conducted by the British Iron and Steel Research Association, used three posters that reminded workers to hook cable slings. The posters were displayed in six plants over a period of six weeks. A seventh plant was used as a control. Tallies made in the six plants before and after display of the posters showed about an 8% increase in compliance with the rule. The seventh plant, in which the posters were not used, showed a very slight decrease in compliance.

  Although use of the posters merely supplemented previous training, the plants that originally had the lowest rates of compliance showed some of the best gains. Using the three test posters separately, on a biweekly basis, proved slightly more effective than simultaneous use of all three during the entire six-week test.

- In a survey conducted by a prominent casualty insurance company, over 200 employees were interviewed in depth on the effectiveness of safety posters, films, and leaflets. Results indicated that all the media were instrumental in bringing workers to a higher level of safety awareness and that all were effective in sustaining that awareness. Employees preferred posters to leaflets, although they acknowledged the value of leaflets for more detailed coverage of safety issues.

- A survey of Council members indicated that about three-fourths of the nearly 800 respondents use a variety of poster subjects with one-third preferring cartoons and all-industry posters. Horror or shocker posters were least preferred.

Fifty-four percent of the respondents used posters to influence general attitudes; 27% to cover special operations; 18% to meet special or seasonal problems; and 14% to promote off-the-job safety.

Sixty-nine percent of the respondents in the Council's poster survey preferred posters of 8 × 11 in. and 17 × 23 in. sizes (21.5 × 28 cm and 43 × 58 cm). Of these two sizes, the smaller was preferred by a six-to-one margin over the larger.

Frequently, an unusually striking photo or overly elaborate (and expensive) artwork actually detracts from the safety message. This is not to say that "eye-popping" illustrations should never be used; they may serve to attract attention to the more conservative, serious messages on other posters. However, be sure photos are related to the message being presented.

Types of Posters
Industrial posters available from the National Safety Council, insurance companies, associations, and other sources fall into three broad categories: general, special industry, and special hazard.

- General posters are concerned with such subjects as risky behavior, disregarding safety rules, forgetting to replace guards, and other human failures.
- Special industry posters, as the term indicates, have application only in specific industries, such as mining or logging.
- Special hazard posters, for example on lifting, ladders, or the storage or handling of flammable liquids, are useful in every industry where the particular hazards are encountered. In some cases, a hazard is so serious that a special poster is developed based on the severity and not the frequency of exposure.

The National Safety Council carries more than 800 different safety posters in stock at any one time. These range from pocket-sized, pressure-sensitive stickers to billboard-sized jumbo posters.

Subject matter is roughly in proportion to the frequency rate of certain types of accidents. For example, more posters are concerned with materials handling than with chemicals and gases. (Call the National Safety Council for illustrations of the many posters available.)

Other Locations
Posters and stickers can be mounted on delivery trucks, buses, industrial trucks, mail trucks and carts, in elevators, and even on doors. Pocket cards or plastic pocket protectors, such as those available from the National Safety Council, might be called "walking safety posters."

Other Materials
Safety messages need not be limited to printing and artwork. They can be very effective when used in illuminated or changeable signs. One company paints a safety message on a plywood welding screen.

Homemade Posters
A company can develop its own posters to deal with special hazards not covered by posters available from outside sources. Even the smallest company can make an occasional special poster inexpensively—using colored paper, crayons, or felt marking pens—to call attention to a special hazard, to commemorate the winning of a safety award, or to point up a problem not likely to be covered by a commercial poster. (See Chapter 24, Media, for details.)

Safety personnel or supervisors can make effective posters using photographs of local conditions or accidents, even if the situations must be posed. A common type of homemade poster is the "testimonial," featuring a photo of an employee and a close-up of damaged safety glasses or safety shoes. The poster carries a brief statement explaining how this equipment protected the wearer.

Homemade posters on new processes, new guards, or new rules personalize the safety program and augment even the best selection of commercial posters. An excellent source of this type of poster is an employee contest, which is easily administered at a nominal cost.

Changing and Mounting Posters
No specific rule can be given for how often posters should be changed because of varying definitions of the term "poster." Some types probably should be mounted permanently. For example, a poster on artificial respiration can be kept in the first-aid room, or one on the use of a certain kind of fire extinguisher can be posted near it.

Most companies change general interest posters at definite intervals, usually weekly. They may rotate them from one area to another or file and reuse them after a year or so.

Management should vary the type of posters displayed. Consecutive posting of several health-related messages, for instance, or of new machinery posters is not desirable unless the company is conducting a special campaign on the topic. It is better to mix poster topics; for example, present an eye safety poster, then a machinery guarding poster, followed by a poster warning against wound infections, and so on.

For maximum effectiveness, posters not only must be selected carefully and changed on a definite schedule, but also displayed attractively in well-lighted locations where they will be seen by the greatest number of people. They should be placed on safety bulletin boards, near time clocks, in cafeterias, and at points of special hazard, such as paint storage rooms, rubbish cans, hazardous machinery, or dangerous intersections. Consider placing safety posters in unique places such as in the washroom above urinals, on the back of stall doors, on mirrors, or by paper towel dispensers.

Bulletin Boards
Bulletin boards should be large enough to allow convenient change of posters and should be placed where employees can see them during break times, such as near drinking fountains. They should be centered at eye level, about 63 in. (1.6 m) from the floor. Place them in a well-lighted location, or provide special lighting. A good size for a bulletin board is about 22 in. wide × 30 in. long (56 × 76 cm).

Boards should be painted attractively and glass-covered. Although a single board in the workplace is usually enough, several panels may be better in lunchrooms

Figure 25-7. Personal protective equipment is the theme of this display at NRPC's Beech Grove, Indiana, maintenance shop. The display is lighted at night and is in Amtrak's impressive 15-foot display case.

or locker rooms. Flashing lights or other eye-catching extras may be suitable for nonproduction areas but are likely to distract and annoy employees in workplaces.

A bulletin board should be used for only one display at a time, but need not carry safety posters exclusively. Any program of mutual interest to company and employees can be displayed on the bulletin boards. In fact, safety posters may have a stronger appeal if they appear along with other displays and announcements.

Bulletin boards in the same company may range from large, enclosed, illuminated boards with special sections for posters, safety bulletins, and other messages to a number of small frames or other inexpensive poster mounts installed at strategic points.

Poster frames to which clip-on literature racks can be added are available from the National Safety Council. This arrangement allows the company to conveniently distribute leaflets and other pickup literature that supports the safety message.

DISPLAYS AND EXHIBITS

The company can also use displays and exhibits to promote off-the-job safety as well as workplace safety. For

example, many companies try to educate their employees to understand that a safe vacation is a happy one.

Personal protective devices, tools, and pieces of fire-fighting equipment can be used to make up display or exhibits, with or without corresponding posters (Figures 25-7 and 25-8). Another good eye-catcher is a seasonal exhibit combining a Council poster and a safety display featuring such items as proper footwear for winter walking or skin and eye protection items in the summer to guard against sunburn and eyestrain. Management can purchase signs with changeable letters, electric tape messages, or unusual lighting for safety topics.

Companies and associations have devised many simple and attractive displays for presenting statistical data to workers. One is a safety clock, whose face is marked off to indicate the frequency of disabling injuries. Twin clocks or dials often are used, one recording the present injury rate and the other the rate for the corresponding period of the past month or year.

One company used large thermometer-like boards, placed at every gate and clock-house. Arrows indicated the present and previous month's records. The comparative standings of departments appeared below the symbols.

Figure 25-8. In this scaled down trailer-mounted house, children learn to escape from fires. They learn to use the proper procedures as they exit the bedroom when the smoke detector goes off (Operation EDITH: Exit Drills In The Home). (Courtesy Springfield, IL, Fire Department.)

An auto race was the theme of another display. Each car represented a department, and the cars moved daily to denote progress being made. Airplanes can be used similarly. Another exhibit featured racehorses participating in a "Safety Derby"; the horses were named after items of personal protective equipment. A display of highway signs with photographs of accidents below them was placed near the plant entrance of another company.

OTHER PROMOTIONAL METHODS

Other methods that can be used effectively to arouse and maintain interest in safety are campaigns, safety stunts, courses and demonstrations, publications, public address systems, and suggestion systems.

Campaigns

Campaigns focus attention on one specific accident problem. They are additions to, and not substitutes for, continuous accident prevention efforts the year round.

Campaigns may promote home safety, vacation safety, fire safety, or the use of safety equipment such as safety shoes. Companies can sponsor a "Clean-Up Week" or run a "Stop Accidents" campaign to promote the development of safe attitudes both on and off the job.

The National Safety Council's nationwide campaigns include posters, films, booklets, and specialty items to promote and maintain employees' interest. In addition, the Council has developed special materials in collaboration with trade associations, often to support special campaigns such as "Stop Shock" and "Fight Falls." Information on current campaign material is found in Council catalogs and other publications.

Many large corporations also have conducted extensive campaigns to promote safety on and off the job. Much of their safety awareness material is aimed at employee families, local citizens, and even groups outside the United States.

Companies must plan suitable publicity for the campaign from kickoff to conclusion, similar to that discussed earlier in this chapter for safety contests. Signs, flags, desktop symbols, and other items can be used to dramatize the campaign. To wind it up, schedule a special event such as giving each employee an inexpensive novelty item, free coffee for a week, or a free breakfast or dinner.

Many of the same promotional ideas that help to maintain interest in contests also can be effective in special campaigns—for example, a first-aid drill or a demonstration of cardiopulmonary resuscitation (CPR). Some companies use safety parades, exhibits of unsafe and safe tools and equipment, pledge cards, and other such features.

Timeliness may be an important factor in the way employees respond to a campaign. Successful safety campaigns have been linked to elections, the World Series, the football season, Thanksgiving, and other special events. See the National Safety Council Data Sheet 616, *Posters, Bulletin Boards, and Safety Displays,* for more ideas.

Unique Safety Ideas

Unique safety ideas or "stunts" capitalize on all the effective aspects of showmanship. Companies can develop them as separate devices for maintaining interest or use them to supplement contests and campaigns. *Safety and Health* and other publications regularly give details on various stunts.

Most companies agree that constructive stunts help inspire employees to high standards while stunts that ridicule usually do more harm than good. Moreover, employees and supervisors who are the objects of ridicule may have just cause to blame management for not setting up safe procedures or providing safe facilities and equipment.

Safety stunts can involve an entire company, a department, a small group, or just the individual. A stunt may be humorous, novel, or dramatic, and occasionally even shocking.

A simple stunt is often the most effective. A pivoted hammer, mounted over a pair of safety glasses in a display case, can be operated by a string to demonstrate the impact resistance of the glasses. To dramatize the importance of eye protection, the "let's pretend" stunt can be used. Several volunteers are blindfolded and then asked to eat, write, and move around.

Stunts developed for the company safety program often serve equally well at company open houses or safety picnics and in community safety projects as well. Such stunts, when supported by visual aids, signs, and printed material, demonstrate the company's interest in accident prevention. They also give employees a chance to participate in programs that help create safer attitudes on the job.

Some firms use a somewhat unique poster or card as an attention getter. For example, safety cards handed out by employees can alert co-workers that they have been risking an accident by their behavior.

Courses and Demonstrations

Most safety and health professionals agree that courses in first aid, lifesaving, water safety, civil defense, and disaster control have bonus values that help prevent work injuries. The worker who has completed a course in first aid and has learned to do CPR will be more aware of the hazards of electric shock and more likely to help maintain electrical equipment in safe condition. Likewise, the employee who learns how to stop arterial bleeding better appreciates the serious consequences of using an unguarded saw, power press, or cutter.

Home study and extension courses also serve to stimulate and maintain interest by giving the employee a better understanding of hazards involved in the job. Most safety training courses are designed specifically to improve the attitudes of both supervisors and employees. Using appropriate videotapes and other visual aids will enhance the effectiveness of the courses. (See Chapter 24, Media.)

Public participation in courses involving employees promotes community goodwill. Many industrial safety people are doing an excellent job of promoting safety and fire prevention by arranging or teaching courses on these subjects for the Boy Scouts and Explorer Scouts, the Girl Scouts, Camp Fire groups, Junior Achievement companies, and other youth or school groups (Figures 25-9 and 25-10).

The National Safety Council's "Defensive Driving Courses" provide an excellent means of promoting good employee and public relations. They stimulate safer attitudes both on the job as well as off.

Fire equipment and fire safety demonstrations have a practical value beyond that of teaching employees how to react in an emergency. The fact that the equipment is provided for their use reminds them of management's concern about their welfare. Moreover, the demonstrations make employees more aware of the dangers of fire and point up the need for obeying fire prevention rules.

Such demonstrations are easily arranged through local fire departments or fire equipment distributors. Many companies conduct their own demonstrations, using extinguishers that require recharging, or "not-in-service" extinguishers kept specifically for this purpose.

Publications

Reports

Safety program progress reports should be written in an interesting and concise manner for superiors and supervisors. Visuals can be used effectively where appropriate. (See Chapter 24, Media.)

The safety and health professional can present the cost of accidents and, perhaps, the cost of prevention in terms significant to management. These costs include medical and compensation costs, production losses, sales losses, increased maintenance costs, and the less tangible, but perhaps more important, hidden costs involved in administrative problems and in impaired public, customer, or employee relations. Reports need not be dull. Photographs, for example, can pinpoint a company's major sources of disabling work accidents.

A statement of accident losses and safety achievements may be included in the company's annual report or in a special annual or monthly safety report issued to top executives and supervisors. If departmental accident losses,

Figure 25-9. Management, employees, and their families participate in a safety fair. Before the children can slide down this authentic firefighters' pole, they are given a safety question to answer.

like incidence rates, can be charged on an equitable basis, such as "per hundred thousand dollars of sales" or "per one thousand employee-days of production," the comparative standings of departments and improvement in departments or units are easy to evaluate. (See the discussion in Chapter 8, Injury and Illness Record Keeping and Incidence Rates.)

The fact that the company records and publicizes such information is in itself an incentive to supervisors. It reminds everyone concerned that accident costs are just as much an integral part of profit and loss as production, sales, maintenance, distribution, and advertising.

Special charts, graphs, and statistical reports also can visually show facts about accidents. One chart can track the number of lost time injuries, others the number of days lost, injury causes, accident causes, or type of injuries. It cannot be too strongly emphasized that unless such charts are kept up to date they can do more harm than good. An outmoded chart indicates management is losing interest in maintaining a strong safety program.

Annual Reports

In recent years, companies have worked hard to make their annual reports to stockholders interesting and clearly

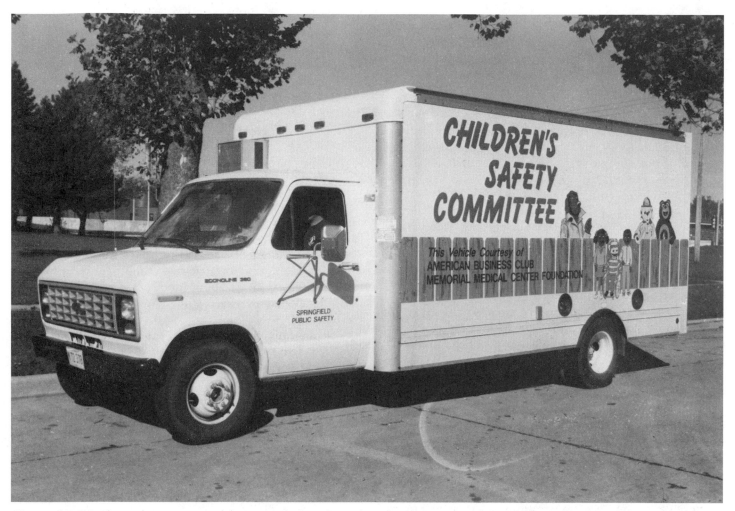

Figure 25-10. This truck contains a mobile puppet show and sound system. The current repertoire consists of four shows: Burn Prevention, Fire Prevention, Substance Abuse, and Safety on Wheels.

understandable. In many cases, annual reports are also distributed to employees so that they can become better acquainted with the company's purposes and problems. A section on aims and accomplishments in accident prevention attracts employee interest and further serves to emphasize the interest of management in the safety of its employees.

Newsletters

Monthly or weekly newsletters or videos are especially important as a means of maintaining interest. They keep employees and supervisors informed, particularly in decentralized or field operations where bulletin boards are not practical. Newsletters can give detailed information on standings in a safety contest and publicize unusual accidents or serious hazards. They can help explain safety rules, remind employees of safe work practices, and support the safety program in general. If workers can serve as "reporters" or help to produce such a newsletter, so much the better (see details later in this chapter). A case history of a particularly unusual or spectacular accident can sometimes be featured.

Booklets, leaflets, and personalized messages take many forms: safety rule booklets, NSC's "New Employee Booklets," special "one-shot" leaflets, monthly publications such as the National Safety Council's *Today's Supervisor* magazine and its *Safe Worker and Safe Driver* for employees, and letters from management. The content of an employee rule booklet, except for material involving company policy, may be developed with the help of safety committees or selected workers to stimulate interest in the topic and help to ensure compliance with the rules.

Larger companies may have their own editors and artists, even their own printing facilities. Smaller companies, however, also can issue attractive booklets, leaflets, and personalized messages, and at negligible expense by using local consultants and printers or their desktop publishing systems.

The National Safety Council, trade associations, and professional organizations publish a wide variety of booklets and leaflets that are authoritative, attractive, and relatively inexpensive. They cover a wide range of subjects—material handling, first aid, housekeeping, fire prevention, vacation safety, safe driving, and the like. Such materials, carefully selected and regularly distributed, effectively supplement company-prepared publications.

Letters commending meritorious service, signed by the manager and addressed to individuals, also make an excellent impression upon workers.

Safety calendars, published by the National Safety Council, together with Christmas letters from the manager have a direct appeal that reaches the workers' homes. Such mailings should include all employees. *Family Safety and Health,* a quarterly magazine, is sent to more than two million homes by organizations who are interested in the welfare of both employees and their families.

Buttons, blotters, book matches, pencils, and other small novelties all conveying a safety message also may be used. For example, packets of silicone tissues for cleaning glasses, imprinted with brief messages or safety rules, serve to remind employees of the rules and to encourage proper use of safety equipment.

Public Address Systems

Public address systems often are used to broadcast announcements and page employees. Many companies have taken advantage of these installations to broadcast safety information as well.

Such messages should be planned carefully. Employees might readily lose interest in long speeches or too-frequent safety reminders. When the public address system is used for broadcasting music, safety announcements can be made between numbers.

Suggestion Systems

Because accident prevention is closely associated with efficient operation, many suggestions not only help to prevent accidents, but also lower production costs, improve manufacturing conditions and methods, and change the outlook of workers.

If a company does not have a general suggestion system, it is probably better not to establish one for safety suggestions alone. Setting safety apart from ordinary operating procedures may deemphasize its importance.

Getting good suggestions is essential and must be encouraged by all levels of management. Company posters, contests, campaigns, merchandise incentives, direct mail, printed handouts, personal appeals, supervisory training, safety clubs, and press releases can motivate employees to submit safety suggestions.

Many effective suggestion systems offer money for ideas, and companies may pay considerable sums for employee suggestions. One firm awards more than $10 million annually, but considers the money well spent on valuable suggestions.

To merit an award, a safety suggestion, like a production suggestion, should be substantial and practical, and offer a real solution to a problem. For example, suggestions worthy of awards include changing a method or material, guarding a hazard, or inventing a safety tool or device. Suggestions that would not merit awards include erecting a sign, cautioning workers, or publishing slogans.

Safety suggestions generally are regarded as a highly desirable way to avoid safety grievances. Improved employee interest and personal involvement are additional benefits.

Suggestion Awards

It is easy to measure the monetary value of suggestions that result in greater efficiency, lower material cost, decreased labor cost, or reduced waste. Usually awards for suggestions in these categories are in proportion to the savings derived by the company. In other instances safety suggestions may have a monetary value but be harder to evaluate. As a result, payments for suggestions that contribute to the welfare of the employees but result in no direct savings to the company are most often estimates or composite judgments. Some firms have developed guidelines that consider such factors as the degree of hazard, originality, extent of application, etc. One company has also developed an award guide based upon disability cost experience.

If management downplays "safety awards" and gives more weight to other awards, employees will believe that the company regards safety as an unimportant sideline. Payment for a safety suggestion must be equal to awards given for other suggestions—it should be based upon its real worth if it can be determined. If a suggested safety device enables an operation to be run at a speed that would be dangerous without the device, a saving may be measured. If a number of accidents have occurred in an operation and a suggested device will eliminate them, the cost of those accidents can be projected and a saving calculated.

Awards should reflect the merit of the suggestions. Most companies award cash and/or bonds. Some award merchandise, all-expense-paid trips, company stock, certificates of merit, medals, gifts for the suggester's spouse and family, or attendance at a recognition luncheon.

Some companies exclude superintendents, supervisors, designers, methods and systems personnel, and other supervisory or technical personnel from receiving awards, so that the other workers will have someone to whom they can go for assistance. Some companies feel supervisory personnel should not be excluded; several firms have separate plans and award schedules for salaried, supervisory, technical, and management personnel.

Suggestion Committee

If necessary, a special subcommittee can be set up to determine the monetary value of safety suggestions so that employees will be rewarded for them exactly as they would be for other money-saving suggestions. However, some firms believe such special treatment sets safety ideas apart from other ideas.

Many established suggestion systems now in operation are producing excellent results in monetary and "people" savings. No company should consider a suggestion plan without first studying carefully the plans now in existence.

Boxes and Forms

Suggestion boxes should be attractive and well placed, and stocked with special blank submission forms. Commercial suggestion forms also are available. To increase the interest of the employees and establish a spirit of cooperation and importance, it is essential that management acknowledge and resolve all suggestions promptly.

Recognition Organizations

Some organizations recognize people in the United States and Canada who have eliminated serious injuries, or who have minimized them by using certain articles of personal protective equipment. The award provides an excellent opportunity for publicity. Three of these organizations are:

Wise Owl Club

Founded in 1947, this club honors industrial employees and students who have saved their eyesight by wearing eye protection. Address inquiries to Wise Owl Club, National Society to Prevent Blindness, Schaumburg, IL 60173.

The Golden Shoe Club

Awards are made to employees who have avoided serious injury because they were wearing safety shoes. Address inquiries to Golden Shoe Club, 2001 Walton Road, P.O. Box 36, St. Louis, MO 63166.

The Turtle Club

Founded in 1946, the Turtle Club recognizes and honors workers who have escaped serious injury because they were wearing a hard hat at the time of an industrial accident. Although OSHA mandated the use of head protection in many U.S. industries in 1971, the Club's goal—to save lives and prevent injuries through promoting the use and acceptance of hard hats—and its purpose—to help business understand the benefits of industrial head protection—remain important today (Figure 25-11). The Club's sponsor is E.D. Bullard. Address inquiries to Mr. E.D. Bullard, International Sponsor, The Turtle Club, P.O. Box 9707, San Rafael, CA 94912–9707.

CAMPAIGNS

The preceding sections covered internal publicity within a company or organization. The remainder of this chapter discusses how to influence the way a company looks to people on the outside. Favorable publicity is an unmistakable bonus to a good safety program. Yet many organizations often disregard it.

Any company likes to have someone—especially a prospective customer—say, "I like what I hear about this company. I understand that it really takes care of its employees. So I figure it must treat its customers right; therefore, I'll be treated right." One good way for a company to get a reputation for taking care of its employees is to be known as a safe place to work. Surprisingly, an amazing number of companies do little or nothing to let

their public—customers, stockholders, the community—know that the safety and health of their employees are important to them. This chapter presents a simple and sensible formula for letting people know that "at my company, the health and safety of the workers are important."

Most companies have a professional public relations (PR) department to handle the public communication program. In smaller companies, the safety and health professional may have to handle public communication. In both cases, the information in the following pages should prove useful. The safety professional should be aware of overall company policies and programs and know when to turn to specialists and when to ask for creative help.

PUBLIC RELATIONS

Public relations is the "management function which evaluates public attitudes, identifies the policies and procedures of an individual or an organization with the public interest, and plans and executes a program of action to earn public understanding and acceptance," according to the magazine *Public Relations News*. Abraham Lincoln knew about PR. Speaking at Ottawa, Illinois, in 1858, Lincoln said: "Public sentiment is everything. With public sentiment nothing can fail; without it, nothing can succeed. He who molds public sentiment goes deeper than he who enacts statutes or pronounces decisions. He makes statutes or decisions possible or impossible to execute."

Lincoln's classic quotation on public sentiment can be traced to Jean-Jacques Rousseau, the 18th-century French philosopher who is generally credited with developing the term *public opinion* as we use it today. At its simplest, PR is the promotion of good public opinion about a person, a company, a government, or other entity. Thus, every

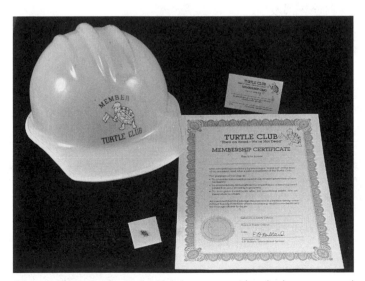

Figure 25-11. The Turtle Club honors people who have escaped serious injury because they were wearing a hard hat. Displayed here are the items new members receive: a hard hat with the club insignia, a membership certificate, a wallet card, and a lapel pin.

employee, every activity, every facility of a company contributes in many ways to the overall opinion that people outside the company have about that company. This is true PR.

Anyone concerned with accident prevention—safety and health professional, supervisor, member of the plant safety and health committee, or officer of the school or community safety council—should realize that any time they communicate with someone outside the committee, department, or company, they are involved in PR. Even an event such as a company-sponsored family picnic can strengthen PR.

To be successful, a PR program must be backed up by an effective management team and a quality organization. Public relations can then reflect the nature of that organization. A strong safety program merits good publicity, but false publicity or publicity based upon inflated facts or specious statistics will be recognized for what it is—company propaganda—and can do more harm than good.

The basis of a successful PR program is a successful management. Such a management team makes sure that staff and employees produce good products safely and efficiently, that they cooperate with each other and with the customers, and that all give the best and friendliest service humanly possible at all times.

The plain fact is that poor PR is costing individuals and organizations in this country millions of dollars each year. The remedy is simple: a better understanding and use of fundamental PR on the part of everyone—and a sincere effort to put it into practical use. For lack of good PR, many a worthy cause has failed to get the support it deserves, and many an organization has failed to gain the business it needs. A good PR program need not cost a great deal of money. But it is worth all the time, effort, and reasonable expense allocated to support its activities.

THE VOICE OF SAFETY

In any genuine, effective PR program, emphasizing safety can be a real help. In fact, it is hard to imagine a PR program where sincere and effective concern for the protection of employees from accidents is not a top priority.

If a company does not have a safety program, it misses vital opportunities for good PR and dramatically increases the chance for adverse public attitudes, not to mention the risk of an accident. A safety program, properly managed and communicated to internal and external audiences, can help offset or at least mitigate adverse news should accidents occur. In addition, the safety program is long-range and sustained, whereas "news" is here today and replaced by other news tomorrow. The time to implement a safety PR program is not when trouble strikes. The program must be in place and functioning, and the public must know about it.

The safety and health professional should not only welcome publicity for safety efforts but should energetically seek it.

Working within the Company

In evaluating a company's PR function, one can begin by asking two questions:

1. Does the company have a PR department?
2. Is there an employee publication in the company?

If both answers are "yes," the safety professional should get in touch with both these units before doing anything about publicizing the safety program. This step is important. It not only ensures professional skill and consistency of efforts to publicize safety activities, but it will save confusion, avoid duplication, and possibly prevent conflict between safety personnel and the PR department.

Why does such an obvious step have to be mentioned? The reason is that too often there is little, if any, communication between a safety and health professional and the PR department and publications editor. Communication between the safety and health professional and the PR staff and editor is indispensable, for these three must work together, or safety will not receive the attention it deserves and needs.

The safety and health professional should explain to the PR staff and the editor, if this has not been done before, that positive safety motivation and accident reduction at the company are priorities and that their communications help and advice are needed to reach these goals. They need to understand the necessity of employee and public acceptance of the safety program and how much the safety and health professional is depending upon them to help earn employee acceptance.

Safety activities contain a wealth of real news. It may be that everyone in the safety business has assumed that safety activities and programs by their very nature must be on the dull side. In recent times, however, more writers and others have begun to recognize that safety can be made interesting. It takes the combined efforts of safety and health professionals, PR people, and editors to do the job.

Sixteen Ways to Make Safety News

Public relations information, to deliver an effective message, need not always be red-hot news. However, it must have an element of spot news, human interest, or self-help. These qualities will give it feature value for the media. Activity—real, honest, legitimate activity—makes news. Of course, dramatic circumstances such as a rescue or near-miss make news. But until they turn up, the best stories are those that describe a company's constructive and creative efforts to promote and maintain a worker health and safety program.

It might be useful to list some of the items and activities that can make safety news in an organization and that editors and PR staff ought to know.

1. No-accident records for the entire company or for any one unit—in terms of either days or worker-hours.

2. Improved safety records for the company or for any one unit, even if no prolonged no-accident period is involved.

3. An interplant safety contest, or an intercompany contest—especially if anyone has dreamed up an unusual angle or prize.

4. Any unusual safety record for safety performance by an officer or employee of the company—either in length of time or character of the job done.

5. Innovations in safety programs that will prevent accidents. An invention, too, has special news value if the company has been plagued with accidents that the new gadget may prevent.

6. An unusual or highly valuable safety suggestion by an employee.

7. Safety conventions or meetings, either those held by the company or those held elsewhere, to which company representatives will go. A digest of such meetings should be publicized.

8. Other special safety events besides conventions—a safety banquet, a safety training course, fire and first aid demonstration, a special meeting, or an award ceremony.

9. Some unusual event intended to get the employees to take their safety training home to their families, or something the company is doing directly with workers' families to promote around-the-clock safety. Open-house tours, local water safety shows, or public showings of safety films are examples.

10. Some pronouncement or statement by the president or other high officer of the company on some unusual or new safety device or company safety service, such as free inspection of employees' cars.

11. A speech by the head of the company or the safety and health professional at a local, regional, state, or national safety convention or conference. The editors and PR staff should have advance copies of it. The person making the speech should be sure to say something worthy of public attention.

12. The company's annual report is the foundation for corporate communications. Stockholders do read these reports. A paragraph or two on the good safety record for the past year will go a long way in achieving sound publicity within the corporate family, as well as to inform the analysts, who recommend stocks, what the company is doing beyond its financial performance.

13. Any act of heroism by someone in the company. This is a sure-fire story for local papers as well as company publications. Maybe this type of news is not pure safety, but news media regard it as part of safety, and it can always be tied in with an indirect safety message.

14. A survey or study of some phase of accident prevention in the company. If the investigators discover that married people who own their own homes are safer than their single counterparts, they have provided a ready-made story.

15. Election or appointment of a company officer or safety professional to an important post as a volunteer officer of the National Safety Council, American Society of Safety Engineers, Board of Certified Safety Professionals, local safety organization, or governmental agency.

16. A company or employee winning an award in a National Safety Council award program or contest. Winners, not losers, are publicized.

A good rule of thumb is to stress the positive, rather than the negative side of an event. For example, instead of a story that says "single people are less safe," it could say that "a company study shows a need for special safety efforts by singles." Another way to stress the positive is to report "Last year 290 employees worked without an occupational injury. Only 10 did not."

SOME BASICS OF PUBLICITY

If a company does not have a PR department, this fact need not prevent its getting information into local papers or on the air. Without PR personnel, the task may be more difficult, because PR people are more experienced and know the media game better than the safety and health professional does. However, even in the absence of a company PR department, the safety and health professional can get publicity for safety activities by going directly to the newspapers, magazines, and radio and TV. It helps to solicit and secure PR advice from local safety organizations, or even business or trade associations with such service. The safety and health professional should always keep the company management informed of what is going on at all times.

The safety and health professional should not pretend to be a PR expert. Instead, he or she should be sincere in approaching editors and TV/radio program managers with sound, worthwhile information and not worry about whether it is packaged perfectly for media acceptance. A good story is a good story. Most editors and reporters are glad to have the material and often will use it. News value and reader interest are what count the most.

Editors and program people are usually not difficult to approach, provided the safety and health professional is courteous and friendly and admits to a lack of specialized knowledge of PR techniques. Naturally, no editor or anyone else likes to have someone come charging in and claim to be doing a big favor by delivering the story the world has been waiting for.

In general, the common sense and salesmanship needed for success in the safety field are enough to enable the safety professional to present his or her case clearly and effectively to the media. Nevertheless, the person should be willing to accept advice from PR professionals on how best to tell the story. The safety and health professional needs to be careful about being quoted out of context.

Select the Audience

"Who must be reached with safety information, and is it of interest to them?" These seem like fundamental questions, but many companies never think to answer them. Because publicity cannot reach everyone, the company must choose its audience carefully, especially if the budget is tight or time is limited. In that case, it is logical to assume that—in addition to the in-plant (company) audience—the company would prefer to reach people who might buy the product or help the company in some other direct and profitable manner.

Use Humor and Human Interest

It is worthwhile to try to brighten safety, to make it positive, rather than ponderous and dreary. It is even possible to evoke a chuckle now and then.

Editors are familiar with the solemn pronouncement that "Safety is a serious subject, and must be taken seriously—safety is no laughing matter." No one can argue with that position. But does it follow, therefore, that no one can put into safety some of the same techniques, the same sales appeal, the same sparkle that are used so successfully to sell commercial products? If those techniques can sell shampoo or a personal care product or an automobile, is it unreasonable to expect they can also sell safety?

Or how about a cartoon treatment? This just may brighten what might otherwise be a slightly dull and drab presentation. The safety and health professional should not be too disturbed if someone points out that a cartoon has shown a negative aspect of a safety message—that is, what not to do. It may well have done just that. This is the very thing that gives a cartoon its punch. A pratfall cartoon will draw attention to the slippery, icy sidewalk in a way that cannot be shown by a person walking and not falling. However, care should be taken to avoid ridiculing or negatively portraying victims of incidents or the negative results of incidents.

It is even possible to get a child or baby into the act, or even a faithful, shaggy dog, in order to get the spark that lifts safety activities out of a dreary, impersonal rut and presents them in terms of human interest.

Names Are News

Remember that facts and figures about injuries and their frequency and severity are made more interesting by injecting human interest elements into safety news. Because human interest means people, the safety and health professional should include more information on *people* and safety and less on things and safety. Mention which managers or employees accomplished safety goals, contributed ideas to the program, and other such activities.

Friendly Rivalry

Safety awards, safety records, safety contests, safety inventions, and gimmicks—these are only a few of the many things that make good safety news. For example, if the company has reached a new injury-free record in its industry, it is headed for headlines. The editor, the PR department, and senior management must be kept informed of safety progress all along the way. They will help arouse public interest in the performance and also stimulate greater interest and greater effort among the employees themselves.

An award loses its motivational and publicity value if kept a secret. The safety professional must publicize the event. Photos of award presentations are sometimes used, but they should be interesting and even unusual to attract special attention. Editors dislike the typical "grip and grin" award photos, so be inventive in trying to get a picture worthy of publication. For example, a representative from the organization conferring the award on the organization could be shown handing the award to a foreman or worker in the workplace setting.

In some instances, top safety awards have been accepted by some companies as if they were a "dime a dozen." On the other hand, companies that have made similar awards the occasion for a major bell-ringing celebration have seen their "safety stock" rise sharply among employees.

A public utility company in Michigan, for example, made so much of its award-winning safety records of the various units throughout the state that an outsider might have thought the company had won the World Series. At one celebration, more than 1,000 employees, from top management on down, jammed the closed-off street in front of company headquarters for presentations, followed by an all-company picnic. This celebration got coverage in the papers and on the air throughout the state of Michigan. Here was public relations that any organization would welcome. Safety had made news.

Techniques

These pointers are offered to the safety and health professional who wants to make the most of PR opportunities:

1. Be honest in what you say. Never exaggerate.
2. Deliver what you promise. If you say to the media that something is going to happen, make certain it happens—as you said it would when you said it would. This often calls for a "run-through" in advance.
3. If for any reason there is a change in plans from what you have announced, notify the papers, radio, and TV stations at once.
4. Be scrupulously accurate with all names, dates, places, and other facts. There is no such thing as being too careful in this respect. If an editor misspells names, the only thing to do is to resubmit the names correctly spelled and hope for the best.
5. Offer ideas but do not ask for specific space or time. Complaining about your company PR department, or complaining that the local paper or station has treated your company shabbily will not only be

fruitless but will create a strained relationship between you and the press.

6. Do not alert your PR department or publications editor (or put out anything yourself) unless you have real news or features to offer. You must not issue material just to be issuing it. Be reasonable with the amount of material you send out. You can wear out your welcome.

7. Tip off your local or industry association, safety councils, and your PR department (or if you do not have one, the papers or radio and TV stations) to anything worthwhile you run across that might make an interesting news story. Do so even if it has no relation to you or your company; the media will appreciate it and be more inclined to view your own materials favorably.

8. Above all, do things that make news. Almost every routine safety item can, with a little extra effort by the safety and health professional and the editor, become a more readable, more constructive piece. News will be published only if something is being done for safety that makes news. News can always be heightened by intelligent, imaginative treatment, but the story must be worth telling. Advertising space can even be purchased for special items.

9. Use good sense and an honest approach if the news is bad. Prove to press representatives and the public that you can roll with the punches. Know your organization's policy and procedure to follow in all situations.

Communications by the Safety Office

If a safety and health professional must handle his or her own publicity with the local media, here are some tips. Many stories have died because someone failed to observe them.

1. Editors and news directors are busy people. Unless a situation is an emergency, media people prefer to receive possible story ideas in writing. Send them a brief release or a letter outlining what you have to offer. Include facts and by all means make your material interesting. Timing is vital: Don't notify the media today of something you are planning for tomorrow. Never suggest that a reporter or news crew be sent out—this is the editor's or news director's province, not yours.

2. Generally, the person to write to is the city or business editor of the paper or the news director of the radio or TV station. Of course, if an item is specifically written for a certain columnist or commentator, it is better to make direct contact. If it applies only to a specialized area (finance or sports, for example), it should be brought to the attention of that editor. Know the publication, whether it is general news or a trade journal. If you are working with TV or radio, don't waste their time on information

or items they'll never use. (Example: broadcast media seldom mention personnel changes.)

3. Write not for your boss, but for your reader or listener. Answer objectively the questions: who, what, when, where, how, and why? Do not load your releases with propaganda for the company. There is no surer way to kill your positive relations with the media.

4. Make your releases as professional in style, appearance, and general quality as you possibly can.

5. In writing a release, be brief and to the point. Printed space is limited and costly. Try to keep the piece to one page. Publications receive thousands of releases each month. These are skimmed, and only the best get into print. Likewise, remember that TV and radio broadcasts usually are measured in seconds, not minutes. Keep your material short and factual.

6. If you are sending a picture with the release, the caption should be typed on a piece of plain white paper and taped to the back or bottom of the photo. Never use paper clips, and never write with a pen or pencil on the back of the picture. Either of these will likely damage the photo and make it difficult to reproduce clearly. Think visuals when working with TV—what can they show that will help to tell or illustrate your story?

7. Leave script writing to the professionals, but check facts. If a radio or TV station requests material, send them the facts, figures, and whatever narrative is necessary. The people at the station will put it into the proper form.

Hints for TV Interviews

The safety and health professional must often be the spokesperson for his or her company, not only for newspaper coverage, but also for radio and television. Although getting the facts correct may be adequate for a newspaper interview, a television or radio interview reflects more of the company than merely what the facts show. It projects a company image through the company spokesperson. If the safety and health professional does not feel that he or she projects a good image over the radio or television, then someone who does should be picked from the safety department or from the PR department.

Here are some tips that will help the safety and health professional give a better radio or TV appearance.

1. Remember that you are being interviewed for your knowledge, not for your personality, entertainment value, or good looks. Be yourself. Don't put on a special voice or worry how the lavaliere microphone looks with your clothes. Don't wave your hands or touch your face or hair. Don't jingle coins or play with jewelry. Keep your hands down.

2. If your TV appearance is preplanned, dress conservatively. Avoid loud clothing or busy patterns that can affect the camera or overpower your message.

3. Review with the interviewer in advance what areas will be discussed. You can steer away from areas you cannot discuss, and you can get help on questions that you might not be able to answer. If you don't know the answer to a question, say so. But add that you will check and get back with the answer in a specific time period.

4. Be natural and cordial. Smile, if the situation is friendly. Be sure to maintain eye contact with the interviewer or camera. Do not memorize a statement and rattle it off. You will waste everyone's time. It is better to be well-prepared and let the on-air material be a question and answer or discussion between you and the interviewer. Don't rush, and don't try to cram a lot of information into a TV slot. Keep it short and light. Avoid statistics.

Handling an Accident Story

In any PR program, it is just as important to know what *not* to do as to know what to do. In fact, it can be even more important.

The foremost warning is this: do not cover up bad news. Good media relations are of utmost importance. It is at such a time that a sound PR program pays off.

Every safety and health professional hopes the day will never come when a serious accident damages a company's safety record, but it happens. In some instances, the repercussions of how the accident reporting and news coverage are handled have been even more unfortunate for the company than the accident itself.

Here is an example of how not to handle a press representative: In a midwestern city some time ago, two workers were killed by a crane. This company enjoyed a first-rate relationship with the newspapers and radio and TV people in that city. It had worked hard at safety and at PR. It was good to its employees and had a fine reputation for honest and open communication.

On this particular occasion, however, someone in the company's top management became anxious over the deaths. As a result, when a reporter came out to the plant to get what was to her paper a routine story of the accident, she ran into a maze of censorship.

First, the safety and health professional turned the reporter over to the personnel manager. This person switched her to the general manager, who gave her some "double talk" about the accident and told her the company doctor was the one to see. The doctor said the safety and health professional was the person to talk to.

By this time the reporter's anger and suspicions were rising. What had started out as a routine assignment now had become a challenge to discover what everyone in management appeared to be covering up.

It is important to remember that a reporter cannot lose in a contest like this. Since the workers had been killed, the coroner would have all the facts. If they had been badly hurt, one of the hospitals would have the information. If

the workers had not been killed or hurt badly, then there would have been no story in the first place.

The reporter obtained the facts regarding the workers' deaths from the coroner's office. She then wrote a story that was as harsh toward the company as it could be without committing libel. The story was edited, headed, set in type, and lay in the composing room, awaiting its turn to get into the paper.

Ordinarily, newspapers set more news in type than they can print. Each day dozens of items get left out—the "overset," as it is called in newspaper parlance. Under normal circumstances, the story of the accident might well have ended up as "overset" or, if printed, might not have received much attention. However, the coverup and runaround the reporter had received at the plant changed all that. As a result, the story was marked "must" when sent to the composing room and ended up as front-page news.

When there is bad news to report, the safety and health professional must back up the PR department 100% to help get out the news as quickly as if the tidings were all in the company's favor. Unless directed to deal personally with the media, the safety professional should stay in the background and provide the proper information to management and public relations.

Along with the bad news, the safety professional's material should mention the good things. It may be that this is the first accident in months or years, that the company has a safety record far better than the national average for its type of operation, and that it has won a number of safety awards. Such counterbalancing information will help to take the sting out of the story. Reporters are usually willing to include these facts, too.

It is not only good policy but a wise precaution to be honest and fair to press representatives whenever there is news—whether pleasant or unpleasant. This principle is vital to a good PR program and will pay considerable dividends in terms of getting fair treatment from the media.

It would be wise for a safety and health professional to anticipate that some day he or she may have to serve as a company spokesperson at an accident or disaster scene. It is imperative to follow company policy and procedure when acting as a company spokesperson. This needs to be planned ahead of time.

News media can actually help during a big emergency. Families, friends, and neighbors will be clamoring for news and the media can get it to them fast.

WORKING WITH COMPANY PUBLICATIONS

If the safety and health professional thinks the company publication has neglected safety, it is time to correct the situation. He or she should ask the editor how more news value and human interest can be worked into safety stories. With some editorial assistance, the safety program can be just as newsworthy as other stories. The safety professional should offer to provide details and descriptions of events. The idea is to give the editor plenty of good, current information. The editor is just as eager as anyone to publish interesting news and features, and will go more than halfway to think up ways to put news value and reader interest into safety doings.

Here is an example of how the safety and health professional and the editor can team up to make a routine safety happening more newsworthy. Suppose one of the employees, Oscar B—, reaches his 25th anniversary of steady work without a day's lost time due to an injury or occupational illness. This achievement probably entitles Oscar to a button or badge or some other award. The PR-conscious safety and health professional can ask, "Instead of just pinning this button on Oscar with a hearty handclasp and a few words of commendation, why not make a real event out of it? Take the occasion to tell Oscar—and all the world—that this plant places top priority on recognizing the contribution employees make to a safer, better workplace. Oscar has made such a contribution through his personal example of safe practices over the years."

As a result, the company publications editor does not merely publish a picture of Oscar and his award along with a two-line caption. The editor finds Oscar, sits down with him over a cup of coffee, and asks him a few questions about his career, about his opinions on safety "way back yonder" and now, and about any ideas he may have for making things even safer at the plant.

This example is only one idea of what can happen when the safety and health professional and the editor of the company publication get together to do a more imaginative and energetic job of publicizing the safety program. For instance, they can develop a system to be used when one department of the company wins an interdepartmental safety contest. Instead of merely recording the results of the contest, the editor can dig into the program of the winning department, interview the people responsible for its success, and perhaps come up with a magazine story that will give every department in the organization some hints on how to improve its safety activities.

Producing a Publication

Materials, such as safety newsletters, instruction cards, bulletins, broadsides, booklets, and manuals for communicating safety rules, information, and ideas in print, require careful planning and preparation. Among steps to be taken in planning both internal (to a company audience) and external (to the general public or other out-of-company groups) publications are:

1. Clearly define the objectives of the publication. Consider the type of audience to be reached by those objectives.
2. Determine how general or how restricted the message is to be.
3. Decide what form of publication will best convey the message.

4. Estimate cost of preparing and printing the publication in whatever forms, sizes, and quantities needed. An expenditure for a new publication must, of course, be provided for in the budget, whether or not the item is produced in-house.

If the objective is to give the worker specific rules to follow in doing a job safely and efficiently, an instruction card may be suitable. To stimulate general safety-consciousness, a broadside (single-sheet printed on one side) may be effective. If a series of short reminders, for example, on fire prevention, is needed, posters may be the answer. To treat a topic of general interest, such as methods of materials handling, a leaflet can be the best choice. Here, posters or leaflets from the National Safety Council, insurance company, or other organizations may be more effective, and more economical than in-house materials. Manuals may be required for highly technical jobs or for more thorough coverage of a plant's safety policies and rules. Even a company-wide (or plant-wide) public address system, closed-circuit TV network, or videotapes would be appropriate.

When the safety professional and editor are deciding on the form of the publication, they should keep in mind that there is a direct relationship between the appearance of a printed piece and the degree of interest it arouses. Most readers will react unfavorably to a bulletin, newsletter, or booklet with text in small type, few or no illustrations, narrow margins, and long paragraphs.

Reasonably large type (10 point or 12 point), selected to fit the size of the page and, of course, to accommodate the volume of material, will make the material more readable. The Council newsletters, published by various sections of the Industrial Division of the National Safety Council, are set in 10-point type in about 2-in.-wide columns. In addition, judicious use of white space and variety in size and placement of illustrations help make a publication both pleasing to the eye and easy to read. In safety, as in other fields, ideas conveyed in print are best received and absorbed if they are well organized and attractively presented.

Illustrations break up the text and help to convey information to the reader. Photographs that show action described in the text add realism to instructional materials such as manuals. Human interest photos are desirable in newsletters. Line drawings and sketches are valuable to clarify technical points on instruction cards, in manuals, and in other training materials. Awards can also be publicized.

If the printing process permits reproduction of photos and other illustrations, pictures of award winners, safety devices, and safe and unsafe practices can be used. To avoid embarrassing or ridiculing employees who have been injured or caught in an unsafe procedure, their features can be blocked out or the pictures can be posed (and so identified) to reproduce the event. Or cartoon illustrations can be made up depicting the unsafe procedures from the photograph.

Some state laws forbid publication of a person's photograph without their written permission. As a result, the editor should obtain a signed release from every person who appears in recognizable form in any picture. Often having a new employee sign a photo release is part of the employment routine. Asking for a photo release is just good manners. Details on illustrations are given in Chapter 24, Media.

Preparation of Material

Once the objectives, scope, and form of a publication are determined, the person preparing it should make an outline of the subject or subjects to be covered. For most types of material, the outline need not be elaborate, but it should be logical and complete, showing how each topic is a part of the overall plan.

Before gathering material, the writer should study the audience for whom the message is intended in order to get some idea of the readers' knowledge and level of comprehension. In the interests of accuracy, completeness, and balance, material should be gathered from several sources—including articles, books, and especially supervisors, workers, and others in the company who have had experience in the matters to be treated. To ensure technical accuracy, it may be necessary to solicit help from specialists in specific areas.

No matter what the form of the publication, the writer should keep in mind certain basic rules of good writing. To get ideas across quickly and easily, use short sentences, simple words, and brief paragraphs.

In a piece of some length, such as a booklet, a system of headings, kept as informal as possible, will both arouse the reader's interest and guide his or her thinking. In a piece designed to instruct, numbered lists of job steps, for instance, will prove helpful. In any case, the writer should follow closely the line of logical thought developed in the outline.

Copy should be written in a positive, constructive style. When the nature of the material and the form of publication permit, a friendly—but never condescending—tone can be used effectively. Personal references and names, as in a newsletter, will increase readership. Humor tied to the message and pitched to the employees' sense of what is funny can add a great deal to some types of publications. For instance, cartoon illustrations and a light touch may be particularly effective in a rule booklet. The proposed publication's readability can be gauged by having a few people from the intended audience test-read the piece for understanding.

Production of Publications

For the technical details of printing, the advertising department or experts in the publishing field can be consulted. In terms of layout and typography, readability should be the first consideration.

How the piece is to be used will determine its size, paper, binding, cover, and similar details. If the materials are to be filed or if revised pages will be inserted, loose-leaf binders may be used. The in-company or outside editor or printer who will handle the job should be asked for technical advice.

Getting Ideas

Everyone in the PR business runs out of ideas now and then. Anyone suffering from this affliction should not hesitate to call on others for help. Employee publications do not compete with one another, so ideas can be borrowed freely from them. Some national agencies produce and supply safety materials that may serve as inspiration. (See Appendix 1, Sources of Help.)

The National Safety Council publishes in *Safety and Health* a "Safetyclips" page, which contains stories and illustrations for editors of employee publications. Council newsletters and other publications contain a wealth of interesting and informative material. Council poster miniatures and other materials usually are released for general use if the customary publication credit is given.

Volunteering to serve as an editor of a National Safety Council section newsletter is good practice.

SUMMARY

- Companies must work to maintain management and employee interest in safety and health programs because accident prevention depends on the desire of people to work safely. Thus, the major objective of a program designed to maintain safety awareness is to involve employees in accident prevention.
- Indications of the need for such a program include an increased rate of accidents and injuries, deteriorating housekeeping, and decreasing interest on the part of supervisors in enforcing safety and health regulations and procedures.
- A well-designed program to maintain interest in safety and health issues attempts to "sell" safety to employees. It must be based on management and worker needs and opinions. When developing the program and its related activities, the safety professional will need to consider company policy and experience, budget and facilities, types of operations, types of employees, basic human interest factors for promotional activities, and other considerations. Activities and promotional efforts should emphasize a positive, constructive approach to safety and be changed often to catch employees' attention.
- Safety and health professionals are responsible for assisting line management in developing and coordinating the program and for encouraging the support of management, supervisors, and employees. Safety and health directors also work with local, community, state, and national groups interested in promoting

accident prevention. They must be able to communicate their ideas clearly and effectively and to educate supervisors to maintain high levels of safety performance in their work areas.

- The ranking managers play a key role in any program designed to create and maintain interest in safety. Under existing laws, they are directly accountable to management and society for their employees' safety. Line managers, including supervisors, must (1) set a good example for their workers, (2) teach safe practices and procedures to employees, (3) enforce the rules fairly and consistently, and (4) keep abreast of the latest developments in safety and health issues.

- Programs designed to create and maintain interest in safety can include safety committees and observers along with quality and safety circles. Committees should consist of management and worker representatives, and membership should be rotated periodically. Quality and safety circles involve employees in hazard recognition and control. These groups meet regularly to discuss safety and health-related issues.

- Safety and health meetings can be conducted for managers, supervisors, employees, or other groups within a company. These gatherings can be of top executives, mass meetings of all employees, departmental meetings, or small group meetings. The gatherings must be carefully planned to focus on specific problems and solutions, or their effectiveness will be diluted.

- Various promotional activities such as contests can help to stimulate worker interest in safety and health policies and procedures. Contests should be undertaken only after management has established a sound safety program in the company—they must not be viewed as a substitute for such a program.

- To be successful, contests must be planned and conducted by a committee representing all competing groups; competing groups must be composed of natural units; methods of grading must be simple and fair; awards must be worth winning; and the contest should be well publicized and promoted. Types of contests include injury-rate contests, Council contests and awards, associate contests, intergroup competitions, personalized contests (raffles, sweepstakes), noninjury-rate contests, community and family contests, and miscellaneous contests.

- The value of contest awards lies in their appeal to basic human interest factors such as pride, need for recognition, urge to compete, and desire for financial gain. Awards must be meaningful to workers to be properly motivating. Many companies allow employees to select their own awards. Award presentations should be staged to provide maximum recognition for the recipients and maximum publicity for the company.

- Posters and displays can reach a large audience with brief, simple messages designed to convey information, to change attitudes, and to change behavior. Properly used, posters and displays have great value in safety programs. They are used to remind employees to work safely, suggest behavior patterns that prevent accidents, inspire interest in safety efforts, foster attitudes that accident-prevention is a skill, and remind employees of specific hazards.

- There are three types of posters: general posters concerned with broad safety topics, special industry posters that apply only to specific industries, and special hazard posters emphasizing particular hazards.

- Posters and displays should be set up where employees are most likely to see them. They should be prominently visible and changed often to maintain worker interest in their messages. Companies can use professionally designed posters or ask managers and employees to design their own. Posters can be mounted on bulletin boards or hung in their own frames.

- Other promotional methods include various campaigns to promote home safety, vacation safety, fire safety, and the like. These campaigns can be timed to coincide with elections, sports events, holidays, and so on. Unique safety ideas or stunts can be effective if they are not overdone. Courses and demonstrations on safety and health topics serve the dual purpose of stimulating interest in safety and teaching workers valuable skills and techniques. Finally, various publications such as reports, annual reports, newsletters, and booklets, leaflets, and personalized messages can help generate interest in safety among workers.

- Company public address systems and suggestion systems are also effective ways to maintain workers' interest. Public address messages should be brief and well timed, or employees tend to tune them out. Suggestion systems must be carefully planned and administered to reward good ideas and demonstrate the company's interest in safety and health policies.

- The Wise Owl Club, Golden Shoe Club, and Turtle Club are three of the most well-known safety recognition organizations. These groups offer assistance to companies that wish to improve their safety record or who want to recognize workers who have escaped serious injury through the use of personal protective equipment.

- Many companies do little or nothing to publicize their internal safety and health programs and activities. Yet the best way for a company to develop a reputation for taking care of employees is to be known as a safe place to work.

- Public relations is defined as the management function that evaluates public attitudes, identifies the firm's policies and activities with the public interest, and brings company news to the public's attention. Most organizations have either a professional public relations department or delegate public relations responsibilities to the safety director.

- To be successful, a public relations program must be actively supported by top management, a strong safety program, and newsworthy stories supplied by the safety professional. Lack of a safety program and poor public relations efforts can cost companies millions of dollars each year in lost productivity and potential business.

- The safety professional can conduct publicity work within a company by coordinating efforts with the public relations department and with the employee publications team to publicize the safety program. The safety professional can point out how the PR director and publications editor can help to gain employee acceptance and support of the safety program.

- Safety activities contain a wealth of real news. Items such as no-accident records, interplant safety contests, safety suggestions by employees, safety award presentations, and the like can be the focus of a company media event.

- In companies without public relations departments, the safety and health director can still obtain publicity for the firm's safety activities. Stories should be selected for a specific audience, contain human interest and appropriate humor, include graphics (cartoons, photographs, posters, etc.), and focus on people, not things.

- Safety professionals who want to make the most of public relations opportunities must (1) be honest in what they say, (2) deliver what they promise, (3) notify the media of any changes immediately, (4) supply accurate information, (5) offer ideas, not demand space or time, (6) feature only real news, (7) tip off local media to any worthwhile story, (8) give stories news value, and (9) be forthright about bad news (accidents, injuries, etc.).

- In working with local media, the safety and health professional should keep in mind the following tips: (1) keep press releases brief, timely, accurate, and complete; (2) be familiar with the publication or radio/TV station and know which editor or news director to contact; (3) tailor press releases for the audience; (4) include captions with photographs; and (5) leave script writing to the professionals but make sure all facts are accurate.

- When the safety and health professional must give a radio or TV interview, the individual should (1) be natural and refrain from distracting gestures or adopting an "official" tone, (2) dress conservatively, (3) review in advance what topics will be covered, and (4) respond naturally to the interviewer's questions without attempting to cram information into the time allotted.

- When a serious accident occurs at a plant or company, the public relations and safety professional must work together to provide accurate, honest information about the incident. Bad news about the accident should be balanced by pointing out the company's overall safety record or safety achievements.

- Company publications represent excellent opportunities to publicize safety programs, activities, and issues. In planning to start internal and/or external publications, the company should take the following steps: (1) clearly define the objectives and purpose of the publication, (2) determine how general or restricted the message will be, (3) decide what form of publication is best, and (4) estimate costs of preparing and printing the publication.

- Eye-catching illustrations and concisely written, lively copy add interest to publications. The more audience-based the publication is, the more readers are likely to accept the messages they deliver. Companies should also seek the advice and expertise of professionals in the public relations field to help ensure the success of their publications.

REFERENCES

Alliance of American Insurers, 1501 Woodfield Road, Schaumburg, IL 60195. *Fire Prevention and Control*, A/V.

Culligan MJ and Crewe D. *Getting Back to the Basics of Public Relations and Publicity*. New York: Crown Publishers, 1982.

Forrestal D. *Public Relations Handbook*, 2nd ed. Chicago: The Dartnell Corp., 1979.

Indiana Labor and Management Council, Inc., 2780 Waterfront Parkway, Indianapolis, IN 46214. *Worker Involvement in Hazard Control Booklet*, 1985.

Lesly P. Lesly's *Public Relations Handbook*. Report. Englewood Cliffs, NJ: Prentice-Hall, 1983.

Moore HF and Kalupa FB. *Public Relations Principles, Cases and Problems*. Homewood, IL: Richard D. Irwin, 1985.

National Safety Council, 1121 Spring Lake Drive, Itasca, IL 60143.

Accident Facts (annual).

Catalog and Poster Directories.

Family Safety & Health Magazine.

Occupational Safety and Health Data Sheets.

Nonprojected Visual Aids, 12304–0564, 1993.

Photography for the Safety Professional, 12304–0619, 1991.

Posters, Bulletin Boards, and Safety Displays, 12304–0616, 1991.

Writing, Publishing, and Administering Employee Safety Regulations, 12304–0664, 1991.

Section "Newsletters."

Today's Supervisor Magazine.

Safety and Health Magazine.

101 More Ideas that Worked.

Safedriver Magazine.

Safeworker Magazine.

Nolte LW and Wilcox DL, eds. *Fundamentals of Public Relations: Professional Guidelines, Concepts, and Integrations,* 2nd ed. Elmsford, NY: Pergamon Press, Inc., 1979.

Parkhurst W. *How to Get Publicity.* New York: Times Books, 1985.

REVIEW QUESTIONS

1. What is the main objective of a safety awareness program?

2. What are five factors a safety and health professional needs to consider when developing a safety awareness program?
   a.
   b.
   c.
   d.
   e.

3. List three indications of the need for a safety awareness program.
   a.
   b.
   c.

4. Which of the following can help to stimulate worker interest in safety and health policies and procedures?
   a. Contests
   b. Unique safety ideas
   c. Posters and displays
   d. Campaigns
   e. Publications
   f. All of the above

5. List five ways to make a contest successful.
   a.
   b.
   c.
   d.
   e.

6. Name the three broad categories of posters.
   a.
   b.
   c.

7. When working with local media, what five tips should the safety and health professional keep in mind?
   a.
   b.
   c.
   d.
   e.

8. What are the four steps the company should take in planning to start internal and/or external publications?
   a.
   b.
   c.
   d.

9. When a serious accident occurs at a facility or company, who must work together to provide information about the incident?
   a. Top management and line managers
   b. Line managers and supervisors
   c. Public relations and the safety professional

10. To be successful, a public relations program must be actively supported by which of the following?
    a. A strong safety program
    b. Newsworthy stories supplied by the safety professional
    c. Top management
    d. All of the above
    e. Only a and c

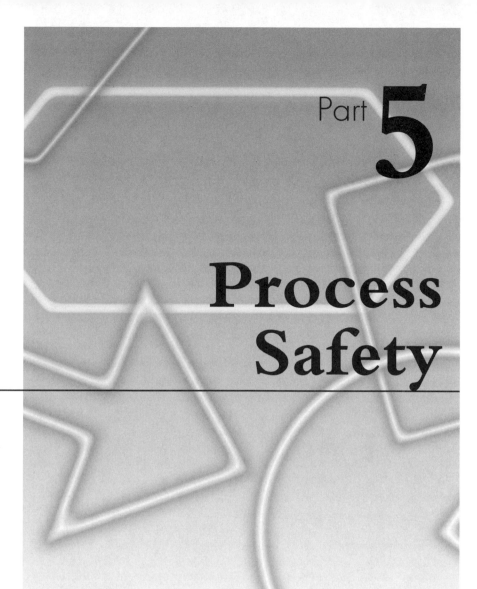

Part **5**

Process Safety

INTRODUCTION

Part five describes the various regulations and activities designed to protect employees, the public, and the environment from the consequences of potential accidents involving hazardous materials. Process safety management represents a major regulatory initiative in the 1990s and illustrates the importance of community involvement and relations. Safety, health, and environmental management now includes not only activities and processes inside facility boundaries but also those extending into the surrounding community.

26

Process Safety Management

Releases of flammable liquids, vapors, and gases or chemicals have probably occurred for as long as there have been processes that use temperature and pressure to produce hazardous and toxic materials. Beginning in the early 1980s, a number of serious major incidents occurred involving highly hazardous chemicals. These incidents resulted in considerable numbers of fatalities and injuries and significant property losses (Tables 26-A and 26-B). They spurred government agencies, labor organizations, and industry associations throughout the world to develop and implement codes, regulations, procedures, and safe work practices aimed at eliminating or reducing these undesirable events. This chapter covers the following topics:

- regulatory, industry, and labor efforts to control the processing of toxic and hazardous substances
- government involvement in regulating the manufacture of toxic and hazardous substances
- components of an effective process safety management program
- OSHA standards for process safety management
- conducting efficient, confidential process safety audits.

HISTORICAL PERSPECTIVE

The term "process safety" is most commonly used to describe various regulations and activities designed to protect employees, the public, and the environment from the consequences of major chemical accidents involving highly hazardous materials. In practice, the term has many definitions. According to the Chemical Manufacturers Association (CMA),

Process safety is the control of hazards which are caused by poor operation or malfunction of the processes used to convert raw materials into finished

Table 26-A. Examples of Serious Hazard Release Incidents (1984–1991)*

Year	Location	Incident	Fatalities	Injuries
1984	Bhopal, India	Toxic chemical release	2,000+	20,000+
1984	Mexico City	LPG release & explosion	542	many
1985	West Virginia	Toxic chemical release	0	135
1988	Louisiana	Vapor cloud explosion	7	unknown
1989	Texas	Vapor cloud explosion & fire	23	132
1990	Texas	Wastewater tank explosion	17	unknown
1991	Louisiana	Explosion and fire	8	128

* Such incidents led to the development and passage of the OSHA PSM standard.

products, which may lead to the unplanned release of hazardous material.

OSHA has defined process safety management as

the proactive identification, evaluation, and mitigation or prevention of chemical releases that could occur as a result of failures in processes, procedures, or equipment.

Companies must consider all of the following in the systematic identification and evaluation of hazards: the process design, technology, changes in the process, materials and changes in materials, operations and maintenance practices and procedures, training, emergency preparedness, and other elements impacting upon the process. Such an analysis will help to determine whether these items have the potential to cause a catastrophic incident in the workplace and surrounding community.

INDUSTRY AND LABOR PROCESS SAFETY INVOLVEMENT

During the 1980s, the petroleum, petrochemical, and chemical industries recognized that process safety technology without process safety management would not prevent catastrophic incidents. A number of industry associations, such as the American Institute of Chemical Engineers, the American Petroleum Institute, and the Chemical Manufacturers Association, initiated programs to develop and provide process safety management guidelines for use by their members who process hazardous materials. (See also the References section at the end of this chapter.)

Center for Chemical Process Safety

The Center for Chemical Process Safety (CCPS) was formed by the American Institute of Chemical Engineers (AIChE) in 1985 to promote the improvement of process safety management techniques among those who store, handle, process, and use hazardous materials. The following documents are examples of some materials developed and issued by the CCPS to promote process safety management systems:

- *Guidelines for Hazard Evaluation Procedures*. This manual describes a selection of hazard evaluation procedures used by chemical and petrochemical companies to identify and evaluate process hazards.
- *A Challenge to Commitment*. This brochure introduces an overview and presents an outline for the technical management of chemical process safety, characterized by the identification and brief description of each of the CCPS's 12 distinct and essential elements of process safety management.
- *Guidelines for Technical Management of Chemical Process Safety*. This book provides the framework and detailed components of the CCPS Chemical Process

Table 26-B. Major Worldwide Process Industry Explosions and Fires, 1965–1995 (Losses in Millions of Dollars Adjusted to 1995 Values)

Location Facility		Number of Incidents and Property Loss ($ MM)				
		Refineries	Petrochem.	Gas Plants	Misc.	Totals
North & South America	No.	29	26	2	1	58
	Amt.	$ 1,930	$ 2,010	$ 55	$ 25	$ 4,020
Europe	No.	7	6		1	14
	Amt.	$ 650	$ 460		$ 45	$ 1,155
Asia & Australia	No.	3	2	1		6
	Amt.	$ 440	$ 100	$ 65		$ 605
Middle East & Africa	No.	2		2	1	5
	Amt.	$ 180		$ 245	$ 15	$ 440
Total No.		41	34	5	3	83
$ Amount		$ 3,200	$ 2,570	$ 365	$ 85	$ 6,220

Source: David Mahoney, ed. *Large Property Damage Losses in the Hydrocarbon-Chemical Industries, A Thirty-Year Review*, 16th ed. Chicago: M&M Protection Services, 1995.

Safety Management System. It discusses the various alternatives for implementation and provides detailed information to help managers understand each of the 12 basic elements and components of process safety management.

* *Plant Guidelines for Technical Management of Chemical Process Safety*. This document provides specific and detailed technical guidance for the implementation of the 12 elements of the process safety management system within process facilities.

American Petroleum Institute

Beginning in 1990, the American Petroleum Institute (API) initiated an industry-wide program titled Strategies for Today's Environmental Partnership (STEP), aimed at addressing public concerns by improving industry's environmental, health, and safety performance. One of the seven strategic elements of the STEP Program covers operating and process safety. The following documents are examples of some of the materials that have been developed as a result of the STEP Program to help prevent or minimize the consequences of catastrophic releases of toxic or explosive materials:

* *API Guiding Environmental Principles and Management Practices*. This synopsis of API's Recommended Practice 9000 discusses the seven elements of the STEP Program.
* *Management of Process Hazards* (RP 750). This recommended practice addresses the management of process hazards in design, construction, start-up operations, inspection, maintenance and facility modifications. It applies specifically to refineries, petrochemical facilities, and major processing facilities that use, produce, process, or store flammable liquids and toxic substances in quantities above certain hazardous levels.

* *Management of Hazards Associated with Location of Process Plant Buildings* (RP 752). This recommended practice, codeveloped by API and CMA, provides a methodology for assessing and evaluating the hazards associated with the location of process buildings.
* *Management Practices, Self-Assessment Process, and Resource Materials* (RP 9000). This recommended practice is intended to act as a bridge between API's STEP Program and individual company activities to improve environmental, health, and safety performance. The resource materials and self-assessment process provide users with a means of measuring progress in implementing process safety management practices.

Chemical Manufacturers Association

In 1988, the CMA initiated its Responsible Care® Program in answer to public concerns about the manufacture and use of chemicals. The first of the 10 elements of Responsible Care®, "The Guiding Principle," outlines each member company's commitment to environmental, health, and safety responsibility in managing chemicals. The "Six Codes of Management Practices" describes virtually every aspect of chemical manufacturing, transporting, and handling. Examples of some of the other CMA documents that have been developed covering process safety issues include the following:

* *Process Safety Management (Control of Acute Hazards)*. This report contains results of a survey of chemical company process safety policies and methods available for hazard identification, assessment, and control. It describes the steps involved in designing, commissioning, and operating process facilities as

well as the role of supervisors and managers in process safety activities.

- *A Resource Guide for the Process Safety Code of Management Practice.* This document covers the CMA process safety code, describes the code in relation to other Responsible Care® elements, and provides information on improving safety performance.
- *Fixed Equipment Inspection Guide.* This manual provides a systematic approach to developing inspection functions integrating technology and skill, which are based on planned, preventive actions.

Other Initiatives

Additional examples of industry associations which have developed materials and programs providing guidance to their members on process safety management include, but are not limited to, the following:

- Organizations Resource Counselors' (ORC) report, *Process Hazards Management of Substances with Catastrophic Potential*
- National Petroleum Refiners Association (NPRA) program, *Building Environmental Stewardship Tools (BEST)*
- Synthetic Organic Chemical Manufacturers Association (SOCMA) program, *Chemical Process Operator Certification Training*
- United Nations International Labour Organization (ILO), *Code of Practice on the Prevention of Major Accident Hazards*
- International Chamber of Commerce (ICC), *Charter for Sustainable Development.*

GOVERNMENT PROCESS SAFETY INVOLVEMENT

In 1985, the Environmental Protection Agency (EPA) responded to public concern over the hazards of chemicals by launching its Chemical Emergency Preparedness Program (CEPP). This evolved into the Emergency Planning and Community Right to Know Act of 1986 (EPCRA), known as SARA Title III, which required companies to develop emergency preparedness, recognition, knowledge, and inventory of hazardous chemicals and reporting of toxic releases.

About the same time, OSHA promulgated its Hazard Communication Standard, which required manufacturers and users to identify hazardous materials in the workplace and inform employees and consumers of the hazards presented by their manufacture, use, storage, and handling. The Clean Air Act Amendments (CAAA) passed in 1990, mandated that both the EPA and OSHA develop regulations for chemical safety management to prevent catastrophic accidents. In response to this mandate, OSHA proposed standards to control process safety within the workplace, and regulations were issued by EPA covering releases primarily affecting areas outside of the workplace.

Occupational Safety and Health Administration

OSHA's process safety management (PSM) standard evolved from a number of previous industry and government guidance documents and initiatives, including OSHA's Hazard Communication Standard and two OSHA directives, *Systems Safety Evaluation* and *PETROSEP*.

- Instruction CPL 2-2.45, *Systems Safety Evaluation of Operations with Catastrophic Potential.* This 1988 directive provided guidance for systems safety programs and inspections in workplaces focusing on chemical-related operations with potential for fires, explosions, and hazardous releases. The directive was superseded by OSHA's PSM standard.
- OSHA *Petrochemical Industries (PETROSEP) Compliance Directive.* This program, initiated in 1990 and canceled in 1992, targeted large petroleum refining and petrochemical processing facilities to determine whether management systems governing safety and health procedures for operations, maintenance, and contractor activities were in place and were able to control risks and prevent disasters. In 1992 PETROSEP was replaced by OSHA's PSM standard.
- 29 *CFR* 1910.119, *Process Safety Management of Highly Hazardous Chemicals.* The major objective of this 1992 PSM standard is to prevent releases of hazardous chemicals that could expose employees and others to danger.

The PSM standard covers a wide range of manufacturing industries involved in handling, storing, processing, or transporting flammable liquids and gases in quantities of 10,000 pounds or more and specific toxic and reactive chemicals in regulated quantities. Exempted from the PSM standard are retail facilities; oil and gas well drilling or servicing operations; normally unoccupied remote facilities; hydrocarbons used solely as fuel in the workplace; and flammable liquids stored in atmospheric tanks or transferred at temperatures below their normal boiling point without cooling, which are not connected to a process. All facilities in the United States that store, use, manufacture, transport, or handle flammable liquids and highly hazardous chemicals should be familiar with the requirements of the PSM standard and whether it applies to their workplaces.

A key provision of the PSM standard requires covered facilities to gather written information concerning process technology, process equipment, and chemical hazards. This information is used to identify the processes that pose the greatest risks in the workplace. Employers then establish priorities for conducting process hazard analyses based on risk and improvement and conduct these analyses within prescribed time limits. The PSM standard requires that at least 25% of the covered processes were to be evaluated by May 26, 1994, with an additional 25% evaluated each following year. All process hazard analyses were to be completed no later than May 1997.

The PSM standard recommends that companies consider one of six commonly used methodologies (discussed later in this chapter) to conduct process hazard analyses. The PSM standard also has specific requirements regarding the experience and qualifications of the persons who are to conduct the process hazard analysis.

Other requirements of the PSM standard cover operating procedures; employee participation; training; contractors and contractor employees; pre-start-up safety reviews; mechanical integrity; hot work permit system; management of change; accident and incident investigation; emergency preparedness; compliance audits; and confidentiality agreements. Information concerning OSHA requirements for process safety management in the United States can be found in the following two booklets: OSHA 3132, *Process Safety Management*; and OSHA 3133, *Process Safety Management Guidelines for Compliance*. A single free copy of each booklet may be obtained by writing the OSHA Publications Office, Room N3101, Washington, DC, 20210; or by calling 202/523–9667.

Environmental Protection Agency

The EPA's process safety management regulations and risk management program evolved from the requirements of the Clean Air Act Amendments (CAAA) of 1990, which led to the following initiatives:

- 40 *CFR* 68 Sections 301 (r) and 304, Clean Air Act (CAA) Provisions for Process Safety Management. The CAA contains provisions requiring EPA to establish a list of substances and thresholds and promulgate accident prevention rules. The CAA mandated the inclusion of 16 specific chemicals, and EPA added 61 more chemicals to the list (as of 1995). The program requirements established by EPA to prevent releases affecting areas outside the facility are similar to those contained in the OSHA PSM standard covering the workplace and include:

  - reviewing and documenting the facility's chemicals, processes and procedures
  - conducting detailed process hazard analyses to identify hazards, assess the likelihood of releases, and evaluate the consequences of such releases
  - developing standard operating procedures and training employees on procedures
  - implementing preventive maintenance and management of change in operations programs
  - conducting reviews prior to initial start-up and prior to start-up following a modification
  - conducting periodic safety audits to assure compliance.

- 40 *CFR* 68 Section 112 (r), Risk Management Programs for Chemical Accidental Release Prevention. Section 112 (r) of the CAA requires facilities that store, handle, manufacture, distribute, or use more than the threshold quantity of a listed substance in a process to develop risk management plans (RMPs) to identify hazards, maintain safe facilities, and minimize the consequences of accidental releases of listed hazardous substances, regardless of quantity. These RMPs are to be registered with the EPA and made available to state and local emergency response agencies and the public.

The EPA's risk management plans regulation requires facilities to conduct hazard assessments that define the off-site impacts of potential releases, including worst-case scenarios. Facilities must also document a five-year history of releases and develop and implement protection programs which build upon OSHA's PSM standard. Information on EPA process safety requirements can be found in Bulletin EPA 510, *Managing Chemicals Safely,* which is available by writing to the EPA Chemical Emergency Preparedness and Prevention Office, OS-120, 401 M. Street, SW, Washington, DC 20460 or calling 202/260–8600.

PROCESS SAFETY MANAGEMENT PROGRAM

A process safety management program should be among the top priorities in a company's organization. All employees must be encouraged to adopt such a program as their own and to ensure that new employees understand its importance. Initially, management must identify the basic requirements for a safety management program, assign accountability for implementing and monitoring it, and continue to look for ways to improve the company's risk-management efforts.

Basic Process Safety Management Requirements

Process safety management is an integral part of a company's overall facility safety program. An effective process safety management program requires the leadership, support, and involvement of top management, facility management, supervisors, engineers, employees, contractors, and contractor employees. Much of the process safety management program relies upon access to good records and documentation. Therefore, management of information (company memory) is one of the key factors to a good program.

When developing a process safety management program, management should consider several important items, including:

- incident prevention objectives
- existing employer and contractor process safety management programs
- use of internal resources versus outside consultants

Process Safety Accountability

Because process safety management is the basis for all other safety efforts within the process facility, the company needs to establish accountability for all its management systems. This step requires that the company clearly define management and supervisory responsibility and accountability and establish specific goals and objectives in order for the program to work.

Components of process safety management accountability identified by the CCPS include:

- interdependent continuity of operations, systems, and organization
- control of process quality, deviations, exceptions, and alternate methods
- management and supervisory accessibility and communications
- company expectations, compliance audits, and measuring performance.

Process Risk Management

Prior to program implementation, top management should establish both long-term and short-term goals and objectives for each of the elements of its process safety management program. Long-term goals should be based on the regulatory requirements for program development and implementation and management's desired level of compliance. Short-term objectives should be based on the hazard analyses conducted within each facility for each specific unit, process, or material; the determination of its hazard potential; the degree of risk that management is willing to accept; and the timeliness and economic feasibility of instituting necessary changes.

Companies should consider the following items when developing process risk management parameters:

- hazard identification and potential impact
- risk analysis to determine degree of risk reduction
- residual risk management capabilities
- available emergency process management and shutdown procedures.

ELEMENTS OF THE PROCESS SAFETY MANAGEMENT PROGRAM

All process safety management programs should cover the same basic program requirements, although the number of elements may vary depending on the criteria used. Whether a company uses OSHA's PSM standard, EPA's risk management program, API's *Recommended Practice 750*, or another company, association, or government source document, certain basic requirements should be included in every process safety management program (Table 26-C).

Process Safety Information

Process safety information is used to determine which processes, materials, and equipment are critical to safety efforts. Process safety information includes compiling all available written information concerning process technology, process equipment, hazardous materials, and chemical

Table 26-C. Comparison of Selected Government and Industry Process Safety Management Program Elements

Program Elements	OSHA PSM Std	EPA RPM	API RP 750	CCPS Elements
Process Safety Information	x	x	x	x
Employee Involvement	x			x
Process Hazard Analysis	x	x	x	
Management of Change	x	x	x	x
Operating Procedures	x	x	x	
Safe Work Practices and Permits	x		x	
Employee Information and Training	x	x	x	x
Contractor Personnel	x			
Pre-Start-up Safety Reviews	x	x	x	x
Design and Quality Assurance	x		x	x
Maintenance and Mechanical Integrity	x	x	x	x
Emergency Planning, Preparedness, and Response	x	*		
Compliance Audits	x	x	x	x
Process Incident Investigation	x	x	x	x
Standards and Regulations				x
Trade Secrets	x			

Note: * EPA covers emergency response separate from RMP.

hazards before conducting a process hazard analysis. This information may exist in various locations throughout the facility, may be in a central location, or may remain at the design or supply source. Other critical information includes documentation of capital project reviews and design criteria. The following items are particularly important.

Chemical Information

This includes the chemical and physical properties, reactivity and corrosive data, and thermal and chemical stability of the highly hazardous materials in the process, as well as the hazardous effects of inadvertently mixing incompatible materials. Material Safety Data Sheets (MSDSs) or other sources of information should be available covering each chemical and material, including its toxicity and permissible exposure limits, so that safety professionals can evaluate the potential hazard.

Management must have information on the maximum intended inventory of highly hazardous chemicals within the process or facility at any one time and their typical usage to determine if company or regulatory quantity limits for listed chemicals are exceeded. Chemical information also includes data that may be needed to conduct environmental hazard assessments of toxic and flammable releases.

Process Technology Information
This includes block-flow diagrams and/or simple process-flow diagrams. The safety professional should provide a description of the chemistry of each specific process, including safe upper and lower limits for temperatures, pressures, flows, compositions, and, where available, process design material and energy balances. In addition, the safety professional should evaluate consequences of deviations, including the effects on employee safety and health. Whenever processes or materials are changed, the information should be updated and reevaluated.

Process Equipment and Mechanical Design Information
This includes all documentation covering the design codes employed and an evaluation as to whether company equipment complies with recognized engineering practices. There should be written documentation verifying that existing equipment designed and constructed in accordance with codes, standards, and practices that are no longer in general use has been maintained, operated, inspected, and tested to ensure its continued safe operation. Whenever equipment or systems are changed, the company should update and reevaluate information on materials of construction, piping and instrument diagrams, relief system design, electrical classification, ventilation design, and safety systems (shutdowns, interlocks, detectors, alarms, etc.). If original design information is not available, the necessary information may be reconstructed from equipment and inspection records.

Employee Involvement
Management should consider developing and implementing a program that enables employees to participate in process safety analyses and other elements of the process safety management program. This program may give employees and contractor employees (or their representatives) access to all process safety information, incident investigations, and process hazard analyses. The following are examples of elements that may be included in employee involvement programs:

- safe work procedures and activities
- operator/process and operator/equipment interface
- administrative controls versus engineering controls
- written operating procedures.

These programs should provide for consultation with employees (and their representatives) as required by applicable regulations, such as the OSHA PSM standard.

Conducting Process Hazard Analyses
After process safety information is compiled, the company should conduct a thorough and systematic multidisciplinary process hazard analysis, appropriate to the complexity of the process to identify, evaluate, and control the hazards of all processes. Persons performing the process analyses must have experience in engineering and general process operations. The analysis team should include at least one person who is thoroughly familiar with the process being analyzed and one person who is competent in the hazard analysis methodology being used.

Each facility or company should determine and document the priority order they will use to begin conducting process hazard analyses, based on the following criteria:

- extent of the process hazards
- number of potentially affected employees
- operating history of the process
- age of the process.

Some of the more frequently used methods of conducting process safety analyses include:

Fault Tree Analysis and Event Tree Analysis
Both methods are formal deductive techniques used to estimate the quantitative likelihood of events occurring. Fault tree analysis works backward from a defined incident to identify and display the combination of operational errors and/or equipment failures involved in the incident. Event tree analysis (the reverse of fault tree analysis), works forward from specific events, or sequences of events, to pinpoint those that could result in hazards and to calculate the likelihood of an event occurring.

HAZOP Study
The hazard and operability (HAZOP) study method is commonly used in the chemical and petroleum industries. HAZOP requires a multidisciplinary team, guided by an experienced leader. The team uses specific guide words (such as "no," "increase," "decrease," "reverse,") that are systematically applied to identify the consequences of deviations from design intent for various processes and operations.

Failure Mode and Effect Analysis
This method is used to correlate each system or unit of equipment with (1) its potential failure modes, (2) the effect of each potential failure on the system or unit, and (3) how critical each failure could be to the integrity of the system. The failure modes and the effects are then ranked according to criticality to determine which ones are most likely to occur and possibly cause a serious incident.

What If?
This method works by asking a series of questions to review potential hazard scenarios and possible consequences. What if is a good method to use when analyzing proposed changes to materials, processes, equipment, or facilities.

Checklist

This method is similar to the what if method, except that a predeveloped checklist is used that is specific to an operation, process, or unit. Companies often use the checklist method when conducting pre-start-up reviews at the time of initial construction or after completing a major turnaround or addition to the process or facility.

What If?/Checklist

This method is a combination of the what if and checklist methods and is typically used when a company analyzes units that are identical in construction, materials, and process.

No matter which method of process hazard analysis is used, all analyses should provide the following information:

- process location, siting, and hazards of the process
- identification of any prior incident with potential catastrophic consequences
- engineering and administrative controls applicable to the hazards
- interrelationships of controls and appropriate application of detection methodology to provide early warnings
- consequences of human factors, facility siting, and failure of the controls
- consequences of safety and health effects on workers within areas of potential failure.

Management should develop a schedule for acting on process hazard analyses findings and recommendations. This schedule may include timetables for planned actions, documentation of actions to be taken (or not taken), and notification of employees involved. Process hazard analyses, updates, and revalidation, including documentation of resolution of recommendations, should be maintained for the life of each process. A good process safety management practice is to review, update, and reevaluate process hazard analyses at least every five years, even if no changes to the processes, equipment, or materials have occurred.

Management-of-Change Programs

Facilities should develop and implement a program to revise process safety information, procedures, and practices as necessary. The program should include a system of management authorization and written documentation for all changes to materials, chemicals, technology, equipment, procedures, personnel, or facilities that affect each process.

The management-of-change program should consider and document the following information, as a minimum:

- change of process technology
- changes in facility, equipment, or materials

- management-of-change personnel and organizational and personnel changes
- temporary changes, variances, and permanent changes
- enhancement of process safety knowledge, including:

  - technical basis for proposed change
  - impact of change on safety and health
  - modifications to operating procedures and safe work practices
  - modifications required to other processes
  - time required for the change
  - authorization requirements for the proposed change
  - updating documentation relating to process information, operating procedures, and safety practices
  - required training or education due to change

- management of subtle change (anything that is not a replacement in kind)
- nonroutine changes.

Prior to the start-up of the process or affected part of the process, managers and supervisors should make sure that employees and maintenance and contractor personnel involved in the process are informed of the changes. They should also provide employees with updated operating procedures, process safety information, safe work practices, and training, as needed.

Operating Procedures

Process facilities should develop and implement written operating instructions and appropriate work practices and procedures that enable employees safely to conduct activities involved in every process. Consistent with the management-of-change program, management should review and update or amend operating instructions and procedures as changes occur, and certify them annually for completeness and accuracy. Operating instructions should specify detailed procedures and cover the process unit's operating limits, including the following three areas:

- consequences of deviation
- steps to avoid or to correct deviation
- functions of safety systems related to operating limits.

Operating instructions should be clear, precise, and understandable, and incorporate existing safe work practices such as hot work permits, lockout/tagout, personal protection, fire prevention, and confined-space entry. Operating instructions should be accessible to all employees and contractor employees involved in the process and cover the following areas, as a minimum:

- initial start-up and start-up after turnaround
- normal start-up, normal and temporary operations, and normal shutdown

- emergency operations and emergency shutdown
- start-up after emergency and after temporary operations
- conditions under which emergency shutdown is required and assignment of shutdown responsibilities to qualified operators
- nonroutine work
- operator/process and operator/equipment interface
- administrative controls versus automated controls.

Safe Work Practices and Permits

Facilities should develop and implement a program that uses hot work and safe work permits to control all work conducted in or near process areas. These permits, as a minimum, should include fire prevention and protection, lockout/tagout, safe work practices, personal protection, emergency response, and confined-space entry requirements applicable to the work, nature of work, materials involved, and object or equipment on which the work is to be performed. Supervisors, employees, and contractor personnel should be trained in the requirements of the permit program, including permit issuance and expiration and appropriate safety, materials handling, and fire protection and prevention measures. Facilities may consider retaining copies of permits on file until the work is completed to provide information in the event that an incident occurs during the course of the work.

Examples of work to be considered for inclusion in a permit program include, but are not limited to:

- hot work
- lockout/tagout of electrical, mechanical, pneumatic energy and pressure
- confined-space entry
- opening process vessels, equipment, lines, and the like (Figure 26-1)
- control of entry into process areas by nonassigned personnel.

Facilities should develop and implement safe work practices that apply to both employees and contractor personnel. These practices should cover appropriate safety and health considerations to control potential hazards during process operations including, but not limited to, the following:

- properties and hazards of chemicals used in the process
- engineering, administrative, and personal protective equipment controls to prevent exposures
- measures to be taken in event of physical contact or exposure
- quality control of raw materials and inventory control of hazardous chemicals
- safety (interlock, suppression, detection, etc.) system functions
- special or unique hazards in the workplace.

Figure 26-1. Following safe procedures and wearing protective equipment, this worker checks a process vessel.

Employee Information and Training

Management should develop and implement a formal process safety training program covering all incumbent, reassigned, and new supervisors and employees. Employers may initially certify that experienced employees are qualified by knowledge, skill, and experience to conduct operating procedures safely. However, requirements for refresher, remedial, and skills-improvement training for each employee involved in a particular process should be reviewed regularly (for example, every three years or sooner, if necessary). The review schedule will be determined by management in consultation with the employees involved. The company should keep detailed training documentation on file, including identity of the employee, date of training, and means of verifying the employee's knowledge, understanding, and skills. (See also Chapter 23, Safety and Health Training.)

Management should consider the following areas, as a minimum, when implementing the training program:

- required skills, knowledge, and qualifications of process employees
- design of process operating and maintenance procedures
- selection and development of process-related training programs
- measuring and documenting employee performance and effectiveness
- selection and training of instructors knowledgeable in the process.

Training should be provided for all operating and maintenance supervisors and personnel, both employee and contractor. Initial and refresher process training programs should include the following, as a minimum:

- overview of process operations and process hazards
- availability and suitability of materials and spare parts for the processes in which they are to be used
- process operating procedures
- process emergency and shutdown procedures
- safety and health hazards related to the process
- facility and process area safe work practices and procedures.

Contractor Personnel

Management should regard contractor personnel performing maintenance, repair, turnaround, major renovation, or specialty work on a process unit or within a process area as temporary employees of the company. Employers should develop and implement procedures to control the entry, presence, and exit of contractor personnel in covered process areas. Contractor personnel working in process areas must be fully aware of the hazards, processes, and operating and safety procedures and equipment in the area. Employers and contractors should periodically evaluate contractor personnel performance to ensure that contractor employees are trained and qualified and follow all safety rules and procedures. As a minimum requirement, all contractors and contractor personnel working in the process area should be informed and aware of the following:

- potential fire, explosion, and toxic release hazards related to their work
- plant safety procedures and contractor safe work practices
- emergency plan and contractor personnel actions
- controls for contractor personnel entry, exit, and presence in process areas.

When selecting contractors, employers should obtain and evaluate their safety records, performance, and programs. To provide safety information for future evaluation, employers may consider maintaining separate contractor injury and illness logs for all work conducted in process areas. Contractor employers should assure management of the following for all contractor personnel:

- Contractor employees are trained to perform work safely, and know and follow the safety rules of the facility.
- They are instructed in the hazards related to the job and process, and in applicable provisions of the emergency plan.
- They are instructed to advise facility management of any hazards found or created as a result of contractor work.

- Contractors document names of their employees who have been trained, dates of training, and means used to verify employees' understanding and knowledge of the training.

(See also Chapter 21, Contractor and Nonemployee Safety.)

Pre-Start-up Safety Reviews

Process facility management should institute a program to ensure that pre-start-up process safety reviews are conducted (1) before start-up of new process facilities; (2) before introduction of hazardous chemicals into new facilities; (3) following a major turnaround; or (4) where facilities have had significant process modifications.

The pre-start-up safety reviews, as a minimum, should make sure the following items have been accomplished:

- Construction, materials, and equipment are in accordance with design criteria.
- Process systems and hardware, including testing of computer control logic, have been inspected, tested, and certified.
- Alarms and instruments, relief and safety devices, signal systems, and fire protection and prevention systems have been inspected, tested, and certified.
- Safety, fire prevention, and emergency response procedures have been developed and reviewed, put in place, and are appropriate and adequate for the job-related hazards.
- Necessary start-up procedures are in place and proper actions have been taken to ensure safe start-up and operation.
- A process hazard analysis has been performed and all recommendations addressed, implemented or resolved, and actions documented.
- All required initial and/or refresher operator and maintenance personnel training is completed, including emergency response, process hazards and health hazards.
- All written procedures, including operating procedures (normal and upset), operating manuals, equipment procedures, and maintenance procedures are completed and in place, have been reviewed, and are appropriate and accurate for the job-related hazards.
- Management-of-change requirements have been met for new processes and modifications to existing processes.

Design and Quality Assurances

When management contemplates new processes or major changes to existing processes, they need to establish a formal design and quality assurance program. Review teams, similar to the teams that perform process hazard analyses, should be established and directed to provide process safety design reviews before and during construction (prior to the pre-start-up review). Depending upon the construction, modifications, and complexity of the project, the appropriate times to perform these design and quality assurance reviews may vary from facility to facility and from process to process. There are, however, some obvious interim review periods common to all projects. For example, the first design control review for a capital project would be conducted just before plans and specifications are issued as "final design drawings," and typically would cover the following areas:

- plot plan, siting, spacing, electrical classification, drainage, and so on
- hazards analysis and process chemistry design review
- project management
- process equipment and mechanical equipment design and integrity
- piping and instrument drawings (P&IDs)
- reliability engineering, alarms, interlocks, reliefs, and safety devices
- materials of construction and compatibility.

Another review is normally conducted just before the start of construction. It would cover such areas as the following:

- demolition and excavation procedures
- control of raw materials
- control of construction personnel and equipment entry into facility and site
- fabrication, construction, and installation procedures and inspection.

One or more reviews may be conducted during the course of construction or modification to verify that the following areas are in accordance with design specifications and facility requirements:

- materials of construction provided and used as specified
- proper assembly and welding techniques, inspections, verifications, and certifications provided
- chemical and occupational health hazards considered during construction
- physical, mechanical, and operational safety hazards considered during construction and facility safety practices followed
- interim protective and emergency response systems provided and working
- facility permit program is in place and used during construction.

Maintenance and Mechanical Integrity

Process facilities should develop and implement a program with written procedures to maintain the ongoing integrity of process equipment. This program would ensure and document the periodic inspection, testing, performance maintenance, corrective action, and quality assurance of process-related equipment. Management should make sure that the mechanical integrity of equipment and materials is reviewed and certified and deficiencies corrected prior to start-up, or provisions made for appropriate safety measures. Mechanical integrity requirements apply to the following equipment and systems:

- pressure vessels and storage tanks
- emergency shutdown systems
- relief and vent systems and devices
- process safeguards such as controls, interlocks, sensors, alarms, and so on
- pumps and piping systems (including components such as valves)
- quality assurance, materials of construction, and reliability engineering
- maintenance and preventive maintenance programs.

The company should develop and implement an inspection and testing program that complies with regulatory requirements or recognized good engineering practices. New and replacement equipment should be inspected during fabrication and installation to verify that it meets design requirements and is suitable for the process application intended. The program also should cover in-spection and testing of maintenance materials, spare parts, and equipment to ensure proper installation and adequacy for the process application involved. The acceptance criteria and frequency of inspections and tests should conform with manufacturers' recommendations, good en-gineering practices, regulatory requirements, industry practices, facility policies, or prior experience.

Inspections and test results should be documented and include the following information, as a minimum:

- date of inspection or test
- name of person performing test
- identification of equipment
- description of work performed
- acceptance limits or criteria and results of test or inspection
- steps required and taken to correct or mitigate deficiencies outside acceptable limits.

Emergency Planning, Preparedness, and Response

Management should develop and implement an emergency preparedness and response plan covering the entire process facility. The company should also make sure that all employees and contractor employees are trained or educated in emergency notification, response, and evacuation procedures. An emergency control center should be established that includes the appropriate process safety information and adequate means of communication should there be an emergency in any process unit within the facility or a hazardous substance release affecting areas outside the facility.

The emergency preparedness plan should comply with applicable company and regulatory requirements and include the following minimum requirements:

- distinctive employee and/or community alarm system
- preferred method of internal reporting of fires, spills, releases, and emergencies
- requirements and methods of reporting process-related incidents to appropriate government agencies
- evacuation, emergency escape procedures, and route assignments
- procedures to account for all personnel
- emergency response capabilities including public safety, contractors, and mutual aid
- employee and contractor employee rescue and emergency duties
- procedures for handling small releases of hazardous chemicals
- hazardous waste emergency response (HAZWOPER) procedures.

Compliance Audits

Process facilities should develop and implement a process safety management audit program. The program would be used to measure facility performance and ensure compliance with internal and external (regulatory and industry) process safety management requirements. Management should establish audit objectives and set a schedule to audit process units periodically. The frequency of audits should be determined by management; however, for compliance with the OSHA PSM standard, employers must certify that all process operations have been audited at least every three years.

When establishing a compliance audit program, facility management should be prepared to accomplish the following:

- Establish goals, schedules, and methods of verification of findings prior to the audit.
- Determine the methodology (or format) to be used in conducting the audit, and develop appropriate checklists or audit report forms.
- Prepare to certify compliance with government, industry, and company requirements.
- Assign a knowledgeable audit team (internal and/or external expertise).
- Promptly address all findings and recommendations, and document actions taken.
- Keep a copy of at least the most recent compliance audit report on file (OSHA requires the last two reports to be maintained on file).

Because the objectives and scope of audits can vary, the compliance audit team should include at least one person knowledgeable in the process being audited, one person with applicable regulatory and standards expertise, and other persons with the skills and qualifications necessary for conducting the audit. Management may decide to include one or more outside experts on the audit team due to lack of facility personnel or expertise or because of regulatory requirements. (See also the section on Process Safety Audits later in this chapter.)

Process Incident Investigation

Management should establish a system to (1) thoroughly investigate and analyze incidents and near-misses associated with process systems, (2) promptly address and resolve findings, and (3) review recommendations with employees and contractors whose jobs are relevant to the incident findings. Any incident (or near-miss) resulting in a catastrophic hazardous chemical release (or potential release) should be thoroughly investigated as soon as possible by a team trained in investigation techniques. This team should include at least one person knowledgeable in the process operation involved, a contractor employee if the work involved a contractor, and others with appropriate knowledge and experience.

A detailed report of each investigation should be prepared and maintained on file (OSHA requires records to be kept for five years) together with documentation of corrective actions (or reasons for no action). The investigation report should include at least the following information:

- date of incident or near-miss
- date investigation began
- description of incident
- contributing factors
- human error assessments
- recommendations, follow-up, and actions resulting from the investigation.

(See also Chapter 3, Loss Control Programs, and Chapter 7, Accident Investigation, Analysis, and Costs.)

Standards and Regulations

Process facilities are subject to two distinct and separate forms of standards and regulations, external and internal.

External

External codes, standards, and regulations applicable to the design, operation, and protection of process facilities and employees typically include government regulations and association and industry standards and practices.

Internal

Internal policies, guidelines, and procedures are those developed or adopted by the company or facility to complement the external requirements and to cover distinct or unique processes. Internal standards should be reviewed periodically and changed when necessary, in accordance with the facility's management-of-change system.

Confidentiality and Trade Secrets

Process facility management needs to provide process information, without regard to possible trade secrets, to persons with the following responsibilities:

- gathering and compiling process safety information
- conducting process hazard analyses
- developing maintenance, operating, and safe work procedures
- assisting in incident (near-miss) investigations
- developing emergency plans and response
- conducting compliance audits.

The OSHA PSM standard provides for confidentiality agreements to protect trade secrets. The standard states that nothing shall prevent employers from requiring that persons who receive information about the processes sign agreements not to disclose the information.

ENFORCEMENT OF OSHA PROCESS SAFETY MANAGEMENT STANDARD

Following the promulgation of OSHA's standard 29 *CFR 1910.119, Process Safety Management of Highly Hazardous Chemicals,* OSHA issued Instruction CPL 2-2.45. This regulation establishes uniform policies and procedures for the enforcement of the PSM standard within the United States. Four types of process safety compliance inspections may be conducted to determine whether a facility is covered by the standard and/or to assess the facility's compliance with the standard:

- **Inspections Resulting from Response to Accidents and Catastrophes.** Unprogrammed inspections, especially those triggered by a serious incident or by a complaint, can occur anytime. OSHA will follow normal procedures in responding to serious events of potential catastrophic significance. A program quality verification (PQV) inspection may be recommended as a result of OSHA's response. (See the Program Quality Verification Inspections section later in this chapter.)
- **Unprogrammed Process Safety Management-Related Inspections.** Any investigation conducted as a result of a complaint or referral will include a determination as to whether or not the facility is covered by the PSM standard. A PQV inspection may be recommended if major deficiencies exist. OSHA, however, will not wait for a company to conduct a PQV to follow through on obvious process safety violations.
- **Programmed General Industry Inspections.** Whenever any other type of programmed general industry safety and health inspection is conducted,

OSHA will determine if the facility should be added to the PQV inspection list.

- **Program Quality Verification Inspections.** Candidates for PQV inspections will be determined by criteria that include the facility's accident/incident history; facility age; number of affected employees; known catastrophic potential of chemicals used in the facility's processes; and past EPA, OSHA, and local fire department experience. A facility may be deleted from the list if its PSM systems have been audited during the past five years; if it is included in a corporate settlement agreement covering process safety; if it is a voluntary protection program (VPP) participant; or if it was previously screened and determined that a PQV inspection would not be conducted. Because of limited resources, OSHA may be able to conduct only a few PQV inspections each year in each region.

Program quality verification inspections are used to evaluate procedures used by facilities and their contractors to manage hazards associated with highly dangerous chemicals. These inspections have the following three goals:

- evaluate the employer's and contractors' PSM programs
- compare the quality of the programs to acceptable industry practices
- verify effective implementation of the programs.

PROCESS SAFETY AUDITS

The two basic principles of conducting self-evaluation audits are (1) gather all relevant documentation covering process safety management requirements at a specific facility and (2) determine the program's implementation and effectiveness by following up on their application to one or more selected processes. Facility management should develop a report of the audit findings and recommendations and should maintain documentation noting how deficiencies were corrected or mitigated and, if not, reasons why no corrective action was taken.

Elements in Process Safety Audits

The following items should be considered, as a minimum, when conducting process safety audits.

Orientation

A detailed overview of the facility—including block diagrams indicating chemicals, materials, equipment, and processes—should be given to the review team. During the orientation, noncompany or nonfacility team members should become familiar with the facility's requirements for personal protective equipment, respiratory protection, alarm systems, and emergency response procedures.

Process Safety Management Program Overview
A person capable of explaining the company's process safety management program should address the audit team. He or she should cover how each of the elements of the program are implemented, who is responsible for the implementation, and what records are used to verify implementation.

Preliminary Walkaround
A walkaround should be considered to give the audit team a basic overview of the facility; to look for potential hazards; and to solicit preliminary input from supervisors, employees, and contractor employee representatives.

Documentation Review
The audit team should have unlimited access to or be provided with copies of general safety, fire protection, health, and process safety-related documents, including, but not limited to, the following:

- accident, injury, and illness records (OSHA logs) for past three years for facility and contractors working in processes
- written plan of action for implementing employee participation
- written current and accurate operating procedures, maintenance procedures, process integrity schedules, and process safety information for the units selected to be reviewed for verification of the program
- documented prioritization, rationale, and method used to conduct process hazard analyses
- records for initial, updated, and refresher training and skills evaluations for supervisors, employees, and contractor employees working in processes to be reviewed
- pre-start-up safety review criteria and documentation for any new and modified processes in the units to be reviewed
- hot work, safe work, and entry permit programs and permits for units to be reviewed
- written management-of-change procedures, including work orders and contractor work notices for the units to be reviewed
- incident investigation reports, actions recommended, and results for the units to be reviewed
- written emergency action plan, including emergency shutdown and nonroutine start-up; documentation of known hazards; and employee and contractor emergency response actions
- the two most recent compliance audit reports for the units to be reviewed, including findings, recommendations, and actions taken
- information related to contractor employee safety performance and programs, including contractor employee safe work practices in process areas and methods of informing contractor employees of potential hazards

- facility's written evaluation of contractor employee performance toward PSM requirements in the units to be reviewed
- for the units to be reviewed, a list of unique hazards presented by contractor's work and contractor's list of unique workplace hazards reported to the employer.

Review of Selected Process Units
The company may conduct a supplemental detailed audit to verify the effectiveness of the process safety program within one or more process units. Selection of the units to be reviewed is based on the following criteria:

- factors observed and employee and contractor employee representative input during the preliminary walkaround
- incident reports, age of process units, and other historical information
- units whose process hazard analyses are completed or prioritized for completion
- nature and quantity of materials and chemicals involved in the process
- units with current hot work, replacement, turnaround, and maintenance activities
- number of employees and contractor employees working in the process or activity.

Self-Evaluation Process Safety Audit Checklist
A company may develop a process safety checklist tailored specifically to the facility and/or process units. The checklist would cover such areas as the following:

- prior incidents and near-misses (in the facility or unit)
- process chemistry hazards
- process unit construction
- process unit documentation
- process unit controls
- process unit maintenance, repair, testing, and inspection
- process unit operations
- process-related training and employee involvement
- process management of change
- process emergency procedures.

Confidentiality of Self-Evaluation Audit Reports
Maintaining confidentiality of the process safety audit report is always a concern to management. Many companies that conduct normal, routine internal process safety audits have questioned whether they should continue to do so for fear that agencies would use the results against them in a civil or criminal proceeding. Reports, recommendations, and resulting actions (or no actions) may be subject to review by OSHA or others in the event of an inspection or incident investigation.

In an effort to mitigate this situation, a number of states have passed (or are considering) audit privilege legislation that protects internal audits from regulators and third parties. Internal audit privilege legislation typically contains these requirements:

- Companies who decide to assert the confidentiality of their process safety audits bear the burden of proof and must actively (legally) assert the privilege, which may be waived by act or omission or otherwise forfeited. Waiver can be expressed or implied, such as when a supposedly privileged internal process safety report is widely distributed internally or among third parties.
- The internal process safety audit report must have attributes common to all internal safety audits conducted by the company. It should be marked or titled as an internal safety audit document. In some states, it must also be labeled privileged.
- Violations discovered during the internal process safety audit must be voluntarily disclosed and promptly addressed. Where this has been done, some states will provide a company with immunity from civil penalties or criminal prosecution and may allow that the report cannot be used by regulators for enforcement.

In some instances, common law may also provide protection from involuntary disclosure of internal process safety audit reports. A degree of confidentiality may be maintained by conducting the self-evaluation audits at the request of the company attorney. In addition, team members and those with access to the report and its findings must follow the rigorous requirements of maintaining attorney-client confidentiality and work-product privilege. Common law concepts that may be applicable include:

- **Attorney-Client Privilege.** Covers confidential legal advice between companies and their attorneys.
- **Work Product Privilege.** Protects work products, such as process safety reports, that have been prepared at the direction of an attorney in anticipation of litigation.
- **Self-Evaluation Privilege.** A new concept recognizing that a company should be able to examine its own process safety compliance confidentially without threat of disclosure.

SUMMARY

- Process safety describes various regulations and activities designed to protect employees, the public, and the environment from the consequences of major chemical accidents involving highly hazardous materials.
- A number of petroleum, petrochemical, and chemical associations recognize the need for safety management of chemical processing operations. These organizations have expended considerable time and effort to promote safety guidelines, practices, and regulations in this area.

- The EPA created CEPP and EPCRA in response to public pressure over the hazardous release and transport of chemicals in communities. OSHA promulgated its Hazard Communication Standard at the same time, along with the Clean Air Act Amendments and process safety management standard.
- Process safety management should be an integral part of a company's overall safety program. To establish a process safety management program, management should establish specific goals and conduct process risk assessments.
- Elements of an effective process safety management program include process safety information, conducting process hazard analyses, establishing management-of-change programs, developing operating procedures, establishing safe work practices and permits, conducting employee training, running pre-start-up safety reviews, and other administrative, regulatory, and operational elements.
- Four types of process safety compliance inspections can be conducted to determine whether a facility is covered by the OSHA process safety management standard: inspections in response to accidents and catastrophes, unprogrammed process safety management inspections, programmed general industry inspections, and program quality verification inspections.
- Two basic principles of conducting self-evaluation audits of process safety management are (1) gather all relevant documentation and (2) evaluate the program's application to one or more selected processes.
- Elements in a process safety audit include orientation, overview, preliminary walkaround, documentation review, and review of selected process units.

REFERENCES

API Guiding Environmental Principles and Management Practices. Washington, DC: American Petroleum Institute, December 1993.

Fixed Equipment Inspection Guide, Washington, DC: Chemical Manufacturers Association, September 1991.

Guidelines for Technical Management of Chemical Process Safety. New York: Center for Chemical Process Safety of the American Institute of Chemical Engineers, 1989.

Guidelines for Hazard Evaluation Procedures. New York: Center for Chemical Process Safety of the American Institute of Chemical Engineers, 1985.

Management of Process Hazards (RP 750). Washington, DC: American Petroleum Institute, 1990.

Management of Hazards Associated with Location of Process Plant Buildings (RP 752). Washington, DC: American Petroleum Institute, 1995.

Management Practices, Self-Assessment Process, and Resource Materials (RP 9000). Washington, DC: American Petroleum Institute, 1992.

Managing Chemicals Safely. Washington, DC: United States Environmental Protection Agency, March 1992.

Plant Guidelines for Technical Management of Chemical Process Safety. New York: Center for Chemical Process Safety of the American Institute of Chemical Engineers, 1992.

Process Safety Management (Control of Acute Hazards). Washington, DC: Chemical Manufacturers Association, May 1985.

29 CFR Part 1910, Process Safety Management of Highly Hazardous Chemicals, Explosives and Blasting Agents, Final Rule. Washington, DC: Occupational Safety and Health Administration, Department of Labor, May 1992.

A Resource Guide for the Process Safety Code of Management Practice. Washington, DC: Chemical Manufacturers Association, September 1991.

REVIEW QUESTIONS

1. According to OSHA, what is process safety management?
2. What two things must manufacturers do, according to the Hazard Communication Standard?
   a.
   b.
3. What is the main difference between the OSHA and EPA chemical safety management regulations?
4. The PSM standard applies to the handling, storage, processing, and transport of flammable gases and liquids in quantities of _____.
   a. 500 gallons or more
   b. 1,000 gallons or more
   c. 10,000 gallons or more
   d. none of the above
5. The three purposes of a risk management program are:
   a.
   b.
   c.
6. Which two people (i.e., what proficiencies) must be present for a hazard analysis team to be successful?
   a.
   b.
7. When deciding the priority order for hazard process analysis, the company should take what four criteria into account?
   a.
   b.
   c.
   d.
8. What is the difference between fault tree and even tree analysis?
9. What does HAZOP stand for?
10. Briefly define failure mode and effects analysis.
11. When is the 'What if?' method of process hazard analysis appropriate?

12. Operating procedures need to be certified for completeness and accuracy every _____.
    a. six months
    b. year
    c. two years
    d. five years
13. List five of the eight areas that operating instructions must address.
    a.
    b.
    c.
    d.
    e.
14. List four times that a pre-start-up process safety review should be conducted.
    a.
    b.
    c.
    d.
15. List three of six items of information that inspection and test result documentation should contain.
    a.
    b.
    c.
16. What does HAZWOPER stand for?
17. What is the purpose of a process safety management audit program?
18. How long does OSHA require accident investigation records be kept?
    a. one year
    b. two years
    c. five years
    d. seven years
19. What are the four types of OSHA process safety compliance inspections?
    a.
    b.
    c.
    d.
20. What are the three goals of PQV inspections?
    a.
    b.
    c.
21. What are two basic principles of conducting self-evaluation audits?
    a.
    b.
22. Mr. Smith is the head of the audit team in charge of reviewing documentation for your area. As a manager, you should provide him with access to which compliance audit reports for the area?
    a. all reports
    b. the most recent
    c. the two most recent
    d. none

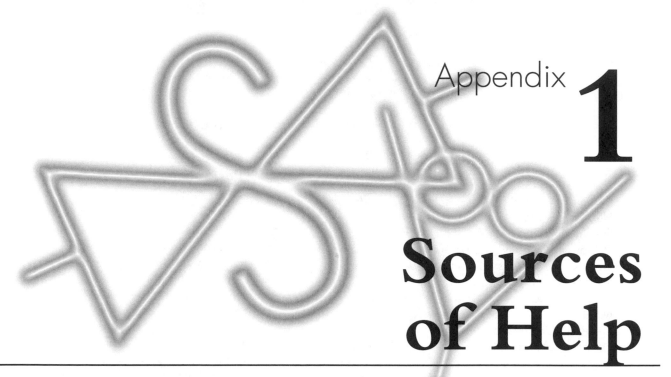

Appendix

1

Sources of Help

Petroleum Institute • American Pulpwood Association • American Road and Transportation Builders • American Trucking Associations, Inc. • American Water Works Association • American Welding Society • Associated Builders and Contractors • Associated General Contractors of America, Inc. • Association of American Railroads • Chemical Manufacturers Association, Inc. • Compressed Gas Association, Inc. • Edison Electric Institute • Graphic Arts Technical Foundation • Human Factors and Ergonomics Society • Illuminating Engineering Society of North America • Industrial Safety Equipment Association, Inc. • Institute of Makers of Explosives • International Association of Drilling Contractors • Laser Institute of America • National Association of Manufacturers • Metal Casting Society • National Propane Gas Association • National Petroleum Refiners Association • National Restaurant Association • National Rural Electric Cooperative Association • New York Shipping Association, Inc. • Portland Cement Association • Power Tool Institute, Inc. • Printing Industries of America, Inc. • Scaffolding, Shoring, and Forming Institute, Inc. • Steel Plate Fabricators Association, Inc.

This appendix offers resources for safety and health professionals who often need specialized or current unpublished information. Although professional societies and trade associations are excellent sources of help, their charters of responsibility are varied.

For particularly difficult safety problems, the safety and health professional may need to contact several sources before finding an effective solution. For example, governmental agencies can help with information about regulations or applicable standards. Insurance companies or their associations may offer assistance through their knowledge of similar problems. The trade association in the industry may have developed materials and aids in solving the problems.

The National Safety Council, through its resources and membership, can provide added information and advice to help safety and health professionals and their companies develop effective countermeasures to current safety problems.

SERVICE ORGANIZATIONS

National Safety Council
1121 Spring Lake Drive
Itasca, IL 60143–3201

The National Safety Council—a nongovernmental, not-for-profit, and nonpolitical organization--is the largest facility in the world devoting its entire efforts to the prevention of accidents and illnesses and the mitigation of injuries and economic losses. The Council's staff members work as a team with more than 2,000 volunteer officers, directors, and members of various divisions and committees to develop and maintain accident prevention material and programs in specific areas of safety and health. These areas include industrial, traffic, home, recreational, and public concerns. Council headquarters' facilities contain one of the largest safety reference libraries in the world.

At the headquarters, a staff of more than 250, about half of whom are engineers, editors, statisticians, writers, educators, data processors, librarians, and other specialists, carry out the Council's major activities. In addition to its main office and Distribution Center, the Council has regional offices in Redwood City, CA; Syracuse, NY; and Atlanta, GA, as well as a Public Policy Office and Environmental Health Center in Washington, DC.

The Council's member services are offered through the following divisions: Business and Industry, Utilities, Construction, Campus Safety, Educational Resources, Labor Community, Youth, Motor Transportation, Highway Safety, and Agriculture.

Business and Industry Division
Because identifying and handling industry's safety and health problems often requires specialized knowledge, the Council has organized its occupational health and safety activities into sections guided by the Business and Industry Division. Each section is administered by its own executive committee, elected from its own membership. The industrial membership of the Council is organized according to the following sections, each devoted to providing special help for all facets of the industrial sector indicated:

- Aerospace—Missile and aircraft manufacture, related components
- Automotive, Tooling, Metalworking, and Associated Industries—Machining, fabrication, assembly, manufacturing
- Cement, Quarry, and Mineral Aggregates—Quarrying, processing, and marketing
- Chemical—Manufacturing chemical compounds and substances
- Food and Beverage—Dairies, brewers, confectioners, distillers, canners and freezers, grain handling and processing, meat processing, restaurants, and related operations
- Forest Industries—Logging, pulp and paper, plywood, and related products
- Health Care—Hospital, patient, employee, visitor safety
- International Air Transport—Ground terminal safety, affecting personnel and equipment
- Marine—Crew, passenger, vessel safety, stevedoring, shipbuilding, repair, cargo safety
- Metals—Foundries, ferrous and nonferrous manufacturing, fabricating, other metal products
- Mining—Underground and open pit coal operations, and metals and minerals extraction and processing
- Petroleum—Exploration, drilling, production, pipeline, marketing, retail
- Power Press, Forging, and Metal Fabricating—Metal stamping and forming, and forging
- Printing and Publishing—Newspaper and commercial publishing
- Public Employee—Employees and governments from local to national levels and associated activities
- Research, Development, and Emerging Technologies—Laboratory safety
- Retail, Trades, and Services—Food service, hotels, motels, mercantile, automotive, leather, financial institutions, warehouses, offices, recreational facilities

- Rubber and Plastics—Manufacturers of rubber plastic products
- Textile and Apparel—Manufacturing and fabrication, natural and synthetic fibers, ginning and finished goods

Assigning Council Business and Industry Division members to sections gives them greater involvement with organizations and professionals in the same topic area. Many members volunteer their time and expertise to assist in accomplishing the Council's mission by serving on committees and in other leadership capacities that deal with their particular safety and health specialties and interests.

The Business and Industry Division meets twice a year to review current occupational safety and health problems and to determine, on a national scale, the best procedures to follow to provide better safety materials and programs to industry. The division consists primarily of industrial members of the Council and comprises the section general chairmen, vice-general chairmen, and other members-at-large drawn from business and industry member organizations, professional and trade associations, and other groups.

Committees focus primarily on improving the Council's services through new technical materials and visual aids. They also carefully plan each National Safety Council program and conduct ongoing recruitment of members. Special committees are often assigned to work on problems unique to an industry and for which no program materials exist.

Program Materials
The following Council publications have proven to be particularly useful for industrial and off-the-job safety programs. (Unless otherwise stated, they are published monthly.)

- *Family Safety & Health* (quarterly)
- *Today's Supervisor*
- *Safety & Health*
- *OSHA Update*
- *Product Safety Up-to-Date*
- Section Newsletters (one for each of the 21 sections—six issues a year)
- *Safe Driver* (issued in three editions—Truck, Passenger Car, and Bus)
- *Safe Worker*
- *Traffic Safety* (six issues a year)

In addition, the following statistical materials are also available from the Council:

- *Accident Facts* (annually)
- *Section Contest Bulletins* (quarterly)
- *Work Injury and Illness Rates* (annually)

Technical Materials

The following is a partial list of technical manuals; see the current Council General Materials Catalog for a complete listing:

- *Accident Prevention Manual for Business & Industry* (3 volumes)
    - *Administration & Programs*
    - *Engineering & Technology*
    - *Environmental Management*
- *Aviation Ground Operations Handbook*
- *Fundamentals of Industrial Hygiene*
- *Lockout/Tagout: The Process of Controlling Hazardous Energy*
- *Motor Fleet Safety Manual*
- *Occupational Health & Safety*
- *Product Safety Management Guidelines*
- *Power Press Safety Manual*
- *Safeguarding Concepts Illustrated*

Training and motivational materials include the following (see Poster Catalog for a complete listing)

- Banners
- Booklets
- Calendars
- Films
- Posters
- Safety and health slides
- Supervisory training pamphlets
- Videotapes

The National Safety Council sponsors the annual Congress and Exposition, one of the largest safety conventions held anywhere. Nearly 200 general or specialized sessions, workshops, and clinics are held, covering the full range of safety and health topics.

The annual meetings or special business meetings of a dozen allied organizations and associations are also held concurrent with the congress, greatly enhancing the exchange of views and information in safety and health.

Library and Statistics

Special services available through the Council to support industrial safety programs include a library and statistical support. With a collection of more than 120,000 documents, of which 95,000 are indexed on an in-house computerized database, the National Safety Council's library is one of the most comprehensive safety and occupational health libraries in the world. The library is available to the general public. It networks with other databases for safety and health information.

The Council's statisticians are a recognized source of highly reliable, accurate, and authoritative data within the safety community. Equipped with data processing and research tools, they study various types of accident data in the continuing search for clues on the causes of accidents.

Safety Training Institute

The National Safety Council's Safety Training Institute (STI), founded in 1946, offers training of all types for industry managers, safety professionals, and others from all levels of management who supervise workplace activities. More than 24,000 students have completed courses. Classes are held at the STI's facilities, in the Council's regional offices, and at onsite locations provided by various organizations.

Course offerings cover such general subjects as Principles of Occupational Safety and Health, Safety Training Methods, and Safety Management Techniques, as well as specific topics that include industrial hygiene, hazard communication, ergonomics, product safety, chemical organizations, laboratory, hospital, advanced hospital safety, and advanced safety concepts. The National Safety Council awards Continuing Education Units (CEUs) for its STI courses. STI students may also earn the Council's Advanced Safety Certificate by completing certain course tracks.

Further information about STI is available from the annual Safety Training Institute Course Catalog & Schedule. STI also offers two Home Study Courses: Supervising for Safety, and Protecting Workers' Lives. These courses are recognized by the National Home Study Council.

Two supervisor training courses can be purchased from STI that have been designed for presentation by in-house training personnel. These programs include the Management Development Program, which is oriented to the new supervisor, and Supervisors' Development Program for individuals with six months or more of experience. (Complete information on these and other courses can be found in the National Safety Council's General Materials Catalog.)

The Forklift Truck Operators Training Course is an eight-hour program designed to help a company comply with training regulations. The course can be presented in the standard form or tailored to fit an individual company's facilities and needs.

A number of these courses offered by STI are also presented by local safety councils. STI can tailor courses and seminars to customer needs and present them at customer locations. Information on costs, scheduling, and specifics can be obtained by contacting the Safety Training Institute.

Safety and Health Management Services (Consulting)

The Council's consulting services are available for loss control, motor fleet, traffic, industrial hygiene, environmental, community, product, motivation, agricultural, and research and statistical services. The loss control consulting services can provide an analysis, recommendation, and structure for a safe and healthful working environment and can conduct follow-up audits. Information can be obtained upon request.

Labor Division

This division and its member unions and government labor agencies represent labor and its safety and health viewpoints in many of the Council's areas of work. Specific problems requiring labor's review or consensus input are developed and approved by the Executive Committee.

Speakers Bureau

The Speakers Bureau provides nationally known speakers who cover a wide range of safety, health, and environmental subjects. Speakers are ideal for seminars and conferences.

First Aid Institute

This institute offers programs, certification of instructors, and materials for teaching first aid and cardiopulmonary resuscitation (CPR) to employees. The courses of several levels meet requirements of governmental agencies and other organizations.

Environmental Health Center

The National Safety Council's Environmental Health Center, established in 1988, serves as a communications link between members of the public concerned about environmental risks and private and public risk managers. The Center's goals are to develop accurate and objective information on environmental and public health risks and disseminate this information to the public.

Local Safety Councils

Nearly 100 local safety councils throughout the United States and Canada have been chartered by the National Safety Council. These councils work under the leadership of public-spirited citizens, commercial and industrial interests, responsible official agencies, and other important groups. They are nonprofit, self-supporting organizations whose purpose is to reduce accidents and illnesses through prevention education.

Accredited councils operate under the guidance of a full-time executive and staff and receive support services from the National Safety Council. (A list of accredited safety councils can be obtained from the Safety Council Relations Department.) These organizations give assistance to local safety engineers and others concerned with occupational safety and health. Safety professionals, in turn, render substantial service through participation in the local council's work.

Many of the local councils offer the following services:

- act as a clearinghouse of information on safety and health problems, and maintain a library of audio and visual and other communication aids

- provide a forum for exchange of experience through regularly scheduled meetings of supervisory personnel

- sponsor a variety of safety and health courses, including many presently being offered by the National Safety Council's Safety Training Institute

- conduct annual, area-wide safety and health conferences

- on request, provide assistance with safety problems and programs

- stimulate interest for and assist in the development of accident prevention programs for all employers

- conduct safety contests, with awards for outstanding safety records.

The scope and extent of activities of each council depend, of course, upon local conditions and available resources.

American National Red Cross
431 18th Street, NW
Washington, DC 20006

The American Red Cross, through its more than 2,600 chapters, offers courses in first aid, cardiopulmonary resuscitation (CPR), swimming, lifesaving, canoeing, and sailing.

Experience in industry shows that first aid and safety training helps reduce accidents—both on and off the job—by helping personnel understand accident causes and effects, by improving attitudes toward safety, and by preparing individuals to give appropriate emergency care to accident victims.

Arrangements for first-aid training can be made through local Red Cross chapters. Most industries prefer to select key personnel to receive training as volunteer instructors who, in turn, can conduct classes for fellow employees. Others may wish to arrange for employee training by instructors provided through the Red Cross.

Texts, instructor's manuals, charts, and other teaching materials and visual aids, such as films and posters, are available through local chapters.

Industrial Health Foundation, Inc.
34 Penn Circle W
Pittsburgh, PA 15206

The Foundation, a nonprofit research association of industries, advocates industrial health programs, improved working conditions, and better human relations. The Foundation maintains a staff of physicians, chemists, engineers, biochemists, and medical technicians. Its activities fall into three major categories:

- to provide professional assistance to member companies in the study of industrial health hazards and their control

- to help companies develop health programs as an essential part of industrial organization
- to contribute to the technical advancement of industrial medicine and hygiene by educational programs and publications.

Activities are classified as follows:

Medical

- Organization and administrative practices
- Opinions on doubtful x-ray pictures
- Surveys of health problems
- Specific industrial medical problems
- Epidemiology studies

Chemistry, Toxicology, Industrial Hygiene

- Field studies—facility or industry basis
- Toxicity of chemicals, physical agents, or processes
- Sampling and analytical procedures

Engineering

- Ventilating systems
- Exhaust hoods

Education

- Training courses in occupational health and safety for:
- Physicians
- Nurses
- Industrial hygienists and engineers
- Symposia on special subjects of current interest in these fields

The foundation holds an annual meeting of members, conferences of member company specialists, and special conferences on problems common in a particular industry. The following publications are issued:

- *Industrial Hygiene Digest,* monthly (abstracts)
- Bibliographies on current interest subjects
- Technical bulletins
- Proceedings of symposia

**Prevent Blindness America
(Formerly known as the National Society to Prevent Blindness)
500 East Remington Road
Schaumburg, IL 60173**

Prevent Blindness America is the oldest national voluntary health organization that works to prevent blindness and preserve sight through community service programs, public and professional education, and research. PBA's occupational health and safety program is guided by a professional advisory council, comprised of experts in the fields of industry, education, medicine, nursing, and accident prevention.

Some of the organization's major activities include:

- Sponsors the Wise Owl Program—an eye-safety incentive program for industrial, military, municipal, and educational organizations. The program provides information for developing eye health and safety programs and for recognizing those who saved their sight by wearing proper protective eyewear.
- Promotes state-wide eye safety for all school and college laboratory and shop students, and their teachers and visitors.
- Provides counseling to school administrators and teachers in establishing and implementing eye safety programs.
- Participates as a member of national standards organizations on committees concerning vision, eye protection, and illumination.
- Recommends the use of protective eyewear for individuals with monocular vision, and for hazardous situations in occupational and educational environments, around the home, and during sports activities.
- Develops and promotes position statements on health and safety issues including: contact lenses, fireworks, protective eyewear, etc.
- Recommends regular eye examinations for all age groups for the early detection and treatment of eye and vision disorders.
- Offers literature and audiovisual materials on a wide range of eye health and safety topics.
- Sponsors the National Center for Sight, a toll-free hotline that provides information and referral services for a wide range of eye health and safety topics.

STANDARDS AND SPECIFICATIONS GROUPS

American National Standards Institute
11 West 42nd Street
New York, NY 10036

The American National Standards Institute (ANSI) coordinates and administers the federated voluntary standardization system in the United States. It also represents the nation in international standardization efforts through the International Organization for Standardization (IOS), the International Electrotechnical Commission (IEC), and the Pacific Area Standards Congress (PASC).

ANSI is a federation of over 1,200 national trade, technical, professional, labor, and consumer organizations, government agencies, and individual companies. It coordinates the standards development efforts of these groups.

When its Board of Standards Review determines that a national consensus exists in favor of a particular standard the groups produce, ANSI approves it as an American National Standard. A catalog of safety standards is issued annually and is distributed without charge.

Many American National Standards, as well as other national consensus standards, have become more important since passage of the Williams-Steiger Occupational Safety and Health Act of 1970. The Occupational Safety and Health administration has stated a definite preference for basing its regulations on consensus standards that have proved their value and practicality on the job.

Under ANSI procedures, the responsibility for the management of specific standards projects is divided according to subject matter and assigned to an ANSI Safety and Health Standards Board. Standards dealing with safety fall under the jurisdiction of the Standards Management Board. Many American National Standards on safety and health are developed by ANSI-accredited Standards Committees.

American Society for Testing and Materials
1916 Race Street
Philadelphia, PA 19103

American Society for Testing and Materials (ASTM) is the world's largest source of voluntary consensus standards for materials, products, systems, and services. There are currently more than 9,000 ASTM standards.

ASTM's 35,000 members include engineers, scientists, researchers, educators, testing experts, companies, associations and research institutes, governmental agencies, and departments (federal, state, and municipal), educational institutions, consumers, and libraries.

ASTM standards are published for such categories as:

- Business supplies
- Cementitious, ceramic, and masonry materials
- Chemicals and products
- Construction
- Electronics
- Energy
- Environmental effects
- Ferrous metals
- Medical devices
- Nonferrous metals
- Occupational safety and health
- Protective equipment for sports
- Security systems
- Transportation systems

ASTM sponsors committees on geothermal resources and energy, quality control, food service equipment, and protective coatings for power generation facilities. The committee on consumer product safety has helped develop standards to help protect the public by reducing risks associated with consumer products such as cigarette lighters, bathtubs, shower structures, children's furniture, trampolines, and nonpowered guns.

ASTM standards are of interest to the safety professional because they identify areas of hazards and establish guidelines for safe performance. The Society also publishes standards for atmospheric sampling and analysis, fire tests of materials and construction, methods of testing building construction, nondestructive testing, fatigue testing, radiation effects, pavement skid resistance, protective equipment for electrical workers, and others.

FIRE PROTECTION ORGANIZATIONS

Factory Mutual Engineering Organization
1151 Boston-Providence Turnpike
Norwood, MA 02062

The Factory Mutual Engineering Organization (FMEO) is a world leader in loss control training, engineering, and research. FMEO is an outgrowth of the philosophy of positive protection, which states that the good risk through rate (premium) reduction is preferable to merely allowing the good risks to help pay for the bad.

FMEO is owned by the following companies: Allendale Mutual Insurance (Johnson, RI), Arkwright (Waltham, MA), Protection Mutual Insurance (Park Ridge, IL).

Factory Mutual Engineering and Research provides loss prevention services for policyholders. Its goal is to protect properties and production facilities from fire, explosion, and other manmade or natural disasters for which coverage is provided. Engineering services include evaluation of hazards and protection through property inspections by Factory Mutual consultants located in major industrial centers. Research services involve the evaluation of fire protection devices and equipment for approval and the development of recommendations based on tests and loss experience for the prevention of loss. The Test Center in Rhode Island provides full-scale fire testing, simulating industrial conditions with regard to height, weight, and protection of major industrial storage occupancies.

The Factory Mutual Training Resource Center provides loss control training services to policyholders and renders assistance for companies' special technical problems. The source of these services is the Factory Mutual Training Resource Center.

Training courses include Practicing Property Conservation, Boiler and Machinery Prevention Maintenance (both developed specifically for insureds), and Designing for Firesafety and Hazard Control (open to all architects and design professionals). These courses are taught by Factory Mutual engineers, research scientists, and training experts.

FMEO's publications include the *Record, Approval Guide, Handbook of Property Conservation, Loss Prevention Data Books,* and *Factory Mutual Engineering and Research*

Property Loss Control Catalog. The *Record* is an internationally recognized bimonthly management magazine dealing with property conservation. The *Approval Guide* lists industrial fire protection equipment and services tested and approved by Factory Mutual Research Corporation. The *Handbook of Property Conservation* and the *Loss Prevention Data Books* cover recommended practices for protection against fire and related hazards.

National Fire Protection Association
Batterymarch Park
Quincy, MA 02269

The National Fire Protection Association (NFPA) is the clearinghouse for information on the subject of fire protection, fire prevention, and firefighting. It is a nonprofit technical and educational organization with a membership of some 65,000 companies and individuals.

The technical standards issued as a result of NFPA committee work are widely accepted by federal, state, and municipal governments as the basis of legislation and used as the basis of good practices. More than 50 serve as OSHA regulations. More than 300 codes and standards are currently issued by NFPA, which are constantly reviewed and updated. Issued as separate booklets, many of these codes and standards supply authoritative guidance to safety engineers. Representative subjects include:

- Air Conditioning and Ventilating Systems
- Brigades
- Flammable and Combustible Liquids Code
- Hazardous Chemicals Data
- Industrial Fire Loss Prevention
- Life Safety Code and Handbook
- Lighting Protection Code
- National Electrical Code and Handbook
- Portable Fire Extinguishers
- Powered Industrial Trucks
- Prevention of Dust Explosions in Industrial Plants
- Protection of Records
- Safeguarding Building Construction Operations
- Sprinkler Systems Organization and Training of Private Fire Brigades
- Storage and Handling of Liquefied Petroleum Gases
- Truck Fire Protection

The standards are also published as the *National Fire Codes* in 12 volumes. Other publications of interest to safety professionals concern hazardous materials, fire safety in health care facilities, and public fire safety. In addition, *The Fire Protection Handbook* is an authoritative encyclopedia on fire and its control.

NFPA is a sponsor of National Fire Prevention Week held every year in October in remembrance of the Great Chicago Fire.

Underwriters Laboratories Inc.
333 Pfingsten Road
Northbrook, IL 60062

Underwriters Laboratories, a not-for-profit organization, maintains laboratories for the examination and testing of devices, systems, and materials to determine their compliance with safety standards.

UL publishes annual directories of manufacturers whose products have met the criteria outlined in appropriate standards and whose products are covered under UL's Follow-Up Services program. These directories are:

- "Appliances, Equipment, Construction Materials, and Components Evaluated in Accordance with International Publications"
- "Automotive, Burglary Protection and Mechanical Equipment"
- "Building Materials"
- "Electrical Appliance and Utilization Equipment"
- "Fire Protection Equipment"
- "Fire Resistance"
- "Gas and Oil Equipment"
- "General Information from Electrical Construction Materials and Hazardous Location Equipment"
- "Hazardous Location Equipment"
- "Marine"
- "Recognized Component"

UL Follow-Up field representatives make unannounced visits to production sites where UL-listed products are made to determine whether the manufacturer's quality control program is in compliance with UL safety standards. Safety professionals often specify the UL mark when they purchase fire, electrical, and other equipment that falls in categories covered under the services of Underwriters Laboratories.

Engineers should be aware, however, that UL listings apply only within the scope of its standards for safety and may have no bearing on performance or other factors not involved in the UL investigation. The markings and instructions provided with the products should tell a safety professional about the function of the device or material. UL tests are conducted under conditions of installation and use that conform to the appropriate standards of the NFPA or other applicable codes. Any departure from these standards by the user may affect the performance qualifications found by Underwriters Laboratories.

INSURANCE ASSOCIATIONS

In addition to the insurance associations listed under Fire Protection Organizations, there are a number of insurance federations with accident prevention departments that produce technical information available to safety people.

Alliance of American Insurers
1501 Woodfield Road
Schaumburg, IL 60195

The Alliance of Insurers is a national organization of leading property-casualty insurance companies. Its membership includes more than 150 companies that safeguard the value of lives and property by providing protection against mishaps in workplaces and losses from fires, traffic accidents, and other perils. Alliance member companies have a tradition of loss prevention that dates from the organization of the first American mutual insurance company in 1752.

Through its Loss Control Department, the Alliance makes a concerted effort to reduce accidents, fires, and other loss-producing incidents. Under the guidance of its loss control advisory committee, effective safety engineering, industrial hygiene, fire protection engineering, and other loss prevention principles are promoted. Major activities include disseminating information on safety subjects, conducting specialized training courses for member company personnel, sponsoring research, cooperating in the development of safety standards, developing visual aids, and publishing technical and promotional safety literature. A catalog of safety material is available.

American Insurance Services Group, Inc.
Engineering and Safety Service
85 John Street
New York, NY 10038

The Engineering and Safety Service of American Insurance Services Group, Inc., is dedicated to providing information, education, and consultation to the technical and loss control personnel of its participating property-casualty insurance companies.

Engineering and Safety Service's stock in trade is the largest and most beneficial information, gathered and prepared in a variety of ways for loss control professionals. The focus is always on safety and risk reduction in industrial, commercial, and residential settings within the following technical fields: occupational safety and health, pollution control, product safety, special and chemical hazards, fire protection, crime prevention, environmental science, building technology, industrial hygiene, construction hazards, commercial fleets, and boilers and machinery. In addition, Engineering and Safety Service represents the insurance industry on numerous national standards-making committees of the American National Standards Institute, National Fire Protection Association, and others.

The Engineering and Safety Service develops and publishes a wide variety of safety-related materials that it makes available to the public, in addition to those it produces for its subscriber companies.

PROFESSIONAL SOCIETIES

American Society of Safety Engineers
1800 East Oakton
Des Plaines, IL 60018

The American Society of Safety Engineers is the only organization of individual safety professionals dedicated to advancing the safety profession and to fostering the well-being and professional development of its members.

To fulfill its purpose, the Society has the following objectives:

- promote the growth and development of the profession
- establish and maintain standards for the profession
- develop and disseminate material that will carry out the purpose of the Society
- promote and develop educational programs for obtaining the knowledge required to perform the functions of a safety professional
- promote and conduct research in areas that further the purpose and objectives of the Society
- provide forums for the interchange of professional knowledge among its members
- provide for liaison with related disciplines

An annual professional development conference is conducted for members.

The Society is actively pursuing its Professional Development Programs including the development of curricula for safety professionals, accreditation of degree programs, presentation of member education courses and publications, and definition of research needs and communications to keep safety practitioners current.

The Society has established the American Society for Safety Engineers Foundation because of its concern for the need for greater research efforts in the prevention of accidents and injuries. The Society believes that increased research will improve accident and injury control techniques. The group was also instrumental in establishing a corporation known as the Board of Certified Safety Professionals of America. (See BCSP later in this Appendix.)

Members of the Society receive the monthly *Professional Safety* as part of their membership. (Nonmembers may also subscribe.) The Society also publishes other technical or specialized information.

Chapters engage in a number of activities designed to enhance the professional competence of their members. Most hold monthly meetings featuring speakers, demonstrations, workshops, and discussions designed to help members keep abreast of developments in their professional field. A list of chapters grouped by states and listed by name and by location can be obtained from the national office.

American Association of Occupational Health Nurses, Inc.
50 Lenox Pointe
Atlanta, GA 30324

As the national professional organization for the registered nurse working in business and industry, the association serves as an advocate for occupational health nurses, establishes health care and provides educational programs for nurses in this special field. *AAOHN Journal*, a scientific peer-reviewed journal, and *AAOHN News* are published monthly.

The annual meeting is held in conjunction with the American Occupational Health Conference, which AAOHN cosponsors with the American College of Occupational and Environmental Medicine. Publications that address safety and health issues in the workplace are also available.

American Board of Industrial Hygiene
4600 West Saginaw, Suite 101
Lansing, MI 48917

This specialty board certifies properly qualified industrial hygienists. It seeks to encourage the study, improve the practice, elevate the standards, and issue certificates to qualified applicants.

American Chemical Society
1155 16th Street, NW
Washington, DC 20036

This society is devoted to the science of chemistry in all its branches, the promotion of research, the improvement of the qualifications and usefulness of chemists, and the distribution of chemical knowledge.

Articles on safety appear in the bimonthly publication *Chemical Health & Safety* or the weekly publication *Chemical and Engineering News*. Digests of papers dealing with aspects of industrial hygiene appear monthly in *Chemical Abstracts*. The environment is discussed in *Environmental Science and Technology*.

The society has a Committee on Chemical Safety and a Division of Chemical Health and Safety as well as a Chemical Health and Safety Referral Service accessible by telephone.

American Conference of Governmental Industrial Hygienists
1330 Kemper Meadow Drive
Cincinnati, OH 45240

ACGIH is an organization devoted to the administrative and technical aspects of worker health and safety protection. For over 50 years, ACGIH has provided leadership, educational, and investigative opportunities to professional and technical personnel representing governmental agencies or educational institutions involved in occupational health and safety activities.

ACGIH publishes a number of well-known texts, including the *Threshold Limit Values/Biological Exposure Indices*. The Conference also publishes *Applied Occupational and Environmental Hygiene*, the monthly journal that combines peer-reviewed papers with wide-ranging columns and articles to offer information and guidance to the practicing professional. An ACGIH Publications catalog listing over 200 texts in occupational safety and health is available.

American Industrial Hygiene Association
2700 Prosperity Avenue
Fairfax, VA 22031

AIHA is an organization of professionals dedicated to the prevention of workplace-related illness or injury. With more than 11,000 members, AIHA is the largest international association serving the needs of occupational and environmental scientists and engineers in industry, government, labor, academic institutions, and independent organizations. Founded in 1939, the purposes of AIHA are to promote the field of industrial hygiene, to provide education and training, to provide a forum for the exchange of ideas and information, and to represent the interests of industrial hygienists and those they serve.

AIHA offers its members and other interested professionals a wide range of products and services on a variety of topics within the occupational health and safety field. AIHA publishes the monthly *AIHA Journal* and *The Synergist*, a monthly informational newsletter. AIHA holds continuing education seminars on the latest topics in the field. AIHA offers an employment service for those professionals seeking new positions and for companies looking for employees. AIHA has laboratory programs to assist laboratories in maintaining high quality standards, including accreditation for every qualified industrial hygiene laboratory in the world.

AIHA sponsors the American Industrial Hygiene Conference (AIHC). Catalogs for continuing education, laboratory programs, and publications are available. Information is available for all AIHA products and services.

American Institute of Chemical Engineers
345 East 47th Street
New York, NY 10017

Founded in 1908, the AIChE is a nonprofit society dedicated to advancing the chemical engineering profession and professional standards of 54,000 chemical engineers through publications, technical meetings, continuing education, and research. The AIChE's Safety and Health division holds annual Loss Prevention and Ammonia Safety symposia and publishes proceedings. The Institute's large continuing education program conducts short courses in many engineering safety subjects several times each year. The AIChE Design Institutes for Physical Property Data (DIPPD) and for Emergency Relief Systems (DIERS) Users Group offers corporations and government agencies the opportunity to join in the cooperative sponsorship of safety-related research that they could not afford to do individually. Symposia proceedings, DIERS and DIPPD research material, and other safety publications are listed in AIChE's annually updated publications catalogue.

The nonprofit Center for Chemical Process Safety, established in 1985 by the American Institute of Chemical Engineers and sponsored by over 50 corporations, gives leadership and increased focus on measures to help improve process safety. CCPS collects, organizes, and publishes the latest in scientific and engineering practices as guidelines for preventing and mitigating major incidents that involve the release of potentially hazardous materials.

CCPS sponsors international technical symposia, usually with Proceedings, where technical papers disseminate and encourage the use of safe engineering and operating practices. Teaching material, developed by CCPS, is being used at colleges to teach the concepts of Safety, Health, and Loss Prevention in Chemical Processes to undergraduate engineers. Research is also carried out to advance the state-of-the-art in measures to prevent and mitigate major events.

Both AIChE and CCPS publications, meetings, and continuing education courses are available to AIChE members and nonmembers.

American Medical Association
515 North State Street
Chicago, IL 60610

The American Medical Association, with a current membership of about 300,000 physicians, has a long record of involvement in public health.

The Department of Environmental, Public, and Occupational Health (DEPOH) was organized in 1970 as a combination of AMA's Departments of Occupational Health and Environmental Health. The expertise of the staff covers the health effects of air, water, chemical, and physical stresses; communicable diseases; population growth; injuries; epidemiology; preventive medicine; sports medicine; and problems of the aged and the handicapped.

The department is principally an authoritative source of information for inquiries about environmental, public, and occupational health. In addition, it plans appropriate conferences and courses, prepares authoritative publications for physicians and the public, carries out special assignments and studies, and acts as the AMA liaison in federal health programs and legislation.

Since 1939, AMA has sponsored and organized annual congresses on occupational health and national conferences on the medical aspects of sports. Proceedings of the conferences are published. The AMA continues to have an active voice in the regulatory process.

The *Journal of the American Medical Association* (JAMA) is published weekly and occasionally contains articles on some aspect of occupational health.

American College of Occupational and Environmental Medicine
55 West Seegers Road
Arlington Heights, IL 60005

This organization fosters the study and discussion of problems peculiar to the practice of occupational medicine, encourages the development of methods adapted to the conservation and improvement of health among workers, and promotes a more general understanding of the purpose and results of employee medical care. The official publication of the association is the monthly *Journal of Occupational Medicine.*

An annual American Occupational Health Conference is held in collaboration with the American Association of Occupational Health Nurses.

American Public Health Association
1015 15th Street NW
Washington, DC 20005

The American Public Health Association is a multidisciplinary, professional association of 32,000 health workers. Through its two periodicals, *The American Journal of Public Health* and *The Nation's Health,* and other publications, APHA disseminates health and safety information to those responsible for state and community health-service programs. The program area interest group is devoted to injury control and emergency health services, and to occupational health and safety.

American Society for Industrial Security
1655 North Fort Myer Drive, Suite 1200
Arlington, VA 22209

The American Society for Industrial Security is an international professional society of more than 25,000 industrial security executives in both the private and public sector. Committee activities that would be of interest to the safety professional are safeguarding proprietary information, physical security, terrorist activities, disaster management, fire prevention and safety, and investigations.

The Society publishes a monthly magazine, *Security Management,* to which safety professionals may subscribe, and issues a newsletter to its members bimonthly. The Society certifies security professionals through its Certified Protection Professional (CPP) program. Other member services include the use of the security library and job placement. The Society also produces numerous workshops and educational programs and an annual seminar open to both members and nonmembers.

American Society of Mechanical Engineers
345 East 47th Street
New York, NY 10017

This society encourages research, prepares papers and publications, sponsors meetings for the dissemination of information, and develops standards and codes under the supervision of its Policy Board.

The society developed the following safety codes under the procedures meeting the criteria of the American National Standards Institute:

- *Safety Code for Elevators*
- *Safety Code for Mechanical Power—Transmission Apparatus*
- *Safety Code for Conveyors, Cableways, and Related Equipment*
- *Safety Code for Cranes, Derricks, and Hoists*
- *Safety Code for Manlifts*
- *Safety Code for Powered Industrial Trucks*
- *Safety Code for Aerial Passenger Tramways*
- *Safety Code for Mechanical Packing*
- *Safety Code for Garage Equipment*
- *Safety Code for Pressure Piping*
- *Safety Standards for Compressor Systems*

The ASME Boiler and Pressure Vessel Committee is responsible for the formation and revision of the *ASME Boiler and Pressure Vessel Code*. The society publishes the monthly periodical *Applied Mechanical Review*.

American Society for Training and Development
Box 1433, 1630 King Street
Alexandria, VA 22313

Members of this professional society are engaged in the training and development of business, industrial, and government personnel. ASTD holds a major annual conference, provides an information service for members, and publishes *Technical & Skill Training* magazine.

Board of Certified Hazard Control Management
8009 Carita Court
Bethesda, MD 20817

Founded in 1976, the Board evaluates and certifies the capabilities of practitioners engaged primarily in the administration of safety and health programs. Levels of certification are senior and master, with master being the highest attainable status indicating that the individual possesses the skill and knowledge necessary to effectively manage comprehensive safety and health programs.

The Board offers advice and assistance to those who wish to improve their status in the profession by acquiring skills in administration and combining them with technical safety abilities. It establishes curricula in conjunction with colleges, universities, and other training institutions to better prepare hazard control managers for their duties. It publishes *Hazard Control Manager*.

Board of Certified Safety Professionals
208 Burwash Avenue
Savoy, IL 61874–9510

The Board of Certified Safety Professionals (BCSP) was organized as a peer certification board in 1969 with the purpose of certifying practitioners in the safety profession. The specific functions of the Board are to evaluate the academic and professional experience qualifications of safety professionals, to administer examinations, and to issue certificates of qualification to those professionals who meet the Board's criteria and successfully pass its examinations.

Flight Safety Foundation, Inc.
2200 Wilson Boulevard
Arlington, VA 22201

The Flight Safety Foundation is an international membership organization dedicated solely to improving the safety of flight. Nonprofit and independent, it serves the public interest by helping to develop and implement programs, policies, and procedures that find ways and means of eliminating accident-inducing factors, by in-depth appraisals of problems in flight and ground safety, and by developing possible solutions to those problems. As a vehicle of information interchange, it cooperates with all other organizations and individuals in the field of aviation safety in educating all segments of the aviation community in the principles of accident prevention.

Perhaps most well-known of the FSF's activities are its annual safety seminars, held in successive years in various parts of the world. The Corporate Aviation Safety seminar is held annually in North America. The programs feature leaders of industry, operator and user experts, as well as government and university researchers.

National Safety Management Society
12 Pickens Lane
Weaverville, NC 28787

The National Safety Management Society, founded in 1968, is a nonprofit corporation that seeks to expand and promote the role of safety management as an integral component of total management by developing and perfecting effective methods of improving control of accident losses. Membership is open to those having management responsibilities related to loss control.

System Safety Society
Five Export Drive, Suite A
Sterling, VA 20164

The System Safety Society is a nonprofit organization of professionals dedicated to safety of products and activities by the effective implementation of the system safety concept. The objectives include:

- advance the state-of-the-art of system safety
- contribute to a meaningful management and technological understanding of system safety

- disseminate newly developed knowledge to all interested groups and individuals
- further the development of the professionals engaged in system safety

Through its local chapters, committees, executive council, publications, and meetings, the society provides many opportunities for interested members to participate in a variety of activities compatible with society objectives. In addition to its operating committees, society activities include publication of *Hazard Prevention,* the official society journal published quarterly.

International System Safety Conferences are sponsored biennially and Proceedings are available.

Veterans of Safety
c/o Dr. Robert Baldwin
Central Missouri State University
Humphries Building
Warrenburg, MO 64093

Membership numbers 1,800 safety engineers with 15 or more years of professional safety experience. Founded in 1941, the objective of Veterans of Safety is to promote safety in all fields. Activities include:

- Safety Town USA, to educate preschool and elementary school children in pedestrian safety
- Most Precious Cargo Program, to improve school bus safety
- Unified Emergency Telephone Numbers Program, to establish nationwide uniformity in emergency telephone numbers to contact fire, police, and medical aid
- Safety and Health Hall of Fame

The organization also gives annual awards for best technical safety papers published and maintains placement service.

TRADE ASSOCIATIONS

American Forest and Paper Association
1111 19th Street, NW
Washington, DC 20036

Member companies include those that grow, harvest, and process wood and wood fiber; manufacture pulp, paper and paperboard; and produce solid wood products. Publishes brochures, booklets, periodicals, and technical papers.

American Foundrymen's Society
505 State Street
Des Plaines, IL 60016–8399

The American Foundrymen's Society is the only technical society that serves the interests of the foundry industry. It disseminates information on all phases of foundry operations, including environmental (air, water, and waste) and

occupational safety and health. Services provided include in-plant visitations by AFS environmental staff for environmental facility reviews, safety and health reviews, and industrial hygiene surveys.

The AFS Lester B. Knight Analytical Laboratory is accredited by the American Industrial Hygiene Association. Analytical services include Leachate analysis of heavy metals and phenol using TCLP and industrial hygiene analysis of heavy metals, silica, and organics. Industrial hygiene pump calibration and rental is also offered.

American Gas Association
1515 Wilson Boulevard
Arlington, VA 22209

The association, through its Accident Prevention Committee, serves as a clearinghouse and in an advisory capacity to persons responsible for employee safety and to safety departments of its member companies. Its purposes are to study accident causes, recommend corrective measures, prepare manuals, and disseminate information to the gas industry that will help reduce employee injuries, motor vehicle accidents, and accidents involving the public. The committee meets several times a year and conducts a Safety and Health Workshop for Supervisors each year. It also provides speakers for regional gas associations and other gas industry meetings.

The committee has project groups that address all aspects of employee safety, including awards and statistics, distribution and utilization, education, motor vehicles, gas transmission, and promotional and advisory.

Published material includes suggested safe practices manuals and quarterly and annual reports on the industry's accident experience and programs. A publication catalog is available.

American Iron and Steel Institute
1101 17th Street, NW
Washington, DC 20036–4700

AISI represents companies accounting for raw steel production in the United States. Its Occupational Health and Safety Committee investigates matters relating to the working environment of employees in the iron and steel industry to enhance the health, hygiene, and safety of steel industry personnel, both in and out of the workplace. This includes recognition and evaluation of occupational health and safety hazards with respect to existing and new technology. The Committee also develops industry information and data that may be used for a variety of purposes, and participates in the development of pertinent national and international consensus standards.

American Mining Congress
1920 N Street, NW
Washington, DC 20036–1012

Congress membership is from coal, metal, and nonmetal mining companies. This association has a Coal Mine

Safety Committee, a Metal-Nonmetal Mine Safety Committee, and an Occupational Health Committee. *The American Mining Congress Journal* regularly carries articles and news about safety and health.

American Petroleum Institute
1220 L Street, NW
Washington, DC 20005

The objective of the Committee on Safety and Fire Protection of the American Petroleum Institute is to reduce the incidence of accidental occurrences and fire hazards to employees and the public and to property. To attain this objective, the Committee on Safety and Fire Protection:

- provides statistical reports, pamphlets, data sheets, and other publications to assist the industry in the prevention of accidents and the prevention, control, and extinguishment of fires
- provides a means for the development and exchange of information on accident prevention and fire protection to be used for education and training in the industry
- provides a forum for discussion and exchange of information concerning safe practices and the science and technology of fire protection and safety engineering
- promotes research and development in the fields of accident prevention and fire protection for the benefit of the petroleum industry as a whole
- maintains contact and cooperates with association and code writing bodies such as NFPA and ANSI

American Pulpwood Association
1025 Vermont Avenue, NW, Suite 1020
Washington, DC 20005

This association fosters study, discussion, and action programs to guide and help the pulpwood industry in growing and harvesting pulpwood raw material for the pulp and paper industry. The group conducts safety and training programs and provides training guides, notebooks, safety alerts, and technical releases that describe items of personal protective equipment, safe working procedures, and other pertinent accident control items.

American Road and Transportation Builders
1010 Massachusetts Avenue, NW
Washington, DC 20001

ARTBA is a national federation of public and private organizations concerned with transportation construction issues. Its members are involved in highway design, construction, management, signing, lighting, and other areas related to road safety.

American Trucking Associations, Inc.
2200 Mill Road
Alexandria, VA 22314

American Trucking Associations is the national federation of the trucking industry. The Safety Management Council is the official safety organization of the national trucking industry. It is responsible to the ATA Executive Committee through the ATA Safety and Engineering Committee. The ATA Safety Department serves as technical advisors to the Council and supports the Council's activities and programs.

Membership in the Safety Management Council is composed of professional truck fleet safety personnel in the field of motor carrier safety and/or personnel work and persons from allied fields and industries who are concerned with motor carrier safety activities. The Council's technical committees provide the trucking industry with information by working on such problems as employee selection, workers' compensation, training and supervision, accident investigation and reporting, transportation of hazardous materials, federal motor carrier safety regulations, drug testing, physical qualifications, and injury control. In addition to regional and national meetings, the Council also sponsors a wide range of continuing professional education programs for truck and workplace safety.

Publications, instructional videos, safety award programs, and trucking safety magazines are available through the association.

American Water Works Association
6666 West Quincy Avenue
Denver, CO 80235

The Association, through its Loss Control Committee, develops loss control programs for the water utility industry. Programs presently available include:

- collecting, analyzing, and compiling annual statistics in both employee and motor vehicle accidents
- a safety award program
- an audiovisual library
- one- and two-day loss control seminars for supervisors
- loss control audit programs
- a safety poster program

In addition to these programs, the Loss Control Committee has three subcommittees on accident prevention, health maintenance, and risk management that study and disseminate information that will assist the water utility industry in increasing its concepts of total loss control.

Regional safety meetings are held with AWWA section safety chairmen to promote safety throughout the water utility industry. These meetings are designed to provide an exchange of ideas and experiences to enhance water utility safety.

American Welding Society
P.O. Box 351040
Miami, FL 33135

The society is devoted to the proper and safe use of welding by industry. Through its Safety and Health Committee, the society coordinates safe practices in welding by promoting new and revising existing standards.

The society sponsors American National Standards Institute committee Z49, whose publication, *Safety in Welding and Cutting,* is the authoritative standard in this field. It deals with the protection of workers from accidents, occupational diseases, and fires arising out of the installation, operation, and maintenance of electric and gas welding and cutting equipment.

Articles on safety in welding appear in the official publication of the society, the *Welding Journal.*

Associated Builders and Contractors
1300 North 17th Street
Rosalyn, VA 22209

The organization consists of construction contractors, subcontractors, suppliers, and associates. Their purpose is to foster the principles of rewarding construction workers and management on the basis of merit. They sponsor management education programs, craft training, and apprenticeship and skill training programs. A publications catalog is available.

Associated General Contractors of America, Inc.
1957 E. Street, NW
Washington, DC 20006

The Associated General Contractors of America (AGC) is a construction trade association representing more than 32,500 firms, including 8,000 of America's leading general contracting companies. These companies are responsible for the employment of more than 3,500,000 employees. The member construction contractors perform more than 80% of America's contract construction of commercial buildings, highways, bridges, heavy industrial, and municipal utilities.

AGC has actively sought to improve safety performance in the construction workplace since its founding. In 1927, AGC first published its Manual of Accident Prevention in Construction and revised it in 1990.

Throughout the years, AGC and its Safety and Health Committee have diligently sought to keep AGC's members on the cutting edge of safety. AGC has safety programs and publications, all designed to enhance safety in the construction industry.

Association of American Railroads
American Railroads Building
50 F Street, NW
Washington, DC 20001

All Class I railroads (those with an annual revenue in excess of $92 million) are members of the AAR. The following divisions and committees of the association are concerned with safety:

- Communication and Signal Section
- Engineering Division
- Mechanical Division
- Medical Section
- Operating Rules Committee
- Police and Security Section
- Safety Section
- State Rail Programs Division

The Safety Section holds annual meetings; it issues a monthly newsletter, produces posters, and publishes pamphlets on railroad safety.

Chemical Manufacturers Association, Inc.
1300 Wilson Boulevard
Arlington, VA 22209

CMA develops and implements policies, programs, and services that benefit the industry and the public. CMA member companies operate worldwide and account for 90 percent of the domestic chemical industry's productive capacity of basic industrial chemicals.

CMA launched the Responsible Care® Initiative in 1988 to enhance its health, safety, and environmental performance and improve the public's perception of that performance. The industry is implementing the Process Safety Code of Responsible Care®, which is designed to prevent fires, explosions and accidental chemical releases. The Employee Health and Safety (EH&S) Code of Responsible Care® is intended to protect and promote the health and safety of people working at or visiting member company work sites.

Compressed Gas Association, Inc.
1235 Jefferson Davis Highway
Arlington, VA 22202

The Compressed Gas Association is dedicated to the development and promotion of technical and safety standards and safe practices in the compressed gas industry in the United States, Canada, and other countries worldwide. More than 200 member companies work together through the committee system to create technical specifications, safety standards, and training and educational materials; to cooperate with governmental agencies in formulating responsible regulations and standards; and to promote compliance with these regulations and standards in the workplace.

The CGA publishes more than 100 technical standards, many of which are formally recognized by U.S. government agencies, as well as safety bulletins, audiovisual safety training programs, a monthly newsletter, *Compressions,* and the *Handbook of Compressed Gases.* This handbook contains complete descriptions of the widely used industrial gases and sets forth the recognized safe methods for handling, storing, and transporting them. In addition, safety and regulatory alerts and position statements are produced when warranted. A complete catalog is available on request.

The CGA and its Canadian Division hold annual meetings each year in January and September, respectively.

Edison Electric Institute
701 Pennsylvania Avenue, NW
Washington, DC 20004–2696

The Edison Electric Institute is the association of the nation's investor-owned electric utilities. Its members serve 99.6% of the customers serviced by the investor-owned segment of the industry.

The Safety and Industrial Health Committee is dedicated to improving working conditions through development of safe work practices. The Committee meets twice a year and compiles reports and publications on subjects of interest to safety professionals in the utility industry.

Graphic Arts Technical Foundation
4615 Forbes Avenue
Pittsburgh, PA 15213

This is a member-supported, nonprofit, scientific, technical, and educational organization serving the international graphic communications industries since 1924. A publications and services catalog is available on all aspects of the graphic arts industry.

Human Factors and Ergonomics Society
P.O. Box 1369
Santa Monica, CA 90406–1369

HFES is a society of psychologists, engineers, physiologists, and other related scientists who are concerned with the use of human factors and ergonomics in the development of systems and devices of all kinds. *Human Factors* is the official journal.

Illuminating Engineering Society
of North America
345 East 47th Street
New York, NY 10017

The society is the scientific and engineering stimulus in the field of lighting. The work of the Industrial Lighting Committee and its numerous subcommittees for various specific industries should be of particular interest to industrial safety professionals as well as many other projects, such as street and highway, aviation, and office lighting. Through the society, safety personnel can obtain reference material on all phases of lighting, including authoritative treatises on nomenclature, testing, and measurement procedures.

The society publishes a monthly magazine, *Lighting Design and Applications;* a quarterly, *Journal of the Illuminating Engineering Society;* and the *IES Lighting Handbook,* a reference guide. In addition, the society has other publications on specific lighting areas such as mining, roadway, office, and emergency lighting.

Industrial Safety Equipment Association, Inc.
1901 North Moore Street
Arlington, VA 22209

The Industrial Safety Equipment Association (ISEA) is a nonprofit organization of manufacturers of personal protective products for industrial environments. Since its inception in 1934, ISEA has been dedicated to the safety of workers who rely on protective equipment, and to the welfare of the safety equipment industry. Member companies are located in the United States and abroad.

ISEA is dedicated to fostering public interest in safety and encouraging the development and use of proper equipment to deal with industrial hazards in the following ways:

- through its member manufacturers who participate in developing consensus standards for product performance and use
- by representing the safety equipment industry before governmental agencies
- by collecting and disseminating information to the general public

One of ISEA's major efforts is helping to create standards that assure the high performance of protective equipment. In addition, ISEA continually evaluates existing standards to ensure industrial environments are as safe as they can feasibly be. ISEA provides manufacturer input in standards development through official representation to such organizations as the American National Standards Institute, National Fire Protection Association, National Society for the Prevention of Blindness, and the American Society for Testing and Materials.

Institute of Makers of Explosives
1120 19th Street, NW
Washington, DC 20036

The Institute is the safety association of the commercial explosives industry in the United States and Canada. Founded in 1913, IME is a nonprofit, incorporated association whose primary concern is safety and its application to the manufacture, transportation, storage, handling, and use of commercial explosive materials used in blasting and other essential operations. The member companies of IME produce over 85% of the commercial explosive materials consumed annually in the United States or some four billion pounds.

International Association of Drilling Contractors
P.O. Box 4287
Houston, TX 77210

This association works to improve oil well drilling contracting operations as a whole and to increase the value of oil well drilling as an integral part of the petroleum industry. The association holds an annual safety clinic and has standing and special safety committees of contractor representatives to study current problems.

Safety meetings for tool pushers, drillers, crew members, and safety directors are sponsored by the associations; they are conducted throughout the country in locations where these people normally reside.

A Supervisory Accident Prevention Training Program has been developed to instruct drillers and tool pushers in how to establish and maintain effective accident prevention programs. Six to eight professional safety instructors personally conduct these programs anywhere in the world where 18 to 25 people wish to enroll. Many other schools of either two-day or five-day duration are available through the association.

Safety award certificates, cards, safety hat decals, and plaques are given to member personnel and rigs that have completed one or more years without a disabling injury.

The group has produced safety manuals for the industry, inspection reports, color codes, safety signs, studies on protective clothing, and other publications. They have produced color films, film strips, and slides on specific drilling rig safety practices. The association also produces safety posters keyed to the hazards of the drilling industry.

Laser Institute of America
12424 Research Parkway, Suite 130
Orlando, FL 32826

The Laser Institute of America is a nonprofit membership society devoted to education and the advancement, promotion, and safe application of laser technology. The LIA is the secretariat of ANSI Z136.1. The association conducts laser safety training courses and publishes a catalog of materials. Its main publication is *The Journal of Laser Applications.*

National Association of Manufacturers
1331 Pennsylvania Avenue, NW, Suite 1500N
Washington, DC 20004

The NAM safety and health activities are carried on under the aegis of its Risk Management Committee which has a dual function: (a) promoting sound health and safety policies and programs in industry; and (b) working with the federal government to assure that present regulation of health and safety practices in industry and proposals for new regulations and legislation are realistic from industry's viewpoint.

Metal Casting Society
455 State Street
Des Plaines, IL 60016

The Metal Casting Society is a trade association that represents gray, ferrous, and nonferrous foundries in the United States and Canada. Founded in 1975 through a merger of the Gray and Ductile Iron Founders' Society and the Malleable Founders' Society, it is a nonprofit, voluntary membership organization with administrative

headquarters in Des Plaines, IL. It is governed by an elected board of directors which is assisted by 11 standing committees in the formulation of programs and policy to promote the progress of its members and the industry.

National Propane Gas Association
1600 Eisenhower Lane
Lisle, IL 60532

Founded in 1931, the association is a nonprofit, cooperative group of producers and distributors of liquefied petroleum gas (LP-gas), manufacturers of LP-gas equipment, and manufacturers and marketers of LP-gas appliances. NPGA promotes technical information and industry standards in its special field.

Its Safety Committee develops and maintains educational programs to train the public and industry in the safe handling and use of LP-gas and in safe practices for the installation and maintenance of equipment and appliances.

An Educational Committee working closely with the Safety Committee arranges training schools and conferences for dealers and distributors. The association distributes informational, technical, and legislative bulletins and publishes a weekly newsletter for members. It holds an annual meeting and sectional meetings with a definite portion of each program devoted to safety.

The association publishes a *Safety Handbook,* training guides, audiovisual training programs, and distributes consumer education leaflets for members to help educate its customers.

National Petroleum Refiners Association
1899 L Street, NW, Suite 1000
Washington, DC 20036

The NPRA, as a national trade association of petroleum refiners and petrochemical manufacturers, gathers and disseminates industry information and statistics and provides an effective channel of communications among its members and with other associations, government, and the public.

Among the several technical meetings that are sponsored by NPRA is the National Safety Conference, which is a two-day session focusing on current safety issues. The trade group also holds one-day Fire and Accident Prevention Group meetings in each of five geographic areas.

The association also prepares and distributes an annual summary of industry statistics dealing with occupational injuries and illnesses. The group sponsors a comprehensive safety awards program in an effort to promote safety in plant operations.

National Restaurant Association
1200 17th Street, NW
Washington, DC 20036

The National Restaurant Association, through its Technical Services, Public Health, and Safety Department, carries on a program to reduce accidents and hazards that affect the safety of food service employees and patrons. The association conducts research to substantiate industry positions on DOE regulations and on OSHA. The NRA prepares and distributes educational materials to the membership, including self-inspection guidelines on general safety concerns, OSHA requirements, and fire protection, as well as posters and audiovisual programs. These materials are also available for purchase.

Safety-related information appears in the association's *Washington Weekly* report and monthly magazine, *Restaurants USA*.

National Rural Electric Cooperative Association
1800 Massachusetts Avenue, NW
Washington, DC 20036

The association, through its Retirement Safety and Insurance Department, promotes loss control among its members. Following is a brief review of its safety programs:

- Rural Electric Safety Accreditation Program (since 1965). The Program appraises the loss control activities of its members through field audits and review of documentation relating to loss control management.
- Rural Electric Accident Control Today (REACT). The association offers a variety of safety literature for its members to make available to their consumers. The material addresses specific exposures characteristic of electric utilities.
- Loss Control Resources Center. This program helps members exchange safety ideas and policies.
- Loss Control Fund. Funds are provided to selected state associations to assist with their loss control activities based on their member's participation in the association-sponsored property and casualty insurance program.

Loss control articles are published in the association's *Rural Electrification* and *Management Quarterly* magazines.

New York Shipping Association, Inc.
2 World Trade Center
New York, NY 10048

New York Shipping Association is composed of American and foreign flag ocean carriers and contracting stevedores, marine terminal operators, and other employers of waterfront labor within the bi-state Port of New York and New Jersey.

Safety by NYSA is supervised by a director who is appointed by the association president. The safety director maintains contact with federal, state, and other agencies involved with industrial safety and health regulations. In this regard, the safety director disseminates relevant data to the member companies of NYSA to assist them in reducing accidents on piers and at marine facilities. The director also coordinates industry activity and maintains liaison with the various stevedoring and marine terminal companies who operate their own company safety and health programs.

Portland Cement Association
5420 Old Orchard Road
Skokie, IL 60077

PCA, devoted to research, educational, and promotional activities to extend and improve the use of portland cement, is supported by more than 60 U.S., Canadian, and Mexican member companies that operate more than 130 cement manufacturing plants. Occupational safety and health have been considered important by the association since its formation in 1916.

The Washington Affairs Office works closely with an occupational safety and health committee of member company representatives to provide activities, services, and materials that are responsive to the needs of those companies. Knowledge and experience are shared through meetings and conferences.

PCA expresses the views and opinions of its member companies to governmental organizations regarding proposed legislation and regulations and other issues affecting the cement industry.

An injury/illness reporting program enables the association to accumulate data and identify significant causal and circumstantial factors. This information is used to define the nature and extent of cement industry injury/illness experience and is disseminated to member companies.

Power Tool Institute, Inc.
1300 Sumner Avenue
Cleveland, OH 44155-2851

The purposes of the Institute are:

- to promote the common business interests of the power tool industry
- to represent the industry before government
- to educate the public as to the usefulness and importance of power tools
- to encourage high standards of safety and quality control in the manufacture of power tools
- to prepare and distribute information about safe use of power tools

Printing Industries of America, Inc.
100 Dangerfield Road
Alexandria, VA 22314

Printing Industries of America, an association of printers' organizations, actively sponsors the development of safety in the graphic arts through its affiliated local organizations and through its participation in the Graphic Arts Technical Foundation (described earlier).

Scaffolding, Shoring, and Forming Institute, Inc.
c/o Thomas Associates, Inc.
1300 Sumner Avenue
Cleveland, OH 44115-2851

The institute has a deep interest in safety; members try to do everything possible to improve this situation in the construction industry. Publishes numerous guidelines dealing with scaffolding and shoring safety.

Steel Plate Fabricators Association, Inc.
3158 Des Plaines Avenue
Des Plaines, IL 60018

The association has an active safety committee that prepares publications on safety for member companies and their employees.

EMERGENCY INFORMATION BY PHONE

It is not always possible to prepare a plan of action that anticipates every emergency. The following source has 24-hour-a-day phone service for emergencies involving hazardous chemicals.

CHEMTREC

Emergency information about hazardous chemicals involved in transportation accidents can be obtained 24 hours a day. It is the Chemical Transportation Emergency Center (CHEMTREC), and it can be reached by a nationwide telephone number—800/424-9300. The Area Code 800 WATS line permits the caller to dial the station-to-station number without charge.

CHEMTREC provides the caller with response/action information for the product or products and tells what to do in case of spills, leaks, fires, and exposures. This informs the caller of the hazards, if any, and provides sufficient information to take immediate first steps in controlling the emergency. CHEMTREC is strictly an emergency operation provided for fire, police, and other emergency services. It is not a source of general chemical information of a nonemergency nature.

ELECTRONIC SOURCES AVAILABLE THROUGH COMPUTER ACCESS

There are more than 4,500 on-line databases available to computer users. To help users identify databases of particular interest, the latest edition of the source listed below can prove useful:

Gale Directory of Databases
Gale Research Inc.
835 Penobscot Building
Detroit, MI 48226–4094

Governmental Information Sources

The U.S. Government collects volumes of information on almost every subject imaginable. Frequently, one can obtain searches and cited documents free of charge; also, many of the staff are experts who can offer guidance as well as answer specific questions, recommend other sources, and provide printed materials.

The governmental agency that provides the most databases that are of interest to safety and health professionals is the National Library of Medicine, Bethesda, MD 800/638-8480. The library offers the following databases:

- MEDLARS. MEDLARS on-line network is the most useful database for the safety professional who needs to know more about industrial hygiene and health subjects. This network consists of approximately 20 bibliographical databases covering worldwide literature in the health sciences. Those of special value are described below. For more information, contact:

  National Library of Medicine
  MEDLARS Management Section
  8600 Rockville Pike
  Bethesda, MD 20209

- AVLINE. AVLINE (Audio Visuals on-Line) contains references to 21,000 audiovisual instructional packages in the health sciences. All of these materials are professionally reviewed for technical quality, currency, accuracy of subject content, and educational design. AVLINE enables teachers, students, librarians, researchers, practitioners, and other health science professionals to retrieve citations which aid in evaluating audiovisual materials with maximum specificity.

- CANCERLIT. CANCERLIT (Cancer Literature) is the National Cancer Institute's on-line database dealing with all aspects of cancer. The database contains more than 1,000,000 citations dealing with cancer; the sources of this database include more than 3,500 U.S. and foreign journals, as well as books and other reference sources.

- CHEMLINE. CHEMLINE (Chemical Dictionary Online) is the National Library of Medicine's on-line, interactive chemical dictionary file created by the Specialized Information Services in collaboration with Chemical Abstracts Service (CAS). It provides a mechanism whereby more than 1.5 million chemical substance names, representing nearly 1 million unique substances, can be searched and retrieved on-line.

  This file contains CAS Registry Numbers; molecular formulas; preferred chemical index nomenclature; generic and trivial names derived from the CAS Registry Nomenclature File; and a locator designation which points to other files in the NLM system containing information on that particular chemical substance. For a limited number of records in the file, there are Medical Subject Headings (MeSH) terms and Wiswesser Line Notations (WLN). In addition, where applicable, each Registry Number record in CHEMLINE contains ring information including—number of rings within a ring system, ring sizes, ring elemental composition, and component line formulas.

- MEDLINE. MEDLINE (Medical Literature Analysis and Retrieval System on-Line) is a database maintained by the National Library of Medicine; it contains references to over 6.8 million citations from 3,900 biomedical journals. It is designed to help health professionals find out easily and quickly what has been published recently on any specific biomedical subject.

- TOXLINE. TOXLINE (Toxicology Information on-Line) is the National Library of Medicine's extensive collection of computerized toxicology information containing references to published human and animal toxicity studies, effects of environmental chemicals and pollutants, adverse drug reactions, and analytical methodology.
- NIOSHTIC. NIOSHTIC, NIOSH's on-line computerized research and reference database, contains bibliographic abstracts of approximately 188,000 documents in many subject areas, such as toxicology, occupational medicine, industrial hygiene, and personal protective equipment. Additional details can be obtained from:

  National Institute for Occupational Safety
  and Health
  Technical Information Branch
  4676 Columbia Parkway
  Cincinnati, OH 45226

- RTECS. As an on-line database, RTECS (Registry of Toxic Effects of Chemical Substances) is available as a real time, interactive computer database that permits the user to search the RTECS for special data or subsets of data and to compile RTECS subfiles tailored to their particular needs. Updates to these systems are provided on a quarterly basis. The RTECS is available through NIOSH.

PRIVATELY OWNED INFORMATION SOURCES

Two nongovernmental organizations make their databases available to their members.

HAZARDLINE

The HAZARDLINE database provides regulatory, handling, identification, and emergency care information for over 90,000 hazardous substances. The information is gathered from regulations issued from state and federal agencies, court decisions, books, and journal articles, in order to assemble a comprehensive record for each substance.

For more information contact:
Occupational Health Services, Inc.
11 W. 42nd Street
New York, NY 10036

Safety on the Internet

The Internet is a network of networks that can provide the user with a link to the world. A user can use E-mail, transfer files, remote log-in to other computer systems, access mailing lists and news, and participate in discussion groups. The Internet provides valuable information concerning safety, health, and environmental issues.

Getting Connected

Getting connected to the Internet requires at least a 486 personal computer, a 14400 baud modem, and a hard drive. It is advisable to buy a hard drive with the most capacity available.

Everyone who accesses the Internet must establish an Internet address. Access to the Internet can be available through a company specializing in providing Internet communications access or an employer who has an electronic networking account. In the United States, CompuServe, American Online, and Prodigy are popular Internet providers. The user pays the provider a monthly fee for an electronic account. The software programs necessary to permit a computer to send data via a modem to the Internet are either furnished by the Internet provider or can be bought from computer equipment and software retailers.

Surfing the Net

Considering the size of the Internet, the number of computers connected to it, and the volumes of data available to the user, it's easy to understand how people can get lost surfing around for information. Gopher, Veronica, World Wide Web (WWW), and search engines are just a few of the tools that help the user surf through the massive information available on the Internet.

The Gopher is a good tool for a new Internet user. Exploration using Gopher navigation is based on an inquiry about a subject and does not depend on knowledge of computer addresses or locations. Once connected to a Gopher site, the user is presented with menus that make it easy to use. Veronica is a keyword search tool that assists navigation of gopher sites. More information regarding Gopher and Veronica can be found at the following Internet addresses:

Gopher: ftp://pit-manager.mit.edu/pub
      ftp://boombox.micro.umn.edu/pub/gopher
Veronica: ftp://cs.dal.ca

The WWW can link documents at one location with files at another. The Web is a worldwide network of sites, each of which has a "home page" that can serve as a gateway to other Web sites. The Web allows navigation through a point and click interface, using the computer's mouse. Software available for browsing the Web include Mosaic for Windows, Mosaic for Macintosh, WinWeb, MacWeb, and Netscape. Information on the Web can be downloaded or printed directly to a PC. Additional information regarding the WWW can be found at the following Internet addresses:

ftp://info.cern.ch
ftp://ftp.ncsa.uiuc.edu
gopher://fatty.law.cornell.edu

If a user is searching for information on the WWW, search engines can provide a listing of Internet sites that may contain the information you are seeking. Some useful search engines include:

LYCOS: http://lycos.cs.cmu.edu/
NETSEARCH: http://www.ais.net:80/netsearch/
INTERNET SLEUTH: http://www.intbc.com/sleuth/

Safety, Health, and Environmental Issues

For information concerning safety, health, and environmental issues, the Internet provides up-to-date national and international information from both government and industry. Some sites to access for safety, health, and environmental information include:

- American National Standards Institute (ANSI): http://www.ansi.org
- American Red Cross: http://www.crossnet.org
- National Safety Council: http://www.nsc.org/nsc
- U.S. Environmental Protection Agency: http://www.epa.gov
- U.S. Federal Register: http://gopher.nara.gov:70/1/register
- U.S. Government Agency Server Index: http://www.eit.com/web/www.servers/government.html
- U.S. Occupational Safety and Health Administration: http://www.osha.gov/index.html

Some general safety and health reference lists on the Internet can be found at:

- http://www.sas.ab.ca/biz/christie/
- gopher://www.sas.ab.ca/businesses/christie/safelist.html
- http://www.ccohs.ca

U.S. GOVERNMENT AGENCIES

An overwhelming amount of safety information is available from the federal government concerning all aspects of safety and health, environmental problems, pollution, statistical data, and other industry problems.

Because of the constant change in government agency activities and frequent reorganizations, it is recommended that the reader consult the *United States Government Organization Manual,* published by the Government Printing Office, Washington, DC 20402. It can be found in most libraries.

DEPARTMENTS AND BUREAUS IN THE STATES AND POSSESSIONS

Safety professionals need to have a good working knowledge of the state agencies responsible for the enforcement of safety and health laws. They should contact the proper groups in their state's labor departments or other pertinent agencies and find out how the various boards, divisions, and services function.

Codes and laws vary widely in the different states and provinces, and those persons who have safety jurisdiction in plants in a number of places must understand these differences.

In many cases, the standards set up by the code may serve only as a minimum, and safety professionals will want to compare them with American National Standards or other regulations to establish more rigid rules for their own plants. They should also know the jurisdiction rights of the factory inspectors, so that they can better understand the job inspectors have to do and how they can help them in the performance of their duties.

Health and Hygiene Services

Departments or boards of health and industrial hygiene services are integral parts of the organization of each of the states, the District of Columbia, and the autonomous territories of the United States.

It is to these organizations that safety professionals must look for their state's specific standards and recommendations on such points as occupational health, food and health engineering, disease control, water pollution, and other facets of the overall field of industrial hygiene.

Industrial hygiene units usually function full time or, in several states, on a limited basis. In addition to the units that operate under state health departments, a number of other industrial hygiene units are run by municipalities or other local authorities.

In addition to direct industrial hygiene services, these state units are able to bring to industry a more or less complete health program by integrating their work with that of other divisions in the state government, such as sanitation and infectious disease control.

OSHA On-Site Consultation Program

To contact the OSHA coordinator call or write:

Directorate of Federal-State Operations
Office of Consultation Programs
Room N3476
200 Constitution Avenue, NW
Washington, DC 20210
Phone: (202) 523-8902

The names and addresses of such state, commonwealth or territorial agencies with which the safety professional may need to communicate are listed in the U.S. Department of Labor Bulletin "OSHA Onsite Consultation Project Directory." The bulletin is revised periodically and is available from the Occupational Safety and Health Administration, Washington, DC 20210.

CANADIAN DEPARTMENTS, ASSOCIATIONS, AND BOARDS

In all provinces of Canada there is a Workmen's Compensation Board or Commission. Some of these handle accident prevention directly. In other provinces there are provisions similar to Section 110 of the Quebec Workmen's Compensation Act, which stipulates

that industries included in any of the classes under Schedule I may form themselves into an Association for accident prevention and formulate rules for that purpose. Further, the Workmen's Compensation Commission, if satisfied that an Association so formed sufficiently represents the employers in the industries included in the class, may make a special grant toward the expense of any such Association.

It is under these provisions that the various safety associations were organized and are functioning. In some provinces, accident prevention is directly assumed by the board itself by establishing a safety department.

Furthermore, all provinces have legal safety requirements which are administered by the Department of Highways, Department of Labor, and the Department of Mines. These sources can be contacted by writing to the deputy minister of the department located in the capital of each province.

Governmental Agencies

The Occupational Safety and Health Branch of Labour Canada is responsible for the implementation and administration of the Canada Labor Code Part IV and pursuant regulations which became effective January 1, 1968, and deals primarily with Occupational Safety and Health. It is also responsible for the development of Occupational Safety and Health regulations, procedures, and standards for regulating all work places under federal jurisdiction. Address: Director, Occupational Safety and Health Branch, Labour Canada, Ottawa, Ontario, Canada KIA 0J2.

Accident Prevention Associations

Canada Safety Council
#6, 2750 Stevenage Drive
Ottawa, Ontario K1G 5N2

Provincial associations
Alberta Safety Council
201-10526 Jasper Avenue
Edmonton T5J 1Z7

British Columbia Safety Council
8589 Baxter Place
Burnaby, V5A 4T2

Manitoba Safety Council
#700.213 Notre Dame Avenue
Winnipeg, R3B 1N3

Newfoundland Safety Council
93 Water Street
P.O. Box 5123
St. John's A1C 5V5

Saskatchewan Safety Council
140 4th Avenue East
Regina S4N 4Z4

Industrial Accident Prevention Association of Ontario
250 Yonge Street
Toronto M5B 2N4

Included in this association are ten class associations:

- Woodworkers Accident Prevention Association
- Ceramics & Stone Accident Prevention Association
- Metal Trades Accident Prevention Association
- Chemical Industries Accident Prevention Association
- Grain, Feed & Fertilizer Accident Prevention Association
- Food Products Accident Prevention Association
- Leather, Rubber & Tanners Accident Prevention Association
- Textile & Allied Industries Accident Prevention Association
- Printing Trades Accident Prevention Association
- Ontario Retail Accident Prevention Association

Ontario Safety League
21 Four Seasons Place
Etobicoke, Ontario M9B 6J8

Mines Accident Prevention Association of Manitoba
305 Broadway
Winnipeg R3C 3J7

Transportation Safety Association of Ontario (Inc.)
220 Traders Blvd. East
Mississauga, Ontario L4Z 1W7

Quebec Occupational Health & Safety Research Institute (IRSST)
505 boul de Maisonneuve ouest
Montreal H3A 3C2

INTERNATIONAL SAFETY ORGANIZATIONS

Inter-American Safety Council
(Consejo Interamericano de Seguridad)
33 Park Place
Englewood, NJ 07631

The Inter-American Safety Council was founded and incorporated in 1938 as a noncommercial, nonpolitical, and nonprofit educational association for the prevention of accidents. It is the Spanish and Portuguese language counterpart of the National Safety Council.

The Council is the first and only association of its kind rendering services to all industries and agencies in the Latin American countries, Spain, and Portugal. The objectives are to prevent accidents and to reduce the number and severity of accidents in every activity, both on the job and off the job. The services that the Council provides for its members are paid by membership dues and sales of the Council's monthly publications and other educational materials. All of its work is done from the headquarters in New Jersey.

Membership is open to all industries, organizations, institutions, or other groups with two or more employees, interested in accident prevention in Latin America and Spain. More than 3,200 plants or work locations in 20 countries are members or are using the materials and services of the Council. In addition, over 300 universities, technical schools, public libraries, and the like receive the monthly publications free of charge.

Services available to members include monthly publications, annual contest, special awards, consultation, statistical service, reproduction and translation rights, and participation in the election of Council officers. In addition, the Council acts as a clearing house of accident prevention materials available in the United States. A catalog is available from the organization.

The Council's monthly publications include two magazines and safety posters and also a quarterly publication:

Noticias de Seguridad (Safety News)

El Supervisor (The Supervisor)

Safety posters in sizes 8 × 11 and 17 × 22 in.

Usted y su familia (You and Your Family)

In addition, the Council publishes translations of publications, films, safety slides, training programs, and other materials of the National Safety Council and other accident prevention organizations.

The Royal Society for the Prevention of Accidents Cannon House, The Priory Queensway Birmingham, England B4 6BS

Founded in 1916, RoSPA, Europe's largest safety organization, works to prevent accidents at home, at work, on the road, and at leisure. It produces safety education materials for schools and runs an occupational safety training center and extensive driver training programs. It is frequently consulted by government departments and other organizations on technical matters.

Occupational safety and health training is offered worldwide. The RoSPA International Health and Safety Exhibition at the National Exhibition Center is the largest annual safety exhibition in Europe. There is an award scheme rewarding progressive and long-term safety achievements by companies.

Three monthly journals are produced: *Occupational Safety and Health, Safety Representative*, and *RoSPA Bulletin* cover all aspects of the workplace safety scene. The library is the largest safety library in Europe. RoSPA is an independent charity under the patronage of HM The Queen.

International Association of Industrial Accident Boards and Commissions
1575 Aviation Center Parkway, Suite 512 Daytona Beach, FL 32114

The IAIABC, founded in 1914, is a professional association of workers' compensation specialists from the private and public sector. Its purpose is the improvement of workers' compensation administration. The IAIABC active membership includes the public worker's compensation agencies in the United States, Canada, Puerto Rico, and Australia. The associate membership consists of private sector worker's compensation administrators, physicians, attorneys, labor representatives, medical and rehabilitation providers, insurers, employers, trade associations, and other groups and individuals involved with workers' compensation programs from the United States, Canada, Guam, South Africa, Australia, and Zimbabwe.

International Labor Organization
International Labor Office, CH1211 Geneva 22, Switzerland

The International Labor Organization (ILO), a specialized agency associated with the United Nations, was created by the Treaty of Versailles in 1919 as part of the League of Nations. Its purpose is to improve labor conditions, raise living standards, and promote economic and social stability as the foundation for lasting peace throughout the world. To this purpose, one of ILO's functions is "the protection of the worker against sickness, disease, and injury arising out of his employment."

The organization consists of about 140 member countries, including the United States (which joined in 1934). ILO functions through an annual conference of member states, a governing body, advisory committees, and a permanent office, the International Labor Office. ILO is distinctive from all other international agencies because it is tripartite--that is, the conference, the governing body, and some of the committees are composed of representatives of governments, employers, and workers.

In the field of safety and health, the International Labor Office maintains a permanent international staff of medical doctors, engineers, and industrial hygienists. Assistance in specific fields is given by panels of consultants, drawn from all parts of the world to act in an advisory capacity and to discuss problems, draft regulations, or render help in emergencies.

The United States has several members on the panels and has been represented on all the temporary expert committees and special conferences.

The main tasks of the ILO in the field of occupational safety and health are:

- International instruments. These include conventions and recommendations, and also model safety codes and codes of practice. An *Encyclopedia of Occupational Health and Safety* has been prepared in English and French, to succeed *Occupation and Health*, which was published in 1930. This is designed to provide guidance to a wide range of people concerned with health, safety, and welfare at work. Although problems are reviewed from an international angle, special account is taken of the needs of developing countries.

- the compilation of technical studies
- the publication of medical and technical studies
- direct assistance to governments by furnishing experts, drafting regulations, supplying information, etc.
- collaboration with other international organizations, the World Health Organization, and the International Organization for Standardization
- assistance to national safety and health organizations, research centers, employers' associations, trade unions, etc., in different countries

In general, the organization works to keep in touch with the safety and health movement throughout the world and assist the movement in any way possible.

International Organization for Standardization
1 rue de Varembe
Case Postale 56
CH-1211, Geneva
Switzerland

A multinational federation of national standard bodies united to promote standardization worldwide. Develops and publishes international standards to help in the exchange of goods and services and foster mutual cooperation in intellectual, scientific, technological, and economic areas of endeavor. Offers a variety of publications on standardization including the "ISO Catalogue" on ISO standards.

Pan American Health Organization
525 23rd Street, NW
Washington, DC 20037

Originally established as the International Sanitary Bureau in 1902, the Pan American Health Organization serves as the regional office for the World Health Organization for the Americas. The purposes of the PAHO are to promote and coordinate the efforts of the countries of the western hemisphere to combat disease, lengthen life, and promote the physical and mental health of the people.

Programs encompass technical collaboration with governments in the field of public health, including such subjects as sanitary engineering and environmental sanitation, eradication or control of communicable diseases, and maternal and child health.

World Health Organization
20 Avenue Appia
CH-1211 Geneva 27
Switzerland

Created in 1948, the World Health Organization (WHO) is a specialized agency of the United Nations with primary responsibility for international health matters and public health. Through this organization, the health professions of its 166 member states exchange their knowledge and experience with the aim of making possible the attainment by all citizens of the world by the year 2000 of a level of health that will permit them to lead a socially and economically productive life.

Through technical cooperation with its member states, WHO promotes the development of comprehensive health services, the prevention and control of diseases, the improvement of environmental conditions, the development of health manpower, the coordination and development of biomedical and health services research, and the planning and implementation of health programs.

WHO also plays a major role in establishing international standards for biological substances, pesticides, and pharmaceuticals; formulating environmental health criteria; recommending international nonproprietary names for drugs; administering the International Health Regulations; revising the International Classification of Diseases, Injuries, and Causes of Death; and collecting and disseminating health statistical information.

Authoritative information on the various fields covered by WHO is to be found in its many scientific and technical publications. Of particular interest to health and safety professionals is the Environmental Health Criteria series, prepared by the International Programme on Chemical Safety. Each book in the series reviews all available information on a selected chemical, environmental pollutant, or method for testing toxicity and carcinogenicity, in order to provide guidance on prevention of health hazards and the setting of safe exposure limits. In 1987, WHO launched a new series of Health and Safety Guides, based on the Criteria series and designed to enhance workers' awareness of precautions needed in the handling of individual potentially dangerous chemicals. Catalogs of new and current publications are available on request.

EDUCATIONAL INSTITUTIONS

Many colleges and universities offer formal courses in industrial safety. In a publication *Safety and Related Degree Programs*, compiled by the American Society of Safety Engineers, accredited four-year colleges and universities are listed that offer a degree program with concentration on industrial safety. Also listed are those that offer one or more credit courses as an elective within engineering or education curricula. Courses in water safety, first aid and safety, firefighting, and driver or traffic education are not listed. Write to the ASSE for a copy of this publication. (Address is listed under Professional Societies discussed earlier in this appendix.)

BIBLIOGRAPHY OF SAFETY AND HEALTH PERIODICALS

Safety

Accident Analysis and Prevention (quarterly)
Pergamon Press Inc.
c/o Elsevier Science Inc.
660 White Plains Road
Tarrytown, NY 10591

Canadian Occupational Safety (bimonthly)
Royal Life Center

277 Lakeshore Road East
Oakville, Ontario L6J 6J3
Canada

Chemical Health & Safety (bimonthly)
American Chemical Society
1155 16th Street NW
Washington, DC 20036

Hazard Prevention (quarterly)
System Safety Society
5 Export Drive, Suite A
Sterling, VA 22170

Health and Safety at Work (monthly)
Drayton Bridge House
3 High Street
West Drayton, Middlesex UB7 7QT
England

Human Factors (quarterly)
The Human Factors and Ergonomics Society
P.O. Box 1369
Santa Monica, CA 90406

Job Safety & Health Quarterly (quarterly)
Superintendent of Documents
U.S. Government Printing Office
Washington, DC 20402

Occupational Hazards (monthly)
Penton Publishing Inc.
1100 Superior Avenue
Cleveland, OH 44114

Occupational Health & Safety (monthly)
Stevens Publishing
3630 I-35
Waco, TX 76706

Occupational Health & Safety Canada (bimonthly)
1450 Don Mills Road
Don Mills, Ontario M3B 2X7

Occupational Safety and Health (monthly)
The Royal Society for the Prevention of Accidents
Cannon House, The Priory
Queensway
Birmingham B4 6BS
England

Professional Safety (monthly)
American Society of Safety Engineers
1800 East Oakton Street
Des Plaines, IL 60018

Safety Science
Elsevier Science B.V.
P.O. Box 211, 1000 AE
Amsterdam, The Netherlands
Industrial Hygiene and Medicine

Archives of Environmental Health (bimonthly)
Heldref Publications
4000 Albemarle Street, NW
Washington, DC 20016

Journal of the American Medical Association (weekly)
American Medical Association
535 North Dearborn Street
Chicago, IL 60610

American Industrial Hygiene Association Journal (monthly)
American Industrial Hygiene Association
345 White Pond Drive
P.O. Box 8390
Akron, OH 44320

American Journal of Public Health (monthly)
American Public Health Association
1015 15th Street, NW
Washington, DC 20005

American Journal of Industrial Medicine (quarterly)
Wiley-Liss
41 East 11th Street
New York, NY 10003

Applied Occupational and Environmental Hygiene (monthly)
6500 Glenway Avenue, Bldg. D-7
Cincinnati, OH 45211-4438

Applied Ergonomics (bimonthly)
Butterworth-Heinemann Ltd.
P.O. Box 63
Westbury House, Bury Street
Gilford, Surrey GU2 5BH
England

Occupational and Environmental Medicine (monthly)
British Medical Association
Tavistock Square
London WC1H 9JR
England

CIS Abstracts (8 times a year)
International Occupational Safety and Health Information Center (CIS)
International Labor Office
1211 Geneva 22
Switzerland

Industrial Hygiene Digest (monthly)
Industrial Health Foundation
34 Penn Circle West
Pittsburgh, PA 15206

Journal of Occupational and Environmental Medicine (monthly)
Williams & Wilkins
428 Preston Street
Baltimore, MD 21202-3993

AAOHN Journal
American Association of Occupational Health Nurses
50 Lenox Pointe
Atlanta, GA 30324

Work and Stress (quarterly)
Taylor & Francis Ltd.
4 John Street
London WC1N 2ET
England

Environmental Health Perspectives (monthly)
Superintendent of Documents

U.S. Government Printing Office
Washington, DC 20402

Environmental Research (8 times a year)
Academic Press Inc.
6277 Sea Harbor Drive
Orlando, FL 32887-4900

Fire
Fire Technology (quarterly)
National Fire Protection Association
Batterymarch Park
Quincy, MA 02269

Fire Engineering (monthly)
Park 80 West, Plaza Two
Saddle Brook, NJ 07663

NFPA Journal (bimonthly)
National Fire Protection Association
Batterymarch Park
Quincy, MA 02269

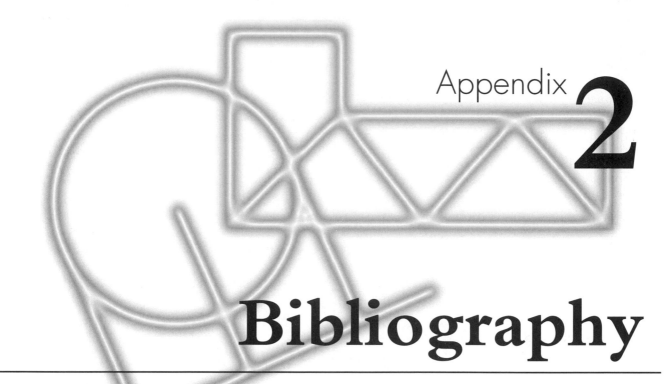

Bibliography

The reference material cited in this Bibliography was selected to provide safety and health professionals with sources of information that are likely to prove most useful in coping with problems of worker health protection and hazard assessment. This compilation is not to be viewed as a comprehensive coverage of the abundant literature on this subject, nor is any endorsement implied. The reference books are listed according to the following outline:

- General principles
- Risk assessment
- Sampling methods
- Toxicology
- Medical
- Dermatitis
- Chemical
- Pollution and hazardous waste
- Control
- Encyclopedias and handbooks
- Safety management
- Emergency
- Safety training
- Accident investigation and analysis
- Product safety
- Fire
- Loss control
- OSHA
- Computer and applications

GENERAL PRINCIPLES

Best's Safety Directory. Oldwick, NJ: A. M. Best Company. Published annually.

Burgess WA. *Recognition of Health and Hazards in Industry,* 2nd ed. New York: John Wiley & Sons, Inc., 1995.

Clayton GD and Clayton FE, eds. *Patty's Industrial Hygiene and Toxicology.* Vol. 1: *General Principles.* New York: John Wiley & Sons, Inc., 1991.

Cralley LJ and Cralley LV, eds. *Patty's Industrial Hygiene and Toxicology.* Vol 3A: *Theory and Rationale of Industrial Hygiene Practice,* 3rd ed. (in two parts: A and B), New York: John Wiley & Sons, Inc., 1993.

Hazard Communication Standard Inspection Manual, 2nd ed. Rockville, MD: Government Institutes, Inc., 1989.

LaDou JL, ed. *Occupational Health and Safety,* 2nd ed. Itasca, IL: National Safety Council, 1993.

Leveson N. *Safeware.* New York: Addison-Wesley Publishing Company, 1995.

Lowry GG and Lowry RC. *Handbook of Hazard Communication and OSHA Requirements.* Chelsea, MI: Lewis Publishers, Inc., 1985.

O'Donnel MP and Harris JS, eds. *Health Promotion in the Workplace,* 2nd ed. Albany, NY: Delmar Publishers, Inc., 1993.

Parmeggiani L, ed. *Encyclopedia of Occupational Health and Safety,* 3rd rev. ed. New York: McGraw-Hill Book Co., 1989.

Waldo AB. *Chemical Hazard Communication Guidebook,* 2nd ed. New York: Executive Enterprises, 1991.

RISK ASSESSMENT—INDUSTRIAL HYGIENE

Andelman JB and Underhill DW. *Health Effects from Hazardous Waste Sites.* Chelsea, MI: Lewis Publishers, Inc., 1987.

Hallenbeck WH. *Quantitative Risk Assessment for Environmental and Occupational Health,* 2nd ed. Chelsea, MI: Lewis Publishers, Inc., 1993.

Threshold Limit Values and Biological Exposure Indices. Cincinnati: ACGIH. Published annually.

SAMPLING METHODS—INDUSTRIAL HYGIENE

American Conference of Governmental Industrial Hygienists. *Air Sampling Instruments—For Evaluation of Atmospheric Contaminants,* 8th ed. Cincinnati: ACGIH, 1995.

TOXICOLOGY

Clayton DB, Krewski D, and Munro I. *Toxicological Risk Assessment: Biological & Statistical Criteria.* Boca Raton, FL: CRC Press, Inc., 1985.

Clayton GD and Clayton FE, eds. *Patty's Industrial Hygiene and Toxicology.* Volume 2A, 2B, and 2C: *Toxicology.* New York: John Wiley & Sons, Inc., 1993.

Documentation of the Threshold Limit Values for Substances in Workroom Air, rev. ed. Cincinnati: ACGIH.

Johnson BL. *Advances in Neurobehavioral Toxicology Applications in Environmental and Occupational Health.* Chelsea, MI: Lewis Publishers, 1990.

Milman HA and Weisburger ED, eds. *Handbook of Carcinogen Testing.* Park Ridge, NJ: Noyes Data Corp., 1985.

Neely WB and Blau GE. *Environmental Exposure from Chemicals.* Boca Raton, FL: CRC Press, Inc., 1985.

MEDICAL

Alderman MH and Hanley MJ, eds. *Clinical Medicine for the Occupational Physician.* New York: Marcel Dekker, Inc., 1982.

Cataldo MF and Coates TJ. *Health and Industry: A Behavior Medicine Perspective.* New York: John Wiley & Sons, Inc., 1986.

Karpilow C. *Occupational Medicine in the International Workplace.* New York: Van Nostrand Reinhold, 1991.

Preventing Illness and Injury in the Workplace. Washington, DC: Office of Technology Assessment, 1985.

Safety Guide for Health Care Institutions, 5th ed. Itasca, IL: National Safety Council and the American Hospital Association, 1994.

Zenz C. *Occupational Medicine,* 3rd ed. St. Louis, MO: Mosby-Yearbook, Inc., 1994.

DERMATOLOGICAL

Adams RM. *Occupational Skin Disease,* 2nd ed. Orlando, FL: W. B. Saunders, 1989.

Kryter KD. *The Effects of Noise on Man,* 2nd ed. San Diego, CA: Academic Press, 1985.

Polk C. *CRC Handbook of Biological Effects of Electromagnetic Fields.* Boca Raton, FL: CRC Press, Inc., 1986.

Sliney D and Wolbarsht ML. *Safety with Lasers and Other Optical Sources: A Comprehensive Handbook.* New York: Plenum Publishing Corp., 1980.

ERGONOMICS

Astrand PO and Rodahl K. *Textbook of Work Physiology,* 3rd ed. New York: McGraw-Hill Book Co., 1986.

Chaffin DB and Anderson JB. *Occupational Biomechanics,* 2nd ed. New York: John Wiley & Sons, Inc., 1991.

Eastman Kodak Company. *Ergonomic Design for People at Work.* Vol. 1. New York: Van Nostrand Reinhold, 1989.

Grandjean E. *Fitting the Task to the Man—An Ergonomic Approach,* 4th ed. London, England: Taylor & Francis, Ltd., 1988.

Konz S. *Work Design: Industrial Ergonomics,* 3rd ed. Worthington, OH: Publishing Horizons, 1990.

McCormick EJ and Sander MS. *Human Factors in Engineering and Design.* 7th ed. New York: McGraw-Hill, 1993.

MacLeod D. *The Ergonomics Edge: Improving Safety, Quality, and Productivity.* New York: Van Nostrand Reinhold, 1995.

Noro K. *Participatory Ergonomics.* Philadelphia, PA: Taylor & Francis, 1991.

Salvendy G, ed. *Handbook of Human Factors.* New York: John Wiley & Sons, Inc., 1987.

CHEMICAL

Compressed Gas Association. *Handbook of Compressed Gases,* 3rd ed. New York: Van Nostrand Reinhold, 1990.

Fawcett HL and Wood WS. *Safety and Accident Prevention in Chemical Operations,* 2nd ed. New York: Interscience Publishers, 1982.

Hathaway G. *Chemical Hazards in the Workplace,* 3rd ed. New York: Van Nostrand Reinhold, 1991.

POLLUTION AND HAZARDOUS WASTE

Dawson GW and Mercer BW. *Hazardous Waste Management.* New York: John Wiley & Sons, Inc., 1986.

Gammage RB and Kaye SV. *Indoor Air and Human Health.* Chelsea, MI: Lewis Publishers, Inc., 1985.

Godish T. *Air Quality,* 2nd ed. Chelsea, MI: Lewis Publishers, Inc., 1991.

Hines AL. *Indoor Air Quality and Control.* Englewood Cliffs, NJ: Prentice-Hall, 1993.

Levine SP and Martin WF. *Protecting Personnel at Hazardous Waste Sites,* 2nd ed. Stoneham, MA: Butterworth Publishers, 1994.

Lioy PJ and Daisey JM, eds. *Toxic Air Pollution.* Chelsea, MI: Lewis Publishers, Inc., 1987.

Management of Toxic and Hazardous Wastes. Chelsea, MI: Lewis Publishers, Inc., 1985.

Martin WF. *Hazardous Waste Handbook for Health and Safety.* Stoneham, MA: Butterworth Publishers, 1987.

Robinson WD. *The Solid Waste Handbook, A Practical Guide.* New York: John Wiley & Sons, Inc., 1986.

Wagner TP. *The Complete Guide to the Hazardous Waste Regulations,* 2nd. ed. New York: Van Nostrand Reinhold, 1991.

CONTROL

Borup B. *Pollution Control for the Petrochemicals Industry.* Chelsea, MI: Lewis Publishers, Inc., 1987.

Cralley LJ and Cralley LV, eds. *Industrial Hygiene Aspects of Plant Operations* (in three volumes). Melbourne, FL: Krieger Publishing Co., 1986.

Guidelines for Process Safety Fundamentals in General Plant Operations. New York: American Institute of Chemical Engineers, 1994.

Heinsohn PA. *Biosafety References Manual,* 2nd ed. Fairfax, VA: American Industrial Hygiene Association, 1995.

Industrial Ventilation, A Manual of Recommended Practice, 22nd ed. Lansing, MI: Committee on Industrial Ventilation, ACGIH, 1995.

LaDou JL, ed. *Occupational Health and Safety,* 2nd ed. Itasca, IL: National Safety Council, 1994.

McDermott H. *Handbook of Ventilation for Contaminant Control,* 2nd ed. Stoneham, MA: Butterworth-Heineman, 1985.

Pepitone DA, ed. *Safe Storage of Laboratory Chemicals,* 2nd ed. New York: John Wiley & Sons, Inc., 1991.

Rajhans G and Blackwell DS. *Practical Guide to Respirator Usage in Industry.* Stoneham, MA: Butterworth-Heineman, 1985.

ENCYCLOPEDIAS AND HANDBOOKS

King R and Hudson R. *Construction Hazard and Safety Handbook.* Stoneham, MA: Butterworth-Heineman, 1985.

Lowry GG and Lowry RC. *Handbook of Hazard Communication and OSHA Requirements.* Chelsea, MI: Lewis Publishers, Inc., 1985.

Parmeggiani L, ed. *Encyclopedia of Occupational Health and Safety,* 3rd rev. ed. (two volumes) Geneva, Switzerland: International Labour Office, 1989.

Ridley J. *Safety at Work,* 4th ed. Stoneham, MA: Butterworth Publishers, 1995.

Sax NI. *Dangerous Properties of Industrial Materials,* 9th ed. New York: Van Nostrand Reinhold, 1993.

Sax NI and Lewis RJ Sr, eds. *Rapid Guide to Hazardous Chemicals in the Workplace,* 3rd ed. New York: Van Nostrand Reinhold, 1994.

NIOSH PUBLICATIONS

The National Institute for Occupational Safety and Health (NIOSH) has published many useful publications in the field of industrial hygiene. Consult the current "Publications Catalog," listing all NIOSH publications in print and their prices, for further information. The NIOSH publications can be obtained by requesting single copies from:

National Institute for Occupational Safety and Health
Division of Technical Services
Publications Dissemination
4676 Columbia Parkway
Cincinnati, OH 45226

Many of these publications are also available from:

Superintendent of Documents
U. S. Government Printing Office
Washington, DC 20402

Some NIOSH publications can also be obtained from:

National Technical Information Services (NTIS)
Springfield, VA 22161

CRITERIA DOCUMENTS

The NIOSH is responsible for providing relevant data from which valid criteria for effective standards can be derived. Recommended standards for occupational exposure, which are the result of this work, are based on the health effects of exposure.

The single most comprehensive source of information on a particular material will probably be found in the NIOSH "Criteria Document" for that substance. The Table of Contents for a Criteria Document is as follows:

I. Recommendations for an Occupational Exposure Standard

Section 1—Environmental (workplace air)
Section 2—Medical
Section 3—Labeling and posting
Section 4—Personal protective equipment and clothing

SAFETY MANAGEMENT

Asfahl CR. *Industrial Safety and Health Management,* 2nd ed. Englewood Cliffs, NJ: Prentice-Hall, 1990.

Goetsch DL. *Occupational Safety and Health in the Age of High Technology,* 2nd ed. Englewood Cliffs, NJ: Prentice-Hall, 1996.

Grimaldi J. *Safety Management,* 5th ed. Burr Ridge, IL: Irwin, 1989.

Hammer W. *Occupational Safety Management and Engineering,* 4th ed. Englewood Cliffs, NJ: Prentice-Hall, 1989.

University of Medicine and Dentistry of New Jersey, Department of Environment and Community Medicine Staff. *Health and Safety in a Small Industry: A Practical Guide for Managers.* Chelsea, MI: Lewis Publishers, Inc., 1989.

Hoover et al. *Health Safety and Environmental Control.* New York: Van Nostrand Reinhold, 1989.

Kase DW. *Safety Auditing: A Management Tool.* New York: Van Nostrand Reinhold, 1990.

Kavianien HR and Wentz CA. Jr. *Occupational and Environment Safety Engineering and Management.* New York: Van Nostrand Reinhold, 1990.

Krause TR. *Behavior-Based Safety Process.* New York: Van Nostrand Reinhold, 1990.

MacCollum DV. *Construction Safety Planning.* New York: Van Nostrand Reinhold, 1995.

McSween TE. *Values-Based Safety Process: Improving Your Safety Culture with a Behavior Approach.* New York: Van Nostrand Reinhold, 1995.

Marshall G. *Safety Engineering,* 2nd ed. Des Plaines, IL: American Society of Safety Engineers, 1994.

Occupational Health Services: A Practical Approach. Chicago American Medical Association, 1990.

Petersen D. *Safety Objectives,* 2nd ed. New York: Van Nostrand Reinhold, 1996.

Petersen D. *Techniques of Safety Management: A Systems Approach,* 3rd ed. Goshen, NY: ALORAY, 1989.

Slote L, ed. *Handbook of Occupational Safety & Health.* New York: John Wiley & Sons, 1987.

Tompkins NC. *How to Write a Company Safety Manual.* Boston, MA: Standard Publishing Corp., 1993.

Vincoli JW. *Basic Guide to System Safety.* New York: Van Nostrand Reinhold, 1993.

EMERGENCY

Himmelfarb AB. *A Guide to Product Failures and Accidents.* Lancaster, PA: Technomic, 1985.

Kelly B. *Industrial Emergency Procedures.* New York: Van Nostrand Reinhold, 1989.

SAFETY TRAINING

Baldwin DA. *Safety and Environmental Training: Using Compliance to Improve Your Company.* New York: Van Nostrand Reinhold, 1992.

Hendrick K. *Systematic Safety Training.* New York: Marcel Dekker, Inc., 1990.

ReVelle JB. *Safety Training Methods,* 2nd ed. New York: Wiley, 1995.

ACCIDENT INVESTIGATION AND ANALYSIS

Ferry S. *Readings in Accident Investigation.* Springfield, IL: Charles Thomas, 1984.

Hendrick K and Benner J. *Investigating Accidents with STEP.* New York: Marcel Dekker, 1987.

Vincoli JW. *Basic Guide to Accident Investigation and Loss Control.* New York: Van Nostrand Reinhold, 1994.

PRODUCT SAFETY

Hammer W. *Product Safety and Management Engineering,* 2nd ed. Des Plaines, IL: American Society of Safety Engineers, 1993.

Selden R. *Product Safety Engineering for Managers.* Englewood Cliffs, NJ: Prentice-Hall, 1984.

FIRE

Cote A and Linville J, eds. *Fire Protection Handbook,* 17th ed. Quincy, MA: National Fire Protection Association, 1991.

Cote AE. *Industrial Fire Hazards Handbook.* Quincy, MA: National Fire Protection Association, 1990.

LOSS CONTROL

Bird FE. *Practical Loss Control Leadership: the Conservation of People, Property, and Profits.* Loganville, GA: International Loss Control Institute, 1992.

Brisbin RE. *Loss Control for the Small to Medium Size Business.* New York: Van Nostrand Reinhold, 1989.

OSHA

Government & Industry Standards Crossreference. Neenah, WI: J. J. Keller & Associates, 1994.

Mintz BW. *OSHA History, Law, and Policy.* Washington, DC: The Bureau of National Affairs, 1984.

OSHA Systems Safety Inspection Guide. Rockville, MD: Government Institutes, Inc., 1989.

Wang CC. *OSHA Compliance and Management Handbook.* Park Ridge, NJ: Noyes Publications, 1993.

COMPUTERS AND APPLICATIONS

Brauer RL. *Directory of Safety-Related Computer Resources.* Des Plaines, IL: American Society of Engineers, 1994.

Dobins D, ed. *Microcomputer Applications in Occupational Health and Safety.* Chelsea, MI: Lewis Publishers, Inc., 1987.

Appendix 3

Answers to Review Questions

1 HISTORICAL PERSPECTIVES

1. The six reasons for preventing injuries and occupational illnesses as given in the text:
   a. Needless destruction of life and health is morally unjustified.
   b. Failure to take necessary precautions against predictable accidents and occupational illnesses makes management and workers morally responsible for those accidents and occupational illnesses.
   c. Accidents and occupational illnesses severely limit efficiency and productivity.
   d. Accidents and occupational illnesses produce far-reaching social harm.
   e. The safety movement has demonstrated that its techniques are effective in reducing accident rates and promoting efficiency.
   f. Recent state and federal legislation mandates management responsibility to provide a safe, healthful workplace.
2. The four components of the factory system that emerged at that time:
   a. substitution of mechanical energy for animal sources of power, particularly steam power through the combustion of coal
   b. substitution of machines for human skills and strength
   c. invention of new methods for transforming raw materials into finished goods, particularly in iron and steel production and industrial chemicals
   d. organization of work into large units, such as factories or forges or mills. This made possible direct supervision of the manufacturing process and an efficient division of labor.
3. The textile industry introduced the factory system in the United States.
4. Fellow servant rule—Employer was not liable for injury to an employee that resulted from negligence of a fellow employee.
5. Contributory negligence—Employer was not liable if the employee was injured due to his own negligence.
6. Assumption of risk—Employer was not liable because the employee took the job with full knowledge of the risks and hazards involved.
7. a. Theodore Roosevelt
8. b. Wisconsin
9. The railroad industry was the first to realize that the actions of people were important in creating accident situations.
10. c. 1912
11. engineering, education, and enforcement
12. List the three trends in safety work and the safety profession that have emerged from the prevention and control measure concepts of the 1950s.

a. First, more emphasis is being placed on analyzing the loss potential of any organization or projected activity.

b. Second, industry is developing more factual, unbiased, and objective information about loss-producing problems and accident causation to help those who are ultimately responsible for worker health and safety make sound decisions.

c. Third, management is making greater use of the safety and health professional's knowledge and assistance in developing safe products.

13. List the necessary elements for a joint safety and health activity to be successful in any workplace.

a. A sincere commitment to the safety and health effort must be displayed by both labor and management leaders in the workplace.

b. Specific roles and responsiblities should be defined for all committe members.

c. An effective communication link with feedback to workers from committee members must be established between the committe members and all employees in the workplace.

d. Measureable, realistic goals and objectives should be established for committee activity.

14. a. Walsh-Healy Act

15. a. one or more employees
b. businesses affected by interstate commerce

16. List four factors that have spurred the drive for international standardization of health and safety regulations.

a. Emerging global markets have intensified the need for international standardization.

b. Worldwide technological innovations result in changes to industrial methods and organizations that threaten worker and consumer safety.

c. The rapid pace of change in science and technology is outstripping standards development in most countries.

d. Developing countries' efforts to industrialize means they may downplay safety and health regulations in favor of rapid economic growth.

17. List five employer obligations relating to the safety and health of workers stated in the Framework Directive (89/391/EEC) of ISO 9000 and ISO 14000.

a. Taking the measures necessary for their safety and health, bearing in mind technical progress

b. Evaluating hazards and instructing workers accordingly

c. Setting up protection and prevention services—for example, the precision of safety and health practitioners—within the workplace, possibly by enlisting competent external services or persons

d. Organization of first aid

e. Evacuation of workers in the event of serious damage.

18. List three employee obligations relating to the safety and health of workers stated in the Framework Directive (89/391/EEC) of ISO 9000 and ISO 14000.

a. Correct use of machinery

b. Correct use of protective equipment

c. The need to report defects in equipment, defects in procedures, and potentially dangerous situations, such as near-miss accidents.

19. The death calendar used by the Russell Sage Foundation in Allegheny County, Pennsylvania, made it clear that the accident and death rate was serious and gave the safety movement a much-needed boost.

20. Since World War II, the growth of safety procedures and policies intensified, particularly as the federal government began encouraging its contractors to adopt safe work practices.

21. The purpose of an occupational safety and health committee is to make the workplace a safer, healthier environment. An independent committee can be important for several reasons:

• It allows representatives of the union to meet and consider the safety of the workplace, without any interference from management.

• It gives the union its own forum to discuss and set priorities and strategies for dealing with workplace hazards.

• It allows the committee members to gain appropriate expertise in researching hazards and seeking effective solutions. (Note that there is an important difference between gaining expertise and becoming an expert.)

• It can be used to monitor the performance of a joint union-management committee, if one exists.

• If a joint committee runs into roadblocks or becomes ineffective and the union side withdraws, the union will already have a structure for handling safety and health concerns.

22. Experience has shown that certain elements are necessary for a joint safety and health activity to be successful in any workplace:

• A sincere commitment to the safety and health effort must be displayed by both labor and management leaders in the workplace.

• Specific roles and responsibilities should be defined for all committee members.

• An effective communication link must be established between the committee members and all employees in the workplace. This includes feedback to workers from committee members.

• Measurable, realistic goals and objectives should be established for committee activity.

23. Management must address serious emerging issues in worker health and safety law. These issues include off-the-job safety and ways to deal with the

special problem of employees who are at risk in the work environment because of physical condition, language problems, or particular susceptibility to injury or disease. Another issue is the burgeoning paperwork required to comply with OSHA and other agency record-keeping regulations, with the Medical Access Standard, and with the Hazard Communication Standard. Industry accepts almost without question the concept of financial responsibility for work injuries. Not all of industry, however, is convinced of the cost effectiveness of government regulation of safety procedures.

2 REGULATORY HISTORY

1. The purpose of the Williams-Steiger Act or OSHAct (Public Law No. 91-596 found in 29 *United States Code* (*USC*) §§651-678) is "to assure so far as possible every working man and woman in the Nation safe and healthful working conditions and to preserve our human resources." The OSHAct is regarded by many as landmark legislation because it goes beyond the present workplace and considers long-term health hazards in the working environment of the future.

2. Prior to the 1960s only a few federal laws directed any attention to occupational safety and health. Several pieces of legislation passed by the Congress during the 1960s focused industry attention on occupational safety and health. Each of these federal laws was applicable only to a limited number of employers. The laws were either directed at those who had obtained federal contracts, or they targeted a specific industry. Even collectively, all the federal safety legislation passed prior to 1970 was not applicable to most employers or employees. There was little attempt to establish the omnibus coverage that is a central feature of the OSHAct.

3. Labor's position was based on the following:
   a. In general, states had inadequate safety and health standards, inadequate enforcement procedures, inadequate staff with respect to quality and quantity, and inadequate budgets.
   b. In the late 1960s, approximately 14,300 employees were killed annually on or in connection with their job and more than 2.2 million employees suffered a disabling injury each year as a result of work-related accidents. The injury/death toll was considered by most to be unacceptably high.
   c. The nation's work-injury rates in most industries were increasing throughout the 1960s. Because the trend was moving in the wrong direction, proponents of federal intervention felt that national legislation would help to reverse this trend.

4. The two organizations that are vested with the administration and enforcement of the OSHA Act are
   a. The Secretary of Labor and Assistant Secretary of Labor for OSHA
   b. The Occupational Safety and Health Review Commission (OSHRC) as an appellate agency. With respect to the enforcement process, the Secretary of Labor, through the Assistant Secretary, performs the investigation and prosecution aspects, and the OSHRC performs the administrative adjudication portion, with possible appeal through the courts.

5. The Occupational Safety and Health Review Commission is a quasi-judicial board of three members appointed by the president and confirmed by the Senate. The principal function of the Commission is to adjudicate cases when an enforcement action taken by OSHA against an employer is contested by the employer, the employees, or their representatives.

6. The primary functions of NIOSH are:
   a. to develop and establish recommended occupational safety and health standards
   b. to conduct research experiments and demonstrations related to occupational safety and health
   c. to develop educational programs to provide an adequate supply of qualified personnel to carry out the purposes of the OSHAct.

7. The five main technical services provided by NIOSH are:
   a. Hazard evaluation—Onsite evaluations of potentially toxic substances used or found on the job.
   b. Technical information—Detailed technical information concerning health or safety conditions at workplaces, such as the possible hazards of working with specific solvents, and guidelines for use of protective equipment.
   c. Accident prevention—Technical assistance for controlling on-the-job injuries, including the evaluation of special problems and recommendations for corrective action.
   d. Industrial hygiene—Technical assistance in the areas of engineering and industrial hygiene, including the evaluation of special health-related problems in the workplace and recommendations for control measures.
   e. Medical service—Assistance in solving occupational medical and nursing problems in the workplace, including assessment of existing medically related needs and development of recommended means for meeting such needs.

8. The final responsibility for compliance rests with the employer.

9. The two voluntary programs that have been established are Star and Merit. OSHA's Star Initiative, the agency's most demanding and prestigious voluntary protection program, is for work sites with outstanding workplace safety and health systems.

10. Federal safety standards are published in the *Federal Register* prior to their being brought up for formal consideration to be enacted into law.

11. The two types of variances that OSHA can grant are:
   a. interpretations of standards
   b. clarifications of standards

12. Employers must report to OSHA within eight hours after an accident occurs that is fatal to one or more employees or that results in the inpatient hospitalization of three or more employees.

13. b. No

14. The five categories of violations:
   a. willful
   b. serious
   c. repeat
   d. other than serious
   e. de minimis (very minor)

15. The four steps a compliance safety and health officer must take to determine that a violation is serious are to evaluate:
   a. The type of accident or health hazard exposure which the violated standard or the general duty clause is designed to prevent.
   b. The type of injury or illness which could reasonably be expected to result from the type of accident or health hazard exposure identified in Step 1.
   c. Whether the types of injury or illness identified in Step 2 could include death or a form of serious physical harm.
   d. Whether the employer knew, or with the exercise of reasonable diligence, could have known of the presence of the hazardous condition.

16. In contesting an OSHA action, the employer must notify the Area Office which initiated the action that the employer is to contest the case. This must be done within 15 working days after receiving OSHA's notice of proposed penalty; it should be sent by certified mail. If the employer does not contest within the required 15 working days, the citation and proposed assessment of penalties are deemed to be a final order of OSHRC and are not subject to review by any court or agency. Employers should then request an informal conference with the Area Director or the Area Director's representative. Many times such informal sessions will resolve questions and issues, thus avoiding the formal contested case proceedings.

17. Employee medical records should be retained by employers for 30 years from the termination date of the employee.

18. The basic purpose of the HCS is to establish uniform requirements to make sure that the hazards of all chemicals produced, imported, or used within the United States are evaluated.

19. The quality of a hazard communication program depends on the accuracy of the initial hazard assessment.

20. The producer of the MSDS is responsible for the information on it and must ensure that all sheets are up to date.

21. With some exceptions, OSHAct applies to every employer in all 50 states and U.S. possessions who has one or more employees and who is engaged in a business affecting commerce. Specifically excluded from coverage are all federal, state, and local government employees. Employees of states and political subdivisions of the states are excluded from the federal OSHAct. However, states with approved state plans are required to provide coverage for these public employees. Public employees in states without approved plans are not covered by the OSHAct in any manner. The OSHAct also does not apply to those operations in which a federal agency (and state agencies acting under the Atomic Energy Act of 1954), other than the DOL, already has the authority to prescribe or enforce standards or regulations affecting occupational safety or health and is performing that function. Also excluded from the OSHAct are operators and miners covered by the U.S. Mine Safety and Health Act of 1977.

22. Employers have the general duty to furnish each employee with employment and places of employment free from recognized hazards causing or likely to cause death or serious physical harm (this is commonly known as the "general duty clause") and the specific duty of complying with safety and health standards promulgated under the Act. Each employee has the duty to comply with the safety and health standards and with all rules, regulations, and orders that apply to employee actions and conduct on the job.

23. The purpose of OSHA's Voluntary Protection Program (VPP) is to emphasize the importance of, encourage the improvement of, and recognize excellence in employer-provided, site-specific occupational safety and health programs.

24. A firm's accident record must meet stringent criteria. In addition, the work site must have an effective, comprehensive safety and health program, including:
   • Management commitment and accountability— The agency has found that top-flight safety and health programs have strong corporate commitment and accountability, including written, clearly defined assignment of responsibility for worker protection at every level of management.

- Hazard assessment—This means a comprehensive inventory of potential safety and health hazards and includes periodic review and updating, especially when processes are changed or new substances are introduced. Effective hazard assessment also includes an effective mechanism for inviting and responding to worker notices of possible hazards.
- Safety rules and enforcement—Star participants have their own policies and procedures that go beyond the protection specified by OSHAct standards.
- Employee training—In addition to formal training, many participants have regular safety meetings and toolbox meetings where safety procedures are reviewed.
- Self-evaluation—Comprehensive safety and health program audits ensure that the program continues to be an effective system for communicating to employees the company's commitment to safety and health and efficient procedures for responding to employee concerns and suggestions.

25. It provides the opportunity to discuss the issues raised.
26. The regulations pertaining to state plans for the development and enforcement of state standards are codified in Title 29 *CFR*, Chapter XVII, Part 1902. The basic criterion for approval of state plans is that the plan must be at least as effective as the federal program.
27. The OSHAct has given employees a significant role to play in occupational safety and health matters. It has raised occupational safety and health issues to a higher priority in business management. It has given new status and responsibility to professionals working in the occupational safety and health field. And, the Act has bestowed a new status to nationally recognized organizations that develop industry standards. The OSHAct has encouraged greater training for professionals in occupational safety and health. The OSHAct also gave new emphasis to the product safety discipline.

3 LOSS CONTROL PROGRAMS

1. The benefits of hazard analysis are:
   a. It forces the conductors of the analysis to view each operation as a part of a system.
   b. It identifies hazardous conditions and potential accidents.
   c. It provides information so effective control measures can be established.
   d. It can determine the level of knowledge, skill, and physical requirements workers need to perform specific tasks.
   e. It can discover and eliminate unsafe procedures, techniques, motions, positions, and actions.
2. The purpose of ranking hazards by risk is to figure out which hazards are the worst. By doing so, a consistent guide for corrective action will be established. The ranking will specify which hazardous conditions warrant immediate action, which have secondary priority, and which ones can be addressed in the future.
3. Hazardous conditions can be either eliminated or controlled:
   a. At the source (substitute a less harmful agent for the one causing the problem, e.g., toxic to nontoxic, flammable to nonflammable).
   b. Along its path (e.g., install machine guards to prevent unwanted contact by workers; put up protective curtains to prevent sparks and welding arc flash; install an exhaust system to remove toxic vapors from breathing zones of workers).
   c. At the worker (e.g., employ automated or remote control options; provide a system of worker rotation or reschedule operations when there are few workers in the plant; provide personal protective equipment).

4 SAFETY, HEALTH, AND ENVIRONMENTAL AUDITING

1. A methodical examination of a facility's procedures and practices that verifies whether they comply with legal requirements and internal policies and evaluates whether they conform to good safety, health, and environmental practices.
2. Any six of the following:
   a. To determine and document compliance status
   b. To improve overall safety, health, and environmental performance at operating facilities
   c. To asses facility management
   d. To increase the overall level of safety, health, and environmental awareness
   e. To accelerate the overall development of S/H/E management control systems
   f. To improve the safety, health, and environmental risk management system
   g. To protect the company from potential liabilities
   h. To develop a basis for optimizing safety, health, and environmental resources
   i. To assess facility management's ability to achieve S/H/E goals.
3. a. Organizational boundaries—address which of the company's operations are included in the auditing program, such as manufacturing, R & D, and distribution.
   b. Geographical boundaries—address how far or wide the program applies, such as state, province, regional, national, or international.

c. Locational boundaries—address what territory is included in a specific audit, such as activities within the facility boundary, off-site manufacturing or packaging, off-site waste disposal, local residences, and a nearby river or lake if there is a potential for environmental damage.

d. Functional areas—define which subject areas are included, such as air and water pollution control, solid and hazardous waste management, employee safety, industrial hygiene, occupational medicine, fire and loss prevention, process safety, and product safety.

e. Compliance boundaries—define the standards against which the facility is measured, such as federal, provincial, regional, and local laws and regulations; corporate or division policies, procedures, standards, and guidelines; local facility operating procedures; or standards established by an outside group.

4. a. Audit planning
   b. Understand management systems
   c. Assess internal controls
   d. Gather audit evidence
   e. Evaluate audit evidence
   f. Report audit findings
   g. Audit follow-up

5. a. Audit protocols—represent plans of how the auditor is to accomplish the objectives of the audit. They list the audit procedures that are to be performed to gain evidence about safety, health, and environmental practices. They also provide the basis for assigning specific tasks to individual members of the audit team, for comparing what was accomplished with what was planned, and for summarizing and recording the work accomplished. When well-designed, audit protocols can be used to train inexperienced auditors and reduce the amount of supervision required by the audit team leader, as well as help build consistency into the audit.

   b. Working papers—document the work performed, the techniques used, and the conclusions reached by the auditors. These papers help the auditor achieve the audit objectives and provide reasonable assurance that an adequate audit was performed consistent with audit program goals. Working-papers should include documentation of compliance or noncompliance.

6. c

7. d

8. e

9. Any five of the following:
   a. Increased growth
   b. Broader scope
   c. Increased rigor and depth of review
   d. Increased effectiveness of field resources

e. Increased emphasis on basic skills
f. Expanded nature of reporting
g. Emergence of S/H/E auditing standards

5 WORKERS' COMPENSATION

1. a. Loss of earnings
   b. Additional expenses

2. An effective loss control program benefits the entire economy by assisting in keeping workers from being injured on the job, and by reducing direct losses to the worker and the worker's family. In addition, an effective loss control program will mitigate society's losses from taxes that the injured worker would have paid, products they would have purchased, and public assistance benefits the family may have needed.

3. The Longshoremen and Harbor Workers' Compensation Act.

4. a. Provide adequate, equitable, prompt, and sure income and medical benefits to work-related accident victims, or income benefits to their dependents, regardless of fault.

   b. Provide a single remedy and reduce court delays, costs, and workloads arising of the personal-injury litigation.

   c. Relieve public and private chanties of financial drains due to uncompensated industrial accidents.

   d. Eliminate payment fees to attorneys and witnesses as well as time-consuming trials and appeals.

   e. Encourage maximum employer interest in safety and rehabilitation through an appropriate experience-rating mechanism.

   f. Promote frank study of accident causes and not faults, reducing preventable accidents and human suffering.

5. A compulsory law requires each employer to accept the provisions and provide for specific benefits, i.e. required compliance with the law.

6. Under quid pro quo of workers' compensation law, employers were required to accept responsibility for injuries arising out of and in the course of employment without regard to fault.

7. a. Expansion of the dual capacity doctrine
   b. The international tort exception

8. a. Loss of income
   b. Medical payments
   c. Rehabilitation

9. 66 percent

10. a. Workers in farming, domestic service, casual employment, charitable or religious organizations
    b. Employees covered under other types of compensation, such as FECA, Jones Act, and FELA

11. a. Temporary total
    b. Permanent total
    c. Permanent partial
    d. Temporary partial
12. a. To prevent accidents
    b. To control costs
    c. To respond to accidents promptly and efficiently
13. To replace the wages lost by workers who are disabled due to job-related injury or illness
14. The Federal Vocational Rehabilitation Act
15. a. Uncertainty about whether an accident arose out of and in the course of employment
    b. The extent of the disability
16. Vocational rehabilitation prepares the injured worker for a new occupation or for ways of continuing in an old one. Vocational rehabilitation is assigned when medical treatment fails to restore the worker to the job held when the individual was injured.
17. a. Whole-person
    b. Wage-loss
    c. Loss of wage-earning capacity
18. a. Expansion of the dual doctrine—injury resulted from the employer's product that is available to consumers.
    b. Employer commits an international tort

6 IDENTIFYING HAZARDS

1. Hazard analysis is an orderly process used to acquire specific hazard and failure data pertinent to a given system for the purpose of eliminating or controlling hazards.
2. a. Inductive
   b. Deductive
3. a. What is the quantity and quality of information desired?
   b. What information is already available?
   c. What is the cost of setting up and conducting analyses?
   d. How much time is available before decisions must be made and action taken?
   e. How many people are available to assist in the hazard analysis, and what are their qualifications?
4. a. Frequency of accidents
   b. Potential for injury
   c. Severity of injury
   d. New or altered equipment, processes, and operations
   e. Excessive material waste or damage to equipment
5. JSA is a procedure used to review job methods and uncover hazards that may have been overlooked in the layout of the facility or building and in the design of the machinery, equipment, tools, workstations, and processes; secondly, to uncover hazards that may have developed after production started; thirdly, to uncover hazards that resulted from changes in work procedures or personnel.
6. a. Breaking the job down into successive steps or activities and observe how these actions are performed
   b. Identifying the hazards and potential accident causes—a critical step since only an identified problem can be eliminated
   c. Developing solutions—recommending safe job procedures to eliminate the hazards and prevent the potential accidents
7. To detect potential hazards so they can be corrected before an accident occurs
8. a. Continuous inspections—informal inspections that do not conform to a set schedule, plan, or checklist, and are conducted by employees, supervisors, and maintenance personnel as part of their job responsibilities
   b. Interval inspections—planned inspections at specific intervals that are deliberate, thorough, and systematic, and are conducted by safety professionals, certified or licensed inspectors, outside investigators, and government inspectors
9. c
10. Building codes, building inspection books, guides to building and facility maintenance, NFPA publications; other publications of the National Safety Council; the *Federal Register* and the *Code of Federal Regulations (CFR)*, Title 29, parts 1900–1950 give OSHA regulations; the Occupational Safety and Health Administration; the Mine Safety and Health Administration (MSHA); the Federal Aviation Administration (FAA); the Nuclear Regulatory Commission (NRC); and the Environmental Protection Agency (EPA)
11. a. The loss severity potential of the problem
    b. The potential for injury to employees
    c. How quickly the item or part can become unsafe
    d. The history of failures
12. a. Personal monitoring—measures the airborne concentrations of contaminants by placing the measuring device as closely as possible to the site at which the contaminant enters the human body.
    b. Environmental monitoring—measures contaminant concentrations in the workroom in the general area adjacent to the worker's usual workstation.
    c. Biological monitoring—measures changes in composition of body fluid, tissues, or expired air to detect the level of contaminant absorption.
    d. Medical monitoring—medical personnel examine workers to see their physiological and psychological response to a contaminant.
13. b

14. TLVs represent an exposure level of airborne concentrations of substances under which most people can work, day after day, without adverse effect. The term TLV refers specifically to limits published by the American Conference of Governmental Industrial Hygienists (ACGIH), and are reviewed and updated annually.

15. PELs represent the legal maximum level of contaminants in the workplace air. The General Industry OSHA Standards currently list about 600 substances with established exposure limits in subpart Z, "Toxic and Hazardous Substances," Sections 1910.1000 through 1910.1500.

16. a. Determines direct causes
    b. Uncovers indirect causes
    c. Prevents similar accidents
    d. Documents facts
    e. Provides information on costs
    f. Promotes safety

17. Immediate, on-the-scene accident investigation provides the most accurate and useful information. The longer the delay in examining the accident scene and interviewing the victim(s) and witnesses, the greater the possibility of obtaining erroneous or incomplete information. The accident scene changes, memories fade, and people discuss what happened with each other.

7 ACCIDENT INVESTIGATION, ANALYSIS, AND COSTS

1. The six (6) fundamental activities are needed for a successful incident prevention program:
   a. study of all working areas to detect and eliminate or control the physical or environmental hazards that contribute to incidents
   b. study of all operating methods and practices and administrative controls
   c. education, instruction, training, and enforcement of procedures to minimize the human factors that contribute to incidents
   d. thorough investigation and causal analysis of every incident resulting in at least a lost-work-day injury to determine contributing circumstances
   e. implementation of programs to change or control the hazardous conditions, procedures, and practices found in the preceding activities
   f. program follow-up and evaluation to ensure that the programs achieve the desired control

2. The primary purpose of an incident investigation is to prevent future incidents. As such, the investigation or analysis must produce factual information leading to corrective actions that prevent or reduce the number of incidents.

3. All incidents should be investigated, regardless of severity of injury or amount of property damage. The extent of the investigation depends on the outcome or potential outcome of the incident. An incident involving only first aid or minor property damage is not investigated as thoroughly as one resulting in death or extensive property damage, that is, unless the potential outcome could have been disabling injury or death.

4. Depending on the nature of the incident and other conditions, the investigation is usually made by the supervisor, perhaps assisted by a fellow worker familiar with the process involved, the safety and health professional or inspector, the employee health professional, the joint safety and health committee, the general safety committee, or an engineer from the insurance company.

5. Each investigation should be conducted as soon after the incident as possible

6. A company can learn the following from incident and accident causation analysis:
   a. Management can identify and locate the principal sources of incidents by determining, from actual experience, the methods, materials, machines, and tools most frequently involved in incidents, and the jobs most likely to produce injuries.
   b. Investigations may disclose the nature and size of the incident problems in departments and among occupations.
   c. Results will indicate the need for engineering revision by identifying the principal hazards associated with various types of equipment and materials.
   d. The investigation can disclose inefficiencies in operating processes and procedures where, for example, poor layout contributes to incidents, or where outdated, physically overtaxing methods or procedure can be avoided.
   e. An incident report will disclose the unsafe practices that need to be corrected by training employees or changing work methods.
   f. The report also will enable supervisors to put their safety work efforts to the best use by giving them information about the principal hazards and unsafe practices in their departments.
   g. Investigation results permit an objective evaluation of the progress of a safety program by noting in continuing analyses the effect of corrective actions, educational techniques, and other methods adopted to prevent injuries.

7. The minimum data that should be collected for each incident includes:
   a. data about employer characteristics
   b. employee characteristics
   c. characteristics of the injury

d. narrative description and accident sequence

e. characteristics of the equipment associated with the incident

f. characteristics of the task being performed when the incident happened

g. time factors

h. task and activity factors

i. supervision information

j. causal factors

k. corrective actions taken immediately after the incident to prevent a recurrence, including interim or temporary actions

8. The three (3) basic steps in a systematic approach to selecting corrective actions are (1) all major actions are considered, (2) the analyst does not stop with familiar and favorite corrective actions, and (3) each corrective action chosen for implementation is carefully thought out.

9. ANSI Z16.2-1995 recommends that general and specific classifications be used for most key facts; for an analysis to be of maximum usefulness, classifications must be set up to encompass the situations pertinent to a particular organization. It also recommends numerical codes for the nature of injury or illness, part of body affected, source of injury or illness, event or exposure, secondary source of injury or illness, occupation, and industry.

10. Inadequate policies, procedures, or management systems might be suggested by thorough analysis of groups of incident investigation reports that were not evident when studying individual cases.

11. The two (2) general categories of work incidents, for the purpose of cost analysis, are (1) incidents resulting in work injuries or illnesses and (2) incidents causing property damage or interfering with production. The inclusion of the no-injury incidents makes "work incidents" roughly synonymous with the type of occurrences a safety department strives to prevent.

8 INJURY AND ILLNESS RECORD KEEPING AND INCIDENCE RATES

1. Safety personnel must maintain records because:
   a. It is required by law and by their management.
   b. It is useful to an effective safety program.

2. The seven ways a good record-keeping system can help the safety professional are:
   a. It provides the means to evaluate accident problems objectively and measures overall progress and effectiveness of the company safety program.
   b. It identifies high incident rate units, plants, or departments and problem areas so extra effort can be made in those areas.

c. It provides data for an analysis of incidents pointing to specific causes or circumstances.

d. It creates interest in safety among supervisors by providing them with information about their department's incident experience.

e. It provides supervisors and safety committees with hard facts about their safety problems so their efforts can be concentrated.

f. It measures the effectiveness of individual countermeasures and determines if specific programs are doing the job they were designed to do.

g. It assists management in performance evaluation.

3. c

4. d

5. c

6. OSHA Form No. 200, Log and Summary of Occupational Injuries and Illnesses serves as the annual summary report that is used by any company subject to the OSHAct.

7. Every company subject to the OSHAct is required to post its annual summary by February 1st of each year for 30 days.

8. The OSHA record-keeping requirements are found in the OSHAct and Title 29 of the *Code of Federal Regulations*, Part 1904.

9. The following employers and individuals who do not have to keep OSHA injury and illness records:
   a. Self-employed individuals
   b. Partners with no employees
   c. Employers of domestics in the employers' private residence for the purpose of housekeeping or child care, or both
   d. Employers engaged in religious activities concerning the conduct of religious services or rites

10. Under the OSHAct, all work-related illnesses must be recorded, while injuries are recorded only when:
    a. They require medical treatment (other than first aid) or involve the loss of consciousness.
    b. They involve restriction of work or motion.
    c. They require a transfer to another job.

11. 8 hours

12. Incidence Rate of Recordable Cases =

$$\frac{\text{No. of injuries and illnesses} \times 200{,}000}{\text{Total hours worked by all employees during period covered}}$$

or

$$\frac{\text{No. of lost workdays} \times 200{,}000}{\text{Total hours worked by all employees during period covered}}$$

13. Off-the-job disabling injuries have far exceeded on-the-job disabling injuries. Any unscheduled absence of employees can cause production slowdowns and delays, costly retraining and replacement, or costly overtime by remaining employees.

14. c

9 COMPUTERS AND INFORMATION MANAGEMENT

1. Any five of the following:
   a. Better availability of data
   b. Improved decision making
   c. Duplication eliminated
   d. Improved communication and quality of data
   e. Standardized data
   f. Improved accuracy
   g. Improved analytical capabilities
   h. Reduced cost
2. a. Develop a thorough understanding of the organization's current and evolving operations and identify realistic needs for data management. Consider reengineering work processes to eliminate activities that add little value to the management process.
   b. Identify potential vendors, review software features and functionality, then select hardware to meet those needs.
   c. Purchase existing software and/or hardware or develop a customized system based upon the identified needs.
   d. Set realistic goals for start-up implementation and for training personnel.
   e. Monitor implementation process, usage, and work processes to evaluate and improve these items.
3. a
4. a. How the system will be used?
   b. Who will use it?
   c. What features are necessary to satisfy the users' immediate and future need?
5. Management should address the need for future integration when first designing or purchasing a system.
6. a. Cost
   b. Convenience
7. Programs that fit an organization's exact needs
8. d
9. a. Establish realistic and achievable goals
   b. Provide on-the-job training to supplement formal programs
   c. Introduce training programs at the right time
   d. Periodically reinforce user learning
10. a. Completeness
    b. Reliability of system hardware and software
    c. User acceptance
    d. Costs
    e. Improved availability of information
    f. New capabilities
    g. Flexibility

10 OCCUPATIONAL HEALTH PROGRAMS

1. Any five of the following:
   a. Maintenance of a healthful work environment
   b. Health examinations
   c. Diagnosis and treatment for occupational injuries and illnesses
   d. Case management services
   e. Immunization programs
   f. Confidential health records
   g. Health promotion, education, and counseling
   h. Open communication between the health personnel and the employee's personal physician
2. a. To protect employees against health hazards in their work environment.
   b. To facilitate placement and ensure the suitability of individuals according to their physical capacities, mental abilities, and emotional makeup in work that they can perform with an acceptable degree of efficiency and without endangering their own health and safety or that of their fellow employees.
   c. To assure adequate health care and rehabilitation of the occupationally injured.
   d. To encourage personal health maintenance.
3. Contact of a chemical with the worker's skin.
4. a. Worker education
   b. Improved management techniques
5. c.
6. a. Emergency treatment—given for immediate, life-threatening conditions; provided by the first aid staff until proper medical care can be given.
   b. Prompt attention—treatment of minor injuries such as cuts, scratches, bruises, and burns that do not require medical attention.
7. To reduce the risk of infection, disability, and missed diagnosis.
8. Any five of the following:
   a. Properly trained and designated first-aid personnel on every shift
   b. Instructions for contacting an ambulance or rescue squad
   c. Posted method for transporting ill or injured employees
   d. Posted instructions for calling physician and notifying the hospital that a patient is en route
   e. Approved first-aid unit and supplies
   f. First-aid manual
   g. List of reactions to chemicals and routes of exposure
   h. Adequate first-aid record system and follow-up.
9. MSHA
10. Restoring disabled workers to their former earning power and occupation as completely and rapidly as possible.
11. The primary purpose of a preplacement examination program is to aid in the selection and appropriate placement of workers.
12. Light duty is an adaptation of the worker's original job to reduce the worker's tasks. Limited duty is

the placement of the worker in a new job that is appropriate to the worker's capabilities.

13. The primary purpose of exit examinations is to document the status of the health of the employee leaving the organization.

14. a. Selection, training, and supervision of auxiliary nursing and other personnel
    b. Transportation and caring for the injured
    c. Transfer of the seriously injured to hospitals
    d. Coordination of these plans with the safety department, security, police, road patrols, fire departments, and other interested community groups

15. c

16. Any five of the following:
    a. Job placement.
    b. Establishing health standards.
    c. Health maintenance programs.
    d. Treatment and rehabilitation.
    e. Workers' compensation cases.
    f. Epidemiologic studies.
    g. Helping management with program evaluation and improvement.
    h. Establishing a health profile of each worker.

17. Wellness is defined as a way of life that promotes a state of health.

18. b

11 INDUSTRIAL HYGIENE PROGRAM

1. Industrial hygiene is the science and art devoted to the recognition, evaluation, and control of environmental factors or stresses arising in and from the workplace. These factors or stresses may cause sickness, injury, or significant discomfort and inefficiency among workers or citizens in the community.

2. e

3. a. Substitution of harmful or toxic materials with less dangerous ones
   b. Changing work processes to eliminate or minimize work exposure
   c. Installation of exhaust ventilation systems
   d. Good housekeeping/appropriate waste disposal methods
   e. Provision of proper personal protective equipment

4. a. Anticipating and recognizing health hazards that arise from work operations processes by the industrial hygiene professional
   b. Evaluating and measuring the magnitude of the hazard by the industrial hygiene professional
   c. Control of the hazard
   d. Commitment and support of the industrial hygienist by management
   e. Recognition and trust of the industrial hygienist by the workers in the facility

5. a. Chemical
   b. Physical
   c. Biological
   d. Ergonomic

6. e

7. Any of the following:
   a. Industrial hygienist
   b. Safety professional
   c. Occupational health nurse
   d. Occupational health physician
   e. Employees
   f. Senior and line management

8. b

9. a. Dermatologic
   b. Allergic
   c. Respiratory
   d. Psychological

10. c

11. a. Gastrointestinal problems
    b. Sleep disorders
    c. Psychological problems

12. Toxicity is the capacity of a chemical to harm or injure a living organism by other than mechanical means.

13. a. Inhalation
    b. Skin absorption
    c. Injection
    d. Ingestion

12 ENVIRONMENTAL MANAGEMENT

1. ISO 14010 is the general principles for environmental auditing, while ISO 14011 is a specific auditing protocol.

2. A gap analysis is a determination of where the company is relative to the requirements and development of a strategy to ensure ISO 14001 conformance.

3. a. legislative and regulatory requirements
   b. identification of significant environmental aspects
   c. examination of existing environmental management practices and procedures
   d. assessment of feedback from previous environmental incidents

4. The purpose is to create a publicly available registry of environmental effects that have been verified by a third party.

13 ERGONOMICS PROGRAMS

1. f

2. a. To reduce the physical and mental stress associated with a given job
   b. To increase the comfort, health, and safety of a work environment
   c. To increase productivity

d. To reduce human errors associated with a task

e. To improve the quality of work life

3. a. Physical demands—lifting, pushing pulling, reaching, exerting force to perform a task, the effort required to do the job, repetitive tasks; walking, standing, or sitting all day

   b. Environmental demands—vibration, temperature, humidity, noise, lighting levels, work organization, pace, shift schedule, need for overtime

   c. Mental demands—information needed to perform a particular task, mental calculations or computations, short-term memory demands, information processing, decision making

4. a. Management commitment

   b. Case management process

   c. Training and education

   d. Workplace improvement process

5. a. Workplace characteristics and accessories

   b. Physiological demands

   c. Physical demands

   d. Environmental demands

   e. Design of displays, control, and dials

6. Workers become bored when performing below their abilities. Boredom can lead to lapses in concentration and attention, resulting in product defects, quality problems, and accidents.

7. a. Heights of object handled

   b. Location of work material

   c. Clearance required and accessibility of work

   d. Design of accessories

8. A neutral posture is one in which the back is naturally curved, the head is held erect, the shoulders are relaxed, the upper arms are close to the sides of the body, the wrists are straight, the elbows are bent at about 90 degrees, and the feet are well supported.

9. b

10. a

14 EMPLOYEE ASSISTANCE PROGRAMS

1. a. Health problems

   b. Marital difficulties

   c. Financial problems

   d. Alcoholism

   e. Drug abuse

   f. Legal problems

   g. Emotional impairment

   h. Stress

2. a

3. EAP assistance is provided through referrals to outside counseling services in the community, treatment services provided by the EAP itself, or by a wide variety of training and educational programs available through the EAP. Services are available to both the employee and to members of the employee's immediate family.

4. Confidentiality—employees need to feel that their employment is not threatened by participation in the EAP and that their problems will remain private.

5. a. A professional assesses the client's problem and recommends treatment.

   b. The assessment and recommendations are recorded by the professional in confidential EAP files.

   c. Treatment or other services will be rendered.

   d. If the client is an employee who has taken time off work for treatment, the EAP may assist with reintegration to the work force as soon as practical.

   e. EPA staff will then conduct follow-up sessions with the client to support his or her recovery.

6. a. Morale improves as disruptive employees are helped.

   b. Management spends less time and resources on discipline.

   c. Valued employees with personal problems remain with the company rather than resigning.

   d. Hiring and training costs are lower.

   e. Safety improves and liability declines.

7. e

8. a. Type of organization or industry

   b. Number of worksites

   c. Type of work/jobs

   d. Size of workforce and demographics (sex, age, ethnicity, education, special needs)

   e. Major employee problems (from benefits and workers' compensation data)

   f. Risk management issues (from safety, medical, insurance data)

   g. Management and labor indentification of problem issues

   h. Regulatory requirements of government agencies

   i. Resources available to EAP from the corporation

9. a. Internal—services delivered by professionals employed by the organization

   b. External—services delivered by a contracted vendor

   c. Union-based—services delivered by trained union personnel to union members

   d. Consortium—a group of smaller companies banded together to jointly contract with an EAP

   e. Blended—any combination of the above

10. e

11. b

12. Any five of the following:

    a. Treatment services

    b. Managed care

    c. Treatment follow-up and monitoring of recovery

    d. Substance abuse professional services and other regulatory compliance

e. Management of workplace threat and violence

f. Critical incident stress debriefing

g. Work/family services

h. Employee training

i. Management consultation and organization development

13. a. Regularly scheduled calls or client visits to the EAP

b. Regular reports to EAP staff from a professional aftercare provider

c. Random drug or alcohol testing coordinated by the EAP

14. e

15 EMERGENCY PREPAREDNESS

1. b

2. Any eight of the following:

a. Fire and explosion

b. Floods

c. Hurricanes and tornados

d. Earthquakes

e. Civil strife and sabotage

f. Work accidents and rumors

g. Shutdowns

h. Wartime emergencies

i. Hazardous materials

j. Radioactive materials

k. Weather-related emergencies

3. Any five of the following:

a. Historical knowledge and records of accidents

b. Fire statistics

c. National Weather Service logs

d. U.S. Geological Survey studies

e. Location of rail lines

f. Location of airports

g. Local U.S. Hazard Vulnerability Community Analysis documents

4. d

5. Any six of the following:

a. Chain of command

b. Training

c. Hazardous materials/spills emergencies (HAZMAT)

d. Command headquarters

e. Uniform incident command system (ICS)

f. Emergency equipment

g. Alarm systems

h. Fire brigades

i. Facility protection and security

j. Emergency medical services (EMS)

k. Warden service and evacuation

l. Transportation

6. Disaster training is essential in keeping a disaster-control plan functioning in good working order. Simulated disaster drills help key people and employees respond to emergencies with greater confidence and effectiveness. Feedback from training drills helps to improve emergency management plans.

7. The U.S. OSHA regulation entitled Hazardous Waste Operations and Emergency Response: Final Rule

8. The function of the HAZMAT team is to control and stabilize actual or potential leaks or spills of hazardous substances requiring possible close approach to the substances.

9. e

16 PRODUCT SAFETY MANAGEMENT

1. d

2. a. Making a commitment to spend money and resources to save money in the long term

b. Choosing a program coordinator

c. Choosing a program auditor

3. a. Engineering

b. Manufacturing

c. Service

d. Legal

e. Purchasing

f. Human resources

g. Risk management/insurance

4. Quality assurance refers to actions taken by management to ensure that manufactured, assembled, and fabricated products conform to design or engineering requirements and to customer acceptance and satisfaction.

5. a. Record keeping

b. Field-information system

c. Product recall or field modification

d. Independent testing

6. a. Complete and accurate records can be convincing evidence in a court of law.

b. Complete and accurate records can enable a company to identify and locate products that will be indicated in a product recall or field-modification program.

7. d

8. a. The auditor's opinion of the company's overall PSM program

b. The auditor's recommendations to correct deficiencies detected during the audit

c. The importance of each recommendation and the possible consequences if the company fails to comply

d. An estimation of a realistic time frame for compliance with the recommendations

9. a. Who is the customer?

b. What types of product are involved?

c. What is the problem?

d. How is the product being used?

10. The manufacturer of a product is responsible for ensuring that the product is both reliable and safe for the customer to use.

17 RETAIL/SERVICE/WAREHOUSE FACILITIES

1. b
2. a. Policy statement
   b. Employee manual
   c. Safety committee
   d. Safety audits
   e. Training
   f. Drills/practices
   g. Investigation and action
   h. Corporate safety culture
3. give specific instructions
4. The employer is responsible for providing employees with a safe place of employment free from recognized hazards that are causing or likely to cause death or serious physical harm to the employees.
5. OSHA 200 provides a summary of all occupational injuries and illnesses. OSHA 101 provides detailed information on each of the occurences listed in the OSHA 200 log.
6. a. exits sufficient to permit prompt escape in case of fire or other emergency
   b. a building that is constructed, arranged, equipped, maintained, and operated to avoid undue danger to its occupants
   c. exits that are of the proper type, location, and number for the building
   d. exits that are accessible at all times, when it is occupied
   e. exits that are clearly visible and marked in a conspicuous manner (lights, signs)
   f. adequate and reliable emergency lighting for all exits
   g. fire alarms in a building of such size, arrangement, or occupancy that individuals may not be immediately aware of the fire
   h. training and drills in emergency exit procedures
   i. at least two accessible exits that remote from each other in each building
7. A standard that provides, to the workers, the identities and hazards, and appropriate protective measures for the chemicals to which they are exposed.
8. a. written hazard communication program
   b. labels and warning on containers of chemicals
   c. Material Safety Data Sheets available for review by employees
   d. information and training for potentially exposed employees.
9. To safeguard employees from hazardous energy during equipment maintenance, tagout procedures require that all control surfaces be tagged to instruct all parties not to open or operate such controls until the tag is removed.
10. c
11. a
12. a
13. meet or exceed
14. a. employee responsibilities and discipline
    b. safety rules
    c. accident/incident reporting
    d. general fire protection
    e. emergency procedures
    f. security alarms and inventory controls
15. b
16. material handling
17. The study of people at work and the use of various methods to create a good fit between the employee and the job
18. a. identify the problems
    b. study the physical demands of the job
    c. create a written action plan
    d. maintain the effort
19. a
20. a. crime-related violence
    b. employment-related violence

18 TRANSPORTATION SAFETY PROGRAMS

1. a. Driver error
   b. Vehicle failure
2. a
3. a. A written safety policy—developed, supported, and enforced by management
   b. A person designated to create and administer the safety program and to advise management
   c. A driver safety program, including driver selection procedures, driver training, and safety-motivating activities; proper supervision and implementation are mandatory for success
   d. An efficient system for collision investigation, reporting, and analysis; determination and application of appropriate corrective action; and follow-up procedures to help prevent future collisions
   e. A vehicle preventive maintenance program
4. a. The applicant's driving record
   b. The applicant's personal traits
5. a. To prevent collisions and delays
   b. To minimize the number of vehicles down for repair
   c. To stabilize the work load of the maintenance department
   d. To save money by preventing excessive wear, breakdown of equipment, and unscheduled down time
6. e
7. U.S. Department of Transportation (DOT) rule 49 *CFR*, Part 40

8. a. The Federal Aviation Administration (FAA)
   b. The National Transportation Safety Board (NTSB)
9. To ensure compliance with regulations and federal mandates
10. a. Training videos give emergency response organizations an understanding of an aircraft's specific procedures and assists them in tailoring response activities to various aircraft.
   b. By providing speeches and audiovisual presentations to emergency response organizations, safety department personnel help promote a better understanding of their operations and enhance teamwork and goodwill.
   c. By using informational briefing cards and videos specific to a certain type of aircraft, safety department personnel can assist in training airport rescue fire-fighting personnel. This information is especially helpful at fire stations in rural areas and developing countries.

19 OFFICE SAFETY

1. c
2. Any five of the following:
   a. Stairways, exits, and doors
   b. Lighting
   c. Ventilation
   d. Electrical equipment/outlets
   e. Equipment placement
   f. Floors
   g. Parking lots and sidewalks
   h. Aisles
   i. Storage of materials
3. d
4. a. Select floor finishes that are durable, maintenance free, and slip-resistant.
   b. Select carpeting instead of bare floors.
   c. Repair defective tiles, boards, or carpet immediately.
   d. Use slip-resistant floor wax.
   e. At lobby and elevator entrances, use floor mats and runners.
5. Ergonomics is the science of optimizing a system of designing for the capabilities and limitations of the human interacting with it.
6. a. Visual demands
   b. Reaches required
   c. Muscular strength exerted to perform the task
7. a. The screen should be placed at or slightly below seated eye height.
   b. The screen should be at a comfortable distance that suits the visual acuity of the operator so that he or she does not have to lean forward.
   c. The screen should be far enough away so that the operator does not have to move his or her head

to read the whole screen. The operator should be able to scan the screen by simply moving the eyes.
   d. The screen should be placed directly in front of the operator to minimize twisting of the trunk or neck.
8. b

20 LABORATORY SAFETY

1. *L*ight *a*mplification by *s*timulated *e*mission of *r*adiation
2. d
3. a. Eyes
   b. Skin
4. Lasers generally use high-voltage electrical power as a source of excitation for the active medium. As a result, electric shock is by far the greatest hazard associated with lasers; additional hazards are heat, toxic by-products, noise, dust, and explosions of high-pressure flash lamps
5. These controls require no action by the individual worker to be effective
6. e
7. a. Program document
   b. Top management support
   c. Laser safety officer
   d. Inventory control
   e. Medical surveillance
   f. Safety training
   g. Inspections of hazard controls
8. a. Reducing the individual's time of exposure
   b. Moving farther from the source of radiation
   c. Shielding the individual
   d. Limiting the amount of radioactive material used
9. c

21 CONTRACTOR AND NONEMPLOYEE SAFETY

1. The employer must obtain and evaluate information regarding the contractor's safety performance and programs.
2. a. Strong employer management
   b. Effective coordination of job tasks
   c. Employer emphasis on safety
   d. Strong interpersonal skills of supervisors
   e. Safe work environment in the employer's facility.
3. Does not
4. To determine equitable premiums
5. Good safety record
6. Before work begins
7. a. Work site safety requirements, including safety manuals and standards related to the proposed work
   b. Detailed outline of the safety responsibilities of the employer, outside management, and outside workers

c. Any special hazards at the work site, and review of hazardous materials and relevant MSDSs

d. Training requirements for outside personnel

e. Schedule to review safety auditing, performance, and training programs after the first few weeks of the job

8. Liability associated with a dangerous condition that is generally a threat to children, such as a swimming pool

9. Enclosed by fence or curb

10. Daily

11. Top and bottom of the landing

12. a. Day, month, year, and time of inspection

b. Observations by mechanics or inspectors

c. All breakdowns, including causes and corrective actions

13. Every three years

14. Incipient, smoldering, flame, heat

15. a. Size of the building makes personnel evacuation impossible or impractical.

b. Part or most of the building is inaccessible by the fire department aerial equipment.

c. Any fire in the building must be attacked from within due to the building height.

d. Building has the potential for stack effect.

16. Smoke and toxic fumes can be drawn into the exhaust or return air duct and distribute to all floors and areas served.

17. a. Tell employees to all security and/or management.

b. Arrange for two or more key personnel to assume assigned positions at entrances and other key points.

18. a. Advise employees to avoid comments or antagonism and to answer all questions courteously.

b. Direct employees to stop selling, lock their drawers and remain as calm as they can.

c. Arrange for frequent cash pickups in areas still doing business.

d. Key employees take their assigned places.

e. Employees do not try to apprehend looters or thieves.

19. a. Unmarked glass doors, panels and windows

b. Parking lots

20. ph test

22 MOTIVATION

1. Motivation entails moving people to action that supports or achieves desired goals.

2. Any five of the following:

a. Individual differences

b. Motivation

c. Emotions

d. Stress

e. Attitudes

f. Behaviors

g. Learning processes

3. a. Positive reinforcement—such as personal recognition or performance awards—is more efficient in achieving higher levels of safety performance than forms of disciplinary action focused on eliminating unwanted behaviors.

b. A reinforcement should be associated with a behavior as soon as possible for the effect to be greatest.

4. a. The present situation—is the employee rushed, stressed, fatigued, or in poor health?

b. Past experiences—were accidents avoided in the past; what amount of training did the employee receive?

c. Workplace and methods of design—were job procedures and the work setting designed to promote safety and healthy behaviors?

5. Abraham Maslow developed a hierarchy of needs depicting how certain needs have a higher priority than other needs in people's lives.

6. c

7. a. Theory X—assumes the worker is uninterested and unmotivated to work

b. Theory Y—assumes the worker has the potential to be interested and motivated to work

8. a

9. a. Feeling

b. Knowing

c. Action

10. Direct personal experiences from the past

11. Any four of the following:

a. Organize the job to give each worker a complete and natural unit of work.

b. Provide new and more difficult task to each worker.

c. Allow the worker to perform specialized tasks in order to provide a unique contribution.

d. Increase the authority of the worker.

e. Eliminate unneeded controls on the worker while maintaining accountability.

f. Require increased accountability of the worker.

g. Provide direct feedback through periodic reports to the worker.

12. Behaviors are observable actions that can be measured. Attitudes are not observable, and therefore cannot be measured. It is easier to change behaviors than it is to change attitudes.

13. Any seven of the following:

a. Reinforcement

b. Knowledge of results

c. Practice

d. Meaningfulness

e. Selective learning

f. Frequency

g. Recall

h. Primacy

i. Intensity

j. Transfer of training

23 SAFETY AND HEALTH TRAINING

1. a. Reinforcement of the operational goals of the organization

   b. Improved performance

   c. Fewer incidents/accidents

   d. Reduced costs

2. Any seven of the following:

   a. Task procedures

   b. Material safety data sheets

   c. Flowcharts

   d. Checklists

   e. Diagrams

   f. Troubleshooting guides

   g. Decision tables

   h. Reference manuals

   i. Help desks or hotlines

   j. Reward systems

   k. Improved physical work environments

   l. Improved work processes

3. Performance-based training is implemented to solve a specific, on-the-job problem or to encourage a specific behavioral change, and can be evaluated by analyzing a worker's performance.

4. a. Assess the knowledge and skills workers need to develop.

   b. Design the objectives based on the needs assessment.

   c. Develop training materials.

   d. Deliver the training.

   e. Evaluate training based on feedback from those involved in the training.

5. b

6. a. Adults need to know why they are learning a particular topic or skill so they can apply learning to immediate, real-life challenges.

   b. Adults need to apply experience to all new learning.

   c. Adults need to be in control of their learning.

   d. Adults need to know that what they learn will make them more effective and successful.

7. a. On-the-job training

   b. Group methods

   c. Individual methods

8. e

9. Computer-assisted training allows an individual to receive information by reading or watching a video presentation and then respond to situations and questions.

24 MEDIA

1. A medium is a channel of communication. It can be anything that carries information between a source and a receiver, such as television, films, diagrams, printed materials, computers, and trainers. The purpose of media is to facilitate communication.

2. Training should train workers in specific competencies and document that such training has been effective.

3. Media can help support training objectives

   a. when areas of weaknesses are identified through accident investigations

   b. whenever personnel changes occur

   c. To enhance effective communication of safety and health issues to higher levels of management.

4. Some of the roles a trainer uses include expert, mentor, and facilitator.

5. As a facilitator, a trainer is a resource person or guide who helps to match individual student needs with appropriate materials.

6. To determine whether the cost of a training medium is justified, management must consider what the expenditure will buy, how quickly the material will be outdated, how it will be used, and how many students will view it. Management must always rely on personal judgement to determine if the return on investment (ROI) is appropriate for the organization and its needs.

7. The factors to consider in choosing between professionally developed or home-made materials include:

   a. the number of showings

   b. how many people will be delivering the training

   c. size and composition of the audience(s)

   d. degree of customizing (is this aid for general use, or is the information for a particular department or facility only?)

   e. the importance of the message

8. The basic purpose of media is to communicate a message.

9. In designing color charts or graphs, remember that 10% of the population is red/green color blind or color poor.

10. Use no more than four colors in a visual.

11. The disadvantages of slides are that they can easily become disordered, stick in the tray, be projected upside down, or, if poorly designed, be distracting.

12. A photographer working from heights should use a fall-arrest harness and lifeline.

25 SAFETY AWARENESS PROGRAMS

1. To maintain interest in safety by involving management and employees in accident prevention

2. a. Company policy and experience

   b. Budget and facilities

c. Types of operations

d. Types of employees

e. Basic human interest factors for promotional activities

3. a. An increased rate of injuries, accidents, and near-accidents

b. Deteriorating housekeeping

c. Incomplete or missing accident reports, indicating decreased supervisory interest

4. f

5. a. Contests must be planned and conducted by a committee representing all competing groups.

b. Competing groups must be composed of natural units.

c. Methods of grading must be simple and fair.

d. Awards must be worth winning.

e. Contests should be well publicized and promoted.

6. a. General posters concerned with broad safety topics

b. Special industry posters that apply only to specific industries

c. Special hazard posters emphasizing particular hazards

7. a. Keep press releases brief, timely, accurate, and complete.

b. Be familiar with the publication or radio/TV station and know which editor or news director to contact.

c. Tailor press releases for the audience.

d. Include captions with photographs.

e. Leave script writing to the professionals, but make sure all facts are accurate.

8. a. Clearly define the objectives and purpose of the publication.

b. Determine how general or restricted the message will be.

c. Decide what form of publication is best.

d. Estimate costs of preparing and printing the publication.

9. c

10. d

26 PROCESS SAFETY MANAGEMENT

1. The proactive identification, evaluation, and mitigation or prevention of chemical releases that could occur as a result of failures in processes, procedures, or equipment.

2. a. identify hazardous materials in the workplace

b. inform employees and consumers of the hazards presented by their manufacture, use, storage, and handling

3. The OSHA regulation deals with process safety within the workplace, while the EPA regulation focuses on releases primarily affecting areas outside the workplace.

4. d. (the PSM standard deals with weight of flammable liquids and gases, not volume)

5. a. identify hazards

b. maintain safe facilities

c. minimize the hazards of accidental releases

6. a. someone thoroughly familiar with the process begin analyzed

b. someone competent in the hazard analysis methodology being used

7. a. extent of the process hazards

b. number of potentially affected employees

c. operating history of the process

d. age of the process

8. Fault tree analysis works backwards from the incident to identify the operations errors and/or equipment failures; event tree analysis works forward from events to identify those that could result in hazards and to calculate the probability of an incident.

9. Hazards and Operability study

10. a. potential failure mode of each system or piece of equipment

b. effect of each potential failure on system or unit

c. criticality of each failure to system integrity

11. When analyzing proposed changes to materials, processes, equipment, or facilities.

12. b

13. a. initial startup and startup after turnaround

b. normal startup and shutdown, and normal and temporary operations

c. emergency operation and emergency shutdown

d. startup after emergency and/or temporary operation

e. conditions requiring emergency shutdown and assignment of shutdown responsibilities

f. nonroutine work

g. operator/process and operator/equipment interface

h. administrative controls versus automated controls

14. a. before startup of new process facilities

b. before introduction of hazardous chemicals into a new facility

c. following major turnaround

d. when facilities have had significant process modifications

15. a. date of inspection or test

b. name of person performing test

c. identification of equipment

d. description of work performed

e. acceptance limits/criteria and results

f. steps required/taken to correct/mitigate deficiencies

16. Hazardous waste emergency response procedures

17. To measure facility performance and ensure

compliance with internal and external process safety management requirements.

18. c

19. a. inspection resulting from response to accidents or catastrophes
    b. unprogrammed inspection
    c. programmed general industry inspection
    d. program quality verification inspection

20. a. evaluate employer's and contractor's PSM programs
    b. compare the quality of the programs to acceptable industry practices
    c. verify effective implementation of the programs

21. a. gather all relevant documentation covering process safety management requirements at a specific facility
    b. determine program's implementation and effectiveness by following up on their application to one or more selected processes

22. c

Index